P9-EEL-456

Single- and Multi-Carrier DS-CDMA

Multi-user detection, Space-Time Spreading, Synchronisation, Networking and Standards

Single and Multi-Carrier DS-CDMA

Multi-User Detection, Space-Time Spreading, Synchronisation, Networking and Standards

L. Hanzo, L-L. Yang,
E-L. Kuan and **K. Yen**

All of
University of Southampton, UK

IEEE PRESS
IEEE Communications Society, Sponsor

Telephone 01243 779777
International (+44) 1243 779777

E-mail (for orders and customer service enquiries): cs-books@wiley.co.uk

Visit our Home Page on www.wileyeurope.com or www.wiley.com

Other Wiley Editorial Offices

John Wiley & Sons, Inc., 111 River Street, Hoboken, NJ 07030, USA

Jossey-Bass, 989 Market Street, San Francisco, CA 94103-1741, USA

Wiley-VCH Verlag GmbH, Boschstr. 12, D-69469 Weinheim, Germany

John Wiley & Sons Australia Ltd, 33 Park Road, Milton, Queensland 4064, Australia

John Wiley & Sons (Asia) Pte Ltd, 2 Clementi Loop #02-01, Jin Xing Distripark, Singapore 129809

John Wiley & Sons (Canada) Ltd, 22 Worcester Road, Etobicoke, Rexdale, Ontario, Canada M9W 1L1

Wiley also publishes its books in a variety of electronic formats. Some content that appears in print may not be available in electronic books.

IEEE Communications Society, Sponsor COMMS-S Liason to IEEE Press, Mostafa Hashem Sherif

Library of Congress Cataloging-in-Publication Data

Single-and multi-carrier CDMA : multi-user detection, space-time spreading, synchronisation, networking, and standards / L. Hanzo... [et al.].
p. cm.
Includes bibliographical references and index.
ISBN 0-470-86309-9 (alk. paper)
1. Code division multiple access. 2. Signal processing–Digital techniques. I. Hanzo, Lajos, 1952-
TK5103.452.S56 2003
621.3845—dc21

2003053813

British Library Cataloguing in Publication Data

A catalogue record for this book is available from the British Library

ISBN 0-470-86309-9

Typeset from pdf files supplied by the author.
Printed and bound in Great Britain by Antony Rowe, Chippenham, Wiltshire.
This book is printed on acid-free paper responsibly manufactured from sustainable forestry in which at least two trees are planted for each one used for paper production.

We dedicate this monograph to the numerous contributors of this field,
many of whom are listed in the Author Index

Contents

IV Multi-Carrier CDMA 607

About the Authors

Lajos Hanzo received his degree in electronics in 1976 and his doctorate in 1983. During his career in telecommunications he has held various research and academic posts in Hungary, Germany and the UK. Since 1986 he has been with the Department of Electronics and Computer Science, University of Southampton, UK, where he holds the chair in telecommunications. He co-authored 10 books totalling 8000 pages on mobile radio communications, published about 450 research papers, organised and chaired conference sessions, presented overview lectures and has been awarded a number of distinctions. Currently he heads an academic research team, working on a range of research projects in the field of wireless multimedia communications sponsored by industry, the Engineering and Physical Sciences Research Council (EPSRC) UK, the European IST Programme and the Mobile Virtual Centre of Excellence (VCE), UK. He is an enthusiastic supporter of industrial and academic liaison and he offers a range of industrial courses. Lajos is also an IEEE Distinguished Lecturer of both the Communications as well as the Vehicular Technology Society and a Fellow of the IEE. For further information on research in progress and associated publications please refer to http://www-mobile.ecs.soton.ac.uk

Lie-Liang Yang received his M.Eng, Ph.D. degrees in communications and electronics from Northern Jiaotong University, Beijing, China in 1991 and 1997, respectively, and his B.Eng. degree in communications engineering from Shanghai TieDao University, Shanghai, China in 1988. Since December 1997, he has been with the Communications Research Group at the Department of Electronics and Computer Science of University of Southampton, UK, where he held various research posts as a visiting postdoctoral research fellow, research fellow and senior research fellow. Currently holds an academic post as a lecturer. From June 1997 to December 1997 he was a visiting scientist of the IREE, Academy of Sciences of the Czech Republic. He has been involved in a number of projects funded by the National Sciences Foundations of China, the Grant Agency of the Czech Republic, the Engineering and Physical Sciences Research Council (EPSRC) of UK and the European Union. His research has covered a wide range of areas in communications, which include data network and security, intelligent wireless networking, error control coding, modulation and demodulation, spread-spectrum communications and multiuser detection, pseudo-noise (PN) code synchronisation, smart antennas, adaptive wireless systems, as well as wideband, broadband and ultrawideband code-division multiple-access (CDMA) for advanced wireless mobile communication systems. He has published over 90 papers in various journals and conference proceedings. He is a senior member of the IEEE.

Ee-Lin Kuan received the B.Eng degree in Electronics Engineering with First Class Honours from the University of Southampton, U.K. She was awarded the Ph.D. degree in Mobile Communications at the University of Southampton, U.K. Her current research interests are associated with mobile transceiver designs with emphasis on adaptive CDMA techniques. In this field she published a dozen various contributions. Recently she joined Multiple Access Communications (MAC) Ltd. in Southampton, UK.

Kai Yen received the B.Eng degree in Electronics Engineering with First Class Honours from the Nanyang Technological University, Singapore in 1996. In 2001, he was awarded the Ph.D. degree in Mobile Communications at the University of Southampton, U.K. He is currently working as a Research Scientist at the Institute for Infocomm Research (I^2R), primarily funded by the Agency of Science, Technology and Research (A*STAR), Singapore. As a member of the spread/ spectrum/multiple access group in I^2R, his current research interests are associated with multiuser detection techniques and spectrally efficient methods for future mobile communication systems.

Other Wiley and IEEE Press Books on Related Topics [1]

- R. Steele, L. Hanzo (Ed): *Mobile Radio Communications: Second and Third Generation Cellular and WATM Systems*, John Wiley and IEEE Press, 2nd edition, 1999, ISBN 07 273-1406-8, 1064 pages
- L. Hanzo, W. Webb, and T. Keller, *Single- and Multi-Carrier Quadrature Amplitude Modulation: Principles and Applications for Personal Communications, WLANs and Broadcasting*, John Wiley and IEEE Press, 2000, 739 pages
- L. Hanzo, F.C.A. Somerville, J.P. Woodard: *Voice Compression and Communications: Principles and Applications for Fixed and Wireless Channels*; IEEE Press and John Wiley, 2001, 642 pages
- L. Hanzo, P. Cherriman, J. Streit: *Wireless Video Communications: Second to Third Generation and Beyond*, IEEE Press and John Wiley, 2001, 1093 pages
- L. Hanzo, T.H. Liew, B.L. Yeap: *Turbo Coding, Turbo Equalisation and Space-Time Coding*, John Wiley and IEEE Press, 2002, 751 pages
- J.S. Blogh, L. Hanzo: *Third-Generation Systems and Intelligent Wireless Networking: Smart Antennas and Adaptive Modulation*, John Wiley and IEEE Press, 2002, 408 pages
- L. Hanzo, C.H. Wong, M.S. Yee: *Adaptive wireless transceivers: Turbo-Coded, Turbo-Equalised and Space-Time Coded TDMA, CDMA and OFDM systems*, John Wiley and IEEE Press, 2002, 737 pages
- L. Hanzo, M. Münster, B.J. Choi and T. Keller: *OFDM and MC-CDMA for Broadband Multi-user Communications, WLANs and Broadcasting*, John Wiley - IEEE Press, May 2003, 980 pages

[1] For detailed contents and sample chapters please refer to http://www-mobile.ecs.soton.ac.uk

Acknowledgments

We are indebted to our many colleagues who have enhanced our understanding of the subject, in particular to Prof. Emeritus Raymond Steele. These colleagues and valued friends, too numerous to be mentioned, have influenced our views concerning various aspects of wireless multimedia communications. We thank them for the enlightenment gained from our collaborations on various projects, papers and books. We are grateful to Steve Braithwaite, Jan Brecht, Jon Blogh, Marco Breiling, Marco del Buono, Sheng Chen, Peter Cherriman, Stanley Chia, Byoung Jo Choi, Joseph Cheung, Sheyam Lal Dhomeja, Dirk Didascalou, Lim Dongmin, Stephan Ernst, Peter Fortune, Eddie Green, David Greenwood, Hee Thong How, Thomas Keller, Ee Lin Kuan, W. H. Lam, C. C. Lee, Xiao Lin, Chee Siong Lee, Tong-Hooi Liew, Matthias Münster, Vincent Roger-Marchart, Jason Ng, Michael Ng, M. A. Nofal, Jeff Reeve, Redwan Salami, Clare Somerville, Rob Stedman, David Stewart, Jürgen Streit, Jeff Torrance, Spyros Vlahoyiannatos, William Webb, Stephan Weiss, John Williams, Jason Woodard, Choong Hin Wong, Henry Wong, James Wong, Lie-Liang Yang, Bee-Leong Yeap, Mong-Suan Yee, Kai Yen, Andy Yuen, and many others with whom we enjoyed an association.

We also acknowledge our valuable associations with the Virtual Centre of Excellence (VCE) in Mobile Communications, in particular with its chief executive, Dr Walter Tuttlebee, and other leading members of the VCE, namely Dr Keith Baughan, Prof. Hamid Aghvami, Prof. Ed Candy, Prof. John Dunlop, Prof. Barry Evans, Prof. Peter Grant, Dr Mike Barnard, Prof. Joseph McGeehan, Prof. Steve McLaughlin and many other valued colleagues. Our sincere thanks are also due to the EPSRC, UK for supporting our research. We would also like to thank Dr Joao Da Silva, Dr Jorge Pereira, Dr Bartholome Arroyo, Dr Bernard Barani, Dr Demosthenes Ikonomou, Dr Fabrizio Sestini and other valued colleagues from the Commission of the European Communities, Brussels, Belgium.

We feel particularly indebted to Denise Harvey for her skilful assistance in correcting the final manuscript in LaTeX. Without the kind support of Mark Hammond, Sarah Hinton, Zöe Pinnock and their colleagues at the Wiley editorial office in Chichester, UK this monograph would never have reached the readers. *Finally, our sincere gratitude is due to the numerous authors listed in the Author Index — as well as to those whose work was not cited owing to space limitations — for their contributions to the state of the art, without whom this book would not have materialised.*

Lajos Hanzo, Lie-Liang Yang, Ee-Lin Kuan and Kai Yen
Department of Electronics and Computer Science
University of Southampton

Chapter 1

Introduction to Intelligent Broadband Wireless Transceivers[1]

1.1 Aim of the Book

Against the background of the explosive expansion of the Internet and the continued dramatic increase in demand for high-speed multimedia wireless services, there is an urgent demand for flexible, bandwidth-efficient transceivers. Multi-standards operation is also an important requirement for the future generations of wireless systems. In this introductory chapter we present a versatile broadband multiple access scheme, combining frequency-hopping (FH) with multicarrier DS-CDMA (FH/MC DS-CDMA). The proposed FH/MC DS-CDMA system framework is capable of meeting the requirements of future mobile wireless systems, by supporting the existing second- and third-generation systems, while also introducing more advanced techniques facilitated by the employment of Software Defined Radios (SDR) [1–3]. and with the aid of efficient adaptive base-band algorithms [4, 5]. *This generic system design framework will be used throughout as the unifying backbone of the various intelligent algorithms and systems proposed. No doubt that the concrete solutions proposed constitute only a miniscule fraction of the set of potential solutions to wireless communications problems. Our hope, however, is that the book will be motivational, inspiring further research and leading to more advanced wireless system solutions.*

The design of attractive wireless transceivers hinges on a range of contradictory factors, which are summarized in Figure 1.1. The essence of this illustration is that of contrasting the conflicting design factors of wireless systems from different perspectives. For example, given a certain wireless channel, it is always feasible to design a transmission scheme capable of further increasing the achievable transmission integrity at the cost of invoking 'smarter' and hence more sophisticated signal processing. Alternatively, the transmission integrity may be increased even without increasing the system's complexity, provided that, for example,

[1] *Single- and Multi-Carrier CDMA Multi-User Detection, Space-Time Spreading, Synchronization, Networking and Standards.* L.Hanzo, L-L.Yang, E-L.Kuan and K.Yen,
©2003 John Wiley & Sons, Ltd. ISBN 0-470-86309-9

the turbo channel coding/interleaving delay can be further extended. Naturally, this is only possible in the context of delay-insensitive, i.e. non-interactive data communications, but not in delay-sensitive interactive speech or video systems. It is also realistic to increase the system's integrity by invoking a more robust but lower throughput modulation scheme, as will be demonstrated throughout this book, again often with the added benefit of reducing the system's complexity. Naturally, diverse system solutions accrue when optimizing different features of the system.

By contrast, it is always possible to increase the achievable effective throughput of the system, if the system's target integrity is relaxed, which facilitates the employment of an increased throughput, but a more error-sensitive transmission scheme. As another design alternative to employing more robust modulation schemes when requiring a reduced target BER, the channel coding rate may be reduced, in order to increase the transmission integrity at the cost of sacrificing the effective throughput of the system. as well as increasing the system's complexity. When the fading channel's characteristics and the associated bit error statistics change, different transceiver modes may have to be activated. Furthermore, the channel's impulse response and the associated dispersion vary, as the receiver roams in different environments. The subject of this book provides powerful solutions to these problems, with particular emphasis on various CDMA transceivers, which have found favour in all the recent third-generation wireless systems.

Our intention with the book is thus:

1) *We would like to pay tribute to all researchers, colleagues and valued friends who contributed to the field.* Hence this book is dedicated to them, since without their quest for better CDMA transceiver solutions this book could not have been conceived. They are too numerous to name here, hence they appear in the Author Index.

2) *Since at the time of writing no joint treatment of the subjects covered by this book exists, it is timely to compile the most recent advances in the field. Hence it is our hope that this book will present an adequate portrayal of the community's recent research efforts and spur this innovation process by stimulating further research in the wireless communications research community.*

1.2 Organization and Novel Contributions of the Book

The book is divided into four parts and the level of challenge as well as the amount of knowledge assumed gradually increase towards the end of the book. Below, we present the outline and rationale of the book:

- Part I provides a detailed discourse on multi-user detection, aimed at providing a detailed historical perspective on the topic. Our discussions are embedded in the generic system design framework to be outlined at a later stage in this introductory chapter.

- **Chapter 2:** A basic exposition of the principles of CDMA transmission and reception is presented in this chapter. The conventional CDMA receiver, i.e. the matched filter

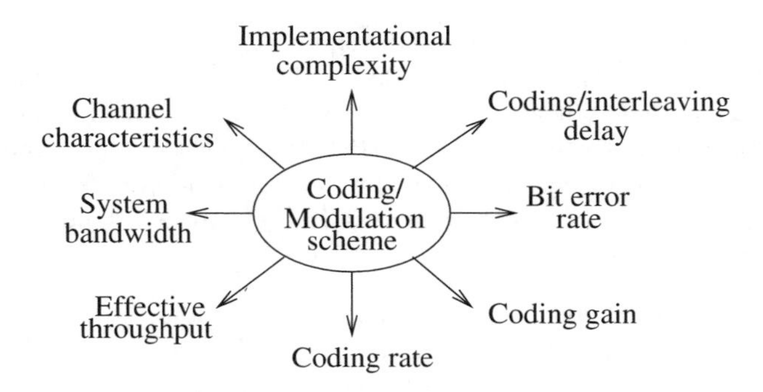

Figure 1.1: Factors affecting the design of wireless transceivers ©John Wiley and IEEE Press, 2002, Hanzo, Liew, Yeap [5]

is investigated and its performance in synchronous Gaussian channels is analyzed. The effects of various spreading sequences on the bit error rate (BER) performance are also considered. Finally, a literature survey of multi-user receivers is discussed and the various receivers are loosely classified.

- **Chapter 3:** Joint detection (JD) receivers, which constitute a class of multi-user receivers are introduced. These multi-user receivers are derived from the well-known single-user equalizers, which are used to equalize signals that have been distorted by inter-symbol interference (ISI) due to multipath channels. The adoption of these single-user equalizers is supported by the fact that the ISI inflicted by a K-path dispersive channel is similar to the multi-user interference due to K users. As a preliminary introduction to joint detection, a brief overview of single-user equalization is presented. This is followed by the portrayal of the theoretical basis of four joint detection receivers, namely the zero-forcing block linear equalizer (ZF-BLE), the minimum mean square error block linear equalizer (MMSE-BLE), zero-forcing block decision feedback equalizer (ZF-BDFE) and the minimum mean square error block decision feedback equalizer (MMSE-BDFE). The implementational complexity of these receivers is estimated and a simple method of reducing the complexity is presented. Finally, the effects of multipath diversity, channel coding and channel estimation errors on the BER performances of the JD receivers are presented and discussed.

- **Chapter 4:** In this chapter, adaptive-rate CDMA techniques are introduced, where the information rate is varied in accordance with the channel quality. The information rate is chosen accordingly, in order to provide the best trade-off between the BER and throughput performance for a given application. The signal to interference plus noise ratio (SINR) at the output of the JD receiver is derived and this parameter is used as the criterion for adapting the information rate. Two different methods of varying the information rate are considered, namely the Adaptive Quadrature Amplitude Modulated (AQAM) JD-CDMA system and the Variable Spreading Factor (VSF) JD-CDMA scheme. AQAM is an adaptive-rate technique, whereby the data modulation mode is chosen according to some criterion related to the channel quality. On the other hand, in VSF transmission, the information rate is varied by adapting the spreading factor of the CDMA codes used, while keeping the chip rate constant. The performance of the proposed AQAM JD-CDMA and VSF JD-CDMA schemes is evaluated and discussed in relation to fixed-rate schemes. A numerical method for obtaining the BER of the JD receiver is presented, which employs the SINR at the receiver output and the associated semi-analytical performance is compared to the simulation results presented. Finally, a performance comparison between the two different adaptive-rate methods is carried out and a combined AQAM-VSF scheme is investigated.

- **Chapter 5:** Another class of multi-user receivers is considered, namely the family of interference cancellation (IC) receivers. These IC receivers are often divided into three categories, namely successive interference cancellation (SIC), parallel interference cancellation (PIC) and hybrid IC, which is a combination of the former two. Here, the SIC and PIC receivers are analyzed in terms of their BER performance and complexity. The performance of these receivers in adaptive-rate CDMA schemes is also investigated and again, the SINR at the output of each receiver is derived and used as the switching criterion in the adaptive schemes. Finally, performance comparisons between the IC and JD receivers are carried out.

- **Chapter 6:** The PIC system investigated in this chapter also benefits from the employment of various channel codecs, such as convolutional coding, as well as Trellis Coded Modulation (TCM) and Turbo Trellis Coded Modulation (TTCM). The latter two schemes achieve an enhanced system performance without any bandwidth expansion. Additionally, the principle of turbo equalization is invoked, where the receiver jointly carries out soft interference cancellation and channel decoding. The attraction of this combined operation is that the channel decoder refrains from making premature 'hard' decisions concerning the transmitted data. Instead, it provides soft information for the PIC scheme and after a few iterations a substantially improved performance may be achieved, which may approach the single-user limit at the cost of a moderate implementational complexity.

- **Chapter 7:** A joint data detector and channel impulse response (CIR) estimator is proposed in this chapter, namely the blind Per-Survivor Processing (PSP) receiver. This receiver performs joint data detection and CIR estimation at the receiver, without the aid of training sequences. The blind PSP receiver is proposed for the bit-synchronous, multipath, multi-user scenario, where a reduced complexity tree-search-based algorithm is employed for data detection and CIR estimation is performed by adaptive Recursive Least Squares (RLS) estimators. The performance of this receiver is investigated for static, slow-fading and fast-fading multipath channels. The PSP-CDMA receiver is also combined with error correction techniques, namely turbo convolutional coding, in order to improve the performance of the system. The performance of the turbo-coded, PSP-CDMA receiver is analyzed for various fading scenarios and the performance of various turbo-coded schemes is compared.

- **Chapter 8:** In this chapter we investigate the performance of single-carrier Wideband CDMA (W-CDMA) schemes. Furthermore, we employ space–time spreading (STS) based transmit diversity, when encountering multipath Nakagami-m fading channels, multi-user interference and background noise. Our analysis and the results will demonstrate that the diversity gains achieved by STS and those owing to multipath diversity accruing as a benefit of the channel's frequency-selectivity are independent. Furthermore, we will show that both the STS-based transmit diversity and the multipath-induced receiver diversity achieved with the aid of the channel's frequency-selectivity appear to have the same order of importance and hence offer similar potential performance benefits.

We will argue, however, that in certain non-dispersive indoor propagation environments the channel may have a single resolvable propagation path and hence no receiver diversity gain may be achieved. Given a fixed chip-rate associated with a constant system bandwidth, such as in the third-generation IMT2000 standard of Chapter 22, in this scenario the number of resolvable multipath components and the associated receiver diversity gain cannot be increased by increasing the chip-rate. We will argue in this chapter that in this case the intelligent system framework proposed at a later stage in this Introduction is expected to configure itself to attain a higher transmit diversity gain.

More specifically, based on the time-varying characteristics of the wireless communication channels, in this chapter an adaptive STS-based transmission scheme is proposed and investigated, which adapts the mode of operation of its STS scheme and its corresponding data rate according to the near-instantaneous frequency-selectivity information fed back from the mobile receiver to the base station's transmitter. Our numerical results show that this adaptive STS scheme is capable of efficiently exploiting the diversity potential provided by the channel's frequency-selectivity, hence significantly improving the effective throughput of W-CDMA systems.

To elaborate a little further, since W-CDMA signals are subjected to frequency-selective fading, the number of resolvable paths at the receiver may vary over a wide range depending on the transmission environment encountered. It is shown that if the channel's grade of frequency-selectivity is sufficiently high, there is no need to employ transmit diversity. Therefore in this chapter an adaptive STS-based transmission scheme is proposed to improve the throughput of W-CDMA systems. Specifically, when the number of resolvable paths is low and hence the resultant BER is higher than the required BER, then a low throughput STS-assisted transmitter mode is activated, which exhibits a high transmit diversity gain. By contrast, when the number of resolvable paths is high and hence the resultant BER is lower than the required BER, then a higher throughput STS-assisted transmitter mode is activated, which has a lower transmit diversity gain. Our numerical results will demonstrate that this adaptive STS-based transmission scheme is capable of significantly improving the effective throughput of W-CDMA systems. Specifically, the studied W-CDMA system's bitrate can be increased by a factor of three at the modest cost of requiring an extra 0.4dB or 1.2dB transmitted power in the context of the investigated urban or suburban areas, respectively.

- **Chapter 9:** Part II of the book commences with Chapter 9, which investigates the potential of genetic algorithm-assisted MUD. In this chapter we continue our discourse by assuming that the reader is familiar with the basic CDMA principles. We commence by giving an overview of GAs, since the concepts of GAs are not widely known in the mobile communications community. We will present the basic functions of a GA in optimizing an objective function, with the aid of a flowchart as well as an example. An insight into why a GA constitutes an efficient function optimizer is also given in the context of the schemata and the schema theorem [6]. Some of the more advanced GA processes are also highlighted here, in order to improve the efficiency of our search for the optimum solution. Finally, a survey of the GA-based CDMA mutiuser detection schemes found in the current literature is conducted.

- **Chapter 10:** In this chapter, we will invoke the GA-assisted multi-user detector in a symbol-synchronous CDMA system over a simple AWGN channel as well as over a single-path Rayleigh fading channel. While this system model is not practical, it can provide us with a better insight into how certain GA operations and parameters behave in the context of our specific application, without considering the effects of the multi-path interference and the asynchronism among the users. We will first define the objective function for our optimization by deriving the correlation metric for the optimum multi-user detector. As mentioned above, the complexity of the optimum multi-user detector is exponential to the number of users and hence it is impractical to implement. Hence we will apply GAs in optimizing the correlation metric, while achieving a reduction in the associated complexity. Through a series of experiments, we will attempt to find the particular GA configuration that is capable of offering a near-optimum bit error probability (BEP) performance at a reduced computational complexity, compared to that of the optimum multi-user detector. Upon determining the GA configuration that we will be adopting for our GA-assisted multi-user detection scheme, we will then investigate its BEP performance in an AWGN channel as well as in a single-path Rayleigh fading channel.

- **Chapter 11:** In Chapter 10 we have assumed that the Channel Impulse Response (CIR) coefficients are perfectly known by the receiver, which allowed us to detect the users' transmitted bits coherently. In practice, these coefficients must be estimated either blindly or with the aid of pilot symbols. By exploiting the capabilities of the GAs in dealing with both binary and floating point variables, we proposed a joint GA-assisted multi-user channel estimation and symbol detection technique in this chapter. Unlike in traditional systems, where the CIR estimation and symbol detection are usually performed by separate but inter-linked algorithms, such as the Kalman filter used for channel estimation [7] and the decorrelator for symbol detection [8], our proposed technique is capable of performing both the channel estimation and symbol detection concurrently using the same GAs. The achievable MSE of the estimated CIR coefficients as well as the BEP performance of our proposed joint GA-assisted multi-user CIR estimator and symbol detector are then evaluated using computer simulations.

- **Chapter 12:** In order to obtain a BEP performance improvement, in this chapter we evaluated the performance of the GA-assisted multi-user detector in conjunction with antenna diversity. More specifically, we investigated the BEP performance of the GA-assisted multi-user detector using two different diversity selection strategies. According to the first strategy, the so-called mating pool is created by selecting the K-bit GA individuals based on the combined figure of merits of the diversity antennas. On the other hand, we can exploit the population-based optimization approach of the GAs and invoke the so-called Pareto optimality [9], in order to aid our search. According to this strategy, the mating pool is comprised of all non-dominated individuals. The BEP performance of the antenna diversity-aided GA-assisted multi-user detector based on these two strategies is evaluated and compared for various fading scenarios.

- **Chapter 13:** The model that we have adopted in the previous GA-assisted MUD chapters was based on a symbol-synchronous CDMA system and multi-user detection is performed at the centralised base station. Unless strict timing control is employed, it is almost impossible to maintain a symbol-synchronous transmission among the users. Hence in Chapter 13 we will propose a GA-assisted multi-user detector for an asynchronous transmission environment. The correlation metric over a finite observation window is derived. The effects of the so-called edge bits, which placed a limitation on the BEP performance, must be taken into account when detection is considered over a finite observation window. In our proposed technique, we can reduce the effects of the edge bits by attempting to estimate the tentative decisions concerning these edge bits, while at the same time detecting the desired bits using the same GA. The BEP performance of our proposed scheme is then compared with that of a similar GA-assisted multi-user detector, where the edge bits are estimated using the conventional single-user matched filter.

- **Chapter 15:** The topic of Part III of the book is the investigation of M-ary CDMA schemes, which is introduced in Chapter 15 in the context of non-coherent detection. In general the transmission of known pilot signals is required for providing a coherent reference in support of coherent demodulation in a fading environment [10]. However, in the uplink, an independent pilot signal/symbol would have to be transmitted by each user to coherently demodulate each signal. This clearly reduces the efficiency of the system. Thus a more practical approach is to use non-coherent detection on the uplink, which does not require a pilot signal for estimating the phase reference [10]. This is the case in the IS-95 CDMA system [11].

- **Chapter 16:** This chapter investigates the achievable performance of a RS coded DS-CDMA system, when M-ary orthogonal signalling is employed in conjunction with a Ratio Statistic Test (RTT) based erasure insertion scheme, when communicating over multipath Rayleigh-fading channels. Two different diversity combining schemes [12], [13] - namely equal-gain combining (EGC) and selection combining (SC) - are considered and their performance is evaluated in the context of the proposed RTT-based erasure insertion scheme. We derive the probability density function (PDF) of the ratio defined in the RTT conditioned on both the correct detection hypothesis (H_1) and erroneous detection hypothesis (H_0) of the received M-ary signals over multipath Rayleigh fading channels. These PDFs are then used to compute the symbol erasure probability and the random symbol error probability after erasure insertion, and finally to compute the codeword decoding error probability. These analytical results allow us to quantify the performance of the system considered using a numerical approach. Furthermore, with the aid of these analytical results, we gain an insight into the basic characteristics of Viterbi's RTT, and determine the optimum decision threshold for practical systems, since the formulae obtained can be evaluated numerically.

- **Chapter 17:** A deficiency of the previously introduced M-ary CDMA systems is that the number of correlators required at the receiver to detect the M-ary symbols is increasing exponentially with the number of bits transmitted. In this chapter a classic number representation system, namely the so-called Residue Number System (RNS) is introduced with the intention of invoking it in the next chapter in the design of novel error correction coded M-ary CDMA system.

- **Chapter 18:** The aim of this chapter is the substantial complexity reduction of the previously discussed M-ary CDMA receivers by decomposing each M-ary message into shorter 'symbols' with the aid of the RNS system. This allows us to use a reduced number of correlators at the receiver. More specifically, our basic approach in this chapter is that the M-ary information symbols are converted into the residue digits of an RNS and all required operations are then carried out in the RNS domain. It is readily seen in this context that each individual M-ary symbol is mapped to a higher number of residue digits. Hence, this mapping operation can be interpreted as breaking up the M-ary symbols into U number of m_u-ary symbols, where U is the number of moduli in the RNS, where m_u, $u = 1, 2, \ldots, U$ represents the moduli of the RNS.

  In a somewhat simplistic, but conceptually feasible manner one could argue then that for the transmission of the U number of reduced-size m_u-ary symbols we now need only $\sum_{u=1}^{U} m_u < M$ correlators at the receiver, which is essentially the aim of this chapter. Hence, in this chapter, we focus our attention on studying the performance of RNS- and Redundant RNS (RRNS) based parallel direct sequence CDMA (DS-CDMA) systems in the context of M-ary orthogonal modulation, when communicating over both Additive White Gaussian Noise (AWGN) channels and multipath fading channels.

- **Chapter 19:** The subject of Part IV of the book is multicarrier CDMA, which is introduced and classified in Chapter 19. It will be shown that there are a number of trade-offs associated with the design and employment of the various multicarrier CDMA scheme proposed in the literature. The BER performance of the various schemes is also characterized, when communicating over frequency selective Rayleigh fading channels. The chapter is concluded with various novel system design proposals invoking frequency hopping for improving the system's achievable performance.

- **Chapter 20:** Following the philosophy of the generic communications system framework proposed in Chapter 1, in Chapter 8 an adaptive STS-aided W-CDMA system was proposed, which was capable of adapting the mode of operation of its STS scheme and its corresponding data rate according to the near-instantaneous frequency-selectivity information fed back from the mobile receiver to the base station's transmitter. The numerical results of Chapter 8 showed that this adaptive STS scheme is capable of efficiently exploiting the diversity potential provided by the channel's frequency-selectivity, hence significantly improving the effective throughput of W-CDMA systems.

  In this chapter our discussions evolve further by initially considering the advantages and disadvantages of single-carrier CDMA, MC-CDMA and MC-DS-CDMA in Section 20.2, with particular attention devoted to communicating in diverse propagation environments exhibiting a time-variant amount of dispersion. More specifically, the benefits and deficiencies of these three systems are analyzed, when aiming for supporting ubiquitous communications over a variety of propagation channels encountered in indoor, open rural, suburban and urban environments. We will demonstrate that, when communicating in such diverse environments, both SC DS-CDMA and MC-CDMA exhibit certain limitations that are hard to circumvent. By contrast, when appropriately selecting the system parameters and using transmit diversity, MC DS-CDMA becomes capable of adapting to such diverse propagation environments at a reasonable detection complexity.

Finally, in the spirit of the generic communications system framework proposed in Chapter 1, a class of generalized MC DS-CDMA schemes is defined and its performance is considered, when communicating over multipath Nakagami-m fading channels. In the generalized MC DS-CDMA schemes considered the spacing between two adjacent subcarriers is a variable, allowing us to gain insight into the effects of the spacing on the BER performance of MC DS-CDMA systems. This generalized MC DS-CDMA scheme includes the subclasses of multitone DS-CDMA and orthogonal MC DS-CDMA as special cases. A unified analytical framework is utilized, which determines the exact average BER of the generalized MC DS-CDMA system transmitting over generalized multipath Nakagami-m fading channels. The optimum spacing of the MC DS-CDMA system required for achieving the minimum BER is investigated and the BER performance of the system having optimum spacing is evaluated. The resultant BER is compared with that of both multitone DS-CDMA and orthogonal MC DS-CDMA.

- **Chapter 21:** Accurate and fast synchronization plays a cardinal role in the efficient utilization of any spread-spectrum system. Typically, the first step in the process of synchronization between the received spread pseudo noise (PN) code (sequence) and the locally generated despread code (sequence) is code acquisition [14–16], which is also referred to as initial synchronization. More explicitly, initial synchronization is constituted by a process of successive decisions, wherein the ultimate goal is to bring the two codes into coarse time alignment, namely within one code-chip interval. Once initial code acquisition has been accomplished, usually a code tracking loop [14] is employed for achieving fine alignment of the two codes and for maintaining their alignment during the whole data transmission process. The aim of this chapter, however, is to focus on the task of initial synchronization in the context of direct-sequence code division multiple-access (DS-CDMA) systems.

  Initial synchronization has been lavishly treated in the literature [15–53] in recent years. Initial code acquisition in DS-CDMA systems is usually achieved with the aid of non-coherent correlation or matched filtering, since prior to despreading the signal-to-noise ratio (SNR) is usually insufficient high for attaining a satisfactory performance with the aid of coherent carrier phase estimators based on carrier-phase tracking loops. Initial code synchronization methods can broadly be classified as serial search acquisition [15, 22, 28, 30, 39, 44] and parallel search acquisition [19, 20, 23–25, 31, 46, 49] techniques. In serial search-based initial code synchronization all potentially possible code phases are searched serially, until synchronization is achieved. More explicitly, in serial search-based code acquisition each reference phase is evaluated by attempting to despread the received signal. If the estimated code phase is correct, successful despreading will take place, which can be detected. If the code phase is incorrect, the received signal will not be successfully despread, and the local reference code replica will be shifted to a new tentative phase for evaluation. By contrast, in parallel search-based code acquisition potentially all of the possible code phases are tested simultaneously. In this chapter both serial and parallel search-based acquisition schemes will be investigated.

It is well known that the near–far interference inflicted upon low-power signals by high-power interfering signals may substantially degrade the system's performance. In the context of DS-CDMA the above-discussed serial and parallel search-based code timing acquisition are as interference-limited and as vulnerable to the near–far problem, as conventional matched-filter-based single-user detection. Recently, a range of timing acquisition algorithms having a high near–far resistance in multi-user environments have also been proposed [17, 18, 21, 37, 54–56]. These timing-acquisition algorithms can be classified as maximum-likelihood synchronization scheme [21, 55], minimum-mean-square-error (MMSE) timing estimation arrangements [17, 37, 54], per-survivor processing (PSP) based blind acquisition techniques [18] and Multiple Signal Classification Algorithm (MUSIC) based timing estimation scheme [54–56]. In this chapter both serial and parallel search-based initial synchronization schemes are considered in the context of both single- and multi-carrier CDMA systems.

- **Chapter 22:** In this chapter we present an overview of the various third-generation terrestrial radio transmission standards recently proposed by ETSI, ARIB, and TIA.

- **Chapter 23:** In this chapter adaptive rate transmissions are investigated in the context of DS-CDMA systems using variable spreading factors (VSF). In the recently established family of adaptive rate transmission schemes [4] the transmission rate is typically adapted in response to the perceived near-instantaneous channel quality. The perceived channel quality is influenced by numerous factors, such as the variation of the number of users supported, which imposes a time-variant MUI load or the fading-induced channel quality fluctuation of the user considered. In Chapter 4 the impact of both of these channel quality factors was considered in the context of multi-user detection aided adaptive modulation as well as variable spreading factor assisted transmissions. It was found that the system performed best, when both the spreading factors and the modulation modes were controlled in a near-instantaneous fashion.

  However, the number of users supported was varied in Chapter 4 on a long-term basis. Hence the average MUI level was constant throughout the simulations. By contrast, in this chapter the individual users' transmission rate is adapted in response to the near-instantaneous MUI fluctuations encountered. Our study will show that by employing the proposed VSF-assisted adaptive rate transmission scheme, the effective throughput may be increased by up to 40%, when compared to that of DS-CDMA systems using constant spreading factors. This increased throughput is achieved without wasting power, without imposing extra interference upon other users and without increasing the BER.

1.3 Introduction to Flexible Transceivers

There is a range of activities in various parts of the globe concerning the standardization, research and development of the third-generation (3G) mobile systems known as the Universal Mobile Telecommunications System (UMTS) in Europe, which was termed as the IMT-2000 system by the International Telecommunications Union (ITU) [57, 58]. This is mainly due to the explosive expansion of the Internet and the continued dramatic increase in demand for all types of advanced wireless multimedia services, including video telephony as well as the more conventional voice and data services. However, advanced high-rate services such as high-resolution interactive video and 'telepresence' services require data rates in excess of 2Mb/s, which are unlikely to be supported by the 3G systems [59–63]. These challenges remain to be solved by future mobile broadband systems (MBS).

The most recent version of the IMT-2000 standard is in fact constituted by a range of five independent standards. These are the UTRA Frequency Division Duplex (FDD) Wideband Code Division Multiple Access (W-CDMA) mode [64], the UTRA Time Division Duplex (TDD) CDMA mode, the Pan-American multi-carrier CDMA configuration mode known as cdma2000 [64], the Pan-American Time Division Multiple Access (TDMA) mode known as UWT-136 and the Digital European Cordless Telecommunications (DECT) [64] mode. It would be desirable to achieve that future systems become part of this standard framework without having to define new standards. The framework proposed in this contribution is capable of satisfying this requirement.

More specifically, these future wireless systems are expected to cater for a range of requirements. Firstly, MBSs are expected to support extremely high bitrate services, while exhibiting different traffic characteristics and satisfying the required quality of service guarantees [59]. The objective is that mobile users become capable of accessing the range of broadband services available for fixed users at data rates upto 155 Mb/s. Multi-standard operation is also an essential requirement. Furthermore, these systems are expected to be highly flexible, supporting multimode and multiband operation as well as global roaming, while achieving the highest possible spectral efficiency. These features have to be sustained under adverse operating conditions, ie for high-speed users, for dynamically fluctuating traffic loads and over hostile propagation channels. These requirements can be conveniently satisfied with the aid of broadband mobile wireless systems based on the concept of adaptive software defined radio (SDR) architectures [1, 2].

In the first part of this chapter a broadband multiple access candidate scheme meeting the above requirements is presented, which is constituted by frequency-hopping (FH) based multicarrier DS-CDMA (FH/MC DS-CDMA) [65–68]. Recent investigations demonstrated that channel-quality controlled rate adaptation is an efficient strategy for attaining the highest possible spectral efficiency in term of b/s/Hz [69–73], while maintaining a certain target integrity. Hence, in the second part of the chapter we consider Adaptive Rate Transmission (ART) schemes associated with supporting both time-variant rate and multirate services. These ART techniques are discussed in the context of the proposed FH/MC DS-CDMA system, arguing that SDRs constitute a convenient framework for their implementation. Therefore, in the final part of this contribution the concept of SDR-assisted broadband FH/MC DS-CDMA is presented and the range of re-configurable parameters are described with the aim of outlining a set of promising research topics. Let us now commence our detailed discourse concerning the proposed FH/MC DS-CDMA system.

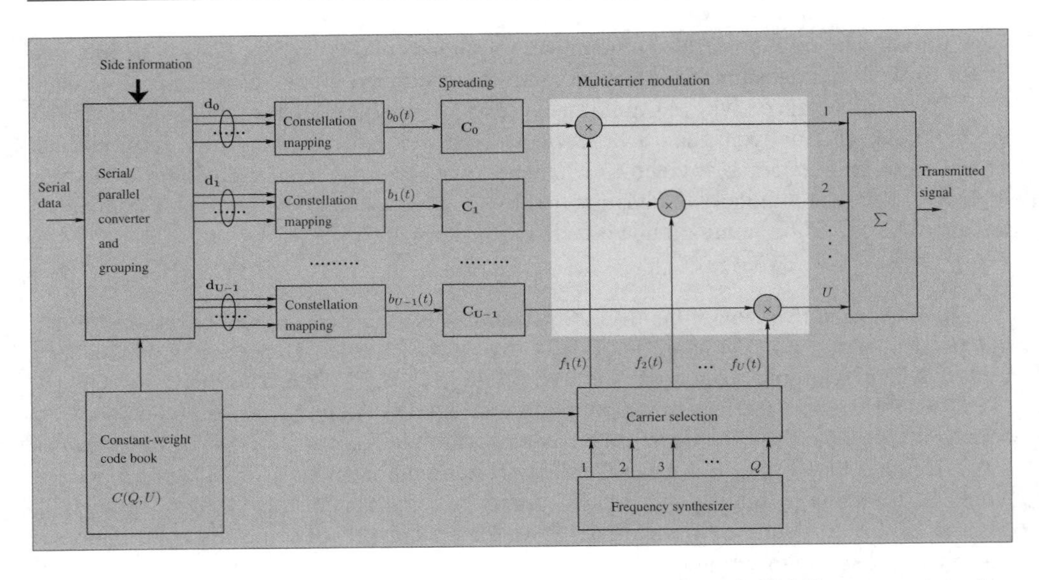

Figure 1.2: Transmitter diagram of the frequency-hopping multicarrier DS-CDMA system using adaptive transmission.

1.4 FH/MC DS-CDMA

The transmitter schematic of the proposed FH/MC DS-CDMA arrangement is depicted in Figure 1.2. Each subcarrier of a user is assigned a pseudo-noise (PN) spreading sequence. These PN sequences can be simultaneously assigned to a number of users, provided that only one user activates the same PN sequence on the same subcarrier. These PN sequences produce narrow-band DS-CDMA signals. In Figure 1.2, C(Q, U) represents a constant-weight code having U number of '1's and $(Q - U)$ number of '0's. Hence the weight of C(Q, U) is U. This code is read from a so-called constant-weight code book, which represents the frequency-hopping patterns. The constant-weight code C(Q, U) plays two different roles. Its first role is that its weight - namely U - determines the number of subcarriers invoked, while its second function is that the positions of the U number of binary '1's determine the selection of a set of U number of subcarrier frequencies from the Q outputs of the frequency synthesizer. Furthermore, in the transmitter 'side-information' reflecting the channel's instantaneous quality might be employed, in order to control its transmission and coding mode, so that the required target throughput and transmission integrity requirements are met.

As shown in Figure 1.2, the original bit stream having a bit duration of T_b is first serial-to-parallel (S-P) converted. Then, these parallel bit streams are grouped and mapped to the potentially time-variant modulation constellations of the U active subcarriers. Let us assume that the number of bits transmitted by a FH/MC DS-CDMA symbol is M, and let us denote the symbol duration of the FH/MC DS-CDMA signal by T_s. Then, if the system is designed for achieving a high processing gain and for mitigating the inter-symbol-interference (ISI) in a constant-rate transmission scheme, the symbol duration can be extended to a multiple of the bit duration, i.e., $T_s = MT_b$. By contrast, if the design aims to support multiple transmission rates or channel-quality matched variable information rates, then a constant bit

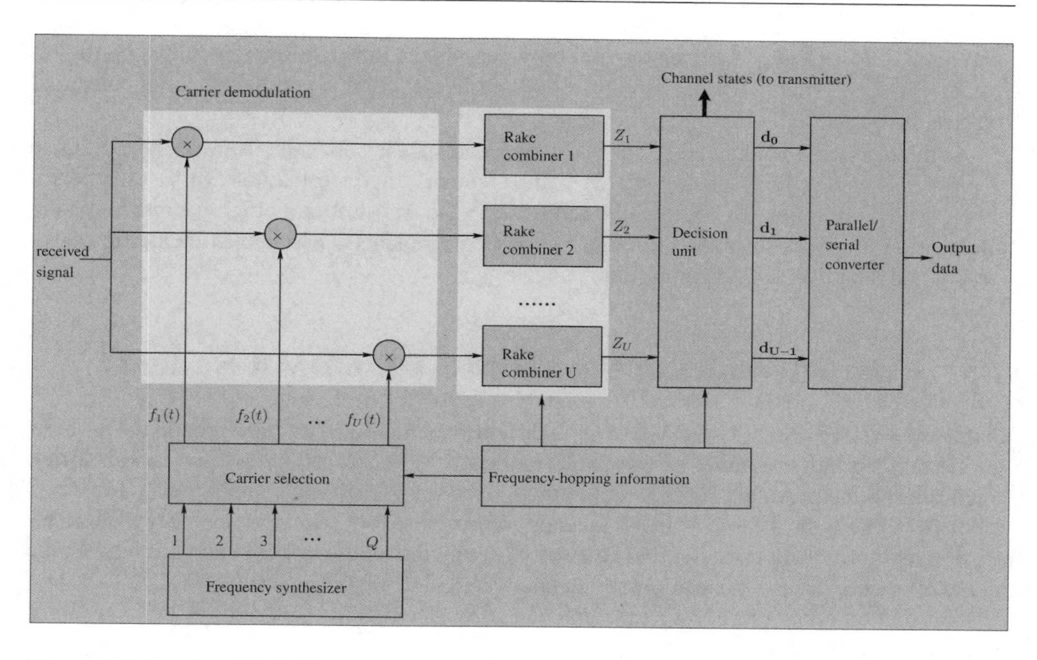

Figure 1.3: Receiver block diagram of the frequency-hopping multicarrier DS-CDMA system using a conventional RAKE receiver.

duration of $T_0 = T_s$ can be employed. Both multirate and variable rate transmissions can be implemented by employing a different number of subcarriers associated with different modulation constellations as well as different spreading gains. As seen in Figure 1.2, after the constellation mapping stage, each branch is DS spread using the assigned PN sequence, and then this spread signal is carrier modulated using one of the active subcarrier frequencies derived from the constant-weight code $C(Q, U)$. Finally, all U active branch signals are multiplexed, in order to form the transmitted signal.

In the FH/MC DS-CDMA receiver of Figure 1.3 the received signal associated with each active subcarrier is detected using, for example, a RAKE combiner. Alternatively, Multi-user Detection (MUD) can be invoked, in order to approach the single-user limit. In contrast to the transmitter side, where only U out of Q subcarriers are transmitted by a user, at the receiver different detector structures might be implemented based on the availability [68] or lack [74] of the FH pattern information. During the FH pattern acquisition stage, which usually happens at the beginning of transmission or during hand-over, tentatively all Q subcarriers can be demodulated. The transmitted information can be detected and the FH patterns can be acquired simultaneously by using blind joint detection algorithms exploiting the characteristics of the constant-weight codes [66, 67]. If, however, the receiver has the explicit knowledge of the FH patterns, then only U subcarriers have to be demodulated. However, if Fast Fourier Transform (FFT) techniques are employed for demodulation - as often is the case in multi-carrier CDMA [75] or OFDM [76] systems, then all Q subcarriers might be demodulated, where the inactive subcarriers only ignore output noise. In the decision unit of Figure 1.3, these noise-output-only branches can be eliminated by exploiting the knowledge of the FH patterns [68,74]. Hence, the decision unit only outputs the information transmitted by the active subcarriers. Finally, the decision unit's output information is parallel-to-serial converted to form the output data.

At the receiver, the channel states associated with all the subchannels might be estimated or predicted using pilot signals [71,73]. This channel state information can be utilized for coherent demodulation. It can also be fed back to the transmitter as highly protected side-information, in order to invoke a range of adaptive transmission schemes including power control and adaptive-rate transmission.

1.5 Characteristics of the FH/MC DS-CDMA Systems

In the proposed FH/MC DS-CDMA system the entire bandwidth of future broadband systems can be divided into a number of sub-bands and each sub-band can be assigned a subcarrier. According to the prevalent service requirements, the set of legitimate subcarriers can be distributed in line with the users' instantaneous information rate requirements. FH techniques are employed for each user, in order to evenly occupy the whole system bandwidth available and to efficiently utilize the available frequency resources. In this respect FH/MC DS-CDMA systems exhibit compatibility with the existing second- and third-generation CDMA systems and hence constitute a highly flexible air interface.

Broadband Wireless Mobile System – To elaborate a little further, our advocated FH/MC DS-CDMA system is a broadband wireless mobile system constituted by multiple narrow-band DS-CDMA sub-systems. Again, FH techniques are employed for each user, in order to evenly occupy the whole system bandwidth and to efficiently utilize the available frequency resources. The constant-weight code-based FH patterns used in the schematic of Figure 1.2 are invoked, in order to control the number of subcarriers invoked, which is kept constant during the FH procedure. In contrast to single-carrier broadband DS-CDMA systems, such as wideband CDMA (W-CDMA) [58] exhibiting a bandwidth in excess of 5 MHz - which inevitably results in extremely high-chip-rate spreading sequences and high-complexity - the proposed FH/MC DS-CDMA system does not have to use high chip-rate DS spreading sequences, since each subcarrier conveys a narrow-band DS-CDMA signal. In contrast to broadband OFDM systems [62] - which have to use a high number of subcarriers and usually result in a high peak-to-average power fluctuation - due to the associated DS spreading, the number of subcarriers of the advocated broadband wireless FH/MC DS-CDMA system may be significantly lower. This potentially mitigates the crest-factor problem. Additionally, with the aid of FH, the peak-to-average power fluctuation of the FH/MC DS-CDMA system might be further decreased. In other words, the FH/MC DS-CDMA system is capable of combining the best features of single-carrier DS-CDMA and OFDM systems, while avoiding many of their individual shortcomings. Finally, in comparison to the FH/MC DS-CDMA system, both broadband single-carrier DS-CDMA systems and broadband OFDM systems are less amenable to interworking with the existing second- and third-generation wireless communication systems. Let us now characterize some of the features of the proposed system in more depth.

Compatibility – The broadband FH/MC DS-CDMA system can be rolled out over the bands

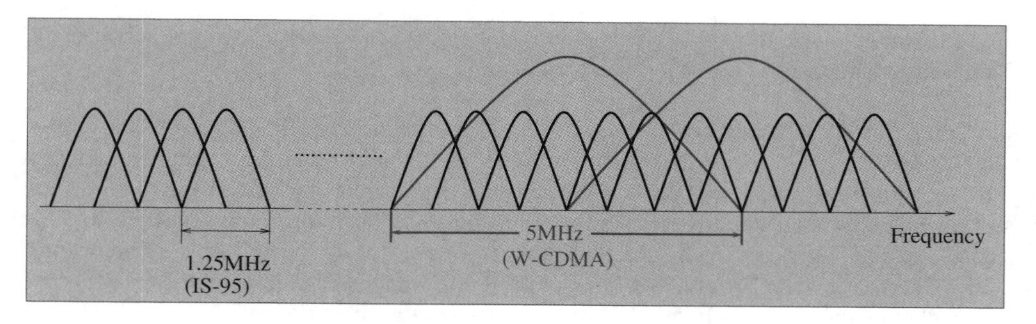

Figure 1.4: Spectrum of FH/MC DS-CDMA signal using subchannel bandwidth of 1.25 MHz and/or 5 MHz.

of the second and third generation mobile wireless systems and/or in the band licensed for future broadband wireless communication systems. In FH/MC DS-CDMA systems the subbands associated with different subcarriers are not required to be of equal bandwidth. Hence existing second and third-generation CDMA systems can be supported using one or more subcarriers. For example, Figure 1.4 shows the spectrum of a frequency hopping, orthogonal multicarrier DS-CDMA signal using a subchannel bandwidth of 1.25 MHz, which constitutes the bandwidth of a DS-CDMA signal in the IS-95 standard [57]. In Figure 1.4 we also show that seven subchannels, each having a bandwidth of 1.25 MHz can be replaced by one subchannel with a bandwidth of 5 MHz ($= 8 \times 1.25/2$ MHz). Hence, the narrow-band IS-95 CDMA system can be supported by a single subcarrier, while the UMTS and IMT-2000 W-CDMA systems might be supported using seven subchannels' bandwidth amalgamated into one W-CDMA channel. Moreover, with the aid of SDRs, FH/MC DS-CDMA is also capable of embracing other existing standards, such as the Time-Division Multiple-Access (TDMA) based Global System of Mobile communications known as GSM [64].

FH Strategy – In FH/MC DS-CDMA systems both slow FH and fast FH techniques can be invoked, depending on the system's design and the state-of-the-art. In slow FH several symbols are transmitted after each frequency hopping, while in fast FH several frequency hops take place in a symbol duration, i.e. each symbol is transmitted using several subcarriers. Moreover, from a networking point of view, random FH, uniform FH and adaptive FH [68] schemes can be utilized, in order to maximize the efficiency of the network. In the context of *random* FH [68], the subcarriers associated with each transmission of a user are determined by the set of pre-assigned FH patterns constituting a group of constant-weight codewords [74]. The active subcarriers are switched from a group of frequencies to another without the knowledge of the FH patterns of the other users. By contrast, for the FH/MC DS-CDMA system using *uniform* FH [68], the FH patterns of all users are determined jointly under the control of the base station (BS), so that each subcarrier is activated by a similar number of users. It can be shown that for the down-link (DL) uniform FH can be readily implemented, since the BS has the knowledge of the FH patterns of all users. However, for implementing uplink (UL) transmissions, the FH patterns to be used must be signalled by the BS to each mobile station (MS), in order to be able to implement uniform FH. Finally, if the near-instantaneous channel quality information is available at the transmitter, advanced adaptive FH can be invoked, where information is only transmitted over a group of subchannels exhibiting a satisfactory Signal to Interference Ratio (SIR).

Implementation of Multicarrier Modulation – The multicarrier modulation block in Figure 1.2 and the multicarrier demodulation block in Figure 1.3 can be implemented using FFT techniques, provided that each of the subchannels occupies the same bandwidth. Since not all of the subcarriers are activated at each transmission in the proposed FH/MC DS-CDMA system, the deactivated subcarriers can be set to 'zero' in the FFT or IFFT algorithm. However, if an unequal bandwidth is associated with the subchannels, multicarrier modulation/demodulation can only be implemented using less efficient conventional, rather than FFT-based carrier modulation/demodulation schemes.

Access Strategy – When a new user attempts to access the channel and commences his/her transmission, a range of different access strategies might be offered by the FH/MC DS-CDMA system, in order to minimize the interference inflicted by the new user to the already active users. Specifically, if there are subchannels, which are not occupied by any other users, or there are subchannels that exhibit a sufficiently high SIR, then the new user can access the network using these passive subchannels or the subchannels exhibiting a high SIR. However, if all the subchannels have been occupied and the SIR of each of the subchannels is insufficiently high, then the new user accesses the network by spreading its transmitted energy evenly across the subchannels. This access scheme imposes the minimum possible impact on the QoS of the users already actively communicating. However, the simplest strategy for a new user to access the network is by randomly selecting one or several subchannels.

Multirate and Variable Rate Services – In FH/MC DS-CDMA systems multirate and variable rate services can be implemented using a variety of approaches. Specifically, the existing techniques, such as employing a variable spreading gain, multiple spreading codes, a variable constellation size, variable-rate Forward Error Correction (FEC) coding, etc. can be invoked to provide multirate and variable rate services. Furthermore, since the proposed FH/MC DS-CDMA systems use constant-weight code-based FH patterns, multirate and variable rate services can also be supported by using constant-weight codes having different weights, i.e., by activating a different number of subcarriers. Note that the above-mentioned techniques can be implemented either separately or jointly in a system.

Diversity – The FH/MC DS-CDMA system includes frequency hopping, multicarrier modulation as well as direct-sequence spreading, hence a variety of diversity schemes and their combinations can be implemented. The possible diversity schemes include the following arrangements.

- If the chip-duration of the spreading sequences is lower than the maximum delay spread of the fading channels, then frequency diversity can be achieved on each of the subcarrier signals.
- Frequency diversity can also be achieved by transmitting the same signal using a number of different subcarriers.
- Time diversity can be achieved by using slow frequency hopping in conjunction with error control coding as well as interleaving.

- Time-frequency diversity can be achieved by using fast frequency hopping techniques, where the same symbol is transmitted using several time slots assigned to different frequencies.
- Spatial diversity can be achieved by using multiple transmit antennas, multiple receiver antennas and polarization.

Initial Synchronization – In our FH/MC DS-CDMA system initial synchronization can be implemented by first accomplishing DS code acquisition. The fixed-weight code book index of the FH pattern used can readily be acquired, once DS code acquisition achieved. During DS code acquisition the training signal supporting the initial synchronization, which is usually the carrier modulated signal without data modulation, can be transmitted using a group of subcarriers. These subcarrier signals can be combined at the receiver using, for example, equal gain combining (EGC) [77]. Hence, frequency diversity can be employed during the DS code acquisition stage of the FH/MC DS-CDMA system's operation, and consequently the initial synchronization performance can be significantly enhanced. Following the DS code acquisition phase, data transmission can be activated and the index of the FH pattern used can be signalled to the receiver using a given set of fixed subchannels. Alternatively, the index of the FH pattern can be acquired blindly from the received signal with the aid of a group of constant-weight codes having a given minimum distance [67].

Interference Resistance – The FH/MC DS-CDMA system concerned can mitigate the effects of inter-symbol interference encountered during high-speed transmissions, and it readily supports partial-band and multi-tone interference suppression. Moreover, the multi-user interference can be suppressed by using multi-user detection techniques [69], potentially approaching the single-user performance.

Advanced Technologies – The FH/MC DS-CDMA system can efficiently amalgamate the techniques of FH, OFDM and DS-CDMA. Simultaneously, a variety of performance enhancement techniques, such as multi-user detection [78], turbo coding [79], adaptive antennas [80], space-time coding and transmitter diversity [81], near-instantaneously adaptive modulation [70], etc, might be introduced.

Flexibility – The future generation broadband mobile wireless systems will aim to support a wide range of services and bit rates. The transmission rates may vary from voice and low-rate messages to very high-rate multimedia services requiring rates in excess of 100Mb/s [59]. The communications environments vary in terms of their grade of mobility, the cellular infrastructure, the required symmetric and asymmetric transmission capacity, and whether indoor, outdoor, urban or rural area propagation scenarios are encountered, etc. Hence flexible air interfaces are required, which are capable of maximizing the area spectrum efficiency expressed in terms of bits/s/Hz/km^2 in a variety of communication environments. Future systems are also expected to support various types of services based on ATM and IP, which require various Quality of Service (QoS). As argued before, FH/MC DS-CDMA systems exhibit a high grade of compatibility with existing systems. These systems also benefit from the employment of FH, MC and DS spreading based diversity-assisted adaptive modulation [76]. In short, FH/MC DS-CDMA systems constitute a high-flexibility air interface.

1.6 Adaptive Rate Transmission

1.6.1 Why Adaptive Rate Transmission?

There are a range of issues, which motivate the application of adaptive rate transmissions (ART) in the broadband mobile wireless communication systems of the near future. The explosive growth of the Internet and the continued dramatic increase in demand for all types of wireless services are fuelling the demand for increasing the user capacity, data rates and the variety of services supported. Typical low-data-rate applications include audio conferencing, voice mail, messaging, email, facsimile, and so on. Medium- to high-data-rate applications encompass file transfer, Internet access, high-speed packet- and circuit-based network access as well as high-quality video conferencing. Furthermore, the broadband wireless systems in the future are also expected to support real-time multimedia services, which provide concurrent video, audio and data services to support advanced interactive applications. Hence, in the future generation mobile wireless communication systems, a wide range of information rates must be provided, in order to support different services, which demand different data rates and different QoS. In short, an important motivation for using ART is to support a variety of services, which we refer to as service-motivated ART (S-ART). However, there is a range of other motivating factors, which are addressed below.

The performance of wireless systems is affected by a number of propagation phenomena: 1) path-loss variation versus distance; 2) random slow shadowing; 3) random multipath fading; 4) Inter-Symbol Interference (ISI), co-channel interference as well as multi-user interference; and 5) background noise. For example, mobile radio links typically exhibit severe multipath fading, which leads to serious degradation in the link's signal-to-noise ratio (SNR) and consequently to a higher bit error rate (BER). Fading compensation techniques such as an increased link budget margin or interleaving with channel coding are typically required to improve the link's performance. However, today's cellular systems are designed for the worst-case channel conditions, typically achieving adequate voice quality over 90-95% of the coverage area for voice users, where the signal to interference plus noise ratio (SINR) is above the designed target [70]. Consequently, the systems designed for the worst-case channel conditions result in poor exploitation of the available channel capacity a good percentage of time. Adapting the transmitter's certain parameters to the time-varying channel conditions leads to better exploitation of the channel capacity available. This ultimately increases the area spectral efficiency expressed in terms of b/s/Hz/km^2. Hence, the second reason for the application of ART is constituted by the time-varying nature of the channel, which we refer to as channel quality motivated ART (C-ART).

1.6.2 What Is Adaptive Rate Transmission?

Broadly speaking, ART in mobile wireless communications implies that the transmission rate at both the base stations and the mobile terminals can be adaptively adjusted according to the instantaneous operational conditions, including the communication environment and service requirements. With the expected employment of SDR-based wireless systems, the concept of ART might also be extended to adaptively controlling the multiple access schemes - including FDMA, TDMA, narrow-band CDMA, W-CDMA and OFDM - as well as the supporting network structures - such as Local Area Networks and Wide Area Networks. In this chapter only C-ART and S-ART are concerned in the context of the proposed FH/MC DS-CDMA scheme. Employing ART in response to different service requests indicates that the transmission rate of the base station and the mobile station can be adapted according to the requirements of the services concerned, as well as to meet their different QoS targets. By contrast, employing ART in response to the time-varying channel quality implies that for a given service supported, the transmission rate of the base station and that of the mobile station can be adaptively controlled in response to their near-instantaneous channel conditions. The main philosophy behind C-ART is the real-time balancing of the link budget through adaptive variation of the symbol rate, modulation constellation size and format, spreading factor, coding rate/scheme, etc, or in fact any combination of these parameters. Thus, by taking advantage of the time-varying nature of the wireless channel and interference conditions, the C-ART schemes can provide a significantly higher average spectral efficiency than their fixed-mode counterparts. This takes place without wasting power, without increasing the co-channel interference, or without increasing the BER. We achieve these desirable properties by transmitting at high speeds under favourable interference/channel conditions and by responding to degrading interference and/or channel conditions through a smooth reduction of the associated data throughput. Procedures that exploit the time-varying nature of the mobile channel are already in place for all the major cellular standards worldwide [70], including IS-95 CDMA, cdma2000 and UMTS W-CDMA [64], IS-136 TDMA, the General Packet Radio Service of GSM and in the Enhanced Data rates for Global Evolution (EDGE) schemes. The rudimentary portrayal of a range of existing and future ART schemes is given below. Note that a system may employ a combination of several ART schemes listed below, in order to achieve the desired data rate, BER or the highest possible area spectrum efficiency.

- **Multiple Spreading Codes** – In terms of S-ART, higher rate services can be supported in CDMA-based systems by assigning a number of codes. For example, in IS-95B each high-speed user can be assigned one to eight Walsh codes, each of which supports a data rate of 9.6kb/s. By contrast, multiple codes cannot be employed in the context of C-ART in order to compensate for channel fading, path-loss and shadowing, **unless they convey the same information and achieve certain diversity gain**. However, if the co-channel interference is low – which usually implies in CDMA-based systems that the number of simultaneously transmitting users is low - then multiple codes can be transmitted by an active user, in order to increase the user's transmission rate.

- **Variable Spreading Factors** – In the context of S-ART higher rate services are supported by using lower spreading factors without increasing the bandwidth required. For example, in UMTS W-CDMA [57] the spreading factors of 4/8/16/32/64/128/256 may be invoked to achieve the corresponding data rates of 1024/512/ 256/128/64/32/16 kb/s. In terms of C-ART, when the SINR experienced is increased, reduced spreading factors are assigned to users for the sake of achieving higher data rates.

- **Variable Rate FEC Codes** – In a S-ART scenario higher rate services can be supported by assigning less powerful, higher rate FEC codes associated with reduced redundancy. In a C-ART scenario, when the SINR improves, a higher-rate FEC code associated with reduced redundancy is assigned, in an effort to achieve a higher data rate.

- **Different FEC Schemes** – The range of coding schemes might entail different classes of FEC codes, code structures, encoding/decoding schemes, puncturing patterns, interleaving depths and patterns, and so on. In the context of S-ART, higher rate services can be supported by coding schemes having a higher coding rate. In the context of C-ART, usually an appropriate coding scheme is selected, in order to maximize the spectrum efficiency. The FEC schemes concerned may entail block or convolutional codes, block or convolutional constituent code-based turbo codes, trellis codes, turbo trellis codes, etc. The implementational complexity and error correction capability of these codes can be varied as a function of the coding rate, code constraint length, the number of turbo decoding iterations, the puncturing pattern, etc. A rule of thumb is that the coding rate is increased towards unity, as the channel quality improves, in order to increase the system's effective throughput.

- **Variable Constellation Size** – In S-ART schemes higher rate services can be supported by transmitting symbols having higher constellation sizes. For example, an adaptive modem may employ BPSK, QPSK, 16QAM and 64QAM constellations [76], which corresponds to 1, 2, 4 and 6 bits per symbol. The highest data rate provided by the 64QAM constellation is a factor six higher than that provided by employing the BPSK constellation. In C-ART scenarios, when the SINR increases, a higher number of bits per symbol associated with a higher order constellation is transmitted for increasing the system's throughput.

- **Multiple Time Slots** – In a S-ART scenario higher rate services can also be supported by assigning a corresponding number of time slots. A multiple time slot based adaptive rate scheme is used in GPRS-136 (1-3 time slots/20 ms) and in Enhanced GPRS (EGPRS) (1-8 time slots/4.615GSM frame) in order to achieve increased data rates. In the context of C-ART, multiple time slots associated with interleaving or frequency hopping can be implemented for achieving time diversity gain. Hence, C-ART can be supported by assigning a high number of time slots for the compensation of severe channel fading at the cost of tolerating a low data throughput. By contrast, assigning a low number of time slots over benign non-fading channels allows us to achieve a high throughput.

- **Multiple Bands** – In the context of S-ART higher rate services can also be supported by assigning a higher number of frequency bands. For example, in UMTS W-CDMA [57] two 5 MHz bands may be assigned to a user, in order to support the highest data rate of 2 Mb/s (= $2\times$ 1024kb/s), which is obtained by using a spreading factor of 4 on each sub-band signal. In the context of C-ART associated with multiple bands, frequency-hopping associated with time-variant redundancy and/or variable rate FEC coding schemes, or frequency diversity techniques might be invoked, in order to increase the spectrum efficiency of the system. For example, in C-ART schemes associated with double-band assisted frequency diversity, if the channel quality is low, the same signal can be transmitted in two frequency bands for the sake of maintaining a diversity order of two. However, if the channel quality is sufficiently high, two independent streams of information can be transmitted in these bands and consequently the throughput can be doubled.

- **Multiple Transmit Antennas** – Employing multiple transmit antennas based on space-time coding [81] is a novel method of communicating over wireless channels, which was also adapted for use in the 3rd-generation mobile wireless systems. ART can also be implemented using multiple transmit antennas associated with different space-time codes. In S-ART schemes higher rate services can be supported by a higher number of transmit antennas associated with appropriate space-time codes. In terms of C-ART schemes, multiple transmit antennas can be invoked for achieving a high diversity gain. Therefore, when the channel quality expressed in terms of the SINR is sufficiently high, the diversity gain can be decreased. Consequently, two or more symbols can be transmitted in each signalling interval and each stream is transmitted by only a fraction of the transmit antennas associated with the appropriate space-time codes. Hence the throughput is increased. However, when the channel quality is poor, all the transmit antennas can be invoked for transmitting one stream of data, hence achieving the highest possible transmit diversity gain of the system, while decreasing the throughput.

Above we have summarized the philosophy of a number of ART schemes, which can be employed in wireless communication systems. An S-ART scheme requires a certain level of transmitted power, in order to achieve the required QoS. Specifically, a relatively high transmitted power is necessary to support high-data rate services, and a relatively low transmitted power is required for offering low-data rate services. Hence, a side-effect of a S-ART scheme supporting high data rate services is the concomitant reduction in the number of users supported due to the increased interference or/and increased bandwidth. By contrast, a cardinal requirement of a C-ART scheme is the accurate channel quality estimation or prediction at the receiver as well as the provision of a reliable side-information feedback between the channel quality estimator or predictor of the receiver and the remote transmitter [71, 73], where the modulation/coding mode requested by the receiver is activated. The parameters capable of reflecting the channel quality may include BER, SINR, transmission frame error rate, received power, path loss, Automatic Repeat Request (ARQ) status, etc. A C-ART scheme typically operates under the constraint of constant transmit power and constant bandwidth. Hence, without wasting power and bandwidth, or without increasing the co-channel interference compromising the BER performance, C-ART schemes are capable of providing a significantly increased average spectral efficiency by taking advantage of the time-varying nature of the wireless channel, when compared to fixed-mode transmissions.

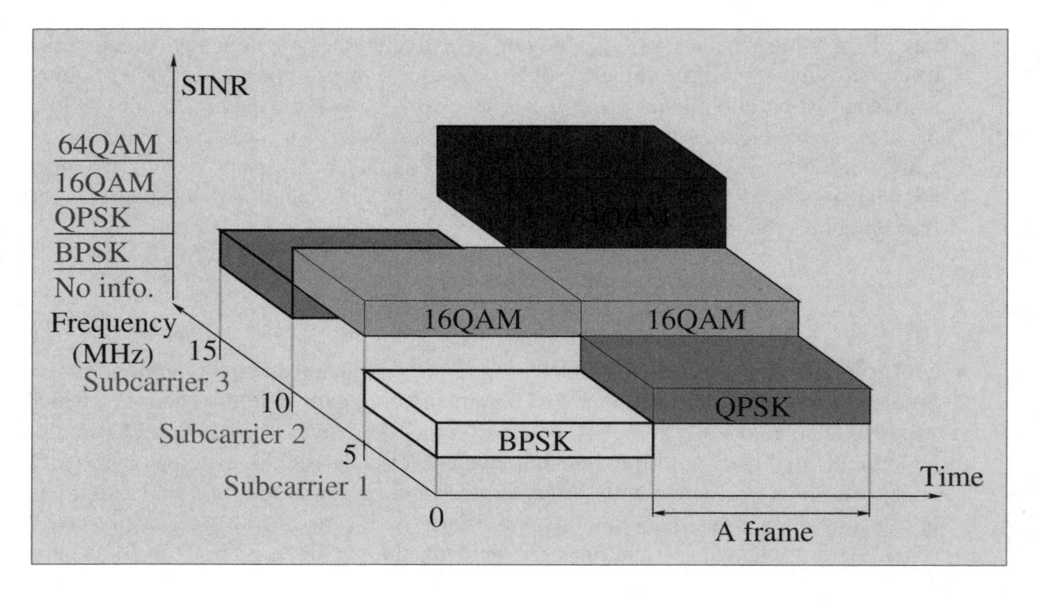

Figure 1.5: A frame structure of burst-by-burst adaptive modulation in multicarrier DS-CDMA systems.

1.6.3 Adaptive Rate Transmission in FH/MC DS-CDMA Systems

Above we have discussed a range of ART schemes, which were controlled by the prevalent service requirements and the channel quality. Future broadband mobile wireless systems are expected to provide a wide range of services characterized by highly different data rates, while achieving the highest possible spectrum efficiency. The proposed FH/MC DS-CDMA-based broadband mobile wireless system constitutes an attractive candidate system, since it is capable of efficiently utilizing the available frequency resources, as we discussed previously, and simultaneously achieving a high grade of flexibility by employing ART techniques. More explicitly, the FH/MC DS-CDMA system can provide a wide range of data rates by combining the various ART schemes discussed above. At the same time, for any given service, the FH/MC DS-CDMA system may also invoke a range of adaptation schemes, in order to achieve the highest possible spectrum efficiency in various propagation environments, such as indoor, outdoor, urban, rural scenarios at low to high speeds. Again, the system is expected to support different services at a variety of QoS, including voice mail, messaging, email, file transfer, Internet access, high-speed packet- and circuit-based network access, real-time multimedia services, and so on. As an example, a channel-quality motivated burst-by-burst ART assisted FH/MC DS-CDMA system is shown in Figure 1.5, where we assume that the number of subcarriers is three, the bandwidth of each subchannel is 5 MHz and the channel quality metric is the SINR. In response to the SINR experienced, the transmitter may transmit a frame of symbols selected from the set of BPSK, QPSK, 16QAM or 64QAM constellations, or simply curtail transmitting information, if the SINR is too low.

1.7 Software-Defined Radio-Assisted FH/MC DS-CDMA

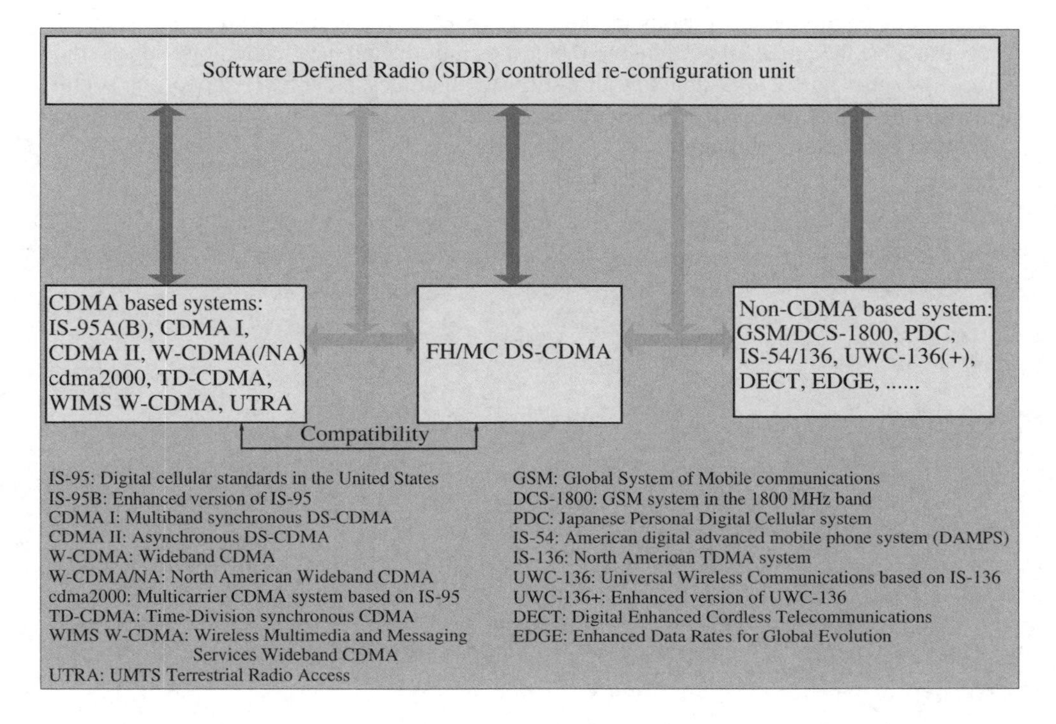

Figure 1.6: Software defined radio assisted FH/MC DS-CDMA and its re-configuration modes.

The range of existing wireless communication systems is shown in Figure 1.6. Different legacy systems will continue to coexist, unless ITU - by some miracle - succeeds in harmonizing all the forthcoming standards under a joint framework, while at the same time ensuring compatibility with the existing standards. In the absence of the 'perfect' standard, the only solution is employing multiband, multimode, multi-standard transceivers based on the concept of Software Defined Radios (SDR) [1–3].

In SDRs the digitization of the received signal is performed at some stage downstream from the antenna, typically after wideband filtering, low-noise amplification, and down-conversion to a lower frequency. The reverse processes are invoked by the transmitter. In contrast to most wireless communication systems which employ Digital Signal Processing (DSP) only at baseband, SDRs are expected to implement the DSP functions at an intermediate frequency (IF) band. An SDR defines all aspects of the air interface, including RF channel access and waveform synthesis in software. In SDRs wide-band analog-to-digital and digital-to-analog converters (ADC and DAC) transform each RF service band from digital and analogue forms at IF. The wideband digitized receiver stream of bandwidth W_s accommodates all subscriber channels, each of which has a bandwidth of $W_c(W_c << W_s)$. Thanks to using programmable DSP chips at both the intermediate frequency as well as at baseband, SDRs are sufficient to efficiently support multiband and multi-standard communications. A SDR employs one or more reconfigurable processors embedded in a real-time multiprocessing fabric, permitting flexible reprogramming and reconfiguration using software downloaded for example with the aid of a signalling channel from the base-station. Hence, SDRs offer an elegant solution to accommodating various modulation formats, coding and radio access schemes. They also have the potential of reducing the cost of introducing new technology superseding legacy systems, and are amenable to future software upgrades, potentially supporting sophisticated future signal processing functions such as array processing, multi-user detection and as yet unknown coding techniques.

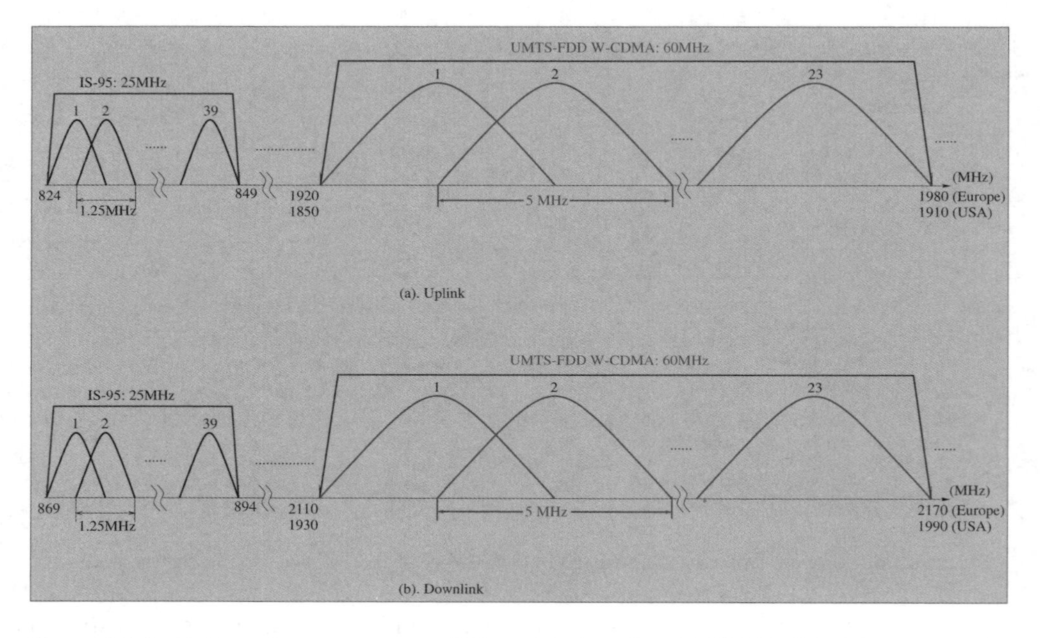

Figure 1.7: Exhibition of spectrum compatibility of the broadband FH/MC DS-CDMA system with IS-95 and UMTS-FDD wideband CDMA systems.

FH/MC DS-CDMA systems will be designed, in order to provide the maximum grade of compatibility with the existing CDMA based systems, such as IS-95 and W-CDMA based systems. For example, the frequency bands of the IS-95 CDMA system in North America are 824-849 MHz (uplink) and 869-894 MHz (downlink), respectively. The corresponding frequency bands for the UMTS-FDD wideband CDMA system are 1850-1910 MHz (uplink) and 1930-1990 MHz (downlink) in North America, while 1920-1980 MHz (uplink) and 2110-2170 MHz (downlink) in Europe. In order to ensure compatibility with these systems, the proposed FH/MC DS-CDMA system's spectrum can be assigned according to Figure 1.7. Particularly, in the frequency band of IS-95, 39 orthogonal subcarriers are assigned, each having a bandwidth of 1.25 MHz, while in the frequency band of UMTS-FDD W-CDMA, 23 orthogonal subcarriers are allocated, each with a bandwidth of 5 MHz. The multicarrier modulation used in the FH/MC DS-CDMA system obeying the above spectrum allocation can be readily implemented using two IFFT sub-systems at the transmitter and two FFT sub-systems at the receiver, where a pair of IFFT-FFT sub-systems carries out modulation/demodulation in the IS-95 band, while another pair of IFFT-FFT sub-systems transmits and receives in the

UMTS-FDD band. If the chip rate for the 1.25 MHz bandwidth subcarriers is 1.2288M chips per second (cps) and for the 5 MHz bandwidth subcarriers is 3.84 Mcps, then the FH/MC DS-CDMA system will be compatible with both the IS-95 and the UMTS-FDD W-CDMA systems.

However, the terminals of future broadband mobile wireless systems are expected not only to support multimode and multi-standard communications, but also to possess the highest possible grade of flexibility, while achieving a high spectrum efficiency. Hence, these systems are expected to be capable of software re-configuration both between different standards as well as within a specific standard. In contrast to re-configuration between different standards invoked, mainly for the sake of compatibility, the objective of re-configuration within a specific standard is to support a variety of services at the highest possible spectrum efficiency. The SDR-assisted broadband FH/MC DS-CDMA system is operated under the control of the software re-configuration unit shown in Figure 1.6. The set of reconfigured parameters of the broadband FH/MC DS-CDMA system may include:

- **Services**: Data rate, QoS, real-time or non-real-time transmission, encryption/decryption schemes and parameters;
- **Error Control**: CRC, FEC codes, coding/decoding schemes, coding rate, number of turbo decoding steps, interleaving depth and pattern;
- **Modulation**: Modulation schemes, signal constellation, partial response filtering;
- **PN Sequence**: Spreading sequences (codes), chip rate, chip waveform, spreading factor, PN acquisition and tracking schemes;
- **Frequency Hopping**: FH schemes (slow, fast, random, uniform and adaptive), FH patterns, weight of constant-weight codes;
- **Detection**: Detection schemes (coherent or non-coherent, etc.) and algorithms (maximum likelihood sequence detection (MLSD) or minimum mean square estimation (MMSE), etc.), parameters associated with space/time as well as frequency diversity, beam-forming, diversity combining schemes, equalization schemes as well as the related parameters (such as the number of turbo equalization iterations, etc.) and channel quality estimation algorithms, parameters;
- **Others**: Subchannel bandwidth, power control parameters.

In the context of different standards - in addition to the parameters listed above - the transceiver parameters that must be re-configurable have to include the clock rate, the radio frequency (RF) bands and air interface modes.

1.8 Chapter Summary and Conclusion

We have presented a flexible broadband mobile wireless communication system based on FH/MC DS-CDMA and reviewed a variety of existing as well as a range of forthcoming techniques, which might be required for developing broadband mobile wireless systems exhibiting a high flexibility and high communications efficiency. We argued that this broadband

FH/MC DS-CDMA system exhibits a high grade of compatibility with the existing CDMA based systems, since it is capable of incorporating a wide range of techniques developed for the second- and third-generation mobile systems. At the time of writing, research is well under way towards the SDR-based implementation of a range of existing systems. It is expected that these efforts will soon encompass a generic scheme not too different from the FH/MC DS-CDMA scheme advocated here. The rest of this book will provide an overview of various communications techniques, which may be accommodated by the system framework introduced in this chapter.

Part I

Multi-User Detection for Adaptive Single-Carrier CDMA

List of Symbols

General notation

- Capital letters in boldface represent matrices, for instance $\mathbf{X}$.
- Column vectors are presented in lowercase letters, for instance $\mathbf{x}$.
- The vector, $\mathbf{x}^{(k)}$, represents the vector $\mathbf{x}$ of the user k.
- The i-th element of the vector $\mathbf{x}$, is represented as $\mathbf{x}_i$.
- The matrix $\mathbf{R}$ is used to represent covariance matrices. For example, $\mathbf{R}_x$ represents the covariance matrix of the vector $\mathbf{x}$ and $\mathbf{R}_x = E[\mathbf{x}\mathbf{x}^H]$.
- The asterisk (*) superscript is used to indicate complex conjugation. Therefore $\mathbf{X}^*$ is used to denote the complex conjugate matrix of $\mathbf{X}$ and a^* represents the complex conjugate of the variable a.
- The notation $\mathbf{X}^T$ implies the transpose matrix of $\mathbf{X}$.
- The notation $\mathbf{X}^H$ is used to represent the complex conjugate and transpose matrix of $\mathbf{X}$, i.e. $\mathbf{X}^H = \mathbf{X}^{*T}$.
- The notation $\mathbf{X}^{-1}$ corresponds to the inverse matrix of $\mathbf{X}$.
- The notation diag($\mathbf{X}$) is used to represent a diagonal matrix containing only the diagonal elements of the matrix $\mathbf{X}$ and $\overline{\text{diag}}(\mathbf{X})$ represents a matrix with zero diagonal elements and containing all the elements of $\mathbf{X}$ except the diagonal elements.
- The notation $[\mathbf{X}]_{ij}$ identifies the element in the i-th row and j-th column of the matrix $\mathbf{X}$ and the notation x_i is used to represent the i-th element of the column vector $\mathbf{x}$.

Special symbols

γ_o	Signal to interference and noise ratio (SINR) at the output of a receiver
ω_c	Carrier frequency
σ^2	Represents the variance of a variable, usually of noise
$\mathbf{0}$	Matrix where all the elements in it are zero.
$\mathbf{0}_{[X \times Y]}$	An all-zero matrix with the dimensions of $X \times Y$.
$\mathbf{A}$	System matrix in a JD-CDMA system
A	Amplitude of a signal
$\mathbf{b}$	Vector representing the discretized combined impulse response of one user in a JD-CDMA system, where $\mathbf{b} = \mathbf{c} * \mathbf{h}$. The term $\mathbf{b}^{(k)}$ represents the vector for user k.
$\mathbf{c}$	Vector representing the spreading sequence of one user in a JD-CDMA system. The term $\mathbf{c}^{(k)}$ represents the vector for user k.
$c_k(t)$	Continous spreading signal of user k.
$\mathbf{D}$	A diagonal matrix obtained after Cholesky decomposition of the cross-correlation matrix in a JD receiver. The elements on the main diagonal are real (non-complex) values.
$\mathbf{d}$	Vector representing the transmitted data symbols of one user in a JD-CDMA system. The term $\mathbf{d}^{(k)}$ represents the vector for user k.
$d_k(t)$	Continous data signal of user k.
$E[x]$	The expectation operator, where this function returns the expectation of the variable x.
E_b	Bit energy
E_b/N_0	Ratio of bit energy to noise power spectral density.
E_s	Symbol energy
E_s/N_0	Ratio of symbol energy to noise power spectral density.
f_d	Doppler frequency
$\mathbf{h}$	Vector representing the discretized channel impulse response vector of one user in a JD-CDMA system. The term $\mathbf{h}^{(k)}$ represents the vector for user k.
$\mathbf{I}$	Identity matrix which is a square matrix where all the elements on the main diagonal have values of 1 and all other elements have values of 0.
$\mathbf{I}_m$	Identity matrix with the dimensions of $(m \times m)$.

K	Number of users in a CDMA system
L	The number of paths or taps in the channel impulse response.
N	Number of information symbols in one transmission burst
$N_0/2$	Double-sided power spectral density of white noise.
$\mathbf{n}$	Vector representing the noise samples corrupting the received signal.
$O(x)$	Represents the order of complexity for a particular operation or algorithm, and the complexity is in the order of x.
P	Power of a signal
Q	Length of a spreading sequence in chips
$\mathbf{R}$	Covariance matrix, where $\mathbf{R}_n$ indicates the covariance matrix of the vector $\mathbf{n}$
R_{jk}	The correlation between two sequences, represented by j and k.
T_c	Duration of one chip
t_i	Switching threshold in the adaptive-rate schemes, where i represents the mode index.
$\mathbf{U}$	An upper triangular matrix obtained after Cholesky decomposition of the cross-correlation matrix in a JD receiver. All the elements on the main diagonal have values of 1.
$\text{Var}[x]$	The variance of the variable x.
W	The length of the channel impulse response in number of chips.
$\mathbf{y}$	Vector representing the CDMA composite received signal at the receiver.
(n, k, t)	Parameters used to specify a block FEC code, where the encoder takes in k input symbols and codes them into n output symbols. The decoder has the ability to correct up to t errors.
$(x \oslash y)$	A mathematical function that returns the remainder that is left after dividing x by y.
$(x \oplus y)$	A mathematical function that returns the modulo-2 addition of x and y.
$\lceil x \rceil$	A function that returns the nearest integer larger than or equal to x.
$\lfloor x \rfloor$	A function that returns the nearest integer smaller than or equal to x.

Chapter 2

CDMA Overview[1]

2.1 Introduction to Code Division Multiple Access

Spread spectrum communications have long been used in military systems Omura [82]. The distinguishing feature of spread spectrum communications is that the transmission bandwidth is significantly higher than that required by the information rate, resulting in the "spread" terminology. Typically, a pseudo-random code is used to 'spread' the information signal to the allocated frequency bandwidth.

Multiple access systems designed for mobile communications have traditionally employed Time Division Multiple Access (TDMA) [64] and Frequency Division Multiple Access (FDMA) [64] techniques. In FDMA, the allocated spectrum is divided into frequency slots, while in TDMA the time-domain transmission frame is periodically divided into time slots. Thus each user accesses its own slot in either the time-domain or frequency-domain and hence they are thus orthogonal to each other either in time or frequency, respectively. Viterbi Qualcomm Inc. was one of the first to propose using Code Division Multiple Access (CDMA) [83] for civilian mobile communications, which eventually led to the North American IS-95 standard [84, 85]. CDMA is a multiple access technique that allows multiple users to transmit independent information within the same bandwidth simultaneously. Each user is assigned a pseudo-random code that is either orthogonal to the codes of all the other users or the code possesses appropriate cross-correlation properties that minimize the multiple access interference (MAI). This code is superimposed on the information signal, making the signal appear noise-like to all the other users. Only the intended receiver has a replica of the same code and uses it to extract the information signal. This then allows the sharing of the same spectrum by multiple users without causing excessive MAI. It also ensures message privacy, since only the intended user is able to "decode" the signal. This code is also known as a spreading code, since it spreads the bandwidth of the original data signal into a much higher bandwidth before transmission. Therefore the term Spread Spectrum Multiple Access

[1] *Single- and Multi-Carrier CDMA Multi-User Detection, Space-Time Spreading, Synchronization, Networking and Standards.* L.Hanzo, L-L.Yang, E-L.Kuan and K.Yen,
 ISBN 0-470-86309-9

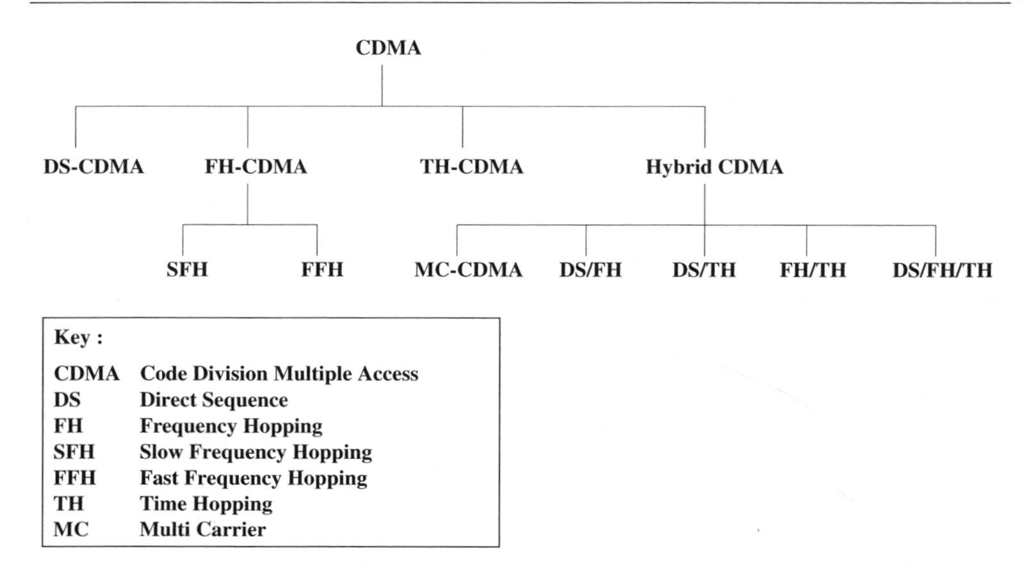

Figure 2.1: Classification of CDMA schemes according to the modulation method used to obtain the spread spectrum signal

(SSMA) is also used interchangeably with the term CDMA. These aspects are explained in further detail later in this chapter.

Figure 2.1 shows a classification tree of the various types of CDMA techniques. These techniques differ from each other in the way that the information signal is transformed to produce a high-bandwidth spread signal. In Direct Sequence CDMA (DS-CDMA), a pseudo-random sequence having a higher bandwidth than the information signal is used to modulate the information signal directly. The resultant signal has a significantly higher bandwidth than the original signal. In Frequency Hopping CDMA (FH-CDMA), the transmission bandwidth is divided into frequency sub-bands, where the bandwidth of each sub-band is equal to the bandwidth of the information signal. A pseudo-random code is then used to select the sub-band, in which the information signal is transmitted and this sub-band changes periodically according to the code. There are two sub-categories of FH-CDMA, namely Slow Frequency Hopping (SFH) [86] and Fast Frequency Hopping (FFH) [86]. In FFH, the frequency sub-band used to transmit one bit of information is changed multiple times within the bit-duration, while in SFH, the sub-band is changed only after multiple bits have been transmitted. Time Hopping CDMA (TH-CDMA) [86] transmits the information signal in short bursts. These short bursts render the transmission bandwidth high. The start of each burst for one user is determined by a pseudo-random code. Hybrid CDMA [86] encompasses a group of techniques that combine two or more of the other techniques already described. One of these hybrid techniques is known as Multi-Carrier CDMA (MC-CDMA) [86]. There are many variations of this technique, which were summarized, for example, by Prasad [75, 86], but their common characteristic is that a spreading code is used to spread a user's signal either in the time or frequency domain, and that more than one carrier frequency is used for transmission. This discusses DS-CDMA only. For more detailed information on the other multiple access schemes, the reader is referred to the excellent monographs by Viterbi [83], Prasad [86], Glisic and Lepännen [87], as well as Glisic and Vucetic [88].

CDMA as a multiple access scheme has advantages and disadvantages over the more traditionally employed FDMA and TDMA. Baier [89] has provided an insightful comparison of CDMA, FDMA and TDMA. If all three multiple access schemes are compared under the hypothetical conditions of a Gaussian channel and all the users are perfectly orthogonal to each other, then all three schemes are equivalent with respect to Shannon's channel capacity [90]. However, mobile communications are usually conducted over radio channels that are more hostile than the ideal Gaussian channel and these radio channels lead to performance differences among the three multiple access schemes. In comparison to FDMA and TDMA systems, CDMA systems suffer more severe multiple access interference (MAI) due to the often imperfect cross-correlation properties of the spreading codes used. The user signals in FDMA and TDMA are inherently orthogonal to each other due to the orthogonal frequency and time slots used. However, FDMA and TDMA are primarily dependent on the availability of bandwidth, which is a costly resource, and the capacities of both these schemes are therefore bandwidth-limited. CDMA, on the other hand, is only interference-limited. In second-generation CDMA systems, specifically, in the Qualcomm IS-95 CDMA standard [11, 83], the multiple access interference (MAI) is treated as noise. We note, however, that upon exploiting the extra knowledge of the users' spreading sequences and their associated channel impulse responses (CIR), the MAI can be substantially reduced, resulting in corresponding user capacity gains.

One of the effects of the mobile radio channel is dispersive multipath propagation. This is due to the reflections and scattering of the signal by objects present in the propagation environment. The multipath channel destroys the orthogonality between spreading codes, thus further increasing the cross-correlation between the users. The multipath spreading of the signal results in inter-symbol interference (ISI), which severely degrades the performance of the system. However, an advantage of CDMA schemes is that the multipath propagation can be exploited to give multipath diversity gains. With the aid of multipath receivers, namely RAKE receivers [91, 92], the correlation properties of the spreading codes can be exploited and the energy from the different paths can be coherently combined, in order to provide a multipath diversity gain. A detrimental effect of the mobile channel is its time-varying property, more commonly known as fading. This time-varying property is caused by the movements of mobile stations and other objects in the propagation environment. This results in the received signal having a large variance about its mean, again resulting in a degraded performance. These effects will be explained in more detail in our further discussions. Other advantages of CDMA include the privilege of privacy through the use of unique spreading codes for different users and diversity techniques to combat multipath fading [93, 94]. Some of these diversity techniques include multipath diversity combining [95], cell-site antenna diversity combining [95] and "soft-handoff" combining, where signals from two or more cell sites are combined [95]. Another advantage of CDMA is that it is well suited to support variable bit rate transmission and data rates can be chosen individually for each user, as proposed for example by Baier [96]; Ottosson and Svensson [97]; Ramakrishna and Holtzman [98]. Further discussions on variable bit rate and adaptive-rate transmissions are provided in Chapter 4.

A disadvantage of CDMA is that multiple access interference (MAI) is the main limiting factor imposed by the cross-correlation between spreading codes. Additionally, even when the spreading codes are designed to be perfectly orthogonal, the wideband mobile chan-

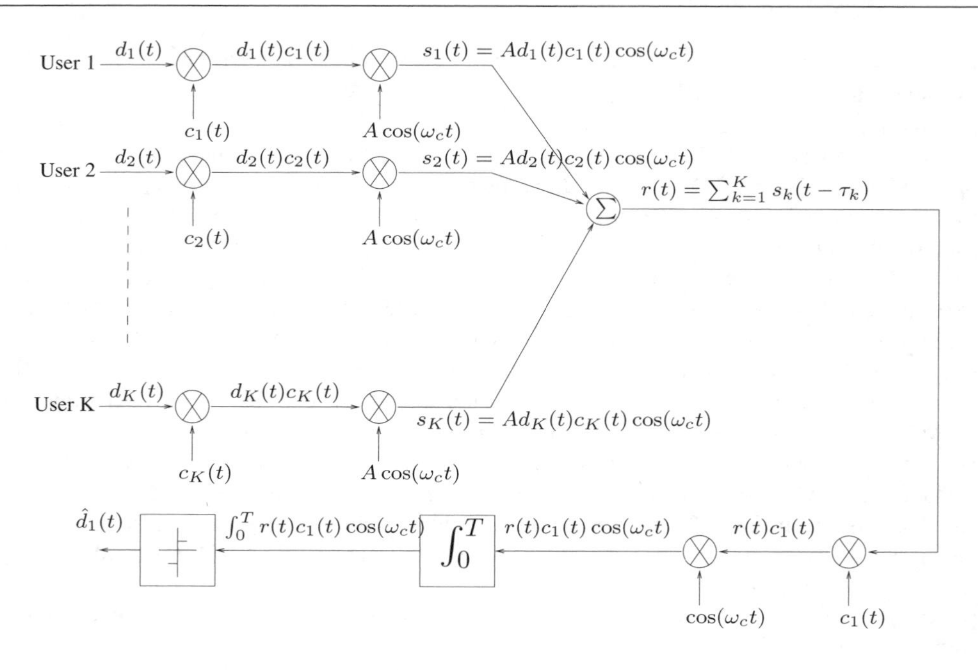

Figure 2.2: Block diagram of a simple asynchronous DS-CDMA system in a noiseless channel. The signals from all the K users arrive at the receiver with different propagation delays, τ_k. Only the receiver for user 1 is shown. The despreading and demodulation processes shown in the figure assume perfect synchronization.

nel destroys this orthogonality between the codes. Traditionally, powerful error-correcting codes [99] are used to overcome this problem. Other interference reduction techniques used in CDMA are voice activity-based bit rate control [93,94] and cell-site sectorization [93,94]. Current research on CDMA receivers includes multi-user detection [78] that exploits the available information about the spreading sequences and CIR estimates, in order to reduce the MAI. An introduction to the basic principles of these receivers was given, for example by Prasad [86] and Verdú [78]. Multi-user receivers are dealt with in Chapters 3 to 5. Furthermore, due to the time-varying nature of the mobile radio channel, the powers of the received signals vary widely among the users. This results in the phenomenon commonly termed as the "near–far" effect, where the stronger signals "swamp" the weaker signals. Therefore, stringent power control [93] is required to ensure that the signals from all the different users arrive at the receiver with relatively similar strengths.

Following the above brief introduction to CDMA, let us now consider the basics of this multiple access technique and the foundations of conventional DS-CDMA receivers in a little more depth.

2.2 DS-CDMA Transmission Model

This section explains the basic principles of operation in a DS-CDMA scheme. For a more detailed discussion, the reader is referred to the monographs by Viterbi [83], Prasad [86], Glisic and Leppännen [87] as well as Glisic and Vucetic [88]. The block diagram of a simple asynchronous CDMA modem in a noiseless channel is shown in Figure 2.2. This system supports K users, each transmitting its own information. The users are identified by $k = 1, 2, \ldots, K$. The modulation scheme used is Binary Phase Shift Keying (BPSK). Each user's data signal is denoted by $d_k(t)$ and each user is assigned a unique pseudo-random code also known as a spreading code denoted by $c_k(t)$. There are two classes of spreading codes in general, binary and complex. For simplicity, the following discussion considers only binary codes. Each spreading code consists of Q pulses, commonly known as chips. Spreading codes are discussed in more detail in Section 2.6.

In this discussion, the wanted signal is the signal of user $k = 1$ and all the other $(K - 1)$ signals are considered to be interfering signals.

2.3 DS-CDMA Transmitter

At the transmitter of user k, each data bit of user k is first multiplied by the spreading code $c_k(t)$. This causes the spectrum of the information signal to be spread across the allocated bandwidth. Next, the signal is modulated onto its carrier before it is transmitted. The transmitted signal is given by:

$$s_k(t) = Ad_k(t)c_k(t)\cos(\omega_c t), \tag{2.1}$$

where ω_c is the carrier frequency in rad/sec and A is the amplitude of the carrier signal. Let us now consider the recovery of the DS-CDMA signal at the receiver.

2.4 DS-CDMA Receiver

At the receiver, the composite of all the K user signals is received, consisting of the transmitted signal from user 1 and the other $(K - 1)$ interfering signals. Ignoring the noise, the received signal is given by:

$$r(t) = \sum_{k=1}^{K} s_k(t - \tau_k), \tag{2.2}$$

where τ_k is the propagation delay from the transmitter to the receiver of the k-th user. Let us now concentrate on the recovery of the information signal.

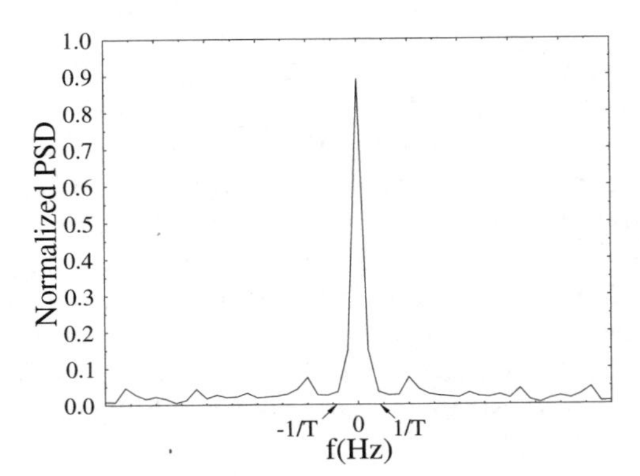

Figure 2.3: The power spectral density (PSD) plot of a despread signal in the presence of interfering users, normalized to the total power in the signal. The received signal consists of the superposition of the signals from more than one user. The PSD values are normalized to the PSD value at frequency 0. The despread signal has a bandwidth of approximately $1/T$ compared to the other signals, which remain spread and have bandwidths of $1/T_c$, where T_c is the duration of one chip in the spreading sequence. The ratio of T/T_c is 128.

2.4.1 Recovery of the Information Signal

Let us first consider the simplest case, where $K = 1$, yielding:

$$\begin{aligned} r(t) &= s_1(t - \tau_1) && (2.3) \\ &= Ad_1(t - \tau_1)c_1(t - \tau_1)\cos(\omega_c t + \theta'), && (2.4) \end{aligned}$$

where the propagation delay-induced carrier shift is given by $\theta' = -\omega_c \tau_1$.

In order to recover the original information of user 1, the received signal is despread by multiplying the received signal with a synchronized replica of the spreading code of user 1, as follows:

$$\begin{aligned} \hat{s}_1(t) &= r(t)c_1(t - \tau') && (2.5) \\ &= Ad_1(t - \tau_1)c_1(t - \tau_1)c_1(t - \tau')\cos(\omega_c t + \theta'), && (2.6) \end{aligned}$$

where τ' is the delay estimate. This despreads the spread signal of user 1 back to its original bandwidth, as illustrated in Figure 2.3.

In order to demodulate the signal, it is then multiplied by the carrier and passed through a correlator followed by a thresholding device, as demonstrated in Figure 2.2. At the correlator output, we then have:

$$z_1 = \int_{t_1}^{t_1+T} \hat{s}_1(t)\cos(\omega_c t + \theta' + \varphi)dt, \tag{2.7}$$

where t_1 is the time of commencement for the data bit, T is the period of one data bit and φ is the phase synchronization error.

If we assume that the replica of the spreading code at the receiver is perfectly synchronized to the code used to spread the signal at the transmitter, then $\tau_1 = \tau'$ and we can set $\tau_1 = 0$, yielding:

$$\hat{s}_1(t) = Ad_1(t)c_1^2(t)\cos(\omega_c t). \tag{2.8}$$

Substituting Equation 2.8 into Equation 2.7 and assuming that there is no phase synchronization error, i.e. $\varphi = 0$, we then have :

$$\begin{aligned} z_1 &= \int_{t_1}^{t_1+T} Ad_1(t)c_1^2(t)\cos^2(\omega_c t) & (2.9)\\ &= \frac{A}{2}\int_{t_1}^{t_1+T} d_1(t)c_1^2(t)[\cos 2(\omega_c t) + 1]\, dt. & (2.10)\end{aligned}$$

The high frequency term $\cos 2(\omega_c t)$ tends to zero after the correlator-based receiver, since the frequency ω_c is significantly higher than that of $d_1(t)c_1^2(t)$, resulting in an equal number of positive and negative terms in the integral, yielding:

$$z_1 = \frac{A}{2}\int_{t_1}^{t_1+T} d_1(t)c_1^2(t)\, dt. \tag{2.11}$$

Since it is assumed that there is no timing error, $d_1(t)$ is BPSK demodulated, giving $d_1(t) = \pm 1$, and $\int_0^T c_1^2(t) = T$, hence we have:

$$z_1 = \pm\frac{AT}{2}. \tag{2.12}$$

The above derivation assumes that there is accurate phase and timing synchronization. If these assumptions are invalid, then we arrive at:

$$\begin{aligned} z_1 &= \int_{t_1}^{t_1+T} Ad_1(t-\tau_1)c_1(t-\tau_1)c_1(t-\tau')\cos(\omega_c t + \theta')\cos(\omega_c t + \theta' + \varphi)dt \\ &= \int_{t_1}^{t_1+T} \frac{A}{2}d_1(t-\tau_1)c_1(t-\tau_1)c_1(t-\tau')\cos(\varphi)dt, & (2.13)\end{aligned}$$

where we used the identity that $\cos A \cos B = \frac{1}{2}[\cos(A+B)+\cos(A-B)]$ and we exploited again that the high-frequency term of $\frac{1}{2}\cos[2(\omega_c t + \theta') + \varphi]$ tends to 0 after the correlator. Since $d_1(t - \tau_1) = \pm 1$, the following ensues:

$$z_1 = \pm\frac{A}{2}\cos(\varphi)\int_{t_1}^{t_1+T} c_1(t-\tau_1)c_1(t-\tau')dt \tag{2.14}$$

$$= \pm\frac{AT}{2}\cos(\varphi)R_{cc}(\tau_1 - \tau'), \tag{2.15}$$

where $R_{cc}(\tau_1 - \tau')$ is the auto-correlation of the spreading code $c_1(t)$ given by:

$$R_{cc}(\tau_1 - \tau') = \int_0^T c_1(\tau_1)c_1(\tau'). \tag{2.16}$$

Therefore $|z_1|$ is maximum, when $(\tau_1 - \tau') = 0$ and $\varphi = 0$. Synchronization errors will result in lower $|z_1|$ values, leading to lower signal-to-noise ratios (SNR) and hence higher bit-error-rates (BER). When inaccurate synchronization occurs, we have $\tau_1 \neq \tau'$ and hence the high values of out-of-phase autocorrelation contributions due to $R_{cc}(\tau_1 - \tau')$ result in high BERs. Therefore, the spreading codes have to be carefully designed in order to have low out-of-phase autocorrelation values, an issue to be addressed in Section 2.6 in greater depth.

2.4.2 Recovery of the Information Signal in Multiple Access Interference

In the case where there are several users, i.e. $K > 1$, the received signal is given by:

$$r(t) = \sum_{k=1}^{K} s_k(t - \tau_k). \tag{2.17}$$

After multiplying the received signal, $r(t)$, with the spreading code, $c_1(t)$, the information signal of user 1, $d_1(t)$, is despread into its original bandwidth. However, the signals of all the other $K-1$ users remain spread, since $c_1(t)$ is orthogonal to the other spreading codes, resulting in:

$$\hat{s}_1(t) = \sum_{k=1}^{K} s_k(t - \tau_k)c_1(t - \tau_1'). \tag{2.18}$$

Next the signal is demodulated and passed through a correlator as well as the thresholding device, as seen in Figure 2.2. The output of the correlator is formulated as:

$$\begin{aligned} z_1(t) &= \int_{t_1}^{t_1+T} \sum_{k=1}^{K} s_k(t-\tau_k)c_1(t-\tau_1')\cos(\omega_c t+\theta')dt \quad (2.19)\\ &= \int_{t_1}^{t_1+T} \Big[s_1(t-\tau_1)c_1(t-\tau_1') + s_2(t-\tau_2)c_2(t-\tau_1') + \ldots \\ &\quad + s_K(t-\tau_K)c_1(t-\tau_1')\Big]\cos(\omega_c t+\theta')\,dt. \quad (2.20) \end{aligned}$$

In a synchronous system, the start of each data bit of a user coincides with the start of the data bits of all the other users. Assuming that there are no phase or timing synchronization errors, i.e. $\tau_k = 0$ and $\theta' = 0$, we have:

$$\begin{aligned} z_1(t) &= \int_{t_1}^{t_1+T} \sum_{k=1}^{K} s_k(t)c_1(t)\cos(\omega_c t)dt \quad (2.21)\\ &= \int_{t_1}^{t_1+T} \sum_{k=1}^{K} Ad_k(t)c_k(t)c_1(t)\cos^2(\omega_c t)dt \quad (2.22)\\ &= \frac{A}{2}\int_{t_1}^{t_1+T} \sum_{k=1}^{K} d_k(t)c_k(t)c_1(t)dt \quad (2.23)\\ &= \frac{A}{2}\int_{t_1}^{t_1+T} \Big[d_1(t)c_1^2(t) + d_2(t)c_2(t)c_1(t) + \ldots + d_K(t)c_K(t)c_1(t)\Big]dt, \end{aligned}$$

where again, the high frequency term of $\cos(2\omega_c t)$ integrates to zero.

If the spreading sequences are designed such that they constitute an orthogonal set, where

$$\int_0^T c_i(t)c_j(t)dt = \begin{cases} T & \text{for } i = j \\ 0 & \text{for } i \neq j \end{cases}, \quad (2.24)$$

then all the interference terms integrate to zero, leaving only the data bits of the wanted user, namely user 1. Since $d_1(t) = \pm 1$ and $\int_0^T c_1^2(t) = T$, we have:

$$z_1 = \pm\frac{AT}{2}, \quad (2.25)$$

as seen for a single user in Equation 2.12.

Thus, under the assumption of perfect synchronization, the data bits of user 1 can be recovered despite the other interfering users. In practice, it is difficult to design a large set of spreading codes that are perfectly orthogonal, since two different demands have to be satisfied. Firstly, zero out-of-phase correlation is required, i.e. $\int_0^T c_i(t-\tau)c_i(t) = 0$, for $\tau \neq 0$. Secondly, Equation 2.24 has to be satisfied. In addition, a range of other practical code design constraints have to be satisfied, as we will see in Section 2.6. The result is usually a compromise by approximating the ideal conditions, where the spreading codes are not perfectly uncorrelated with each other, but the cross-correlation values are kept as small as possible. These cross-correlations result in multiple access interference (MAI) that degrades the performance of the system. These correlations are given by:

$$\int_0^T c_i(t)c_j(t)dt = \begin{cases} T & \text{for } i = j \\ R_{ij} \neq 0 & \text{for } i \neq j \end{cases} \tag{2.26}$$

Considering Equation 2.24 again in conjunction with the correlation constraints of Equation 2.26, we have:

$$\begin{aligned} z_1(t) &= \frac{A}{2}\int_{t_1}^{t_1+T}\Big[d_1(t)c_1^2(t) + d_2(t)c_2(t)c_1(t) + d_K(t)c_K(t)c_1(t)\Big]dt && (2.27) \\ &= \frac{A}{2}\Big[\pm T \pm TR_{12} + \ldots \pm TR_{1K}\Big]. && (2.28) \end{aligned}$$

The first term is the wanted data and all the other terms contribute to the MAI.

Let us now consider the corresponding operations in the frequency domain.

2.5 Frequency-Domain Representation

The spreading and despreading of the information signal are more easily represented in the frequency domain. If the information signal, $d_k(t)$, is BPSK-modulated, having an amplitude of ± 1 and a bit rate of $1/T$ bits/sec, then the power spectral density (PSD) of the information signal is given by:

$$S(f) = T\text{sinc}^2(fT). \tag{2.29}$$

This is shown in Figure 2.4(a). The frequencies of the first nulls of the spectrum are at $\pm 1/T$, giving an approximate bandwidth of $1/T$.

The spreading signal, $c_k(t)$, has a chip rate of $1/T_c$. If the chip amplitudes are also ± 1, then the spreading signal has a PSD of:

$$S_c(f) = T_c\,\text{sinc}^2(fT_c). \tag{2.30}$$

If the spread time-domain signal is given by the product $d_k(t)c_k(t)$, the resulting frequency-domain PSD is that of the corresponding convolved spectra, which is shown in Figure 2.4(b). Since the T_c/T ratio is typically high, the PSD of Equation 2.29 is a 'near-Dirac' impulse and hence the PSD of Figure 2.4(b) is similar to $S_c(f)$ in Equation 2.30. Therefore, for conventional receivers, the bandwidth of the original information signal is spread by the ratio of T/T_c. The ratio of the spread bandwidth to the information bandwidth is known as the processing gain . Historically, the term processing gain was used in jamming systems, which employed the principles of spread spectrum to protect a signal against jamming interference from an unknown or hostile source. The term processing gain was used to

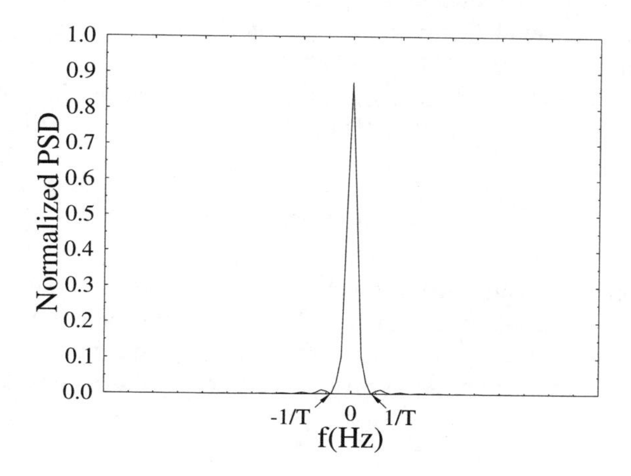

(a) PSD of the information signal before spreading. The first nulls are at $\pm 1/T$, where T is the bit rate of the signal.

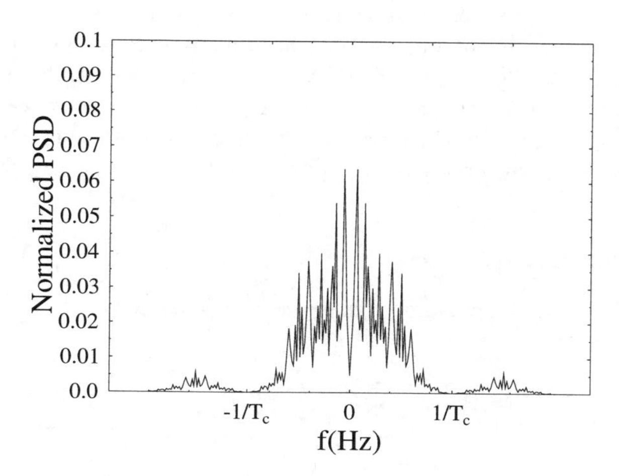

(b) PSD of the information signal after spreading. The first PSD nulls are at $\pm 1/T_c$, where T_c is the chip rate of the signal.

Figure 2.4: Power spectral density (PSD) plot of the signal before and after spreading, normalized to the total power in the signal. The PSD values are normalized to the PSD value at frequency 0. Note that the frequency scales in (a) and (b) are different for simple graphical representation reasons, since $T/T_c = 128$.

quantify the factor of advantage of a signal that had been spread over the jammer interference. In DS-CDMA systems, the spread bandwidth is approximately the spreading code chip rate, $1/T_c$ and the information bandwidth is approximately $1/T$. This thus gives a processing gain of T/T_c. The term spreading factor is also used interchangeably with processing gain to convey a similar meaning.

If at the receiver, the spread signal, $d_k(t)c_k(t)$, is multiplied by the spreading signal, $c_k(t)$, again, the spread signal is then despread into its original bandwidth and will have a PSD of:

$$S(f) = T\text{sinc}^2(fT), \tag{2.31}$$

while all the other signals remain spread. Figure 2.3 shows the PSD of a despread signal in the presence of interfering users, which corresponds to the convolution of the received signal spectrum with that of the spreading code for an arbitrary T/T_c ratio of 128.

In the next section, a few examples of spreading sequences will be considered.

2.6 Spreading Sequences

In Section 2.4.2, we noted that the multiple access interference (MAI) of Equation 2.28 is largely dependent on the cross-correlation (CCL) between the spreading code of the wanted user and the spreading codes of all the other interfering users. The ideal CCL value is zero for all values of $t \neq 0$. Ideally, the out-of-phase auto-correlation (ACL) of the spreading sequences should also be zero, in order to tolerate inaccurate synchronization of the spreading sequences. Therefore, for the conventional CDMA receiver, the design of the set of spreading sequences to be used in a CDMA system is very important as to its performance. The design is usually based on a set of criteria which will be the topic of the forthcoming sections. Let us commence by considering the correlation properties of the spreading sequences.

2.6.1 Correlation of Sequences

The periodic ACL of a complex spreading sequence, $c^{(k)}$, is defined as:

$$R_{kk}(q) = \sum_{i=1}^{Q} c_i^{(k)} c_{(q+i)\oslash Q}^{(k)*}, \tag{2.32}$$

for $q = -(Q-1), \ldots, Q$; where the expression $[(q+i) \oslash Q]$ represents the remainder after dividing $(q+i)$ by Q, $c^{(k)*}$ is the complex conjugate of $c^{(k)}$ and Q is the length of the sequence. The periodic CCL of two different sequences, $c^{(j)}$ and $c^{(k)}$ is defined as:

$$R_{jk}(q) = \sum_{i=1}^{Q} c_i^{(j)} c_{q+i}^{(k)*}, \tag{2.33}$$

for $q = -(Q-1), \ldots, Q$.

A user's signal that has been spread has to appear noise-like to all the other users. This can be achieved by using a random spreading code, having an infinite repetition time. However, if the code is truly random, the original data signal cannot be recovered at the receiver by correlation techniques. For the spreading code to be useful, it has to be deterministic, but known only to the transmitter and the intended receiver. However, it still has to appear like random noise to non-intended receivers. Therefore, the spreading codes are generally known as pseudo-random or pseudo-noise (PN) sequences. These codes have a length of Q and are repeated periodically with a period of Q. Each spreading code set has a family size of K, i.e. there are K codes in the set. This will be explained later in the following sections. There are basically two classes of spreading sequences, namely the binary and the non-binary sequences also known as polyphase sequences [100]. The latter are complex-valued sequences.

Traditionally, the measure used to compare different code sets was defined as R_{max}, where:

$$R_{max} = \max |R_a, R_c|, \tag{2.34}$$

and

$$\begin{aligned} R_a &= \max |R_{kk}(q)| \quad \text{for } 1 \leqslant k \leqslant K; \quad -(Q-1) < q \leqslant Q \;\; q \neq 0 \\ R_c &= \max |R_{jk}(q)| \quad \text{for } 1 \leqslant j, k \leqslant K, \;\; j \neq k; \quad -(Q-1) < q \leqslant Q. \end{aligned} \tag{2.35}$$

The value of the periodic CCL or out-of-phase periodic ACL that has the highest magnitude is taken to be the R_{max} of a particular code set. A low out-of-phase periodic ACL allows easier code acquisition or synchronization, while a low periodic CCL reduces the MAI. A measure of "goodness" of R_{max} is how well it compares with the well-known Welch [82] or Sidelnikov [82] bounds. For a set of K sequences having a period of Q, the Welch bound is defined as [82]:

$$R_{max} \geqslant Q\sqrt{\frac{K-1}{KQ-1}}. \tag{2.36}$$

This value is dependent on the number, K, of codes in the set, and the length, Q, of the codes. Therefore, for a code set to be optimum, according to Welch, it had to achieve the minimum value given by Equation 2.36.

For the same set of sequences, the Sidelnikov bound is defined as [82]:

$$R_{max} > (2Q-2)^{\frac{1}{2}}. \tag{2.37}$$

A code set is optimum according to these criteria, if its performance measure, R_{max}, approaches the Welch or Sidelnikov lower bounds, given by Equations 2.36 and 2.37, respectively. Some numerical examples will be given in Sections 2.6.3 and 2.6.4 in the context of Gold sequences and Kasami sequences, respectively.

The criterion of family size, K, is important because large families of good codes enable more users to be accommodated in the same bandwidth, thus increasing the user capacity of the system. Since a CDMA system is interference-limited, as will be shown in Section 2.7, the larger the code set having low CCL values, the higher the capacity. The difficulty in designing spreading codes lies in achieving a large family size having good CCL values for a specific value of spreading code length, Q. These issues will be demonstrated in Sections 2.6.2, 2.6.3 and 2.6.4 using various examples, namely the m-sequences [82], Gold codes [101] and Kasami sequences [102], respectively.

2.6.2 *m*-Sequences

An important class of binary sequences is the binary maximal-length shift register sequences, commonly known as m-sequences [82]. Sequences can be generated using the well-known linear generator polynomials of degree m, where:

$$g(x) = g_m x^m + g_{m-1} x^{m-1} + g_{m-2} x^{m-2} + \ldots + g_1 x + g_0. \tag{2.38}$$

In order to generate m-sequences, the generator polynomial, $g(x)$, must be from the class of polynomials known as primitive polynomials, which implies in simple terms, that $g(x)$ cannot be factorized into lower-order polynomials. For a more detailed discussion on primitive polynomials, the reader is referred to BlahutBlahut's excellent monograph on error correction coding [99], where generator polynomials are used extensively. The m-sequence generated has a period of $Q = 2^m - 1$, where m is the degree of the generator polynomial. By contrast, a sequence generated by a non-primitive generator polynomial, $g(x)$, may have a period of less than $2^m - 1$ and hence this sequence is not an m-sequence.

One of the properties of m-sequences is that the ACL of these sequences can be calculated as:

$$\begin{aligned} R_{kk}(0) &= Q \\ R_{kk}(q) &= -1 \quad \text{for } q \neq 0, \end{aligned} \tag{2.39}$$

which accrues from the simple binary nature of the sequences. Explicitly, when a Q-chip sequence is perfectly aligned with itself, corresponding to a shift of $q = 0$ chip intervals, there are Q number of $+1$ terms in the autocorrelation sum of Equation 2.32. By contrast, for $q \neq 0$, the summation always consists of one extra -1 value compared to the number of $+1$ values. These ACL properties are near-ideal for code acquisition or synchronization, where the perfectly aligned condition of $q = 0$ between the received and locally stored sequences has to be detected. As an additional constraint, in multi-user communications, a large set of spreading sequences exhibiting low CCL values is needed. The m-sequences do not satisfy this requirement, since some m-sequence pairs have large CCL values. Let us now consider the family of Gold codes.

2.6.3 Gold Sequences

One of the best-known binary sequences having relatively good correlation values is the Gold sequence set [101]. A set of Gold sequences is constructed from a preferred pair of m-sequences, $\underline{x}$ and $\underline{y}$, having identical length Q. In 1967, Gold proved that these preferred pairs of m-sequences of length Q have only three possible CCL values, which are shown in

Equations 2.41, 2.42 and 2.43. The period of the Gold sequences generated by $\underline{x}$ and $\underline{y}$ is also Q. Each Gold sequence in a set is generated by a modulo-2 sum of $\underline{x}$ and cyclic shifts of $\underline{y}$. The set also includes the m-sequences $\underline{x}$ and $\underline{y}$. The entire set of Gold sequences having a period of Q is given by:

$$S_g = \left\{ \underline{x},\ \underline{y},\ \underline{x} \oplus \underline{y},\ \underline{x} \oplus T^{-1}\underline{y},\ \underline{x} \oplus T^{-2}\underline{y},\ \ldots,\ \underline{x} \oplus T^{-(Q-1)}\underline{y} \right\} \tag{2.40}$$

where $T^{-q}\underline{y}$ for $q = 0, 1, \ldots, Q-1$, represents a cyclic shift of $\underline{y}$ by q chip intervals; and the symbol $\oplus$ represents modulo-2 addition.

The family size of a Gold sequence set having a period of Q is $K = Q + 2$. A property of Gold sequences, as stated earlier, is that the CCLs and out-of-phase ACLs have only three possible values, which are given by:

$$R_{kk}(q) = \begin{cases} Q & \text{for } q = 0 \\ \{-1,\ -t(m),\ t(m)-2\} & \text{for } q \neq 0 \end{cases} \tag{2.41}$$

$$R_{jk}(q) = \{-1,\ -t(m),\ t(m)-2\} \quad \text{for all } q \text{ and } j \neq k, \tag{2.42}$$

where

$$\begin{aligned} t(m) &= 2^{(m+1)/2} + 1 \quad m \text{ is odd} \\ t(m) &= 2^{(m+2)/2} + 1 \quad m \text{ is even.} \end{aligned} \tag{2.43}$$

Since Gold sequences are constructed from preferred pairs of m-sequences, the ACL is given by Q, as seen in Equation 2.39. The detailed derivation of the CCL values and out-of-phase ACL values was given by Simon *et al.* [82]. In order to compute the correlation values, the binary bits 0 and 1 are mapped to $+1$ and -1, respectively. From the correlation values, it can be seen that for a set of Gold sequences, $R_{max} = t(m)$. Consider a Gold sequence set of period $Q = 63$, corresponding to $m = 6$. Then from Equation 2.43 we have:

$$R_{max} = t(m) = 2^{(6+2)/2} + 1 = 17. \tag{2.44}$$

The Welch bound of Equation 2.36 for this Gold sequence set, which has a family size of $K = Q + 2 = 65$, is computed as follows:

$$R_{max} \geqslant 63\sqrt{\frac{65-1}{63(65)-1}} \approx 7.88. \tag{2.45}$$

Note that the Welch bound is as low as 46% of $|R_{max}| = 17$ in Equation 2.44, computed for our example of a code set of $Q = 63$ and $K = 65$. Two examples of Gold sequences are shown in Figure 2.5.

The periodic CCL of these two sequences was calculated using Equation 2.33 and the CCL is plotted in Figure 2.6. The binary bits 0 and 1 were mapped to $+1$ and -1, respectively. It can be seen from Figure 2.6 that the CCL has three values, i.e. -1, -17 and $+15$. This set of spreading sequences yields comparatively high CCLs at the indices, where the spikes are

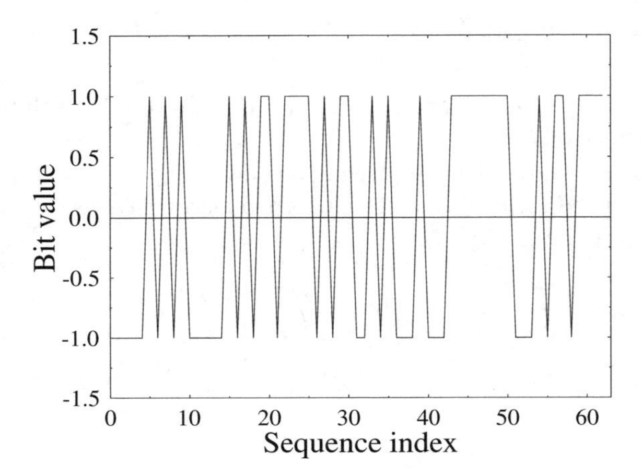

(a) A Gold sequence from the code set of $Q = 63$.

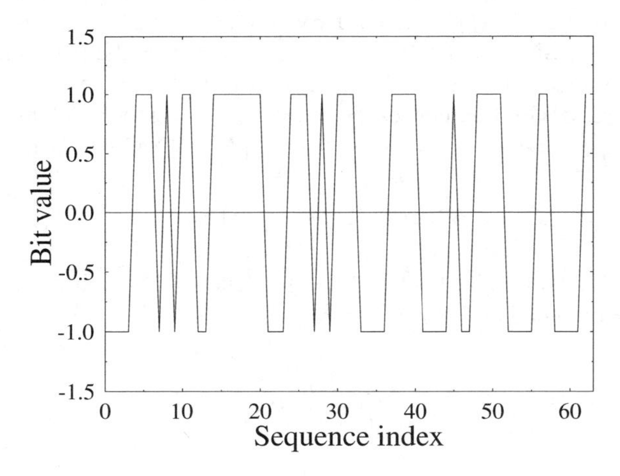

(b) A different Gold sequence from code set of $Q = 63$.

Figure 2.5: Examples of Gold sequences

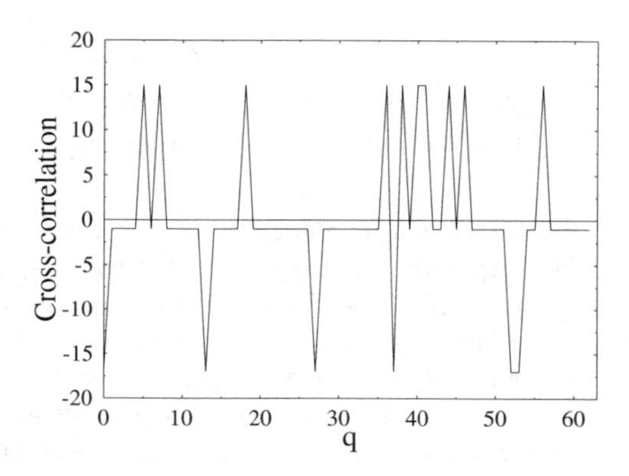

Figure 2.6: Cross-correlation values between two pairs of Gold sequences versus q as specified in Equation 2.33. The binary bits 0 and 1 were mapped to $+1$ and -1 respectively to compute the correlation values. The CCL values are -1, -17 or 15.

located in the figure.

Following the above brief notes on Gold codes, let us now consider Kasami sequences.

2.6.4 Kasami Sequences

A set of binary sequences that are near-optimum with respect to the Welch bound is the Kasami sequence family [102]. Each sequence set is generated from one m-sequence, $\underline{x}$, which has a period of Q. The m-sequence $\underline{x}$ is decimated by sampling it periodically. Then, a second sequence, $\underline{y}$, is formed by concatenating the decimated sequence repeatedly, until a sequence of length Q is obtained. The entire Kasami set is then generated by the modulo-2 addition of the sequence $\underline{x}$ and the sequences generated by cyclic shifts of $\underline{y}$.

The generator polynomial of $\underline{x}$ is of degree m and $\underline{x}$ has a period of $Q = 2^m - 1$. Kasami sequences can only be generated for even values of m. The decimation factor used to decimate $\underline{x}$ is given by $s(m) = 2^{m/2} + 1$. The decimated sequence, $\underline{y}$, has a shorter period of $(2^m - 1)/s(m) = 2^{m/2} - 1$. Hence, the set of Kasami sequences is given by:

$$S_k = \left\{ \underline{x},\ \underline{x} \oplus \underline{y},\ \underline{x} \oplus T^{-1}\underline{y},\ \underline{x} \oplus T^{-2}\underline{y},\ \ldots,\ \underline{x} \oplus T^{-(2^{m/2}-2)}\underline{y} \right\}. \tag{2.46}$$

The correlation values of a Kasami sequence set are also ternary, similar to those of the Gold sequences, but the values are lower for sequences of the same length, which are given by:

$$R_{kk}(q) = \begin{cases} Q & \text{for } q = 0 \\ \{-1,\ -s(m),\ s(m)-2\} & \text{for } q \neq 0 \end{cases} \tag{2.47}$$

$$R_{jk}(q) = \{-1,\ -s(m),\ s(m)-2\} \quad \text{for all } q \text{ and } j \neq k. \tag{2.48}$$

The number of Kasami sequences is now reduced to $K = 2^{m/2}$, which is lower than that of the m-sequences. More explicitly, the usage of a decimated sequence reduces the number of sequences in a set, compared to the Gold sequence set, having the same period. For example, a Gold sequence set having a period of $Q = 63$ has a family size of $K = 65$, while Kasami sequences of the same length have a set size of $K = 8$. However, due to the reduced set size, the high CCL sequences can be eliminated and hence the Kasami sequences have lower CCL values, which is essentially due to the usage of a decimated sequence. This is explained in more depth by Simon *et al.* [82]. These lower CCL values result in Kasami sequences being nearer to achieving the optimum Welch lower bound compared to Gold sequences of the same length. The Welch bound of Equation 2.36 for the Kasami set of $Q = 63$ and $K = 8$ is given by:

$$R_{max} \geqslant 63\sqrt{\frac{8-1}{63(8)-1}} \approx 7.43, \tag{2.49}$$

while

$$\mid s(m) \mid = 2^{(6/2)} + 1 = 9. \tag{2.50}$$

Therefore, the optimum Welch lower bound is now 83% of R_{max}, compared to the previous 46% for the equivalent Gold sequence set. Having considered a range of spreading sequences, let us now concentrate our discussions on the performance of the DS-CDMA system.

2.7 DS-CDMA System Performance

In this section, a model for a DS-CDMA system using BPSK modulation is presented and a theoretical analysis of its performance is obtained. The model shown in Figure 2.7 is similar to the model shown in Figure 2.2, the difference being that the noise signal $n(t)$ is now added to the received signal. From this model, an expression for the bit-error-rate (BER) performance of the system is derived. The conditions under which this model is presented are:

- BPSK modulation is used for the information signal. The carrier signal has a frequency of ω_c and amplitude of $A = \sqrt{P}$, where P is the average power of the signal. This gives an energy per bit of $E_b = A^2 T = PT$.

- The length of the spreading code used is Q.

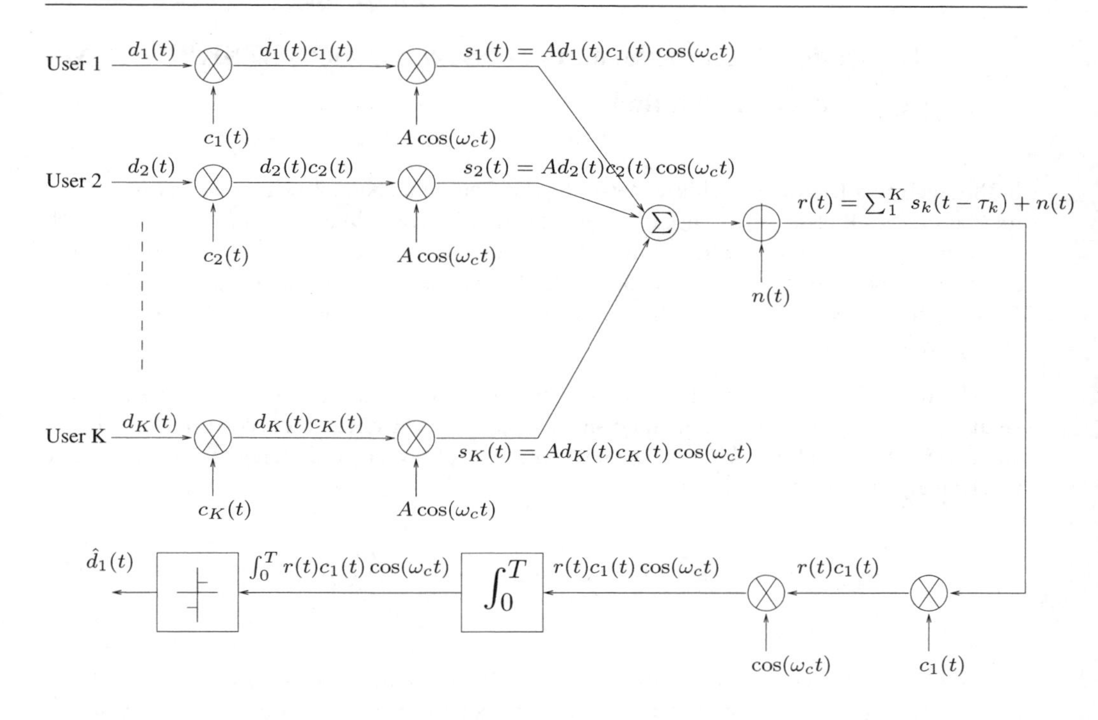

Figure 2.7: Block diagram of a simple asynchronous DS-CDMA system transmitting over a Gaussian channel. The signals from all the K users arrive at the receiver with different propagation delays, τ_k. Only the receiver for user 1 is shown. The despreading and demodulation processes shown in the figure assume perfect synchronization.

- Each information/data bit has a period of T and each chip has a period of T_c, where $T = QT_c$

- The pulse shape is rectangular and has an amplitude of ± 1 for the information signal. The pulse shape for each chip of the spreading code is also rectangular, having an amplitude of ± 1.

- The total number of users in the system is represented by K and each user is identified by the subscript k.

- All K users are transmitting at the same bit rate.

- There is perfect power control for all K users.

- This system operates in a single cell environment.

2.7.1 Theoretical BER Performance of Asynchronous BPSK/DS-CDMA over Gaussian Channels

In this section a BER expression is derived for a Gaussian channel, where it is assumed that for each user, all errors are caused by zero-mean Additive White Gaussian Noise (AWGN) at the receiver and multiple access interference (MAI) from the other $(K-1)$ users. The derivation is performed at the baseband level, which greatly simplifies the analysis. However, the derivation is still valid for a bandpass system, because baseband and bandpass systems are equivalent [92].

The desired user is user 1 and all the other $(K-1)$ users are interferers. The received signal, $r(t)$, is a sum of the transmitted signals from all K users and it is corrupted by Gaussian noise. The signal of each user arrives at a different propagation delay, given by τ_k. The received signal is formulated as:

$$r(t) = \sum_{k=1}^{K} \sqrt{P} d_k(t-\tau_k) c_k(t-\tau_k) + n(t). \tag{2.51}$$

This signal is then despread with a replica of the spreading code of user 1. A correlator-based receiver is used to obtain the associated decision statistic, z_1, from which a decision is made concerning the bit transmitted. Therefore, we have:

$$z_1 = \int_0^T r(t) c_1(t-\tau_1-\tau')dt, \tag{2.52}$$

where τ' is the code synchronization error, which degrades the demodulator's correlation properties; $\theta = -\omega_c \tau_1$, which is the delay-induced carrier-phase term; and $\varphi = -\omega_c \tau'$ is the demodulator's phase error.

If perfect synchronization is assumed for user 1, then we can set $\tau_1 = \tau' = 0$. Substituting Equation 2.51 into Equation 2.52 leads to:

$$\begin{aligned} z_1 &= \int_0^T r(t) c_1(t)\, dt && (2.53) \\ &= \int_0^T \left[\sqrt{P} d_1(t) c_1^2(t) + \sum_{k=2}^{K} \sqrt{P} d_k(t-\tau_k) c_k(t-\tau_k) c_1(t) + n(t) c_1(t)\right] dt \\ &= D_1 + I + \eta, && (2.54) \end{aligned}$$

where the corresponding terms are defined below, namely, D_1 is the bit transmitted by user 1:

$$D_1 = \int_0^T \sqrt{P} d_1(t) c_1^2(t)\, dt, \tag{2.55}$$

and taking into account that $c_1^2(t) = 1$ and $d_1(t) = \pm 1$, this yields:

$$D_1 = \pm\sqrt{P}T. \tag{2.56}$$

The term η in Equation 2.54 is the component due to the AWGN, $n(t)$, which corresponds to the despread and demodulated term of:

$$\eta = \int_0^T n(t)c_1(t)\,dt. \tag{2.57}$$

Since $n(t)$ is the zero-mean AWGN having a variance of $N_0/2 = \sigma^2$, η is a Gaussian variable with zero mean and variance Var$[\eta]$, which is derived as:

$$\begin{aligned} \text{Var}[\eta] &= E[\eta^2] && (2.58)\\ &= E\Big[\int_0^T n(t)c_1(t)\,dt \int_0^T n(u)c_1(u)\,du\Big] && (2.59)\\ &= \int_0^T \int_0^T E[n(t)n(u)]c_1(t)c_1(u)\,dt\,du. && (2.60) \end{aligned}$$

But $E[n(t)n(u)]$ is the autocorrelation of $n(t)$, where:

$$E[n(t)n(u)] = \frac{N_0}{2}\delta(t-u). \tag{2.61}$$

Therefore, the variance Var$[\eta]$ becomes:

$$\begin{aligned} \text{Var}[\eta] &= \frac{N_0}{2}\int_0^T \int_0^T \delta(t-u)c_1(t)c_1(u)\,dtdu && (2.62)\\ &= \frac{N_0}{2}\int_0^T c_1^2(u)\,du. && (2.63) \end{aligned}$$

Since $\int_0^T c_1^2(u) = T$, we have:

$$\text{Var}[\eta] = \frac{N_0 T}{2}. \tag{2.64}$$

The middle term, I, in Equation 2.54 is the MAI component from all the other $(K-1)$ users, which is given by:

$$I = \sum_{k=2}^{K} \sqrt{P}\int_0^T d_k(t-\tau_k)c_k(t-\tau_k)c_1(t)\,dt. \tag{2.65}$$

From the central limit theorem [90], the summation of $(K-1)$ independent random variables can be modelled by the Gaussian distribution and hence in our analysis, I is approximated by a Gaussian random variable. By using the Gaussian approximation method [103] and assuming that there is perfect power control, the variance of I was derived by Pursley as follows:

$$\begin{aligned} \text{Var}[I] &= \frac{QT_c^2}{3} \sum_{k=2}^{K} P \\ &= \frac{QT_c^2}{3}(K-1)P. \end{aligned} \tag{2.66}$$

Substituting $T = QT_c$, we arrive at:

$$\text{Var}[I] = \frac{T^2}{3Q}(K-1)P. \tag{2.67}$$

Therefore, the equivalent SNR at the output of the correlator receiver is:

$$\text{SNR}_o = \frac{D_1^2}{\text{Var}[\eta] + \text{Var}[I]}. \tag{2.68}$$

Substituting Equations 2.56, 2.64 and 2.66 into Equation 2.68 and employing the expression $E_b = PT$, yields:

$$\text{SNR}_o = \left[\frac{PT^2}{\frac{N_0T}{2} + \frac{T^2}{3Q}(K-1)P}\right] \tag{2.69}$$

$$= \left[\frac{\frac{N_0T}{2}}{PT^2} + \frac{\frac{T^2}{3Q}(K-1)P}{PT^2}\right]^{-1} \tag{2.70}$$

$$= \left[\frac{N_0}{2E_b} + \frac{K-1}{3Q}\right]^{-1}. \tag{2.71}$$

The second term of $(K-1)/(3Q)$ in the brackets in Equation 2.71 represents the MAI that causes an SNR degradation resulting in a degraded SNR performance for a particular value of E_b/N_0. It can be seen that this degradation depends on the number of users, K, and the sequence length, Q. An increase in K or a decrease in Q would degrade the performance because it would increase the cross-correlation (CCL) between the received signals from all the users.

By assuming that the MAI has a Gaussian PDF, the BER of a BPSK-modulated system

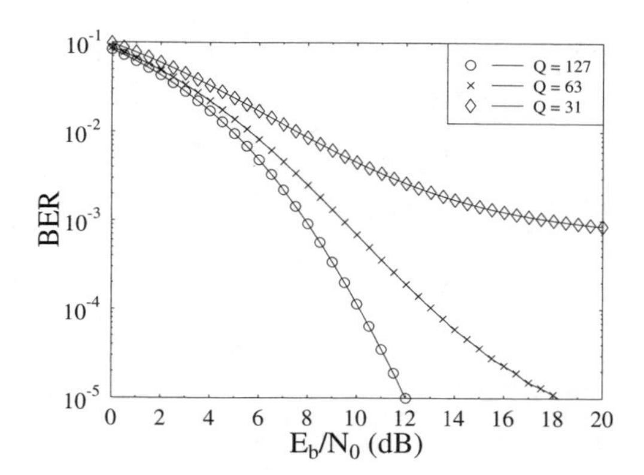

Figure 2.8: BER versus E_b/N_0 plots for an asynchronous CDMA system in a Gaussian channel. The number of users is fixed at $K = 10$, but varying values of sequence lengths, Q, are used. These graphs were plotted using Equation 2.72. As the sequence length, Q, decreases, the BER performance degrades.

in a Gaussian channel is given as BER $= Q(\sqrt{\text{SNR}_o})$, leading to:

$$\begin{aligned} \text{BER} &= Q(\sqrt{\text{SNR}_o}) \\ &= Q\left(\left[\frac{N_0}{2E_b} + \frac{K-1}{3Q}\right]^{-\frac{1}{2}}\right), \end{aligned} \tag{2.72}$$

where $Q(x)$ is the Gaussian Q-function [90].

According to Equation 2.72, if there is only one user, then $K = 1$ and the equation simplifies to BER $= Q(\sqrt{2E_b/N_0})$, which is the same as the theoretical BER performance of a BPSK system.

Figure 2.8 shows a plot of three BER versus E_b/N_0 curves. These curves were plotted using Equation 2.72 for $K = 10$ users and for sequence lengths of $Q = 31$, 63 and 127. The curves show that as the sequence length, Q, decreased, the BER performance degraded.

Figure 2.9 shows another plot of three BER versus E_b/N_0 curves. These curves were also plotted using Equation 2.72, but for $Q = 63$ and for $K = 10$, 20 and 30. The curves show that as the number of users, K, increased, the BER performance degraded. This was due to the increase in MAI.

Having characterized the system performance for K asynchronous users over a Gaussian channel, let us now consider synchronous users.

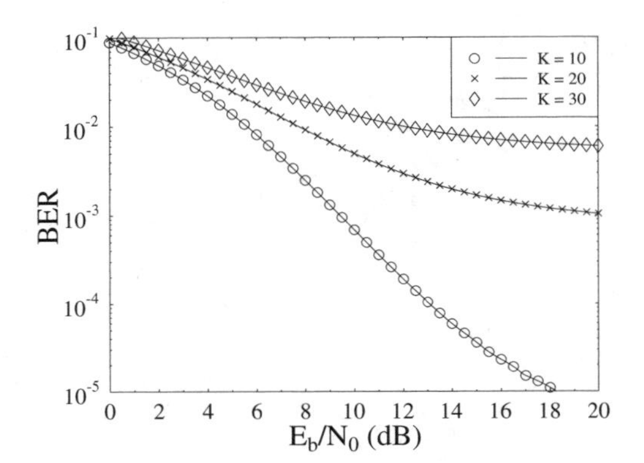

Figure 2.9: BER versus E_b/N_0 plots for an asynchronous CDMA system in a Gaussian channel. The sequence length is fixed at $Q = 63$ chips, but varying numbers of users, K, are used. These graphs were plotted using Equation 2.72. As the number of users, K, increases, the BER performance degrades.

2.7.2 Theoretical BER Performance of Bit-Synchronous BPSK/DS-CDMA Systems over Gaussian Channels

As stated above, the BER performance expression given in Equation 2.72 is valid for a system, where the users are asynchronous with respect to each other. Simulation studies are often conducted in a synchronous environment due to limited resources since then no over-sampling of the signals is necessary. This is especially true for studies that combine Time Division Multiple Access (TDMA) and CDMA transmission. The users are allocated time slots and all the users in the same time slot transmit simultaneously, leading to synchronous reception at the receiver. Therefore, the BER performance of a perfectly synchronous system is analysed next.

Again, the output of the integrator in Figure 2.7 is often referred to as the decision statistic, z_1, and its expression was given in Equation 2.54 as:

$$z_1 = D_1 + I + \eta,$$

where D_1 is the component representing the original bit transmitted by user 1. The expression for D_1 was given in Equation 2.56 as:

$$D_1 = \pm\sqrt{P}T.$$

The variance of the noise component η was previously derived as:

$$\text{Var}[\eta] = \frac{N_0 T}{2}. \tag{2.73}$$

Finally, I is the interference component contributed by all the other $(K-1)$ users, which is given by:

$$I = \int_0^T \sum_{k=2}^{K} \sqrt{P} d_k(t) c_k(t) c_1(t)\, dt. \tag{2.74}$$

The interference, I_k, from the k-th user is represented as:

$$\begin{aligned} I_k &= \int_0^T \sqrt{P} d_k(t) c_k(t) c_1(t)\, dt & (2.75) \\ &= \pm\sqrt{P} \int_0^T c_k(t) c_1(t)\, dt. & (2.76) \end{aligned}$$

The term $\int_0^T c_k(t)c_1(t)\, dt$ represents the normalized periodic in-phase cross-correlation between the spreading codes of user 1 and user k, which can be rewritten as:

$$\int_0^T c_k(t) c_1(t)\, dt = R_{k,1} T. \tag{2.77}$$

Substituting this into Equation 2.76 leads to:

$$I_k = \sqrt{P} R_{k,1} T. \tag{2.78}$$

If we assume that the power of I, represented by S_I, is the sum of the powers of all the $K-1$ interfering users, then we have:

$$\begin{aligned} S_I &= \sum_{k=2}^{K} I_k^2 & (2.79) \\ &= \sum_{k=2}^{K} [\sqrt{P} R_{k,1} T]^2 & (2.80) \\ &= PT^2 \sum_{k=2}^{K} R_{k,1}^2. & (2.81) \end{aligned}$$

The signal-to-noise ratio is given as:

$$\text{SNR}_o = \frac{D_1^2}{\text{Var}[\eta] + S_I}. \tag{2.82}$$

Q	MAI value
31	0.0593
63	0.0587
127	0.0055

Table 2.1: MAI term of Equation 2.85 for various values of Q and for $K = 10$.

Combining Equations 2.56, 2.64 and 2.81 into Equation 2.82, we arrive at:

$$\text{SNR}_o = \frac{PT^2}{\frac{N_0 T}{2} + PT^2 \sum_{k=2}^{K} R_{k,1}^2} \tag{2.83}$$

$$= \left[\frac{\frac{N_0 T}{2}}{PT^2} + \frac{PT^2 \sum_{k=2}^{K} R_{k,1}^2}{PT^2} \right]^{-1}. \tag{2.84}$$

Since the amplitude of each user's signal is $\sqrt{P}$, the energy per bit is $E_b = PT$. This leads to:

$$\text{SNR}_o = \left[\frac{1}{(2E_b/N_0)} + \sum_{k=2}^{K} R_{k,1}^2 \right]^{-1}. \tag{2.85}$$

The second term in Equation 2.85 represents the SNR degradation due to MAI. This term depends directly on the CCL between the spreading codes, rather than indirectly as in Equation 2.71. This is because the signals from all the users are synchronized with each other and therefore the CCL values of the spreading codes determine the amount of excess interference at the output of the integrator in Figure 2.7.

Assuming that the combined noise and interference components have a Gaussian distribution, the BER performance is given as:

$$\text{BER} = Q(\sqrt{\text{SNR}_o}) \tag{2.86}$$

$$= Q\left(\left[\frac{1}{(2E_b/N_0)} + \sum_{k=2}^{K} R_{k,1}^2 \right]^{-\frac{1}{2}} \right), \tag{2.87}$$

where $Q(x)$ is the Gaussian Q-function [90].

Figure 2.10 shows a plot of three BER versus E_b/N_0 curves. These curves were plotted using Equation 2.87 for $K = 10$ and for Gold sequences of $Q = 31$, 63 and 127. In general, the curves show that as the sequence length, Q, decreased, the BER performance degraded. However, there is virtually no difference between the curves for $Q = 31$ and $Q = 63$. This is because the MAI term in Equation 2.85 has very similar values for both of these sequence lengths, as shown in Table 2.1.

Figure 2.11 portrays a plot of three BER versus E_b/N_0 curves using Equation 2.87 for $Q = 63$ and for $K = 10$, 20 and 30. In harmony with our expectations, as the number of

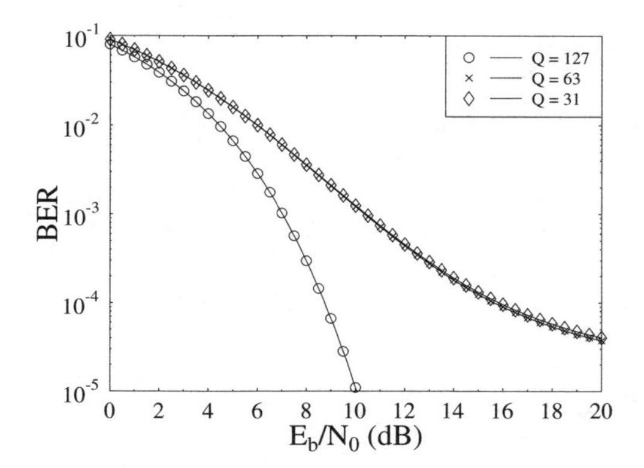

Figure 2.10: BER versus E_b/N_0 plots for a synchronous CDMA system in a Gaussian channel. Gold sequences are used with varying sequence lengths, Q, but the total number of users in the system is fixed at $K = 10$. These graphs were plotted using Equation 2.87. As the sequence length, Q, decreases, the BER performance degrades.

K	MAI value
10	0.0587
20	0.0612
30	0.0637

Table 2.2: MAI term of Equation 2.85 for various values of K and for $Q = 63$.

users, K, increased, the BER performance degraded. However, the performance degradation was small compared to the performance degradation shown in Figure 2.9. In a synchronous system, the BER performance depends directly on the CCL values between the spreading codes, which indirectly depends on the number of users. If the CCL values are small, then the increase in MAI is small, leading to a small degradation in performance. Table 2.2 shows the corresponding MAI values, i.e. the second term in Equation 2.85 for $K = 10$, 20 and 30.

Having considered the theoretical performance of a DS-CDMA system, let us now turn our attention to the results of our simulation studies.

2.8 Simulation Results and Discussion

Let us now study the performance of a BPSK-modulated DS-CDMA system, quantifying:

- the effect of increasing the number of users in the system on the BER performance,
- the effect of spreading sequence length on the BER performance of the system,

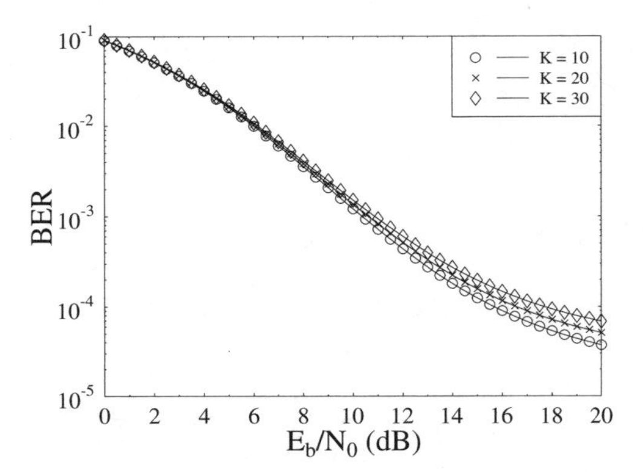

Figure 2.11: BER versus E_b/N_0 plots for a synchronous CDMA system in a Gaussian channel. Gold sequences are used with the sequence length fixed at $Q = 63$ chips, but varying numbers of users, K, are used. These graphs were plotted using Equation 2.87. As the number of users, K, increases, the BER performance degrades.

- the effect of various spreading codes having different cross-correlation values on the BER performance.

The simulation model used is shown in Figure 2.12. At the transmitter, the data bits were generated by a pseudo-random bit generator for all the K users. Each user was assigned a unique spreading code and the data bits of each user were spread using this code. Both the data bits and chips were binary. The spread signals were then summed synchronously, whereby the first chip of each bit for each user coincided with the first chip of each bit of all the other users.

At the receiver, Gaussian noise was added to the received signal. The noisy signal was then despread using a synchronous replica of the spreading code of the wanted user, i.e. it was assumed that there were no timing errors. The despread signal was fed to a so-called correlation receiver, where it was then integrated over one bit period. The output of the receiver was used to make a decision concerning the transmitted bit. The experiments were conducted under the following conditions:

- The simulations were performed at baseband, because the performance of baseband systems is equivalent to that of bandpass systems [92], but simulations carried out at baseband are more economical on computing resources.

- Perfect synchronization of the spreading codes was assumed at the receiver.

- Perfect power control was assumed.

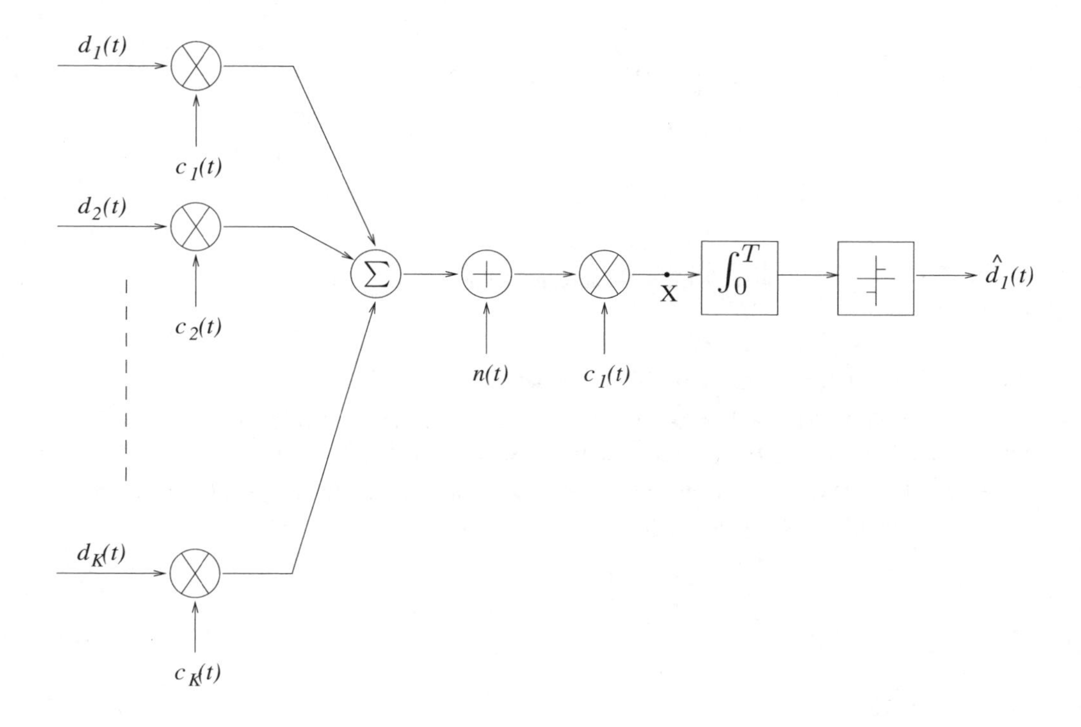

Figure 2.12: Simulation model of a synchronous BPSK DS-CDMA modem in a Gaussian channel

2.8.1 Estimation of E_b/N_0

The E_b/N_0 value at point X of Figure 2.13, in a communications system is estimated by first obtaining the average signal power, $\overline{S}$, and the average noise power, $\overline{N}$. The average signal-to-noise ratio, $\overline{\text{SNR}}$, is calculated as:

$$\overline{\text{SNR}} = \frac{\text{Signal power}}{\text{Noise power}} = \frac{\overline{S}}{\overline{N}}. \tag{2.88}$$

The energy per bit, E_b, is defined as $E_b = \overline{S}T$, while the noise power spectral density, N_0, is defined as $N_0 = \overline{N}/W_b$, where W_b is the bandwidth of the noisy signal at the point where the $\overline{\text{SNR}}$ is measured. Therefore, the relationship between E_b/N_0 and $\overline{\text{SNR}}$ is:

$$\frac{E_b}{N_0} = \frac{\overline{S} \times T}{\overline{N}/W_b} \tag{2.89}$$

$$= \frac{\overline{S}}{\overline{N}} \times \frac{W_b}{R} \tag{2.90}$$

$$= \overline{\text{SNR}} \times \frac{W_b}{R}, \tag{2.91}$$

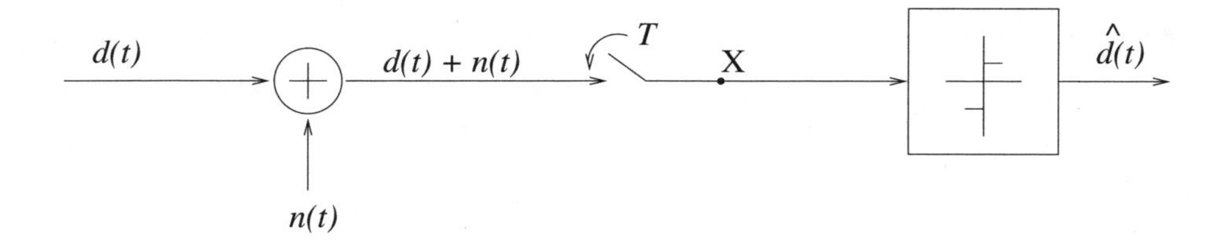

Figure 2.13: Simulation model of a BPSK modem

where T is the bit period and $R = 1/T$ is the data bit rate.

For a simple BPSK modem as shown in Figure 2.13, the bandwidth, W_b, at point X, is equal to the data bit rate, R. Therefore, at point X, we have $E_b/N_0 = \overline{\text{SNR}}$.

For the DS-CDMA system shown in Figure 2.12, at point X, the bandwidth requirement was extended due to the spreading of the signal by a factor of:

$$\frac{W_b}{R} = \frac{T}{T_c} \tag{2.92}$$

$$= Q. \tag{2.93}$$

Hence, we have:

$$\frac{E_b}{N_0} = \overline{\text{SNR}} \times \frac{T}{T_c} \tag{2.94}$$

$$= \overline{\text{SNR}} \times Q, \tag{2.95}$$

where T is the bit period and T_c is the period of one chip in the spreading code.

2.8.2 Simulated DS-CDMA BER Performance over Gaussian Channels for Synchronous Users

Our initial simulations were carried out in a synchronous environment, where all the users were perfectly synchronized at the receiver. The simulation results were compared with the analytical ones given in Equation 2.72, which exploited the Gaussian approximation and in Equation 2.87, which modelled a perfectly synchronous system. The graphs are presented in Figure 2.14, showing the BER versus E_b/N_0 curves for a synchronous BPSK/DS-CDMA system using Gold codes of length, $Q = 63$. Figures 2.14(a) and 2.14(b) have different number of users in the system, namely $K = 31$ and $K = 63$, respectively.

From the figures it can be seen that the simulation curves match the theoretical BER curves of the synchronous model quite well. Let us now consider explicitly the effect of the MAI due to different number of users.

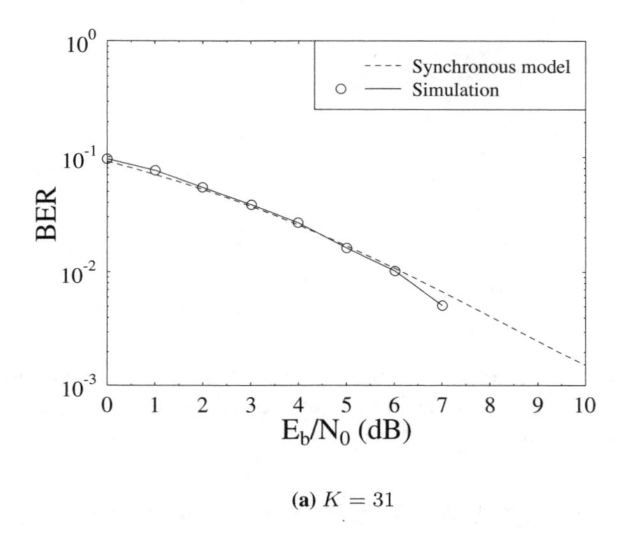

(a) $K = 31$

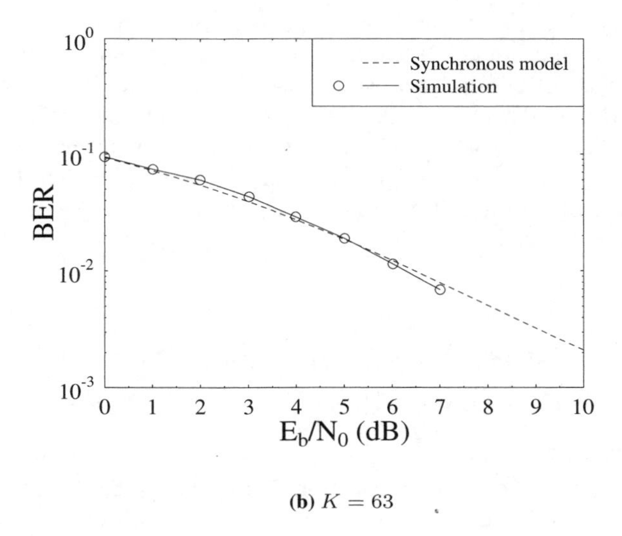

(b) $K = 63$

Figure 2.14: Simulated BER versus E_b/N_0 curves for a synchronous BPSK/DS-CDMA system using Gold codes of $Q = 63$ and supporting $K = 31$ users as well as $K = 63$ users as specified. The curve for the synchronous model was obtained from Equation 2.87.

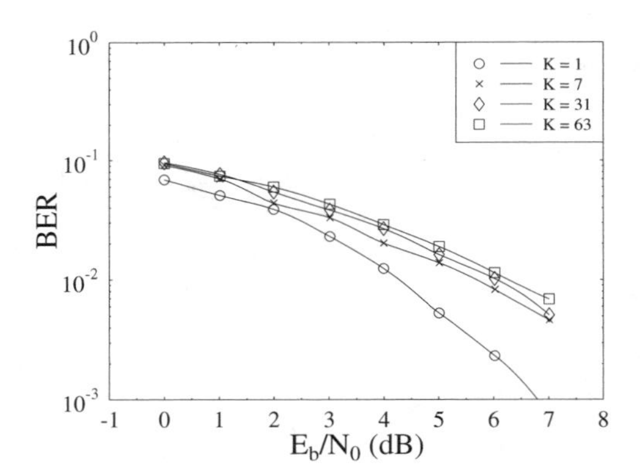

Figure 2.15: Comparison of simulation-based BER versus E_b/N_0 curves for a synchronous BPSK/DS-CDMA system for various numbers of users in the system, K. The spreading codes used were Gold codes having a length of $Q = 63$.

2.8.3 Simulated DS-CDMA BER Performance Versus the Number of Users over Gaussian Channels

Figure 2.15 shows the plot of BER versus E_b/N_0 curves for synchronous DS-CDMA simulations using different values of K. It was expected that as the number of users, K, in the system increased, the BER performance would also suffer due to the increase in multiple access interference (MAI). This was evident in the simulation results using Gold codes, as shown in Figure 2.15. However, it should be noted that the performance differences for $K = 7$, 31 and 63 are small compared to the difference between $K = 1$ and $K = 7$. This is due to the fact that the MAI term of Equation 2.87 has similar values for $K = 7$, 31 and 63, as explained in Section 2.7.2. We continue our elaborations by investigating the effect of the spreading code length.

2.8.4 Simulated DS-CDMA BER Performance Versus Spreading Code Length over Gaussian Channels

Figure 2.16 shows the BER versus E_b/N_0 performance for $K = 7$ users, in conjunction with different Gold code lengths, Q. Increasing Q from 7 to 31 and to 63 decreased the BER for a given E_b/N_0. However, there was only a slight difference between the BER curves for $Q = 31$ and $Q = 63$. This can be explained by examining the maximum in-phase cross-correlation (CCL) values, $R_{max}(0)$ of Equation 2.34, for the different values of Q in Table 2.3, where $Q = 31$ and $Q = 63$ had rather similar CCL values. Only the in-phase CCL values were considered, since the simulations were conducted in a synchronous environment, where the received signals of the different users were bit-synchronous.

From Table 2.3, $R_{max}(0)$ for $Q = 7$ is approximately three times higher than that of

Q	CCL ($R_{max}(0)$)
7	0.714
31	0.226
63	0.238
127	0.133

Table 2.3: Maximum normalized in-phase CCL values of Equation 2.34 for Gold codes of various lengths

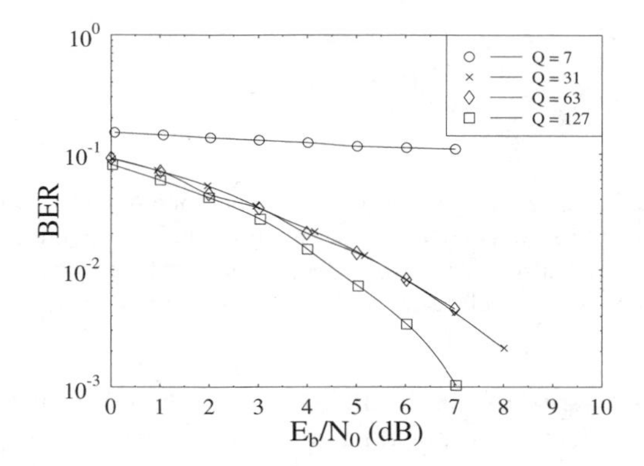

Figure 2.16: Comparison of BER versus E_b/N_0 curves for a synchronous BPSK/DS-CDMA system and for various Gold code lengths, Q, while supporting $K = 7$ users.

$Q = 31$ and $Q = 63$, which accounts for the BER curves of $Q = 31$ and $Q = 63$ showing lower BERs. However, again, the $R_{max}(0)$ values for $Q = 31$ and $Q = 63$ are similar, which results in the similarity between the BER curves despite their different code lengths. Again, for $Q = 127$, the $R_{max}(0)$ value is half as high as that of $Q = 63$, resulting in a better BER performance for $Q = 127$. Therefore, the performance of the system is dependent on the $R_{max}(0)$ values of Equation 2.34 and not specifically on Q. Let us now study the effect of various spreading codes.

2.8.5 Simulated DS-CDMA BER Performance for Spreading Code Sets Having Different Cross-Correlation Values over Gaussian Channels

Figure 2.17 offers a BER versus E_b/N_0 comparison for two different code sets, namely Gold codes with $Q = 63$ and Walsh codes for $Q = 64$. The number of users, $K = 15$ is the same for both code sets. The Walsh codes are perfectly orthogonal to each other. Therefore, in a perfectly synchronous environment, the BER curve matches that of the theoretical single-

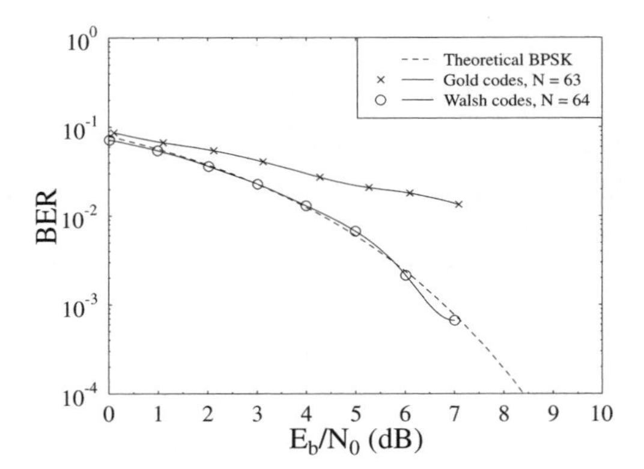

Figure 2.17: Simulated BER versus E_b/N_0 curves for synchronous BPSK/DS-CDMA supporting $K = 15$ users in the system, and comparing different types of codes, namely Gold codes and Walsh codes. Walsh codes have perfect cross-correlation properties, i.e. the codes are perfectly orthogonal to each other, while Gold codes are not perfectly orthogonal codes.

user BPSK curve. In comparison to the Gold codes, they perform much better. However, Walsh codes are not very effective as spreading codes, since they do not spread the spectrum of the signal sufficiently widely in order to cater for high multipath diversity gains, an issue to be treated in our forthcoming discussions throughout the thesis. The power spectral density (PSD) plot of a data signal spread by a Walsh code of length $Q = 64$ and the PSD of that spread by a Gold code of length $Q = 63$ are shown in Figures 2.18(a) and 2.18(b), respectively.

2.9 Discussion

In the preceding sections of this chapter, a basic portrayal of the conventional CDMA transmitter and receiver was provided. A few examples of binary sequences were presented and the effects of their cross-correlation values on the BPSK CDMA BER performance were examined. The BER performance for an asynchronous CDMA system utilizing the Gaussian approximation was presented along with a modified version for the bit-synchronous CDMA system. The theoretical and simulation performances of bit-synchronous CDMA systems in Gaussian channels employing the conventional correlation receiver were compared. It was concluded that for these simple CDMA systems, the BER performance was dependent on the cross-correlation between the spreading codes of the users and the total multiple access interference in the system. Many of our early assumptions were impractical. For instance, synchronization and timing errors at the receiver will increase the correlation between spreading codes and increase the MAI. Furthermore, synchronous transmission is difficult to achieve practically for the up-link, since the transmissions from different mobiles will commence and terminate at different times. In a wideband channel the orthogonality between spreading

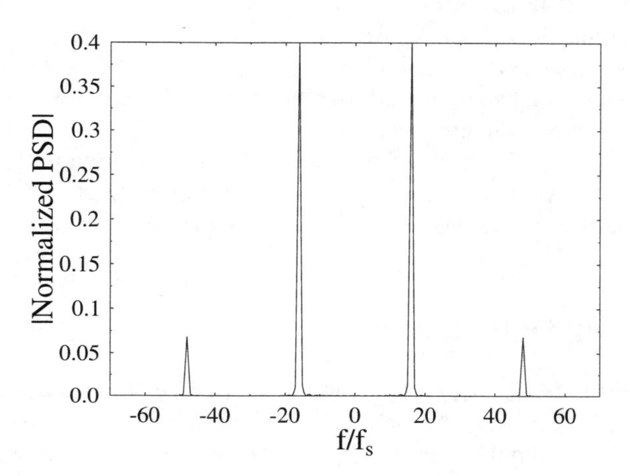

(a) PSD plot of a signal that has been spread with a Walsh code of length $Q = 64$

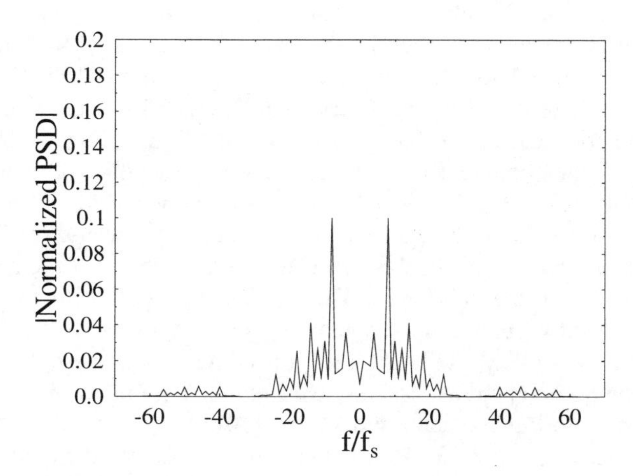

(b) PSD plot of a signal that has been spread with a Gold code of length $Q = 63$

Figure 2.18: Power spectral density plots of a signal that has been spread with two different codes, in order to show the difference in bandwidth spread.

codes will be destroyed and this increases the cross-correlation between users and degrades the BER performance. Other propagation phenomena such as fast fading, shadowing and path loss affect the power of the received signal making it impossible to achieve perfect power control. Imperfect power control has a very significant effect, because a strong signal from one source can completely obscure a weak signal from another source. This will render the BER performance for some users much worse than the average. Finally, the simulations were carried out in a single cell environment. In practice, interference arises both from users within the same cell and also from other cells.

Having gained a basic knowledge of the theoretical and simulation performance of the conventional BPSK/DS-CDMA receiver in a Gaussian environment, let us turn our attention to the non-conventional multi-user detection techniques in the forthcoming section.

2.10 Multi-user Detection

Multiple access communications using DS-CDMA is interference-limited due to the multiple access interference (MAI) generated by the users transmitting within the same bandwidth simultaneously. The signals from the users are separated by means of spreading sequences that are unique to each user. These spreading sequences are usually non-orthogonal. Even if they are orthogonal, the asynchronous transmission or the time-varying nature of the mobile radio channel may partially destroy this orthogonality. The non-orthogonal nature of the codes results in MAI, which degrades the performance of the system. The frequency selective mobile radio channel also gives rise to inter-symbol interference (ISI) due to multi-path propagation. This is exacerbated by the fact that the mobile radio channel is time-varying.

Conventional CDMA detectors – such as the matched filter [90, 104] and the RAKE combiner [92] – are optimized for detecting the signal of a single desired user. RAKE combiners exploit the inherent multi-path diversity in CDMA, since they essentially consist of matched filters for each resolvable path of the multipath channel. The outputs of these matched filters are then coherently combined according to a diversity combining technique, such as maximal ratio combining, equal gain combining or selection diversity combining [92]. These conventional single-user detectors are inefficient, because the interference is treated as noise and there is no utilization of the available knowledge of the mobile channel or the spreading sequences of the interferers. The efficiency of these detectors is dependent on the cross-correlation (CCL) between the spreading codes of all the users. The higher the cross-correlation, the higher the MAI. This CCL-induced MAI is exacerbated by the multi-path channel or by asynchronous transmissions. Conventionally, this MAI is reduced by the use of voice-activity monitoring and cell sectorization [10]. The utilization of these conventional receivers results in an interference-limited system and soft hand-over capabilities are required in order to provide an acceptable grade of service [93]. Another weakness of the conventional CDMA detectors is the phenomenon known as the "near–far effect" [10, 93]. For conventional detectors to operate efficiently, the signals from all the users have to arrive at the receiver with approximately the same power. A signal that has a much weaker signal strength compared to the other signals will be "swamped" by the relatively higher powers of the other signals and the quality of the weaker signal at the output of the conventional receiver will be severely degraded. Therefore, stringent power control algorithms are needed to ensure that the signals arrive at relatively similar powers at the receiver, in order to achieve similar

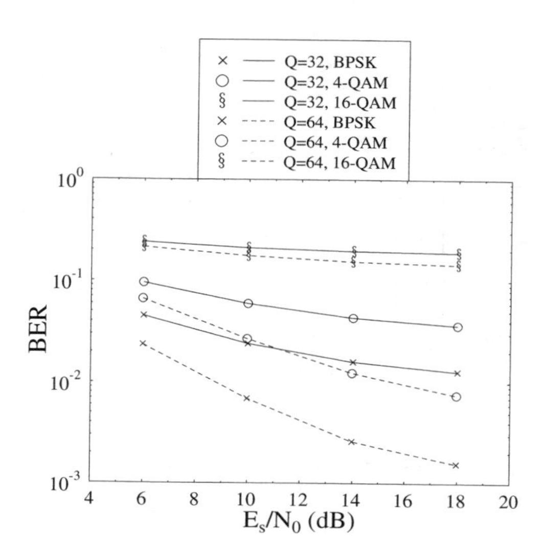

Figure 2.19: BER performance curves for the RAKE receiver with $K = 8$ users using BPSK, and over the seven-path Bad Urban channel with the impulse response shown in Figure 2.20. Different modulation modes, including BPSK, 4-QAM and 16-QAM, were investigated along with the spreading sequence lengths of $Q = 32$ and $Q = 64$.

qualities of service for different users [10]. Using conventional detectors to detect a signal corrupted by MAI, while encountering a hostile channel results in an irreducible BER, even if the E_s/N_0 is increased. This is because at high E_s/N_0 values the errors due to thermal noise are insignificant compared with the errors caused by the MAI and the channel. Therefore, detectors that can reduce or remove the effects of MAI and ISI are needed in order to achieve capacity gains. These detectors also have to be "near–far resistant" in order to avoid the need for stringent power control requirements. The performance of RAKE receivers for a synchronous up-link DS-CDMA system is shown in Figure 2.19 in conjunction with BPSK, 4-QAM and 16-QAM modulation, where an error floor is observed for all the different modulation modes, including BPSK. The simulations were carried out over the COST 207 [105] seven-path Bad Urban channel shown in Figure 2.20. The performance improves, however, with an increase in the spreading sequence length, Q, although the error floor still remains.

The COST 207 [105] channel profiles were developed approximately around the time when the GSM system was standardized, in order to assist, for example, in GSM performance investigations. The third generation wideband CDMA systems [64] have, however, a higher bandwidth and a significantly higher chip rate than the bit rate of the GSM system [64], thus leading to a higher multipath resolution and a larger number of resolvable multipath components. Nonetheless, in our performance investigations the COST 207 channels were adopted for the sake of the comparability of our results, since these channels are widely utilized in the CDMA research community.

In order to mitigate the problem of MAI, Verdú [106] proposed and analyzed the optimum multi-user detector for asynchronous Gaussian multiple access channels. The optimum

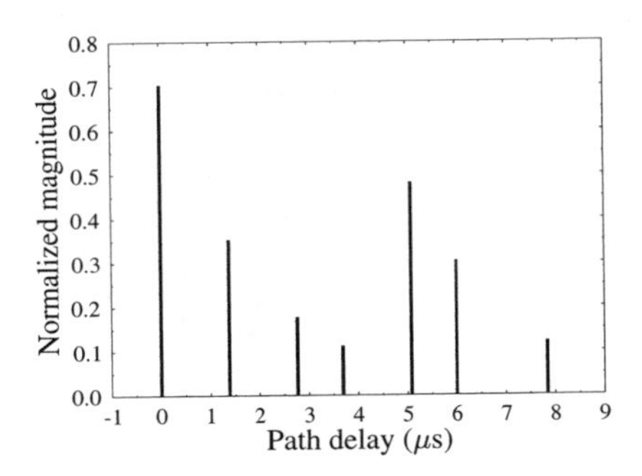

Figure 2.20: Normalized channel impulse response for the COST 207 [105] seven-path Bad Urban channel.

detector searches all the possible bit sequences in order to find the sequence that maximizes the correlation metric given by [78]:

$$\Omega(\mathbf{r}, \mathbf{d}) = 2\mathbf{d}^T\mathbf{r} - \mathbf{d}^T\mathbf{R}\mathbf{d}, \tag{2.96}$$

where the elements in vector $\mathbf{r}$ represent the cross-correlation of the received signal with each of the user spreading sequences, the vector $\mathbf{d}$ consists of the bits transmitted by the users and the matrix $\mathbf{R}$ is the correlation matrix of the spreading sequences. This optimum detector significantly outperforms the conventional detector and it is near–far resistant, but unfortunately its complexity grows exponentially in the order of $O(2^{NK})$, where N is the number of overlapping asynchronous bits considered in the detector window and K is the number of interfering users. In order to reduce the complexity of the receiver and yet provide acceptable BER performances, significant research efforts have been invested in the field of sub-optimal CDMA multi-user receivers [78]. Multi-user detection exploits the base station's knowledge of the spreading sequences and that of the estimated channel impulse response (CIR) in order to remove the MAI. These multi-user detectors can be categorized in a number of ways, such as linear versus non-linear, adaptive versus non-adaptive algorithms or burst transmission versus continuous transmission regimes. Excellent summaries of some of these sub-optimum detectors can be found in the monographs by Prasad [86], Glisic and Vucetic [88] and Verdú [78]. Other MAI-mitigating techniques include the employment of adaptive antenna arrays in order to suppress the level of MAI at the receiver. Research efforts invested in this area include, among others, research carried out by Thompson, Grant and Mulgrew [107, 108]; Naguib and Paulraj [109]; Godara [110]; as well as Kohno, Imai, Hatori and Pasupathy [111]. However, the area of adaptive antenna arrays is beyond the scope of this thesis and the reader is referred to the references cited for further discussion. In the next section, a brief survey of the sub-optimal multi-user receivers will be presented.

2.10.1 Survey of Multi-user Receivers

Following the seminal work by Verdú [106], numerous sub-optimum multi-user detectors have been proposed for a variety of channels, data modulation schemes and transmission formats Lupas and Verdú [112] initially suggested a sub-optimum linear detector for symbol-synchronous transmissions and further developed it for asynchronous transmissions in a Gaussian channel [113]. This linear detector inverted the CCL matrix, which was constructed from the spreading codes of the users and was termed the decorrelating detector. It was shown that this decorrelator exhibited the same degree of near–far resistance as the optimum multi-user detector. A further sub-optimum multi-user detector investigated was the minimum mean square error (MMSE) detector, where a biased version of the CCL matrix was inverted in order to optimize the receiver obeying the MMSE criterion Zvonar and Brady [114] proposed a multi-user detector for a synchronous CDMA system designed for a frequency-selective Rayleigh fading channel. This approach also used a bank of matched filters followed by a whitening filter, but maximal ratio combining was used to combine the resulting signals. The decorrelating detector of [113] was further developed for differentially-encoded coherent multi-user detection in flat fading channels by Zvonar *et al.* [115]. Zvonar also amalgamated the decorrelating detector with diversity combining, in order to achieve performance improvements in frequency selective fading channels [116]. A multi-user detector jointly performing decorrelating channel estimation and data detection was investigated by Kawahara and Matsumoto [117]. Path-by-path decorrelators were employed for each user in order to obtain the input signals required for channel estimation and the channel estimates as well as the outputs of a matched filter bank were fed into a decorrelator for demodulating the data. A variant of this idea was also presented by Hosseinian, Fattouche and Sesay [118], where training sequences and a decorrelating scheme were used for determining the channel estimate matrix. This matrix was then used in a decorrelating decision feedback scheme for obtaining the data estimates, Juntti, Aazhang and Lilleberg [119] proposed iterative schemes for obtaining the decorrelator and linear MMSE detectors in order to reduce the complexity Sung and Chen [120] advocated using a sequential estimator for minimizing the mean square error. The cross-correlations between the spreading codes and estimates of the faded amplitude of the received signal of each user were needed in order to obtain estimates of the transmitted data of each user. Duel-Hallen [121] proposed a decorrelating decision-feedback detector for removing the MAI from a synchronous system in a Gaussian channel. The outputs from a bank of filters matched to the spreading codes of the users were passed through a whitening filter. This filter was obtained by decomposing the CCL matrix of the user spreading codes through the Cholesky decomposition [122] technique. The results showed that MAI could be removed from each user successively, assuming that there was no error propagation. However, estimates of the received signal strengths of the users were needed, because the users had to be ranked in order of decreasing signal strengths so that the more reliable estimates were obtained first. The decorrelating feedback detector was improved by Wei and Schlegel [123] with a sub-optimum variant of the Viterbi algorithm, where only those metrics which were most likely were retained. Decorrelating decision feedback detection was improved with the assistance of soft-decision convolutional coding by Hafeez and Stark [124]. Soft decisions from a Viterbi decoder were fed back into the filter for signal cancellation.

The effect of MAI on the desired signal is similar to the impact of multipath propagation-induced ISI inflicted upon the same signal. Each user in a K-user system suffers from MAI due to the other $(K-1)$ users. This MAI can also be viewed as a single-user signal perturbed by ISI from $(K-1)$ paths in a multipath channel. Therefore, classic equalization techniques [90, 125] used to mitigate the effects of ISI can be modified for multi-user detection and these types of multi-user detectors can be classified as joint detection receivers. These joint detection receivers were developed for burst-based rather than continuous transmission. The concept of joint detection for the up-link was proposed by Klein and Baier [126] for synchronous burst transmission, where the performance of a zero-forcing block linear equalizer (ZF-BLE) was investigated for frequency-selective channels. Other joint detection schemes for up-link situations were also proposed by Jung, Blanz, Nasshan, Steil, Baier and Klein, such as the minimum mean-square error block linear equalizer (MMSE-BLE) [127–130], the zero-forcing block decision feedback equalizer (ZF-BDFE) [129, 130] and the minimum mean-square error block decision feedback equalizer (MMSE-BDFE) [129,130]. These joint-detection receivers were also combined with coherent receiver antenna diversity (CRAD) techniques [128–131] and turbo coding [132, 133] for performance improvement. Joint detection receivers were also proposed for downlink scenarios by Nasshan, Steil, Klein and Jung [134, 135]. Channel estimates were required for the joint detection receivers and some channel estimation algorithms were proposed by Steiner and Jung [136] for employment in conjunction with joint detection. Werner [137] extended the joint detection receiver by combining ZF-BLE and MMSE-BLE techniques with a multistage decision using soft inputs to a Viterbi decoder.

Interference cancellation (IC) schemes constitute another variant of multi-user detection and they can broadly be divided into three categories: parallel cancellation, successive cancellation and a hybrid of both. Varanasi and Aazhang [138] proposed a multistage detector for an asynchronous system, where the outputs from a matched filter bank were fed into a detector that performed MAI cancellation using a multistage algorithm. At each stage in the detector, the estimates of all the other users from the previous stage were used for reconstructing an estimate of the MAI and this estimate was then subtracted from the interfered signal representing the wanted bit. The computational complexity of this detector was linear with respect to the number of users, K. Varanasi further modified the parallel cancellation scheme in order to create a parallel group detection scheme for Gaussian channels [139] and later developed it further for frequency-selective slow Rayleigh fading channels [140]. In this scheme, K users were divided into P groups and each group was demodulated in parallel using a group detector. Yoon, Kohno and Imai [141] then extended the applicability of the multistage interference cancellation detector to a multipath, slowly fading channel. At each cancellation stage, hard decisions generated by the previous stage were used for reconstructing the signal of each user and for cancelling its contribution from the composite signal. The effects of CIR estimation errors on the performance of the cancellation scheme were also considered. A multi-user receiver that integrated MAI rejection and channel decoding was investigated by Giallorenzi and Wilson [142]. The MAI was cancelled via a multistage cancellation scheme and soft-outputs were fed from the Viterbi decoder of each user to each stage for improving the performance.

The parallel interference cancellation (PIC) receiver [138] was also modified for employment in multi-carrier modulation [143] by Sanada and Nakagawa, where convolutional coding was used in order to obtain improved estimates of the data for each user at the initial stage and these estimates were then utilized for interference cancellation in the following stages. The employment of convolutional coding improved the performance by 1.5 dB.

Latva-aho, Juntti and Heikkilä [144] enhanced the performance of the parallel interference cancellation receiver by feeding back channel estimates to the signal reconstruction stage of the multistage receiver and proposed an algorithm for mitigating error propagation. Dahlhaus, Jarosch, Fleury and Heddergott [145] combined multistage detection with channel estimation techniques utilizing the outputs of antenna arrays. The channel estimates obtained were fed back into the multistage detector in order to refine the data estimates. An advanced parallel cancellation receiver was also proposed by Divsalar, Simon and Raphaeli [146]. At each cancellation stage, only partial cancellation was carried out by weighting the regenerated signals with a less than unity scaling factor. At each following stage, the weights were increased based on the assumption that the estimates became increasingly accurate.

A simple successive interference cancellation (SIC) scheme was analyzed by Patel and Holtzman [147]. The received signals were ranked according to their correlation values, which were obtained by utilizing the correlations between the received signal and the spreading codes of the users. The transmitted information of the strongest user was estimated enabling the transmitted signal to be reconstructed and subtracted from the received signal. This was repeated for the next strongest user, where the reconstructed signal of this second user was cancelled from the composite signal remaining after the first cancellation. The interference cancellation was carried out successively for all the other users, until eventually only the signal of the weakest user remained. It was shown that the SIC receiver improved the BER and the system's user capacity over that of the conventional matched filter for the Gaussian, narrowband Rayleigh and dispersive Rayleigh channels. Multipath diversity was also exploited by combining the SIC receiver with the RAKE correlator [147]. Soong and Krzymien [148] extended the SIC receiver by using reference symbols in order to aid the CIR estimation. The performance of the receiver was investigated in flat and frequency-selective Rayleigh fading channels, as well as in emulated multi-cell scenarios. A soft-decision based adaptive SIC scheme was proposed by Hui and Letaief [149] where soft decisions were used in the cancellation stage and if the decision statistic did not satisfy a certain threshold, no data estimation was carried out for that particular data bit, in order to reduce error propagation.

Hybrid SIC and PIC schemes were proposed by Oon, Li and Steele [150, 151], where SIC was first performed on the received signal, followed by a multistage PIC arrangement. This work was then extended to an adaptive hybrid scheme for flat Rayleigh fading channels [152]. In this scheme, successive cancellation was performed for a fraction of the users, while the remaining users' signals were processed via a sub-parallel cancellation stage. Finally, multistage parallel cancellation was invoked. The number of serial and sub-parallel cancellations performed was varied adaptively according to BER estimates. Sawahashi, Miki, Andoh and Higuchi [153] proposed a pilot symbol-assisted multistage hybrid successive-parallel cancellation scheme. At each stage, data estimation was carried out successively for all the users, commencing with the user having the strongest signal and ending with the weakest signal. For each user, interference from other users was regenerated using the estimates of the current stage for the stronger users and the estimates of the previous stage for the weaker users. Channel estimates were obtained for each user by employing pilot symbols and a recursive estimation algorithm. Another hybrid successive and parallel interference cancellation receiver was proposed by Sun, Rasmussen, Sugimoto and Lim [154], where the users to be detected were split into a number of groups. Within each group, PIC was performed on the signals of these users belonging to the group. Between the separate groups, SIC was employed. This had the advantage of a reduced delay and improved performance compared to the SIC receiver. A further variant of the hybrid cancellation scheme was constituted by the combination of MMSE detectors with SIC receivers, as proposed by Cho and Lee [155]. Single-user MMSE detectors were used to obtain estimates of the data symbols, which were then fed back into the SIC stages. An adaptive interference cancellation scheme was investigated by Agashe and Woerner [156] for a multicellular scenario, where interference cancellation was performed for both in-cell interferers and out-of-cell interferers. It was shown that cancelling the estimated interference from users having weak signals actually degraded the performance, since the estimates were inaccurate. The adaptive scheme exercised interference cancellation in a discriminating manner, using only the estimates from users having strong received signals. Therefore signal power estimation was needed and the threshold for signal cancellation was adapted accordingly.

Several tree-search detection [157–159] receivers have been proposed in the literature, in order to reduce the complexity of the original maximum likelihood detection scheme proposed by Verdú [106]. Specifically, Rasmussen, Lim and Aulin [157] investigated a tree-search detection algorithm, where a recursive, additive metric was developed in order to reduce the search complexity. Reduced tree-search algorithms, such as the well-known M-algorithms [160] and T-algorithms [160] were used by Wei, Rasmussen and Wyrwas [158] in order to reduce the complexity incurred by the optimum multi-user detector. Motivated by the M-algorithm, at every node of the Viterbi search algorithm, only M paths were retained, depending on certain criteria such as the highest-metric M number of paths. Alternatively, all the paths that were within a fixed threshold, T, compared to the ideal metric were retained. At the decision node, the path having the highest metric was chosen as the most likely transmitted sequence. Maximal-ratio combining was also used in conjunction with the reduced tree-search algorithms and the combining detectors outperformed the "non-combining" detectors. The T-algorithm was combined with soft-input Viterbi detectors for channel-coded CDMA multi-user detection in the work carried out by Nasiri-Kenari, Sylvester and Rushforth [159]. The recursive tree-search detector generated soft-outputs, which were fed into single-user Viterbi channel decoders, in order to generate the bit estimates.

Multi-user projection receivers were proposed by Schlegel, Roy, Alexander and Jiang [161] and by Alexander, Rasmussen and Schlegel [162]. These receivers reduced the MAI by projecting the received signal onto a space which was orthogonal to the unwanted MAI, where the wanted signal was separable from the MAI.

In all the multi-user receiver schemes discussed earlier, all the required parameters except for the transmitted data estimates were assumed to be known at the receiver. In order to remove this constraint while reducing the complexity, adaptive receiver structures have been proposed [163]. An excellent summary of these adaptive receivers has been provided by Woodward and Vucetic [164]. Several adaptive algorithms have been introduced for approximating the MMSE receivers, such as the Least Mean Squares (LMS) [125] algorithm, the Recursive Least Squares (RLS) algorithm [125] and the Kalman filter [125]. Xie, Short and Rushforth [165] showed that the adaptive MMSE approach could be applied to multi-user receiver structures with a concomitant reduction in complexity. In the adaptive receivers employed for asynchronous transmission by Rapajic and Vucetic [163], training sequences were employed, in order to obtain the estimates of the parameters required. Lim, Rasmussen and Sugimoto introduced a multi-user receiver for an asynchronous flat-fading channel based on the Kalman filter [166], which compared favourably with the finite impulse response MMSE detector. An adaptive decision feedback joint detection scheme was investigated by Seite and Tardivel [167], where the least mean squares (LMS) algorithm was used to update the filter coefficients, in order to minimize the mean square error of the data estimates. New adaptive filter architectures for the downlink DS-CDMA receivers were suggested by Spangenberg, Cruickshank, McLaughlin, Povey and Grant [168], where an adaptive algorithm was employed in order to estimate the CIR, and this estimated CIR was then used by a channel equalizer. The output of the channel equalizer was finally processed by a fixed multi-user detector in order to provide the data estimates of the desired user.

The novel class of multi-user detectors, referred to as "blind" detectors, does not require explicit knowledge of the spreading codes and CIRs of the multi-user interferers. These detectors do not require the transmission of training sequences or parameter estimates for their operation. Instead, the parameters are estimated "blindly" according to certain criteria, hence the term "blind" detection. RAKE-type blind receivers have been proposed, for example by Povey, Grant and Pringle [169] for fast-fading mobile channels, where decision-directed channel estimators were used for estimating the multipath components and the output of the RAKE fingers was combined employing various signal combining methods. Liu and Li [170] also proposed a RAKE-type receiver for frequency-selective fading channels. In [170], a weight vector was utilized for each RAKE finger which was calculated based on maximizing the signal-to-interference-plus-noise ratio (SINR) at the output of each RAKE finger. Xie, Rushforth, Short and Moon [171] proposed an approximate Maximum Likelihood Sequence Estimation (MLSE) solution known as the per-survivor processing (PSP) type algorithm, which combined a tree-search algorithm for data detection with the aid of the Recursive Least Squares (RLS) adaptive algorithm used for channel amplitude and phase estimation. The PSP algorithm was first proposed by Seshadri [172]; as well as by Raheli, Polydoros and Tzou [173, 174] for blind equalization in single-user ISI-contaminated channels. Xie, Rushforth, Short and Moon extended their earlier work [171] in order to include the estimation of user-delays along with channel- and data-estimation [175]. Iltis and Mailaender [176] combined the PSP algorithm with the Kalman filter, in order to adaptively estimate the amplitudes and delays of the CDMA users. In other blind detection schemes, Mitra and Poor compared the application of neural networks and LMS filters for obtaining data estimates of the CDMA users [177]. In contrast to other multi-user detectors, which required the knowledge of the spreading codes of all the users, only the spreading code of the desired user was needed for this adaptive receiver [177]. An adaptive decorrelating detector was also developed by Mitra and Poor [178], which was used to determine the spreading code of a new user entering the system. Blind equalization was combined with multi-user detection for slowly fading channels in the work published by Wang and Poor [179]. Only the spreading sequence of the desired user was needed and a zero-forcing as well as an MMSE detector were developed for data detection. As a further solution, a sub-space approach to blind multi-user detection was also proposed by Wang and Poor [180], where only the spreading sequence and the delay of the desired user were known at the receiver. Based on this knowledge, a blind subspace tracking algorithm was developed for estimating the data of the desired user. Further blind adaptive algorithms were developed by Honig, Madhow and Verdú [181], Mandayam and Aazhang [182], as well as by Ulukus and Yates [183]. In [181], the applicability of two adaptive algorithms to the multi-user detection problem was investigated, namely that of the stochastic gradient algorithm and the least squares algorithm; while in [183] an adaptive detector that converged to the decorrelator was analyzed.

The employment of the Kalman filter for adaptive data, CIR and delay estimation was advocated, for example, by Lim and Rasmussen [184], demonstrating that the Kalman filter gave a good performance and exhibited a high grade of flexibility. However, the Kalman filter required reliable initial delay estimates in order to initialize the algorithm. Miguez and Castedo [185] modified the well-known constant modulus approach [186, 187] to blind equalization for ISI-contaminated channels in the context of multi-user interference suppression. Fukawa and Suzuki [188] proposed an orthogonalizing matched filtering detector, which consisted of a bank of despreading filters and a signal combiner. One of the despreading filters was matched to the desired spreading sequence, while the other despreading sequences were arbitrarily chosen such that the impulse responses of the filters were linearly independent of each other. The filter outputs were adaptively weighted in the complex domain with the criterion that the average output power of the combiner was minimized. A constraint was imposed on the combining process such that the combiner's response to the desired user's signal was kept constant. In another design, an iterative scheme used to maximize the log-likelihood function was the basis of the research by Fawer and Aazhang [189]. RAKE correlators were employed for exploiting the multipath diversity and the outputs of the correlators were fed to an iterative scheme for joint channel estimation and data detection using the Gauss-Seidel [190] algorithm.

Several hybrid multi-user receiver structures have also been proposed recently [191–194]. Bar-Ness [191] advocated the hybrid multi-user detector that consisted of a decorrelator for detecting asynchronous users, followed by a maximum-SNR data combiner, an adaptive canceller and another data combiner. The decorrelator matrix was adaptively determined. A multi-user receiver employing an iterative hybrid genetic algorithm as a search technique has also been proposed by Yen *et al.* [192]. The search-space for the most likely sequence was limited to a certain population of sequences and the sequences were updated at each iteration according to certain probabilistic, aptly termed genetic operations, known as *reproduction*, *cross-over* or *mutation* operations. Commencing with a population of tentative decisions, the best n sequences were selected as so-called "parent" sequences according to a fitness criterion in order to generate the "offspring" for the next generation of sequence estimates. The offspring of sequence estimates were generated by employing a uniform "cross-over process" where the bits between two parent sequences were exchanged according to a random mask and a certain probability value. Finally, "mutation" was performed where the value of a bit is flipped according to a certain probability. In order to prevent the loss of "high-fitness" individuals that were not selected as parents, the worst offspring was replaced by the best non-parent individual of the earlier generation.

Neural network-type receivers have also been proposed as CDMA receivers [195, 196]. Tanner and Cruickshank proposed a non-linear receiver that exploited neural-network structures and employed pattern recognition techniques for data detection [195]. This work [195] was extended to a reduced complexity neural network receiver for the downlink scenario [196].

Other novel techniques employed for mitigating the multipath fading effects inflicted upon multiple users include joint transmitter-receiver optimization proposed by Jang, Vojčić and Pickholtz [193, 194]. In these schemes, transmitter precoding was carried out, such that the mean squared errors of the signals at all the receivers were minimized. This required the knowledge of the CIRs of all the user channels and the assumption was made that the channel fading was sufficiently slow, such that the channel prediction could be employed reliably.

Recently, there has been significant interest in iterative detection schemes, where chan-

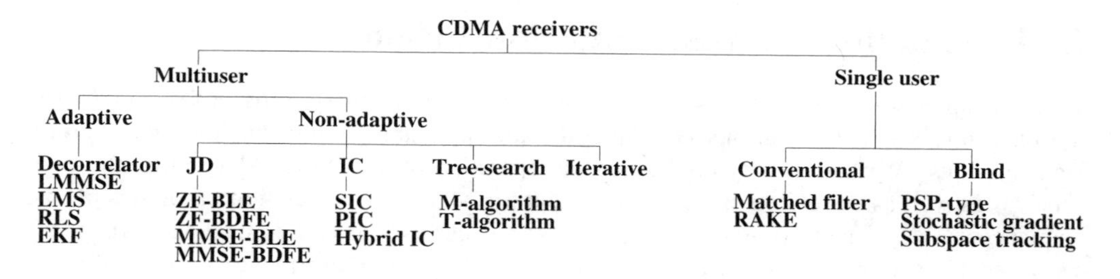

Figure 2.21: Classification of CDMA detectors

nel coding was exploited in conjunction with multi-user detection, in order to obtain a high BER performance. The spreading of the channel-coded symbols and their corruption by the wideband channel were viewed as a serially concatenated code structure, where the CDMA channel was viewed as the inner code and the single user convolutional codes made up the outer codes. After processing the received signal in a bank of matched filters or orthogonalizing whitening matched filter, the matched filter outputs were processed using the turbo-style iterative decoding (TEQ) [197] process. In this process, a multi-user decoder was used to produce confidence measures which were used as soft inputs to the single-user channel decoders. These single-user decoders then provided similar confidence metrics for the multi-user detector. This iterative process continued, until no further performance improvement was recorded.

Giallorenzi and Wilson [198] presented the maximum likelihood solution for the asynchronous CDMA channel, where the user data was encoded with the aid of convolutional codes. Near-single-user performance was achieved for the two-user case in conjunction with fixed-spreading codes. The decoder employed was based on the Viterbi algorithm, where the number of states increased exponentially with the product of the number of users and the constraint length of the convolutional codes. Later, a suboptimal modification of this technique was proposed [142], where the MAI was cancelled via multistage cancellation and the soft outputs from the Viterbi algorithm were fed to each stage for improving the performance. Following this, several proposals of iterative multi-user detection for channel-coded signals have been presented [199–204]. For example, Alexander, Astenstorfer, Schlegel and Reed [201, 203] proposed the multi-user maximum a-posteriori (MAP) detectors for the decoding of the inner CDMA channel code and single-user MAP decoders for the outer convolutional codes. A reduced complexity solution employing the M-algorithm [160] was also suggested which resulted in a complexity that increased linearly – rather than exponentially, as in [198] – with the number of users [202]. Wang and Poor [204] employed a soft-output multi-user detector for the inner channel code, which combined soft interference cancellation and instantaneous linear MMSE filtering, in order to reduce the complexity. These iterative receiver structures showed considerable promise and near-single-user performance was achieved at high SNRs.

Figure 2.21 portrays the classification of most of the CDMA detectors that have been discussed previously. All the acronyms for the detectors have been defined in the text. Examples of the different classes of detectors are also included.

2.11 Chapter Summary and Conclusion

In this chapter we commenced our discussions with a rudimentary introduction to CDMA systems and classified the various techniques that are applicable to spreading the signal before transmissions. We then briefly considered the operation of the DS-CDMA transmitter and receiver as well as the effects of the channel and the MUI. The correlation properties of m-sequences, Gold-sequences as well as Kasami-sequences were reviewed next and a range of basic results were presented for characterizing the achievable performance as a function of the various system parameters. The chapter was conscluded by a brief overview of various multi-user detectors.

In conclusion, multi-user detectors reduce the error floor due to MAI and this translates into user capacity gains for the system. If the performance of multi-user detectors is independent of the spreading codes used, then the codes can be chosen in order to optimize the system's spectral efficiency. These multi-user detectors are also near–far resistant to a certain extent and this results in less stringent power control requirements.

On the other hand, multi-user detectors are more complex than conventional detectors. Coherent detectors require knowledge of the CIR estimates, which means that a channel estimator is needed in the receiver and training sequences have to be included in the bursts that are transmitted. These multi-user detectors also exhibit an inherent latency, which results in a delayed reception.

Multi-user detection is more suitable for up-link receivers due to its increased complexity. A hand-held mobile receiver has to be compact and lightweight, rendering multi-user detection impractical for the downlink. Recent research into blind receivers has shown that data detection for the desired user – without using the spreading sequences and channel estimates of other users – is possible, as discussed in Section 2.10.1, hence using these detectors for downlink receivers may become a reality. However, the inherent latency in convergence and the increased complexity may still remain a limiting factor in these downlink receivers. less harmful to the mobile station, because all the signals of all the users in one transmission burst are generated at the same signal strength from the base station. The near–far problem becomes serious only when the interference from other cells is high. To mitigate the problem of MAI for the downlink, measures such as using spreading codes with good cross-correlation and auto-correlation properties can be employed. The cross-correlation property is less likely to be destroyed, as downlink transmissions can be made synchronous, especially in hybrid TDMA/CDMA systems such as the ones in the FRAMES proposal [205]. To combat fading, RAKE receivers can be employed, in order to exploit multi-path diversity.

Chapter 3

Joint Detection of CDMA Signals[1]

The joint detection receivers are derivatives of the well-known single-user equalizers, which are used to equalize signals that have been distorted by inter-symbol interference (ISI) due to multipath channels. Therefore, as a precursor to joint detection, a brief introduction to single-user equalization is presented in Section 3.1. This is followed by the portrayal of the theoretical basis of joint detection CDMA receivers and the complexity estimation of four joint detection algorithms in Sections 3.3 and 3.4. Finally, the BER performance of the JD receivers for various scenarios is characterized and discussed in Sections 3.5 and 3.6.

3.1 Basic Equalizer Theory

Single user transmissions over a mobile channel suffer from ISI due to multipath propagation. In order to remove the linear distortion caused by ISI, equalization techniques are used, which are well documented in the literature [90, 92, 206]. Therefore, only a brief introduction to some linear equalizers is presented next, in order to set the scene for the multiuser detection schemes, which invoke these schemes.

Figure 3.1 shows the magnitudes of the channel at instances of x_n for each symbol, y_j, in a transmission burst, where $n = 1, 2, \ldots$ and $j = 1, 2, \ldots$. The x-axis represents the instances where the channel impulse response (CIR) taps were recorded, the y-axis represents the symbol index for each symbol transmitted through the channel and the z-axis represents magnitude. If we consider the x and z axes for a specific value of y_n on the y-axis, the values on the x-axis represent the values of the faded CIR taps for one received symbol at consecutive recording instances, x_n. By contrast, if we consider now the cross-section for a fixed value of x_n, the y-axis represents the consecutive received symbol indices in the transmission burst at the recording instance of x_n.

[1] *Single- and Multi-Carrier CDMA Multi-User Detection, Space-Time Spreading, Synchronization, Networking and Standards.* L.Hanzo, L-L.Yang, E-L.Kuan and K.Yen,
 ISBN 0-470-86309-9

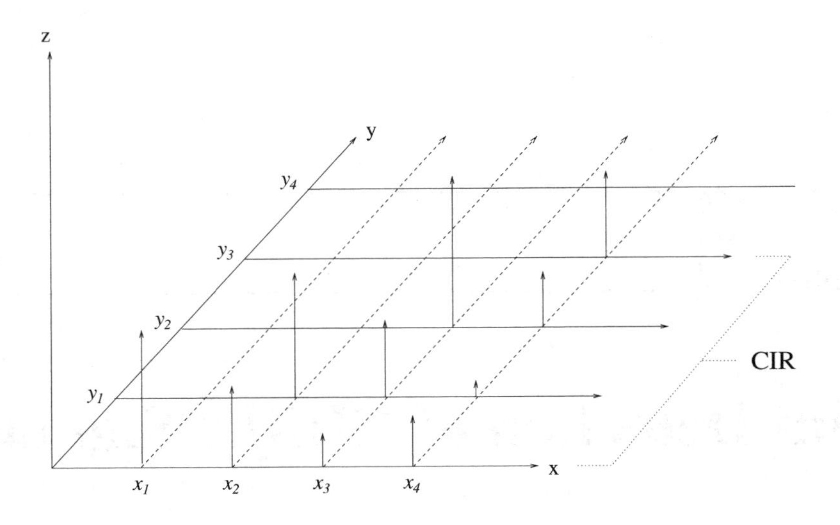

Figure 3.1: Stylized three-dimensional representation of the magnitudes of a mobile channel at instances of x_n for each symbol, y_j, in a transmission burst, where $n = 1, 2, \ldots$ and $j = 1, 2, \ldots$. The CIR at instance x_n is given by the CIR profile parallel to the y-axis at x_n. At x_4, for example, the CIR affects four consecutive received symbols, namely y_0, y_1, y_2, y_3.

If we assume that the CIR is known or can be accurately estimated at the receiver, then the received signal can be equalized. By referring to Figure 3.1, the channel estimates for a fixed sampling instance of x_n can be obtained by taking a cross-section of the graph across the x-axis at x_n. Linear equalizers such as the zero-forcing (ZF) equalizer and the minimum mean square error (MMSE) equalizer can also be viewed as linear filters with an infinite number of taps. Ignoring noise for the moment, let the received signal be the convolution of the transmitted signal d_n and the CIR, b_n, giving:

$$r_i = \sum_{j=-\infty}^{\infty} d_j b_{i-j}. \tag{3.1}$$

In order to recover the transmitted signal, an equalizer with an impulse response f_n is employed and the output of the equalizer is given by:

$$\begin{aligned} \hat{d}_n &= \sum_{i=-\infty}^{\infty} r_i f_{n-i} && (3.2) \\ &= \sum_{i=-\infty}^{\infty} \sum_{j=-\infty}^{\infty} d_j b_{i-j} f_{n-i} && (3.3) \\ &= \sum_{j=-\infty}^{\infty} d_j \sum_{i=-\infty}^{\infty} f_{n-i} b_{i-j}. && (3.4) \end{aligned}$$

Letting $n - i = v$, leads to:

$$\hat{d}_n = \sum_{j=-\infty}^{\infty} d_j \sum_{v=-\infty}^{\infty} f_v b_{n-j-v}. \tag{3.5}$$

If the convolution of b_n and f_n, is represented as:

$$g_n = \sum_{v=-\infty}^{\infty} f_v b_{n-v}, \tag{3.6}$$

then the data estimates at the output of the equalizer can be expressed as:

$$\hat{d}_n = \sum_{j=-\infty}^{\infty} d_j g_{n-j} \tag{3.7}$$

$$= d_n g_0 + \sum_{j \neq n} d_j g_{n-j}. \tag{3.8}$$

3.1.1 Zero-Forcing Equalizer

In our discussion on zero-forcing equalizers, the approach of Messerschmitt [206] will be followed. The zero-forcing criterion [206] constrains the signal component at the output of the equalizer to be free of ISI. Referring to Equation 3.8, this implies that

$$d_n g_0 = d_n \quad \text{and} \quad \sum_{j \neq n} d_j g_{n-j} = 0, \tag{3.9}$$

which then leads to

$$g_n = \sum_{j=-\infty}^{\infty} f_j b_{n-j} = \begin{cases} 1 & (n = 0) \\ 0 & (n \neq 0). \end{cases} \tag{3.10}$$

The z-transform of Equation 3.10 is given by:

$$G(z) = F(z)B(z) = 1, \tag{3.11}$$

$$F(z) = \frac{1}{B(z)}, \tag{3.12}$$

where $G(z)$, $F(z)$ and $B(z)$ are the z-transforms of g_n, f_n and b_n, respectively. This shows that the zero-forcing equalizer is constituted by the inverse filter of the channel, which has the transfer function, $B(z)$. Denoting $D(z)$ and $N(z)$ to be the z-transforms of the transmitted signal and the additive noise respectively, the z-transform of the received signal can be

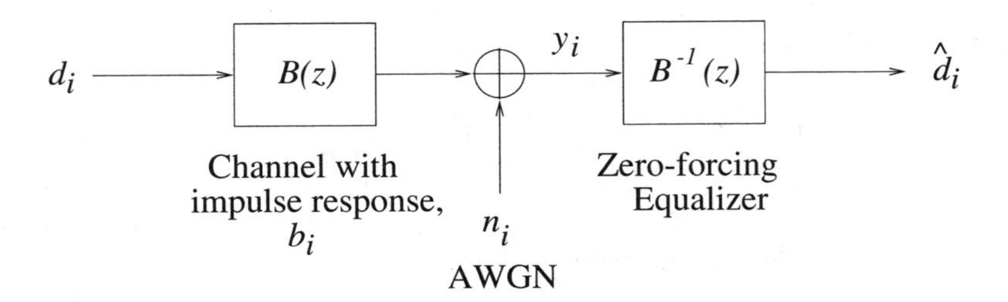

Figure 3.2: Block diagram of a simple transmission scheme using a zero-forcing equalizer.

represented by $R(z)$, where

$$R(z) = D(z)B(z) + N(z). \tag{3.13}$$

The z-transform of the equalizer output will be

$$\begin{aligned} \hat{D}(z) &= F(z)R(z) && (3.14) \\ &= \frac{R(z)}{B(z)} && (3.15) \\ &= D(z) + \frac{N(z)}{B(z)}. && (3.16) \end{aligned}$$

From Equation 3.16, it can be seen that the output signal is free of ISI. However, the noise component is enhanced by the inverse of the transfer function of the channel. This has a disastrous effect on the output of the equalizer, in terms of noise amplification in the frequency domain at frequencies where the transfer function of the channel was severely faded.

Figure 3.2 shows the block diagram of a simple transmission scheme communicating over a channel having an impulse response b_i, which is equalized by a zero-forcing equalizer (ZFE). The noise enhancement at the output of the ZFE is more explicitly observed by scrutinizing the receiver in the z-domain. By utilizing Equation 3.16, the z-transform of the error, $E(z)$, is given by:

$$\begin{aligned} E(z) &= \hat{D}(z) - D(z) && (3.17) \\ &= \frac{N(z)}{B(z)}. && (3.18) \end{aligned}$$

The power associated with the mean squared error at the output of the ZFE is obtained by calculating the integral of the power spectral density around the unit circle, where $z = e^{j\omega}$, and is given by:

$$P = \int_{-\pi}^{\pi} |E(z = e^{j\omega})|^2 d\omega. \tag{3.19}$$

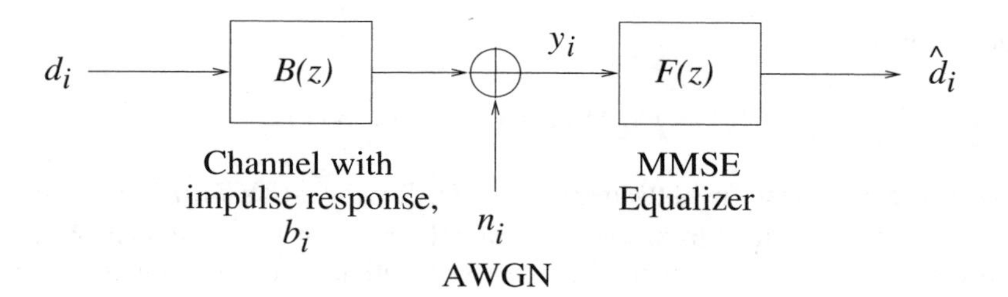

Figure 3.3: Block diagram of a simple transmission scheme employing an MMSE equalizer.

The term $|E(z)|^2$ represents the power spectral density (PSD) of the mean square error at the output of the ZFE and using Equation 3.18 it can be expanded into:

$$|E(z)|^2 = \frac{N(z)N^*(z)}{B(z)B^*(z)}. \tag{3.20}$$

If we assume that the noise input to the receiver is Gaussian and has a double-sided noise power spectral density of $N_0/2$, then $N(z)N^*(z) = N_0/2$ and therefore:

$$|E(z)|^2 = \frac{N_0}{2|B(z)|^2}. \tag{3.21}$$

From Equation 3.21, it can be seen that the MSE is dependent on the inverse of the z-domain transfer function of the channel. At frequencies where the channel transfer function is severely attenuated, the equalizer attempts to boost the signal power by increasing the gain at that frequency. Unfortunately, this also results in a proportional increase in noise power at the same frequency. Therefore, the output of the ZFE suffers from SNR degradation. As another alternative, let us now consider the minimum mean square error linear equalizer.

3.1.2 Minimum Mean Square Error Equalizer

Minimum mean square error (MMSE) equalizers have been considered in depth in [206] and a similar approach is followed here. Upon invoking the MMSE criterion [206], the equalizer tap coefficients are calculated in order to minimize the MSE at the output of the equalizer, where the MSE is defined as:

$$e_k^2 = E[|d_k - \hat{d}_k|^2], \tag{3.22}$$

where the function $E[x]$ indicates the expected value of x. Figure 3.3 shows the system's schematic using an MMSE equalizer. In order to show that the MMSE equalizer minimizes the MSE, let us consider the equalizer in the z-domain [206]. In Figure 3.3, $B(z)$ is the channel transfer function and $F(z)$ is the transfer function of the equalizer. The output of the equalizer is given by:

$$\hat{D}(z) = F(z)B(z)D(z) + F(z)N(z), \tag{3.23}$$

where $D(z)$ is the z-transform of the data bits d_i, $\hat{D}(z)$ is the z-transform of the data estimates $\hat{d}_i$ and $N(z)$ is the z-transform of the noise samples n_i. For the remainder of this section, the 'z' is dropped from the z-transform for ease of notation. The mean squared error, $|E|^2$, is expressed as [206]:

$$\begin{aligned}
|E|^2 &= |\hat{D} - D|^2 && (3.24)\\
&= (\hat{D} - D)\,(\hat{D} - D)^* && (3.25)\\
&= [D(FB-1) + FN]\,[D(FB-1) + FN]^* && (3.26)\\
&= DD^*(FB-1)(F^*B^*-1) + FNF^*N^* && (3.27)\\
&= |D|^2(FB-1)(F^*B^*-1) + |N|^2|F|^2, && (3.28)
\end{aligned}$$

where the terms $FN[D(FB-1)]^*$ and $F^*N^*[D(FB-1)]$ have been dropped, because the noise samples are assumed to be uncorrelated with the other terms. Since D is the z-transform of the data bits that have a constant variance, $|D|^2$ can be written as σ_D^2, which is the power of the data signal. It is also assumed that the noise at the receiver input is Gaussian with a double-sided noise power spectral density (PSD) of $N_0/2$. This leads to [206]:

$$\begin{aligned}
|E|^2 &= \sigma_D^2\Big(|F|^2|B|^2 - FB - F^*B^* + 1\Big) + \frac{N_0}{2}|F|^2 && (3.29)\\
&= \sigma_D^2|F|^2|B|^2 + \frac{N_0}{2}|F|^2 - \sigma_D^2 FB - \sigma_D^2 F^*B^* + \sigma_D^2 && (3.30)\\
&= \Big(\sigma_D^2|B|^2 + \frac{N_0}{2}\Big)\Big[|F|^2 - \Big(\frac{\sigma_D^2 FB + \sigma_D^2 F^*B^*}{\sigma_D^2|B|^2 + \frac{N_0}{2}}\Big)\Big] + \sigma_D^2 && (3.31)\\
&= K_e\Big[\Big(F - \frac{\sigma_D^2 B^*}{K_e}\Big)\Big(F^* - \frac{\sigma_D^2 B}{K_e}\Big) - \frac{\sigma_D^4|B|^2}{K_e^2}\Big] + \sigma_D^2 && (3.32)\\
&= K_e\Big|F - \frac{\sigma_D^2 B^*}{K_e}\Big|^2 - \frac{\sigma_D^4|B|^2}{K_e} + \sigma_D^2, && (3.33)
\end{aligned}$$

where $K_e = \sigma_D^2|B|^2 + N_0/2$ is a constant value. In order to find an equalizer transfer function, F, that minimizes $|E|^2$, the following equation has to be solved:

$$\Big|F - \frac{\sigma_D^2 B^*}{K_e}\Big|^2 = 0, \tag{3.34}$$

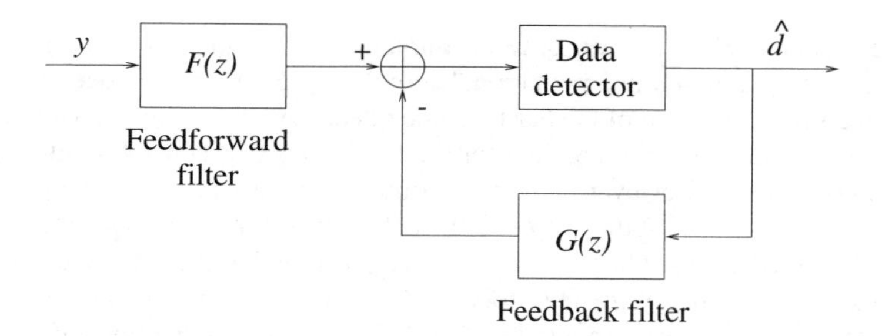

Figure 3.4: Block diagram of a decision feedback equalizer.

since the other terms are not dependent on the transfer function of the equalizer. The MMSE equalizer is thus derived as:

$$F = \frac{\sigma_D^2 B^*}{K_e} \tag{3.35}$$

$$= \frac{\sigma_D^2 B^*}{\sigma_D^2 |B|^2 + N_0/2}. \tag{3.36}$$

Upon substituting Equation 3.34 into Equation 3.33, the MSE at the output of the MMSE equalizer is given by:

$$|E|^2 = \sigma_D^2 - \frac{\sigma_D^4 |B|^2}{K_e}. \tag{3.37}$$

Substituting with $K_e = \sigma_D^2 |B|^2 + N_0/2$ from above, leads to:

$$|E|^2 = \frac{N_0}{2|B|^2 + N_0/\sigma_D^2}. \tag{3.38}$$

Comparing the mean squared error at the output of the ZFE which was given in Equation 3.21, and that of the MMSE equalizer in Equation 3.38, it can be seen that the MSE at the output of the MMSE equalizer has an extra positive term of N_0/σ_D^2 in the denominator. Therefore, the MSE at the output of the MMSE equalizer is always lower or equal to the mean squared error at the output of the ZFE. Let us now briefly consider the family of decision feedback equalizers [206].

3.1.3 Decision Feedback Equalizers

The decision feedback equalizer (DFE) [206] can be separated into two components, a feedforward filter and a feedback filter. A schematic of a general DFE is depicted in Figure 3.4. The philosophy of the DFE is two-fold. First, it aims to reduce the filter-order of the ZFE since in Equations 3.10 and 3.12, as well as Figure 3.2, it becomes explicit that the inverse filter of the channel, $B^{-1}(z)$, can only be implemented as an Infinite Impulse Response (IIR) filter, requiring a high implementational complexity. Second, provided that there are no transmission errors, the output of the hard-decision detector delivers the transmitted data bits, which can provide valuable training data for the DFE. Hence a reduced-length feed-forward filter can be used, which, however, does not entirely eliminate the ISI. Instead, the feedback filter uses the data estimates at the output of the data detector in order to subtract the ISI from the output of the feed-forward filter, such that the input signal to the data detector has less ISI than the signal at the output of the feed-forward filter. If it is assumed that the data estimates fed into the feedback filter are correct, then the DFE is superior to the linear equalizers, since the noise enhancement is reduced. One way of explaining this would be to say that if the data estimates are correct, then the noise has been eliminated and there is no noise enhancement in the feedback loop. However, if the data estimates are incorrect, these errors will propagate through to future decisions and this problem is known as error propagation.

There are two basic DFEs, the ZF-DFE and the MMSE-DFE. Analogous to its linear counterpart, the coefficients of the feedback filter for the ZF-DFE are calculated so that the ISI at the output of the feed-forward filter is eliminated and the input signal to the data detector is free of ISI [92]. Let us now focus our attention on CDMA multiuser detection equalizers.

3.1.4 Equalizer Modifications for Joint Detection

By concatenating the data symbols of each CDMA user successively, as though they were transmitted by one user, we can visualize the similarities between ISI and MAI and apply the principles of conventional TDMA-oriented channel equalization [92] to multiuser detection. The concatenated data frame is thus:

$$\mathbf{d} = d_1^1, d_2^1, \ldots, d_N^1, d_1^2, d_2^2, \ldots, d_N^2, \ldots, d_1^K, \ldots d_N^K, \tag{3.39}$$

where $\mathbf{d}$ represents the data frame and d_n^k represents the n-th symbol of the k-th user for $n = 1, 2, \ldots N$ and $k = 1, 2, \ldots K$.

Figure 3.5 is a modification of Figure 3.1, where the CIR magnitudes of multiple users transmitting simultaneously on the uplink are shown, and both ISI and MAI are visualized. If each user in a K–user CDMA system transmits N symbols per transmission burst, then the indices y_1 to y_N represent the symbols of user $k = 1$, the indices y_{N+1} to y_{2N} represent the symbols of user $k = 2$ and so on. Comparing Figures 3.1 and 3.5, both graphs are similar and again, a cross-section across the x-axis of Figure 3.5 will be similar to its counterpart in Figure 3.5. By referring to Figure 3.5, the CIR at instant x_n is related to the cross-section of the graph across the x-axis at x_n, while the instances x_n can be interpreted as consecutive symbol transmission instances, when a different CIR is experienced. With this information, a single user equalizer can be modified to equalize the multiuser signals by treating the signals from multiple users as inter-symbol interference. This is the principle upon which joint detection is based. The matrix $\mathbf{A}$ in Figure 3.9, which is described later in Section 3.2, is a matrix representation of the three dimensional graph in Figure 3.5. Let us now consider a system model of the uplink of a synchronous CDMA burst transmission system.

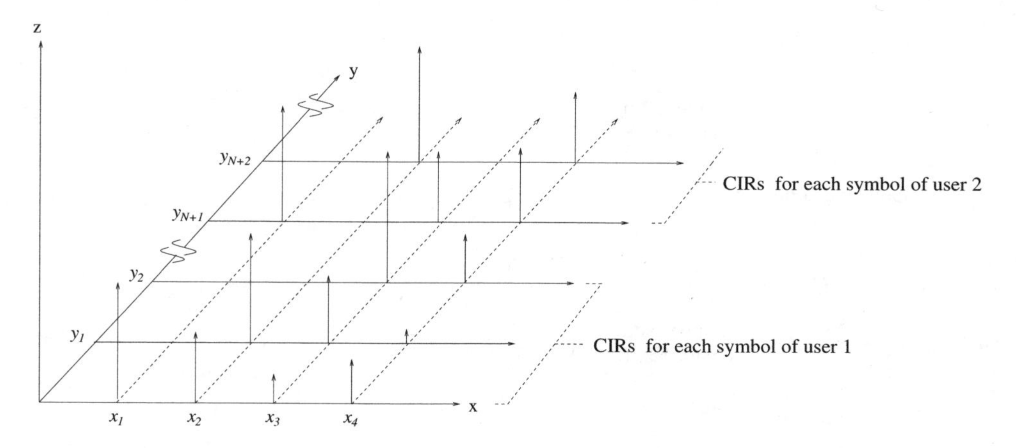

Figure 3.5: Stylized three-dimensional representation of the magnitudes of the CIRs for different users transmitting simultaneously on the uplink, at instances of x_n for each symbol, y_j, where $n = 1, 2, \ldots$ and $j = 1, 2, \ldots$. The transmitted symbols, $y_1, \ldots, y_N$, are transmitted by user 1, the symbols $y_{N+1}, \ldots, y_{2N}$ by user 2 and so on. By concatenating the transmitted symbols of all the users as one long burst transmitted by a single user, the CIR at instance x_n is given by the CIR profile parallel to the y-axis at x_n. At x_4, for example, the CIR affects four symbols, namely $y_0, y_1, y_2, y_{N+1}, y_{N+2}$.

3.2 System Model

The notations used in the following sections are as follows:

- Capital letters in boldface represent matrices, for instance $\mathbf{X}$.
- Column vectors are presented in lowercase letters, for instance $\mathbf{x}$.
- The vector, $\mathbf{x}^{(k)}$, represents the vector $\mathbf{x}$ of user k.
- The matrix $\mathbf{R}$ is used to represent covariance matrices. For example, $\mathbf{R}_x$ represents the covariance matrix of the vector $\mathbf{x}$, where $\mathbf{R}_x = E[\mathbf{x}\mathbf{x}^H]$.
- The asterisk (*) superscript is used to indicate complex conjugation. Therefore $\mathbf{X}^*$ is used to denote the complex conjugate matrix of $\mathbf{X}$ and a^* represents the complex conjugate of the variable a.
- The notation $\mathbf{X}^T$ implies the transpose matrix of $\mathbf{X}$.
- The notation $\mathbf{X}^H$ is used to represent the complex conjugate and transpose matrix of $\mathbf{X}$, i.e. $\mathbf{X}^H = \mathbf{X}^{*T}$.
- The notation $\mathbf{X}^{-1}$ corresponds to the inverse matrix of $\mathbf{X}$.
- The notation diag$(\mathbf{X})$ is used to represent a diagonal matrix containing only the elements on the main diagonal of the matrix $\mathbf{X}$. The complement of this is denoted by $\overline{\text{diag}}(\mathbf{X}) = \mathbf{X} - \text{diag}(\mathbf{X})$ and it represents a matrix containing all the elements of $\mathbf{X}$,

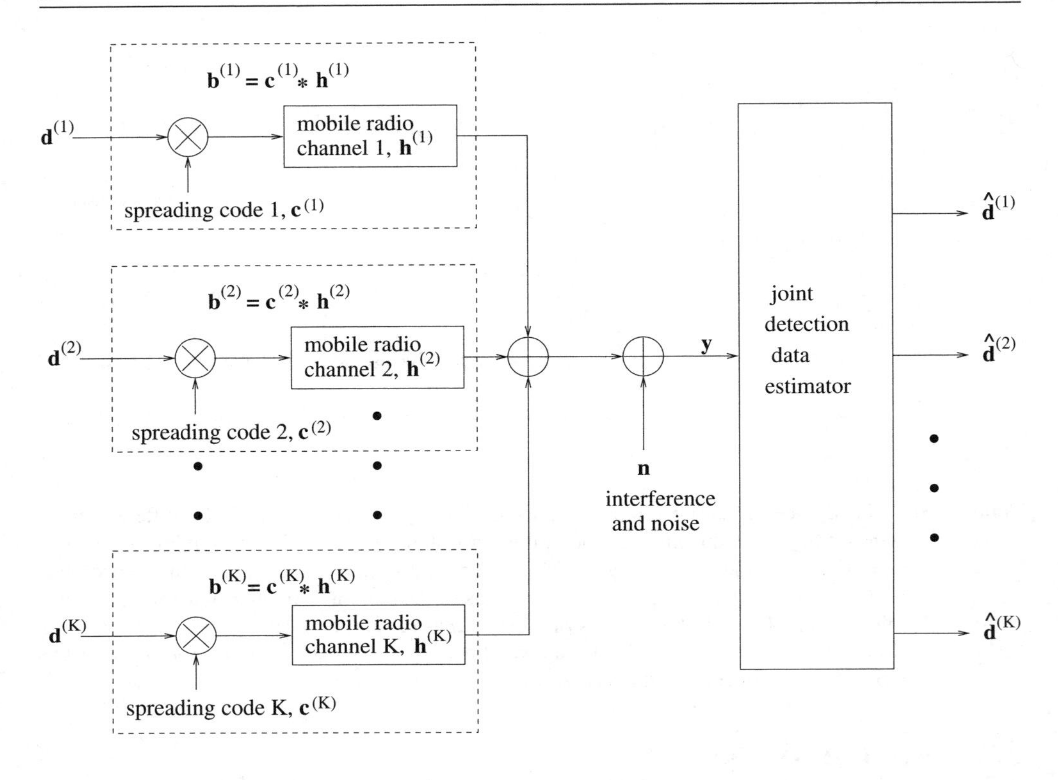

Figure 3.6: System model of a synchronous CDMA system on the uplink using joint detection.

except the ones on the main diagonal, where the elements on the main diagonal have values of zero.

- The notation $[\mathbf{X}]_{ij}$ identifies the element in the i-th row and j-th column of the matrix $\mathbf{X}$ and the notation x_i is used to represent the i-th element of the column vector $\mathbf{x}$.

- The matrix $\mathbf{I}$ is the identity matrix and the notation $\mathbf{I}_m$ is used to denote an identity matrix with the dimensions of $(m \times m)$.

Figure 3.6 depicts the block diagram representation of a synchronous system model for uplink transmission. Transmission is in bursts and all the users are synchronous with each other. The list of the symbols used is given as:

- K for the total number of users,

- N is the number of data symbols transmitted by each user in one transmission burst,

- Q represents the number of chips in each spreading sequence,

- W denotes the length of the wideband CIR, where W is assumed to be an integer multiple of the number of chip intervals, T_c.

- L indicates the number of multipath components or taps in the wideband CIR.

The following analysis is performed at the baseband level in order to simplify the notation. However, this does not infringe on the validity of the derivation, since baseband and bandpass systems can be considered equivalent [92]. Let us consider the transmission of one symbol of one user, $d_n^{(k)}$. The superscript (k) will be dropped temporarily, since only one user is considered. At the baseband, the signal transmitted by the k-th user is given by,

$$x_n(i) = d_n c(i), \text{ for } i = 1, \dots, Q, \tag{3.40}$$

where $c(i)$ is the i-th chip of the user spreading code. After corruption by the mobile channel, the received signal is expressed as a convolution of $x_n(i)$ and the CIR, $h_n(i)$:

$$\begin{aligned} r_n(i) &= x_n(i) * h_n(i) && (3.41)\\ &= \sum_{j=1}^{Q} x_n(j) h_n(i-j+1) && (3.42)\\ &= \sum_{j=1}^{Q} d_n c(j) h_n(i-j+1) && (3.43)\\ &= d_n \sum_{j=1}^{Q} c(j) h_n(i-j+1) && (3.44)\\ &= d_n b_n(i), \text{ for } i = 1, \dots, Q+W-1, && (3.45) \end{aligned}$$

where $h_n(i)$ denotes the CIR for the n-th symbol and $b_n(i) = \sum_{j=1}^{Q} c(j) h_n(i-j+1)$ represents the combined impulse response, which is the convolution of the spreading code and the CIR. Since the CIR has a delay spread of W, inter-symbol interference (ISI) will occur between the N symbols in a transmission burst.

In order to introduce more compact mathematical expressions, matrix notation will be employed. The transmitted data symbol sequence of the k-th user is represented by a vector as:

$$\begin{aligned} &\mathbf{d}^{(k)} = (d_1^{(k)}, d_2^{(k)}, \dots, d_n^{(k)}, \dots, d_N^{(k)})^T, \\ &\text{for } k = 1, \dots, K; \;\; n = 1, \dots, N, \end{aligned} \tag{3.46}$$

where k is the user index and n is the symbol index. There are N data symbols per transmission burst and each data symbol is generated using an m-ary modulation scheme [92].

The Q-chip spreading sequence vector of the k-th user is expressed as:

$$\begin{aligned} &\mathbf{c}^{(k)} = (c_1^{(k)}, c_2^{(k)}, \dots, c_q^{(k)}, \dots, c_Q^{(k)})^T, \\ &\text{for } k = 1, \dots, K; \;\; q = 1, \dots, Q. \end{aligned} \tag{3.47}$$

The CIR for the n-th data symbol of the k-th user is represented as:

$$\mathbf{h}_n^{(k)} = (h_n^{(k)}(1), h_n^{(k)}(2), \ldots, h_n^{(k)}(w), \ldots, h_n^{(k)}(W))^T, \quad (3.48)$$
$$\text{for } k = 1, \ldots, K; \; w = 1, \ldots, W,$$

consisting of W complex samples $h_n^{(k)}(w)$ taken at the chip rate of $1/T_c$.

The combined impulse response, $\mathbf{b}_n^{(k)}$, due to the spreading sequence and the CIR is defined by the convolution of $\mathbf{c}^{(k)}$ and $\mathbf{h}_n^{(k)}$, which is represented as:

$$\mathbf{b}_n^{(k)} = (b_n^{(k)}(1), b_n^{(k)}(2), \ldots, b_n^{(k)}(l), \ldots, b_n^{(k)}(Q+W-1))^T \quad (3.49)$$
$$= \mathbf{c}^{(k)} * \mathbf{h}_n^{(k)}, \quad (3.50)$$
$$\text{for } k = 1 \ldots K; \; n = 1, \ldots N.$$

In order to represent the ISI due to the N symbols and the dispersive combined impulse responses, the discretized received signal, $\mathbf{r}^{(k)}$, of user k can be expressed as the product of a matrix $\mathbf{A}^{(k)}$ and its data vector $\mathbf{d}^{(k)}$, where:

$$\mathbf{r}^{(k)} = \mathbf{A}^{(k)} \mathbf{d}^{(k)}. \quad (3.51)$$

The i-th element of the received signal vector $\mathbf{r}^{(k)}$ is:

$$r_i^{(k)} = \sum_{n=1}^{N} [\mathbf{A}^{(k)}]_{in} d_n^{(k)}, \quad \text{for } i = 1, \ldots, NQ+W-1. \quad (3.52)$$

The matrix $\mathbf{A}^{(k)}$ is the so-called system matrix of the k-th user and it is constructed from the combined impulse responses of Equation 3.50. It represents the effect of the combined impulse responses on each data symbol $d_n^{(k)}$ in the data vector, $\mathbf{d}^{(k)}$. Each column in the matrix $\mathbf{A}$ indexed by n contains the combined impulse response, $\mathbf{b}_n^{(k)}$ that affects the n-th symbol of the data vector. However, since the data symbols are spread by the Q-chip spreading sequences, they are transmitted Q chips apart from each other. Hence the start of the combined impulse response, $\mathbf{b}_n^{(k)}$, for each column is offset by Q rows from the start of $\mathbf{b}_{n-1}^{(k)}$ in the preceding column. Therefore, the element in the $[(n-1)Q+l]$-th row and the n-th column of $\mathbf{A}^{(k)}$ is the l-th element of the combined impulse response, $\mathbf{b}_n^{(k)}$, for $l = 1, \ldots, Q+W-1$. All other elements in the column are zero-valued. For instance, when

$n = 1$, we have:

$$
\begin{array}{lcl}
[\mathbf{A}^{(k)}]_{1,1} & = & b_1^{(k)}(1) \\
[\mathbf{A}^{(k)}]_{2,1} & = & b_1^{(k)}(2) \\
 & \vdots & \\
[\mathbf{A}^{(k)}]_{(Q+W-1),1} & = & b_1^{(k)}(Q+W-1) \\
[\mathbf{A}^{(k)}]_{(Q+W),1} & = & 0 \\
 & \vdots & \\
[\mathbf{A}^{(k)}]_{(NQ+W-1),1} & = & 0.
\end{array}
\tag{3.53}
$$

For a general value of n, we arrive at:

$$
\begin{array}{lcl}
[\mathbf{A}^{(k)}]_{1,n} & = & 0 \\
 & \vdots & \\
[\mathbf{A}^{(k)}]_{(n-1)Q,n} & = & 0 \\
[\mathbf{A}^{(k)}]_{(n-1)Q+1,n} & = & b_n^{(k)}(1) \\
[\mathbf{A}^{(k)}]_{(n-1)Q+2,n} & = & b_n^{(k)}(2) \\
 & \vdots & \\
[\mathbf{A}^{(k)}]_{nQ+W-1,n} & = & b_n^{(k)}(Q+W-1) \\
[\mathbf{A}^{(k)}]_{nQ+W,n} & = & 0 \\
 & \vdots & \\
[\mathbf{A}^{(k)}]_{(NQ+W-1),n} & = & 0.
\end{array}
\tag{3.54}
$$

This can be represented in a more compact form as:

$$
[\mathbf{A}^{(k)}]_{in} = \begin{cases} b_n^{(k)}(l) & \text{for } i = (n-1)Q + l; \;\; n = 1, \ldots, N; \;\; l = 1, \ldots, Q + W - 1 \\ 0 & \text{otherwise.} \end{cases}
\tag{3.55}
$$

The pictorial representation of Equation 3.51 is shown in Figure 3.7, where $Q = 4$, $W = 2$ and $N = 3$. As it can be seen from the diagram, in each column of matrix $\mathbf{A}^{(k)}$ – where a box with an asterisk marks a non-zero element – the vector $\mathbf{b}_n^{(k)}$ starts at an offset of $Q = 4$ rows below its preceding column, except for the first column, which starts at the first row. The total number of elements in the vector $\mathbf{b}_n^{(k)}$ is $(Q + W - 1) = 5$. The total number of columns in the matrix $\mathbf{A}^{(k)}$ equals the number of symbols in the data vector, $\mathbf{d}^{(k)}$, i.e. N. Finally, the received signal vector product, $\mathbf{r}^{(k)}$ in Equation 3.51, has a total of $(NQ + W - 1) = 13$ elements due to ISI from the multipath channel, as opposed to $NQ = 12$ elements in a narrowband channel.

The joint detection receiver aims to detect the symbols of all the users jointly by utilizing the information available on the spreading sequences and channel estimates of all the users. Therefore, as seen in Figure 3.8, the data symbols of all K users can be viewed as the transmitted data sequence of a single user, by concatenating all the data sequences. The overall

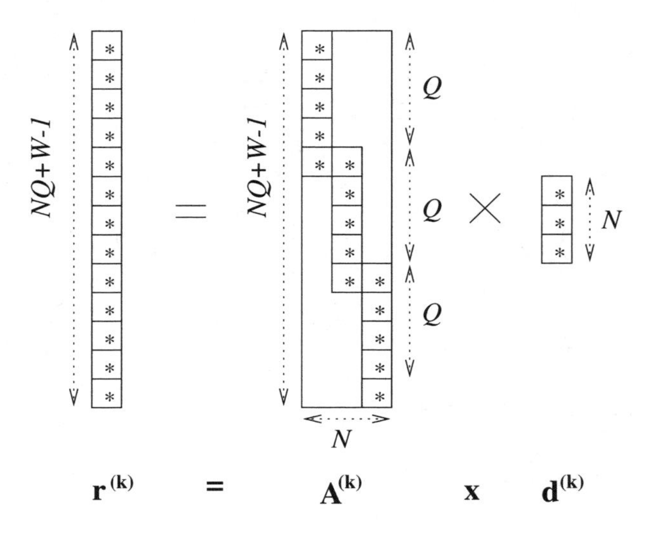

Figure 3.7: Stylized structure of Equation 3.51 representing the received signal vector of a wideband channel, where $Q = 4$, $W = 2$ and $N = 3$. The column vectors in the matrix $\mathbf{A}^{(k)}$ are the combined impulse response vectors, $\mathbf{b}_n^{(k)}$ of Equation 3.50. A box with an asterisk in it represents a non-zero element, and the remaining notation is as follows: K represents the total number of users, N denotes the number of data symbols transmitted by each user, Q represents the number of chips in each spreading sequence, and W indicates the length of the wideband CIR.

transmitted sequence can be rewritten as:

$$\mathbf{d} = (\mathbf{d}^{(1)T}, \mathbf{d}^{(2)T}, \ldots, \mathbf{d}^{(K)T})^T \quad (3.56)$$
$$= (d_1, d_2, \ldots, d_{KN})^T, \quad (3.57)$$

where $d_j = d_n^{(k)}$ for $j = n + N.(k-1)$, $k = 1, 2, \ldots, K$ and $n = 1, 2, \ldots, N$.

The system matrix for the overall system can be constructed by appending the $\mathbf{A}^{(k)}$ matrix of each of the K users column-wise, whereby:

$$\mathbf{A} = (\mathbf{A}^{(1)}, \mathbf{A}^{(2)}, \ldots, \mathbf{A}^{(k)}, \ldots, \mathbf{A}^{(K)}). \quad (3.58)$$

The construction of matrix $\mathbf{A}$ from the system matrices of the K users is depicted in Figure 3.8. Therefore, the discretized received composite signal can be represented in matrix form as:

$$\mathbf{y} = \mathbf{A}\mathbf{d} + \mathbf{n}, \quad (3.59)$$
$$\mathbf{y} = (y_1, y_2, \cdots, y_{NQ+W-1})^T,$$

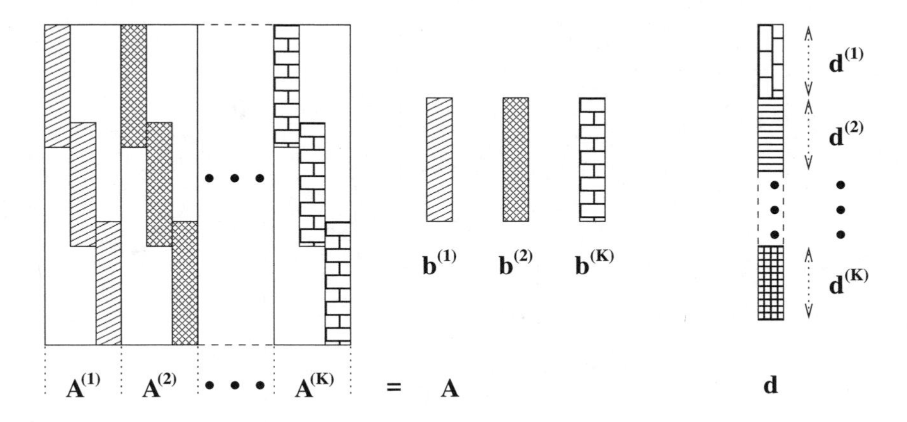

Figure 3.8: The construction of matrix $\mathbf{A}$ from the individual system matrices, $\mathbf{A}^{(k)}$ seen in Figure 3.7, and the data vector $\mathbf{d}$ from the concatenation of data vectors, $\mathbf{d}^{(k)}$, of all K users.

where $\mathbf{n} = (n_1, n_2, \ldots, n_{NQ+W-1})^T$, is the noise sequence, which has a covariance matrix of $\mathbf{R}_n = E[\mathbf{n}.\mathbf{n}^H]$. The composite signal vector $\mathbf{y}$ has $(NQ + W - 1)$ elements for a data burst of length N symbols. Upon multiplying the matrix $\mathbf{A}$ with the vector $\mathbf{d}$ seen in Figure 3.8, we obtain the MAI- and ISI-contaminated received symbols according to Equation 3.59.

Taken as a whole, the system matrix, $\mathbf{A}$, can be constructed from the combined response vectors, $\mathbf{b}_n^{(k)}$ of all the K users, in order to depict the effect of the system's response on the data vector of Equation 3.56. The dimensions of the matrix are $(NQ + W - 1) \times KN$. Figure 3.9 shows an example of the matrix, $\mathbf{A}$, for an N-bit long data burst. For ease of representation, we assumed that the channel length, W, for each user is the same and that it remains constant throughout the data burst. We have also assumed that the channel experiences slow fading and that the fading is almost constant across the data burst. Therefore, the combined response vector for each transmitted symbol of user k is represented by $\mathbf{b}^{(k)}$, where $\mathbf{b}^{(k)} = \mathbf{b}_1^{(k)}\mathbf{b}_2^{(k)} = \ldots = \mathbf{b}_N^{(k)}$. Focusing our attention on Figure 3.9, the elements in the j-th column of the matrix constitute the combined response vector that affects the j-th data symbol in the transmitted data vector $\mathbf{d}$. Therefore, columns $j = 1$ to N of matrix $\mathbf{A}$ correspond to symbols $m = 1$ to N of vector $\mathbf{d}$, which are also the data symbols of user $k = 1$. The next N columns correspond to the next N symbols of data vector $\mathbf{d}$, which are the data symbols of user $k = 2$ and so on.

For user k, each successive response vector, $\mathbf{b}^{(k)}$, is placed at an offset of Q rows from the preceding vector, as shown in Figure 3.9. For example, the combined response vector in column 1 of matrix $\mathbf{A}$ is $\mathbf{b}^{(1)}$ and it starts at row 1 of the matrix because that column corresponds to the first symbol of user $k = 1$. In column 2, the combined response vector is also $\mathbf{b}^{(1)}$, but it is offset from the start of the vector in column 1 by Q rows. This is because the data symbol corresponding to this matrix column is transmitted Q chips later. This is repeated until the columns $j = 1, \ldots, N$ contain the combined response vectors that affect all the data symbols of user $k = 1$. The next column of $j = N + 1$ in the matrix $\mathbf{A}$ contains the combined impulse response vector that affects the data symbol, $d_{N+1} = d_1^{(2)}$, which is the first data symbol of user $k = 2$. In this column, the combined response vector for user

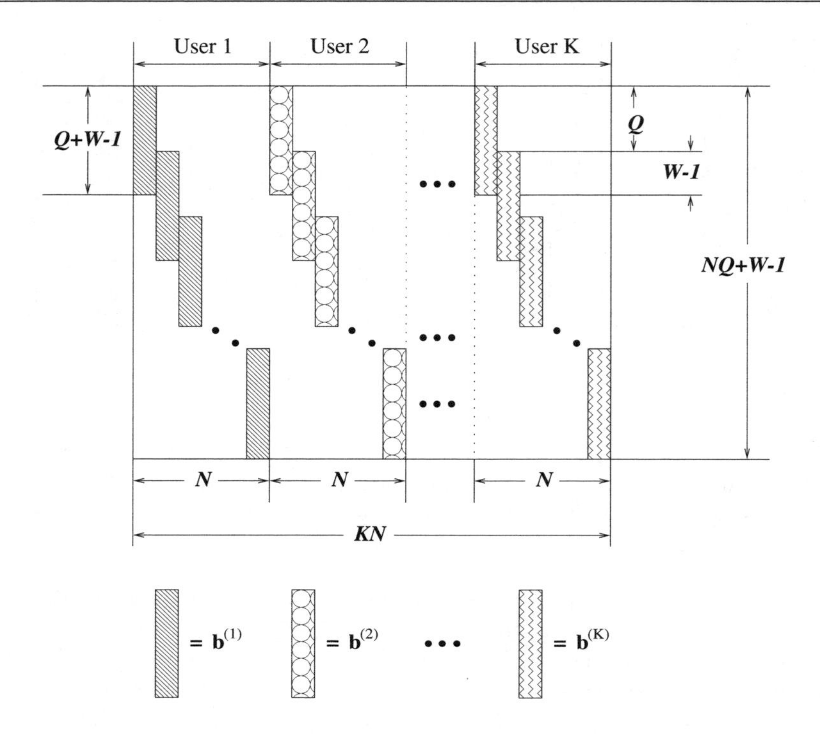

Figure 3.9: Stylized structure of the system matrix $\mathbf{A}$, where $\mathbf{b}^{(1)}$, $\mathbf{b}^{(2)}$ and $\mathbf{b}^{(K)}$ are column vectors representing the combined impulse responses of users 1, 2 and K, respectively in Equation 3.50. The notation is as follows: K represents the total number of users, N denotes the number of data symbols transmitted by each user, Q represents the number of chips in each spreading sequence, and W indicates the length of the wideband CIR.

$k = 2$, $\mathbf{b}^{(2)}$, is used and the vector starts at row 1 of the matrix because it is the first symbol of this user. The response matrix, $\mathbf{b}^{(2)}$ is then placed into columns $j = N+1, \ldots, 2N$ of the matrix $\mathbf{A}$, with the same offsets for each successive vector, as was carried out for user 1. This process is repeated for all the other users until the system matrix is completely constructed.

The mathematical representation of matrix $\mathbf{A}$ in general can be written as:

$$[\mathbf{A}]_{ij} = \begin{cases} b_n^{(k)}(l) & \text{for } k = 1, \ldots, K;\ \ n = 1, \ldots, N; \\ & l = 1, \ldots, Q + W - 1 \\ 0 & \text{otherwise}, \end{cases} \tag{3.60}$$
$$\text{for } i = 1, \ldots, NQ + W - 1, \quad j = 1, \ldots, KN,$$

where $i = Q(n-1) + l$ and $j = n + N(k-1)$.

Figure 3.10 shows the stylized structure of Equation 3.59 for a specific example. In the figure, a system with $K = 2$ users is depicted. Each user transmits $N = 3$ symbols per transmission burst, and each symbol is spread with a signature sequence of length $Q = 3$ chips. The channel for each user has a dispersion length of $W = 3$ chips. The blocked

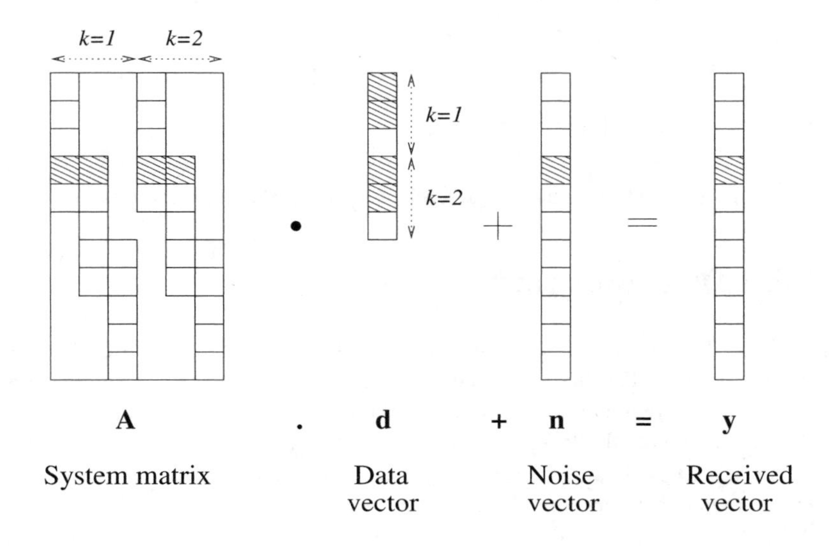

Figure 3.10: Stylized structure of the matrix equation $\mathbf{y} = \mathbf{Ad} + \mathbf{n}$ for a $K = 2$–user system. Each user transmits $N = 3$ symbols per transmission burst, and each symbol is spread with a signature sequence of length $Q = 3$ chips. The channel for each user has a dispersion length of $W = 3$ chips.

segments in the figure represent the combination of elements that result in the element y_4, which is obtained from Equation 3.59 by:

$$y_4 = \sum_{i=1}^{KN=6} [\mathbf{A}]_{4,i} d_i + n_4 \tag{3.61}$$

$$= [\mathbf{A}]_{4,1} d_1 + [\mathbf{A}]_{4,2} d_2 + [\mathbf{A}]_{4,4} d_4 + [\mathbf{A}]_{4,5} d_5 + n_4 \tag{3.62}$$

Given the above transmission regime, the basic concept of joint detection is centred around processing the received composite signal vector, $\mathbf{y}$, in order to determine the transmitted data vector, $\mathbf{d}$. This concept is encapsulated in the following set of equations:

$$\hat{\mathbf{y}} = \mathbf{S}\hat{\mathbf{d}} = \mathbf{My}, \tag{3.63}$$

where $\mathbf{S}$ is a square matrix with dimensions $(KN \times KN)$ and the matrix $\mathbf{M}$ is a $[KN \times (NQ + W - 1)]$-matrix. These two matrices determine the type of joint detection algorithm, as will become explicit during our further discourse. The schematic in Figure 3.11 shows the receiver structure represented by this equation.

Having considered the system model for synchronous DS-CDMA in the uplink, let us now concentrate on the joint detection techniques.

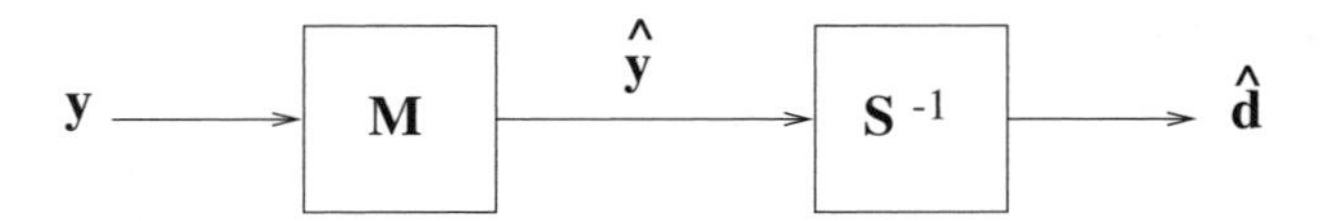

Figure 3.11: Structure of the receiver represented in Equation 3.63.

3.3 Joint Detection Techniques

The conventional detector for DS-CDMA systems is the matched filter, which was previously invoked in Section 2.7. The first data detection scheme that will be considered here is the well-known whitening matched filter (WMF). All the JD schemes that are presented next can be viewed as extensions to the WMF, as will be shown below.

The WMF [127, 129] is an extension of the conventional data estimation technique that uses a bank of matched filters, one for each user. A pre-whitening filter, which will be described in Section 3.3.1, is used to decorrelate the noise in the received samples, before passing the received samples through a matched filter. Hence the name "decorrelator" or "whitening matched filter". A matched filter [92, 206] maximizes the SNR at a fixed sampling point for a particular data estimate, as will be discussed in Section 3.3.2. This technique treats ISI and MAI as noise and it is therefore inefficient. However, this approach is introduced as a prelude to the other techniques to be discussed, since all the other JD receivers can be viewed as extensions of the WMF. Before the WMF is discussed, the whitening filter and matched filter are explored in more depth in the following section.

3.3.1 Whitening Filter

The filter that outputs white uncorrelated Gaussian noise samples, when fed with correlated non-white Gaussian noise at the input is termed a whitening filter. In communications systems, it is quite common to use a whitening filter in order to process the received signal prior to its detection. This is because after passing the received signal through a whitening filter, the detection problem using the output signal becomes one of detecting a signal in the presence of white noise. Solutions for detecting these signals are well known [104]. The basic concept of the whitening filter is explained next.

Let n_i represent the noise samples and $N(z)$ be the z-transform of n_i. The z-transform of the autocorrelation function of the noise samples is given by $R_n(z) = N(z)N^*(z)$, which corresponds to the power spectral density of the noise samples. If $R_n(z)$ satisfies the Paley-Wiener criterion [104], which in simple terms implies that $R_n(z)$ can be factored into a product of two terms as follows:

$$R_n(z) = L(z)L^*(z), \tag{3.64}$$

then the noise whitening filter has the z-transform of $1/L(z)$. Let the output of the whitening filter be n'_i, when n_i is the input. In order to show that the filter $1/L(z)$ whitens the noise, let us consider $N'(z) = N(z)/L(z)$. The z-transform of the autocorrelation of n'_i is given by

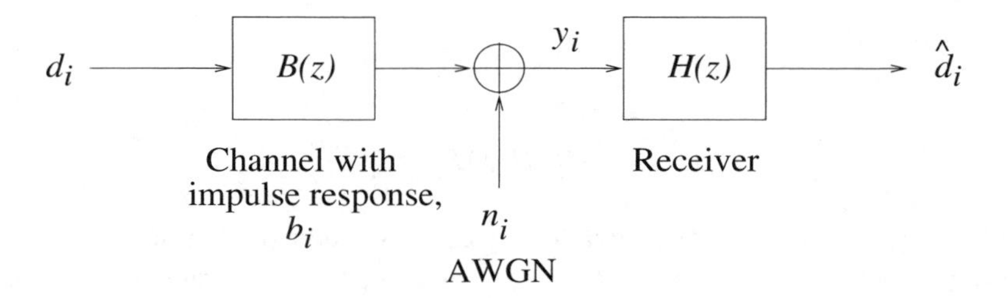

Figure 3.12: Block diagram of a simple transmission scheme.

the following product:

$$R_{n'}(z) = \frac{N(z)}{L(z)} \times \frac{N^*(z)}{L^*(z)} \quad (3.65)$$
$$= \frac{R_n(z)}{R_n(z)} \quad (3.66)$$
$$= 1. \quad (3.67)$$

In the time domain, we have $R_{n'} = \delta_i$, which shows that the output noise samples are uncorrelated. In matrix form $R_n(z) = L(z)L^*(z)$ is represented as $\mathbf{R}_n = \mathbf{L}\mathbf{L}^H$, where $\mathbf{L}$ and $\mathbf{L}^H$ are obtained using the well-known Cholesky decomposition [122]. The whitening filter is then represented as $\mathbf{L}^{-1}$. In other words, if the autocorrelation of the non-white noise samples is known, then the Cholesky decomposition generates the lower triangular matrix, $\mathbf{L}^{-1}$, containing the z-domain transfer function samples of the whitening filter. Let us now briefly consider the principles of matched filtering.

3.3.2 Matched Filter

The matched filter is a filter that maximizes the SNR at the required sampling instant at its output for a given received waveform [92]. The block diagram of a simple transmitter-receiver is shown in Figure 3.12. The received signal at the input of the filter is $y_i = (b_i * d_i) + n_i$, where $*$ in this case denotes convolution. The received signal y_i is the result of convolution between the data samples, d_i, and the channel response, b_i, plus the corruption due to the additive white Gaussian noise (AWGN), n_i. Following, for example, Schwarz's approach [206], the z-transforms of y_i, the data estimates $\hat{d}_i$ and the channel impulse response, b_i are represented by $Y(z)$, $\hat{D}(z)$ and $B(z)$, respectively, leading to:

$$Y(z) = B(z)D(z) + N(z) \quad (3.68)$$
$$\hat{D}(z) = H(z)B(z)D(z) + H(z)N(z), \quad (3.69)$$

where $X(z)$ denotes the z-transform of x_i.

The derived signal component of $\hat{D}(z)$ is $H(z)B(z)D(z)$, and if we assumed that the data samples, d_i are uncorrelated with a variance of σ_D^2, then its power spectral density, Σ

can be expressed as:

$$\Sigma = |D(z)|^2 |H(z)B^*(z)|^2 \tag{3.70}$$
$$= \sigma_D^2 |H(z)B^*(z)|^2 \tag{3.71}$$

where $H(z)B^*(z)$ is the power spectral density due to the convolution of b_i and h_i.

The noise component of $\hat{D}(z)$ is $H(z)N(z)$, and its power spectral density, Γ is:

$$\Gamma = H^*(z)H(z)N^*(z)N(z) \tag{3.72}$$
$$= |H(z)|^2 |N(z)|^2 \tag{3.73}$$
$$= \frac{N_0}{2} |H(z)|^2, \tag{3.74}$$

where $N_0/2$ is the double-sided power spectral density of white Gaussian noise.

Therefore the SNR is given by:

$$\text{SNR} = \frac{\Sigma}{\Gamma} \tag{3.75}$$
$$= \frac{\sigma_D^2 \, |H(z)B^*(z)|^2}{\frac{N_0}{2} \, |H(z)|^2}. \tag{3.76}$$

According to Schwarz's Inequality [206] of:

$$|H(z)B^*(z)|^2 \leqslant |H(z)|^2 \, |B(z)|^2, \tag{3.77}$$

we have:

$$\text{SNR} = \frac{\sigma_D^2 \, |H(z)B^*(z)|^2}{\frac{N_0}{2} \, |H(z)|^2} \leqslant \frac{\sigma_D^2 \, |H(z)|^2 \, |B(z)|^2}{\frac{N_0}{2} \, |H(z)|^2}. \tag{3.78}$$

Therefore, in order to maximize the SNR, $|H(z)B^*(z)|^2$ must be equal to $|H(z)|^2 \, |B(z)|^2$ and this will hold if and only if:

$$H(z) = cB^*(z), \tag{3.79}$$

where $c \neq 0$ is a scalar constant. In matrix form $H(z) = B^*(z)$ is represented as $\mathbf{H} = \mathbf{B}^{*T} = \mathbf{B}^H$, explicitly stating the condition of maximizing the SNR at the output of the matched filter.

3.3.3 Whitening Matched Filter

The WMF has been the topic of a range of studies [90, 104, 206] and here we follow the approach similar to that of Messerschmitt [206]. From Equation 3.59, the received vector is given by $\mathbf{y} = \mathbf{Ad} + \mathbf{n}$. The covariance matrix of $\mathbf{n}$ is represented by $\mathbf{R}_n$, which can be decomposed using Cholesky decomposition [122] into:

$$\mathbf{R}_n = \mathbf{L}\mathbf{L}^H, \tag{3.80}$$

where $\mathbf{L}$ is a lower triangular matrix having real-valued elements on its main diagonal and zero elements above its main diagonal. In order to decorrelate or pre-whiten the noise, the z-domain coefficients of the whitening filter are the elements of $\mathbf{L}^{-1}$, as argued in Section 3.3.1.

The output of the whitening filter is given as:

$$\begin{aligned} \mathbf{y}' &= \mathbf{L}^{-1}\mathbf{y} && (3.81) \\ &= \mathbf{L}^{-1}\mathbf{A}\mathbf{d} + \mathbf{L}^{-1}\mathbf{n}. && (3.82) \end{aligned}$$

Since the signal component at the output of the whitening filter is $\mathbf{L}^{-1}\mathbf{A}\mathbf{d}$, the matched filter solution for the resultant output of the whitening filter would therefore be:

$$\begin{aligned} \text{MF} &= (\mathbf{L}^{-1}\mathbf{A})^H && (3.83) \\ &= \mathbf{A}^H(\mathbf{L}^{-1})^H, && (3.84) \end{aligned}$$

as demonstrated in Equation 3.79 of Section 3.3.2. The resultant signal at the output of the matched filter is given by:

$$\begin{aligned} \hat{\mathbf{d}}_{\text{WMF}} = \mathbf{y}'' &= MF \cdot \mathbf{y}' && (3.85) \\ &= \mathbf{A}^H(\mathbf{L}^{-1})^H\mathbf{L}^{-1}\mathbf{y} && (3.86) \\ &= \mathbf{A}^H(\mathbf{L}\mathbf{L}^H)^{-1}\mathbf{y} && (3.87) \\ &= \mathbf{A}^H\mathbf{R}_n^{-1}\mathbf{y}, && (3.88) \end{aligned}$$

as stated by Klein *et al.* [127].

Substituting Equation 3.59 into Equation 3.88 gives:

$$\begin{aligned} \hat{\mathbf{d}}_{\text{WMF}} &= \mathbf{A}^H\mathbf{R}_n^{-1}\mathbf{A}\mathbf{d} + \mathbf{A}^H\mathbf{R}_n^{-1}\mathbf{n} && (3.89) \\ &= \underbrace{\text{diag}(\mathbf{A}^H\mathbf{R}_n^{-1}\mathbf{A})\mathbf{d}}_{\text{desired symbols}} \\ &\quad + \underbrace{\overline{\text{diag}}(\mathbf{A}^H\mathbf{R}_n^{-1}\mathbf{A})\mathbf{d}}_{\text{ISI and MAI}} + \underbrace{\mathbf{A}^H\mathbf{R}_n^{-1}\mathbf{n}}_{\text{noise}}, && (3.90) \end{aligned}$$

where the term $[\text{diag}(\mathbf{A}^H\mathbf{R}_n^{-1}\mathbf{A})\mathbf{d}]$ constitutes the useful part of the output and allows the recovery of the transmitted data $\mathbf{d}$, while $\overline{\text{diag}}(\mathbf{A}^H\mathbf{R}_n^{-1}\mathbf{A})\mathbf{d}$ indicates the off-diagonal values constituting MAI and ISI.

By using Equation 3.88, the i-th element of the vector $\mathbf{y}''$ can be rewritten as:

$$\begin{aligned}
y_i'' &= \sum_j [\mathbf{A}^H \mathbf{R}_n^{-1}]_{ij} y_j && (3.91)\\
&= \sum_j \sum_l [\mathbf{A}^H]_{il} [\mathbf{R}_n^{-1}]_{lj} y_j && (3.92)\\
&= \sum_j \sum_l [\mathbf{A}^H]_{il} [\mathbf{R}_n^{-1}]_{lj} (\sum_m [\mathbf{A}]_{jm} d_m + n_j) && (3.93)\\
&= \sum_j \sum_l \sum_m [\mathbf{A}^H]_{il} [\mathbf{A}]_{jm} [\mathbf{R}_n^{-1}]_{lj} d_m + z_i && (3.94)\\
&= \sum_j \left| [\mathbf{A}]_{ji} \right|^2 [\mathbf{R}_n^{-1}]_{jj} d_i + \sum_j \sum_{m, m \neq i} [\mathbf{A}^H]_{ij} [\mathbf{A}]_{jm} [\mathbf{R}_n^{-1}]_{jj} d_m + \\
&\quad \sum_j \sum_{l, l \neq j} \sum_m [\mathbf{A}^H]_{il} [\mathbf{A}]_{jm} [\mathbf{R}_n^{-1}]_{lj} d_m + z_i, && (3.95)
\end{aligned}$$

where $z_i = \sum_j \sum_l [\mathbf{A}^H]_{il} [\mathbf{R}_n^{-1}]_{lj} n_j$ is the noise component of the i-th symbol. The first term of Equation 3.95 contains the desired data symbol scaled with the maximum SNR of $\sum_j \left| [\mathbf{A}]_{ji} \right|^2 [\mathbf{R}_n^{-1}]_{jj}$. The second and third terms represent the residual interference at the output of the WMF.

If it is assumed that the vector $\mathbf{n}$ consists of uncorrelated noise samples that are zero mean Gaussian variables with a variance of σ^2, then its covariance matrix $\mathbf{R}_n = \sigma^2 \mathbf{I}_{(NQ+W-1)}$, where $\mathbf{I}_{(NQ+W-1)}$ is the identity matrix with the dimensions of $[(NQ + W - 1) \times (NQ + W - 1)]$. Using this in Equation 3.88 gives:

$$\begin{aligned}
\hat{\mathbf{d}}_{\text{WMF}} \Big|_{\mathbf{R}_n = \sigma^2 \mathbf{I}_{(NQ+W-1)}} &= \mathbf{A}^H (\frac{1}{\sigma^2} \mathbf{I}_{(NQ+W-1)}) \mathbf{y} && (3.96)\\
&= \frac{1}{\sigma^2} \mathbf{A}^H \mathbf{y}. && (3.97)
\end{aligned}$$

The maximal ratio combining (MRC) RAKE [92] receiver which is used for multipath CDMA reception is a type of WMF, where the equation for the MRC-RAKE receiver can be expressed as:

$$\hat{\mathbf{d}}_{\text{MRC-RAKE}} = \mathbf{A}^H \mathbf{y}. \qquad (3.98)$$

Having analyzed the WMF let us now focus our attention on the joint detection schemes.

3.3.4 Zero Forcing Block Linear Equalizer

The joint detection scheme that is termed the zero-forcing block linear equalizer (ZF-BLE) is similar to the zero-forcing linear equalizer presented in Section 3.1.1. The derivation that follows uses matrix notation and is derived from the principles of maximum likelihood sequence estimation (MLSE) [104, 125], where the terminology is justified at a later stage. MLSE operates by estimating the vector $\hat{\mathbf{d}}$, based on the highest probability vector, given the received vector $\mathbf{y}$. The most likely values of $\hat{\mathbf{d}}$, are those for which the so-called conditional joint probability density function, $p(\mathbf{y}|\mathbf{d})$, [104] is maximum. From Equation 3.59, if the received vector $\mathbf{y}$ is assumed to be Gaussian distributed, where $\mathbf{n}$ is a zero mean multivariate Gaussian random vector with covariance matrix $\mathbf{R}_n$, then the joint probability density function function of $\mathbf{y}$ conditioned on $\mathbf{d}$ is given as [104]:

$$p(\mathbf{y}|\mathbf{d}) = \frac{1}{(2\pi)^{(NQ+w-1)/2}|\mathbf{R}_n|^{1/2}} \exp\left[-\frac{1}{2}(\mathbf{y}-\mathbf{A}\mathbf{d})^H\mathbf{R}_n^{-1}(\mathbf{y}-\mathbf{A}\mathbf{d})\right] \quad (3.99)$$

The assumption that the received vector is Gaussian distributed is valid for a high number of users due to the Central Limit Theorem [90]. In order to obtain the maximum of $p(\mathbf{y}|\mathbf{d})$, the following quadratic form $Q(\mathbf{d})$ has to be minimized, which is simply in the exponent in Equation 3.99:

$$Q(\mathbf{d}) = (\mathbf{y}-\mathbf{A}\mathbf{d})^H\mathbf{R}_n^{-1}(\mathbf{y}-\mathbf{A}\mathbf{d}). \quad (3.100)$$

From Equation 3.100, it can be seen that the vector $\hat{\mathbf{d}}$ that minimizes $Q(\mathbf{d})$, satisfies the following equation [104]:

$$\mathbf{y} = \mathbf{A}\hat{\mathbf{d}} \quad (3.101)$$

$$\therefore \hat{\mathbf{d}} = \mathbf{A}^{-1}\mathbf{y}, \quad (3.102)$$

in which case the exponent in Equation 3.99 becomes zero.

One of the identities for an inverse matrix is given by [104]:

$$\mathbf{J}^{-1} = (\mathbf{J}^H\mathbf{\Lambda}^{-1}\mathbf{J})^{-1}\mathbf{J}^H\mathbf{\Lambda}^{-1}, \quad (3.103)$$

where $\mathbf{\Lambda}$ is a square matrix and $\mathbf{J}$ is a matrix that has the same number of rows as $\mathbf{\Lambda}$. The identity holds true, only if $(\mathbf{J}^H\mathbf{\Lambda}^{-1}\mathbf{J})$ has an inverse.

By using Equation 3.102 and the identity in Equation 3.103, the solution which gives the ZF-BLE estimator is:

$$\hat{\mathbf{d}}_{\text{ZF-BLE}} = (\mathbf{A}^H\mathbf{R}_n^{-1}\mathbf{A})^{-1}\mathbf{A}^H\mathbf{R}_n^{-1}\mathbf{y}. \quad (3.104)$$

Equations 3.102 and 3.104 are equivalent. However, the implementation of Equation 3.104 in matrix form is easier, since $(\mathbf{A}^H\mathbf{R}_n^{-1}\mathbf{A})$ is a Hermitian matrix and efficient techniques for solving Equation 3.104 are known [99].

Upon substituting $\mathbf{y} = \mathbf{A}\mathbf{d} + \mathbf{n}$ from Equation 3.59 into Equation 3.104, we arrive at:

$$\hat{\mathbf{d}}_{\text{ZF-BLE}} = \underbrace{\mathbf{d}}_{\text{desired symbols}} + \underbrace{(\mathbf{A}^H\mathbf{R}_n^{-1}\mathbf{A})^{-1}\mathbf{A}^H\mathbf{R}_n^{-1}\mathbf{n}}_{\text{noise}}. \quad (3.105)$$

From Equation 3.105, it can be seen that the output of the ZF-BLE contains only the de-

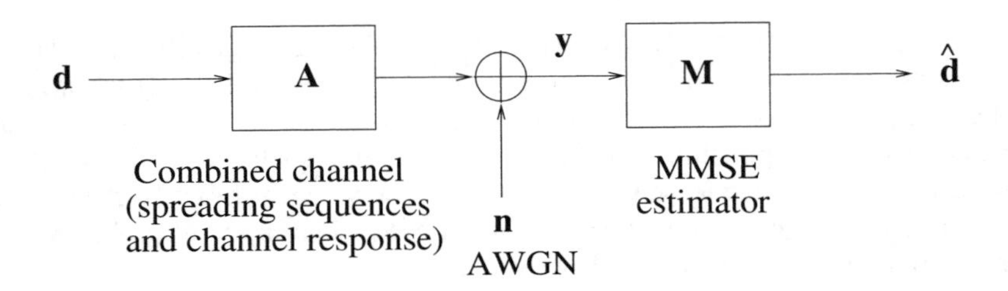

Figure 3.13: Block diagram of a simple transmission arrangement using an MMSE estimator.

sired symbols and a noise term. In contrast to the second term of the WMF solution of Equation 3.90, when using the ZF-BLE, there is no residual ISI or MAI in Equation 3.105. Therefore, this data estimator is termed "zero-forcing", since it forces the ISI and MAI to zero. This type of detector is also often referred to as a decorrelating detector, since it decorrelates the MAI and ISI. However, the removal of ISI and MAI is performed at the expense of noise enhancement. The noise at the output of the estimator is increased compared to the noise at the input of the estimator, hence resulting in SNR degradation. This is justified later on in this section. Comparing Equations 3.63 and 3.104, we have $\mathbf{M} = (\mathbf{A}^H \mathbf{R}_n^{-1} \mathbf{A})^{-1} \mathbf{A}^H \mathbf{R}_n^{-1}$ and $\mathbf{S} = \mathbf{I}_{KN}$. Substituting Equation 3.88 into Equation 3.104 yields:

$$\begin{aligned} \hat{\mathbf{d}}_{\text{ZF-BLE}} &= (\mathbf{A}^H \mathbf{R}_n^{-1} \mathbf{A})^{-1} \mathbf{A}^H \mathbf{R}_n^{-1} \mathbf{y} \\ &= (\mathbf{A}^H \mathbf{R}_n^{-1} \mathbf{A})^{-1} \hat{\mathbf{d}}_{\text{WMF}}. \end{aligned} \tag{3.106}$$

This shows that the ZF-BLE detector is an extension of the WMF, i.e. it takes the output of the WMF and decorrelates the MAI and ISI.

If it is assumed that the vector $\mathbf{n}$ consists of noise samples that are zero mean Gaussian variables having a variance of σ^2, then its covariance matrix is $\mathbf{R}_n = \sigma^2 \mathbf{I}_{(NQ+W-1)}$, where $\mathbf{I}_{(NQ+W-1)}$ is the identity matrix with the dimensions of $[(NQ+W-1) \times (NQ+W-1)]$. Using this in Equation 3.104 yields:

$$\begin{aligned} \hat{\mathbf{d}}_{\text{ZF-BLE}}\Big|_{\mathbf{R}_n=\sigma^2\mathbf{I}_{(NQ+W-1)}} &= \left(\mathbf{A}^H (\frac{1}{\sigma^2}\mathbf{I}_{(NQ+W-1)})\mathbf{A}\right)^{-1} \mathbf{A}^H (\frac{1}{\sigma^2}\mathbf{I}_{(NQ+W-1)})\, \mathbf{y} \\ &= (\mathbf{A}^H \mathbf{A})^{-1} \mathbf{A}^H \mathbf{y} \end{aligned} \tag{3.107}$$

3.3.5 Minimum Mean Square Error Block Linear Equalizer

The topic of the MMSE-BLE has been studied for example in [127–130], which constitute the basis of our discussions here. The block diagram depicting the transmission schematic when using a minimum mean square error (MMSE) estimator, is shown in Figure 3.13. Analogous to the MMSE linear equalizer for a single user, the principle behind MMSE estimation is the minimization of the error between the data estimate, $\hat{\mathbf{d}}$, and the actual data, $\mathbf{d}$, hence it minimizes the effects of both MAI, ISI and noise. Explicitly, the MMSE estimator minimizes the simple quadratic form [127]:

$$Q(\hat{\mathbf{d}}) = E[(\mathbf{d}-\hat{\mathbf{d}})^H(\mathbf{d}-\hat{\mathbf{d}})]. \tag{3.108}$$

Upon invoking the well-known Orthogonality Principle [207], in order to minimize the mean squared error (MSE), the error vector $\mathbf{e} = \mathbf{d} - \hat{\mathbf{d}}$ has to be set orthogonal by the MMSE equalizer to the estimator's input vector $\mathbf{y}$. This implies that:

$$E[(\mathbf{d}-\hat{\mathbf{d}})\mathbf{y}^H] = \mathbf{0}, \tag{3.109}$$

where $\mathbf{0}$ is a matrix with all the elements being zero-valued. If we let $\hat{\mathbf{d}} = \mathbf{M}\mathbf{y}$, where $\mathbf{M}$ is a linear estimator, then,

$$\begin{aligned}
E[(\mathbf{d}-\mathbf{M}\mathbf{y})\mathbf{y}^H] &= \mathbf{0} && (3.110)\\
E[(\mathbf{d}\mathbf{y}^H - \mathbf{M}\mathbf{y}\mathbf{y}^H)] &= \mathbf{0} && (3.111)\\
E[(\mathbf{d}\mathbf{y}^H)] - \mathbf{M}\,E[\mathbf{y}\mathbf{y}^H] &= \mathbf{0} && (3.112)\\
\mathbf{R}_{dy} - \mathbf{M}\mathbf{R}_y &= \mathbf{0} && (3.113)\\
\therefore \mathbf{M} &= \mathbf{R}_{dy}\mathbf{R}_y^{-1}, && (3.114)
\end{aligned}$$

where $\mathbf{R}_{dy} = E[\mathbf{d}\mathbf{y}^H]$ and $\mathbf{R}_y = E[\mathbf{y}\mathbf{y}^H]$.

For the special case of Equation 3.59, i.e. when $\mathbf{y} = \mathbf{A}\mathbf{d} + \mathbf{n}$, we have [104]:

$$\begin{aligned}
\mathbf{R}_{dy} &= E[\mathbf{d}(\mathbf{A}\mathbf{d}+\mathbf{n})^H] && (3.115)\\
&= E[\mathbf{d}\mathbf{d}^H\mathbf{A}^H + \mathbf{d}\mathbf{n}^H], && (3.116)
\end{aligned}$$

and assuming that the transmitted data vector, $\mathbf{d}$, and the noise vector, $\mathbf{n}$, are uncorrelated with each other, i.e. $E[\mathbf{d}\mathbf{n}^H] = 0$, we arrive at:

$$\begin{aligned}
\mathbf{R}_{dy} &= E[\mathbf{d}\mathbf{d}^H\mathbf{A}^H] && (3.117)\\
&= \mathbf{R}_d\mathbf{A}^H, && (3.118)
\end{aligned}$$

where $\mathbf{R}_d = E[\mathbf{d}\mathbf{d}^H]$. Furthermore, the covariance matrix, $\mathbf{R}_y$, of the received vector $\mathbf{y}$ in Equation 3.114 is given by:

$$\begin{aligned}
\mathbf{R}_y &= E[(\mathbf{A}\mathbf{d}+\mathbf{n})(\mathbf{A}\mathbf{d}+\mathbf{n})^H] && (3.119)\\
&= E[\mathbf{A}\mathbf{d}\mathbf{d}^H\mathbf{A}^H + \mathbf{n}\mathbf{n}^H] && (3.120)\\
&= \mathbf{A}\mathbf{R}_d\mathbf{A}^H + \mathbf{R}_n. && (3.121)
\end{aligned}$$

Substituting Equations 3.118 and 3.121 into Equation 3.114, we get:

$$\begin{aligned}
\mathbf{M} &= \mathbf{R}_d\mathbf{A}^H(\mathbf{A}\mathbf{R}_d\mathbf{A}^H + \mathbf{R}_n)^{-1} && (3.122)\\
&= (\mathbf{A}^H\mathbf{R}_n^{-1}\mathbf{A} + \mathbf{R}_d^{-1})^{-1}\mathbf{A}^H\mathbf{R}_n^{-1}. && (3.123)
\end{aligned}$$

In order to have a better understanding of the MMSE estimator, let us rearrange Equation 3.122 to give:

$$\mathbf{R}_d\mathbf{A}^H = \mathbf{M}(\mathbf{A}\mathbf{R}_d\mathbf{A}^H + \mathbf{R}_n). \tag{3.124}$$

Expanding the matrix term on the left-hand-side of Equation 3.124 leads to:

$$[\mathbf{R}_d\mathbf{A}^H]_{pq} = \sum_{r=1}^{KN}[\mathbf{R}_d]_{pr}[\mathbf{A}^H]_{rq} \tag{3.125}$$

$$= \sum_{r=1}^{KN}\sigma_{pr}^2 A_{qr}^*, \quad \text{for } p = 1,\ldots,KN; q = 1,\ldots,NQ+W-1 \tag{3.126}$$

where σ_{pr}^2 is the covariance between the p-th and q-th data symbols of vector $\mathbf{d}$, while the notation A_{qr}^* represents the complex conjugate of the element in the q-th row and r-th column of matrix $\mathbf{A}$.

Taking the matrix expression $(\mathbf{A}\mathbf{R}_d\mathbf{A}^H + \mathbf{R}_n)$ from the right-hand-side of Equation 3.124 and expanding it, gives us:

$$[\mathbf{A}\mathbf{R}_d\mathbf{A}^H + \mathbf{R}_n]_{st} = \sum_{u=1}^{KN} A_{su}\sum_{v=1}^{KN}\sigma_{uv}^2 A_{vt}^* + \eta_{st}^2 \tag{3.127}$$

$$= \sum_{u=1}^{KN}\sum_{v=1}^{KN}\sigma_{uv}^2 A_{su}A_{vt}^* + \eta_{st}^2 \tag{3.128}$$

$$\text{for } s = 1,\ldots,NQ+W-1; t = 1,\ldots,NQ+W-1.$$

The term η_{st}^2 represents the noise covariance between the s-th and t-th element of the noise vector $\mathbf{n}$. From Equation 3.128, we can see that the term $\sum_{v=1}^{KN}\sigma_{uv}^2 A_{su}A_{vt}^*$ represents the composite interference in the received signal at the sampling instant of u.

Therefore, by utilizing Equations 3.126 and 3.128, Equation 3.124 can be rewritten as:

$$[\mathbf{R}_d\mathbf{A}^H]_{pq} = \sum_{s=1}^{NQ+W-1} M_{ps}[\mathbf{A}\mathbf{R}_d\mathbf{A}^H + \mathbf{R}_n]_{sq} \tag{3.129}$$

$$\sum_{r=1}^{KN}\sigma_{pr}^2 A_{qr}^* = \sum_{s=1}^{NQ+W-1} M_{ps}\sum_{u=1}^{KN}\sum_{v=1}^{KN}\sigma_{uv}^2 A_{su}A_{vq}^* + \eta_{sq}^2, \tag{3.130}$$

$$\text{for } p = 1,\ldots,KN; \quad \text{and } q = 1,\ldots,NQ+W-1.$$

Considering Equation 3.130, we can observe that it consists of KN sets of $(NQ+W-1)$

linear equations each. Solving each set of equations simultaneously leads to the estimator coefficients, M_{ps} for $p = 1, \ldots, KN$ and $s = 1, \ldots, NQ + W - 1$, which correspond to the MMSE criterion in Equation 3.114.

Comparing the MMSE multiuser detector in Equation 3.122 with the single-user MMSE equalizer of Equation 3.36, we can see that the two expressions are analogous to each other. The single-user MMSE equalizer is expressed as $C = (\sigma_D^2 B^*)/(\sigma_D^2 |B|^2 + N_0/2)$, where B represents the channel transfer function, σ_D^2 is the signal power, and $N_0/2$ is the double-sided power spectral density of additive white Gaussian noise. The numerator expression, $(\sigma_D^2 B^*)$, is analogous to $\mathbf{R}_d \mathbf{A}^H$, since $\mathbf{R}_d$ is the covariance matrix of the data symbols and $\mathbf{A}^H$ is the Hermitian of the system matrix. Similarly, in the denominator the expressions of $(\sigma_D^2 |B|^2)$ and $N_0/2$ are analogous to the matrices $(\mathbf{A}\mathbf{R}_d\mathbf{A}^H)$ and $\mathbf{R}_n$, respectively in Equation 3.122.

Substituting the MMSE detector expression of Equation 3.123 into the Equation $\hat{\mathbf{d}} = \mathbf{M}\mathbf{y}$, we arrive at:

$$\hat{\mathbf{d}}_{\text{MMSE-BLE}} = (\mathbf{A}^H \mathbf{R}_n^{-1} \mathbf{A} + \mathbf{R}_d^{-1})^{-1} \mathbf{A}^H \mathbf{R}_n^{-1} \mathbf{y}. \tag{3.131}$$

Substituting Equation 3.59 into Equation 3.131, we get:

$$\begin{aligned}
\hat{\mathbf{d}}_{\text{MMSE-BLE}} &= [(\mathbf{A}^H \mathbf{R}_n^{-1} \mathbf{A} + \mathbf{R}_d^{-1})^{-1} \mathbf{A}^H \mathbf{R}_n^{-1}] \mathbf{A}\mathbf{d} + \\
&\quad (\mathbf{A}^H \mathbf{R}_n^{-1} \mathbf{A} + \mathbf{R}_d^{-1})^{-1} \mathbf{A}^H \mathbf{R}_n^{-1} \mathbf{n} \qquad (3.132) \\
&= [\mathbf{I}_{KN} + (\mathbf{R}_d \mathbf{A}^H \mathbf{R}_n^{-1} \mathbf{A})^{-1}]^{-1} \mathbf{d} + \\
&\quad (\mathbf{A}^H \mathbf{R}_n^{-1} \mathbf{A} + \mathbf{R}_d^{-1})^{-1} \mathbf{A}^H \mathbf{R}_n^{-1} \mathbf{n} \qquad (3.133) \\
&= \underbrace{\text{diag}\Big([\mathbf{I}_{KN} + (\mathbf{R}_d \mathbf{A}^H \mathbf{R}_n^{-1} \mathbf{A})^{-1}]^{-1} \Big) \mathbf{d}}_{\text{desired symbols}} + \\
&\quad \underbrace{\overline{\text{diag}}\Big([\mathbf{I}_{KN} + (\mathbf{R}_d \mathbf{A}^H \mathbf{R}_n^{-1} \mathbf{A})^{-1}]^{-1} \Big) \mathbf{d}}_{\text{ISI and MAI}} + \\
&\quad \underbrace{(\mathbf{A}^H \mathbf{R}_n^{-1} \mathbf{A} + \mathbf{R}_d^{-1})^{-1} \mathbf{A}^H \mathbf{R}_n^{-1} \mathbf{n}}_{\text{noise}}. \qquad (3.134)
\end{aligned}$$

Upon comparing Equation 3.134 to Equation 3.105 for the ZF-BLE, it can be seen that unlike the ZF-BLE, ISI and MAI are still present in the data estimator's output in the MMSE-BLE. This is because this technique does not seek to eliminate the ISI and MAI at the cost of noise amplification, but rather it attempts to achieve a balance between the different types of data corruption in order to minimize the mean squared estimation error in Equation 3.108 [206].

Upon reformulating Equation 3.131, we have:

$$\begin{aligned}
\hat{\mathbf{d}}_{\text{MMSE-BLE}} &= (\mathbf{A}^H\mathbf{R}_n^{-1}\mathbf{A}+\mathbf{R}_d^{-1})^{-1}\mathbf{A}^H\mathbf{R}_n^{-1}\mathbf{y} && (3.135)\\
&= [(\mathbf{A}^H\mathbf{R}_n^{-1}\mathbf{A})(\mathbf{I}_{KN}+(\mathbf{R}_d\mathbf{A}^H\mathbf{R}_n^{-1}\mathbf{A})^{-1})]^{-1}\mathbf{A}^H\mathbf{R}_n^{-1}\mathbf{y} && (3.136)\\
&= (\mathbf{I}_{KN}+(\mathbf{R}_d\mathbf{A}^H\mathbf{R}_n^{-1}\mathbf{A})^{-1})^{-1}(\mathbf{A}^H\mathbf{R}_n^{-1}\mathbf{A})^{-1}\mathbf{A}^H\mathbf{R}_n^{-1}\mathbf{y}. && (3.137)
\end{aligned}$$

Substituting $\hat{\mathbf{d}}_{\text{ZF-BLE}}$ from Equation 3.104 into Equation 3.137, we then arrive at:

$$\begin{aligned}
\hat{\mathbf{d}}_{\text{MMSE-BLE}} &= [\mathbf{I}_{KN}+(\mathbf{R}_d\mathbf{A}^H\mathbf{R}_n^{-1}\mathbf{A})^{-1}]^{-1}\hat{\mathbf{d}}_{\text{ZF-BLE}} && (3.138)\\
&= \mathbf{W}_o\hat{\mathbf{d}}_{\text{ZF-BLE}}, && (3.139)
\end{aligned}$$

where $\mathbf{W}_o = [\mathbf{I}_{KN}+(\mathbf{R}_d\mathbf{A}^H\mathbf{R}_n^{-1}\mathbf{A})^{-1}]^{-1}$. Since $\hat{\mathbf{d}}_{\text{ZF-BLE}}$ is obtained from $\hat{\mathbf{d}}_{\text{WMF}}$, as shown in Equation 3.106, $\hat{\mathbf{d}}_{\text{MMSE-BLE}}$ is also indirectly derived from $\hat{\mathbf{d}}_{\text{WMF}}$. Therefore, the MMSE-BLE can also be viewed as an extension of the WMF.

A Wiener filter [125] is a linear filter that produces an MMSE estimate of the quantity wanted. Therefore, $\mathbf{W}_o$ is termed as a Wiener filter, since it can be viewed as filtering the output of the ZF-BLE and producing an MMSE estimate of the data vector, $\mathbf{d}$.

For the special case, where $\mathbf{R}_n = \sigma^2\mathbf{I}_{(NQ+W-1)}$ and $\mathbf{R}_d = \mathbf{I}_{KN}$, i.e. the noise samples are uncorrelated and have a variance of σ^2, while the transmitted data symbols are also uncorrelated and normalized to a variance of one, then from Equation 3.131, we arrive at:

$$\hat{\mathbf{d}}_{\text{MMSE-BLE}}\bigg|_{\mathbf{R}_n=\sigma^2\mathbf{I}_{(NQ+W-1)},\ \mathbf{R}_d=\mathbf{I}_{KN}} \qquad (3.140)$$

$$\begin{aligned}
&= \left(\mathbf{A}^H(\frac{1}{\sigma^2}\mathbf{I}_{(NQ+W-1)})\mathbf{A}+\mathbf{I}_{KN}\right)^{-1}\mathbf{A}^H(\frac{1}{\sigma^2}\mathbf{I}_{(NQ+W-1)})\mathbf{y} && (3.141)\\
&= [\mathbf{A}^H\mathbf{A}+\sigma^2\mathbf{I}_{KN}]^{-1}\mathbf{A}^H\mathbf{y}. && (3.142)
\end{aligned}$$

Let us now continue by considering the family of non-linear joint detection schemes, which incorporate the non-linear hard-decision stage in the equalization loop, commencing with the zero forcing block decision feedback equalizer.

3.3.6 Zero Forcing Block Decision Feedback Equalizer

The zero forcing block decision feedback (ZF-BDFE) estimation technique [129, 130] introduces a non-linearity into the system by feeding back previous estimates of data symbols in order to remove the MAI. This technique is similar to the decorrelating decision-feedback detector proposed by Duel-Hallen [121].

Observing Equations 3.88 and 3.104, the output of the ZF-BLE is given by the output of the WMF in Equation 3.88, multiplied by the matrix $(\mathbf{A}^H\mathbf{R}_n^{-1}\mathbf{A})^{-1}$. The Cholesky decomposition of $(\mathbf{A}^H\mathbf{R}_n^{-1}\mathbf{A})$ gives:

$$\mathbf{A}^H\mathbf{R}_n^{-1}\mathbf{A} = (\mathbf{DU})^H\mathbf{DU}, \qquad (3.143)$$

where $\mathbf{U}$ is an upper triangular matrix, where all the elements on its main diagonal have the value of one and $\mathbf{D}$ is a diagonal matrix having real-valued elements. Applying the matrix $((\mathbf{DU})^H)^{-1}$ to the output of the WMF, $\hat{\mathbf{d}}_{\text{WMF}}$ in Equation 3.89 leads to:

$$((\mathbf{DU})^H)^{-1}\hat{\mathbf{d}}_{\text{WMF}} = ((\mathbf{DU})^H)^{-1}\mathbf{A}^H\mathbf{R}_n^{-1}\mathbf{A}\mathbf{d} + ((\mathbf{DU})^H)^{-1}\mathbf{A}^H\mathbf{R}_n^{-1}\mathbf{n}. \quad (3.144)$$

Upon invoking Equation 3.143 and recognizing that $\mathbf{D}^H = \mathbf{D}$, we have:

$$\begin{aligned} ((\mathbf{DU})^H)^{-1}\hat{\mathbf{d}}_{\text{WMF}} &= \mathbf{DUd} + ((\mathbf{DU})^H)^{-1}\mathbf{A}^H\mathbf{R}_n^{-1}\mathbf{n} && (3.145)\\ &= \mathbf{DUd} + \mathbf{D}^{-1}(\mathbf{U}^H)^{-1}\mathbf{A}^H\mathbf{R}_n^{-1}\mathbf{n}. && (3.146) \end{aligned}$$

Since $\mathbf{D}$ is a diagonal matrix having real-valued elements, it can be treated as a scaling factor and therefore it can be removed by multiplying both sides of Equation 3.146 with its inverse, giving the output vector $\hat{\mathbf{y}}$ as:

$$\begin{aligned} \hat{\mathbf{y}} &= \mathbf{D}^{-1}((\mathbf{DU})^H)^{-1}\hat{\mathbf{d}}_{\text{WMF}} && (3.147)\\ &= \mathbf{D}^{-1}[\mathbf{DUd} + \mathbf{D}^{-1}(\mathbf{U}^H)^{-1}\mathbf{A}^H\mathbf{R}_n^{-1}\mathbf{n}] && (3.148)\\ &= \mathbf{Ud} + \mathbf{D}^{-1}\mathbf{D}^{-1}(\mathbf{U}^H)^{-1}\mathbf{A}^H\mathbf{R}_n^{-1}\mathbf{n} && (3.149)\\ &= \mathbf{d} + (\mathbf{U} - \mathbf{I}_{KN})\mathbf{d} + \mathbf{z}, && (3.150) \end{aligned}$$

where $\mathbf{I}_{KN}$ is an identity matrix that has the same dimensions as $\mathbf{U}$, i.e. $(KN \times KN)$ and

$$\mathbf{z} = \mathbf{D}^{-1}\mathbf{D}^{-1}(\mathbf{U}^H)^{-1}\mathbf{A}^H\mathbf{R}_n^{-1}\mathbf{n}. \quad (3.151)$$

The matrix of $\mathbf{D}^{-1}((\mathbf{DU})^H)^{-1}$ in Equation 3.147 is analogous to the feed-forward filter in single-user decision feedback equalizers, which removes some of the interference present in the composite received signal. Equation 3.150 now gives rise to the ZF-BDFE. Since $\mathbf{U}$ is an upper triangular matrix with ones on its main diagonal, $(\mathbf{U} - \mathbf{I}_{\mathbf{KN}})$ is an upper triangular matrix with zeroes on its main diagonal. Therefore, from Equation 3.150 the i-th element of vector $\hat{\mathbf{y}}$ is:

$$\hat{y}_i = d_i + \sum_{j=i+1}^{J} [\mathbf{U} - \mathbf{I}_{KN}]_{i,j}\, d_j + z_i, \quad (3.152)$$

where $J = KN$. The second term, $(\sum_{j=i+1}^{J}[\mathbf{U} - \mathbf{I}_{KN}]_{i,j}\, d_j)$ is the ISI and MAI term. The summation is only from $j = i+1$ to J, since $(\mathbf{U} - \mathbf{I}_{KN})$ is an upper triangular matrix having zeroes on its main diagonal. For $i = J$, we have:

$$\hat{y}_J = d_J + z_J. \quad (3.153)$$

In other words, for $i = J$, the expression in Equation 3.153 does not contain a MAI or ISI term. Therefore, the i-th data estimate is $Q(\hat{y}_J)$, where $Q\{.\}$ is the thresholding or quantization operator performed in a threshold detector. For $i = 1, \ldots, J-1$, if the estimates of the data are obtained in decreasing order of i, then for each $\hat{y}_i$, the data estimates $\hat{d}_{i+1}, \ldots, \hat{d}_J$ have already been obtained. Therefore, the data estimate, $\hat{d}_i$ is given by:

$$\hat{d}_i = d_i + \sum_{j=i+1}^{J} [\mathbf{U} - \mathbf{I}_{KN}]_{i,j} \, (d_j - \hat{d}_j) + z_i. \tag{3.154}$$

Assuming that there is no error propagation, i.e. the data estimates are correct, the ISI and MAI term, $(\sum_{j=i+1}^{J} [\mathbf{U} - \mathbf{I}_{KN}]_{i,j} \, d_j)$ in Equation 3.152 can be cancelled out to obtain data estimates free of ISI and MAI. Without error propagation, i.e. if $d_i = \hat{d}_i$, the second term of Equation 3.154 is eliminated and we have:

$$\hat{d}_i = d_i + z_i. \tag{3.155}$$

If $\hat{d}_i$ of Equation 3.154 is fed into a demodulator or threshold detector, the data estimate will be obtained. All ISI and MAI is completely eliminated, thus leading to the terminology of "zero-forcing block decision feedback equalizer".

Upon rearranging Equation 3.152, the input to the threshold detector can be represented as:

$$\begin{aligned} t_i &= \hat{y}_i - \sum_{j=i+1}^{J=KN} [\mathbf{U} - \mathbf{I}_{KN}]_{i,j} \hat{d}_j & (3.156) \\ &= \hat{y}_i - \sum_{j=i+1}^{J=KN} ([\mathbf{U}]_{ij} - [\mathbf{I}_{KN}]_{ij}) \hat{d}_j & (3.157) \\ &= \hat{y}_i - \sum_{j=i+1}^{J=KN} [\mathbf{U}]_{ij} \hat{d}_j, & (3.158) \end{aligned}$$

where $[\mathbf{I}_{KN}]_{ij} = 0$ for $i \neq j$. Using this expression, the structure of the ZF-BDFE can be depicted schematically as shown in Figure 3.14. The composite received vector, $\mathbf{y}$, is processed through the WMF, which is represented by the matrix $(\mathbf{A}^H \mathbf{R}_n^{-1})$ and this operation is expressed in Equation 3.88. The output of the WMF is then passed through the feed-forward filter which is made up of the lower triangular matrix of $((\mathbf{DU})^H)^{-1}$ and the scaling factor matrix of $\mathbf{D}^{-1}$. These two operations were expressed in Equation 3.147. The output of the feed-forward filter is then processed in order to obtain the data estimates, $\hat{\mathbf{d}}_j$, where the estimates are obtained in the order of $j = J, J-1, \ldots, 1$. As the data estimates, $\hat{\mathbf{d}}_j$, are obtained, they are fed back into the receiver, where they are multiplied by the elements in the upper triangular matrix, $[\mathbf{U}]_{ij} = U_{ij}$, as shown in Equation 3.158.

The performance of this estimation technique can be degraded by error propagation in the feedback loop, where erroneous decisions affect the following decisions. This problem can be mitigated by using a technique referred to as channel sorting [127], where the vector $\mathbf{d}$ in Equation 3.56 and the system matrix $\mathbf{A}$ in Equation 3.60 are reordered so that decisions are performed on the more reliable symbols first.

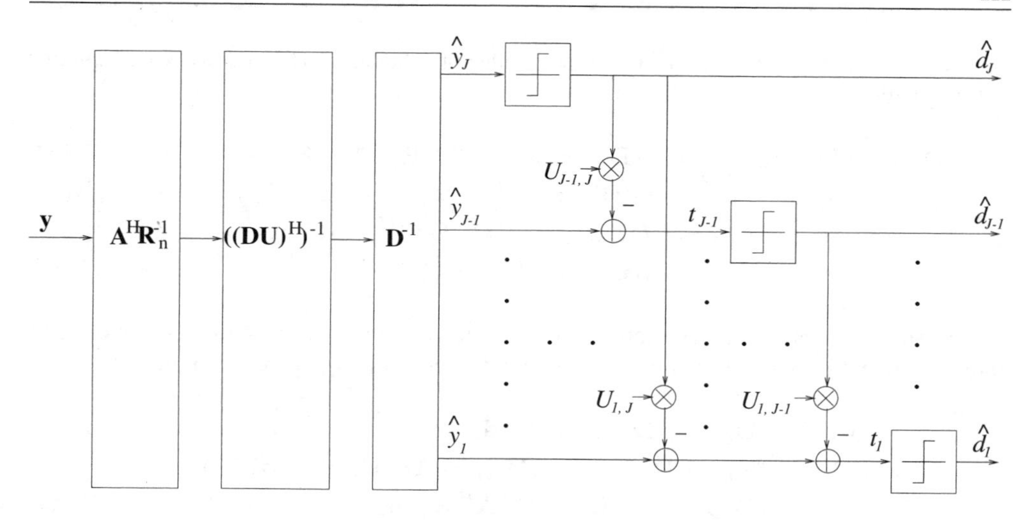

Figure 3.14: Structure of the ZF-BDFE-based receiver. The composite received vector, $\mathbf{y}$, is processed through the WMF, which is represented by the matrix $(\mathbf{A}^H\mathbf{R}_n^{-1})$ and this operation is expressed in Equation 3.88. The output of the WMF is then passed through the feed-forward filter which is made up of the lower triangular matrix of $((\mathbf{DU})^H)^{-1}$ and the scaling factor matrix of $\mathbf{D}^{-1}$. These two operations were expressed in Equation 3.147. The output of the feed-forward filter is then processed in order to obtain the data estimates, $\hat{\mathbf{d}}_j$, where the estimates are obtained in the order of $j = J, J-1, \ldots, 1$. As the data estimates, $\hat{\mathbf{d}}_j$, are obtained, they are fed back into the receiver, where they are multiplied by the elements in the upper triangular matrix, $[\mathbf{U}]_{ij} = U_{ij}$, as shown in Equation 3.158.

3.3.7 Minimum Mean Square Error Block Decision Feedback Equalizer

The structure of the MMSE-BDFE is the same as that of the ZF-BDFE in Figure 3.14 except that Cholesky decomposition [122] is performed on the matrix $(\mathbf{A}^H\mathbf{R}_n^{-1}\mathbf{A} + \mathbf{R}_d^{-1})$ [127], rather than on $(\mathbf{A}^H\mathbf{R}_n^{-1}\mathbf{A})$ as we have seen in Equation 3.143 for the ZF-BDFE. From Equation 3.131 for the MMSE-BLE, we have:

$$\hat{\mathbf{d}}_{\text{MMSE-BLE}} = (\mathbf{A}^H\mathbf{R}_n^{-1}\mathbf{A} + \mathbf{R}_d^{-1})^{-1}\mathbf{A}^H\mathbf{R}_n^{-1}\mathbf{y}. \tag{3.159}$$

By taking a Cholesky decomposition of the matrix $(\mathbf{A}^H\mathbf{R}_n^{-1}\mathbf{A} + \mathbf{R}_d^{-1})$, we obtain:

$$\mathbf{A}^H\mathbf{R}_n^{-1}\mathbf{A} + \mathbf{R}_d^{-1} = (\mathbf{D}_m\mathbf{U}_m)^H\mathbf{D}_m\mathbf{U}_m. \tag{3.160}$$

By rearranging Equation 3.89 for the output of the WMF, we obtain:

$$\begin{aligned}\hat{\mathbf{d}}_{\text{WMF}} &= \mathbf{A}^H\mathbf{R}_n^{-1}\mathbf{A}\mathbf{d} + \mathbf{A}^H\mathbf{R}_n^{-1}\mathbf{n} &(3.161)\\ &= (\mathbf{A}^H\mathbf{R}_n^{-1}\mathbf{A} + \mathbf{R}_d^{-1})\mathbf{d} - \mathbf{R}_d^{-1}\mathbf{d} + \mathbf{A}^H\mathbf{R}_n^{-1}\mathbf{n}. &(3.162)\end{aligned}$$

Applying the matrix $((\mathbf{D}_m\mathbf{U}_m)^H)^{-1}$ to both sides of Equation 3.162 and invoking Equation 3.160, yields:

$$\begin{aligned}
((\mathbf{D}_m\mathbf{U}_m)^H)^{-1}\hat{\mathbf{d}}_{\text{WMF}} &= ((\mathbf{D}_m\mathbf{U}_m)^H)^{-1}(\mathbf{A}^H\mathbf{R}_n^{-1}\mathbf{A} + \mathbf{R}_d^{-1})\mathbf{d} \quad (3.163)\\
&\quad -((\mathbf{D}_m\mathbf{U}_m)^H)^{-1}\mathbf{R}_d^{-1}\mathbf{d} + ((\mathbf{D}_m\mathbf{U}_m)^H)^{-1}\mathbf{A}^H\mathbf{R}_n^{-1}\mathbf{n}\\
&= \mathbf{D}_m\mathbf{U}_m\mathbf{d} - ((\mathbf{D}_m\mathbf{U}_m)^H)^{-1}\mathbf{R}_d^{-1}\mathbf{d} +\\
&\quad ((\mathbf{D}_m\mathbf{U}_m)^H)^{-1}\mathbf{A}^H\mathbf{R}_n^{-1}\mathbf{n}. \quad (3.164)
\end{aligned}$$

Again, since $\mathbf{D}_m$ is a diagonal matrix, it can be treated as a scaling factor and removed by multiplying both sides of Equation 3.164 with its inverse, giving the resultant vector of $\hat{\mathbf{y}}$ as:

$$\begin{aligned}
\hat{\mathbf{y}} &= (\mathbf{D}_m)^{-1}((\mathbf{D}_m\mathbf{U}_m)^H)^{-1}\hat{\mathbf{d}}_{\text{WMF}} \quad (3.165)\\
&= (\mathbf{D}_m)^{-1}\mathbf{D}_m\mathbf{U}_m\mathbf{d} - (\mathbf{D}_m)^{-1}((\mathbf{D}_m\mathbf{U}_m)^H)^{-1}\mathbf{R}_d^{-1}\mathbf{d} +\\
&\quad (\mathbf{D}_m)^{-1}((\mathbf{D}_m\mathbf{U}_m)^H)^{-1}\mathbf{A}^H\mathbf{R}_n^{-1}\mathbf{n} \quad (3.166)\\
&= \mathbf{U}_m\mathbf{d} - \mathbf{e} + \mathbf{z} \quad (3.167)\\
&= \mathbf{d} + (\mathbf{U}_m - \mathbf{I}_{KN})\mathbf{d} - \mathbf{e} + \mathbf{z}, \quad (3.168)
\end{aligned}$$

where

$$\mathbf{e} = (\mathbf{D}_m)^{-1}((\mathbf{D}_m\mathbf{U}_m)^H)^{-1}\mathbf{R}_d^{-1}\mathbf{d}, \quad (3.169)$$

and

$$\mathbf{z} = (\mathbf{D}_m)^{-1}((\mathbf{D}_m\mathbf{U}_m)^H)^{-1}\mathbf{A}^H\mathbf{R}_n^{-1}\mathbf{n}. \quad (3.170)$$

Therefore, for the MMSE-BDFE, the received vector, $\mathbf{y}$, is processed through a WMF, followed by the feed-forward filter that is represented by the matrix $(\mathbf{D}_m)^{-1}((\mathbf{D}_m\mathbf{U}_m)^H)^{-1}$. The output of the feed-forward filter is given as:

$$\begin{aligned}
\hat{\mathbf{y}} &= (\mathbf{D}_m)^{-1}((\mathbf{D}_m\mathbf{U}_m)^H)^{-1}\mathbf{A}^H\mathbf{R}_n^{-1}\mathbf{y} \quad (3.171)\\
&= \mathbf{M}\mathbf{y}, \quad (3.172)
\end{aligned}$$

where

$$\mathbf{M} = (\mathbf{D}_m)^{-1}((\mathbf{D}_m\mathbf{U}_m)^H)^{-1}\mathbf{A}^H\mathbf{R}_n^{-1}. \quad (3.173)$$

The matrix $\mathbf{M}$ represents the combination of the WMF and feed-forward filter. Analogously to Equation 3.152 for the ZF-BDFE, Equation 3.168 can be expanded into:

$$\hat{d}_i = d_i + \sum_{j=i+1}^{J} [\mathbf{U}_m - \mathbf{I}_{KN}]_{i,j}\,(d_j - \hat{d}_j) - e_i + z_i. \quad (3.174)$$

If the data estimates, $\hat{d}_j$ are accurately estimated, i.e. $\hat{d}_j = d_j$, Equation 3.174 becomes:

$$\hat{d}_i = d_i - e_i + z_i, \qquad (3.175)$$

giving an MMSE estimate of the data symbol. For the MMSE-BDFE, the feedback operator, $\mathbf{S}$ can be derived from Equation 3.174 as:

$$\mathbf{S} = \mathbf{U}_m - \mathbf{I}_{KN}, \qquad (3.176)$$

where $\mathbf{S}$ is an upper diagonal matrix and all the elements on its main diagonal have values of zero. If the data symbols are estimated in descending order of j, the previously estimated data symbols of $\hat{d}_j$ are fed back into the system via this feedback operator, resulting in the threshold detector or data demodulator input of t_i, given as:

$$\begin{aligned} t_i &= \hat{y}_i - \sum_{j=i+1}^{J=KN} [\mathbf{U}_m - \mathbf{I}_{KN}]_{ij} \hat{d}_j & (3.177)\\ &= \hat{y}_i - \sum_{j=i+1}^{J=KN} ([\mathbf{U}_m]_{ij} - [\mathbf{I}_{KN}]_{ij}) \hat{d}_j & (3.178)\\ &= \hat{y}_i - \sum_{j=i+1}^{J=KN} [\mathbf{U}_m]_{ij} \hat{d}_j, & (3.179) \end{aligned}$$

where $[\mathbf{I}_{KN}]_{ij} = 0$ for $i \neq j$. The schematic for the MMSE-BDFE is the same as that of the ZF-BDFE shown in Figure 3.14 except that the elements in the matrices $\mathbf{D}$ and $\mathbf{U}$ are now different, since $\mathbf{D}$ and $\mathbf{U}$ were derived by decomposing Equation 3.143, rather than Equation 3.160.

Having reviewed a variety of detection techniques, let us now consider their complexity. We note, however, that the forthcoming detailed complexity analysis is summarized in Table 3.1 and hence readers interested only in the final results can proceed to Table 3.1, followed by Section 3.5.

3.4 Complexity Calculations

This section analyzes the complexity of each of the four joint detection schemes described in Section 3.3. The complexity calculations are presented in terms of the number of multiplications and the number of additions needed to solve the matrix equations that describe the joint detection schemes. In the following calculations, the variables of Section 3.3 will be used, where

- The number of paths in the CIR is represented by L and the total channel dispersion is given by W chips.
- The number of users is represented by K, the spreading sequence length is Q chips, and each user transmits N symbols per burst.

- $\mathbf{A}$ is a matrix having dimensions of $[(NQ+W-1) \times (KN)]$. Let us set $T = (NQ+W-1)$ and $U = KN$. Therefore, $\mathbf{A}$ is a $(T \times U)$ matrix.
- $\mathbf{y}$ is a column vector having the dimensions of $[(NQ+W-1) \times 1]$ i.e. $(T \times 1)$.
- $\mathbf{R}_n^{-1}$ is a square matrix with the dimensions of $[(NQ+W-1) \times (NQ+W-1)]$. From this point onwards, we assume that $\mathbf{R}_n^{-1}$ is always known and the complexity calculation for obtaining $\mathbf{R}_n^{-1}$ is not considered. This is because all the four JD schemes require this covariance matrix and therefore all four schemes will have the same overhead pertaining to $\mathbf{R}_n^{-1}$.

The number of multiplications and additions needed to perform certain matrix operations are given as follows:

- Multiplication of a $(B \times C)$ matrix with a $(C \times D)$ matrix requires $B(C-1)D$ additions and BCD multiplications. However, in order to simplify the complexity calculations, the number of additions has been approximated as BCD, with the assumption that C is sufficiently large such that $C - 1 \approx C$.
- Multiplication of a $(B \times C)$ matrix with a $(C \times 1)$ vector requires approximately BC additions and BC multiplications.
- Cholesky decomposition of a Hermitian matrix having the dimensions of $(C \times C)$ requires approximately $C^3/6$ additions and $C^3/6$ multiplications [190]. An approximate Cholesky factorization has been proposed by Karimi and Anderson [208], in order to reduce the complexity. However, in this discussion, the complexity of full Cholesky factorization will be used.
- The solution of a linear equation, $\mathbf{Ax} = \mathbf{b}$ for $\mathbf{x}$, where $\mathbf{b}$ and $\mathbf{x}$ are column vectors with dimensions of $(C \times 1)$ and $\mathbf{A}$ is a Hermitian matrix with the dimensions of $(C \times C)$, requires Cholesky decomposition of $\mathbf{A}$, followed by an additional C^2 additions and C^2 multiplications.

3.4.1 Whitening Matched Filter

The complexity of the WMF is evaluated first, since it is a precursor for all the other schemes. From Equation 3.88, the solution for the WMF is given as:

$$\hat{\mathbf{d}}_{\text{WMF}} = \mathbf{A}^H \mathbf{R}_n^{-1} \mathbf{y}.$$

Therefore, the complexity calculations are:

Matrix operation	Additions	Multiplications
$\mathbf{y}' = \mathbf{R}_n^{-1}\mathbf{y}$	T^2	T^2
$\hat{\mathbf{d}}_{\text{WMF}} = \mathbf{A}^H\mathbf{y}'$	UT	UT

The total number of addition and multiplication operations is hence given by:

$$\Sigma_{\text{WMF}} = 2(UT + T^2). \quad (3.180)$$

The number of detected symbols is equal to $U = KN$. Therefore, the total number of operations per detected symbol is:

$$\bar{\Sigma}_{\text{WMF}} = 2\Big(\frac{UT + T^2}{U}\Big) \tag{3.181}$$

$$= 2\Big(T + \frac{T^2}{U}\Big) \tag{3.182}$$

3.4.2 Zero Forcing Block Linear Equalizer

From Equation 3.104, the estimated data for the ZF-BLE is given as:

$$\hat{\mathbf{d}}_{\text{ZF-BLE}} = (\mathbf{A}^H\mathbf{R}_n^{-1}\mathbf{A})^{-1}\mathbf{A}^H\mathbf{R}_n^{-1}\mathbf{y}.$$

In order to construct the system matrix $\mathbf{A}$, the combined impulse responses, $\mathbf{b}_n^{(k)}$, of all K users and all N symbols have to be calculated. For a spreading sequence of Q chips and a CIR having L paths and dispersion of W chips, the complexity required to obtain $\mathbf{b}_n^{(k)}$ for one symbol of one user is $L(Q + W - 1)$ additions and QL multiplications. For a total of K users and N symbols per user, the total complexity required to construct the matrix $\mathbf{A}$ is $KNL(Q + W - 1)$ additions and $KNQL$ multiplications. Employing $U = KN$ and substituting V for $(Q+W-1)$ gives $KNL(Q+W-1) = ULV$ additions and $KNQL = UQL$ multiplications. Hence the associated complexity calculations for the implementation of the ZF-BLE are:

Matrix operation	Additions	Multiplications
Construction of $\mathbf{A}$	ULV	UQL
$\mathbf{y}' = \mathbf{A}^H\mathbf{R}_n^{-1}\mathbf{y}$	$UT + T^2$	$UT + T^2$
$\mathbf{S} = (\mathbf{A}^H\mathbf{R}_n^{-1}\mathbf{A})$	$UT^2 + U^2T$	$UT^2 + U^2T$
Cholesky decomposition of $\mathbf{S}$	$U^3/6$	$U^3/6$
Solve for $\hat{\mathbf{d}}$: $\mathbf{S}\hat{\mathbf{d}} = \mathbf{y}'$	U^2	U^2

The total number of addition and multiplication operations is therefore:

$$\Sigma_{\text{ZF-BLE}} = ULV + UQL + 2\Big[UT + T^2 + UT^2 + U^2T + \frac{U^3}{6} + U^2\Big]. \tag{3.183}$$

Again, the number of detected symbols is $U = KN$, hence, the total number of operations per detected symbol is

$$\bar{\Sigma}_{\text{ZF-BLE}} = LV + QL + 2\Big[T + \frac{T^2}{U} + T^2 + UT + \frac{U^2}{6} + U\Big]. \tag{3.184}$$

For the simplified version of the ZF-BLE shown in Equation 3.107, where it is assumed that $\mathbf{R}_n = \sigma^2\mathbf{I}$, the complexity is reduced. Here, the structure of the matrix $\mathbf{A}$ can be exploited, since the simplified ZF-BLE detector is now given by the expression:

$$\hat{\mathbf{d}}_{\text{ZF-BLE}} = (\mathbf{A}^H\mathbf{A})^{-1}\mathbf{A}^H\mathbf{y}. \tag{3.185}$$

Based on the structure of the matrix $\mathbf{A}$ as shown in Figure 3.9, we note that the matrix is a sparse matrix and the grade of sparsity depends on the spread of the combined impulse response, $\mathbf{b}_n^{(k)}$. The column vector, $\mathbf{b}_n^{(k)}$ has a length of $(Q+W-1)$ chips, if it is assumed that all the users employ the same spreading sequence length and have the same channel delay spread. If we introduce the short-hand $\mathbf{J} = \mathbf{A}^H\mathbf{A}$ in Equation 3.185, then:

$$[\mathbf{J}]_{i,j} = \sum_{m=1}^{NQ+W-1} [\mathbf{A}^H]_{i,m} \times [\mathbf{A}]_{m,j}, \tag{3.186}$$

where $i = 1, \ldots, KN$ and $j = 1, \ldots, KN$. The elements $[\mathbf{J}]_{i,j}$ will have non-zero values only when $[\mathbf{A}^H]_{i,m} \neq 0$ **and** $[\mathbf{A}]_{m,j} \neq 0$. Therefore, the multiplications and additions involved in Equation 3.186 only have to be carried out when the corresponding elements in both matrices $\mathbf{A}^H$ and $\mathbf{A}$ are non-zero. For example, in Figure 3.9, columns 3 to N do not overlap with column 1. Nor do columns $(N+3)$ to $2N$, $(2N+3)$ to $3N$, $((K-1)N+3)$ to KN. Therefore:

$$[\mathbf{J}]_{1,j} = 0,$$

for $j = 3, \ldots, N; (N+3), \ldots, 2N; (2N+3), \ldots, 3N; ((K-1)N+3), \ldots, KN.$

This procedure can be repeated for all the elements in $\mathbf{J}$ by identifying the overlapping columns of matrix $\mathbf{A}$, which thus reduces the complexity of the multiplication of the matrix $\mathbf{A}^H$ with $\mathbf{A}$.

Let us consider a system matrix $\mathbf{A}$ for K users, where $Q+W-1 \leqslant 2Q$. For user 1, there are K columns where there is a full overlap of all $(Q+W-1)$ elements in $\mathbf{b}_1^{(1)}$ with $\mathbf{b}_1^{(k)}$, for $k = 1, \ldots, K$, leading to $K(Q+W-1)$ number of multiplications and additions. There are also K columns, where there is a partial overlap between $\mathbf{b}_1^{(1)}$ and $\mathbf{b}_2^{(k)}$, for $k = 1, \ldots, K$, leading to $K(Q+W-1-Q) = K(W-1)$ number of multiplications and additions. Since there are K users, the total number of multiplications and additions, M is given by:

$$M = K[K(Q+W-1) + K(W-1)] \tag{3.187}$$

$$= K^2[Q + 2(W-1)]. \tag{3.188}$$

Let us now consider a general matrix $\mathbf{A}$ and commence by assuming that each column in the matrix overlaps with all the columns. Therefore, the total number of multiplications and additions, is given by:

$$M = K^2N^2(Q+W-1). \tag{3.189}$$

However, since only some of the columns actually do overlap to produce non-zero elements of $[\mathbf{J}]_{i,j}$, let the ratio of overlap be represented by ρ:

$$\rho = \frac{\left\lceil \frac{(Q+W-1)}{Q} \right\rceil}{N}, \tag{3.190}$$

where $\lceil (Q+W-1)/Q \rceil$ gives us the number of symbols that overlap due to the multi-path channel. This leads to an approximation of the complexity, where the total number of multiplications and additions is given by:

$$\begin{aligned} M &\approx \rho K^2 N^2 (Q+W-1) & (3.191)\\ &= \rho U^2 V, & (3.192) \end{aligned}$$

where $U = KN$ and $V = Q + W - 1$.

Using the same approach, the reduced complexity for the matched filtering expression of $\mathbf{y}' = \mathbf{A}^H \mathbf{y}$ in Equation 3.185 can be obtained. Each row in the matrix $\mathbf{A}^H$ contains only $(Q+W-1)$ non-zero elements. Therefore, for the multiplication of one row in $\mathbf{A}^H$ with the column vector $\mathbf{y}$ requires $(Q+W-1)$ multiplication and addition operations. The reduced complexity for the matrix operation of $\mathbf{y}' = \mathbf{A}^H \mathbf{y}$ is then given by:

$$\begin{aligned} G &= KN(Q+W-1) & (3.193)\\ &= UV. & (3.194) \end{aligned}$$

Employing Equations 3.192 and 3.194, the complexity calculations for the simplified ZF-BLE can now be written as:

Matrix operation	**Additions**	**Multiplications**
Construction of $\mathbf{A}$	ULV	UQL
$\mathbf{y}' = \mathbf{A}^H \mathbf{y}$	UV	UV
$\mathbf{S} = (\mathbf{A}^H \mathbf{A})$	$\rho U^2 V$	$\rho U^2 V$
Cholesky decomposition of $\mathbf{S}$	$U^3/6$	$U^3/6$
Solve for $\hat{\mathbf{d}} : \mathbf{S}\hat{\mathbf{d}} = \mathbf{y}'$	U^2	U^2

The total number of addition and multiplication operations per detected symbol for the simplified version of the ZF-BLE specified by Equation 3.185 is given by:

$$\begin{aligned} \bar{\Sigma}_{\text{ZF-BLE}} &= \frac{1}{U}\Big[ULV + UQL\Big] + \frac{2}{U}\Big[2UQL + UV + \rho U^2 V + \frac{U^3}{6} + U^2\Big] & (3.195)\\ &= LV + QL + 2\Big[V + \rho UV + \frac{U^2}{6} + U\Big], & (3.196) \end{aligned}$$

where $U = KN$ and $V = Q + W - 1$.

3.4.3 Minimum Mean Square Error Block Linear Equalizer

From Equation 3.131, the equalizer output for the MMSE-BLE is given by:

$$\hat{\mathbf{d}}_{\text{MMSE-BLE}} = (\mathbf{A}^H\mathbf{R}_n^{-1}\mathbf{A} + \mathbf{R}_d^{-1})^{-1}\mathbf{A}^H\mathbf{R}_n^{-1}\mathbf{y},$$

while the complexity calculations ensue as follows:

Matrix operation	Additions	Multiplications
Construction of $\mathbf{A}$	ULV	UQL
$\mathbf{y}' = \mathbf{A}^H\mathbf{R}_n^{-1}\mathbf{y}$	$UT + T^2$	$UT + T^2$
$\mathbf{S} = (\mathbf{A}^H\mathbf{R}_n^{-1}\mathbf{A})$	$UT^2 + U^2T$	$UT^2 + U^2T$
$\mathbf{S}' = \mathbf{S} + \mathbf{R}_d^{-1}$	U^2	$-$
Cholesky decomposition of $\mathbf{S}'$	$U^3/6$	$U^3/6$
Solve for $\hat{\mathbf{d}}$: $\mathbf{S}'\hat{\mathbf{d}} = \mathbf{y}'$	U^2	U^2

The required number of addition and multiplication operations therefore amounts to:

$$\Sigma_{\text{MMSE-BLE}} = ULV + UQL + 2\Big[UT + T^2 + UT^2 + U^2T + \frac{U^3}{6} + U^2\Big] + U^2. \tag{3.197}$$

The number of symbols detected is again $U = KN$. Therefore, the number of addition and multiplication operations per detected symbol is

$$\bar{\Sigma}_{\text{MMSE-BLE}} = LV + QL + 2\Big[T + \frac{T^2}{U} + T^2 + UT + \frac{U^2}{6} + U\Big] + U. \tag{3.198}$$

The simplified version of the MMSE-BLE given in Equation 3.142 can be employed to provide reduced complexity calculations following the approach outlined for the ZF-BLE previously. Upon repeating Equation 3.142 for convenience, the reduced complexity MMSE-BLE is given as:

$$\hat{\mathbf{d}}_{\text{MMSE-BLE}} = (\mathbf{A}^H\mathbf{A} + \sigma^2\mathbf{I}_{KN})^{-1}\mathbf{A}^H\mathbf{y}. \tag{3.199}$$

Employing Equations 3.192 and 3.194, the reduced complexity calculations for the MMSE-BLE can be formulated as:

Matrix operation	Additions	Multiplications
Construction of $\mathbf{A}$	ULV	UQL
$\mathbf{y}' = \mathbf{A}^H\mathbf{y}$	UV	UV
$\mathbf{S} = (\mathbf{A}^H\mathbf{A})$	ρU^2V	ρU^2V
$\mathbf{S}' = \mathbf{S} + \sigma^2\mathbf{I}$	U	$-$
Cholesky decomposition of $\mathbf{S}$	$U^3/6$	$U^3/6$
Solve for $\hat{\mathbf{d}}$: $\mathbf{S}\hat{\mathbf{d}} = \mathbf{y}'$	U^2	U^2

The total number of addition and multiplication operations per detected symbol for the simplified version of the MMSE-BLE is given by:

$$\bar{\Sigma}_{\text{MMSE-BLE}} = \frac{1}{U}(ULV + UQL) + \frac{2}{U}\Big[UV + \rho U^2 V + \frac{U^3}{6} + U^2\Big] + 1 \quad (3.200)$$

$$= LV + QL + 2\Big[V + \rho UV + \frac{U^2}{6} + U\Big] + 1, \quad (3.201)$$

where $U = KN$ and $V = Q + W - 1$. Apart from the extra addition operation in Equation 3.201, the complexity of the simplified MMSE-BLE is the same as that of the simplified ZF-BLE in Equation 3.196.

3.4.4 Zero Forcing Block Decision Feedback Equalizer

From Equations 3.147 and 3.158, the formulae that describe the ZF-BDFE are given by:

$$\hat{\mathbf{y}} = \mathbf{D}^{-1}((\mathbf{DU})^H)^{-1}\mathbf{A}^H\mathbf{R}_n^{-1}\mathbf{y}$$

and

$$t_i = \hat{y}_i - \sum_{j=i+1}^{J=KN} [\mathbf{U}]_{ij}\hat{d}_j,$$

where

$$\mathbf{A}^H\mathbf{R}_n^{-1}\mathbf{A} = (\mathbf{DU})^H\mathbf{DU},$$

leading to the following complexity calculations:

Matrix operation	Additions	Multiplications
Construction of $\mathbf{A}$	ULV	UQL
$\mathbf{r} = \mathbf{A}^H\mathbf{R}_n^{-1}\mathbf{y}$	$UT + T^2$	$UT + T^2$
$\mathbf{S} = (\mathbf{A}^H\mathbf{R}_n^{-1}\mathbf{A})$	$UT^2 + U^2T$	$UT^2 + U^2T$
Cholesky decomposition of $\mathbf{S}$	$U^3/6$	$U^3/6$
Solve for $\mathbf{y}'$: $\mathbf{D}^{-1}(\mathbf{DU})^H\mathbf{y}' = \mathbf{r}$	$U^2/2$	$U^2/2$
Feedback operation: $t_i = \hat{y}_i - \sum_{j=i+1}^{J=KN}[\mathbf{U}]_{ij}\hat{d}_j$	$U^2/2$	$U^2/2$

Therefore the required number of addition and multiplication operations is:

$$\Sigma_{\text{ZF-BDFE}} = ULV + UQL + 2\Big[UT + T^2 + UT^2 + U^2T + \frac{U^3}{6} + U^2\Big]. \quad (3.202)$$

Since the number of detected symbols is $U = KN$, the number of addition and multiplication operations per detected symbol is

$$\bar{\Sigma}_{\text{ZF-BDFE}} = LV + QL + 2\Big[T + \frac{T^2}{U} + T^2 + UT + \frac{U^2}{6} + U\Big]. \quad (3.203)$$

With the aid of Equations 3.192 and 3.194, the simplified complexity calculations for the ZF-BDFE can now be broken down into:

Matrix operation	Additions	Multiplications
Construction of $\mathbf{A}$	ULV	UQL
$\mathbf{r} = \mathbf{A}^H\mathbf{y}$	UV	UV
$\mathbf{S} = (\mathbf{A}^H\mathbf{A})$	$\rho U^2 V$	$\rho U^2 V$
Cholesky decomposition of $\mathbf{S}$	$U^3/6$	$U^3/6$
Solve for $\mathbf{y}' : \mathbf{D}^{-1}(\mathbf{DU})^H\mathbf{y}' = \mathbf{r}$	$U^2/2$	$U^2/2$
Feedback operation: $t_i = \hat{y}_i - \sum_{j=i+1}^{J=KN}[\mathbf{U}]_{ij}\hat{d}_j$	$U^2/2$	$U^2/2$

The total number of addition and multiplication operations per detected symbol for the reduced complexity version of the ZF-BDFE is therefore

$$\begin{aligned}\bar{\Sigma}_{\text{ZF-BDFE}} &= \frac{1}{U}(ULV + UQL) + \\ &\quad \frac{2}{U}\Big[UV + \rho U^2 V + \frac{U^3}{6} + U^2\Big] \quad (3.204)\\ &= LV + QL + 2\Big[V + \rho UV + \frac{U^2}{6} + U\Big], \quad (3.205)\end{aligned}$$

where $U = KN$ and $V = Q + W - 1$. This is also the same expression as that of the reduced complexity for the ZF-BLE given in Equation 3.196.

Finally, let us consider the complexity of the MMSE-BDFE.

3.4.5 Minimum Mean Square Error Block Decision Feedback Equalizer

The implementation of the MMSE-BDFE is similar to that of the ZF-BDFE, except that Cholesky decomposition is carried out on the matrix $(\mathbf{A}^H\mathbf{R}_n^{-1}\mathbf{A} + \mathbf{R}_d^{-1})$ instead of $(\mathbf{A}^H\mathbf{R}_n^{-1}\mathbf{A})$. Using the results of Equation 3.202 for the ZF-BDFE and including an extra U^2 additions for the matrix addition of $(\mathbf{A}^H\mathbf{R}_n^{-1}\mathbf{A} + \mathbf{R}_d^{-1})$, the total number of addition and multiplication operations for the MMSE-BDFE becomes:

$$\begin{aligned}\Sigma_{\text{MMSE-BDFE}} &= ULV + UQL + \\ &\quad 2\Big[UT + T^2 + UT^2 + U^2T + \frac{U^3}{6} + U^2\Big] + U^2. \quad (3.206)\end{aligned}$$

As before, the number of detected symbols is $U = KN$. Therefore, the number of addition and multiplication operations per detected symbol is

$$\bar{\Sigma}_{\text{MMSE-BDFE}} = LV + QL + 2\Big[T + \frac{T^2}{U} + T^2 + UT + \frac{U^2}{6} + U\Big] + U. \quad (3.207)$$

Here again, analogous to the similarity between the simplified ZF-BLE and the simplified MMSE-BLE, the complexity per detected symbol for the simplified MMSE-BDFE is the same as the complexity of the simplified ZF-BDFE given in Equation 3.205.

Table 3.1 summarizes the complexity calculations per detected symbol for all the four JD schemes. It can be seen that the four schemes have similar complexities and that both of the MMSE schemes have an extra U number of additions per detected symbol. In the table, $T = (NQ + W - 1)$ and $U = KN$. For a system supporting $K = 10$ users, transmitting $N = 20$ symbols per burst, spreading sequences of $Q = 31$ chips in length and assuming an impulse response spread of $W = 10$ and $L = 4$ for a four-path typical urban channel as defined in COST 207 [105], we have:

$$T = NQ + W - 1 = (20 \times 31) + 10 - 1 = 629,$$

$$U = KN = 10 \times 20 = 200,$$

and

$$V = Q + W - 1 = 31 + 10 - 1 = 40.$$

Therefore, for the ZF-BLE and the ZF-BDFE, the number of addition and multiplication operations per detected symbol is

$$\begin{aligned}\bar{\Sigma} &= LV + QL + 2 \times \Big[T + \frac{T^2}{U} + T^2 + UT + \frac{U^2}{6} + U\Big] \\ &= 4(40) + 31(4) + \Big[629 + \frac{(629)^2}{200} + (629)^2 + (629 \times 200) + \frac{(200)^2}{6} + 200\Big] \\ &\approx 1.06 \times 10^6. \end{aligned} \quad (3.208)$$

For the MMSE-BLE and the MMSE-BDFE, the number of addition and multiplication operations per detected symbol is also approximately 1.06×10^6.

If the simplified versions of the joint detectors are employed, the number of combined multiplications and additions per detected symbol is reduced. For $V = Q + W - 1 = 40$ and assuming that $\rho = (\lceil (Q + W - 1)/Q \rceil)/N = 0.1$, the number of addition and multiplication

JD receiver	Full complexity	Reduced complexity
ZF-BLE	$LV+QL+$ $2(T+T^2/U+T^2+UT+U^2/6+U)$	$LV+QL+$ $2(V+\rho UV+U^2/6+U)$
ZF-BDFE	$LV+QL+$ $2(T+T^2/U+T^2+UT+U^2/6+U)$	$LV+QL+$ $2(V+\rho UV+U^2/6+U)$
MMSE-BLE	$LV+QL+U+$ $2(T+T^2/U+T^2+UT+U^2/6+U)$	$LV+QL+$ $2(V+\rho UV+U^2/6+U)$
MMSE-BDFE	$LV+QL+U+$ $2(T+T^2/U+T^2+UT+U^2/6+U)$	$LV+QL+$ $2(V+\rho UV+U^2/6+U)$

Table 3.1: Total number of addition and multiplication operations per detected symbol for the four joint detection receivers studied, where $U = KN, T = NQ + W - 1$ and $V = Q + W - 1$.

operations per detected symbol is given by:

$$\begin{aligned}\bar{\Sigma} &= LV + QL + 2\Big[V + \rho UV + \frac{U^2}{6} + U\Big] \\ &= 4(40) + 31(4) + 2\Big[40 + (0.1)(200)(40) + \frac{(200)^2}{6} + (200)^2\Big] \\ &\approx 14614. \qquad (3.209)\end{aligned}$$

From Table 3.1, it can be seen that for the reduced complexity JD receivers, the complexity per detected symbol is in the order of $O(U^2)$, where $U = KN$. This complexity is incurred by the Cholesky decomposition required to calculated the inverse of the cross-correlation matrix $(\mathbf{A}^H\mathbf{A})$ or the biased cross-correlation matrix $(\mathbf{A}^H\mathbf{A} + \sigma^2\mathbf{I}_{KN})$. Several reduced complexity versions of calculating this inverse have been proposed, including an approximate Cholesky factorization that involves rearranging the columns of the system matrix and exploiting the so-called band-diagonal nature of the matrix proposed by Karimi and Anderson [208]. Iterative solutions have also been proposed including the method of conjugate gradients and the Jacobian method advocated by Karimi [209]. Benvenuto and Sostrato [210] proposed a reduced complexity matrix inversion solution by partitioning the cross-correlation matrix into circulant submatrices and employing the Fast Fourier Transform (FFT) in order to decompose these submatrices. Upon invoking these algorithms, the complexity of the JD receivers can be further reduced. Let us now provide a brief critical analysis of the advantages and disadvantages of JD receivers.

3.5 Advantages and Disadvantages of Joint Detection Receivers

The advantages can be summarized as follows:

- JD combats both MAI and ISI, therefore channel equalization is inherent in a JD receiver, hence the receiver is near–far resistant and the output signal of the receiver does not require further equalization.
- The complexity and the implementation of the JD algorithms are independent of the modulation mode used for transmission [126]. This provides a simple approach to supporting variable data rates, for example, by changing the modulation mode without needing to modify the JD receiver. However, as expected, the performance of the JD receivers usually degrades as the number of modulation constellation points is increased, since the distance between the constellation points is reduced. However, adaptive techniques can be invoked in order to implement adaptive detection of CDMA signals as will be exemplified in Chapter 4.
- JD can be readily combined with coherent receiver antenna diversity (CRAD), in order to mitigate the effects of fading by achieving diversity gain. CRAD reduces the variation in SNR at the receiver, resulting in a less variable BER fluctuation and in a more constant quality of service.
- All the users' signals are detected simultaneously, which makes JD especially useful for base station receivers.

The disadvantages can be highlighted as follows:

- The JD schemes reviewed were designed based on the assumption that burst synchronization and chip synchronization could be achieved. However, the JD algorithms can be extended to account for asynchronous transmission by increasing the size of the system matrix, if the relative delays of the users are known. This further increases the complexity.
- Channel estimates are required by the joint detector, which implies that the transmitted signals must include training sequences or a midamble for channel estimation. This reduces the data rate and capacity of the system. Furthermore, at the receiver channel estimators must be employed, in order to produce the estimates needed by the joint detector, which increases the complexity of the system.

Let us now quantify the performance of the JD receivers studied.

3.6 Experimental Work on Joint Detection Receivers

3.6.1 Experimental Conditions

This section characterizes the BER performance of the four JD schemes presented in Section 3.3. A synchronous DS-CDMA model was used for conducting simulations, investigating

Parameter	Value
Doppler frequency	80 Hz
Spreading ratio, Q	16
Chip rate	2.167 MBaud
Frame burst structure	FRAMES Mode 1 Spread burst 1 [205] in Figure 3.17
Burst duration	577 μs
Number of QAM symbols per JD block, N	28
Modulation mode	4-QAM
Number of CDMA users, K	8

Table 3.2: Simulation parameters for JD-CDMA systems.

the performance of the ZF-BLE, MMSE-BLE, ZF-BDFE and MMSE-BDFE detectors. The simulation parameters for the investigations carried out are presented in Table 3.2 and the experiments were carried out under the following assumptions:

- Transmission was implemented in bursts and the data symbols were assumed to be uncorrelated and the statistically wide-sense stationary, i.e. $\mathbf{R}_d = \mathbf{I}_{KN}$.

- The uplink was considered, thus the channel impulse response, $\mathbf{h}_n^{(k)}$ of each user was different and independent. The channel impulse responses were derived from the COST 207 channel model for a rural area (RA), a typical urban area (TU) and a bad urban area (BU) [105]. The normalized channel impulse responses are shown in Figures 3.15, 3.18 and 3.20. Each path in the channel impulse response was faded independently using Rayleigh fading. The signal level variations due to path loss and shadowing were assumed to be eliminated by power control. The COST 207 [105] channel profiles were developed for GSM performance investigations. The third generation wideband CDMA systems [64] have a higher bandwidth and chip rate than the GSM system [64], thus leading to a higher multipath resolution and a larger number of resolvable multipaths. However, in our investigations, the COST 207 channels were adopted for the sake of the comparability of our results, since these channels are widely used in the CDMA research community.

- The noise, $\mathbf{n}$ is assumed to be additive white Gaussian noise (AWGN) having zero mean and a covariance matrix of $\sigma^2 \mathbf{I}_{(NQ+W-1)}$, where σ^2 is the noise variance.

- Due to the high complexity of the joint detection receiver, the sample size used in our investigations was limited to a duration of approximately $200t_D$, where $t_D = 1/2f_D$, and f_D is the Doppler frequency spread of the channel. This is justified by the results concluded by Bello [211], who showed that when the symbol rate, f_s is much higher than f_D, i.e. $f_s/f_d >> 1$, then the sample size required to estimate the BER can be approximated by small sample sizes.

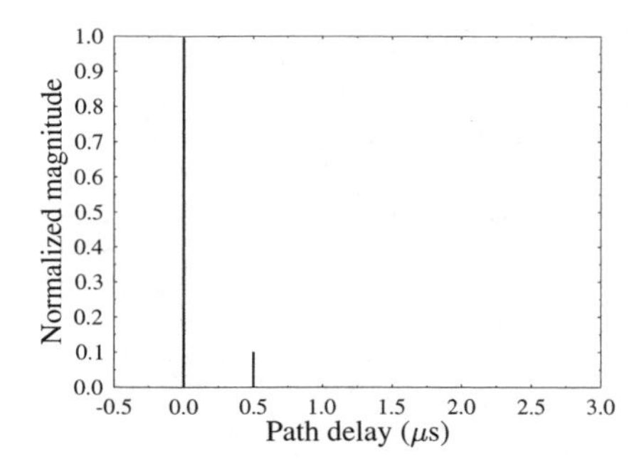

Figure 3.15: Normalized channel impulse response for the COST 207 [105] **two-path Rural Area channel**.

3.6.2 Comparison of Joint Detection Receivers in Different Channels

Figure 3.16 compares the BER performance of all four JD schemes using the simulation conditions described previously. The transmission burst format of the FRAMES Mode 1 (FMA1) spread speech/data burst 1 [205] was adopted. The burst structures of the spread speech/data bursts 1 and 2 are shown in Figure 3.17. Random sequences of $Q = 16$ chips were used as signature sequences for each user. For each scheme, $K = 8$ users were admitted to the system. Each path in the rural area channel was faded independently using Rayleigh fading with a Doppler frequency of $f_d = 80$ Hz and the Baud rate was 2.167 MBaud. From the plot, it can be seen that all four schemes outperform the conventional WMF. With the WMF, the BER remains near-constant as the SNR per bit increases. Since the errors produced are due to the effects of MAI and ISI, increasing the SNR does not alleviate the problem. In contrast to the WMF, for all four JD schemes the BER decreases as the SNR per bit increases. When comparing the JD schemes, it can be seen that the MMSE-BLE performs slightly better than the ZF-BLE and the MMSE-BDFE performs better than the ZF-BDFE. However, the disparity in performance is not obvious, because the channel did not introduce severe ISI effects, and all four schemes were capable of mitigating the interference. Finally, the BDFE schemes performed slightly better than the linear schemes.

The four JD schemes were also utilized in simulations over the Typical Urban channel model of Figure 3.18. All other parameters were similar to the investigations conducted over the Rural Area channel. The results obtained from these simulations are shown in Figure 3.19. Here again, it can be seen that all four schemes significantly improve the BER compared to the WMF. The performance difference between the four JD algorithms is more obvious for this channel, where the ISI is higher due to an increased number of paths compared to the Rural Area channel of Figure 3.15.

Finally, the four joint detection schemes were tested in simulations using the COST 207 [105] seven path Bad Urban channel profile of Figure 3.20. Again, the simulation conditions of Table 3.2 were used. The results for the simulations are presented in Figure 3.21. Here, the difference in performance between the four joint detection schemes is more significant, since there is a higher number of dispersive paths in the channel, leading to higher ISI. Having

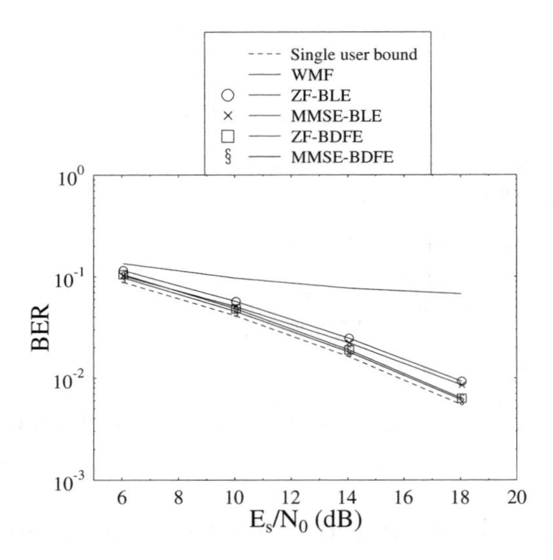

Figure 3.16: BER performance curves for the WMF, ZF-BLE, MMSE-BLE, ZF-BDFE and MMSE-BDFE with $K = 8$ users and for a **Rural Area channel** with the impulse response shown in Figure 3.15, using the parameters of Table 3.2.

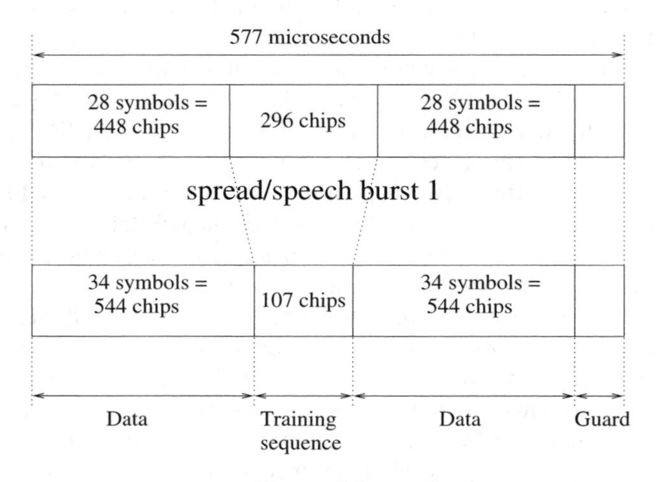

Figure 3.17: Transmission burst structures of the FMA1 spread speech/data bursts 1 and 2 of the FRAMES proposal [205].

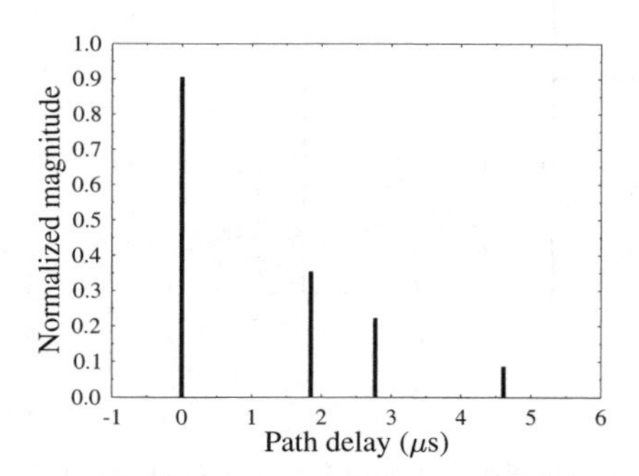

Figure 3.18: Normalized channel impulse response for the COST 207 [105] **four-path Typical Urban channel**.

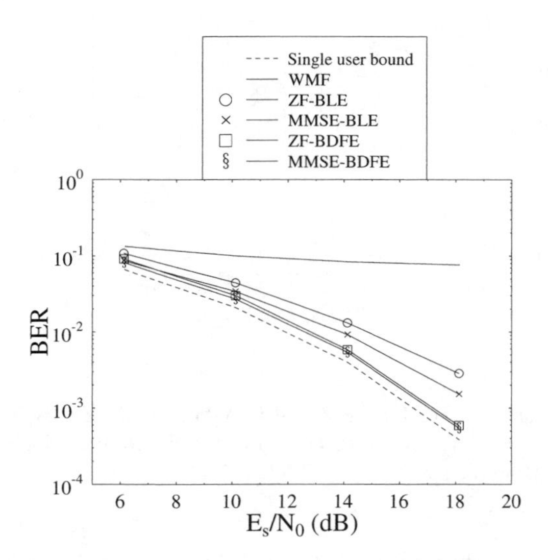

Figure 3.19: BER performance curves for the WMF, ZF-BLE, MMSE-BLE, ZF-BDFE and MMSE-BDFE with $K = 8$ users and for a **Typical Urban channel** with the impulse response shown in Figure 3.18, using the parameters of Table 3.2.

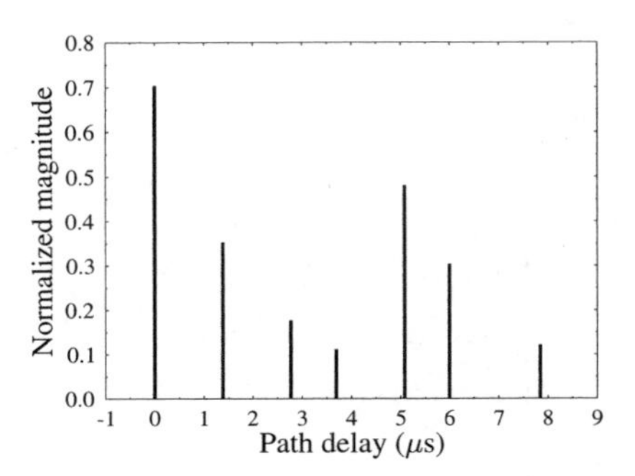

Figure 3.20: Normalized channel impulse response for the COST 207 [105] **seven-path Bad Urban channel**.

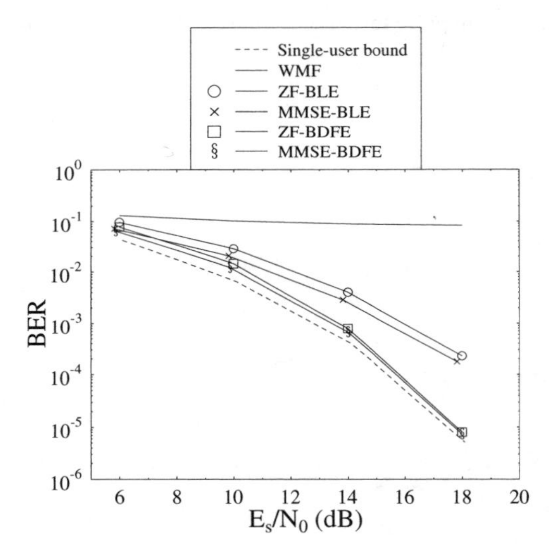

Figure 3.21: BER performance curves for the ZF-BLE, MMSE-BLE, ZF-BDFE and MMSE-BDFE with $K = 8$ users and for a **Bad Urban channel** with the impulse response shown in Figure 3.20, using the parameters of Table 3.2.

investigated the BER performance of the four JD schemes in three different channels, let us now consider the effect of multipath diversity on the performance of these receivers.

3.6.3 Effect of Multipath Diversity

The effect of multipath diversity on the performance of the joint detector is discussed in this section. Figure 3.22 shows the performance comparisons for the **MMSE-BDFE** supporting

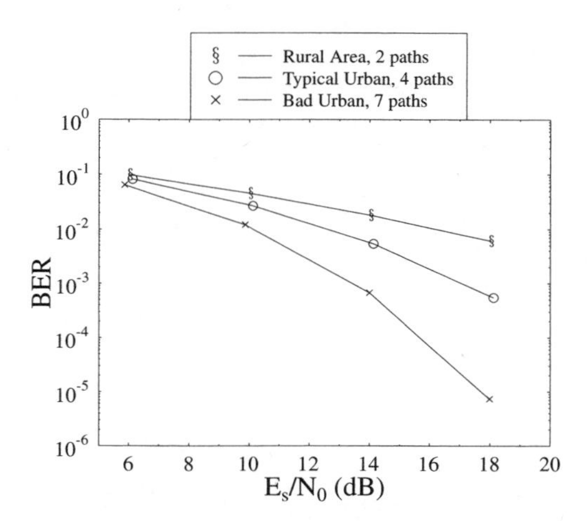

Figure 3.22: BER performance comparisons of the **MMSE-BDFE** supporting $K = 8$ users and for three different CIRs with different orders of multipath diversity. The CIRs are for the Rural Area (two path), Typical Urban (four path) and Bad Urban (seven path) channels. The remaining system parameters are summarized in Table 3.2.

$K = 8$ users and for three different channel impulse responses (CIR) with different orders of multipath diversity. The CIRs are those of the Rural Area (two path), Typical Urban (four path) and Bad Urban (seven path) channels, where the corresponding CIR profiles are depicted in Figures 3.15, 3.18 and 3.20, respectively. It can be observed from Figure 3.22 that with an increasing order of multipath diversity, the BER performance of the MMSE-BDFE also improves. At the BER of 10^{-2}, the performance gain due to the highly dispersive seven-path Bad Urban channel was approximately 2.5 dB over the Typical Urban channel, and approximately 6 dB over the Rural Area channel. Due to the higher order of multipath diversity, the joint detector was capable of compensating for the fading of the mobile channel and improve the BER performance. Let us now investigate the effect of increasing the number of users on the JD performance.

3.6.4 Effect of Increasing the Number of Users, K

Figure 3.23 shows the variation in BER versus the number of CDMA users, parametrized according to the E_s/N_0 values encountered. The transmissions were simulated over a Bad Urban channel using pseudo-random sequences of $Q = 16$ chips in length and the MMSE-BDFE was used as the joint detection receiver. From the results depicted, it can be observed that the BER performance degraded gracefully, as the number of users increased. This phenomenon is known as the soft degradation or soft capacity effect of CDMA [127]. The joint detection receiver is capable of exploiting the knowledge of the spreading sequences and channel impulse responses of all the users in order to mitigate the MAI and hence the BER performance degrades gradually instead of suffering a sharp drop-off in quality. It should also be noted that when the number of users, K, exceeds a certain number, typically half the

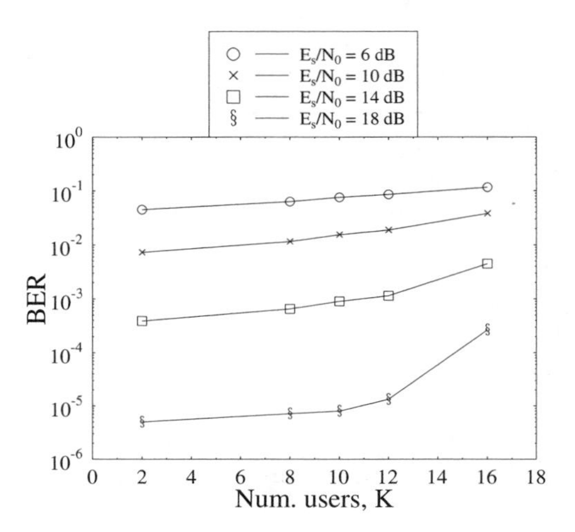

Figure 3.23: The effect of increasing the load, K, on the BER performance of the **MMSE-BDFE** over a Bad Urban channel using the impulse response shown in Figure 3.20. The remaining system parameters are summarized in Table 3.2.

value of Q, the BER suffers a steep increase. This is due to the fact that the higher number of users results in a higher noise enhancement, thus increasing the BER.

3.6.5 Joint Detection and Channel Coding

The effect of combining JD equalizers with channel coding was also investigated. Two types of channel coding were considered, Bose-Chauduri-Hocquenghem (BCH) codes [64] and turbo convolutional codes [79, 197]. BCH codes constitute a class of binary block codes, where each BCH code is characterized by the parameters (n, k, t), encoding k input bits into n output bits. The error-correcting capability of the codes is given by the parameter t, which implies that the code is capable of correcting up to t errors within a block of n number of bits. The rate of the code is defined as $R = k/n$.

Turbo convolutional coding is a powerful method of channel coding, which has been reported to produce excellent results [79, 197]. The turbo encoder takes in information bits and generates output bits by employing two convolutional codes as the component codes as well as a turbo interleaver. In their investigations, Barbulescu and Pietrobon [212] concluded that the optimum performance from the turbo coding scheme is achieved when a large turbo interleaver is employed; a good interleaver structure would be a square-shaped interleaver matrix with an odd number of rows and columns. Previous work invoking turbo coding and joint detection was carried out for example by Jung *et al.* [132, 133], where the ZF-BLE was combined with turbo coding in order to achieve performance improvement, using a 12×16 block interleaver in the turbo encoder.

Interleaving is a technique used to disperse and randomize burst errors, so that channel coding, which is capable of combating random errors, can function more effectively. Input bits or symbols to an interleaver are rearranged so that they are no longer in the consecutive order they were in before interleaving. Therefore, when burst errors occur in the channel, these errors are spread after deinterleaving and can be corrected using error correction codes designed for correcting randomly distributed errors. At the receiver, a deinterleaver arranges the received bits or symbols back into their original order.

The gain in using channel coding in conjunction with RAKE receivers [92] for seven-path Bad Urban channels was investigated. Two types of channel codes were used, namely BCH codes and turbo codes. The uncoded transmission packet size was fixed at 198 bits. When BCH codes were employed, a range of codes was used, namely codes with the (n, k, t) combinations of (255,131,18), (127,64,10), (15,7,2), (11,11,0), arranged to give a total of $255 + 127 + 15 + 11 = 408$ bits for all the codes after channel coding. The total number of input bits required was $131 + 64 + 7 + 11 = 213$ bits. Therefore, an extra number of 15 known bits were added to the original packet size of 198 uncoded bits in order to make up the required packet size. The overall coding rate was approximately 1/2. When turbo codes were utilized for channel coding, a 14×14 turbo block interleaver was employed and the component codes were half-rate convolutional codes with constraint lengths of 3. After appropriate puncturing, this resulted in a half-rate turbo-coded output. The decoder employed the well-known Soft Output Viterbi Algorithm (SOVA) [213] algorithm with 8 iterations for decoding. The Euclidean distances between the soft outputs of the RAKE receiver and the legitimate modulation constellation points were used to calculate the so-called log likelihood ratios (LLR) which were then used as soft inputs to the turbo decoder. A simple block channel interleaver was employed with the dimensions of 15×15.

Figure 3.24 shows the uncoded and coded BER performance of a RAKE receiver [92] utilizing BCH codes or turbo codes. From this figure, it can be observed that both coding schemes offered performance improvements over the uncoded arrangement and the turbo codes resulted in the best BER performance. However, even with the use of turbo codes, the error floor still persisted due to the inaccuracy of the initial symbol estimates input to the respective channel decoders.

The performance of joint detection in combination with channel coding was investigated by employing the MMSE-BDFE. The BCH codes and turbo codes employed previously in the RAKE receiver were used also in the MMSE-BDFE-based system. Figures 3.25 and 3.26 show the performance improvements due to channel coding over the uncoded BER of the MMSE-BDFE for $K = 2$ and $K = 8$ users, respectively. Here, there was no error floor due to the more reliable symbol estimates and soft outputs provided by the MMSE-BDFE. Furthermore, the performance gains attained were higher when the MMSE-BDFE was used than in the case of the RAKE receiver. It can also be observed that the employment of turbo codes provided a superior performance compared to the BCH codes.

Having analyzed the effects of BCH and turbo channel coding, let us now quantify the effects of channel estimation errors on the performance of the JD receiver.

3.6.6 Effect of Channel Estimation Errors

Joint detection algorithms require the knowledge of the CIRs of all the users, in order to equalize the signals and to generate the data estimates for each user. Here, the impact of channel estimation errors on the performance of the MMSE-BDFE is assessed. Following the approach adopted by Yoon, Kohno and Imai [141], the channel estimation errors for both

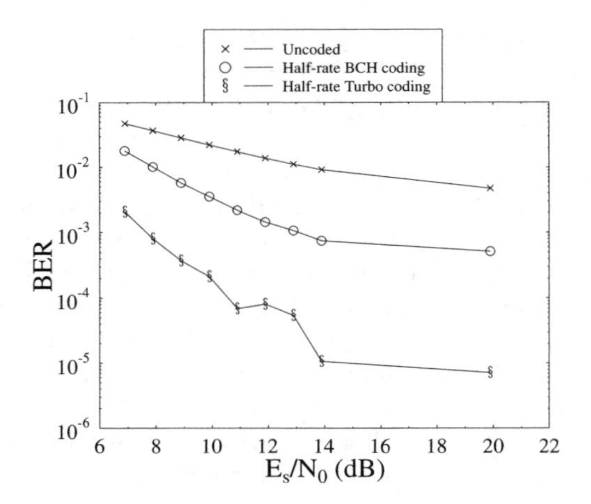

Figure 3.24: Average uncoded, half-rate BCH-coded and half-rate turbo-coded BER versus E_s/N_0 for an uplink synchronous CDMA system supporting $K = 2$ users, over the seven-path Bad Urban channel of Figure 3.20. Pseudo-random codes of length $Q = 16$ chips were utilized for each user and the **RAKE** receiver was employed for detecting the symbols and generating soft outputs for the turbo decoder. The turbo codec used a 14×14 square interleaver and the SOVA algorithm with 8 iterations for decoding. Channel interleaving was employed using a block interleaver with the dimensions of 15×15. The remaining system parameters are summarized in Table 3.2.

the phase and amplitude are modelled as independent and identically distributed zero mean Gaussian random variables. If the actual complex channel impulse at the l-th path of the k-th user is represented by $\alpha_{k,l}e^{j\theta_{k,l}}$, then the estimated amplitude, $\hat{\alpha}_{k,l}$, and phase $\hat{\theta}_{k,l}$ is given by:

$$\hat{\alpha}_{k,l} = \alpha_{k,l} + \delta\alpha_{k,l} \tag{3.210}$$

$$\hat{\theta}_{k,l} = \theta_{k,l} + \delta\theta_{k,l}, \tag{3.211}$$

where $\delta\alpha_{k,l}$ and $\delta\theta_{k,l}$ are the amplitude and phase estimation errors, respectively, for the l-th path of the k-th user. The respective variances of the estimation errors are defined as [141]:

$$\sigma^2_{\delta\alpha_{k,l}} \triangleq E[(\delta^2 e)] \tag{3.212}$$

$$\sigma^2_{\delta\theta_{k,l}} \triangleq \pi^2 E[(\delta^2 e)], \tag{3.213}$$

where $E[(\delta^2 e)]$ is the mean squared error of the channel estimates. The percentage of errors in the channel estimates can thus be defined as the standard deviation $\times$ 100%, which is equivalent to $\sigma_{\delta\alpha_{k,l}} \times 100\%$. As an example, for a channel estimate error standard deviation of -13 dB, this corresponds to errors within $\pm 10^{-1.3} \times 100\% = \pm 5\%$ of the actual channel value.

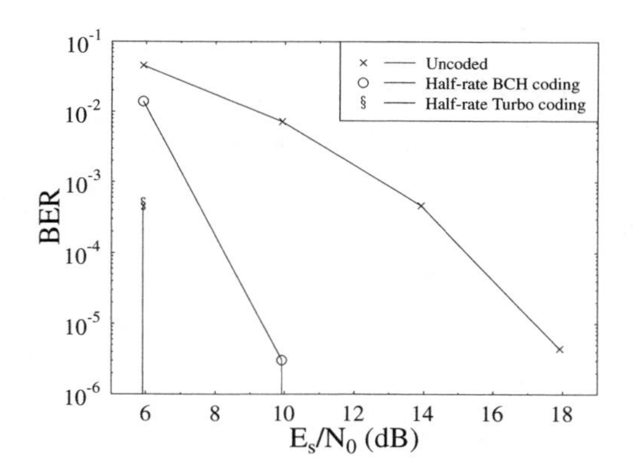

Figure 3.25: Average uncoded, half-rate BCH-coded and half-rate turbo-coded BER versus E_s/N_0 for an uplink synchronous JD-CDMA system supporting ***K*=2** users, over the seven-path Bad Urban channel of Figure 3.20. Pseudo-random codes of length $Q = 16$ chips were utilized for each user and the JD receiver used was the **MMSE-BDFE**, which generated soft outputs to the turbo decoder. The turbo codec used a 14×14 square interleaver and the SOVA algorithm with 8 iterations for decoding. Channel interleaving was employed using a block interleaver with the dimensions of 15×15. The remaining system parameters are summarized in Table 3.2. The vertical BER curve of the turbo-coded system indicates that for $E_s/N_0 > 6$ dB, there the errors were all corrected by the channel decoder.

The BER performance of the MMSE-BDFE parametrized by different values of channel estimation errors is compared in Figure 3.27. The mobile propagation channel simulated was the seven-path Bad Urban COST 207 channel of Figure 3.20 with a Doppler frequency of 80 Hz. The channel estimation error indicated was inflicted on each individual path in the CIR profile. For example, for a channel estimation error of 5%, the amplitude and phase of each path in the CIR were corrupted by independent zero mean Gaussian random variables having a standard deviation of 0.05, corresponding to a mean squared error of 0.0025. From Figure 3.27, the BER can be seen to degrade progressively, as the error increases from 0% to 14%. The E_s/N_0 performance loss in dB due to channel estimation errors is tabulated for two different BER values as follows:

BER	**Error percentage**			
	3%	5%	10%	14%
1%	0 dB	0 dB	1 dB	4 dB
0.1%	0 dB	1 dB	3.5 dB	>5 dB

The E_s/N_0 performance loss increased as the BER reduced and the table shows that the performance loss was more than 3 dB for a BER of 0.1% at a channel estimation error of $\pm 10\%$ per path. Therefore, in order to maintain a loss of no more than 3 dB in performance, the channel estimation errors must not exceed 10%. These results show that the performance of the MMSE-BDFE is dependent on the accuracy of the channel estimation algorithm. Significant research efforts have been invested in the area of CDMA channel estimation, including

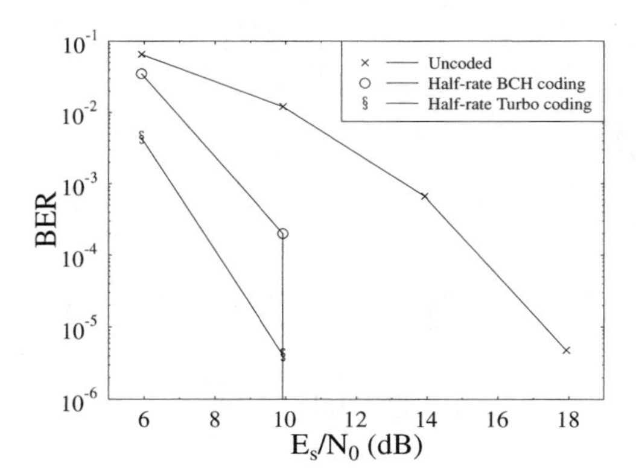

Figure 3.26: Average uncoded, half-rate BCH-coded and half-rate turbo-coded BER versus E_s/N_0 for an uplink synchronous JD-CDMA system supporting ***K*=8** users, over the seven-path Bad Urban channel of Figure 3.20. Pseudo-random codes of length $Q = 16$ chips were utilized for each user and the JD receiver used was the **MMSE-BDFE**, which generated soft outputs to the turbo decoder. The turbo codec used a 14×14 square interleaver and the SOVA algorithm with 8 iterations for decoding. Channel interleaving was employed using a block interleaver with the dimensions of 15×15. The remaining system parameters are summarized in Table 3.2.

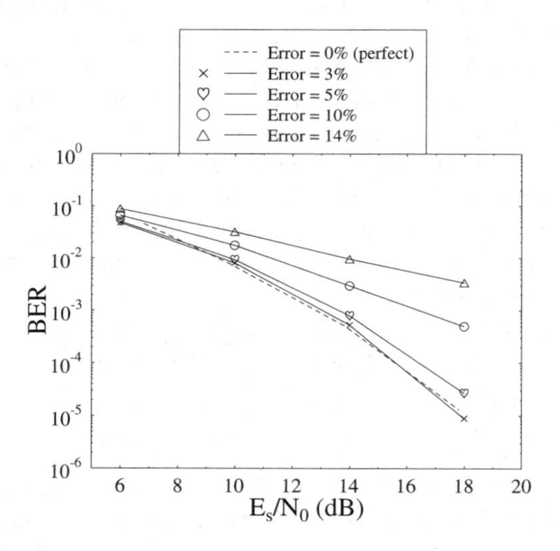

Figure 3.27: Impact of channel estimation errors on the BER performance of the MMSE-BDFE.

work carried out for example by Steiner and Jung [136], as well as Papproth and Kaleh [214], who designed optimum and sub-optimum channel estimation algorithms for CDMA systems applying joint detection algorithms. The optimum channel estimator was found to be the maximum likelihood estimation algorithm, while the sub-optimum algorithm was realized by using a correlator or Fast Fourier Transforms (FFT) in conjunction with specially designed midamble codes. Soong and Krzymien [148] inserted reference symbols at periodic intervals into the transmitted bursts in order to estimate the channel. In the third-generation UTRA standard [215], continuous user pilots are proposed in order to provide high-quality channel estimates for supporting JD receivers. In order not to interrupt the organizational flow of the thesis, where the next chapter is focused on adaptive-rate schemes in conjunction with JD-CDMA receivers, the discussion on CIR estimation is deferred to Chapter 7, which is concentrated on blind PSP receivers for joint data and CIR estimation. For our subsequent investigations in Chapters 4 and 5, perfect CIR estimation was assumed.

Having considered the performance of the JD receivers in various scenarios, let us now summarize the findings of this chapter.

3.7 Chapter Summary and Conclusion

In this chapter, the theoretical background and the derivation of four joint detection (JD) algorithms were presented. The JD receivers were shown to be modified variants of the single user equalizers developed for wideband single-user transmission. The implementational complexities of the four JD algorithms were evaluated and compared in Section 3.4, arriving at the conclusion that all four schemes resulted in a similar complexity. Reduced complexity versions of the JD algorithms were also proposed, which exploited the sparsity of the system matrix. The performance of the JD receivers was compared for different channels and different number of users in Sections 3.6.3 and 3.6.4, respectively. Channel coding in combination with the JD receivers was also investigated and it was ascertained that higher performance gains could be achieved using the JD receivers compared to the conventional RAKE receiver approach due to the more reliable estimates fed into the channel decoders from the output of the JD receiver. Finally, the impact of channel estimation errors on the BER performances was assessed in Section 3.6.6. It was shown that the performance of the receivers was sensitive to channel estimation errors and an estimation error of $\pm 10\%$ per channel path caused a performance loss of 3 dB at a BER of 0.1%.

Having gained an insight into the theoretical basis of JD receivers and having evaluated their performance in fixed-mode schemes, let us now turn our attention to the combination of JD receivers in conjunction with adaptive-rate systems.

Chapter 4

Adaptive-Rate Joint Detection-Aided CDMA[1]

4.1 Introduction

Mobile radio signals are subject to propagation path loss as well as slow fading and fast fading. Due to the nature of the fading channel, transmission errors occur in bursts when the channel exhibits deep fades or when there is a sudden surge of multiple access interference (MAI) or inter-symbol interference (ISI). In mobile communications systems power control techniques [64] are used to mitigate the effects of path loss and slow fading. However, in order to counteract the problem of fast fading and co-channel interference, agile power control algorithms are required [10]. Another technique that can be used to overcome the problems due to the time-variant fluctuations of the channel is adaptive-rate transmission [216, 217], where the information rate is varied according to the quality of the channel, rather than according to user requirements. When the channel quality is low, a lower information rate is chosen in order to reduce the number of errors. Conversely, when the channel quality is high, a higher information rate is used to increase the throughput of the system. More explicitly, this method is similar to multi-rate transmission [97], except that in this case, the transmission rate is modified according to the channel quality instead of the service required by the mobile user. Various methods of multi-rate transmission have been proposed in the research literature. Next we will briefly discuss some of the current research on multi-rate transmission schemes, before focusing our attention on adaptive-rate systems.

Ottosson and Svensson [97] compared various multi-rate systems, including multi spreading factor (SF) based, multi-code and multimode-modulation schemes. According to the multi-code philosophy, the SF is kept constant for all users, but multiple spreading codes are

[1] *Single- and Multi-Carrier CDMA Multi-User Detection, Space-Time Spreading, Synchronization, Networking and Standards.* L.Hanzo, L-L.Yang, E-L.Kuan and K.Yen,
 ISBN 0-470-86309-9

assigned to users requiring higher bit rates. Multiple data rates can also be supported by a multiple SF scheme, where the chip rate is kept constant, but the data rates are varied by changing the SF of the spreading codes assigned to the users; the lower the SF, the higher the supported data rate. Performance comparisons for both of these schemes have been carried out by Ottosson and Svensson [97]; as well as by Ramakrishna and Holtzman [98], demonstrating that both schemes achieved a similar performance. Adachi, Ohno, Higashi, Dohi and Okumura proposed the employment of multi-code CDMA in conjunction with pilot symbol-assisted channel estimation, RAKE reception and antenna diversity for providing multi-rate capabilities [218, 219]. Multi-level modulation schemes were also investigated by Ottosson and Svensson [97], where higher rate users were assigned higher order modulation modes transmitting several bits per symbol. However, it was concluded that the performance experienced by users requiring higher rates was significantly worse than that experienced by the lower rate users. The use of M-ary orthogonal modulation in providing variable rate transmission was investigated by Schotten, Elders-Boll and Busboom [220]. According to this method, each user was assigned an orthogonal sequence set, where the number of sequences, M, in the set was dependent on the data rate required – the higher the rate required, the larger the sequence set. Each sequence in the set was mapped to a particular combination of $b = (\log_2 M)$ bits to be transmitted. The M-ary sequence was then spread with the aid of a spreading code of a constant SF before transmission. It was found [220] that the performance of the system depended not only on the multiple access interference (MAI), but also on the Euclidean distance between the sequences in the M-ary sequence set.

Saquib and Yates [221] investigated the employment of the decorrelating detector in conjunction with the multiple SF scheme and proposed a modified decorrelating detector, which utilized soft decisions and maximal ratio combining in order to detect the bits of the different rate users. The performance comparison of various multi-user detectors for a multiple SF transmission scheme was presented, for example, by Juntti [222] where the detectors compared were the decorrelator, the parallel interference cancellation (PIC) receiver and the group serial interference cancellation (GSIC) receiver. It was concluded that the GSIC and decorrelator performed better than the PIC receiver but all the interference cancellation schemes including the GSIC, exhibited an error floor at high SNRs due to error propagation.

The transmission rate of each user can also be adapted according to the channel quality, in order to mitigate the effects of channel quality fluctuations. Kim [217] analyzed the performance of two different methods of combating the variations in the mobile channel, which were the adaptation of the transmitter power to compensate for the channel quality variations or the switching of the information rate, in order to suit the channel conditions. Using a RAKE receiver [92], it was demonstrated that rate adaptation provided a higher average information rate, than power adaptation for a given average transmit power and a given BER [217]. Abeta, Sampei and Morinaga [223] have conducted investigations into an adaptive packet transmission-based CDMA scheme, where the transmission rate is modified by varying the channel code rate and the processing gain of the CDMA user, employing the carrier to interference plus noise ratio (CINR) as the switching parameter. When the channel quality was favourable, the transmission bit rate was increased and conversely, the bit rate was reduced when the channel quality dropped. The overall packet transmission rate was kept constant by transmitting in shorter bursts, when the transmission bit rate was high, and lengthening the burst when the bit rate was low. This resulted in a decrease in interference power, which translated to an increase in system capacity. Hashimoto, Sampei and Morinaga [224] extended this work also to demonstrate that the proposed system was capable of achieving a higher capacity with a smaller hand-off margin and lower average transmitter power. In these schemes, the conventional RAKE receiver [92] was used for the detection of the data symbols. A variable-rate CDMA scheme – where the transmission rate was modified by varying the error code rate and correspondingly, the M-ary modulation constellations – was investigated by Lau and Maric [225]. As the channel code rate was increased, the dimensionality of the M-ary modulation scheme was increased correspondingly. Another adaptive system was proposed by Tateesh, Atungsiri and Kondoz [226], where the rates of the speech and channel codecs were varied adaptively [226]. In their adaptive system, the gross transmitted bit rate was kept constant, but the speech codec and channel codec rates were varied according to the channel quality. When the channel quality was low, a lower rate speech codec was used, resulting in increased redundancy and thus a more powerful channel code could be employed. This resulted in an overall coding gain, although the speech quality dropped with decreasing speech rate. A variable rate data transmission scheme was proposed by Okumura and Adachi [227], where the fluctuating transmission rate was mapped to discontinuous transmission, in order to reduce the interference to the other users when there was no transmission. The transmission rate was detected blindly at the receiver with the help of cyclic redundancy check decoding and RAKE receivers were employed for coherent reception, where pilot-symbol-assisted channel estimation channel estimation was performed. Adaptive CDMA systems are also useful for employment in arbitrary propagation environments or in hand-over scenarios, such as those encountered, when a mobile user roams from an indoor to an outdoor environment or in a so-called 'birth-death' scenario, where the number of transmitting users changes frequently [168].

In our work, we also propose to vary the information rate in accordance with the channel quality. However, in comparison to conventional power control techniques - which may disadvantage other users in an effort to maintain the quality of the links considered - the proposed technique does not disadvantage other users and increases the network capacity [228]. The instantaneous channel quality can be estimated at the receiver and the chosen information rate can then be communicated to the transmitter via explicit signalling in a closed-loop scheme. Conversely, in an open-loop scheme, by assuming reciprocity in the uplink and downlink channels of Wideband Time Division Duplex (TDD) CDMA systems, the information rate for the downlink transmission is chosen according to the channel quality estimate related to the uplink and vice versa. Channel reciprocity issues in TDD-CDMA systems have been investigated by Miya *et al.* [229], Kato *et al.* [230] and Jeong *et al.* [231]. In our investigations the COST 207 channel models developed for the Global System of Mobile Communications known as GSM were employed. Although due to the orthogonal spreading sequences and due to the resultant high chip-rate the number of resolvable multipath components is higher in our CDMA system than in GSM, nonetheless we used the COST 207 models, since these are widely used in the community.

In this chapter, two different methods of varying the information rate are considered, namely the Adaptive Quadrature Amplitude Modulated (AQAM)/JD-CDMA system and the Variable Spreading Factor (VSF)/JD-CDMA scheme. AQAM is an adaptive-rate technique, whereby the data modulation mode is chosen according to some criterion related to the channel quality. On the other hand, in VSF transmission, the information rate is varied by adapting the spreading factor of the CDMA codes used, while keeping the chip rate constant. Further elaborations on these two methods will be given in subsequent sections. The remainder of

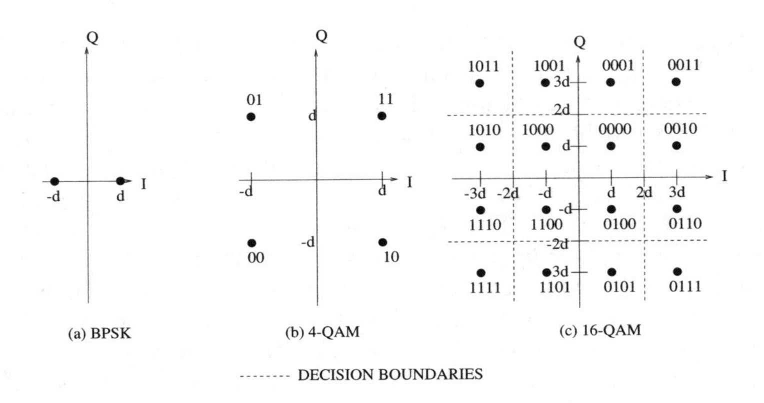

Figure 4.1: QAM constellation points for BPSK, 4-QAM and 16-QAM.

this chapter is structured as follows. Section 4.2 describes the AQAM scheme while Section 4.3 discusses the employment of AQAM in conjunction with JD-CDMA. The performance of the proposed AQAM/JD-CDMA scheme is evaluated and discussed in Section 4.3.5. The VSF/JD-CDMA scheme is studied and evaluated in Section 4.4. A performance comparison between the two different adaptive-rate methods is presented in Section 4.5 and a combined AQAM-VSF scheme is investigated in Section 4.6. Finally, our conclusions are summarized.

4.2 Adaptive Quadrature Amplitude Modulation

In Quadrature Amplitude Modulation (QAM) [92], n bits are grouped to form a signalling symbol and $m = 2^n$ different symbols fully convey all combinations of the n bits. These m symbols are arranged in a phasor constellation seen in Figure 4.1, in order to form an m-QAM scheme. The modulation modes of BPSK (2-QAM), 4-QAM and 16-QAM convey 1, 2 and 4 bits per symbol, respectively and Figure 4.1 shows the arrangement of the phasor constellation points for the three modes. The BER performance of the three QAM modes for a single user in a non-dispersive Gaussian channel is presented in Figure 4.2. For a given channel Signal to Noise Ratio (SNR), the BER performance degrades, as the number of constellation points increases, since the average distance between adjacent constellation points is reduced, which increases the probability of demodulation errors.

Burst-by-burst Adaptive Quadrature Amplitude Modulation (AQAM) [92] is a technique that attempts to increase the average throughput of the system by switching between modulation modes depending on the instantaneous state or quality of the channel. When the channel quality is favourable, a modulation mode having a high number of constellation points is used to transmit as many bits per symbol as possible, in order to increase the throughput. Conversely, when the channel is hostile, the modulation mode is switched to one using a low number of constellation points, in order to reduce the error probability and to maintain a certain adjustable target BER. Figure 4.3 shows the stylized amplitude variation in a fading channel and the switching of the modulation modes in a four-mode AQAM system, where both the BER and the throughput increase when switching from Mode 1 to 4.

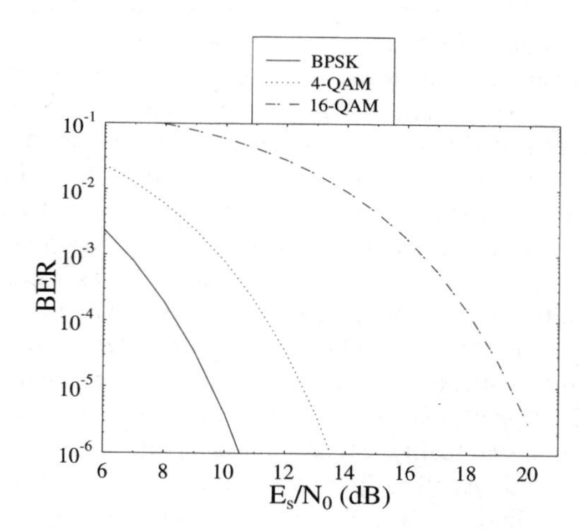

Figure 4.2: BER performance curves for BPSK, 4-QAM and 16-QAM in a non-dispersive Gaussian channel for a single-user system

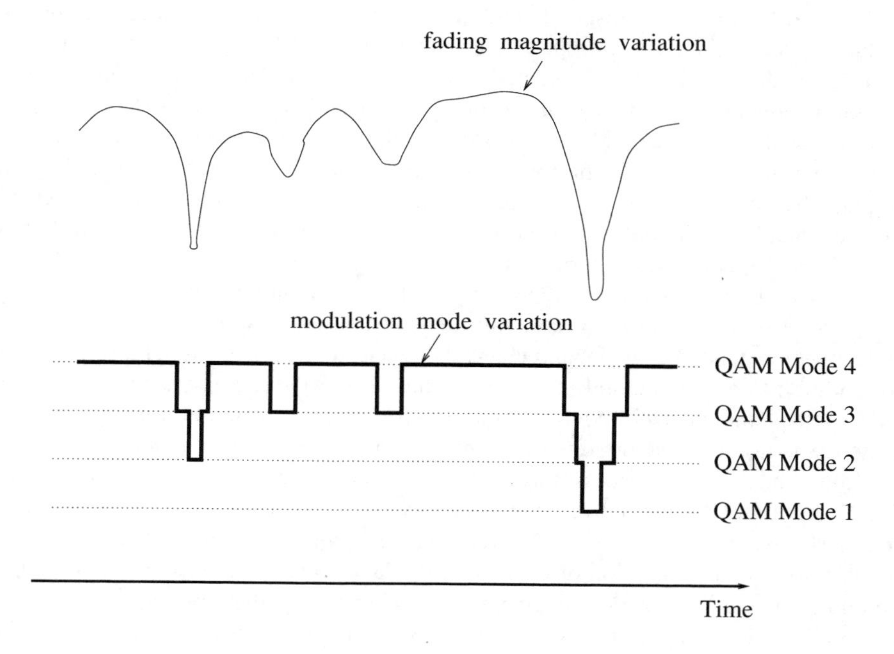

Figure 4.3: Basic concept of a four-mode AQAM transmission in a narrowband channel. The variation of the modulation mode follows the fading variation of the channel over time.

In order to determine the best choice of modulation mode in terms of the required trade-off between BER and throughput, the near-instantaneous quality of the channel has to be estimated. This parameter can be estimated at the receiver and the chosen modulation mode is then communicated using explicit signalling to the transmitter in a closed-loop scheme. In other words, the receiver instructs the transmitter as to the choice of modulation mode, in order to satisfy the prevalent BER and bits per symbol (BPS) throughput trade-off. This closed-loop scheme is depicted schematically in Figure 4.4(a). By contrast, in the open-loop scheme of Figure 4.4(b), the channel quality estimation can be carried out at the transmitter itself based on the receiver's perception of the channel quality during the last received packet. Then the transmitter explicitly informs the remote receiver as to the modem mode used in the packet and modulation mode detection would then be performed at the receiver. This scheme performs most successfully in situations where the fading in the channel varies slowly in comparison to the packet transmission rate. Channel estimation is less accurate in a fast-fading channel and this lag in the quality estimation renders the choice of modulation mode less appropriate for the channel. Another approach to this issue would be for the receiver to detect the modulation mode blindly [232, 233], as shown in Figure 4.4(c).

As stated earlier, a metric corresponding to the near-instantaneous channel quality is required in order to adapt the AQAM modes. Some examples of these metrics include the carrier-to-interference (C/I) ratio of the channel [234], the SNR of the channel [92, 216], the received signal strength indicator (RSSI) [235], the mean square error (MSE) at the output of the receiver and the BER of the system [236]. The most accurate metric is the BER of the system, since this metric corresponds directly to the system's performance, irrespective of the actual source of the channel impairment. However, the BER is dependent on the AQAM mode employed, and cannot be estimated directly for most receivers. For a system that incorporates channel coding, such as turbo coding, the log-likelihood ratios (LLR) at the input and output of the turbo decoder can also be employed as the adaptation metric. AQAM systems were first proposed for narrowband channels and the research in this field includes work published by Webb and Steele [92, 216], Sampei, Komaki and Morinaga [234] Goldsmith and Chua [237]; as well as Torrance [235]. Webb *et al.* [92, 216] employed Star QAM and the channel quality was determined by measuring the received signal strength and the instantaneous BER. Sampei *et al.* [234] switched the modulation modes by estimating the signal to co-channel interference ratio and the expected delay spread of the channel. This work has been extended to wideband channels, where the received signal also suffers from ISI in addition to amplitude and phase distortions due to the fading channel. In wideband AQAM systems the channel SNR, C/I ratio or RSSI metrics cannot be readily estimated or predicted, among other factors due to the multipath nature of the channel or as a result of the so-called 'birth-death' processes associated with the sudden appearance or disappearance of users. Additionally, the above simple metrics do not provide accurate measures of the system performance at the output of the receiver employed, since the effects of the CIR are not considered in the estimation of these metrics. Wong *et al.* [238] proposed a joint adaptive modulation and equalization scheme, where a channel equalizer was used to mitigate the effects of ISI inflicted upon the signal. The CIR estimate was used to calculate the Signal to residual ISI plus Noise Ratio (SINR) at the output of the channel equalizer and this SINR value was used to switch the modulation modes. This was a more appropriate switching parameter than the received signal level, since it was a reliable indicator of the performance that could be achieved after equalization. Let us now consider the feasibility of combining

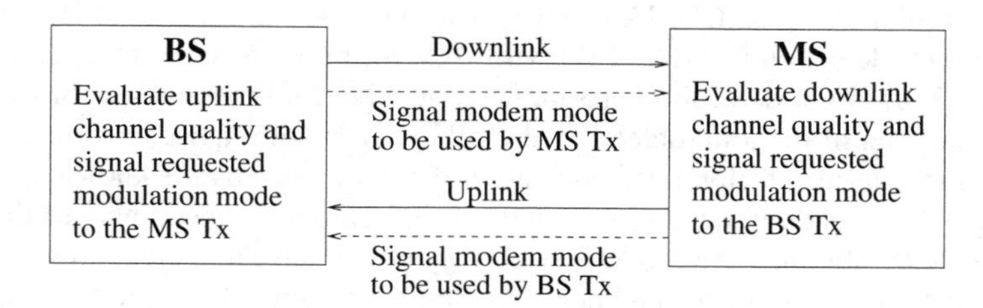

(a) Closed-loop modulation mode signalling from receiver to transmitter

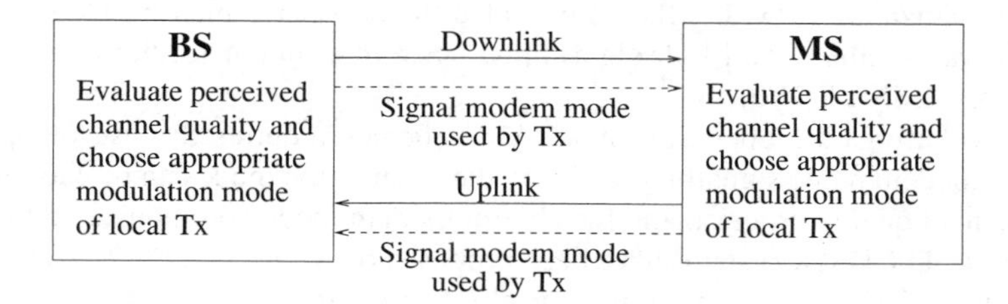

(b) Open-loop modulation mode signalling from transmitter to receiver

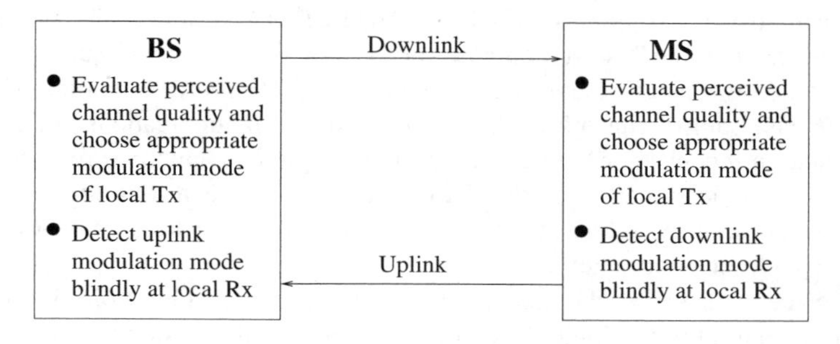

(c) Blind modulation mode detection at the receiver

Figure 4.4: Three different methods of modulation mode signalling for the adaptive schemes, where BS represents the Base Station, MS denotes the Mobile Station, the transmitter is represented by Tx and the receiver is denoted by Rx.

AQAM with JD-CDMA.

4.3 Joint Detection-Assisted AQAM-Based CDMA

The theory of joint detection CDMA receivers was discussed in Chapter 3. Here we propose to combine joint detection CDMA [127] with AQAM, by modifying the approach used by Wong *et al.* [238]. Joint detection is particularly suitable for combining with AQAM, since the implementation of the joint detection algorithms does not require any knowledge of the modulation mode used. The joint detection algorithm requires only the knowledge of the CIR estimates and the spreading sequences of all the users. Therefore, the joint detection receivers are suitable for combining with AQAM, since they do not have to be reconfigured each time the modulation mode is switched and hence the implementational complexity is independent of the modulation mode used. In Chapter 3, four basic joint detection algorithms were considered [127], and it was concluded that the Minimum Mean Square Error Block Decision Feedback Equalization (MMSE-BDFE) provided the best performance, while the detection complexity was similar for all four algorithms. Therefore, in our further investigations, we have used the MMSE-BDFE.

As stated before, the operation of AQAM schemes requires the use of open-loop or closed-loop modem mode signalling, whereby the modulation mode can be chosen according to some channel quality criterion and the chosen modem mode communicated to the remote transceiver. TDD-CDMA systems have been proposed by Miya *et al.* [229], Kato *et al.* [230] and Jeong *et al.* [231]. They showed that the close correlation between the downlink and uplink channel quality could be used to implement transmission and reception space diversity at the base station, as well as open-loop power control. This leads us to conclude that this correlation can also be exploited by duplex AQAM/CDMA systems. The uplink transmission can be utilized by the base station receiver to estimate the channel quality for downlink transmissions due to this close correlation, leading to a "best-effort" choice of modulation mode for the base station transmitter. Channel quality estimation can also be performed by the mobile station receiver for the next uplink transmission. Again, we refer to this regime as an open-loop mode signalling regime. In contrast to the open-loop regime, the receiver perception of the required mode for the remote transmitter can be signalled to the remote transmitter in a closed-loop scheme.

In joint detection systems, the Signal to residual Interference plus Noise Ratio (SINR) for each user at the output of the joint detector can be calculated by using the channel estimates and the spreading sequences of all the users [127]. The derivation of the corresponding SINR expression is provided later in Section 4.3.2. Initially, preliminary investigations were carried out, in order to gauge the potential of the system, where a two-user JD-CDMA system was simulated and the SINR values at the output of the MMSE-BDFE for the two different users over the same period of time were calculated. The simulation parameters are tabulated in Table 4.1. Figure 4.5 shows the variation of output SINR and the number of errors per transmission burst for the different users over the time frame considered. The figure shows that not only does the output SINR vary significantly over time, but the variation in output SINR is different for the different users. This suggests potential advantages in combining AQAM with JD-CDMA and the AQAM/JD-CDMA scheme will be explored in later sections. It can also be seen in the figure that the variation in the number of errors in each burst follows

Parameter	Value
Channel type	COST 207 Bad Urban
Number of paths in channel	7
Doppler frequency	80 Hz
Spreading ratio	16
Spreading sequence	Pseudo-random
Chip rate	2.167 MBaud
Frame burst structure	FRAMES Mode 1 Spread burst 1 [205]
Burst duration	577 μs
Receiver type	MMSE-BDFE

Table 4.1: The simulation parameters employed in a 4-QAM JD-CDMA transmission scheme, where the SINR at the output of the MMSE-BDFE was estimated and compared to the number of errors corresponding to the SINR estimated.

Switching criterion	Modulation mode
$\gamma_o(k) < t_1$	V_1
$t_1 \leqslant \gamma_o(k) < t_2$	V_2
.	.
.	.
.	.
$t_M \leqslant \gamma_o(k)$	V_M

Table 4.2: The general rules employed for switching the modulation modes in an AQAM/JD-CDMA system. The choice of modulation modes are denoted by V_m, where the total number of modulation modes is M and $m = 1, 2, \ldots, M$. The modulation modes with the lowest and highest number of constellation points are V_1 and V_M, respectively. The SINR at the output of the MMSE-BDFE is represented by $\gamma_o(k)$ and the values $(t_1, \cdots, t_M)$ represent the switching thresholds, where $t_1 < t_2 < \cdots < t_M$.

that of the output SINR. Consequently, the output SINR of each user was chosen as the switching criterion for our AQAM/JD-CDMA system.

4.3.1 System Model

In the AQAM/JD-CDMA system, the output SINR value – $\gamma_o(k)$, for $k = 1, \ldots, K$ – of all the K users is estimated by employing the estimated CIRs and spreading sequences of all the users. After $\gamma_o(k)$ is calculated, the modulation mode is chosen accordingly and communicated to the transmitter. Let us designate the choice of modulation modes by V_m, where the total number of modulation modes is M and $m = 1, 2, \ldots, M$. The modulation mode having the lowest number of constellation points is V_1 and the one with the highest is V_M. The rules used to switch the modulation modes are tabulated in Table 4.2, where $\gamma_o(k)$ is the SINR of the k-th user at the output of the MMSE-BDFE and the values $(t_1, \cdots, t_M)$ represent the switching thresholds, where $t_1 < t_2 < \cdots < t_M$.

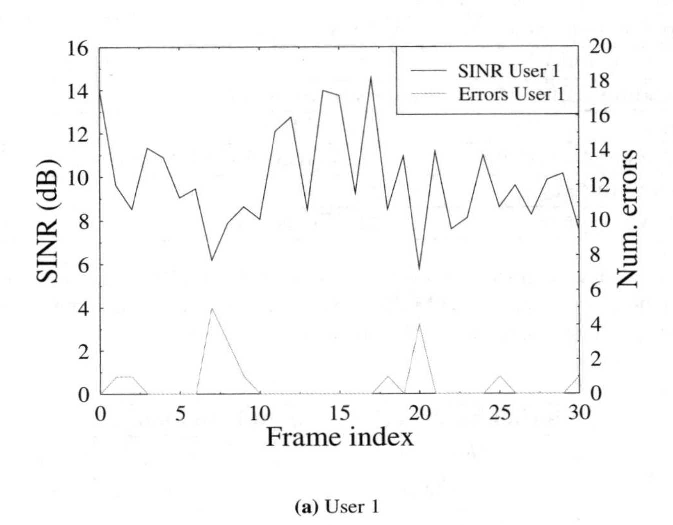

(a) User 1

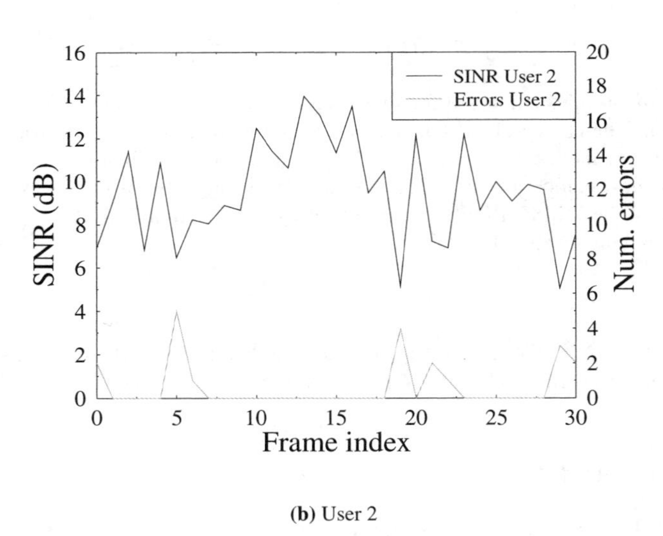

(b) User 2

Figure 4.5: Comparison of the SINR at the output of the MMSE-BDFE for different users over the same period of time. The number of errors corresponding to the SINR is also shown. There were 2 users in the system, transmitting at E_s/N_0=10dB and the 4-QAM modulation mode was used. The remaining system parameters were summarized in Table 4.1, while the CIR was plotted in Figure 3.20.

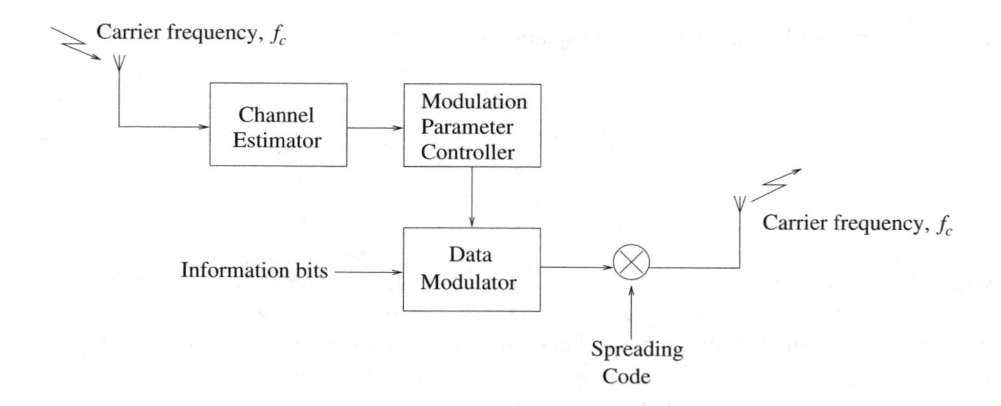

Figure 4.6: The schematic of the transmitter of an AQAM/CDMA system. In this transmitter, a TDD transmission scheme is assumed. Channel estimates are obtained by assuming close correlation between the uplink and downlink channels, which are used to measure the quality of the channel. This quality measure is passed to the modulation parameter controller which selects the modulation mode according to the thresholds set. The data bits are mapped to QAM symbols according to the chosen modulation mode, spread with the spreading code and modulated on to the carrier.

The schematic of the transmitter is shown in Figure 4.6. The data bits are mapped to their respective symbols according to the modulation mode chosen. The QAM symbols are then spread with the spreading code assigned to the user, modulated on to the carrier and transmitted. At the receiver, the despread and joint-detected QAM symbols have to be demodulated according to the modulation mode used for transmission. If the choice of modulation mode was made at the transmitter, the modulation mode would either have to be conveyed to the receiver by using explicit signalling symbols or the receiver would have to perform blind detection of the modulation mode. In our investigations, perfect detection of the modulation modes was assumed.

In the next section, the derivation of the switching criterion, $\gamma_o(k)$, is presented. A general expression for all four joint detection algorithms is presented first, followed by the specific derivation in the context of the MMSE-BDFE.

4.3.2 Output SINR for the Joint Detection Receiver

From Equation 3.63 in Section 3.2 and using the same notation as in Chapter 3, the equation is modified for a linear receiver, by eliminating the feedback matrix $\mathbf{S}$ of Figure 3.11, yielding:

$$\hat{\mathbf{d}} = \mathbf{M}\mathbf{y}, \tag{4.1}$$

where $\hat{\mathbf{d}}$ is the data estimate vector at the output of the linear receiver, $\mathbf{M}$ is the matrix representing the linear receiver and $\mathbf{y}$ is the vector for the composite received sequence defined in Equation 3.59. Again, the feedback matrix, $\mathbf{S}$, in Equation 3.63 is set to the identity matrix,

$\mathbf{I}_{KN}$. Substituting Equation 3.59 into Equation 4.1 gives:

$$\hat{\mathbf{d}} = \mathbf{MAd} + \mathbf{Mn} \quad (4.2)$$
$$= \mathbf{Fd} + \mathbf{Mn} \quad (4.3)$$
$$= \underbrace{\text{diag}(\mathbf{F})\mathbf{d}}_{\text{desired symbols}} + \underbrace{\overline{\text{diag}}(\mathbf{F})\mathbf{d}}_{\text{interference}} + \underbrace{\mathbf{Mn}}_{\text{noise}}, \quad (4.4)$$

where $\mathbf{F} = \mathbf{MA}$, and $\mathbf{d}$ represents the vector of the actual data symbols transmitted. As before, the notation diag$(\mathbf{X})$ represents a diagonal matrix, where the elements on the main diagonal are obtained from the main diagonal of the matrix $\mathbf{X}$, and the complement of this is denoted by $\overline{\text{diag}}(\mathbf{X}) = \mathbf{X} - \text{diag}(\mathbf{X})$. The term diag$(\mathbf{F})\mathbf{d}$ constitutes the useful part of the output and allows the recovery of the transmitted data, while $\overline{\text{diag}}(\mathbf{F})\mathbf{d}$ indicates the off-diagonal values constituting interference.

Using Equation 4.4, the SINR for the n-th symbol of the k-th user can be defined as:

$$\gamma_o(j) = \frac{\overbrace{E\left\{\left|[\text{diag}(\mathbf{F})\mathbf{d}]_j\right|^2\right\}}^{\text{desired signal}}}{\underbrace{E\left\{\left|[\overline{\text{diag}}(\mathbf{F})\mathbf{d}]_j\right|^2\right\}}_{\text{interference}} + \underbrace{E\left\{\left|[\mathbf{Mn}]_j\right|^2\right\}}_{\text{noise}}}, \quad (4.5)$$

where $j = n + N(k-1)$; the notation $[\mathbf{x}]_j$ denotes the j-th element in the column vector $\mathbf{x}$; and $E\{\cdot\}$ is the expectation operator.

In the numerator,

$$E\left\{\left|[\text{diag}(\mathbf{F})\mathbf{d}]_j\right|^2\right\} = \left|[\mathbf{F}]_{j,j}\right|^2 E\{|d_j|^2\}, \quad (4.6)$$

where $d_j = [\mathbf{d}]_j$ and $[\mathbf{F}]_{j,j}$ represents the element in the j-th row and j-th column of the matrix $\mathbf{F}$.

Using the definition of $[\overline{\text{diag}}(\mathbf{F}) = \mathbf{F} - \text{diag}(\mathbf{F})]$ in the interference term of the denominator gives:

$$E\left\{\left|[\overline{\text{diag}}(\mathbf{F})\mathbf{d}]_j\right|^2\right\} = E\left\{\left|[(\mathbf{F} - \text{diag}(\mathbf{F}))\mathbf{d}]_j\right|^2\right\} \quad (4.7)$$
$$= E\left\{\left|[\mathbf{Fd}]_j - [\text{diag}(\mathbf{F})\mathbf{d}]_j\right|^2\right\} \quad (4.8)$$
$$= E\left\{[\mathbf{Fd}(\mathbf{Fd})^H]_{j,j}\right\} - 2 \cdot E\left\{[\mathbf{Fd}(\text{diag}(\mathbf{F})\mathbf{d})^H]_{j,j}\right\} + E\left\{\left|[\text{diag}(\mathbf{F})\mathbf{d}]_j\right|^2\right\} \quad (4.9)$$
$$= [\mathbf{FR}_d\mathbf{F}^H]_{j,j} - 2 \cdot \text{Re}\left\{[\mathbf{FR}_d]_{j,j}[\mathbf{F}]^*_{j,j}\right\} + \left|[\mathbf{F}]_{j,j}\right|^2 E\{|d_j|^2\}, \quad (4.10)$$

where the notation $\text{Re}\{x\}$ represents the real part of the complex value x.

The covariance matrix of the noise term is given by:

$$E\left\{\left|[\mathbf{Mn}]_j\right|^2\right\} = [\mathbf{Mn}(\mathbf{Mn})^H]_{j,j} \tag{4.11}$$

$$= [\mathbf{MR}_n\mathbf{M}^H]_{j,j}. \tag{4.12}$$

By substituting Equations 4.6, 4.10 and 4.12 into Equation 4.5, we obtain:

$$\gamma_o(j) = \frac{\overbrace{\left|[\mathbf{F}]_{j,j}\right|^2 E\{|d_j|^2\}}^{\text{desired signal}}}{\underbrace{[\mathbf{FR}_d\mathbf{F}^H]_{j,j} - 2\text{Re}\left\{[\mathbf{FR}_d]_{j,j}[\mathbf{F}]^*_{j,j}\right\} + \left|[\mathbf{F}]_{j,j}\right|^2 E\{|d_j|^2\}}_{\text{interference}} + \underbrace{[\mathbf{MR}_n\mathbf{M}^H]_{j,j}}_{\text{noise}}} \tag{4.13}$$

where Equation 4.13 is the general expression for the SINR at the output of either of the two linear receivers, namely the ZF-BLE and MMSE-BLE.

We can further simplify the expression in Equation 4.13 by making two assumptions. First, the noise values are assumed to be uncorrelated and have a variance of σ^2, giving $\mathbf{R}_n = \sigma^2\mathbf{I}_{(NQ+W-1)}$. Second, the data symbols are also assumed to be uncorrelated and have a variance of g_j^2. This results in $\mathbf{R}_d = \mathbf{D}_g$, where $\mathbf{D}_g$ is a diagonal matrix having real values on its main diagonal and $[\mathbf{D}_g]_{j,j} = E\{|d_j|^2\} = g_j^2$. With these assumptions, the SINR at the output of the linear joint detection receiver is:

$$\gamma_o(j)\Big|_{\mathbf{R}_n=\sigma^2\mathbf{I}_{(NQ+W-1)},\mathbf{R}_d=\mathbf{D}_g}$$

$$= \frac{g_j^2\left|[\mathbf{F}]_{j,j}\right|^2}{[\mathbf{FD}_g\mathbf{F}^H]_{j,j} - 2\text{Re}\left\{g_j^2[\mathbf{F}]_{j,j}[\mathbf{F}]^*_{j,j}\right\} + g_j^2\left|[\mathbf{F}]_{j,j}\right|^2 + \sigma^2[\mathbf{MM}^H]_{j,j}} \tag{4.14}$$

$$= \frac{g_j^2\left|[\mathbf{F}]_{j,j}\right|^2}{[\mathbf{FD}_g\mathbf{F}^H]_{j,j} - 2g_j^2\left|[\mathbf{F}]_{j,j}\right|^2 + g_j^2\left|[\mathbf{F}]_{j,j}\right|^2 + \sigma^2[\mathbf{MM}^H]_{j,j}} \tag{4.15}$$

$$= \frac{g_j^2\left|[\mathbf{F}]_{j,j}\right|^2}{[\mathbf{FD}_g\mathbf{F}^H]_{j,j} - g_j^2\left|[\mathbf{F}]_{j,j}\right|^2 + \sigma^2[\mathbf{MM}^H]_{j,j}} \tag{4.16}$$

$$= \frac{g_j^2\left|[\mathbf{F}]_{j,j}\right|^2}{\sum_{l=1,l\neq j}^{KN} g_l^2\left|[\mathbf{F}]_{j,l}\right|^2 + \sigma^2\sum_{q=1}^{NQ+W-1}\left|\mathbf{M}_{jq}\right|^2}, \tag{4.17}$$

where $j = n+N(k-1)$. For a system supporting K users, each transmitting N data symbols per burst, employing spreading sequences of length Q, and transmitting over a channel of length W chips, the matrix $\mathbf{M}$ has dimensions of $[KN \times (NQ + W - 1)]$ and the matrix $\mathbf{F}$ is a square matrix having KN number of columns.

In Equation 3.63 the general expression representing the decision feedback joint detection receivers was formulated as:

$$\mathbf{S}\hat{\mathbf{d}} = \mathbf{M}\mathbf{y}, \tag{4.18}$$

where $\mathbf{S}$ is the feedback operator. Therefore, the input to the threshold detector or data demodulator can be expressed as:

$$\begin{aligned} \mathbf{t} &= \mathbf{M}\mathbf{y} - \mathbf{S}\hat{\mathbf{d}} && (4.19) \\ &= \mathbf{M}\mathbf{A}\mathbf{d} + \mathbf{M}\mathbf{n} - \mathbf{S}\hat{\mathbf{d}}, && (4.20) \end{aligned}$$

where we have used the expression $\mathbf{y} = \mathbf{A}\mathbf{d} + \mathbf{n}$ from Equation 3.59.

If we assume that no erroneous decisions occurred, i.e. $\hat{\mathbf{d}} = \mathbf{d}$, this leads to:

$$\begin{aligned} \mathbf{t} &= (\mathbf{M}\mathbf{A} - \mathbf{S})\mathbf{d} + \mathbf{M}\mathbf{n} && (4.21) \\ &= \mathbf{P}\mathbf{d} + \mathbf{M}\mathbf{n} && (4.22) \\ &= \underbrace{\text{diag}(\mathbf{P})\mathbf{d}}_{\text{desired symbols}} + \underbrace{\overline{\text{diag}}(\mathbf{P})\mathbf{d}}_{\text{interference}} + \underbrace{\mathbf{M}\mathbf{n}}_{\text{noise}}, && (4.23) \end{aligned}$$

where we used the shorthand

$$\mathbf{P} = \mathbf{M}\mathbf{A} - \mathbf{S}. \tag{4.24}$$

Equation 4.23 is analogous to Equation 4.4 except that $\mathbf{F} = \mathbf{M}\mathbf{A}$, while $\mathbf{P} = \mathbf{M}\mathbf{A} - \mathbf{S}$, which allows us to use the output SINR expression of Equation 4.13. By replacing the matrix $\mathbf{F}$ in Equation 4.13 with the matrix $\mathbf{P}$ for the decision feedback equalizers, we obtain an expression for the SINR at the output of the decision feedback equalizers, which is given by:

$$\gamma_{o,\text{bdfe}}(j) = \frac{\overbrace{\left|[\mathbf{P}]_{j,j}\right|^2 E\{|d_j|^2\}}^{\text{desired signal}}}{\underbrace{[\mathbf{P}\mathbf{R}_d\mathbf{P}^H]_{j,j} - 2\text{Re}\left\{[\mathbf{P}\mathbf{R}_d]_{j,j}[\mathbf{P}]^*_{j,j}\right\} + \left|[\mathbf{P}]_{j,j}\right|^2 E\{|d_j|^2\}}_{\text{interference}} + \underbrace{[\mathbf{M}\mathbf{R}_n\mathbf{M}^H]_{j,j}}_{\text{noise}}}, \tag{4.25}$$

where $j = n + N(k-1)$.

In Sections 3.3.6 and 3.3.7, it was shown that the feedback operator, $\mathbf{S}$, was constituted by an upper triangular matrix where all the elements on its main diagonal had values of zero.

Therefore, based on Equation 4.24, the elements on the main diagonal of matrix $\mathbf{P}$ can be simplified to:

$$[\mathbf{P}]_{j,j} = [\mathbf{MA}]_{j,j} = [\mathbf{F}]_{j,j}. \tag{4.26}$$

By substituting this into Equation 4.25, we arrive at a general expression for the SINR at the output of the MMSE-BDFE, given by:

$$\gamma_{o,\text{bdfe}}(j) = \frac{\overbrace{\left|[\mathbf{F}]_{j,j}\right|^2 E\{|d_j|^2\}}^{\text{desired signal}}}{\underbrace{[\mathbf{PR}_d\mathbf{P}^H]_{j,j} - 2\text{Re}\Big\{[\mathbf{PR}_d]_{j,j}[\mathbf{F}]^*_{j,j}\Big\} + \left|[\mathbf{F}]_{j,j}\right|^2 E\{|d_j|^2\}}_{\text{interference}} + \underbrace{[\mathbf{MR}_n\mathbf{M}^H]_{j,j}}_{\text{noise}}}. \tag{4.27}$$

Having derived the general expression for the output SINR of a JD receiver, let us now consider the SINR at the output of the MMSE-BDFE specifically.

4.3.3 Output SINR of the MMSE-BDFE

The performance of the four joint detection schemes, namely that of the ZF-BLE, MMSE-BLE, ZF-BDFE and MMSE-BDFE, was compared in Chapter 3. It was concluded that the MMSE-BDFE had the best performance. Therefore, we shall now derive the output SINR expression for the MMSE-BDFE, $\gamma_{o,\text{MMSE-BDFE}}(j)$, by using Equations 3.160, 3.173 and 3.176 from Section 3.3.7 along with Equation 4.27.

Commencing from Equation 3.173 and substituting the expression for the matrix $\mathbf{F} = \mathbf{MA}$ as well as taking into account Equation 3.160, we arrive at:

$$\begin{aligned}
\mathbf{F} &= \mathbf{D}_m^{-1}((\mathbf{D}_m\mathbf{U}_m)^H)^{-1}\mathbf{A}^H\mathbf{R}_n^{-1}\mathbf{A} && (4.28)\\
&= \mathbf{D}_m^{-1}((\mathbf{D}_m\mathbf{U}_m)^H)^{-1}((\mathbf{D}_m\mathbf{U}_m)^H\mathbf{D}_m\mathbf{U}_m - \mathbf{R}_d^{-1}) && (4.29)\\
&= \mathbf{U}_m - \mathbf{D}_m^{-2}(\mathbf{U}_m^H)^{-1}\mathbf{R}_d^{-1}, && (4.30)
\end{aligned}$$

where $\mathbf{D}_m^{-2} = \mathbf{D}_m^{-1}\mathbf{D}_m^{-1}$.

By using Equations 4.30 and 3.176, the matrix $\mathbf{P} = \mathbf{MA} - \mathbf{S} = \mathbf{F} - \mathbf{S}$ can be expressed as:

$$\begin{aligned}
\mathbf{P} &= \mathbf{U}_m - \mathbf{D}_m^{-2}(\mathbf{U}_m^H)^{-1}\mathbf{R}_d^{-1} - (\mathbf{U}_m - \mathbf{I}_{KN}) && (4.31)\\
&= \mathbf{I}_{KN} - \mathbf{D}_m^{-2}(\mathbf{U}_m^H)^{-1}\mathbf{R}_d^{-1} && (4.32)
\end{aligned}$$

Some of the identities that we will be exploiting include:

- $\mathbf{U}_m$ is an upper triangular matrix having the value of unity for all the elements on its main diagonal, $[\mathbf{U}_m]_{j,j} = [\mathbf{U}_m]^H_{j,j} = [\mathbf{U}_m]^*_{j,j} = 1$.

- $[(\mathbf{U}_m\mathbf{R}_d)^H]_{j,j} = [\mathbf{U}_m\mathbf{R}_d]_{j,j}$, since the matrix $(\mathbf{U}_m\mathbf{R}_d)$ is an upper triangular matrix with real values on its main diagonal. Therefore, the elements on the main diagonal of the matrix and those of its Hermitian matrix are the same.

- For the covariance matrices, we have $\mathbf{R}_d = \mathbf{R}_d^H$ and $\mathbf{R}_n = \mathbf{R}_n^H$.

- $[\mathbf{D}_m\mathbf{X}]_{j,j} = [\mathbf{D}_m]_{j,j}[\mathbf{X}]_{j,j}$, where $\mathbf{D}_m$ is a diagonal matrix having real values on its main diagonal and $\mathbf{X}$ is a square matrix.

- Finally, $[\mathbf{R}_d]_{j,j} = E\{|d_j|^2\}$.

Upon substituting Equation 4.30 in the numerator of Equation 4.27, the variance of the desired signal can be rewritten as:

$$\begin{aligned}
\left|[\mathbf{F}]_{j,j}\right|^2 E\{|d_j|^2\} &= \left|\left[\mathbf{U}_m - \mathbf{D}_m^{-2}(\mathbf{U}_m^H)^{-1}\mathbf{R}_d^{-1}\right]_{j,j}\right|^2 E\{|d_j|^2\} && (4.33)\\
&= \left|[\mathbf{U}_m]_{j,j} - [\mathbf{D}_m^{-2}(\mathbf{U}_m^H)^{-1}\mathbf{R}_d^{-1}]_{j,j}\right|^2 E\{|d_j|^2\} && (4.34)\\
&= \left|1 - [\mathbf{D}_m]_{j,j}^{-2}[((\mathbf{U}_m\mathbf{R}_d)^{-1})^H]_{j,j}\right|^2 E\{|d_j|^2\} && (4.35)\\
&= [\mathbf{D}_m]_{j,j}^{-4}\left|[\mathbf{D}_m]_{j,j}^2 - [\mathbf{Y}]_{j,j}\right|^2 E\{|d_j|^2\}, && (4.36)
\end{aligned}$$

where we used the identity $[(\mathbf{U}_m\mathbf{R}_d)^H]_{j,j} = [\mathbf{U}_m\mathbf{R}_d]_{j,j}$; and $\mathbf{Y} = (\mathbf{U}_m\mathbf{R}_d)^{-1}$.

The first term of the interference variance in the denominator of Equation 4.27 can be expanded by utilizing Equation 4.32:

$$\begin{aligned}
&[\mathbf{P}\mathbf{R}_d\mathbf{P}^H]_{j,j}\\
&= \left[(\mathbf{I}_{KN} - \mathbf{D}_m^{-2}(\mathbf{U}_m^H)^{-1}\mathbf{R}_d^{-1})\mathbf{R}_d(\mathbf{I}_{KN} - \mathbf{D}_m^{-2}(\mathbf{U}_m^H)^{-1}\mathbf{R}_d^{-1})^H\right]_{j,j} && (4.37)\\
&= \left[\mathbf{R}_d - \mathbf{D}_m^{-2}(\mathbf{U}_m^H)^{-1} - \mathbf{U}_m^{-1}\mathbf{D}_m^{-2} + \mathbf{D}_m^{-2}(\mathbf{U}_m\mathbf{R}_d\mathbf{U}_m^H)^{-1}\mathbf{D}_m^{-2}\right]_{j,j} && (4.38)\\
&= [\mathbf{R}_d]_{j,j} - [\mathbf{D}_m]_{j,j}^{-2}[(\mathbf{U}_m^H)^{-1}]_{j,j} - [\mathbf{U}_m^{-1}]_{j,j}[\mathbf{D}_m]_{j,j}^{-2} + [\mathbf{D}_m]_{j,j}^{-4}[\mathbf{X}^{-1}]_{j,j} && (4.39)\\
&= E\{|d_j|^2\} - 2\cdot[\mathbf{D}_m]_{j,j}^{-2}[\mathbf{U}_m^{-1}]_{j,j} + [\mathbf{D}_m]_{j,j}^{-4}[\mathbf{X}^{-1}]_{j,j} && (4.40)\\
&= E\{|d_j|^2\} - 2\cdot[\mathbf{D}_m]_{j,j}^{-2} + [\mathbf{D}_m]_{j,j}^{-4}[\mathbf{X}^{-1}]_{j,j}, && (4.41)
\end{aligned}$$

where $\mathbf{X} = \mathbf{U}_m\mathbf{R}_d\mathbf{U}_m^H$.

The second term of the interference variance in the denominator of Equation 4.27 can be expanded by utilizing Equations 4.30 and 4.32, yielding:

$$
\begin{aligned}
& 2 \cdot \mathrm{Re}\Big\{ [\mathbf{PR}_d]_{j,j} [\mathbf{F}]^*_{j,j} \Big\} \\
&= 2 \cdot \mathrm{Re}\Big\{ \Big[(\mathbf{I}_{KN} - \mathbf{D}_m^{-2} (\mathbf{U}_m^H)^{-1} \mathbf{R}_d^{-1}) \mathbf{R}_d \Big]_{j,j} \Big[\mathbf{U}_m - \mathbf{D}_m^{-2} (\mathbf{U}_m^H)^{-1} \mathbf{R}_d^{-1} \Big]^*_{j,j} \Big\} && (4.42) \\
&= 2 \cdot \mathrm{Re}\Big\{ \Big[\mathbf{R}_d - \mathbf{D}_m^{-2} (\mathbf{U}_m^H)^{-1} \Big]_{j,j} \Big[\mathbf{U}_m - \mathbf{D}_m^{-2} (\mathbf{U}_m^H)^{-1} \mathbf{R}_d^{-1} \Big]^*_{j,j} \Big\} && (4.43) \\
&= 2 \cdot \mathrm{Re}\Big\{ [\mathbf{R}_d]_{j,j} [\mathbf{U}_m]^*_{j,j} - [\mathbf{D}_m^{-2} (\mathbf{U}_m^H)^{-1}]_{j,j} [\mathbf{U}_m]^*_{j,j} - [\mathbf{R}_d]_{j,j} [\mathbf{D}_m^{-2} (\mathbf{U}_m^H)^{-1} \mathbf{R}_d^{-1}]^*_{j,j} + \\
& \quad [\mathbf{D}_m^{-2} (\mathbf{U}_m^H)^{-1}]_{j,j} [\mathbf{D}_m^{-2} (\mathbf{U}_m^H)^{-1} \mathbf{R}_d^{-1}]^*_{j,j} \Big\} && (4.44)
\end{aligned}
$$

By recognizing the identities of $[\mathbf{R}_d]_{j,j} = E\{|d_j|^2\}$, $[\mathbf{U}_m]_{j,j} = 1$ and $[\mathbf{D}_m^{-2}(\mathbf{U}_m^H)^{-1}]_{j,j} = [\mathbf{D}_m]^{-2}_{j,j}$, Equation 4.44 can be simplified to:

$$
\begin{aligned}
& 2 \cdot \mathrm{Re}\Big\{ [\mathbf{PR}_d]_{j,j} [\mathbf{F}]^*_{j,j} \Big\} \\
&= 2 \cdot E\{|d_j|^2\} - 2 \cdot [\mathbf{D}_m]^{-2}_{j,j} - 2 \cdot \mathrm{Re}\Big\{ [\mathbf{D}_m]^{-2}_{j,j} [\mathbf{Y}]^*_{j,j} \Big\} E\{|d_j|^2\} + \\
& \quad 2 \cdot \mathrm{Re}\Big\{ [\mathbf{D}_m]^{-4}_{j,j} [\mathbf{U}_m^{-1}]_{j,j} [\mathbf{Y}]^*_{j,j} \Big\} && (4.45) \\
&= 2 \cdot E\{|d_j|^2\} - 2 \cdot [\mathbf{D}_m]^{-2}_{j,j} - 2 \cdot [\mathbf{D}_m]^{-2}_{j,j} \mathrm{Re}\{[\mathbf{Y}]_{j,j}\} \, E\{|d_j|^2\} + \\
& \quad 2 \cdot [\mathbf{D}_m]^{-4}_{j,j} \mathrm{Re}\{[\mathbf{Y}]^*_{j,j}\}, && (4.46)
\end{aligned}
$$

where $\mathbf{Y} = (\mathbf{U}_m \mathbf{R}_d)^{-1}$.

The third term of the interference variance in the denominator of Equation 4.27 is the same expression as the variance of the desired signal in Equation 4.36. Combining Equations 4.36, 4.41 and 4.46 for all three terms of the interference variance and recognizing that $[\mathbf{Y}]_{j,j} = \mathrm{Re}\{[\mathbf{Y}]_{j,j}\}$ leads to the expression:

$$
\begin{aligned}
& [\mathbf{PR}_d \mathbf{P}^H]_{j,j} - 2 \cdot \mathrm{Re}\Big\{ [\mathbf{PR}_d]_{j,j} [\mathbf{F}]^*_{j,j} \Big\} + \Big| [\mathbf{F}]_{j,j} \Big|^2 E\{|d_j|^2\} \\
&= [\mathbf{D}_m]^{-4}_{j,j} [\mathbf{X}^{-1}]_{j,j} - 2 [\mathbf{D}_m]^{-4}_{j,j} \mathrm{Re}\Big\{ [\mathbf{Y}]_{j,j} \Big\} + [\mathbf{D}_m]^{-4}_{j,j} \Big| [\mathbf{Y}]_{j,j} \Big|^2 E\{|d_j|^2\} && (4.47)
\end{aligned}
$$

With the aid of the expression $\mathbf{M} = \mathbf{D}_m^{-1}((\mathbf{D}_m \mathbf{U}_m)^H)^{-1} \mathbf{A}^H \mathbf{R}_n^{-1}$ in Equation 3.173, the noise term in the denominator of Equation 4.27 can be rewritten as:

$$
\begin{aligned}
[\mathbf{MR}_n \mathbf{M}^H]_{j,j} &= \Big[\mathbf{D}_m^{-1} ((\mathbf{D}_m \mathbf{U}_m)^H)^{-1} \mathbf{A}^H \mathbf{R}_n^{-1} \mathbf{R}_n \mathbf{R}_n^{-1} \mathbf{A} (\mathbf{D}_m \mathbf{U}_m)^{-1} \mathbf{D}_m^{-1} \Big]_{j,j} && (4.48) \\
&= \Big[(\mathbf{U}_m - \mathbf{D}_m^{-2} (\mathbf{U}_m^H)^{-1} \mathbf{R}_d^{-1}) \mathbf{U}_m^{-1} \mathbf{D}_m^{-2} \Big]_{j,j} && (4.49) \\
&= [\mathbf{D}_m]^{-2}_{j,j} - [\mathbf{D}_m^{-2} (\mathbf{U}_m \mathbf{R}_d \mathbf{U}_m^H)^{-1} \mathbf{D}_m^{-2}]_{j,j} && (4.50) \\
&= [\mathbf{D}_m]^{-2}_{j,j} - [\mathbf{D}_m]^{-4}_{j,j} [\mathbf{X}^{-1}]_{j,j}, && (4.51)
\end{aligned}
$$

where we have employed the shorthand notation of $\mathbf{X} = \mathbf{U}_m \mathbf{R}_d \mathbf{U}_m^H$ used previously.

By substituting Equations 4.36, 4.47 and 4.51 into Equation 4.27, we obtain an expression for the SINR at the output of the MMSE-BDFE, which can be simplified to:

$$\gamma_{o,\text{MMSE-BDFE}}(j)$$

$$= \frac{[\mathbf{D}_m]_{j,j}^{-4} \left| [\mathbf{D}_m]_{j,j}^2 - [\mathbf{Y}]_{j,j} \right|^2 E\{|d_j|^2\}}{[\mathbf{D}_m]_{j,j}^{-4} \left| [\mathbf{Y}]_{j,j} \right|^2 E\{|d_j|^2\} - 2 \cdot [\mathbf{D}_m]_{j,j}^{-4} \cdot \text{Re}\left\{ [\mathbf{Y}]_{j,j} \right\} + [\mathbf{D}_m]_{j,j}^{-2}}$$

$$= \frac{\overbrace{\left| [\mathbf{D}_m]_{j,j}^2 - [(\mathbf{U}_m \mathbf{R}_d)^{-1}]_{j,j} \right|^2 \cdot E\left\{ \left| d_j \right|^2 \right\}}^{\text{desired signal}}}{\underbrace{\left| [(\mathbf{U}_m \mathbf{R}_d)^{-1}]_{j,j} \right|^2 E\left\{ \left| d_j \right|^2 \right\} - 2 \cdot \text{Re}\left\{ [(\mathbf{U}_m \mathbf{R}_d)^{-1}]_{j,j} \right\} + [\mathbf{D}_m]_{j,j}^2}_{\text{interference + noise}}}, \quad (4.52)$$

where $j = n + N(k-1)$.

The expression in Equation 4.52 can be simplified by stipulating the same assumptions as in obtaining Equation 4.17, namely that $\mathbf{R}_d = \mathbf{D}_g$, where $\mathbf{D}_g$ is a diagonal matrix having real values on its main diagonal. Furthermore, $[\mathbf{D}_g]_{j,j} = E\{|d_j|^2\} = g_j^2$, where g_j^2 is the variance of the transmitted symbol. Using these assumptions, the matrix $(\mathbf{U}_m \mathbf{R}_d)^{-1}$ can be expressed as:

$$[(\mathbf{U}_m \mathbf{R}_d)^{-1}]_{j,j} = [\mathbf{D}_g^{-1} \mathbf{U}_m^{-1}]_{j,j} \quad (4.53)$$

$$= [\mathbf{D}_g]_{j,j}^{-1} [\mathbf{U}_m^{-1}]_{j,j} \quad (4.54)$$

$$= \frac{1}{g_j^2} \times 1 \quad (4.55)$$

$$= \frac{1}{g_j^2}. \quad (4.56)$$

With the aid of this expression, the SINR at the output of the MMSE-BDFE can be simplified to:

$$\gamma_{o,\text{MMSE-BDFE}}(j) = \frac{\left| [\mathbf{D}_m]_{j,j}^2 - \frac{1}{g_j^2} \right|^2 g_j^2}{\frac{1}{g_j^4} g_j^2 - 2 \frac{1}{g_j^2} + [\mathbf{D}_m]_{j,j}^2} \quad (4.57)$$

$$= g_j^2 [\mathbf{D}_m]_{j,j}^2 - 1, \qquad j = n + N(k-1), \quad (4.58)$$

under the assumption of uncorrelated noise and data.

4.3.4 Numerical Analysis of the JD-CDMA System Performance

A numerical solution to the BER performance of the JD-CDMA system can be obtained semi-analytically by employing the SINR at the output of the joint detector. Following the approach given by Klein *et al.* [127], the semi-analytical average BER of a fixed-mode JD-CDMA system for a given E_s/N_0 value can be calculated as follows:

$$P^b(E_s/N_0) = \int_0^{\infty} P^b_m(\gamma_o) \cdot p(\gamma_o)\, d\gamma_o, \tag{4.59}$$

where $P^b(E_s/N_0)$ is the BER performance corresponding to the E_s/N_0 value specified; γ_o is the SINR at the output of the joint detector; and $P^b_m(\gamma_o)$ is the theoretical BER performance corresponding to γ_o over a Gaussian channel for the chosen fixed modulation scheme. Equation 4.59 simply corresponds to weighting the error probability, $p(\gamma_o)$, at the encountered SINR, namely γ_o, by the probability of encountering this specific SINR, and then averaging this value over the range of γ_o values. The closed-form solutions for the BER performances over Gaussian channels of the various modulation schemes – BPSK, 4-QAM, 16-QAM and 64-QAM – used in the proposed AQAM scheme were outlined by Webb *et al.* [92] and the corresponding BER expressions for BPSK, 4-QAM and 16-QAM are given below:

$$P^b_{\text{BPSK}}(\gamma) = Q(\sqrt{2\gamma}\,) \tag{4.60}$$

$$P^b_{\text{4-QAM}}(\gamma) = Q(\sqrt{\gamma}\,) \tag{4.61}$$

$$P^b_{\text{16-QAM}}(\gamma) = \frac{1}{2}\left[Q\left(\sqrt{\frac{\gamma}{5}}\right) + Q\left(\sqrt{3\cdot\frac{\gamma}{5}}\right)\right]. \tag{4.62}$$

The notation $p(\gamma_o)$ represents the probability distribution function (PDF) of the output SINR, γ_o, for the given E_s/N_0 value. The instantaneous γ_o values are calculated for a particular set of simulation conditions by employing Equation 4.52 for the MMSE-BDFE or Equation 4.27 for the joint detector in general and the discretized PDF of the SINR γ_o is constructed from these γ_o values to the accuracy required. The PDF is then substituted into Equation 4.59, in order to obtain the numerical BER performance. The integration is approximated by the trapezoidal rule [190] with the integration limits set to the lowest and highest encountered values of γ_o, respectively.

The BER expression in 4.59 can be modified to obtain the average BER and bits-per-symbol (BPS) throughput performance of an AQAM/JD-CDMA system following the approach given by Wong *et al.* [238]. Utilizing the SINR at the output of the joint detector, γ_o, and assuming perfect SINR estimation and perfect channel estimation, the upper bound performance of a four-mode AQAM scheme that includes BPSK, 4-QAM and 16-QAM and 64-QAM can be obtained as follows:

$$P^b(E_s/N_0) = B^{-1} \cdot \begin{bmatrix} 1 \cdot \int_0^{t_1} P^b_{\text{BPSK}}(\gamma_o) \cdot p(\gamma_o)\, d\gamma_o + \\ 2 \cdot \int_{t_1}^{t_2} P^b_{\text{4-QAM}}(\gamma_o) \cdot p(\gamma_o)\, d\gamma_o + \\ 4 \cdot \int_{t_2}^{t_3} P^b_{\text{16-QAM}}(\gamma_o) \cdot p(\gamma_o)\, d\gamma_o + \\ 6 \cdot \int_{t_3}^{\infty} P^b_{\text{64-QAM}}(\gamma_o) \cdot p(\gamma_o)\, d\gamma_o \end{bmatrix}, \tag{4.63}$$

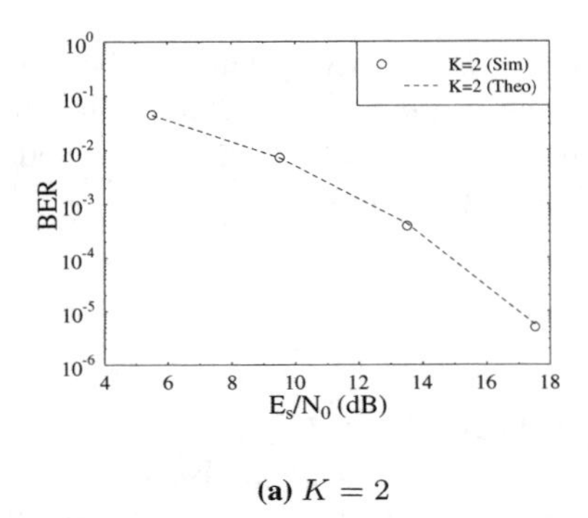

(a) $K = 2$

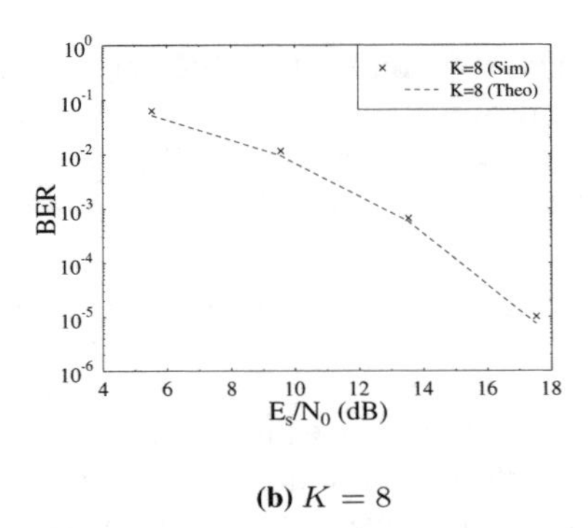

(b) $K = 8$

Figure 4.7: Comparison of the numerical and simulation performances of the JD-CDMA system employing 4-QAM as the modulation mode and the MMSE-BDFE as the JD receiver. The comparison was carried out for the seven-path Bad Urban channel of Figure 3.20 and for $K = 2$, as well as $K = 8$.

where B is the average number of bits-per-symbol (BPS) transmitted, which is given by:

$$B = \begin{bmatrix} 1 \cdot \int_0^{t_1} p(\gamma_o)\, d\gamma_o + \\ 2 \cdot \int_{t_1}^{t_2} p(\gamma_o)\, d\gamma_o + \\ 4 \cdot \int_{t_2}^{t_3} p(\gamma_o)\, d\gamma_o + \\ 6 \cdot \int_{t_3}^{\infty} p(\gamma_o)\, d\gamma_o \end{bmatrix}. \tag{4.64}$$

The symbols t_1, t_2 and t_3 are the switching thresholds for 4-QAM, 16-QAM and 64-QAM, respectively. For a system having a lower number of modes, the expressions in Equations 4.63 and 4.64 can be modified accordingly by including only the terms that involve the modulation mode choices in the AQAM scheme.

Figure 4.7 shows the numerical- and simulation-based BER performance comparison for a fixed-mode JD-CDMA system for $K = 2$ and $K = 8$ users, employing 4-QAM as the modulation mode. The γ_o values were obtained for the COST 207 seven-path Bad Urban channel [105] shown in Figure 3.20. From the performance curves, it can be seen that the expression in Equation 4.59 provides an accurate measure of the upper-bound performance of the joint detection scheme.

4.3.5 Simulation Parameters

In this section, the results of the AQAM/JD-CDMA simulations are presented. The simulation parameters used are listed in Table 4.3. The channel profile used was the COST 207 Bad Urban channel [105] consisting of seven-paths, where each path was faded independently at a Doppler frequency of 80 Hz. This channel profile is depicted in Figure 3.20.

In our simulations, perfect channel estimation and perfect SINR estimation was assumed. This issue will be addressed in terms of more practical assumptions in subsequent sections,

Parameter	Value
Doppler frequency	80 Hz
Spreading ratio	16
Spreading sequence	Pseudo-random
Chip rate	2.167 MBaud
Frame burst structure	FRAMES Mode 1 Spread burst 1 [205]
Burst duration	577 μs
Number of QAM symbols per JD block, N	28
Receiver type	MMSE-BDFE

Table 4.3: Simulation parameters for the proposed AQAM/JD-CDMA systems

but currently we wish to obtain the upper bound performance of the AQAM/JD-CDMA system. In order to use the simplified version of the SINR expression at the output of the MMSE-BDFE given in Equation 4.58, under the assumption that both the data symbols and the noise would be uncorrelated, i.e. that $\mathbf{R}_d = \mathbf{D}_g$ and $\mathbf{R}_n = \sigma^2\mathbf{I}_{(NQ+W-1)}$, where σ^2 represents the noise variance and g_j^2 represents the variance of the data symbols, i.e. $[\mathbf{D}_g]_{j,j} = g_j^2 = E\{|d_n^{(k)}|^2\}$ for $j = n + N(k-1)$.

Since the modulation mode in AQAM/JD-CDMA systems is likely to change many times over the course of a simulation run, the transmitted power of the k-th user is maintained at the required level by multiplying the QAM symbol with the required signal gain. The results presented in the following figures are plotted against E_s/N_0, where E_s/N_0 represents the ratio of the energy per QAM symbol to the noise power spectral density.

4.3.6 Performance of a Two-User, Twin-Mode AQAM/JD-CDMA System

A twin-mode AQAM/JD-CDMA system supporting $K = 2$ users was simulated employing the parameters in Table 4.3. The modulation mode of the data symbols was switched between $m_1 =$ BPSK and $m_2 =$ 4-QAM. By examining the performance of fixed-mode QAM systems presented in Figure 4.8, the E_s/N_0 required to achieve a target BER of 1% is approximately 6.5 dB for BPSK and approximately 9.5 dB for 4-QAM. Since 4-QAM conveys two bits per symbol, the performance of the two modes in terms of E_b/N_0 is identical, since the difference between E_s/N_0 and E_b/N_0 is 3 dB for 4-QAM. In order to achieve a target BER of 1% using a twin-mode AQAM system, the E_s/N_0 switching threshold between the two modulation modes was set to $t_1 = 10$ dB. This system was simulated and the rules for the switching of modulation modes are tabulated in Table 4.4.

Due to employing time-variant modulation modes, the performance results are plotted against E_s/N_0 on the horizontal axis, where E_s represents the energy per QAM symbol. The value of E_s/N_0 can be converted to the corresponding E_b/N_0 value upon dividing E_s/N_0 by the number of bits per QAM symbol. The performance of the AQAM schemes was evaluated by analyzing the BER and the throughput expressed in terms of the average number of bits per symbol (BPS) transmitted. The BER and BPS throughput performance of the simulated

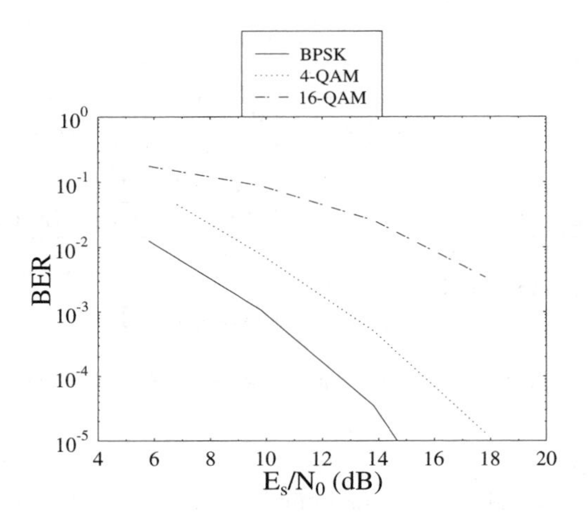

Figure 4.8: BER performance for a fixed-mode QAM/JD-CDMA system supporting $K = 2$ users over the seven-path Bad Urban channel of Figure 3.20. The QAM modes used for the three performance curves were BPSK, 4-QAM and 16-QAM. The simulation parameters are listed in Table 4.3.

Switching criterion	Modulation mode
$\gamma_o(k) < 10$ dB	BPSK
8 dB $\leqslant \gamma_o(k)$	4-QAM

Table 4.4: The switching rules for a twin-mode AQAM/JD-CDMA system, where $\gamma_o(k)$ is the SINR of the k-th user at the output of the MMSE-BDFE. The output SINR was calculated by using Equation 4.58 and $\gamma_o(k) = \frac{1}{N}\sum_{n=1}^{N} \gamma_{o,\text{MMSE-BDFE}}(j)$, for $j = n + N(k-1)$.

system is presented in Figure 4.9. The BER curve has to be read by referring to the y-axis on the left of the figure, while the throughput curve is interpreted by referring to the right y-axis that is labelled BPS. At low E_s/N_0 values the BER of the AQAM scheme mirrored that of BPSK. This is because, at low E_s/N_0 values, the threshold of $t_1 = 10$ dB was seldom reached, and the BPSK modulation mode was the predominant mode. As E_s/N_0 was increased, more frames were transmitted in the 4-QAM mode, since the threshold of 10 dB was achieved more often. This resulted in the BER curve of the AQAM system gradually moving towards the BER curve of 4-QAM. The throughput performance curve is also plotted in the same figure. As E_s/N_0 increased, more and more frames were transmitted using 4-QAM and at $E_s/N_0 = 14$ dB the average throughput increased to approximately 1.9 bits per symbol, which was close to the 2 bits per symbol offered by fixed-mode 4-QAM. The BER at this stage was approximately 6×10^{-5}, which was lower than that achieved by the 4-QAM system. When the channel quality was low, the modulation mode was switched to BPSK which had a lower error probability. This decreased the number of errors received and thus lowered the BER, compared to the 4-QAM mode.

Figure 4.10 shows the modem mode Probability Distribution Functions (PDF) of the

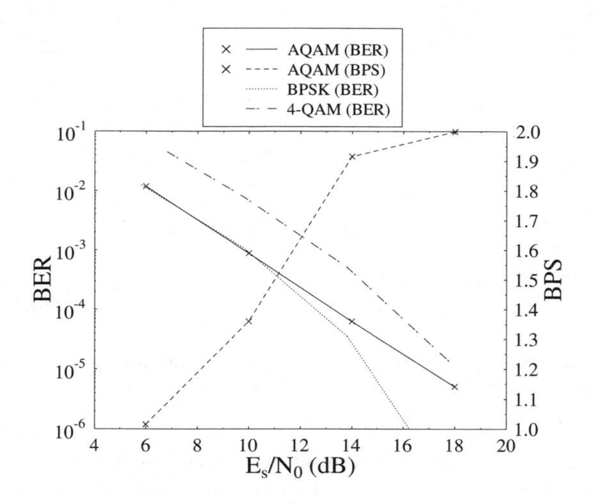

Figure 4.9: BER and throughput (bits per symbol/BPS) performance for an AQAM/JD-CDMA system supporting $K = 2$ users and employing two modulation modes, namely BPSK and 4-QAM. The switching threshold was set to $t_1 = 10$ dB and the simulation parameters are listed in Table 4.3, while the channel model is depicted in Figure 3.20.

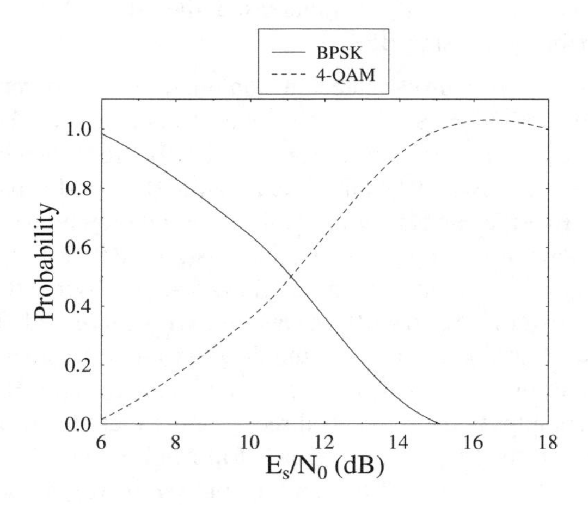

Figure 4.10: The probability of each modulation mode being chosen for transmission in a twin-mode (BPSK, 4-QAM), two-user AQAM/JD-CDMA system employing $t_1 = 10$ dB over the channel model of Figure 3.20 using the parameters of Table 4.3.

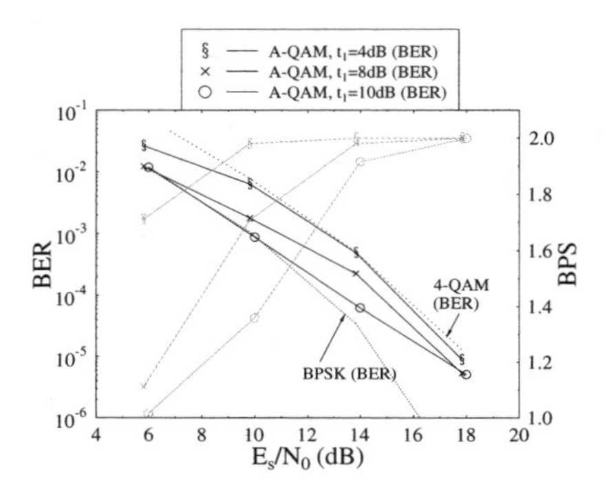

(a) BER comparisons are in bold, BPS comparisons are in grey.

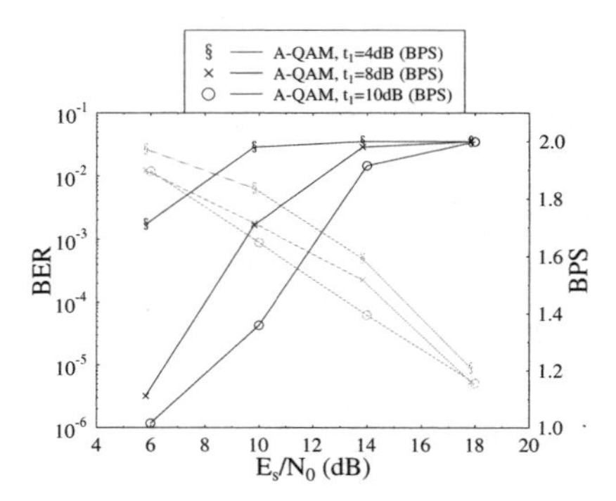

(b) BPS comparisons are in bold, BER comparisons are in grey.

Figure 4.11: The average BER and BPS performance versus average E_s/N_0 of a combined JD-CDMA/AQAM system for $K = 2$ users. Two modulation modes were used, namely BPSK and 4-QAM. The simulation parameters are listed in Table 4.3. The performance associated with different switching thresholds is compared and these thresholds were set to $t_1 = 4$, 8 and 10 dB, respectively. The BER curves for the fixed modulation modes of BPSK, 4-QAM and 16-QAM for $K = 2$ users are also shown. The simulations were conducted over the seven-path Bad Urban channel model of Figure 3.20.

AQAM scheme versus E_s/N_0. As can be seen from the figure, at low E_s/N_0 values, BPSK was the predominant mode, but as the E_s/N_0 increased, the 4-QAM mode was chosen more often, reducing the probability of using BPSK.

The AQAM system was also investigated in conjunction with other threshold values, namely using $t_1 = 4$ dB and $t_1 = 8$ dB for the sake of comparison. The results of these simulations, as well as the performance curves for $t_1 = 10$ dB, are shown in Figure 4.11. The same performance curves are shown twice but for clarity, the BER performance is emphasized in Figure 4.11(a) and the BER performances of the fixed modulation modes of BPSK, 4-QAM and 16-QAM are also shown for comparison. By contrast, the BPS throughput performance is highlighted in Figure 4.11(b). Comparing the BER curves, we observe that as the switching threshold t_1 was reduced from 10 dB to 4 dB, the average BER increased. This was expected, since as the switching threshold was lowered, the 4-QAM mode was activated more often, thus increasing the number of errors. This also led to an increase in the BPS throughput and we can see from the throughput curves of all three systems that at low E_s/N_0 values, the $t_1 = 4$ dB–system achieved the highest throughput, followed by the $t_1 = 8$ dB–system and the $t_1 = 10$ dB–system. Comparing all three performance curves, it can be seen that the $t_1 = 8$ dB–system appeared to achieve the best performance trade-off. The BER is below the target rate of 1% in the range of 6 to 18 dB, and the throughput approaches the maximum of 2 BPS offered by a 4-QAM system at $E_s/N_0 = 14$ dB. It is possible to optimize the switching thresholds for the AQAM system by utilizing, for example, the Powell Multi-dimensions Line Minimization [190] technique in conjunction with a cost function designed for a certain target BER and BPS performance. This technique was proposed by Wong *et al* [238] for

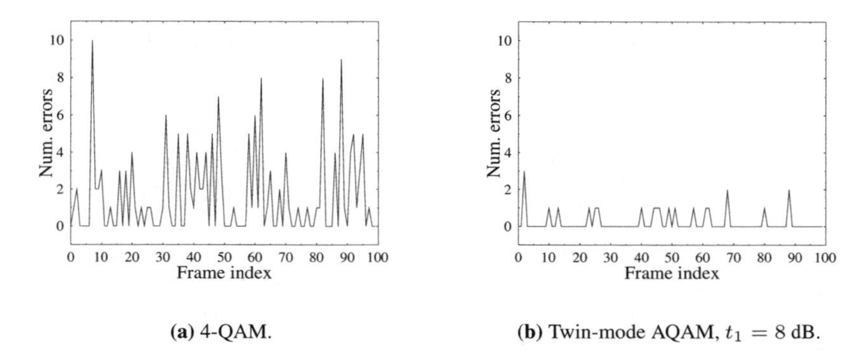

(a) 4-QAM.

(b) Twin-mode AQAM, $t_1 = 8$ dB.

Figure 4.12: The comparison of the number of errors per frame versus frame index for a 4-QAM/JD-CDMA system and for a twin-mode AQAM/JD-CDMA system with $t_1 = 8$ dB, at $E_s/N_0 = 10$ dB. The simulations were conducted over the seven-path Bad Urban channel model of Figure 3.20.

wideband single-user AQAM systems. However, this technique requires the knowledge of the Probability Distribution Function (PDF) of the SINR at the output of the receiver and relied on the existence of an analytical formula describing the associated BER performance. For JD-CDMA systems, the output SINR is dependent on the channel conditions as well as on the number of users in the system and an analytical solution to the BER performance of JD-CDMA systems is difficult to obtain. However, we will show in Section 4.3.10 that the employment of thresholds optimized for single-user wideband AQAM systems is also feasible in the context of AQAM/JD-CDMA systems. Here we emphasize furthermore that apart from its improved performance, the proposed AQAM/JD-CDMA scheme exhibits a high flexibility, always attaining the highest possible BPS throughput that can be supported under the constraint of the prevalent channel quality and target BER. It also allows us to observe one of the consequences of information theory, namely that the higher the tolerable BER, the higher the throughput. Finally, due to the AQAM mode switching regime, the transceiver exhibits a less bursty error distribution than conventional fixed-mode modems, as can be seen in Figure 4.12, where the error variation for a 4-QAM/JD-CDMA system is compared with that of an AQAM/JD-CDMA system. Hence AQAM/JD-CDMA is more amenable to channel coding and may attain higher channel coding gain than fixed-mode transceivers.

Let us now consider the twin-mode AQAM performance for systems supporting different number of users.

4.3.7 Performance Comparisons of a Twin-Mode AQAM/JD-CDMA System for Various Number of Users

In this section the findings of Section 4.3.6 were repeated, but the number of users supported by the system was varied, using K = 1, 2 and 8. The switching threshold was set to t_1 =

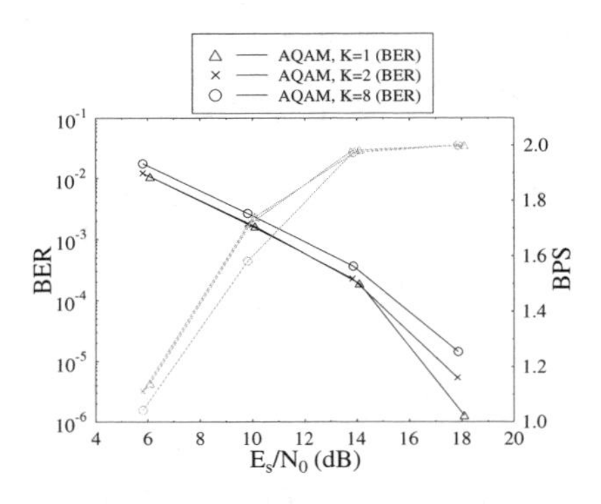

(a) BER comparisons are in bold, BPS comparisons are in grey.

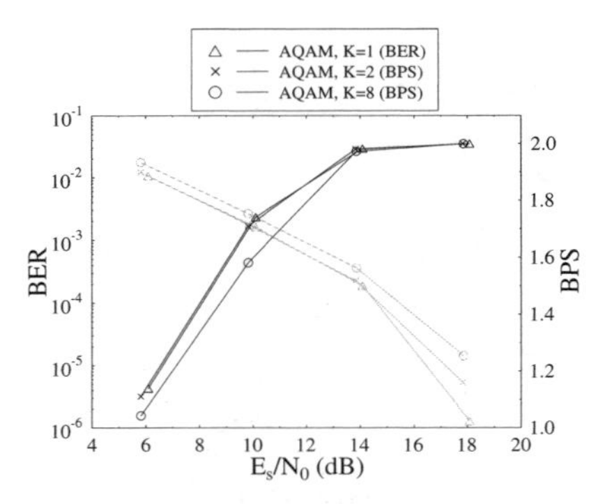

(b) BPS comparisons are in bold, BER comparisons are in grey.

Figure 4.13: BER and throughput (bits per symbol/BPS) performance comparisons for $K = 1$, 2 and 8 users in an AQAM/JD-CDMA system. The modulation mode was switched between BPSK and 4-QAM. The rest of the simulation parameters are listed in Table 4.3 and the simulations were conducted over the seven-path Bad Urban channel model of Figure 3.20.

8 dB for all three values of K. The results of these simulations are presented in Figure 4.13. From the figure, it can be seen that the performance curves for the different number of users matched each other quite closely, with only a slight performance degradation, when K was increased from 1 to 8. This was because the MMSE-BDFE was capable of reducing the MAI by utilizing all the available information regarding the spreading sequences and channel estimates. The slight performance degradation for $K = 8$ users was due to the higher interference from an increased number of users causing the SINR at the output of the MMSE-BDFE to increase slightly.

In order to ascertain the accuracy of the numerical BER and BPS expressions of the AQAM system given in Equations 4.63 and 4.64, the numerical performance of the twin-mode AQAM scheme employing $t_1 = 8$ dB was calculated for $K = 2$ and $K = 8$ users and compared with the simulation-based performance in Figure 4.14. From this figure, it can be seen that the numerical and simulation performance curves matched each other quite closely and the expressions provided a reasonable accuracy in estimating the upper-bound performance of the AQAM scheme.

Having studied the effect of the number of users, let us now consider a triple-mode AQAM system.

4.3.8 Performance of a Two-User, Triple-Mode AQAM/JD-CDMA System

In this section the twin-mode AQAM system is extended to a triple-mode system by adding the 16-QAM mode as one of the modulation choices. The rules for switching the modulation modes are tabulated in Table 4.5.

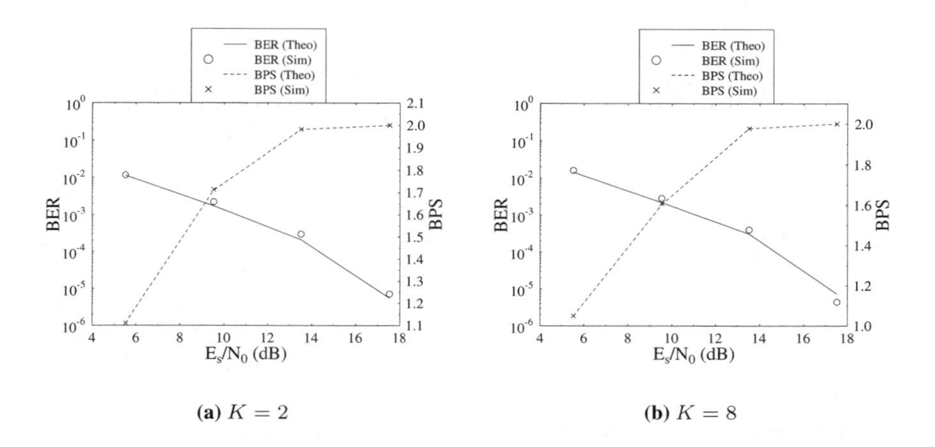

Figure 4.14: Comparison of the numerical- and simulation-based performance of the twin-mode AQAM/JD-CDMA system, where the modulation mode was switched between BPSK and 4-QAM using a switching threshold of $t_1 = 8$ dB. The MMSE-BDFE was employed as the JD receiver and the comparison was carried out for the seven-path Bad Urban channel of Figure 3.20 while supporting $K = 2$ and 8 users.

Switching criterion	Modulation mode
$\gamma_o(k) < t_1$	BPSK
$t_1 \leqslant \gamma_o(k) < t_2$	4-QAM
$t_2 \leqslant \gamma_o(k)$	16-QAM

Table 4.5: The switching rules for a triple-mode AQAM/JD-CDMA system, where $\gamma_o(k)$ is the SINR of the k-th user at the output of the MMSE-BDFE. The output SINR was calculated by using Equation 4.58 and $\gamma_o(k) = \frac{1}{N}\sum_{n=1}^{N}\gamma_{o,\text{MMSE-BDFE}}(j)$, for $j = n + N(k-1)$.

Considering the fixed-mode BER curves in Figure 4.8, the E_s/N_0 values required to achieve a target BER of 1% are approximately 9.5 dB and 16 dB for 4-QAM and 16-QAM, respectively. Therefore, the initial switching thresholds were set to $t_1 = 10$ dB and $t_2 = 16$ dB, for the simulated triple-mode AQAM system. In addition, an "aggressive" system was contrived, which aimed for a higher target BPS throughout, where the corresponding threshold values were chosen to be $t_1 = 8$ dB and $t_2 = 12$ dB. Below our performance results for the triple-mode AQAM/JD-CDMA system supporting $K = 2$ users are presented.

The results of the triple-mode simulations are shown in Figure 4.15 along with the BER performance curves for various fixed-QAM/JD-CDMA systems using the same simulation parameters. Upon comparing the BER performance of the triple-mode AQAM arrangement with that of the twin-mode AQAM system in Figure 4.11, similar characteristics were observed. At low values of E_s/N_0, the BPSK mode was chosen with a higher probability, as demonstrated by the modem mode Probability Distribution Function (PDF) of Figure 4.16, thus reducing the throughput. However, as E_s/N_0 increased, the channel quality improved, thus allowing the 4-QAM and 16-QAM modes to be activated more often, which resulted in

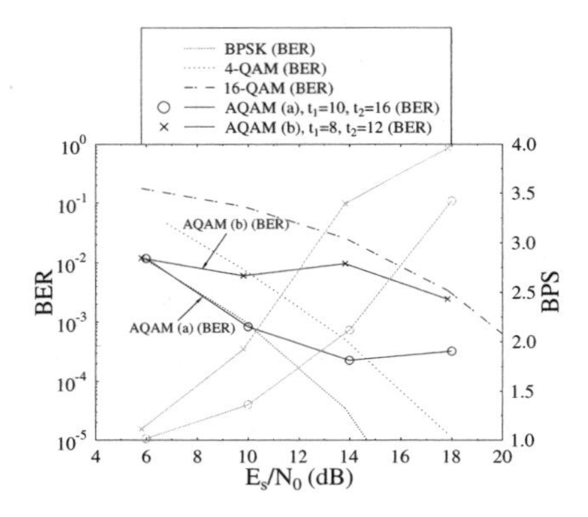

(a) BER comparisons are in bold, BPS comparisons are in grey.

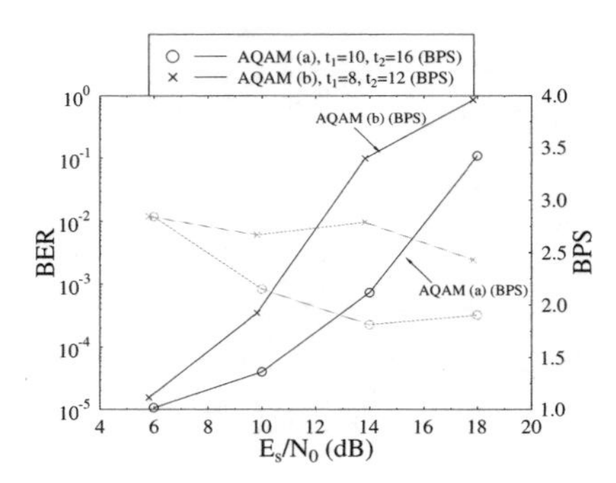

(b) BPS comparisons are in bold, BER comparisons are in grey.

Figure 4.15: The average BER and BPS performance versus average E_s/N_0 of an AQAM/JD-CDMA system for $K = 2$ users. Three modulation modes were used, namely BPSK, 4-QAM and 16-QAM. Two different sets of thresholds were evaluated, as shown in the legends. The BER curves for the same system but using fixed modulation modes of BPSK, 4-QAM and 16-QAM for $K = 2$ users are also shown. The simulation parameters are listed in Table 4.3 and the simulations were conducted over the seven-path Bad Urban channel model of Figure 3.20.

an increased average BPS performance. Both of these AQAM systems were capable of maintaining the target BER of 1% or better. However, the more "aggressive" thresholds of $t_1 = 8$ dB and $t_2 = 12$ dB guaranteed a higher BPS throughput performance, achieving a throughput of 3.5 BPS at $E_s/N_0 = 14$ dB. At this stage, the BER of the AQAM scheme was still lower than the BER achieved by the 16-QAM system. This was because when the channel quality dropped, the modulation mode was switched to a more robust one that provided a lower error probability, thus reducing the number of errors. We can see from the results presented that lowering the threshold values degraded the BER performance, but improved the average throughput of the system. As expected, this is because the higher order modulation modes were chosen with increasing probability, thus allowing more bits to be transmitted. By the same token, this caused the degradation in terms of the BER performance due to the higher BER of the higher throughput – but more vulnerable – higher order modulation modes. However, it should be noted that the BER kept to the level of 1% as targeted. The modem mode PDFs for both sets of switching thresholds are plotted in Figure 4.16. Specifically, Figure 4.16(a) shows the PDFs for the more conservative thresholds of $t_1 = 10$ dB, $t_2 = 16$ dB, while the PDFs of the more aggressive, $t_1 = 8$ dB, $t_2 = 12$ dB–system, are shown in Figure 4.16(b)

Following the BER and BPS study of a two-user, triple-mode AQAM/JD-CDMA system, let us now quantify the effects of different numbers of users.

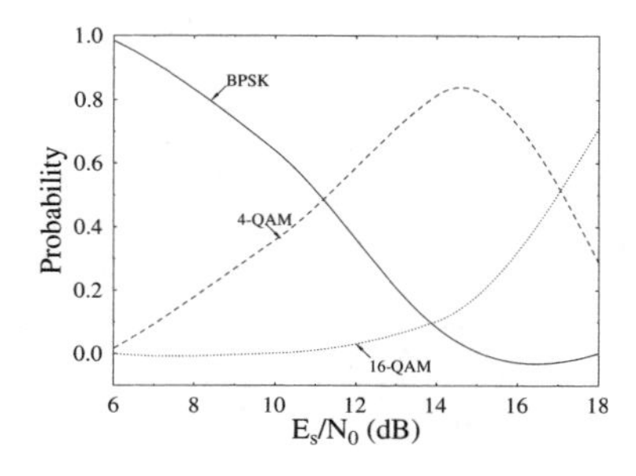

(a) Conservative system : $t_1 = 10$ dB, $t_2 = 16$ dB

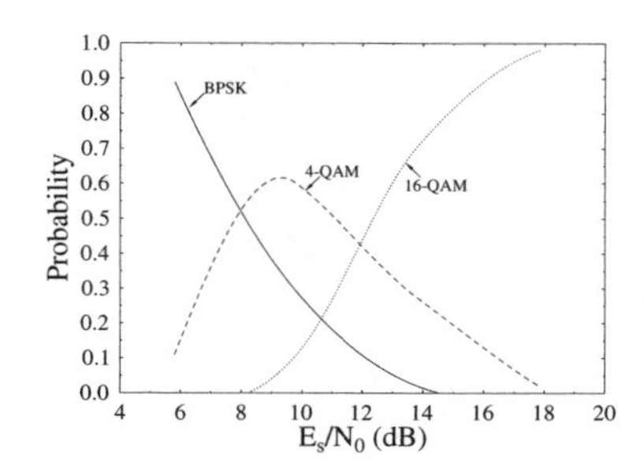

(b) "Aggressive" system : $t_1 = 8$ dB, $t_2 = 12$ dB

Figure 4.16: The PDF of each modulation mode being chosen for transmission in a triple-mode (BPSK, 4-QAM, 16-QAM), two-user AQAM/JD-CDMA system, using the simulation parameters of Table 4.3 and the CIR of Figure 3.20.

4.3.9 Performance Comparisons of a Triple-Mode AQAM/JD-CDMA System for Various Number of Users

The investigations of Section 4.3.8 were also repeated for different numbers of users, i.e. for $K = 1$, 2 and 8. The graphs in Figure 4.17 show the resultant performance curves. The results are similar to those presented for the twin-mode system in Figure 4.13, where the performance curves for the different number of users are quite similar, exhibiting a degradation of less than 1 dB for $K = 8$ users. This is because the MMSE-BDFE exploited the available knowledge on the spreading sequences and channel estimates of all the users, in order to remove a large portion of the MAI.

Figure 4.18 shows the comparison between the numerical and simulation-based performance for the triple-mode AQAM system employing $t_1 = 8$ and $t_2 = 12$ as the switching thresholds. The numerical expressions for the BER and BPS performance were given by Equations 4.63 and 4.64, respectively. From this figure, it can be seen that the numerical expressions provided a fairly accurate estimation of the upper-bound performance of the triple-mode AQAM scheme.

4.3.10 Performance of a Five-Mode AQAM/JD-CDMA System

In this section we extend our previous AQAM systems to a five-mode system, where the 64-QAM mode was used in addition to BPSK, 4-QAM and 16-QAM and a transmission blocking mode (NOTX) was also incorporated. The blocking mode was utilized, when the channel quality was extremely low and hence in this mode no data transmission was carried out. Instead, known dummy symbols were transmitted, in order to assist in channel estimation for the next transmission burst. These known symbols were discarded at the receiver, in order to reduce the complexity of the system. The switching rules for this five-mode scheme are tabulated in Table 4.6.

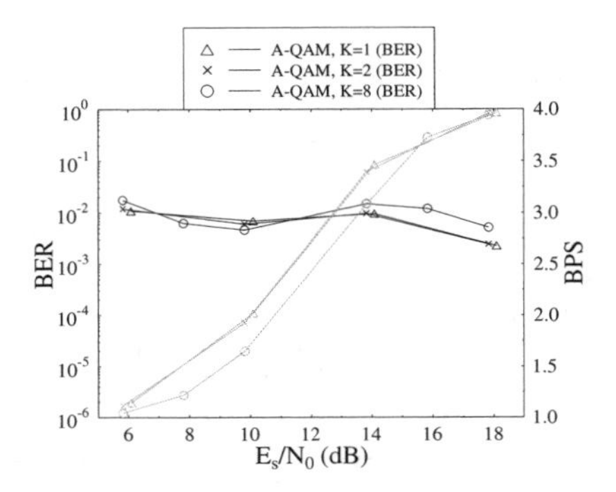

(a) BER comparisons are in bold, BPS comparisons are in grey.

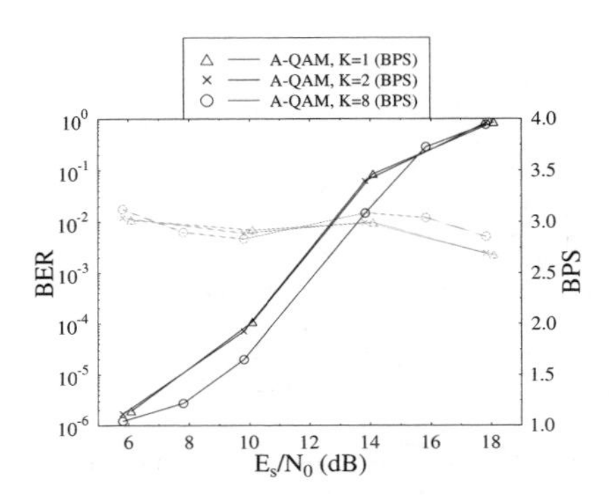

(b) BPS comparisons are in bold, BER comparisons are in grey.

Figure 4.17: BER and throughput (bits per symbol/BPS) performance comparisons for $K = 1$, 2 and 8 users in a triple-mode AQAM/JD-CDMA system. The modulation modes used for transmission were BPSK, 4-QAM or 16-QAM. The simulation parameters are listed in Table 4.3 and the simulations were conducted over the seven-path Bad Urban channel model of Figure 3.20.

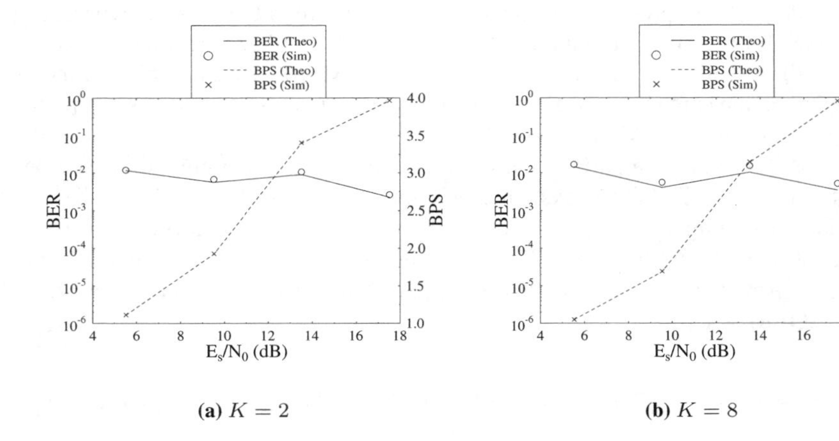

(a) $K = 2$

(b) $K = 8$

Figure 4.18: Comparison of the numerical and simulation performance of the twin-mode AQAM/JD-CDMA system, where the modulation mode was chosen to be BPSK, 4-QAM or 16-QAM with the switching thresholds of $t_1 = 8$ dB and $t_2 = 12$ dB. The MMSE-BDFE was employed as the JD receiver and the comparison was carried out for the seven-path Bad Urban channel of Figure 3.20 with two different loads of users. The rest of the simulation parameters are listed in Table 4.3.

Switching criterion	Modulation mode
$\gamma_o(k) < t_0$	NOTX
$t_0 \leqslant \gamma_o(k) < t_1$	BPSK
$t_1 \leqslant \gamma_o(k) < t_2$	4-QAM
$t_2 \leqslant \gamma_o(k) < t_3$	16-QAM
$t_3 \leqslant \gamma_o(k)$	64-QAM

Table 4.6: The switching rules for a general five-mode AQAM/JD-CDMA system, where the notation NOTX denotes transmission blocking. The output SINR, $\gamma_o(k)$, was calculated by using Equation 4.58 and $\gamma_o(k) = \frac{1}{N}\sum_{n=1}^{N}\gamma_{o,\text{MMSE-BDFE}}(j)$, for $j = n + N(k-1)$. The values $(t_1, \cdots, t_M)$ represent the switching thresholds, where $t_1 < t_2 < \cdots < t_M$.

Switching criterion	Modulation mode
$\gamma_o(k) < 8.25$ dB	NOTX
8.25 dB $\leqslant \gamma_o(k) <$ 10.46 dB	BPSK
10.46 dB $\leqslant \gamma_o(k) <$ 16.8 dB	4-QAM
16.8 dB $\leqslant \gamma_o(k) <$ 23.76 dB	16-QAM
23.76 dB $\leqslant \gamma_o(k)$	64-QAM

Table 4.7: The switching rules for a five-mode AQAM/JD-CDMA system, where $\gamma_o(k)$ is the SINR of the k-th user at the output of the MMSE-BDFE. The notation NOTX denotes transmission blocking.

For this scheme the target BER was set to 0.01%, which is more acceptable for data transmission, instead of the previous 1%. In previous sections it was shown that the performance of the two-user system was similar to the single-user system. The single-user joint detector is actually similar to the decision feedback equalizer [90,92,206] used for conventional TDMA transmission. In the research literature AQAM systems for wideband TDMA transmission have been proposed [238] and the switching thresholds for a five-mode system have been optimized using the Powell Multi-dimensions Line Minimization [190] technique, in order to provide the best possible throughput, while maintaining a target BER of 0.01% for a single-user AQAM scheme in a wideband channel. Therefore, these switching levels have been adopted in our investigations, since the performance of the joint detector is similar to that of the single-user bound. Using these switching thresholds, namely $t_0 = 8.25$, $t_1 = 10.46$, $t_2 = 16.8$, and $t_3 = 23.76$, the rules for switching transmission modes are tabulated in Table 4.7.

The BER and BPS throughput performance of the two-user, five-mode AQAM/JD-CDMA system is presented in Figure 4.19. The channel profile used in this investigation is the COST 207 Bad Urban seven-path channel shown in Figure 3.20. The BER curves for the four fixed constituent modulation modes used are also represented in the same graph. At $E_s/N_0 \leqslant 14$ dB, the BER is less than the BER of fixed-mode BPSK. This is because transmission blocking was used and this reduced the number of bit errors, which occurred when the mobile channel quality was particularly low. It can be seen from the BPS curve that for $E_s/N_0 < 12$ dB, the throughput was less than 1 BPS, since the total number of information bits transmitted was less than the total number of symbols available for transmis-

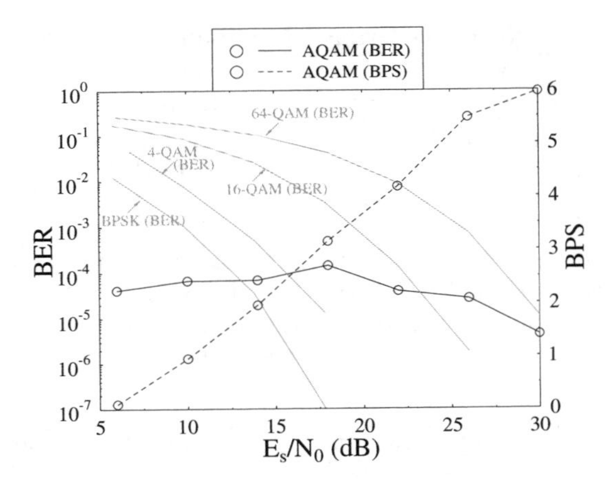

Figure 4.19: BER and BPS performance of a five-mode AQAM/JD-CDMA system supporting $K = 2$ users over the **seven-path Bad Urban channel** of Figure 3.20. The transmission modes were switched between no transmission (NOTX), BPSK, 4-QAM, 16-QAM and 64-QAM. The target BER was set to 0.01% and the switching thresholds were optimized, yielding $t_0 = 8.25$ dB, $t_1 = 10.46$ dB, $t_2 = 16.8$ dB, and $t_3 = 23.76$ dB. The rest of the simulation parameters are listed in Table 4.3.

sion. As the E_s/N_0 value increased, the output SINR exceeded the thresholds required for switching to increasingly higher order modulation modes and this increased the BPS throughput correspondingly. Due to this switching mechanism, a near-constant BER of 0.01% was maintained. Considering the BPS curve, the E_s/N_0 value required to achieve a throughput of 1 BPS was approximately 12 dB. For the fixed-mode BPSK system, the BER of 0.01% was achieved at $E_s/N_0 = 14$ dB. Therefore, at 1 BPS, the SNR performance gain of the AQAM system over fixed-mode BPSK was approximately 2 dB. Similarly, for 2 and 4 BPS, the SNR performance gains of the AQAM system were also in the range of 0.5 to 2 dB over the fixed-mode schemes of 4-QAM and 16-QAM, respectively.

The five-mode AQAM transmission scheme was also investigated for other COST 207 channels, namely the Typical Urban four-path channel of Figure 3.18 and the Rural Area two-path channel of Figure 3.15. The corresponding performance results are shown in Figure 4.20 for the Typical Urban channel and Figure 4.21 for the Rural Area channel. The same performance trends as for the Bad Urban channel can be observed from these two figures. At low E_S/N_0 values, the transmission blocking mode was used predominantly, but as E_s/N_0 increased, the higher order modulation modes were used more often, leading to increasingly higher BPS throughput values. Throughout these E_s/N_0 values, the target BER of 0.01% was maintained. For the Typical Urban channel, as E_s/N_0 increased beyond 20 dB, the BER started to drop and eventually followed the BER curve of the fixed-mode 64-QAM system. This drop-off E_s/N_0 value was higher for the Rural Area channel, at an SNR of approximately 30 dB. The SNR performance gain for the Typical Urban channel was in the range of 4 to 6 dB over the fixed-mode schemes at BPS values of 1, 2 and 4. This gain was even more remarkable for the Rural Area channel and it was in the range of 11 to 14 dB. The

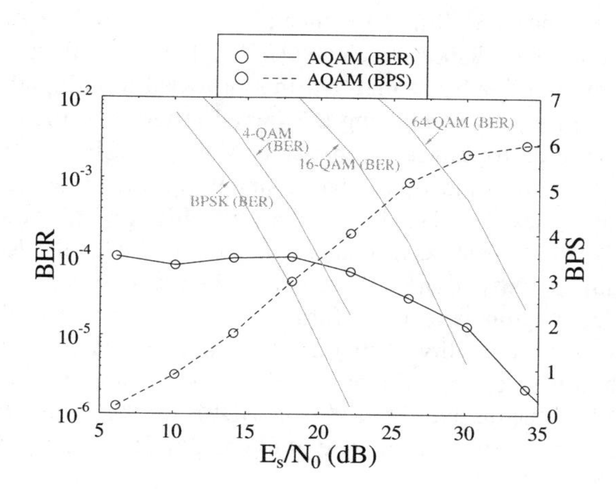

Figure 4.20: BER and BPS performance of a five-mode AQAM/JD-CDMA system for $K = 2$ users over the **four-path Typical Urban channel** of Figure 3.18. The transmission modes were switched between no transmission (NOTX), BPSK, 4-QAM, 16-QAM and 64-QAM. The target BER was set to 0.01% and the switching thresholds were optimized, yielding $t_0 = 8.25$ dB, $t_1 = 10.46$ dB, $t_2 = 16.8$ dB, and $t_3 = 23.76$ dB. The rest of the simulation parameters are listed in Table 4.3.

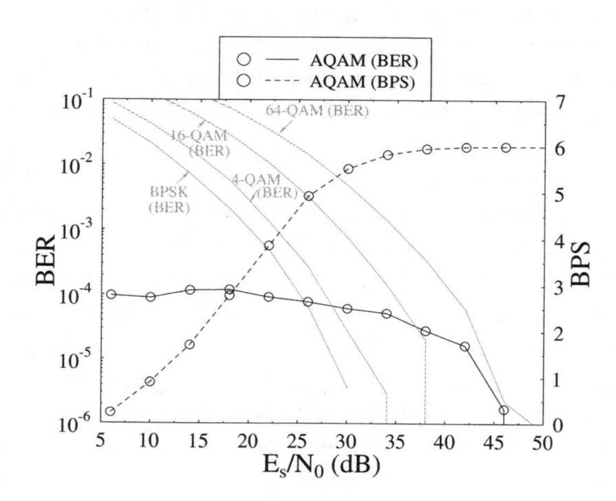

Figure 4.21: BER and BPS performance of a five-mode AQAM/JD-CDMA system for $K = 2$ users over the **two-path Rural Area channel** of Figure 3.15. The transmission modes were switched between no transmission (NOTX), BPSK, 4-QAM, 16-QAM and 64-QAM. The target BER was set to 0.01% and the switching thresholds were optimized, yielding $t_0 = 8.25$ dB, $t_1 = 10.46$ dB, $t_2 = 16.8$ dB, and $t_3 = 23.76$ dB. The rest of the simulation parameters are listed in Table 4.3.

reason for the improvement in SNR performance gain was due to the reduction in multipath diversity provided by the less dispersive channels. For the Bad Urban channel, there were seven CIR paths in the multipath channel and this provided a multipath diversity gain for the joint detection algorithm, thus reducing the spread in terms of the output SINR for the MMSE-BDFE. From the performance comparisons of the MMSE-BDFE in Figure 3.22 for the Rural Area, Typical Urban and Bad Urban channels, it was observed that the best BER performance was achieved over the Bad Urban channel, which provided the highest multipath diversity. Conversely, the lowest performance was obtained over the Rural Area channel, which had a multipath diversity in the order of two. The diversity gain provided by the Bad Urban channel enabled the joint detector to further compensate for the fading in the channel, but it also led to a lower probability of modulation mode switching in the AQAM scheme and the AQAM system was less capable of exploiting the variability of the mobile channel. The PDFs of the SINR at the output of the MMSE-BDFE for the three different channels are shown in Figure 4.22. The mean and variance of the output SINR at $E_s/N_0 = 14$ dB for the three channels are tabulated as:

Channel type	SINR Mean	SINR Standard Deviation
Rural Area (two-path)	14 dB	13.7 dB
Typical Urban (four-path)	14 dB	12.9 dB
Bad Urban (seven-path)	14 dB	11.1 dB

From the PDFs of Figure 4.22 as well as from the SINR standard deviation values, it can be observed that a high order of multipath diversity provides a lower standard deviation, thus leading to a lower SNR performance gain for the AQAM scheme. The AQAM scheme was capable of achieving higher performance gains in mobile channels exhibiting lower orders of multipath diversity, since it could compensate for the lack of multipath diversity. This enabled the AQAM/JD-CDMA system to achieve a certain BPS throughput at similar E_s/N_0 values for different channels with varying orders of multipath diversity. For example, in the five-mode AQAM scheme, the E_s/N_0 required to achieve a BER of 0.01% and the BPS throughput of 2 BPS is approximately 15 dB for the Rural Area (two-paths) channel of Figure 3.15 and for the Typical Urban (four-paths) channel of Figure 3.18. For attaining the same BER and BPS performance, the required E_s/N_0 value was approximately 14 dB for the Bad Urban (seven-paths) channel of Figure 3.20. In the fixed-mode 4-QAM scheme the E_s/N_0 values required to achieve the BER of 0.01% are 28 dB, 21 dB and 16.5 dB over the Rural Area, Typical Urban and Bad Urban channels, respectively. Figure 4.23 shows the SNR performance gain in dB for the AQAM scheme over the fixed-mode schemes versus the channel diversity order. Explicitly, the SNR performance gain is obtained by calculating the difference in E_s/N_0 required to achieve the same BPS throughput for a BER of 0.01% over the CIRs of Figures 3.15, 3.18 and 3.20.

The five-mode AQAM investigations were repeated for a higher system load of $K = 8$ users and the performance comparisons between two and eight users are shown in Figures 4.24, 4.25 and 4.26 for the three different channels. From these figures, it can be observed that increasing the load in the system from 2 users to 8 users caused only a slight degradation in throughput performance due to the capability of the JD receiver to mitigate MAI.

Having studied the upper-bound performance of AQAM/JD-CDMA by assuming perfect SINR estimation, let us now investigate the effects of SINR estimation errors.

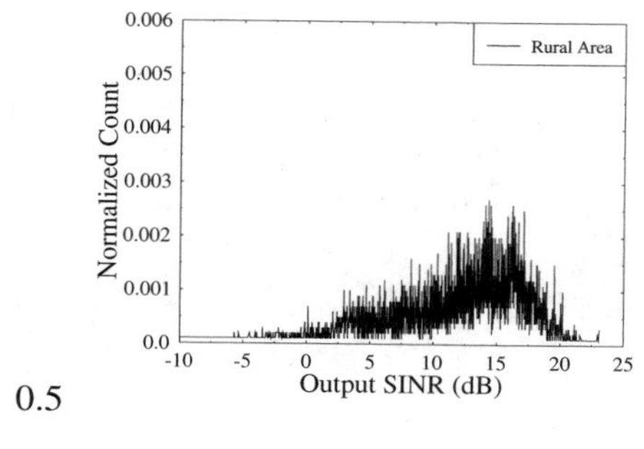

(a) Rural Area channel of Figure 3.15

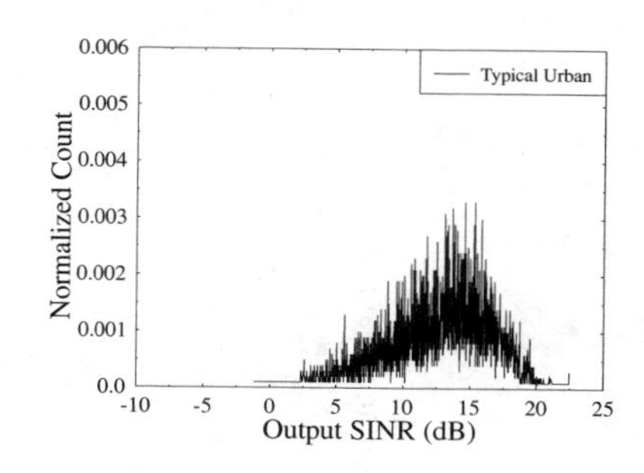

(b) Typical Urban channel of Figure 3.18

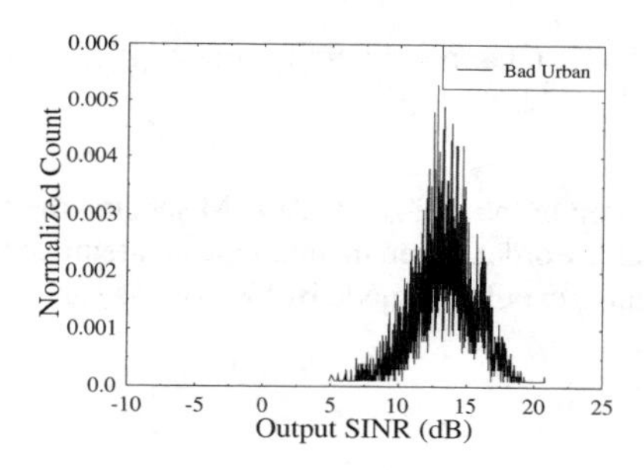

(c) Bad Urban channel of Figure 3.20

Figure 4.22: PDF of the SINR at the output of the MMSE-BDFE for CIRs of varying orders of multipath diversity and $E_s/N_0 = 14$ dB. The rest of the simulation parameters are listed in Table 4.3.

4.3.11 Effect of Imperfect SINR Estimation on AQAM Performance

In all of our previous simulations perfect SINR estimation was assumed. However, in realistic scenarios perfect SINR estimation is impractical because the instantaneous CIR used in the SINR estimation is different from the CIR in the next transmission burst, resulting in an SINR estimation latency. Therefore, an investigation into the effects of imperfect SINR estimation on the AQAM system was conducted and the results are presented in this section. Instead of assuming perfect SINR estimation, the output SINR is calculated at the output of the joint detection receiver using the spreading sequences and the CIR estimates of the received burst. This SINR estimate is then passed on to the transmitter and used as the modem mode selection criterion for the next frame to be transmitted. The accuracy of the SINR estimate is dependent upon the rate at which the mobile channel varies. If the channel variations are slow, then this imperfect SINR estimate will have a minor impact on the selection of modulation modes. Conversely, if the channel varies rapidly, then the SINR estimates become more obsolete by the time they are invoked for modem mode selection during the next transmission burst, causing a performance degradation.

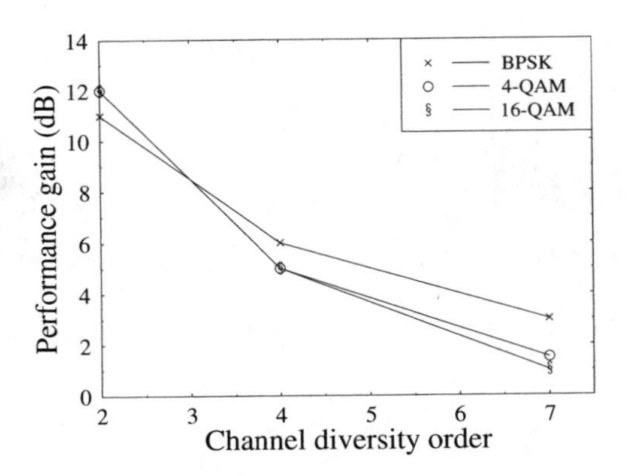

Figure 4.23: SNR performance gain in the five-mode AQAM scheme over fixed-mode QAM schemes versus channel diversity order, when maintaining an identical BER performance of 0.01%. The performance gains over fixed-mode BPSK, 4-QAM and 16-QAM schemes are shown.

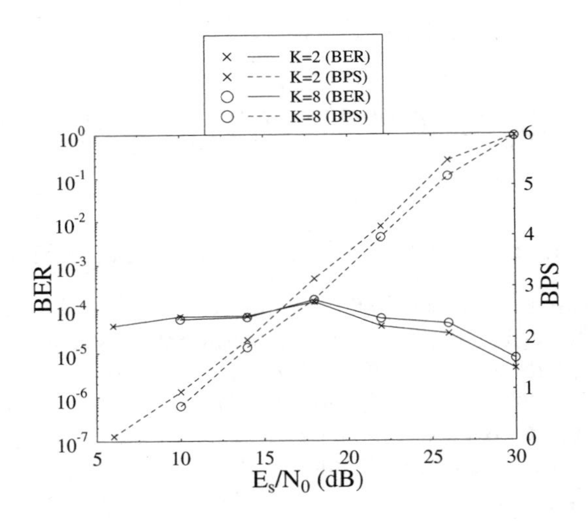

Figure 4.24: BER and BPS performance comparison of the five-mode AQAM scheme for $K = 2$ and $K = 8$ users over the seven-path **Bad Urban channel** of Figure 3.20. The rest of the simulation parameters are listed in Table 4.3 and the switching thresholds were set to $t_0 = 8.25$ dB, $t_1 = 10.46$ dB, $t_2 = 16.8$ dB, and $t_3 = 23.76$ dB.

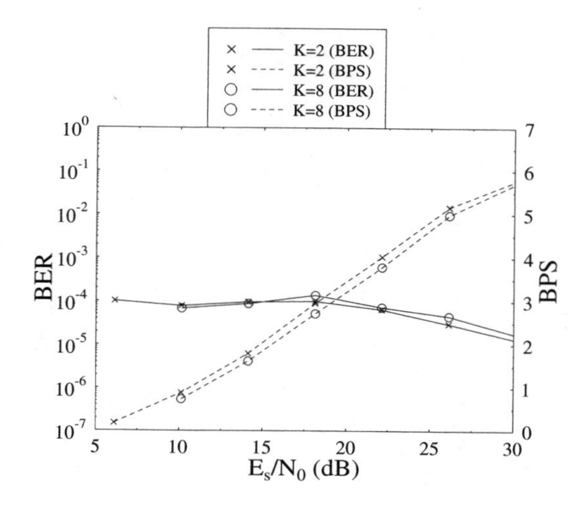

Figure 4.25: BER and BPS performance comparison of the five-mode AQAM scheme for $K = 2$ and $K = 8$ users over the **Typical Urban channel** of Figure 3.18. The rest of the simulation parameters are listed in Table 4.3 and the switching thresholds were set to $t_0 = 8.25$ dB, $t_1 = 10.46$ dB, $t_2 = 16.8$ dB, and $t_3 = 23.76$ dB.

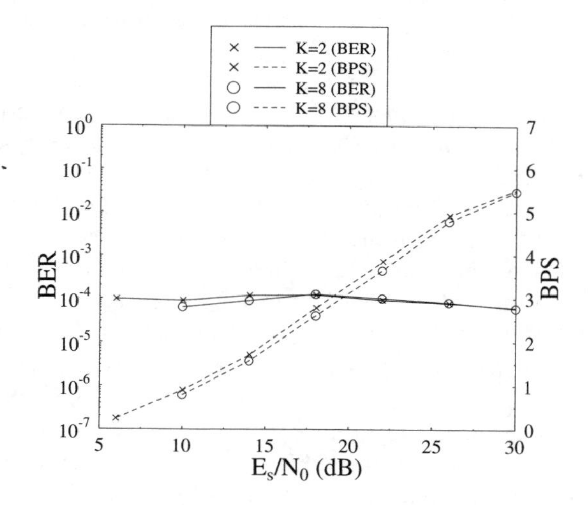

Figure 4.26: BER and BPS performance comparison of the five-mode AQAM scheme for $K = 2$ and $K = 8$ users over the **Rural Area channel** of Figure 3.15. The rest of the simulation parameters are listed in Table 4.3 and the switching thresholds were set to $t_0 = 8.25$ dB, $t_1 = 10.46$ dB, $t_2 = 16.8$ dB, and $t_3 = 23.76$ dB.

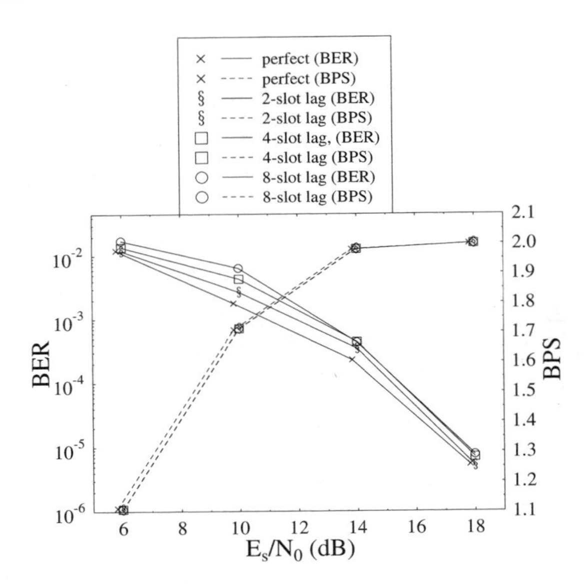

Figure 4.27: BER and throughput performance comparisons for a twin-mode AQAM system using perfect and imperfect SINR estimation over the seven-path Bad Urban channel of Figure 3.20. The Doppler frequency was set at **80 Hz**, and the performance was compared for latency values of two time slots (1.15 ms), four time slots (2.3 ms) and eight time slots (4.6 ms). The switching thresholds for all the performance curves were kept at the same value of $t_1 = 8$ dB and the rest of the simulation parameters are listed in Table 4.3.

Figure 4.27 shows the effects of imperfect SINR estimation on a twin-mode AQAM/JD-CDMA system that switches between BPSK and 4-QAM, where the Doppler frequency is 80 Hz. In our simulations, the TDMA frame structure of the FRAMES Mode 1 proposal [205] was used, where there were eight transmission time slots per TDMA frame and each time slot lasted for a period of 577 μs. Considering a worst-case scenario, where a user transmits in one time slot per TDMA frame, the latency in SINR estimation is eight time slots. The performance curves for various degrees of latency are compared in Figure 4.27, where the latency is varied from two time slots (1.154 ms) to a maximum of eight time slots (4.6 ms).

Comparing the perfect and imperfect SINR estimation results in Figure 4.27, we can see that for the same switching threshold of $t_1 = 8$ dB in both systems, the imperfect SINR estimation results in a performance degradation, as expected. As the latency was increased from two time slots to eight time slots, the BER performance suffered an increasing degradation, while the BPS throughput remained more or less the same. The BPS throughput was unaffected by the latency, since the threshold was kept at $t_1 = 8$ dB for all the simulations, since the output SINR values remained 'statistically speaking' constant, resulting in the same AQAM mode choices averaged over time. However, the degradation in BER performance was caused by the inaccurate choice of AQAM mode and was particularly severe, when the channel was hostile, exhibiting a low E_s/N_0.

In an effort to improve the performance, a higher, more conservative switching threshold

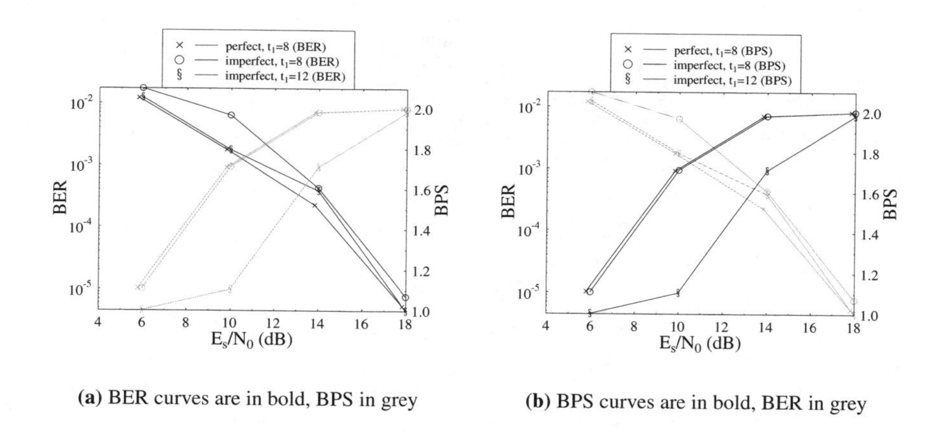

(a) BER curves are in bold, BPS in grey

(b) BPS curves are in bold, BER in grey

Figure 4.28: BER and throughput performance comparisons for a twin-mode AQAM system assisted by perfect and imperfect SINR estimation over the seven-path Bad Urban channel of Figure 3.20 at a Doppler frequency of **80 Hz**. The latency is eight time slots (4.6 ms). The switching threshold was changed from $t_1 = 8$ dB to 12 dB and the rest of the simulation parameters are listed in Table 4.3.

of $t_1 = 12$ dB was tested, in order to ascertain its effect on the imperfect system. The performance of $t_1 = 8$ dB and $t_1 = 12$ dB for a latency of 8 time slots was plotted in Figure 4.28. When t_1 was increased to 12 dB, the BER performance improved, eventually matching the performance of the perfect SINR estimation-based system. However, the increase in t_1 caused a corresponding decrease in throughput, and for $t_1 = 12$ dB the throughput achieved was approximately 1.7 bits per symbol at $E_s/N_0 = 14$ dB, compared with approximately 2 BPS for $t_1 = 8$ dB.

The effect of the channel fading rate on the accuracy of the SINR estimation was also examined by reducing the Doppler frequency to 26.7 Hz. For the same switching threshold of $t_1 = 8$ dB, the effect of latency on this more slowly fading scenario was investigated by varying the latency from four time slots to eight time slots. Figure 4.29 shows the performance comparisons of the perfect and imperfect SINR estimation based systems for the slower fading channel. From the figure, we can see that the latencies of four and eight time slots did not inflict any degradation in BER performance, since the fading rate was sufficiently slow, such that the AQAM scheme was capable of compensating for the inaccuracies in SINR estimation.

The effects of imperfect SINR estimation were also quantified for the triple-mode AQAM system, where the modulation modes were BPSK, 4-QAM or 16-QAM. Here again, the higher fading rate with a Doppler frequency of 80 Hz was employed and the latency values investigated were a lag of two time slots (1.15 ms), four time slots (2.3 ms) and eight time slots (4.6 ms). The performance results in Figure 4.30 show the same trends as for the twin-mode AQAM system, where the BER performance suffered increasing degradation as the latency values were increased. In an attempt to achieve the target BER of 1%, the switching thresholds were increased to $t_1 = 10$ dB and $t_2 = 14$ dB from the original set of $t_1 = 8$ dB and $t_2 = 12$ dB. The results of these investigations are shown in Figure 4.31. Upon using

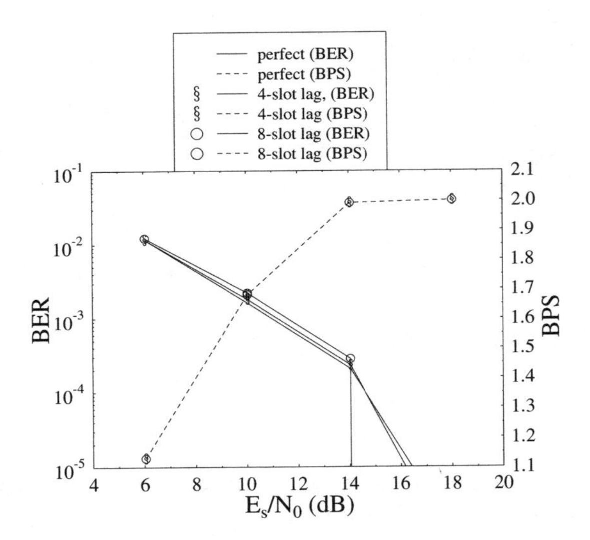

Figure 4.29: BER and throughput performance comparisons for a twin-mode AQAM system using perfect and imperfect SINR estimation over the seven-path Bad Urban channel of Figure 3.20 at a lower Doppler frequency of **26.7 Hz**. The performance is compared for latency values of four time slots (2.3 ms) and eight timeslots (4.6 ms). The switching threshold for all the performance curves was kept at the same value of $t_1 = 8$ dB and the rest of the simulation parameters are listed in Table 4.3.

the higher thresholds, the AQAM system was capable of maintaining the target BER of 1% at the expense of a degradation in BPS throughput performance.

The performance of the triple-mode AQAM system in a more slowly fading channel was also investigated and the results shown in Figure 4.32 are the associated performance curves for a Doppler frequency of 26.7 Hz. Again, it can be observed that in this scenario the latency in SINR estimation did not have a significant effect on the BER or BPS throughput performance, since the channel quality fluctuated less rapidly and the AQAM scheme was capable of coping with less obsolete SINR estimates.

It can be seen from the results presented in this section that the AQAM scheme is more suited for employment in a slowly fading channel. Although the AQAM scheme can also be used in a channel exhibiting a higher fading rate, the switching thresholds have to be increased, in order to maintain a target BER and this causes a degradation in BPS throughput. However, the AQAM scheme is still capable of outperforming the fixed-mode QAM schemes. Another method that can be used to improve the accuracy of the SINR estimation is the employment of channel prediction techniques. Recent research into mobile channel modelling and adaptive long-range channel prediction by Eyceoz, Duel-Hallen and Hallen [239] have shown that the mobile channel can be reliably predicted. The predicted CIR can then be used for SINR estimation.

Let us now summarize our findings for AQAM/JD-CDMA in the next section.

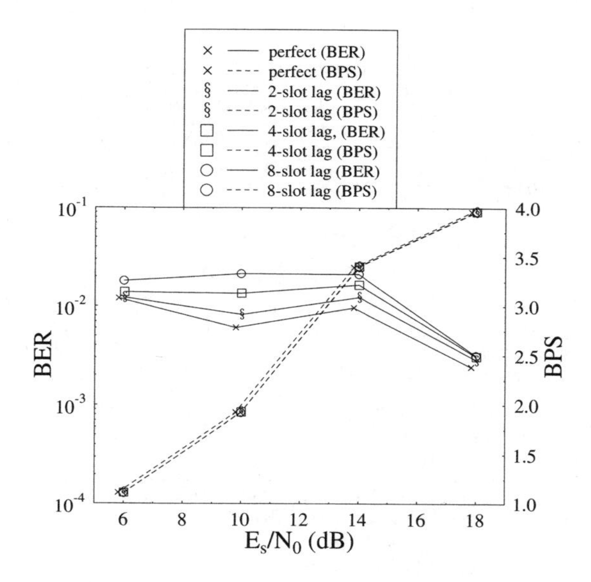

Figure 4.30: BER and throughput performance comparisons for a triple-mode AQAM system invoking perfect and imperfect SINR estimation over the seven-path Bad Urban channel of Figure 3.20 at a Doppler frequency of **80 Hz**. The performance is compared for latency values of two time slots (1.15 ms), four time slots (2.3 ms) and eight time slots (4.6 ms). The switching thresholds for all the performance curves were kept at the same value of $t_1 = 8$ dB and $t_2 = 12$ dB and the rest of the simulation parameters are listed in Table 4.3.

4.3.12 Discussion

AQAM provides a simple technique for increasing the throughput of the system by exploiting the variability of the mobile channel quality. The results presented in the previous sections have demonstrated that combining AQAM with JD-CDMA provides substantial gains in terms of both BER and BPS throughput performance terms. The availability of the various modulation modes and transmission blocking capabilities allow the system to choose the modulation mode that constitutes the best trade-off in terms of BER and BPS performance based on the instantaneous channel conditions. The system is capable of maximizing the throughput, when the channel is of high quality and to reduce the number of errors, when the channel is hostile. The AQAM scheme is capable of achieving higher performance gains in a mobile channel having a lower order of multipath diversity. This allows the joint detector to compensate for a lack of multipath diversity, thus enabling a certain BPS throughput to be achieved at the same E_s/N_0 value for different channels having varying orders of multipath diversity. The proposed AQAM/JD-CDMA scheme exhibits a high flexibility always attaining the highest possible BPS throughput that can be supported under the constraint of the prevalent channel quality and target BER. Due to the modem-mode switching regime the transceiver exhibits a less bursty error distribution than fixed-mode modems and hence AQAM/JD-CDMA may attain higher channel coding gain than fixed-mode transceivers.

As stated previously, the joint detection algorithm does not require any knowledge of the

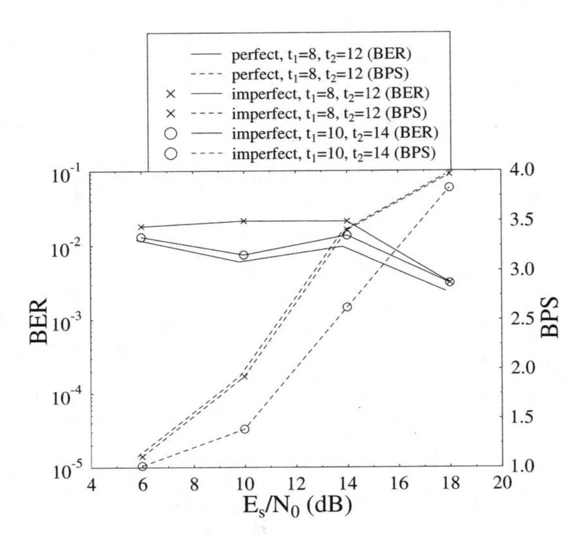

Figure 4.31: BER and throughput performance comparisons for a triple-mode AQAM system assisted by perfect and imperfect SINR estimation over the seven-path Bad Urban channel of Figure 3.20 at a Doppler frequency of **80 Hz**. The symbols t_1 and t_2 represent the switching thresholds from BPSK to 4-QAM and from 4-QAM to 16-QAM respectively and the rest of the simulation parameters are listed in Table 4.3.

modulation mode used, which makes it suitable for combination with AQAM. No changes have to be made to the joint detection receiver, in order to accommodate the time-variant modulation modes and this allows the complexity of the joint detection receiver to remain constant, irrespective of the modulation modes encountered.

One disadvantage of this system is that it requires channel estimates for future frames. In a slowly-fading channel, the channel estimation is capable of tracking the channel fluctuations fairly closely. However, a slowly varying channel also implies that the channel will stay in a certain state for relatively long periods of time and this may impose unacceptable delays upon interactive speech or video communications, when the channel is unfavourable. This situation is partly mitigated in a multipath channel, since CDMA is capable of utilizing the multipaths to provide multipath diversity gain, thus alleviating the problem. Another disadvantage of AQAM/JD-CDMA is that the throughput gains could be slightly reduced, since the modulation mode used has to be signalled from the transmitter to the receiver. Another approach to this problem would be for the receiver to detect the modulation mode blindly [232, 233], resulting in an increase in the complexity of the system.

As a comparative basis for our AQAM/JD-CDMA systems, let us consider the family of variable spreading factor-based adaptive CDMA systems in the next section.

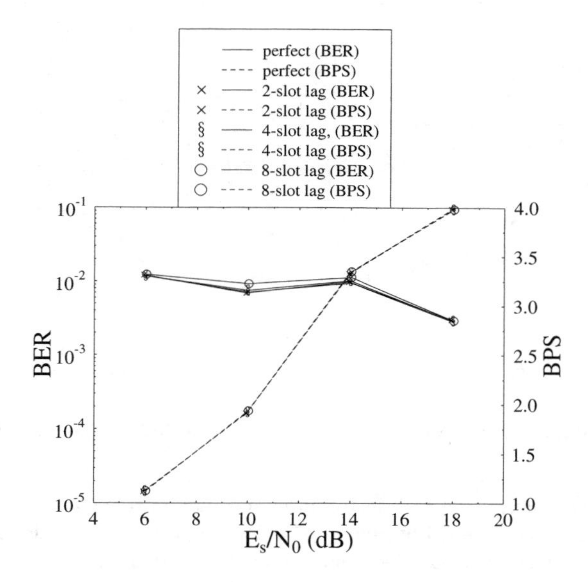

Figure 4.32: BER and throughput performance comparisons for a triple-mode AQAM system employing perfect and imperfect SINR estimation over the seven-path Bad Urban channel of Figure 3.20 at a lower Doppler frequency of **26.7 Hz**. The performance is compared for latency values of two time slots (1.15 ms), four time slots (2.3 ms) and eight time slots (4.6 ms). The switching thresholds for all the performance curves were kept at the same values of $t_1 = 8$ dB and $t_2 = 12$ dB and the rest of the simulation parameters are listed in Table 4.3.

4.4 Variable Spreading Factor-Based Adaptive-Rate CDMA

Multi-rate transmission systems using spreading sequences having different processing gains have been proposed in the literature among others by Adachi, Sawahashi and Okawa [240]; Ottosson and Svensson [97]; Ramakrishna and Holtzman [98]; Saquib and Yates [221]; as well as by Johansson and Svensson [241]. In the FRAMES FMA2 Wideband CDMA proposal for UMTS [242], different bit rates are accommodated by supporting variable spreading factor (VSF) [240] and multi-code-based operations. In this section, we discuss the employment of VSF codes in adaptive-rate CDMA systems. In our proposed VSF system, the chip rate of the CDMA users is kept constant throughout the transmission, while the bit rate is varied by using spreading codes exhibiting different spreading factors over the course of transmission. For example, by keeping the chip rate constant, the number of bits transmitted in the same period of time upon using a spreading code of length $Q = 16$ is twice the number of bits transmitted upon using a spreading code of length $Q = 32$.

Kim [217] demonstrated that rate adaptation provided a higher average information rate, than power adaptation for a given average transmit power and a given BER. Therefore, we employ here a VSF scheme in order to provide rate adaptation. In the VSF system the chip rate of the CDMA users is kept constant throughout the transmission. The bit rate is varied by using spreading codes having different spreading factors over the course of transmission.

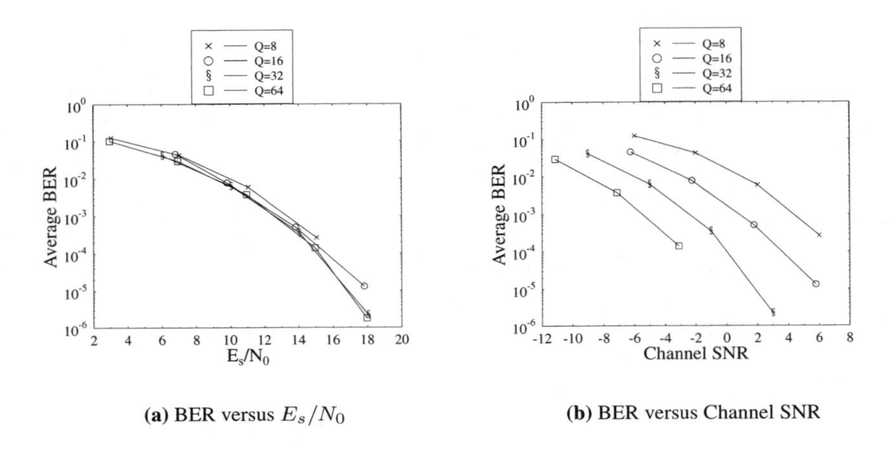

(a) BER versus E_s/N_0 **(b)** BER versus Channel SNR

Figure 4.33: BER performance comparisons for different values of spreading factor, Q over the seven-path Bad Urban channel of Figure 3.20. Figure (a) plots the BER performance comparisons with respect to E_s/N_0, while the curves in figure (b) are plotted with respect to channel SNR, where channel SNR $\times\ Q = E_s/N_0$.

When the channel quality is high, a low Q is used in order to increase the throughput and conversely, when the channel conditions are hostile, a high Q is employed for maintaining a target BER performance. Figure 4.33(a) shows the BER versus E_s/N_0 curves for 4-QAM JD-CDMA transmissions in conjunction with different spreading factors (SF), represented by the notation Q. These comparisons imply that the length of the spreading sequence, or the value of Q, does not affect the BER performance of the joint detector. Therefore, no performance gain is obtained when comparing the adaptive-rate scheme to a fixed-rate scheme. However, if we replot these BER curves with respect to channel SNR (CSNR) instead of E_s/N_0, where CSNR $\times\ Q = E_s/N_0$, then the difference in BER performance can be observed due to the different values of Q. This is shown in Figure 4.33(b). Therefore, in order to accommodate or indeed to exploit the channel fluctuations, when Q is varied, the E_s/N_0 is also varied accordingly while maintaining a constant CSNR by using the expression $E_s/N_0 = \text{CSNR} \times Q$. When the channel quality is high and a low value of Q is employed, the transmitted power – which is proportional to E_s – is also reduced. For example, let us consider a triple-mode VSF scheme at $E_s/N_0 = 10$ dB, where Q is switched between $Q = 64$, 32 and 16. Of all the three modes, the CSNR is lowest when $Q = 64$ is employed, since CSNR $= 10$ dB $- 10\log_{10}(64)$ dB ≈ -8 dB. When $Q = 32$ is employed, the CSNR is now approximately -5 dB. However, in order to maintain the CSNR at -8 dB, the E_s is multiplied by a factor of $1/f$, such that the resultant CSNR value becomes -8 dB. For example, for $Q = 32$, the factor f has the value of 2, such that we have again CSNR$' = -5$ dB $- 10\log_{10} 2 \approx -8$ dB. In the VSF schemes, the employment of the mode associated with the highest spreading factor provides the lowest possible value of CSNR, and hence requires the lowest value of transmitted power, i.e. the lowest E_s value. Therefore, in our VSF/JD-CDMA scheme, the CSNR is always maintained at the value associated with the highest spreading factor. In short, the VSF scheme is exploiting the variability in mobile channel quality by adapting

Switching criterion	Spreading code
$\gamma_o(k) < t_1$	$\mathbf{c}_1^{(k)}$, Q_h
$t_1 \leqslant \gamma_o(k) < t_2$	$\mathbf{c}_2^{(k)}$, $2^{-1}Q_h$
.	.
.	.
.	.
$t_M \leqslant \gamma_o(k)$	$\mathbf{c}_M^{(k)}$, $2^{-(M-1)}Q_h$

Table 4.8: The general switching rules for a VSF/JD-CDMA system, where $\gamma_o(k)$ is the SINR of the k-th user at the output of the MMSE-BDFE. The values of $t_1, \cdots, t_M$ represent the switching thresholds for the different modulation modes, where $t_1 < t_2 < \cdots < t_M$.

the spreading factor according to channel quality without increasing the transmitted power above that of the lowest possible value associated with the highest spreading factor. When the spreading factor is decreased due to an improved channel quality, E_s is proportionally reduced.

4.4.1 System Model

Each user in the VSF-CDMA system is assigned M number of legitimate spreading codes having different lengths, $Q_1, Q_2, \dots, Q_M$. Analogously to the AQAM/JD-CDMA system proposed in Section 4.3, the SINR at the output of the MMSE-BDFE will be estimated and used as a criterion for choosing the spreading code to be used for transmission. Generally, when the channel quality is favourable, a code with a low spreading factor will be used, in order to increase the throughput. Conversely, when the channel is hostile, a code with a high spreading factor will be used, in order to minimize the number of errors inflicted. Since the BER performance at the output of the joint detector is independent of the spreading sequence length as we have seen in Figure 4.33, the spreading sequence with the highest spreading factor will always be used in the SINR calculations.

In order to have a system that can accommodate a large number of spreading codes having different lengths, the simplest choice of spreading codes would be those with lengths of $Q = 2^r$, where $r = 1, 2, 3, \cdots$. Let us denote the set of spreading codes assigned to the k-th user by $\{\mathbf{c}_1^{(k)}, \mathbf{c}_2^{(k)}, \cdots, \mathbf{c}_M^{(k)}\}$, where $\mathbf{c}_1^{(k)}$ is the spreading code having the highest spreading factor and the code $\mathbf{c}_M^{(k)}$ has the lowest spreading factor. If the highest spreading factor is denoted by Q_h, then the rules used to choose the spreading code are tabulated in Table 4.8. In the proposed VSF systems, each user maintains a constant level of total transmission power in order to ensure that the interference seen by other users remains relatively constant.

4.4.2 Simulation Results

The performance of a triple-mode VSF/JD-CDMA system was evaluated using the simulation parameters listed in Table 4.9. By employing the spread burst 1 format of FMA1 [205], where the number of chips used for transmission is 896 chips and the chip rate is 2.167 Mchips/s, the total period of data transmission is approximately 413 μs. When the chosen modulation

Parameter	Value
Doppler frequency	80 Hz
Chip rate	2.167 MBaud
Frame burst structure	FRAMES Mode 1 Spread burst 1 [205]
Frame duration	4.615 ms
Burst duration	577 μs
Data modulation mode	4-QAM
Receiver type	MMSE-BDFE

Table 4.9: Simulation parameters for VSF/JD-CDMA systems

Switching criterion	Spreading code
$\gamma_o(k) < t_1$	$Q_1 = 64$, $f_1 = 1$
$t_1 \leqslant \gamma_o(k) < t_2$	$Q_2 = 32$, $f_2 = 2$
$t_2 \leqslant \gamma_o(k)$	$Q_3 = 16$, $f_3 = 4$

Table 4.10: The switching rules for a triple-mode VSF/JD-CDMA system employing 4-QAM, where $\gamma_o(k)$ represents the SINR at the output of the MMSE-BDFE for the k-th user. The values t_1 and t_2 represent the switching thresholds, where $t_1 < t_2$. The symbols Q_1, Q_2 and Q_3 are the spreading factors for the three modes and f_1, f_2 and f_3 are the corresponding power factors that determine the E_s/N_0 value for each mode.

mode is 4-QAM and the highest spreading factor of $Q = 64$ is used, the total number of 4-QAM symbols transmitted is $896/64 = 14$. Since each 4-QAM symbol represents two bits, this gives the minimum throughput value of approximately $[14 \times 2]/[413 \times 10^{-6}] = 68$ kbits/s. Similarly, the maximum throughput is approximately $[(896/16) \times 2]/[413 \times 10^{-6}] =$ 271 kbits/s, when the spreading factor of $Q = 16$ is used. Since the wideband channel may destroy the orthogonality of the spreading codes, the benefit of using spreading codes exhibiting good correlation properties often erodes. Hence we opted for using pseudo-random spreading sequences. The SINR at the output of the MMSE-BDFE, $\gamma_o(k)$, was estimated at the receiver, and it was used to choose the spreading code most suitable for transmission, given the current interference and channel conditions. The chosen spreading code was then used by the transmitter. The rules used to switch the spreading codes are tabulated in Table 4.10.

Different switching thresholds were evaluated, in order to determine their effect on the performance of the system. The transmission rate was varied adaptively by switching between the spreading factors of $Q_1 = 64$, $Q_2 = 32$ and $Q_3 = 16$. Two different sets of threshold values were compared with the aim of achieving an average BER of 1%. Specifically, for System A we employed $t_1 = 10$ dB, $t_2 = 13$ dB; and for System B: $t_1 = 8$ dB, $t_2 = 11$ dB. The performance curves versus average E_s/N_0 of these two VSF schemes are presented in Figure 4.34. In this figure, the throughput performance is normalized to the minimum throughput of 68 kbits/s. Therefore, the right y-axis denotes the normalized throughput values in the range of 1 to 4.

At low E_s/N_0 values, the $Q_1 = 64$-chip spreading code was chosen predominantly,

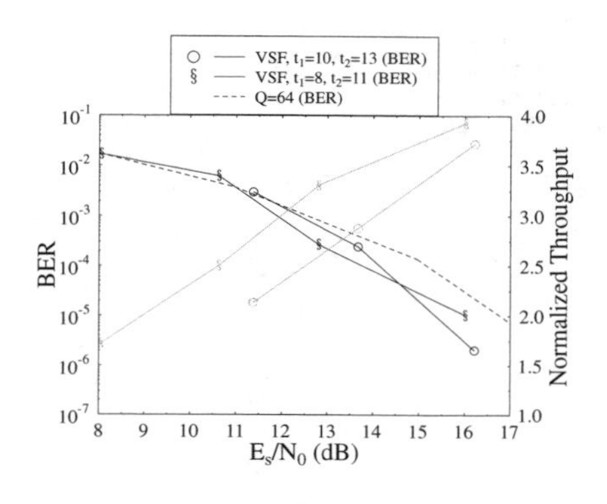

(a) BER is in bold, throughput is in grey

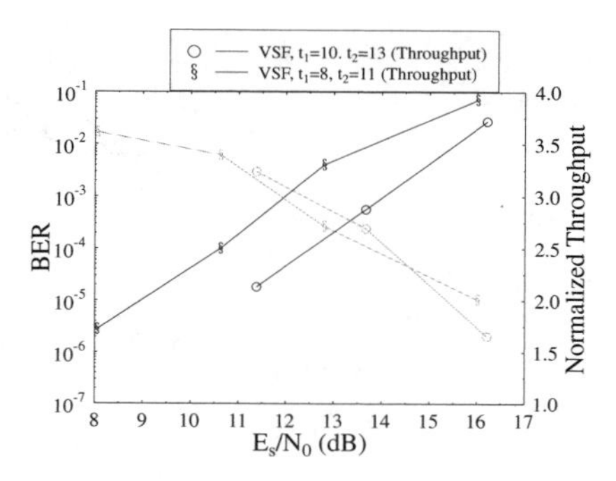

(b) Throughput is in bold, BER is in grey

Figure 4.34: The average BER and throughput performance versus average E_s/N_0 of a VSF/JD-CDMA system for $K = 2$ users over the **Bad Urban seven-path channel** of Figure 3.20. The rest of the simulation parameters are listed in Table 4.9 and perfect SINR estimation was assumed. The transmission rate was varied adaptively by switching between codes having different spreading factors of $Q_1 = 64$, $Q_2 = 32$ and $Q_3 = 16$. The switching thresholds used are shown in the legends. The throughput values were normalized to the minimum throughput of 68 kbits/s, which was also the throughput of the fixed-rate $Q = 64$-chip system.

causing the BER to mimic the performance of the fixed-rate, $Q = 64$-chip system at the same E_s/N_0 values. As the transmission power was increased, the spreading codes having the lower spreading factors of $Q_2 = 32$ and $Q_3 = 16$ were chosen with a higher probability, thus improving the throughput performance. At the same time, with the increasing use of these lower spreading factors, the average E_s/N_0 values were also decreasing due to the constraint of a constant, lowest transmitted power. Therefore, as the throughput performance improved, the BER performance also began to diverge from the BER performance of the $Q = 64$-chip fixed spreading factor scheme. From the figure, it can be observed that for System A, at a BER of 0.01%, there was a performance gain of 1 dB with respect to the $Q = 64$-chip fixed-rate benchmarker. This gain increased to approximately 2 dB at a BER of 0.001% and this gain continued to improve as the SNR was increased. The VSF/JD-CDMA system was capable of providing a throughput equivalent to the $Q_2 = 32$ and $Q_3 = 16$ fixed-rate schemes as E_s/N_0 increased, while using the lowest transmitted power, namely that of the $Q_1 = 64$ scheme. This was because the adaptive system had the option of choosing a higher spreading factor, when the channel was hostile, thus lowering the number of errors in the burst. Conversely, when the channel quality was high, the lower spreading factors could be chosen, thus increasing the total number of bits transmitted and thereby further lowering the average BER.

The performance of the triple-mode VSF scheme was also investigated over the Typical Urban four- path channel of Figure 3.18 and the Rural Area two-path channel of Figure 3.15. The results of these investigations are plotted in Figures 4.35 and 4.36 for the Typical Urban and Rural Area channels, respectively. It can be observed that the performance gain is higher

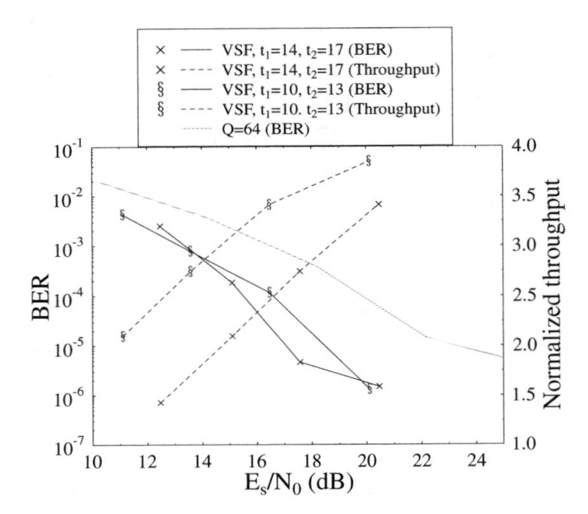

Figure 4.35: The average BER and throughput performance versus average E_s/N_0 of a VSF/JD-CDMA system for $K = 2$ users over the **Typical Urban four-path channel** of Figure 3.18. The rest of the simulation parameters are listed in Table 4.9 and perfect SINR estimation was assumed. The transmission rate was varied adaptively by switching between codes associated with different spreading factors of $Q_1 = 64$, $Q_2 = 32$ and $Q_3 = 16$. The switching thresholds used are shown in the legends. The throughput values were normalized to the minimum throughput of 68 kbits/s, which was also the rate of the fixed-mode $Q = 64$ system.

for channels exhibiting a lower order of diversity. For the Typical Urban channel, the SNR performance gain is approximately 3 dB at a BER of 0.01% and 4 dB at a BER of 0.001%. Similarly for the Rural Area channel, the SNR gains are approximately 7 and 8 dB for BERs of 0.01% and 0.001% respectively. At these BER values, the normalized throughput is in the range of 3.4 to 3.6 BPS, which is superior to the throughput of the fixed-rate $Q = 64$-chip spreading factor scheme.

Figure 4.37 shows the effect of increasing the number of users on the performance of the triple-mode VSF scheme. As can be observed from the figure, increasing the number of users from $K = 1$ to 8 did not result in a significant performance degradation in BER or throughput terms. The joint detection algorithm exploited the available knowledge on spreading sequences and CIRs, in order to mitigate the multiple access interference.

4.5 Comparison of AQAM/JD-CDMA and VSF/JD-CDMA

In the previous sections, we have invoked two different methods of adaptively varying the bit rate of the system, namely AQAM and VSF-based JD-CDMA. Figure 4.38 presents our performance comparison between the two systems. In order to make a fair comparison, the combination of spreading factors and modulation modes was chosen such that the minimum and

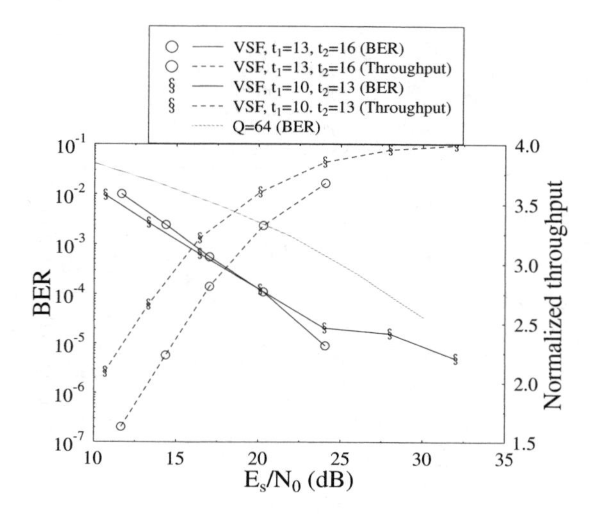

Figure 4.36: The average BER and throughput performance versus average E_s/N_0 of a VSF/JD-CDMA system for $K = 2$ users over the **Rural Area two-path channel** of Figure 3.15. The rest of the simulation parameters are listed in Table 4.9 and perfect SINR estimation was assumed. The transmission rate was varied adaptively by switching between codes having different spreading factors of $Q_1 = 64$, $Q_2 = 32$ and $Q_3 = 16$. The switching thresholds used are shown in the legends. The throughput values were normalized to the minimum throughput of 68 kbits/s, which was also the rate of the fixed-mode $Q = 64$ system.

maximum throughput of both systems were the same. Specifically, the chosen AQAM/JD-CDMA system was a triple-mode scheme, which switched modulation modes among BPSK, 4-QAM and 16-QAM, while the spreading sequences had a spreading factor of $Q = 32$. The VSF systems used a modulation mode of 4-QAM and had a choice of three spreading factors, namely $Q_1 = 64$, $Q_2 = 32$ and $Q_3 = 16$. Therefore, using the FMA1 CDMA spread burst 1 that has 896 chips per burst [205], the minimum and maximum number of bits transmitted per burst were 28 and 112, respectively, corresponding to the minimum and maximum throughput of 68 and 271 kbits/s. The switching SINR thresholds of $t_1 = 8$ dB and $t_2 = 12$ dB were chosen for the AQAM/JD-CDMA scheme in order to obtain a target BER of 1%. For the VSF systems, two different sets of thresholds were compared. A more conservative SINR threshold set of $t_1 = 10$ dB and $t_2 = 13$ dB, as well as a more aggressive threshold set of $t_1 = 8$ dB and $t_2 = 10$ dB. In order to provide a comparison of the throughput values, the throughput values of the three systems compared were normalized to the throughput of a fixed-rate system employing $Q = 64$ and 4-QAM as the modulation mode.

From the results presented in Figure 4.38, we can see that the AQAM system maintained a fairly constant BER of 1%, while the VSF systems showed a gradual drop in BER, as the average E_s/N_0 increased. The constant BER performance of the AQAM system was due to the use of 16-QAM, which yielded a high throughput, but exhibited an inferior BER performance due to the closer distance of the modem constellation points. For the VSF

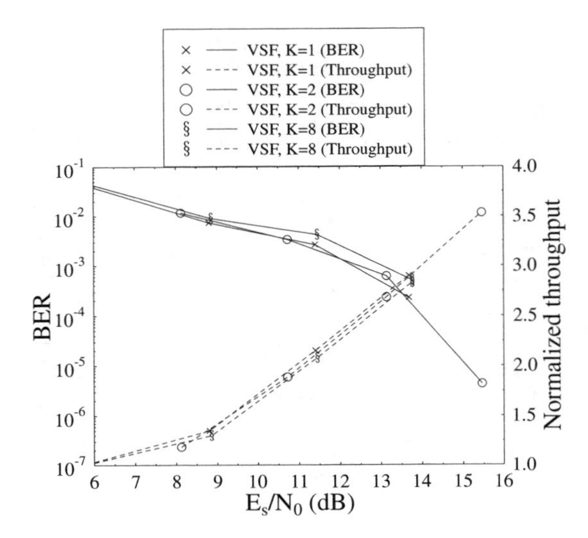

Figure 4.37: The average BER and throughput performance versus average E_s/N_0 of a VSF/JD-CDMA system for $K = 1, 2$ and 8 users over the Bad Urban seven-path channel of Figure 3.20. The rest of the simulation parameters are listed in Table 4.9 and perfect SINR estimation was assumed. The transmission rate was varied adaptively by switching between codes having different spreading factors of $Q_1 = 64$, $Q_2 = 32$ and $Q_3 = 16$. The switching thresholds used are shown in the legends. The throughput values were normalized to the minimum throughput of 68 kbits/s, which was also the rate of the fixed-mode $Q = 64$ system.

schemes, this gradual drop in BER was due to the fact that the BER performance of the joint detector is independent of the spreading factor Q. Therefore, as the E_s/N_0 increased and the lower spreading factors were used increasingly, the BER continued to follow the "waterfall-shaped" drop of the $Q = 32$ and $Q = 16$ fixed-rate BER curves. When comparing the associated throughput performance, the more aggressive VSF system resulted in a higher throughput than the AQAM system. The more conservative VSF/JD-CDMA system had a consistently lower throughput than both of the other systems. By comparing the aggressive VSF scheme and the AQAM scheme, it can be observed that the VSF scheme outperformed the AQAM in both BER and throughput terms.

In comparing the complexities of the three systems, let us use the complexity of the fixed-rate 4-QAM, $Q = 32$ scheme as a benchmark. The complexity is on the order of $O(KN)^3$ per transmitted burst for the Cholesky decomposition segment of the algorithm in Equation 3.160, where K represents the number of users and N represents the number of symbols per transmission burst. The complexity of the AQAM/JD-CDMA system was the same as that of the benchmark, since the choice of modulation mode does not have any effect on the complexity of the JD algorithm. The VSF schemes had the same complexity as the benchmark, when $Q_2 = 32$ was used but the complexity decreased by a factor of eight, when $Q_1 = 64$ was used and increased by a factor of eight when $Q_3 = 16$ was used. This was because the employment of $Q_1 = 64$ decreased the number of transmitted symbols, N,

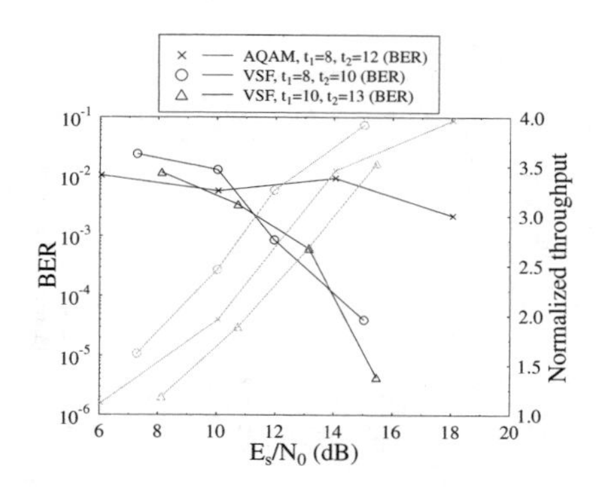

(a) BER comparisons are in bold, throughput comparisons are in grey.

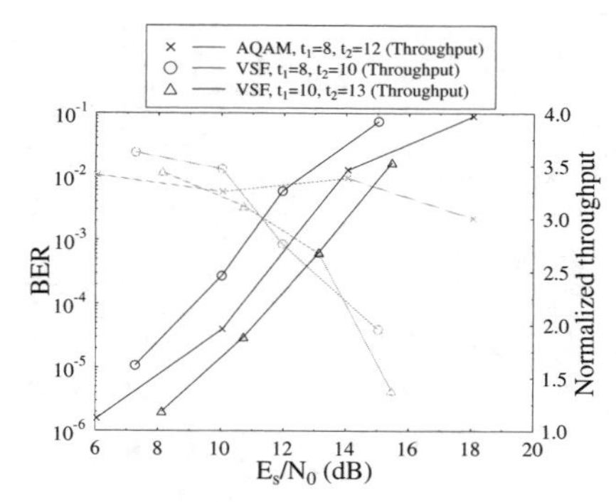

(b) Throughput comparisons are in bold, BER comparisons are in grey.

Figure 4.38: Comparison of the BER and throughput performance of **triple-mode VSF** and **triple-mode AQAM/JD-CDMA** for $K = 2$ users. In the AQAM system, the spreading sequences were of $Q = 32$ chips in length and the modulation mode was switched between BPSK, 4-QAM and 16-QAM. The two VSF systems used 4-QAM and chose the appropriate spreading factor from the set $Q_1 = 64$, $Q_2 = 32$ and $Q_3 = 16$. The throughput values were normalized to the minimum throughput of 68 kbits/s, which was also the rate of the fixed-mode $Q = 64$ system. The simulations were conducted over the seven-path Bad Urban channel model of Figure 3.20, where perfect SINR estimation was assumed. The rest of the simulation parameters are listed in Table 4.9.

by a factor of two, while the shorter spreading factor of $Q_3 = 16$ doubled the value of N. However, in the VSF/JD-CDMA systems, the configuration of the algorithm had to be altered whenever the spreading code of any of the users was changed, thus increasing the complexity.

By contrast, AQAM/JD-CDMA provides a relatively simple method of adaptively varying the bit rate, in order to exploit and accommodate the time-varying nature of the mobile channel without disadvantaging other users. The change in modulation mode does not affect the operation of the joint detection algorithm, thus there is no complexity penalty. However, the employment of higher order modulation modes such as 16-QAM increases the BER due to the reduced distance between the modulation constellation points. Hence, the employment of VSFs results in a better BER performance. Having compared the VSF and AQAM-based JD-CDMA systems, let us now attempt to combine their advantages.

4.6 Combining AQAM and VSF with JD-CDMA

In order to amalgamate the benefits of both AQAM and VSF in conjunction with JD-CDMA, a combined AQAM and VSF based adaptive-rate JD-CDMA system was simulated and its performance was analyzed. Ue, Sampei and Morinaga [243] carried out investigations on a combined symbol-rate and modulation-level controlled adaptive modulation system for TDMA/TDD schemes in narrowband channels. In our system, the SINR at the output of the MMSE-BDFE was estimated as before, for both the AQAM/JD-CDMA and the VSF/JD-

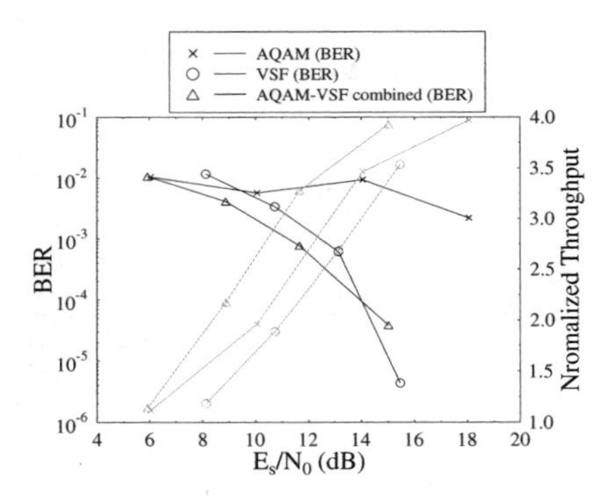

(a) BER comparisons are in bold, throughput comparisons are in grey.

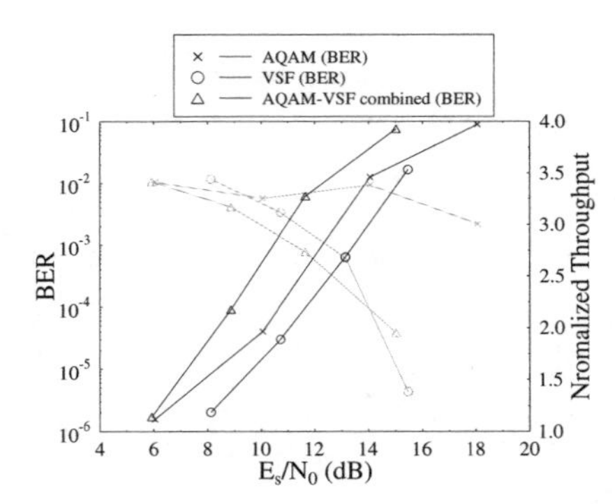

(b) Throughput comparisons are in bold, BER comparisons are in grey.

Figure 4.39: BER and throughput performance of a **combined AQAM and VSF adaptive-rate JD-CDMA system** over the seven-path Bad Urban channel model of Figure 3.20. The rest of the simulation parameters are listed in Table 4.9. The bit rate was varied adaptively by using both AQAM and a variable spreading factor. The modulation modes were switched between BPSK with $Q = 32$; 4-QAM with $Q = 32$; and 4-QAM with $Q = 16$. The switching thresholds were set to $t_1 = 8$ dB and $t_2 = 10$ dB. Also shown are the performance results for the triple-mode AQAM (BPSK, 4-QAM, 16-QAM) scheme employing $Q = 32$, and the triple-mode VSF scheme utilizing 4-QAM as the modulation mode. The throughput values were normalized to the minimum throughput of 68 kbits/s, which was also the throughput of the mode utilizing BPSK and the spreading factor of $Q = 32$.

Switching criterion	Mode
$\gamma_o(k) < 8$ dB	Mode 1: $Q = 32$ and BPSK
8 dB $\leqslant \gamma_o(k) < 10$ dB	Mode 2: $Q = 32$ and 4-QAM
10 dB $\leqslant \gamma_o(k)$	Mode 3: $Q = 16$ and 4-QAM

Table 4.11: The switching rules for a triple-mode VSF/JD-CDMA system, where $\gamma_o(k)$ is the SINR of the k-th user at the output of the MMSE-BDFE. The switching thresholds – which were chosen for maintaining a target BER of 1% – were set to $t_1 = 8$ dB and $t_2 = 10$ dB.

CDMA systems. Three different transmission modes were chosen for the combined AQAM-VSF/JD-CDMA system, which were Mode 1 – BPSK and $Q = 32$; Mode 2 – 4-QAM and $Q = 32$; and, finally, Mode 3 – 4-QAM and $Q = 16$. The rules used for switching the transmission formats are tabulated in Table 4.11. The results of the associated simulations are presented in Figure 4.39. Also shown for comparison are the triple-mode AQAM/JD-CDMA scheme using $Q = 32$ and the triple-mode VSF/JD-CDMA scheme using 4-QAM that were characterized in Section 4.5.

Considering the BER curves of Figure 4.39, we can see that the combined scheme outperformed both the triple-mode AQAM scheme and the VSF scheme. The target BER of 1% was achieved and as E_s/N_0 increased, the BER also gradually decreased. In BPS throughput terms, the combined scheme also outperformed the other two individual schemes.

Let us now summarize the conclusions of this chapter.

4.7 Chapter Summary and Conclusion

Two methods of adaptively varying the transmission bit rate, namely adapting the modulation mode (AQAM) or varying the spreading factor (VSF) of the spreading code were proposed. The bit rate was varied in accordance with the quality of the channel. When the channel was of high quality, a higher number of bits was transmitted, in order to increase the throughput, while the bit rate was lowered in order to maintain the target BER when the channel was hostile. This allowed the system to accommodate and exploit the time-varying nature of the mobile channel. The metric used for switching the bit rate was the SINR at the output of the multi-user receiver. Our simulation results showed that the adaptive-rate schemes were capable of maintaining a target BER, while providing an increased throughput.

For the AQAM schemes, the modulation mode was switched according to the instantaneous channel conditions. Up to four modulation modes were utilized, which were BPSK, 4-QAM, 16-QAM and 64-QAM. The inclusion of a transmission blocking mode – where no information bits were transmitted – enabled the AQAM scheme to maintain target BERs, which were lower than the BER provided by BPSK schemes at low E_s/N_0 values. However, this was achieved at the expense of a low throughput. The AQAM/JD-CDMA system was capable of maximizing the throughput when the channel quality was favourable, and to reduce the number of errors when the channel was hostile. The AQAM scheme was also capable of improving the performance over channels having a lower order of multipath diversity. The lower the degree of multipath diversity, the higher the performance gain achieved, when compared to fixed-mode schemes over the same channel. The joint detector compensated for the lack of multipath diversity, thus enabling a certain BPS throughput to be achieved at similar E_s/N_0 values for different channels exhibiting varying orders of multipath diversity. It was also shown that varying the number of users from $K = 1$ to 8 for a system that utilized a spreading factor of $Q = 16$, did not significantly degrade the BER and BPS performance of the AQAM/JD-CDMA scheme.

In Section 4.4, it was shown that for VSF/JD-CDMA systems, the BER of fixed-rate schemes was independent of the spreading factor, Q. Therefore, when the channel conditions were good, the shorter spreading factors were utilized without increasing the power of the transmitted burst. It was shown that the VSF/JD-CDMA scheme was capable of outperforming the fixed-rate $Q = 64$ scheme at the same transmitted power in terms of BPS throughput and BER.

The two methods of varying the bit rate were also compared, where the minimum and maximum throughput values available were the same for both methods. It was shown that the VSF-based method outperformed the AQAM/JD-CDMA scheme. However, it was an advantage of the AQAM system that the complexity of the JD remained constant despite the change in modulation mode, while the complexity of the VSF receiver varied according to the SF used. Furthermore, the receiver had to be reconfigured for different spreading codes each time the spreading code was changed in the VSF scheme. It was also demonstrated that a combined AQAM-VSF/JD-CDMA system, where the bit rate was varied by switching both the modulation mode and the spreading factor, resulted in improved BER and throughput

AQAM			VSF			AQAM-VSF		
E_s/N_0	BER	NTP	E_s/N_0	BER	NTP	E_s/N_0	BER	NTP
6	1%	1.1	8	1%	1.2	6	1%	1.1
18	0.1%	4	15.5	0.0001%	3.5	15	0.001%	3.9

Table 4.12: Performance summary of the **triple-mode AQAM**, **triple-mode VSF** and **triple-mode AQAM-VSF combined** schemes for the **JD-CDMA**. The simulations were conducted over the seven-path Bad Urban channel model of Figure 3.20 and the rest of the simulation parameters are listed in Table 4.9. The number of users in the system is $K = 2$. The E_s/N_0 values in the table are the values at which the each system achieves the BER stated. The notation "NTP" represents the normalized throughput given by the ratio of the adaptive scheme's throughput to the throughput of the fixed-rate scheme utilizing 4-QAM and $Q = 64$, giving a minimum normalized throughput of 1 and a maximum normalized throughput of 4 when 4-QAM and $Q = 16$ or 16-QAM and $Q = 32$ are employed.

performance over the purely AQAM or VSF JD-CDMA systems. Table 4.12 summarizes the performance comparisons for the triple-mode AQAM, triple-mode VSF and the triple-mode combined AQAM-VSF/JD-CDMA schemes, where all three schemes are seen to achieve the target BER of 1% at $E_s/N_0 \leqslant 8$ dB. As the average E_s/N_0 increased, the combined AQAM-VSF/JD-CDMA scheme outperformed the purely VSF-based scheme in terms of throughput, while still achieving a BER of 0.001%, which was better than the 0.1% BER offered by the AQAM/JD-CDMA scheme.

In our investigations of adaptive rate JD-CDMA systems, the effects of diversity techniques such as antenna diversity [95] and space-time coding [244, 245] were not considered. The employment of antenna diversity improves the performance of the receiver by providing a diversity gain. It was shown in Section 4.3.10 that the performance gain of the AQAM system over fixed-mode systems decreased, as the order of multipath diversity increased. Therefore, it is expected that with the utilization of antenna diversity, the performance gains of adaptive rate systems over fixed-rate systems will also decrease. The complexity of combining diversity arrangements with adaptive rate schemes will increase due to the need for obtaining multiple channel quality metrics for the multiple channels. Furthermore, the optimization of the adaptation thresholds for a multi-user, multi-channel system will not be a trivial process. Additionally, the employment of antenna diversity incurs increased cost due to the need for multiple antennas and the processing of multiple versions of the received signal. Nonetheless, adaptive schemes remain attractive in terms of being able to communicate always at the highest possible BPS throughput, irrespective of the nature and severity of the channel impairments encountered, whilst satisfying the target integrity requirements, which are typically stipulated in terms of the BER.

Having considered the family of high-complexity JD-CDMA schemes, let us now focus our attention in the next chapter on the class of lower-complexity compromise schemes constituted by adaptive interference cancellation-based arrangements.

Chapter 5

Adaptive-Rate Interference Cancellation-Assisted CDMA [1]

In Section 2.10, a variety of multi-user detectors was discussed. A specific variant of multi-user detection is based on interference cancellation (IC). IC receivers attempt to remove the multiple access interference (MAI) by reconstructing the original transmitted signals of one or more users and cancelling the interference imposed by these reconstructed signals on the composite received signal [138, 147]. The resultant signal is then processed using the same procedure, in order to obtain the data estimates for the remaining users and to cancel their interference effects in the received multi-user signal, until all the users' signals are detected. Generally, the composite received multi-user signal is processed through a first stage that consists of either a bank of matched filters or RAKE receivers. After this stage, initial estimates of the information symbols are obtained and used for signal reconstruction and MAI cancellation in the subsequent stages. Interference cancellation techniques are often divided into two categories, namely successive interference cancellation (SIC) [147] and parallel interference cancellation (PIC) [138]. Hybrid cancellation schemes have also been proposed, where PIC and SIC are combined in a variety of ways [152–154, 246, 247]. Several methods of refining the data estimates and rendering them more reliable have also been proposed, such as re-estimating the CIR at each cancellation stage [153], integrating forward error correction methods with the interference cancellation [142, 143] and using soft decisions in order to de-emphasize less reliable decisions [146]. In comparison to joint detection (JD) schemes, interference cancellation algorithms exhibit a complexity that increases linearly with the number of users. Let us now consider the family of SIC techniques in a little more depth.

[1] *Single- and Multi-Carrier CDMA Multi-User Detection, Space-Time Spreading, Synchronization, Networking and Standards.* L.Hanzo, L-L.Yang, E-L.Kuan and K.Yen,
 ISBN 0-470-86309-9

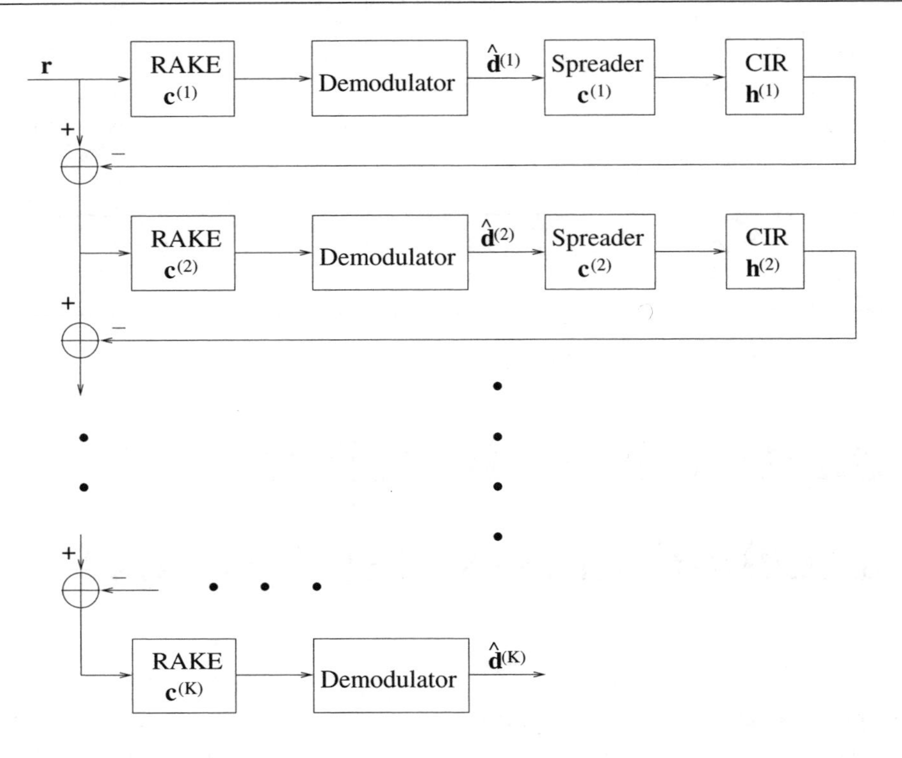

Figure 5.1: Schematic of the successive interference cancellation (SIC) receiver for K users. The users' signals have been ranked, where user 1's signal was received at the highest power, while user K's signal at the lowest power. In the order of ranking, the data estimates of each user are obtained and the received signal of each user is reconstructed and cancelled from the received composite signal, $\mathbf{r}$.

5.1 Successive Interference Cancellation

SIC [147] receivers rank the users according to a certain criterion, by estimating the received signal quality, for example, or the auto-correlation value for each user. This is carried out in order to determine the reliability of the initial data estimates for each user. The schematic of the SIC receiver is depicted in Figure 5.1, where all the K users have been ranked, with the strongest user being labelled 1 and the weakest user labelled K. After signal ranking, the received composite signal is processed through a matched filter or RAKE receiver for obtaining the initial data estimates of user 1. The received signal of this user is then reconstructed using the initial estimates of the transmitted data, the CIR and the spreading sequence of this user and this estimated received signal is subtracted from the composite received signal. The remaining signal is then processed through the matched filter bank or RAKE receiver of user 2 in order to obtain the data estimates for this user. Employing these estimates of the transmitted data, as well as the CIR and spreading sequence of this second user, the corresponding received signal is reconstructed and subtracted again from the remaining composite multi-user signal that has already had the strongest user's signal cancelled from it. This is

Operation	Multiplications	Additions
Data detection (RAKE receiver)	$NQL + NL$	$NQL + NL$
Signal reconstruction	$NQ + NQL$	$NQ + W - 1$
Interference cancellation	$-$	$NQ + W - 1$
Total	$2NQL + NQ + NL$	$NQL + 2(NQ + W - 1) + NL$

Table 5.1: Total number of addition and multiplication operations required for data detection, signal reconstruction and interference cancellation in the SIC receiver, where N represents the number of symbols per IC block, Q denotes the spreading sequence length, W indicates the total CIR dispersion length and finally, L represents the number of paths in the CIR.

repeated, until the data estimates of all the users have been obtained. Early research into SIC receivers includes, for example, work carried out by Patel and Holtzman [147], and Soong and Krzymien [148].

5.1.1 Complexity Calculations

In comparison to high-complexity JD-CDMA schemes, SIC algorithms have a lower complexity, which increases linearly with the number of users. In order to carry out a comparison with JD schemes, let us consider a burst-based cancellation receiver, where N represents the number of data symbols per transmitted burst, Q is the length of the spreading sequence for each user in terms of chips, L is the number of resolvable paths at the RAKE receiver and K is the total number of users. For each user, data detection using a RAKE receiver requires NQL multiplications and additions, i.e. NQ multiplications and additions per resolvable path, since there are NQ chips per transmitted burst and L multipath components are combined coherently. The diversity combining of the L paths requires L multiplications and additions per symbol. Therefore, the total number of multiplications and additions required for diversity combining of N symbols is NL. In total, the detection of the initial data estimates requires $NQL + NL$ multiplications and additions per burst. The reconstruction of the individual received signal of each user involves re-spreading of the data estimates, which leads to NQ multiplications. This is followed by a convolution with the L-path channel impulse response (CIR) and this requires NQL multiplications and $(NQ + W - 1)$ additions, where W is the total dispersion of the multipath channel expressed in chips. The interference cancellation of one reconstructed signal requires a total of $(NQ+W-1)$ subtractions/additions. The number of operations needed for each process is summarized in Table 5.1.

From Table 5.1, the total number of mathematical operations required for one stage of data detection, signal reconstruction and interference cancellation is:

$$T = 2NQL + NQ + NL + NQL + 2(NQ + W - 1) + NL \quad (5.1)$$

$$= [3NQL + NQ + 2NL + 2(NQ + W - 1)]. \quad (5.2)$$

For the SIC receiver supporting K users, a total of $(K-1)$ cancellations have to be performed, since interference cancellation is not required for the last detected user. Hence, for the SIC receiver the total number of operations amounts to $(K - 1)T$ addition and multiplication

Parameter	Value
Channel type	COST 207 Bad Urban
Number of paths in channel	7
Doppler frequency	80 Hz
Chip rate	2.167 MBaud
Frame burst structure	FRAMES Mode 1 Spread burst 1 [205]
Frame duration	4.615 ms
Burst duration	577 μs
Receiver type	SIC receiver

Table 5.2: Simulation parameters for SIC systems

operations.

For a system supporting $K = 10$ users, $N = 20$ symbols, $Q = 31$ chips and assuming an impulse response spread of $L = 4$ and $W = 10$ for the four-path Typical Urban channel of Figure 3.18 as defined in COST 207 [105], we have:

$$\begin{aligned} T &= 3NQL + NQ + 2NL + 2(NQ + W - 1) \\ &= 3(20)(31)(4) + 20(31) + 2(20)(4) + 2[20(31) + 10 - 1] \\ &= 9478. \end{aligned}$$

Therefore, the number of operations required for the SIC receiver per detected symbol is

$$\begin{aligned} \Sigma &= \frac{(K-1)T}{KN} \\ &= \frac{9(9478)}{10(20)} \\ &\approx 427. \end{aligned} \tag{5.3}$$

Comparing the complexity of the SIC receiver in Equation 5.3 with the complexity of 14614 for the JD receiver in Equation 3.209 employing the same parameters, it can be seen that the complexity of the SIC receiver is much lower than that of the JD receiver.

Having analyzed the implementational complexity of the SIC receiver, let us now consider its performance.

5.1.2 BER Performance of SIC-CDMA

A CDMA system utilizing an SIC receiver was simulated, in order to evaluate its performance. The simulation parameters are listed in Table 5.2.

The graphs in Figure 5.2 show the effects of the spreading factor, Q, on the BER performance for a $K = 2$-user system. Both users employed a fixed data modulation mode of 4-QAM and the chip rate was constant for all the various values of Q. As can be seen from Figure 5.2, the BER performance improves as the spreading sequence length increases. This

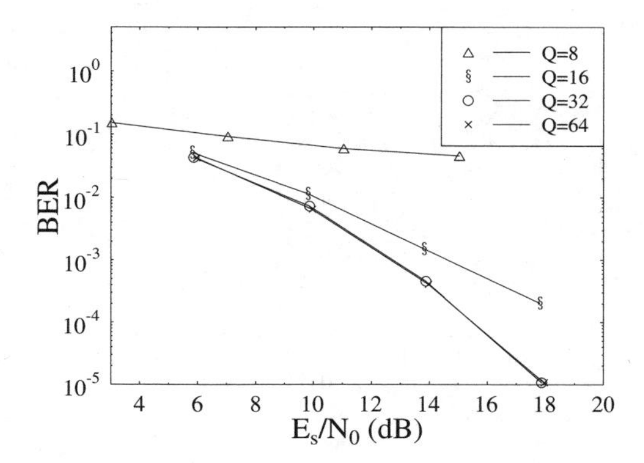

Figure 5.2: The effect of spreading sequence length, Q, on the BER performance of an SIC receiver in a two-user, **4-QAM**-based CDMA system is demonstrated here. The simulations were conducted over the COST 207 **seven-path Bad Urban channel** of Figure 3.20. The rest of the simulation parameters are listed in Table 5.2 and perfect channel estimation was assumed.

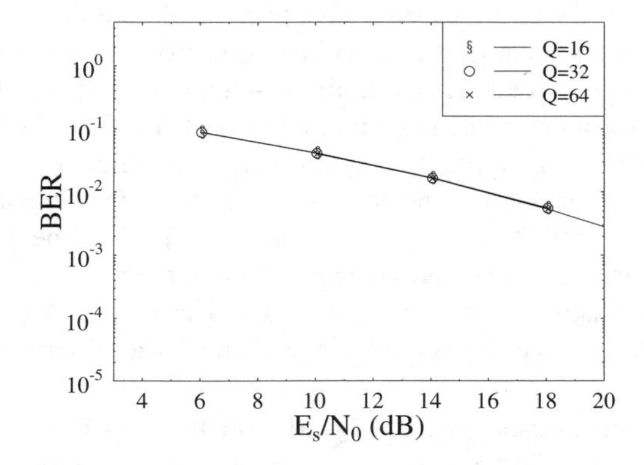

Figure 5.3: The effect of spreading sequence length, Q, on the BER performance of an SIC receiver in a two-user, **4-QAM**-based CDMA system is demonstrated here. The simulations were conducted over the COST 207 **two-path Rural Area channel** of Figure 3.15. The rest of the simulation parameters are listed in Table 5.2 and perfect channel estimation was assumed.

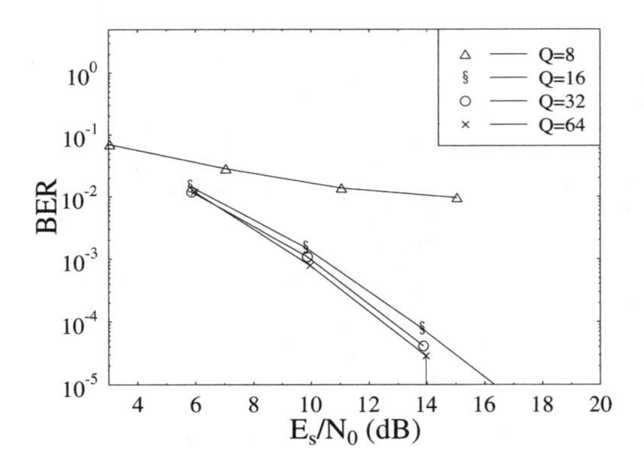

Figure 5.4: The effect of spreading sequence length, Q, on the BER performance of an SIC receiver in a two-user, **BPSK**-based CDMA system is demonstrated here. The simulations were conducted over the COST 207 seven-path Bad Urban channel of Figure 3.20. The rest of the simulation parameters are listed in Table 5.2 and perfect channel estimation was assumed.

performance improvement reduces as the spreading sequence length increases from $Q = 8$ to $Q = 32$, until finally the BER performance for $Q = 64$ becomes similar to that of $Q = 32$. The COST 207 Bad Urban channel [105] of Figure 3.20 used in our simulation had seven-paths and a delay spread of 7.8 μs, which is approximately equivalent to 17 chips at a chip rate of 2.167 Mchips/s. As Q was increased from 8 to 64 while keeping the chip rate at 2.167 Mchips/s, the duration of one symbol was also increased from 8 chips to 64 chips. Therefore, the symbol duration increased relative to the 17-chip delay spread of the channel, thus reducing the inter-symbol interference (ISI) and this translated to an improvement in BER performance. Beyond $Q = 32$ chips, the symbol period was much higher than the delay spread of the channel and the ISI was no longer a significant factor. Hence there was no further performance improvement, as shown in Figure 5.2. This is evident also in Figure 5.3, which shows the variation in BER performance parametrized with Q for transmissions over the two-path Rural Area channel shown in Figure 3.15. The BER performance did not vary upon increasing in Q, since the delay spread of the channel was already small compared to Q.

The same characteristics observed for 4-QAM over the Bad Urban channel of Figure 3.20 were also exhibited when BPSK was used as the data modulation mode, as depicted in Figure 5.4. For BPSK, the performance difference between $Q = 16$, 32 and 64 was less than that observed for 4-QAM. This is because the BPSK modulation mode has a larger distance between its modulation constellation points than 4-QAM, giving rise to reduced error propagation in the interference cancellation stage.

Increasing the number of users in the system increases the multiple access interference (MAI). This results in a degradation of the BER performance, as can be seen from the BER comparisons for $K = 2$ and $K = 8$ users in Figure 5.5. The initial data estimates used for

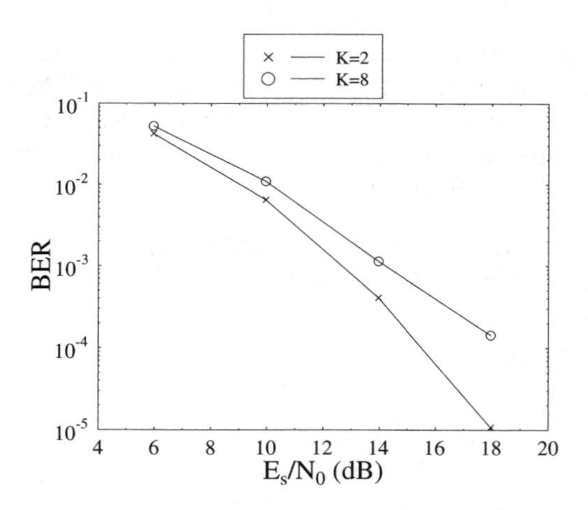

Figure 5.5: BER performance comparisons of $K = 2$ and $K = 8$ users, employing the SIC receiver. A two-user CDMA system was simulated with pseudo-random spreading sequences of length $Q = 64$ and 4-QAM was used for data modulation. The rest of the simulation parameters are listed in Table 5.2 and perfect channel estimation was assumed.

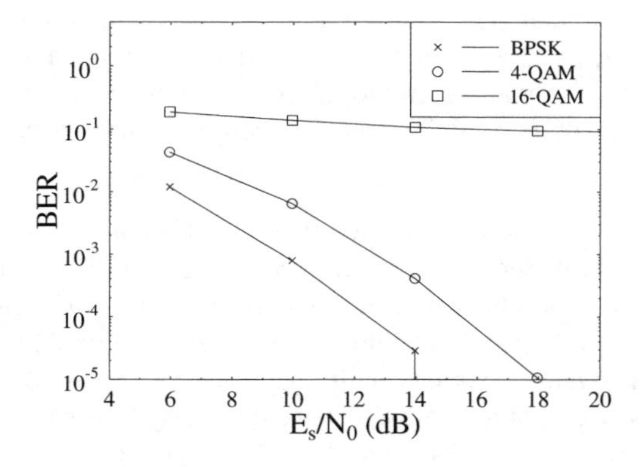

Figure 5.6: BER performance comparisons of the SIC receiver for three different data modulation modes. A two-user CDMA system was simulated with pseudo-random spreading sequences of length $Q = 64$. The rest of the simulation parameters are listed in Table 5.2.

cancellation are provided by the RAKE receiver. The performance of the RAKE receiver is interference-limited, therefore the degradation of the output of the RAKE receiver leads to error propagation in the cancellation stages, thus further degrading the performance.

Finally, the graphs in Figure 5.6 demonstrate the BER performance of the SIC-CDMA system for the three fixed modulation modes of BPSK, 4-QAM and 16-QAM. As expected, the BER performance suffers from increasing degradations as the number of constellation points in the modulation modes increases. The system using 16-QAM as the modulation mode suffers from an error floor, since the relatively closer distance between the modulation constellation points leads to inaccurate initial data estimates and these inaccurate estimates are further propagated in the cancellation stages.

Having analyzed the performance of fixed-rate SIC-CDMA, let us now turn our attention to combining adaptive-rate schemes with the SIC receiver.

5.1.3 Adaptive-Rate SIC-CDMA

Multi-rate transmission involving interference cancellation receivers have previously been investigated among others by Johansson and Svensson [241, 248]; and Juntti [222]. Typically, multiple users transmitting at different bit rates are supported in the same CDMA system invoking multiple codes or different spreading factors. SIC schemes and multi-stage cancellation schemes were used at the receiver for mitigating the MAI [222, 241, 248]. Moreover, the bit rate of the users was dictated by user requirements. In Chapter 4, we characterized the MMSE-BDFE JD receiver, in combination with adaptive-rate transmission schemes, where the bit rate was varied adaptively according to the prevalent channel conditions. In this section, we consider the employment of SIC receivers in conjunction with the same adaptive-rate transmission schemes.

In the adaptive-rate JD-CDMA systems investigated in Chapter 4, a measure of reception quality was required in order to determine the information rate of the next transmission burst. This channel quality measure or metric was chosen to be the Signal to Interference plus Noise Ratio (SINR) at the output of the MMSE-BDFE for JD-CDMA, as explained in Section 4.3.3. Similarly, for the SIC receiver, a channel quality measure was required, and this was again chosen to be the SINR at the output of the SIC receiver. The SINR estimate was derived by modifying the approach presented by Patel and Holtzmann [147].

Let us consider the up-link scenario for a synchronous CDMA system supporting K users. The received composite signal due to K users and L CIR paths, can be expressed as:

$$y_i = \sum_{k=1}^{K} \sum_{n=1}^{N} d_n^{(k)} b_{i-(n-1)Q}^{(k)} + n_i, \tag{5.4}$$

where $d_n^{(k)}$ represents the n-th data symbol transmitted by the k-th user; b_u represents the u-th sample, of the combined impulse response of the k-th user based on the CIR and the spreading sequence; and n_i represents the i-th noise sample. After processing the received signal by a RAKE receiver once, the n-th decision variable for the first user, $k = 1$, is given by [147]:

$$
\begin{aligned}
z_n^{(1)} &= \sum_{i=1}^{Q+W-1} y_{(n-1)Q+i}[b_i^{(1)}]^* && (5.5)\\
&= \sum_{i=1}^{Q+W-1} d_n^{(1)} b_i^{(1)}[b_i^{(1)}]^* + \sum_{j=2}^{K}\sum_{i=1}^{Q+W-1} d_n^{(j)} b_i^{(j)}[b_i^{(1)}]^* + \sum_{i=1}^{Q+W-1} n_i[b_i^{(1)}]^* && (5.6)\\
&= d_n^{(1)} I_1 + \sum_{j=2}^{K} d_n^{(j)} I_{j,1} + \eta_1 && (5.7)\\
&= d_n^{(1)} I_1 + C_1, && (5.8)
\end{aligned}
$$

where $I_{j,k} = \sum_{i=1}^{Q+W-1} b_i^{(j)}[b_i^{(k)}]^*$ represents the elements of the cross-correlation between the combined impulse responses of user j and user k; $I_k = I_{k,k}$; and $\eta_k = \sum_{i=1}^{Q+W-1} n_i[b_i^{(k)}]^*$. Furthermore, the variable C_1 represents the interference plus noise for user 1, where $C_1 = \sum_{j=2}^{K} d_n^{(j)} I_{j,1} + \eta_1$.

The decision variable $z_n^{(1)}$ is used to regenerate the received signal of user 1 and the regenerated received signal is then cancelled or deducted from the received composite signal. After processing the resulting signal by a RAKE receiver, the n-th decision variable of the second user, $k = 2$, is given by [147]:

$$
\begin{aligned}
z_n^{(2)} &= d_n^{(2)} I_2 + \sum_{j=3}^{K} d_n^{(j)} I_{j,2} - C_1 I_{1,2} + \eta_2 && (5.9)\\
&= d_n^{(2)} I_2 + C_2, && (5.10)
\end{aligned}
$$

where the term $C_1 I_{1,2}$ represents the imperfect cancellation due to data estimation errors in the first cancellation round. The three terms $(\sum_{j=3}^{K} d_n^{(j)} I_{j,2} - C_1 I_{1,2} + \eta_2)$ symbolize the total interference plus noise corrupting the desired term of $d_n^{(2)} I_2$ and these terms are represented collectively by the variable C_2.

We can deduce that after $(k-1)$ cancellation rounds, the decision variable for the k-th user is given by [147]:

$$
z_n^{(k)} = d_n^{(k)} I_k + C_k, \qquad (5.11)
$$

where

$$
C_k = \sum_{j=k+1}^{K} d_n^{(j)} I_{j,k} - \sum_{j=1}^{k-1} C_j I_{j,k} + \eta_k. \qquad (5.12)
$$

Parameter	Value
Doppler frequency	80 Hz
Spreading ratio	64
Spreading sequence	Pseudo-random
Chip rate	2.167 MBaud
Frame burst structure	FRAMES Mode 1 Spread burst 1 [205]
Burst duration	577 μs
Receiver type	SIC

Table 5.3: Simulation parameters for AQAM/SIC-CDMA systems

The SINR at the output of the SIC receiver is defined by [147]:

$$\gamma_o(n,k) = \frac{\text{Var}[d_n^{(k)} \cdot I_k]}{\text{Var}[C_k|d_n^{(k)}]}. \tag{5.13}$$

The variance of the desired term is conditioned on $d_n^{(k)}$ and is given by:

$$\text{Var}[d_n^{(k)} \cdot I_k] = |d_n^{(k)}|^2 |I_k|^2. \tag{5.14}$$

The variance of C_k conditioned on $d_n^{(k)}$ is as follows:

$$\text{Var}[C_k|d_n^{(k)}] = \sum_{j=k+1}^{K} |d_n^{(j)}|^2 |I_{j,k}|^2 + \sum_{j=1}^{k-1} \text{Var}[C_j|d_n^{(j)}] |I_{j,k}|^2 + \text{Var}[\eta_k], \tag{5.15}$$

where $\text{Var}[\eta_k] = \sigma^2$ is the noise variance. Finally, the SINR at the output of the SIC receiver is:

$$\gamma(n,k) = \frac{|d_n^{(k)}|^2 \, |I_k|^2}{\sum_{j=k+1}^{K} |d_n^{(j)}|^2 \, |I_{j,k}|^2 + \sum_{j=1}^{k-1} \text{Var}[C_j|d_n^{(j)}] \, |I_{j,k}|^2 + \text{Var}[\eta_k]}. \tag{5.16}$$

5.1.4 AQAM/SIC-CDMA

In Section 4.2 the concept of Adaptive Quadrature Amplitude Modulation (AQAM) was introduced and the improved BER and throughput performance of AQAM transmission over fixed modulation-based transmissions in JD-CDMA systems were demonstrated in Section 4.3.5. Here, the performance of SIC-assisted AQAM-CDMA is characterized. The simulation parameters used were the same as those chosen for the JD-CDMA simulations in Section 4.3.5, and are listed again in Table 5.3. Perfect CIR estimation was assumed and RAKE receivers were used for providing the initial data estimates.

Figure 5.7 presents the twin-mode AQAM performance curves for two different AQAM systems, each utilizing a different spreading factor, Q. From the figure, we can see that for

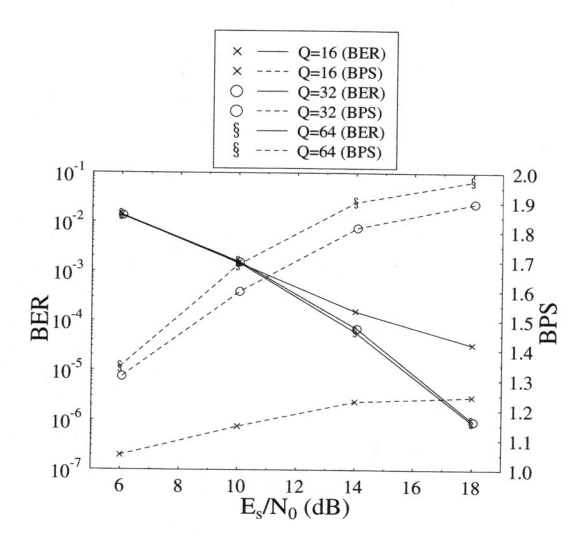

Figure 5.7: Performance of a two-user, twin-mode AQAM/SIC-CDMA system transmitting over the seven-path Bad Urban channel of Figure 3.20, where the modulation modes were switched between BPSK and 4-QAM. The BER and BPS performance curves are shown for **three different values of spreading factor,** Q. The rest of the simulation parameters are listed in Table 5.3 and perfect channel estimation was assumed.

Switching criterion	Modulation mode
$\gamma_o(k) < t_1$	BPSK
$t_1 \leqslant \gamma_o(k)$	4-QAM

Table 5.4: The switching rules for a twin-mode AQAM/SIC-CDMA system, where t_1 is the switching threshold and $\gamma_o(k)$ is the SINR of the k-th user at the output of the SIC receiver, which was calculated by using Equation 5.16 and $\gamma_o(k) = \frac{1}{N}\sum_{n=1}^{N}\gamma(n,k)$.

a given E_s/N_0 value, the AQAM scheme that employs $Q = 64$ as the spreading factor performs the best in BER and throughput terms. At $E_s/N_0 = 18$ dB, there is an approximately 0.7 BPS throughput performance difference between the $Q = 64$ system and $Q = 16$ system. This is because the $Q = 64$ scenario benefitted from a higher spreading factor and hence it invoked the higher throughput 4-QAM mode more frequently. Additionally, the system employing $Q = 64$ provided the best BER performance. Based on these results, in subsequent investigations into AQAM/SIC-CDMA the spreading factor of $Q = 64$ was employed.

Figure 5.8 presents the BER and BPS throughput performance of a two-user, twin-mode AQAM system. The modulation mode of each user was switched between BPSK and 4-QAM, depending on the SINR of the user at the output of the SIC receiver. The switching rules used are tabulated in Table 5.4. Two sets of simulations were conducted, using two different threshold values of $t_1 = 8$ dB and $t_1 = 12$ dB, for comparison. For both thresholds, the BER performance was better than the 4-QAM BER performance and worse than the

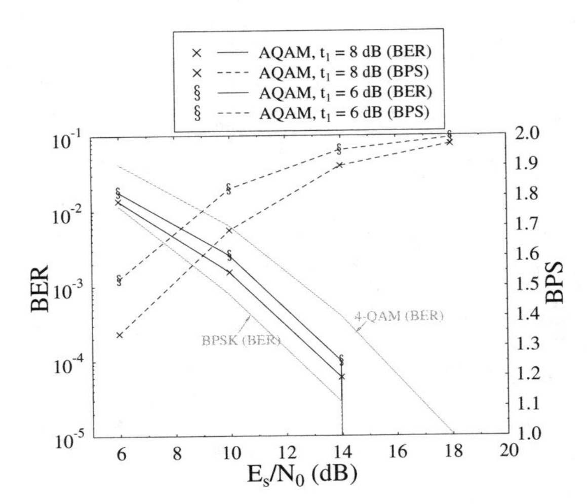

Figure 5.8: **Twin-mode AQAM/SIC-CDMA** system transmitting over the seven-path Bad Urban channel of Figure 3.20, where the modulation modes were switched between BPSK and 4-QAM, using Q = 64-chip spreading sequences. The switching threshold is represented by t_1. The rest of the simulation parameters are listed in Table 5.3 and perfect channel estimation was assumed. The BER and BPS performance curves are shown for two different threshold values. The BER performance of the constituent fixed-mode systems is also shown for comparison.

Switching criterion	Modulation mode
$\gamma_o(k) < t_1$	BPSK
$t_1 \leqslant \gamma_o(k) < t_2$	4-QAM
$t_2 \leqslant \gamma_o(k)$	16-QAM

Table 5.5: The switching rules for a triple-mode AQAM/SIC-CDMA system, where $\gamma_o(k)$ is the SINR of the k-th user at the output of the SIC receiver, which was calculated by using Equation 5.16 and $\gamma_o(k) = \frac{1}{N}\sum_{n=1}^{N}\gamma(n,k)$. The switching thresholds are represented by t_1 and t_2, where $t_1 < t_2$.

BPSK BER. This is because, when the channel conditions were unfavourable, the more robust modulation mode of BPSK was used, thus improving the BER performance. It should also be noted that as E_s/N_0 increased, the throughput of both systems increased and approached the 2 BPS offered by the 4-QAM system, while maintaining a BER lower than that of 4-QAM. From the figure, we can also see that the employment of a higher threshold of $t_1 = 8$ dB resulted in a better BER performance but yielded a lower throughput compared to the $t_1 = 6$ dB–system. This is due to the fact that a higher SINR had to be achieved by each user, before a higher-order modulation mode could be used, thus improving the BER performance, but reducing the average number of data bits per symbol.

A triple-mode AQAM system was also investigated, where the modulation modes were

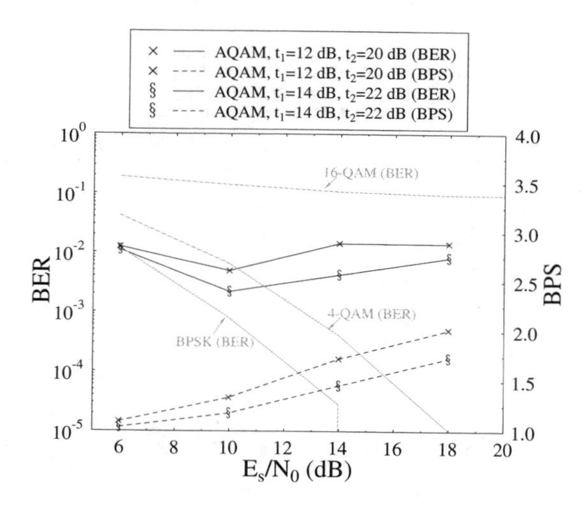

Figure 5.9: Triple-mode AQAM/SIC-CDMA system transmitting over the seven-path Bad Urban channel of Figure 3.20, where the modulation modes were switched between BPSK, 4-QAM and 16-QAM, using $Q = 64$-chip spreading sequences. The switching thresholds are represented by t_1 and t_2. The rest of the simulation parameters are listed in Table 5.3 and perfect channel estimation was assumed. The BER and BPS performance curves are shown for two different set of thresholds. The BER performance of the constituent fixed-mode systems is also shown for comparison.

switched between BPSK, 4-QAM and 16-QAM, according to the rules tabulated in Table 5.5. The results of these simulations are shown in Figure 5.9. The first set of thresholds chosen was $t_1 = 12$ dB and $t_2 = 20$ dB. The BER performance was maintained at a level of approximately 1% for $E_s/N_0 < 10$ dB, but increased to a level above 1% as E_s/N_0 increased. This was due to the fact that 16-QAM was one of the modulation modes used. The BER performance of the 16-QAM mode could not dip below 1% and this degraded the BER performance of the AQAM system. Even with an increase in threshold settings to $t_1 = 14$ dB and $t_2 = 22$ dB, the BER performance improved only slightly. The BPS throughput performances remained below 2 BPS and both schemes were unable to match the maximum BPS of 4 offered by the fixed-mode 16-QAM system without exceeding the target BER of 1%.

Comparing the twin-mode and triple-mode AQAM/SIC-CDMA schemes, it can be concluded that the inclusion of 16-QAM does not offer potential performance gains, since the inaccurate data estimates generated by 16-QAM modulation mode results in error propagation, which the SIC receiver is unable to resolve. However, the twin-mode AQAM schemes showed performance gains over the fixed-mode schemes and further comparisons will therefore be concentrated on the twin-mode schemes.

The performance of the twin-mode AQAM/SIC-CDMA scheme was also investigated for the four-path Typical Urban channel of Figure 3.18 and the two-path Rural Area channel of Figure 3.15. The resulting BER and BPS curves are presented in Figures 5.10 and 5.11 for the Typical Urban and Rural Area channels, respectively. From these figures, it can be seen that

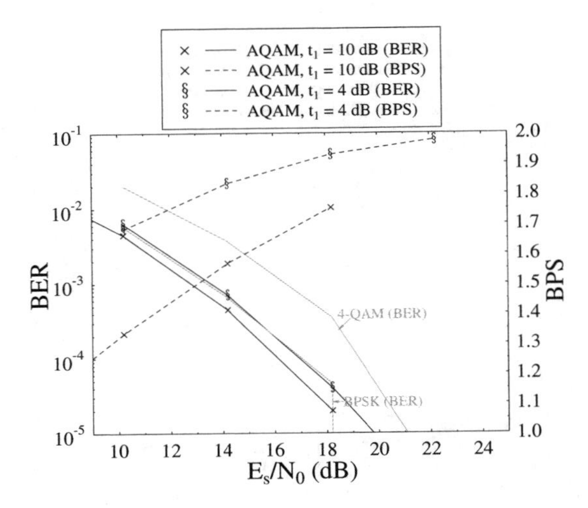

Figure 5.10: Twin-mode AQAM/SIC-CDMA system transmitting over the **four-path Typical Urban channel** of Figure 3.18, where the modulation modes are switched between BPSK and 4-QAM. The switching threshold is represented by t_1. The rest of the simulation parameters are listed in Table 5.3 and perfect channel estimation was assumed. The BER and BPS performance curves are shown for two different threshold values. The BER performance of the constituent fixed-mode QAM systems is also shown for comparison.

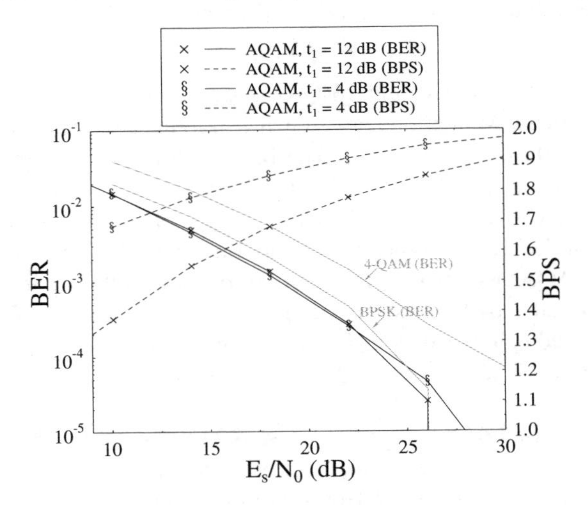

Figure 5.11: Twin-mode AQAM/SIC-CDMA system transmitting over the **two-path Rural Area channel** of Figure 3.15, where the modulation modes are switched between BPSK and 4-QAM. The switching threshold is represented by t_1. The rest of the simulation parameters are listed in Table 5.3 and perfect channel estimation was assumed. The BER and BPS performance curves are shown for two different threshold values. The BER performance of the constituent fixed-mode QAM systems is also shown for comparison.

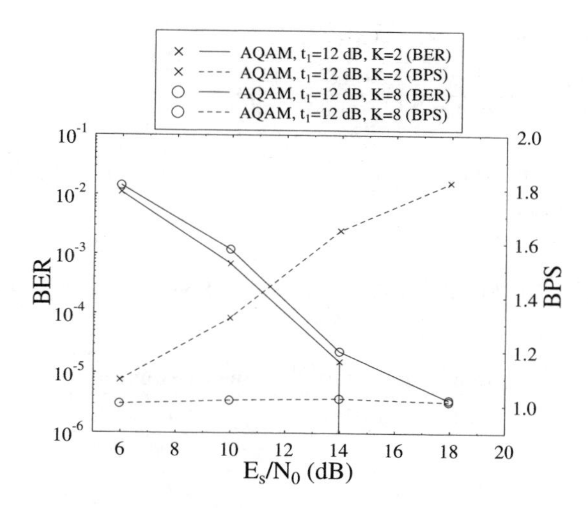

Figure 5.12: Twin-mode AQAM/SIC-CDMA system transmitting over the **seven-path Bad Urban** channel of Figure 3.20, where the modulation modes are switched between BPSK and 4-QAM. The rest of the simulation parameters are listed in Table 5.3 and perfect channel estimation was assumed. The performance of two different systems, one supporting $K = 2$ users and the other supporting $K = 8$ users is compared. The switching threshold is kept the same for both systems.

the AQAM/SIC-CDMA schemes outperformed the fixed-mode schemes in BER performance terms. For the Typical Urban channel, the AQAM/SIC-CDMA BER performance was similar to that of the fixed-mode BPSK scheme, while the BPS throughput performance achieved was higher than the 1 BPS offered by fixed-mode BPSK. Similarly, for the Rural Area channel, the AQAM/SIC-CDMA scheme outperformed the fixed-mode BPSK scheme in BER and BPS throughput terms. The reason for this is that the AQAM/SIC-CDMA scheme was capable of switching to a more robust modulation mode when the channel quality was low, thus lowering the number of errors inflicted. When the channel quality was high, a higher-order modulation mode was employed, which increased the total number of bits transmitted. These two factors in combination assisted in reducing the average BER.

The performance comparison of the twin-mode AQAM/SIC-ACDMA scheme for $K = 2$ and $K = 8$ users is presented in Figure 5.12. From the figure, we can see that although both systems were capable of achieving the same BER performance, the throughput of the $K = 8$ system was almost constant at 1 BPS. This is due to the increase in MAI, resulting in the output SINR values being lower than the required switching threshold.

Having evaluated the performance of AQAM schemes in conjunction with SIC-CDMA, let us now focus our attention on another adaptive-rate scheme, namely the VSF/SIC-CDMA system.

Parameter	Value
Doppler frequency	80 Hz
Spreading sequence	Pseudo-random
Chip rate	2.167 MBaud
Frame burst structure	FRAMES Mode 1 Spread burst 1 [205]
Burst duration	577 μs
Receiver type	SIC

Table 5.6: Simulation parameters for VSF/SIC-CDMA systems

Switching criterion	Transmission mode
$\gamma_o(k) < t_1$	$\mathbf{c}_1^{(k)}, \; Q_1 = 64$
$t_1 \leqslant \gamma_o(k) < t_2$	$\mathbf{c}_2^{(k)}, \; Q_2 = 32$
$t_2 \leqslant \gamma_o(k)$	$\mathbf{c}_3^{(k)}, \; Q_3 = 16$

Table 5.7: The switching rules for a VSF/SIC-CDMA system, where $\gamma_o(k)$ is the SINR of the k-th user at the output of the SIC receiver, which was calculated by using Equation 5.16 and $\gamma_o(k) = \frac{1}{N}\sum_{n=1}^{N}\gamma(n,k)$. The switching thresholds are represented by t_1 and t_2, where $t_1 < t_2$.

5.1.5 VSF/SIC-CDMA

In Section 5.1.2 we showed the effect of the spreading sequence length, Q, on the BER performance of the SIC receiver. As a general rule, the BER performance improves as the length of the spreading sequence increases relative to the delay spread of the propagation channel. This property can be exploited in a variable spreading factor (VSF) CDMA system, where the spreading sequence length is increased when the mobile channel is hostile in order to reduce the number of errors, and decreased when channel conditions are favourable in order to improve the throughput of the system. By keeping the chip period constant, the data throughput can be varied upon changing the spreading factors of the codes used. Previously, for the VSF/JD-CDMA schemes the transmission power was kept constant, regardless of the spreading factor chosen for transmission, since the BER performance of the joint detector was not affected by the spreading factor. However, for the VSF/SIC-CDMA schemes, the transmitter power was increased when a lower spreading factor was used and decreased when a higher spreading factor was employed in order to keep the E_s/N_0 value constant. This is because the BER performance of the SIC receiver is dependent on the spreading factor and the E_s/N_0 value has to be kept constant, in order to exploit the full potential of the VSF/SIC-CDMA scheme.

In this section, we present the simulation results for such an adaptive system, where the SIC receiver is used to detect the data. There were three modes of transmission, i.e. $Q = 64$, $Q = 32$ and $Q = 16$. The modulation mode of 4-QAM was used for $K = 2$ users. The relevant simulation parameters are listed in Table 5.3. In order to calculate the SINR at the output of the SIC receiver, the spreading code having the highest spreading factor was used to reduce the complexity of the SINR estimation. Three different sets of switching thresholds

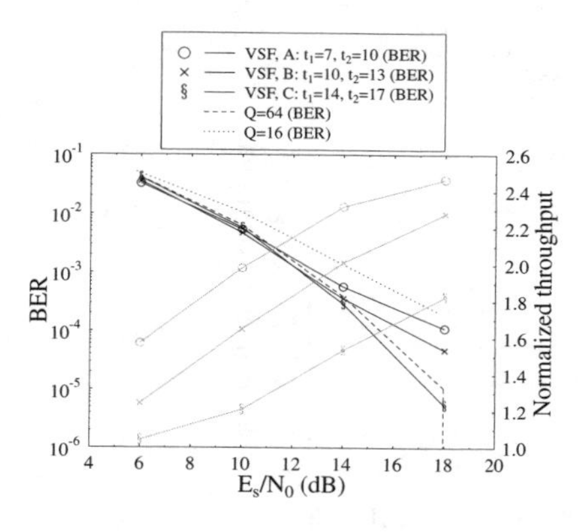

(a) BER comparisons are in bold, throughput comparisons are in grey.

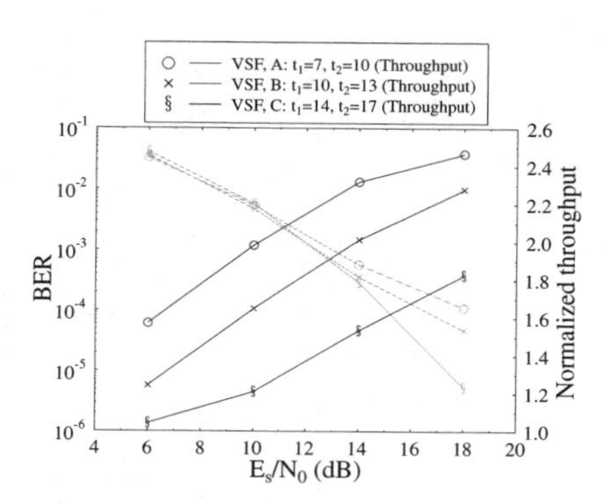

(b) Throughput comparisons are in bold, BER comparisons are in grey.

Figure 5.13: BER and normalized throughput performance of a triple-mode VSF/SIC-CDMA system transmitting over the seven-path Bad Urban channel of Figure 3.20. This is a two-user system, and the data modulation mode is 4-QAM. The spreading factor is switched between $Q_1 = 64$, $Q_2 = 32$ and $Q_3 = 16$. Three different sets of switching thresholds are compared. The rest of the simulation parameters are listed in Table 5.6 and perfect channel estimation was assumed.

were investigated, namely System A: $t_1 = 7$ dB, $t_2 = 10$ dB; System B: $t_1 = 10$ dB, $t_2 = 13$ dB; and System C: $t_1 = 14$ dB, $t_2 = 17$ dB. The switching rules employed are tabulated in Table 5.7. Perfect channel estimation was assumed and RAKE receivers were used to provide the initial data estimates.

The results of these simulations are shown in Figure 5.13. The throughput performance is normalized to the minimum throughput of the scheme, which is when $Q_1 = 64$ is chosen as the spreading factor, thus giving a minimum normalized throughput of 1 and a maximum normalized throughput of 4. From the results presented in Figure 5.13 we observe that at $E_s/N_0 < 14$ dB the BER performances of all three systems were similar to that of the fixed-rate $Q = 64$ scheme. At these E_s/N_0 values the VSF scheme had an option of switching to a higher throughput rate when the channel quality was high. This increased the total number of bits transmitted and thus reduced the average BER. As E_s/N_0 was increased, the BER performance of the more "aggressive" Systems A and B gradually approached that of the fixed-rate $Q = 16$ scheme, since the lower spreading factors were chosen with an increasing probability. The most conservative system, System C, was still capable of mimicking the performance of the fixed-rate $Q = 64$ BER, since the switching thresholds were the highest of the three systems, and the lower spreading factors were not chosen as often. By comparing the normalized throughput performance of all three systems, it can be observed that the most aggressive system resulted in the highest throughput, and as the switching thresholds were increased, a drop in throughput also followed.

Figure 5.14 shows the performance comparisons for a triple-mode VSF/SIC-CDMA system supporting $K = 2$ and $K = 8$ users. The BER and normalized throughput performance

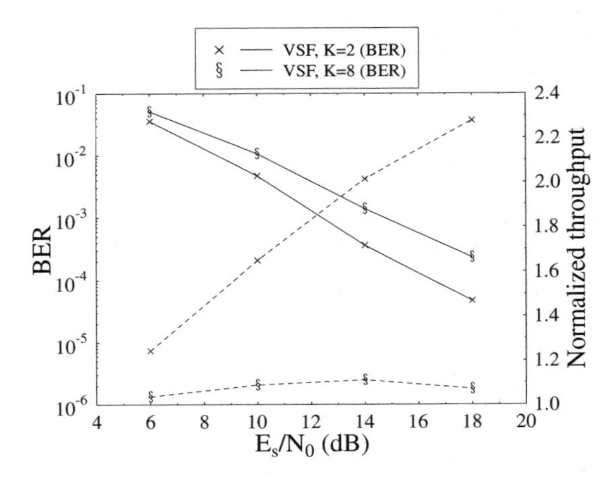

Figure 5.14: BER and normalized throughput performance comparisons for triple-mode VSF/SIC-CDMA schemes supporting $K = 2$ and $K = 8$ users. The modulation mode was chosen to be 4-QAM. The rest of the simulation parameters are listed in Table 5.6 and the seven-path Bad Urban channel of Figure 3.20 was employed, where perfect channel estimation was assumed.

of the $K = 8$–user system is degraded compared to the two-user system. This is due to the increase in MAI inflicted by the extra users and due to the RAKE receivers, which were unable to provide accurate data estimates, leading to error propagation in the cancellation stage.

Having investigated the SIC-CDMA system for fixed-rate and adaptive-rate schemes, let us now focus our attention on another type of IC receiver, the parallel interference cancellation (PIC) receiver.

5.2 Parallel Interference Cancellation

Parallel interference cancellation (PIC) algorithms usually consist of multiple stages of interference cancellation. Figure 5.15 shows a single cancellation stage for one of the users. In each cancellation stage, the signal of each user is reconstructed by invoking the data estimates from the previous cancellation stage. Then, for each user, the reconstructed signals of all the other users are subtracted from the received composite signal, and the resulting signal is processed by a matched filter or RAKE receiver, in order to obtain a new set of data estimates for this user. This reconstruction, cancellation and re-estimation stage is repeated as many times, as required or can be afforded by the system. The cancellation is carried out in parallel for all the users, hence the term *parallel interference cancellation.* Research into PIC receivers include, for example, the work previously carried out by Varanasi and Aazhang [138]; Yoon, Kohno and Imai [141]; as well as Divsalar, Simon and Raphaeli [146].

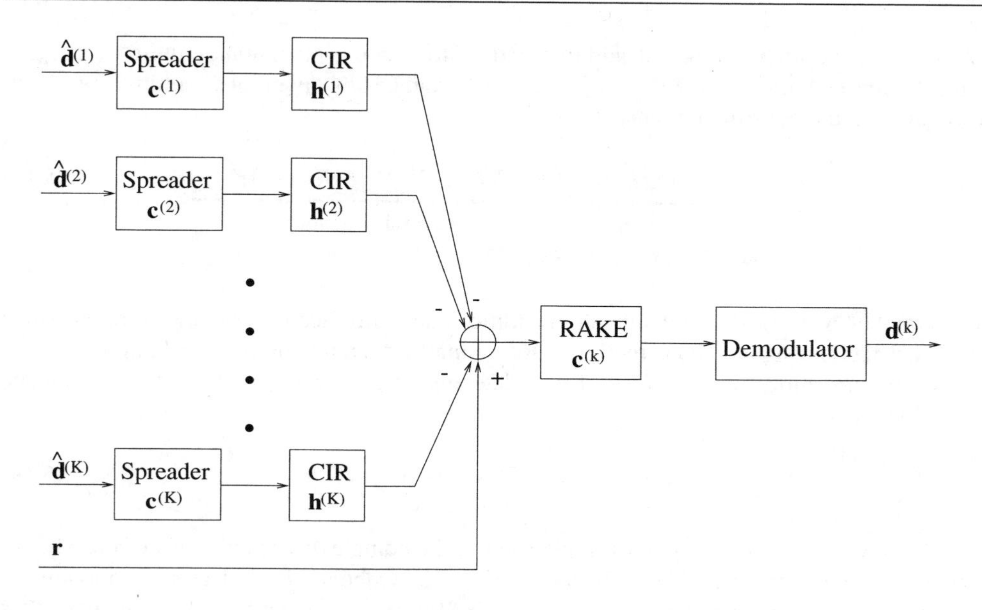

Figure 5.15: A schematic of a single cancellation stage for user K in the parallel interference cancellation (PIC) receiver for K users. The data estimates, $\mathbf{d}^{(1)}, \ldots, \mathbf{d}^{(K)}$ from the other $(K-1)$ users were obtained from the previous cancellation stage and the received signal of each user other than the K-th one is reconstructed and cancelled from the received signal, $\mathbf{r}$.

Operation	Multiplications	Additions
Data detection (RAKE receiver)	$NQL + NL$	$NQL + NL$
Signal reconstruction	$NQ + NQL$	$NQ + W - 1$
Interference cancellation	$-$	$NQ + W - 1$
Total	$2NQL + NQ + NL$	$NQL + 2(NQ + W - 1) + NL$

Table 5.8: Total number of addition and multiplication operations for a single cancellation stage in the PIC receiver, where N represents the number of symbols per IC block, Q denotes the spreading sequence length, W indicates the total CIR dispersion length and, finally, L represents the number of paths in the CIR.

5.2.1 Complexity Calculations

Following the approach of Section 5.1.1 for the SIC receiver, the complexity calculations for the PIC receiver can be similarly conducted. The breakdown of the number of operations required for each process in the interference cancellation receiver is shown in Table 5.8.

For the PIC receiver, signal reconstruction and cancellation of all the other users' signals are required for each user. Thus, for each user, a total of $(K-1)$ cancellations have to be performed, leading to $(K-1)(NQ+W-1)$ operations for signal cancellation, since $(NQ+W-1)$ is the total length of the received signal for one user. Therefore, for a total of K users, the total number of addition and multiplication operations required for signal cancellation is $K(K-1)(NQ+W-1)$. The number of operations required for signal estimation and reconstruction only is:

$$\begin{aligned} U &= \underbrace{2(NQL+NL)}_{\text{data detection}} + \underbrace{NQ+NQL+NQ+W-1}_{\text{signal reconstruction}} \qquad (5.17) \\ &= 3NQL+2NL+2NQ+W-1. \qquad (5.18) \end{aligned}$$

For a total of K users, the total number of addition and multiplication operations for one stage of data detection, signal reconstruction is KU. Finally, the total number of addition and multiplication operations for one stage of data detection, signal reconstruction and interference cancellation is:

$$T = K(K-1)(NQ+W-1) + KU. \qquad (5.19)$$

Employing the same system example used in the complexity calculations of the SIC receiver, a system supporting $K = 10$ users, $N = 20$ symbols, $Q = 31$ chips and assuming an impulse response spread of $L = 4$ and $W = 10$ for a four-path typical urban channel as defined in COST 207 [105], we have:

$$\begin{aligned} U &= 3NQL+2NL+2NQ+W-1 \\ &= 3(20)(31)(4)+2(20)(4)+2(20)(31)+10-1 \\ &= 8849. \end{aligned}$$

Utilizing the value of $U = 8849$ calculated above, the final number of operations required for the PIC receiver to detect one symbol is:

$$\begin{aligned} \Sigma &= \frac{K(K-1)(NQ+W-1)+KU}{KN} \\ &= \frac{(K-1)(NQ+W-1)+U}{N} \\ &= \frac{9[20(31)+10-1]+8849}{20} \\ &\approx 726. \end{aligned}$$

Comparing the number of operations required for the SIC and PIC receivers, the complexity of the PIC receiver is approximately twice that of the SIC receiver. This is because in PIC, for each user the signals of all the other users are reconstructed and cancelled from the received composite signal. In comparison, in SIC each user's signal is only cancelled once from the composite signal, resulting in a lower complexity. When the complexity of both the SIC and PIC receivers are compared to the 7300 combined multiplications and additions required by the simplified JD algorithms, as shown in Section 3.4, we can see that the IC receivers have approximately an order of magnitude lower complexity than the JD schemes. The complexity of the SIC receiver increases linearly with the number of users and the num-

Parameter	Value
Doppler frequency	80 Hz
Chip rate	2.167 MBaud
Spreading factor	$Q = 64$
Frame burst structure	FRAMES Mode 1 Spread burst 1 [205]
Frame duration	4.615 ms
Burst duration	577 μs
Receiver type	PIC receiver

Table 5.9: Simulation parameters for PIC systems

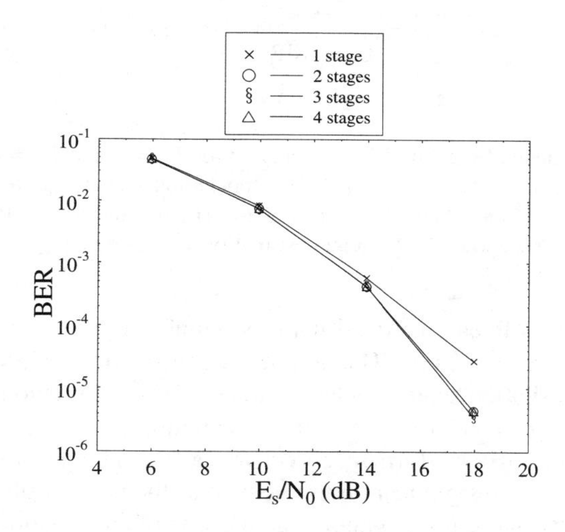

Figure 5.16: BER performance of the PIC-CDMA receiver for different number of cancellation stages. The number of users in the system was $K = 8$, the modulation mode was set to 4-QAM and the propagation channel used was the COST 207 seven-path Bad Urban channel of Figure 3.20. The rest of the simulation parameters are listed in Table 5.9 and perfect channel estimation was assumed.

ber of symbols per transmission burst while the complexity of the PIC receiver is of the order of $O(K^2)$. In joint detection, the complexity is of the order of $O(K^3)$ and $O(N^3)$. This relatively low complexity of the IC receivers renders them an attractive option for implementing multi-user CDMA receivers. Let us now consider the performance of the PIC receiver in the next section.

5.2.2 BER Performance of PIC-CDMA

The BER performance of the PIC receiver is evaluated in this section. The simulation parameters for the investigations are listed in Table 5.9.

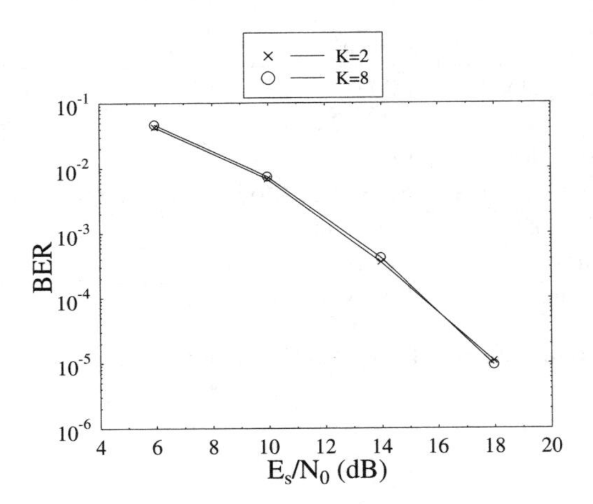

Figure 5.17: BER performance of the PIC-CDMA receiver for $K = 2$ and $K = 8$ users. The modulation mode was set to 4-QAM and the propagation channel used was the COST 207 seven-path Bad Urban channel of Figure 3.20. The rest of the simulation parameters are listed in Table 5.9 and perfect channel estimation was assumed.

The improvement in BER parametrized with the number of cancellation stages is presented in Figure 5.16 for $K = 8$ users. The number of cancellation stages is increased from one to four. At each cancellation stage, the data estimates from the previous stage are used to reconstruct the received signals of all the users for cancellation. From the results shown in the figure, we observe that the BER performance improves as the number of stages is increased from one to two but there is no significant improvement as the number of stages is increased to three and four. In subsequent investigations the number of cancellation stages is fixed to two in order to reduce the associated complexity.

The BER performance of the PIC receiver is shown in Figure 5.17 for different numbers of users. Although K was increased from two to eight, the BER performance of the PIC receiver remained similar. This is in contrast to the BER performance of the SIC receiver of Figure 5.5, which degraded when the number of users was increased. The reason for this performance disparity is that the SIC receiver offers the benefits, when there is a large received signal power difference among the different users. When the received signal powers for all the users are similar, the PIC outperforms the SIC, since the interference inflicted by the unwanted users is cancelled from the received composite signal in order to obtain the data estimates of the desired user.

Having briefly investigated the performance of the PIC receiver for fixed-rate CDMA schemes, let us now focus our attention on adaptive-rate PIC-CDMA schemes.

5.2.3 Adaptive-Rate PIC-CDMA

In this section, the performance of the PIC-CDMA receiver with respect to adaptive-rate CDMA schemes is evaluated. Similarly to the SIC-CDMA and JD-CDMA receivers, the metric chosen for switching the information rate is the SINR at the output of the PIC-CDMA receiver. This output SINR expression was derived by modifying the approach of Patel and Holtzman [147] given in Section 5.1.3 for the SIC receiver. Using Equation 5.7 and modifying it for user K in general, we obtain the decision statistic at the output of the RAKE receiver as:

$$z_{n,0}^{(k)} = d_n^{(k)} I_k + \sum_{j=1, j\neq k}^{K} d_n^{(k)} I_{j,k} + \eta_k \quad (5.20)$$

$$= d_n^{(k)} I_k + C_{k,0}, \quad (5.21)$$

where $I_{j,k} = \sum_{i=1}^{Q+W-1} b_i^{(j)} [b_i^{(k)}]^*$; $I_k = I_{k,k}$; and $\eta_k = \sum_{i=1}^{Q+W-1} n_i [b_i^{(k)}]^*$. The variable $C_{k,0}$ represents the interference plus noise for user 1 at the 0-th stage, where $C_{k,0} = \sum_{j=1, j\neq k}^{K} d_n^{(j)} I_{j,k} + \eta_k$.

In the first cancellation stage of Figure 5.15, the decision variable, $z_{n,0}^{(k)}$, is used to regenerate the received signal of user K and the regenerated signal is cancelled from the received composite signal. For user K, after the cancellation of the reconstructed signals of all the other $(K-1)$ users from the composite signal, the resultant signal is obtained as:

$$y_{i,1}^{(k)} = \sum_{u=1}^{K} d_n^{(u)} b_{i-(n-1)Q}^{(u)} + n_i - \sum_{j=1, j\neq k}^{K} z_{n,0}^{(j)} b_{i-(n-1)Q}^{(j)} \quad (5.22)$$

$$= \sum_{u=1}^{K} d_n^{(u)} b_{i-(n-1)Q}^{(u)} + n_i - \sum_{j=1, j\neq k}^{K} [d_n^{(j)} I_j + C_{j,0}] b_{i-(n-1)Q}^{(j)}. \quad (5.23)$$

$$(5.24)$$

If it is assumed that the data estimates of $d_n^{(j)}$ are the same as the transmitted data, $d_n^{(u)}$, and perfect CIR estimates are available, i.e. $d_n^{(u)} b_{i-(n-1)Q}^{(u)} \approx d_n^{(j)} I_j b_{i-(n-1)Q}^{(j)}$ for $j = u$, then this gives:

$$y_{i,1}^{(k)} = d_n^{(k)} b_{i-(n-1)Q}^{(k)} + n_i - \sum_{j=1, j\neq k}^{K} C_{j,0} b_{i-(n-1)Q}^{(j)}. \quad (5.25)$$

After this signal is processed by the RAKE receiver, the resultant decision statistic at the output of the first PIC iteration can be expressed as:

$$z_{n,1}^{(k)} = d_n^{(k)} I_k + \eta_k - \sum_{j=1, j\neq k}^{K} C_{j,0} I_{j,k}. \quad (5.26)$$

Therefore, after one stage of cancellation the SINR at the output of the PIC receiver is defined

Parameter	Value
Doppler frequency	80 Hz
Spreading ratio	64
Spreading sequence	Pseudo-random
Chip rate	2.167 MBaud
Frame burst structure	FRAMES Mode 1 Spread burst 1 [205]
Burst duration	577 μs
Receiver type	PIC, 2 cancellation stages

Table 5.10: Simulation parameters for AQAM/PIC-CDMA systems

as:

$$\gamma_o(n,k) = \frac{\text{Var}[d_n^{(k)} I_k]}{\text{Var}[\eta_k - \sum_{j=1,j\neq k}^{K} C_{j,0} I_{j,k}]}, \tag{5.27}$$

where the notation $\text{Var}[x]$ represents the variance of x.

Substituting Equations 5.14 in the numerator of Equation 5.27 results in:

$$\gamma_o(n,k) = \frac{|d_n^{(k)}|^2 |I_k|^2}{\text{Var}[\eta_k - \sum_{j=1,j\neq k}^{K} C_{j,0} I_{j,k}]} \tag{5.28}$$

$$= \frac{|d_n^{(k)}|^2 \ |I_k|^2}{\text{Var}[\eta_k] + \sum_{j=1,j\neq k}^{K} \text{Var}[C_{j,0}] \ \text{Var}[I_{j,k}]}. \tag{5.29}$$

Based on the SINR expression of Equation 5.29, the AQAM/PIC-CDMA regime can now be constructed in the next section.

5.2.4 AQAM/PIC-CDMA

In this section, the performance of the AQAM scheme in combination with the PIC-CDMA receiver is evaluated. The expression in Equation 5.29 is used to calculate the output SINR of the PIC and this is employed as the switching criterion for the AQAM/PIC-CDMA schemes. The simulation parameters are listed in Table 5.10. The PIC receiver employed two cancellation stages which was deemed sufficient, as evidenced by Figure 5.16.

The performance of our twin-mode AQAM/PIC-CDMA scheme is compared in Figure 5.18 for two different thresholds. Also shown for comparison are the BER curves for fixed-mode BPSK and 4-QAM. Again, it can be observed that increasing the switching threshold resulted in an improvement in BER performance terms, although this was accompanied by a drop in BPS throughput. For $t_1 = 20$ dB, the BER achieved at $E_s/N_0 = 14$ dB was similar to the BER of fixed-mode BPSK, but the throughput was approximately 1.8 BPS, which was higher than the 1 BPS offered by fixed-mode BPSK.

At this stage, the number of users was increased to $K = 8$ for the twin-mode AQAM/PIC-

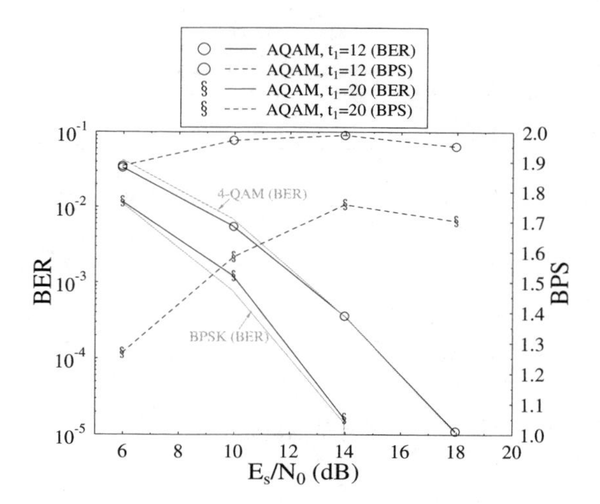

Figure 5.18: BER and BPS performance comparisons for twin-mode AQAM/PIC-CDMA schemes supporting $K = 2$ users and transmitting over the seven-path Bad Urban channel of Figure 3.20. The modulation mode was switched between BPSK and 4-QAM. The rest of the simulation parameters are listed in Table 5.9 and perfect channel estimation was assumed.

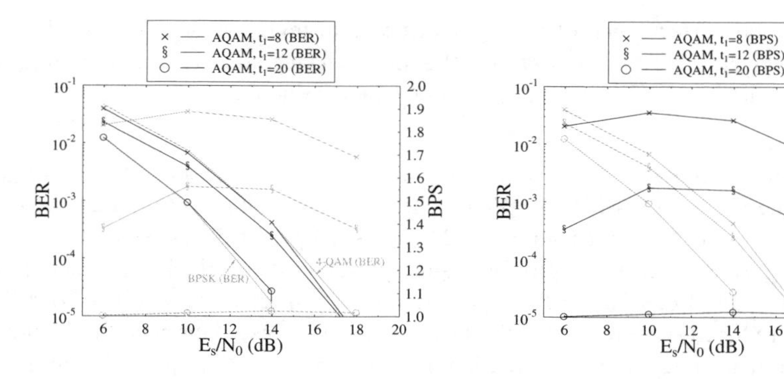

(a) BER comparisons are in bold, BPS comparisons are in grey.

(b) BPS comparisons are in bold, BER comparisons are in grey.

Figure 5.19: BER and BPS performance comparisons for **twin-mode AQAM/PIC-CDMA** schemes supporting $K = 8$ users and transmitting over the seven-path Bad Urban channel of Figure 3.20. The rest of the simulation parameters are listed in Table 5.10 and perfect channel estimation was assumed.

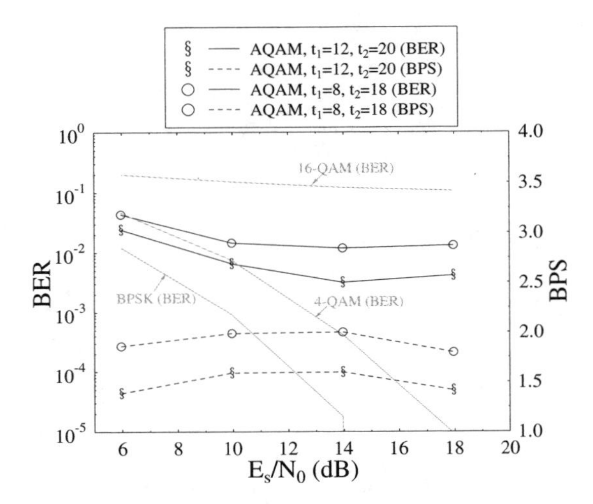

Figure 5.20: BER and BPS performance comparisons for a **triple-mode AQAM/PIC-CDMA** scheme supporting $K = 8$ users and transmitting over the seven-path Bad Urban channel of Figure 3.20. The modulation mode was switched between BPSK, 4-QAM and 16-QAM. The rest of the simulation parameters are listed in Table 5.10 and perfect channel estimation was assumed.

CDMA system. The performance of this system using different values of t_1 is shown in Figure 5.19. Here again, the improvement in BER performance and the corresponding drop in BPS throughput are in accordance with the increasing thresholds.

In order to further enhance the system performance, the twin-mode scheme was then extended to a triple-mode AQAM/PIC-CDMA scheme by including 16-QAM as one of the legitimate modulation mode choices. Figure 5.20 shows the performance of the triple-mode scheme supporting $K = 8$ users. From the results presented here, it can be observed that the inclusion of 16-QAM caused the BER performance to suffer from an error floor. This was because the reduced distance between the modulation constellation points in the 16-QAM mode led to inaccuracy in data estimation, thus leading to error propagation in the PIC stages.

The percentage of errors due to each modulation mode for the triple-mode AQAM/PIC-CDMA scheme is plotted in Figure 5.21. For BPSK and 4-QAM, as E_s/N_0 increased, the number of errors dropped or remained at a very low level. However, for 16-QAM, the error percentage increased as E_s/N_0 increased. This was because 16-QAM was chosen more frequently as the modulation mode and the 16-QAM BER suffered from an error floor when the PIC receiver was employed.

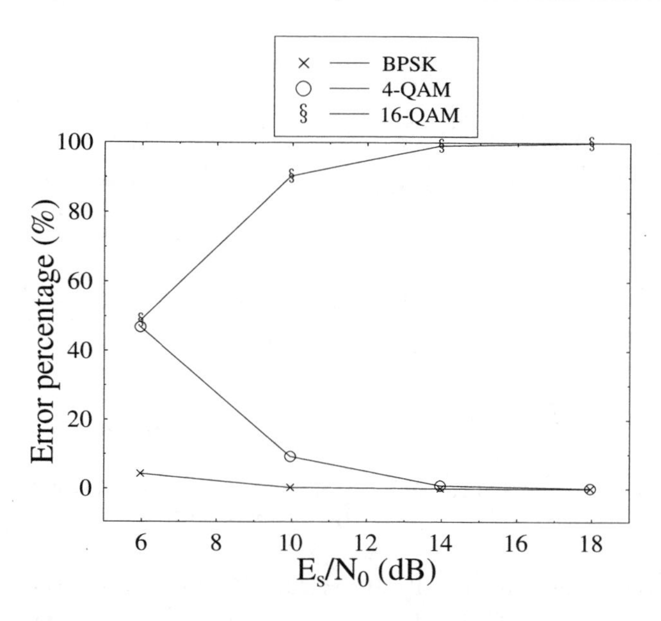

Figure 5.21: Error percentage due to the various modes in the triple-mode AQAM/PIC-CDMA scheme for $K = 8$ users.

5.3 Comparison of JD, SIC and PIC-CDMA Receivers for AQAM Transmission

In this section a comparison between the performance of JD-CDMA, SIC-CDMA and PIC-CDMA systems in the context of AQAM transmission will be discussed. The performance of the SIC receiver is dependent on the accuracy of the initial data estimates provided by the RAKE receivers. As has been observed in Section 5.1.4, the BER performance of the SIC-CDMA receiver improves as the spreading sequence length, Q, increases. In contrast, the value of Q does not have a significant effect on the performance of JD-CDMA receivers. Therefore, in our AQAM/CDMA comparisons, a spreading factor of $Q = 64$ was used instead of $Q = 16$ that was employed in the AQAM/JD-CDMA systems discussed in Chapter 4. The rest of the simulation parameters are summarized in Table 5.11.

Figure 5.22 compares the performance of the JD, the SIC receiver and the PIC receiver for the twin-mode, two-user AQAM-CDMA system. Here, the BER performance of all three receivers was kept as similar as possible and the performance comparison was evaluated on the basis of the BPS throughput. From Figure 5.22 it can be observed that the SIC and PIC receivers were capable of achieving a higher BPS throughput than the JD at low E_s/N_0 values and match it at high E_s/N_0 values. This was a consequence of the fact that the AQAM/CDMA system investigated supported only $K = 2$ users and therefore, the MAI was relatively low.

The performance comparison of the twin-mode, two-user AQAM/CDMA scheme was

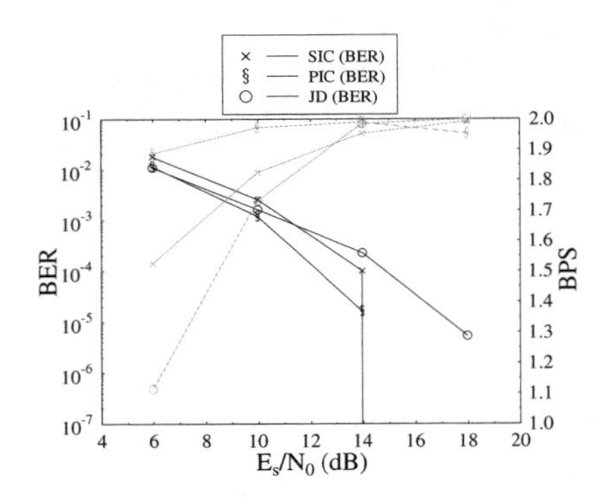

(a) BER comparisons are in bold, BPS comparisons are in grey.

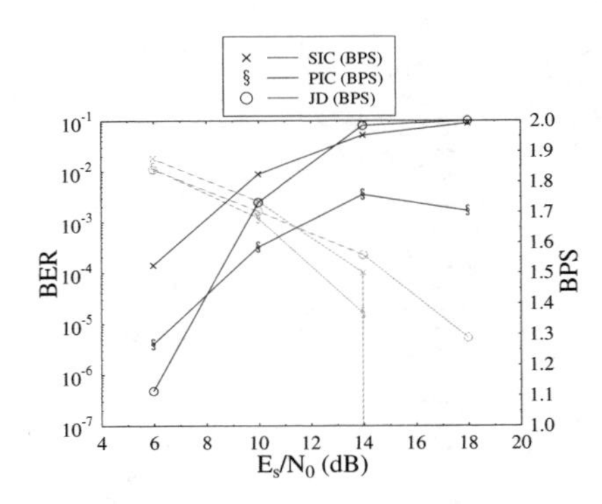

(b) BPS comparisons are in bold, BER comparisons are in grey.

Figure 5.22: Performance comparison of the JD, SIC and PIC CDMA receivers for **twin-mode** (BPSK, 4-QAM) AQAM transmission and $K = 2$ users over the **Bad Urban channel** of Figure 3.20. The rest of the simulation parameters are shown in Table 5.11.

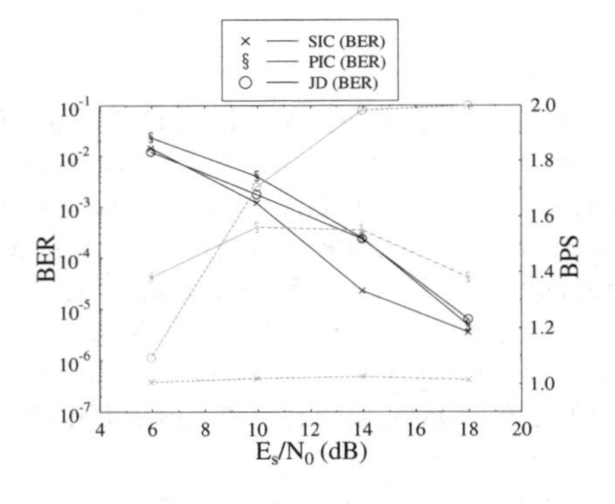

(a) BER comparisons are in bold, BPS comparisons are in grey.

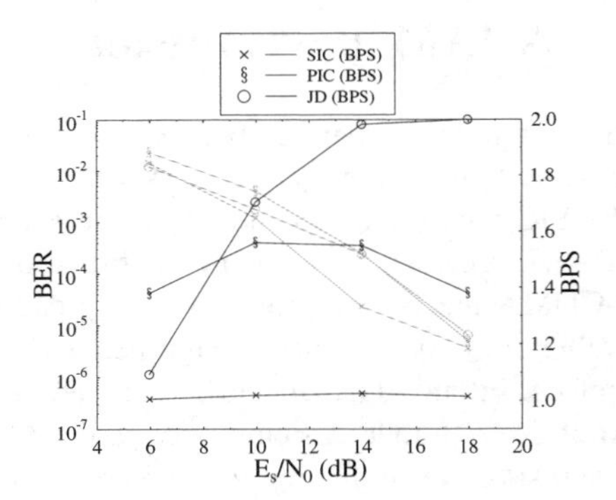

(b) BPS comparisons are in bold, BER comparisons are in grey.

Figure 5.23: Performance comparison of the JD, SIC and PIC CDMA receivers for **twin-mode** (BPSK, 4-QAM) AQAM transmission and $K = 8$ users over the **Bad Urban channel** of Figure 3.20. The rest of the simulation parameters are shown in Table 5.11.

Parameter	Value
Doppler frequency	80 Hz
Spreading ratio	64
Spreading sequence	Pseudo-random
Chip rate	2.167 MBaud
Frame burst structure	FRAMES Mode 1 Spread burst 1 [205]
Burst duration	577 μs

Table 5.11: Simulation parameters for the JD, SIC and PIC AQAM-CDMA systems

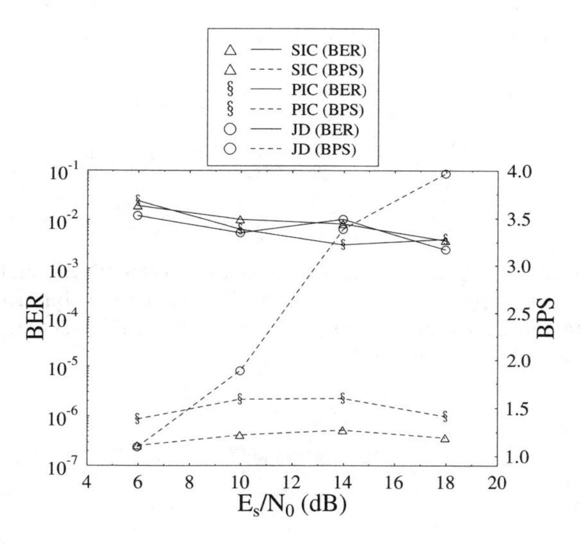

Figure 5.24: BER and BPS performance comparisons for **triple-mode** JD, SIC and PIC AQAM-CDMA schemes supporting $K = 8$ users over the **Bad Urban channel** of Figure 3.20. The modulation mode was chosen to be BPSK, 4-QAM or 16-QAM. The rest of the simulation parameters are tabulated in Table 5.11.

extended to systems supporting $K = 8$ users and the corresponding results are presented in Figure 5.23. Here again, similar BER performance was achieved, but there was a marked difference in BPS throughput terms. The BPS throughput of the JD was the highest, where approximately 1.9 BPS was achieved at $E_s/N_0 = 14$ dB. The PIC receiver outperformed the SIC receiver in BPS throughput terms, where the BPS throughput of the PIC receiver was approximately 1.55 BPS at $E_s/N_0 = 14$ dB compared to the approximately 1.02 BPS achieved by the SIC receiver. The two IC receivers suffered from MAI and they were unable to match the performance of the JD receiver. The PIC receiver outperformed the SIC receiver since two-parallel cancellation stages were implemented and the PIC receiver was capable of cancelling out a higher degree of MAI than the SIC receiver.

The previous performance comparisons between the three multi-user receivers were then

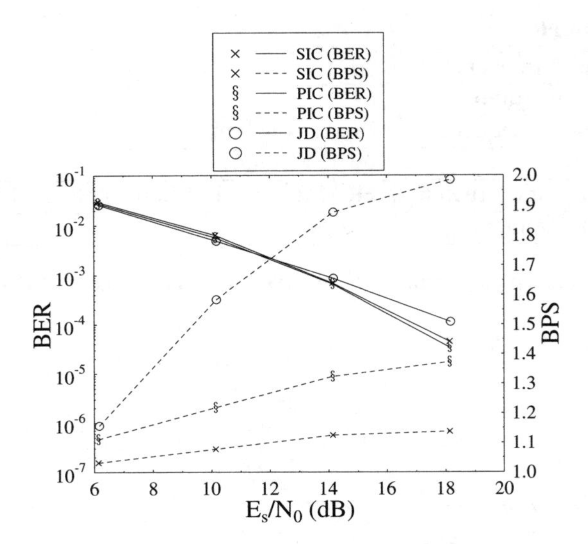

Figure 5.25: BER and BPS performance comparisons for **twin-mode** JD, SIC and PIC AQAM-CDMA schemes supporting $K = 8$ users over the **Typical Urban channel** of Figure 3.18. The modulation mode was chosen to be BPSK or 4-QAM. The rest of the simulation parameters are tabulated in Table 5.11.

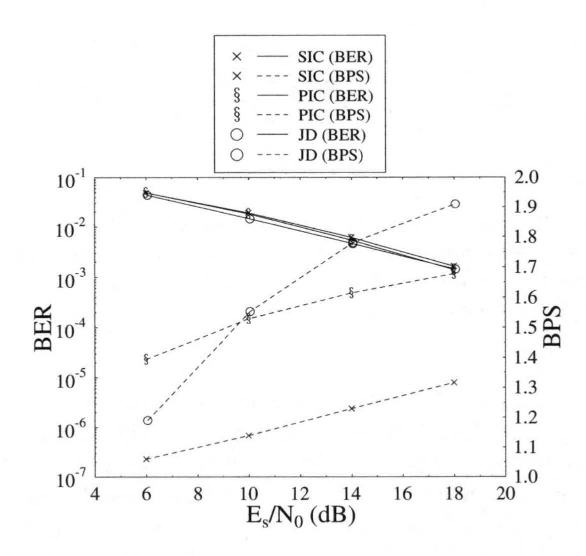

Figure 5.26: BER and BPS performance comparisons for **triple-mode** JD, SIC and PIC AQAM-CDMA schemes supporting $K = 8$ users over the **Rural Area channel** of Figure 3.15. The modulation mode was chosen to be BPSK or 4-QAM. The rest of the simulation parameters are tabulated in Table 5.11.

Parameter	Value
Doppler frequency	80 Hz
Modulation mode	4-QAM
Spreading sequence	Pseudo-random
Chip rate	2.167 MBaud
Frame burst structure	FRAMES Mode 1 Spread burst 1 [205]
Burst duration	577 μs

Table 5.12: Simulation parameters for the JD, SIC and PIC VSF-CDMA systems

extended to triple-mode AQAM systems that supported $K = 8$ users, as portrayed in Figure 5.24. For these systems, we can observe from the results that both the PIC and SIC receivers were unable to match the BPS performance of the JD receiver, when the multi-user receivers achieved similar BERs. This was because the increase in the number of users aggravated the MAI, thus degrading the ability of the RAKE receivers to provide reliable data estimates for interference cancellation. Here, again, the PIC receiver outperformed the SIC receiver in BPS throughput terms, since the PIC receiver had two interference cancellation stages and the data estimates were further refined in these stages.

The twin-mode AQAM scheme comparisons cast in the context of the PIC, SIC and JD receivers were also extended to the Typical Urban channel of Figure 3.18 and Rural Area channel of Figure 3.15. The corresponding performance results are presented in Figures 5.25 and 5.26 for the Typical Urban and Rural Area channels, respectively. The BER performances for all the three receivers were kept as similar as possible and the BPS throughput was compared. Similarly to the previous comparisons in the context of the Bad Urban channel, the BPS throughput of the JD was the highest of all three receivers, followed by the PIC receiver and lastly, the SIC receiver.

Let us now evaluate the performance comparison for the JD, SIC and PIC receivers in VSF-CDMA schemes.

5.4 Comparison of JD, SIC and PIC-CDMA Receivers for VSF Transmission

In this section, the JD, PIC and SIC receivers are compared in the context of adaptive variable spreading factor (VSF) schemes. The spreading factor used was varied adaptively, opting for $Q_1 = 64$, $Q_2 = 32$ or $Q_3 = 16$. Since the BER performance of the IC receivers degrades as the spreading factor decreases, the value of E_s/N_0 was not varied when the spreading factor was switched. Therefore, the transmission power of the signal varied adaptively, in order to maintain a constant value of E_s/N_0. This is in contrast to the previous VSF/JD-CDMA schemes of Section 4.4, where the transmission power was kept constant, in order to maintain a constant interference power inflicted to all the other users. The simulation parameters employed are summarized in Table 5.12.

Figure 5.27 portrays the associated BER and throughput comparisons for the adapt-

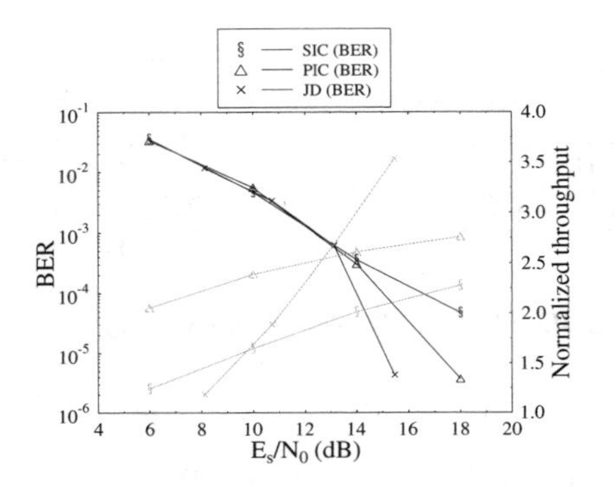

(a) BER comparisons are in bold, throughput comparisons are in grey.

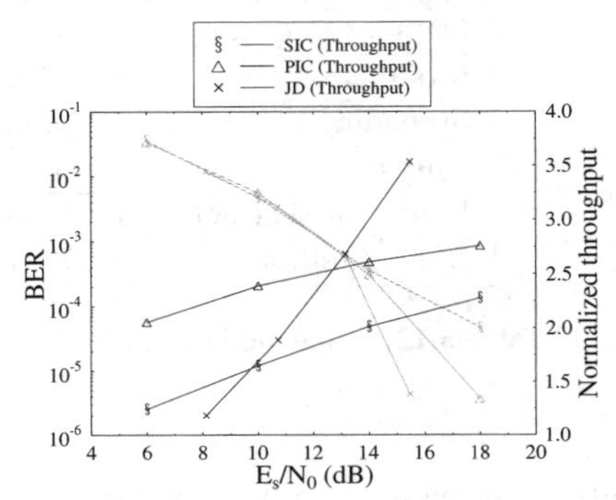

(b) Throughput comparisons are in bold, BER comparisons are in grey.

Figure 5.27: BER and throughput performance comparisons for triple-mode JD, SIC and PIC VSF-CDMA schemes supporting $K = 2$ users over the Bad Urban channel of Figure 3.20. The spreading factor was adaptively varied using $Q_1 = 64$, $Q_2 = 32$ or $Q_3 = 16$; and 4-QAM was used as the data modulation mode. The rest of the simulation parameters are tabulated in Table 5.12.

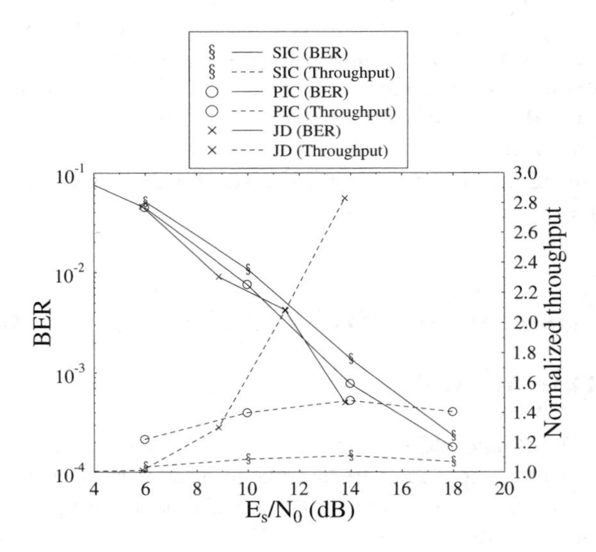

Figure 5.28: BER and throughput performance comparisons for triple-mode JD, SIC and PIC VSF-CDMA schemes supporting $K = 8$ users over the Bad Urban channel of Figure 3.20. The modulation mode was chosen to be 4-QAM. The rest of the simulation parameters are tabulated in Table 5.12.

ive VSF PIC-, SIC- and JD-CDMA schemes using 4-QAM. The minimum and maximum throughput values were approximately 68 kbits/s and 271 kbits/s, respectively. The throughput performance values were normalized to the minimum throughput of 68 kbits/s, thus giving minimum and maximum normalized throughput values of 1 and 4, respectively. We compared the normalized throughput of the JD-CDMA system to that of the PIC and SIC based systems, which showed similar BER performances. At low E_s/N_0 values the PIC and SIC receivers outperformed the JD receiver in throughput terms because the MAI was relatively low in a two-user scenario and the additive noise was the main corrupting influence at low E_s/N_0 values. However, as E_s/N_0 increased, the JD gradually outperformed the two cancellation receivers in both BER and BPS throughput performance terms, achieving a normalized throughput of 3.5 compared to 2.7 and 2 for the PIC and SIC receivers, respectively, at $E_s/N_0 = 16$ dB.

The above performance comparisons cast in the context of the triple-mode VSF schemes using 4-QAM were then extended to systems supporting $K = 8$ users. From the results presented in Figure 5.28, it can be seen that the JD consistently outperformed the PIC and SIC receivers in both BER and throughput performance terms. The VSF/IC-CDMA systems were unable to accommodate the variability in the channel conditions and hence their normalized throughput performance remained in the range of 1 to 1.5, compared to the JD which was capable of achieving a normalized throughput of 2.8 at $E_s/N_0 = 14$ dB.

Let us now finally summarize the conclusions of this chapter.

5.5 Chapter Summary and Conclusion

Successive interference cancellation (SIC) and parallel interference cancellation (PIC) receivers were introduced in this chapter. The performance of the SIC receiver improved as the spreading sequence length, Q and hence the symbol duration of QT_c increased with respect to the delay spread of the propagation channel. Despite this potential improvement, the SIC receiver exhibited an error floor in higher-order modulation modes due to the high error propagation across the cancellation stages. However, the SIC receiver exhibited a complexity that increased linearly with the number of users, K, compared to the complexity of the JD schemes, which were of the order of $O(K^3)$. For the PIC receiver it was shown that the BER performance improved as the number of stages increased. It was shown that no further improvement could be obtained upon invoking more than two stages of interference cancellation.

Both IC receivers were also investigated in the context of AQAM/CDMA schemes, which were outperformed by the JD-based twin-mode AQAM/CDMA schemes for a spreading factor of $Q = 64$ and $K = 8$ users. Both IC receivers were unable to provide good performances in the triple-mode AQAM/CDMA arrangement, since the BER curves exhibited error floors. In the VSF/CDMA systems the employment of variable spreading factors of $Q = 32$ and $Q = 16$ enabled the PIC and SIC receivers to provide a reasonable BER and throughput performance, which was nonetheless inferior to that of the JD. When the number of users in the system was increased, the PIC and SIC receivers were unable to exploit the variability in channel conditions in order to provide a higher information throughput, as opposed to the JD which showed performance gains in both the adaptive-rate AQAM and VSF CDMA schemes. However, the complexity of the IC receivers increased only linearly with the num-

K	JD			SIC			PIC		
	E_s/N_0	BPS	Cmplx	E_s/N_0	BPS	Cmplx	E_s/N_0	BPS	Cmplx
$K = 2$	10	1.82	683	10	1.73	1068	10	1.97	4207
$K = 2$	14	1.95	683	14	1.98	1068	14	1.99	4207
$K = 8$	10	1.7	2641	10	1.02	1869	10	1.55	4605
$K = 8$	14	1.98	2641	14	1.03	1869	14	1.38	4605

Table 5.13: Summary of **twin-mode AQAM-CDMA** results for the JD, SIC and PIC receivers, using a spreading factor of $Q = 64$ and transmitting over the seven-path Bad Urban channel of Figure 3.20. The E_s/N_0 values in the table are the values at which the AQAM-CDMA systems achieved the target BER of 1% or less. The complexity values are in terms of the number of additions plus multiplications required per detected data symbol.

K	JD			SIC			PIC		
	E_s/N_0	BPS	Cmplx	E_s/N_0	BPS	Cmplx	E_s/N_0	BPS	Cmplx
$K = 8$	10	1.88	2641	10	1.02	1869	10	1.99	4605
$K = 8$	14	3.39	2641	14	1.03	1869	14	2	4605

Table 5.14: Summary of **triple-mode AQAM-CDMA** results for the JD, SIC and PIC receivers, using a spreading factor of $Q = 64$ and transmitting over the seven-path Bad Urban channel of Figure 3.20. The E_s/N_0 values in the table are the values at which the AQAM systems achieved the target BER of 1% or less. The complexity values are in terms of the number of additions plus multiplications operations required per detected data symbol.

ber of CDMA users, K, compared to the joint detector, which exhibited a complexity proportional to $O(K^3)$. Tables 5.13, 5.14 and 5.15 summarize our performance 1 comparisons for all three multi-user detectors in terms of E_s/N_0 required to achieve the target BER of 1% or less, the normalized throughput performance and the complexity in terms of the number of operations required per detected symbol.

K	JD			SIC			SIC		
	E_s/N_0	Tp.	Cmplx	E_s/N_0	Tp.	Cmplx	E_s/N_0	Tp.	Cmplx
$K = 2$	10.7	1.89	1496	10	1.64	1068	10	2.39	4207
$K = 2$	13.1	2.68	1496	14	2.01	1068	14	2.61	4207
$K = 8$	11.4	2.07	18298	10	1.08	1869	10	1.4	4605
$K = 8$	13.7	2.83	18298	14	1.11	1869	14	1.48	4605

Table 5.15: Summary of **triple-mode VSF-CDMA** results for the JD, SIC and PIC receivers, where 4-QAM was employed as the modulation mode and transmission was conducted over the Bad Urban channel of Figure 3.20. The number of users in the system was $K = 2$ and $K = 8$. The E_s/N_0 in the table are the values at which the VSF systems achieved the target BER of 1% or less. The notation "Tp." denotes the throughput values normalized to the throughput of the fixed-rate scheme employing $Q = 64$ and 4-QAM. The complexity values indicated are valid for the modem mode that incurred the highest complexity for each receiver.

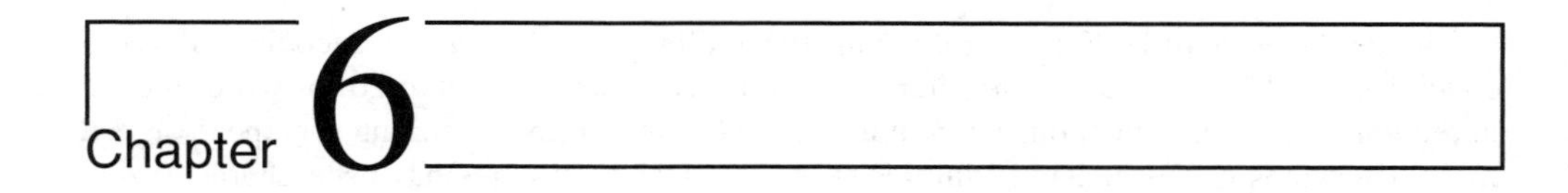

Iterative Parallel Interference Cancellation Using Convolutional Codes, Turbo Codes, LDPC Codes, TCM and TTCM[1]

H. Wei, S. X. Ng and L. Hanzo

6.1 Introduction

As we argued in Chapter 5, Parallel Interference Cancellation (PIC) constitutes a powerful interference rejection technique, which becomes particularly attractive when the CDMA system benefits from accurate power control. This is because in this case all users' signals arrive at the base station at a similar power level and therefore the power-based ranking of the users – as necessitated in Serial Interference Cancellation (SIC) – would become unreliable.

The system also investigated benefits from the employment of various channel codecs, such as convolutional coding, as well as Trellis Coded Modulation (TCM) and Turbo Trellis Coded Modulation (TTCM). The latter two schemes achieve an enhanced system performance without any bandwidth expansion. Additionally, the principle of turbo equalization is invoked, where the receiver jointly carries out soft interference cancellation and channel decoding. The attraction of this combined operation is that the channel decoder refrains from

[1] *Single- and Multi-Carrier CDMA Multi-User Detection, Space-Time Spreading, Synchronization, Networking and Standards.* L.Hanzo, L-L.Yang, E-L.Kuan and K.Yen,
©2003 John Wiley & Sons, Ltd. ISBN 0-470-86309-9

making premature "hard" decisions concerning the transmitted data. Instead, it provides soft information for the PIC scheme and after a few iterations a substantially improved performance may be achieved, which may approach the single-user bound at the cost of a moderate implementational complexity.

To elaborate a little further, following the philosophy of iterative decoding of turbo codes [5, 79, 197] in recent years, iterative multi-user detection captured growing interest in the wireless communications community. In [249] an iterative decoding scheme designed for synchronous Code Division Multiple Access (CDMA) systems has been characterized. The algorithm proposed in [249] has a computational complexity proportional to the order of $O(2^{Kv})$, where K is the number of users supported, and v is the channel codec's memory length. This iterative multi-user detector exhibits near single-user performance. However, its excessive complexity renders its real-time implementations unrealistic, except when the number of users supported is low.

In [250,251] several schemes have been proposed for reducing the implementational complexity imposed. In the set of these schemes, Interference Cancellation (IC) based iterative multi-user detection exhibits the lowest complexity, rendering the complexity proportional to the order of $O(K \cdot 2^v)$.

In [252] an IC-based iterative Multi-user Detector (MUD) was developed, which exhibits a further reduced implementational complexity. The iterative receiver advocated in this chapter obeys a similar structure to that proposed in [252], except for employing different channel coding schemes and modulation schemes. In [250, 251] the proposed iterative multi-user detectors invoked soft estimation of the real-valued BPSK modulated signal. In our approach, we develop a method which is capable of generating soft estimates of the complex transmitted symbol by exploiting the channel coded bits' A Posteriori Probabilities (APP). Our initial investigations described in Section 6.4 are based on employing a convolutional code. In Section 6.3 we also propose a novel symbol probability-based iterative multi-user detection assisted receiver for employment in a Trellis Coded Modulation (TCM) and Turbo Trellis Coded Modulation (TTCM) aided system. Symbol-based *Maximum aposteriori* (MAP) algorithms [5] are employed for both the TCM decoder and the TTCM decoder, which are capable of feeding back the symbol probabilities to the soft-symbol estimation and interference cancellation stages. More specifically, this algorithm relies on utilizing the probabilities of the channel-coded information symbols for symbol estimation, rather than on employing the individual APP of the coded bits.

This chapter is organized as follows: in Section 6.2 we introduce the basic philosophy of the TCM and TTCM schemes as well as that of the symbol-based MAP algorithm. In Section 6.3 we continue our discourse by describing the IC-based iterative multi-user detector scheme. Section 6.4 discusses the performance of the convolutional coding assisted iterative multi-user detector, while Section 6.5 characterizes the performance of the TCM-based iterative multi-user detector. These evaluations are followed in Section 6.6 by the performance study of the TTCM-aided iterative multi-user detector. Similar studies are conducted in Section 6.7 and 6.8 in the context of turbo decoded and LDPC decoded systems. Finally, in Section 6.9 we offer our conclusions and outline the topics of our future research.

6.2 Coded Modulation Schemes

TCM [253] was originally proposed for communicating over Gaussian channels and it was later further developed for applications in mobile communications [254]. Turbo Trellis Coded Modulation (TTCM) [255] is a more recent joint coding and modulation scheme that has a structure similar to that of the family of power-efficient binary turbo codes [5, 79, 197], but employs TCM schemes as component codes. Both TCM and TTCM use symbol-based interleavers and Set-Partitioning (SP)-based signal labelling [5].

6.2.1 TCM-Assisted Turbo Decoder

The TCM-aided turbo MUD operates similarly to the corresponding module of a trained convolutional coding based turbo equalizer [256]. The sole difference is the replacement of the convolutional decoder by a TCM decoder. Hence, a bit-to-symbol converter is placed before the TCM decoder for converting the Log Likelihood Ratio (LLR) values to symbol probabilities, which are necessary for facilitating TCM decoding. Similarly, a symbol-to-bit converter is employed at the output of the TCM decoder for feeding the users' bits to the MUD's input. The calculation of the symbol probabilities from the probabilities of the input bits is based on the following relationship [5]:

$$\begin{aligned} Prob(Symbol = A_i) &= \Pi_j Prob(Bit_j(Symbol) = Bit_j(A_i)), \\ & \quad i = 0, \cdots, Q,\ j = 0, \cdots, K, \end{aligned} \tag{6.1}$$

where Q is the number of symbols in the modulation constellation used, K is the number of bits per symbol, A_i represents the symbols of the modem's constellation and the function $Bit_j(A_i)$ returns the value of the j-th bit of symbol A_i. The assumption implicitly stipulated here is that the bits of a TCM symbol are independent of each other. This, however, is not a valid assumption, since channel coding has deliberately imposed a certain amount of correlation or interdependence on the bits, in order to be able to exploit this redundancy for correcting transmission errors. Hence the symbol probability calculation is somewhat inaccurate. Fortunately the effects of this inaccuracy are gradually mitigated by the iterative detection process, as iterative detection ensues. Similarly to the bit-to-symbol conversion of Equation 6.1, the symbol-to-bit conversion procedure is based on the following relationship:

$$\begin{aligned} Prob(Bit_j = Bit) &= \sum_i Prob[Symbol_i; Bit_j(Symbol_i) = Bit] \\ & \quad i = 0, \cdots, Q,\ j = 0, \cdots, K. \end{aligned} \tag{6.2}$$

This formula is accurate in the sense that it does not require any assumptions concerning the correlation of the input symbols.

It is widely recognized that the LLR is defined as [5]:

$$LLR(Bit) = ln\left[\frac{Prob(Bit = 1)}{Prob(Bit = 0)}\right]. \tag{6.3}$$

Given the LLR value, namely $LLR(Bit)$ of a binary bit, we can calculate the probability of $Bit = +1$ or $Bit = -1$ as follows. Remembering that $Prob(Bit = -1) = 1 - Prob(Bit = +1)$, and taking the exponent of both sides in Equation 6.3 we can write:

$$e^{LLR(Bit)} = \frac{Prob(Bit = +1)}{1 - Prob(Bit = +1)}. \tag{6.4}$$

Hence we have:

$$Prob(Bit = 1) = \frac{e^{LLR(Bit)}}{1 + e^{LLR(Bit)}}, \tag{6.5}$$

and

$$Prob(Bit = -1) = \frac{1}{1 + e^{LLR(Bit)}}. \tag{6.6}$$

Then, assuming that the bits of a symbol are independent of each other, the probability of a $symbol$, which is represented by the bits $Bit^1, \ldots, Bit^n$, can be calculated as in Equation 6.1, where we have $symbol \in (0, \ldots, 2^n - 1)$ for the 2^n-ary modulation scheme used. The probability of the trellis transition from states s' to state s, commonly defined as $\gamma(s', s)$, in the context of the non-binary MAP decoder [5, 79, 197], can be calculated as:

$$\gamma(s', s) = \eta_{Symbol} \cdot \prod_{i=1}^{i=n} Prob(Bit^i), \tag{6.7}$$

where $symbol$ is the trellis transition branch label associated with state s' to state s, and η_{Symbol} is the "a priori" information of the $symbol$. Then the α and β values of the forward and backward recursion involved in the MAP algorithm can be obtained from:

$$\alpha_k(s) = \sum_{s'} \gamma_k(s', s) \cdot \alpha_{k-1}(s') \tag{6.8}$$

$$\beta_{k-1}(s') = \sum_{s} \gamma_k(s', s) \cdot \beta_k(s). \tag{6.9}$$

The number of transitions emerging from a specific trellis state is equal to 2^k, where k is the number of information bits per n-bit modulation symbol. The coding rate used is $R = \frac{k}{n}$, where $k = n - 1$. Therefore the log-MAP decoder used is non-binary, when $k > 1$. By contrast, if $k = 1$, then the number of transitions emerging from a trellis state is equal to $2^1 = 2$, i.e. a binary MAP decoder is used. The *a posteriori* probability (APP) of the

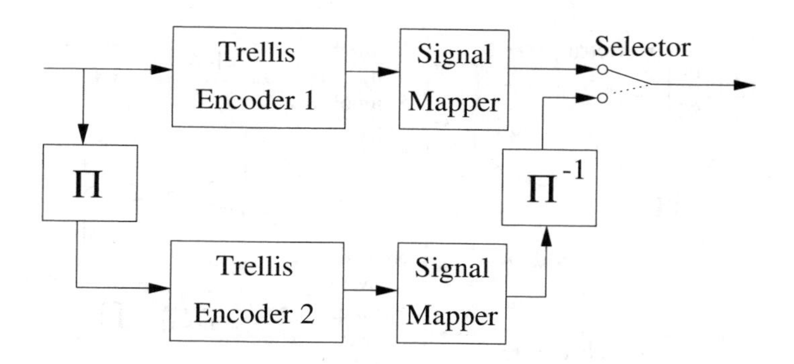

Figure 6.1: Turbo trellis–coded modulation encoder.

information symbol u_t, $u_t \in (0, \ldots, 2^k - 1)$ at time instant t can be computed as [5,79,197]:

$$APP(u_t) = \sum_{s' \rightarrow s, u_t} \alpha_{k-1}(s') \cdot \gamma_k(s', s) \cdot \beta_k(s). \tag{6.10}$$

The final decoded information symbol at instant t is the hard decision-based symbol generated from these APP values. However, we have to feed back the LLR values of all the n number of coded bits of a symbol, rather than just the LLRs of the k information bits, to the PSP equalizer, after improving their reliability by the channel decoder. The APP of the coded symbol x_t, $x_t \in (0, \ldots, 2^n - 1)$ at time instant t can be computed from:

$$APP(x_t) = \sum_{s' \rightarrow s, x_t} \alpha_{k-1}(s') \cdot \gamma_k(s', s) \cdot \beta_k(s), \tag{6.11}$$

while Equation (6.10) formulated the APP of the original encoded information symbol. The probability of bit i assuming a value of binary 1 in a coded symbol x is calculated from:

$$Prob(Bit^i = 1) = \sum_{x=0}^{x=2^n-1} APP(x^i = 1), \tag{6.12}$$

where x^i denotes the binary value at bit position i of the symbol x, $x^i \in (0, 1)$ and in verbal terms the probability of $Bit^i = 1$ is given by the sum of the probabilities of all symbols from the set of $2^n - 1$ number of phasors, which host a binary 1 at bit position i. A similar procedure is invoked for determining $Prob(Bit^i = 0)$ and finally the LLR of the bits can be computed from Equation 6.3.

6.2.2 TTCM-Assisted Turbo Decoder

An extension of the turbo-PSP equalizer employing Turbo Trellis-Coded Modulation (TTCM) has also been considered. TTCM was proposed in [255], where the information bits are transmitted only once, while the parity bits are provided alternatively by the two con-

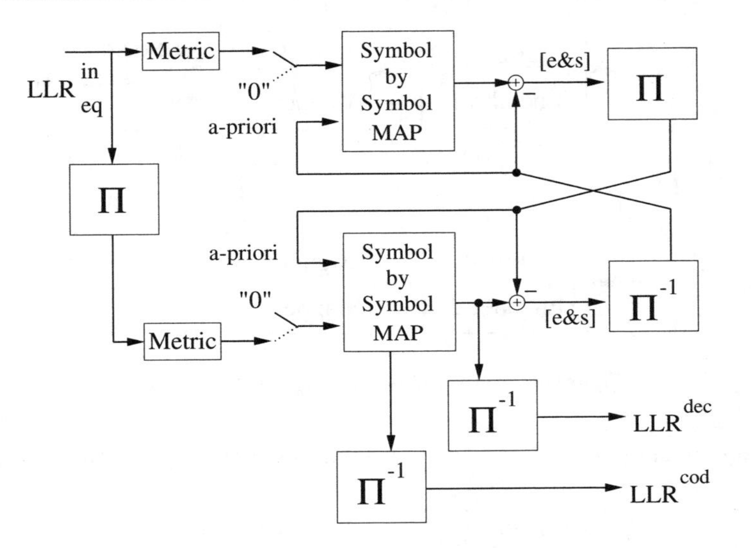

Figure 6.2: TTCM decoder schematic.

stituent TCM encoders [5]. The schematic of the TTCM encoder [5] is shown in Figure 6.1, which comprises two identical TCM encoders linked by the symbol interleaver Π [5,79,197]. The TTCM decoder structure shown in Figure 6.2 is similar to that of binary turbo codes, except for the difference in the nature of the information passed from one decoder to the other. Each decoder alternately processes its corresponding encoder's channel impaired output symbol, and then the other encoder's channel impaired output symbol. The information bits, i.e. the systematic bits, are constituted by the corresponding systematic TCM encoder's output bits received over the channel in both cases. The systematic information and the parity information are transmitted together in the same symbol. Hence, the systematic information component cannot be separated from the extrinsic information, since the noise that affects the parity component of a TTCM symbol also affects the systematic information component. The output of each symbol-based MAP decoder of Figure 6.2 can be split into two components [5,79,197]:

1) the $a\ priori$ component, and

2) the amalgamated i.e. inseparable (extrinsic and systematic) $[e\&s]$ component.

Each decoder of Figure 6.2 has to pass only the $[e\&s]$ component to the other decoder, which is written in parentheses in order to emphasize the inseparability of the extrinsic and systematic components. The reason the *a priori* information component is subtracted from the output of the symbol-based MAP decoder of Figure 6.2 is that this information component was generated by the other constituent decoder and hence it must not be fed back to it. Otherwise the probability estimates of the two decoders become dependent on each other, and this lack of independent estimates prevents them from enhancing the decoder's decision reliability. The LLR_{eq}^{in} output of the equalizer is forwarded to the "metric" calculation block of Figure 6.2, in order to generate a set of 2^n number of symbol reliabilities. The selectors at the input of the symbol-by-symbol MAP decoder of Figure 6.2 select the current symbol's reliabilities from the "metric" calculation block, if the current received symbol corresponds to the specific component decoder concerned. Otherwise depuncturing will be applied, where the reliabilities of the symbols are set to "0" corresponding to the absence of *a priori* information. The "metric" calculation block provides the decoder with the parity and systematic $[p\&s]$ information, and the second input to the symbol by symbol MAP decoder of Figure 6.2 is the *a priori* information acquired from the other component decoder. The MAP decoder then provides the *a posteriori* information constituted by the $(a\ priori + [e\&s])$ components as the output. Then the *a priori* information is subtracted from the *a posteriori* information, again, for the sake of ensuring that information is not used more than once in the other decoder. The resultant $[e\&s]$ information is appropriately interleaved (or de-interleaved) in order to create the *a priori* input of the other constituent decoder. This decoding process will continue iteratively, in order to generate an improved version of the set of symbol reliabilities for the other component decoder. One iteration comprises the decoding of the received symbols by both of the component decoders. Then, the *a posteriori* information of the lower component decoder of Figure 6.2 will be de-interleaved, in order to extract $(n - 1)$ decoded information bits per symbol. On the other hand, the *a posteriori* information of the n coded bits is de-interleaved, in order to convert the LLR^{cod} to symbol probabilities with the aid of equation 6.2 and 6.4. Finally, the symbol probabilities will be fed back to the input of the MUD as shown in Figure 6.4.

6.3 Iterative Parallel Interference Cancellation

6.3.1 The Concept of Interference Cancellation

Recall from Chapter 5 that the family of Interference Cancellation (IC) techniques may be divided into two broad categories, namely successive interference cancellation (SIC) [257] and parallel interference cancellation (PIC) [138]. Both techniques rely on the philosophy that if all users' decision bits have been detected without decision errors, then the multiple access interference (MAI) can be readily recreated by remodulating the detected bits and subtracting them from the received MAI-contaminated signal, provided that we have perfect knowledge of the channel parameters. In comparison to the family of joint detectors (JD) [78], the interference cancellation algorithms exhibit a lower complexity, which increases linearly as a function of the number of users. In this treatise we focus our attention on a powerful PIC scheme. Interference cancellation techniques typically employ "hard" decisions concerning the transmitted bits or symbols. However, hard decision techniques are prone to error propagation effects. Hence, in order to improve the achievable performance, we rely on the soft estimation of the transmitted bits or symbols during the process of interference cancellation.

Figure 6.3 shows the structure of a parallel interference cancellation scheme, where the soft estimates of the transmitted symbols, namely $\hat{b}_k$, which are output by the channel decoder, are utilized for reconstructing the transmitted signal. Then, for the sake of decontaminating the received signal of each user, the reconstructed signals of all the other users are subtracted from the composite multi-user signal, and the resultant signal $\hat{y}_k$ is processed by a matched filer or RAKE receiver as seen in Figure 6.3. These steps of modulated signal reconstruction, interference cancellation and desired signal re-estimation stage are repeated as many times, as the number of affordable iterations employed in the multi-user detector.

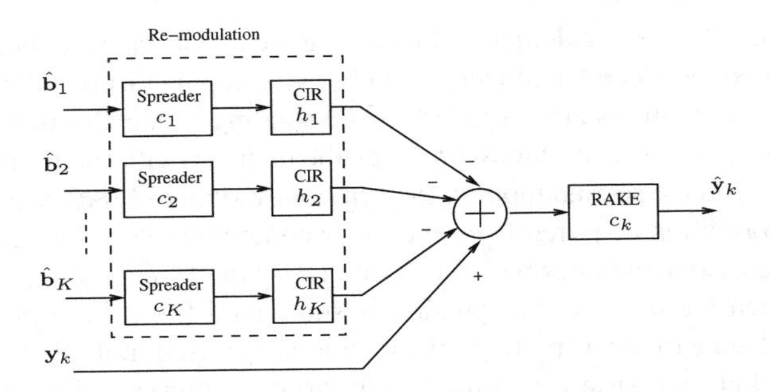

Figure 6.3: Schematic of a single PIC stage.

6.3.2 Estimation of the M-ary Coded Symbol

Let us assume that an M-QAM modulation scheme [92] is employed. Then the M-ary TCM or TTCM-based symbols [5] can be written as: $s_i = u_i + jv_i, i = 1, ...M$. For the TTCM and TCM schemes the k-th user's channel decoder will output all the M-ary soft symbol probabilities $\mathbf{Pr}\{b_k = s_i\}, k = 1, ..., K, i = 1, ..., M$ of the K users supported. Hence, the estimated symbol $\hat{b}_k$ may be obtained as:

$$\hat{b}_k = \sum_{i=1}^{M} \mathbf{Pr}\{b_k = s_i\} \cdot s_i, \tag{6.13}$$

where s_i is a TCM or TTCM symbol. By contrast, if a convolutional code is employed in this system, the k-th user's channel decoder will output the soft APP of the channel coded bits. Hence a transformation between the APP of the channel coded bits and the symbol probabilities is needed. First, we may generate the symbol probabilities $Prob(symbol)$ according to Equation 6.1, again, bearing in mind that this equation assumes the independence of the bits of a symbol, and this inaccuracy is compensated by the iterative detection process. Once we acquired the symbol probabilities, we can estimate the symbols according to Equation 6.13.

6.3.3 Iterative Multi-user Detection

Figure 6.4 shows the schematic of the PIC-assisted iterative multi-user detector, where the Matched Filter's (MF) output is given by: $\mathbf{y} = [y_1, ..., y_K]^T$. The vector $\hat{\mathbf{b}}$ contains the estimated symbols of all the K users, which are represented by the symbol probabilities or by the soft APP of the channel coded bits output by the channel decoder, which can be represented for the K users as: $\hat{\mathbf{b}} = [\hat{b}_1, ..., \hat{b}_K]^T$. Once we acquired a soft-estimate of a symbol, soft MAI cancellation is performed. The soft output signal vector $\hat{\mathbf{y}}$ generated by the interference cancellation process of Figure 6.4 is expressed as: $\hat{\mathbf{y}} = [\hat{y}_1, ..., \hat{y}_K]^T$. The acronyms Π and Π^{-1} denote the channel interleaver and deinterleaver, respectively.

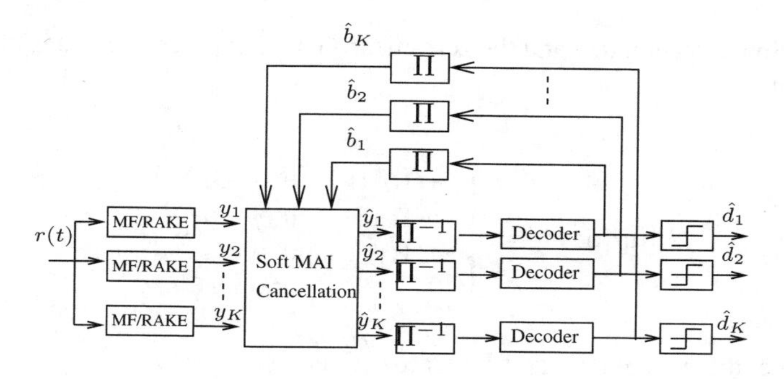

Figure 6.4: Schematic of the iterative multi-user detector

6.3.3.1 Iterative MUD for Synchronous DS-CDMA

Let us now consider a special case, when the synchronous DS-CDMA system communicates over a non-dispersive Additive White Gaussian Noise (AWGN) channel. According to [78], we can express the received signal vector $\mathbf{y} = [y_1, ..., y_K]^T$ containing the output of the MF related to each of the K users seen in Figure 6.3 as:

$$\mathbf{y} = \mathbf{R} \cdot \mathbf{A} \cdot \mathbf{b} + \mathbf{n}, \tag{6.14}$$

where $\mathbf{R}$ is the normalized $K \times K$ dimensional cross-correlation matrix of the user's spreading code. The element in the ith row and jth column of the matrix is denoted as ρ_{ij}, and we have $\rho_{ij} = \rho_{ji}$. Furthermore, the vector $\mathbf{b}$ containing each transmitted symbol of the K users is expressed as $\mathbf{b} = [b_1, ..., b_K]^T$, and $\mathbf{n} = [n_1, ..., n_K]^T$ is the $K \times 1$ dimensional vector of noise samples at the output of the MF, while $\mathbf{A}$ = diag$\{A_1, ..., A_K\}$ is a diagonal matrix, which contains the amplitude of each user's signal.

Let us define $\hat{\mathbf{b}}_k$ as a vector which equals $\hat{\mathbf{b}}$ except that its kth element is zero. Then the signal vector $\hat{\mathbf{y}}_k$ containing the desired signal of user k in its kth position, which was decontaminated from the effects of MAI, can be written as:

$$\hat{\mathbf{y}_k} = \mathbf{R} \cdot \mathbf{A} \cdot (\mathbf{b} - \hat{\mathbf{b}}_k) + \mathbf{n}, \quad k = 1, ..., K. \tag{6.15}$$

Explicitly, once a sufficiently reliable soft estimate $\hat{\mathbf{b}}$ becomes available, the signal vector components hosted by $\hat{\mathbf{y}}$ become free from MAI, which are now contaminated only by the channel noise.

6.3.3.2 Iterative MUD for Asynchronous DS-CDMA

According to [78], in an asynchronous DS-CDMA scenario the K-dimensional received signal vector $\mathbf{y}$ containing the signal components of the K users can also be expressed as:

$$\mathbf{y}[i] = \mathbf{R}^T[1] \cdot \mathbf{A} \cdot \mathbf{b}[i+1] + \mathbf{R}[0] \cdot \mathbf{A} \cdot \mathbf{b}[i] + \mathbf{R}[1] \cdot \mathbf{A} \cdot \mathbf{b}[i-1] + \mathbf{n}[i], \tag{6.16}$$

where i is the time instant index and the zero-mean Gaussian process $\mathbf{n}[i]$ has the crosscorrelation matrix of:

$$E[\mathbf{n}[i]\mathbf{n}^T[j]] = \begin{cases} \sigma^2\mathbf{R}^T[1], & \text{if } j = i+1; \\ \sigma^2\mathbf{R}[0], & \text{if } j = i; \\ \sigma^2\mathbf{R}[1], & \text{if } j = i-1; \\ \mathbf{0}, & otherwise. \end{cases} \tag{6.17}$$

Furthermore, the matrix $\mathbf{R}[1]$ and $\mathbf{R}[0]$ are defined by:

$$\mathbf{R}_{jk}[0] = \begin{cases} 1, & \text{if j = k;} \\ \rho_{jk}, & \text{if } j < k; \\ \rho_{kj}, & \text{if } j > k, \end{cases} \tag{6.18}$$

$$\mathbf{R}_{jk}[1] = \begin{cases} 0, & \text{if } j \geq k; \\ \rho_{kj}, & \text{if } j < k. \end{cases} \tag{6.19}$$

Let us define the vectors $\hat{\mathbf{b}}_k[i+1], \hat{\mathbf{b}}_k[i], \hat{\mathbf{b}}_k[i-1] \;\; k = 1, \ldots, K$ which contain the estimates of the transmitted signal at the time index of $i+1, i$ and $i-1$, where the component corresponding to user k is zero. Therefore, after the soft MAI cancellation process obeying Equation 6.15 is invoked, the signal vector $\hat{\mathbf{y}}_k[i], \;\; k = 1, \ldots, K$ containing in its kth position the signal of the kth user, which was decontaminated from the MAI can be represented as:

$$\begin{aligned} \hat{\mathbf{y}}_k[i] \;\; = \;\; & \mathbf{R}^T[1] \cdot \mathbf{A} \cdot (\mathbf{b}[i+1] - \hat{\mathbf{b}}_k[i+1]) + \mathbf{R}[0] \cdot \mathbf{A} \cdot (\mathbf{b}[i] - \hat{\mathbf{b}}_k[i]) \\ & + \mathbf{R}[1] \cdot \mathbf{A} \cdot (\mathbf{b}[i-1] - \hat{\mathbf{b}}_k[i-1]) + \mathbf{n}[i]. \end{aligned} \tag{6.20}$$

According to Equation 6.20, we can observe that the signal vector $\hat{\mathbf{y}}_k[i]$ contains components, which are free from multiple access interference, provided that the estimated symbols $\hat{\mathbf{b}}_k[i+1], \hat{\mathbf{b}}_k[i], \hat{\mathbf{b}}_k[i-1]$ are reliable. In comparison to the synchronous scenario of Equation 6.15, we can see that this iterative multi-user receiver operating in an asynchronous environment is more prone to error propagation, because its reliable operation requires three consecutive correctly estimated K-dimensional symbol vectors. Even if only one of these three vectors contains errors, this will lead to the incorrectly cancelled interference and to inevitable error propagation between the various user decisions. Hence, as expected, it is more difficult to ensure that the PIC-based iterative multi-user receiver converges in an asynchronous scenario, than in a synchronous system.

When communicating over a dispersive multipath channel, this PIC-assisted multi-user receiver maintains the same structure and requires the same operations as in a non-dispersive channel, except for the remodulation process and for the soft MAI cancellation stage, since all the K users have different Channel Impulsive Responses (CIR), as seen in Figure 6.3.

Having described the operation of the PIC-assisted multi-user detector, let us now consider its performance. We arranged for all these coded schemes to have a similar complexity.

Coded Scheme	memory length m	inner iterations
CC	6	
TCM	6	
TTCM	3	2/4
TC	3	2/4
LDPC		8/16

Table 6.1: The basic simulation parameters of the coding schemes employed by the iterative PIC.

Hence we fixed the non-iterative coded schemes' memory length – namely that of the CC and TCM arrangements – to be $m = 6$. This resulted in a total of 64 trellis states. By contrast, the code memory of the iterative schemes – namely that of the TC and TTCM schemes was adjusted to $m = 3$, resulting in eight trellis states per decoder. Since there are two constituent channel decoders, in case of one iteration we had a total of 16 trellis states. Hence, when aiming for the same total complexity, we could afford a total of four iterations. However, at this stage we set the number of inner iterations to two, while at the final outer iteration stage the number of inner iteration was set to four. This design choice was motivated by the observation that the overall system performance has not improved, when we tentatively opted for employing the total affordable number of inner channel decoder iterations, namely four. This observation was deemed to be the consequence of unnecessarily iterating in the inner channel decoding loop, when the corresponding channel decoder's inner decoding loop was unable to reverse the effects of channel impairments imposed on gravely damaged bits. This phenomenon often even aggravated the situation in case of bits having a high confidence measure, despite their corrupted polarity. When the moderate number of two – rather than four – inner channel decoding iterations was used, the outer loop still had a chance of reversing the polarity of the corresponding high-confidence, but wrong-polarity bits. An additional benefit of this philosophy was that the total receiver complexity was reduced. The corresponding number of inner iterations for the LDPC scheme was set to 8 and 16, respectively, since an LDPC iteration is significantly less complex than a convolutional turbo decoding iteration. All these parameters are listed in Table 6.1.

In our system, a random channel interleaver having a length of $L = 1920$ was adopted, and an $N = 15$-chip m-sequence based spreading code was employed. In the synchronous AWGN environments $K = 15$ users were supported, while in the more interference-contaminated asynchronous AWGN channel a reduced number of $K = 7$ users communicated. Finally, in the two-path equal weight Rayleigh fading channel $K = 10$ users were supported.

6.4 Performance of the Iterative PIC Assisted by Convolutional Decoding

We adopted the LOG-MAP technique [5, 258] as the convolutional channel decoding algorithms, which required a slight modification. Specifically, the procedure outlined in [258]

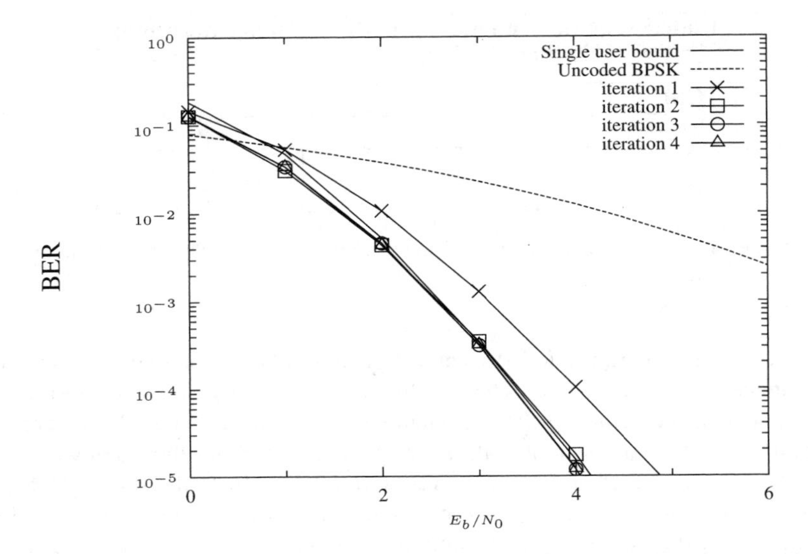

Figure 6.5: The PIC-based iterative multi-user detector's **BER** versus E_b/N_0 performance, when communicating over a **synchronous non-dispersive AWGN** channel, while supporting $K = 15$ users, each encoded with the aid of a $\frac{1}{2}$ rate, $m = 6$ convolutional code. **QPSK** modulation is employed.

only provides the *a posteriori* probabilities of the information bits, but not those of the parity bits. However, in turbo detection algorithms all the bits have to be fed back to the detector's input and hence we additionally need the *a posteriori* probabilities of the parity bits for both soft estimation and for interference cancellation. These can be obtained in the same manner as highlighted in [249] [252]. Apart from this modification, the LOG-MAP algorithm invoked here is the same as that described in [5, 258].

In this section, we study the performance of a convolutional coded DS/CDMA system supporting $K = 15$ users with the aid of an m-sequence-based spreading code having $N = 15$ chips. The low-complexity half rate convolutional code employed in this system has a low memory length of $m = 3$, which is sufficient for employment in an iterative receiver. We investigate the achievable performance for transmission over two types of of channels, an asynchronous two-path uncorrelated Rayleigh fading environment and a synchronous non-dispersive AWGN environment. Figure 6.5 shows the achievable BER performance, when communicating over a synchronous non-dispersive AWGN environment, while Figure 6.7 shows the performance attained, when communicating over an asynchronous two-path uncorrelated Rayleigh fading channel.

From Figures 6.5, 6.6 and 6.7, we observe that the receiver assisted by the low-complexity convolutional decoder is capable of approaching the optimum single-user bound after two iterations. More explicitly, the receiver only obtains insignificant further gains after two iterations, when communicating over a synchronous non-dispersive AWGN channel or over a two-path uncorrelated Rayleigh fading channel having an equal-weight, chip-spaced CIR.

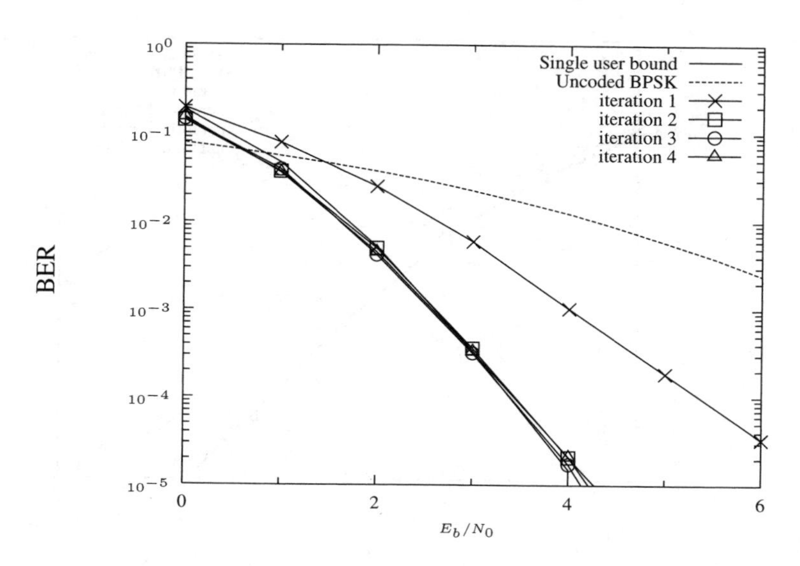

Figure 6.6: The PIC-based iterative multi-user detector's **BER** versus E_b/N_0 performance, when communicating over an **asynchronous non-dispersive AWGN** channel, while supporting $K = 7$ users , each encoded with the aid of a $\frac{1}{2}$ rate, $m = 6$ convolutional code. **QPSK** modulation is employed.

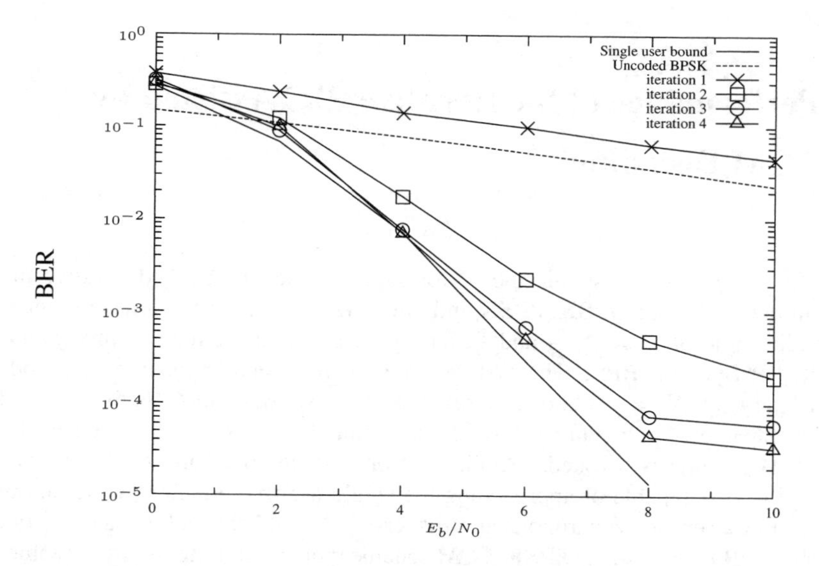

Figure 6.7: The PIC-based iterative multi-user detector's **BER** versus E_b/N_0 performance, when communicating over an **asynchronous two-path uncorrelated Rayleigh** channel having a chip-spaced equal-weight CIR, while supporting $K = 10$ users , each encoded with the aid of a $\frac{1}{2}$ rate, $m = 6$ convolutional code. **QPSK** modulation is employed.

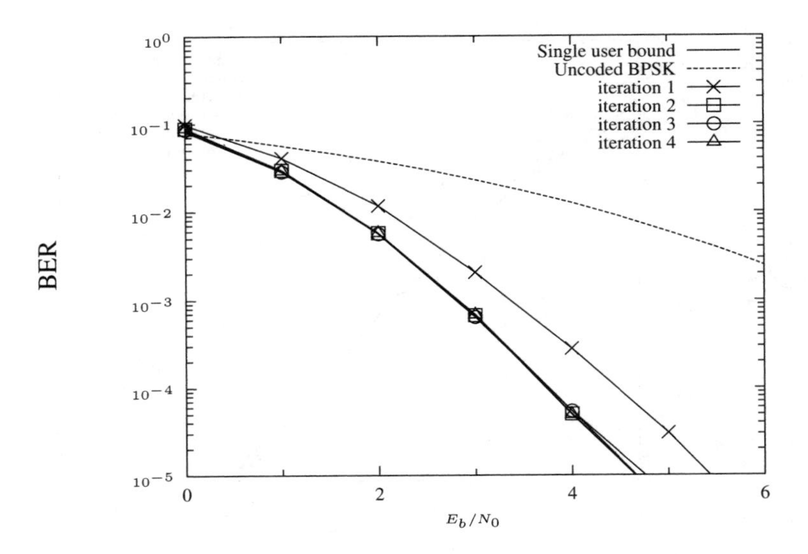

Figure 6.8: The PIC-based iterative multi-user detector's **BER** versus E_b/N_0 performance, when communicating over a **synchronous non-dispersive AWGN** channel, while supporting $K = 15$ users. A $\frac{1}{2}$ rate **QPSK TCM** scheme having an effective throughput of 1 BPS was invoked.

6.5 Performance of the Iterative PIC Assisted by TCM Decoding

In this section we will investigate the performance of the TCM-based iterative multi-user detector in the context of **QPSK**, **8PSK** and **16QAM** schemes. The corresponding schemes have a coding rate of $R = \frac{1}{2}, \frac{2}{3}$ and $\frac{3}{4}$, and a corresponding effective throughput of 1, 2 and 3 Bits Per Symbol (BPS). The system employs m-sequences as spreading codes, each having a length of $N = 15$ chips. From Figures 6.8, 6.9 and 6.10, it is clear that the receiver is capable of achieving the single-user bound after 3 iterations, when the $\frac{1}{2}$ rate **QPSK TCM** scheme is invoked. Similarly, from Figures 6.11 and 6.12 we observe that the receiver is also capable of approaching the single-user bound after two iterations, when communicating over the synchronous non-dispersive AWGN channel using the $\frac{2}{3}$ rate **8PSK TCM** scheme. By contrast, the **8PSK TCM** scheme requires four iterations for achieving the single-user performance, when communicating over an asynchronous non-dispersive AWGN channel. From Figures 6.13 and 6.14, we can see that when a $\frac{3}{4}$ rate **16QAM TCM** scheme is invoked, the receiver also requires two iterations for approaching the single-user bound, when transmitting over a synchronous AWGN channel, while necessitating four iterations, when communicating over an asynchronous AWGN channel.

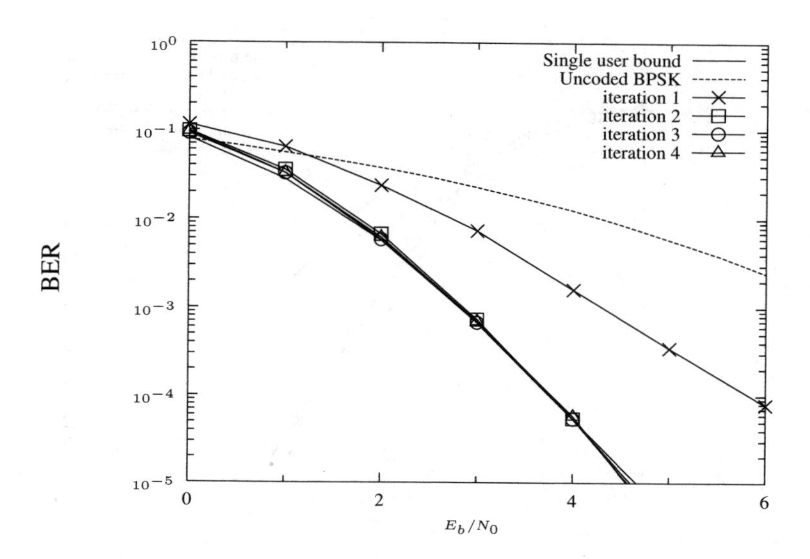

Figure 6.9: The PIC-based iterative multi-user detector's **BER** versus E_b/N_0 performance, when communicating over an **asynchronous non-dispersive AWGN** channel, while supporting $K = 7$ users. A $\frac{1}{2}$ rate **QPSK TCM** scheme having an effective throughput of 1 BPS was invoked.

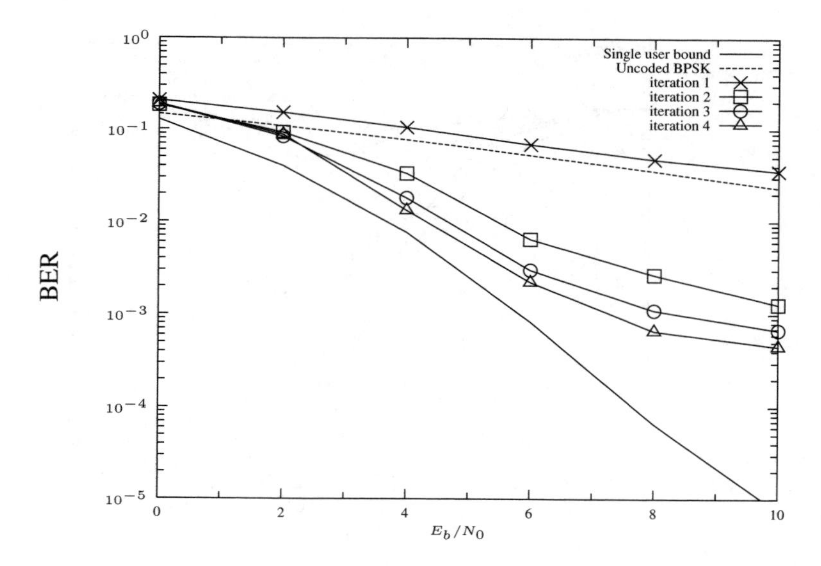

Figure 6.10: The PIC-based iterative multi-user detector's **BER** versus E_b/N_0 performance, when communicating over an **asynchronous two-path equal-weight uncorrelated Rayleigh** channel, while supporting $K = 10$ users. A $\frac{1}{2}$ rate **QPSK TCM** scheme having an effective throughput of 1 BPS was invoked.

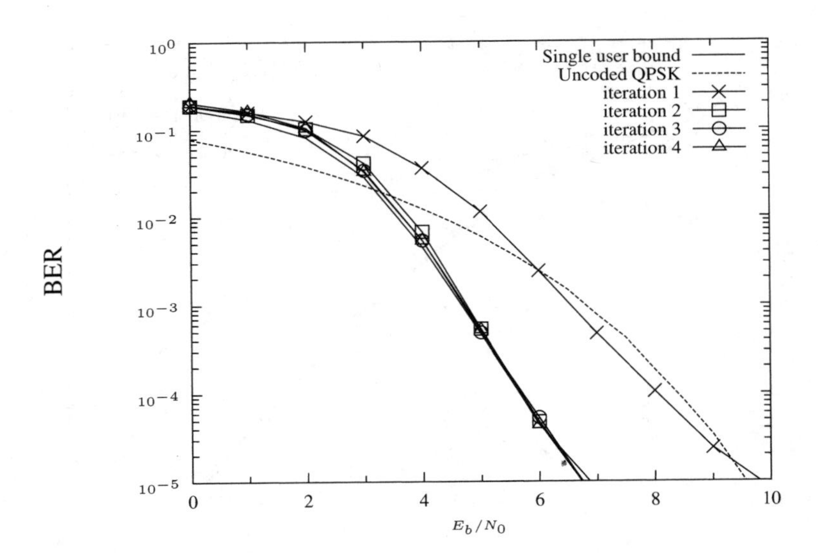

Figure 6.11: The PIC-based iterative multi-user detector's **BER** versus E_b/N_0 performance, when communicating over a **synchronous non-dispersive AWGN** channel, while supporting $K = 7$ users. A $\frac{2}{3}$ rate **8PSK TCM** scheme having an effective throughput of 2 BPS was invoked.

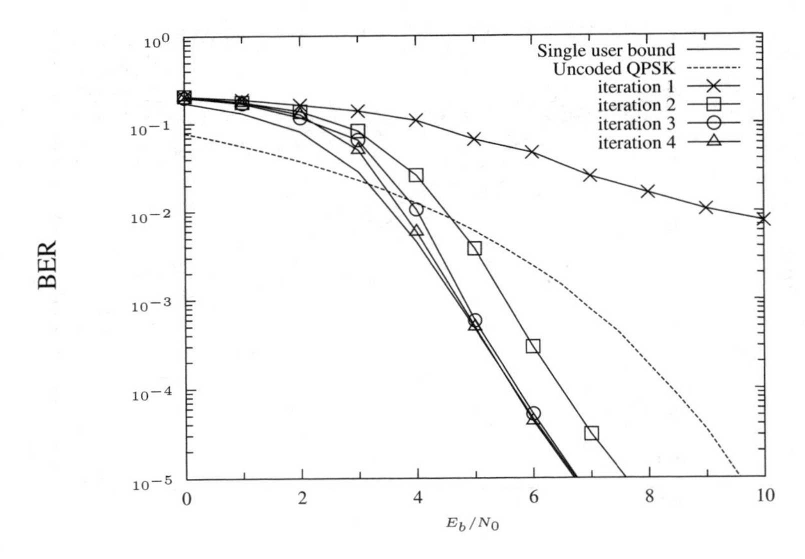

Figure 6.12: The PIC-based iterative multi-user detector's **BER** versus E_b/N_0 performance, when communicating over an **asynchronous non-dispersive AWGN** channel, while supporting $K = 7$ users. A $\frac{2}{3}$ rate **8PSK TCM** scheme having an effective throughput of 2 BPS was invoked.

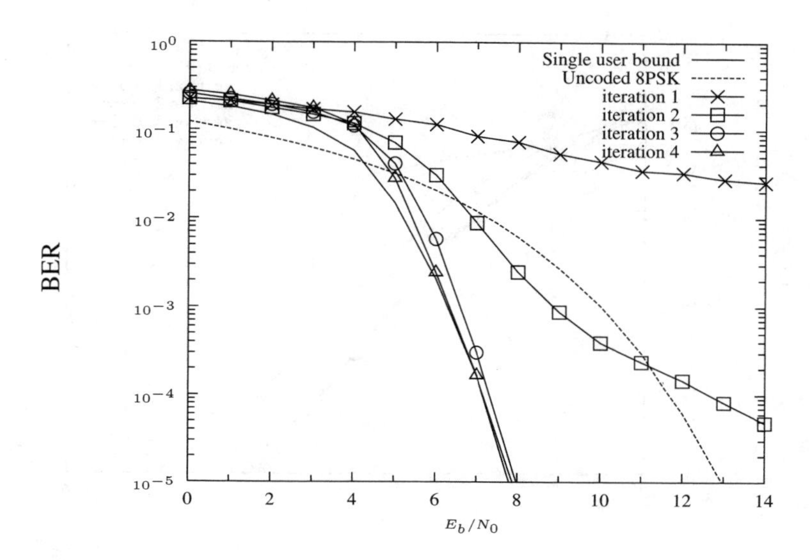

Figure 6.13: The PIC-based iterative multi-user detector's **BER** versus E_b/N_0 performance, when communicating over a **synchronous non-dispersive AWGN** channel, while supporting $K = 7$ users. A $\frac{3}{4}$ rate **16QAM TCM** scheme having an effective throughput of 3 BPS was invoked.

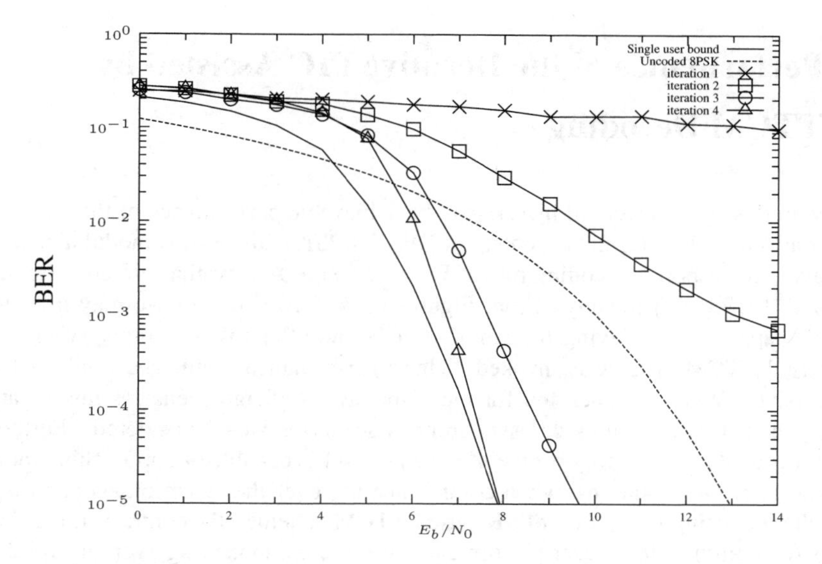

Figure 6.14: The PIC-based iterative multi-user detector's **BER** versus E_b/N_0 performance, when communicating over an **asynchronous non-dispersive AWGN** channel, while supporting $K = 7$ users. A $\frac{3}{4}$ rate **16QAM TCM** scheme having an effective throughput of 3 BPS was invoked.

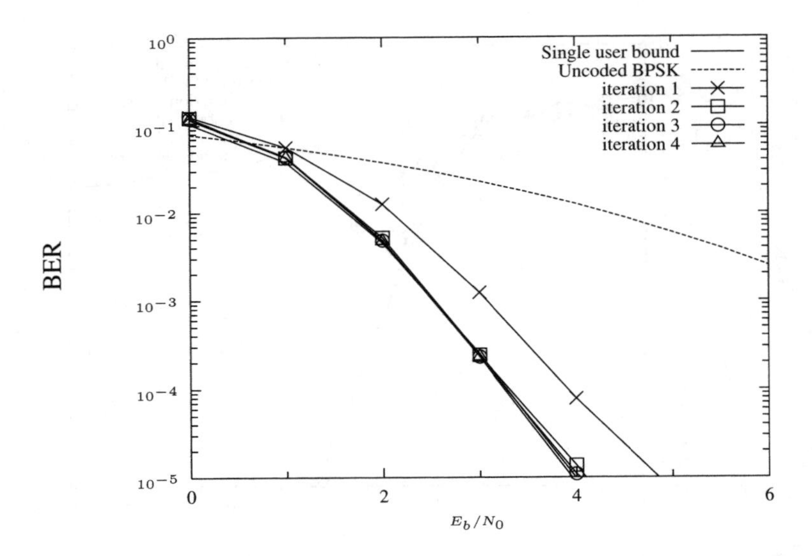

Figure 6.15: The PIC-based iterative multi-user detector's **BER** versus E_b/N_0 performance, when communicating over a **synchronous non-dispersive AWGN** channel, while supporting $K = 15$ users. A $\frac{1}{2}$ rate **QPSK TTCM** scheme having an effective throughput of 1 BPS was invoked.

6.6 Performance of the Iterative PIC Assisted by TTCM Decoding

Similarly, in this section we will investigate the achievable performance of the TTCM-based iterative multi-user detector in the context of QPSK, 8PSK and 16QAM modulation schemes, which have a corresponding coding rate of $\mathbf{R} = \frac{1}{2}, \frac{2}{3}$ and $\frac{3}{4}$, as well as effective throughputs of 1, 2 and 3 BPS, respectively. From Figures 6.15, 6.16 and 6.17 we may infer that the receiver is capable of achieving the single-user bound after two iterations, when a $\frac{1}{2}$ rate **QPSK**-based TTCM scheme is invoked. These performance trends are valid for both the non-dispersive AWGN channel and for the dispersive Rayleigh scenarios investigated, regardless, whether synchronous or asynchronous scenarios were considered. Furthermore, from Figures 6.18 and 6.19 we observe that the receiver is capable of approaching the single-user bound after two iterations, when communicating over the synchronous non-dispersive AWGN channel using the $\frac{2}{3}$-rate **8PSK**-based TTCM scheme. By contrast, it requires four iterations to attain the single-user performance, when communicating over an asynchronous non-dispersive AWGN channel. From Figures 6.20 and 6.21 it becomes clear that when the $\frac{3}{4}$-rate **16QAM** TTCM scheme is invoked, the receiver also requires two iterations for maintaining a near single-user bound, when transmitted over a synchronous AWGN channel. However, the number of required iterations becomes four, when communicating over an asynchronous AWGN channel.

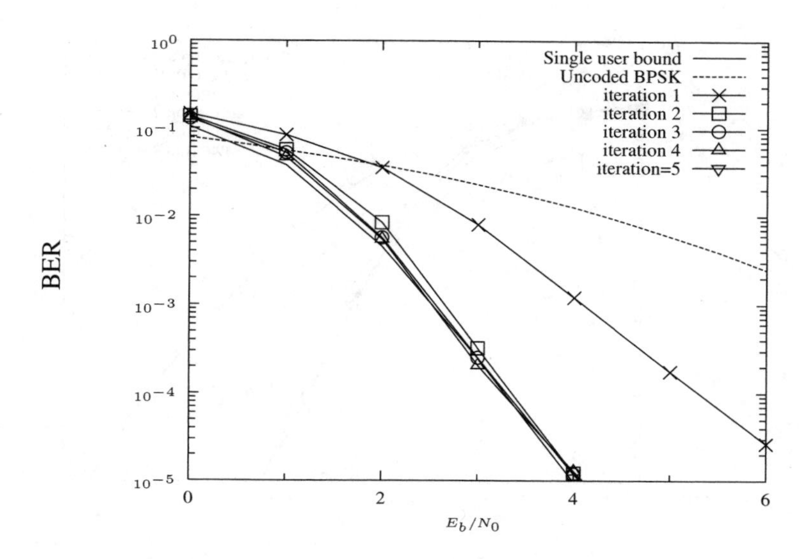

Figure 6.16: The PIC-based iterative multi-user detector's **BER** versus E_b/N_0 performance, when communicating over an **asynchronous non-dispersive AWGN** channel, while supporting $K = 7$ users. A $\frac{1}{2}$ rate **QPSK TTCM** scheme having an effective throughput of 1 BPS was invoked.

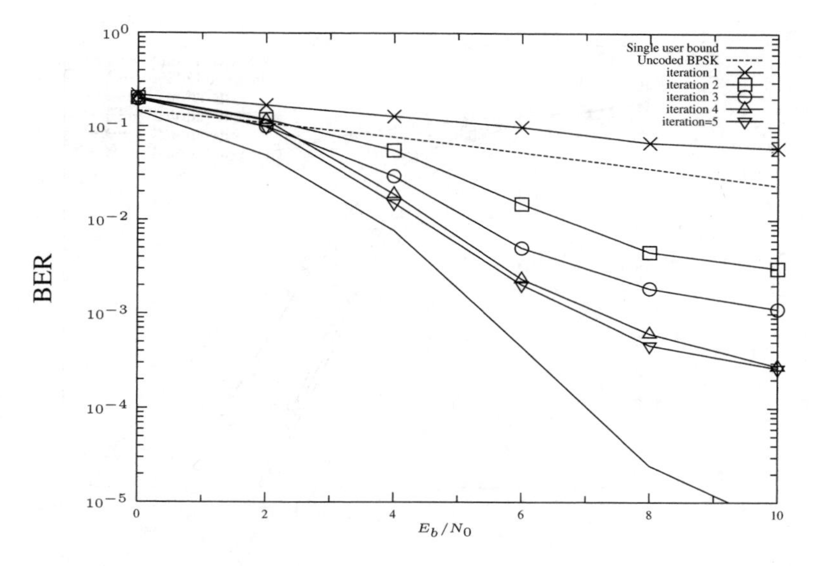

Figure 6.17: The PIC-based iterative multi-user detector's **BER** versus E_b/N_0 performance, when communicating over an **asynchronous two-path equal-weight uncorrelated Rayleigh** channel, while supporting $K = 10$ users. A $\frac{1}{2}$ rate **QPSK TTCM** scheme having an effective throughput of 1 BPS was invoked.

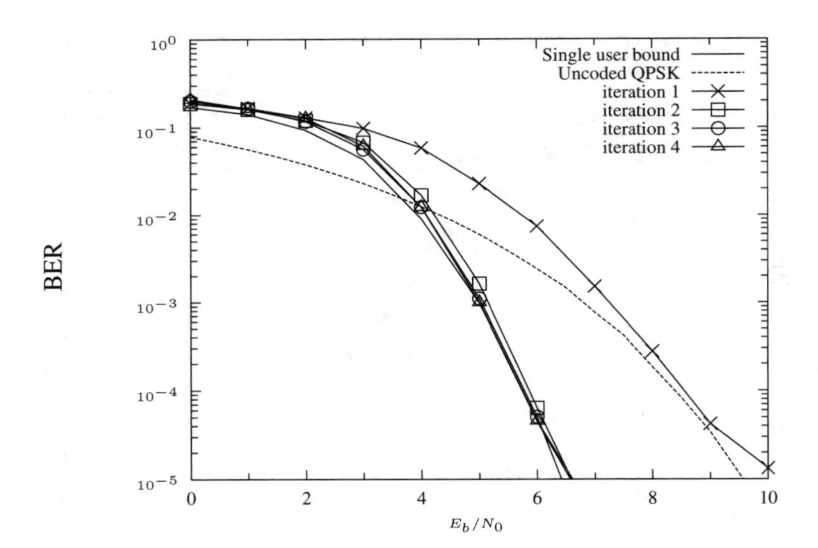

Figure 6.18: The PIC-based iterative multi-user detector's **BER** versus E_b/N_0 performance, when communicating over a **synchronous non-dispersive AWGN** channel, while supporting $K = 7$ users. A $\frac{2}{3}$ rate **8PSK TTCM** scheme having an effective throughput of 2 BPS was invoked.

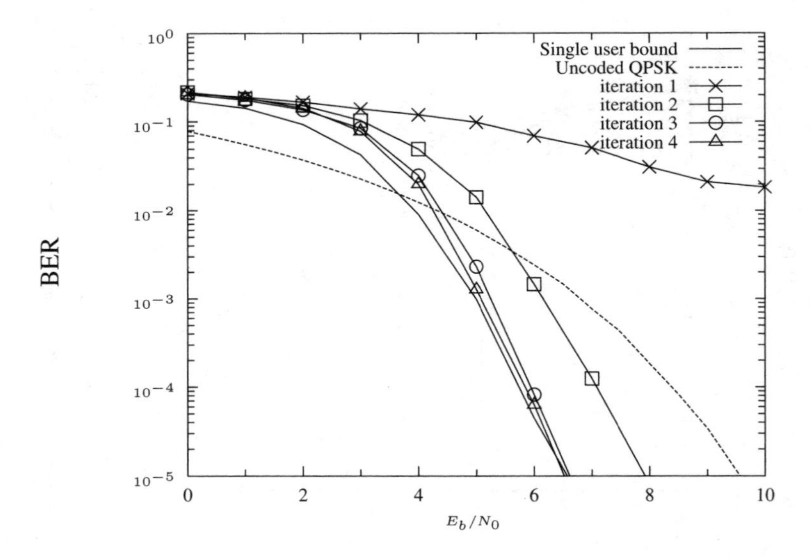

Figure 6.19: The PIC-based iterative multi-user detector's **BER** versus E_b/N_0 performance, when communicating over an **asynchronous non-dispersive AWGN** channel, while supporting $K = 7$ users. A $\frac{2}{3}$ rate **8PSK TTCM** scheme having an effective throughput of 2 BPS was invoked.

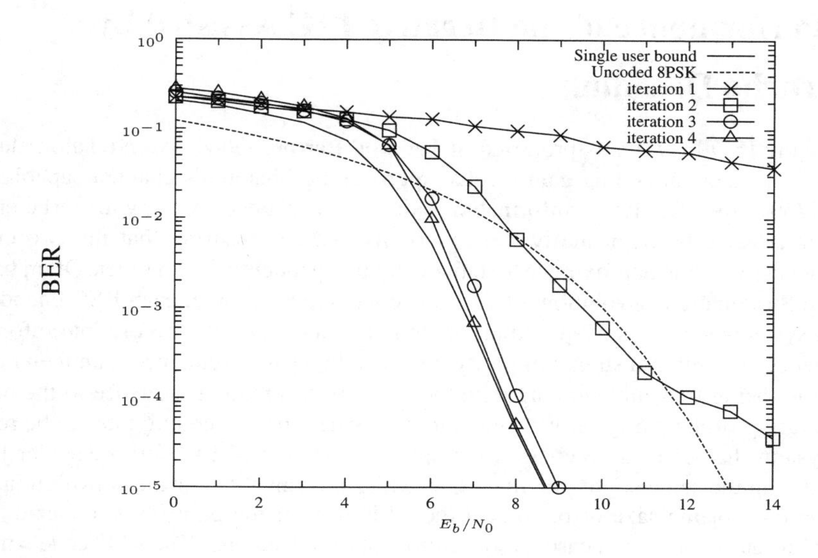

Figure 6.20: The PIC-based iterative multi-user detector's **BER** versus E_b/N_0 performance, when communicating over a **synchronous non-dispersive AWGN** channel, while supporting $K = 7$ users. A $\frac{3}{4}$ rate **16QAM TTCM** scheme having an effective throughput of 3 BPS was invoked.

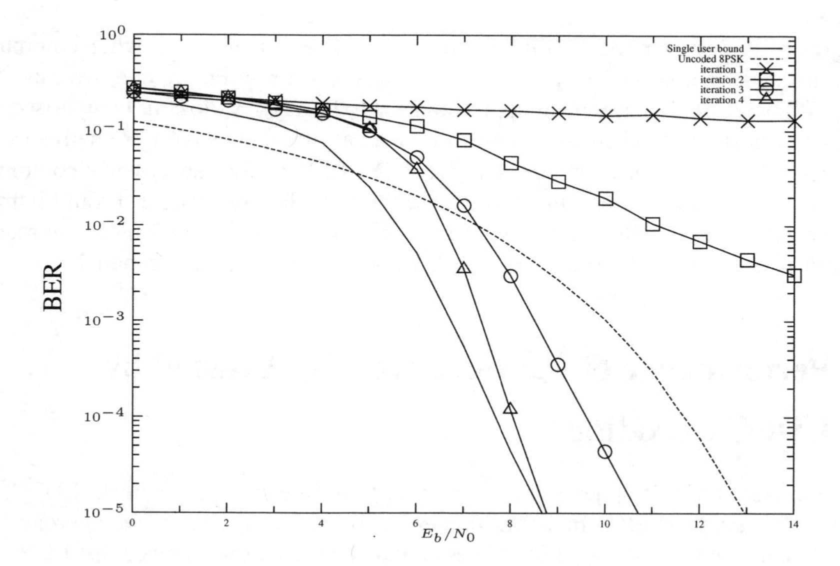

Figure 6.21: The PIC-based iterative multi-user detector's **BER** versus E_b/N_0 performance, when communicating over an **asynchronous non-dispersive AWGN** channel, while supporting $K = 7$ users. A $\frac{3}{4}$ rate **16QAM TTCM** scheme having an effective throughput of 3 BPS was invoked.

6.7 Performance of the Iterative PIC Assisted by Turbo Decoding

Turbo coding [5, 79, 197] was proposed in 1993 by Berrou, Glavieux and Thitimajashima, who reported excellent coding gain results, approaching Shannon's channel capability predictions [259]. Specifically, the information sequence is encoded twice by the turbo encoder, with an interleaver between the two encoders invoked for ensuring that the two encoded data sequences become approximately statistically independent of each other. Often half-rate Recursive Systematic Convolutional (RSC) encoders are used, with each RSC encoder producing a systematically encoded output bit stream containing both the original information bits sequence, as well as a stream of parity bits. The two parity sequences can then be punctured before being transmitted along with the original information sequence to the decoder. The puncturing of the parity information allows a wide range of coding rates to be realized. In our system, half-rate turbo codes are employed. The Log-MAP turbo decoder [5] was employed with the same slight modifications, which was invoked for the convolutional code in Section 6.4, for the sake of producing the APP also for the parity bits, rather than only for the information bits, as in case of conventional turbo decoding. The RSC code's memory length was $m = 3$ and a random channel interleaver length of 1920 bits was employed in our system. The turbo interleaver was a 10×10 dimensional block interleaver. Figure 6.22 shows the achievable BER performance, when communicating over a synchronous non-dispersive AWGN channel, respectively. From Figure 6.22 we can observe that the PIC-based MUD exhibits an approximately 0.5dB E_b/N_0 loss in comparison to the single-user performance at a BER of 10^{-5}.

Figures 6.23 characterizes the PIC-based MUD's BER performance, when communicating over an asynchronous non-dispersive AWGN channel. From Figure 6.23 we can observe that the PIC-based MUD suffers an approximately 0.9dB E_b/N_0 loss in comparison to the single-user performance, when communicating in an asynchronous AWGN environment.

Figures 6.24 characterizes the PIC-based MUD's BER performance, when communicating over an asynchronous two-path uncorrelated Rayleigh fading channel. From Figure 6.23 we can observe that the PIC-based MUD requires more than 2 iterations to approach the single-user performance when communicating over a dispersive fading channel.

6.8 Performance of the Iterative PIC Assisted by LDPC Decoding

In 1963, Gallager [260, 261] devised the family of Low Density Parity Check (LDPC) codes during his PhD study at MIT. In 1981, Tanner [262] suggested a recursive approach to the construction of LDPC codes. In 1995, MacKay and Neal [263,264] showed that LDPC codes are capable of achieving a comparable performance to that of the family of turbo codes. Hence in this section we employ the family of LDPC codes in our system.

When we consider the achievable performance of this system in a synchronous environment, we can infer from Figure 6.25 the system achieves a slightly further gain when the number of outer iterations is higher than 2, and when communicating over an asynchronous

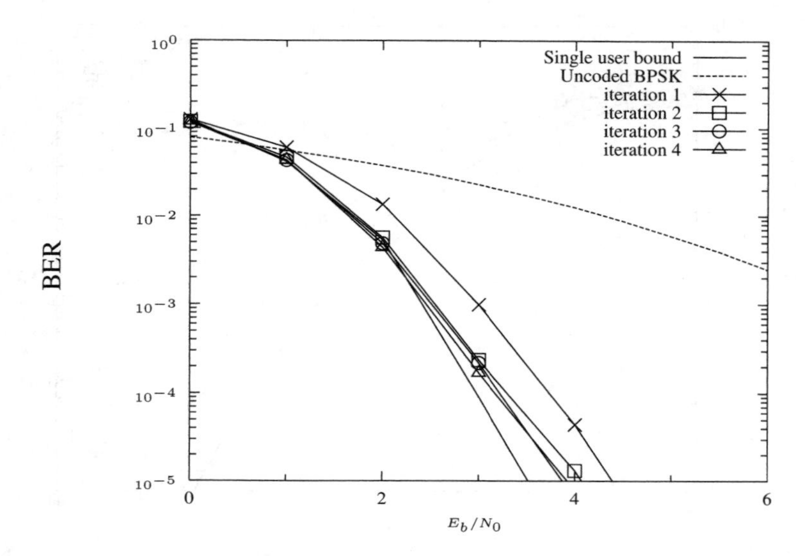

Figure 6.22: The PIC-based iterative multi-user detector's **BER** performance, when communicating over a **synchronous non-dispersive AWGN** channel, while supporting $K = 15$ users, each encoded with the aid of a $\frac{1}{2}$-rate, $m = 3$ **turbo code**. **QPSK** modulation was employed.

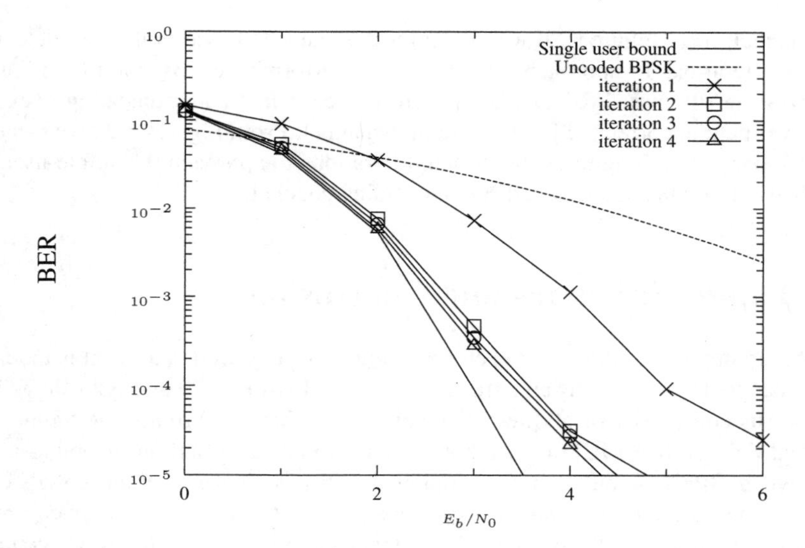

Figure 6.23: The PIC-based iterative multi-user detector's **BER** performance, when communicating over an **asynchronous non-dispersive AWGN** channel, while supporting $K = 7$ users, each encoded with the aid of a $\frac{1}{2}$-rate, $m = 3$ **turbo code**. **QPSK** modulation was employed.

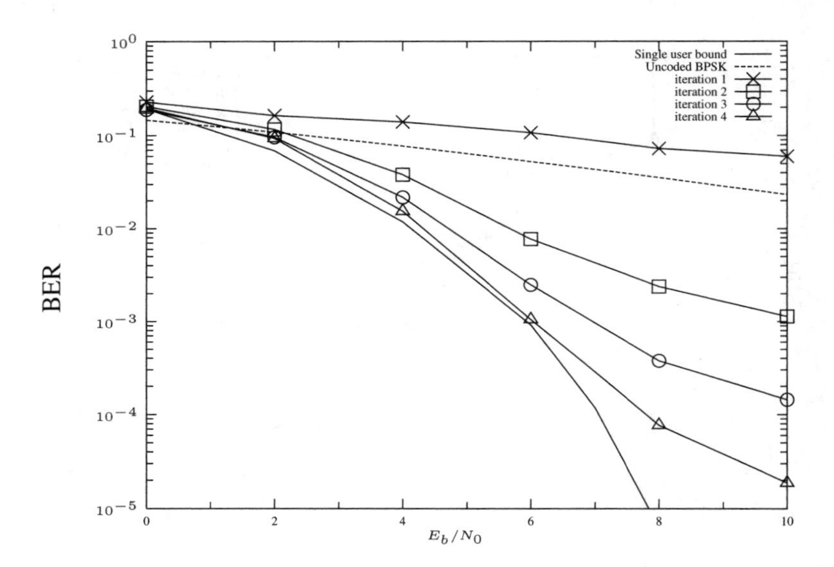

Figure 6.24: The PIC-based iterative multi-user detector's **BER** performance, when communicating over an **asynchronous two-path uncorrelated Rayleigh fading** channel, while supporting $K = 7$ users, each encoded with the aid of a $\frac{1}{2}$-rate, $m = 3$ **turbo code**. **QPSK** modulation was employed.

AWGN channel, from Figure 6.26 we may infer that the PIC-based multi-user detector also requires two iterations to approach the single-user performance. By contrast, Figure 6.24 characterizes the PIC-based MUD's BER performance, when communicating over a asynchronous two-path uncorrelated Rayleigh fading channel. From Figures 6.24 we can observe that the PIC based MUD requires more than 3 iterations to approach the single-user performance when communicating over a dispersive fading channel.

6.9 Chapter Summary and Conclusion

The PIC-based iterative multi-user detector's complexity per decoded user bit is independent of the number of users, which is an attractive property. Figures 6.29 and 6.28 show the coding gain versus complexity of all these channel coded schemes. From these figures we can observe that the iterative MUD is capable of approaching the single-user bound after two iterations when **QPSK** modulation was employed for transmission over an AWGN channel. Figure 6.30 shows the coding gain versus complexity of all these schemes when communicating over a two-path uncorrelated Rayleigh fading channel. From this figure we can observe that the performance of the iterative PIC scheme converges slower when communicating over a Rayleigh fading channel than over an AWGN channel. More specifically, in this scenario, it requires about three or four iterations for approaching the single-user bound at a BER of 10^{-3}. Figures 6.31 and 6.32 show the coding gain versus complexity of TCM and TTCM when **8PSK** and **16QAM** are employed. Form these figures we infer that the iterative PIC

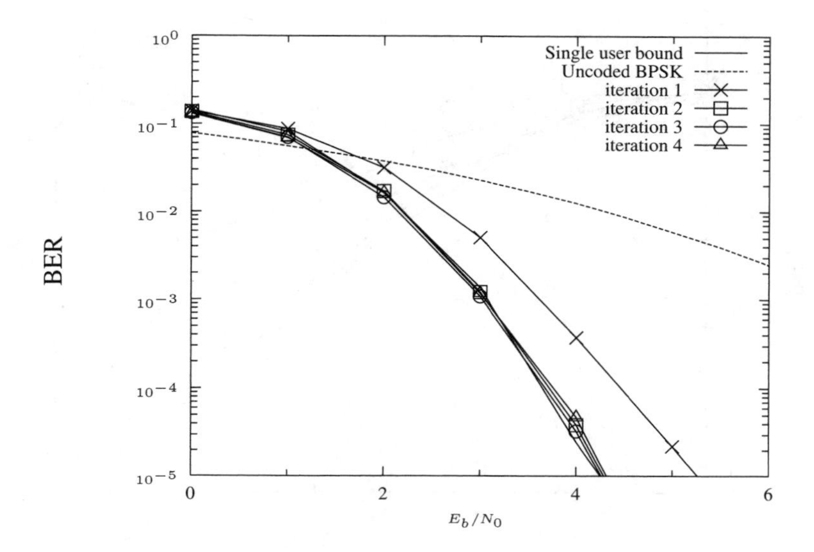

Figure 6.25: The PIC-based iterative multi-user detector's **BER** performance, when communicating over a **synchronous non-dispersive AWGN** channel, while supporting $K = 15$ users. A $\frac{1}{2}$-rate **LDPC** was invoked as well as **QPSK** modulation was employed.

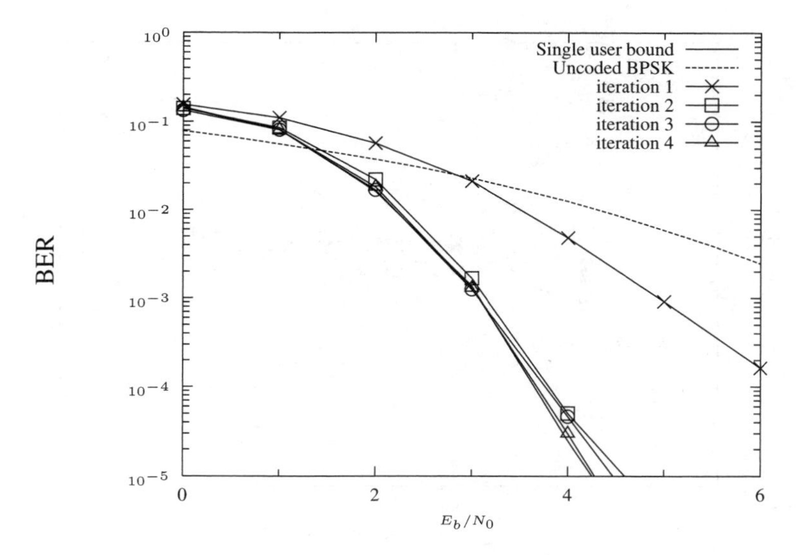

Figure 6.26: The PIC-based iterative multi-user detector's **BER** performance, when communicating over an **asynchronous non-dispersive AWGN** channel, while supporting $K = 7$ users. A $\frac{1}{2}$-rate **LDPC** was invoked as well as **QPSK** modulation was employed.

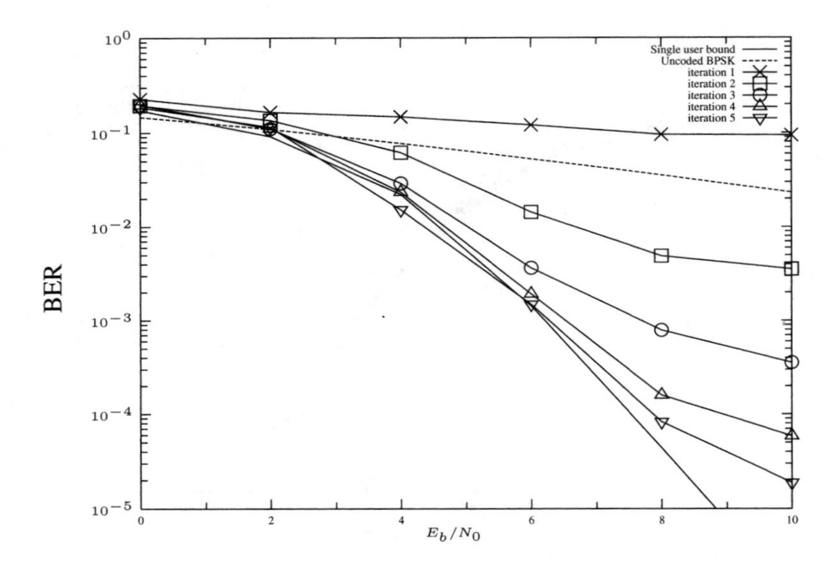

Figure 6.27: The PIC-based iterative multi-user detector's **BER** performance, when communicating over an **asynchronous two-path uncorrelated Rayleigh fading** channel, while supporting $K = 7$ users. A $\frac{1}{2}$-rate **LDPC** was invoked as well as **QPSK** modulation was employed.

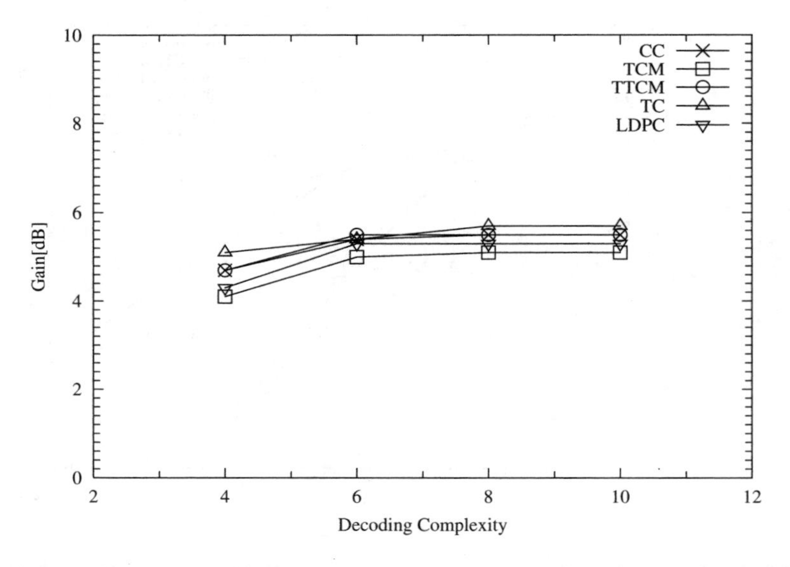

Figure 6.28: The PIC-based iterative multi-user detector's coding gain versus the complexity of all these coding schemes at a BER of 10^{-5}, when communicating over a **synchronous AWGN** channel, while supporting $K = 15$ users, and employing **QPSK** modulation.

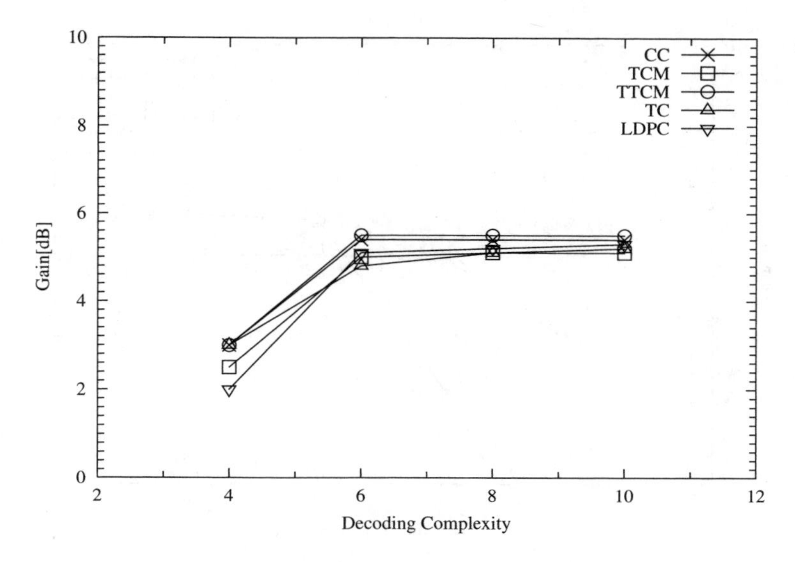

Figure 6.29: The PIC-based iterative multi-user detector's coding gain versus the complexity of different coding schemes at a BER of 10^{-5}, when communicating over an **asynchronous AWGN** channel, while supporting $K = 7$ users, and employing **QPSK** modulation.

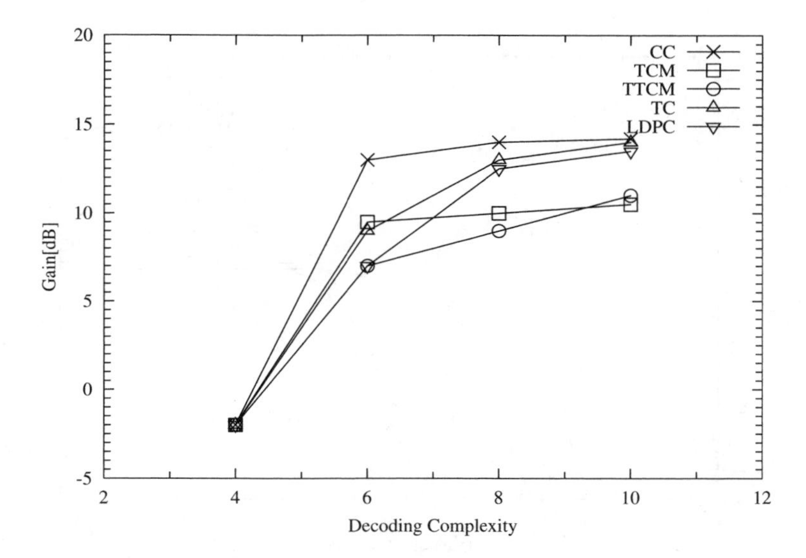

Figure 6.30: The PIC-based iterative multi-user detector's coding gain versus the complexity of different coding schemes when communicating over a **two-path uncorrelated Rayleigh fading channel having equal-weight CIR taps** at a BER of 10^{-3}, while supporting $K = 10$ users, and employing **QPSK** modulation.

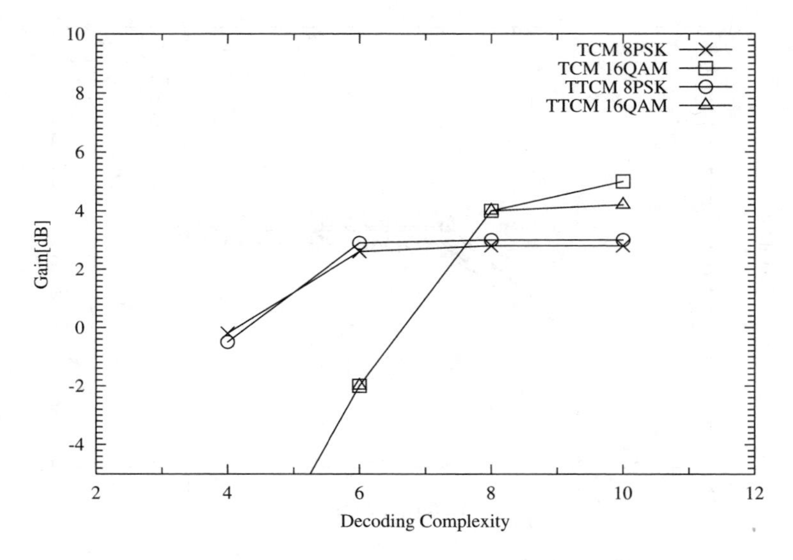

Figure 6.31: The PIC-based iterative multi-user detector's coding gain versus the complexity of coded modulation schemes when communicating over a **synchronous** channel at a BER of 10^{-5}, while supporting $K = 15$ users, and employing **8PSK** and **16QAM** modulation.

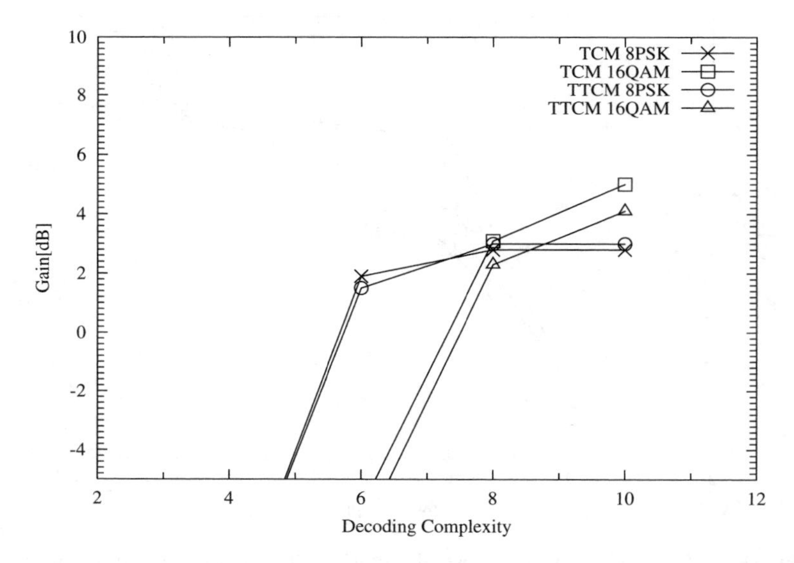

Figure 6.32: The PIC-based iterative multi-user detector's coding gain versus the complexity of coded modulation schemes when communicating over an **asynchronous** channel at a BER of 10^{-5}, while supporting $K = 7$ users, and employing **8PSK** and **16QAM** modulation.

scheme converges slower in the asynchronous environment than in the synchronous environment, since it requires four iterations for approaching the single-user bound instead of two or three iterations.

Chapter 7

Blind Per-Survivor Processing-Aided CDMA [1]

In multi-user detection, the knowledge of the channel impulse responses (CIR) and spreading sequences of all the users is exploited in order to obtain reliable data estimates. However, in order to obtain the CIRs of all the users, training sequences or reference sequences have to be transmitted periodically and multi-user CIR estimation has to be carried out at the receiver. The inclusion of these training sequences reduces the throughput, while requiring extra transmitted power. There is a class of CDMA receivers, collectively known as blind receivers [169–171, 175–177, 179–181], where the receiver estimates the transmitted data without the need for training sequences. For most of these receivers the estimation of the relevant parameters – such as the delays of the users and the CIRs – is integrated with the data estimation algorithm. A brief discussion of some examples of blind receivers was presented in Section 2.10.1. In this chapter, the focus of our discussions is on a specific type of blind receivers, namely on the so-called blind Per-Survivor Processing (PSP) detector [173].

The optimum estimation technique for a sequence that has undergone channel coding and has been corrupted by inter-symbol interference (ISI) and additive noise is the well-known Maximum Likelihood Sequence Estimation (MLSE) [90] strategy. When the parameters of the dispersive channel are perfectly known, the receiver produces the most likely transmitted sequence through the employment of the well-known trellis-search algorithm, namely the Viterbi algorithm [90, 213]. In the scenario where the channel parameters are unknown, an approximation of this technique may be obtained by estimating the required parameters in a so-called "per-survivor" fashion [173]. This implies that a set of parameter estimators is employed for estimating these channel parameters and each estimator is data-aided by a surviving data sequence of the trellis. Again, this approach is termed PSP, which was proposed by , and [173], who demonstrated the applicability of the PSP algorithms in var-

[1] *Single- and Multi-Carrier CDMA Multi-User Detection, Space-Time Spreading, Synchronization, Networking and Standards.* L.Hanzo, L-L.Yang, E-L.Kuan and K.Yen,
 ISBN 0-470-86309-9

ious environments. A reduced complexity blind PSP algorithm based on the well-known M-algorithm [160] for joint data and channel estimation was proposed by [172]. The procedure produced the data sequence and the required channel estimates under the constraint of the overall least squares error. The PSP algorithm in combination with a simplified Kalman filter [125] was employed by and [265] in order to approximate MLSE over fast frequency-selective fading channels.

The references cited previously have been concerned with the application of PSP algorithms in wideband channels for single user transmission. The employment of blind PSP for joint signal detection and channel estimation in a multi-user asynchronous DS-CDMA system was first investigated by , , and [171]. In their work, the data estimator employed a reduced-complexity tree-search algorithm, while a Recursive Least Squares (RLS) estimator [125] was used for estimating the received signal amplitudes of all the users for each surviving data sequence in the trellis. Each user transmitted over a static narrowband channel and amplitude estimation was carried out without the use of a training sequence, although it was assumed that the time delays of all the users were known at the receiver. This work was extended by the same authors to a blind PSP-based multi-user receiver, where channel estimation was carried out without the knowledge of the individual time delays of the users [175]. Previously, several tree-search-based multi-user detectors that exploited perfect channel estimation have been proposed, including the arrangement advocated by , and [158] for multi-user CDMA systems over Rayleigh fading channels. The employment of adaptive algorithms – such as the RLS algorithm – for adaptive symbol and CIR estimation without the aid of a tree-search algorithm has been investigated by , and [166, 184] in the context of multi-user CDMA detection. The combination of Kalman filtering and the M-algorithm in the tree-search was evaluated by and [176] for adaptive joint amplitude and delay estimation of the multi-user signals.

In this chapter, the PSP algorithm proposed by *et al.* [171] for narrowband static channels is extended to a multi-user detector for employment in a wideband synchronous CDMA system. Furthermore, this multi-user detector is combined with turbo convolutional coding in order to improve its performance, where the PSP multi-user detector generates soft outputs as reliability information for the individual turbo decoders of the users.

The notation used in the following sections are the same as those employed in Chapter 3 and which are repeated here for convenience:

- Capital letters in boldface represent matrices, for instance $\mathbf{X}$.
- Column vectors are presented in lowercase letters, for instance $\mathbf{x}$.
- The vector, $\mathbf{x}^{(k)}$, represents the vector $\mathbf{x}$ of user k.
- The asterisk (*) superscript is used to indicate complex conjugation. Therefore $\mathbf{X}^*$ is used to denote the complex conjugate matrix of $\mathbf{X}$ and a^* represents the complex conjugate of the variable a.
- The notation $\mathbf{X}^T$ implies the transpose matrix of $\mathbf{X}$.
- $\mathbf{X}^H$ is used to represent the complex conjugate and transpose matrix of $\mathbf{X}$, i.e. $\mathbf{X}^H = \mathbf{X}^{*T}$, which is also referred to as the Hermitian matrix of $\mathbf{X}$.
- $\mathbf{X}^{-1}$ corresponds to the inverse matrix of $\mathbf{X}$.

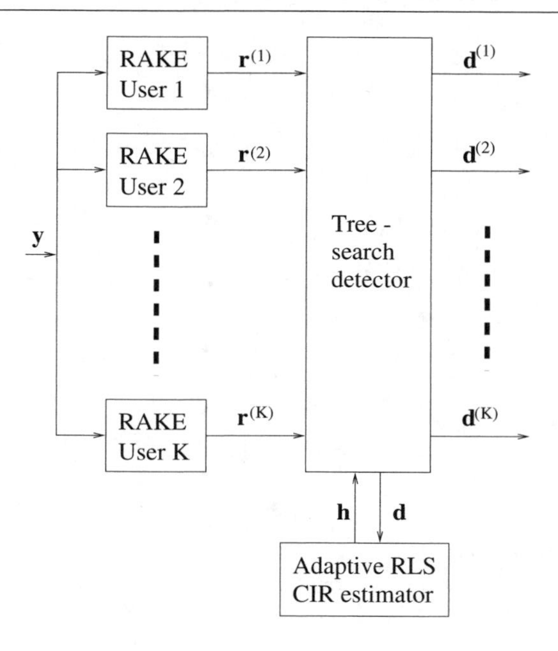

Figure 7.1: Schematic of the multi-user PSP receiver.

- $[\mathbf{X}]_{ij}$ identifies the element in the i-th row and j-th column of the matrix $\mathbf{X}$ and the notation x_i is used to represent the i-th element of the column vector $\mathbf{x}$.
- The matrix $\mathbf{I}$ is the identity matrix.
- The matrix $\mathbf{0}$ is an all-zero matrix.
- The notation $\mathbf{0}_{[X \times Y]}$ represents an all-zero matrix with the dimensions of $X \times Y$.

7.1 Blind PSP for bit-synchronous CDMA data detection and multipath CIR estimation

The schematic of the proposed blind PSP multi-user receiver is shown in Figure 7.1. The receiver consists of a bank of RAKE receivers, one for each CDMA user, followed by a reduced complexity tree-search-based data sequence estimator that is linked to a bank of RLS CIR estimators. The receiver determines the data sequence, $\mathbf{d}$, and the corresponding set of CIR estimates, $\mathbf{h}$, that maximize the correlation metric [78]:

$$\Omega(\mathbf{r}, \mathbf{d}) = 2\mathbf{d}^H \mathbf{\Lambda} \mathbf{r} - \mathbf{d}^H \mathbf{\Lambda} \mathbf{R} \mathbf{\Lambda}^H \mathbf{d}, \tag{7.1}$$

where the elements of the vector $\mathbf{r}$ represent the cross-correlation (CCL) of the received signal with each of the users' spreading sequences, i.e. the output signal of the matched filter.

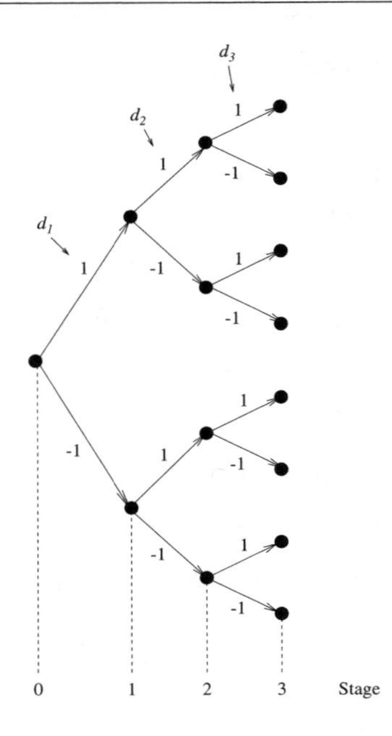

Figure 7.2: Tree-search diagram for a three-user symbol-synchronous system.

Furthermore, the vector $\mathbf{d}$ consists of the bits transmitted by the users; the matrix $\mathbf{R}$ is the CCL matrix of the spreading sequences; and finally the matrix $\mathbf{\Lambda}$ is a diagonal matrix, and its main diagonal elements are the received signal amplitudes due to the transmitted symbols. In the tree-search-based data sequence estimator, the data sequence leading to each surviving path in the tree is applied to an RLS estimator, in order to obtain the CIR estimates associated with that data sequence. After the channel estimates have been obtained, a metric based on the likelihood function in Equation 7.1 is calculated and a corresponding high-probability subset of the paths is remembered, depending on the metric values of the paths. Finally, at the decision stage, the data sequence associated with the surviving path that yields the highest metric in Equation 7.1 is chosen as the most likely transmitted sequence.

7.2 Tree-search Algorithm for Data Sequence Detection

The tree-search-based detection algorithm proposed by Xie *et al* [171] is adopted in our receiver and its implementation is explained next. In order to implement a tree-search-based detection algorithm for CDMA systems, the transmitted symbols of all the CDMA users can be concatenated, in order to generate a single data sequence. In a K-user system, where $d_n^{(k)}$ represents the n-th symbol transmitted by the k-th user, the data sequence of all the users are concatenated, in order to be able to invoke convolutional channel equalization techniques, as was proposed for example in [127]:

$$\mathbf{d} = [d_1^{(1)}, d_1^{(2)}, \ldots, d_1^{(K)}, d_2^{(1)}, d_2^{(2)}, \ldots, d_N^{(1)}, \ldots, d_N^{(K)}]^T \quad (7.2)$$
$$k = 1, \ldots, K; \; n = 1, \ldots, N; \; K, N \in \mathbb{N},$$

where N is the number of symbols transmitted per user, which can also be considered as the detector window length for each user. Figure 7.2 shows a tree-diagram for a $K = 3$-user CDMA system that is symbol-synchronous and BPSK is used as the data modulation mode. At stage i, there exists a set of nodes and associated with each node is a hypothesized data sequence of length i as well as a set of CIR estimates corresponding to that data sequence. The maximum likelihood sequence detector calculates the likelihood metric of Equation 7.1 for each node at stage i and this is carried out exhaustively in order to obtain the most likely sequence. For wideband channels which result in ISI, the detection window has to be extended to cover a longer sequence of symbols, for example $N + \lceil W/Q \rceil$, in order to minimize the error probability, where W is the CIR duration in terms of chips and the expression $\lceil W/Q \rceil$ represents the nearest integer larger than or equal to W/Q. However, the complexity of the estimator rises exponentially with the window length and the number of users, K. In order to implement a more practical algorithm in our receiver, the M-algorithm [160] was chosen as the reduced complexity tree-search algorithm. At each stage, i, only a subset of the nodes was kept for per-survivor processing at the next stage. Specifically, if the number of nodes at the i-th stage is M and the data modulation mode is BPSK, then the procedure for moving from stage i to stage $i + 1$ can be formulated as follows:

1) For each node at stage i, extend the node to two extra nodes by hypothesizing the next data bit in the sequence to be, first, a logical +1 and, then, a logical 0, associated with -1. This results in a total of $2M$ nodes. The corresponding data sequences associated with the nodes are stored for CIR estimation and metric calculation.

2) For each of the new $2M$ nodes, update the CIR estimates with the aid of a Recursive Least Squares (RLS) [125] estimator and then calculate the metric value of Equation 7.6 based on the data bit sequence associated with that node.

3) Out of the total of $2M$ nodes, select the M nodes associated with the highest metric values for the next stage, namely for stage $i + 1$.

At the decision stage of $i = NK$, the bit sequences and CIR estimates associated with the node that has the highest metric constitute the best joint data and CIR estimates of the K users.

7.3 Metric Calculation

For the K-user synchronous CDMA narrowband channel the likelihood function was derived by [78], as stated in Equation 7.1, which can be rewritten as:

$$\Omega = \sum_{n=1}^{N} \sum_{k=1}^{K} d_n^{(k)} \alpha_n^{(k)} \left[2r_n^{(k)} - \sum_{m=1}^{N} \sum_{j=1}^{K} (d_m^{(j)} \cdot \alpha_n^{(k)})^* \cdot R_{kj}(0) \right], \quad (7.3)$$

where $d_n^{(k)}$ represents the n-th symbol of the k-th user; $\alpha_n^{(k)}$ represents the received signal amplitude of the n-th symbol of the k-th user; $r_n^{(k)}$ is the output of the matched filter for the n-th symbol of the k-th user; and $R_{kj}(0)$ denotes the periodic CCL between the spreading sequences of user k and j, respectively. The notation x^* represents the complex conjugate of the variable x. Equation 7.3 was modified by Xie *et al* [171] for the metric calculation.

In our approach, Equation 7.3 was modified to take into account the multipath channel of each user. Let the maximum delay spread of the mobile channels of all the users be represented by W chips. Equation 7.3 can then be modified to include the effects of multipath interference, resulting in:

$$\Omega = \sum_{n=1}^{N}\sum_{k=1}^{K} d_n^{(k)} \sum_{w=1}^{W} (h_n^{(k)}(w))^* \Big[2r_n^{(k)}(w) - \sum_{m=1}^{N}\sum_{j=1}^{K} (d_m^{(j)})^* \sum_{v=1}^{W} h_m^{(j)}(v) \cdot R_{kj}(w-v)\Big], \quad (7.4)$$

where $r_n^{(k)}(w)$ represents the output of the w-th finger of the RAKE receiver for the n-th symbol of the k-th user; and $h_n^{(k)}(w)$ represents the w-th tap of the CIR associated with the n-th symbol of the k-th user. The notation $R_{kj}(w-v)$ represents the CCL between the users' spreading sequences, which is given by:

$$R_{kj}(w-v) = \begin{cases} \sum_{i=1}^{Q-(w-v)} c^{(k)}(i) \cdot c^{(j)*}(i+w-v), & \text{for } w-v \geqslant 0 \\ \sum_{i=1}^{Q+(w-v)} c^{(k)}(i-(w-v)) \cdot c^{(j)*}(i), & \text{for } w-v < 0 \end{cases} \quad (7.5)$$

where $c^{(k)}(i)$ represents the i-th chip of the spreading sequence of the k-th user and the spreading sequence length is Q chips. For slowly fading channels, the CIR can be assumed to be constant over the detector window length and hence $h_n^{(k)}(w)$ can be simplified to $h^{(k)}(w)$.

The metric calculation associated with extending the trellis from stage $I-1$ to stage I in the tree-search algorithm is performed using a modified form of Equation 7.4, given as:

$$m[I] = \Re\Big\{ \sum_{i=1}^{I} d_i \sum_{w=1}^{W} (h^{(\zeta(i))}(w))^* \cdot \Big[2r_i(w) - \sum_{j=1}^{K-1} (d_{i-j})^* \sum_{v=1}^{W} h^{(\zeta(i-j))}(v) \cdot R_{\zeta(i),\zeta(i-j)}(w-v)\Big]\Big\}. \quad (7.6)$$

In the tree-search algorithm, the transmitted data symbols of all the K users are concatenated into one long sequence of KN bit durations as seen in Equation 7.2. The notation d_i in Equation 7.6 refers to the i-th symbol in the data sequence, i.e. $d_1 = d_1^{(1)}$, $d_2 = d_1^{(2)}$ and so on. In order to map the data bit being detected at stage i to the corresponding n-th symbol of the k-th user, two mapping functions are employed, namely $k = \zeta(i)$ and $n = \nu(i)$, where:

$$k = \zeta(i) = [(i-1) \oslash K] + 1, \quad (7.7)$$

and

$$n = \nu(i) = \left\lceil \frac{i}{K} \right\rceil, \tag{7.8}$$

where the expression $(x \oslash y)$ represents the remainder upon dividing x by y; the function $\lceil x \rceil$ returns the nearest integer larger than or equal to x; and K is the total number of CDMA users. As an example, for $K = 2$, we have:

$$\begin{aligned} d_3 &= d_{\nu(3)}^{[\zeta(3)]} \\ &= d_2^{(1)}. \end{aligned} \tag{7.9}$$

Similarly, the notation $r_i(w)$ represents the output of the RAKE receiver at the w-th tap of the $(\zeta(i))$-th user.

In the steady state, the metric $m[I]$ of Equation 7.6 can be updated incrementally by using the metric $m[I-1]$ calculated for the previous stage with the aid of the following equation:

$$\begin{aligned} m[I] &= m[I-1] + \\ & \Re\left\{ d_I \sum_{w=1}^{W} (h^{(\zeta(I))}(w))^* \cdot \right. \\ & \left. \left[2r_I(w) - \sum_{j=1}^{K-1} (d_{I-j})^* \cdot \sum_{v=1}^{W} h^{(\zeta(I-j))}(v) \cdot R_{\zeta(I),\zeta(I-j)}(w-v) \right] \right\}. \end{aligned} \tag{7.10}$$

Having analyzed the data estimation component of the receiver, let us now focus our attention on the CIR estimation algorithm.

7.4 Adaptive Recursive Least Squares Multipath CIR Estimator

The tree-search algorithm described in Section 7.2 requires a CIR estimator for each of the hypothesized data sequences associated with the $2M$ nodes. In our scheme, the adaptive RLS estimator [125] was employed as the multipath CIR estimator following the approach proposed by *et al.* [171] for narrowband channels. The RLS algorithm is a recursive implementation of the least squares method [125], where the algorithm is optimized to obtain the adaptive CIR estimator tap weights that yield the least squares error between the desired CIR and the output of the CIR estimator. At the PSP receiver, the outputs of the RAKE fingers of all the users in Figure 7.1 are chosen as the reference or desired response and the CIR estimates are the estimator tap weights that have to be recursively obtained. Therefore, the error at the output of the adaptive CIR estimator is defined as:

$$\mathbf{e}[i] = \mathbf{r}[i] - \mathbf{T}^H[i] \cdot \mathbf{f}[i]. \tag{7.11}$$

The vector $\mathbf{r}[i]$ in Equation 7.11 contains the RAKE finger outputs in Figure 7.1 for data bit i, where:

$$\mathbf{r}[i] = [\, r_i(1),\, r_i(2), \ldots ,\, r_i(W)\,]^T, \tag{7.12}$$

and

$$r_i(w) = \sum_{q=1}^{Q} y(q+w) \cdot c^{(\zeta(i))}(q). \tag{7.13}$$

The symbol $c^{(\zeta(i))}(q)$ in Equation 7.13 represents the q-th chip of the spreading sequence of user $k = \zeta(i)$.

Furthermore, the CIR estimator tap weights, $\mathbf{f}[i]$, are constituted by the CIR estimates of all the users:

$$\mathbf{f}[i] = [\, h^{(1)}(1),\, h^{(1)}(2), \ldots ,\, h^{(1)}(W),\, h^{(2)}(1),\, h^{(2)}(2), \ldots ,\, h^{(K)}(W)\,]^T \tag{7.14}$$

The matrix $\mathbf{T}[i]$ is a little more difficult to construct. Let us consider the construction of this matrix step-by-step. Let the matrix $\mathbf{\Delta}_i$ be defined for the data bit d_i as a diagonal matrix, where all the elements on the main diagonal are d_i, i.e.:

$$[\mathbf{\Delta}_i]_{w,x} = \begin{cases} d_i & \text{for } w = x, \text{ and } w = 1, \ldots , W \\ 0 & \text{for } w \neq x, \text{ and } w, x = 1, \ldots , W. \end{cases} \tag{7.15}$$

An example of the matrix $\mathbf{\Delta}_i$ is shown below:

$$\mathbf{\Delta}_i = \begin{bmatrix} d_i & 0 & 0 & \ldots & 0 \\ 0 & d_i & 0 & \ldots & 0 \\ \vdots & \vdots & \vdots & \vdots & 0 \\ 0 & 0 & 0 & 0 & d_i \end{bmatrix}. \tag{7.16}$$

Next, the CCL matrix, $\mathbf{\Theta}_{kj}$, between the spreading sequences of user k and user j for the time delay shifts of the CIR is arranged as:

$$\mathbf{\Theta}_{kj} = \begin{bmatrix} R_{k,j}(0) & R_{k,j}(-1) & \ldots & R_{k,j}(-W+1) \\ R_{k,j}(1) & R_{k,j}(0) & \ldots & \ldots \\ \vdots & \vdots & \vdots & \vdots \\ R_{k,j}(W-1) & \ldots & \ldots & R_{k,j}(0) \end{bmatrix}. \tag{7.17}$$

For the data bit of d_i at stage i in the tree-search algorithm, the matrices of $\mathbf{\Sigma}_{\zeta(i)}$, $\mathbf{\Sigma}_{\zeta(i-1)}$,

..., $\mathbf{\Sigma}_{\zeta(i-K+1)}$ can be constructed with the help of Equations 7.15 and 7.17 as:

$$\begin{array}{rcl} \mathbf{\Sigma}_{\zeta(i)} & = & \mathbf{\Delta}_i \cdot \mathbf{\Theta}_{\zeta(i),\zeta(i)} \\ \mathbf{\Sigma}_{\zeta(i-1)} & = & \mathbf{\Delta}_{i-1} \cdot \mathbf{\Theta}_{\zeta(i),\zeta(i-1)} \\ & \vdots & \\ \mathbf{\Sigma}_{\zeta(i-K+1)} & = & \mathbf{\Delta}_{i-K+1} \cdot \mathbf{\Theta}_{\zeta(i),\zeta(i-K+1)}, \end{array} \tag{7.18}$$

where $\zeta(j) \in \{1, 2, \dots, K\}$.

With the aid of Equation 7.18, the input matrix $\mathbf{T}^H[i]$ can be constructed as:

$$\mathbf{T}^H[i] = \begin{bmatrix} \mathbf{\Sigma}_1 & \mathbf{\Sigma}_2 & \dots & \mathbf{\Sigma}_K \end{bmatrix}^*. \tag{7.19}$$

The matrix $\mathbf{T}^H[i]$ is constructed in this manner in order to enable the filter output of $\mathbf{T}^H[i]\mathbf{f}[i]$ in Equation 7.11 to be as similar to the RAKE finger output, $\mathbf{r}[i]$, as possible. The RAKE finger output in Figure 7.1 consists of the CCL of the spreading sequences multiplied by the desired symbol and the CIR response. The error, $\mathbf{e}[i]$, of Equation 7.11 tends to zero, if the CIR estimates, $\mathbf{f}[i]$, are accurate.

In the method of exponentially weighted least squares, the cost function to be minimized is [125]:

$$\mathcal{E}[n] = \sum_{i=1}^{n} \lambda^{n-i} \mathbf{e}^H[i]\mathbf{e}[i], \tag{7.20}$$

where λ is the exponential weighting factor. The negative exponentially decaying weighting function de-emphasizes older error terms in Equation 7.20. The optimum value of the CIR vector, $\mathbf{f}[i]$, for which the cost function of Equation 7.20 attains its minimum values is defined by the matrix equation [125]:

$$\mathbf{\Phi}[n]\mathbf{f}[n] = \mathbf{z}[n], \tag{7.21}$$

where the input sequence correlation matrix, $\mathbf{\Phi}[n]$, is defined by:

$$\mathbf{\Phi}[n] = \sum_{i=1}^{n} \lambda^{n-i} \mathbf{S}[i]\mathbf{T}^H[i] \tag{7.22}$$

$$= \lambda\Big(\sum_{i=1}^{n-1} \lambda^{n-1-i} \mathbf{S}[i]\mathbf{T}^H[i]\Big) + \mathbf{S}[n]\mathbf{T}^H[n] \tag{7.23}$$

$$= \lambda\mathbf{\Phi}[n-1] + \mathbf{S}[n]\mathbf{T}^H[n], \tag{7.24}$$

and

$$\mathbf{S}[n] = \begin{bmatrix} \mathbf{0}_{[W \times (k-1)W]} \\ \mathbf{\Delta}_n \\ \mathbf{0}_{[W \times (K-k)W]} \end{bmatrix}. \tag{7.25}$$

Equation 7.24 constitutes a recursive update formula for $\mathbf{\Phi}[n]$ on the basis of $\mathbf{\Phi}[n-1]$.

The CCL vector $\mathbf{z}[n]$, between the inputs to the CIR estimator constituted by the RAKE receiver outputs, $\mathbf{r}[i]$, and the matrix $\mathbf{S}[i]$, is expressed as:

$$\mathbf{z}[n] = \sum_{i=1}^{n} \lambda^{n-i}\mathbf{S}[i]\mathbf{r}^*[i] \tag{7.26}$$

$$= \lambda\Big(\sum_{i=1}^{n-1} \lambda^{n-1-i}\mathbf{S}[i]\mathbf{r}^*[i]\Big) + \mathbf{S}[n]\mathbf{r}^*[n] \tag{7.27}$$

$$= \lambda\mathbf{z}[n-1] + \mathbf{S}[n]\mathbf{r}^*[n], \tag{7.28}$$

for $k = \zeta(i)$ and the notation $\mathbf{0}_{[X\times Y]}$ represents a matrix having the dimensions of $(X \times Y)$ and all the elements in it have values of zero. Equation 7.28 constitutes a recursive update formula for $\mathbf{z}[n]$.

For a matrix $\mathbf{A}$, where $\mathbf{A} = \mathbf{W} + \mathbf{XYZ}$, its inverse, $\mathbf{A}^{-1}$, can be obtained by employing the matrix inversion lemma [125] that is given here as:

$$\mathbf{A}^{-1} = \mathbf{W}^{-1} - \mathbf{W}^{-1}\mathbf{X}(\mathbf{Y}^{-1} + \mathbf{Z}\mathbf{W}^{-1}\mathbf{X})^{-1}\mathbf{Z}\mathbf{W}^{-1}. \tag{7.29}$$

Commencing from Equation 7.24 and employing the lemma of Equation 7.29, where $\mathbf{A} = \mathbf{\Phi}[n]$, $\mathbf{W} = \lambda\mathbf{\Phi}[n-1]$, $\mathbf{X} = \mathbf{S}[n]$, $\mathbf{Y} = \mathbf{I}$ and $\mathbf{Z} = \mathbf{T}^H[n]$, the inverse of $\mathbf{\Phi}[n]$ can be expressed as:

$$\mathbf{\Phi}^{-1}[n] = \lambda^{-1}\mathbf{\Phi}^{-1}[n-1] - \lambda^{-1}\mathbf{\Phi}^{-1}[n-1]\mathbf{S}[n]\Big(\mathbf{I} + \mathbf{T}^H[n]\lambda^{-1}\mathbf{\Phi}^{-1}[n-1]\mathbf{S}[n]\Big)^{-1}\mathbf{T}^H[n]\lambda^{-1}\mathbf{\Phi}^{-1}[n-1]. \tag{7.30}$$

For convenience, let

$$\mathbf{P}[n] = \mathbf{\Phi}^{-1}[n], \tag{7.31}$$

and

$$\mathbf{g}[n] = \lambda^{-1}\mathbf{P}[n-1]\mathbf{S}[n]\Big(\mathbf{I} + \mathbf{T}^H[n]\lambda^{-1}\mathbf{P}[n-1]\mathbf{S}[n]\Big)^{-1}. \tag{7.32}$$

Equation 7.30 can then be rewritten as:

$$\mathbf{P}[n] = \lambda^{-1}\mathbf{P}[n-1] - \lambda^{-1}\mathbf{g}[n]\mathbf{T}^H[n]\mathbf{P}[n-1]. \tag{7.33}$$

By rearranging Equation 7.32, we obtain:

$$\mathbf{g}[n] = \lambda^{-1}\mathbf{P}[n-1]\mathbf{S}[n] - \lambda^{-1}\mathbf{g}[n]\mathbf{T}^H[n]\mathbf{P}[n-1]\mathbf{S}[n] \quad (7.34)$$

$$\mathbf{g}[n] = \left(\lambda^{-1}\mathbf{P}[n-1] - \lambda^{-1}\mathbf{g}[n]\mathbf{T}^H[n]\mathbf{P}[n-1]\right)\mathbf{S}[n]. \quad (7.35)$$

Substituting Equation 7.33 in Equation 7.35 yields:

$$\mathbf{g}[n] = \mathbf{P}[n]\mathbf{S}[n]. \quad (7.36)$$

In order to obtain the recursive equation for updating the CIR estimates, $\mathbf{f}[i]$, Equation 7.21 is expanded with the aid of Equations 7.28 and Equation 7.31 as:

$$\begin{aligned} \mathbf{f}[n] &= \mathbf{\Phi}^{-1}[n]\mathbf{z}[n] && (7.37)\\ &= \mathbf{P}[n]\mathbf{z}[n] && (7.38)\\ &= \lambda\mathbf{P}[n]\mathbf{z}[n-1] + \mathbf{P}[n]\mathbf{S}[n]\mathbf{r}^*[n]. && (7.39) \end{aligned}$$

Upon substituting $\mathbf{P}[n]$ from Equation 7.33 in the first term only at the right-hand side of Equation 7.39 yields:

$$\begin{aligned} \mathbf{f}[n] &= \mathbf{P}[n-1]\mathbf{z}[n-1] - \lambda^{-1}\mathbf{g}[n]\mathbf{T}^H[n]\mathbf{P}[n-1]\mathbf{z}[n-1] + \mathbf{P}[n]\mathbf{S}[n]\mathbf{r}^*[n]\\ &= \mathbf{f}[n-1] - \lambda^{-1}\mathbf{g}[n]\mathbf{T}^H[n]\mathbf{f}[n-1] + \mathbf{P}[n]\mathbf{S}[n]\mathbf{r}^*[n]. && (7.40) \end{aligned}$$

With the aid of Equation 7.36, this leads to:

$$\begin{aligned} \mathbf{f}[n] &= \mathbf{f}[n-1] + \mathbf{g}[n]\left(\mathbf{r}^*[n] - \lambda^{-1}\mathbf{T}^H[n]\mathbf{f}[n-1]\right) && (7.41)\\ &= \mathbf{f}[n-1] + \mathbf{g}[n]\mathbf{e}[n], && (7.42) \end{aligned}$$

where we used the shorthand of:

$$\mathbf{e}[n] = \mathbf{r}^*[n] - \lambda^{-1}\mathbf{T}^H[n]\mathbf{f}[n-1]. \quad (7.43)$$

Equations 7.32, 7.33, 7.42 and 7.43 constitute the adaptive RLS CIR estimator. Rearranging them in the order in which the RLS estimator carries out these steps gives:

$$\mathbf{e}[n] = \mathbf{r}^*[n] - \lambda^{-1}\mathbf{T}^H[n]\mathbf{f}[n-1], \quad (7.44)$$

$$\mathbf{g}[n] = \lambda^{-1}\mathbf{P}[n-1]\mathbf{S}[n]\left(\mathbf{I} + \lambda^{-1}\mathbf{T}^H[n]\mathbf{P}[n-1]\mathbf{S}[n]\right)^{-1}, \quad (7.45)$$

$$\mathbf{P}[n] = \lambda^{-1}\mathbf{P}[n-1] - \lambda^{-1}\mathbf{g}[n]\mathbf{T}^H[n]\mathbf{P}[n-1], \quad (7.46)$$

$$\mathbf{f}[n] = \mathbf{f}[n-1] + \mathbf{g}[n]\mathbf{e}[n], \quad (7.47)$$

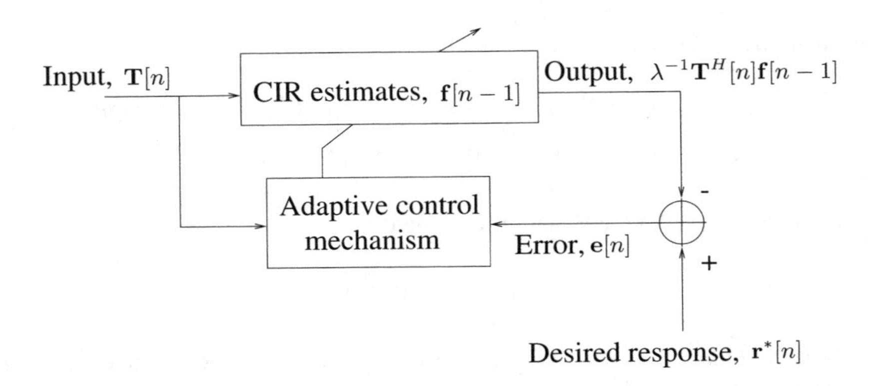

Figure 7.3: Schematic of the adaptive RLS CIR estimator for each surviving path in the PSP receiver.

where $\mathbf{e}[n]$ represents the error between the desired response and the output of the adaptive filter, $\mathbf{g}[n]$ symbolizes the gain for adapting the filter tap weights, $\mathbf{P}[n]$ represents the inverse of the input sequence correlation matrix and, finally, the multipath channel estimates are represented by the filter tap weights, $\mathbf{f}[n]$.

Figure 7.3 depicts the schematic of the adaptive RLS CIR estimator. The error, $\mathbf{e}[n]$, is obtained by comparing the desired response, $\mathbf{r}^*[n]$, with the output of the filter, $\lambda^{-1}\mathbf{T}^H[n]\mathbf{f}[n-1]$, as shown in Equation 7.44. This error is fed into the adaptive weight-control mechanism, which carries out the steps encapsulated by Equations 7.45 and 7.46. Finally, the adaptive filter tap weights are updated from $\mathbf{f}[n-1]$ to $\mathbf{f}[n]$, as shown in Equation 7.47.

Having obtained the expressions for the blind PSP CDMA receiver, let us now investigate its performance.

7.5 Blind PSP Receiver Performance

The performance of the blind PSP receiver assisted by adaptive RLS CIR estimation is presented in this section. In order to resolve any phase ambiguity due to the blind CIR estimation at the receiver, Differential Binary Phase Shift Keying (DBPSK) was used as the data modulation scheme. The spreading sequence length was set to $Q = 31$. The number of retained survivors in the tree-search algorithm was set to $M = 2$, in order to limit the associated complexity. A bank of RLS CIR estimators was implemented, using Equations 7.44, 7.45, 7.46 and 7.47. In order to implement the RLS CIR estimator, the matrices of $\mathbf{f}[0]$ and $\mathbf{P}[0]$ were initialized to $\mathbf{f}[0] = \mathbf{0}$ and $\mathbf{P}[0] = \delta^{-1}\mathbf{I}$, where δ is a small positive constant. The exponential weighting factor, λ was set to 1, in order to implement a static or a slowly fading channel, where the CIR remained constant over a detected burst of $N = 100$ bits. For the CIR estimator, it was assumed that the delays of the CIR paths were known at the receiver, but not their amplitudes and the phases. The determination of the CIR path delays as well as the CIR amplitude and phase estimation were previously investigated by , , and [175].

The convergence of the adaptive RLS CIR estimator is shown in Figure 7.4 for the CIR amplitudes and phases of the two-user CDMA system considered. The channels simulated

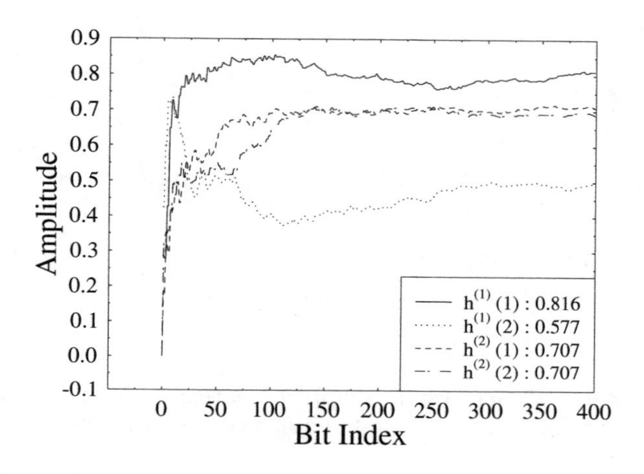

(a) Amplitude

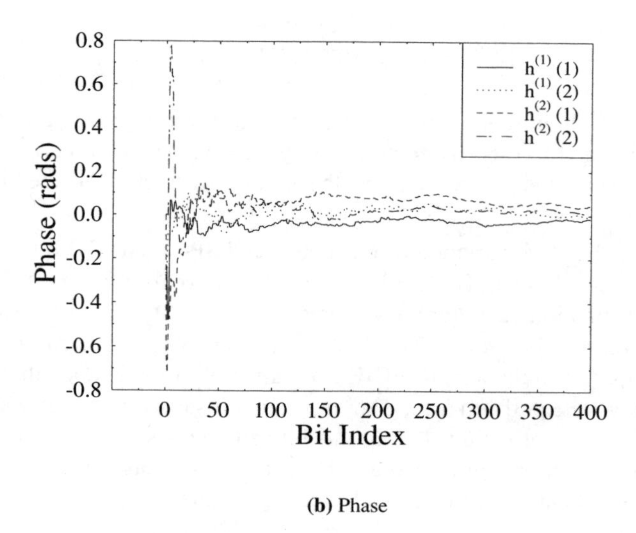

(b) Phase

Figure 7.4: Estimated signal amplitudes and phases for a two-user DBPSK DS-CDMA system where each user transmitted over a static two-path channel at $E_s/N_0 = 6$ dB. The CIRs were $\mathbf{h}^{(1)} = \{0.816, 0.577z^{-1}\}$ and $\mathbf{h}^{(2)} = \{0.707, 0.707z^{-1}\}$, where the notation z^{-1} indicates a delay of 1 chip.

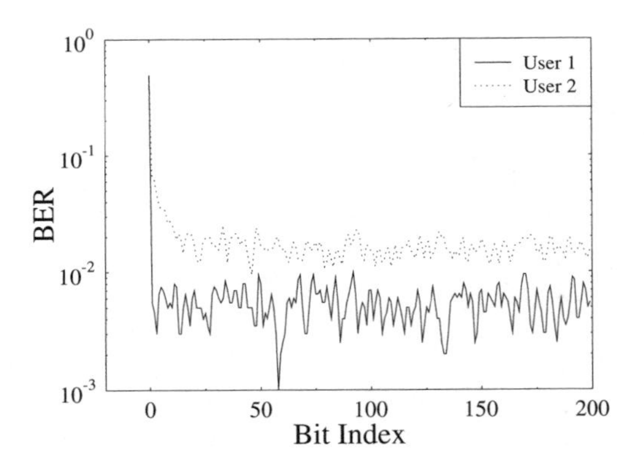

Figure 7.5: Variation in the BER for each individual bit in a received burst for a two-user CDMA system transmitting over the static two-path channels of Equations 7.48 and 7.49, at $E_s/N_0 = 6$ dB.

were static two-path channels, represented by:

$$\mathbf{h}^{(1)} = \{0.816, 0.577z^{-1}\}, \quad (7.48)$$

$$\mathbf{h}^{(2)} = \{0.707, 0.707z^{-1}\}, \quad (7.49)$$

where the notation z^{-1} indicates a delay of 1 chip. From the figures we observe that the estimated CIRs converged to their associated steady-state values within approximately 100 iterations and the steady-state values seen in the figures were close to the actual CIR tap values specified above.

Figure 7.5 shows the BER variation of our two-user PSP-based CDMA system for each individual bit in a received burst. The BER variation here reveals a faster convergence compared to the CIR estimation in Figure 7.4, demonstrating that the steady state BER was achieved after approximately 5 bits. The BER of the first bit was high compared to the BER of the other bits due to the adaptive CIR estimation algorithm, since the estimator had not reached its steady state performance. Therefore, in our subsequent BER calculations the errors in the first bit were not included since this bit can be considered as a training bit. Finally, the BER of the two users converged to different values because user 1 suffered from a more severe multipath channel resulting in a lower BER value.

The BER performance of a two-user CDMA system employing the blind PSP receiver is presented in Figure 7.6. The users transmitted over static two-path channels having the impulse responses of Equations 7.48 and 7.49. The BER performance of two types of sequences was compared, namely that of m-sequences and Gold sequences of Sections 2.6.2 and 2.6.3, respectively. From the results shown, it can be observed that the scheme employing Gold sequences outperformed the system using m-sequences. The former system was capable of achieving zero BER at $E_s/N_0 = 10$ dB, while the latter scheme exhibited an error floor in

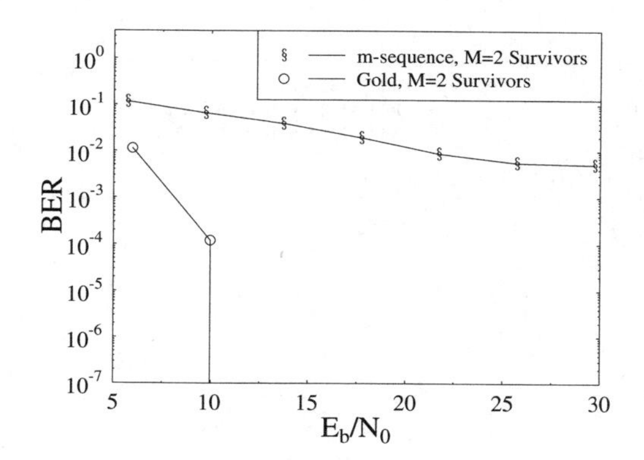

Figure 7.6: BER performance of a two-user CDMA system employing blind PSP receivers in conjunction with adaptive RLS channel estimation. The BER performance for two types of sequences, namely m-sequences and Gold sequences of $Q = 31$ each, are compared. Transmissions were carried out over static two-path channels having the CIRs of Equations 7.48 and 7.49. The vertical part of BER curve for the system employing Gold sequences indicates that for $E_s/N_0 > 10$ dB, all the errors were corrected by the PSP-CDMA receiver.

its BER performance. This is because the m-sequences have large CCL values compared to Gold sequences, as we saw in Section 2.6.3. Since the performance of the system is dependent on the initial estimates generated by the RAKE receiver, the CCL values play an important role in determining the performance of the system.

In the following experiments, the performance of the PSP receiver in a two-path Rayleigh-fading scenario was evaluated. Figure 7.7 shows the BER performance of a two-user CDMA system transmitting over the previous chip-spaced two-path channels that were Rayleigh-faded. Different fading conditions were compared, namely a non-fading channel, a slow-fading channel – where the CIR remained constant over one detection burst length, albeit varied from burst to burst – and a fast-fading channel, where the CIR varied continuously over the entire burst. From the results shown in Figure 7.7, it can be noted that as the fading rate increased, the performance degraded. In both fading scenarios, an error floor exists and this error floor is higher for the fast-fading channel. This is not a surprising conclusion, since the receiver obtains the CIR estimates blindly, without the benefit of training sequences and the accuracy of the CIR estimates in a fading scenario is limited.

In the next section, channel coding is employed in order to improve the BER performance over fading channels.

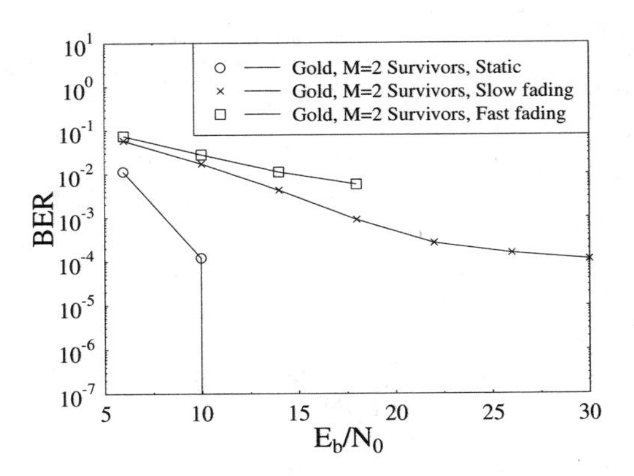

Figure 7.7: BER performance of a two-user CDMA system employing blind PSP receivers with adaptive RLS channel estimation. The BER performance for different fading conditions is compared. The transmission was carried out over Rayleigh-faded two-path channels having the CIRs of Equations 7.48 and 7.49. The vertical part of BER curve for the system where a static channel was employed indicates that for $E_s/N_0 > 10$ dB, all the errors were corrected by the PSP-CDMA receiver.

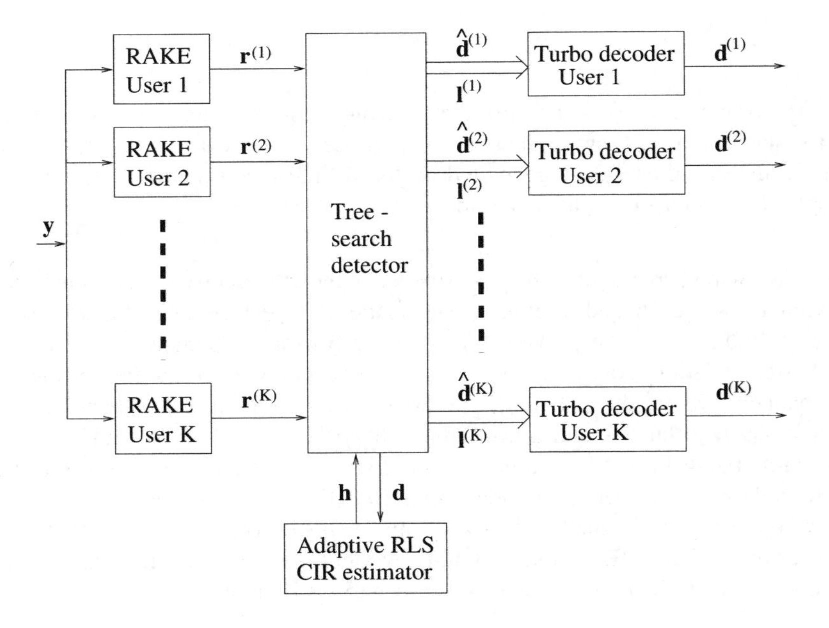

Figure 7.8: Schematic of the multi-user PSP receiver in conjunction with turbo decoding.

7.6 PSP with Turbo Coding

In this section, the approach adopted by , and [159] was employed in the proposed blind PSP receiver. In their work, *et al.* combined a tree-search-based detector with convolutional coding and the Soft Output Viterbi Algorithm (SOVA) [213] for channel decoding. However, their receiver assumed perfect amplitude estimation and no CIR estimation was performed. In this section, the tree-search-based detector is modified, in order to invoke turbo-coding [79, 197] rather than convolutional coding, and to generate soft outputs as reliability values for a Max-Log-MAP turbo decoder [266, 267]. This receiver differs from the receiver proposed by *et al.* in that the CIR estimates were obtained blindly with the aid of adaptive RLS CIR estimators and again, turbo convolutional coding [79, 197] was used as the channel coding method. In the investigations carried out with the aid of turbo convolutional coding, the component codes that constituted the turbo code consisted of two half-rate convolutional codes with the constraint length of 3. The octally represented generator polynomials were 5 and 7. The turbo interleaver employed was a square interleaver where its dimensions depended on the performance requirements. Instead of the SOVA decoder employed by *et al.*, the Max-Log-MAP decoder [266, 267] was utilized for turbo decoding. The Max-Log-MAP algorithm was developed in order to simplify the highly complex MAP algorithm [90], which is the optimal decoding algorithm. Therefore, the Max-Log-MAP decoder is a sub-optimal decoder that has, however, a significantly reduced complexity compared to the MAP and SOVA decoders. Turbo decoding is an iterative decoding process, where invoking eight iterations provides an attractive trade-off between BER performance and decoding complexity. The algorithm for generating the soft inputs to the Max-Log-MAP decoder is presented next.

The soft inputs required by the turbo decoder are in the form of reliability or confidence measures for each bit that is output by the PSP receiver of Figure 7.1. In traversing from stage i to stage $i + 1$ in the tree of Figure 7.2, the following steps are added after Step 2 in the tree-search block of the PSP algorithm described in Section 7.2.

a. Out of the $2M$ nodes that have been extended from the previous stage, consider one extended node, $\eta[i]$, and let its associated, accumulative metric value be represented by μ_i, which is obtained from $\mu_i = m[I = i]$ of Equation 7.6, where the associated bit of d_i has an estimated bit value of δ, for $\delta \in \{1, -1\}$.

b. Let S be the set of all the other extended nodes, where the associated bit value d_i is not equal to δ.

c. From the set S, discard the node that has the same metric value of μ_i as the extended node $\eta[i]$ being considered. This is because, when DBPSK is used to resolve the phase ambiguity, this results in one other node that has the opposite phase estimate, but exhibits the same metric value and which eventually results in the same bit sequence after differential demodulation. Therefore, the node with the same metric value has to be discarded from the reliability calculations.

d. Now, from the remaining nodes in the set S, choose the node associated with the highest metric and let its metric value be represented by μ_s.

e. The reliability metric for $d_i = \delta$ of node $\eta[i]$ is calculated as $l = \delta \times |\mu_i - \mu_s|$.

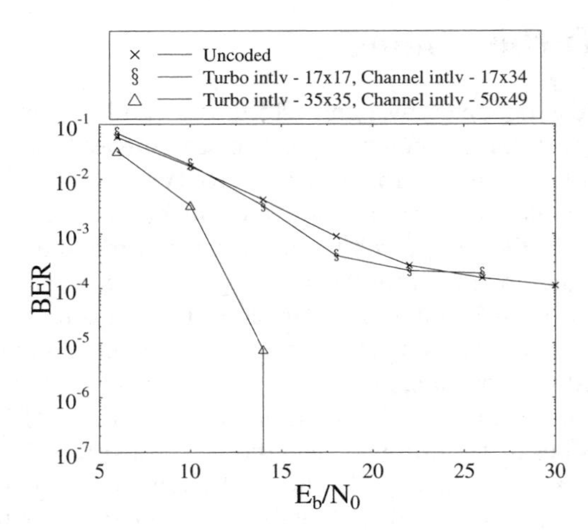

Figure 7.9: BER performance of a two-user CDMA system employing blind PSP receivers in conjunction with adaptive RLS channel estimation. Half-rate, turbo convolutional coding was employed with a constraint length of 3 and using a Max-Log-MAP decoder. Transmission was carried out over two-path Rayleigh-faded channels having the CIRs of Equations 7.48 and 7.49.

f. Repeat steps (a) to (e) for all the other extended nodes, until the soft outputs for those nodes have been generated.

For the performance results presented in this section, the data bits of each user were passed to a half-rate turbo codec having a constraint length of 3, and the component convolutional codes were recursive convolutional codes which were represented by octal generator polynomials of 5 and 7. The turbo-coded output bits were modulated using DBPSK and spread with a Gold sequence of $Q = 31$ chips. Figure 7.8 shows the schematic of the PSP receiver having individual turbo decoders for each user. From the composite received signal, $\mathbf{y}$, the PSP receiver generated the estimated bit sequence and the associated soft reliability values for each bit in the estimated sequence. The estimated bit sequence and associated soft outputs were separated into two sets, namely the bits, $\hat{\mathbf{d}}^{(k)}$, and soft outputs, $\mathbf{l}^{(k)}$, for each user, which were then passed on to the individual turbo decoders of the users. The turbo decoder of each user then employed the Max-Log-MAP algorithm in order to generate the final data bits, $\mathbf{d}^{(k)}$.

Figure 7.9 shows the turbo-decoded BER performance of the turbo-coded PSP receiver for a two-path Rayleigh-faded channel with different turbo interleaver sizes. From the performance results shown, it can be observed that a large turbo interleaver with the square dimensions of (35×35) and a channel interleaver with the dimensions of (50×49) were required, in order to remove the error floor. When the turbo interleaver size was decreased to (17×17) and the channel interleaver size was reduced to (17×34), the error floor persisted.

The BER performance over the Typical Urban four-path channel [105] of Figure 3.18 is shown in Figure 7.10. In these investigations the fading was assumed to be slow and constant over one transmission burst, but varied from burst to burst. Here, the size of the turbo

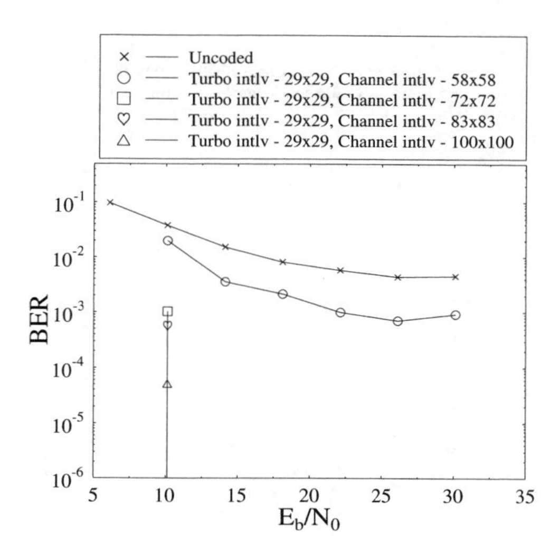

Figure 7.10: BER performance of a two-user CDMA system employing blind PSP receivers in conjunction with adaptive RLS channel estimation. Half-rate, turbo convolutional coding was employed with a constraint length of 3 and using a Max-Log-MAP decoder. Transmissions were carried out over slowly-fading four-path Typical Urban channels [105] of Figure 3.18. The fading rate was set to a Doppler frequency of 0.8 Hz. The vertical BER curves of some of the turbo-coded systems indicates that for $E_s/N_0 > 10$ dB, the errors were all corrected by the turbo decoder.

interleaver was kept constant and had dimensions of (29×29), but the channel interleaver size was varied. By increasing the channel interleaver size, the burst errors were further randomized and this benefitted the channel decoding process. From the results shown, it can be seen that it is very important to randomize the burst errors with the aid of a large interleaver size. For channel interleaver sizes of (72×72), (83×83), and (100×100), the error floor in the BER performance was eliminated.

In the next graph in Figure 7.11, the effect of the turbo interleaver size on the BER performance was investigated for the Typical Urban channel of Figure 3.18, as before. In these investigations the channel interleaver size was maintained at (100×100), while two different turbo interleaver sizes of (29×29) and (35×35) were compared. Here, in conjunction with a large channel interleaver size of (100×100), varying the turbo interleaver size did not have a significant impact on the BER performance and the error floor of the uncoded BER curve was eliminated with the aid of turbo coding.

Finally, the BER performance of the turbo-coded system is evaluated for the fast-fading Typical Urban four-path channel of Figure 3.18. In this scenario, the fading varied continuously over the entire transmitted burst. We observe from the results shown in Figure 7.12 that a large turbo interleaver and a large channel interleaver were required in order to remove the error floor. The dimensions of the turbo and channel interleaver that achieved this were (55×55) and (156×156), respectively, where the increased performance potential was mainly attributed to the increased turbo interleaver size.

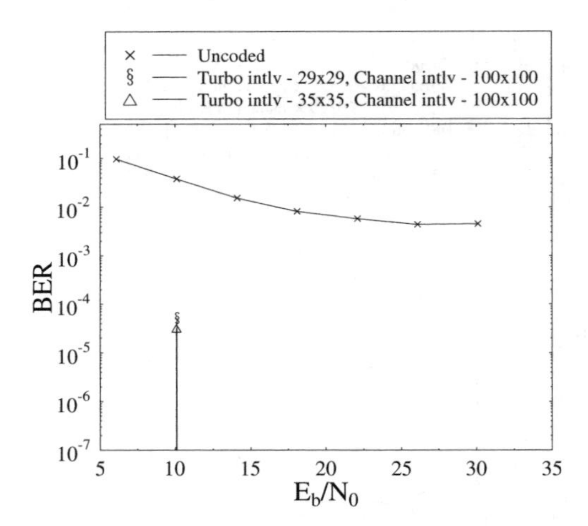

Figure 7.11: BER performance of a two-user CDMA system employing blind PSP receivers in conjunction with adaptive RLS channel estimation. Half-rate, turbo convolutional coding was employed with a constraint length of 3 using a Max-Log-MAP decoder. Transmissions were carried out over the slow-fading, four-path Typical Urban channel [105] of Figure 3.18. The fading rate was set to a Doppler frequency of 0.8 Hz. The vertical BER curves of the turbo-coded systems indicates that for $E_s/N_0 > 10$ dB, the errors were all corrected by the turbo decoder.

7.7 Chapter Summary and Conclusion

In this section, a blind PSP receiver assisted by adaptive RLS CIR estimation was proposed and investigated over a synchronous multipath multi-user channel. The receiver and channel estimators operated reliably over static channels, but the BER performance degraded, resulting in an error floor when fading was inflicted by the channel. This error floor was eliminated, when turbo convolutional coding was employed. The PSP detector was modified, in order to produce soft outputs, which were then fed to the turbo decoder. With the aid of large turbo and channel interleavers, the BER performance of the multi-user interference contaminated fading channels was improved and the error floor was eliminated.

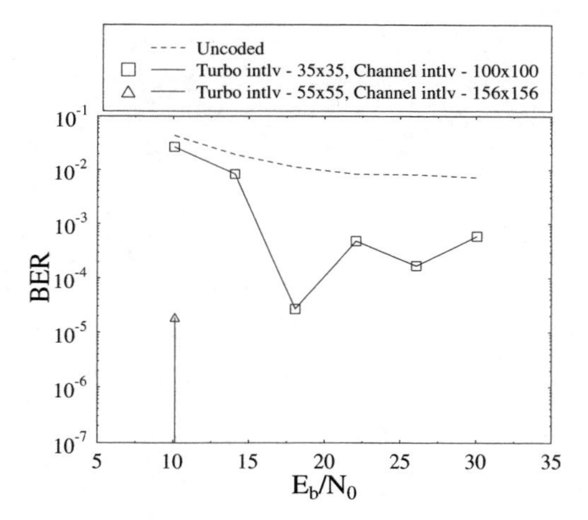

Figure 7.12: BER performance of a two-user CDMA system employing blind PSP receivers in conjunction with adaptive RLS channel estimation. Half-rate, turbo convolutional coding was employed with a constraint length of 3 using a Max-Log-MAP decoder. Transmission was carried out over the Rayleigh-faded, four-path Typical Urban channel [105] as shown in Figure 3.18. The fading rate was set to a Doppler frequency of 80 Hz. The vertical BER curve of the turbo-coded system with a turbo interleaver of size (55×55) indicates that for $E_s/N_0 > 10$ dB, the errors were all corrected by the turbo decoder.

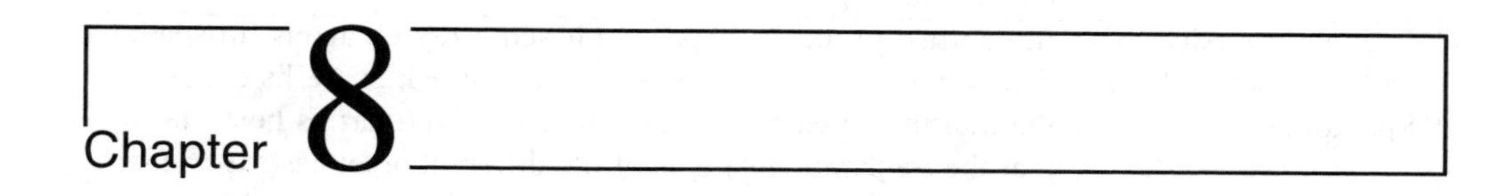

Space-Time Spreading Aided Single-Carrier Wideband CDMA Communicating over Multipath Nakagami Fading Channels

8.1 Introduction

In recent years various smart antenna designs have emerged, which have found application in diverse scenarios, as seen in Table 8.1. The main objective of employing smart antennas is that of combating the effects of multipath fading on the desired signal and suppressing interfering signals, thereby increasing both the performance and capacity of wireless systems [271]. Specifically, in smart antenna assisted systems multiple antennas may be invoked at the transmitter and/or the receiver, where the antennas may be arranged for achieving spatial diversity, directional beamforming or for attaining both diversity and beamforming. In smart antenna systems the achievable performance improvements are usually a function of the antenna spacing and that of the algorithms invoked for processing the signals received by the antenna elements.

In beamforming arrangements [268] typically $\lambda/2$-spaced antenna elements are used for the sake of creating a spatially selective transmitter/receiver beam. Smart antennas using beamforming have widely been employed for mitigating the effects of various interfering signals and for providing beamforming gain. Furthermore, the beamforming arrangement is capable of suppressing co-channel interference, which allows the system to support multiple users within the same bandwidth and/or same time-slot by separating them spatially. This

Beamforming [268]	Typically $\lambda/2$-spaced antenna elements are used for the sake of creating a spatially selective transmitter/receiver beam. Smart antennas using beamforming have been employed for mitigating the effects of cochannel interfering signals and for providing beamforming gain.
Spatial Diversity [5] and Space-Time Spreading	In contrast to the $\lambda/2$-spaced phased array elements, in spatial diversity schemes, such as space-time block or trellis codes [5] the multiple antennas are positioned as far apart as possible, so that the transmitted signals of the different antennas experience independent fading, resulting in the maximum achievable diversity gain.
Space Division Multiple Access [269]	SDMA exploits the unique, user-specific "spatial signature" of the individual users for differentiating amongst them. This allows the system to support multiple users within the same frequency band and/or time slot.
Multiple Input Multiple Output Systems [270]	MIMO systems also employ multiple antennas, but in contrast to SDMA arrangements, not for the sake of supporting multiple users. Instead, they aim for increasing the throughput of a wireless system in terms of the number of bits per symbol that can be transmitted by a given user in a given bandwidth at a given integrity.

Table 8.1: Applications of multiple antennas in wireless communications

spatial separation, however, becomes only feasible, if the corresponding users are separable in terms of the angle of arrival of their beams. These beamforming schemes, which employ appropriately phased antenna array elements that are spaced at distances of $\lambda/2$ typically result in an improved SINR distribution and enhanced network capacity [268].

In contrast to the $\lambda/2$-spaced phased array elements, *in spatial diversity schemes* the multiple antennas are positioned as far apart as possible, using a typical spacing of 10λ [271], so that the transmitted signals of the different antennas experience independent fading, when they reach the receiver. This is because the maximum diversity gain can be achieved, when the signal replicas experience independent fading. Although spatial diversity can be achieved by employing multiple antennas at either the base station, mobile station, or both, it is more cost effective and practical to employ multiple antennas at the base station. A system having multiple receiver antennas has the potential of achieving receiver diversity, while that employing multiple transmit antennas exhibits transmit diversity. Recently, the family of transmit diversity schemes based on space-time coding, either space-time block codes or space-time trellis codes, has received wide attention and has been invoked in the third-generation systems [272]. The aim of using spatial diversity is to provide both transmit as well as receive diversity and hence enhance the system's integrity/robustness. This typically results in a better physical-layer performance and hence a better network-layer performance, hence space-time codes indirectly increase not only the transmission integrity, but also the achievable spectral efficiency.

A third application of smart antennas is often referred to as *Space Division Multiple Ac-*

cess [269] (SDMA), which exploits the unique, user-specific "spatial signature" of the individual users for differentiating amongst them. In simple conceptual terms one could argue that both a conventional CDMA spreading code and the Channel Impulse Response (CIR) affect the transmitted signal similarly - they are namely convolved with it. Hence, provided that the CIR is accurately estimated, it becomes known and certainly unique, although - as opposed to orthogonal Walsh-Hadamad spreading codes, for example - not orthogonal to the other CIRs. Nonetheless, it may be used for uniquely identifying users after channel estimation and hence for supporting several users within the same bandwidth. Provided that a powerful multi-user detector is available, one can support even more users than the number of antennas. Hence this method enhances the achievable spectral efficiency directly.

Finally, Multiple Input Multiple Output (MIMO) systems [270] also employ multiple antennas, but in contrast to SDMA arrangements, not for the sake of supporting multiple users. Instead, they aim for increasing the throughput of a wireless system in terms of the number of bits per symbol that can be transmitted by a given user in a given bandwidth at a given integrity.

In this chapter we investigate the performance of single-carrier Wideband CDMA (W-CDMA) schemes using multiple transmission antennas. Specifically, we study the performance of W-CDMA systems, when employing space-time spreading (STS) based transmit diversity. We assume that we encounter dispersive multipath Nakagami-m fading channels, multi-user interference and background noise. Our analysis and the results will demonstrate that the total diversity order achieved is the product of the diversity order achieved by STS and that owing to multipath diversity accruing as a benefit of the channel's frequency-selectivity. Furthermore, we will show that both the STS-based transmit diversity and the multipath-induced receiver diversity achieved with the aid of the channel's frequency-selectivity appear to have the same order of importance and hence offer similar potential performance benefits.

The frequency-selective frequency-domain transfer function of W-CDMA wireless channels may vary slowly, but often over a wide dynamic range, when roaming in urban and suburban areas [273]. Therefore, the number of resolvable paths encountered at the receiver can be modelled as a random variable distributed over a certain range, depending on the location of the receiver, where the number of resolvable paths varies slowly, as the receiver (mobile station) moves. Consequently, STS schemes designed on the basis of a low number of resolvable paths or based on the premise of encountering a constant number of resolvable paths may not achieve the maximum communication efficiency expressed in terms of the effective throughput. Motivated by the above arguments, in this chapter an adaptive STS-based transmission scheme is proposed and investigated, which adapts the mode of operation of its STS scheme and its corresponding data rate, according to the near-instantaneous frequency-selectivity information fed back from the mobile receiver to the base station's transmitter. The numerical results show that this adaptive STS scheme is capable of efficiently exploiting the diversity potential provided by the channel's frequency-selectivity, hence significantly improving the effective throughput of W-CDMA systems.

The remainder of this chapter is organized as follows. First we provide an introduction to the basic STS principles. Then, in Section 8.3 the W-CDMA system's model using STS and the channel model are described. Section 8.4 considers the detection of STS-based W-CDMA signals. In Section 8.5 we derive the corresponding BER expression and summarize our numerical results, while in Section 8.6 we describe the proposed adaptive STS scheme and investigate its BER performance. Finally, our conclusions are offered in Section 8.7.

8.2 Principle of Space-time Spreading

A transmit diversity scheme using space-time spreading (STS) has been investigated in [274]. The STS arrangement considered constitutes an open-loop transmit diversity scheme, in which the symbols are spread using multiple orthogonal codes, such as Walsh codes. Below we review the STS principle by first considering two transmission antennas and one receiver antenna.

8.2.1 Two Transmission Antennas and One Receiver Antenna

Let the original serial data stream of rate $R_b = 1/T_b$ be split into an even-indexed substream of $\{b_1\}$ and odd-indexed substream of $\{b_2\}$. The data rate of these substreams is hence halved to $R_s = 1/T_s = 1/2T_b$. Let c_k be a G-length orthogonal code assigned to the kth user, where $G = T_b/T_c$ represents the number of chips per bit-duration and T_c is the chip-duration of the orthogonal spreading codes. Since in the STS scheme using two transmission antennas the transmitted signal's symbol duration is $T_s = 2T_b$, two orthogonal code waveforms having a period of $2G$ are required by the STS. These two orthogonal spreading waveforms can be derived from the orthogonal code waveform c_k by concatenating two related versions of the original code, which are expressed as:

$$\begin{aligned} c_{k1} &= [c_k \; c_k] \\ c_{k2} &= [c_k \; -c_k], \end{aligned} \tag{8.1}$$

which can be expressed in matrix form as:

$$\mathbf{c}^T = \begin{bmatrix} c_{k1} \\ c_{k2} \end{bmatrix} = \begin{bmatrix} 1 & 1 \\ 1 & -1 \end{bmatrix} \oplus \begin{bmatrix} c_k \\ c_k \end{bmatrix}, \tag{8.2}$$

where the superscript of $()^T$ represents the transpose of a vector or a matrix and $\oplus$ represents the previously introduced concatenation operation. By observing (8.2) it becomes explicit that the required double-length orthogonal waveforms are obtained by concatenating the originally assigned orthogonal code waveform of c_k with a Hadamard matrix. Since the data rate per substream is halved splitting the data rate into even and odd streams, the original chip rate remains unchanged, as the length of the orthogonal codes is doubled. The overall data rate also retmains unchanged, since the even and odd streams are sent simultaneously. Hence, by using the extended orthogonal codes as described in (8.2), in STS no additional orthogonal codes are required for spreading the signals of two, rather than one antenna.

When using the orthogonal code waveforms of $c_{k1}(t)$ and $c_{k2}(t)$ in the context of STS using two transmission antennas, the resultant transmitted signals of antenna 1 and 2 are given by:

$$s_{k1}(t) = \sqrt{\frac{P}{2}} \left[b_1 c_{k1}(t) + b_2 c_{k2}(t) \right], \tag{8.3}$$

$$s_{k2}(t) = \sqrt{\frac{P}{2}} \left[b_2 c_{k1}(t) - b_1 c_{k2}(t) \right]. \tag{8.4}$$

Let h_1 and h_2 represent the non-dispersive channel impulse responses with respect to the first and second antennas, where h_1 and h_2 assume complex values. The received signal can then be expressed as:

$$\begin{aligned} r(t) &= h_1 s_{k1}(t) + h_2 s_{k2}(t) + n(t) \\ &= \sqrt{\frac{P}{2}} h_1 \left[b_1 c_{k1}(t) + b_2 c_{k2}(t)\right] + \sqrt{\frac{P}{2}} h_2 \left[b_2 c_{k1}(t) - b_1 c_{k2}(t)\right] + n(t), \end{aligned} \quad (8.5)$$

where $n(t)$ represents the additive white Gaussian noise (AWGN) contributed by the background noise sources. Upon correlating the received signal against $c_{k1}(t)$ and $c_{k2}(t)$, we arrive at:

$$d_1 = \sqrt{2PT_b}\left[h_1 b_1 + h_2 b_2\right] + n_1, \quad (8.6)$$

$$d_2 = \sqrt{2PT_b}\left[-h_2 b_1 + h_1 b_2\right] + n_2, \quad (8.7)$$

where

$$n_1 = \int_0^{T_s} n(t) c_{k1}(t) dt, \quad n_2 = \int_0^{T_s} n(t) c_{k2}(t) dt. \quad (8.8)$$

If we introduce the following definitions:

$$\mathbf{d} = \begin{bmatrix} d_1 \\ d_2 \end{bmatrix}, \quad \mathbf{H} = \begin{bmatrix} h_1 & h_2 \\ -h_2 & h_1 \end{bmatrix}, \quad \mathbf{b} = \begin{bmatrix} b_1 \\ b_2 \end{bmatrix}, \quad \mathbf{n} = \begin{bmatrix} n_1 \\ n_2 \end{bmatrix}, \quad (8.9)$$

then (8.6) can be written as:

$$\mathbf{d} = \sqrt{2PT_b}\mathbf{H}\mathbf{b} + \mathbf{n}. \quad (8.10)$$

Let us assume that the receiver has obtained perfect knowledge of h_1 and h_2. Multiplying both sides of (8.10) by $\mathbf{H}^\dagger$, where the operation of $()^\dagger$ denotes complex conjugate transpose, it can be shown that the decision variables $\mathbf{Z} = [Z_1\ Z_2]^T$ corresponding to b_1 and b_2 can be expressed as:

$$\begin{aligned} \mathbf{Z} &= \text{Re}\left\{\mathbf{H}^\dagger \mathbf{d}\right\} \\ &= \sqrt{2PT_b}\text{Re}\left\{\mathbf{H}^\dagger \mathbf{H}\mathbf{b}\right\} + \text{Re}\left\{\mathbf{H}^\dagger \mathbf{n}\right\}, \end{aligned} \quad (8.11)$$

where Re(x) represents the real part of x. Upon extending (8.11) with the aid of (8.9), it yields:

$$\begin{aligned} \mathbf{Z} &= \begin{bmatrix} Z_1 \\ Z_2 \end{bmatrix} = \sqrt{2PT_b}\begin{bmatrix} |h_1|^2 + |h_2|^2 & 0 \\ 0 & |h_1|^2 + |h_2|^2 \end{bmatrix}\begin{bmatrix} b_1 \\ b_2 \end{bmatrix} + \text{Re}\left\{\mathbf{H}^\dagger \mathbf{n}\right\} \\ &= \sqrt{2PT_b}\begin{bmatrix} (|h_1|^2 + |h_2|^2) b_1 \\ (|h_1|^2 + |h_2|^2) b_2 \end{bmatrix} + \text{Re}\left\{\mathbf{H}^\dagger \mathbf{n}\right\}, \end{aligned} \quad (8.12)$$

which shows that the optimum two-fold diversity gain can be achieved compared to the case of using a single-transmission antenna. More explicitly, in (8.12) the two-order of diversity manifests itself in terms of the sum of $(|h_1|^2 + |h_2|^2)$, indicating that even if one of the diversity channels is badly faded, the other one is likely to be providing an independent source of received signal energy.

Following a similar analytical approach to that outlined above in the context of the single-receiver antenna scenario, a STS scheme using two transmission antennas and multiple receiver antennas can also be readily analyzed. Specifically, it can be shown that when N_r number of receiver antennas are employed in addition to the two transmission antennas, the diversity gain achieved is of the order of $2N_r$, as indicated in [274].

8.2.2 Using U Transmission Antennas and One Receiver Antenna

In this section the scenario of employing a $U = 2$ transmission antenna assisted STS scheme is generalized to the case of using U number of transmission antennas, where we assume that $U = 2^i$ and $i \geq 1$ is an integer, since in this case the resultant STS scheme is capable of providing the maximum achievable transmit diversity gain without requiring extra spreading codes [274].

In the context of STS using U number of transmission antennas, U number of data substreams are processed by the STS scheme. Let the original serial data stream of rate $R_b = 1/T_b$ be serial-to-parallel (S/P) converted to U substreams, namely $\{b_1\}, \{b_2\}, \ldots, \{b_U\}$. Consequently, the data rate of each of these substreams is $R_s = 1/T_s = 1/UT_b$ and hence the transmitted symbol duration is $T_s = UT_b$. Again, let c_k be a G-length orthogonal code waveform assigned to the kth user. The U number of orthogonal spreading waveforms required by the STS can be derived from the original orthogonal code waveform c_k by concatenating it with its appropriately inverted version under the control of the $(U \times U)$-dimensional Hadamard matrix of $\mathbf{W}_{U\times U}$, which can be formulated as:

$$\mathbf{c}^T = \begin{bmatrix} c_{k1} \\ c_{k2} \\ \cdots \\ c_{kU} \end{bmatrix} = \mathbf{W}_{U\times U} \oplus \begin{bmatrix} c_k \\ c_k \\ \cdots \\ c_k \end{bmatrix}. \tag{8.13}$$

Based on this set of orthogonal codes, the transmitted STS signals corresponding to employing U number of transmission antennas can be described as:

$$\mathbf{s}_k(t) = \begin{bmatrix} s_{k1}(t) \\ s_{k2}(t) \\ \cdots \\ s_{kU}(t) \end{bmatrix} = \mathbf{c}(t)\mathbf{B}_U(t), \tag{8.14}$$

where a power-related scaling factor has been removed for the sake of simplicity. In (8.14) the vector $\mathbf{c}(t)$ hosts the orthogonal spreading waveforms derived from (8.13), $\mathbf{B}_U(t)$ represents the $U \times U$ dimensional transmitted data matrix created by mapping U input data bits to the U parallel sub-streams according to the specific design rules decribed in [274], so that the maximum possible transmit diversity is achieved, while using relatively low-complexity signal detection algorithms. Specifically, $\mathbf{B}_U(t)$ can be written as:

$$\mathbf{B}_U(t) = \begin{pmatrix} a_{11}b_{11} & a_{12}b_{12} & \dots & a_{1U}b_{1U} \\ a_{21}b_{21} & a_{22}b_{22} & \dots & a_{2U}b_{2U} \\ \vdots & \vdots & \ddots & \vdots \\ a_{U1}b_{U1} & a_{U2}b_{U2} & \dots & a_{UU}b_{UU} \end{pmatrix}(t), \tag{8.15}$$

where the time dependence of the (i, j)th element is indicated at the right of the matrix for simplicity. In (8.15) a_{ij} represents the sign of the element at the ith row and the jth column, which is determined by the STS design rule, while b_{ij} is the data bit assigned to the (i, j)th element, which is one of the U input data bits $\{b_1, b_2, \dots, b_U\}$. Each input data bit of $\{b_1, b_2, \dots, b_U\}$ appears only once in a given row and in a given column. For $U = 2$ and 4, $\mathbf{B}_2(t)$ and $\mathbf{B}_4(t)$ are given by [274]:

$$\mathbf{B}_2(t) = \begin{pmatrix} b_1 & b_2 \\ b_2 & -b_1 \end{pmatrix}(t), \quad \mathbf{B}_4(t) = \begin{pmatrix} b_1 & b_2 & b_3 & b_4 \\ b_2 & -b_1 & b_4 & -b_3 \\ b_3 & -b_4 & -b_1 & b_2 \\ b_4 & b_3 & -b_2 & -b_1 \end{pmatrix}(t). \tag{8.16}$$

Let us assume that the STS signal of (8.14) is transmitted over a wireless channel having the non-dispersive channel impulse responses (CIR) for the U different antennas expressed by:

$$\mathbf{h}^T = [h_1, h_2, \dots, h_U], \tag{8.17}$$

where h_i represents the CIR corresponding to the ith transmission antenna. Then, the received signal can be expressed as:

$$r(t) = \mathbf{c}(t)\mathbf{B}_U(t)\mathbf{h} + n(t), \tag{8.18}$$

where $n(t)$ represents the AWGN. Upon correlating the received signal $r(t)$ with the orthogonal codes $\mathbf{c}^T(t)$, it yields:

$$\mathbf{d} = \begin{bmatrix} d_1 \\ d_2 \\ \dots \\ d_U \end{bmatrix} = \mathbf{B}_U\mathbf{h} + \mathbf{n}, \tag{8.19}$$

where $\mathbf{n}^T = [n_1, n_2, \dots, n_U]$ and

$$n_u = \int_0^{T_s} n(t)c_{ku}(t)dt, \ u = 1, 2, \dots, U. \tag{8.20}$$

It can be shown that [274], for the STS scheme using $U = 2^i$ number of transmission antennas, there exist STS schemes that obeys:

$$\mathbf{B}_U\mathbf{h} = \mathbf{H}\mathbf{b}, \tag{8.21}$$

where $\mathbf{H}$ is an $(U \times U)$-dimensional matrix with its elements derived from the channel gains of $\{h_1, h_2, \dots, h_U\}$, while

$$\mathbf{b} = [b_1, b_2, \dots, b_U]. \tag{8.22}$$

Therefore, according to (8.21), (8.19) can be written as:

$$\mathbf{d} = \mathbf{H}\mathbf{b} + \mathbf{n}. \tag{8.23}$$

The decision variables derived for the U number of data bits are obtained by multiplying both sides of (8.23) with $\mathbf{H}^\dagger$, which can be formulated as:

$$\mathbf{Z} = \begin{bmatrix} Z_1 \\ Z_2 \\ \cdots \\ Z_U \end{bmatrix} = \text{Re}\left\{\mathbf{H}^\dagger\mathbf{H}\right\}\mathbf{b} + \text{Re}\left\{\mathbf{H}^\dagger\mathbf{n}\right\}. \tag{8.24}$$

We note that for the most attractive set of STS schemes corresponding to using $U = 2^i$ transmission antennas, we have $\mathbf{H}^\dagger\mathbf{H} = \mathbf{h}^\dagger\mathbf{h}\mathbf{I}$ [274]. Hence, in this case the decision variables corresponding to the U transmitted bits obey:

$$\begin{aligned} \mathbf{Z} = \begin{bmatrix} Z_1 \\ Z_2 \\ \cdots \\ Z_U \end{bmatrix} &= \text{Re}\left\{\mathbf{h}^\dagger\mathbf{h}\right\}\mathbf{I}\mathbf{b} + \text{Re}\left\{\mathbf{H}^\dagger\mathbf{n}\right\} \\ &= \begin{bmatrix} \sum_{u=1}^{U} |h_u|^2 \times b_1 \\ \sum_{u=1}^{U} |h_u|^2 \times b_2 \\ \cdots \\ \sum_{u=1}^{U} |h_u|^2 \times b_U \end{bmatrix} + \text{Re}\left\{\mathbf{H}^\dagger\mathbf{n}\right\}. \end{aligned} \tag{8.25}$$

It can be shown according to (8.25) that for the most attractive STS schemes associated with using $U = 2^i$ transmission antennas, a Uth-order diversity gain can be achieved, as indicated by the sum of squared terms in (8.25), when U number of transmission antennas are used. In summary, the family of transmit diversity schemes using space-time spreading (STS) exhibits the following characteristics:

- STS-based transmit diversity is an efficient scheme designed for CDMA systems;
- It is an open-loop transmit diversity scheme, which does not require the knowledge of the CIR coefficients at the transmitter;
- For the specific set of CDMA transmitters using $U = 2, 4, 8, \dots$, etc. number of trans-

mission antennas, no extra spreading codes are required for carrying out the spreading of the signal to multiple transmission antennas, and each user's spreading codes used for STS can be derived from a single spreading code assigned to that user;

- STS-based transmit diversity schemes are capable of attaining the maximum achievable transmit diversity gain without requiring an increased transmit power, in comparison to a system using a single transmission antenna.

Following the above rudimentary review of the STS principle, in the forthcoming sections of this chapter we investigate the performance of W-CDMA systems using STS-assisted transmit diversity.

8.3 System Model

8.3.1 Transmitted Signal

The W-CDMA system considered in this chapter consists of U transmitter antennas and one receiver antenna. The transmitter schematic of the kth user and the receiver schematic of the reference user are shown in Figure 8.1, where real-valued data symbols using BPSK modulation and real-valued spreading [274] were assumed. Note that the analysis of this chapter can be extended to W-CDMA systems using U transmitter antennas and more than one receiver antenna, or to W-CDMA systems using complex-valued data symbols as well as complex-valued spreading. As shown in Figure 8.1(a), at the transmitter side the binary input data stream having a bit duration of T_b is serial-to-parallel (S/P) converted to U parallel sub-streams. The new bit duration of each parallel sub-stream, in other words the symbol duration becomes $T_s = UT_b$. After S/P conversion, the U number of parallel bits are direct-sequence spread using the STS schemes proposed in [274] and discussed in Section 8.2 with the aid of U number of orthogonal spreading sequences - for example, Walsh codes - having a period of UG, where $G = T_b/T_c$ represents the number of chips per bit and T_c is the chip-duration of the orthogonal spreading sequences. As seen in Figure 8.1(a), following STS, the U parallel signals to be mapped to the U transmission antennas are scrambled using the kth user's pseudo-noise (PN) sequence $\text{PN}_k(t)$, in order that the transmitted signals become randomized, and to ensure that the orthogonal spreading sequences employed within the STS block of Figure 8.1 can be reused by the other users. Finally, after the PN sequence based scrambling, the U number of parallel signals are carrier modulated and transmitted by the corresponding U number of antennas.

As described above, we have assumed that the number of parallel data sub-streams, the number of orthogonal spreading sequences used by the STS block of Figure 8.1 and the number of transmission antennas is the same, namely U. This specific STS scheme constitutes a specific sub-class of the generic family of STS schemes, where the number of parallel data sub-streams, the number of orthogonal spreading sequences required by STS block and the number of transmission antennas may take different values. The study conducted in [274] has shown that the number of orthogonal spreading sequences required by STS is usually higher, than the number of parallel sub-streams. The STS scheme having an equal number of parallel sub-streams, orthogonal STS-related spreading sequences as well as transmission antennas constitutes an attractive scheme, since this STS scheme is capable of providing

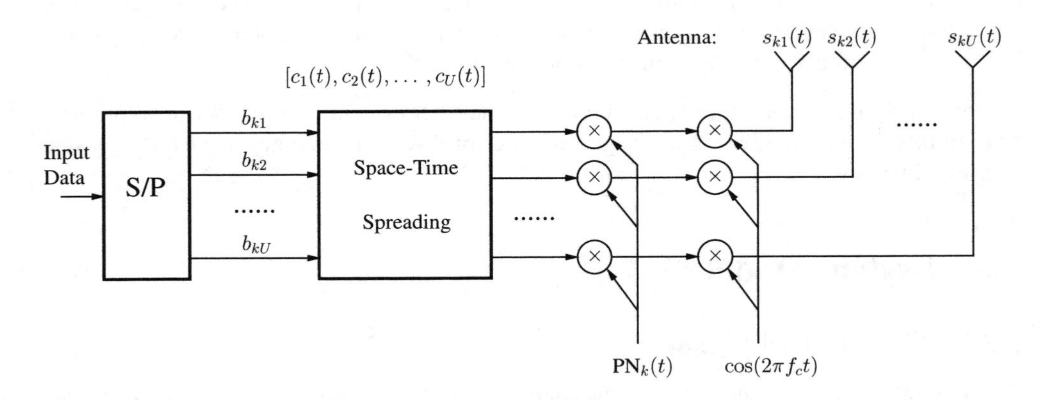

(a) Transmitter

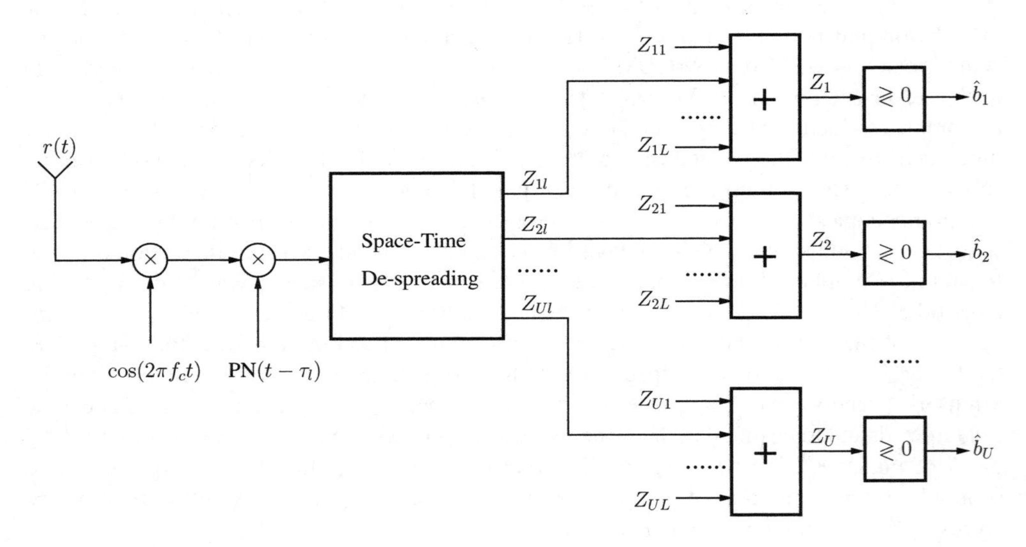

(b) Receiver

Figure 8.1: Transmitter and receiver block diagram of the W-CDMA system using space-time spreading.

maximal transmit diversity without requiring extra STS spreading codes. Note that for the specific values of $U = 2, 4$ the above-mentioned attractive STS schemes have been specified in [274]. In this section, we only investigate these attractive STS schemes.

Based on the philosophy of STS as discussed in [274] and referring to Figure 8.1(a), the transmitted signal of the kth user can be written as:

$$\mathbf{s}_k(t) = \sqrt{\frac{2P}{U^2}}\mathbf{c}(t)\mathbf{B}_U(t) \times \mathrm{PN}_k(t)\cos(2\pi f_c t), \tag{8.26}$$

where P represents each user's transmitted power, which is constant for all users, $\mathbf{s}_k(t) = [s_{k1}(t)\ s_{k2}(t)\ ...\ s_{kU}(t)]$ represents the transmitted signal vector of the U transmission antennas, while $\mathrm{PN}_k(t)$ and f_c represent the DS scrambling-based spreading waveform and the subcarrier frequency, respectively. The scrambling sequence waveform is given by $\mathrm{PN}_k(t) = \sum_{j=-\infty}^{\infty} p_{kj} P_{T_c}(t - jT_c)$, where p_{kj} assumes values of +1 or -1 with equal probability, while $P_{T_c}(t)$ is the rectangular chip waveform, which is defined over the interval $[0, T_c)$. In (8.26) the vector $\mathbf{c}(t) = [c_1(t)\ c_2(t)\ ...\ c_U(t)]$ is constituted by the U number of orthogonal signals assigned for the STS, $c_i(t) = \sum_{j=-\infty}^{\infty} c_{ij} P_{T_c}(t - jT_c),\ i = 1, 2, \ldots, U$ denotes the individual components of the STS-based orthogonal spread signals, where $\{c_{ij}\}$ is an orthogonal sequence of period UG for each index i; $\mathbf{B}_U(t)$ represents the $U \times U$ dimensional transmitted data matrix created by mapping U input data bits to the U parallel sub-streams according to specific design rules [274], so that the maximum possible transmit diversity is achieved, while using relatively low-complexity signal detection algorithms. Specifically, $\mathbf{B}_U(t)$ can be expressed as:

$$\mathbf{B}_U(t) = \begin{pmatrix} a_{11}b_{k,11} & a_{12}b_{k,12} & \cdots & a_{1U}b_{k,1U} \\ a_{21}b_{k,21} & a_{22}b_{k,22} & \cdots & a_{2U}b_{k,2U} \\ \vdots & \vdots & \ddots & \vdots \\ a_{U1}b_{k,U1} & a_{U2}b_{k,U2} & \cdots & a_{UU}b_{k,UU} \end{pmatrix}(t), \tag{8.27}$$

where the time dependence of the (i, j)th element is indicated at the right of the matrix for simplicity. In (8.27) a_{ij} represents the sign of the element at the ith row and the jth column, which is determined by the STS design rule, while $b_{k,ij}$ is the data bit assigned to the (i, j)th element, which is one of the U input data bits $\{b_{k1}, b_{k1}, \ldots, b_{kU}\}$ of user k. Each input data bit of $\{b_{k1}, b_{k2}, \ldots, b_{kU}\}$ appears only once in any given row and in any given column. For $U = 2, 4$, $\mathbf{B}_2(t)$ and $\mathbf{B}_4(t)$ are given in (8.16).

Based on Equations (8.26) and (8.27) the signal transmitted by the uth antenna to the kth user can be explicitly formulated as:

$$s_{ku}(t) = \sqrt{\frac{2P}{U^2}}\left[c_1(t)a_{1u}b_{k,1u}(t) + c_2(t)a_{2u}b_{k,2u}(t) + \ldots + c_U(t)a_{Uu}b_{k,Uu}(t)\right] \times \mathrm{PN}_k(t)\cos(2\pi f_c t),\ u = 1, 2, \ldots, U. \tag{8.28}$$

8.3.2 Channel Model

The U number of parallel sub-signals $\mathbf{s}_k(t) = [s_{k1}(t)\ s_{k2}(t)\ ...\ s_{kU}(t)]$ are transmitted by the U number of antennas over frequency-selective fading channels, where each parallel sub-signal experiences independent frequency-selective Nakagami-m fading. The complex low-pass equivalent representation of the impulse response experienced by the uth parallel sub-signal of user k is given by [90]:

$$h_k^u(t) = \sum_{l=1}^{L} h_{kl}^u \delta(t - \tau_{kl}) \exp\left(j\phi_{kl}^u\right), \tag{8.29}$$

where h_{kl}^u, τ_{kl} and ψ_{kl}^u represent the attenuation factor, delay and phase-shift of the lth multipath component of the channel, respectively, while L is the total number of resolvable multipath components and $\delta(t)$ is the Kronecker Delta-function. We assume that the phases $\{\psi_{kl}^u\}$ in (8.29) are independent identically distributed (iid) random variables uniformly distributed in the interval $[0, 2\pi)$, while the L multipath attenuations $\{h_{kl}^u\}$ in (8.29) are independent Nakagami random variables with a Probability Density Function (PDF) of [275–277]:

$$\begin{aligned} p(h_{kl}^u) &= M(h_{kl}^u, m_{kl}^{(u)}, \Omega_{kl}^u), \\ M(R, m, \Omega) &= \frac{2m^m R^{2m-1}}{\Gamma(m)\Omega^m} e^{(-m/\Omega)R^2}, \end{aligned} \tag{8.30}$$

where $\Gamma(\cdot)$ is the gamma function [90], and $m_{kl}^{(u)}$ is the Nakagami-m fading parameter, which characterizes the severity of the fading over the l-th resolvable path [278] between the uth transmission antenna and user k. Specifically, $m_{kl}^{(u)} = 1$ represents Rayleigh fading, $m_{kl}^{(u)} \to \infty$ corresponds to the conventional Gaussian scenario and $m_{kl}^{(u)} = 1/2$ describes the so-called one-sided Gaussian fading, i.e., the worst-case fading condition. The Rician and log-normal distributions can also be closely approximated by the Nakagami distribution in conjunction with values of $m_{kl}^{(u)} > 1$. The parameter Ω_{kl}^u in (8.30) is the second moment of h_{kl}^u, i.e., we have $\Omega_{kl}^u = E[(\alpha_{kl}^u)^2]$. We assume a negative exponentially decaying multipath intensity profile (MIP) given by $\Omega_{kl}^u = \Omega_{k1}^u e^{-\eta(l-1)}$, $\eta \geq 0$, where Ω_{k1}^u is the average signal strength corresponding to the first resolvable path and η is the rate of average power decay.

We support K asynchronous CDMA users in the system and assume perfect power control. Consequently, when the K users' signals obeying the form of (8.26) are transmitted over the frequency-selective fading channels characterized by (8.29), the received complex low-pass equivalent signal at a given mobile station can be expressed as:

$$R(t) = \sum_{k=1}^{K} \sum_{l=1}^{L} \sqrt{\frac{2P}{U^2}} \mathbf{c}(t - \tau_{kl}) \mathbf{B}_U(t - \tau_{kl}) \mathbf{h}_{kl} \times \text{PN}_k(t - \tau_{kl}) + N(t), \tag{8.31}$$

where $N(t)$ is the complex valued low-pass-equivalent AWGN having a double-sided spectral density of N_0, while

$$\mathbf{h}_{kl} = \begin{pmatrix} h_{kl}^1 \exp(j\psi_{kl}^1) \\ h_{kl}^2 \exp(j\psi_{kl}^2) \\ \cdots \\ h_{kl}^U \exp(j\psi_{kl}^U) \end{pmatrix}, \; k = 1, 2, \ldots, K; \; l = 1, 2, \ldots, L \tag{8.32}$$

represents the channel's complex impulse response in the context of the kth user and the lth resolvable path, where $\psi_{kl}^u = \phi_{kl}^u - 2\pi f_c \tau_{kl}$. Furthermore, in (8.31) we assumed that the signals transmitted by the U number of transmission antennas arrive at the receiver antenna after experiencing the same set of delays. This assumption is justified by the fact that in the frequency band of cellular system the propagation delay differences among the transmission antenna elements is on the order of nanoseconds, while the multipath delays are on the order of microseconds [274], provided that U is a relatively low number.

8.3.3 Receiver Model

Let the first user be the user-of-interest and consider a receiver using space-time de-spreading as well as diversity combining, as shown in Figure 8.1(b), where the subscript of the reference user's signal has been omitted for notational convenience. The receiver of Figure 8.1(b) carries out the inverse processing of Figure 8.1(a), in addition to multipath diversity combining. In Figure 8.1(b) the received signal is first down-converted using the carrier frequency f_c, and then de-scrambled using the DS scrambling sequence of PN$(t - \tau_l)$ in the context of the lth resolvable path, where we assumed that the receiver is capable of achieving near-perfect multipath-delay estimation for the reference user. The de-scrambled signal associated with the lth resolvable path is space-time de-spread using the approach of [274] - which will be further discussed in Section 8.4, in order to obtain U separate variables, $\{Z_{1l}, Z_{2l}, \ldots, Z_{Ul}\}$, corresponding to the U parallel data bits $\{b_1, b_2, \ldots, b_U\}$, respectively. Following space-time de-spreading, a decision variable is formed for each parallel transmitted data bit of $\{b_1, b_2, \ldots, b_U\}$ by combining the corresponding variables associated with the L number of resolvable paths, which can be written as:

$$Z_u = \sum_{l=1}^{L} Z_{ul}, \; u = 1, 2, \ldots, U. \tag{8.33}$$

Finally, the U number of transmitted data bits $\{b_1, b_2, \ldots, b_U\}$ can be decided based on the decision variables $\{Z_u\}_{u=1}^{U}$ using the conventional decision rule of a BPSK scheme.

Above we have described the transmitter model, the channel model as well as the receiver model of W-CDMA using STS. Let us now describe the detection procedure of the W-CDMA scheme using STS.

8.4 Detection of Space-Time Spreading W-CDMA Signals

Let $\mathbf{d}_l = [d_{1l}\ d_{2l}\ \ldots\ d_{Ul}]^T$, $l = 1, 2, \ldots, L$ - where T denotes vector transpose - represent the correlator's output variable vector in the context of the lth ($l = 1, 2, \ldots, L$) resolvable path, where

$$d_{ul} = \int_{\tau_l}^{UT_b+\tau_l} R(t)c_u(t-\tau_l)\text{PN}(t-\tau_l)dt. \tag{8.34}$$

When substituting (8.31) into (8.34), it can be shown that:

$$\begin{aligned} d_{ul} = \sqrt{2P}T_b \left[a_{u1}b_{u1}h_l^1 \exp\left(j\psi_l^1\right) + a_{u2}b_{u2}h_l^2 \exp\left(j\psi_l^2\right) + \ldots \right. \\ \left. \ldots + a_{uU}b_{uU}h_l^U \exp\left(j\psi_l^U\right)\right] + J_u(l),\ u = 1, 2, \ldots, U, \end{aligned} \tag{8.35}$$

where

$$J_u(l) = J_{Su}(l) + J_{Mu}(l) + N_u(l),\ u = 1, 2, \ldots, U \tag{8.36}$$

and $J_{Su}(l)$ is due to the multipath-induced self-interference of the signal-of-interest inflicted upon the lth path signal, where $J_{Su}(l)$ can be expressed as:

$$\begin{aligned} J_{Su}(l) = \sum_{j=1, j\neq l}^{L} \sqrt{\frac{2P}{U^2}} \int_{\tau_l}^{UT_b+\tau_l} \mathbf{c}(t-\tau_j)\mathbf{B}_U(t-\tau_j)\mathbf{h}_j\text{PN}(t-\tau_j) \\ \times c_u(t-\tau_l)\text{PN}(t-\tau_l)dt, \end{aligned} \tag{8.37}$$

$J_{Mu}(l)$ represents the multi-user interference due to the signals transmitted simultaneously by the other users, which can be expressed as:

$$\begin{aligned} J_{Mu}(l) = \sum_{k=2}^{K}\sum_{j=1}^{L} \sqrt{\frac{2P}{U^2}} \int_{\tau_l}^{UT_b+\tau_l} \mathbf{c}(t-\tau_{kj})\mathbf{B}_U(t-\tau_{kj})\mathbf{h}_{kj}\text{PN}_k(t-\tau_{kj}) \\ \times c_u(t-\tau_l)\text{PN}(t-\tau_l)dt, \end{aligned} \tag{8.38}$$

and finally $N_u(l)$ is due to the AWGN, which can be written as:

$$N_u(l) = \int_{\tau_l}^{UT_b+\tau_l} N(t)c_u(t-\tau_l)\text{PN}(t-\tau_l)dt, \tag{8.39}$$

which is a Gaussian distributed variable having zero mean and a variance of $2UN_0T_b$.

Let $\mathbf{J}(l) = [J_1(l)\ J_2(l)\ \ldots\ J_U(l)]^T$. Then, the correlator's output variable vector $\mathbf{d}_l$ can be written as:

$$\mathbf{d}_l = \sqrt{2P}T_b\mathbf{B}_U\mathbf{h}_l + \mathbf{J}(l),\ l = 1, 2, \ldots, L, \tag{8.40}$$

where $\mathbf{B}_U$ is the reference user's $U \times U$-dimensional transmitted data matrix, which is given by (8.27), but ignoring the time dependence, while $\mathbf{h}_l$ is the channel's complex impulse response between the base station and the reference user, as shown in (8.32) in the context of the reference user.

Attractive STS schemes have the property [274] of $\mathbf{B}_U\mathbf{h}_l = \mathbf{H}_U\mathbf{b}$, i.e., Equation (8.40) can be written as:

$$\mathbf{d}_l = \sqrt{2PT_b}\mathbf{H}_U\mathbf{b} + \mathbf{J}(l), \tag{8.41}$$

where $\mathbf{b} = [b_1\ b_2\ \ldots\ b_U]^T$ represents the U number of transmitted data bits, while $\mathbf{H}_U$ is a $U \times U$-dimensional matrix with elements from $\mathbf{h}_l$. Each element of $\mathbf{h}_l$ appears once and only once in a given row and also in a given column of the matrix $\mathbf{H}_U$ [274]. The matrix $\mathbf{H}_U$ can be expressed as:

$$\mathbf{H}_U(l) = \begin{pmatrix} \alpha_{11}(l) & \alpha_{12}(l) & \ldots & \alpha_{1U}(l) \\ \alpha_{21}(l) & \alpha_{22}(l) & \ldots & \alpha_{2U}(l) \\ \vdots & \vdots & \ddots & \vdots \\ \alpha_{U1}(l) & \alpha_{U2}(l) & \ldots & \alpha_{UU}(l) \end{pmatrix}, \tag{8.42}$$

where $\alpha_{ij}(l)$ takes the form of $d_{ij}h_l^m \exp(j\psi_l^m)$, and $d_{ij} \in \{+1, -1\}$ represents the sign of the (i, j)th element of $\mathbf{H}_U$, while $h_l^m \exp(j\psi_l^m)$ belongs to the mth element of $\mathbf{h}_l$. For $U = 2, 4$, with the aid of [274], it can be shown that

$$\mathbf{H}_2(l) = \begin{pmatrix} h_l^1 \exp(j\psi_l^1) & h_l^2 \exp(j\psi_l^2) \\ -h_l^2 \exp(j\psi_l^2) & h_l^1 \exp(j\psi_l^1) \end{pmatrix}, \tag{8.43}$$

$$\mathbf{H}_4(l) = \begin{pmatrix} h_l^1 \exp(j\psi_l^1) & h_l^2 \exp(j\psi_l^2) & h_l^3 \exp(j\psi_l^3) & h_l^4 \exp(j\psi_l^4) \\ -h_l^2 \exp(j\psi_l^2) & h_l^1 \exp(j\psi_l^1) & -h_l^4 \exp(j\psi_l^4) & h_l^3 \exp(j\psi_l^3) \\ -h_l^3 \exp(j\psi_l^3) & h_l^4 \exp(j\psi_l^4) & h_l^1 \exp(j\psi_l^1) & -h_l^2 \exp(j\psi_l^2) \\ -h_l^4 \exp(j\psi_l^4) & -h_l^3 \exp(j\psi_l^3) & h_l^2 \exp(j\psi_l^2) & h_l^1 \exp(j\psi_l^1) \end{pmatrix}. \tag{8.44}$$

With the aid of the analysis in [274], it can be shown that the matrix $\mathbf{H}_U(l)$ has the property of $\text{Re}\left\{\mathbf{H}_U^\dagger(l)\mathbf{H}_U(l)\right\} = \mathbf{h}_l^\dagger\mathbf{h}_l \cdot \mathbf{I}$, where $\dagger$ denotes complex conjugate transpose and $\mathbf{I}$ represents a $U \times U$-dimensional unity matrix. Letting $\mathbf{h}_u(l)$ denote the uth column of $\mathbf{H}_U(l)$, the variable Z_{ul} in (8.33) can be formulated as [274]:

$$Z_{ul} = \text{Re}\left\{\mathbf{h}_u^\dagger(l)\mathbf{d}_l\right\} = \sqrt{2PT_b}b_u \sum_{u=1}^{U} |h_l^u|^2 + \text{Re}\left\{\mathbf{h}_u^\dagger(l)\mathbf{J}(l)\right\},\ u = 1, 2, \ldots, U. \tag{8.45}$$

Finally, according to (8.33) the decision variables associated with the U parallel transmitted data bits $\{b_1, b_2, \ldots, b_U\}$ of the reference user can be expressed as:

$$Z_u = \sqrt{2P T_b} b_u \sum_{l=1}^{L} \sum_{u=1}^{U} |h_l^u|^2 + \sum_{l=1}^{L} \text{Re}\left\{\mathbf{h}_u^{\dagger}(l)\mathbf{J}(l)\right\}, \; u = 1, 2, \ldots, U, \tag{8.46}$$

which shows that the receiver is capable of achieving a diversity order of UL, as indicated by the related sums of the first term.

Above we have analyzed the detection procedure applicable to W-CDMA signals generated using STS. Let us now derive the corresponding BER expression.

8.5 BER Performance

8.5.1 BER Analysis

In this section we derive the BER expression of the STS-assisted W-CDMA system by first analyzing the statistics of the variable Z_u, $u = 1, 2, \ldots, U$ with the aid of the Gaussian approximation [103]. According to (8.46), for a given set of complex channel transfer factor estimates $\{h_l^u\}$, Z_u can be approximated as a Gaussian variable having a mean given by:

$$\text{E}\left[Z_u\right] = \sqrt{2P T_b} b_u \sum_{l=1}^{L} \sum_{u=1}^{U} |h_l^u|^2. \tag{8.47}$$

Based on the assumption that the interferences imposed by the different users, the different paths as well as by the AWGN constitute independent random variables, the variance of Z_u can be expressed as:

$$\begin{aligned} \text{Var}\left[Z_u\right] &= \text{E}\left[\left(\sum_{l=1}^{L} \text{Re}\left\{\mathbf{h}_u^{\dagger}(l)\mathbf{J}(l)\right\}\right)^2\right] \\ &= \sum_{l=1}^{L} \text{E}\left[\left(\text{Re}\left\{\mathbf{h}_u^{\dagger}(l)\mathbf{J}(l)\right\}\right)^2\right] \\ &= \frac{1}{2} \sum_{l=1}^{L} \text{E}\left[\left(\mathbf{h}_u^{\dagger}(l)\mathbf{J}(l)\right)^2\right]. \end{aligned} \tag{8.48}$$

Substituting $\mathbf{h}_u(l)$, which is the uth column of $\mathbf{H}_u(l)$ in (8.42), and $\mathbf{J}(l)$ having elements given by (8.36) into the above equation, it can be shown that for a given set of channel estimates $\{h_l^u\}$, (8.48) can be simplified as:

$$\begin{aligned}\text{Var}[Z_u] &= \frac{1}{2}\sum_{l=1}^{L}\sum_{u=1}^{U}|h_l^u|^2\text{E}\left[(J_u(l))^2\right] \\ &= \frac{1}{2}\sum_{l=1}^{L}\sum_{u=1}^{U}|h_l^u|^2\text{Var}[J_u(l)],\end{aligned} \quad (8.49)$$

where $J_u(l)$ is given by (8.36). In deriving (8.49) we exploited the assumption of $\text{Var}[J_1(l)] = \text{Var}[J_2(l)] = \ldots = \text{Var}[J_U(l)]$.

As shown in (8.36), $J_u(l)$ consists of three terms, namely the AWGN $N_u(l)$ having a variance of $2UN_0T_b$, $J_{Su}(l)$, which is the multipath-induced self-interference inflicted upon the lth path of the user-of-interest and $J_{Mu}(l)$ imposed by the $(K-1)$ interfering users. By careful observation of (8.37), it can be shown that $J_{Su}(l)$ consists of U^2 terms and each term takes the form of $\sum_{j=1,j\neq l}^{L}\sqrt{\frac{2P}{U^2}}\int_{\tau_l}^{UT_b+\tau_l} c_m(t-\tau_j)a_{mn}b_{mn}(t-\tau_j)h_j^n\exp(j\psi_j^n)\text{PN}(t-\tau_j)$ $\times\, c_u(t-\tau_l)\text{PN}(t-\tau_l)dt$. Assuming that $\text{E}[(h_j^n)^2] = \Omega_1 e^{-\eta(j-1)}$, i.e., that $\text{E}[(h_j^n)^2]$ is independent of the index of the transmission antenna, and following the analysis in [276], it can be shown that the above term has a variance of $2\Omega_1E_bT_b[q(L,\eta)-1]/(GU)$, where $q(L,\eta) = (1-e^{-L\eta})/(1-e^{-\eta})$, if $\eta \neq 0$ and $q(L,\eta) = L$, if $\eta = 0$. Consequently, we have $\text{Var}[J_{Su}(l)] = U^2 \times 2\Omega_1E_bT_b[q(L,\eta)-1]/(GU) = 2U\Omega_1E_bT_b[q(L,\eta)-1]/G$. Similarly, the multi-user interference term $J_{Mu}(l)$ of (8.38) also consists of U^2 terms, and each term has the form of $\sum_{k=2}^{K}\sum_{j=1}^{L}\sqrt{\frac{2P}{U^2}}\int_{\tau_l}^{UT_b+\tau_l} c_m(t-\tau_{kj})a_{mn}b_{mn}(t-\tau_{kj})h_{kj}^n\exp(j\psi_{kj}^n)\text{PN}_k(t-\tau_{kj})c_u(t-\tau_l)\text{PN}(t-\tau_l)dt$. Again, with the aid of the analysis in [276], it can be shown that this term has the variance of $(K-1)4\Omega_1E_bT_bq(L,\eta)/(3GU)$, and consequently the variance of $J_{Mu}(l)$ is given by $\text{Var}[J_{Mu}(l)] = (K-1)4U\Omega_1E_bT_bq(L,\eta)/(3G)$. Therefore, the variance of $J_u(l)$ can be written as:

$$\text{Var}[J_u(l)] = 2N_0UT_b + \frac{2U\Omega_1E_bT_b[q(L,\eta)-1]}{G} + \frac{(K-1)4U\Omega_1E_bT_bq(L,\eta)}{3G}, \quad (8.50)$$

and the variance of Z_u for a given set of channel estimates $\{h_l^u\}$ can be expressed as:

$$\begin{aligned}\text{Var}[Z_u] &= \sum_{l=1}^{L}\sum_{u=1}^{U}|h_l^u|^2 \\ &\times\left[N_0UT_b + \frac{U\Omega_1E_bT_b[q(L,\eta)-1]}{G} + \frac{(K-1)2U\Omega_1E_bT_bq(L,\eta)}{3G}\right].\end{aligned} \quad (8.51)$$

Based on Equations (8.47) and (8.51), the BER conditioned on h_l^u for $u = 1, 2, \ldots, U$ and $l = 1, 2, \ldots, L$ can be written as:

$$P_b(E|\{h_l^u\}) = Q\left(\sqrt{\frac{\text{E}^2[Z_u]}{\text{Var}[Z_u]}}\right) = Q\left(\sqrt{2\cdot\sum_{l=1}^{L}\sum_{u=1}^{U}\gamma_{lu}}\right), \quad (8.52)$$

where $Q(x)$ represents the Gaussian Q-function, which can also be represented in its less conventional form as $Q(x) = \frac{1}{\pi}\int_0^{\pi/2} \exp\left(-\frac{x^2}{2\sin^2\theta}\right) d\theta$, where $x \geq 0$ [278, 279]. Furthermore, γ_{lu} in (8.52) is given by:

$$\gamma_{lu} = \overline{\gamma}_c \cdot \frac{(h_l^u)^2}{\Omega_1}, \tag{8.53}$$

and

$$\overline{\gamma}_c = \frac{1}{U}\left[\frac{(2K+1)q(L,\eta)-3}{3G} + \left(\frac{\Omega_1 E_b}{N_0}\right)^{-1}\right]^{-1}. \tag{8.54}$$

The average BER, $P_b(E)$ can be obtained by averaging the conditional BER of (8.52) over the joint PDF of the instantaneous SNR values corresponding to the L multipath components and to the U transmit antennas $\{\gamma_{lu} : l = 1, 2, \ldots, L;\ u = 1, 2, \ldots, U\}$. Since the random variables $\{\gamma_{lu} : l = 1, 2, \ldots, L;\ u = 1, 2, \ldots, U\}$ are assumed to be statistically independent, the average BER can be formulated as [280](23):

$$P_b(E) = \frac{1}{\pi}\int_0^{\pi/2} \prod_{l=1}^{L}\prod_{u=1}^{U} I_{lu}\left(\overline{\gamma}_{lu}, \theta\right) d\theta, \tag{8.55}$$

where

$$I_{lu}\left(\overline{\gamma}_{lu}, \theta\right) = \int_0^{\infty} \exp\left(-\frac{\gamma_{lu}}{\sin^2\theta}\right) p_{\gamma_{lu}}(\gamma_{lu}) d\gamma_{lu}. \tag{8.56}$$

Since $\gamma_{lu} = \overline{\gamma}_c \cdot \frac{(h_l^u)^2}{\Omega_1}$ and h_l^u obeys the Nakagami-m distribution characterized by (8.30), it can be shown that the PDF of γ_{lu} can be formulated as:

$$p_{\gamma_{lu}}(\gamma_{lu}) = \left(\frac{m_l^{(u)}}{\overline{\gamma}_{lu}}\right)^{m_l^{(u)}} \frac{\gamma^{m_l^{(u)}-1}}{\Gamma(m_l^{(u)})} \exp\left(-\frac{m_l^{(u)}\gamma_{lu}}{\overline{\gamma}_{lu}}\right), \quad \gamma_{lu} \geq 0, \tag{8.57}$$

where $\overline{\gamma}_{lu} = \overline{\gamma}_c e^{-\eta(l-1)}$ for $l = 1, 2, \ldots, L$.

Upon substituting (8.57) into (8.56) it can be shown that [278]:

$$I_{lu}\left(\overline{\gamma}_{lu}, \theta\right) = \left(\frac{m_l^{(u)}\sin^2\theta}{\overline{\gamma}_{lu} + m_l^{(u)}\sin^2\theta}\right)^{m_l^{(u)}}. \tag{8.58}$$

Finally, upon substituting (8.58) into (8.55), the average BER of the STS-assisted W-

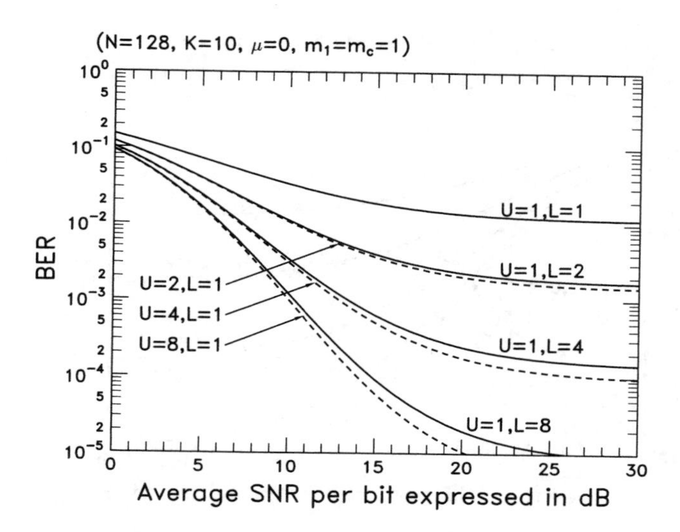

Figure 8.2: BER versus the SNR per bit, E_b/N_0, performance comparison between the space-time spreading based transmit diversity scheme and the conventional RAKE receiver arrangement using only one transmission antenna when communicating over flat-fading (for space-time spreading) and multipath (for RAKE) Rayleigh fading ($m_l = m_c = 1$) channels evaluated from (8.60) by assuming that the average power decay rate was $\eta = 0$.

CDMA system using U transmission antennas can be written as:

$$P_b(E) = \frac{1}{\pi}\int_0^{\pi/2}\prod_{l=1}^{L}\prod_{u=1}^{U}\left(\frac{m_l^{(u)}\sin^2\theta}{\overline{\gamma}_{lu} + m_l^{(u)}\sin^2\theta}\right)^{m_l^{(u)}} d\theta, \tag{8.59}$$

which shows that the diversity order achieved is LU, the product of the diversity due to STS and the diversity contributed by the RAKE receiver. Furthermore, if we assume that $m_l^{(u)}$ is independent of u, i.e., that all of the parallel transmitted sub-signals experience an identical Nakagami fading, then (8.59) can be expressed as:

$$P_b(E) = \frac{1}{\pi}\int_0^{\pi/2}\prod_{l=1}^{L}\left(\frac{m_l\sin^2\theta}{\overline{\gamma}_{lu} + m_l\sin^2\theta}\right)^{Um_l} d\theta. \tag{8.60}$$

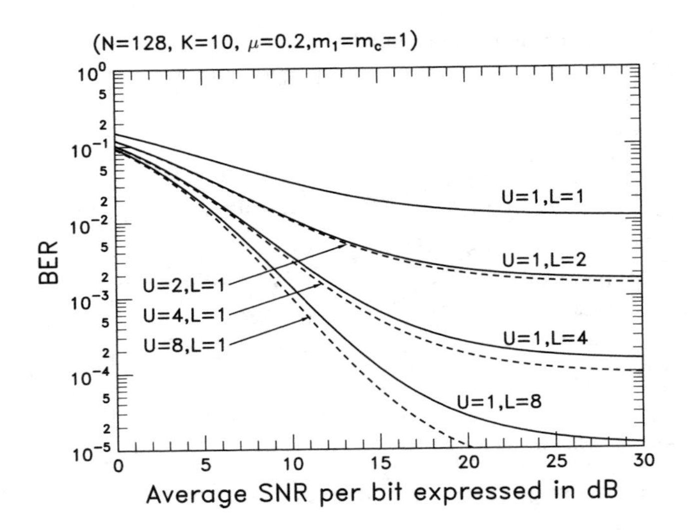

Figure 8.3: BER versus the SNR per bit, E_b/N_0, performance comparison between the space-time spreading based transmit diversity scheme and the conventional RAKE receiver arrangement using only one transmission antenna when communicating over flat-fading (for space-time spreading) and multipath (for RAKE) Rayleigh fading ($m_l = m_c = 1$) channels evaluated from (8.60) by assuming that the average power decay rate was $\eta = 0.2$.

8.5.2 Numerical Results and Discussion

In Figures 8.2 - 8.6 we compare the BER performance of the STS-assisted W-CDMA system transmitting over flat-fading channels and that of the conventional RAKE receiver using only one transmission antenna, but communicating over frequency-selective fading channels. The results in these figures were all evaluated from (8.60) by assuming appropriate parameters, which are explicitly shown in the corresponding figures. In Figures 8.2, 8.3 and 8.4 the BER was drawn against the SNR/bit, namely E_b/N_0, while in Figures 8.5 and 8.6 the BER was drawn against the number of users, K, supported by the system. From the results we observe that for transmission over Rayleigh fading channels ($m_l = 1$), as characterized by Figures 8.2, 8.3 and 8.5, both the STS-based transmit diversity scheme transmitting over the frequency non-selective Rayleigh fading channel and the conventional RAKE receiver scheme communicating over frequency-selective Rayleigh fading channels having the same number of resolvable paths as the number of transmission antennas in the STS-assisted scheme achieved a similar BER performance, with the STS scheme slightly outperforming the conventional RAKE scheme. For transmission over general Nakagami-m fading channels, if the first resolvable path is less severely faded, than the other resolvable paths, such as in Figures 8.4 and 8.6 where $m_1 = 2$ and $m_2 = m_3 = \ldots = m_c = 1$, the STS-based transmit diversity scheme communicating over the frequency non-selective Rayleigh

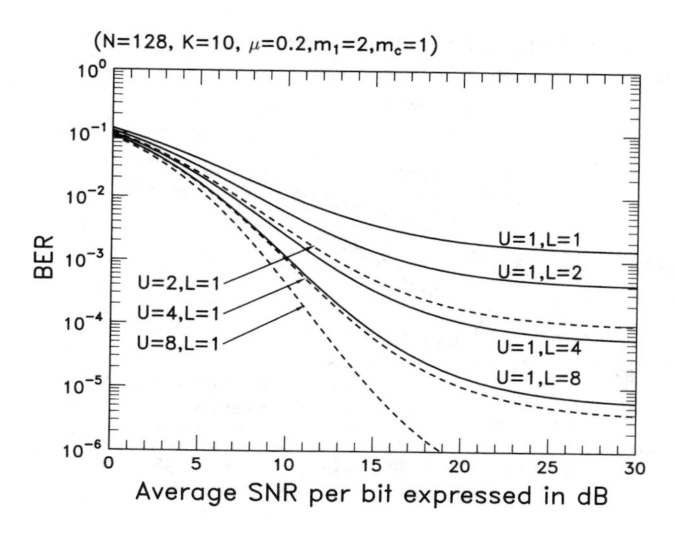

Figure 8.4: BER versus the SNR per bit, E_b/N_0, performance comparison between the space-time spreading based transmit diversity scheme and the conventional RAKE receiver arrangement using only one transmission antenna when communicating over flat-fading (for space-time spreading) and multipath (for RAKE) Nakagami-m fading channels evaluated from (8.60) by assuming that the average power decay rate was $\eta = 0.2$, where $m_1 = 2$ indicates that the first resolvable path constitutes a moderately fading path, while the other resolvable paths experience more severe Rayleigh fading ($m_c = 1$).

fading channel may significantly outperform the corresponding conventional RAKE receiver assisted scheme communicating over frequency-selective Rayleigh fading channels. This is because the STS-based transmit diversity scheme communicated over a single non-dispersive path, which benefited from having a path experiencing moderate fading. However, if the number of resolvable paths is sufficiently high, the conventional RAKE receiver scheme is also capable of achieving a satisfactory BER performance.

Above we assumed that the number of resolvable paths was one, if the STS using more than one antennas was considered. By contrast, the number of resolvable paths was equal to the number of transmit antennas of the corresponding STS-based system, when the conventional RAKE receiver was considered. However, in practical W-CDMA systems the number of resolvable paths only depends on the delay-spread [90] of the channel, and it is independent of the number of transmission antennas. In other words, the number of resolvable paths of each antenna's transmitted signal depends on its transmission environment and the number of resolvable paths dynamically changes, as the mobile traverses through different transmission environments. Specifically, in some scenarios the number of path may be as low as $L = 1$, and in other scenarios it may be as high as $L > 10$. When the number of resolvable paths is as low as $L = 1$ or 2, employing STS-based transmit diversity is particularly valuable. However,

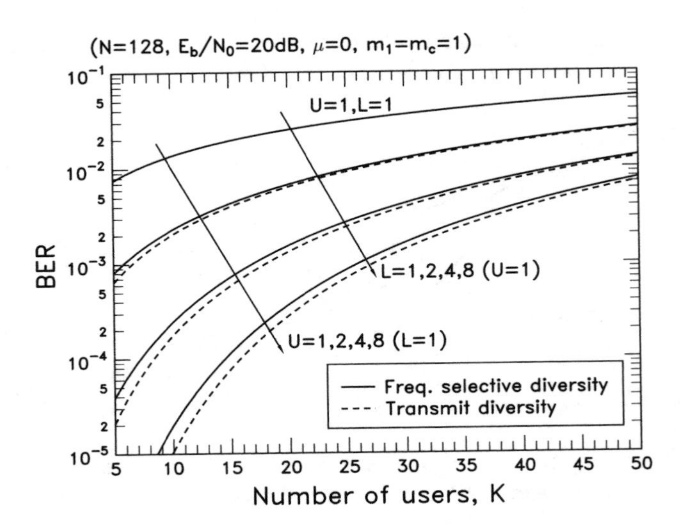

Figure 8.5: BER versus the number of users, K, performance comparison between the space-time spreading based transmit diversity scheme and the conventional RAKE receiver arrangement using only one transmission antenna when communicating over flat-fading (for space-time spreading) and multipath (for RAKE) Rayleigh fading channels evaluated from (8.60) by assuming that the average power decay rate was $\eta = 0$.

when the number of resolvable paths is reasonably high, for example $L > 4$, the employment of STS-based transmit diversity is not necessary. An attractive approach is to adapt the mode of operation of the STS scheme, which is discussed in the following subsection.

8.6 Adaptive Space-time Spreading

The main philosophy behind the proposed adaptive STS scheme is the real-time balancing of the link budget through the adaptive control of the STS-based transmission scheme, in order that the system achieves its maximum throughput, while maintaining the required BER performance. In the context of the STS-assisted W-CDMA system, the delay-spread of the wireless channels, and hence the number of resolvable paths varies slowly over a range spanning from one to dozens of paths. The STS scheme designed based on a low number of resolvable paths, or even based on a relatively high but constant number of resolvable paths, cannot maximize the achievable throughput. For example, if the STS scheme is designed based on a low number of resolvable paths, in order to guarantee a required Quality-of-Service (QoS), the practically achieved QoS may be excessive, when the number of resolvable paths is high, provided that these resolvable paths are efficiently combined. However, if only a low but constant number of resolvable paths is combined, the diversity potential provided by the high

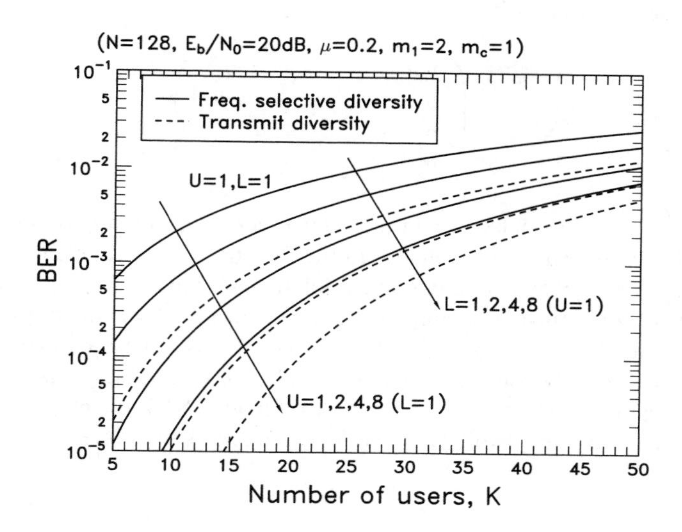

Figure 8.6: BER versus the number of users, K, performance comparison between the space-time spreading based transmit diversity scheme and the conventional RAKE receiver arrangement using only one transmission antenna when communicating over the flat-fading (for space-time spreading) and multipath (for RAKE) Nakagami-m fading channels evaluated from (8.60) by assuming that the average power decay rate was $\eta = 0.2$, where $m_1 = 2$ indicates that the first resolvable path constitutes a moderately fading path, while the other resolvable paths experience more severe Rayleigh fading ($m_c = 1$).

number of resolvable paths is inevitably wasted. A high-efficiency STS-based communication scheme must be capable of combining the transmitted energy, which was scattered over an arbitrary number of resolvable paths, and the mode of operation of the STS scheme can be adaptively controlled according to the receiver's detection performance.

When the number of resolvable paths is low and hence the resultant BER is higher than the required BER, then a low throughput STS-assisted transmitter mode is activated, which exhibits a high transmit diversity gain, as it will be demonstrated below with the aid of an example. By contrast, when the number of resolvable paths is high and hence the resultant BER is lower than the required BER, then a higher throughput STS-assisted transmitter mode is activated, which has a lower transmit diversity gain.

Specifically, the principle of implementing channel-dispersion controlled adaptive rate transmission using adaptive STS may be readily interpreted by referring to the examples provided in Figures 8.7, 8.8 and 8.9. Let the transmitter employ a total of four transmission antennas. If the number of resolvable paths experienced by the receiver is low, the transmitter is instructed by the receiver to employ a STS scheme based on four transmit antennas, using the STS scheme described as [274]:

$$\mathbf{S} = [c_1 \; c_2 \; c_3 \; c_4] \begin{pmatrix} b_1 & b_2 & b_3 & b_4 \\ b_2 & -b_1 & b_4 & -b_3 \\ b_3 & -b_4 & -b_1 & b_2 \\ b_4 & b_3 & -b_2 & -b_1 \end{pmatrix}, \tag{8.61}$$

where $c_1, \; c_2, \; c_3, \; c_4$ are four STS-related orthogonal codes having a period of $4T_b$. The transmitted signal's waveforms corresponding to the four transmission antennas are shown in Figure 8.7, which shows that four data bits are jointly transmitted within a time-duration of $4T_b$ with the aid of four transmission antennas. Hence, this STS scheme transmits $U = 4$ parallel data bits during the interval of $4T_b$ and hence the effective transmission rate becomes $R_b = 4 \times 1/4T_b = 1/T_b$. By contrast, when the number of resolvable paths increases, the transmitter is instructed by the receiver to employ two separate STS schemes, each based on two transmit antennas and sends two STS symbols within a $4T_b$ time-interval, which can be formulated as:

$$\mathbf{S} = \begin{pmatrix} [c_1 \; c_2] \begin{pmatrix} b_1 & b_2 \\ b_2 & -b_1 \end{pmatrix} & [c_3 \; c_4] \begin{pmatrix} b_3 & b_4 \\ b_4 & -b_3 \end{pmatrix} \\ [c_1 \; c_2] \begin{pmatrix} b_5 & b_6 \\ b_6 & -b_5 \end{pmatrix} & [c_3 \; c_4] \begin{pmatrix} b_7 & b_8 \\ b_8 & -b_7 \end{pmatrix} \end{pmatrix}, \tag{8.62}$$

which, again, constitutes four independent 2-antenna based STS schemes $\mathbf{B}_2(t)$ of (8.16), where $c_1, \; c_2, \; c_3, \; c_4$ are the $U = 4$ STS-related orthogonal codes having a period of $2T_b$. The STS operations and the resultant transmitted waveforms of the four antennas are shown in Figure 8.8. Figure 8.8 shows that the first and second antennas jointly transmit two bits within a time-duration of $2T_b$, while the third and fouth antennas jointly transmit another two bits within the same $2T_b$ duration. However, the operation of transmission of antennas 1 and 2 is independent of the actions of antennas 3 and 4. Based on the above four 2-antenna assisted STS schemes shown in Figure 8.8, $U = 4$ parallel data bits are transmitted during the first $2T_b$-duration interval using the STS scheme $\mathbf{B}_2(t)$ of (8.16). Specifically, antennas 1 and 2 are activated with the aid of $c_1, \; c_2$, while activating antennas 3 and 4 using $c_3, \; c_4$. During the following $2T_b$-duration slot another four data bits are transmitted using the same scheme as outlined above. Consequently, the above four 2-antenna based STS schemes transmit a total of eight data bits during two consecutive $2T_b$-duration time-slots having a total duration of $4T_b$, and the effective transmission rate is now doubled to $2R_b$. Furthermore, if the number of resolvable paths is sufficiently high, which results in requiring no transmit diversity at all, then the four transmission antennas can transmit their information independently, as shown in Figure 8.7, and the corresponding transmission mode can be described as:

$$\mathbf{S} = \begin{pmatrix} c_1 b_1 & c_2 b_2 & c_3 b_3 & c_4 b_4 \\ c_1 b_5 & c_2 b_6 & c_3 b_7 & c_4 b_8 \\ c_1 b_9 & c_2 b_{10} & c_3 b_{11} & c_4 b_{12} \\ c_1 b_{13} & c_2 b_{14} & c_3 b_{15} & c_4 b_{16} \end{pmatrix}, \tag{8.63}$$

which implies that each of the 16 bits is transmitted independently using an antenna within

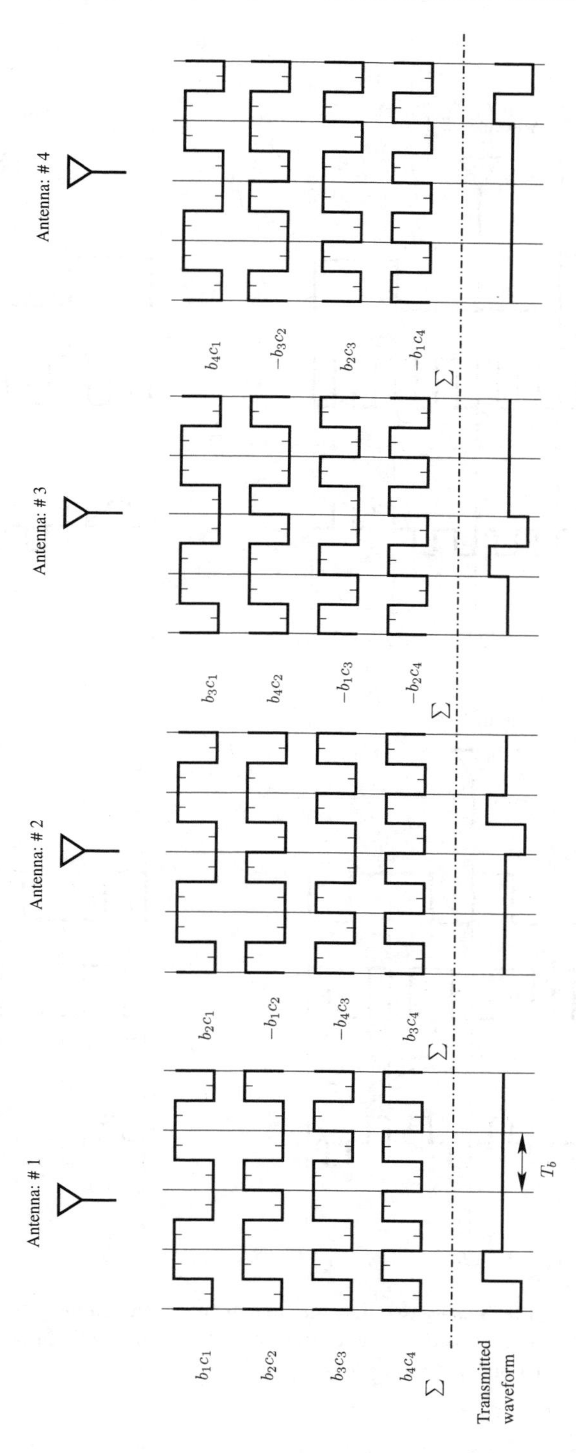

Figure 8.7: Illustration of STS using four transmission antennas tansmitting 4 bits within $4T_b$ duration, where $b_1 = b_2 = b_3 = b_4 = +1$ was assumed.

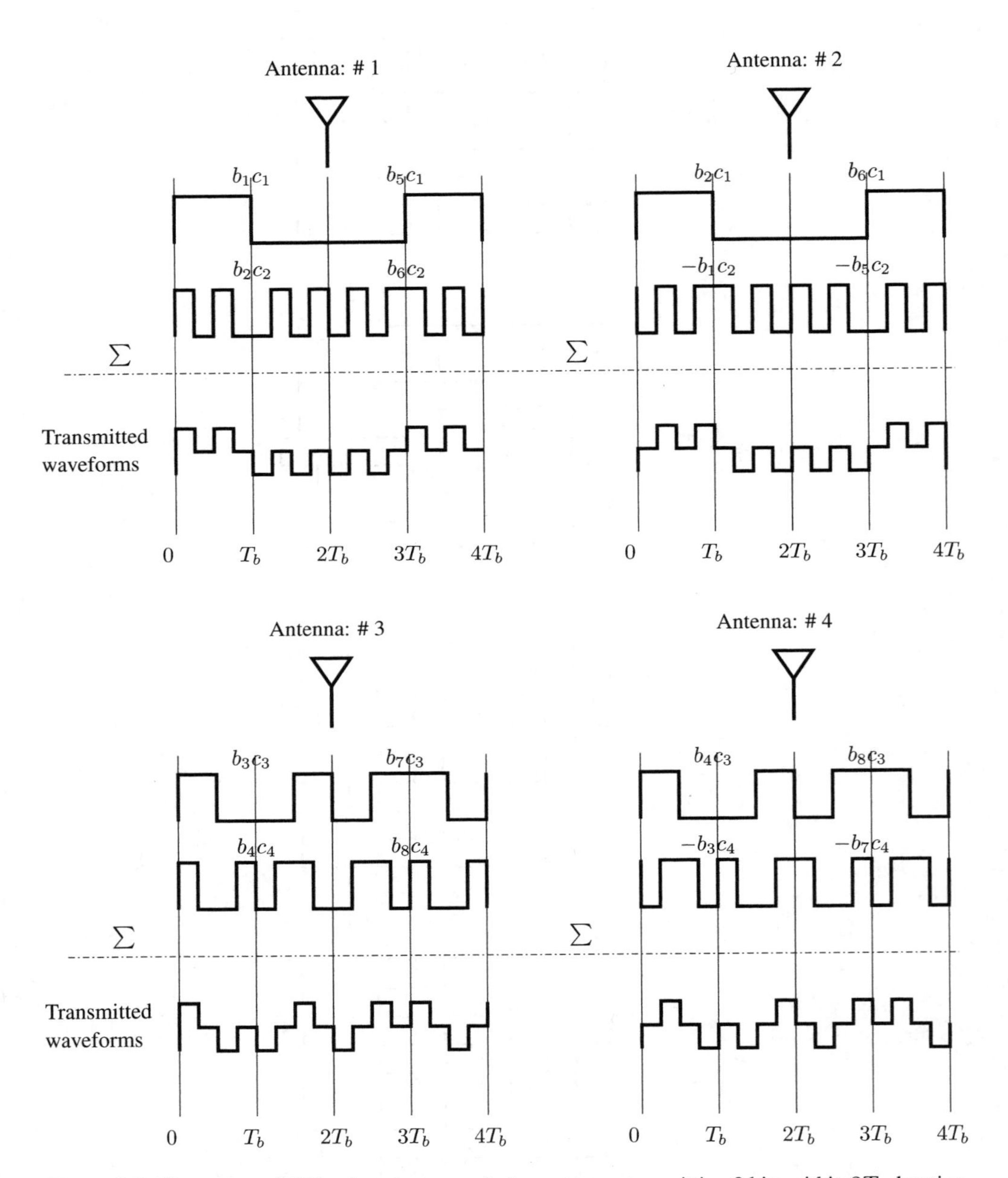

Figure 8.8: Illustration of STS using two transmission antennas tansmitting 2 bits within $2T_b$ duration. Hence, four transmission antennas transmit 8 bits within $4T_b$ duration, where $b_1 = b_2 = b_3 = b_4 = +1$ and $b_5 = b_6 = b_7 = b_8 = -1$ was assumed.

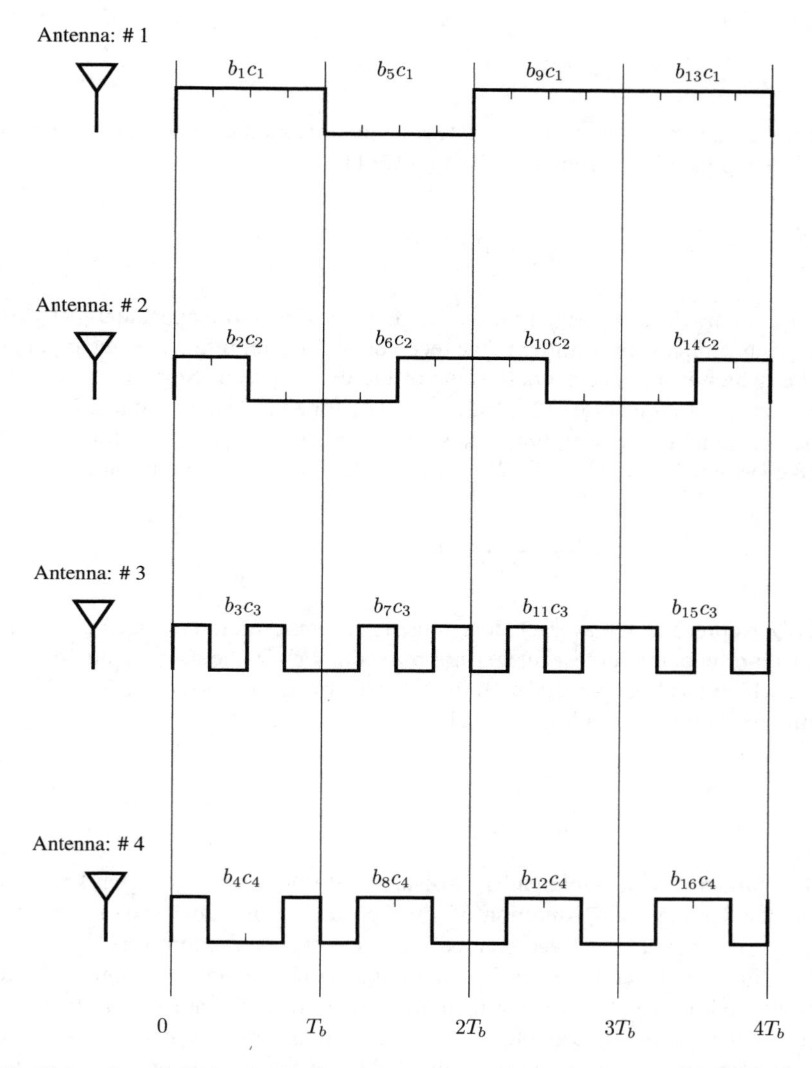

Figure 8.9: Illustration of the transmitted waveforms of the transmission scheme without using STS, i.e., the four transmission antennas transmit their data independently. In this figure we assumed that $b_1 = b_2 = b_3 = b_4 = +1$, $b_5 = b_6 = b_7 = b_8 = -1$, $b_9 = b_{10} = +1$, $b_{11} = b_{12} = -1$, $b_{13} = +1$, $b_{14} = +1$, $b_{15} = +1$, $b_{16} = -1$.

a duration T_b, where c_1, c_2, c_3, c_4 are four orthogonal codes having a period of T_b, each mapped to one antenna. Explicitly, this scheme is capable of transmitting a total of 16 data bits during an interval of $4T_b$, and hence we achieve a transmission rate of $4R_b$.

The PDF of the delay spread in a wireless communication channel can be approximated by a negative exponential distribution given by [281]:

$$f(\tau) = \frac{1}{T_m} \exp\left(-\frac{\tau - \tau_0}{T_m}\right), \ \tau \geq \tau_0, \tag{8.64}$$

where the minimum delay τ_0 is the time required for the signal to propagate directly following the line-of-sight from the transmitter to the receiver, and T_m represents the mean square of the distribution, which is also the average value of the delay spread. Some typical examples of T_m in different environments are [273]: $T_m < 0.1\mu s$ for an indoor environment, $T_m < 0.2\mu s$ for an open rural area, $T_m \approx 0.5\mu s$ for a suburban area and $T_m \approx 3\mu s$ for a typical urban area. In (8.64) we let $\tau_r = (\tau - \tau_0)/T_c$. Then the PDF of τ_r can be formulated as:

$$f(\tau_r) = \frac{1}{T_m/T_c} \exp\left(-\frac{\tau_r}{T_m/T_c}\right), \ \tau_r \geq 0, \tag{8.65}$$

where T_m/T_c represents the average delay-spread to chip-duration ratio, and $\lfloor T_m/T_c \rfloor + 1$ - where $\lfloor x \rfloor$ represents the largest integer not exceeding x - is the average number of resolvable paths, which has been widely used in the performance analysis of DS-CDMA systems transmitting over multipath fading channels.

Let the number of resolvable paths associated with the reference signal be L_r. For DS-CDMA signals having a chip-duration of T_c, the number of near-instantaneous resolvable paths $L_r = \lfloor (\tau - \tau_0)/T_c \rfloor + 1$ can be modelled as a discrete random variable, which varies slowly depending on the communication environment encountered. For a given BER, let the maximum throughput conditioned on the number of resolvable paths L_r be $B(L_r)$. Ideally, assuming that the receiver is capable of combining an arbitrary number of resolvable paths and that the transmitter has the perfect knowledge of the number of resolvable path with the aid of a feedback channel, and that the feedback delay is negligible, the unconditional throughput, B, using adaptive STS can be written as:

$$B = \sum_{L_r=1}^{\infty} P(L_r) \cdot B(L_r), \tag{8.66}$$

where $P(L_r)$ is the probability that there are L_r resolvable paths at the receiver. With the aid of (8.65), this probability can be approximated as:

$$\begin{aligned} P(L_r) &= \int_{\max\{0,L_r-1-0.5\}}^{L_r-1+0.5} f(\tau_r)d\tau_r \\ &= \int_{\max\{0,L_r-1-0.5\}}^{L_r-1+0.5} \frac{1}{T_m/T_c}\exp\left(-\frac{\tau_r}{T_m/T_c}\right)d\tau_r \\ &= \exp\left(-\frac{\max\{0,L_r-1-0.5\}}{T_m/T_c}\right)-\exp\left(-\frac{L_r-1+0.5}{T_m/T_c}\right), \end{aligned} \quad (8.67)$$

where $[L_r-1-0.5, L_r-1+0.5]$ is the normalized delay-spread range having L_r resolvable paths. In (8.66) $B(L_r)$ represents the maximum possible throughput conditioned on having L_r number of resolvable paths. For example, for the proposed adaptive STS scheme using four-antenna based STS, two-antenna based STS as well as conventional single antenna based transmission, as characterized in (8.61), (8.62) and (8.63), $B(L_r)$ may achieve values of R_b, $2R_b$ or $4R_b$, respectively, depending on the specific number of resolvable paths encountered.

Figures 8.10 and 8.11 show the throughput versus SNR/bit performance of the STS-assisted W-CDMA system using a maximum of four antennas. Depending on the number of resolvable paths at the receiver and on the corresponding achievable BER performance, the transmitter may activate one of the transmission schemes described by (8.61), (8.62) and (8.63). In our related investigations, the target BER was set to 0.01. Specifically, if a sufficiently high number of resolvable paths is encountered by the receiver, which results in a BER of less than 0.01 for the scheme described by (8.63), then the transmitter supports a bit-rate of $4R_b$. If the number of resolvable paths is in a range, where the BER using the scheme described by (8.63) is higher than 0.01, but that of the STS scheme described by (8.62) is lower than 0.01, then the transmitter transmits at a rate of $2R_b$. Finally, if the number of resolvable paths is in a range, where the BER using the STS scheme described by (8.62) is higher than 0.01, but that described by (8.61) is lower than 0.01, then the transmitter transmits at a rate of R_b. Otherwise, if the number of resolvable paths is too low, which results in BER > 0.01 for the STS scheme described by (8.61), then the transmitter simply disables transmissions. In the context of Figure 8.10 we assumed that the number of users was $K = 1$, and that the fading associated with each resolvable path obeyed the Rayleigh distribution ($m = 1$). By contrast, in Figure 8.11 we assumed that the number of users was $K = 10$, and that the fading associated with the first resolvable path obeyed the Nakagami-m distribution in conjunction with $m = 2$, while the fading of the other resolvable paths obeyed the Rayleigh distribution ($m_c = 1$).

¿From the results of Figures 8.10 and 8.11 we observe that with the aid of the adaptive STS scheme, the system's effective throughput is significantly increased, if the average delay-spread of the channel is sufficiently high, or in other words, if the number of resolvable paths varies over a sufficiently wide range. Let us highlight the significance of this observation in more detail. Using $T_m = 0.5\mu s$ and $3\mu s$ as examples and by observing Figure 8.10 we find that the SNR/bit required for transmitting at the data rate of R_b is about 5.2dB for $T_m = 0.5\mu s$ and 4.6dB for $T_m = 3\mu s$. Similarly, the SNR/bit required for supporting the data rate of $3R_b$ is about 6.4dB for $T_m = 0.5\mu s$ and 5dB for $T_m = 3\mu s$. Hence, the adaptive STS-assisted W-CDMA system increased the achievable transmission rate by a factor of three, while requiring only a modest transmitted power increase of about 1.2dB for

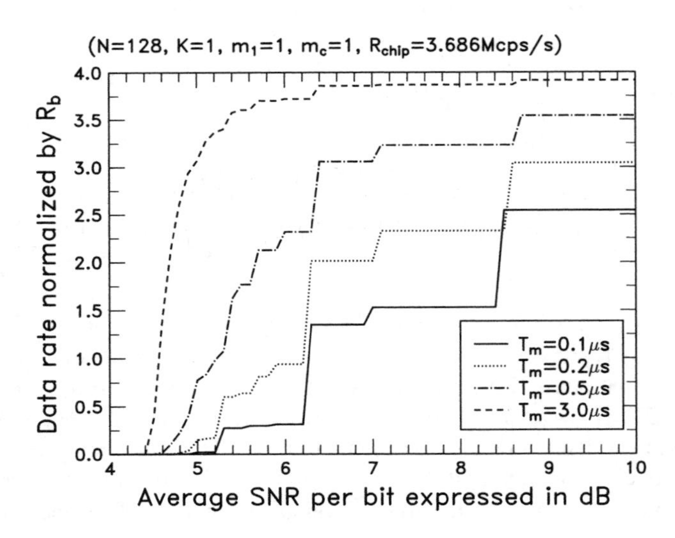

Figure 8.10: Normalized throughput versus the SNR per bit, E_b/N_0, performance of the adaptive space-time spreading assisted W-CDMA system using 4-antenna based STS of (8.61), the 2-antenna aided STS of (8.62) and the conventional single-antenna scheme for transmission over four typical wireless channels experiencing Rayleigh fading ($m = 1$). The target BER of the reference user is 0.01 and there are no interference users, i.e., $K = 1$.

$T_m = 0.5\mu s$ and 0.4dB for $T_m = 3\mu s$. Similar results can also be observed in Figure 8.11, where an extra 0.4dB or 1.2dB transmitted power is required for achieving a data rate of $3R_b$ instead of R_b. However, if the number of resolvable paths varies over a relatively low range, the required increase of the transmitted power becomes higher. For example, for the case of $T_m = 0.1\mu s$ in Figure 8.10 and Figure 8.11 an extra 2.2dB (Figure 8.10) or 1.2dB (Figure 8.11) transmitted power must invested, in order to achieve a data rate of $2R_b$ instead of R_b. In this scenario, due to the associated extra complexity of the adaptive STS-assisted scheme required by the channel dispersion estimation and feedback, and due to the control channel requirement of the dispersion feedback, the adaptive STS-aided scheme might not constitute a more attractive alternative.

In the context of the third-generation (3G) W-CDMA systems, since a chip rate of $R_c \geq 3.686\ Mcps/s$ has been employed, the number of resolvable paths may vary over a wide range in urban and suburban areas. Hence, using adaptive STS-aided transmissions constitutes a promising option, when supporting high data rate services.

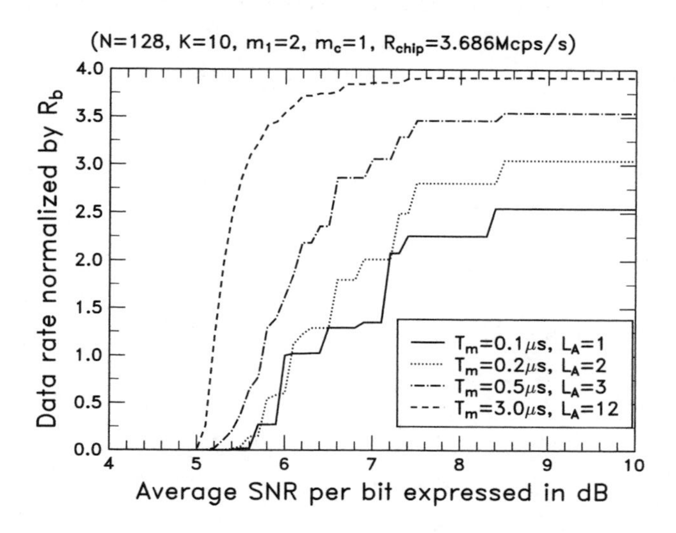

Figure 8.11: Normalized throughput versus the SNR per bit, E_b/N_0, performance of the adaptive space-time spreading assisted W-CDMA system using the 4-antenna based STS of (8.61), the 2-antenna aided STS of (8.62) and the conventional single-antenna scheme for transmission over four typical wireless channels obeying the Nakagami-m distribution ($m_1 = 2,\ m_c = 1$). The target BER of the reference user is 0.01, while the interfering users communicate using the 4-antenna based STS of (8.61) and each interfering signal has an average of L_A number of resolvable paths.

8.7 Chapter Summary and Conclusion

In this chapter the performance of STS-assisted W-CDMA systems has been investigated, when multipath Nakagami-m fading, multi-user interference and background noise induced impairments are considered. The analysis and numerical results demonstrated that the total diversity order achieved is the product of the transmit diversity order and the frequency-selective diversity order. Furthermore, both the transmit diversity and the frequency-selective diversity have a similar influence on the BER performance of the W-CDMA systems considered. Since W-CDMA signals typically experience high dispersion and hence high-dynamic frequency-selective fading in both urban and suburban areas, determining the number of transmission antennas on the premise of assuming a low number of resolvable paths would result in wasteful 'over-engineering' of the system. Based on the above scenarios, an adaptive STS transmission scheme was proposed, which adapts its STS configuration according to the frequency-selectivity information fed back from the mobile receivers. The numerical results show that by efficiently exploiting the channel's frequency-selectivity, the proposed adaptive STS scheme is capable of significantly improving the throughput of W-CDMA systems. For W-CDMA systems transmitting at a data rate of $3R_b$ instead of R_b, only an extra of 0.4dB and

1.2dB transmitted power is required in the urban and suburban areas considered, respectively.

Part II

Genetic Algorithm Assisted Multiuser Detection

List of Symbols in Part II of the Book

$a_k(t)$	kth user's signature sequence.
$\boldsymbol{a}$	Signature sequence vector.
$\alpha_k(t)$	Rayleigh distributed amplitude.
$\hat{b}_k^{(m)}$	Detected mth data bit of the kth user at the receiver.
$b_k^{(m)}$	mth data bit of the kth user.
$\boldsymbol{b}$	Data bit vector.
$\boldsymbol{C}$	CIR coeffcients diagonal matrix.
$\delta(H)$	Defining length of schema H.
Δ	Mutation size for real-valued variables.
$\Gamma_\tau(t)$	Rectangular pulse of unity amplitude from $0 \leq t < \tau$ and zero otherwise.
$h_k(t)$	Complex lowpass channel impulse response associated with the kth user's signal.
H	Schema notation.
K	Number of users in the system.
$\Lambda(\boldsymbol{X})$	Objective function where $\boldsymbol{X}$ is a vector containing the decision variables to be optimised.
λ_{max}	Maximum mutation size for real-valued variables.
L	Number of multipaths
L_d	Number of diversity antennas
$m(H, y)$	Number of instances corresponding to the schema H at the yth generation.
M	Number of data bits transmitted in a packet.
$n(t)$	Zero-mean complex additive white Gaussian noise with

independent real and imaginary components, each having a double-sided power spectral density of $N_0/2$ W/Hz.

$\boldsymbol{n}$ Gaussian noise vector.

N_c Spreading factor, or the number of chips in one data bit duration T_b.

$o(H)$ Order to schema H.

$\Omega(\boldsymbol{b})$ Correlation metric.

p_i Probability of selection corresponding to the ith individual.

p_m Probability of mutation.

$p_s(H)$ Probability of survival of schema H.

$\phi_k(t)$ Channel phase uniformly distributed between $[0, 2\pi)$.

P Population size.

$r(t)$ Received signal.

ρ_{lk} Cross-correlation of the lth user's and the kth user's signature sequence.

$\boldsymbol{R}$ User signature sequence cross-correlation matrix.

$\hat{s}_k(t)$ Equivalent lowpass representation of the kth user transmitted CDMA signal.

$s_k(t)$ Channel impaired signal of the kth user.

$\tau_{k,l}$ Random delay corresponding to user k over the lth path.

τ_{N1}, τ_{N2} Observation window boundary

T Mating pool size.

T_b Data bit duration.

T_c Chip duration.

ξ_k kth user's signal energy per bit.

$\boldsymbol{\xi}$ Signal energy per bit diagonal matrix.

z_l Matched filter output associated with the lth user.

$\boldsymbol{Z}$ Matched filter output vector.

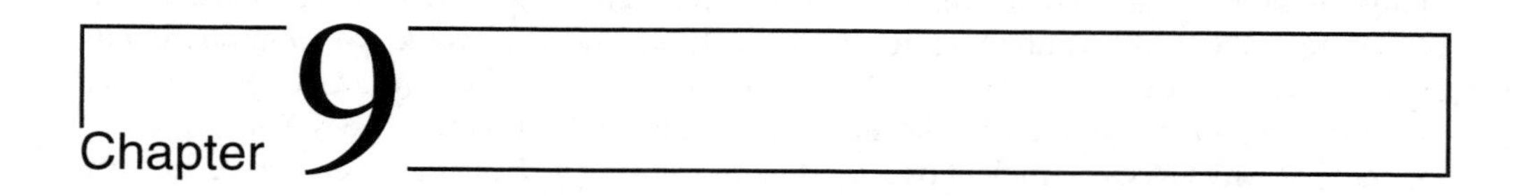

Overview of Genetic Algorithms used for Multi-user Detection[1]

"I have called this principle, by which each slight variation, if useful, is preserved, by the term of Natural Selection." *Charles Darwin, On the Origin of Species [282]*

Darwin's theory of evolution by natural selection, or survival of the fittest, has fascinated many scientists. In particular, the idea that the concept of evolution could be used as an optimization tool for engineering problems was conceived by a group of independent scientists [6, 283–285] in the 1960s and during the 1970s. Their pioneering ideas eventually led to the creation of a new discipline in engineering, known as *Evolutionary Computation* [286]. The concept behind evolutionary computation was to evolve a population (more than one) of candidate solutions to a given problem, using operators stimulated by natural genetic variations and natural selection.

The implementations of evolutionary algorithms can be classified into three strongly related, but independently developed methodologies :

- Evolutionary Strategies (ESs) [283, 284].
- Evolutionary Programming (EP) [285].
- Genetic Algorithms [6].

[1]*Single- and Multi-Carrier CDMA Multi-User Detection, Space-Time Spreading, Synchronization, Networking and Standards.* L.Hanzo, L-L.Yang, E-L.Kuan and K.Yen,
 ISBN 0-470-86309-9

Although the conceptual framework in mimicking the process of natural evolution with the aid of these methods is similar, each of these methods implements the associated algorithms in a different manner.

In this chapter, we are only concerned with the family of GAs and their application as an optimization tool to the problem of multi-user detection in CDMA. Hence, only the concepts of GAs [6, 9, 287, 288] will be highlighted in our further discourse. We will commence with an introduction to GAs in Section 9.1. Following this, we will show how a GA operates as an optimization tool in practice with the aid of an example in Section 9.2. We will then proceed to highlight why GAs are efficient optimization tools by identifying the resemblance between human search traits and the GA in Section 9.3. The derivation of the fundamental theorem of GAs is also included in this section. The various elements that constitute a GA are then highlighted in Section 9.4. A survey of GA-assisted CDMA multi-user detection schemes found in the current literature is presented in Section 9.5. Finally, Section 9.6 concludes this chapter. It should be stressed here that the GAs we will be describing in this chapter constitute only a small portion of the entire GA literature. Furthermore, the GAs that are used for solving our optimization problem are modified, in order to suit our applications and hence the GAs are slightly different from those commonly found in the GA literature. The interested readers might like to consult references [6, 9, 287] for a more detailed discourse on GAs.

9.1 An Introduction to Genetic Algorithms

The origins of GAs [6, 9, 287, 289, 290] can be traced back to the 1960s, when Holland [6] and his students undertook the task of studying the phenomenon of adaptation, as it occurs in nature and then imported these adaptive mechanisms into artificial systems. The results of these studies were published in the seminal monograph by Holland [6] in 1975. Since then, the level of interest in GAs has been growing rapidly and has attracted the attention of numerous scientists, including Goldberg [9], Mühlenbein [291] and Grefenstette [292], just to mention a few.

Although GAs have been used in countless applications, such as machine learning and modelling adaptive processes, by far the largest application of GAs is in the domain of function optimization. In contrast to traditional search methods, such as the method of steepest descent, GAs can be invoked in robust global search and optimization procedures that do not require the knowledge of the objective function's derivatives or any gradient-related information concerning the search space. Hence, non-differentiable functions as well as functions with multiple local optima represent classes of problems, where GAs can be efficiently applied [289].

The basic approach of a GA employed in optimizing a specific problem defined by an objective function is simple. The flowchart of a GA is shown in Figure 9.1. First, an initial population consisting of P number of so-called *individuals* is created in the 'Initialization' block, where P is known as the population size. Each individual represents a legitimate solution to the given optimization problem. An individual can be considered as a vector consisting of the decision variables to be optimized, as shown in Figure 9.2. Here, we will regard the leftmost decision variable in an l-length vector as the 1st decision variable, while the rightmost decision variable is referred to as the lth decision variable. Traditionally, the individuals

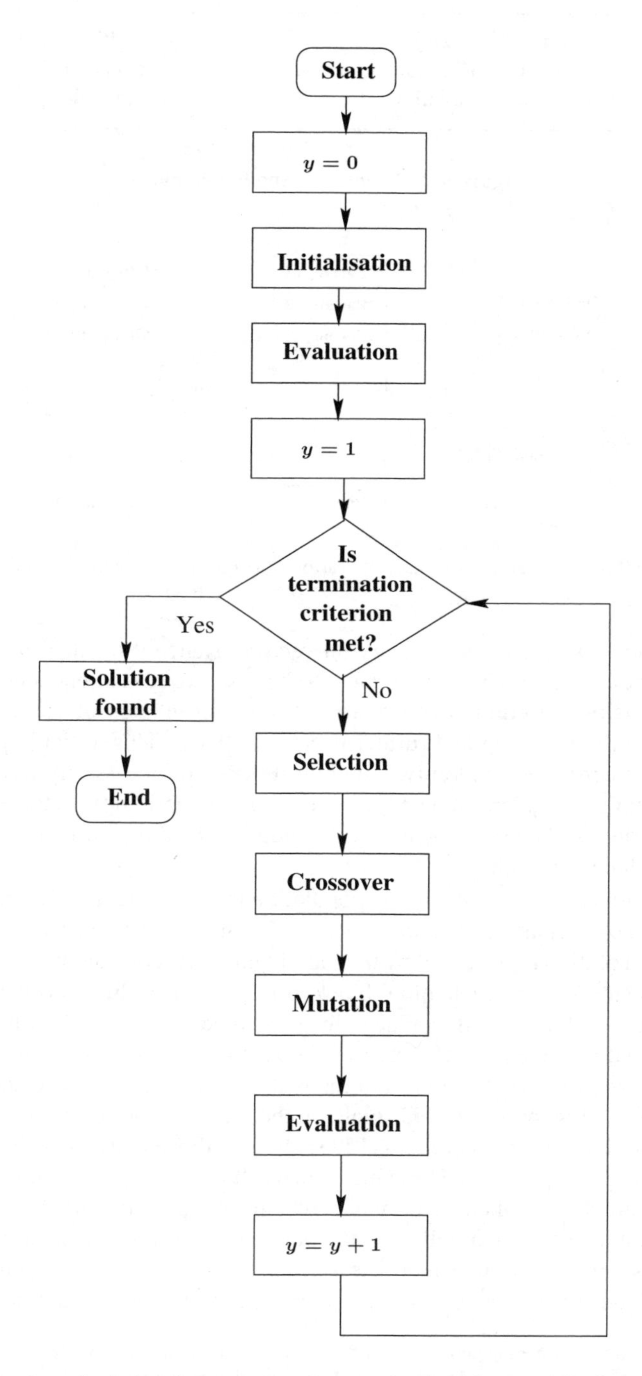

Figure 9.1: A flowchart depicting the structure of a generic algorithm used for function optimization.

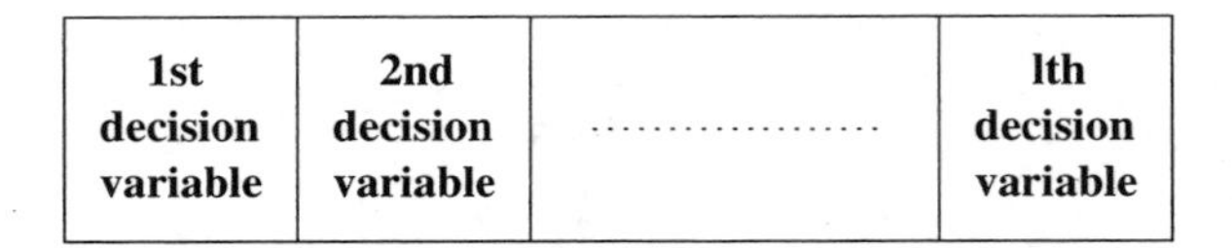

Figure 9.2: A typical l-length individual.

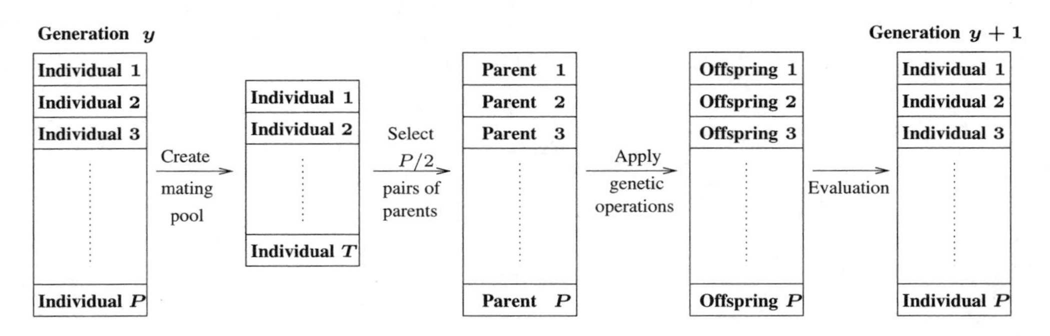

Figure 9.3: An example of a GA operation during a single cycle or generation.

in a GA population take the form of binary bit vectors. Hence if the decision variables to be optimized are not binary in nature, they have to be discretized and encoded to a bit vector, analogously to analogue-to-digital conversion. The representation of the decision variables as an individual will be highlighted further in Section 9.4.1. This initial population of individuals is usually generated randomly, although it does not necessarily have to be random specifically. If explicit *a priori* knowledge concerning the optimum vector is available, then this knowledge can also be used to generate the individuals of the initial population, in order to bias and expedite the search.

Associated with each individual in the population there is a figure of merit, or more commonly known in GA parlance as the *fitness* value. The fitness value is evaluated by substituting the candidate solution represented by the individual under consideration into the objective function, as indicated by the 'Evaluation' block of Figure 9.1. Individuals having the T number of highest fitness values are then placed in a so-called *mating pool*,[2] where $2 \leq T \leq P$. Using a kind of natural selection scheme together with the genetically-inspired operators of *cross-over* and *mutation*, the individuals in the mating pool are then evolved to a new population, as depicted in Figure 9.3. Based solely on the fitness values of these individuals in the current mating pool, the *selection* process [293] chooses those individuals in the mating pool that will be allowed to reproduce. These individuals chosen from the mating pool, referred to as *parents* in Figure 9.3, are then used by the cross-over and mutation operations, in order to generate new individuals, which will form the new population for the next iteration. The selection process is invoked for improving the average fitness value of the population by giving individuals of higher fitness values a higher probability to be reproduced in the new popula-

[2]Note that the definition of mating pool here is different from that found in the GA literature [9, 290]. Here the mating pool consists of T individuals associated with the highest fitness values in the current population, whereas in [9, 290] the mating pool is where the selected parents of the previous population are placed. We adopted our definition in this dissertation, in order to aid our description of GAs, which are modified to suit our specific application.

tion. Hence it focuses the search on the promising regions in the search space, which might contain the optimum solution. Numerous selection schemes have been proposed in the GA literature. Some of the more common selection methods will be highlighted in Section 9.4.2. However, the selection process does not alter the individuals. If the optimum solution is to be found, new individuals must be generated. The task of generating new individuals, using the individuals chosen by the selection process is accomplished by the cross-over operation.

The *cross-over* [287] operation is a process in which arbitrary decision variables are exchanged between a pair of selected parents, mimicking the biological recombination process between two single-chromosome organisms. Hence, the cross-over operation creates two new individuals, known as *offspring* in GA parlance, as portrayed in Figure 9.3, which have a high probability of having better fitness values than their parents. In order to generate P number of new offspring, $P/2$ number of cross-over operations are required. A new pair of parents is selected from the mating pool for each cross-over operation. The newly created offspring will form the basis of the new population.

During the *mutation* [287] operation, each decision variable in the offspring is perturbed, with a probability of p_m, by either a predetermined or a random value. This allows new areas in the search space to be explored. The mutation probability of a decision variable is usually very small, in the region of 0.1-0.01 [9]. However, the mutation operation is necessary in a GA, in order to prevent the phenomenon of so-called *premature convergence*. Premature convergence refers to the loss of population diversity before the optimum solution has been found.

Each cycle of selection, cross-over, mutation and evaluation constitutes a so-called *generation* in the execution of a GA, as depicted in Figure 9.3. This cycle will continue until some termination criterion is met, as shown in Figure 9.1. Generally, if this cycle is executed for many generations, the population will eventually converge on a set of individuals, in which the individual that corresponds to the highest fitness value is deemed to be the optimum or near-optimum solution.

GAs are not guaranteed to find the optimal solution and their effectiveness is determined largely by the population size P, i.e. the number of individuals in a population. However, because GAs operate on the basis of a population of candidate solutions simultaneously, the probability of finding a suboptimal solution is reduced in comparison to the methods that iterate from point to point, such as the method of steepest descent. Hence, the size of the population in a GA is a major factor in determining the accuracy of convergence [294]. As the population size increases, the GA has a better chance of finding the global optimum solution, but the computational cost also increases as a function of the population size. Apart from the population size, a GA's performance will also depend substantially on other factors, such as the choice of the selection method, the type of genetic operations employed, the parameter settings, for example, the value of T and p_m, as well as the particular iteration termination criterion. We will be highlighting the effects of each of these factors in our further discourse. Let us now consider an example of how GAs search for the optimum solution during a single generation, as exemplified by Figure 9.3.

	Individuals					Fitness f_i	Mapped fitness $f'_i = f_i + 10$	Selection probability[4] $p_i = f'_i / \sum_j^T f'_j$
A_1	1	1	1	-1	1	9.06	19.06	0.3985
A_2	1	1	-1	-1	-1	-1.056	8.944	0.1870
A_3	-1	1	-1	-1	-1	-6.368	3.632	0.07594
A_4	1	-1	-1	1	1	6.192	16.192	0.3385
							$\sum_j^P f'_j = 47.828$	

Table 9.1: An example of the initial population consisting of $P = 4$ individuals, where the associated fitness values are evaluated according to the objective function given by Equation (9.1).

9.2 Genetic Algorithms at Work

Consider an optimization problem, where the objective function is given by[3] :

$$\Lambda(\boldsymbol{b}) = 2\boldsymbol{b}\boldsymbol{y} - \boldsymbol{b}\boldsymbol{R}\boldsymbol{b}^T. \tag{9.1}$$

Assuming that the following information is available :

$$\boldsymbol{y} = \begin{bmatrix} 1.328 \\ -2.183 \\ 0.044 \\ -2.856 \\ 2.485 \end{bmatrix}; \quad \boldsymbol{R} = \begin{bmatrix} 1 & 0.1 & 0.2 & 0.3 & 0.4 \\ 0.1 & 1 & 0.4 & 0.2 & 0.3 \\ 0.2 & 0.4 & 1 & 0.5 & 0.1 \\ 0.3 & 0.2 & 0.5 & 1 & 0.6 \\ 0.4 & 0.3 & 0.1 & 0.6 & 1 \end{bmatrix},$$

the goal of this optimization process is to find the decision variable vector $\boldsymbol{b}$, which consists of $l = 5$ antipodal bits, that maximizes the objective function of Equation (9.1).

Since the candidate solutions in this case are in the form of vectors consisting of $l = 5$ antipodal bits, no conversion of the individuals to this format is required. GAs commence their search by generating an initial population of individuals at random. For this example, we adopted a population of size $P = 4$. Each bit of the $l = 5$ bits of an individual can be constructed by tossing an unbiased coin. These randomly generated individuals and their associated fitness values are shown in Table 9.1. For simplicity, we will assume that the mating pool size is equal to the population size, i.e. $T = P$. In this case, all the individuals of Table 9.1 will be given a chance to reproduce. The next step of the GA is the selection of parents, in order to create new offspring. A common selection method used in GAs is the so-called *fitness-proportionate selection* [9], in which the probability of selection p_i of the ith individual is equal to its fitness value f_i divided by the total fitness value of the mating pool. This method requires the fitness values to be positive for all combinations of $\boldsymbol{b}$ because a negative fitness value would yield a negative probability of selection. According to our objective function of Equation (9.1), this requirement is not met, since certain combinations of $\boldsymbol{b}$ exhibit negative fitness values, as shown in Table 9.1. Hence a mapping function must be invoked,

[3]Notice that the objective function of Equation (9.1) is identical to the correlation metric $\Omega(\boldsymbol{b})$ of Equation (10.23) for the optimum multi-user detection in CDMA [106], which will be shown in Chapter 10.

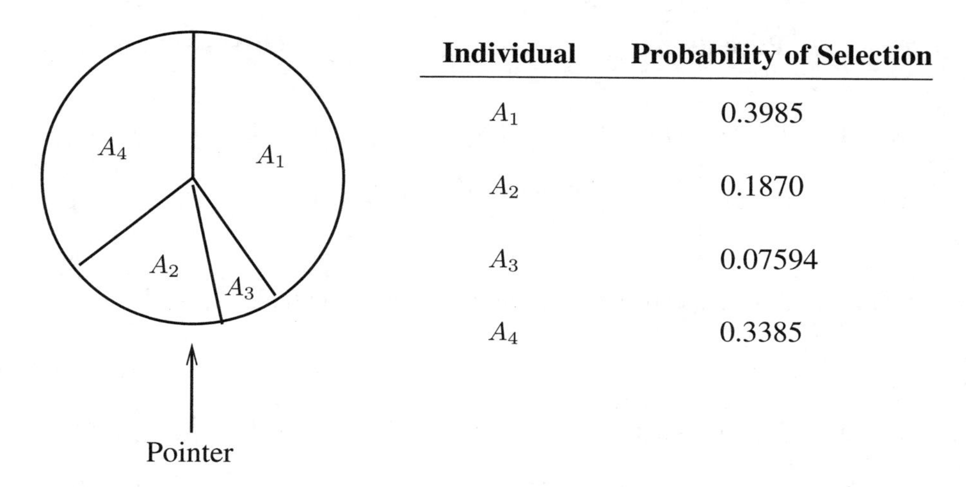

Individual	Probability of Selection
A_1	0.3985
A_2	0.1870
A_3	0.07594
A_4	0.3385

Figure 9.4: An implementation of the fitness-proportionate selection scheme using a roulette wheel, whereby each individual of Table 9.1 is allocated a slice of the wheel proportional in area to the individual's probability of selection.

in order to ensure that the fitness values for all combinations of $\boldsymbol{b}$ become positive. A simple mapping function, referred to as *offset mapping* [9], is to add a constant positive value to the fitness values of all the individuals, as indicated in Table 9.1. Summing the mapped fitness values f_i' over all four individuals, we obtained a total mapped fitness value of 47.828 and an average mapped fitness value of 11.957 for the initial population of Table 9.1. The probability of selection p_i for each individual is calculated in proportion to their individual fitnesses and the values are as listed in Table 9.1. We can see that individuals having higher fitness values are allocated a higher probability of selection. Note that the probability of selection p_i of an individual is defined with respect to the average fitness of the current population. Hence, the probability of selection of the same individual would be different in a different population.

A simple method of implementing the fitness-proportionate selection scheme is the so-called *roulette wheel sampling* [6], whereby each individual is allocated a slice of a circular roulette wheel proportional in area to the individual's probability of selection. An example of a roulette wheel is shown in Figure 9.4 for the population of Table 9.1. When the roulette wheel is spun and the pointer comes to standstill on one of the wedge-shaped slices, the corresponding individual will be selected as a parent. Each spin of the roulette wheel yields a new parent. Hence we can see here that by using this implementation, individuals with higher probability of selection have a higher chance of being selected as parents.

Once a pair of parents is selected, the cross-over operation is then applied to this pair of parents, as depicted in Figure 9.3. A number of variants of the cross-over operations were proposed and the simplest form is the so-called *single-point cross-over* [9]. In a single-point cross-over operation, a so-called *cross-over point* x is arbitrarily selected between the range $[1, l-1]$, where l is the length of the individual. Two offspring are created by swapping all decision variables beyond this cross-over point between the parents, i.e. from the $(x+1)$st decision variable to the lth decision variable. For example, let us assume that the individuals A_1 and A_4 were selected as parents from our initial population of Table 9.1. The resulting

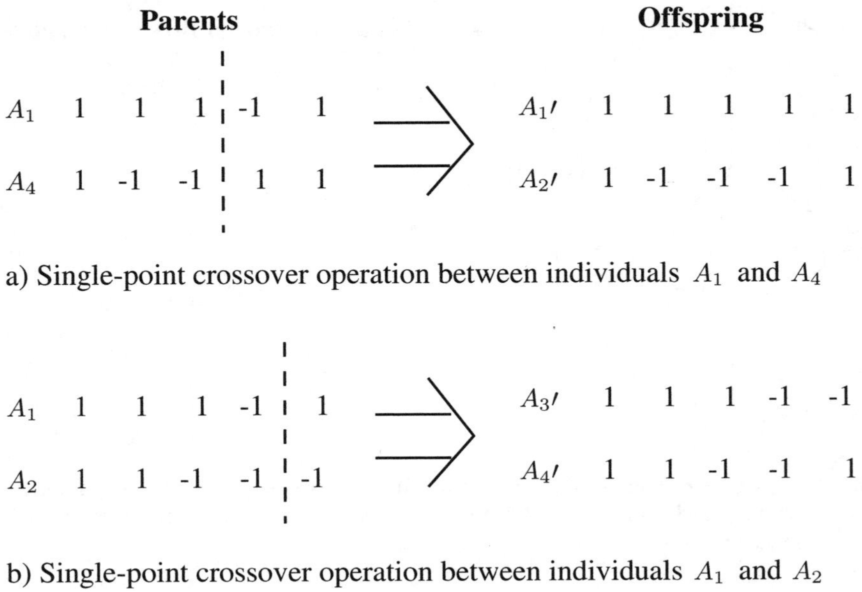

	Parents							Offspring				
A_1	1	1	1 ¦	-1	1	⇒	$A_1\prime$	1	1	1	1	1
A_4	1	-1	-1 ¦	1	1	⇒	$A_2\prime$	1	-1	-1	-1	1

a) Single-point crossover operation between individuals A_1 and A_4

A_1	1	1	1	-1 ¦	1	⇒	$A_3\prime$	1	1	1	-1	-1
A_2	1	1	-1	-1 ¦	-1	⇒	$A_4\prime$	1	1	-1	-1	1

b) Single-point crossover operation between individuals A_1 and A_2

Figure 9.5: Examples of the single-point cross-over operation between the pairs of selected individuals of Table 9.1, where the vertical dashed line represents the cross-over point.

	Individuals					Fitness f_i	Mapped fitness $f_i' = f_i + 10$
$A_1\prime$	1	1	1	1	1	-2.364	8.364
$A_2\prime$	1	-1	-1	-1	1	17.616	27.616
$A_3\prime$	1	1	1	-1	-1	-0.88	9.88
$A_4\prime$	1	1	-1	-1	1	8.884	18.884
							$\sum_j^P f_j' = 64.744$

Table 9.2: An example of the new population created by the single-point cross-over operation between pairs of selected individuals of Table 9.1, as shown in Figure 9.5. Their associated fitness values are evaluated according to the objective function given by Equation (9.1).

cross-over between these two parents, as illustrated in Figure 9.5a with the cross-over point indicated by the vertical dashed line, yields two new offspring, $A_1\prime$ and $A_2\prime$. The cross-over operation is then repeated for another newly selected pair of individuals, for example, individuals A_1 and A_2 of Table 9.1, with a newly generated cross-over point, in order to create two more new offspring, namely $A_3\prime$ and $A_4\prime$, as shown in Figure 9.5b. These two pairs of newly created offspring with their associated fitness values shown in Table 9.2, will form the new population for the next generation. From Table 9.2, we can clearly see that both the maximum and the average mapped fitness values have improved during the transition to the new population over that of the previous population of Table 9.1. The average mapped fitness value of the population has improved from 11.957 to 16.186 during the lifetime of a generation. Similarly, the maximum mapped fitness value has increased from 19.06 to 27.616

during the same generation. Note that the fitness values of Table 9.2 are calculated before the mutation operation is invoked. The conventional procedure [9], however, would be to perform the mutation operation on the offspring before their fitness values are evaluated, as depicted in Figure 9.1. On the other hand, as mentioned in the previous section, the probability of mutation is fairly small. Hence, on average the mutation operation will not affect the average fitness value of the population significantly, even though the maximum fitness value may be different due to the probability that offspring $A_{2'}$ may be mutated. Despite the fact that the mutation of individuals may occasionally cause the loss of important information learnt from the previous generation, there are instances when the mutation operation is useful. Considering Table 9.2, we note that due to the bias of the selection process, the leftmost bit of all the offspring that is produced is a logical '1'. Further cross-over operations between any pairs of these offspring in the subsequent generation will be unable to change the state of the leftmost bit. On the other hand, it is possible that the leftmost bit of the optimum solution is in fact a logical '0' represented by '-1'. Hence in this case, with the aid of cross-overs alone, we will be unable to find the optimum solution. This effect is more significant during the latter stages of the GA, when the individuals in the population began to resemble each other more closely due to the effects of selection and cross-over. Although the probability is small, the mutation operation ensures that new information is injected and hence provides an opportunity to sample the unexplored regions of the search space. In order to prevent the individuals in the population becoming self-similar too rapidly, potentially resulting in a premature convergence, it is important to maintain a high diversity of individuals in the population, especially during the initial phase of a search.

With the aid of this example, we have shown how by granting the individuals associated with high fitness values a higher probability of reproduction, reminiscent of the natural selection theory, then, by using the simple probabilistic operations of cross-over and mutation in the context of these individuals, new individuals are produced, which in general will have a higher fitness values than their parents. Yet, this example raises even more questions than answers. How can it be possible that better solutions may be found by simply exchanging decision variables between two individuals in a random manner? What will ensure that the average fitness value of the population will be improved, even after a single generation? We will attempt to provide the answers to these questions in the next section.

9.3 How do GAs Operate?

From our brief conceptual introduction of the family of GAs given in Section 9.1, it may appear still somewhat far-fetched that simple GA operations such as the selection of individuals based purely on their associated fitness values, on the partial random exchange of decision variables between individuals and on the random perturbation of a few decision variables lends itself to solving even the most complex optimization problems. Even with the aid of our example in Section 9.2, it is still non-trivial as to how GAs can efficiently be used to search for the optimum solution. In this section, we will attempt to further augment the key concepts of GAs.

The fundamental concept of GAs can readily be justified by first observing the way humans perform a search for the optimum solution based purely on human intuition, given a set of candidate solutions. This approach is plausible, since the notion of GAs is in effect based on natural adaptation.

9.3.1 Optimization from a Human's Perspective

Let us consider again the optimization example highlighted in Section 9.2. Upon observing the four individuals given in Table 9.1, we will notice certain similarities in specific segments of the individuals. Furthermore, certain bit sequences are associated with a high fitness value. For example, observe that the three individuals A_1, A_2 and A_4, which have a higher fitness value compared to individual A_3, have a common decision variable, since their leftmost bit is a logical '1'. Hence, it is highly probable that the optimum solution contains a logical '1' in its leftmost bit position. We also noticed that the individuals ending with a logical '1' seem to be associated with high fitness values. This fact is gleaned by comparing individuals A_1 and A_4 against the individuals A_2 and A_3. Hence, it is also highly probable that the optimum solution contains a logical '1' at its rightmost bit position. However, at this stage, we are uncertain about the value of the middle three bits, since these three bits of the two fittest individuals A_1 and A_4 differ from each other. In order to resolve this dilemma, we may exchange some decision variables between these two individuals that are associated with a high fitness value, with the hope that the resulting individual will have an even higher fitness value. This course of action is analogous to our behaviour of exchanging and combining good ideas from different sources. For example, techniques from the field of GAs and CDMA multi-user detection may be exchanged, with the hope that a more likely solution can be found for the transmitted CDMA signals. Hence, in order to continue our search for the optimum solution, we will generate four new individuals (offspring) using the information that is available to us at this moment as a guide. We may then end up with a new population that resembles that shown in Table 9.2.

In summary, we have followed this course of action in order to seek similarities among individuals in the population and then sought a causal relationship between these similarities while aiming for high fitness values [9]. The idea is that it is these similarities that confer high fitness values on the individuals. In our next optimization step we attempt to amalgamate these similarities in the hope of creating a new individual that will exhibit an even higher fitness value. This is accomplished in GAs by the selection and cross-over mechanisms, which bestow them with their ability to converge.

There will come a time when no more information can be gleaned from the individuals in a population, when all the individuals in the population become similar. In order to ascertain that the individual associated with the maximum fitness value in this population is the optimum solution, we will attempt to change the state of certain bits and then check the associated effects on the fitness through a series of trial-and-error steps. In other words, we safeguard our tentative solution against premature convergence. This is executed in GAs with the aid of the mutation operation. However, in order to refrain from changing the state of the bits of an individual aggressively, the mutation operation in GAs is usually carried out with a low probability. Again, this is analogous to our behaviour, when combining the ideas of GAs and CDMA multi-user detection. After combining these two ideas, we tend to modify the resulting solution, in order to gauge whether the solution can be improved further. This will usually involve a series of trial-and-error steps.

So far, we have compared the approach of GAs with certain human search traits and have identified the resemblance between the way humans and GAs perform a search in their quest for the optimum solution. Let us now highlight the theoretical aspects of the optimization from a GA's perspective.

9.3.2 Optimization from a GA's Perspective

We have seen from the previous section that we can search for the optimum solution more effectively, if we exploit important similarities among highly fit individuals. Hence the focus here is no longer on individuals alone, but rather on their similarities. Holland [6] introduced the notion of a so-called *schema* (plural, *schemata*), in order to explain how GAs search for regions of high fitness. A schema H is a similarity template, defined over the alphabet $\{0, 1, *\}$, where 0 and 1 are referred to as *defined bits*, while $*$ denotes a *don't care* symbol. The *order* of a schema $o(H)$ is determined by the number of defined bits in that schema. By contrast, the *defining length* of a schema $\delta(H)$ is the distance between its leftmost and rightmost defined bits, including only one end of the interval in the distance calculation. An individual is said to be an *instance* of a particular schema, if at every position in that schema, a 1 matches a 1 in the individual, a 0 matches a 0, or a $*$ matches either. Using our example in Section 9.2, by replacing the logical '-1' with a logical '0', individuals A_1 and A_4 of the initial population of Table 9.1 are instances of the schema 1***1, while the individuals A_2 and A_3 are instances of the schema *1000. These two schemata have an order of 2 and 4, respectively, and have a defining length of 4 and 3, respectively. Hence we see that the definition of schemata provides us with a better representation of the similarities between the individuals in a population and simplifies the analysis of the GAs.

From a different perspective, schemata can be considered to represent hyperplanes in the search space [290]. Perhaps the best way of visualizing schemata as hyperplanes is to consider an $l = 3$-dimensional search space, where each candidate solution contains 3 bits, as shown in Figure 9.6 [9]. In this case, the search space takes the form of a cube, where the corners of the cube represent the legitimate individuals, which constitute schemata of order 3. The edges of the cube are schemata of order 2, as illustrated in Figure 9.6, while the planes of the cube are schemata of order 1. The whole space is represented by the schema of order 0.

From Figure 9.6, we can readily see that each individual representing a candidate solution of length l contains 2^l schemata, or hyperplanes. Hence, a population of size P will contain between 2^l – when all the individuals in the population are identical – and $P \times 2^l$ schemata. In other words, there are in fact $P \times 2^l$ number of similarities present in a population of size P, which the GA can exploit. The explicit fitness value of an individual evaluated from the objective function implicitly gives some information about the average fitness value of the 2^l different schemata of which the individual concerned is an instance. Hence the explicit fitness evaluation of a population of P individuals, at a given generation, also implies an implicit evaluation of the estimated average fitness value of a much higher number of schemata. This simultaneous evaluation of a high number of schemata in a population of P individuals is referred to as *implicit parallelism* [6]. It should be stressed here that the implicit average fitness value of a schema evaluated in this case is only an estimation, since the instances evaluated in a finite-size population constitute only a small sample of all possible instances. For example, consider the schema 1**** of length $l = 5$. In order to obtain its actual average fitness value, we have to evaluate the fitness of 16 different individuals, i.e. that of all $l = 5$-bit individuals that contain a logical '1' in the leftmost bit. However, in a practical GA, the

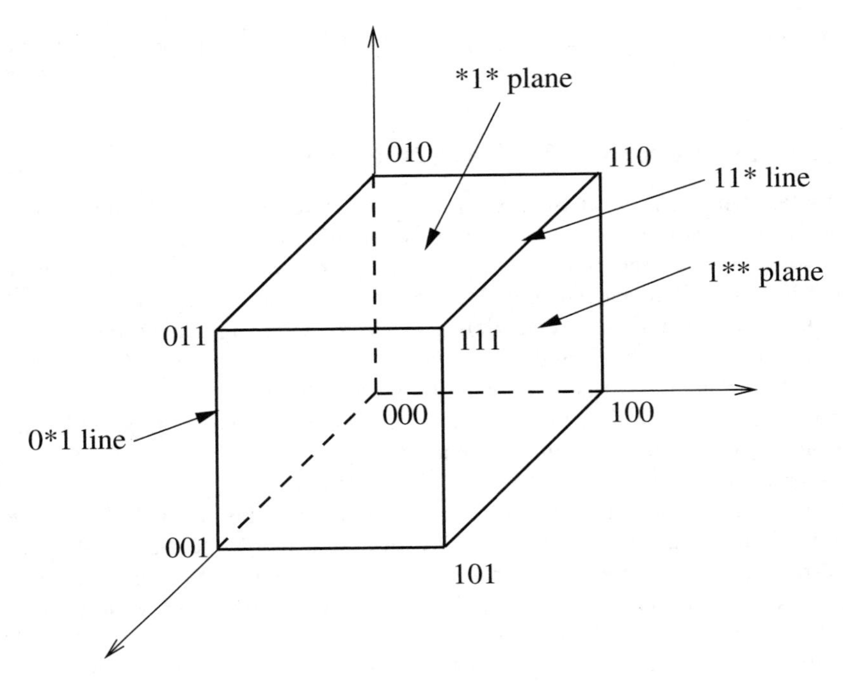

Figure 9.6: Visualization of schemata as hyperplanes in the $l = 3$-dimensional search space [9].

population size invoked will be typically less than 16, when searching for a 5-bit optimum solution.

Having demonstrated that a high number of schemata is present in a given population and their estimated average fitness values are implicitly estimated by the GAs, we will now consider the growth-rate and decaying-rate of these schemata in response to selection, cross-over and mutation.

9.3.2.1 Effects of Selection on the Schemata

Let us assume that there are $m(H, y)$ number of instances corresponding to the schema H present in a population of size P at generation y. The estimated average fitness value of the schema H, considering its $m(H, y)$ number of instances at generation y is denoted as $f(H, y)$. Following our example given in Section 9.2, the probability of selection p_i as a parent for an individual A_i is equal to its fitness value f_i divided by the total fitness value of the population[5], i.e. $p_i = f_i / \sum_j^P f_j$. Hence the expected number of offspring associated with individual A_i in the next generation $y + 1$, ignoring the effects of cross-over and mutation, is equal to $A_i \times P \times p_i$ or $A_i \times f_i / \bar{f}$, where $\bar{f} = (\sum_j^P f_j)/P$ is the average fitness value of the population. Similarly, we can express the expected growth rate of the schema H in terms of its estimated average fitness value $f(H, y)$ and the population's average fitness value $\bar{f}$

[5]Again, here we assumed that the mating pool size T is equal to the population size P.

as [9] :

$$m(H, y+1) = m(H, y)\frac{f(H, y)}{\bar{f}}, \tag{9.2}$$

where $m(H, y{+}1)$ is the expected number of instances of the schema H in the next generation $y + 1$. According to Equation (9.2), due to the effects of the selection process alone, the number of schemata having average fitness values above the population's average fitness value is expected to increase in the next generation. At the same time, the number of schemata having average fitness values lower than the population's average fitness value is expected to decrease. Furthermore, when the population is evolving over many generations, the estimate of a schema's average fitness should, in principle, become more and more accurate, since the GA is sampling more and more instances of that schema.

9.3.2.2 Effects of Cross-over on the Schemata

In order to observe the effects of cross-over on the schemata, let us first consider the following two schemata of length $l = 5$ [289] :

$$\begin{aligned} H_1 &= 11 * * * \\ H_2 &= 1 * * * 1. \end{aligned}$$

Note that the individual A_1 of Table 9.1 is an instance of the schemata H_1 and H_2 and they have a defining length $\delta(H)$ of 1 and 4, respectively. We assume that the single-point cross-over operation is invoked, as highlighted in Section 9.2. Recall that the cross-over point is randomly generated in the range $[1, l-1]$, inclusively. Hence we can readily see that the probability that the defined bits of schema H_1 will be separated during the cross-over operation is only $1/(l-1)$. Specifically this will happen if the cross-over point is located at position 1. On the other hand, the probability that the defined bits of schema H_2 will be separated during the cross-over operation is $(l-1)/(l-1) = 1$. In other words, schema H_2 will be 'destroyed' by the cross-over operation, regardless of where the cross-over point is located. Based on this observation, we can see that the probability $p_s(H)$ that a schema H will survive during the cross-over operation is dependent on its defining length $\delta(H)$. If the generated cross-over point is beyond the defining length of a schema, this schema will remain intact. However, we should also note that even if the cross-over point is within the defining length of a schema, there is still a finite probability that the schema will survive. This will happen, if both the parents selected for the cross-over are instances of that schema. Hence, we can express a lower bound of $p_s(H)$ as [9] :

$$p_s(H) \geq 1 - \frac{\delta(H)}{l-1}. \tag{9.3}$$

Now if we combine the effects of both the selection and cross-over on the schema H, we can expect the number of instances of that schema in the next generation $y+1$ to be [9] :

$$m(H, y+1) \geq m(H, y) \cdot \frac{f(H, y)}{\bar{f}} \left[1 - \frac{\delta(H)}{l-1}\right]. \tag{9.4}$$

According to Equation (9.4), schemata having above-average fitness values and short defining lengths are expected to increase their number of instances in the subsequent generation. Let us next consider the effects of mutation on the schemata.

9.3.2.3 Effects of Mutation on the Schemata

It was mentioned in Section 9.1 that mutation is the random alteration of each bit with a probability of p_m. Hence, in order for a schema H to survive, all its associated defined bits must themselves survive. Recall that the order of a schema $o(H)$ is defined by the number of defined bits it contains. Therefore, the probability that a schema H of order $o(H)$ will survive during the mutation operation is $(1-p_m)^{o(H)}$. In short, the probability of survival under the effects of mutation is higher for lower-order schemata.

Now we can formulate the expected growth-rate of the schemata by combining the effects of selection, cross-over and mutation, which is given as [9] :

$$m(H, y+1) \geq m(H, y) \cdot \frac{f(H, y)}{\bar{f}} \left[1 - \frac{\delta(H)}{l-1}\right] (1-p_m)^{o(H)}. \tag{9.5}$$

Equation (9.5), as formulated by Holland [6], is known as the *Schema Theorem* [6,9], which is the fundamental theorem behind the concepts of GAs. The schema theorem, according to Equation (9.5) states that short, low-order, above-average schemata will increase in their number of instances in the subsequent generation. These schemata are known in GA parlance as *building blocks* and the supposition that this is indeed the nature of the process behind GAs is known as the *building block hypothesis* [287]. According to the building block hypothesis, the GA initially biases its search towards higher fitness values in certain low-order schemata and converges on this part of the search space. Over time, it gradually biases its search towards higher-order schemata by combining information from low-order schemata with the aid of cross-overs and eventually converges on a small region of the search space that exhibits a high fitness value.

However, Holland's schema theorem has its critiques. As seen in Equation (9.5), the schema theorem only makes predictions concerning the expected number of instances of schemata from one generation to the next. Unfortunately it does not provide predictions about the quality of the solution that the GA can deliver over many generations, or about the speed at which the GA will converge, or indeed provide an exact picture of the GAs' behaviour. Hence intensive research has been carried out in order to provide a more exact mathematical analysis concerning the behaviour of GAs [295–297]. However, the portrayal of this analysis is beyond the scope of this dissertation. Nevertheless, the schema theorem of Equation (9.5) provides a fundamental stepping stone towards a better understanding of why GAs work. In the next section, we will highlight the various elements of GAs in more detail.

9.4 Elements of Genetic Algorithms

9.4.1 Representation

Representation refers to the way candidate solutions are represented by individuals. Traditionally, as defined by Holland [6], the individuals are represented in the form of bit vectors, in which each vector is comprised of a combination of zeros and ones. The strong preference for using binary representations of solutions in GAs was justified by Holland [6] according to the schema theory, as highlighted in Section 9.3. It is claimed [9] that GAs are well suited to handle pseudo-Boolean and combinatorial optimization problems. For optimization problems involving nonbinary or real-valued decision variables, these decision variables have to be encoded into binary-valued bit vectors, in order to perform the genetic operations. Similarly, these bit vectors must be converted back to their original real-valued form, in order to evaluate their associated fitness values from the objective function.

9.4.1.1 Binary Encoding

There are several potential encoding schemes for mapping nonbinary decision variables to binary-valued bit vectors. The so-called *binary encoding* [287] is the simplest and most commonly used encoding scheme. Encoding of nonbinary integers is straightforward. For example, 4 and 12 can be represented as 100 and 1100, respectively. For real-valued decision variables, the number of bits invoked will determine the resolution of the encoding. Suppose a real-valued decision variable x, where $a \leq x \leq b$, is to be encoded to an n-bit vector. First we can convert x to a nonbinary integer y according to [298] :

$$y = \left\lfloor \frac{b-a}{2^n} \times x \right\rfloor . \tag{9.6}$$

We can then encode the integer y according to any nonbinary integer encoding. Binary encoding has the drawback that in some cases all the bits must be changed in order to increase a number by 1. For example, the bit pattern 001 translates to 3 in decimal, but 4 is represented by 100. This can make it implementationally difficult for an individual that is close to an optimum solution to move even closer to the optimum with the aid of the cross-over and mutation operation [290].

9.4.1.2 Gray Encoding

In order to overcome this drawback, a different encoding scheme, namely *Gray coding* [287] was proposed. Gray codes have the property that incrementing or decrementing any integer number by 1 always involves only a one-bit change. The mapping function from the binary coded n-bit vector to a Gray coded n-bit vector is given by [298] :

$$g_k = \begin{cases} b_1 & \text{if } k = 1 \\ b_{k+1} \oplus b_k & \text{if } k > 1 \end{cases} , \tag{9.7}$$

where g_k and b_k are the kth Gray code bit and binary code bit, respectively, for $k = 1, \ldots, n$ and $\oplus$ denotes a modulo 2 addition. The conversion from Gray coding to binary coding is given by [298] :

$$b_k = \sum_{i=1}^{k} g_i. \tag{9.8}$$

In practice, Gray-coded representations are often more successful for real-valued parameter function optimization applications than binary-coded representations.

The issue of the representation of nonbinary or real-valued decision variables in terms of bit vectors is still open to debate. If real-valued decision variables are represented in their original form, the search space is continuous and will have an infinite number of search points. Coding these decision variables into bit vectors discretizes the search space and reduces its size. On the other hand, analoguously to the analogue-to-digital conversion process in digital systems, this conversion results in a *quantization error*, where the accuracy of the decision variables is determined by the number of bits used to represent the real-valued decision variables. Furthermore, an additional complexity and delay are incurred, since the real-valued decision variables have to be binary encoded and decoded for each generation in order to perform the required genetic operations and to evaluate their corresponding fitness values, respectively. Moreover, comparisons [298, 299] have shown that GAs representing the real-valued decision variables in their original form, using *real-valued coding* exhibit a better performance, than those converting the decision variables to bit vectors. Hence in this treatise we will be using the real-valued representation of the decision variables, as will be highlighted in Chapter 11. Let us now review the range of selection schemes in the next section.

9.4.2 Selection

There are numerous ways in which a new population can be created from the previous population. However, regardless of what method is used, it is imperative that individuals having higher fitness values in a given mating pool must be given a better chance of reproducing offspring in the subsequent generation, than the lower-fitness individuals in the same mating pool. Otherwise the GA will be unable to take advantage of the presence of high-quality individuals in the population and to efficiently search for the optimum solution. The task of choosing these individuals for reproduction is performed by the selection process [293]. The type of selection scheme used predetermines the convergence characteristics of the GA. A strongly selective scheme implies that suboptimal, but highly fit individuals will dominate the population, reducing the diversity needed for further change and progress and hence may lead to premature convergence, without exploring the entire search space. On the other hand, a weakly selective scheme will result in a slower convergence rate due to the presence of poor quality, low-fitness individuals. Numerous selection schemes have been proposed in the GA literature. We will highlight some of the more commonly used regimes below. Here we will assume that the selection process is invoked in the mating pool, which contains T number of individuals associated with the highest fitness values in a given population.

9.4.2.1 Fitness-proportionate Selection

In *fitness-proportionate* selection, as invoked in our example in Section 9.2, the probability of selection p_i of the ith individual is defined as :

$$p_i = \frac{f_i}{\sum_j^T f_j}, \tag{9.9}$$

where f_i is the fitness value associated with the ith individual. However, the fitness-proportionate selection scheme has several deficiencies. Based on Equation (9.9), we can see that if there is only a small percentage of individuals with relatively high fitness values in a mating pool, then these individuals will be assigned with a high probability of selection compared to the other individuals in the mating pool. Hence the offspring produced in this case are fairly similar in the subsequent generation. This may lead to a premature convergence, since the search space has not been sufficiently well explored. This phenomenon typically occurs during the early stages of a GA's operation, when the initial population is randomly generated and the fitness distribution of the mating pool happens to be non-uniform at the beginning. Hence a small number of individuals have a tendency of dominating the selection process.

Furthermore, if the fitness value distribution of the mating pool is fairly uniform, i.e. the fitness value of each individual is fairly close to one another, then all the individuals in the mating pool will have an approximately equal probability of selection. Hence all solutions will have a similar chance of being assigned to the mating pool and hence producing offspring.

9.4.2.2 Sigma Scaling

The *sigma scaling* selection scheme [300] was proposed in order to make the GA less susceptible to premature convergence. Under this scheme, the probability of selection of the ith individual is a function of several variables, namely that of its fitness value, the mating pool mean fitness and the mating pool fitness' standard deviation, as given by [287] :

$$p_i = \begin{cases} \frac{1}{T}\left(1 + \frac{f_i - \bar{f}}{2\sigma}\right) & \text{if } \sigma \neq 0 \\ \frac{1}{T} & \text{if } \sigma = 0, \end{cases} \tag{9.10}$$

where f_i is the fitness value of the ith individual, $\bar{f}$ is the mean fitness of the mating pool and σ is the standard deviation of the mating pool's fitness values. According to Equation (9.10), it is possible that the value of p_i is negative. In the context of maximization, p_i will be set to a small value (eg. $0.1/T$ in [300]) since a negative p_i calculated from Equation (9.10) implies that the corresponding individual has a fitness value significantly lower than the population's average fitness value. As specified by Equation (9.10), we can see that if the standard deviation of the mating pool fitness values is high, individuals having high fitness values will not be assigned a high probability of selection. Hence the individuals having lower fitness values are given a fair chance of reproducing. On the other hand, if the individuals in the mating pool are similar, resulting in a low standard deviation, then the individuals exhibiting higher fitness values will be assigned a higher probability of selection.

9.4.2.3 Linear Ranking Selection

The *linear ranking* selection scheme [301] is an alternative method of preventing premature convergence. According to this method, the individuals in the mating pool are ranked according to their associated fitness values, such that the rank T is assigned to the individual associated with the highest fitness value in the mating pool, while the rank 1 is assigned to the individual exhibiting the lowest fitness value in the mating pool. Similarly, the remaining individuals in the mating pool are ranked accordingly. The ith individual will then be assigned its probability of selection p_i, based on its specific ranking $rank_i$ in the mating pool, as given by [302]:

$$p_i = \frac{1}{T}\left[\eta^- + \left(\eta^+ - \eta^-\right)\frac{rank_i - 1}{T-1}\right], \tag{9.11}$$

where $\frac{\eta^-}{T}$ is the probability of selection assigned to the individual associated with the lowest fitness value and $\frac{\eta^+}{T}$ the probability of selection assigned to the individual having the highest fitness value. If the mating pool size T always remains the same from generation to generation, the conditions $\eta^+ = 2 - \eta^-$ and $\eta^- \geq 0$ must be fulfilled.

Hence in this selection scheme, we can see that each individual's probability of selection is determined by its rank in the mating pool and it is independent of the fitness value distribution of the mating pool. However, this scheme suffers from a slow convergence rate since the probability of selection of an individual is determined regardless of its relative fitness value in the mating pool.

9.4.2.4 Tournament Selection

According to the *tournament selection* scheme [303], t number of individuals are chosen randomly from the mating pool, where $t < T$ is referred to as the *tournament size*. The individual associated with the highest fitness value out of these t preferred individuals will be selected as a parent. This process is repeated for another t set of individuals, in order to form a pair of parents for the cross-over operation.

Again, the probability of selection of each individual is independent of the fitness distribution of the mating pool according to the tournament selection scheme.

9.4.2.5 Incest Prevention

According to the previously discussed selection regimes it is possible that two identical individuals are selected. The offspring resulting from the cross-over operation between these two so-called *incest* individuals will also be identical. While this will ensure that some individuals will be transferred from one generation to the next generation, this method does not promote diversity and may also lead to premature convergence [287].

Hence an alternative technique, known as *incest prevention* has been proposed [304], which only allows different individuals to be selected for the cross-over operation. This

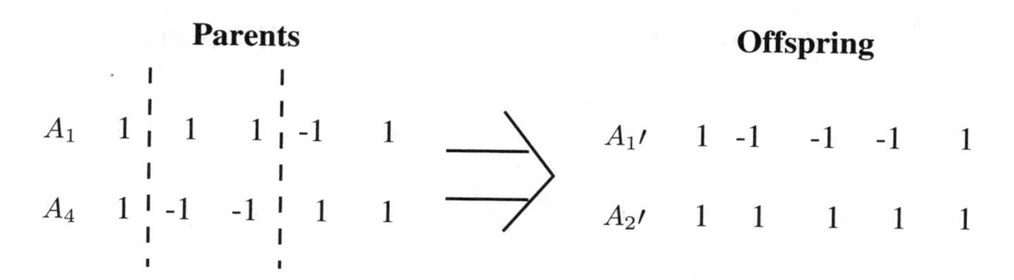

Figure 9.7: Example of the double-point cross-over operation between the individuals A_1 and A_4 of Table 9.1, where the vertical dashed lines represent the cross-over points, in order to produce the offspring $A_1\prime$ and $A_2\prime$.

technique will ensure that sufficient diversity is maintained from generation to generation and the likelihood of a premature convergence is reduced. In our application, we will invoke the above-mentioned incest prevention scheme by ensuring that all individuals in the mating pool are dissimilar.

We will compare the performance of each of the above-mentioned selection schemes in the context of GAs in Chapter 10. More importantly, we will study the effects of the mating pool size T and that of incest prevention on the performance of GAs. Let us now consider the family of cross-over operations.

9.4.3 Cross-over

9.4.3.1 Single-point Cross-over

The simplest form of a cross-over operation is the *single-point cross-over*, which we have highlighted in our example in Section 9.2. The single-point cross-over was also used by Holland [6] in deriving the schema theorem of Equation (9.5). However, the single-point cross-over has several shortcomings. First, as we have observed in Equation (9.3), single-point cross-over is more likely to destroy schemata of long defining lengths. In other words, schemata that can be created or destroyed by the single-point cross-over depend strongly on the location of the bits in the individual. Second, single-point cross-over cannot combine all possible schemata [287]. For example, instances of schemata 1***1 and **11* cannot be combined to form an instance of 1*111. In order to mitigate these shortcomings, two other cross-over operations were introduced in GAs, namely the double-point cross-over [287] and the uniform cross-over [305]. These two operations will be highlighted below.

9.4.3.2 Double-point Cross-over

A *double-point cross-over* [287] operation uses two randomly chosen cross-over points. Decision variables that fall between these cross-over points are then exchanged between the parents. Figure 9.7 illustrates an example of the double-point cross-over between individuals A_1 and A_4 of Table 9.1. Double-point cross-over is less likely to destroy schemata having a high defining length and can combine more schemata than the single-point cross-over [287]. However, there are still certain schemata that the double-point cross-over cannot combine.

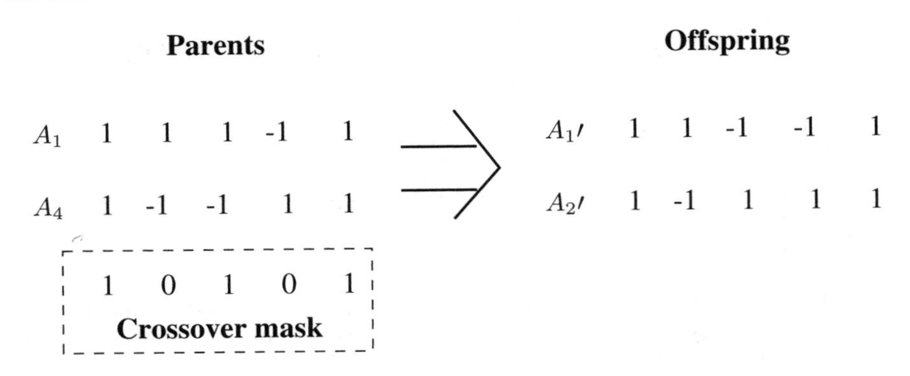

Figure 9.8: Example of the uniform cross-over operation between the individuals A_1 and A_4 of Table 9.1 with the aid of a cross-over mask, in order to produce the offspring $A_1\prime$ and $A_2\prime$.

9.4.3.3 Uniform Cross-over

In a *uniform cross-over* operation [305], a so-called *cross-over mask* is invoked instead of the cross-over point. The cross-over mask is a vector consisting of randomly generated 1s and 0s of equal probability, having a length equal to that of the individuals. Bits are exchanged between the selected pair of parents at locations corresponding to a 1 in the cross-over mask. An illustration of the uniform cross-over operation is shown in Figure 9.8. While it was shown in [306] that the uniform cross-over operation has a higher probability of destroying a schema, it is also capable of creating new schemata.

Intensive research efforts have been invested in quantifying and comparing the usefulness of these cross-over operations. However, the results did not give a definitive guidance on when to use a specific type of cross-over operation, since their effectiveness is very much dependent on the problem in which they are used. Hence the general consensus is that various cross-overs have to be tested and possibly modified, in order to determine what type of cross-over operation is most suitable for solving the problem at hand. Let us next consider the potential mutation operations proposed in the literature [287].

9.4.4 Mutation

The various selection methods and cross-over operations we have highlighted so far are applicable to binary or real-valued, but binary-coded as well as real-valued individuals. In other words, the procedures in carrying out these selection and cross-over operations are the same, regardless of the type of coding used for mapping legitimate solutions to individuals. However, the mutation operation will be different for a binary-coded individual and a real-valued individual. Recall from Section 9.4.1 that a binary-coded individual consists of bit decision variables, while a real-valued individual consists of real-valued decision variables. After the production of the offspring with the aid of the cross-over operation, each decision variable of the offspring will be mutated with a probability of p_m. The mutation operation is invoked, in order to ensure that sufficient diversity is maintained in the population so as to protect it against premature convergence. Numerous studies have been carried out, in order to determine the optimum value of p_m. A high probability of mutation may prevent the survival of schemata of high fitness values and hence may lead to suboptimal solutions. On the other hand, a low probability of mutation may result in premature convergence to suboptimum solutions due to the lack of diversity in the population. Schaffer *et al.* [307] suggested that the value of p_m should lie in the range of $[0.005, 0.01]$, Grefenstette [308] recommended the choice of $p_m \approx 0.01$, while Bäck [309] claimed that $p_m = 1/l$, where l is the length of the individual, is the most useful choice for unimodal functions. Adaptive mutation rates that change during the search process have also been proposed by Bäck [310]. We will be evaluating the effects of the value of p_m in the context of our specific optimization problem in Chapter 10. Let us now highlight the effects of the mutation operation on a binary decision variable, which will be followed by a discourse on the mutation of a real-valued decision variable.

9.4.4.1 Mutation of Binary Decision Variables

There are only two possible values for each binary decision variable hosted by an individual. Hence, when mutation is invoked for a particular bit, the value of the bit is toggled to the other possible value. For example, a bit of logical '1' is changed to a logical '0' and vice versa.

9.4.4.2 Mutation of Real-valued Decision Variables

The mutation of real-valued decision variables [298] is slightly more complicated, since each decision variable can assume an infinite number of possible values. Due to the associated granularity of representing the individuals, it is impossible to obtain the exact value that conforms to the optimum solution. Hence we can only strive for achieving a value that is as close to the optimum value as possible.

When a decision variable x is picked for mutation, the direction of mutation is chosen randomly with equal probability. Then a real-valued *mutation size* Δ is randomly generated, whose value ranges between $[0, \Delta_{max}]$, where Δ_{max} is the maximum mutation range. This value is usually pre-determined, in order to ensure that the value of x after mutation does not exceed the maximum and minimum limits specified by the problem. The value of x is then increased or decreased accordingly by a magnitude prescribed by the mutation size as $x = x \pm \Delta$. The mutation operation on real-valued decision variables will be further elaborated on in Section 11.3.3.

9.4.5 Elitism

We mentioned in Section 9.3 that the cross-over and mutation operations are capable of destroying a schema. This also implies that an individual associated with a high fitness value may be lost from one generation to the next. A good example of this scenario can be found by considering Table 9.1 and Table 9.2, which characterize the population before and after the cross-over operation was invoked, respectively. Notice that the individual A_1 of Table 9.1, which has a corresponding fitness value of 9.06 did not appear in the new population of Table 9.2. While the individual A_1 will never qualify as the optimum solution, since the individual A'_2 was found to have an even higher fitness value, it should be given a chance to be exploited further, since it has the second highest fitness value so far.

Hence, in order to ensure that high-merit individuals are not lost from one generation to the next, the best or a few of the best individuals are copied into the forthcoming generation, replacing the worst offspring of the new population. This technique is known as *elitism* [287].

9.4.6 Termination Criterion

The exact structure of the search space is often unknown in optimization problems. Hence in search algorithms, with the exception of an exhaustive search, it is typically infeasible to ensure that the optimum solution can be found. There are numerous ways of determining the termination criterion for GAs. The GA-assisted search can be terminated, if there are no further improvements in the maximum fitness value after several consecutive generations. In this case, the time required for the GA to reach a decision is uncertain. On the other hand, if the structure of the search space is time-invariant, then it is possible to set a threshold, such that the GA-assisted search is terminated, once the fitness value of an individual is found to exceed this threshold. Unfortunately neither of these termination criteria can be applied to GA-assisted CDMA multi-user detection, since typically a fixed implementational complexity is required and also the search space is time-variant due to the noise and fading imposed by the transmission channel. Hence in our application, we will terminate the GA-assisted search at the Yth generation and the individual associated with the highest fitness value at this point will be the detected solution. Hence the value of Y must be carefully determined, in order to ensure that a high-probability solution is obtained. Furthermore, by specifying the exact number of generations, the computational complexity of the GA can be determined. In this case, the upper bound limit on the number of objective function evaluations is equivalent to $P \times Y$. This figure is derived by assuming that we calculate the fitness values of all the individuals in the population at every generation. If we can store the fitness values associated to each different individual in memory, then if identical individuals are created in the subsequent generations, their associated fitness values can be directly accessed from memory and do not need to be calculated again. This will lead to a significant reduction in the computational complexity.

9.5 Survey of Genetic Algorithm-Assisted CDMA Multi-user Detection

Despite establishing itself as a useful optimization tool in numerous scientific as well as non-scientific applications, the employment of GAs in the area of mobile communications, especially at the physical layer, has been extremely rare. There were only a handful of proposals in the current literature that invoked GAs in CDMA multi-user detection. Below we shall give a brief review of these proposals by outlining the GA configurations invoked.

The earliest notion of a GA-based CDMA multi-user detection scheme was suggested by Juntti *et al.* [311] in 1997. In their contribution, the performance of the GA-based multi-user detector was studied by computer simulations based on a synchronous 20-user CDMA system. Random signature sequences with a spreading factor of 31 were used. The interfering users were assumed to have a power 3 dB higher than the desired user. It was not explicitly stated in the contribution to which selection scheme was invoked nor the probability of mutation and the population size. The single-point cross-over operation was used. Three different approaches of generating the initial population of individuals were simulated, which are as follows :

1. All the individuals of the initial population were randomly generated.

2. Some individuals of the initial population were randomly generated, while the remaining individuals were based on the hard decisions made at the output of the conventional matched filter detector.

3. Some individuals of the initial population were randomly generated, some were based on the hard decisions made by the conventional matched filter detector, while the remaining individuals were based on the hard decisions made by the decorrelating detector.

A performance comparison was made between these three different approaches. While no results were explicitly shown, it was concluded that at high signal-to-noise ratios (SNRs), the bit error probability (BEP) associated with the first approach exhibited a residual value. On the other hand, if the initial population contains some good guesses of the likely solution, as provided by the hard decisions at the output of the matched filter or the decorrelating detector, then the performance of the GA-based multi-user detector is close to that of the single-user system. Hence, it was concluded that using GAs alone cannot provide a robust multi-user detection performance or a high near–far resistance.

In the following year, a GA-based multi-user detection scheme was proposed by Wang *et al.* [312] for an asynchronous CDMA system communicating over an AWGN channel. The users' bits were detected sequentially in conjunction with a modified Viterbi algorithm. A so-called *window mapping* technique was invoked, in order to ensure that all legitimate solutions have positive fitness values. The fitness-proportionate selection scheme and the uniform cross-over operation were used. Furthermore, the elitism strategy was employed for ensuring that some of the best individuals from the previous population were copied into the new population. As we mentioned in Section 9.1, the population size and the rate of convergence, and therefore the computational complexity, are proportional to the size of the search space. Hence, in order to reduce the size of the search space, Wang *et al.* used a threshold value for estimating the BEP based on its corresponding matched filter output. If the matched filter output is above the threshold value, then the corresponding bit can be directly determined based on the hard decision provided by its matched filter output. Consequently, this bit will not be involved in the GA optimization. The performance of the GA-based multi-user detector was also compared against that of the decorrelating detector and the MMSE detector when supporting 20 users. The signature sequences were based on 31-chip Gold codes. A population size of 20 was used in the simulations and the GA was terminated after 6 generations. The probability of mutation was not explicitly stated. It was shown that the performance of the GA-based multi-user detector is very close to that of the MMSE detector and compares favourably with that of the decorrelating detector.

More recently, Ergün *et al.* [313, 314] proposed a hybrid approach that employs a GA and a multistage detector (MSD) for the multi-user detection. Three implementations were evaluated :

	Juntti *et al.* [311]	Wang *et al.* [312]	Ergün *et al.* [313,314]
Selection method	Not specified	Fitness-proportionate	Fitness-proportionate
Cross-over operation	Single-point	Uniform	Single-point
Mutation operation		Standard binary mutation	
Elitism	No	Yes	Yes
Incest prevention	No	No	No
Population size P	Not specified	20 for $K = 20$	30 for $K = 10$
Mating pool size T	Not specified	P	P
Probability of mutation p_m	Not specified	Small	0.05

Table 9.3: A summary of the configuration of the GA used in [311–314] for the application in CDMA multi-user detection.

1. The GA was used for detection. The computational complexity per bit is $O(PY)$, where P and Y denote the population size and the number of generations, respectively.

2. The GA was used as the first stage of the MSD in order to provide a good initial point for the successive stages of the MSD. The computational complexity per bit is $O(K^2)$, where K is the number of users. This approach will be referred to using the acronym C-GA/MSD.

3. The MSD was embedded into the GA as a genetic operator in order to further improve the fitness of the population at each generation. The computational complexity per bit is $O(K^2) + PY \cdot O(K)$. This approach will be referred to using the acronym E-GA/MSD.

Again, the window mapping function was invoked, in order to ensure that the fitness values for all legitimate solutions are positive. The fitness-proportionate selection scheme was employed with a mating pool size of $T = P$, where P is the population size. A single-point cross-over and the elitism strategy were also used. The probability of mutation was set to a value of 0.05. Simulations were performed for a 10-user CDMA system over an AWGN channel and Gold codes with a spreading factor of 15 were used as the signature sequences. Performance results were shown for both synchronous and asynchronous scenarios. In the synchronous case, the results showed that the E-GA/MSD converges to the optimum performance at an SNR of about 6dB after 7 generations for a population size of $P = 30$. On the other hand, the GA and C-GA/MSD techniques attained the optimum performance only after 50 generations. This leads to an excessive computational complexity. Furthermore, the E-GA/MSD approach is capable of achieving a near-optimal performance for all near–far ratios. In an asynchronous transmission scenario using a very short packet size of 4, it was shown that neither the GA nor the C-GA/MSD approaches attained the optimum performance, even after 50 generations for a population size of $P = 50$ at the SNR of 6 dB. On the other hand, the E-GA/MSD approach was capable of attaining the optimal performance after about 10 generations.

A summary of the GA configuration utilized in these proposals is listed in Table 9.3. Based on the limited number of proposals found in the literature, the general conclusion [311, 313, 314] was that multi-user detection based on GAs alone was not attractive due to its slow convergence, even when a high population size was invoked, and hence a high computational complexity was invested. In order to increase its convergence rate, it was proposed that other forms of detection ought to be used in conjunction with the GA [313, 314]. The size of the search space was reduced by the proposition in [312]. Nevertheless, more research must be carried out, in order to establish the feasibility of GA-assisted multi-user detection schemes, which is the aim of this chapter. A different GA-assisted multi-user detection scheme was recently proposed by Abedi [315], in which the optimum filter coefficients at the detector were acquired by the GA in order to detect the users' transmitted bits. Due to its contrasting approach, this method will not be compared against our proposed GA-assisted scheme.

9.6 Chapter Summary and Conclusions

In this chapter we have presented a brief overview of GAs. Specifically, we introduced the terminologies and procedures of GAs in Section 9.1. We then proceeded with an example in Section 9.2, which demonstrated how a GA operates in practice. Section 9.3 augmented the rationale of using GAs by first identifying the resemblance between the way humans and GAs performed a search for the optimum solution. These discussions were followed by the derivation of the schema theorem, which is the fundamental theorem of GAs. Following that, we reviewed some of the more commonly used GA-based operations, such as selection schemes, cross-over and mutation operations as well as implementation strategies in Section 9.4. Last but not least, a survey of the GA-based multi-user detection schemes found in the literature was conducted in Section 9.5.

GAs have successfully been applied in many function optimization problems, as justified by the countless references found in the GA literature. It is this potential of the GAs in solving complex optimization problems that provided the motivation for this chapter.

Finally, it should be stressed here that no GA configuration exists that is universally applicable to the solution of every optimization problem. As seen in Section 9.4, there are many ways of implementing a GA, using different combinations of selection schemes, cross-overs and mutation operations. There is no definite theoretical justification as to which combination gives the optimum performance, since different combinations work best for different problems. The best way of identifying the specific combination of operations that is most suitable for the problem at hand is to critically appraise and adapt these combinations to the problem. On this note, let us proceed with the application of GAs in the context of CDMA multi-user detection.

Chapter 10

Genetic Algorithm-Assisted Multi-user Detection for Synchronous CDMA[1]

10.1 Introduction

In this chapter, we will apply a GA-assisted scheme as a suboptimal multi-user detection technique in bit-synchronous CDMA systems over single-path Rayleigh fading channels. This provides a simple model for investigating the feasibility of applying GAs in CDMA multi-user detection as well as for determining the GA's configuration, in order to obtain a satisfactory performance. Based on the results obtained in this chapter, the GA-assisted CDMA multi-user detector will then be subsequently extended to an asynchronous system model incorporating multipath Rayleigh fading channels in Chapter 13.

This chapter is organized as follows. We will first highlight our system model used in this chapter in Section 10.2. The notations defined here will also be used in the subsequent chapters. An equivalent discrete-time system model is also highlighted in Section 10.3. We will then derive the optimum multi-user detector based on the Maximum Likelihood (ML) criterion for the system model adopted in this chapter in Section 10.4, which can be seen to have a computational complexity exponentially proportional to the number of users. GA-assisted multi-user detectors are then developed through a series of experiments, in order to find the GA configuration that is best suited for our application and the results will be shown in Section 10.5. Finally, using this GA configuration, the BEP performance of the GA-assisted multi-user detector based on our system model is assessed by simulations in

[1] *Single- and Multi-Carrier CDMA Multi-User Detection, Space-Time Spreading, Synchronization, Networking and Standards.* L.Hanzo, L-L.Yang, E-L.Kuan and K.Yen,
 ISBN 0-470-86309-9

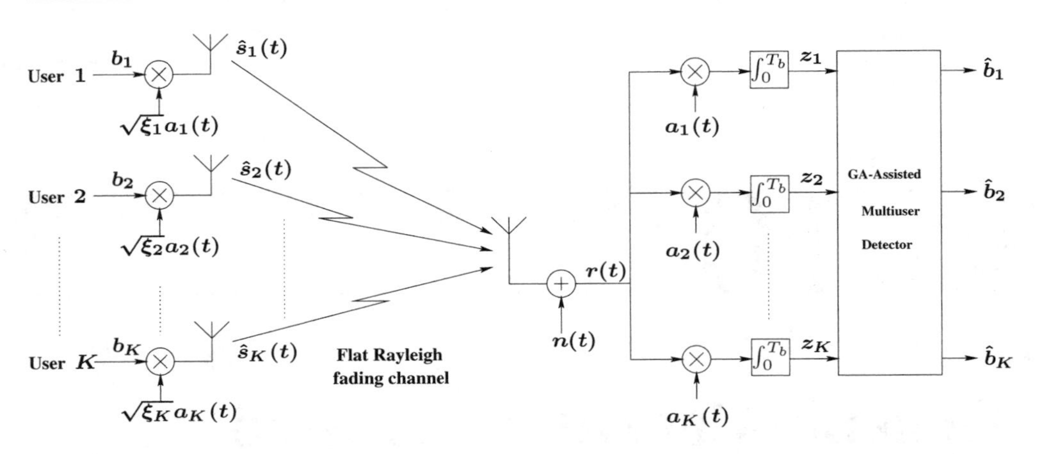

Figure 10.1: Block diagram of the K-user synchronous CDMA system model in a flat Rayleigh fading channel.

Section 10.6. The summary of this chapter is given in Section 10.7.

Before we commence our in-depth discourse, a few observations are made regarding our mathematical notations used in this chapter. Vectors and matrices are represented in boldface, while $(\cdot)^T$ and $(\cdot)^*$ denote the transpose matrix and the conjugate matrix of $(\cdot)$, respectively. Hermitian matrices, defined as the complex conjugate transpose of the matrices, are denoted as $(\cdot)^H$. Furthermore, diag$(\cdot)$ represents a diagonal matrix, where the diagonal elements correspond to the vector $(\cdot)$.

10.2 Synchronous CDMA System Model

We consider a bit-synchronous CDMA system as illustrated in Figure 10.1, where K users simultaneously transmit data packets of equal length to a single receiver. In this chapter we will adopt the Binary Phase Shift Keying (BPSK) modulation technique for all the transmissions. The transmitted signal of the kth user can be expressed in an equivalent lowpass representation as :

$$\hat{s}_k(t) = \sqrt{\xi_k} \sum_{m=0}^{M-1} b_k^{(m)} a_k(t - mT_b), \quad \forall k = 1, \ldots, K \tag{10.1}$$

where ξ_k is the kth user's signal energy per bit, $b_k^{(m)} \in \{+1, -1\}$ denotes the mth data bit of the kth user, $a_k(t)$ is the kth user's signature sequence, T_b is the data bit duration and M is the number of data bits transmitted in a packet. When considering a synchronous system experiencing no multipath interference, it is sufficient to observe the signal over a single bit duration T_b, since there is no interference inflicted by symbols outside this duration. Hence without loss of generality, we can omit the superscript (m) from all our equations in this chapter.

The kth user's signature sequence $a_k(t)$ may be written as :

$$a_k(t) = \sum_{h=0}^{N_c-1} a_k^{(h)} \Gamma_{T_c}(t - hT_c), \quad 0 \leq t < T_b, \quad \forall k = 1, \ldots, K \tag{10.2}$$

where T_c is the chip duration, $a_k^{(h)} \in \{+1, -1\}$ denotes the hth chip, N_c is the spreading factor, which refers to the number of chips per data bit duration T_b such that $N_c = T_b/T_c$ and $\Gamma_{T_c}(t)$ is the chip pulse shape. In practical applications, $\Gamma_\tau(t)$ has a bandlimited waveform, such as a raised cosine Nyquist pulse. However, for the sake of simplicity in our analysis and simulation, we will assume that $\Gamma_\tau(t)$ is a rectangular pulse throughout this dissertation, which is defined as :

$$\Gamma_\tau(t) = \begin{cases} 1, & 0 \leq t < \tau \\ 0, & \text{otherwise.} \end{cases} \tag{10.3}$$

Without loss of generality, we assume that the signature sequence $a_k(t)$ of all K users has unit energy, as given by :

$$\int_0^{T_b} a_k^2(t)dt = 1, \quad \forall k = 1, \ldots, K. \tag{10.4}$$

Each user's signal $\hat{s}_k(t)$ is assumed to propagate over a single-path frequency-nonselective slowly Rayleigh fading channel, as shown in Figure 10.1 and the fading of each path is statistically independent for all users. The complex lowpass channel impulse response (CIR) for the link between the kth user's transmitter and the receiver, as shown in Figure 10.1, can be written as :

$$h_k(t) = \alpha_k(t) e^{j\phi_k(t)} \delta(t), \quad \forall k = 1, \ldots, K \tag{10.5}$$

where the amplitude $\alpha_k(t)$ is a Rayleigh distributed random variable and the phase $\phi_k(t)$ is uniformly distributed between $[0, 2\pi)$.

Hence, when the kth user's spread spectrum signal $\hat{s}_k(t)$ given by Equation (10.1) propagates through a slowly Rayleigh fading channel having an impulse response given by Equation (10.5), the resulting output signal $s_k(t)$ over a single bit duration can be written as :

$$s_k(t) = \sqrt{\xi_k} \alpha_k b_k a_k(t) e^{j\phi_k}, \quad \forall k = 1, \ldots, K \tag{10.6}$$

Upon combining Equation (10.6) for all K users, the received signal at the receiver, which is denoted by $r(t)$ in Figure 10.1, can be written as :

$$r(t) = \sum_{k=1}^{K} s_k(t) + n(t), \tag{10.7}$$

where $n(t)$ is the zero-mean complex Additive White Gaussian Noise (AWGN) with independent real and imaginary components, each having a double-sided power spectral density of $\sigma^2 = N_0/2$ W/Hz.

At the receiver, the output of a bank of filters matched to the corresponding set of the users' signature sequences is sampled at the end of the bit interval. The output of the lth user's matched filter, denoted as z_l in Figure 10.1, can be written as :

$$\begin{aligned} z_l &= \int_0^{T_b} r(t)a_l(t)dt \\ &= \int_0^{T_b} \sum_{k=1}^{K} \sqrt{\xi_k}\alpha_k b_k a_k(t)e^{j\phi_k} a_l(t)dt + \int_0^{T_b} n(t)a_l(t)dt \\ &= \underbrace{\sqrt{\xi_l}\alpha_l b_l e^{j\phi_l}}_{\text{Desired signal}} + \underbrace{\sum_{\substack{k=1 \\ k\neq l}}^{K} \sqrt{\xi_k}\alpha_k b_k \rho_{lk} e^{j\phi_k}}_{\text{Multiple Access Interference}} + \underbrace{n_l}_{\text{Noise}}, \end{aligned} \tag{10.8}$$

where ρ_{lk} is the cross-correlation of the lth user's and the kth user's signature sequence, as given by :

$$\rho_{lk} = \int_0^{T_b} a_l(t)a_k(t)dt, \tag{10.9}$$

and

$$n_l = \int_0^{T_b} n(t)a_l(t)dt. \tag{10.10}$$

As seen in Equation (10.8), apart from the Gaussian noise n_l, the desired signal is interfered by signals transmitted by the other users. This interference due to the other users' signals is also known as Multiple Access Interference (MAI).

Assuming that the receiver has perfect knowledge of the lth user's CIR coefficients $\alpha_l e^{j\phi_l}$, the detected bit $\hat{b}_{l,MF}$ of the lth user based on the conventional coherent single-user detector will be given by the sign of the matched filter output in Equation (10.8) as :

$$\hat{b}_{l,MF} = \text{sgn}\left[\Re\left(z_l \alpha_l e^{-j\phi_l}\right)\right]. \tag{10.11}$$

Multiplication by $\alpha_l e^{-j\phi_l}$ is necessary for coherent detection, because the phase rotation introduced by the channel has to be removed. By approximating the MAI as a Gaussian distributed random variable by virtue of the central limit theorem [316–318], the Bit Error

Probability (BEP) of the desired user can be shown to be given by [319] :

$$P_l = \frac{1}{2}\left(1 - \sqrt{\frac{\xi_l}{N_0/2 + \sum_k \xi_k \rho_{lk}}}\right). \qquad (10.12)$$

Hence from Equation (10.12), we can see that unless the signature sequences of the interfering users are orthogonal to that of the desired user, yielding $\rho_{lk} = 0$ for $k = 1, \ldots, K, k \neq l$, the BEP performance of the desired user will be inferior to that achieved in a single-user environment in conjunction with a single-user matched filter. Furthermore, since the BEP performance will deteriorate in conjunction with an increasing number of users, the conventional single-user detector is highly vulnerable to near–far effects [320].

10.3 Discrete-Time Synchronous CDMA Model

For our application, it is more convenient to express the associated signals in discrete-time format. Invoking Equation (10.6) describing the transmitted signal of each user, the sum of the transmitted signals of all users can be expressed in vector notation as :

$$\begin{aligned} s(t) &= \sum_{k=1}^{K} \hat{s}_k(t) \\ &= \boldsymbol{aC\xi b}, \end{aligned} \qquad (10.13)$$

where

$$\begin{aligned} \boldsymbol{a} &= [a_1(t), \ldots, a_K(t)] \\ \boldsymbol{C} &= \text{diag}\left[\alpha_1 e^{j\phi_1}, \ldots, \alpha_K e^{j\phi_K}\right] \\ \boldsymbol{\xi} &= \text{diag}\left[\sqrt{\xi_1}, \ldots, \sqrt{\xi_K}\right] \\ \boldsymbol{b} &= [b_1, \ldots, b_K]^T. \end{aligned} \qquad (10.14)$$

Hence the received signal of Equation (10.7) can be written as :

$$r(t) = s(t) + n(t). \qquad (10.15)$$

Based on Equations (10.13) and (10.15), the output vector $\boldsymbol{Z}$ of the bank of matched filters portrayed in Figure 10.1 can be formulated as :

$$\begin{aligned} \boldsymbol{Z} &= [z_1, \ldots, z_K]^T \\ &= \boldsymbol{RC\xi b} + \boldsymbol{n}, \end{aligned} \qquad (10.16)$$

where

$$\boldsymbol{R} = \begin{bmatrix} 1 & \rho_{12} & \cdots & \rho_{1K} \\ \rho_{21} & 1 & \cdots & \rho_{2K} \\ \vdots & \vdots & \vdots & \vdots \\ \rho_{K1} & \rho_{K2} & \cdots & 1 \end{bmatrix} \tag{10.17}$$

is the $K \times K$ dimensional user signature sequence cross-correlation matrix having elements given by Equation (10.9) and

$$\boldsymbol{n} = [n_1, \ldots, n_K]^T$$

is a zero-mean Gaussian noise vector with a covariance matrix $\boldsymbol{R}_n = 0.5N_0\boldsymbol{R}$. Based on this discrete-time model, we will next derive the optimum multi-user detector based on the maximum likelihood criterion for the synchronous CDMA system considered [321].

10.4 Optimum Multi-user Detector for Synchronous CDMA Systems

In this section we will derive the joint optimum decision rule for a K-user CDMA system based on the synchronous system model highlighted in Section 10.2. Specifically, we want to maximize the probability of jointly correct decisions of the K users supported by the system based on the received signal $r(t)$ of Equation (10.15).

From Equation (10.14) we note that there are $m = 2^K$ possible combinations of $\boldsymbol{b}$. We shall denote the ith combination as $\boldsymbol{b}_i$ and the combined transmit signal of all users in Equation (10.13) corresponding to the ith combination as $\boldsymbol{b}_i \leftrightarrow s_i(t)$.

Based on the above notations, we can express the joint maximum *a posteriori* probability (MAP) criterion as [90] :

$$\hat{\boldsymbol{b}} = \arg\left\{\max_{\boldsymbol{b}_i}\left[P\left(s_i(t)|r(t)\right)\right]\right\}, \tag{10.18}$$

where $\hat{\boldsymbol{b}}$ denotes the detected bit combination. Using Bayes' theorem, the *a posteriori* probability expression of Equation (10.18) can be written as [90] :

$$P\left(s_i(t)|r(t)\right) = \frac{p\left(r(t)|s_i(t)\right)P(s_i(t))}{p(r(t))}, \tag{10.19}$$

where $p\left(r(t)|s_i(t)\right)$ is the conditional joint probability density function (pdf) of the received signal $r(t)$ in Equation (10.15), $P(s_i(t))$ is the *a priori* probability of the signal containing the ith bit combination and $p(r(t))$ is the pdf of the received signal. Since the transmitted data bits of the K users are independent, the *a priori* probability $P(s_i(t)) = 1/2^K$ is equal for all $m = 2^K$ bit combinations. Furthermore, the received signal pdf $p(r(t))$ is independent of which of the $m = 2^K$ bit combinations is transmitted. Consequently, the decision rule based on finding the signal that maximizes $P(s_i(t)|r(t))$ is equivalent to finding the signal that maximizes $p(r(t)|s_i(t))$. This decision criterion based on the maximum of $p(r(t)|s_i(t))$ is termed as the Maximum Likelihood (ML) criterion and $p(r(t)|s_i(t))$ is referred to as a *likelihood function* [90].

According to Equation (10.7), the received signal $r(t)$ is a Gaussian distributed random variable having a mean equal to that of $s(t)$ given by Equation (10.13). Hence, it can be shown that the likelihood function $p(r(t)|s_i(t))$ is given by [319] :

$$\begin{aligned} p(\boldsymbol{Z}|\boldsymbol{s}) &= \exp\left(-\frac{1}{2\sigma^2}\int_0^{T_b}|r(t)-s(t)|^2\,dt\right) \\ &= \exp\left(-\frac{1}{2\sigma^2}\int_0^{T_b}\left|r(t)-\sum_{k=1}^{K}\sqrt{\xi_k}\alpha_k b_k a_k(t)e^{j\phi_k}\right|^2 dt\right). \end{aligned} \quad (10.20)$$

Taking the natural logarithm of the likelihood function of Equation (10.20), the resulting so-called *Log-Likelihood Function* (LLF) can be written as :

$$\begin{aligned} \ln p(\boldsymbol{Z}|\boldsymbol{s}) &= -\frac{1}{2\sigma^2}\left\{\int_0^{T_b}|r(t)|^2\,dt + \int_0^{T_b}\left|\sum_{k=1}^{K}\sqrt{\xi_k}\alpha_k b_k a_k(t)e^{j\phi_k}\right|^2 dt\right. \\ &\quad \left. -2\Re\left[\int_0^{T_b} r(t)\sum_{k=1}^{K}\sqrt{\xi_k}\alpha_k b_k a_k(t)e^{-j\phi_k}dt\right]\right\}. \end{aligned} \quad (10.21)$$

The term $|r(t)|^2$ is common to all decision metrics, and hence it can be ignored during the optimization. Similarly, the constant term $1/2\sigma^2$ will not influence the maximization. Thus we can express the log-likelihood function of Equation (10.21) in the form of a correlation metric as [90] :

$$\begin{aligned} \Omega(\boldsymbol{b}) &= 2\Re\left[\int_0^{T_b} r(t)\sum_{k=1}^{K}\sqrt{\xi_k}\alpha_k b_k a_k(t)e^{-j\phi_k}dt\right] - \int_0^{T_b}\left|\sum_{k=1}^{K}\sqrt{\xi_k}\alpha_k b_k a_k(t)e^{j\phi_k}\right|^2 dt \\ &= 2\Re\left[\sum_{k=1}^{K}\sqrt{\xi_k}\alpha_k b_k e^{-j\phi_k}z_k\right] - \sum_{l=1}^{K}\sum_{k=1}^{K}\sqrt{\xi_l\xi_k}b_l b_k \alpha_l \alpha_k e^{j\phi_l}e^{-j\phi_k}\rho_{lk}, \end{aligned} \quad (10.22)$$

where z_k and ρ_{lk} are given by Equation (10.8) and Equation (10.9), respectively. Employing our discrete-time model highlighted in Section 10.3, the correlation metric of Equation (10.22) can be expressed in vector notation as [319] :

$$\Omega(\boldsymbol{b}) = 2\Re\left[\boldsymbol{b}^T\boldsymbol{\xi}\boldsymbol{C}^*\boldsymbol{Z}\right] - \boldsymbol{b}^T\boldsymbol{\xi}\boldsymbol{C}\boldsymbol{R}\boldsymbol{C}^*\boldsymbol{\xi}\boldsymbol{b}. \quad (10.23)$$

Hence the decision rule for the optimum CDMA multi-user detection scheme based on the maximum likelihood criterion is to choose the specific bit combination $\boldsymbol{b}$, which maximizes the correlation metric of Equation (10.23). Hence,

$$\hat{\boldsymbol{b}} = \arg\left\{\max_{\boldsymbol{b}}\left[\Omega\left(\boldsymbol{b}\right)\right]\right\}. \tag{10.24}$$

The maximization of Equation (10.23) is a combinatorial optimization problem, which requires an exhaustive search for each of the $m = 2^K$ combination of $\boldsymbol{b}$, in order to find the one that maximizes the correlation metric of Equation (10.23). Explicitly, since there are $m = 2^K$ possible combinations of $\boldsymbol{b}$, the optimum multi-user detection has a complexity that grows exponentially with the number of users K.

We mentioned in Chapter 9 that GAs have been known to solve combinatorial optimization problems efficiently in many other applications [9]. Hence, in this chapter, we will investigate the feasibility of invoking GAs in dealing with the CDMA multi-user detection optimization problem as governed by Equation (10.23).

10.5 Experimental Results

As we mentioned in Section 9.1 of Chapter 9, a GA's performance is dependent on numerous factors, such as the population size P, the choice of the selection method, the genetic operation employed, the specific parameter settings as well as the particular termination criterion used. In this section, we will attempt to find an appropriate GA set-up and parameter configurations that are best suited for our optimization problem.

Our objective function is defined by the correlation metric of Equation (10.23). Here, the legitimate solutions are the $m = 2^K$ possible combinations of the K-bit vector $\boldsymbol{b}$. Hence, each individual will take the form of a K-bit vector corresponding to the K users' bits during a single bit interval. We will denote the pth individual here as $\tilde{\boldsymbol{b}}_p(y) = \left[\tilde{b}_{p,1}(y), \ldots, \tilde{b}_{p,K}(y)\right]$, where y denotes the yth generation. Our goal is to find the specific individual that corresponds to the highest fitness value. However, we note that the fitness values corresponding to certain combinations of $\boldsymbol{b}$ evaluated from the correlation metric of Equation (10.23) will be negative. However, we mentioned in Section 9.4.2 that some selection schemes can only operate with the aid of positive fitness values. Hence, in order to ensure that the fitness values are positive for all combinations of $\boldsymbol{b}$, we modify the correlation metric of Equation (10.23) according to [311] :

$$\exp\left\{\Omega\left(\boldsymbol{b}\right)\right\} = \exp\left\{2\Re\left[\boldsymbol{b}^T\boldsymbol{\xi}\boldsymbol{C}^*\boldsymbol{Z}\right] - \boldsymbol{b}^T\boldsymbol{\xi}\boldsymbol{C}\boldsymbol{R}\boldsymbol{C}^*\boldsymbol{\xi}\boldsymbol{b}\right\}. \tag{10.25}$$

Our performance metric is the average Bit Error Probability (BEP) evaluated over the course of several generations. In the context of CDMA multi-user detection the three most important criteria to be satisfied by an efficient detection scheme are its BEP performance, its detection time as well as its computational complexity. The detection time of the GA is governed by the number of generations Y required, in order to obtain a reliable decision. We also mentioned in Section 9.4.6 that the computational complexity of the GA, in the context of the total number of objective function evaluations, is related to $P \times Y$. On the other

Parameter	Value
Spreading factor N_c	31
Modulation mode	BPSK
Number of CDMA users, K	10 (20 for Figure 10.4)
SNR per bit ξ_k/N_0	9 dB for $k = 1, \ldots, K$

Table 10.1: Simulation parameters for the experiments of Figures 10.2-10.9.

hand, it is well known that the convergence accuracy of the GA is mainly determined by the population size P, as alluded to in Section 9.1. Hence, in this section the purpose of our study is to find GA configurations that achieve a satisfactory BEP performance at the expense of an acceptable computational complexity within a reasonable time. Since our GA-assisted multi-user detector is based on optimizing the modified correlation metric of Equation (10.25), the computational complexity is deemed to be acceptable, if there is a significant amount of reduction in comparison to the optimum multi-user detector, which requires $m = 2^K$ objective function evaluations, in order to reach a decision, as highlighted in Section 10.4. In order to evaluate the average BEP performance of the GA-assisted multi-user detectors, randomly generated signature sequences will be used in our simulations. The simulation parameters used for our investigations in this section are presented in Table 10.1 and the following assumptions are stipulated :

- We will assume that perfect power control is invoked by all users such that, on average, their signals arrive at the receiver with the same power.
- Initially only the AWGN channel is invoked, such that $\alpha_k = 1.0$ and $\phi_k = 0$ for $k = 1, \ldots, K$, i.e. there is no fading.

10.5.1 Effects of the Population Size

Let us commence our experiments by investigating the effects of the population size P on the convergence rate of the GA. Since at this moment we have no knowledge of which configuration of the GA is best suited for our optimization problem, we will adopt the most commonly used GA configuration found in the literature for our initial simulations. This configuration is tabulated in Table 10.2, which follows the flowchart of Figure 9.1. Basically, the individuals, which represent the candidate solutions of $\boldsymbol{b}$, are randomly created during the initialization phase of the GA. Upon evaluating their associated fitness values based on the modified objective function of Equation (10.25), pairs of individuals found in the mating pool are selected for cross-over and mutation operations, in order to produce offspring. The associated fitness values of these offspring are then evaluated and these offspring will form the new population of the next generation. The processes of selection, cross-over, mutation and evaluation are repeated for a total of $Y - 1$ generations.[2] Based on this configuration, the BEP performance of the GA-assisted multi-user detector was evaluated and the results, which showed the achievable BEP at the end of each generation, are displayed in Figure 10.2. Note that the BEP at each generation, with the exception of the 0th generation, is derived by identifying

[2]The 0th generation only consists of initialization and evaluation.

Set-up/Parameter	**Method/Value**
Individual initialization method	Random
Selection method	Fitness-Proportionate
Cross-over operation	Single-point
Mutation operation	Standard binary mutation
Elitism	No
Incest Prevention	No
Population size P	Given in Figure 10.2
Mating pool size T	Population size P
Probability of mutation p_m	0.01

Table 10.2: Configuration of the GA used to obtain the results of Figure 10.2. Explicit description of the fitness-proportionate selection scheme and the single-point cross-over operation can be found in Section 9.4.2 and Section 9.4.3, respectively.

the offspring associated with the highest fitness value among all the offspring created at that generation. The BEP at the 0th generation is derived by identifying the specific individual that exhibits the highest fitness value after the random initialization. It is seen from the figure that the BEP performance of the GA-assisted multi-user detector improved by increasing the population size. However, the computational cost also increases as a function of the population size, as highlighted in Section 9.4.6. In order to maintain a moderate computational complexity, we shall adopt a fixed population size of $P = 30$ for all our simulations in this section. Upon closer inspection of Figure 10.2, we will notice that the BEP performance of the GA-assisted multi-user detector based on the configuration of Table 10.2 is far from promising. Furthermore, the convergence rate of the GA is very slow. Let us now study whether the performance can be improved by varying some of the GA parameters and the GA configuration, commencing with the probability of mutation.

10.5.2 Effects of the Probability of Mutation

As we mentioned in Section 9.4.4, the rate of mutation plays an important role in determining the quality of convergence of a GA. A high probability of mutation p_m may disrupt schemata of potentially high fitness values and hence may lead to suboptimal solutions, while a low probability of mutation may result in premature convergence due to the lack of diversity in the population. This assertion is supported by Figure 10.3, which shows the achievable BEP performance of the GA-assisted multi-user detector over $Y = 20$ generations for various values of p_m. The configuration of the GA implemented for this simulation study is listed in Table 10.3, which is similar to the one given in Table 10.2 of Section 10.5.1. Considering the results shown in Figure 10.3, we will immediately notice that the BEP performance has improved significantly over the rather poor results obtained in the previous section for $p_m > 0.01$. Furthermore, according to Figure 10.3, $p_m = 0.1$ appears to give the best performance. All the other values of p_m have a slower convergence rate, leading to solutions far from the optimal one. On the other hand, the value of p_m is very much dependent on the length of the individual, as exemplified in Figure 10.4 for $K = 20$ users. Here, each individual

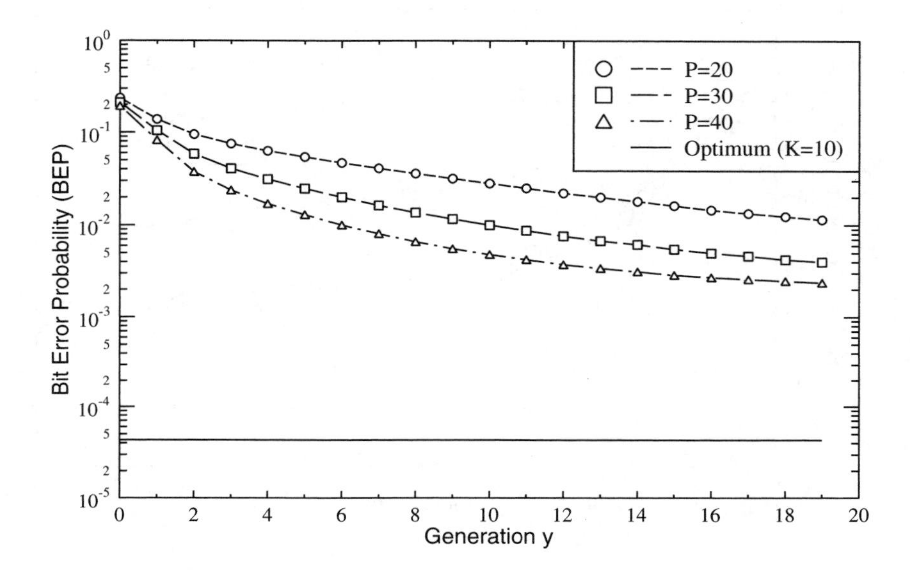

Figure 10.2: The bit error probability performance with respect to the number of generations of the GA-assisted multi-user detector for various population sizes and $K = 10$ users. The configuration of the GA used to obtain these results is tabulated in Table 10.2, while the simulation parameters are listed in Table 10.1.

Set-up/Parameter	Method/Value
Individual initialization method	Random
Selection method	Fitness-Proportionate
Cross-over operation	Single-point
Mutation operation	Standard binary mutation
Elitism	No
Incest Prevention	No
Population size P	30
Mating pool size T	Population size P
Probability of mutation p_m	Given in Figure 10.3 and Figure 10.4.

Table 10.3: Configuration of the GA used to obtain the results of Figures 10.3 and 10.4. Explicit description of the fitness-proportionate selection scheme and the single-point cross-over operation can be found in Section 9.4.2 and Section 9.4.3, respectively.

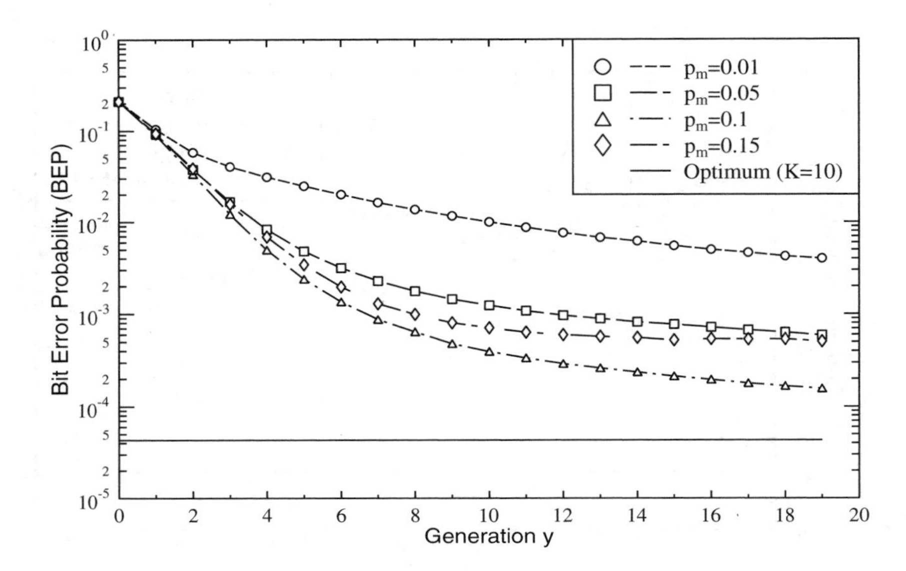

Figure 10.3: The bit error probability performance with respect to the number of generations of the GA-assisted multi-user detector for various probability of mutation values p_m and for $K = 10$ users. The configuration of the GA is specified in Table 10.3, while the simulation parameters are listed in Table 10.1.

will consist of 20-bit variables to be optimized. In this case, $p_m = 0.07$ gives the best performance.[3] Since our simulations performed in this chapter are based on a CDMA system supporting $K = 10 - 20$ users, we will adopt a probability of mutation $p_m = 0.1$ for all our subsequent simulations, since this value was shown in Figures 10.3 and 10.4 to give a good BEP performance for this user population range. However, further investigations concerning the suitable value of p_m must be performed for a higher number of users. Let us now consider whether we can further improve the achievable BEP performance by using different cross-over operations.

10.5.3 Effects of the Choice of Cross-over Operation

In this section, we will investigate, whether the choice of the cross-over operation will have an effect on the convergence rate of the GA. Three types of cross-over operations are investigated, namely the single-point cross-over, the double-point cross-over and the uniform cross-over, which were highlighted in Section 9.4.3. The configuration of the GA is characterized by Table 10.4 and the associated results are shown in Figure 10.5. Judging from the

[3]The poor BEP shown in Figure 10.4 is due to the inadequate population size in handling a sizeable search space for $K = 20$.

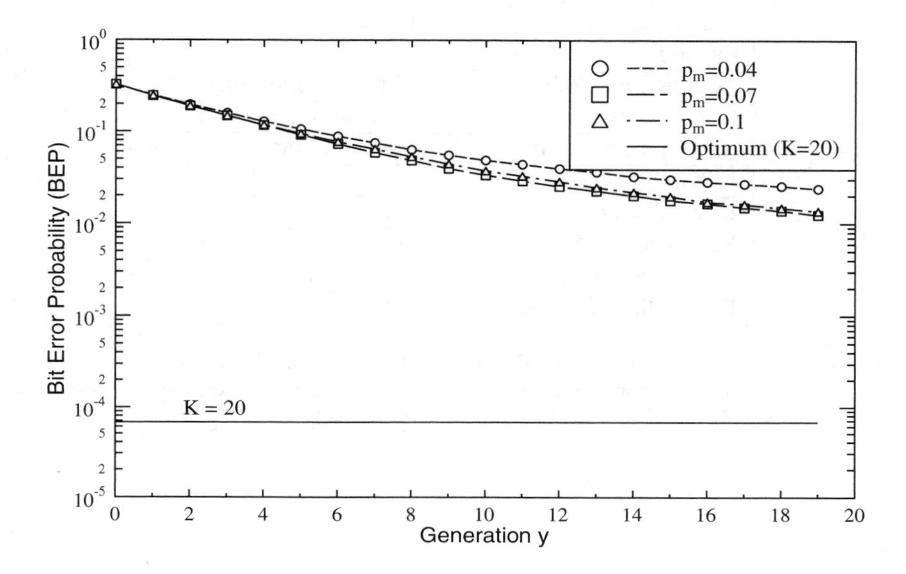

Figure 10.4: The bit error probability performance with respect to the number of generations of the GA-assisted multi-user detector for various probability of mutation value p_m and for $\boldsymbol{K} = \mathbf{20}$ users. The configuration of the GA is specified in Table 10.3, while the simulation parameters are listed in Table 10.1.

results displayed in Figure 10.5, there is no significant performance disparity among the different cross-over operations. Nonetheless, the GA employing the uniform cross-over can be seen to exhibit a slightly faster convergence rate, than that using the single-point and double-point cross-over. This may be due to the fact that for the uniform cross-over operation, every bit of the individual has an equal probability of being exchanged, unlike in the single-point cross-over or the double-point cross-over, where the leftmost and the rightmost bits have a lower probability of being exchanged. Hence, we shall be adopting the uniform cross-over operation for all our subsequent simulations.

10.5.4 Effects of Incest Prevention and Elitism

Let us now investigate the effects of invoking the incest prevention and the elitism strategy, as featured in Section 9.4. In the case of the incest prevention strategy, we will ensure that the individuals in the mating pool are not identical. Hence, the mating pool size $T \leq P$ will not be fixed, because it depends on the number of non-identical individuals in the population. As for the elitism strategy, we will only replace the offspring having the lowest fitness value in the new population with the individual corresponding to the highest fitness value in the old population. The configuration of the GA for this investigation is specified by Table 10.5

Set-up/Parameter	Method/Value
Individual initialization method	Random
Selection method	Fitness-Proportionate
Cross-over operation	Given in Figure 10.5
Mutation operation	Standard binary mutation
Elitism	No
Incest Prevention	No
Population size P	30
Mating pool size T	Population size P
Probability of mutation p_m	0.1

Table 10.4: Configuration of the GA used to obtain the results of Figure 10.5. Explicit description of the fitness-proportionate selection scheme and the various cross-over operations can be found in Section 9.4.2 and Section 9.4.3, respectively.

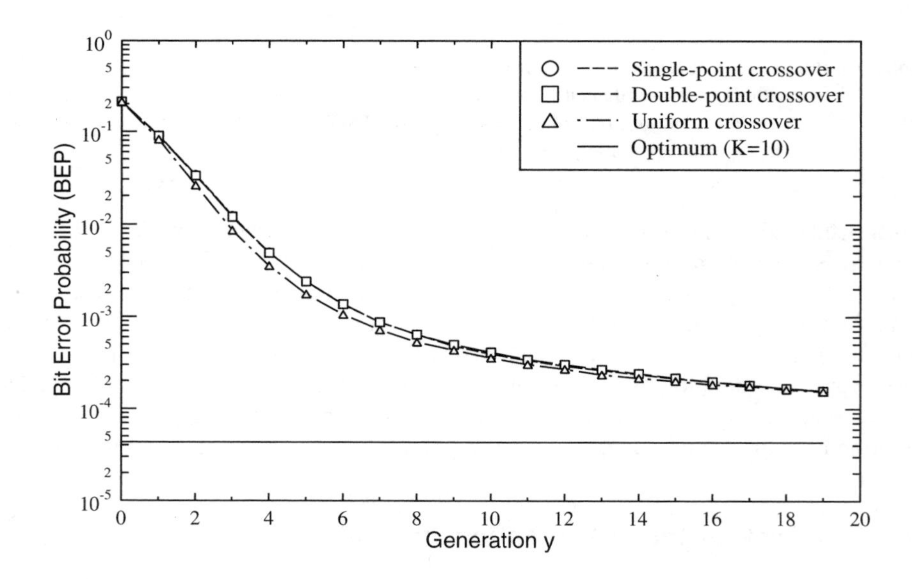

Figure 10.5: The bit error probability performance with respect to the number of generations of the GA-assisted multi-user detector employing the single-point cross-over, double-point cross-over and the uniform cross-over for $K = 10$ users. The configuration of the GA is specified in Table 10.4, while the simulation parameters are listed in Table 10.1.

Set-up/Parameter	Method/Value
Individual initialization method	Random
Selection method	Fitness-Proportionate
Cross-over operation	Uniform
Mutation operation	Standard binary mutation
Elitism	Given in Figure 10.6
Incest Prevention	Given in Figure 10.6
Population size P	30
Mating pool size T	– Population size P if incest prevention is not invoked – $\leq P$ if incest prevention is invoked
Probability of mutation p_m	0.1

Table 10.5: Configuration of the GA used to obtain the results of Figure 10.6. Explicit description of the fitness-proportionate selection scheme and the incest prevention strategy can be found in Section 9.4.2 while the uniform cross-over operation and the elitism strategy can be found in Section 9.4.3 and Section 9.4.5, respectively.

and the associated results are shown in Figure 10.6. A welcome improvement that can be gleaned from Figure 10.6 is that the GA-assisted multi-user detector has finally managed to achieve the optimum performance for $K = 10$, based on the configuration of Table 10.5 in conjunction with the incest prevention and elitism strategies. Furthermore, we can see that the optimum performance is attained only if both strategies are invoked. This can be explained as follows. First, the incest prevention strategy will always ensure that a high diversity of individuals is maintained in the population, since only non-identical individuals are allowed to mate. Hence, the offspring that are produced by the cross-over and mutation operations will have a high probability that they are not identical to their parents. This will ensure that new areas in the search space will be explored, which is always a good trait from an optimization point of view. On the other hand, this will also obliterate the parents associated with high fitness values, since the offspring will constitute the new population. This is undesirable especially, if the parent is actually the optimum solution. Hence the elitism strategy can be invoked in order to counteract this effect. Since in our optimization problem we are interested in finding only one specific individual that gives the highest fitness value and not a set of likely individuals, the elitism strategy is required to keep track of the individual having the highest fitness value found during the course of evolution. Hence by combining these two strategies, a fast convergence rate and a good performance can be achieved. Now that we know that the GA-assisted multi-user detector is capable of attaining the optimum performance within 20 generations, as shown in Figure 10.6, let us now consider whether the detector is capable of achieving this level of performance at a faster convergence rate by invoking various selection schemes.

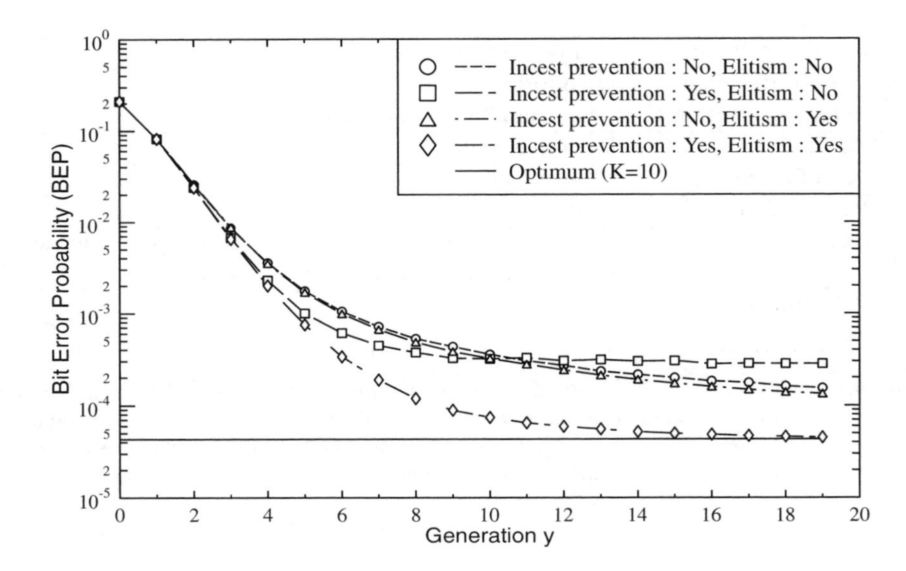

Figure 10.6: The bit error probability performance with respect to the number of generations of the GA-assisted multi-user detector employing the incest prevention strategy and/or elitism strategy, as featured in Section 9.4. The configuration of the GA is specified in Table 10.5, while the simulation parameters are listed in Table 10.1 for $K = 10$ users.

10.5.5 Effects of the Choice of Selection Schemes

In this section, we will attempt to identify the specific selection scheme for our GA-assisted multi-user detector that is capable of offering a fast convergence rate, while maintaining the same level of BEP performance that was attained in Figure 10.6. The selection schemes that were reviewed in Section 9.4.2, namely the fitness-proportionate selection, the sigma scaling selection, the linear ranking selection and the tournament selection, will be investigated. The configuration of the GA is listed in Table 10.6. For the linear ranking selection scheme, we set η^+ and η^- in Equation (9.11) to 1.9 and 0.1, respectively so as to place a higher emphasis on the individuals exhibiting higher fitness values. As for the tournament selection scheme, $t = 5$ individuals are selected from the population randomly with equal probability and the individual that corresponds to the highest fitness value within this group of t individuals will be chosen as the parent. Finally, for sigma scaling selection, if the probability of selection p_i corresponding to the ith individual is a negative value when calculated according to Equation (9.10), then we will set this p_i value to 0.0 and discard the associated individual from the selection process. The BEP results are shown in Figure 10.7.

As we can see, GAs utilizing the fitness-proportionate selection scheme gave the best performance. On the other hand, GAs using either the sigma scaling selection scheme or the

Set-up/Parameter	Method/Value
Individual initialization method	Random
Selection method	Given in Figure 10.7
Cross-over operation	Uniform cross-over
Mutation operation	Standard binary mutation
Elitism	Yes
Incest Prevention	Yes
Population size P	30
Mating pool size T	$T \leq P$ depending on the total number of non-identical individuals
Probability of mutation p_m	0.1

Table 10.6: Configuration of the GA used to obtain the results of Figure 10.7. Explicit description of the various selection schemes and the uniform cross-over operation can be found in Section 9.4.2 and Section 9.4.3, respectively.

linear ranking selection scheme exhibited a slow convergence rate. A plausible explanation is due to the fact that the mating pool size T spanned over all non-identical individuals. Hence the fitness value variance of the mating pool was high. As a result, individuals having high fitness values are not given sufficient priority to be selected as a parent in the case of the sigma scaling selection. Similarly, because of the linearity of Equation (9.11), the higher-rank individuals are not assigned a high probability of selection.

A feasible way of overcoming these shortcomings is to reduce the size T of the mating pool, such that $T \ll P$. This implies that only the $T \ll P$ number of non-identical individuals that are associated with the highest fitness values in the current population will be placed in the mating pool. If the number of non-identical individuals in the population happens to be less than T, then the value of T is set to be equivalent to the number of available non-identical individuals, in order to prevent incest mating. We set $T = 10$ and when using the GA configuration given by Table 10.7, the corresponding simulation results are shown in Figure 10.8. Now we can see that the achievable BEP performance of the GAs employing either the sigma scaling selection scheme or the linear ranking selection scheme has improved significantly. In particular, the GA-assisted MUD employing the sigma scaling selection scheme has an almost identical performance to that using the fitness-proportionate selection scheme. From these results we conclude that the mating pool size T plays a significant part in determining the convergence rate of the GA using a particular type of selection scheme. More specifically, for the sigma scaling selection and the linear ranking selection, the value of T must be set appropriately. We also have to determine the best value of t for the tournament selection scheme. On the other hand, the results obtained in Figure 10.8 for the fitness-proportionate selection scheme are similar to those shown in Figure 10.7, where no specific mating pool size T constraint was imposed, using T equal to the number of non-identical individuals in the population during the particular generation. Furthermore, the fitness-proportionate selection scheme does not involve any external parameters for it to work and judging from Figure 10.7 and Figure 10.8, GAs utilizing the fitness-proportionate selection scheme gave the best performance from the range of selection schemes considered. Hence, in order to re-

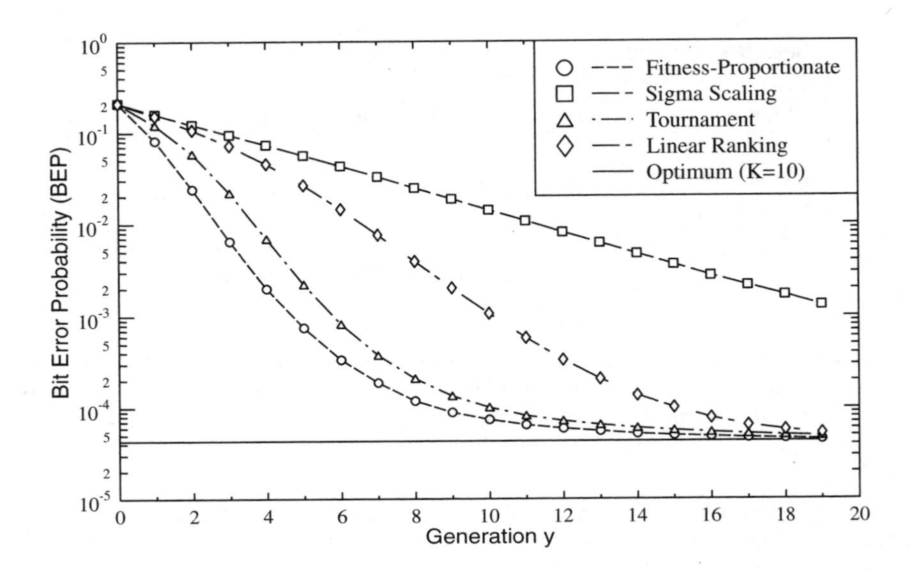

Figure 10.7: The bit error probability performance with respect to the number of generations of the GA-assisted multi-user detector employing various selection schemes. The configuration of the GA is specified in Table 10.6, while the simulation parameters are listed in Table 10.1 for $K = 10$ users.

duce the number of parameters to be optimized for the GAs to perform reliably, we will only consider the fitness-proportionate selection scheme hereafter. In most cases, the mating pool size T will also be set according to the number of non-identical individuals in the population. However, as we will see in our further discussion, there are certain situations where a specific value of T must be set, in particular, when the population contains many non-identical individuals.

10.5.6 Effects of a Biased Generated Population

In Section 9.1, we mentioned that instead of randomly creating the initial population of individuals at the commencement of a GA-assisted search, we can invoke any useful information that is available to create the initial population of individuals, in order to create our search at the beginning. In our case, we can use the hard decisions offered by the matched filter outputs $\boldsymbol{Z}$ of Equation (10.16), in order to create our search. We shall denote these hard decisions here as :

$$\hat{\boldsymbol{b}}_{MF} = \left[\hat{b}_{1,MF}, \hat{b}_{2,MF}, \dots, \hat{b}_{K,MF}\right] \tag{10.26}$$

Set-up/Parameter	Method/Value
Individual initialization method	Random
Selection method	Given in Figure 10.8
Cross-over operation	Uniform cross-over
Mutation operation	Standard binary mutation
Elitism	Yes
Incest Prevention	Yes
Population size P	30
Mating pool size T	$T \leq 10$ depending on the number of non-identical individuals
Probability of mutation p_m	0.1

Table 10.7: Configuration of the GA used to obtain the results of Figure 10.8. Explicit description of the various selection schemes and the uniform cross-over operation can be found in Section 9.4.2 and Section 9.4.3, respectively.

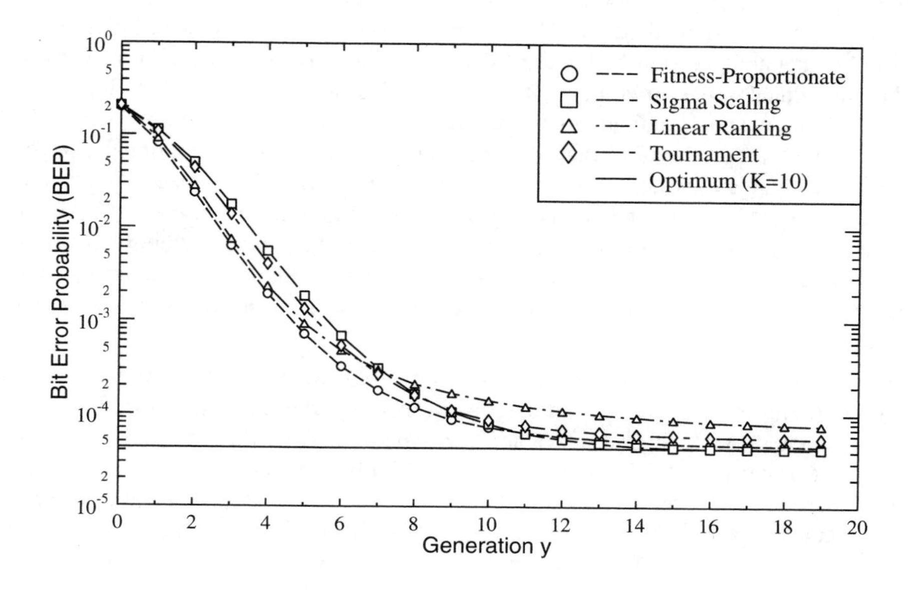

Figure 10.8: The bit error probability performance with respect to the number of generations of the GA-assisted multi-user detector employing various selection schemes. The configuration of the GA is specified in Table 10.7, while the simulation parameters are listed in Table 10.1.

Set-up/Parameter	Method/Value
Individual initialization method	Given in Figure 10.9
Selection method	Fitness-proportionate
Cross-over operation	Uniform cross-over
Mutation operation	Standard binary mutation
Elitism	Yes
Incest Prevention	Yes
Population size P	30
Mating pool size T	$T \leq P$ depending on the number of non-identical individuals
Probability of mutation p_m	0.1

Table 10.8: Configuration of the GA used to obtain the results of Figure 10.9. Explicit description of the fitness-proportionate selection scheme and the uniform cross-over operation can be found in Section 9.4.2 and Section 9.4.3, respectively.

where $\hat{b}_{k,MF}$ for $k = 1, \ldots, K$ is given by Equation (10.11). Two methods of biasing are proposed for our investigations. First, we will assign the hard decisions of Equation (10.26) to only one individual. The remaining $P - 1$ individuals will be randomly generated. This will ensure that a high diversity of individuals are present in the population at the beginning. We shall refer to this method as M1. For our second method, we will assign a different randomly 'mutated' version of the hard decision vector $\hat{\boldsymbol{b}}_{MF}$ of Equation (10.26) to each of the individuals in the initial population. We shall adopt the same probability of mutation as p_m. In this way, the individuals in the initial population will be almost identical. Note that we cannot assign the same hard decision vector $\hat{\boldsymbol{b}}_{MF}$ to all the individuals, since incest prevention is invoked, which will not allow identical individuals to mate. We shall refer to this method as M2. Using the GA configuration listed in Table 10.8, the simulation results are shown in Figure 10.9. As the figure suggests, method M2 gives a better performance in terms of a faster convergence rate due to a good initial population of individuals. This fact conforms to the results obtained in [311]. Hence we shall adopt method M2 of initializing the initial population.

Based on the results gathered from our simulations, the final GA configuration that we will be utilizing for most of our simulations in this dissertation, unless specified otherwise in the associated plots, is given in Table 10.9. The associated flowchart is depicted in Figure 10.10. Further useful information can be gleaned by comparing our GA configuration to the GA-assisted multi-user detectors of [311–314,322] in the literature, as given by Table 9.3. Notice that a low probability of mutation p_m as well as no incest prevention strategy were invoked in these proposals [311–314]. On the other hand, according to our results summarized in Section 10.5.2 and Section 10.5.4, the effects of the value of p_m and those of the incest prevention strategy can have a significant impact on the convergence rate and hence also on the BEP performance of the GA-assisted multi-user detector. As shown in Figure 10.9, our proposed GA-assisted multi-user detector is capable of reaching a near-optimal BEP performance within $Y = 10$ generations with the aid of a population size of $P = 30$ for $K = 10$ users over an AWGN channel at an SNR of 9 dB. This constitutes a total of $P \times Y = 300$

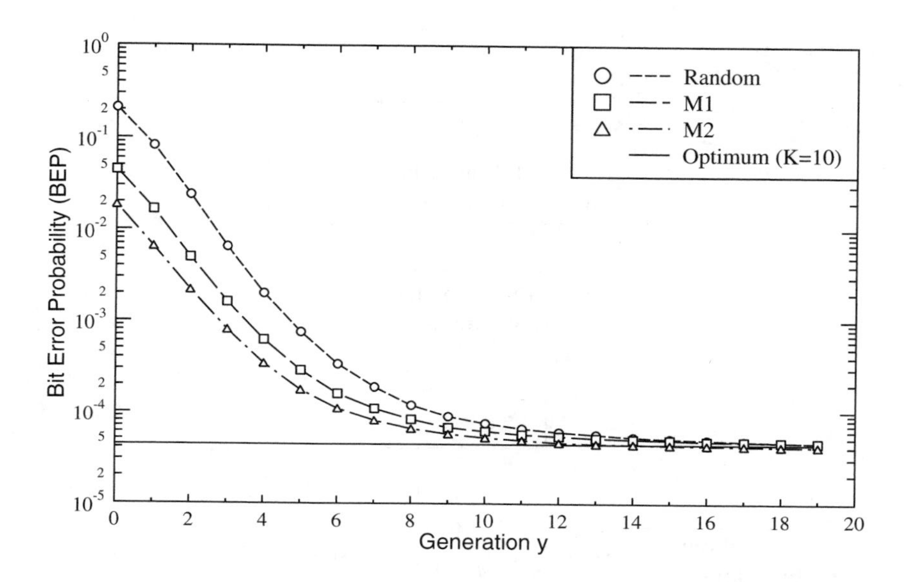

Figure 10.9: The bit error probability performance with respect to the number of generations of the GA-assisted multi-user detector employing different methods of initialising the individuals at the beginning. The configuration of the GA is specified in Table 10.8, while the simulation parameters are listed in Table 10.1 for $K = 10$ users.

number of correlation metric evaluations according to Equation (10.25). In fact, as we mentioned in Section 9.4.6, this number was derived based on the fact that the fitness value is calculated for every individual in the population at every generation. However, in reality, certain individuals will reappear over the course of the evolution. Hence, the fitness values of these individuals need not be recalculated, if they are stored in the memory. Based on our simulations for $P = 30$, $Y = 10$ and $K = 10$, we found that the average number of unique K-bit combinations that were evaluated by the GA for a single bit interval was ≈ 89. Comparing this number to that of the optimum multi-user detector, which requires $2^{10} = 1024$ correlation metric evaluations for every $\boldsymbol{b}$ combination, our proposed GA-assisted multi-user detector is capable of attaining a significantly reduced computational complexity and yet delivering a near-optimum BEP performance up to a specific SNR value. Hence the implementation of our proposed GA-assisted multi-user detector is feasible in practical terms and offers an alternative to the implementation of the optimum multi-user detector.

Finally, it should be stressed that we have only explored a fraction of the numerous possible configurations of GAs, a fact noted at the beginning of Chapter 9. Furthermore, the settings of certain parameters were not investigated extensively, such as the probability of mutation p_m, the ideal mutation pool size T and the population size P. These values will depend very much on the number of users K and the desired quality of detection. There are

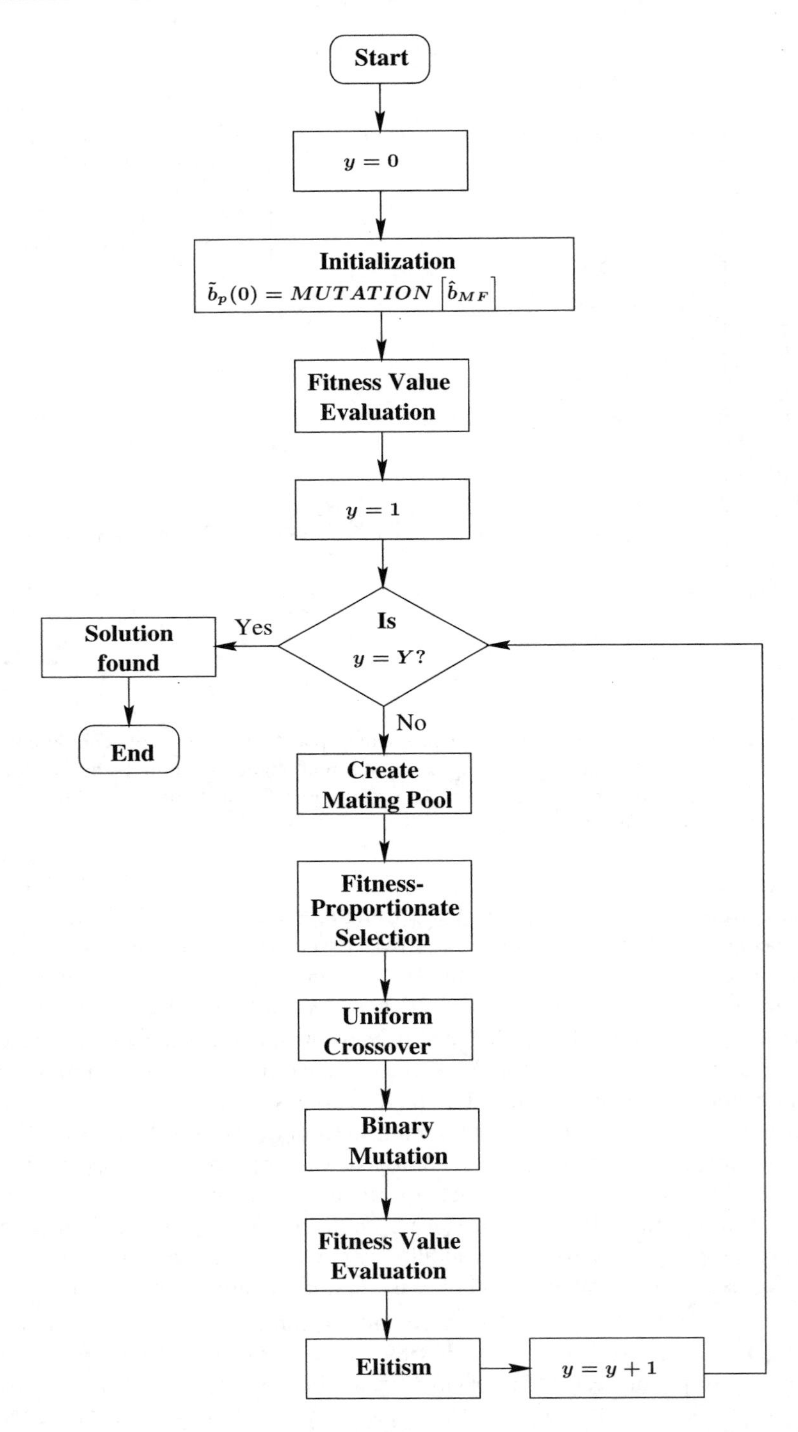

Figure 10.10: A flowchart depicting the structure of the genetic algorithm adopted for our GA-assisted multi-user detection technique, which is a specific version of Figure 9.1.

Set-up/Parameter	Method/Value
Individual initialization method	Mutation of $\hat{\boldsymbol{b}}_{MF}$ of Equation (10.26)
Selection method	Fitness-proportionate
Cross-over operation	Uniform cross-over
Mutation operation	Standard binary mutation
Elitism	Yes
Incest Prevention	Yes
Population size P	30
Mating pool size T	$T \leq P$ depending on the number of non-identical individuals
Probability of mutation p_m	0.1
Termination generation Y	10

Table 10.9: Configuration of the GA that will be used in this chapter hereafter, unless otherwise specified.

also many variants of GAs that have not been studied or indeed even highlighted in this chapter. Hence we made no claims about the optimality of the GAs used in this chapter for the application in CDMA multi-user detection. Hence it is possible that different GA variants, which were not covered in this chapter, or a different set of parameters may give an even better performance. However, we will show with the aid of the following simulation results of this chapter as well as in subsequent chapters that the GA configuration of Table 10.9 together with the set of GA parameters that we have adopted is capable of offering a satisfactory trade-off between computational complexity, detection delay and an acceptable BEP performance. Using the GA configuration of Table 10.9, let us now consider the BEP performance of the GA-assisted CDMA multi-user detector in both an AWGN channel as well as in a non-dispersive-Rayleigh fading channel.

10.6 Simulation Results

10.6.1 AWGN Channel

All the results in this section were based on evaluating the BEP performance of a bit-synchronous K-user CDMA system over an AWGN channel. The signature sequences were randomly generated 31-chip per bit sequences and the transmit bit energy ξ_k was assumed to be equal for all users.

Figure 10.11 shows the average BEP as a function of the SNR per bit for the GA-assisted multi-user detector for various population sizes P and different number of generations Y for $K = 10$. The values of P and Y are assigned such that the maximum number of times the objective function of Equation (10.25) was evaluated upon detecting the bit vector $\hat{\boldsymbol{b}}$ during a bit interval is approximately 300. The optimum performance of the multi-user detector utilizing an exhaustive search for $K = 10$ is also shown. In this case, the optimum multi-user detector has to compute the objective function of Equation (10.25) $2^{10} = 1024$

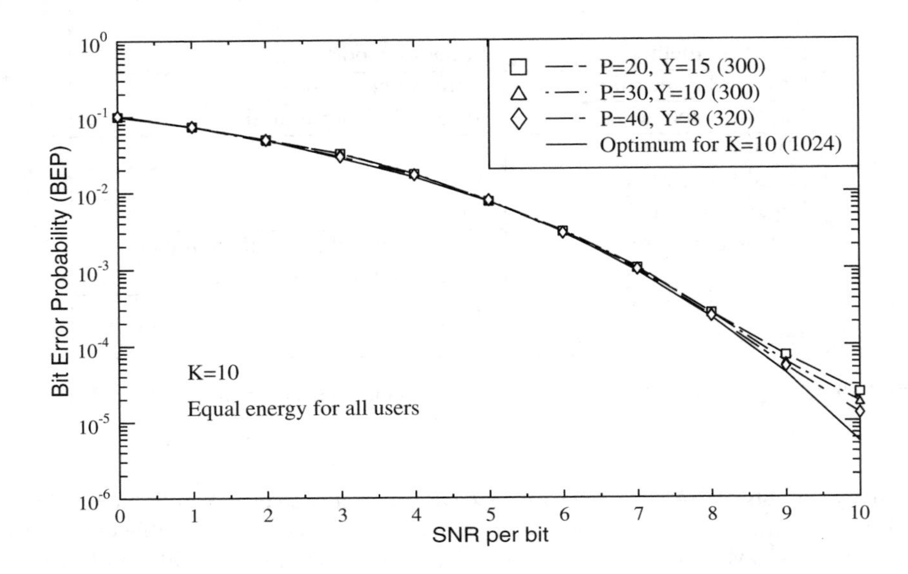

Figure 10.11: The bit error probability performance of the GA-assisted multi-user detector as a function of the SNR per bit with population size of $P = 20, 30, 40$ using binary random signature sequences of length $N_c = 31$ for $K = 10$ users. The GA configuration used is listed in Table 10.9. The values in round brackets in the legend denote the maximum number of times the objective function of Equation (10.25) was evaluated upon detecting the bit vector $\hat{\boldsymbol{b}}$ during a bit interval.

times, which corresponds to every possible combination of $\boldsymbol{b}$. Upon observing Figure 10.11, we notice that the GA-assisted multi-user detector is capable of achieving a near-optimum performance up to an SNR of 8 dB at a lower computational complexity than that of the optimum multi-user detector. Specifically, the minimum reduction in the computational complexity offered by the GA-assisted multi-user detector over its optimum counterpart is $[(1024 - 320)/1024] \times 100 = 68.75\%$. As mentioned previously in Section 10.5, this reduction in the computational complexity is expected to be higher if the repeated fitness value evaluation of the same individual is circumvented. From Figure 10.11, we can see that the BEP performance curves began to flatten at an SNR of 10 dB. This is due to the inadequate population size and/or number of generations required for the GA to converge to the optimum performance. Recall that in Section 9.1 we stated that the GAs are not guaranteed to find the optimal solution, unless a sufficiently large population size and an appropriate number of generations are guaranteed. On the other hand, it is not a prerequisite that the optimum performance must be achieved for every SNR value. The integrity of detection required is usually dependent on the type of service the detection scheme is intended for. For example, a speech signal may tolerate a relatively high BEP of 10^{-3} but no latency, while a data sig-

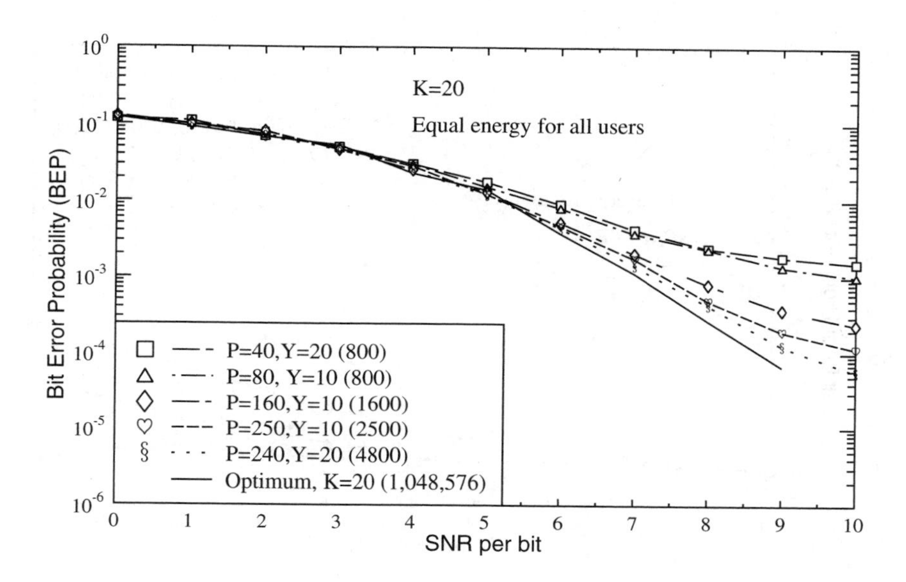

Figure 10.12: The bit error probability performance of the GA-assisted multi-user detector as a function of the SNR per bit with various population sizes P and number of generations Y for $\boldsymbol{K = 20}$ using binary random signature sequences of length $N_c = 31$. The GA configuration used is listed in Table 10.9. The values in round brackets in the legend denote the number of times the objective function of Equation (10.25) was evaluated upon detecting the bit vector $\hat{\boldsymbol{b}}$ during a bit interval.

nal may require a BEP performance below 10^{-6}, but it can tolerate a higher detection delay. Hence we can immediately see that the GA-assisted multi-user detector is capable of offering this trade-off by simply adjusting the values of P and Y.

Let us now consider what happens, if the number of users K is increased to 20. In the context of the optimum multi-user detector, an exhaustive search would required a staggering $2^{20} = 1,048,576$ number of objective function evaluations, in order to obtain the optimum solution. The BEP performance of the GA-assisted multi-user detector for various values of P and Y is shown in Figure 10.12. Again, we see that by simply expanding the population size P and by extending the number of generations Y, the near-optimal BEP performance of the GA-assisted multi-user detector found for $K = 10$ can be maintained, when the number of users is increased to $K = 20$. While this will increase the associated computational complexity of the GA-assisted multi-user detector, the maximum number of objective function evaluations given by $P \times Y$ is still significantly lower than that required by the optimum detector.

Figure 10.13 shows the average BEP performance of the GA-assisted multi-user detector as a function of the number of users K for an SNR value of 7 dB. As we can see, for a given

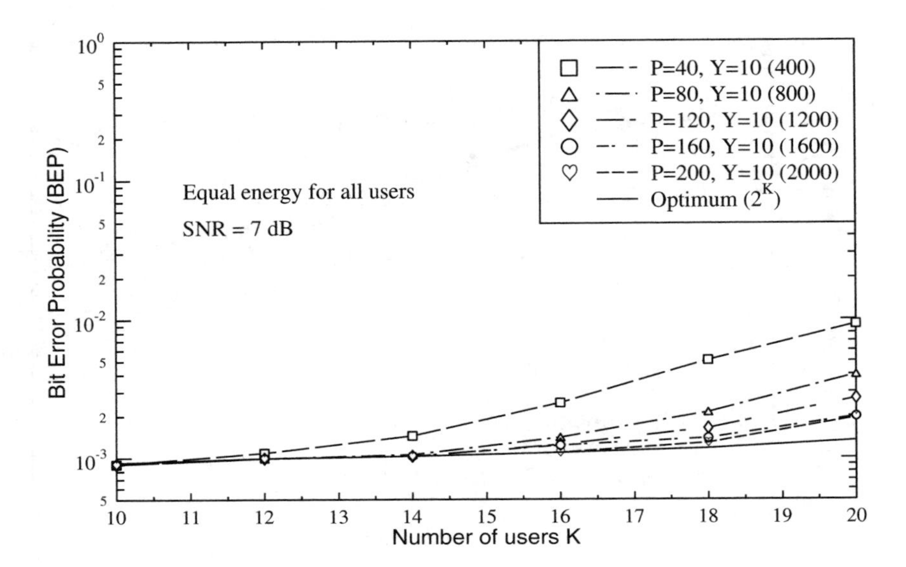

Figure 10.13: The bit error probability performance of the GA-assisted multi-user detector as a function of the number of K users with population size of $P = 40, 80, 120, 160, 200$ and $Y = 10$ using binary random signature sequences of length $N_c = 31$. The GA configuration used is listed in Table 10.9. The values in round brackets in the legend denotes the number of times the objective function of Equation (10.25) was evaluated upon detecting the K-bit vector $\hat{\boldsymbol{b}}$ during a bit interval.

number of generations Y, the population size P must be increased, in order to maintain the same level of BEP performance as K is increased. However, the required increase in the population size P is not exponentially proportional to the number of users. Furthermore, there is a gradual degradation in the BEP performance for a given population size P and termination generation Y, as the number of users K increased.

10.6.2 Single-path Rayleigh Fading Channel

In this section, the BEP performance of the GA-assisted multi-user detector is evaluated for a bit-synchronous K-user CDMA system over a non-dispersive Rayleigh fading channel. The signature sequences were randomly generated $N_c = 31$-chip per bit sequences and the users' CIR coefficients $\alpha_k e^{j\phi_k}$ were assumed to be known at the receiver. With the exception of Figure 10.17, the transmit bit energy ξ_k was assumed to be equal for all users.

Figure 10.14 shows the BEP performance of the GA-assisted multi-user detector for different number of generations Y and for different population sizes P. The optimum performance for $K = 10$ users was also plotted for comparison. As can be seen from the figure,

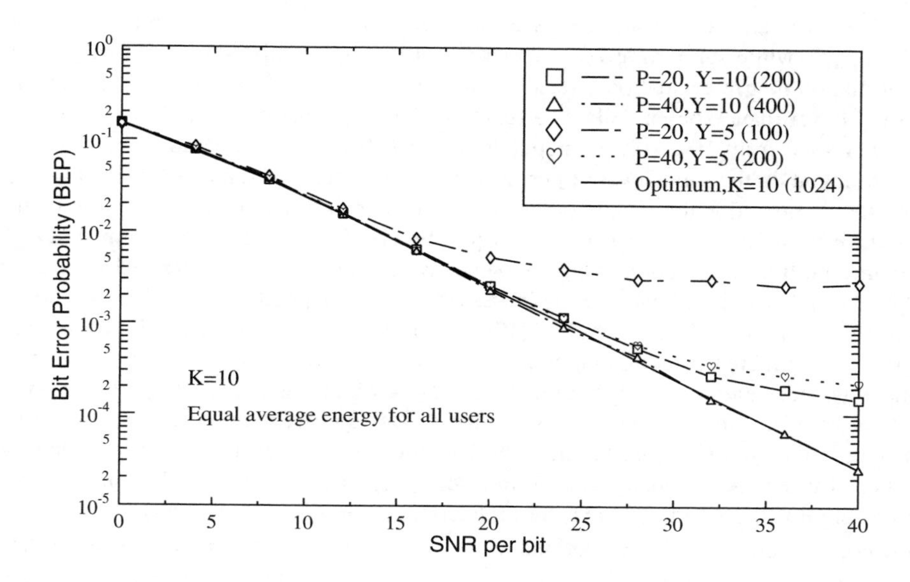

Figure 10.14: The bit error probability performance of the GA-assisted multi-user detector as a function of the SNR per bit for $K = 10$ users with various combinations of P and Y in conjunction with perfect channel estimation using binary random signature sequences of length $N_c = 31$. The GA configuration used is listed in Table 10.9. The values in round brackets in the legend denote the maximum number of times the objective function of Equation (10.25) was evaluated upon detecting the bit vector $\hat{\boldsymbol{b}}$ during a bit interval.

the combination of $P = 40$ and $Y = 10$ – which constitutes a maximum of $40 \times 10 = 400$ number of objective function evaluations according to Equation (10.23) – was capable of achieving a near-optimal BEP performance. For SNR values beyond 40 dB, the system exhibited an error floor due to the performance limitations of the GA in conjunction with the given P and Y values studied. At lower values of Y and P, the error floor occurred at a lower SNR value. For instance, at $Y = 10$ and $P = 20$, which requires a maximum of 200 number of objective function evaluations according to Equation (10.23), the error floor occurred at an SNR value of about 32 dB, while for SNR values up to 24 dB, the detector exhibited near-optimum BEP performance. Hence, again it was shown here that the GA-assisted multi-user detector was capable of offering a trade-off between computational complexity and the optimum BEP performance.

In order to show that the computational complexity of the GA is not exponentially dependent on the number of users K, the BEP performance was evaluated in Figure 10.15 for various number of users, employing $P = 40, 80, 120, 160, 200$ in conjunction with $Y = 10$. The results are shown in Figure 10.15. At $P = 40$ and $Y = 10$, we can see that the BEP performance gradually degrades upon increasing the number of users, due to the limited population size P, which was too small for adequately exploring a significantly larger search space. As the population size P is increased, the BEP improves. For a population size of $P = 160$, we can see that the GA-assisted detector is capable of attaining a near-optimal performance, while supporting $K = 20$ users. More importantly, we noted that the number of correlation metric evaluations, seen within the brackets in the legend of Figure 10.15 increases slower than exponentially as a function of the number of users. For example, when K is increased from 10 to 16, the population size P has to be increased from 40 to 120, in order to maintain the same level of performance. This constituted a factor of $1200/400 = 3$ increased computational complexity, when K was increased from 10 to 16, while maintaining a near-optimum BEP performance. By contrast, the computational complexity of the optimum multi-user detection using an exhaustive search would be increased by a factor of $2^{16}/2^{10} = 64$. Similarly, when K is increased to 20, a population size of $P = 160$ is sufficient for attaining the same level of BEP performance. This constituted only a factor of $1600/400 = 4$ increased computational complexity. Furthermore, in contrast to the reduced-complexity tree-search type algorithms [323, 324] – which can also achieve a near-optimum BEP performance at a complexity lower than that of the optimum detector – the detection time required by our GA-based multi-user detector to reach a decision is independent of the number of users. Additionally, for the tree-search algorithms a noise whitening filter is required. Figure 10.16 portrays the achievable complexity reduction factor of the GA-assisted multi-user detector, which was defined as $\frac{2^K}{P \times Y}$. Specifically, the numerator quantifies the number of correlation metric evaluations required by the optimum multi-user detector, while the denominator indicates the number of correlation metric evaluations required by the GA-assisted multi-user detector, in order to attain the optimum performance at an SNR value of 24 dB. This figure was extracted from the results obtained in Figure 10.15. As seen from the figure, the complexity reduction offered by the GA-assisted multi-user detector over the optimum detector becomes more significant, as the number of users is increased.

Figure 10.17 shows the near–far resistance of the proposed GA-assisted multi-user detector in conjunction with perfect CIR estimation. The average received bit energy ξ_1 of the desired user remained unchanged, while the energies of all other users ξ_k for $k = 2, \ldots, K$ were either 6 dB or 10 dB higher than that of the desired user. We can see that the GA-assisted multi-user detector was near–far resistant.

10.7 Chapter Summary and Conclusion

In this chapter, our model of a bit-synchronous CDMA system communicating over a single-tap Rayleigh fading channel was presented in Section 10.2 and its equivalent discrete representation was considered in Section 10.3. Based on this model, the optimum multi-user detector based on the maximum likelihood criterion was derived in Section 10.4. It was shown that the correlation metric of Equation (10.23) for the optimum multi-user detection scheme is cast in the form of a combinatorial optimization function and its computational complexity is exponentially proportional to the number of users. Thus, its implementation becomes impractical, when there is a high number of users. A GA-assisted multi-user detector was proposed in this chapter, in order to circumvent the above-mentioned complexity problem. According to the results obtained from other similar GA-assisted multi-user detector proposals [311–314] found in the literature, which were summarized in Section 10.5.1,

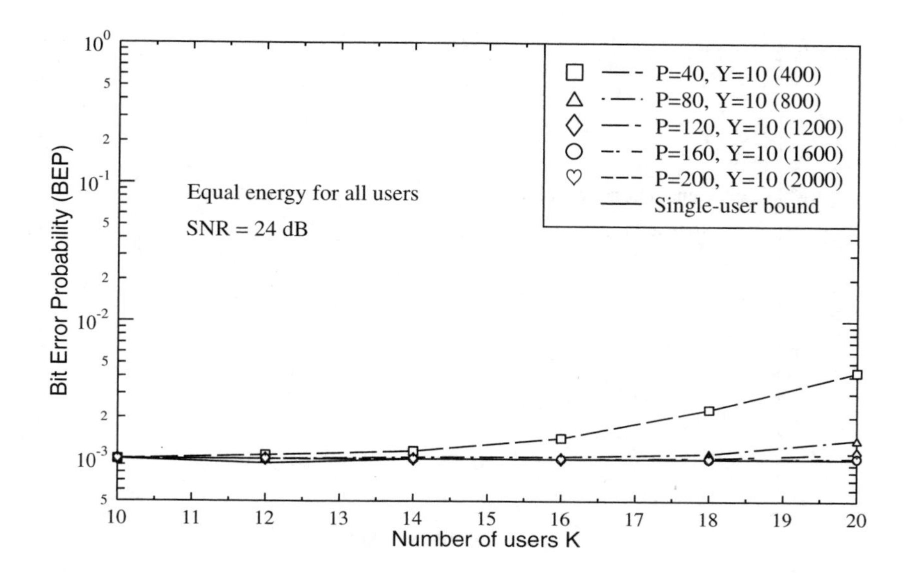

Figure 10.15: The bit error probability performance of the GA-assisted multi-user detector as a function of the number of K users with population size of $P = 40, 80, 120, 160, 200$ and $Y = 10$ in conjunction with perfect channel estimation using binary random signature sequences of length $N_c = 31$. The GA configuration used is listed in Table 10.9. The values in round brackets in the legend denote the number of times the objective function of Equation (10.25) was evaluated upon detecting the K-bit vector $\hat{\boldsymbol{b}}$ during a bit interval.

traditional GAs generally have a slow convergence rate, rendering them unsuitable for real-time data detection. In order to mitigate this impediment, we conducted a series of experiments presented in Section 10.5, for finding a particular GA configuration, from the family of techniques highlighted in Chapter 9, which can offer the best trade-off between the detection delay, computational complexity and BEP performance. Based on the results obtained from our experiments, the GA configuration that we have adopted in our GA-assisted multi-user detector was given in Table 10.9. The notable differences between our favoured GA configuration and those utilized in [311–314], as shown in Table 9.3, are the probability of mutation p_m and the employment of the incest prevention strategy. As suggested by Figure 10.3 and Figure 10.6, these two features can have a significant impact on the convergence rate and hence on the achievable BEP performance of the GA-assisted multi-user detector. For our advocated GA configuration, the incest prevention strategy and a relatively high value of p_m were invoked, since a faster convergence and an improved BEP performance can be achieved, as illustrated in Figure 10.3 and in Figure 10.6. Last but not least, the BEP performance of the GA-assisted multi-user detector was assessed in Section 10.6 for a bit-synchronous CDMA

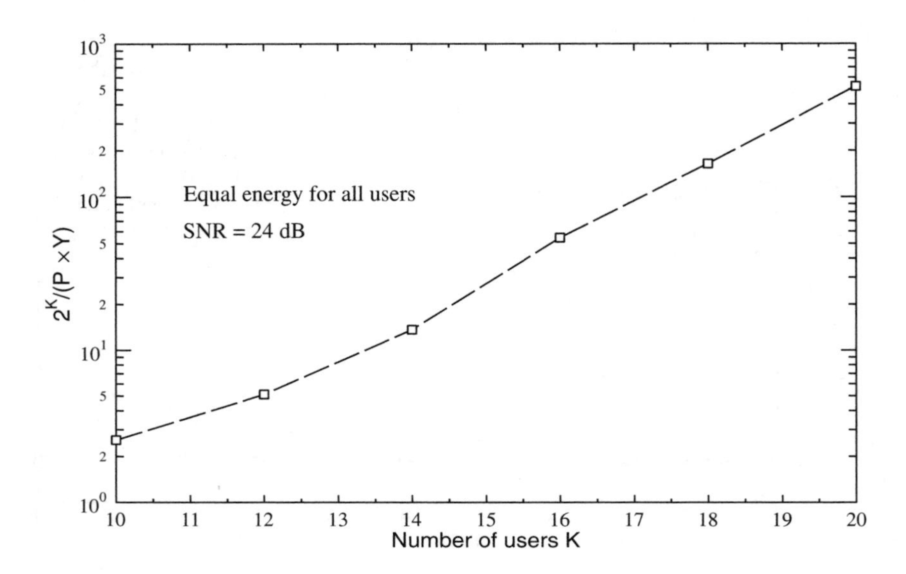

Figure 10.16: The complexity reduction factor between the optimum multi-user detector and the GA-assisted multi-user detector according to $\frac{2^K}{P \times Y}$, where the numerator denotes the number of correlation metric evaluations required by the optimum multi-user detector, while the denominator denotes that required by the GA-assisted multi-user detector, in order to attain the optimum performance at an SNR value of 24 dB, based on the results of Figure 10.15.

system for transmission over an AWGN channel as well as a single-path Rayleigh fading channel. We have shown that with the aid of a sufficiently high population size P and for a reasonable number of generations Y, the BEP performance of the GA-assisted multi-user detector approaches that of the optimum multi-user detector at the cost of a significantly lower computational complexity. Our results for a single-tap Rayleigh fading channel were obtained based on the perfect knowledge of each user's CIR coefficients at the receiver. In reality, these CIR coefficients have to be estimated by some means. In our next chapter, we will show that it is possible to extend the GA-assisted multi-user detector introduced in this chapter, such that the users' CIR coefficients can be estimated concurrently with the data detection without the assistance of any pilot signals.

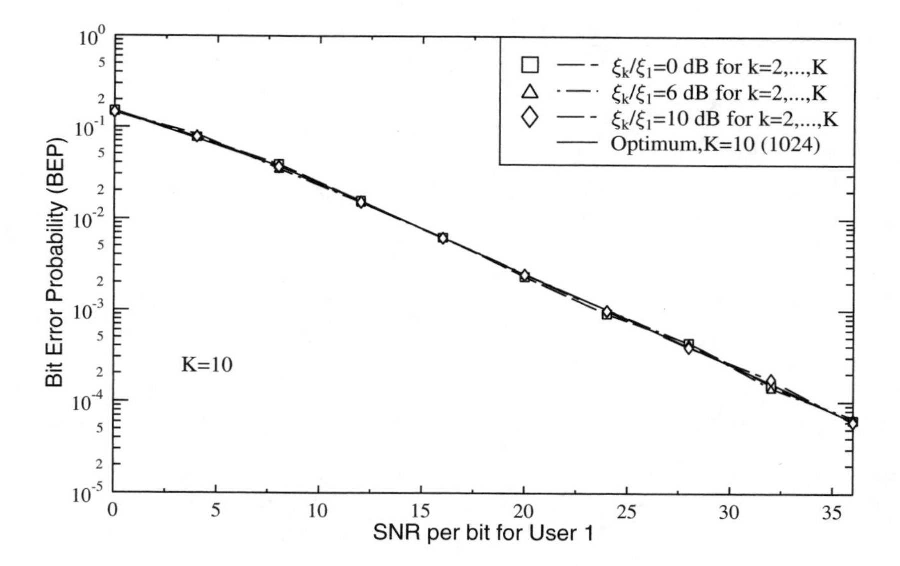

Figure 10.17: The bit error probability performance of the GA-assisted multi-user detector for $K = 10$ users with ξ_k/ξ_1 at 0 dB, 6 dB and 10 dB for $k = 2, \ldots, K$ in conjunction with perfect channel estimation using binary random signature sequences of length 31. The GA configuration used is listed in Table 10.9. The values in round brackets in the legend denote the number of times the objective function of Equation (10.25) was evaluated upon detecting the K-bit vector $\hat{\boldsymbol{b}}$ during a bit interval.

Chapter 11

Joint Genetic Algorithm-Assisted Channel Estimation and Multi-user Detection[1]

11.1 Introduction

Our work in the previous chapter was based on the assumption that perfect channel estimation is available at the receiver, in order to perform coherent detection of the received signals. Conventionally, the fading CIR coefficients of Equation (10.5) are usually estimated using a pilot signal, as for example on the downlink of the IS-95 system [325], or employing a sequence of pilot symbols [326], in order to facilitate coherent detection. However this technique becomes inefficient on the uplink, since an independent pilot signal is required for each user in order to estimate the independent fading CIR coefficients experienced by each user's signal. Nonetheless, in order to support multi-user detection, this approach was used in the third-generation UTRA system [327]. According to our proposal, the associated inefficiency can be eliminated by invoking joint channel and data estimation, which is the topic of this chapter.

The notion of joint multi-user symbol detection and channel estimation was addressed for example in [8, 328–330]. In [328], symbol detection was accomplished using a tree-search algorithm, while the users' complex signal amplitudes were estimated using recursive least-squares techniques. In [329], Gauss-Seidel iterations were applied, in order to solve the joint symbol detection and channel estimation problem. The channel estimation was performed using the Expectation Maximization (EM) algorithm, while a multistage detection

[1]*Single- and Multi-Carrier CDMA Multi-User Detection, Space-Time Spreading, Synchronization, Networking and Standards.* L.Hanzo, L-L.Yang, E-L.Kuan and K.Yen,
©2003 John Wiley & Sons, Ltd. ISBN 0-470-86309-9

algorithm was used for detecting the data packets. In [330], joint multi-user detection and channel estimation were performed using two different types of decorrelators in conjunction with a channel estimator. A path-by-path decorrelator was used to provide noisy channel information for the channel estimator, while a channel-matched decorrelator decides on the symbols transmitted and these decisions were fed back to the channel estimator as reference signals. In [8], a decorrelator and a Kalman filter were used for symbol detection and channel estimation, respectively. A so-called per-survivor approach was also proposed in [8], which used a bank of Kalman filters for channel estimation. We note that in all the proposed methods mentioned above [8, 328–330], the symbol detection and channel estimation were performed using two separate but interlinked techniques, which potentially incurs additional complexity.

In this chapter, we present a novel approach to the problem of joint symbol detection and channel estimation in DS/CDMA for transmissions over flat-fading channels based on a GA-assisted innovation. In Chapter 10, GAs were invoked, in order to detect a particular combination of the users' transmitted bits $\boldsymbol{b}$ that maximizes the objective function of Equation (10.23). Hence the search space was discrete, having a finite number of search points that is exponentially dependent on the number of users. However, in the context of joint CIR estimation and symbol detection solely by the GAs – as considered in this chapter – the search space is continuous having an infinite number of possible points, simply because the fading attenuation and phase trajectories are continuous. A GA-based channel estimation technique was previously proposed in [331], which employed the Viterbi algorithm for data detection in a single-user receiver. We will show in Section 11.3 that the CIR estimation can be performed jointly with the symbol detection using the same GAs simultaneously, without incurring any additional computational complexity. Hence, unlike the research presented in the previous chapter, the CDMA multi-user detector proposed here takes into account the channel estimation error. Furthermore, in contrast to Kalman filter-based CIR estimation [7], which is CIR-dependent, no knowledge of the CIR is required for our proposed estimator. Since the CIR estimation can be conducted without explicit training sequences or decision feedback, our proposed detector is capable of offering a potentially higher throughput and a shorter detection delay than that of explicitly trained CDMA multi-user detectors.

This chapter is organized as follows. Section 11.2 describes the system model used in this chapter, which is only slightly different from the model we detailed in Section 10.2 of the previous chapter. This modification is introduced, in order to take into account the correlation of the CIR coefficients between consecutive bit intervals. Section 11.3 describes the GAs used to implement our proposed joint multi-user channel estimator and symbol detector. The GA configuration used in this chapter will be slightly different from the one listed in Table 10.9, since floating point or real-valued variables are involved here. Our simulation results are presented in Section 11.4, while Section 11.5 concludes the chapter.

11.2 System Model

We will again adopt the symbol-synchronous CDMA system model highlighted in Section 10.2, where K users transmit data packets over a single-path frequency-nonselective Rayleigh fading channel, as depicted in Figure 11.1.

The channel impulse response (CIR) of each user is as given by Equation (10.5). However, in this chapter we will assume that the CIR coefficients $c_k = \alpha_k e^{j\phi_k}$ for $k = 1, \ldots, K$

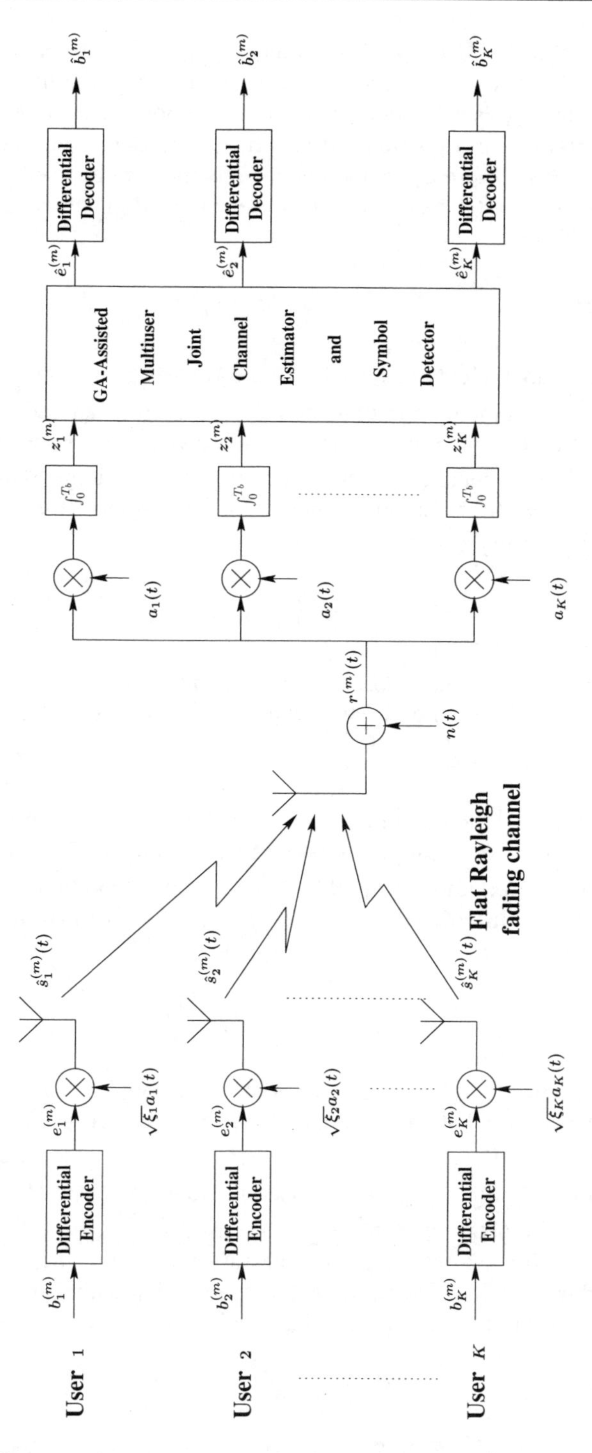

Figure 11.1: Block diagram of the K-user synchronous CDMA system communicating over a flat Rayleigh fading channel using differential encoding.

are varied over the duration T_f of the M-bit transmission frame according to the maximum Doppler frequency f_d. For mobile terrestrial communications, the maximum Doppler frequency is related to the user's velocity v and the transmission wavelength λ by $f_d = v/\lambda$ [8]. Hence, the superscript (m) in Equation (10.1), which denotes the mth bit interval, has to be included in our analysis. There are numerous models that can be used to describe the fading channel characteristics, for example, Jakes' model [319] or a first-order Gauss-Markov model, as given by [332]:

$$c_k^{(m+1)} = ac_k^{(m)} + \nu_k^{(m)}, \tag{11.1}$$

where $c_k^{(m)}$ is the CIR coefficient associated with the kth user during the mth symbol interval, $a = \exp(-2\pi f_d T_b)$, T_b is the bit duration and $\nu_k^{(m)}$ is a zero-mean white Gaussian variable. Kalman filter-based CIR estimation requires exact knowledge of a and that of the variance $\nu_k^{(i)}$ of the kth user, which must be acquired with the aid of training symbols and has to be updated frequently over a time-variant channel [333, 334]. Alternatively, we can express the relation between $c_k^{(m)}$ and $c_k^{(m+1)}$ as:

$$c_k^{(m+1)} = c_k^{(m)} + \Delta_k^{(m)}, \tag{11.2}$$

where $\Delta_k^{(m)}$ is a random variable whose value is dependent on a and $\nu_k^{(m)}$. Note that Equation (11.2) is analogous to the Least Mean Square (LMS)-based CIR estimator [335]), in which $\Delta_k^{(m)} = \mu b_k^{(m)*}\left[c_k^{(m)} b_k^{(m)} + n_k^{(m)} - \hat{c}_k^{(m)} b_k^{(m)}\right]$ (see Equation (8) of [335]), where μ is the step size.

In the context of joint channel estimation and symbol detection using Phase Shift Keying (PSK), there is always the inherent problem of a phase ambiguity of π in the estimated CIR coefficients. In order to overcome this problem, the incoming antipodal data bit stream $b_k^{(m)}$, $m = 0, \ldots, M-1$, is differentially encoded [336, 337], as shown in Figure 11.1. Then according to [319] we have:

$$e_k^{(m)} = e_k^{(m-1)} b_k^{(m)}, \tag{11.3}$$

where $e_k^{(m)}$ for $m = 0, \ldots, M-1$ constitutes the differentially encoded bit stream. From Equation (11.3), we can see that the differentially encoded bit stays at the same logical value as the previous differentially encoded bit, when $b_k^{(m)} = 1$ and vice versa. The differential encoder is initialized with $e_k^{(-1)} = 1$, which is known to the receiver. Hence we will transmit the pilot symbol $e_k^{(-1)}$ together with the encoded bits $e_k^{(m)}$, $m = 0, \ldots, M-1$. However, in Section 11.3 we will show that it is not necessary to transmit the pilot symbol, for reasons to be highlighted at a later stage. Hence in conjunction with differential encoding, the transmitted signal $\hat{s}_k(t)$ of the kth user, as illustrated in Figure 11.1, becomes:

$$\hat{s}_k(t) = \sqrt{\xi_k} \sum_{m=-1}^{M-1} e_k^{(m)} a_k(t - mT_b), \quad \forall k = 1, \ldots, K \tag{11.4}$$

where $(M+1)$ is the number of data bits in a frame transmitted by each user, ξ_k is the bit energy of the kth user and $a_k(t)$ is the signature sequence of the kth user, as given by Equation (10.2).

At the receiver, upon detecting the differentially encoded bits $\hat{e}_k^{(m)}$ of Figure 11.1, the actual data bits are obtained according to:

$$\hat{b}_k^{(m)} = \hat{e}_k^{(m-1)} \hat{e}_k^{(m)} \quad \text{for} \quad m = 0, \ldots, M-1. \tag{11.5}$$

Hence, if the CIR coefficients are estimated with a phase offset of π for at least two consecutive bits, such that $e^{-j\phi_k^{(m-1)}}$ and $e^{-j\phi_k^{(m)}}$ are detected instead of the actual values of $e^{j\phi_k^{(m-1)}}$ and $e^{j\phi_k^{(m)}}$, respectively, then the sign of the differentially encoded detected bits $\hat{e}_k^{(m-1)}$ and $\hat{e}_k^{(m)}$ will be the opposite of the actual transmitted differentially encoded bit, since again their phase is shifted by π. In this case it is assumed that the bits are detected 'correctly'. Differential decoding can cancel out this CIR-estimation-induced phase shift according to Equation (11.5). Therefore we can see that differential encoding is important, if the channel estimation is not aided by pilot signals/symbols. On the other hand, systems utilizing differential encoding suffer from a 3 dB SNR loss compared to coherently detected BPSK modulation, since a detection error in a differentially encoded bit $e_k^{(m)}$ results in two errors of the decoded bits, as can be deduced from Equation (11.5).

In this chapter, we are interested in determining the unknown differentially encoded data bits $e_k^{(m)}$ as well as the CIR coefficients $c_k^{(m)}$ for $m = 0, \ldots, M-1$ and $k = 1, \ldots, K$ at the receiver, in order to perform coherent detection of the received signals. Following the analysis conducted in Sections 10.2–10.4, it can be shown that the correlation metric conditioned on the matrix $\boldsymbol{C}^{(m)}$ containing the CIR coefficients in its diagonal and on the vector $\boldsymbol{e}^{(m)}$ incorporating the data bits is given by [323, 329]:

$$\Omega\left(\boldsymbol{e}^{(m)}, \boldsymbol{C}^{(m)}\right) = 2\Re\left[\boldsymbol{e}^{(m)^T} \boldsymbol{\xi} \boldsymbol{C}^{(m)^*} \boldsymbol{Z}^{(m)}\right] - \boldsymbol{e}^{(m)^T} \boldsymbol{\xi} \boldsymbol{C}^{(m)} \boldsymbol{R} \boldsymbol{C}^{(m)^*} \boldsymbol{\xi} \boldsymbol{e}^{(m)}, \tag{11.6}$$

where $\boldsymbol{e}^{(m)} = \left[e_1^{(m)}, \ldots, e_K^{(m)}\right]^T$, $\boldsymbol{C}^{(m)}$ and $\boldsymbol{\xi}$ are as given by Equation (10.14), with the inclusion of the superscript (m) for denoting the mth signalling interval. Furthermore, $\boldsymbol{Z}^{(m)} = \left[z_1^{(m)}, \ldots, z_K^{(m)}\right]^T$ denotes the matched filter outputs at the mth signalling interval, as shown in Figure 11.1, whose elements are given by Equation (10.8) with b_k replaced by $e_k^{(m)}$ for $k = 1, \ldots, K$. The $K \times K$-dimensional signature sequence cross-correlation matrix $\boldsymbol{R}$ is identical to that given by Equation (10.17). The decision rule for the optimum joint multi-user channel estimation and symbol detection scheme is to choose the channel coefficient matrix $\hat{\boldsymbol{C}}^{(m)}$ and the differentially encoded symbol vector $\hat{\mathbf{e}}^{(m)}$, which maximizes the correlation metric given in Equation (11.6), under a constraint on the channel coefficient matrix $\boldsymbol{C}^{(m)}$ as imposed by Equation (11.2). Hence,

$$\begin{aligned}\left(\hat{\boldsymbol{e}}^{(m)}, \hat{\boldsymbol{C}}^{(m)}\right) &= \arg\left\{\max_{\boldsymbol{e}^{(m)}, \boldsymbol{C}^{(m)}} \left[\Omega\left(\boldsymbol{e}^{(m)}, \boldsymbol{C}^{(m)}\right)\right]\right\} \\ \text{subject to} \qquad & \boldsymbol{C}^{(m)} = \boldsymbol{C}^{(m-1)} + \boldsymbol{\Delta}^{(m-1)}. \end{aligned} \tag{11.7}$$

According to Equation (11.7), when both the CIR coefficients and the transmitted data sequence are unknown, their optimal estimates can, in fact, be jointly obtained by optimizing the correlation metric. Equation (11.7) constitutes a global optimization problem, which is non-linear, since it entails taking the maximum of $\Omega\left(\boldsymbol{e}^{(m)}, \boldsymbol{C}^{(m)}\right)$. Furthermore, Equation (11.7) cannot be solved using a conventional linear optimization approach or employing an exhaustive tree search, because the actual values of the channel coefficient matrix $\boldsymbol{C}^{(m)}$ in Equation (11.6) are unknown, unless a separate CIR estimator is incorporated, as in [328].

11.3 Joint GA-assisted Multi-user Channel Estimation and Symbol Detection

In this section, GAs are developed, in order to solve the joint CIR estimation and symbol detection optimization problem, where the required objective function is defined by the correlation metric in Equation (11.6). Again, we will take the exponential of Equation (11.6), in order to ensure that all the fitness values calculated from this equation become positive. Hence the modified objective function becomes:

$$\exp\left\{\Omega\left(\boldsymbol{e}^{(m)}, \boldsymbol{C}^{(m)}\right)\right\} = \exp\left\{2\Re\left[\boldsymbol{e}^{(m)^T}\boldsymbol{\xi}\boldsymbol{C}^{(m)^*}\boldsymbol{Z}^{(m)}\right] - \boldsymbol{e}^{(m)^T}\boldsymbol{\xi}\boldsymbol{C}^{(m)}\boldsymbol{R}\boldsymbol{C}^{(m)^*}\boldsymbol{\xi}\boldsymbol{e}^{(m)}\right\}. \tag{11.8}$$

In this case, we are interested in determining the CIR matrix $\boldsymbol{C}^{(m)}$ for $m = -1, \ldots, M-1$ and the data vector $\boldsymbol{e}^{(m)}$ for $m = 0, \ldots, M-1$ that maximize the modified objective function of Equation (11.8). Hence, each individual of the GA will consist of two K-variable vectors, namely the CIR coefficient vector $\tilde{\boldsymbol{C}}_p^{(m)}(y) = \left[\tilde{c}_{p,1}^{(m)}(y), \tilde{c}_{p,2}^{(m)}(y), \ldots, \tilde{c}_{p,K}^{(m)}(y)\right]$, which is composed of continuous-valued real and imaginary parts, and the data vector $\tilde{\boldsymbol{e}}_p^{(m)}(y) = \left[\tilde{e}_{p,1}^{(m)}(y), \tilde{e}_{p,2}^{(m)}(y), \ldots, \tilde{e}_{p,K}^{(m)}(y)\right]$, which is composed of the binary-valued antipodal bits of the K users at instant m. The parameters m, y and p denote the mth signalling interval, the yth generation and the pth individual, respectively. Each individual is associated with a certain fitness value. The fitness value of the pth individual, denoted by $f_p\left[\tilde{\boldsymbol{C}}_p^{(m)}(y), \tilde{\boldsymbol{e}}_p^{(m)}(y)\right]$ is computed by substituting its corresponding CIR coefficient vector estimate $\tilde{\boldsymbol{C}}_p^{(m)}(y)$ and data vector estimate $\tilde{\boldsymbol{e}}_p^{(m)}(y)$ into the objective function of Equation (11.8). The individual that corresponds to the highest-fitness value at the end of the evolution is finally chosen as the descriptor of the estimated users' CIR coefficients $\hat{\boldsymbol{C}}^{(m)}$

and the transmitted encoded bits $\hat{\mathbf{e}}^{(m)}$.

11.3.1 Initialization

In the previous chapter, we were only interested in detecting the users' transmitted bits per signalling interval. Since the transmitted bits are independent of each other in consecutive bit intervals, the initialization of the GA, as invoked in Chapter 10 is also independent for consecutive signalling intervals. On the other hand, in this chapter, the CIR coefficients to be estimated are correlated for consecutive bit intervals according to Equation (11.2). Hence we can use the CIR coefficients estimated during the previous bit interval for initializing the CIR coefficient vector associated with each individual of the population in the current bit interval. More explicitly, assuming that the current signalling interval is the mth interval, we have:

$$\begin{aligned}\tilde{\boldsymbol{C}}_1^{(m)}(0) &= \hat{\boldsymbol{C}}^{(m-1)} \\ \tilde{\boldsymbol{C}}_p^{(m)}(0) &= MUTATION\left[\hat{\boldsymbol{C}}^{(m-1)}\right] \quad \text{for } p = 2, \ldots P,\end{aligned} \tag{11.9}$$

where $\hat{\boldsymbol{C}}^{(m-1)}$ consists of the K users' estimated CIR coefficients accruing from the previous signalling interval. The mutation operation applied to the pth individual for $p = 2, \ldots, P$ is necessary, in order to diversify the initial population as well as to provide dissimilar individuals for preventing incest mating, as highlighted in Section 9.4.2. The mutation process of the CIR coefficients will be detailed further in Section 11.3.3.

Recall from our discourse in the previous section that at $m = -1$ the users' data has to be a known bit, since the transmitted bit sequences are differentially encoded. We also mentioned that we will transmit this known symbol, even though it is demonstrated in Section 11.2 that it is not absolutely necessary. The reason for transmitting it is that we can use this pilot symbol, in order to assist in estimating the CIR coefficients $\hat{\boldsymbol{C}}^{(-1)}$. Hence based on this known data bit, the CIR coefficient vectors $\tilde{\boldsymbol{C}}_p^{(0)}(0)$ associated with the P individuals of the initial population can be assigned by estimating the CIR coefficients with the aid of the output of the matched filters according to:

$$\begin{aligned}\tilde{\boldsymbol{C}}_1^{(-1)}(0) &= \boldsymbol{Z}^{(-1)^T}\operatorname{diag}\left[\frac{1}{\sqrt{\xi_1}}, \frac{1}{\sqrt{\xi_2}}, \ldots, \frac{1}{\sqrt{\xi_K}}\right] \\ \tilde{\boldsymbol{C}}_p^{(-1)}(0) &= MUTATION\left[\tilde{\boldsymbol{C}}_1^{(-1)}(0)\right] \quad \text{for } p = 2, \ldots, P\end{aligned} \tag{11.10}$$

where $\boldsymbol{Z}^{(-1)}$ is given by Equation (10.16) with $b_k^{(-1)}$ replaced by $e_k^{(-1)} = 1$ for $k = 1, \ldots, K$. The estimated CIR coefficients will be contaminated by the MAI, as it is evidently seen from Equation (10.8), and hence will be inaccurate. Nevertheless, these CIR coefficient estimates will provide a good foundation for the GA to evolve from. Again, for the individuals $\tilde{\boldsymbol{C}}_p^{(-1)}(0)$, where $p = 2, \ldots, P$, the associated CIR coefficient vectors are randomly mutated versions of $\tilde{\boldsymbol{C}}_1^{(-1)}(0)$ for the reasons highlighted above.

Similarly to the previous chapter, the K-bit data vector associated with each individual is generated at the beginning by randomly mutating the hard decisions generated from the matched filter outputs, as was highlighted in Section 10.5.6 for the method M2. However, in this case, the CIR coefficients for the current signalling interval are unknown. Since the CIR coefficients are highly correlated between consecutive signalling intervals, as indicated by Equation (11.2), we can use the estimated CIR coefficients $\hat{C}^{(m-1)}$ generated during the previous signalling interval for coherently detecting the hard decisions derived from the matched filter outputs during the current signalling interval, in order to obtain an adequately biased initial population. Hence we have:

$$\tilde{e}_p^{(m)}(0) = MUTATION\left\{\text{sgn}\left[\Re\left(Z^{(m)^T}\hat{C}^{(m-1)^*}\right)\right]\right\} \quad \text{for } p = 1, \ldots, P \qquad (11.11)$$

After the initialization of the individuals in the population, the GA is then invoked, in order to find the individual that yields the highest fitness value according to Equation (11.8). A flowchart depicting the structure of the proposed GA used to jointly estimate the users' CIR coefficients $\hat{C}^{(m)}$ and to detect the transmitted differentially encoded bits $\hat{e}^{(m)}$ at the mth signalling interval is shown in Figure 11.2. As we can see, the GA that is adopted here is similar to that employed in the previous chapter, as shown in Figure 10.10. However, due to the continuous-valued nature of the CIR coefficients, there will be an infinite number of possible solutions for the CIR coefficient vector that the GAs will handle. Hence, the GA configuration specified in Table 10.9 that is used to search for the optimum CIR coefficient vector in this chapter is slightly different from that used for finding the optimum data vector. Specifically, the mutation operation designed for real-valued decision variables, as highlighted in Section 9.4.4, will be adopted for determining the CIR coefficient vector associated with each individual, which requires the determination of the maximum mutation size λ_{max}. Furthermore, the mating pool size T will have a significant impact on the CIR estimation performance, since we are unlikely to come across identical individuals in a given population due to the infinite number of possible solutions of the continuous-valued CIR coefficient vectors. Because of this, all the individuals in the population are likely to be considered potential parents. However, due to the high number of potential parents in the mating pool, insufficient selection emphasis might be placed on any individual, since each continuous-valued CIR tap estimate is likely to be encountered only once, despite having a range of similar continuous-valued CIR taps.

11.3.2 Effects of the Mating Pool Size

In this section, we will investigate the effects of the mating pool size T on the performance of the CIR estimator section of our proposed GA-assisted multi-user detector. Hence we shall assume that the received data bits are known. In other words, the data vectors $\tilde{e}_p^{(m)}(y)$ associated with all the individuals $p = 1, \ldots, P$ and $y = 0, \ldots, Y-1$ in the population are equal to the transmitted data vector $e^{(m)}$. In order to quantify the channel estimator's performance, the Mean Squared Error (MSE) between the true value and the estimated value of the channel's attenuation was obtained. Summaries of the various parameters and the GA configuration that are used in our simulations in this section are shown in Table 11.1 and Table 11.2, respectively.

Figure 11.3 shows the achievable CIR estimation MSE over 20 generations at the end of

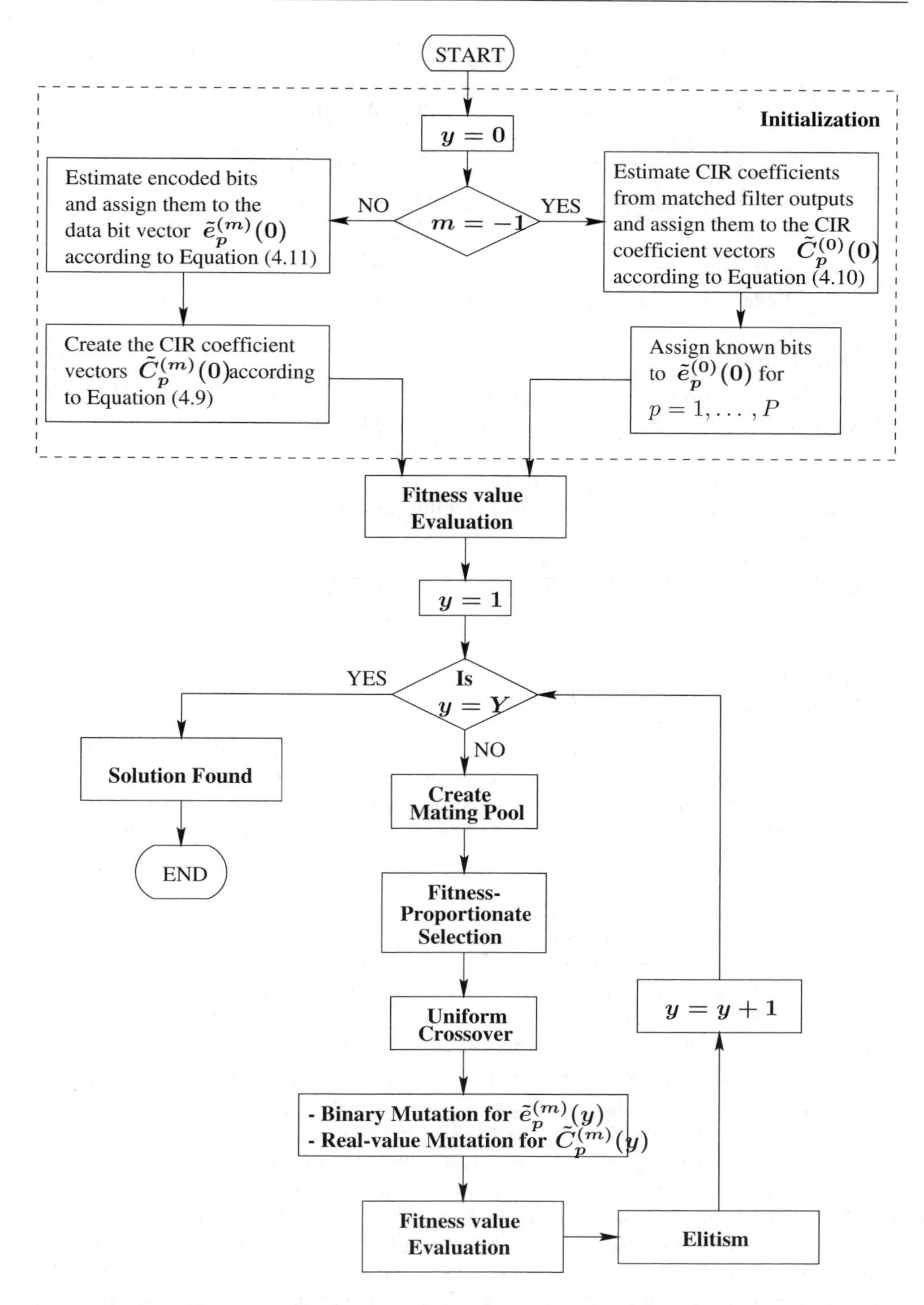

Figure 11.2: A flowchart depicting the structure of the proposed genetic algorithm used to jointly estimate the users' CIR coefficients and to detect the transmitted differentially encoded bits at the mth signalling interval.

Parameter	Value
Spreading factor N_c	31
Modulation mode	Differential Encoded BPSK
Number of CDMA users K	10
SNR per bit ξ_k/N_0	Infinity for $k = 1, \ldots, K$
Doppler frequency f_d	200 Hz
Data rate R_b	64 kbps
Packet size M	100

Table 11.1: Simulation parameters for the experiments of Figures 11.3-11.4.

Set-up/Parameter	Method/Value
Individual initialization method	According to Equation (11.9) for $m = 0, \ldots, M-1$ and Equation (11.10) for $m = -1$
Selection method	Proportionate-Fitness
Cross-over operation	Uniform cross-over
Mutation operation	Standard floating point mutation
Elitism	Yes
Incest prevention	Yes
Population size P	Given in the associated plots
Number of generations Y	Given in the associated plots
Mating pool size T	Given in the associated plots
Probability of mutation p_m	0.1
Mutation size λ_{max}	0.1

Table 11.2: Configuration of the GA used to obtain the results of Figure 11.3 and Figure 11.4. Explicit description of the fitness-proportionate selection scheme, the uniform cross-over operation and the floating point mutation operation can be found in Section 9.4.2, Section 9.4.3 and Section 9.4.4, respectively. The definition of the mutation size will be given in Section 11.3.3.

the $M = 100$-bit data packet for various mating pool sizes T with a population size of 40 individuals. As can be seen, when $T = P = 40$, the lowest achievable MSE of the GA-assisted CIR estimator is relatively high, in the region of 0.02. Furthermore, the convergence rate is relatively low. As the mating pool size is reduced, an MSE improvement can be observed, reaching values as low as 0.0001 for $T = 10$ and 5.

Figure 11.4 shows the achievable MSE for various population sizes P in conjunction with different mating pool sizes T. As can be seen, the value of T has a significant impact on the achievable MSE. Specifically, the MSE becomes higher as T increases. The MSE is less sensitive to small values of T. This is justified with the aid of the following simple example. Let us assume that $T = 4$, in which case the lowest possible probability of selection associated with the individual having the highest fitness value is 0.25. On the other hand, in case of $T = 10$, the lowest possible probability of selection of the same individual becomes 0.1. Hence we can see that the emphasis placed on the best individual is lower for the latter case, which resulted in a slower convergence. This phenomenon will be investigated in more

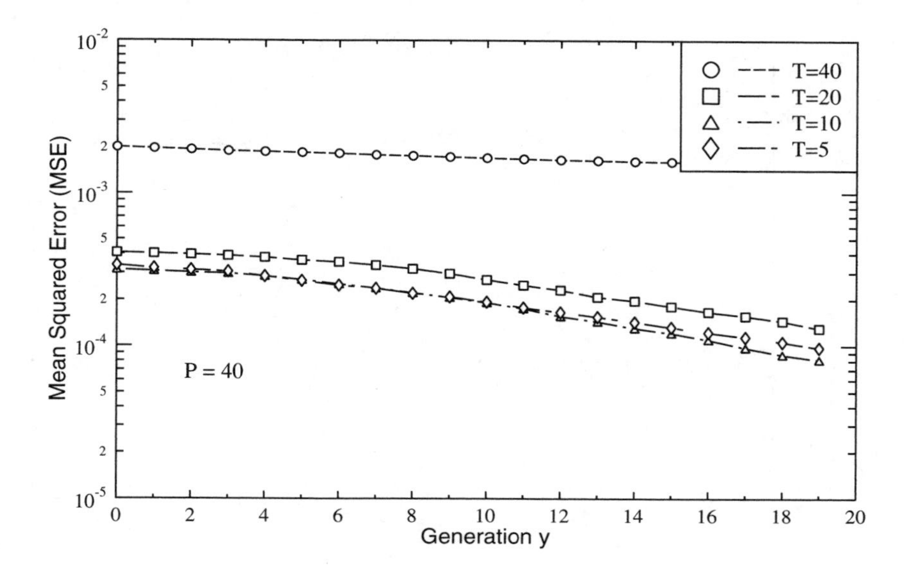

Figure 11.3: The mean squared error performance of the GA-assisted CIR estimator as a function of the number of generations y for a population size of $P = 40$ in conjunction with various mating pool sizes T using binary random signature sequences of length $N_c = 31$. The GA configuration and the simulation parameters used are listed in Table 11.2 and Table 11.1, respectively.

depth in the context of Figure 13.6 in conjunction with asynchronous GA-assisted multi-user detector. Hence in our case, we will adopt a mating pool size of $T = 5$ for our simulations in this chapter.

11.3.3 Effects of the Mutation Size

The mutation operation that is adopted to alter the value of the estimated CIR coefficients is conducted as follows. When a complex-valued variable associated with the CIR coefficient vector is picked for mutation, the direction of mutation is chosen randomly with equal probability for both the real and imaginary part of the CIR coefficient. Then a real-valued random mutation size $\hat{\Delta}_k^{(m)}(y)$ is generated, in the range of $[0, \lambda_{max}]$. The value of both the real and imaginary part of the CIR coefficient is then increased or decreased accordingly by a

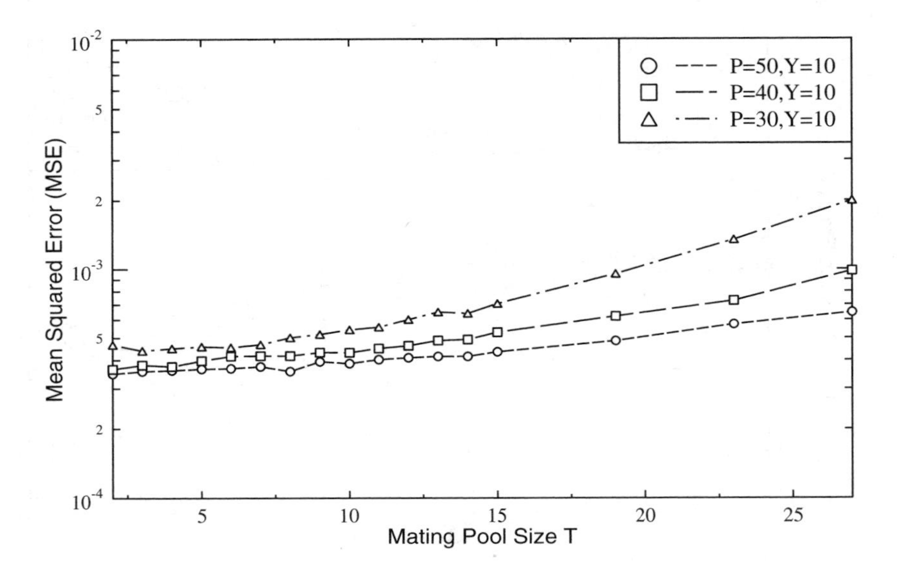

Figure 11.4: The mean squared error performance of the GA-assisted CIR estimator as a function of the mating pool size T in conjunction with various population sizes P using binary random signature sequences of length $N_c = 31$. The GA configuration and the simulation parameters used are listed in Table 11.2 and Table 11.1, respectively.

magnitude prescribed by the mutation size:

$$\begin{aligned} \Re\left[\tilde{c}_{p,k}^{(m)}(y)\right] &= \Re\left[\tilde{c}_{p,k}^{(m)}(y-1)\right] \pm \hat{\Delta}_k^{(m)}(y-1) \\ \Im\left[\tilde{c}_{p,k}^{(m)}(y)\right] &= \Im\left[\tilde{c}_{p,k}^{(m)}(y-1)\right] \pm \hat{\Delta}_k^{(m)}(y-1). \end{aligned} \quad (11.12)$$

Notice that a limit is imposed on the value of $\hat{\Delta}_k^{(m)}(y)$ by the parameter λ_{max}, in order to ensure that the associated phase ambiguity becomes significantly less than π. This is to ensure that the phase ambiguity will not change from the $(m-1)$th symbol to the mth symbol, unless the phase is near zero, as we shall see in the context of one of our simulation results in Section 11.4. On the other hand, the value of λ_{max} should be sufficiently high – especially for high Doppler shifts – in order to track the evolving CIR coefficients from one symbol to the next. Hence in this section, we will attempt to determine the appropriate value of λ_{max} that is acceptable for our application. Again, we will consider the MSE performance of the CIR estimator section of our proposed detector in conjunction with known transmitted bits. The simulation parameters and the GA configuration used for our study in this section are given in Table 11.3 and Table 11.4, respectively.

Parameter	Value
Spreading factor N_c	31
Modulation mode	Differential Encoded BPSK
Number of CDMA users K	10
SNR per bit ξ_k/N_0	Given in the associated plots
Doppler frequency f_d	200 Hz
Data rate R_b	64 kbps
Packet size M	200

Table 11.3: Simulation parameters for the experiments of Figures 11.5-11.6.

Set-up/Parameter	Method/Value
Individual initialization method	According to Equation (11.9) for $m = 0, \ldots, M-1$ and Equation (11.10) for $m = -1$
Selection method	Proportionate-Fitness
Cross-over operation	Uniform cross-over
Mutation operation	Standard floating point mutation
Elitism	Yes
Incest prevention	Yes
Population size P	40
Number of generations Y	10
Mating pool size T	5
Probability of mutation p_m	0.1
Mutation size λ_{max}	Given in the associated plots

Table 11.4: Configuration of the GA used for obtaining the results of Figure 11.5 and Figure 11.6. Explicit description of the fitness-proportionate selection scheme, the uniform cross-over operation and the floating point mutation operation can be found in Section 9.4.2, Section 9.4.3 and Section 9.4.4, respectively.

In Figure 11.5 we examined the effects of different λ_{max} values on the lowest achievable CIR estimation MSE for various Doppler frequencies f_d and SNR values. In order to characterize the worst-case scenarios in terms of the vehicular speeds in Figure 11.5, we opted for using extremely high Doppler frequencies. Explicitly, we found that the GA-assisted multi-user detector MSE reached its best possible value even for a Doppler frequency of $f_d = 600$ Hz at the SNR value investigated. From the figure we can see that the value of λ_{max} can have a significant impact on the lowest achievable CIR estimation MSE for different f_d values. In an effort to quantify the worst-case performance of the algorithm, we tested it in high-speed scenarios, such as for example a vehicular speed of 114 km/h – 342 km/h at 1.9 GHz carrier frequency. This resulted in a Doppler frequency of 200-600 Hz. For example, when $f_d = 200$ Hz, $\lambda_{max} \approx 0.04$ gives the optimal MSE for all SNR values. However, for the same value of $\lambda_{max} = 0.04$, the MSE for $f_d = 600$ Hz becomes excessive due to the fact that a low λ_{max} value is incapable of tracking the rapidly changing CIR coefficients between symbols. Moreover, we can see that the achievable MSE is more sensitive to lower values of λ_{max} for the various Doppler frequencies assumed. On the other hand, only a slight degradation in the

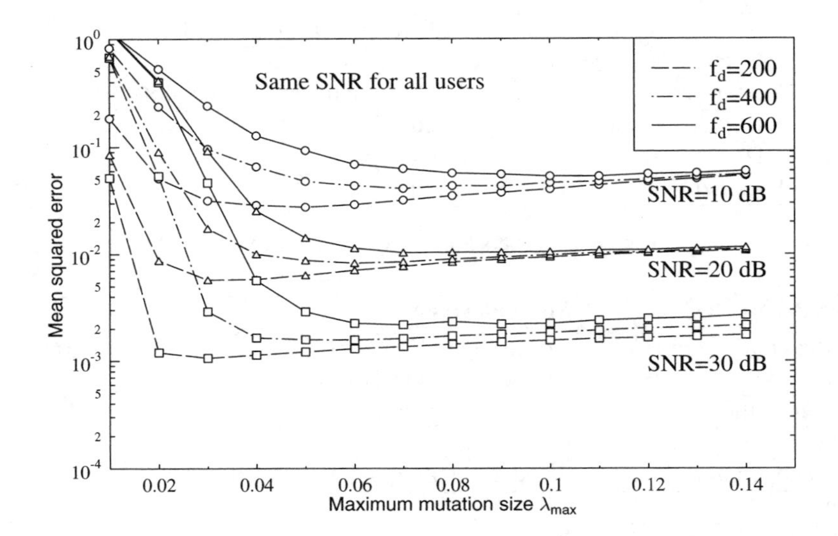

Figure 11.5: Average mean squared channel estimation error after convergence in a $K = 10$-user synchronous CDMA system transmitting known bits and having equal averaged received bit energy for all users over a narrowband Rayleigh-fading channel at various Doppler frequencies f_d. The GA configuration and the simulation parameters used are listed in Table 11.4 and Table 11.3, respectively.

achievable MSE is observed, as λ_{max} increases to a higher value.

Figure 11.6 compares the average MSE performance versus symbol index of the GA-assisted CIR estimator in conjunction with known bits for different SNR values and for $\lambda_{max} = 0.05$ as well as for $\lambda_{max} = 0.1$ measured over a frame of 200 known bits. Averaging over 200 transmitted frames was carried out with equal average received bit energy for all users. It is seen in the figure that GAs using $\lambda_{max} = 0.05$ can achieve a lower MSE, than in conjunction with $\lambda_{max} = 0.1$. However, the former suffered from a longer convergence period.

Based on the results obtained in Figure 11.5 and in Figure 11.6, we decided to adopt $\lambda_{max} = 0.1$ for our simulations hereafter, since this value resulted in a fairly consistent MSE over an f_d range of 200 Hz to 600 Hz as well as ensuring a fast convergence rate. Finally, due to its moderate value it avoided the phase-ambiguity problem.

11.4 Simulation Results

In this section our simulation results are presented in order to characterize the performance of the proposed joint multi-user CIR estimator and symbol detector. Summaries of the various

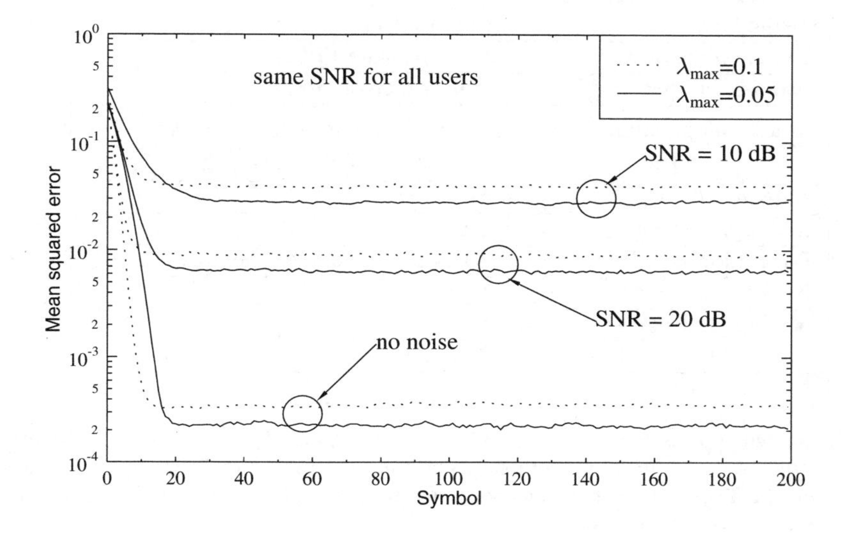

Figure 11.6: Average mean squared CIR estimation error versus transmitted symbol index in a $K = 10$-user synchronous CDMA system for a frame of 200 known bits. Averaging over 200 transmitted frames was carried out with equal averaged received bit energy for all users over a narrowband Rayleigh-fading channel at $f_d = 200$ Hz. The GA configuration and the simulation parameters used are listed in Table 11.4 and Table 11.3, respectively.

Parameter	Value
Spreading factor N_c	31
Modulation mode	Differential Encoded BPSK
Number of CDMA users K	10
Doppler frequency f_d	200 Hz
Data rate R_b	64 kbps
Packet size M	640

Table 11.5: Simulation parameters for the experiments of Figures 11.7-11.12.

parameters and the GA configuration that are used in our simulations are shown in Table 11.5 and Table 11.6, respectively.

Before we examine the BEP performance of the proposed detector, let us first consider the tracking capability of the GA-assisted CIR estimator both in conjunction with known and unknown bits, as characterized in Figure 11.7 and Figure 11.8, respectively. Specifically, a snap-shot of the estimated real and imaginary components of the CIR coefficient of a user is compared with its corresponding true value. Notice the mirror image of the estimated components after about 400 symbols with respect to the zero level of the y-axis in Figure 11.8,

Set-up/Parameter	Method/Value
Individual initialization method	According to the flowchart of Figure 11.2
Selection method	Proportionate-Fitness
Cross-over operation	Uniform cross-over
Mutation operation	- Floating point mutation for $\tilde{C}_p^{(m)}(y)$ - Binary mutation for $\tilde{e}_p^{(m)}(y)$
Elitism	Yes
Incest prevention	Yes
Population size P	40, unless specified otherwise
Number of generations Y	10
Mating pool size T	5
Probability of mutation p_m	0.1
Mutation size λ_{max}	0.1

Table 11.6: Configuration of the GA used to obtain the results of Figures 11.7-11.12. Explicit description of the fitness-proportionate selection scheme, the uniform cross-over operation and the floating point mutation operation can be found in Section 9.4.2, Section 9.4.3 and Section 9.4.4, respectively.

when the transmitted bits and the CIR are jointly estimated by the proposed GA-assisted multi-user detector. This will result in the phase ambiguity we have mentioned previously due to the 180° change in the CIR's phase potentially changing the sign of the estimated bit in the case of BPSK. More explicitly, this change in the phase is caused by the mutation process in an attempt to estimate the desired CIR coefficients. The bits in this ambiguity region will be detected in error, unless differential encoding and decoding are invoked. However, on the whole, the GA-assisted CIR estimator was capable of tracking the channel variations closely, regardless of whether the bits were known or unknown.

Figure 11.9 compares the MSE of the proposed CIR estimator to that of the conventional correlation-type estimator [7]. The single-user bound using the Linear Minimum Mean Squared Error (LMMSE) CIR estimator given in [338] was also plotted for comparison. It can be seen that our proposed GA-assisted CIR estimator exhibited a significantly lower MSE value than that of the conventional CIR estimator and its associated BEP performance was not far from the single-user bound.

Figure 11.10 shows the BEP performance of the proposed GA-assisted joint CIR estimator and symbol detector for spreading factors of $N_c = 31$ and $N_c = 127$. The BEP performance of the GA-assisted symbol detector using imperfect CIR estimation having a MSE of 0.01 and 0.001 is also shown. Furthermore, we plotted in the figure the differentially coded single user bound in conjunction with perfect CIR estimation, which is given by [90]:

$$P_2 = \frac{1}{2}\left(1 - \frac{\bar{\gamma}_c}{1+\bar{\gamma}_c}\right), \tag{11.13}$$

as well as the differentially decoded BEP performance of the proposed GA-assisted symbol detector using perfect CIR estimation. As can be observed, the joint CIR and data detector ex-

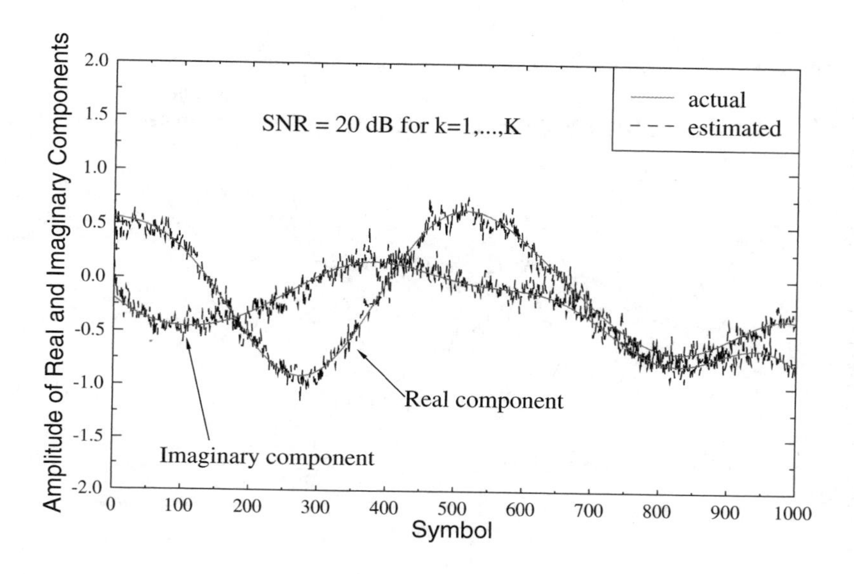

Figure 11.7: A snap-shot of the estimated real and imaginary components of the CIR coefficients in conjunction with known transmitted bits corresponding to one user compared to its true value for a narrowband Rayleigh-fading channel at $f_d = 200$ Hz, where the GA configuration and the simulation parameters used are listed in Table 11.6 and Table 11.5, respectively.

hibited an error floor due to the imperfect CIR estimation and the MSE of the CIR estimation was somewhere between 0.01 and 0.001, which conforms to our results obtained previously in Figure 11.6. The error floor phenomenon can also be observed in the context of other multi-user detectors suffering from CIR estimation errors [332]. For the sake of comparison, the joint symbol detection and CIR estimation using a decorrelator and an ideal Kalman filter shown in [332] achieved a BEP of 10^{-3} for $K = 10$ users and for a processing gain of 127. As shown in Figure 11.10, our proposed joint data and CIR detector attains a BEP performance close to 10^{-3}. Furthermore, it should be noted that our BEP is calculated over the entire length of the transmitted bit sequence, i.e. from the zeroth symbol to the $(M-1)$th symbol, rather than after the initial convergence. Hence the bit errors observed during the acquisition of the CIR estimates were also taken into account. The CIR estimation error induced BEP floor can also be observed for a single-user transmission scenario, i.e. for $K = 1$ using the matched filter-based coherent receiver, as shown in Figure 11.11. It can be seen that the BEP performance of both the matched filter detector supporting $K = 1$ user and that of the GA-assisted multi-user detector for $K = 10$ is almost identical for CIR estimation MSE values of 0.01 and 0.001. This shows that the GA-assisted multi-user detector is operating near its optimum performance.

Figure 11.12 characterizes the BEP performance of the proposed GA-assisted joint multi-user CIR estimator and symbol detector for different population sizes P. As can be seen from

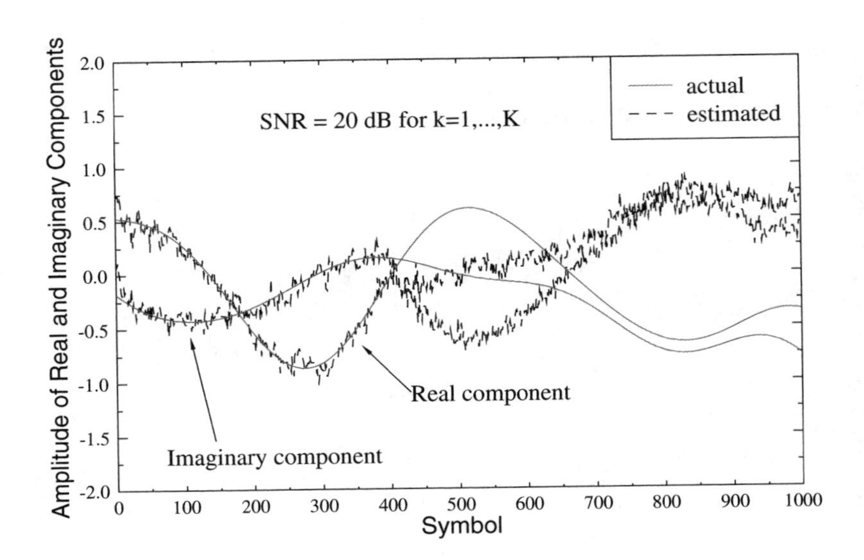

Figure 11.8: A snap-shot of the estimated real and imaginary components of the CIR coefficients in conjunction with unknown transmitted bits corresponding to one user compared to its true value for a narrowband Rayleigh-fading channel at $f_d = 200$ Hz, where the GA configuration and the simulation parameters used are listed in Table 11.6 and Table 11.5, respectively.

the figure, no significant BEP performance improvement can be achieved by increasing the population size. This is an expected observation, since in Figure 11.4 the achievable MSE was observed to be almost the same for the different population sizes studied.

11.5 Chapter Summary and Conclusion

In this chapter, GAs were developed in order to jointly estimate the CIR coefficients as well as the transmitted bits simultaneously for all users in a symbol-synchronous CDMA system based on the ML decision rule. The system model used in this chapter was highlighted in Section 11.2. Differential encoding was invoked, in order to circumvent the phase ambiguity problem, when the CIR coefficients were estimated without the aid of pilot symbols. The GA-assisted joint multi-user CIR estimator and symbol detector were introduced in Section 11.3. Because of the continuous nature of the CIR coefficients as well as due to their correlation between consecutive bit intervals, the configuration of the GA used in this chapter is slightly different from that employed in Chapter 10. In particular, investigations were carried out in Section 11.3.2 and Section 11.3.3, in order to determine the ideal mating pool size T and the best possible mutation size λ_{max} for our application, respectively. The BEP performance of the GA-assisted joint multi-user CIR estimation and symbol detection scheme was then

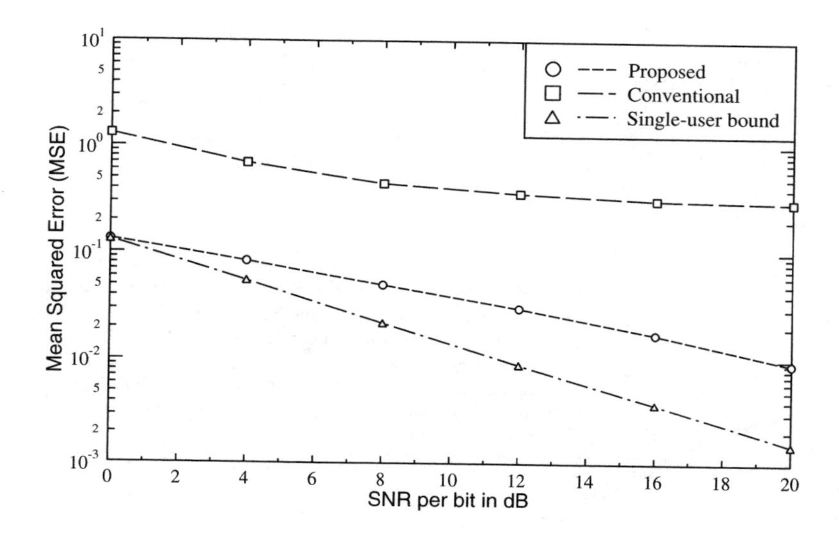

Figure 11.9: Average mean squared CIR estimation error in a $K = 10$-user synchronous CDMA system with known transmitted bits compared to that of a conventional correlator-type CIR estimator in a $K = 10$ user system and to that of a single-user LMMSE estimator over a narrowband Rayleigh-fading channels at $f_d = 200$ Hz. The GA configuration and the simulation parameters used are listed in Table 11.6 and Table 11.5, respectively.

examined using computer simulations in Section 11.4.

Our results showed that as a channel estimator, the GA was capable of tracking the variations of the fading channel, while achieving a channel gain estimation MSE as low as 10^{-3} in a noiseless channel with a Doppler frequency $f_d = 200$ Hz. Upon exploiting its capabilities as a channel estimator as well as a symbol detector, as seen from the previous chapter, the proposed joint channel estimator and symbol detector can achieve a BEP as low as 2×10^{-3} at a SNR value of 30 dB in a 10-user CDMA environment without channel coding or diversity. An error floor was observed beyond SNR = 30 dB due to the imperfect channel estimation. Furthermore, since the channel estimation and symbol detection are performed simultaneously, no pilot symbols or decision feedback are necessary, which results in a higher throughput and shorter detection time than that of explicitly trained CDMA multi-user detectors.

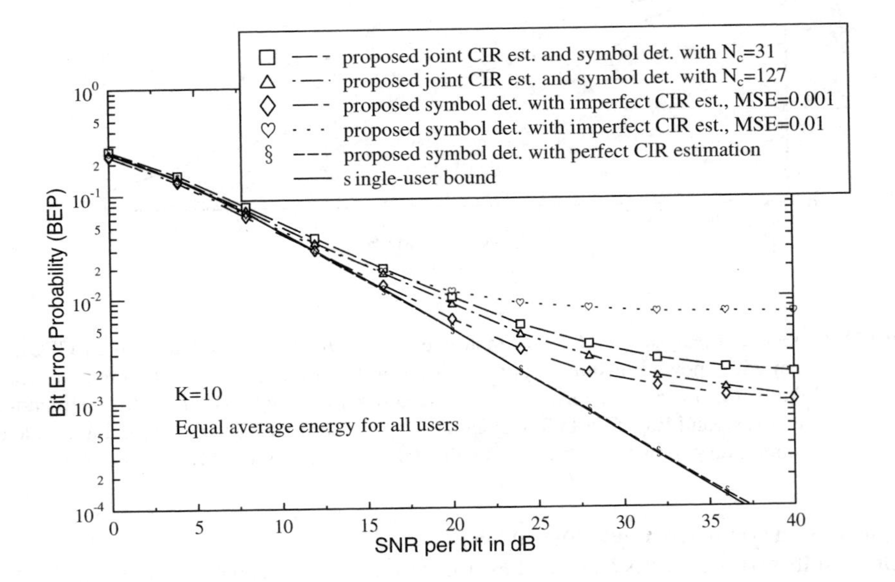

Figure 11.10: BEP performance of the proposed GA-assisted joint CIR estimator and symbol detector for $K = 10$ users over narrowband Rayleigh-fading channels at $f_d = 200$ Hz after the differential decoder. Results were shown for spreading factors of $N_c = 31$ and $N_c = 127$. Also shown are the BEP performances of the GA-assisted data detector with imperfect CIR estimation for MSE values of 0.01 and 0.001. The GA configuration and the simulation parameters used are listed in Table 11.6 and Table 11.5, respectively.

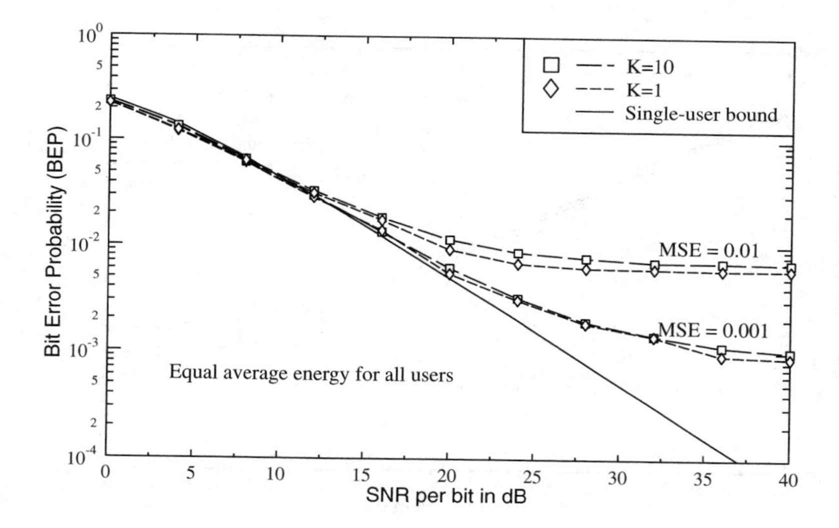

Figure 11.11: BEP performance comparison between the proposed GA-assisted CIR and symbol detector for $K = 10$ users and the matched filter for $K = 1$ over narrowband Rayleigh-fading channels at $f_d = 200$ Hz after the differential decoder using imperfect CIR estimation with MSE values of 0.01 and 0.001. Results were shown for a spreading factor of $N_c = 31$.

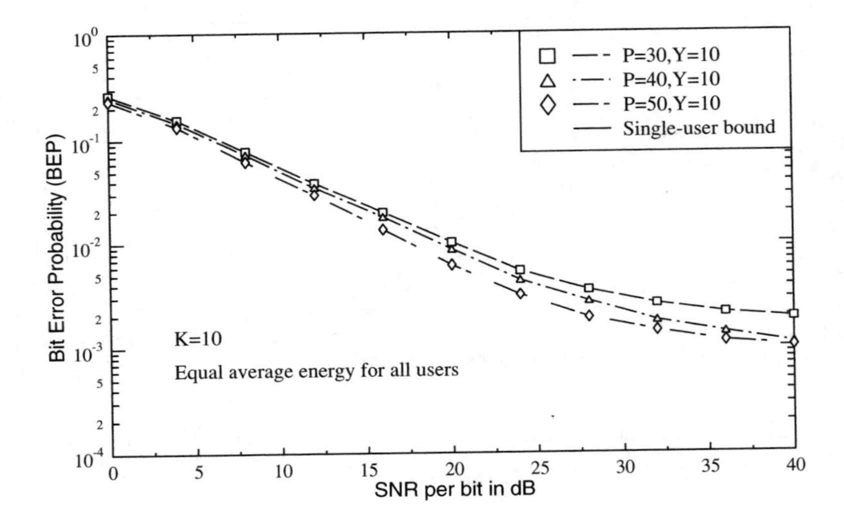

Figure 11.12: BEP performance of the proposed GA-assisted joint CIR estimator and symbol detector for $K = 10$ users with various population sizes P over narrowband Rayleigh-fading channels at $f_d = 200$ Hz after the differential decoder. The GA configuration and the simulation parameters used are listed in Table 11.6 and Table 11.5, respectively.

Chapter 12

Genetic Algorithm-Assisted, Antenna Diversity-Aided Multi-user Detection[1]

12.1 Introduction

It is well known that the hostile effects of fading constitute a major limitation of the system performance, which can be mitigated by diversity techniques [339, 340]. A commonly used diversity technique is receiver *antenna diversity* [90]. The distance between the receiving antennas is expected to be higher than half the wavelength, such that the signals received by the antennas become uncorrelated, experiencing sufficiently different path loss, fading and shadowing conditions [339]. Antenna diversity in conjunction with CDMA has been investigated for example in [341–343].

In this chapter, we present a novel approach to the problem of multi-user detection in DS/CDMA over flat Rayleigh-fading channels assisted by antenna diversity based on the GA-assisted multi-user detector developed in Chapter 10. The antennas are assumed to be sufficiently far apart, such that the received signals at the antennas are faded independently, resulting in an independent correlation metric obeying Equation (10.23) for each antenna. This poses a problem to the optimization process due to the fact that while a specific bit sequence $\boldsymbol{b}$ may optimize the correlation metric of one antenna, the same bit sequence may not necessarily optimize the correlation metric of the other antennas. In order to resolve this dilemma, two different strategies of creating the mating pool are considered. In our first approach, all the non-identical individuals in a given population of the GA are picked for the

[1] *Single- and Multi-Carrier CDMA Multi-User Detection, Space-Time Spreading, Synchronization, Networking and Standards.* L.Hanzo, L-L.Yang, E-L.Kuan and K.Yen,
 ISBN 0-470-86309-9

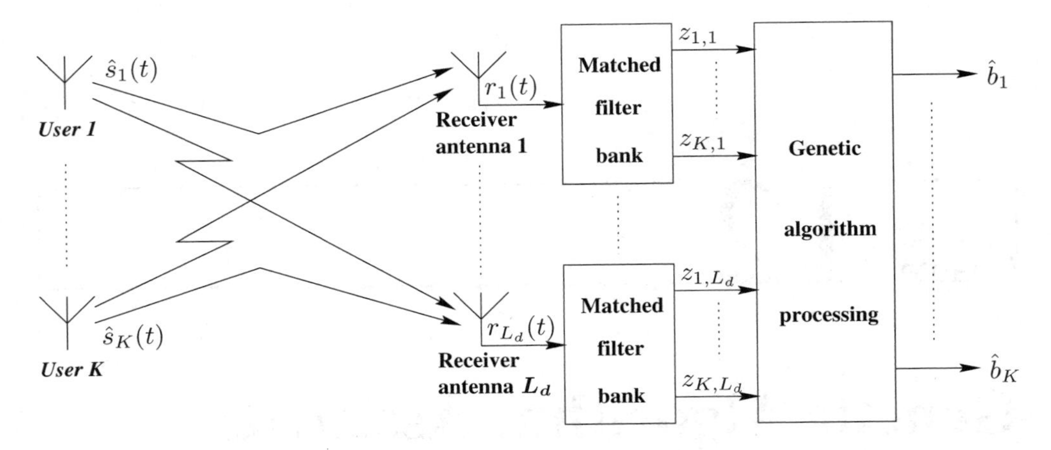

Figure 12.1: Block diagram of the K-user system model incorporating L_d-antenna diversity.

mating pool. This approach is identical to that adopted so far in Chapter 10 and Chapter 11. According to our second strategy, the individuals in a given population associated with the GA are picked for the mating pool based on the concept of the so-called *Pareto optimality* [9], which uses the information from the antennas independently.

This chapter is organized as follows. Section 12.2 describes our system model, which again is assumed to be a K-user symbol-synchronous CDMA system communicating over uncorrelated non-frequency-selective Rayleigh fading channels and receiving using L_d number of antennas. We note, however, that the proposed GA-assisted multi-user detector can also be applied to asynchronous systems transmitting over frequency-selective Rayleigh fading channels using multiple receiving antennas. This can be achieved by simply modifying the correlation metric of Equation (10.23), as will be presented in Chapter 13. The GA-assisted joint multi-user CIR estimator and symbol detector proposed in the previous chapter can also be applied in the context of multiple receiver antennas. Section 12.3 describes the GAs used for implementing our proposed detector in conjunction with diversity reception invoking two different strategies of creating the mating pool, as mentioned previously. Our simulation results are presented in Section 12.4, where the BEP performance of the GA-assisted multi-user detector using two diversity antennas will be investigated under the assumption of perfect CIR estimation. The BEP performance of the GA-assisted joint multi-user CIR estimator and symbol detector proposed in Chapter 11 using two diversity antennas will also be evaluated. Finally, Section 12.5 concludes this chapter.

12.2 System Model

The system model used in this chapter is depicted in Figure 12.1, where the transmitted signals of the K users are received at the base station over L_d receiver antennas. The transmitted signal $\hat{s}_k(t)$, $k = 1, \ldots, K$, of each user is given by Equation (10.1) of Chapter 10. We assumed that the L_d antennas are sufficiently separated spatially, such that the received signals of the K users at each antenna are statistically independent. Hence we can express the received signal at the ith antenna as:

$$r_i(t) = \sum_{k=1}^{K} s_{k,i}(t) + n_i(t), \tag{12.1}$$

with

$$s_{k,i}(t) = \sqrt{\xi_k}\alpha_{k,i} b_k a_k(t) e^{j\phi_{k,i}}, \quad \forall k = 1, \ldots, K \tag{12.2}$$

where ξ_k, b_k and $a_k(t)$ correspond to the bit energy, the transmitted bit and the signature sequence associated with the kth user, respectively. Furthermore, $\alpha_{k,i}$ and $\phi_{k,i}$ describe the channel attenuation and phase for the link between the kth user and the ith antenna, which was given by Equation (10.5) for a single antenna. The path amplitudes are normalized, such that $\sum_{i=1}^{L_d} E\left[\alpha_{k,i}^2\right] = 1$ for $k = 1, \ldots, K$. Following the analysis carried out in Section 10.2, the output $\boldsymbol{Z}_i$ of the matched filter bank at the ith diversity antenna, as portrayed in Figure 12.1, is given by the vector:

$$\boldsymbol{Z}_i = [z_{1,i}, z_{2,i}, \ldots, z_{K,i}]^T = \boldsymbol{R}\boldsymbol{\xi}\boldsymbol{C}_i\boldsymbol{b} + \boldsymbol{n}_i, \tag{12.3}$$

where

$$\begin{aligned}
\boldsymbol{C}_i &= \operatorname{diag}\left[\alpha_{1,i}e^{j\phi_{1,i}}, \ldots, \alpha_{K,i}e^{j\phi_{K,i}}\right] \\
\boldsymbol{\xi} &= \operatorname{diag}\left[\sqrt{\xi_1}, \ldots, \sqrt{\xi_K}\right] \\
\boldsymbol{b} &= [b_1, \ldots, b_K]^T \\
\boldsymbol{n}_i &= [n_{1,i}, \ldots, n_{K,i}]^T
\end{aligned} \tag{12.4}$$

and $\boldsymbol{R}$ is a $K \times K$-dimensional user signature sequence cross-correlation matrix, as given by Equation (10.17). Hence based on the observation vector $\boldsymbol{Z}_i$ given in Equation (12.3), we can express the correlation metric corresponding to the ith antenna as [90]:

$$\Omega_i(\boldsymbol{b}) = 2\Re\left[\boldsymbol{b}^T\boldsymbol{\xi}\boldsymbol{C}_i^*\boldsymbol{Z}_i\right] - \boldsymbol{b}^T\boldsymbol{\xi}\boldsymbol{C}_i\boldsymbol{R}\boldsymbol{C}_i^*\boldsymbol{\xi}\boldsymbol{b} \quad \forall i = 1, \ldots, L_d \tag{12.5}$$

The decision rule for the optimum multi-user detector associated with the ith antenna is to choose the specific bit vector $\hat{\boldsymbol{b}}$, which maximizes the correlation metric given in Equation (12.5). Hence, the estimated transmitted bit vector of the K users is given by:

$$\hat{\boldsymbol{b}} = \arg\left\{\max_{\boldsymbol{b}} \Omega_i(\boldsymbol{b})\right\}. \tag{12.6}$$

Since the channel characteristics for each antenna are statistically independent, we have typically $\Omega_i(\boldsymbol{b}) \neq \Omega_{j\neq i}(\boldsymbol{b})$ for the correlation metrics of the L_d diversity antennas. In certain scenarios such as during deep fades, the above inequality implies that:

$$\arg\left\{\max_{\boldsymbol{b}}\left[\Omega_i\left(\boldsymbol{b}\right)\right]\right\} = \hat{\boldsymbol{b}} \neq \arg\left\{\max_{\boldsymbol{b}}\left[\Omega_{j\neq i}\left(\boldsymbol{b}\right)\right]\right\} \quad \forall i = 1, \ldots, L_d \tag{12.7}$$

In other words, there may not exist a single solution $\hat{\boldsymbol{b}}$, which is the best with respect to all the L_d correlation metrics. This creates a so-called *optimization conflict* [344], since the optimization of the L_d correlation metrics may sometimes lead to two or more possible solutions and any one of them is an acceptable solution. Nevertheless, for optimum detection, the correlation metrics corresponding to the L_d number of diversity antennas are combined according to [329]:

$$\begin{aligned}\Omega\left(\boldsymbol{b}\right) &= \sum_{i=1}^{L_d}\Omega_i\left(\boldsymbol{b}\right)\\ &= 2\Re\left\{\boldsymbol{b}^T\vec{\boldsymbol{C}}^H\vec{\boldsymbol{\xi}}\vec{\boldsymbol{Z}}\right\} - \boldsymbol{b}^T\vec{\boldsymbol{C}}^H\vec{\boldsymbol{\xi}}\vec{\boldsymbol{R}}\vec{\boldsymbol{\xi}}\vec{\boldsymbol{C}}\boldsymbol{b},\end{aligned} \tag{12.8}$$

where $\vec{\boldsymbol{Z}} = \left[z_{1,1}, \ldots, z_{1,L_d}, \ldots, z_{K,1}, \ldots, z_{K,L_d}\right]^T$, $\vec{\boldsymbol{\xi}} = \text{diag}\left[\sqrt{\xi_1}\boldsymbol{I}, \ldots, \sqrt{\xi_K}\boldsymbol{I}\right]$ with $\boldsymbol{I}$ being a unity vector of length L_d. Furthermore, $(\cdot)^H$ denotes a Hermitian matrix and $\vec{\boldsymbol{C}} = \text{diag}\left[(\alpha_{1,1}e^{j\theta_{1,1}}, \ldots, \alpha_{1,L_d}e^{j\theta_{1,L_d}})^T, \ldots, (\alpha_{K,1}e^{j\theta_{K,1}}, \ldots, \alpha_{K,L_d}e^{j\theta_{K,L_d}})^T\right]$. The decision rule is then to find the estimated transmitted bit vector $\hat{\boldsymbol{b}}$ that maximizes $\Omega\left(\boldsymbol{b}\right)$ in Equation (12.8).

In the next section we will highlight the philosophy of our GA-assisted diversity-aided multi-user detector with emphasis on the strategies invoked in creating the mating pool, in order to detect the users' transmitted bits.

12.3 GA-Assisted Diversity-Aided Multi-user Detection

The flowchart of the GA invoked in this chapter is depicted in Figure 12.2. Apart from the specific approach used in creating the mating pool, the structure of the GA invoked here is identical to the one highlighted in Chapter 10. Similarly to Chapter 10, there are P number of individuals in a population, where the pth individual is represented by a K-bit vector as $\tilde{\boldsymbol{b}}_p(y) = \left[\tilde{b}_{p,1}(y), \ldots, \tilde{b}_{p,K}(y)\right]$ and y denotes the generation index. The individuals during the initialization phase of Figure 12.2 are generated based on the maximal ratio combining [90] of the matched filter outputs corresponding to all the antennas. Hence we have [90]:

$$\begin{aligned}\tilde{\boldsymbol{b}}_1(0) &= \text{sgn}\left[\Re\left(\sum_{i=1}^{L_d}\boldsymbol{Z}_i\boldsymbol{C}_i^*\right)\right]\\ \tilde{\boldsymbol{b}}_p(0) &= MUTATION\left[\tilde{\boldsymbol{b}}_1(0)\right] \quad \text{for } p = 2, \ldots, P\end{aligned} \tag{12.9}$$

In a system consisting of L_d receiving antennas, each individual is associated with L_d

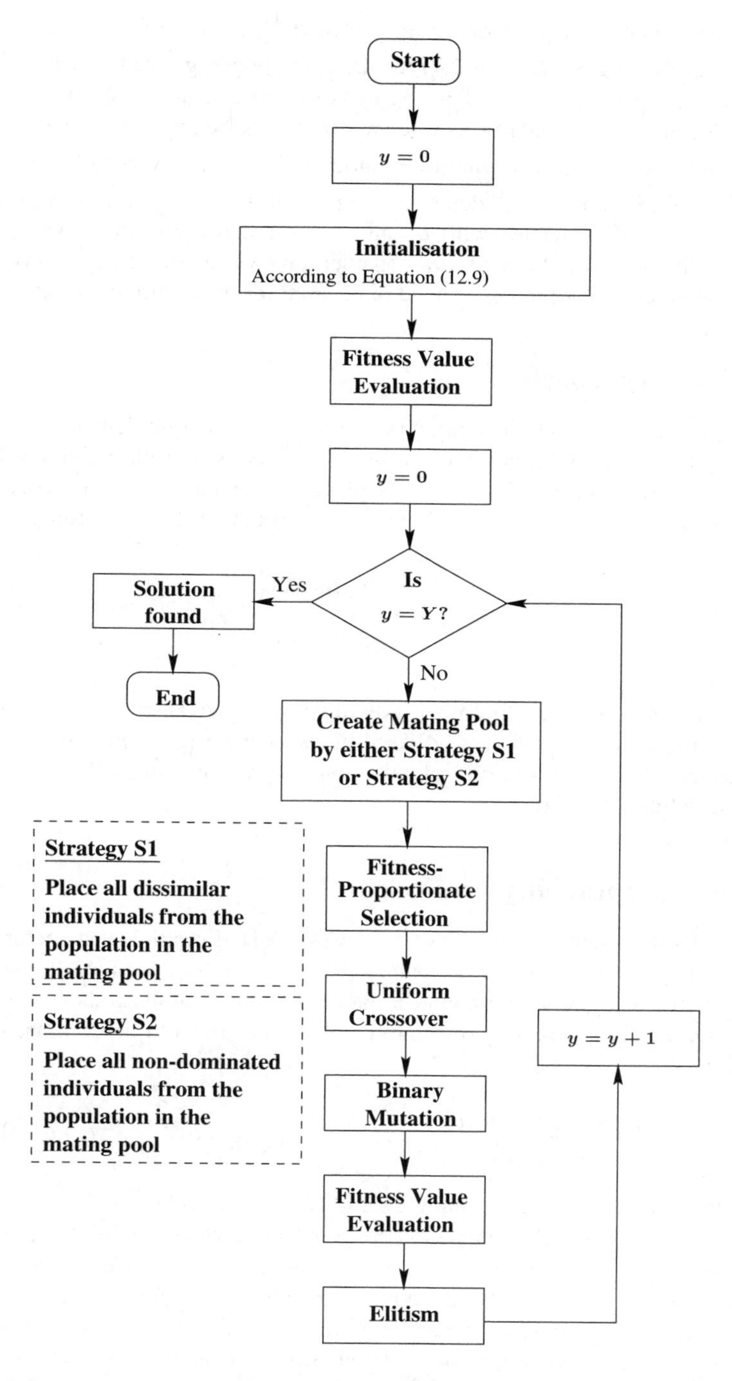

Figure 12.2: A flowchart depicting the structure of a genetic algorithm used for function optimization.

number of antenna-specific figures of merit denoted as $\Omega_i\left(\tilde{\boldsymbol{b}}_p(y)\right)$ for $i = 1, \ldots, L_d$, which are derived by evaluating Equation (12.7) for the corresponding antenna, where $\boldsymbol{b}$ is defined by the individual. We shall refer to these figures of merit as antenna-specific fitness values. We also introduced an additional fitness value referred to as the diversity-specific fitness value $\Omega\left(\tilde{\boldsymbol{b}}_p(y)\right)$, which is derived according to Equation (12.8). The diversity-based fitness value associated with each individual will determine its probability of selection. Hence in summary, each individual will be associated with L_d number of antenna-specific fitness values and a diversity-based fitness value. We will now consider how we can make use of this information in order to create the mating pool and to aid our search for the optimum K-bit vector $\boldsymbol{b}$.

12.3.1 Direct Approach

The direct approach of creating the mating pool is similar to that implemented in Chapter 10. Basically, only the diversity-based fitness value associated with each individual will be considered here. All dissimilar individuals will be placed in the mating pool and their probability of selection is computed following the philosophy of Equation (9.9) according to:

$$p_i = \frac{\exp\left[\Omega\left(\tilde{\boldsymbol{b}}_i(y)\right)\right]}{\sum_{j=1}^{T}\exp\left[\Omega\left(\tilde{\boldsymbol{b}}_j(y)\right)\right]}, \tag{12.10}$$

where $T \leq P$ is the number of dissimilar individuals in a given population. Again, we have considered the exponent of the diversity-based fitness value associated with each individual, in order to ensure that the probability of selection becomes positive. We shall refer to this direct approach strategy as S1.

12.3.2 Pareto Optimality Approach

Our second individual-selection strategy of the GA-assisted multi-user detector is based on the concept of the so-called *Pareto optimality* [9]. This strategy favours the so-called *non-dominated* individuals by retaining them for the mating pool and discards the so-called *dominated* individuals. Then the pth K-bit individual is considered to be dominated by the qth individual iff [345]:

$$\forall i \in \{1,2\} : \Omega_i\left(\tilde{\boldsymbol{b}}_q^{(y)}\right) \geq \Omega_i\left(\tilde{\boldsymbol{b}}_p^{(y)}\right) \wedge \exists j \in \{1,2\} : \Omega_j\left(\tilde{\boldsymbol{b}}_q^{(y)}\right) > \Omega_j\left(\tilde{\boldsymbol{b}}_p^{(y)}\right). \tag{12.11}$$

In verbal terms this implies that an individual is considered to be dominated if there exists another individual in which all the antenna-specific fitness values are higher than that of the dominated individual. If an individual is not dominated in the sense of Equation. (12.11) by any other K-bit individuals in the population, then by definition it is considered to be non-dominated. The non-dominated individuals are also known as Pareto-optimal individuals [344]. Since GAs work with a population of candidate solutions, a number of Pareto-optimal individuals may be captured using GAs. According to our second individual-selection strategy, all the non-dominated K-bit individuals are selected and placed in the

Parameter	Value
Spreading factor N_c	31
Modulation mode	BPSK
Number of CDMA users K	10
Number of diversity antennas L_d	2

Table 12.1: Simulation parameters for the experiments of Figures 12.3-12.6.

mating pool. The probability of selection of these individuals in the mating pool is then computed according to Equation (12.10) using their corresponding diversity-based figure of merit, where T in this case denotes the number of Pareto-optimal individuals in a given population. If there is only one non-dominated individual in a given population, then the next set of non-dominated individuals in the population will be found and placed in the mating pool, together with the ultimate non-dominated individual.

Observe that this strategy uses the information provided by the L_d antennas independently, in order to decide which individuals are placed in the mating pool. By contrast, the direct approach of Section 12.3.1 based its decisions on only the diversity-based fitness values. We shall refer to the *Pareto Optimality* approach as S2. Note that the Pareto optimality concept can only be applied to GAs or population-based algorithms, since non-dominated individuals can only be identified if more than one candidate solutions are evaluated at a time.

From a detection point of view, the concept of Pareto optimality does not give the most likely transmitted bit sequence. Upon termination of the GA, if there is only one Pareto-optimal individual in the final population, then this solution will be deemed as the detected bit sequence $\hat{\boldsymbol{b}}$. On the other hand, there may exist a number of Pareto-optimal individuals in the final population. In this case, we will adopt the optimum criteria according to Equation. (12.8) and then the individual that corresponds to the highest diversity-specific fitness value will be the detected bit sequence $\hat{\boldsymbol{b}}$.

12.4 Simulation Results

In this section our computer simulation results are presented, in order to characterize the BEP performance of the GA-assisted multi-user detector in conjunction with L_d number of received antennas employing both strategies of creating the mating pool, which were highlighted in Section 12.3. All the results in this chapter were based on evaluating the BEP performance of a bit-synchronous K-user CDMA system using L_dth-order antenna diversity reception over Rayleigh fading channels, where the signals of the diversity channels were uncorrelated with each other. The spreading factor was $N_c = 31$ and the signature sequences were randomly generated. The results shown in Figures 12.3–12.4 were based on the asumption that perfect CIR estimation is invoked at each antenna, while in Figures 12.5-12.6, imperfect CIR estimation was assumed. Summaries of the simulation parameters and the GA configuration invoked is listed in Table 12.1 and Table 12.2, respectively.

Specifically, Figure 12.3 shows the BEP performance over a narrow-band Rayleigh channel against the average SNR per bit for the GA-assisted $K = 10$-user detector employing

Set-up/Parameter	Method/Value
Individual initialization method	According to Equation (12.10)
Selection method	Proportionate-Fitness
Cross-over operation	Uniform cross-over
Mutation operation	Standard binary mutation
Elitism	Yes
Incest prevention	Yes
Population size P	Given in the associated plots
Number of generations Y	10
Mating pool size T	– Strategy S1: All dissimilar individuals in the population – Strategy S2: All non-dominated individuals in the population
Probability of mutation p_m	0.1

Table 12.2: Configuration of the GA used to obtain the results of Figure 12.3 and Figure 12.6. Explicit description of the fitness-proportionate selection scheme, the uniform cross-over operation and the floating point mutation operation can be found in Section 9.4.2, Section 9.4.3 and Section 9.4.4, respectively.

both strategy S1 and S2 assuming equal average received energy at the $L_d = 2$ antennas, i.e. for $E\left[\alpha_{k,1}^2\right] = E\left[\alpha_{k,2}^2\right] = 0.5$. Perfect power control and CIR estimation were assumed. The number in parentheses denotes the maximum number of times the correlation metric of Equation (12.8) is evaluated by the GA-assisted multi-user detector. Again, this complexity figure is compared to the complexity of the optimum multi-user detector, which requires 2^K correlation metric evaluations. The single-user bound, which assumed an equal average received energy at both antennas, was computed using [90]:

$$P_2 = \left[\frac{1}{2}(1-\mu)\right]^{L_d} \sum_{k=0}^{L_d-1} \binom{L_d - 1 + k}{k} \left[\frac{1}{2}(1+\mu)\right]^k, \tag{12.12}$$

where $\mu = \sqrt{\frac{\bar{\gamma}_k}{1+\bar{\gamma}_k}}$ and $\bar{\gamma}_k$ is the average SNR per bit of the kth user. An error floor is observed in Figure 12.3. Again, this is due to the limitations of the GAs for a given population size P and for a number of generations Y. However, the BEP performance improved, when the population size P was increased from $P = 10$ to $P = 16$. However, this also increased the computational complexity. Hence the value of P can be selected, in order to find a trade-off between computational complexity and performance. More importantly, we can see from Figure 12.3 that the GA employing strategy S2 performs better, exhibiting a lower error floor, as compared to employing strategy S1. Nevertheless, both strategies were capable of matching the single-user bound performance up to SNRs of $\bar{\gamma}_k = 16$ dB and $\bar{\gamma}_k = 20$ dB for $P = 10$ and $P = 16$, respectively.

We then investigated the BEP performance of the GA-based multi-user detector employing both selection strategy S1 and S2 in conjunction with unequal average received energy

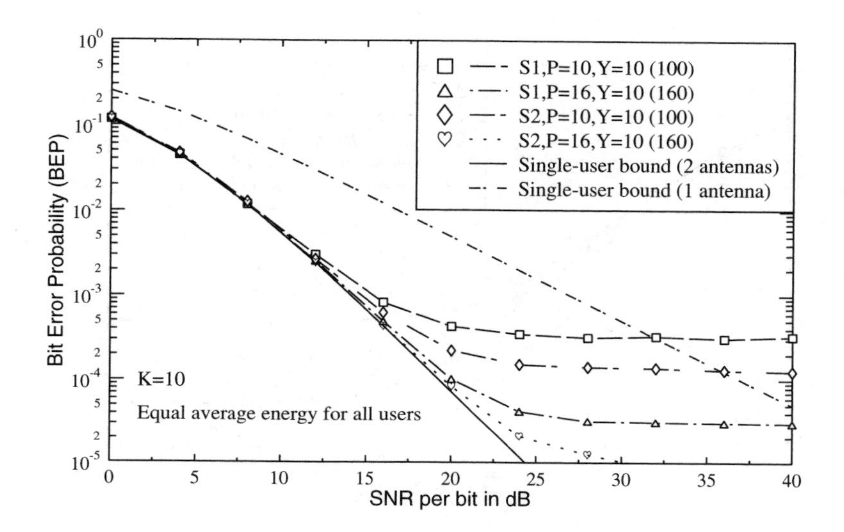

Figure 12.3: BEP performance of the proposed GA-assisted multi-user detector over narrow-band Rayleigh channels employing strategies S1 and S2 in creating the mating pool with population sizes of $P = 10, 16$ using binary random signature sequences of length $N_c = 31$ and supporting $K = 10$ users. The average received energy at the antennas was assumed to be equal, i.e. for $E\left[\alpha_{k,1}^2\right] = E\left[\alpha_{k,2}^2\right] = 0.5$. The GA configuration and the simulation parameters used are listed in Table 12.2 and Table 12.1, respectively.

at the two antennas, setting $E\left[\alpha_{k,1}^2\right] = 0.8$ and $E\left[\alpha_{k,1}^2\right] = 0.2$. Perfect power control and CIR estimation were assumed again. The associated results are shown in Figure 12.4 in comparison to the single-user bound given by Equation (12.12). Again, we can see that GAs invoking strategy S2 exhibit a lower BEP compared to strategy S1.

Figure 12.5 and Figure 12.6 portray the BEP performance of the GA-assisted multi-user detector in the context of imperfect CIR estimation having a CIR estimation MSE of 0.01 and 0.001, respectively. Perfect power control is assumed with equal average received energy at both antennas. In Figure 12.5 we can see that there is no significant difference in the achievable BEP performance between S1 and S2 at $P = 16$. This is due to the high MSE of the CIR estimation, which limits the performance, as also highlighted in [346] in the context of conventional CDMA detectors. The BEP for a single-user transmission scenario using a matched filter and maximal ratio combining in conjunction with a CIR estimation MSE of 0.01 constitutes the lower bound, as shown in Figure 12.5. At the lower CIR estimation MSE of 0.001, we can see from Figure 12.6 that the BEP was lower and the detector was capable of matching the single-user bound up to an SNR of about 20 dB. We can also see from Figure 12.3, which assumed perfect CIR estimation, and from Figure 12.6, which assumed a CIR estimation MSE of 0.001 for both $P = 10$ and $P = 16$ that the error floors in both

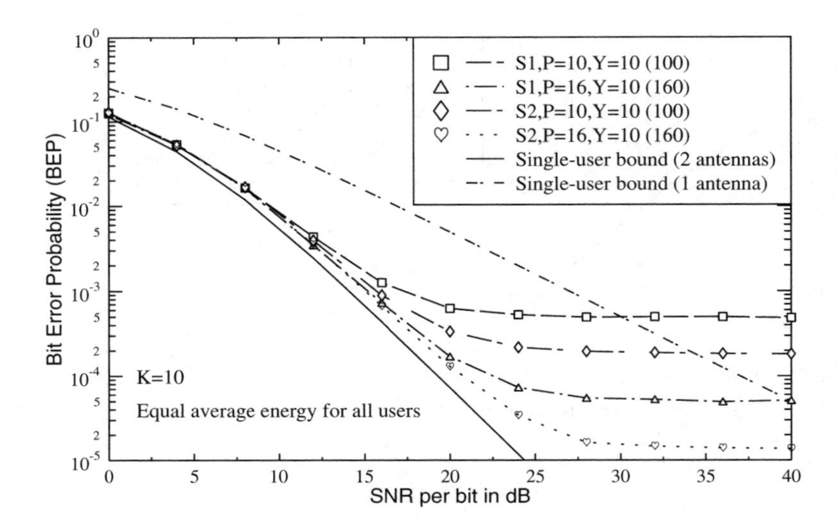

Figure 12.4: BEP performance of the proposed GA-assisted multi-user detector over narrow-band Rayleigh channels employing strategies S1 and S2 in creating the mating pool with population sizes of $P = 10, 16$ using binary random signature sequences of length $N_c = 31$ and supporting $K = 10$ users. The average received energy at the antennas was assumed to be unequal with $E\left[\alpha_{k,1}^2\right] = 0.8$ and $E\left[\alpha_{k,2}^2\right] = 0.2$. The GA configuration and the simulation parameters used are listed in Table 12.2 and Table 12.1, respectively.

figures occur at the same BEP. Hence the BEP floor was deemed to be due to the limitations of the GAs and not the CIR estimation error, since the single-user bound is much lower.

Finally, the BEP performance of the joint GA-assisted multi-user CIR estimation and symbol detection scheme introduced in Chapter 11 was evaluated in conjunction with two diversity antennas. Apart from the creation of the mating pool, which follows the strategies highlighted in Section 12.3, the entire detection process is the same as that implemented in Chapter 11. In this case, the antenna-specific fitness values corresponding to each individual are evaluated according to Equation (11.6) associated with each antenna and the diversity-based fitness value of each individual is obtained by combining its corresponding antenna-specific fitness values. The GA configuration used for this simulation is characterized in Table 12.3. Note that for strategy S1, only $T = 5$ non-identical individuals associated with the highest fitness values in the population were placed in the mating pool. The reason for this course of action was highlighted in Section 11.3.2. The BEP achievable performance is shown in Figure 12.7.

First, we compared the performance gain achieved by utilizing two antennas instead of one without increasing the computational complexity. This is represented in Figure 12.7 by the curves corresponding to $P = 40, Y = 10, L_d = 1$ for a single antenna and to $P = 20, Y = 10, L_d = 2$ for two antennas. We can see that there is a significant BEP perfor-

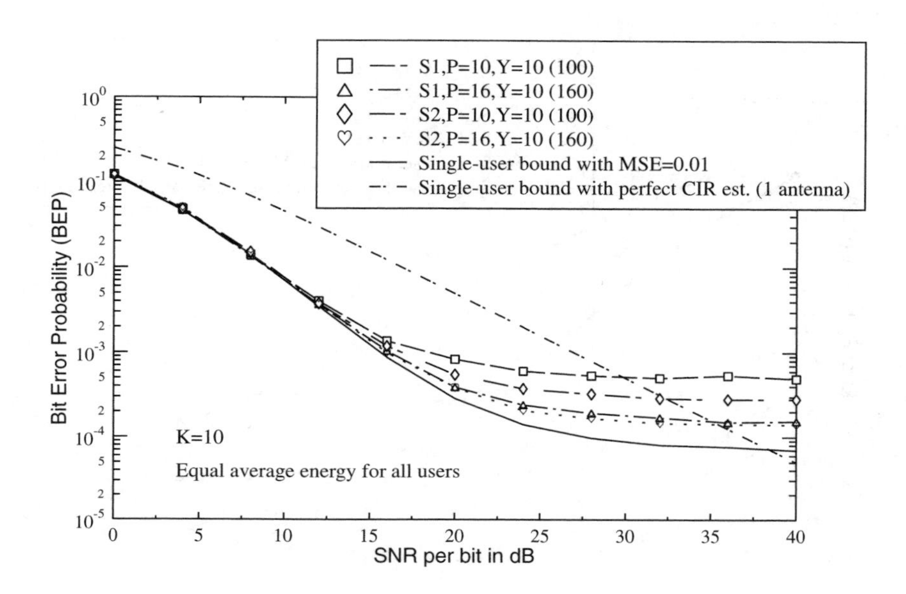

Figure 12.5: BEP performance of the proposed GA-assisted multi-user detector over narrow-band Rayleigh channels employing strategies S1 and S2 in creating the mating pool with population sizes of $P = 10, 16$ using binary random signature sequences of length $N_c = 31$ and supporting $K = 10$ users. Both antennas were assumed to exhibit a CIR estimation error of 0.01. The GA configuration and the simulation parameters used are listed in Table 12.2 and Table 12.1, respectively.

mance improvement for the twin-antenna-assisted system. However, there is no performance difference between S1 and S2, since the BEP is limited by the CIR estimation error. We can reduce the CIR estimation error by increasing the population size P, as seen previously in Figure 11.4, which in turn will reduce the BEP. This is explicitly shown in Figure 12.7, where there is a BEP improvement, when the population size is increased from $P = 20$ to $P = 40$.

12.5 Chapter Summary and Conclusion

In this chapter, we developed a GA-assisted multi-user detector for a symbol synchronous CDMA system incorporating L_d number of diversity antennas. These antennas are expected to be separated by a distance higher than half the wavelength, so that the received signal at each antenna transmitted from any of the users becomes uncorrelated. However, the GA's figure of merit that is obtained from the correlation metrics associated with each antenna is typically different. As a result, there may not exist a particular bit sequence, which is the best with respect to all the antennas' correlation metric.

We have resolved this optimization conflict to our advantage by selecting only the so-called non-dominated individuals of a given population for the mating pool. This process

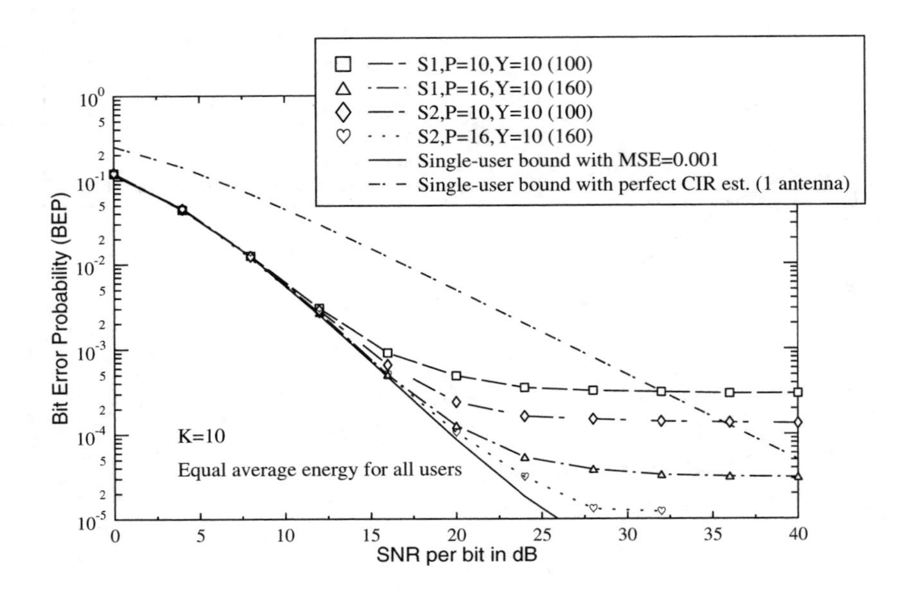

Figure 12.6: BEP performance of the proposed GA-assisted multi-user detector over narrow-band Rayleigh channels employing strategies S1 and S2 in creating the mating pool with population sizes of $P = 10, 16$ using binary random signature sequences of length $N_c = 31$ and supporting $K = 10$ users. Both antennas were assumed to exhibit a CIR estimation error of 0.001. The GA configuration and the simulation parameters used are listed in Table 12.2 and Table 12.1, respectively.

was based on exploiting the Pareto optimality. The creation of the mating pool based on Pareto optimality was referred to here as Strategy S2. This strategy was then compared with the direct approach, which was used previously in Chapter 10 and Chapter 11, whereby all non-identical individuals in a given population were selected for the mating pool. This direct approach was referred to here as Strategy S1.

We have shown that GAs employing Strategy S2 in creating the mating pool always exhibit a lower BEP compared to those employing Strategy S1. We have also shown that the BEP performance can be improved by increasing the population size. Finally, we showed that a significant BEP performance gain can be achieved by the joint GA-assisted CIR estimator and symbol detector, when utilizing two receiving antennas instead of a single antenna without increasing the computational complexity.

Set-up/Parameter	Method/Value
Individual initialization method	According to the flowchart of Figure 11.2 and Equation (12.9)
Selection method	Proportionate-Fitness
Cross-over operation	Uniform cross-over
Mutation operation	- Floating point mutation for $\tilde{\boldsymbol{C}}_p^{(m)}(y)$ - Binary mutation for $\tilde{\boldsymbol{e}}_p^{(m)}(y)$
Elitism	Yes
Incest prevention	Yes
Population size P	Given in Figure 12.7
Number of generations Y	10
Mating pool size T	– Strategy S1: $T = 5$ dissimilar individuals associated with the highest diversity-specific fitness values in the population – Strategy S2: All non-dominated individuals in the population
Probability of mutation p_m	0.1
Mutation size λ_{max}	0.1

Table 12.3: Configuration of the GA used to obtain the results of Figure 12.7. Explicit description of the fitness-proportionate selection scheme, the uniform cross-over operation and the floating point mutation operation can be found in Section 9.4.2, Section 9.4.3 and Section 9.4.4, respectively.

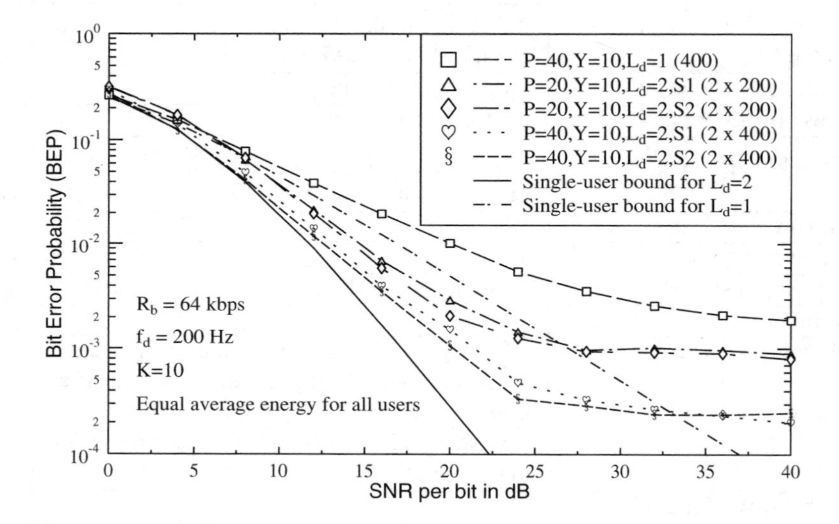

Figure 12.7: BEP performance of the proposed GA-assisted joint channel estimator and symbol detector for $K = 10$ users over narrowband Rayleigh-fading channels at $f_d = 200$ Hz after the differential decoder. Results were shown for spreading factors of $N_c = 31$. The GA configuration and the simulation parameters used are listed in Table 12.3 and Table 11.5, respectively.

Chapter 13

Genetic Algorithm-Assisted Multi-user Detection for Asynchronous CDMA [1]

13.1 Introduction

So far, we have assumed that all the users transmit their signals synchronously. In order to accomplish this symbol-synchronization, a form of closed-loop timing control would be required between the base station's receiver and all the mobile users' transmitters [319]. In practice symbol-synchronous CDMA reception is not easy to implement. However, one of the advantages of CDMA over the more traditional Frequency Division Multiple Access (FDMA) and Time Division Multiple Access (TDMA) is its capability of supporting uncoordinated uplink signal transmission. Hence it is possible to allow the users to transmit their signals in an asynchronous manner.

In an asynchronous DS-CDMA system, every bit of each user is interfered with by two bits of every other user in the system, which are overlapping with the bit of interest, assuming an identical channel bit rate for all users. Hence the multi-user detector must have knowledge of these two overlapping bits, in order to efficiently detect the desired bit. Conventional multi-user detectors, such as the decorrelator [347], operate on the entire length M of the users' bit sequence at once. This results in a long detection delay as well as in a significant receiver complexity, when M is high. Several methods [348–352] have been proposed in order to reduce the detection delay and the receiver complexity in asynchronous DS-CDMA systems. The simplest way is to periodically cease transmission for a fixed time interval for

[1] *Single- and Multi-Carrier CDMA Multi-User Detection, Space-Time Spreading, Synchronization, Networking and Standards.* L.Hanzo, L-L.Yang, E-L.Kuan and K.Yen,
 ISBN 0-470-86309-9

all users [323, 348]. This will effectively break the continuous transmissions into frames and hence reduce the complexity of the multi-user detector. However, this method still requires synchronisation among the users, although not as strictly, as in symbol-synchronous transmissions. Furthermore, this method will degrade the bandwidth efficiency of the system. In the proposal by Xie *et al.* [349], the detection observation window is truncated, such that only a portion of the bit sequence length M is considered by the detector at a time. In [349] the bits that coincide with the window's edge, referred to as the *edge bits* in this chapter, are tentatively estimated employing the conventional single-user correlator. The desired bits within the truncated observation window are then detected using conventional multi-user detection techniques. The overall performance of this technique is largely dependent on the estimation reliability of the edge bits by the single-user correlator, which degrades as K increases. In order to reduce the effects of the edge bits the adjacent subsequences input to the detector can be arranged to overlap. Wijayasuriya *et al.* [350] proposed a technique, where the edge bits are predicted using previously detected bits with the aid of convolutional decoding, although other channel codecs can also be used. Juntti *et al.* [351] proposed a finite-memory-length detector, referred to as a Finite Impulse Response (FIR) detector, in order to reduce the high memory length associated with traditional multi-user detectors employed in asynchronous CDMA systems. In the contribution by Shen *et al.* [352], the edge bits are estimated using a modified decorrelator. These proposals [349, 350, 352] demonstrated that maintaining a low edge bit error probability is essential, in order to attain a high overall bit error rate performance.

Using a similar approach to that in [349,350,352] we proposed a multi-user detector for an asynchronous DS-CDMA system transmitting over L-path Rayleigh fading channels based on a GA. In order to reduce the complexity of the detector, as well as to decrease the detection time, the observed window is truncated such that it encompasses at most one complete symbol interval of all users in any detection window. Let us assume that we are interested in detecting the ith bit of all users. Then the edge bits will be the $(i-1)$st bits and the $(i+1)$st bits of all interfering users, referred to in this chapter as the start edge bits (SEB) and the end edge bits (EEB), respectively. The SEBs have been detected in the previous observed window and hence they are known to the receiver. Two different strategies are adopted, in order to estimate the EEBs of all users. In our first strategy, the EEBs are estimated employing the conventional single-user correlator, a technique similar to that in [349]. GAs are then developed, in order to estimate the desired ith bits of all users. In our second strategy, we extend the same GAs in order to simultaneously improve the EEB error probability (EBEP). In contrast to the previously proposed techniques [350, 352], the EEB and the desired bits in the latter strategy are estimated simultaneously using the same process. This results in minimal detection delay and no additional computation is required for predicting the EEB.

The performance of the proposed multi-user detector is examined by computer simulations, whereby the measure of interest is the desired bit error probability (DBEP). We will investigate the effects of the ambiguity of the edge bits on the DBEP. The improvement in the EBEP using our second strategy, i.e. the GA-based estimation, over that of our first strategy employing the single-user correlator based edge-bit predictor is also shown. Furthermore, we will evaluate the effects of varying the GA parameters on the DBEP performance, in order to strike a balance between detection complexity and performance. Our simulation results showed that the DBEP performance corresponding to the first detection strategy is limited by the high EBEP. On the other hand, upon using GAs for improving the accuracy of the edge

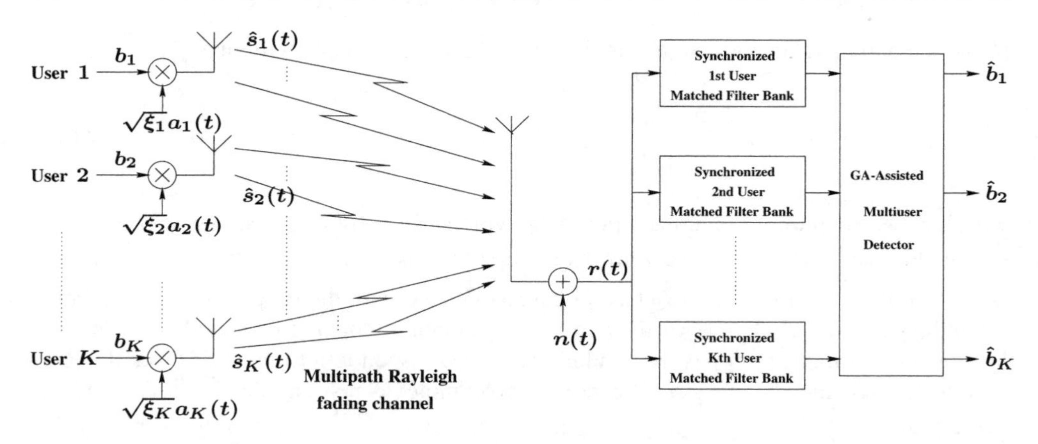

Figure 13.1: Block diagram of the K-user asynchronous CDMA system model in a multipath Rayleigh fading channel.

bits, our proposed multi-user detector can achieve a near-optimum DBEP performance, while imposing a lower complexity compared to that of the optimum multi-user detector [106].

The remainder of this chapter is organized as follows. Section 13.2 describes our asynchronous CDMA system communicating over multipath Rayleigh fading channels. The correlation metric required for the optimization process in conjunction with an asynchronous CDMA system is also developed. Section 13.3 describes the GAs used for implementing our proposed multi-user detector. The structure of the GAs will be slightly different from that invoked in Chapter 10, since the SEBs and the EEBs have to be taken into acount. Our simulation results are presented in Section 13.4, while Section 13.5 concludes the chapter.

13.2 Asynchronous CDMA System Model

We consider Binary Phase Shift Keying (BPSK) transmissions over a common AWGN channel shared by K asynchronous users employing DS-CDMA, as illustrated in Figure 13.1. The signal of each user is assumed to be propagating over L independent slowly Rayleigh fading paths to the base station's receiver. The complex lowpass impulse response of the channel for the kth user over the mth symbol interval of the M-bit transmission burst can be expressed as:

$$h_k^{(m)}(t) = \sum_{l=1}^{L} \alpha_{k,l}^{(m)} \exp(j\theta_{k,l}^{(m)})\delta(t - \tau'_{k,l}), \tag{13.1}$$

where $\alpha_{k,l}^{(m)}$, $\tau'_{k,l}$ and $\theta_{k,l}^{(m)}$ are the lth path gain, propagation delay and phase, respectively.

Assuming ideal lowpass receiver filtering for removing the high-frequency noise components, the baseband received signal as shown in Figure 13.1 can be written as:

$$r(t) = \sum_{m=0}^{M-1} \sum_{k=1}^{K} \sum_{l=1}^{L} \sqrt{\xi_k} b_k^{(m)} c_{k,l}^{(m)} a_k(t - mT_b - \tau_{k,l}) + n(t), \tag{13.2}$$

where M is the number of transmitted data symbols in a frame, ξ_k is the energy per bit of the kth user and $b_k^{(m)}$ is the mth data symbol of the kth user. Furthermore, $c_{k,l}^{(m)} = \alpha_{k,l}^{(m)} \exp(j\theta_{k,l}^{(m)})$ is the complex channel gain associated with the lth path of the kth user at the mth symbol interval, $a_k(t)$ is the normalized signature sequence of the kth user, as given by Equation (10.2) and $\tau_{k,l}$ is the random delay[2] corresponding to user k. Over Rayleigh fading channels the channel gain is a zero mean complex Gaussian random variable, where the amplitude $\alpha_{k,l}^{(m)}$ is Rayleigh distributed and the phase $\theta_{k,l}^{(m)}$ is uniformly distributed between $[0, 2\pi)$. For simplicity and without loss of generality, we assumed an ordering of the random delays $\tau_{k,l}$ such that $0 = \tau_{1,1} < \tau_{1,2} < \ldots < \tau_{1,L} < \tau_{2,1} < \ldots < \tau_{K,L} < T_b$. We also assumed that the energies, channel gains and random delays of all users are known to the receiver and that the channel gain is normalized so that the average signal energy levels at the output and input of the channel are the same, which is formulated as:

$$E\left[\sum_{l=1}^{L} \left|c_{k,l}^{(m)}\right|^2\right] = 1, \quad \text{for } k = 1, 2, \ldots, K. \tag{13.3}$$

The channel noise $n(t)$ is modelled by a zero mean, complex white Gaussian process exhibiting independent real and imaginary components, each having a double-sided power spectral density of $N_0/2$.

Again, we can represent the received signal due to the M-bit transmission burst by using the vector notation as:

$$r(t) = \sum_{m=0}^{M-1} \boldsymbol{a}^T(t - mT_b)\boldsymbol{\xi}\boldsymbol{c}^{(m)}\boldsymbol{b}^{(m)} + n(t), \tag{13.4}$$

where $\boldsymbol{a}(t) = [a_1(t - \tau_{1,1}), \ldots, a_1(t - \tau_{1,L}), \ldots, a_K(t - \tau_{K,L})]^T$ is the K users' signature sequence vector, $\boldsymbol{\xi} = \text{diag}\left[\sqrt{\xi_1}\boldsymbol{I}, \sqrt{\xi_2}\boldsymbol{I}, \ldots, \sqrt{\xi_K}\boldsymbol{I}\right]$ is a $KL \times KL$-dimensional diagonal matrix containing the energy of the K users, while $\boldsymbol{I}$ is an $L \times L$-dimensional identity matrix, $\boldsymbol{c}^{(m)} = \text{diag}\left[c_{1,1}^{(m)}, \ldots, c_{1,L}^{(m)}, \ldots, c_{K,L}^{(m)}\right]$ is the $KL \times KL$ diagonal CIR matrix of the K users for the L-path Rayleigh channels, $\boldsymbol{b}^{(m)} = \left[\boldsymbol{b}_1^{(m)}, \boldsymbol{b}_2^{(m)}, \ldots, \boldsymbol{b}_K^{(m)}\right]^T$ is the $KL \times 1$ data vector of the K users transmitting over their respective L-path channels, where $\boldsymbol{b}_k^{(m)}$ is the kth $1 \times L$ user bit vector.

We can define the $KL \times KL$-dimensional cross-correlation matrix $\boldsymbol{R}(m)$ of the signature

[2]The random delay $\tau_{k,l}$ takes into account the asynchronous nature of the transmission as well as the propagation delay $\tau'_{k,l}$ given in Equation (13.1).

sequences, such that the (p, q)th element is given by:

$$\rho_{p,q}(m) = \int_{-\infty}^{+\infty} a_{k_p}(t - \tau_{k_p,l_p}) a_{k_q}(t + mT_b - \tau_{k_q,l_q}) dt, \tag{13.5}$$

where $k_p = \lceil \frac{p}{L} \rceil$, $k_q = \lceil \frac{q}{L} \rceil$, $l_p = p - \lfloor \frac{p-1}{L} \rfloor \cdot L$ and $l_q = q - \lfloor \frac{q-1}{L} \rfloor \cdot L$. Since the modulating signals are time-limited, $\boldsymbol{R}(m) = \boldsymbol{0}\ \forall |m| > 1$ and $\boldsymbol{R}(-1) = \boldsymbol{R}^T(1)$. Note that $\boldsymbol{R}(1)$ is an upper triangular matrix having a zero diagonal.

The front end of the receiver illustrated in Figure 13.1 consists of a bank of KL filters, matched to the signature sequences of the K users transmitting over their L-path channels. Assuming perfect synchronization for each individual user, which transmit asynchronously with respect to each other, the output of the kth user's matched filter corresponding to the lth path sampled at the end of the ith symbol interval is given as [353]:

$$\begin{aligned} z_{k,l}^{(i)} &= \int_{-\infty}^{+\infty} r(t) a_k(t - iT_b - \tau_{k,l}) dt \\ &= \sum_{j=(k-1)L+l+1}^{KL} \rho_{(k-1)L+l,j}(1) \sqrt{\xi_{k_j}} c_{k_j,l_j}^{(i-1)} b_{k_j}^{(i-1)} \\ &\quad + \sum_{j=1}^{(k-1)L+l-1} \rho_{(k-1)L+l,j}(-1) \sqrt{\xi_{k_j}} c_{k_j,l_j}^{(i+1)} b_{k_j}^{(i+1)} \\ &\quad + \sum_{j=1}^{KL} \rho_{(k-1)L+l,j}(0) \sqrt{\xi_{k_j}} c_{k_j,l_j}^{(i)} b_{k_j}^{(i)} + n_{k,l}^{(i)}. \end{aligned} \tag{13.6}$$

Using vector notation, the output $\boldsymbol{z}^{(i)}$ of the matched filter bank at the ith symbol interval can be written as:

$$\begin{aligned} \boldsymbol{z}^{(i)} &= \left[z_{1,1}^{(i)}, \ldots, z_{1,L}^{(i)}, \ldots, z_{K,L}^{(i)} \right]^T \\ &= \boldsymbol{R}(1) \boldsymbol{\xi} \boldsymbol{c}^{(i-1)} \boldsymbol{b}^{(i-1)} + \boldsymbol{R}(0) \boldsymbol{\xi} \boldsymbol{c}^{(i)} \boldsymbol{b}^{(i)} + \boldsymbol{R}^T(1) \boldsymbol{\xi} \boldsymbol{c}^{(i+1)} \boldsymbol{b}^{(i+1)} + \boldsymbol{n}^{(i)}. \end{aligned} \tag{13.7}$$

From Equation (13.7) given by the first and third terms we can see the presence of the interference contributed by the edge bits. Hence any joint decision made on the ith bits of the K users has to take into account the decisions on either the $(i-1)$st bit or the $(i+1)$st bit of each user, as shown in Figure 13.2.

Let us first assume that the receiver has explicit knowledge of the SEB and EEB of all the

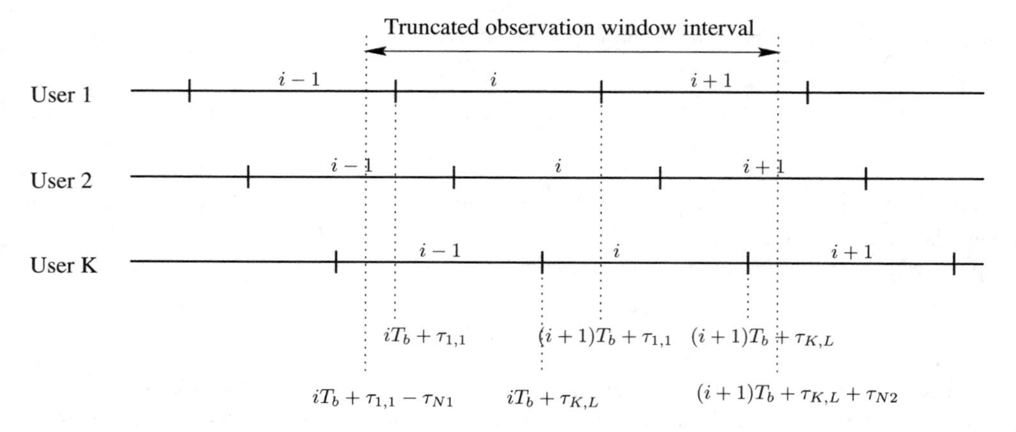

Figure 13.2: Received sequences of an asynchronous DS-CDMA system assuming a non-dispersive channel associated with $L = 1$.

users. Let us also introduce:

$$\boldsymbol{R}'(0) = \begin{bmatrix} \rho'_{1,1} & \cdots & \rho'_{1,L} & \cdots & \rho'_{1,KL} \\ \rho'_{2,1} & \cdots & \rho'_{2,L} & \cdots & \rho'_{2,KL} \\ \vdots & \vdots & & \vdots & \\ \rho'_{KL,1} & \cdots & \rho'_{KL,L} & \cdots & \rho'_{KL,KL} \end{bmatrix} \tag{13.8}$$

and

$$\boldsymbol{R}''(0) = \begin{bmatrix} \rho''_{1,1} & \cdots & \rho''_{1,L} & \cdots & \rho''_{1,KL} \\ \rho''_{2,1} & \cdots & \rho''_{2,L} & \cdots & \rho''_{2,KL} \\ \vdots & \vdots & & \vdots & \\ \rho''_{KL,1} & \cdots & \rho''_{KL,L} & \cdots & \rho''_{KL,KL} \end{bmatrix}, \tag{13.9}$$

where the (p, q)th element is given by:

$$\rho'_{p,q} = \int_{\tau_{1,1}+T_b-\tau_{N1}}^{\tau_{k_p,l_p}+T_b} a_{k_p}(t-\tau_{k_p,l_p})a_{k_q}(t-\tau_{k_q,l_q})dt$$
$$\rho''_{p,q} = \int_{\tau_{k_p,l_p}+T_b}^{\tau_{K,L}+T_b+\tau_{N2}} a_{k_p}(t-\tau_{k_p,l_p})a_{k_q}(t-\tau_{k_q,l_q})dt,$$

which will be used at a later stage in Equation (13.11) and Equation (13.12). The truncated observation window duration is governed by τ_{N1} and τ_{N2}, where $0 \leq \tau_{N1}, \tau_{N2} < \tau_{1,1} - \tau_{K,L} + T_b$. As illustrated by Figure 13.2, the truncated observation window interval can span from the most recently received $(i-1)$st bit of the Kth user to the end of the first received $(i+1)$st bit of the 1st user, i.e. $[(i-1)T_b + \tau_{K,L}, (i+2)T_b + \tau_{1,1}]$. In this way, the decisions made on the desired ith bits of the K users will only depend on either the $(i-1)$st or $(i+1)$st bits of all users.

Let us from now on consider the simplified scenario of non-dispersive channels associated with $L = 1$. Based on the observation vector $\boldsymbol{z}^{(i)}$ given in Equation (13.7) and then following the analysis carried out in Section 10.4 in the context of a synchronous CDMA system, it can be shown that the correlation metric required for detecting the ith bit of all users within the truncated observation window, given that the K-dimensional vectors $\boldsymbol{b}^{(i-1)}$ and $\boldsymbol{b}^{(i+1)}$ are known to the receiver, can be written as:

$$\Omega\left(\boldsymbol{b}^{(i)}\right) = 2\Re\left\{\boldsymbol{B}^T\boldsymbol{C}^*\boldsymbol{W}\boldsymbol{Z}\right\} - \boldsymbol{B}^T\boldsymbol{C}\boldsymbol{W}\boldsymbol{R}\boldsymbol{W}\boldsymbol{C}^*\boldsymbol{B}, \tag{13.10}$$

where

$$\begin{aligned}
\boldsymbol{B} &= \left[\boldsymbol{b}^{(i-1)^T}, \boldsymbol{b}^{(i)^T}, \boldsymbol{b}^{(i+1)^T}\right]^T \\
\boldsymbol{C} &= \operatorname{diag}\left[\boldsymbol{c}^{(i-1)}, \boldsymbol{c}^{(i)}, \boldsymbol{c}^{(i+1)}\right] \\
\boldsymbol{W} &= \operatorname{diag}\left[\boldsymbol{\xi}, \boldsymbol{\xi}, \boldsymbol{\xi}\right] \\
\boldsymbol{Z} &= \left[\boldsymbol{z}^{(i-1)'}, \boldsymbol{z}^{(i)}, \boldsymbol{z}^{(i+1)''}\right]^T \\
\boldsymbol{R} &= \begin{bmatrix} \boldsymbol{R}'(0) & \boldsymbol{R}^T(1) & \boldsymbol{0} \\ \boldsymbol{R}(1) & \boldsymbol{R}(0) & \boldsymbol{R}^T(1) \\ \boldsymbol{0} & \boldsymbol{R}(1) & \boldsymbol{R}''(0) \end{bmatrix}.
\end{aligned}$$

The vectors $\boldsymbol{z}^{(i-1)'}$ and $\boldsymbol{z}^{(i+1)''}$ represent the correlations of the partial matched filter outputs at instances $[iT_b + \tau_{1,1} - \tau_{N1}, iT_b + \tau_{k,l}]$ and $[(i+1)T_b + \tau_{k,l}, (i+1)T_b + \tau_{K,L} + \tau_{N2}]$, respectively, for $k = 1, 2, \ldots, K$, which are given by:

$$\boldsymbol{z}^{(i-1)'} = \boldsymbol{R}'(0)\boldsymbol{w}\boldsymbol{c}^{(i-1)}\boldsymbol{b}^{(i-1)} + \boldsymbol{R}^T(1)\boldsymbol{w}\boldsymbol{c}^{(i)}\boldsymbol{b}^{(i)} + \boldsymbol{n}^{(i-1)'} \tag{13.11}$$

$$\boldsymbol{z}^{(i+1)''} = \boldsymbol{R}(1)\boldsymbol{w}\boldsymbol{c}^{(i)}\boldsymbol{b}^{(i)} + \boldsymbol{R}''(0)\boldsymbol{w}\boldsymbol{c}^{(i+1)}\boldsymbol{b}^{(i+1)} + \boldsymbol{n}^{(i+1)''}. \tag{13.12}$$

The optimum decision concerning the K-dimensional user bit-related vector $\boldsymbol{b}^{(i)}$, provided that $\boldsymbol{b}^{(i-1)}$ and $\boldsymbol{b}^{(i+1)}$ are known to the receiver, is formulated as $\hat{\boldsymbol{b}}^{(i)} = \left[\hat{\boldsymbol{b}}_1^{(i)}, \hat{\boldsymbol{b}}_2^{(i)}, \ldots, \hat{\boldsymbol{b}}_K^{(i)}\right]^T$, which maximizes the correlation metric given in Equation (13.10). However, in practice the receiver is oblivious of the EEB-related K-dimensional vectors $\boldsymbol{b}^{(i+1)}$ during the detection of $\boldsymbol{b}^{(i)}$, unless they are pilot bits. On the other hand, the SEBs $\boldsymbol{b}^{(i-1)}$ can be derived from the previous detection process and if the DBEP of the receiver is sufficiently low, it can be assumed that the SEBs $\boldsymbol{b}^{(i-1)}$ are perfectly known. Hence, in order to optimize the decision concerning $\boldsymbol{b}^{(i)}$, it is imperative that the EEBs are estimated as reliably as possible. One way of estimating the EEBs is by taking a hard decision based on their maximum ratio combined correlator outputs [349]. This can be written as:

$$\tilde{\boldsymbol{b}}_{MF}^{(i+1)} = \text{sgn}\left\{\text{diag}\left[\boldsymbol{I}_L, \ldots, \boldsymbol{I}_L\right]\left[\boldsymbol{c}^{(i+1)^*}\boldsymbol{\xi}\boldsymbol{z}^{(i+1)''}\right]\right\}, \tag{13.13}$$

where $\boldsymbol{I}_L$ is a $1 \times L$ unity vector and the K-dimensional vector $\tilde{\boldsymbol{b}}_{MF}^{(i+1)}$ denotes the tentative decisions concerning the EEBs based on the hard decision of the correlator. In this chapter, this approach of detecting the EEBs is denoted as Strategy 1 or S1. GAs are then invoked in order to estimate the current bits by optimizing the correlation metric of Equation (13.10) with respect to the K-dimensional vector $\boldsymbol{b}^{(i)}$, yielding:

$$\hat{\boldsymbol{b}}^{(i)} = \arg\left\{\max_{\boldsymbol{b}^{(i)}}\left[\Omega\left(\boldsymbol{b}^{(i)}\right)\right]\right\}. \tag{13.14}$$

However, due to the presence of MAI, as shown in Equation (13.12) and Equation (13.13), the EBEP is high, especially in a worst-case single-path scenario, where no diversity gain is achieved. This high EBEP will have a significant detrimental impact on the overall performance of the detector, as we shall see in Section 13.4. Hence, in order to reduce the EBEP, we invoke the proposed GA for improving the tentative decision accuracy of the EEBs $\boldsymbol{b}^{(i+1)}$, and at the same time we optimize the correlation metric in order to detect $\boldsymbol{b}^{(i)}$. In this case, the correlation metric is expressed as:

$$\Omega\left(\boldsymbol{b}^{(i)}, \boldsymbol{b}^{(i+1)}\right) = 2\Re\left\{\boldsymbol{B}^T\boldsymbol{C}^*\boldsymbol{W}\boldsymbol{Z}\right\} - \boldsymbol{B}^T\boldsymbol{C}\boldsymbol{W}\boldsymbol{R}\boldsymbol{W}\boldsymbol{C}^*\boldsymbol{B}, \tag{13.15}$$

since the desired K-dimensional bit vector $\boldsymbol{b}^{(i)}$ and the K-dimensional EEB vector $\boldsymbol{b}^{(i+1)}$ now jointly constitute the decision variables. Again, this approach of detecting the EEBs based on GAs is denoted here as Strategy 2 or S2. Hence, the estimated transmitted bit vector $\hat{\boldsymbol{b}}^{(i)}$ of the K users can be found by optimizing Equation (13.10) with respect to the desired bits $\boldsymbol{b}^{(i)}$ and the EEBs $\boldsymbol{b}^{(i+1)}$, yielding:

$$\hat{\boldsymbol{b}}^{(i)}, \tilde{\boldsymbol{b}}_{GA}^{(i+1)} = \arg\left\{\max_{\boldsymbol{b}^{(i)}, \boldsymbol{b}^{(i+1)}}\left[\Omega\left(\boldsymbol{b}^{(i)}, \boldsymbol{b}^{(i+1)}\right)\right]\right\}, \tag{13.16}$$

where $\tilde{\boldsymbol{b}}_{GA}^{(i+1)}$ denotes the tentative decisions concerning the EEBs based on GA-assisted optimization. In the next section we will further discuss the philosophy of our GA-assisted multi-user detector used for simultaneously estimating both the desired users' bits and the EEBs.

13.3 GA-Assisted Multi-user Detection in Asynchronous CDMA Systems

The flowchart of the GAs invoked for detecting the users transmitted bits in an asynchronous CDMA system is depicted in Figure 13.3.

Apart from the initialization phase, the main structure of the GA is identical to that employed in Figure 10.10 for detecting the users' transmitted bits in the symbol-synchronous CDMA system of Chapter 10. Hence we will be using the fitness-proportionate selection scheme in conjunction with a uniform cross-over operation and a standard binary mutation operation as well as invoking the incest prevention strategy and elitism strategy, as shown in Figure 13.3. However, the structure of the individual adopted here is slightly different, since we have to take into account the SEBs and the EEBs. Each individual will consist of $3 \times K$ antipodal bits. Assuming that the current desired signalling interval is the ith interval, we shall express the pth individual here as $\tilde{\boldsymbol{b}}_p(y) = \left[\hat{\boldsymbol{b}}_{p,SEB}^{(i-1)}, \tilde{\boldsymbol{b}}_p^{(i)}(y), \tilde{\boldsymbol{b}}_{p,EEB}^{(i+1)}(y)\right]$, where $\hat{\boldsymbol{b}}_{p,SEB}^{(i-1)}$, $\tilde{\boldsymbol{b}}_p^{(i)}(y)$ and $\tilde{\boldsymbol{b}}_{p,EEB}^{(i+1)}(y)$ are K-bit vectors which denote the SEBs, the desired bits and the EEBs at the yth generation, respectively. The fitness value associated with each individual, denoted as $f\left[\tilde{\boldsymbol{b}}_p(y)\right]$ for $p = 1, \ldots, P$ is then computed by substituting the corresponding vectors $\hat{\boldsymbol{b}}_{p,SEB}^{(i-1)}$, $\tilde{\boldsymbol{b}}_p^{(i)}(y)$ and $\tilde{\boldsymbol{b}}_{p,EEB}^{(i+1)}(y)$ into the correlation metric of Equation (13.10) and then using it as exponent, in order to obtain positive fitness values. Based on the evaluated fitness value, a new population of P individuals is created for the $(y+1)$st generation with the aid of the various processes, as illustrated in Figure 13.3. The explicit description of each of the process can be found in Chapter 9 and Chapter 10. Upon the GA's termination at the Yth generation, as shown in Figure 13.3, the desired bit vector $\tilde{\boldsymbol{b}}_p^{(i)}(Y)$ of the individual corresponding to the highest fitness value in the population constitutes the detected K users' ith bit associated with the truncated observation window interval considered. In other words, we have $\hat{\boldsymbol{b}}^{(i)} = \tilde{\boldsymbol{b}}_j^{(i)}(Y)$, where $\tilde{\boldsymbol{b}}_j(Y) = \max\left\{f\left[\tilde{\boldsymbol{b}}_1(Y)\right], \ldots, f\left[\tilde{\boldsymbol{b}}_P(Y)\right]\right\}$.

As seen in Figure 13.3, the GA must be initialized for every new truncated observation window before commencing the optimization process. Similarly to the GA initialization invoked in the symbol-synchronous CDMA system of Chapter 10, we can exploit the information already available at the beginning of each detection step, in order to aid and accelerate the optimization. Again, let us assume that the current bit of interest is the ith bit of all users. Hence, the SEB vector will be constituted by the $(i-1)$st bits, while the EEBs are constituted by the $(i+1)$st bits. At this point, the SEBs will have been detected in the previous truncated observation window, when the $(i-1)$st bits were the desired bits, i.e. $\hat{\boldsymbol{b}}^{(i-1)}$ will be known. Therefore, we can assign $\hat{\boldsymbol{b}}_{p,SEB}^{(i-1)} = \hat{\boldsymbol{b}}^{(i-1)}$ for all p. It is well known that the computational complexity – in the context of the population size P and the number of generations Y – of the GA required to attain a specified level of performance increases with the number of variables to be optimized [9]. Hence, in order to reduce the computational complexity of the GA, the SEBs will not be involved in the optimization process, since these SEBs have been detected previously and modification of the SEBs will not affect the performance of the detector significantly. Thus the generation index y is omitted for the SEB vector.

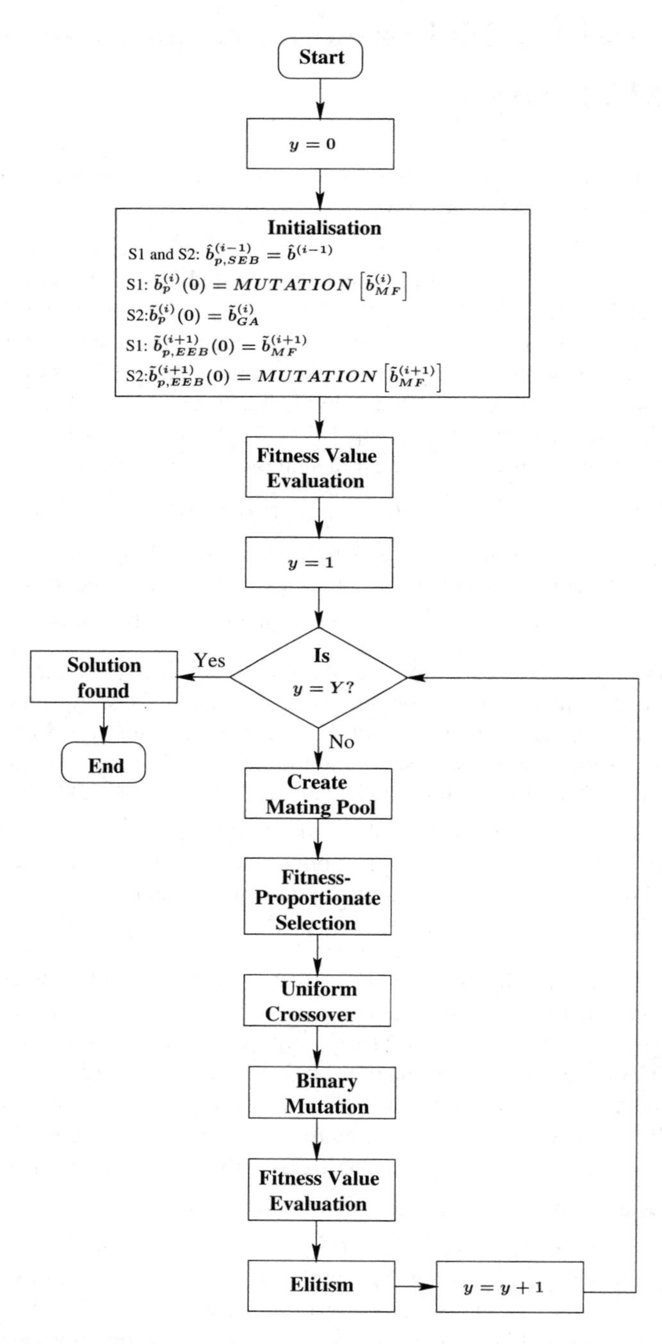

Figure 13.3: A flowchart depicting the structure of the proposed genetic algorithm used for detecting the transmitted users' bit $\boldsymbol{b}^{(i)}$ as well as providing the tentative solutions of $\boldsymbol{b}^{(i+1)}$ during the ith truncated observation window.

13.3.1 Matched Filter-Assisted EEB Estimation

It is now clear that the unknown variables involved in the optimization process consist of the current desired bits as well as the EEBs. As mentioned before, here two methods are investigated in order to provide tentative solutions for the EEBs. **In our first method denoted here as S1**, we invoke the hard decisions obtained from the users' correlator outputs as the tentative EEBs, according to Equation (13.13). Hence:

$$\tilde{\boldsymbol{b}}^{(i+1)}_{p,EEB}(y) = \tilde{\boldsymbol{b}}^{(i+1)}_{MF} \quad \text{for } p = 1, \ldots, P \text{ and } y = 0, \ldots, Y. \tag{13.17}$$

The error probability of these EEBs estimated on the basis of the correlator outputs is generally excessive and therefore it will degrade the performance of the optimization process, as we will see later in Section 13.4. Given these EEBs in this case, only the current desired bits are involved in the GA-assisted optimization process. During the initialization phase of Figure 13.3, the desired bit vector of each individual is a 'mutated' version of the tentative hard decisions. In other words, we will change the state of each bit of $\tilde{\boldsymbol{b}}^{(i)}_{MF}$ with a probability of p_{m1}. Typically, the value of p_{m1} is governed by the BEP that can be achieved by the correlator and a practical choice is to set $p_{m1} = BEP$. In this treatise, we shall set a nominal value of $p_{m1} = 0.1$. Hence:

$$\tilde{\boldsymbol{b}}^{(i)}_{p}(0) = MUTATION\left[\tilde{\boldsymbol{b}}^{(i)}_{MF}\right]_{p_{m1}=0.1} \quad \text{for } p = 1, \ldots, P. \tag{13.18}$$

The mutation process [287] of Equation (13.18) is used for ensuring that the GA has a highly diversified search range at the beginning of its operation as well as for providing dissimilar individuals. Without this mutation process all the individuals at the initialization stage would be identical, which is not allowed by the incest prevention strategy. After initialization, the GA will commence searching for the optimum solution. The main advantage of the strategy S1 is that since only K variables are considered, the population size P required to attain a specified optimization quality associated with a given DBEP can be lower.

13.3.2 GA-Assisted EEB Estimation

In our second method denoted here as S2, the GA is invoked in order to lower the EBEP. At the 0th generation, the unknown EEB $\tilde{\boldsymbol{b}}^{(i+1)}_{p}(0)$, $p = 1, \ldots, P$, can be initially estimated based on the hard decisions of the correlator outputs of Figure 13.1. This is equivalent to using S1 at the 0th generation. Again, these bits are mutated with a probability of $p_{m1} = 0.1$ in order to ensure a diversified search. Hence the mutation process is identical to that of Equation (13.18) used in S1, but it is applied to the edge bits at index $(i + 1)$ yielding:

$$\tilde{\boldsymbol{b}}^{(i+1)}_{p}(0) = MUTATION\left[\tilde{\boldsymbol{b}}^{(i+1)}_{MF}\right]_{p_{m1}=0.1} \quad \text{for } p = 1, \ldots, P. \tag{13.19}$$

Let us now assume that upon termination of the GA-assisted search at the end of every truncated observation window, the error probability of the EEBs will be sufficiently low and hence these bits can be considered the tentative solutions for the GA during initialization, when these EEBs become the desired bits in the next truncated observation window. Hence according to Equation (13.16) we have:

$$\tilde{\boldsymbol{b}}_p^{(i)}(0) = \tilde{\boldsymbol{b}}_{GA}^{(i)} \quad \text{for } p = 1, \ldots, P. \tag{13.20}$$

where $\tilde{\boldsymbol{b}}_{GA}^{(i)}$ is specified by Equation (13.16). After initialization, the GAs are then invoked in order to search for the K users' desired bit vector as well as for the EEB vector that optimizes the correlation metric according to Equation (13.16). The advantage of the S2 strategy is that the optimization performance is not limited by the high EBEP exhibited by the correlator outputs. On the other hand, since there are now $2K$ variables to be optimized, a higher population size P is required in order to attain a specified level of performance.

13.3.3 Complexity Issues

Since our proposed GA-assisted multi-user detector optimizes the correlation metric of Equation (13.10), we will only consider its complexity in terms of the number of correlation metric computations required for the optimization. The optimum multi-user detector using exhaustive search requires 2^K evaluations of the correlation. By contrast, our proposed detector requires a maximum of $Y \times P$ correlation metric evaluations. In fact, the number of such correlation metric evaluations can be reduced by avoiding repeated evaluations of identical individuals, either within the same generation or across the entire iteration process, if the receiver has the necessary memory. We note that since the EEBs of strategy S1 are fixed throughout the iteration process, the number of additions and multiplications per correlation metric evaluation will be a factor KL lower, than that of strategy S2, since they do not have to be re-computed.

Before we present our simulation results, we should note here that the employment of our proposed GA-based multi-user detector is not restricted to joint bit-by-bit detection. The truncated observation window can actually span over several users' bits. In such cases, the individuals of the GA must contain an increased number of bits. However, since there are more unknown bits to be detected, a higher P and more generations must be invoked.

13.4 Simulation Results

In this section, our computer simulation results are presented, in order to characterize the DBEP performance of the GA-assisted multi-user detector employing the two EEB estimation strategies highlighted in the previous section. All the results in this section were based on evaluating the DBEP performance of a chip-asynchronous 10-user CDMA system over both single-path and two-path Rayleigh fading channels. For ease of simulation, the relative delays between the different received signals were arranged in the single-path scenario such that $\tau_{j+1,1} - \tau_{j,1} = \left[\frac{T_c}{8}, T_c\right)$. Hence for a system supporting $K = 10$ users, the maximum delay between the 1st user and the highest-delay Kth user is 10 chips. Since the chip duration is about $\frac{1}{64000} \times \frac{1}{31} \approx 0.5\mu sec$, assuming a bit rate of 64 kbps and a spreading factor of 31, the 10-chip maximum propagation delay difference corresponds to about $5\mu sec$. This scenario would be encountered by two mobiles, where the 1st user is directly at the base station and

Parameter	Value
Spreading factor N_c	31
Modulation mode	BPSK
Number of CDMA users K	10, unless stated otherwise
Number of multipaths L	1 and 2

Table 13.1: Simulation parameters for the experiments of Figures 13.4-13.11.

the Kth user is, for example, directly at the edge of a 1500m-radius propagation cell, where the radio waves' propagation delay becomes $5\mu sec$. Given this maximum propagation delay, the truncated observation window of Figure 13.2 can encompass between 0 to 20 chips of the EEBs and SEBs. Hence in our simulations, we considered an average scenario associated with $t_{N1} = t_{N2} = 10$ chips in Figures 13.4-13.10. The BEP performance difference between the two extreme cases of 0 and 20 chips, respectively, will be shown in Figure 13.11. For the two-path scenario, the relative delays were arranged according to $\tau_{j,2} - \tau_{j,1} = \left[\frac{T_c}{8}, T_c\right)$ and $\tau_{j+1,1} - \tau_{j,2} = \left[\frac{T_c}{8}, T_c\right)$. The two paths were assumed to have equal average received energy, i.e. $E\left[\alpha_{k,1}\right] = E\left[\alpha_{k,2}\right] = 0.5$. The processing gain was $N_c = 31$ and the signature sequences were randomly generated. Perfect power control and CIR estimation were assumed for all the simulations. We also assumed that the first bit $\boldsymbol{b}^{(0)}$ of all the users was known to the receiver. Summaries of the simulation parameters and the GA configuration are given in Table 13.1 and Table 13.2, respectively. Upon observing the GA configuration of Table 13.2, we can see that there are a couple of parameters, which are different from those presented in the previous chapters, namely the mating pool size T and the probability of mutation p_m for S2. The probability of mutation adopted here can be readily justified. As mentioned in Section 13.3.2, there are $2 \times K$ decision variables, comprising the desired bits and the EEBs, to be optimized for S2. In Section 10.5.2, it was shown that the BEP performance of the GA-assisted multi-user detector critically depends on the value of p_m for a specific number of decision variables. From Figure 10.4, we can see that for $K = 20$ users, it is desirable to have $p_m < 0.1$, in order to obtain a lower BEP. Since employing S2 for $K = 10$ users is analogous to a 20-user synchronous CDMA system scenario, we decided to adopt $p_m = 0.05$ for the mutation process. For S1, the value of p_m remains at 0.1, since it involves the same number of decision variables as in the previous chapters. This hypothesis can be extended to the case of $K = 15$ users. The choice of the mating pool size T for S2 will be explained next.

13.4.1 Effects of the Mating Pool Size

Let us first consider the achievable BEP over the course of 10 generations for both S1 and S2 in conjunction with a mating pool size equivalent to the number of non-identical individuals, i.e. $T \leq P$ as well as using $T = 4$. The results are shown in Figure 13.4 for an SNR value of 36 dB over single-path fading Rayleigh fading channels. It can be seen that for GAs employing S1, there is no significant difference in the BEP attained over the course of the evolution for mating pool sizes of $T \leq P$ and $T = 4$. This also proves that our results obtained in the previous chapters are relatively consistent, since the approach we adopted in those chapters is identical to GAs employing S1. On the other hand, for GAs employing S2,

Set-up/Parameter	Method/Value
Individual initialization method	According to Figure 13.3
Selection method	Proportionate-Fitness
Cross-over operation	Uniform cross-over
Mutation operation	Standard binary mutation
Elitism	Yes
Incest prevention	Yes
Population size P	Given in the associated plots
Number of generations Y	10
Mating pool size T	– S1: All dissimilar individuals in the population – S2: Given in the associated plots
Probability of mutation p_{m1} and p_m for $K = 10$	– S1: 0.1 and 0.1, respectively – S2: 0.1 and 0.05, respectively
Probability of mutation p_{m1} and p_m for $K = 15$ of Figure 13.10	– S1: 0.1 and 0.07, respectively – S2: 0.1 and 0.03, respectively

Table 13.2: Configuration of the GA used to obtain the results of Figures 13.4-13.11. Explicit description of the fitness-proportionate selection scheme, the uniform cross-over operation and the floating point mutation operation can be found in Section 9.4.2, Section 9.4.3 and Section 9.4.4, respectively.

a substantial difference in the achievable BEP can be observed between the mating pool sizes of $T \leq P$ and $T = 4$.

This phenomenon can be explained by considering the distribution of the probability of selection p_i corresponding to the individuals having the highest fitness value after the initialization phase of Figure 13.3. The corresponding selection distribution curves for the GAs employing S1 and S2 are shown in Figure 13.5 and Figure 13.6, respectively. From Figure 13.5, the probability of selection corresponding to the individual having the highest fitness value for the GA employing S1 has a similar distribution for both $T \leq P$ and $T = 4$. From this observation, we can conclude that for GAs employing S1, there is only a handful of individuals having relatively high fitness values that dominate the population. Furthermore, since the initial individuals are generated based on the matched filter output according to Equation (13.18), their fitness values are far from optimum and hence the random process of mutation may substantially increase some of these fitness values.

On the other hand, the distribution of the probability of individual selection for GAs employing S2 is different for $T \leq P$ and $T = 4$, as seen from Figure 13.6. In particular, for $T \ll P$ we see that the selection pdf peak is around the region of 0.2. This implies that most of the time the selection probability of the fittest individual is only about 0.2. This implies a weakly selective process, as highlighted in Section 9.4.2, which reduces the convergence rate. The reason for this weakly selective process is because the EEBs now have a significantly lower BEP due to their GA-assisted estimation, which are used for the initialization of the desired bits in the subsequent truncated observation window. Hence the fitness values of all the individuals are relatively high, exhibiting near-optimum values. As a result, a high

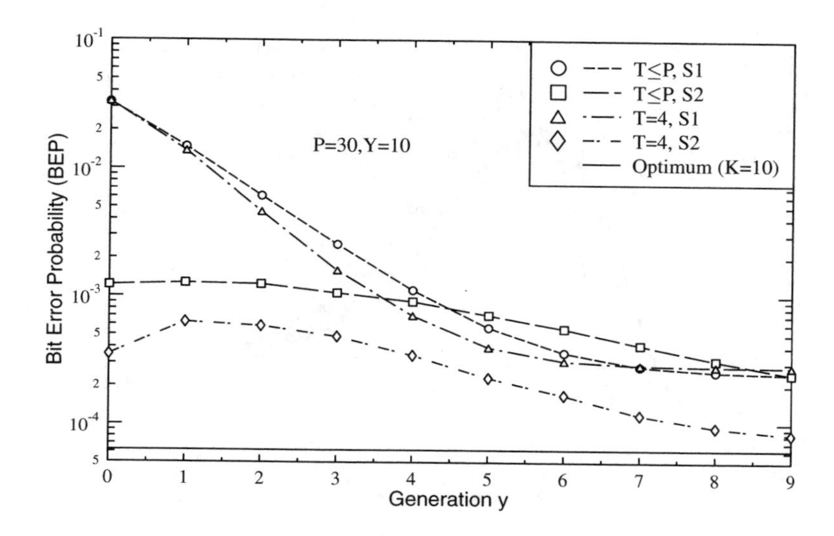

Figure 13.4: The desired bits' error probability with respect to the number of generations for the GA-assisted multi-user detector over narrow-band Rayleigh fading channels employing the EEB detection strategies S1 and S2 with a population size of $P = 30$ using random signature sequences of length $N_c = 31$ and supporting $K = 10$ users at an SNR value of 36 dB. The GA configuration and the simulation parameters used are listed in Table 13.2 and Table 13.1, respectively.

number of individuals will have similar fitness values, which ultimately reduces the probability of selection for each individual. By contrast, a majority of the probability of selections for $T = 4$ centres around the region of 0.3. This provides further evidence of the argument that the majority of the individuals have almost equal fitness values, since the lowest probability of selection for $T = 4$ is 0.25, when these 4 individuals have the same probability of selection. Hence, by using a small mating pool, we can place more emphasis on the individuals exhibiting high fitness values.

Figure 13.7 and Figure 13.8 show the DBEP performance and the EBEP performance, respectively, against the average SNR per bit for the GA-assisted multi-user detector supporting $K = 10$ users, when employing the two EEB estimation strategies. The single-user bound was computed using Equation (12.12), with L_d replaced by L. As Figure 13.7 shows, the DBEP of the GA-assisted multi-user detector employing S1 was inferior compared to that of S2. The error floor observed for S1 in the single-path scenario was caused by the high EBEP, as seen in Figure 13.8. In this case, the GA-assisted multi-user detector was termed as EEB interference-limited. The same can be said for the two-path scenario. On the other hand, we can see from Figure 13.8 that the EBEP upon employing S2 is fairly low. As a result, the performance of the GA-assisted multi-user detector utilizing this strategy was not limited by the EEB errors and hence it was capable of achieving a near-optimum single-user-like

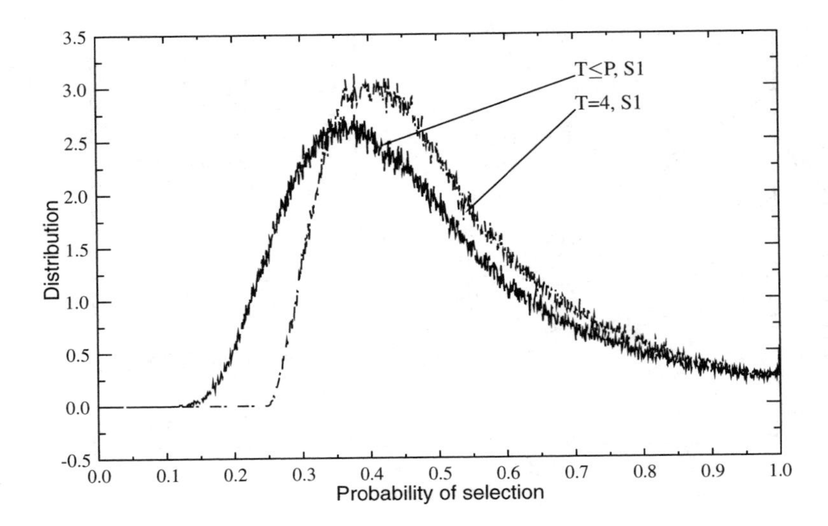

Figure 13.5: Distribution of the probability of selection corresponding to the individual having the highest fitness value at the 0th generation for the GA-assisted multi-user detector over narrow-band Rayleigh fading channels employing the EEB detection strategy S1 with a population size of $P = 30$ using random signature sequences of length $N_c = 31$ and supporting $K = 10$ users at an SNR value of 36 dB. The GA configuration and the simulation parameters used are listed in Table 13.2 and Table 13.1, respectively.

DBEP performance. Furthermore, in comparison to the 'brute-force' ML detector requiring $2^{10} = 1024$ correlation metric evaluations, our proposed multi-user detector is substantially less complex, requiring only a maximum of $10 \times 30 = 300$ correlation metric evaluations, yet performing close to the optimum performance.

The notion of an EEB interference-limited DBEP performance employing strategy S1 is further substantiated in Figure 13.9, which characterizes the DBEP performance of the proposed detector for a population size of $P = 20$. Naturally, we would expect the performance to degrade as compared to Figure 13.7, when the population size P decreases. As seen in the figure for both the one-path and two-path scenarios, the DBEP performance of the proposed detector employing strategy S1 did not show significant degradation in comparison to that associated with $P = 30$, as illustrated in Figure 13.7. This is due to the fact that the EBEP is the same for both $P = 30$ and $P = 20$ and hence the corresponding DBEP performances are limited by the poor reliability of the EEBs. This becomes explicit in comparison to the curve characterizing the scenario using perfect knowledge of the SEBs and EEBs, which exhibited a near-single-user DBEP performance even for $P = 20$. On the other hand, for detectors employing strategy S2, a degradation can be observed for $P = 20$ compared to that for $P = 30$, as shown in Figure 13.7. This is because in this case there are $2K$ variables to be optimized and therefore a higher population size is required in order to achieve optimum performance.

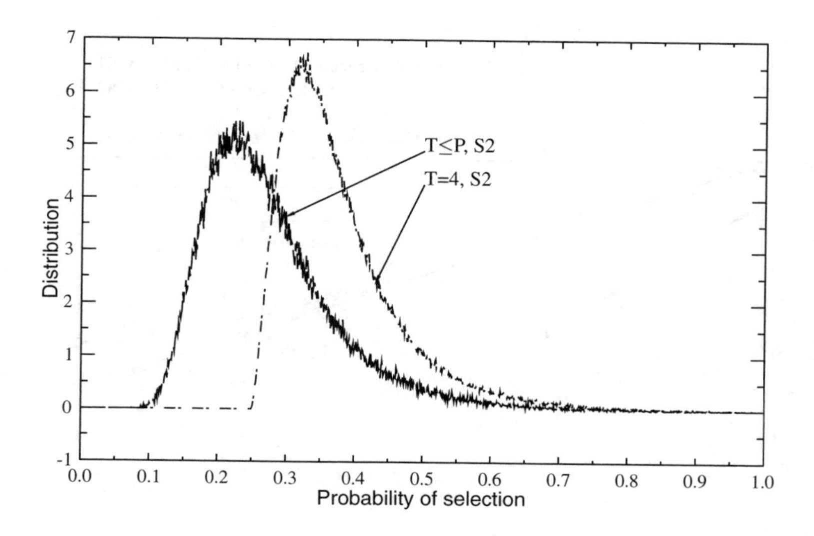

Figure 13.6: Distribution of the probability of selection corresponding to the individual having the highest fitness value at the 0th generation for the GA-assisted multi-user detector over narrow-band Rayleigh fading channels employing the EEB detection strategy S2 with a population size of $P = 30$ using random signature sequences of length $N_c = 31$ and supporting $K = 10$ users at an SNR value of 36 dB. The GA configuration and the simulation parameters used are listed in Table 13.2 and Table 13.1, respectively.

Hence, when $P = 20$, the performance of the detector is degraded. In this case, we referred to the DBEP performance as 'GA-limited'.

Figure 13.10 shows the DBEP performance of our proposed multi-user detector for $K = 15$ users. Because of the higher number of variables to be optimized, we increased the population size P to 40 and 50. We note from the figure that for $P = 40$, the GA employing strategy S1 now exhibits a more significant degradation in terms of its DBEP performance with respect to the single-user bound, than that employing strategy S2. This is due to the fact that as the number of users increases, the EBEP becomes higher. Increasing the population size to 50 does not show any significant improvement using the same strategy, since the performance is limited by the EEB interference. We also note that for $P = 40$ the DBEP performance of GAs employing strategy S2 did not match the single-user bound, even though it outperformed strategy S1. This is due to the limited population size, which was too small for optimizing 2×15 variables. However, by increasing P to 50, the DBEP performance becomes near-optimum. Hence, while achieving a superior performance, the associated additional computational complexity has to be tolerated. An important observation is that when K is increased from 10 to 15 users, a near-optimum DBEP performance can be maintained by increasing the population size P from 30 to 50, while keeping $Y = 10$ and employing strategy S2. This constitutes a factor of 5/3 increase in the number of correlation

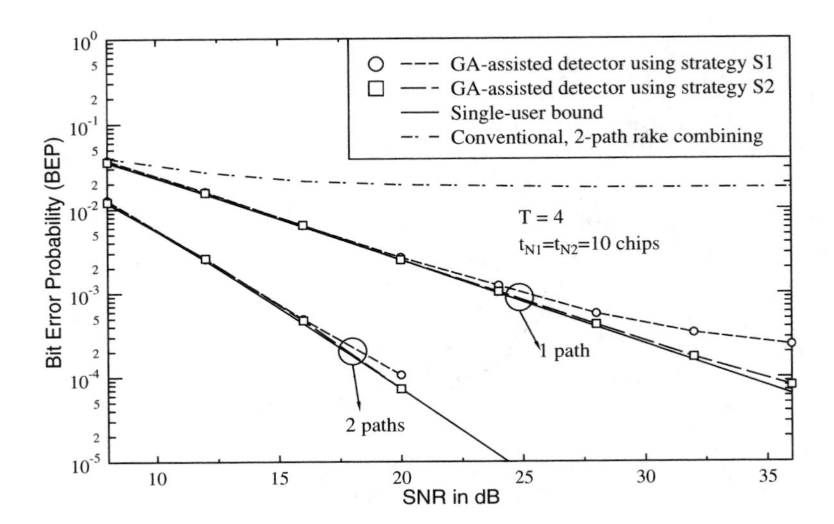

Figure 13.7: The **desired bits'** error probability for the GA-assisted multi-user detector over Rayleigh fading channels employing the EEB detection strategies S1 and S2 with a population size of $P = 30$ using random signature sequences of length 31 and supporting $K = 10$ users. The GA configuration and the simulation parameters used are listed in Table 13.2 and Table 13.1, respectively.

metric computations. On the other hand, the computational complexity of the conventional optimum detector using brute-force optimization is increased by a factor of $2^5 = 32$.

All the simulation results we have seen so far were based on a truncated window size of $t_{N1} = t_{N2} = 10$ chips.[3] This corresponds to a minimum spreading factor of 10 for both the SEBs and EEBs. The effects of the window size on the SEB error probability can be ignored, since these bits have been detected previously. On the other hand, the EEBs have to be tentatively detected based on only the reduced spreading factor. In practice, it is not always possible to set $t_{N1} = t_{N2} = 10$ chips. The worst case would be $t_{N1} = t_{N2} = 0$ chips, while the ideal case would be $t_{N1} = t_{N2} = \tau_{1,1} - \tau_{K,L} + T_b$ chips. We studied the effects of varying the window size on the DBEP performance based on these two settings and the associated results are shown in Figure 13.11.

We can see that for a narrow window size of $t_{N1} = t_{N2} = 0$ chips, the performance of the detectors employing strategy S1 deteriorates more significantly compared to the scenario using a wider window size of $t_{N1} = t_{N2} = 20$ chips, when employing strategy S2. This implies that the DBEP performance of detectors using the EEBs based on the hard decisions of the correlator outputs in Figure 13.1 are more sensitive to the varying window size, than those invoking GA-assisted EEB detection.

[3]We have set $t_{N1} = t_{N2}$ in order to arrive at a symmetric truncated window for ease of simulation.

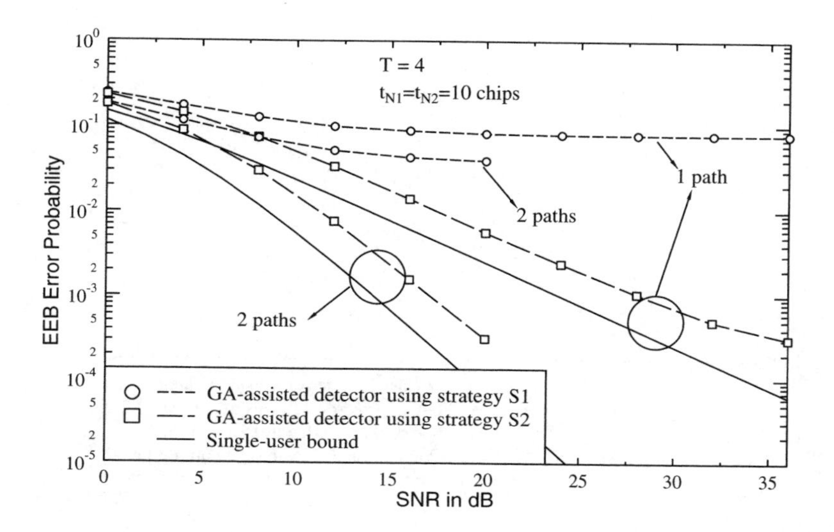

Figure 13.8: The **EEBs'** error probability performance for the GA-assisted multi-user detector over Rayleigh fading channels employing the EEB detection strategies S1 and S2 with a population size of $P = 30$ using random signature sequences of length 31 and supporting $K = 10$ users. The GA configuration and the simulation parameters used are listed in Table 13.2 and Table 13.1, respectively.

13.5 Chapter Summary and Conclusion

In this chapter, we considered an asynchronous CDMA system communicating over a dispersive Rayleigh fading channel, as presented in Section 13.2. The correlation metric based on a truncated window size was also formulated. We then developed a GA-assisted multi-user detector for this model in Section 13.3, in order to search for the particular bit sequence that optimizes the correlation metric. Two strategies were evaluated for providing tentative decisions concerning the EEBs. In the first approach, as highlighted in Section 13.3.1, the EEBs were tentatively detected based on the hard decisions at the correlator outputs. The desired bits within the truncated observation window were detected using GAs. In our second approach presented in Section 13.3.2, GAs were invoked in order to improve the EBEP and at the same time to detect the desired bits within the window. Our simulation results presented in Section 13.4 showed that the DBEP performance of the detectors using the first approach were limited by the high error rate of the EEBs. On the other hand, using the same number of correlation metric evaluations, detectors employing the second approach can achieve a near-optimal DBEP performance at the cost of a lower number of correlation metric evaluation compared to the optimum multi-user detector using a brute-force approach. Furthermore, both the EEBs and the desired bits are detected by the same GAs, resulting in potential complexity savings.

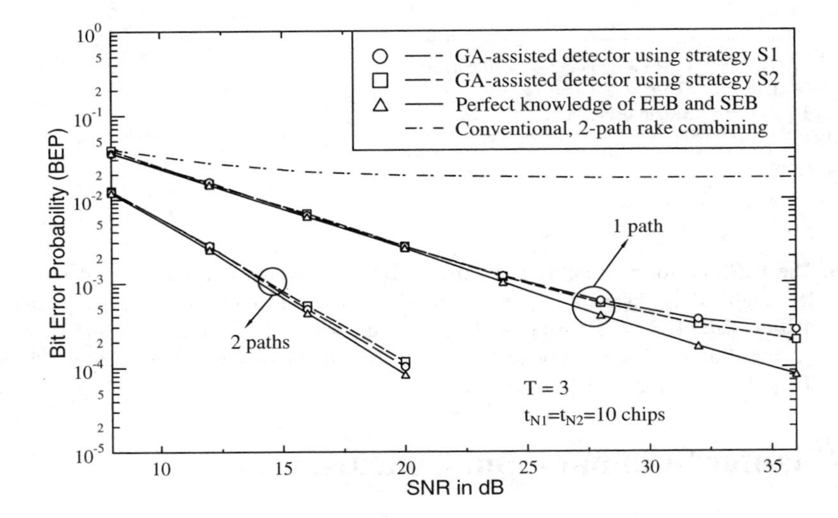

Figure 13.9: The **desired bits'** error probability performance for the GA-assisted multi-user detector over Rayleigh fading channels employing the EEB detection strategies S1 and S2 with a population size of $P = 20$ using random signature sequences of length 31 and supporting $K = 10$ users. The GA configuration and the simulation parameters used are listed in Table 13.2 and Table 13.1, respectively.

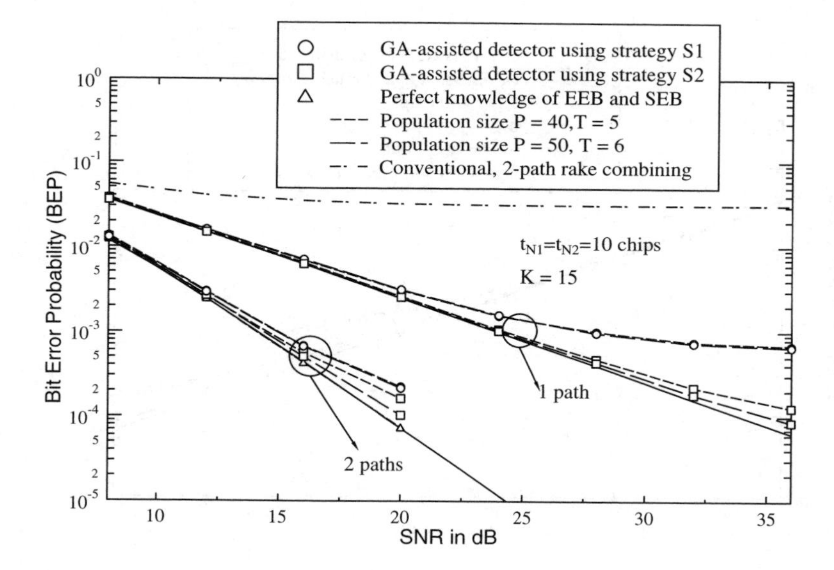

Figure 13.10: The **desired bits'** error probability performance for the GA-assisted multi-user detector over Rayleigh fading channels employing the EEB detection strategies S1 and S2 with population sizes of $P = 40$ and $P = 50$ using random signature sequences of length 31 and supporting $K = 15$ users. The GA configuration and the simulation parameters used are listed in Table 13.2 and Table 13.1, respectively.

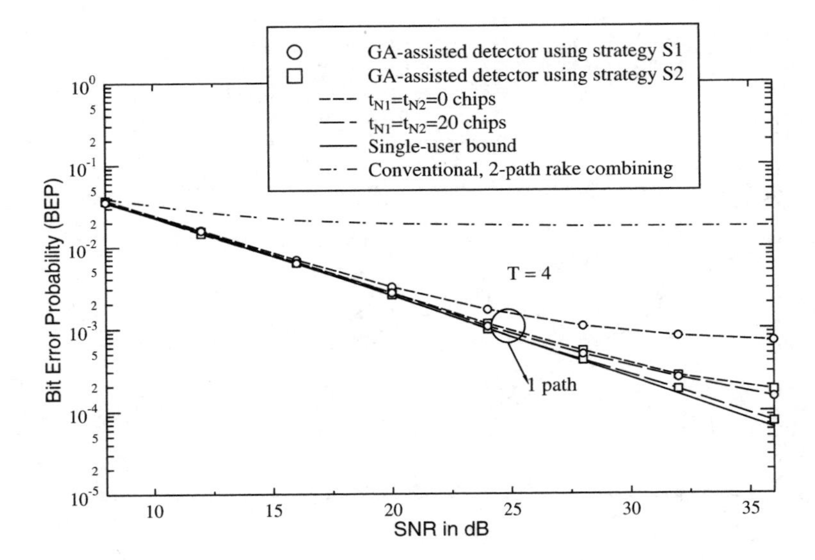

Figure 13.11: The **desired bits'** error probability performance for the GA-assisted multi-user detector over Rayleigh fading channels employing the EEB detection strategies S1 and S2 with a population size of $P = 30$ using random signature sequences of length 31 and supporting $K = 10$ users with various truncated window sizes. The GA configuration and the simulation parameters used are listed in Table 13.2 and Table 13.1, respectively.

Chapter 14

Genetic Algorithm-Assisted Multi-user Detection for Multicarrier CDMA[1]

Hua Wei and L. Hanzo

14.1 Introduction

Multicarrier CDMA (MC-CDMA) [75, 354] is a novel transmission technique, which combines DS-CDMA and Orthogonal Frequency Division Multiplexing (OFDM) [92, 269]. In MC-CDMA systems, instead of applying spreading sequences in the time domain for spreading each bit, we employ spreading sequences in the frequency domain. Hence, we are capable of achieving frequency diversity gain at the cost of a reduced spreading gain. Numerous Multi-user Detection (MUD) schemes have been proposed in the literature [355–357]. For example, the Minimum Mean Square Error (MMSE) MUD has been described, for example, in [355], while an Interference Cancellation (IC) based MUD has been proposed in [356].

In [358], the Maximum Likehood (ML) MUD designed for MC-CDMA had been considered. In this specific MUD, the receiver constructs all the possible combinations of the transmitted signal and employs the estimated channel transfer function for generating all the possible received signals, in order to find the one, which has the smallest Euclidean distance from the received signal. Hence, the ML detection-based MUD designed for MC-CDMA is capable of achieving the optimum performance. However, it requires the calculation of

[1] *Single- and Multi-Carrier CDMA Multi-User Detection, Space-Time Spreading, Synchronization, Networking and Standards.* L.Hanzo, L-L.Yang, E-L.Kuan and K.Yen,
©2003 John Wiley & Sons, Ltd. ISBN 0-470-86309-9

2^K number of possible received signal combinations in conjunction with Binary Phase Shift Keying (BPSK) modulation. In other words, the ML detection-based MUD's complexity will increase exponentially with the number of users K. Hence the complexity imposed will become excessive, when the number of users K is high. Therefore, in this chapter we will invoke Genetic Algorithms (GA) [9, 287] for reducing the complexity of the ML detection-based MUD employed in MC-CDMA systems. The GA-based MUD was first proposed by Juntti *et al.* [311] for a synchronous DS-CDMA system communicating over an Additive White Gaussian Noise (AWGN) channel. Yen *et al.* [359–361] further improved the performance of the GA-based MUD, demonstrating that the performance of the GA-based MUD approaches the single-user performance bound at a significantly lower computational complexity, than that of Verdu's optimum MUD [78]. Again, in this chapter we will employ a GA-assisted MUD scheme as a suboptimal MUD technique applicable to both bit-synchronous and asynchronous MC-CDMA systems communicating over broadband frequency selective fading channels. We assume that each subcarrier obeys independent Rayleigh fading. More explicitly, we will investigate the performance of this specific GA-assisted MUD as a function of the affordable detection complexity.

This chapter is organized as follows. Sections 14.2 and 14.5 describe the operation of the GA-assisted MUD in the context of both synchronous MC-CDMA and asynchronous MC-CDMA, respectively. Sections 14.3 and 14.6 characterize the achievable performance of the GA-assisted MUD, when no channel coding techniques are employed. By contrast, Sections 14.4 and 14.7 characterize the performance of this MUD when turbo codes are invoked for enhancing the achievable performance. Finally, Section 14.8 offers our conclusions and outlines our future work.

14.2 GA-assisted MUD for Synchronous MC-CDMA systems

Initially we consider the bit-synchronous MC-CDMA system illustrated in Figure 14.1. Observe in Figure 14.1 that the ith bit $b_k^{(i)}$ of the kth user is spread to M parallel subcarriers, each conveying one of the M number of N-chip spreading signature sequences $c_{k,m}(t), \quad m = 1, \ldots, M$, each of which spans $(0, T_b)$ and we have $T_b/T_c = N$, where T_b and T_c are the bit duration and chip duration, respectively. Each of the M spreading signatures is mapped to a different subcarrier. In other words, a single-carrier system occupying the same bandwidth as the multicarrier system considered would use a spreading signature having NM chips/bit, and both of these systems have a processing gain of NM. Hence, the transmitted signal of the kth user associated with the mth subcarrier can be expressed in an equivalent lowpass representation as:

$$s_{k,m}(t) = \sqrt{\frac{2E_{bk}}{M}} c_{k,m}(t) b_k^{(i)} e^{jw_m t}, \qquad (14.1)$$

where E_{bk} is the kth user's signal energy per bit, $b_k^{(i)} \in (1, -1), \quad k = 1, \ldots, K$ denotes the ith transmitted bit of the kth user, while the kth user's signature waveform is $c_{k,m}(t), \quad k =$

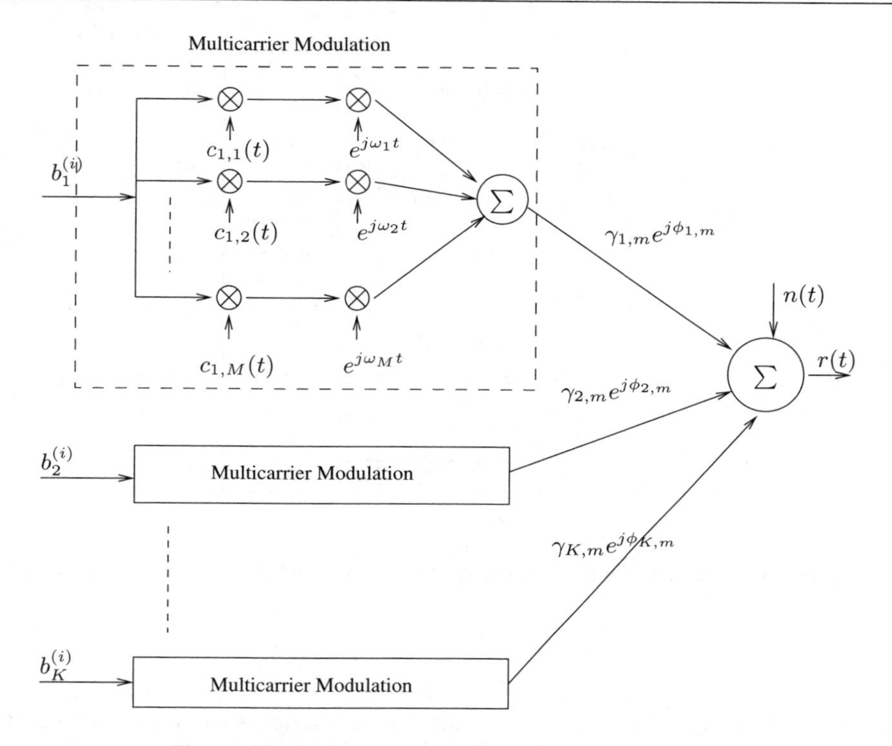

Figure 14.1: The transmitter of a MC-CDMA system

$1,\ldots,K,\ \ m=1,\ldots,M$ on the mth subcarrier, which again has a length of N chips, and can be written as:

$$c_{k,m}(t)=\sum_{n=0}^{N-1}c_{k,m}^{(n)}p(t-nT_c),\quad m=1,\ldots,M,\ \ k=1,\ldots,K, \tag{14.2}$$

where T_c is the chip duration, N is the number of chips per bit associated with each subcarrier and we have $T_b/T_c=N$. Again, the total processing gain is NM, while $p(t)$ is the rectangular chip waveform employed, which can be expressed as:

$$p(t)=\begin{cases}1 & 0\le t<T_c\\ 0 & otherwise.\end{cases} \tag{14.3}$$

Without loss of generality, we assume that the signature waveform $c_{k,m}(t)$ used for spreading the bits to a total of M subcarriers for all the K users has unity energy, which can be written as:

$$\int_0^{T_b}c_{k,m}^2(t)dt=1\quad k=1,\ldots,K,\ \ m=1,\ldots,M. \tag{14.4}$$

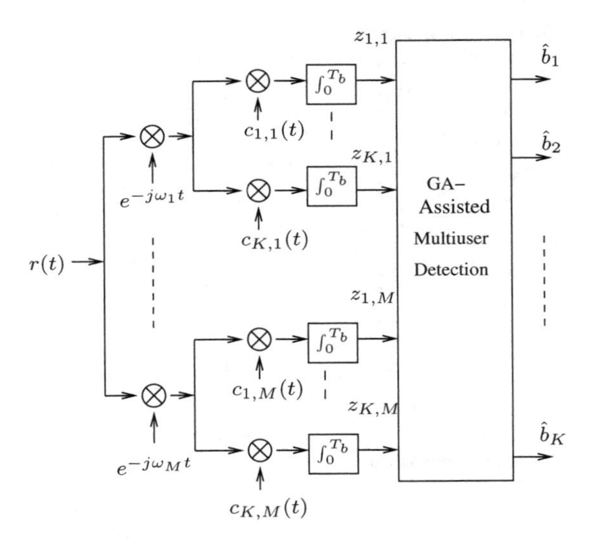

Figure 14.2: Schematic of the GA-assisted MUD-aided MC-CDMA base station receiver

Each user's signal $s_{k,m}(t)$ transmitted on the mth subcarrier is assumed to propagate over an independent non-dispersive single-path Rayleigh fading channel and the fading envelope of each path is statistically independent for all the users. Hence, the single-tap narrowband Channel Impulse Response (CIR) of the kth user on the m subcarrier can be expressed as: $\gamma_{k,m}e^{j\phi_{k,m}}$, where the amplitude $\gamma_{k,m}$ is a Rayleigh distributed random variable, while the phase $\phi_{k,m}$ is uniformly distributed between $[0, 2\pi)$.

Having described the transmitter and the channel, the received signal on the mth subcarrier can be expressed as:

$$r_m(t) = \sum_{i=-\infty}^{\infty} \sum_{k=1}^{K} \sqrt{\frac{2E_{bk}}{M}} c_{k,m}(t - iT_b)\gamma_{k,m} b_k^{(i)} e^{(jw_m t + \phi_{k,m})} + n(t), \qquad (14.5)$$

where K is the number of users supported and $n(t)$ is the Gaussian noise process with a variance of $N_0/2$.

Figure 14.2 shows the schematic of the GA-assisted MUD employed in a synchronous MC-CDMA system. The first step of the receiver's operation is the demodulation of all the subcarrier signals. This is followed by Matched Filtering (MF) for each of the K users and the outputs of the K users' matched filters are input to the GA-based MUD. It is more convenient to express the associated signals in matrix and vectorial formats, when the sum of the transmitted signals of all users can be expressed in vectorial notation as :

$$r_m(t) = \mathbf{C}_m \mathbf{W}_m \mathbf{A}\mathbf{b} + \mathbf{n}, \qquad (14.6)$$

where we have

$$\begin{aligned}
\mathbf{C_m} &= [c_{1,m}(t), \ldots, c_{K,m}(t)] \\
\mathbf{W}_m &= \text{diag}[\gamma_{1,m}e^{j\phi_{1,m}}, \ldots, \gamma_{K,m}e^{j\phi_{K,m}}] \\
\mathbf{A} &= \text{diag}[\sqrt{\tfrac{2E_{b1}}{M}}, \ldots, \sqrt{\tfrac{2E_{bK}}{M}}] \\
\mathbf{b} &= [b_1, \ldots, b_K]^T \\
\mathbf{n} &= [n_1, \ldots, n_K]^T.
\end{aligned} \tag{14.7}$$

Based on Equation 14.6, the output vector $\mathbf{Z}_m$ of the bank of matched filters portrayed in Figure 14.2 can be formulated as [78]:

$$\begin{aligned}
\mathbf{Z}_m &= [z_{1,m}, \ldots, z_{K,m}] \\
&= \mathbf{R}_m\mathbf{W}_m\mathbf{A}\mathbf{b} + \mathbf{n}
\end{aligned} \tag{14.8}$$

where we have

$$\mathbf{R}_m = \begin{bmatrix} \rho_{11}^{(m)} & \rho_{12}^{(m)} & \cdots & \rho_{1K}^{(m)} \\ \rho_{21}^{(m)} & \rho_{22}^{(m)} & \cdots & \rho_{2K}^{(m)} \\ \vdots & \vdots & \vdots & \vdots \\ \rho_{K1}^{(m)} & \rho_{K2}^{(m)} & \cdots & \rho_{KK}^{(m)} \end{bmatrix}. \tag{14.9}$$

and the elements $\rho_{jk}^{(m)}$ of the matrix $\mathbf{R}_m$ are the auto- and cross-correlation of the spreading code, which can be expressed as:

$$\rho_{jk}^{(m)} = \int_0^{T_b} c_{j,m}(t)c_{k,m}(t)dt. \tag{14.10}$$

According to [78, 359], the optimum multi-user detector of the mth subcarrier will maximize the following objective function:

$$\Omega_m(\mathbf{b}) = 2\text{Re}[\mathbf{b}^T\mathbf{A}\mathbf{W}_m^*\mathbf{Z}_m] - \mathbf{b}^T\mathbf{A}\mathbf{W}_m\mathbf{R}_m\mathbf{W}_m^*\mathbf{A}\mathbf{b}, \tag{14.11}$$

where the superscript $*$ indicates the conjugate complex version of a matrix. Therefore, combining the contributions of a total of M parallel subcarriers, the objective function to be maximized in the context of an optimum multi-user detected MC-CDMA system can be expressed as:

$$\begin{aligned}
\Omega(\mathbf{b}) &= \sum_{m=1}^{M} \Omega_m(\mathbf{b}) \\
&= \sum_{m=1}^{M} \{2\text{Re}[\mathbf{b}^T\mathbf{A}\mathbf{W}_m^*\mathbf{Z}_m] - \mathbf{b}^T\mathbf{A}\mathbf{W}_m\mathbf{R}_m\mathbf{W}_m^*\mathbf{A}\mathbf{b}\}.
\end{aligned} \tag{14.12}$$

Hence the decision rule for Verdu's optimum CDMA multi-user detection scheme [78] based on the maximum likelihood criterion is to choose the specific K-user bit combination $\mathbf{b}$, which maximizes the metric of Equation 14.12. Hence, we have to find:

$$\hat{\mathbf{b}} = \arg\left\{\max_{\mathbf{b}}[\Omega(\mathbf{b})]\right\}. \tag{14.13}$$

The maximization of Equation 14.12 is a combinatorial optimization problem, which requires an exhaustive search for each of the $J = 2^K$ combinations of $\mathbf{b}$, in order to find the one that maximizes the metric of Equation 14.12. Explicitly, since in case of binary transmissions there are $J = 2^K$ possible combinations of $\mathbf{b}$, the optimum multi-user detector has a complexity that increases exponentially with the number of users K.

Hence, we invoked GA for finding a solution near to the maximum of the objective function defined by the metric of Equation 14.12 without an exhaustive search. Again, the legitimate solutions are the $J = 2^K$ possible combinations of the K-bit vector $\mathbf{b}$. During the GA's operation, each individual of the GA [359] will take the form of a K-bit vector corresponding to the K users' transmitted bits during a single bit interval, which can be denoted for the pth individual of the GA as $\tilde{\mathbf{b}}_p(y) = [\tilde{b}_{p,1}(y), \ldots, \tilde{b}_{p,K}(y)]$, where $y,\ y = 1, \ldots, Y$ denotes the yth generation, and $p,\ p = 1, \ldots, P$ denotes the pth individual in the mating pool.

We create the initial biased population with the aid of the Maximum Ratio Combining (MRC) based matched filter outputs, which are subjected to hard decision, rather than randomly generating the initial population at the commencement of a GA-assisted search. Explicitly, according to [75], the MRC-combined output vector $\hat{\mathbf{b}}_{MRC}$ of the matched filter output can be expressed as: $\hat{\mathbf{b}}_{MRC} = [\hat{b}_{1,MRC}, \ldots, \hat{b}_{K,MRC}]$, where we have:

$$\hat{b}_{k,MRC} = \sum_{m=1}^{M} z_{k,m}\gamma_{k,m}e^{-j\phi_{k,m}}. \tag{14.14}$$

Having generated $\hat{\mathbf{b}}_{MRC}$, we adopt a 'mutated' version of the hard decision vector $\hat{\mathbf{b}}_{MRC}$ for creating each individual in the initial population, where each bit of the vector $\hat{\mathbf{b}}_{MRC}$ is toggled according to the mutation probability used. Hence, the first individual of the population, namely $\tilde{\mathbf{b}}_p(0)$ can be written as:

$$\tilde{\mathbf{b}}_p(0) \quad = \quad \text{MUTATION}[\hat{\mathbf{b}}_{MRC}]. \tag{14.15}$$

To elaborate a little further, Figure 14.3 shows the flowchart of the GA-assisted MUD, which follows the philosophy of [360]. We will characterize the performance of the GA-assisted MUD in the following section.

14.3 Performance of GA-assisted MUD-aided Synchronous MC-CDMA

The basic parameters of the GA used in our simulations are listed in Table 14.1. From Figures 14.4 and 14.5 we observe that the GA-assisted MUD's performance improves, when the

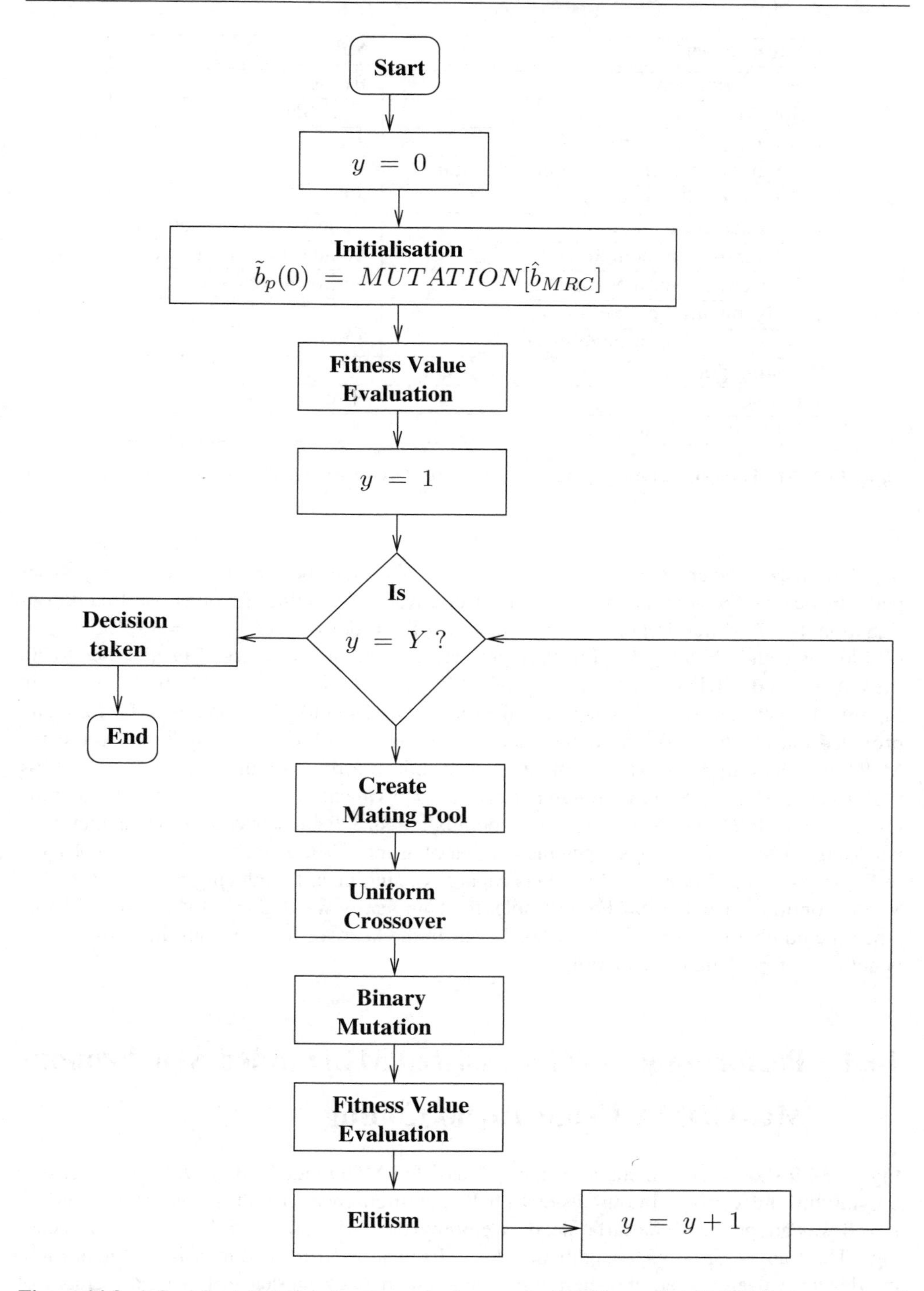

Figure 14.3: A flowchart depicting the structure of a genetic algorithm assisted MUD in the context of the synchronous MC-CDMA base station receiver

Parameters	Value
Modulation scheme	BPSK
Spreading code	WALSH
Number of subcarriers M	4
length of subcarrier spreading signature N	8
Total spreading gain MN	32
GA's selection method	Fitness-proportionate
GA's mutation method	Standard binary mutation
GA's cross-over method	Uniform cross-over
GA's mutation probability p_m	0. 1
GA's cross-over probability p_c	1
Mating pool size T	5
Elitism	Yes
Incest prevention	Yes

Table 14.1: The basic simulation parameters used by the GA-assisted MUD-aided MC-CDMA system

population size P increases. For example, for E_b/N_0 values below 14dB a near-single-user performance can be achieved for $K = 10$ users, when evaluating the objective function of Equation 14.12, which imposes a complexity on the order of $O(P \cdot Y) = O(40 \cdot 10) = O(400)$, as seen in Figure 14.4. Furthermore, when the number of users K is increased to 20, the GA-assisted MUD has a complexity of $O(P \cdot Y) = O(160 \cdot 10) = O(1600)$, as seen in Figure 14.5. We can see in Figure 14.6 that the GA-assisted MUD is also near–far resistant, provided that we perfectly know the channel parameters. More explicitly, the GA-assisted MUD exhibits a high robustness against power control errors. Figure 14.7 shows the BER performance as a function of the number of users K. We can infer from Figure 14.7 that the GA-assisted MUD required a population population size P in excess of 80 for achieving a near-single-user performance, when the number of users K is higher than 14. We can observe in Figure 14.8 that GA-assisted MUD is capable of significantly reducing the complexity of Verdu's optimum MUD [78]. For example, the complexity was reduced by a factor of 1300, when the number of users was $K = 20$. Let us in the next section consider the performance benefits of employing turbo coding.

14.4 Performance of GA-assisted MUD-aided Synchronous MC-CDMA Using Turbo Coding

Figure 14.9 shows the schematic of the GA-assisted MUD-aided MC-CDMA receiver employing turbo decoding. In our system a half-rate turbo encoder with a constraint length of $v = 3$ was adopted and the turbo decoder employed four iterations in the process of decoding. The transmission burst length was $L = 100$ and a 10×10 dimensional block turbo interleaver was employed. Furthermore, a random channel interleaver having a memory of $100 \times 2 \times 4 = 800$ bits was employed in our system. As seen in Figures 14.10 and 14.11,

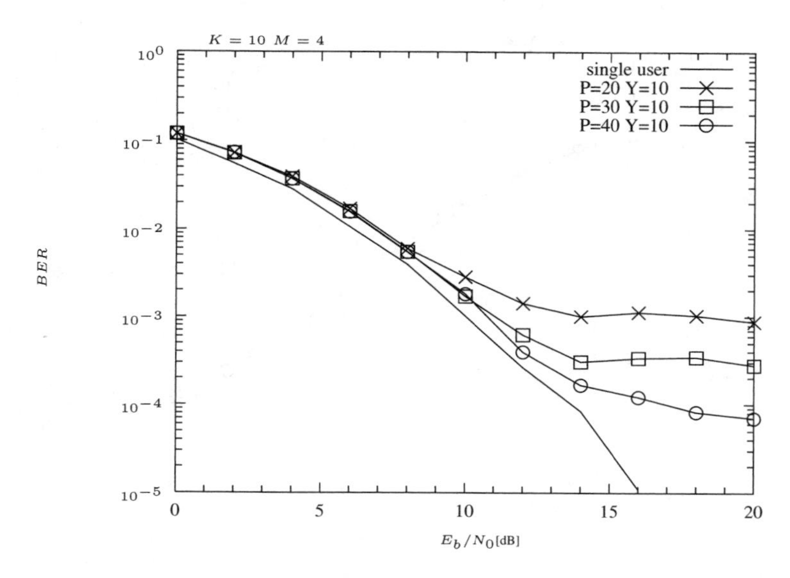

Figure 14.4: BER performance of the GA-assisted MUD designed for a bit-synchronous MC-CDMA system, using a 32-chip Walsh code. The number of users supported was $K = 10$. The number of generations was $Y = 10$ and the population size was $P = 20$, 30 and 40. The number of subcarriers was $M = 4$ and each subcarrier experienced uncorrelated narrowband Rayleigh fading.

when channel coding was used, the GA-assisted MUD was capable of further reducing the complexity required for attaining a BER of 10^{-5}. When the number of users was $K = 10$, the MUD required only $P = 20$ and $Y = 10$ for approaching the single-user bound, resulting in a total complexity of $O(200)$. By contrast, recall from Figure 14.4, where no turbo coding was used that the GA-assisted MUD required a complexity investment of $O(400)$ for approaching the single-user bound, while attaining a BER of 10^{-3}. When the number of users supported was increased to $K = 20$, the MUD required $P = 80$ and $Y = 10$ for approaching the single-user bound, which corresponded to a complexity of $O(800)$. For the sake of comparison recall from Figure 14.5, where no turbo coding was used that the GA-assisted MUD required a complexity of $O(1600)$ for approaching the single-user bound, while aiming for a BER of 10^{-3}. Having studied the less realistic scenario of synchronous GA-assisted MUD-aided MC-CDMA, let us now focus our attention on the more realistic asynchronous uplink scenario.

14.5 GA-assisted MUD-aided Asynchronous MC-CDMA System

Figure 14.12 shows the schematic of GA-assisted MUD-aided asynchronous MC-CDMA base station receiver, which retains the same structure as the synchronous MC-CDMA BS's

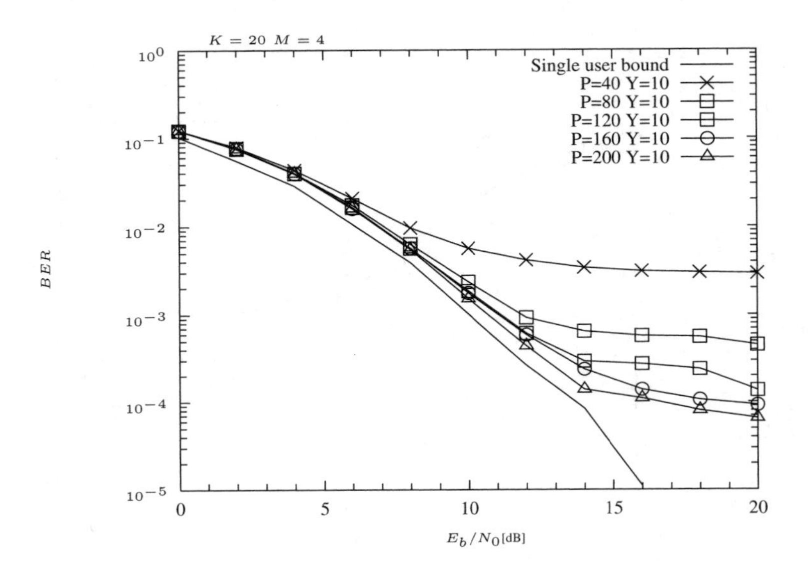

Figure 14.5: BER performance of the GA-assisted MUD designed for a bit-synchronous MC-CDMA system, using a 32-chip Walsh code. The number of users supported was $K = 20$. The number of generations was $Y = 10$ and the population size was $P = 40,\ 80,\ 120,\ 160$ and 200. The number of subcarriers was $M = 4$ and each subcarrier experienced uncorrelated narrowband Rayleigh fading.

receiver. However, the each user's matched filter must be synchronized according to the user's individual delay τ_k. A similar formalism to that of the synchronous MC-CDMA system can be adopted. Therefore, when communicating over asynchronous environments, the received signal can also be expressed similarly to Equation 14.5, namely as:

$$r_m(t) = \sum_{i=-\infty}^{\infty} \sum_{k=1}^{K} \sqrt{\frac{2E_{bk}}{M}} c_{k,m}(t - iT_b - \tau_k)\gamma_{k,m} b_k^{(i)} e^{(j\omega_m t + \phi_{k,m})} + n(t), \tag{14.16}$$

where τ_k is the kth user's delay. Without loss of generality, we assume that we have $0 \leq \tau_1 < \tau_2 \cdots < \tau_K < T_b$.

The output of the kth user's matched filter corresponding to the mth subcarrier sampled at the end of the ith symbol interval is given as:

$$\mathbf{z}_{k,m}^{(i)} = \int_{-\infty}^{\infty} r(t) c_{k,m}(t - iT_b - \tau_k) dt. \tag{14.17}$$

According to [78], the received signal vector $\mathbf{Z}_m$ recorded at the output of the bank of

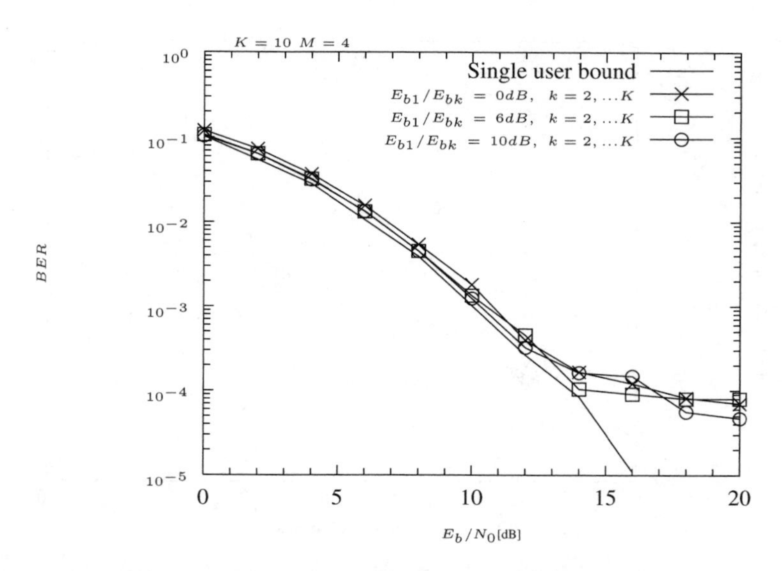

Figure 14.6: BER performance of the GA-assisted MUD designed for a bit-synchronous MC-CDMA system when the power of the interfering users was varied . The number of users supported was $K = 10$. The number of generations was $Y = 10$ and the population size was $P = 40$. The ratio of the reference user to interfering user power was $\frac{E_{b1}}{E_{bk}} = 0,\ 6,\ 10$dB, $k = 2, \ldots, K$, respectively.

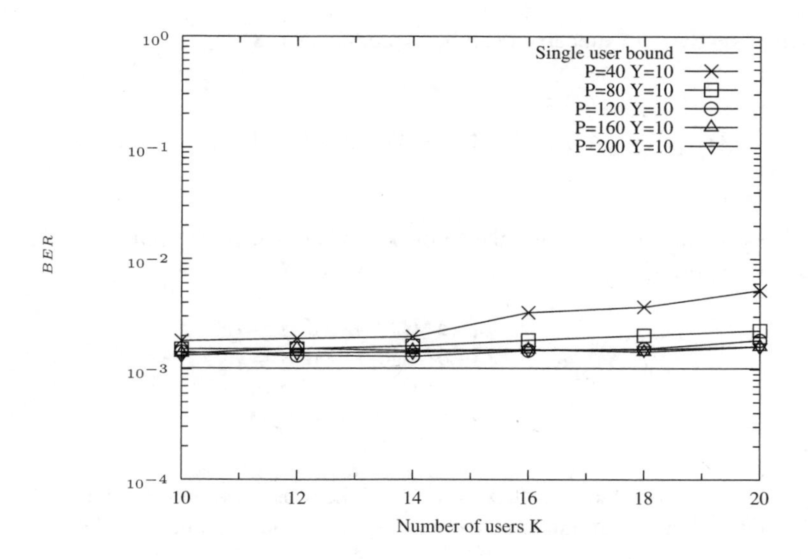

Figure 14.7: BER performance of the GA-assisted MUD as a function of the number of users K for the population sizes of $P = 40,\ 80,\ 120,\ 160,\ 200$, and for $Y = 10$ generations. We had $E_b/N_0 = 10$dB.

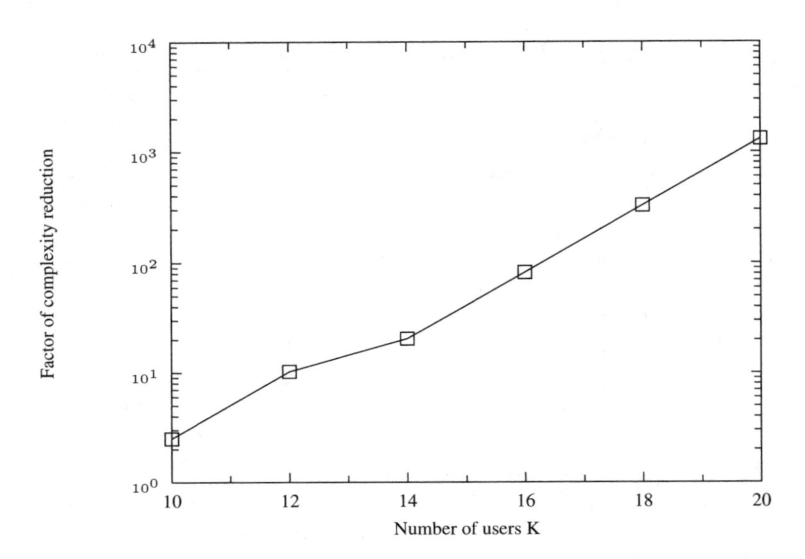

Figure 14.8: The complexity reduction factor of $\frac{2^K}{P \times Y}$ was defined as the ratio of the number of objective function computations required for approaching the single-user bound at a BER of 10^{-3}, when communicating over a **synchronous** environment, where P is the population size, and Y is the number of generations, while K is the number of users supported.

matched filters on the mth subcarrier can be expressed as [78]:

$$\begin{aligned}
\mathbf{Z}_m^{(i)} &= [z_{1,m}, \ldots, z_{K,m}] \\
&= \mathbf{R}_m[1]\mathbf{W}_m\mathbf{A}\mathbf{b}^{(i-1)} + \mathbf{R}_m[0]\mathbf{W}_m\mathbf{A}\mathbf{b}^{(i)} + \mathbf{R}_m^T[1]\mathbf{W}_m\mathbf{A}\mathbf{b}^{(i+1)} + \mathbf{n_m},
\end{aligned} \tag{14.18}$$

where i is the time instant index and the zero-mean Gaussian noise vector $\mathbf{n_m}$ has the cross-correlation matrix of:

$$E[\mathbf{n}[i]\mathbf{n}^T[j]] = \begin{cases} \sigma^2\mathbf{R}^T[1], & \text{if } j = i+1; \\ \sigma^2\mathbf{R}[0], & \text{if } j = i; \\ \sigma^2\mathbf{R}[1], & \text{if } j = i-1; \\ \mathbf{0}, & otherwise. \end{cases} \tag{14.19}$$

The partial cross-correlation matrix $\mathbf{R}_m[1]$ and the cross-correlation matrix $\mathbf{R}_m[0]$ of the spreading codes, when communicating over an asynchronous channel, are defined as [78]:

$$\mathbf{R}_{jk}^{(m)}[0] = \begin{cases} 1, & \text{if } j = k; \\ \rho_{jk}^{(m)}, & \text{if } j < k; \\ \rho_{kj}^{(m)}, & \text{if } j > k, \end{cases} \tag{14.20}$$

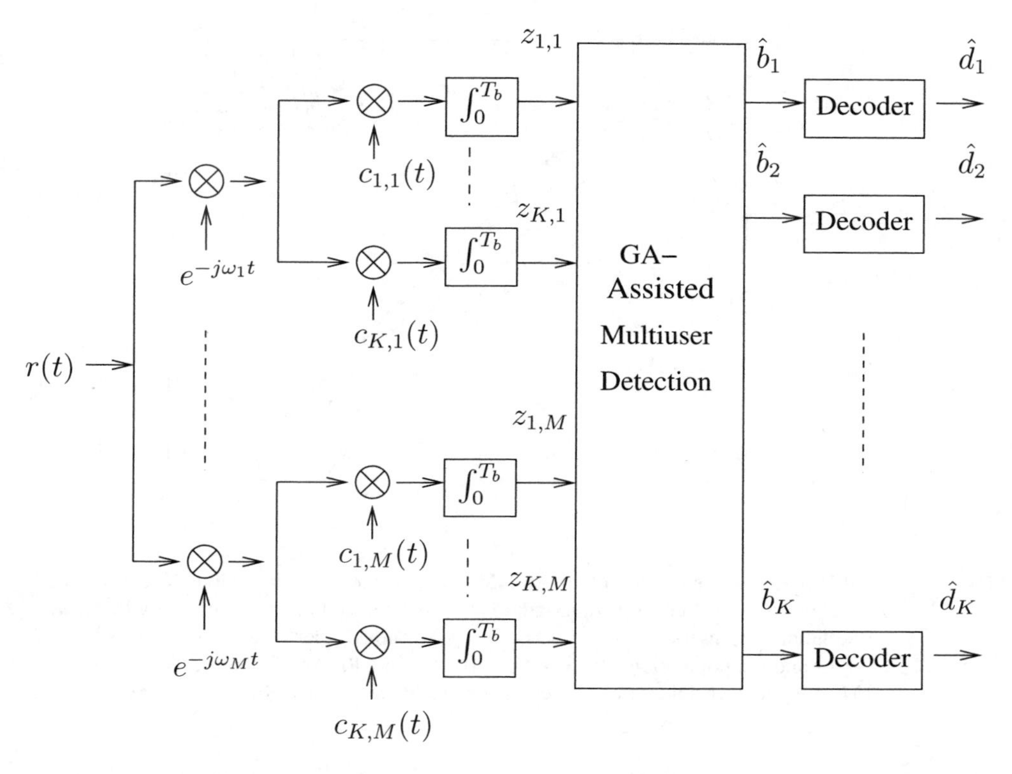

Figure 14.9: Schematic of the GA-assisted MUD-aided MC-CDMA base station receiver employing turbo decoding.

$$\mathbf{R}_{jk}^{(m)}[1] = \begin{cases} 0, & \text{if } j \geq k; \\ \rho_{kj}^{(m)}, & \text{if } j < k. \end{cases} \tag{14.21}$$

where the coefficients $\rho_{kj}^{(m)}$ and $\rho_{kj}^{(m)}$ on the mth subcarrier are the pair of cross-correlations of the spreading codes recorded in the asynchronous CDMA environment, which can be written as [78]:

$$\rho_{jk}^{(m)} = \int_{\tau}^{T_b} c_{j,m}(t)c_{k,m}(t-\tau)dt \tag{14.22}$$

$$\rho_{kj}^{(m)} = \int_{0}^{\tau} c_{j,m}(t)c_{k,m}(t-\tau+T_b)dt, \tag{14.23}$$

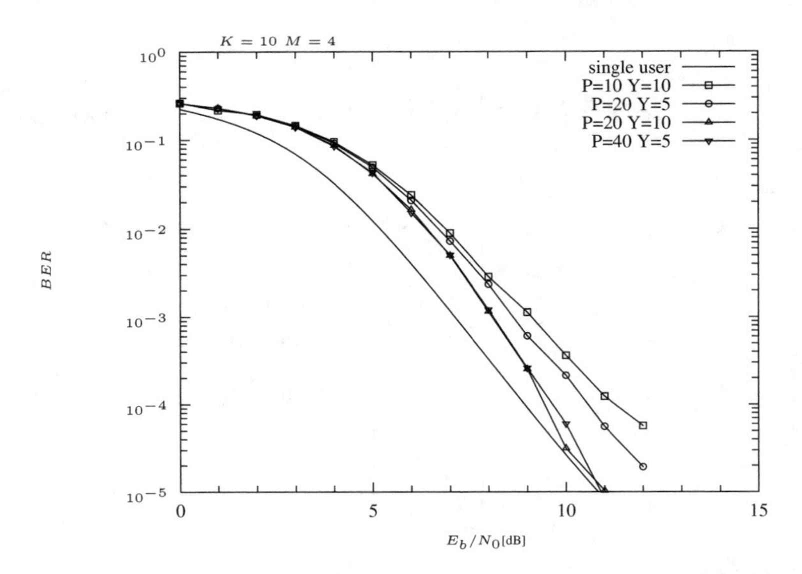

Figure 14.10: BER performance of the GA-assisted MUD designed for synchronous MC-CDMA assisted by $R = \frac{1}{2}$ rate, constraint length $m = 3$ turbo coding using four iterations, when the number of users supported was $K = 10$. The number of generations was $Y = 5$ or 10 and the population size was $P = 10,\ 20$ and 40. The number of subcarriers was $M = 4$ and each subcarrier obeyed uncorrelated narrowband Rayleigh fading.

according to Equation 14.18, the noise sampling vector $\mathbf{n_m}$ can also be expressed as :

$$\begin{aligned} \mathbf{n_m} \;=\; & \mathbf{Z}_m^{(i)} - \mathbf{R}_m[1]\mathbf{W}_m\mathbf{A}\mathbf{b}^{(i-1)} \\ & -\mathbf{R}_m[0]\mathbf{W}_m\mathbf{A}\mathbf{b}^{(i)} - \mathbf{R}_m^T[1]\mathbf{W}_m\mathbf{A}\mathbf{b}^{(i+1)}. \end{aligned} \tag{14.24}$$

Hence, the objective function of the optimum ML detector on the mth subcarrier can be expressed as:

$$\Omega_m(\mathbf{b}^{(i)}) = \arg\{\min_{\mathbf{b}^{(i-1)},\mathbf{b}^{(i)},\mathbf{b}^{(i+1)}} E[\mathbf{n_m} \cdot \mathbf{n_m}^T]\}. \tag{14.25}$$

Hence, according to Equation 14.25, the GA's objective metric for the mth subcarrier, which have to be maximized, can be expressed as:

$$\begin{aligned} \Omega_m(\mathbf{b}^{(i)}) \;=\; & \exp\{- \parallel \mathbf{Z}_m^{(i)} - \mathbf{R}_m[1]\mathbf{W}_m\mathbf{A}\mathbf{b}^{(i-1)} \\ & -\mathbf{R}_m[0]\mathbf{W}_m\mathbf{A}\mathbf{b}^{(i)} - \mathbf{R}_m^T[1]\mathbf{W}_m\mathbf{A}\mathbf{b}^{(i+1)} \parallel^2\}, \end{aligned} \tag{14.26}$$

where $\parallel \cdot \parallel$ denotes the Euclidian norm of a complex quantity expressed for the arbitrary variable $v = a + jb$ as $\parallel v \parallel = \sqrt{a^2 + b^2}$.

Therefore, when combining the signals of the M subcarriers, the modified objective func-

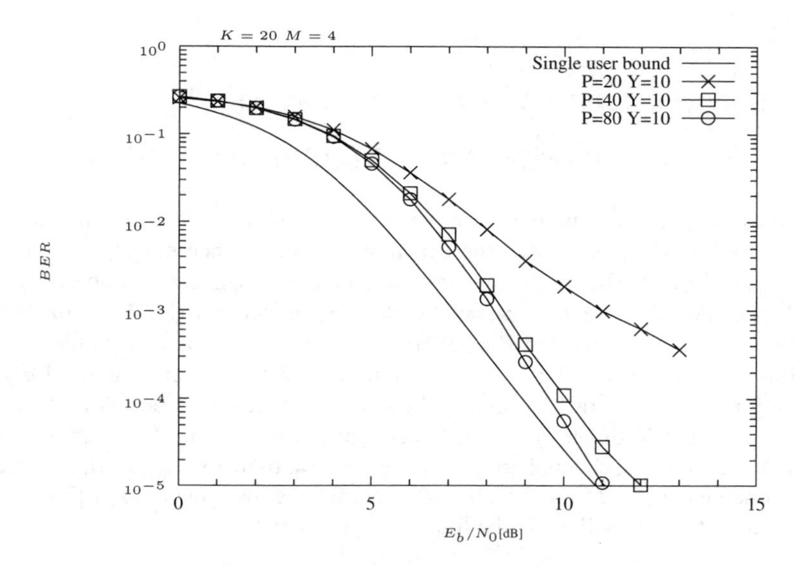

Figure 14.11: BER performance of the GA-assisted MUD designed for synchronous MC-CDMA assisted by $R = \frac{1}{2}$ rate, constraint length $m = 3$ turbo coding using four iterations, when the number of users supported was $K = 20$. The number of generations was $Y = 5$ or 10 and the population size was $P = 10,\ 20$ and 40. The number of subcarriers was $M = 4$ and each subcarrier obeyed uncorrelated narrowband Rayleigh fading.

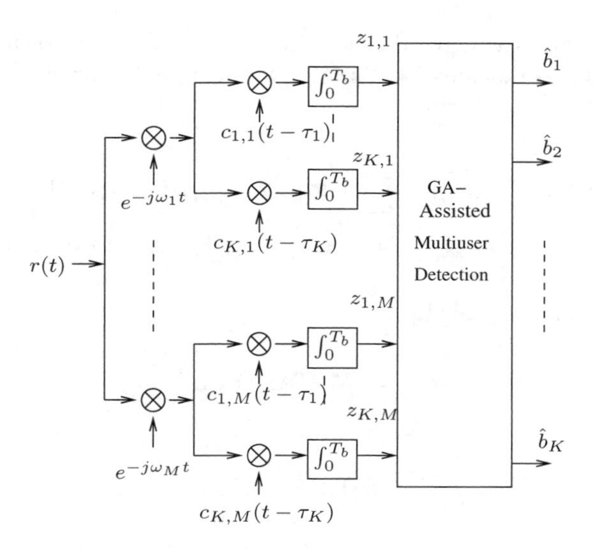

Figure 14.12: Schematic of the GA-assisted MUD-aided asynchronous MC-CDMA base station receiver.

tion becomes:

$$\begin{aligned}\Omega(\mathbf{b}^{(i)}) &= \exp\{-\sum_{m=1}^{M} \| \mathbf{Z}_m^{(i)} - \mathbf{R}_m[1]\mathbf{W}_m\mathbf{A}\mathbf{b}^{(i-1)} \\ &-\mathbf{R}_m[0]\mathbf{W}_m\mathbf{A}\mathbf{b}^{(i)} - \mathbf{R}_m^T[1]\mathbf{W}_m\mathbf{A}\mathbf{b}^{(i+1)} \|^2\}\end{aligned} \quad (14.27)$$

Therefore, for achieving the optimum performance, we have to maximize the metric Ω of Equation 14.27. More explicitly, the optimum decision concerning the K-dimensional Current Estimated Bit (CEB) vector $\mathbf{b}^{(i)}$ will maximize the cross-correlation metric in Equation 14.27, provided that the K-dimensional Start Estimated Bit (SEB) vector $\mathbf{b}^{(i-1)}$ and the K-dimensional End Estimated Bit (EEB) vector $\mathbf{b}^{(i+1)}$ are perfectly known to the receiver. However, in practice the receiver is oblivious of the K-dimensional EEB vectors $\mathbf{b}^{(i+1)}$ during the detection of $\mathbf{b}^{(i)}$, unless these are estimate-based on pilot bits or training bits. Furthermore, the K-dimensional SEB vector $\mathbf{b}^{(i-1)}$ is never perfectly known by the receiver as a consequence of channel errors. Hence we have to invoke appropriate strategies for finding reasonable choices of $(\mathbf{b}^{(i-1)}, \mathbf{b}^{(i)}, \mathbf{b}^{(i+1)})$ for the maximization of Equation 14.27. In the next subsection, we will describe four strategies in detail.

14.5.1 Edge-bit Generation

In the context of all these four edge-bit selection strategies we will employ the hard decision bit vector $\hat{\mathbf{b}}^{(i-1)}$ as the bit vector $\mathbf{b}^{(i-1)}$ in order to reduce the search space of the GA, although this is a suboptimum strategy. Nevertheless, in a low-BER scenario we may assume that the vector $\hat{\mathbf{b}}^{(i-1)}$ is close to a perfect estimate. Let us now describe the four different edge-bit selection strategies in detail.

According to the **first strategy (S1)**, the SEB vector $\mathbf{b}^{(i-1)}$ will be set to the hard decision vector $\hat{\mathbf{b}}^{(i-1)}$. Hence, in Equation 14.27 $\hat{\mathbf{b}}^{(i-1)}$ will replace the vector $\mathbf{b}^{(i-1)}$ and hence we will maximize the following function:

$$\begin{aligned}\Omega(\mathbf{b}^{(i)}, \hat{\mathbf{b}}^{(i+1)}) &= \exp\{-\sum_{m=1}^{M} \| \mathbf{Z}_m^{(i)} - \mathbf{R}_m[1]\mathbf{W}_m\mathbf{A}\hat{\mathbf{b}}^{(i-1)} \\ &-\mathbf{R}_m[0]\mathbf{W}_m\mathbf{A}\mathbf{b}^{(i)} - \mathbf{R}_m^T[1]\mathbf{W}_m\mathbf{A}\mathbf{b}^{(i+1)} \|^2\},\end{aligned} \quad (14.28)$$

and the initial population is given by:

$$\mathbf{b}_p^{(i)}(0) = \text{MUTATION}[\hat{\mathbf{b}}_{MRC}^{(i)}], \quad (14.29)$$

$$\mathbf{b}_p^{(i+1)}(0) = \text{MUTATION}[\hat{\mathbf{b}}_{MRC}^{(i+1)}] \quad p = 1, \ldots, P; \quad (14.30)$$

where the subscript p indicates the individuals' index in the population of the GA, and (0) refers to the initial generation of the GA.

According to the **second strategy S2**, we also employ Equation 14.28 as the function to be maximized. However, in order to obtain a higher grade of diversity for the individuals of the GA invoked for finding the most likely K-dimensional vector $\mathbf{b}^{(i)}$, we randomly initialized the K-dimensional vector $\mathbf{b}^{(i)}$ and hence the initial population associated with the 0th generation can be expressed as:

$$\mathbf{b}_p^{(i)}(0) = \hat{\mathbf{b}}_{GA}^{(i)}, \quad (14.31)$$

$$\mathbf{b}_p^{(i+1)}(0) = \text{MUTATION}[\hat{\mathbf{b}}_{MRC}^{(i+1)}]. \quad (14.32)$$

The philosophical difference between Equations 14.29, 14.30 and Equations 14.31, 14.32 is that the MRC output $\mathbf{b}_{MRC}^{i}$ used in Equation 14.29 for initialization has been replaced by $\mathbf{b}_{GA}^{i}$, which is a randomly initialized K-bit vector, subjected to no mutation.

From Equation 14.28 we can observe that it requires a GA-assisted search for the best individual in a space of 2^{2K} elements, which requires a larger population size P and a higher number of generation Y, than that necessitated by the search space of 2^K elements required by the synchronous system. According to the **third strategy (S3)**, we will reduce the size of the search space having 2^{2K} elements to a search space of 2^K elements. This is achieved by invoking a hard decision both for the vector $\mathbf{b}^{(i+1)}$ and for the vector $\mathbf{b}^{(i-1)}$. Explicitly, according to strategy S3, the vectors $\mathbf{b}^{(i+1)}$ and $\mathbf{b}^{(i-1)}$ are given by:

$$\mathbf{b}^{(i-1)} = \hat{\mathbf{b}}^{(i-1)}, \quad (14.33)$$

$$\mathbf{b}^{(i+1)} = \hat{\mathbf{b}}_{MRC}^{(i+1)}. \quad (14.34)$$

Furthermore, we randomly initialize the population $\mathbf{b}_p^{(i)}(0)$ using no mutation, which can be expressed as:

$$\mathbf{b}_p^{(i)}(0) = \mathbf{b}_{GA}^{(i)}. \quad (14.35)$$

Finally, according to **fourth strategy S4**, the vectors $\mathbf{b}^{(i+1)}$ and $\mathbf{b}^{(i-1)}$ are set to the same value as in the context of strategy S3 expressed in Equation 14.33 and 14.34. However, as in the context of the synchronous MC-CDMA system, we adopted biased initialization for creating the individuals of the initial population, where each bit of the MRC's output vector is toggled according to the mutation probability used. Hence, the initial generation for the vector $\mathbf{b}_p^{(i)}(0)$ can be created according to:

$$\mathbf{b}_p^{(i)}(0) = \text{MUTATION}[\mathbf{b}_{MRC}^{(i)}]. \quad (14.36)$$

Therefore, the objective function to be maximized for the strategies S3 and S4 can be ex-

	$\mathbf{b}^{(i-1)}$	$\mathbf{b}^{(i)}$	$\mathbf{b}^{(i+1)}$	$\tilde{\mathbf{b}}_p^{(i)}(0)$	$\tilde{\mathbf{b}}_p^{(i+1)}(0)$
S1	$\hat{\mathbf{b}}^{(i-1)}$	$\tilde{\mathbf{b}}_p^{(i)}(y)$	$\tilde{\mathbf{b}}_p^{(i+1)}(y)$	biased initialization	biased initialization
S2	$\hat{\mathbf{b}}^{(i-1)}$	$\tilde{\mathbf{b}}_p^{(i)}(y)$	$\tilde{\mathbf{b}}_p^{(i+1)}(y)$	random initialization	biased initialization
S3	$\hat{\mathbf{b}}^{(i-1)}$	$\tilde{\mathbf{b}}_p^{(i)}(y)$	$\hat{\mathbf{b}}_{MRC}^{(i+1)}$	random initialization	
S4	$\hat{\mathbf{b}}^{(i-1)}$	$\tilde{\mathbf{b}}_p^{(i)}(y)$	$\hat{\mathbf{b}}_{MRC}^{(i+1)}$	biased initialization	

Table 14.2: Summary of the four different edge-bit generation strategies characterizing the CEB vector $\mathbf{b}^{(i)}$, SEB vector $\mathbf{b}^{(i-1)}$, EEB vector $\mathbf{b}^{(i+1)}$ generation and the initialization method.

pressed as:

$$\begin{aligned}\Omega(\mathbf{b}^{(i)}) = & \exp\{-\sum_{m=1}^{M} \parallel \mathbf{Z}_m^{(i)} - \mathbf{R}_m[1]\mathbf{W}_m\mathbf{A}\hat{\mathbf{b}}^{(i-1)} \\ & -\mathbf{R}_m[0]\mathbf{W}_m\mathbf{A}\mathbf{b}^{(i)} - \mathbf{R}_m^T[1]\mathbf{W}_m\mathbf{A}\hat{\mathbf{b}}_{MRC}^{(i+1)} \parallel^2\}.\end{aligned} \quad (14.37)$$

Finally, the four different edge-bit generation strategies are summarized in Table 14.2, showing the differences and similarities of these strategies at a glance. The entire process of the GA-aided MUD's operation is depicted in Figure 14.13 for the asynchronous MC-CDMA system considered.

Having described the four different edge-bit generation strategies, let us now investigate their performance in the next section.

14.6 Performance of GA-assisted MUD-aided Asynchronous MC-CDMA

From Figures 14.14 and 14.16 we can observe that strategy S4 is capable of achieving a better performance than the other strategies. By studying Figures 14.14, 14.15, 14.16 and 14.17, we can infer that the biased initialization strategies, namely S1 and S4, outperformed the random initialization strategies of S2 and S3.

Furthermore, the strategy S4 is capable of approaching the near-single-user performance, despite its low complexity, when the number of users K is not higher than 15. However, from Figures 14.18, 14.19 and 14.20, we can observe that for $K = 20$ the GA-assisted MUD is incapable of approaching the single-user performance, because the search space has 2^{60} elements and hence its complexity would become excessive.

Figures 14.21, 14.22 and 14.23 show the performance of the GA-assisted MUD as a function of the number of generations Y. From Figures 14.21, 14.22 and 14.23 we may conclude that the biased initialization strategies tend to approach the single-user performance faster than the random initialization strategies.

From Figures 14.4 and 14.14 we can see that the GA-assisted MUD imposes a similar complexity, when approaching the single-user performance in case of supporting $K = 10$ users communicating either over synchronous or asynchronous environments. However, from

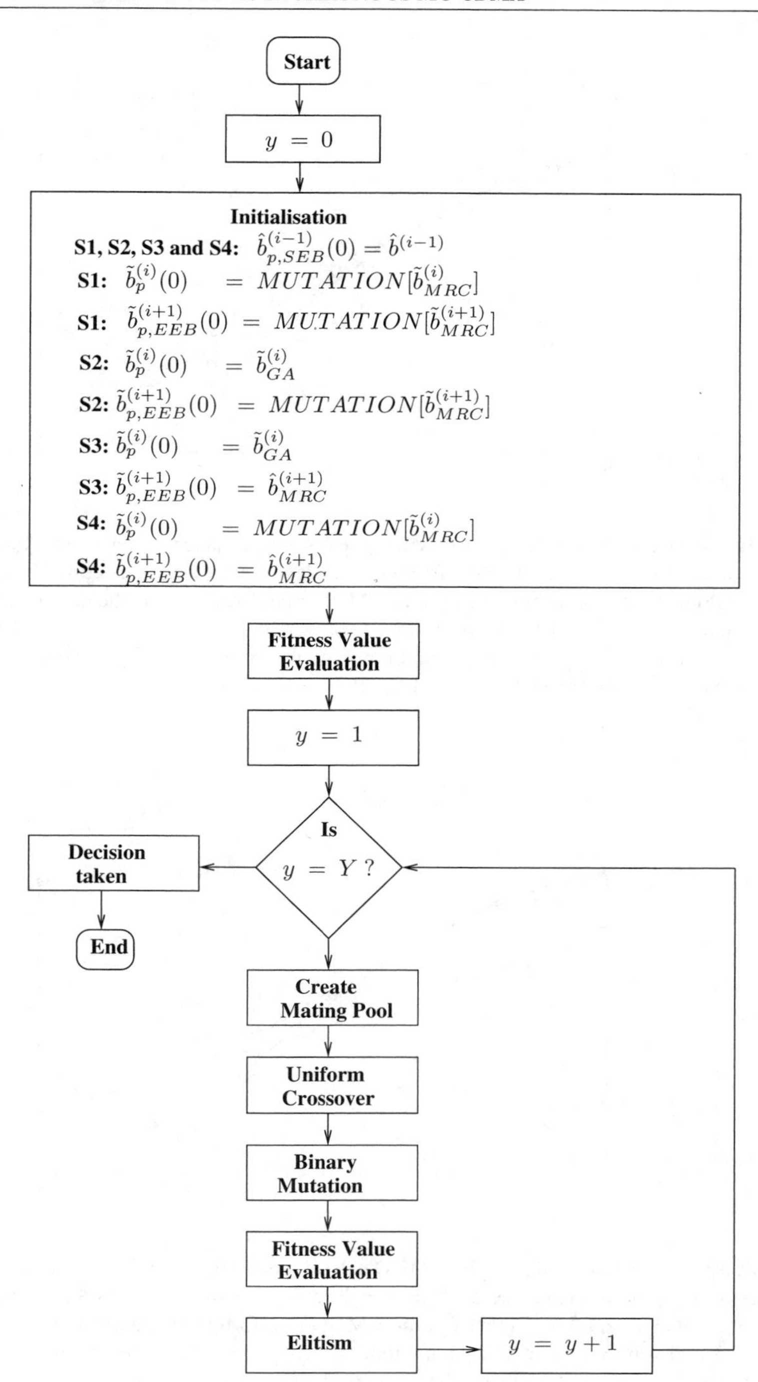

Figure 14.13: A flowchart depicting the structure of a genetic algorithm-assisted MUD-aided MC-CDMA base station receiver.

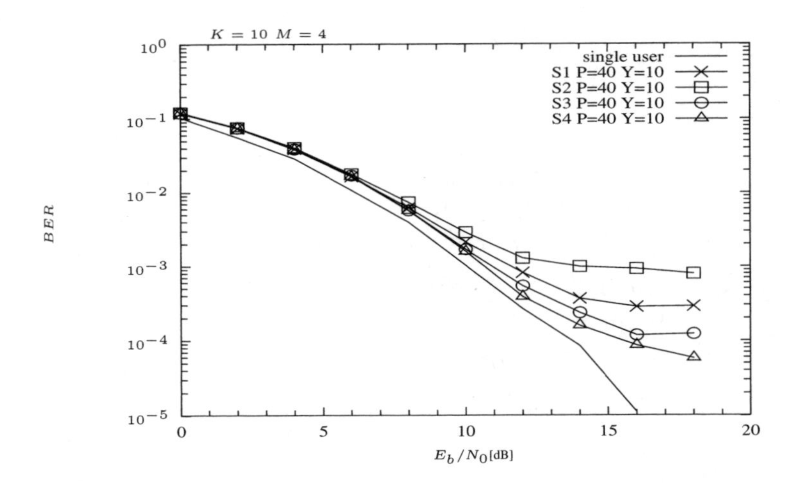

Figure 14.14: The GA-assisted MC-CDMA MUD's BER performance, when communicating over asynchronous environments. The four different GA strategies listed in Table 14.2 were adopted for approaching the optimum MUD's performance and the number of users supported was $K = 10$. The population size was $P = 40$ and the number of generations was $Y = 10$. Furthermore, the number of subcarriers was $M = 4$ and each subcarrier experienced uncorrelated narrowband Rayleigh fading.

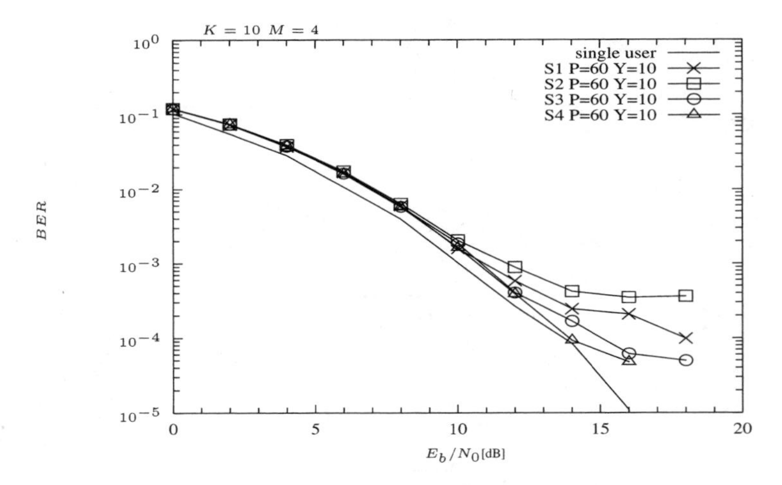

Figure 14.15: The GA-assisted MC-CDMA MUD's BER performance, when communicating over asynchronous environments. The four different GA strategies listed in Table 14.2 were adopted for approaching the optimum MUD's performance and the number of users supported was $K = 10$. The population size was $P = 60$ and the number of generations was $Y = 10$. Furthermore, the number of subcarriers was $M = 4$ and each subcarrier experienced uncorrelated narrowband Rayleigh fading.

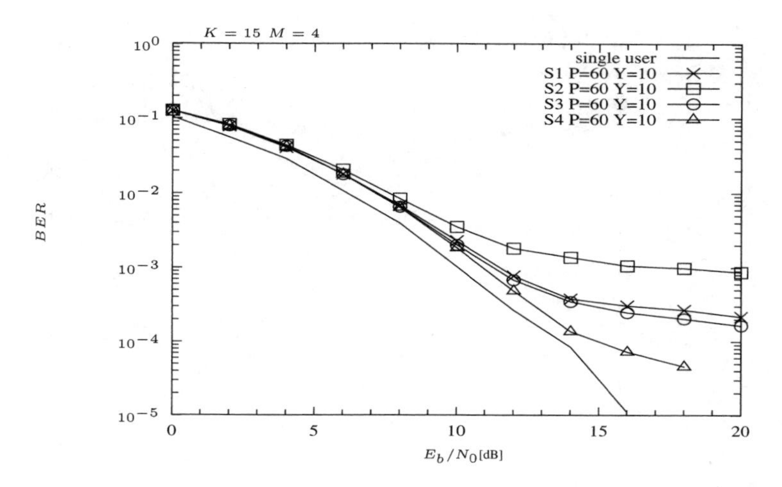

Figure 14.16: The GA-assisted MC-CDMA MUD's BER performance, when communicating over asynchronous environments. The four different GA strategies listed in Table 14.2 were adopted for approaching the optimum MUD's performance and the number of users supported was $K = 15$. The population size was $P = 60$ and the number of generations was $Y = 10$. Furthermore, the number of subcarriers was $M = 4$ and each subcarrier experienced uncorrelated narrowband Rayleigh fading.

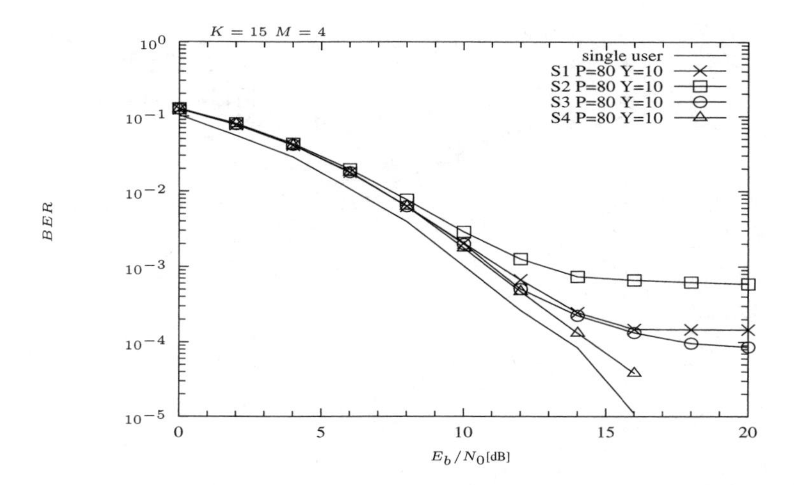

Figure 14.17: The GA-assisted MC-CDMA MUD's BER performance, when communicating over asynchronous environments. The four different GA strategies listed in Table 14.2 were adopted for approaching the optimum MUD's performance and the number of users supported was $K = 15$. The population size was $P = 80$ and the number of generations was $Y = 10$. Furthermore, the number of subcarriers was $M = 4$ and each subcarrier experienced uncorrelated narrowband Rayleigh fading.

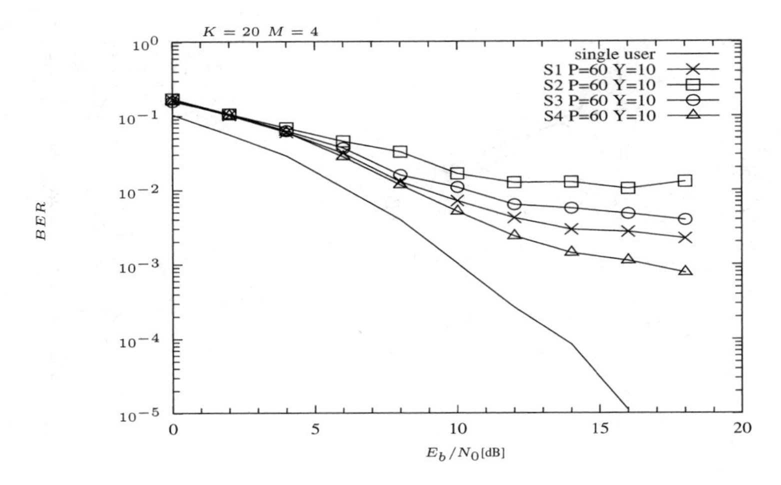

Figure 14.18: The GA-assisted MC-CDMA MUD's BER performance, when communicating over asynchronous environments. The four different GA strategies listed in Table 14.2 were adopted for approaching the optimum MUD's performance and the number of users supported was $K = 20$. The population size was $P = 60$ and the number of generations was $Y = 10$. Furthermore, the number of subcarriers was $M = 4$ and each subcarrier experienced uncorrelated narrowband Rayleigh fading.

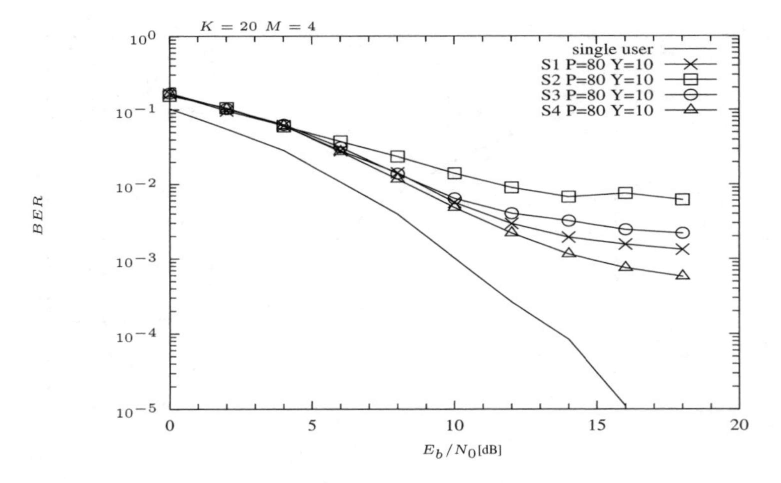

Figure 14.19: The GA-assisted MC-CDMA MUD's BER performance, when communicating over asynchronous environments. The four different GA strategies listed in Table 14.2 were adopted for approaching the optimum MUD's performance and the number of users supported was $K = 20$. The population size was $P = 80$ and the number of generations was $Y = 10$. Furthermore, the number of subcarriers was $M = 4$ and each subcarrier experienced uncorrelated narrowband Rayleigh fading.

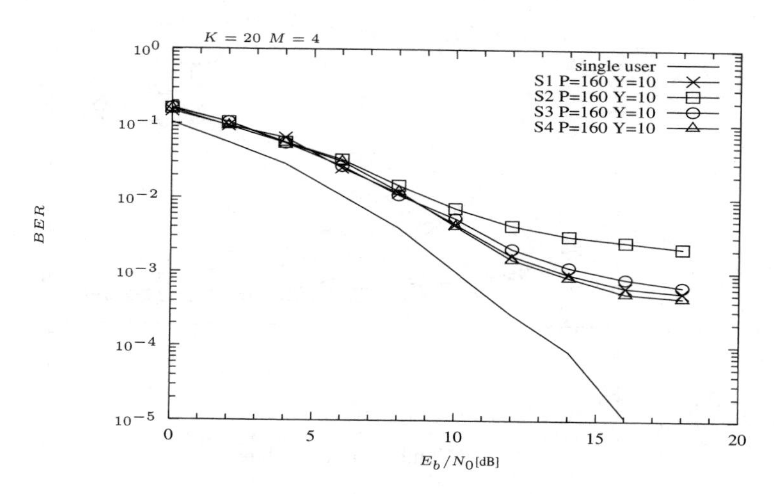

Figure 14.20: The GA-assisted MC-CDMA MUD's BER performance, when communicating over asynchronous environments. The four different GA strategies listed in Table 14.2 were adopted for approaching the optimum MUD's performance and the number of users supported was $K = 20$. The population size was $P = 160$ and the number of generations was $Y = 10$. Furthermore, the number of subcarriers was $M = 4$ and each subcarrier experienced uncorrelated narrowband Rayleigh fading.

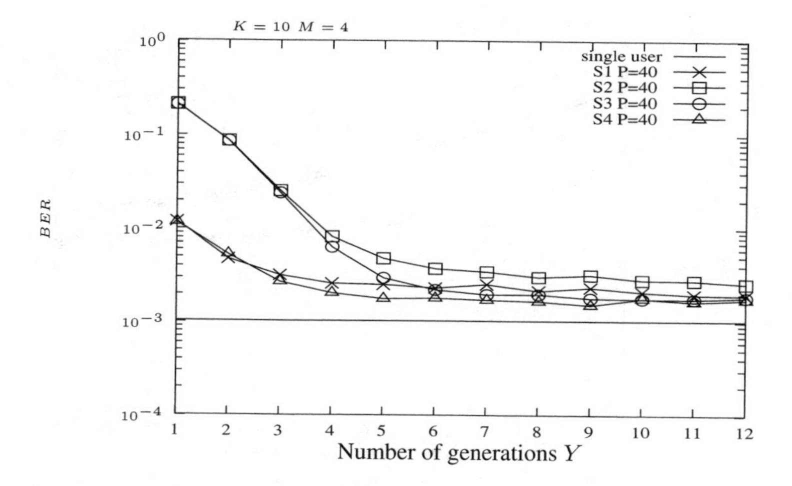

Figure 14.21: The GA-assisted MC-CDMA MUD's BER performance as a function of the number of generations Y, when communicating over asynchronous environments. The four different GA strategies listed in Table 14.2 were adopted for approaching the optimum MUD's performance and the number of users supported was $K = 10$. The population size was $P = 40$, the number of subcarriers was $M = 4$ and each subcarrier experienced uncorrelated narrowband Rayleigh fading.

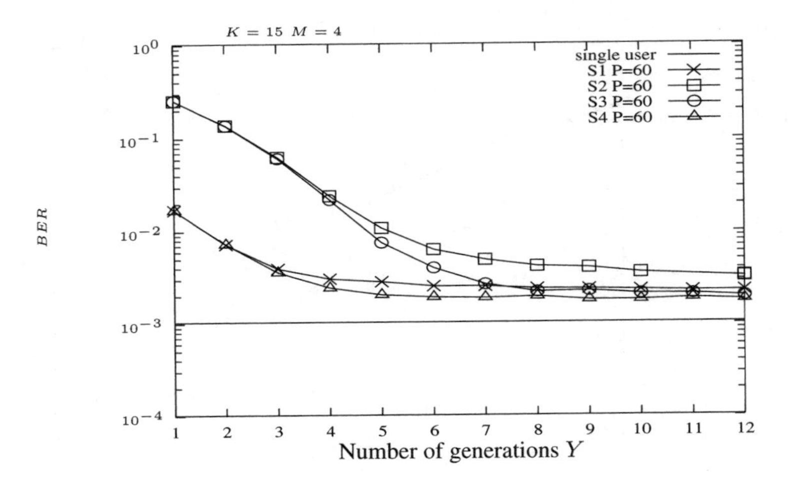

Figure 14.22: The GA-assisted MC-CDMA MUD's BER performance as a function of the number of generations Y, when communicating over asynchronous environments. The four different GA strategies listed in Table 14.2 were adopted for approaching the optimum MUD's performance and the number of users supported was $K = 15$. The population size was $P = 40$, the number of subcarriers was $M = 4$ and each subcarrier experienced uncorrelated narrowband Rayleigh fading.

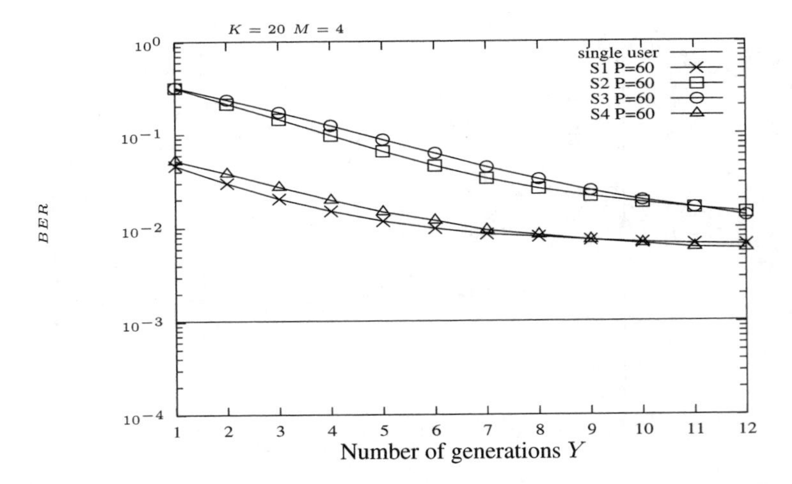

Figure 14.23: The GA-assisted MC-CDMA MUD's BER performance as a function of the number of generations Y, when communicating over asynchronous environments. The four different GA strategies listed in Table 14.2 were adopted for approaching the optimum MUD's performance and the number of users supported was $K = 20$. The population size was $P = 60$, the number of subcarriers was $M = 4$ and each subcarrier experienced uncorrelated narrowband Rayleigh fading.

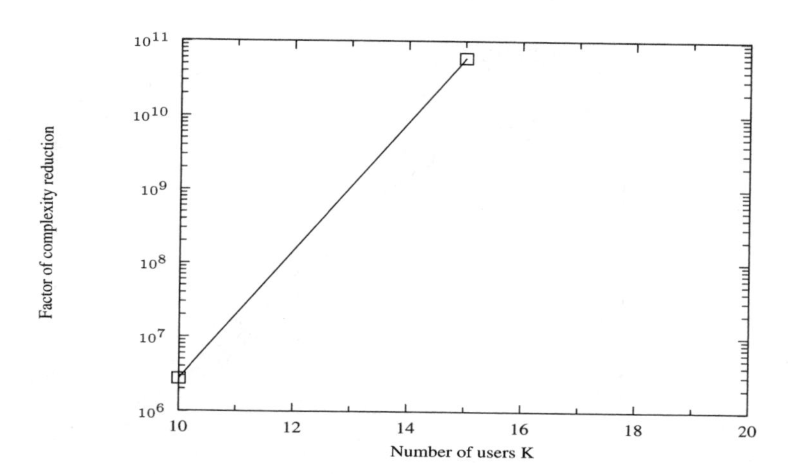

Figure 14.24: The complexity reduction factor of $\frac{2^{3K}}{P \times Y}$ was defined as the ratio of the number of objective function computations required for approaching the single-user bound at a BER of 10^{-3}, in comparison to Verdu's optimum MUD when communicating over an **asynchronous** environment, where P is the population size, and Y is the number of generations, while K is the number of users supported.

Figures 14.5 and 14.20 we can see that the performance of the GA-assisted MUD communicating over the synchronous environment is superior in comparison to that over the asynchronous channel, when the number of users K is increased to 20. This is because the search space hosts 2^{60} elements when communicating in an asynchronous scenario, which imposes an excessive complexity. In this excessive search space the GA-assisted MUD cannot approach the single-user performance and hence an inferior performance is achieved. By contrast, when communicating over the synchronous channel, the search space has 2^{20} elements and hence the GA-assisted MUD is capable of approaching the single-user performance.

From Figure 14.24 we can observe that the GA-assisted MUD is capable of reducing the complexity imposed by a factor of 6×10^{10} in comparison to that of Verdu's optimum MUD, when the number of users K is 15, suggesting that the GA-assisted MUD has the potential of significantly reducing the complexity of the optimum MUD.

14.7 Performance of GA-assisted MUD-aided asynchronous MC-CDMA using turbo coding

In this section, we briefly investigate the performance of the GA-assisted MUD-aided asynchronous MC-CDMA scheme employing turbo decoding. From Figures 14.25 and 14.26 we can observe that in asynchronous environments the GA-assisted MUD employing turbo decoding has a 1.5dB E_b/N_0 loss at a BER of 10^{-4} in comparison to the single-user performance. The GA-assisted MUD's implementation complexity is significantly lower than that of Verdu's optimum MUD [78]. For example, when supporting $K = 10$ users, the complex-

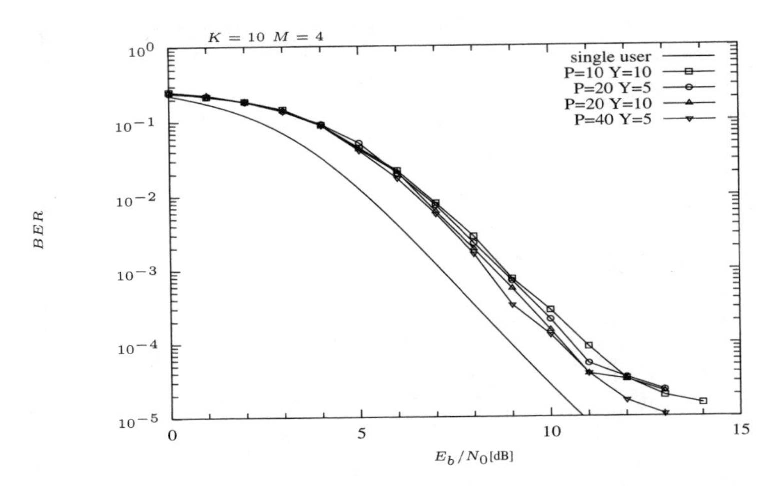

Figure 14.25: BER performance of the GA-assisted MUD designed for asynchronous MC-CDMA assisted by $R = \frac{1}{2}$ rate, constraint length $m = 3$ turbo coding using four iterations, when the number of users supported was $K = 10$. The number of generations was $Y = 5$ or 10 and the population size was $P = 10$, 20 and 40. Furthermore, the number of subcarriers was $M = 4$ and each subcarrier obeyed uncorrelated narrowband Rayleigh fading.

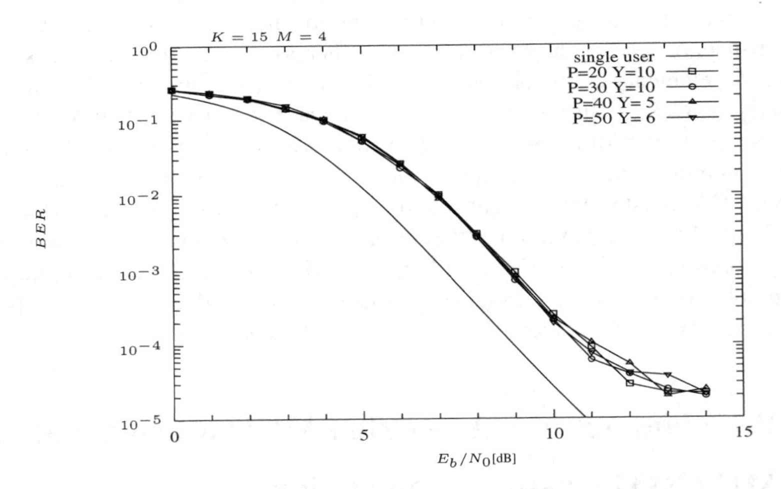

Figure 14.26: BER performance of the GA-assisted MUD designed for asynchronous MC-CDMA assisted by $R = \frac{1}{2}$ rate, constraint length $m = 3$ turbo coding using 4 iterations, when the number of users supported was $K = 15$. The number of generations was $Y = 5, 6$ or 10 and the population size was $P = 20, 30, 40$ and 50. Furthermore, the number of subcarriers was $M = 4$ and each subcarrier obeyed uncorrelated narrowband Rayleigh fading.

ity expressed in terms of the number of objective function evaluations was $O(200)$, while for $K = 15$ the complexity was $O(300)$.

14.8 Chapter Summary and Conclusion

In conclusion, the GA-assisted MUD is capable of significantly reducing the detection complexity in comparison to Verdu's optimum MUD, especially when the number of users supported is higher than $K = 15$. When channel coding techniques are employed, the GA-assisted MUD has the potential of further reducing the complexity imposed. Our future work will comparatively study the performance versus complexity trade-offs of GA versus M-Algorithm [158] based MUDs. Another important area of further study is the employment of multilevel modulation schemes both with and without various trellis-coded error protection schemes.

Part III

M-ary Single-Carrier CDMA

Chapter 15

Non-Coherent M-ary Orthogonal Modulation in CDMA[1]

15.1 Background

In the past, non-coherent reception was typically employed in the uplink for reasons of reduced system complexity, an issue to be further discussed. In general the transmission of known pilot signals is required for providing a coherent reference in support of coherent demodulation in a fading environment [10]. However, in the uplink, an independent pilot signal/symbol would have to be transmitted by each user for coherently demodulating each signal. This clearly reduces the efficiency of the system. Thus a more practical approach is to use non-coherent detection on the uplink, which does not require a pilot signal for estimating the phase reference [10]. This is the case in the IS-95 CDMA system [11].

Differential phase shift keying (DPSK) was analyzed, for example in [362, 363], while non-coherently detected star 16-QAM was discussed, for example in [92], which are techniques that do not require a phase reference for demodulation. M-ary orthogonal signalling based on square-law detection is another attractive technique for non-coherent reception. It was shown in [83, 90] that the BER performance of M-ary orthogonal modulation in narrowband systems is better than that of DPSK for $M \geq 8$.

M-ary orthogonal modulation using 64-ary Walsh functions has been proposed for the uplink transmission in the IS-95 system [11], where accordingly the information is organized in 6-bit 64-ary symbols. M-ary orthogonal modulation in the context of DS-CDMA over an AWGN channel was first proposed and analyzed by Enge and Sarwate [364]. Subsequently, M-ary orthogonal signalling over an indoor fading channel in DS-CDMA was analyzed by Pahlavan and Chase [365]. In a series of recent papers [109, 277, 366, 367], the performance

[1] *Single- and Multi-Carrier CDMA Multi-User Detection, Space-Time Spreading, Synchronization, Networking and Standards.* L.Hanzo, L-L.Yang, E-L.Kuan and K.Yen,
 ISBN 0-470-86309-9

of M-ary orthogonal modulation in DS-CDMA systems using Walsh functions and various demodulation techniques over different transmission environments was studied. The bit error rate (BER) performance of non-coherent M-ary orthogonal modulation in CDMA was studied by Bi in [368] and [369] for an AWGN channel and a multipath Rayleigh fading channel, respectively. Jalloul and Holtzman [366] analyzed the BER performance in a general multipath fading channel, while the investigations by Aalo *et al.* [277] were based on a Nakagami fading channel. In [367], the effects of concatenated signature sequences and Walsh functions on the performance of a DS-CDMA system using M-ary orthogonal signalling were studied. In [109], antenna arrays were used to improve the performance, while in [147], an interference cancellation technique was used to reduce the multi-user interference.

The systems studied by Patel and Holtzman [147] are based on the so-called *quadriphase spreading* concept. However, it was shown in [370,371] that the *performance of a BPSK modulated system using quadriphase spreading is identical to that using biphase spreading, if the relative phase between the desired signal and the interfering signals is uniformly distributed over* $[0, 2\pi)$. *This is usually the case in asynchronous uplink transmissions. However, if the phases of all the users are the same, such that their relative phase is zero, then quadriphase spreading has a 3dB advantage over biphase spreading [371].* Hence, in this section, we will commence by analyzing the performance of M-ary orthogonal modulation using biphase spreading. We will also present novel results concerning the performance of M-ary orthogonal modulation in CDMA. Specifically, in Chapter 18 we will present the original concept of M-ary orthogonal modulation based on the so-called Residue Number System (RNS), which was first proposed by Yang *et al.* [372–378], where the performance analysis was based on a single-user narrowband system. We note that the theoretical results of [372, 373] were confirmed by the corresponding simulations.

Let us commence our discourse with a brief introduction to M-ary orthogonal modulation employed in the context of non-dispersive narrowband systems.

15.2 Introduction to M-ary Orthogonal Modulation

In general, M-ary orthogonal modulation schemes, such as M-ary frequency shift keying (MFSK), are used primarily to improve the BER performance of the system, although at the cost of reducing its bandwidth efficiency, an issue analyzed, for example in [90]. Without delving in the analysis details, it was shown by Proakis [90] that the average symbol error probability of *coherently detected* M-ary orthogonal modulation over AWGN channels is given by:

$$Pr_s(\epsilon) = \frac{1}{\sqrt{2\pi}} \int_{-\infty}^{\infty} \left[1 - \left(\frac{1}{\sqrt{2\pi}} \int_{-\infty}^{y} e^{-\frac{x^2}{2}} dx\right)^{M-1}\right] e^{\left[\frac{1}{2}\left(y - \sqrt{\frac{2\xi_s}{N_0}}\right)^2\right]} dy, \qquad (15.1)$$

where ξ_s and N_0 are the symbol and noise energies, respectively. On the other hand, the average symbol error probability of M-ary orthogonal modulation in a Gaussian channel using *non-coherent reception* is given by [90]:

$$Pr_s(\epsilon) = \sum_{n=1}^{M-1}(-1)^{n+1}\binom{M-1}{n}\frac{1}{n+1}\exp\left[-\frac{n\xi_s}{(n+1)N_0}\right]. \quad (15.2)$$

In a frequency selective Rayleigh slow fading channel, the average symbol error probability using *non-coherent detection* was shown by Proakis to be given by:

$$\begin{aligned} Pr_s(\epsilon) &= 1 - \int_0^{\infty}\frac{1}{(1+\bar{\gamma}_c)^{L_P}(L_P-1)!}y^{L_P-1}\exp\left(-\frac{y}{1+\bar{\gamma}_c}\right) \\ &\quad \times\left(1-e^{-y}\sum_{k=0}^{L_P-1}\frac{y^k}{k!}\right)^{M-1} dy, \end{aligned} \quad (15.3)$$

where $\bar{\gamma}_c$ is the average SNR per diversity channel at the receiver, which is given by $\frac{\xi_s}{N_0}E(\alpha^2)$, and L_P is the order of diversity, while α is the Rayleigh fading amplitude.

The above equations quantify the symbol error rate. It is often desirable to present the performance results in terms of the bit error rate. For M-ary orthogonal signalling, the bit error probability can be derived from the symbol error probability by applying Proakis' equation in [90]:

$$Pr_b(\epsilon) = \frac{2^{\beta-1}}{2^{\beta}-1}Pr_s(\epsilon), \quad (15.4)$$

where $\beta = \log_2 M$ is the number of bits per symbol. We note furthermore that for high values of β we have

$$Pr_b(\epsilon) \approx \frac{1}{2}Pr_s(\epsilon). \quad (15.5)$$

Figures 15.1, 15.2 and 15.3 illustrate the bit error probability against the average SNR per bit using Equations 15.1, 15.2 and 15.3, respectively. *From these figures, it becomes explicit that increasing the value of M in M-ary orthogonal modulation schemes – such as MFSK – improves the BER performance.* It is also clear that, as expected, coherent demodulation has a better performance, than non-coherent demodulation. As noted previously [83, 90], the BER performance of non-coherent M-ary orthogonal modulation becomes better than that of DPSK for $M \geq 8$.

However, this BER performance improvement is achieved at the cost of a bandwidth efficiency degradation. Bandwidth efficiency is defined as the ratio of data rate R_b to channel bandwidth B [379], and is given by:

$$\eta = \frac{R_b}{B}. \quad (15.6)$$

For example, in MFSK, an orthogonal set of M frequency-shifted signals are required for

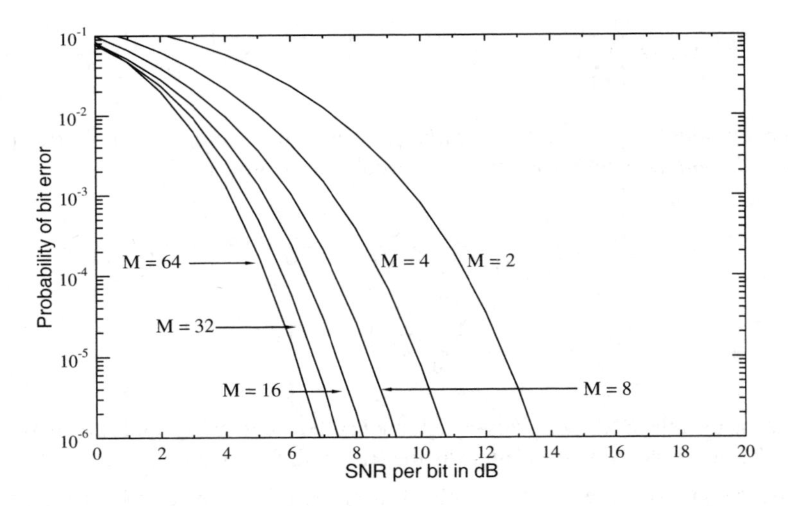

Figure 15.1: Probability of bit error for *coherent detection* of M-ary orthogonal signalling in a non-dispersive Gaussian channel using Equation 15.1

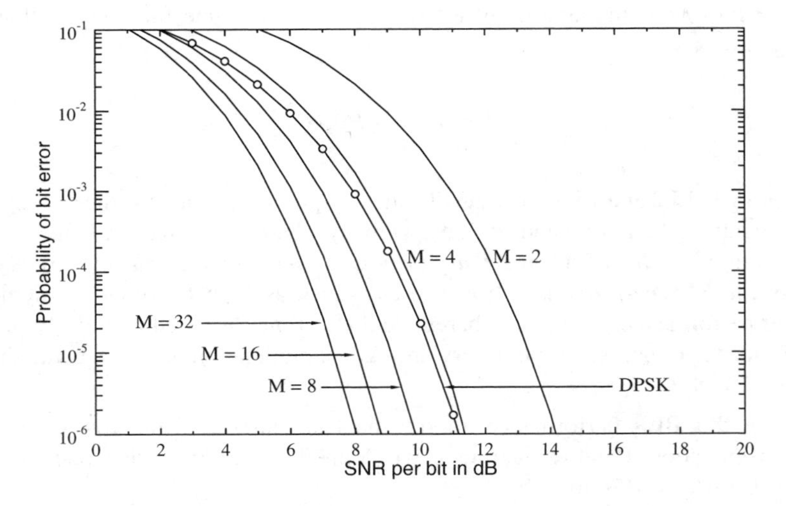

Figure 15.2: Probability of bit error for *non-coherent detection* of M-ary orthogonal signalling in a non-dispersive Gaussian channel using Equation 15.2

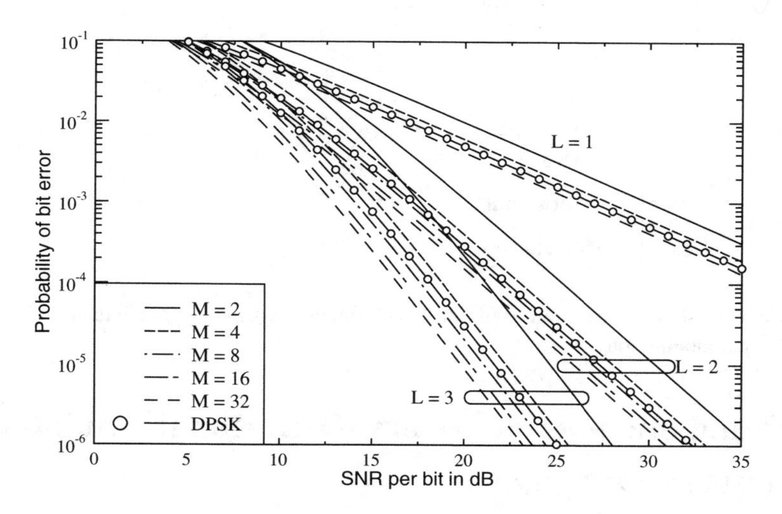

Figure 15.3: Probability of bit error for *non-coherent detection* of M-ary orthogonal signalling *in a non-dispersive Rayleigh fading channel* using Equation 15.3

transmission. In order to maintain the orthogonality, the frequency between adjacent signalling tones is separated by at least $1/2T_s$ Hz, where T_s is the symbol period. Hence for M number of signals, the total channel bandwidth required would be $B = M/2T_s$. Substituting this into Equation 15.6 yields a bandwidth efficiency of:

$$\begin{aligned}\eta &= \frac{2T_s}{T_b M} \\ &= \frac{2\beta}{M},\end{aligned} \tag{15.7}$$

where $R_b = 1/T_b$. Hence, as the value of M increases, the bandwidth efficiency linearly decreases. *However, since $\mathbf{M}$ is typically increased as an integer power of two, in order to convey an integer number of bits, the bandwidth efficiency $\boldsymbol{\eta}$ will decrease exponentially, as the number of bits per symbol is increased. At the same time the BER improvements in Figures 15.1, 15.2 and 15.3 slow down at higher $\mathbf{M}$ values.* Furthermore, increasing the value of M would lead to an increased complexity of the receiver, since more matched filters are required. But considering Figure 15.3, it can be seen that the BER performance of $M = 2$ with three diversity paths is worse than that of $M > 2$ with two diversity paths for SNRs below 16 dB. Hence, increasing the value of M might be a better solution in BER terms, than increasing the order of diversity, if bandwidth efficiency is not the overriding constraint, since both increasing M and the diversity order will increase the complexity of the receiver similarly.

All the results discussed above were based on a narrowband system. We shall now apply M-ary orthogonal modulation in a wideband system, such as CDMA and analyze its performance.

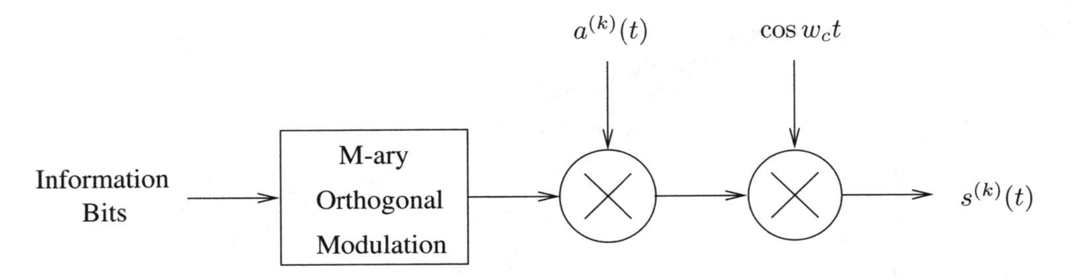

Figure 15.4: The kth user's transmitter for an asynchronous DS-CDMA system using M-ary orthogonal modulation

15.3 Fundamentals of M-ary Orthogonal Modulation in CDMA [277, 366]

15.3.1 The Transmitter Model

The block diagram of the kth user's transmitter is shown in Figure 15.4. While previously we referred to using for example M-ary FSK, as an orthogonal signal set, here each of the K users in the system is assigned an orthogonal sequence set or code consisting of M Walsh functions. The information bits are mapped to one of the M Walsh functions. Although the M orthogonal Walsh-functions can be transmitted in the same bandwidth, the bandwidth expansion is now due to the increased bandwidth requirement of the orthogonal Walsh functions. The associated CDMA bandwidth efficiency will be quantified in Subsection 15.5. The μth Walsh function of the kth user may be written as:

$$V_{\mu}^{(k)}(t) = \sum_{g=0}^{M-1} V_{\mu,g}^{(k)} \Gamma_{T_v}(t - gT_v), \quad 0 \leq t < T_s, \ \mu = 1, 2, \ldots, M \tag{15.8}$$

where $\Gamma_{T_v}(t)$ represents the pulse-shaping function of the chip and $V_{\mu,g}^{(k)} \in \{+1, -1\}, g = 0, 1, \ldots, M-1$ is known as the gth Walsh chip of the μth Walsh function of the kth user, each of which has a duration of T_v. Without loss of generality, we shall assume that $\Gamma_{T_v}(t)$ is a square-shaped pulse of unit amplitude for a duration of T_v and zero elsewhere. The code is used to transmit equiprobable M-ary data symbols at a rate of one symbol every T_s seconds.

A signature sequence $a^{(k)}(t)$, as seen in Figure 15.4, is then used to spread and randomize the assigned Walsh function. Besides providing isolation between users, the pseudo-noise (PN) signature sequence $a^{(k)}(t)$ – which is exemplified by the m-sequences or Gold codes of Chapter 2 also aids the code acquisition process [380–383] at the receiver due to its low cross-correlation. Unfortunately, Walsh functions typically exhibit undesirably high cross-correlation peaks, which results in a high probability of false code acquisition [384] and hence are not attractive for acquisition. Therefore low asynchronous cross-correlations are desirable and a low cross-correlation will also lead to a reduction in the MAI. The signal transmitted by the kth user in order to send the μth data symbol associated with aid of the μth

Walsh-function during the time interval $[0, T_s)$ is

$$\begin{aligned} s^{(k)}(t) &= \text{Re}\left\{\sqrt{2P_s^{(k)}}V_{\mu}^{(k)}(t)a^{(k)}(t)e^{jw_ct}\right\} \\ &= \sqrt{2P_s^{(k)}}V_{\mu}^{(k)}(t)a^{(k)}(t)\cos w_ct, \quad 0 \le t < T_s \end{aligned} \tag{15.9}$$

where P_s is the average signal power and w_c is the angular carrier frequency in Figure 15.4, which is common to all users. The signature sequence of the kth user is represented by

$$a^{(k)}(t) = \sum_{h=0}^{N_c-1} a_h^{(k)}\Gamma_{T_c}(t - hT_c), \quad a_h^{(k)} \in \{+1, -1\}, \quad 0 \le t < T_s \tag{15.10}$$

where $a_h^{(k)}$ is referred to as the hth signature sequence chip of the kth user's signature sequence. Thus, there are M Walsh chips and N_c signature sequence chips in a symbol period of T_s. Since the signature sequences $a^{(k)}(t)$ are used to spread and randomize the orthogonal Walsh functions $V_{\mu}^{(k)}(t)$, where $N_c \ge M$, hence there will be N_c/M signature sequence chips in a single Walsh chip. For simplicity, we will assume that the energy of the Walsh-signature concatenated sequence is given by

$$\int_0^{T_s}\left|V_{\mu}^{(k)}(t)a^{(k)}(t)\right|^2 dt = T_s, \quad k = 1, 2, \ldots, K \text{ and } \mu = 1, 2, \ldots, M. \tag{15.11}$$

It can be shown that for an M-ary system, where $M > 2$, N_c is not equal to the processing gain N. Explicitly, the processing gain N - which is defined as:

$$N = \frac{B_s}{B}, \tag{15.12}$$

where B_s is the bandwidth of the spread spectrum signal while B is the bandwidth of the original information signal - is related to N_c by the following equation:

$$\begin{aligned} N &= \frac{N_c}{\beta} \\ &= \frac{N_c}{\log_2 M}, \end{aligned} \tag{15.13}$$

where $\beta = \log_2 M$ is the number of bits per symbol. Hence for binary communications, i.e. for $M = 2$ we have $\beta = 1$, yielding $N = N_c$, while for $M > 2$ we have $N < N_c$. Let us now consider our channel model.

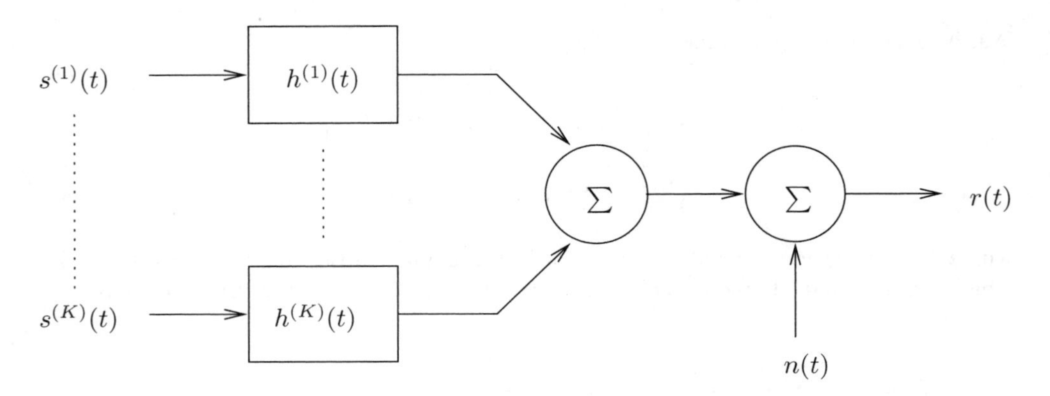

Figure 15.5: The system and channel model block diagram ©IEEE Jalloul and Holtzman, 1994 [366].

15.3.2 Channel Model

The performance analysis in [366] and [277] is based on a general multipath fading and Nakagami fading channel, respectively. Here, we will assume that the channel has Rayleigh distributed fading statistics, similarly to [369]. The complex lowpass equivalent impulse response of the channel associated with the kth user's transmission is given by:

$$\tilde{h}^{(k)}(t) = \sum_{l=1}^{L_P} \left|\alpha_l^{(k)}(t)\right| e^{j\phi_l^{(k)}(t)} \delta(t - \tau_l^{(k)}). \quad (15.14)$$

For simplicity, we will assume that the channel is slowly fading, such that it appears as a time-invariant channel over the symbol duration T_s. Hence, the channel parameters are constant for the entire symbol period and the dependence on t in the amplitude and phase in Equation 15.14 can be omitted, since the analysis is confined to a symbol period. We also assumed that the multipath channel's fading amplitudes $|\alpha_l^{(k)}(t)|, l = 1, 2, \ldots, L_P$, are independent and identically distributed (i.i.d.) random variables and all users have the same number of multipath components. Thus the channel-impaired received signal $y^{(k)}(t)$ of the kth user becomes

$$y^{(k)}(t) = \sum_{l=1}^{L_P} \sqrt{2P_s^{(k)}} \left|\alpha_l^{(k)}\right| V_\mu^{(k)}(t - \tau_l^{(k)}) a^{(k)}(t - \tau_l^{(k)}) \cos(w_c t + \theta_l^{(k)}), \quad (15.15)$$

where $\theta_l^{(k)} = \phi_l^{(k)} - w_c \tau_l^{(k)}$.

15.3.3 The Receiver Model

Let K be the total number of active M-ary CDMA users in the system. Then the received signal at the base station is the sum of all the K users' impaired signals plus noise, as shown in Figure 15.5, which is given by:

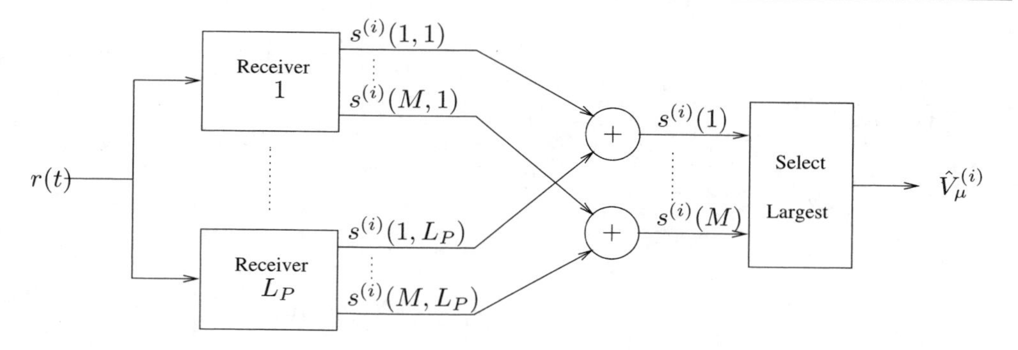

Figure 15.6: The ith user's L_P-branch equal-gain combining RAKE receiver block diagram ©IEEE Jalloul and Holtzman, 1994 [366].

$$r(t) = \sum_{k=1}^{K} y^{(k)}(t) + n(t), \tag{15.16}$$

where $n(t)$ is the AWGN with zero mean and a double-sided power spectral density (PSD) of $N_0/2$. The AWGN may be represented in terms of its low-pass equivalent given by

$$n(t) = n_c(t)\cos w_c t + n_s(t)\sin w_c t, \tag{15.17}$$

where $n_c(t)$ and $n_s(t)$ are the baseband equivalent low-pass Gaussian processes [92].

At the receiver, a RAKE structure invoking equal gain combining is used for demodulation. An illustration of the ith user's receiver associated with L_p diversity paths is shown in Figure 15.6. The block diagram of the l_p-th branch receiver is shown in Figure 15.7. We will concentrate our analysis on the demodulation of the ith user's signal. From Figure 15.7, the output of the low-pass filter of the I-channel is given by

$$\tilde{d}_I(t) = \sum_{k=1}^{K}\sum_{l=1}^{L_P}\sqrt{\frac{P_s^{(k)}}{2}}\left|\alpha_l^{(k)}\right| V_\mu^{(k)}(t-\tau_l^{(k)})a^{(k)}(t-\tau_l^{(k)})\cos\theta_l^{(k)} + \frac{n_c(t)}{2}. \tag{15.18}$$

Similarly for the Q-channel, the output of the low-pass filter is given by

$$\tilde{d}_Q(t) = -\sum_{k=1}^{K}\sum_{l=1}^{L_P}\sqrt{\frac{P_s^{(k)}}{2}}\left|\alpha_l^{(k)}\right| V_\mu^{(k)}(t-\tau_l^{(k)})a^{(k)}(t-\tau_l^{(k)})\sin\theta_l^{(k)} + \frac{n_s(t)}{2}, \tag{15.19}$$

where the - sign is due to the trigonometry relationship. For simplicity, *we will assume that the ith transmitter and the ith receiver are perfectly synchronized,* i.e. $\tau_l^{(i)}, l = 1, 2, \ldots, L_P$ are known. Then, according to Figure 15.7, the output signal of the low-pass filter in both the I and Q channels is multiplied with the signature sequence belonging to the ith user, i.e.

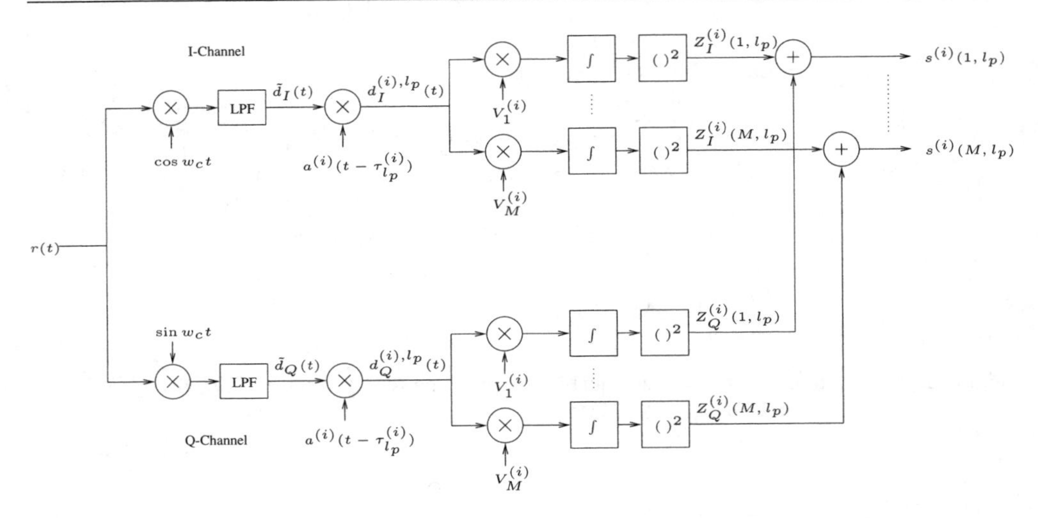

Figure 15.7: The ith user receiver model for the l_p-th RAKE branch ©IEEE Jalloul and Holtzman, 1994 [366]. The associated variables are given by the corresponding equations: $\tilde{d}_I(t)$-Equation 15.18, $\tilde{d}_Q(t)$-Equation 15.19, $d_I^{(i),l_p}(t)$-Equation 15.24, $d_Q^{(i),l_p}(t)$-Equation 15.25, $Z_I^{(i)}(m,l_p)$-Equation 15.26, $Z_Q^{(i)}(m,l_p)$-Equation 15.28

$a^{(i)}(t)$. We will define $d_I^{(i),l_p}(t)$ as

$$d_I^{(i),l_p}(t) = \tilde{d}_I(t)a^{(i)}(t-\tau_{l_p}^{(i)}). \tag{15.20}$$

Then upon substituting Equation 15.18 into Equation 15.20, we obtain [366]:

$$\begin{aligned} d_I^{(i),l_p}(t) &= \\ &\sum_{k=1}^{K}\sum_{l=1}^{L_P}\sqrt{\frac{P_s^{(k)}}{2}}\left|\alpha_l^{(k)}\right| V_\mu^{(k)}(t-\tau_l^{(k)})a^{(k)}(t-\tau_l^{(k)})\cos\theta_l^{(k)}a^{(i)}(t-\tau_{l_p}^{(i)}) \\ &+\frac{n_c(t)}{2}a^{(i)}(t-\tau_{l_p}^{(i)}), \end{aligned} \tag{15.21}$$

yielding:

$$
\begin{aligned}
d_I^{(i),l_p}(t) &= \\
&= \underbrace{\sqrt{\frac{P_s^{(i)}}{2}}\left|\alpha_{l_p}^{(i)}\right| V_\mu^{(i)}(t-\tau_{l_p}^{(i)})a^{(i)}(t-\tau_{l_p}^{(i)})a^{(i)}(t-\tau_{l_p}^{(i)})\cos\theta_{l_p}^{(i)}}_{\text{desired signal}} \\
&\quad \underbrace{+\sum_{\substack{l=1\\ l\neq l_p}}^{L_P}\sqrt{\frac{P_s^{(i)}}{2}}\left|\alpha_l^{(i)}\right| V_\mu^{(i)}(t-\tau_l^{(i)})a^{(i)}(t-\tau_l^{(i)})a^{(i)}(t-\tau_{l_p}^{(i)})\cos\theta_l^{(i)}}_{\text{multipath interference}} \\
&\underbrace{+\sum_{\substack{k=1\\ k\neq i}}^{K}\sum_{l=1}^{L_P}\sqrt{\frac{P_s^{(k)}}{2}}\left|\alpha_l^{(k)}\right| V_\mu^{(k)}(t-\tau_l^{(k)})a^{(k)}(t-\tau_l^{(k)})a^{(i)}(t-\tau_{l_p}^{(i)})\cos\theta_l^{(k)}}_{\text{multi-user interference}} \\
&\quad + \underbrace{\frac{n_c(t)}{2}a^{(i)}(t-\tau_{l_p}^{(i)})}_{\text{white Gaussian noise}} \ .
\end{aligned} \tag{15.22}
$$

In order to simplify the above expression, we will use the following notation:

$$a^{(i,j)}(\tau_l^{(i)},\tau_{l_p}^{(j)}) = a^{(i)}(t-\tau_l^{(i)})a^{(j)}(t-\tau_{l_p}^{(j)}). \tag{15.23}$$

Thus, $d_I^{(i),l_p}(t)$ becomes

$$
\begin{aligned}
d_I^{(i),l_p}(t) &= \sqrt{\frac{P_s^{(i)}}{2}}\left|\alpha_{l_p}^{(i)}\right| V_\mu^{(i)}(t-\tau_{l_p}^{(i)})a^{(i,i)}(\tau_{l_p}^{(i)},\tau_{l_p}^{(i)})\cos\theta_{l_p}^{(i)} \\
&\quad +\sum_{\substack{l=1\\ l\neq l_p}}^{L_P}\sqrt{\frac{P_s^{(i)}}{2}}\left|\alpha_l^{(i)}\right| V_\mu^{(i)}(t-\tau_l^{(i)})a^{(i,i)}(\tau_l^{(i)},\tau_{l_p}^{(i)})\cos\theta_l^{(i)} \\
&\quad +\sum_{\substack{k=1\\ k\neq i}}^{K}\sum_{l=1}^{L_P}\sqrt{\frac{P_s^{(k)}}{2}}\left|\alpha_l^{(k)}\right| V_\mu^{(k)}(t-\tau_l^{(k)})a^{(k,i)}(\tau_l^{(k)},\tau_{l_p}^{(i)})\cos\theta_l^{(k)} \\
&\quad +\frac{n_c(t)}{2}a^{(i)}(t-\tau_{l_p}^{(i)}).
\end{aligned} \tag{15.24}
$$

Similarly,

$$\begin{aligned} d_Q^{(i),l_p}(t) &= -\sqrt{\frac{P_s^{(i)}}{2}}\left|\alpha_{l_p}^{(i)}\right| V_\mu^{(i)}(t-\tau_{l_p}^{(i)})a^{(i,i)}(\tau_{l_p}^{(i)},\tau_{l_p}^{(i)})\sin\theta_{l_p}^{(i)} \\ &\quad -\sum_{\substack{l=1\\ l\neq l_p}}^{L_P}\sqrt{\frac{P_s^{(i)}}{2}}\left|\alpha_l^{(i)}\right| V_\mu^{(i)}(t-\tau_l^{(i)})a^{(i,i)}(\tau_l^{(i)},\tau_{l_p}^{(i)})\sin\theta_l^{(i)} \\ &\quad -\sum_{\substack{k=1\\ k\neq i}}^{K}\sum_{l=1}^{L_P}\sqrt{\frac{P_s^{(k)}}{2}}\left|\alpha_l^{(k)}\right| V_\mu^{(k)}(t-\tau_l^{(k)})a^{(k,i)}(\tau_l^{(k)},\tau_{l_p}^{(i)})\sin\theta_l^{(k)} \\ &\quad +\frac{n_s(t)}{2}a^{(i)}(t-\tau_{l_p}^{(i)}). \end{aligned} \tag{15.25}$$

Both the outputs $d_I^{(i),l_p}(t)$ and $d_Q^{(i),l_p}(t)$ are then passed through a bank of M matched filters, as seen in Figure 15.7, and are correlated with the M possible Walsh functions, $V_\mu^{(i)}(t), \mu = 1, 2, \ldots, M$. Hence, with $d_I^{(i),l_p}(t)$ as the input, the output of the mth correlator of the l_pth branch in Figure 15.7 becomes:

$$Z_I^{(i)}(m,l_p) = \frac{1}{\sqrt{T_s}}\int_{\tau_{l_p}^{(i)}}^{T_s+\tau_{l_p}^{(i)}} d_I^{(i),l_p}(t)V_m^{(i)}(t-\tau_{l_p}^{(i)})dt. \tag{15.26}$$

After substituting Equation 15.24 into Equation 15.26, the output of the mth correlator of the l_pth branch in Figure 15.7 becomes:[2]

$$Z_I^{(i)}(m,l_p) = \tag{15.27}$$
$$\begin{cases} \left|\alpha_{l_p}^{(i)}\right|\sqrt{\frac{\xi_s^{(i)}}{2}}\cos\theta_{l_p}^{(i)} + I_{IM}^{(i,i)}(l_p) + I_{MAI,I}^{(i,k)}(l_p) + N_I^{(i)}(l_p), & m=\mu \\ I_{IM}^{(i,i)}(l_p) + I_{MAI,I}^{(i,k)}(l_p) + N_I^{(i)}(l_p), & m\neq\mu \end{cases}$$

where $\xi_s^{(i)} = \beta \times \xi_b^{(i)}$ is the symbol energy, $\xi_b^{(i)}$ is the bit energy, $I_{IM}^{(i,i)}(l_p)$ is the self-interference inflicted by the M signalling sequences due to multipath propagation, $I_{MAI,I}^{(i,k)}(l_p)$ is the multiple access interference due to other users and $N_I^{(i)}(l_p)$ is the term due to the presence of AWGN in the I channel. The expressions for these interference terms can be found in Section A.1.1.

Similarly, the correlator output $Z_Q^{(i)}(m,l_p)$ is derived the same way and the associated

[2]See Section A.1.1 for derivation.

terms are given by:[3]

$$Z_Q^{(i)}(m, l_p) = \tag{15.28}$$
$$\begin{cases} -\left|\alpha_{l_p}^{(i)}\right| \sqrt{\frac{\xi_s}{2}} \sin\theta_{l_p}^{(i)} - I_{QM}^{(i,i)}(l_p) - I_{MAI,Q}^{(i,k)}(l_p) + N_Q^{(i)}(l_p), & m = \mu \\ -I_{QM}^{(i,i)}(l_p) - I_{MAI,Q}^{(i,k)}(l_p) + N_Q^{(i)}(l_p), & m \neq \mu \end{cases}$$

For the equal gain combining RAKE receiver shown in Figure 15.6, the decision variable of the ith user becomes:

$$s^{(i)}(m) = \sum_{l=1}^{L_P} s^{(i)}(m, l), \quad m = 1, 2, \ldots, M \tag{15.29}$$

where

$$s^{(i)}(m, l) = \left[Z_I^{(i)}(m, l)\right]^2 + \left[Z_Q^{(i)}(m, l)\right]^2. \tag{15.30}$$

The receiver uses the maximum likelihood decision rule by selecting the maximum among the variables $s^{(i)}(m), m = 1, 2, \ldots, M$ in Figure 15.6. The index of the largest decision variable denotes the transmitted M-ary symbol, which allows us to output $\log_2 M$ number of detected bits.

15.4 Performance Analysis

As in references [109, 277, 364–367, 385], we will use the standard Gaussian approximation [103], in order to model the interference. We will further assume that $\theta_l^{(k)}, k = 1, 2, \ldots, K$ is uniformly distributed over the interval $[0, 2\pi]$ and $\tau_l^{(k)}, k = 1, 2, \ldots, K$ and $l = 1, 2, \ldots, L_P$ is uniformly distributed over the interval $[0, T_s]$.

15.4.1 Noise Analysis

The noise term $N_I^{(i)}(l_p)$ in Equation 15.26 is given by:[4]

$$N_I^{(i)}(l_p) = \frac{1}{2\sqrt{T_s}} \int_{\tau_{l_p}^{(i)}}^{T_s + \tau_{l_p}^{(i)}} n_c(t) a^{(i)}(t - \tau_{l_p}^{(i)}) V_m(t - \tau_{l_p}^{(i)}) dt, \tag{15.31}$$

which is a Gaussian distributed random variable with zero mean and variance $N_0/4$. Similarly, $N_Q^{(i)}(l)$ are all zero-mean Gaussian random variables with variance $N_0/4$. Following the above brief notes quantifying the amount of noise encountered, let us now consider the

[3] See Section A.1.2 for derivation.

[4] See Equation A.6.

self-interference inflicted by the M signalling sequences due to dispersive multipath propagation.

15.4.2 Self-interference Analysis

The self-interference, denoted by $I_{IM}^{(i,i)}(l_p)$, is given by:[5]

$$\begin{aligned} I_{IM}^{(i,i)}(l_p) &= \sum_{\substack{l=1 \\ l \neq l_p}}^{L_P} \sqrt{\frac{P_s}{2T_s}} \left|\alpha_l^{(i)}\right| \int_{\tau_{l_p}^{(i)}}^{T_s+\tau_{l_p}^{(i)}} V_\mu^{(i)}(t-\tau_l^{(i)}) V_m^{(i)}(t-\tau_{l_p}^{(i)}) \\ & \times a^{(i,i)}(\tau_l^{(i)}, \tau_{l_p}^{(i)}) \cos\theta_l^{(i)} dt. \end{aligned} \quad (15.32)$$

The term $I_{IM}^{(i,i)}(l_p)$ can also be modelled as a Gaussian random variable. As stated before, we will assume that there are N_c/M signature chips per Walsh chip, so that the sequences can be modelled as random binary sequences. Thus, the term $\int_{\tau_{l_p}^{(i)}}^{T_s+\tau_{l_p}^{(i)}} V_\mu^{(i)}(t-\tau_l^{(i)}) V_m^{(i)}(t-\tau_{l_p}^{(i)}) a^{(i,i)}(\tau_l^{(i)}, \tau_{l_p}^{(i)}) dt$ can be treated as the cross-correlation of two random sequences with a variance $I_{IM}^{(i,i)}(l_p)$ given by [103]:

$$\begin{aligned} \text{Var}\left[I_{IM}^{(i,i)}(l_p)\right] &= \sum_{\substack{l=1 \\ l \neq l_p}}^{L_P} \frac{P_s^{(i)}}{2T_s} \times \frac{1}{T_s} \times \frac{2T_s^3}{3N_c} \times \frac{1}{2} E\left[\left|\alpha_l^{(i)}\right|^2\right] \\ &= \sum_{\substack{l=1 \\ l \neq l_p}}^{L_P} \frac{P_s^{(i)} T_s}{6N_c} E\left[\left|\alpha_l^{(i)}\right|^2\right] \\ &= \frac{\xi_s^{(i)}}{6N_c} \sum_{\substack{l=1 \\ l \neq l_p}}^{L_P} E\left[\left|\alpha_l^{(i)}\right|^2\right]. \end{aligned} \quad (15.33)$$

Similarly, due to symmetry, the terms $I_{QM}^{(i,i)}(l_p)$ have a variance given by Equation 15.33. Let us now focus our attention on quantifying the multi-user interference in the next section.

---

[5]See Equation A.4.

15.4.3 MAI Analysis

The MAI is given by:[6]

$$\begin{aligned} I_{MAI,I}^{(i,k)}(l_p) &= \sum_{\substack{k=1 \\ k\neq i}}^{K} \sum_{l=1}^{L_P} \sqrt{\frac{P_s}{2T_s}} \left|\alpha_l^{(i)}\right| \int_{\tau_{l_p}^{(i)}}^{T_s+\tau_{l_p}^{(i)}} V_\mu^{(k)}(t-\tau_l^{(k)}) V_m^{(i)}(t-\tau_{l_p}^{(i)}) \\ &\times a^{(k,i)}(\tau_l^{(k)}, \tau_{l_p}^{(i)}) \cos\theta_l^{(k)} dt, \end{aligned} \tag{15.34}$$

with a variance, which equals to [103]:

$$\text{Var}\left[I_{MAI,I}^{(i,k)}(l_p)\right] = \frac{\xi_s^{(k)}}{6N_c} \sum_{\substack{k=1 \\ k\neq i}}^{K} \sum_{l=1}^{L_P} E\left[\left|\alpha_l^{(k)}\right|^2\right]. \tag{15.35}$$

The variance of the $I_{MAI,Q}^{(i,k)}(l_p)$ terms is also given by Equation 15.35.

15.4.4 Decision Statistic and Error Probability

For simplicity, we will assume that the received signal energy of all users is equal, implying perfect power control. Hence, $\xi_s^{(k)} = \xi_s$ for $k = 1, 2, \ldots, K$. Having approximated both the self-interference and the MAI as Gaussian noise, the decision variable $s^{(i)}(m, l)$ of Figure 15.6 given in Equation 15.30 becomes a sum of two squared Gaussian random variables with the variance of the self-interference and MAI given by Equations 15.33 and 15.35, yielding:

$$\sigma^2 = \frac{N_0}{4} + \frac{\xi_s}{6N_c} \left\{ \sum_{\substack{k=1 \\ k\neq i}}^{K} \sum_{l=1}^{L_P} E\left[\left|\alpha_l^{(k)}\right|^2\right] + \sum_{\substack{l=1 \\ l\neq l_p}}^{L_P} E\left[\left|\alpha_l^{(i)}\right|^2\right] \right\}. \tag{15.36}$$

We can express the total self-interference plus MAI variance σ^2 in terms of the bit energy and the processing gain of $N = N_c/\beta$ by substituting Equation 15.13 into Equation 15.36,

[6] See Equation A.5.

which yields:

$$\begin{aligned}\sigma^2 &= \frac{N_0}{4} + \frac{\beta\xi_b}{6N_c}\left\{\sum_{\substack{k=1\\k\neq i}}^{K}\sum_{l=1}^{L_P} E\left[\left|\alpha_l^{(k)}\right|^2\right] + \sum_{\substack{l=1\\l\neq l_p}}^{L_P} E\left[\left|\alpha_l^{(i)}\right|^2\right]\right\} \\ &= \frac{N_0}{4} + \frac{\xi_b}{6N}\left\{\sum_{\substack{k=1\\k\neq i}}^{K}\sum_{l=1}^{L_P} E\left[\left|\alpha_l^{(k)}\right|^2\right] + \sum_{\substack{l=1\\l\neq l_p}}^{L_P} E\left[\left|\alpha_l^{(i)}\right|^2\right]\right\}. \end{aligned} \tag{15.37}$$

Thus, $s^{(i)}(m,l)$ in Equation 15.30 is a Chi-square distributed random variable with two degrees of freedom [90]. Assuming that $\left|\alpha_l^{(i)}\right|, l = 1, 2, \ldots, L_P$ are i.i.d. random variables and that all users have the same average signal energy at the receiver as a result of perfect power control, then the total interference variance σ^2 is the same for all branches. According to Equation 15.27, the mean of $s^{(i)}(m,l)$ for the desired signal, i.e. when $m = \mu$, is equal to

$$\left|\alpha_{l_p}^{(i)}\right|\sqrt{\frac{\xi_s}{2}}. \tag{15.38}$$

From Equation 15.29, $s^{(i)}(m)$ is the sum of L_P number of Chi-square distributed random variables $s^{(i)}(m,l), l = 1, 2, \ldots, L_P$. Then, conditioned on $\left|\alpha_{l_p}^{(i)}\right|$, $s^{(i)}(m)$ can be described statistically as a Chi-square distributed random variable with $2L_P$ degrees of freedom. The pdf of $s^{(i)}(m)$ is given by [90]:

$$p_{s^{(i)}(m)}(s_m|c^2) = \frac{1}{2\sigma^2}\left(\frac{s_m}{c^2}\right)^{\frac{L_P-1}{2}} e^{-\frac{c^2+s_m}{2\sigma^2}} I_{L_P-1}\left(\frac{\sqrt{c^2 s_m}}{\sigma^2}\right), \quad m = \mu \tag{15.39}$$

for the desired signal m, while for other values of m in Figure 15.7 we have:

$$p_{s^{(i)}(m)}(s_m) = \frac{1}{2^{L_P}\sigma^{2L_P}(L_P-1)!} s_m^{L_P-1} e^{-\frac{s_m}{2\sigma^2}}, \quad m \neq \mu. \tag{15.40}$$

In Equation 15.39 c^2 is the non-centrality parameter of the Chi-square distribution, which is given by [90]:

$$c^2 = \frac{\xi_s}{2}\sum_{l=1}^{L_P}\left|\alpha_l^{(i)}\right|^2, \tag{15.41}$$

and $I_{L_P}(\cdot)$ is the modified Bessel function of the first kind of the L_P-th order. The conditioning in Equation 15.39 may be removed by using the derivations in Section A.2, yielding:

$$p_{s^{(i)}(m)}(s_m) = \int_0^{\infty} p_{s^{(i)}(m)}(s_m|c^2)p(c^2)dc^2, \quad m = \mu. \tag{15.42}$$

In a Rayleigh fading channel, the unconditional pdf of $p_{s^{(i)}(m)}(s_m)$, when $m = \mu$ is given by:[7]

$$p_{s^{(i)}(m)}(s_m) = \frac{1}{(2\sigma_T^2)^{L_P}(L_P - 1)!} s_m^{L_P - 1} \exp\left(-\frac{s_m}{2\sigma_T^2}\right), \tag{15.43}$$

where $2\sigma_T^2$ is the sum of the variances of the interference-plus-noise variable and the Rayleigh distributed path amplitude variable, which is given by $2\sigma^2 + \frac{\xi_s}{2} E\left[\left|\alpha_l^{(i)}\right|^2\right]$.

In order to derive the symbol error probability, without loss of generality, suppose that the signal $V_1^{(i)}(t)$ was transmitted. The probability that the symbol is detected correctly is given by the joint probability of the decision variable $s^{(i)}(1)$ being larger than all other decision variables, i.e.,:

$$P_c = P\left(s^{(i)}(2) < s^{(i)}(1), s^{(i)}(3) < s^{(i)}(1), \ldots, s^{(i)}(M) < s^{(i)}(1)\right). \tag{15.44}$$

Assuming that $s^{(i)}(m), m = 2, 3, \ldots, M$ are i.i.d. random variables, then Equation 15.44 can be reformulated using the product of the $M - 1$ number of individual probabilities, which is then weighted by the probability of occurence or pdf of $p_{s^{(i)}(1)}(s_1)$ and averaged or integrated over the legitimate range $0, \ldots, \infty$ of s_1, yielding:

$$P_c = \int_0^{\infty} \left[P\left(s^{(i)}(2) < s_1 | s^{(i)}(1) = s_1\right)\right]^{M-1} p_{s^{(i)}(1)}(s_1)ds_1. \tag{15.45}$$

The conditional probability in Equation 15.45 is given by

$$P\left(s^{(i)}(2) < s_1 | s^{(i)}(1) = s_1\right) = \int_0^{s_1} p_{s^{(i)}(2)}(s_2)ds_2 \tag{15.46}$$

[7] See Section A.2 for derivation.

and involving Equation 15.43 yields:

$$\begin{aligned} P\left(s^{(i)}(2) < s_1 | s^{(i)}(1) = s_1\right) &= \int_0^{s_1} \frac{1}{(2\sigma^2)^{L_P}(L_P-1)!} s_2^{L_P-1} \\ &\quad \times \exp\left(-\frac{s_2}{2\sigma^2}\right) ds_2 \\ &= 1 - \exp\left(-\frac{s_1}{2\sigma^2}\right) \sum_{l=0}^{L_P-1} \frac{s_1^l}{(2\sigma^2)^l l!}. \end{aligned} \tag{15.47}$$

Substituting Equations 15.43 and 15.47 back in Equation 15.45, the probability that the symbol will be detected correctly becomes [90]:

$$\begin{aligned} P_c &= \int_0^{\infty} \frac{1}{(2\sigma_T^2)^{L_P}(L_P-1)!} s_1^{L_P-1} \exp\left(-\frac{s_1}{2\sigma_T^2}\right) \\ &\quad \times \left[1 - \exp\left(-\frac{s_1}{2\sigma^2}\right) \sum_{l=0}^{L_P-1} \frac{s_1^l}{(2\sigma^2)^l l!}\right]^{M-1} ds_1 \\ &= \int_0^{\infty} \frac{1}{(2\sigma^2)^{L_P}(1+\bar{\gamma}_c)^{L_P}(L_P-1)!} s_1^{L_P-1} \exp\left(-\frac{s_1}{2\sigma^2(1+\bar{\gamma}_c)}\right) \\ &\quad \times \left[1 - \exp\left(-\frac{s_1}{2\sigma^2}\right) \sum_{l=0}^{L_P-1} \frac{s_1^l}{(2\sigma^2)^l l!}\right]^{M-1} ds_1, \end{aligned} \tag{15.48}$$

where the signal-to-interference-plus-noise ratio (SINR) is given by:

$$\bar{\gamma}_c = \frac{\xi_s}{2} E\left[\left|\alpha_l^{(i)}\right|^2\right] / 2\sigma^2, \tag{15.49}$$

with the total interference variance σ^2 given in Equation 15.37 and $E[x]$ representing the expectation of x. Then the average symbol error probability can be written as:

$$\begin{aligned} Pr_s(\epsilon) &= 1 - P_c \\ &= 1 - \int_0^{\infty} \frac{1}{(1+\bar{\gamma}_c)^{L_P}(L_P-1)!} s_1^{L_P-1} \exp\left(-\frac{s_1}{(1+\bar{\gamma}_c)}\right) \\ &\quad \times \left[1 - e^{-s_1} \sum_{l=0}^{L_P-1} \frac{s_1^l}{l!}\right]^{M-1} ds_1, \end{aligned} \tag{15.50}$$

and the corresponding bit error rate can be found by applying Equation 15.4.

The SINR $\bar{\gamma}_c$ is identical to that given in [277] for quadriphase spreading. Hence, if the transmitted power is the same, there is no difference in performance between a quadriphase spreading system and a biphase spreading system.

Parameter	Definition
γ_b	Signal-to-noise ratio (SNR) per bit defined as $L_P\bar{\gamma}_c/\log_2 M$
N_c	Number of chips per symbol
N	Processing gain
M	Number of orthogonal sequences
L_R	Number of resolvable paths which is the same for all users
L_P	Order of diversity = L_R
β	Number of bits per symbol

Table 15.1: Simulation parameters for the non-coherent M-ary orthogonal modulated DS-CDMA system

15.5 Bandwidth Efficiency in M-ary Modulated CDMA

In contrast to the M-ary orthogonally modulated scenario of Section 15.2, where the MFSK signalling tones occupied a total bandwidth of $B = M/2T_s$, when transmitting at a bit rate of $R_b = 1/T_b$, in CDMA, the M orthogonal Walsh functions are transmitted using the same carrier frequency over a bandwidth determined by the chip rate. Hence *the relative multi-user bandwidth efficiency in CDMA* can be written as:

$$\eta = K\frac{T_c}{T_b}, \tag{15.51}$$

where K is the total number of users transmitting. Since $T_s = N_c T_c$ and $T_s = T_b\beta$, Equation 15.51 becomes:

$$\begin{aligned} \eta &= K\frac{\beta}{N_c} \text{ [in (bits/s)/Hz]} \\ &= K\frac{\log_2 M}{N_c} \\ &= \frac{K}{N}, \end{aligned} \tag{15.52}$$

where $N = N_c/\beta$ is the processing gain. The resulting Equation above shows that the bandwidth efficiency in M-ary modulated CDMA is not inversely proportional to the value of M, as it was the case in Equation 15.7 for a narrowband system. In fact, *we expect the bandwidth efficiency of M-ary modulated CDMA to improve in Equation 15.52 upon increasing M due to the fact that $\beta = \log_2 M$ increases with M.* The only question now to be resolved is, whether the BER performance will also improve with increasing M. This question will be answered in the next section.

15.6 Numerical Simulation Results

For convenience, the various parameters affecting the probability of bit error and bandwidth efficiency are summarized in Table 15.1. We will also assume that the mean powers of the

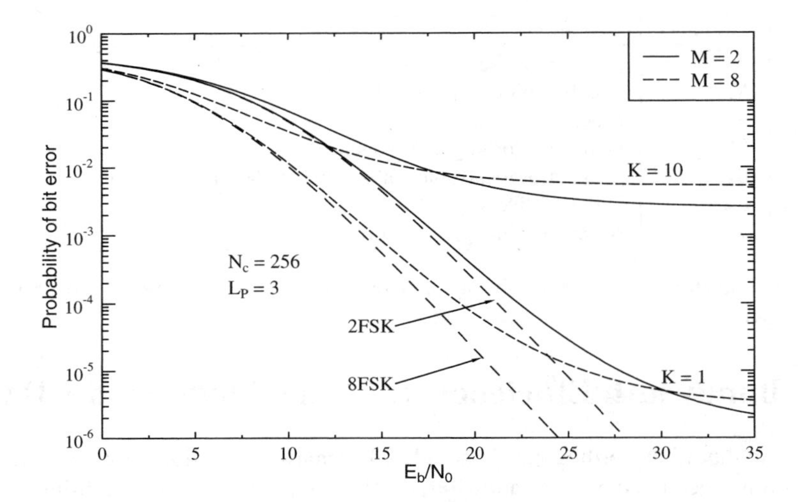

Figure 15.8: BER versus average SNR per bit for the non-coherent M-ary orthogonal modulated DS-CDMA system in a frequency-selective Rayleigh fading channel for $M = 2, 8$ with $K = 1, 10$. The parameters used are $N_c = 256$, $L_P = 3$ with asynchronous transmissions, employing Equation 15.50 for the numerical computations using a three-path, symbol-spaced, uniform-magnitude channel impulse response (CIR) profile, where the tap positions are randomly distributed over the length of the 256-chip spreading code assuming symbol-invariant slow fading.

diversity paths are equal, such that $E\left[\left|\alpha_l^{(i)}\right|^2\right]$ is equal to $1/L_P$ for $l = 1, 2, \ldots, L_P$.

Table 15.2 shows a summary of the experimental conditions related to Figures 15.8 to 15.12. The probability of bit error against the average SNR per bit is shown in Figure 15.8 for $M = 2$ and 8 when $K = 1$ and 10 users communicate with $N_c = 256$ and $L_P = 3$. The bit error rate is compared with that of the conventional narrowband M-ary orthogonal modulation of Section 15.2 denoted by 2FSK and 8FSK. The bit error rate in CDMA is almost identical to that of MFSK for $K = 1$ at lower E_b/N_0 values. At higher E_b/N_0 values, an error floor is observed for all M values in CDMA. This error floor is due to the multipath interference, which increases the total noise variance given in Equation 15.36. This increase in the noise variance is more significant under multi-user transmissions, as indicated by the plot for $K = 10$. This error floor is always present in a CDMA system, regardless of using a binary system or M-ary system because of the interference, as argued in Chapter 2. We can also see from the plot that the BER performance for $M = 8$ becomes worse than that for $M = 2$ above a certain E_b/N_0. This phenomenon can be explained by considering the total variance given by Equation 15.36. Namely, as M increases, the interference-related second term of Equation 15.36 increases due to the fact that a higher symbol energy ξ_s is transmitted, which becomes more explicit in the context of Equation 15.27. Observe in Equation 15.36 that the multipath-induced self-interference and the MAI are linearly proportional to the symbol energy ξ_s. Hence, this increases the overall variance – a phenomenon which becomes more significant at higher E_b/N_0.

This phenomenon is also evidently shown in Figure 15.9, which plots the BER against

Figures	M	L_P	K	N_c	N	E_b/N_0
Figure 15.8: BER versus E_b/N_0 performance comparison of M-ary signalling in a frequency-selective Rayleigh fading channel using orthogonal frequencies and orthogonal codes with constant N_c	2, 8	3	1, 10	256	256, 85.3	N/A
Figure 15.9: BER versus K performance comparison of M-ary signalling in a frequency-selective Rayleigh fading channel using orthogonal codes for different L_P values with constant N_c	2, 8	1, 3, 5	N/A	256	256, 85.3	10 dB
Figure 15.10: BER versus E_b/N_0 performance comparison of M-ary signalling in a frequency-selective Rayleigh fading channel using orthogonal codes for various M values and K = 1 with constant N	2, 4, 8, 16, 32	3	1	256, 512, 768, 1024, 1280	256	N/A
Figure 15.11: BER versus E_b/N_0 performance comparison of M-ary signalling in a frequency-selective Rayleigh fading channel using orthogonal codes for various M values and K = 10 with constant N	2, 4, 8, 16, 32	3	10	256, 512, 768, 1024, 1280	256	N/A
Figure 15.12: BER versus K performance comparison of M-ary signalling in a frequency-selective Rayleigh fading channel using orthogonal codes for different L_P values with constant N	2, 4, 8, 16, 32	3, 5	N/A	256, 512, 768, 1024, 1280	256	10 dB

N/A: Not applicable

Table 15.2: Summary of the experimental conditions for the non-coherent M-ary orthogonal modulated DS-CDMA system related to Figures 15.8 to 15.12 based on Equation 15.50 and assuming symbol-invariant slow fading where the tap positions are randomly distributed over the length of the 256-chip spreading code.

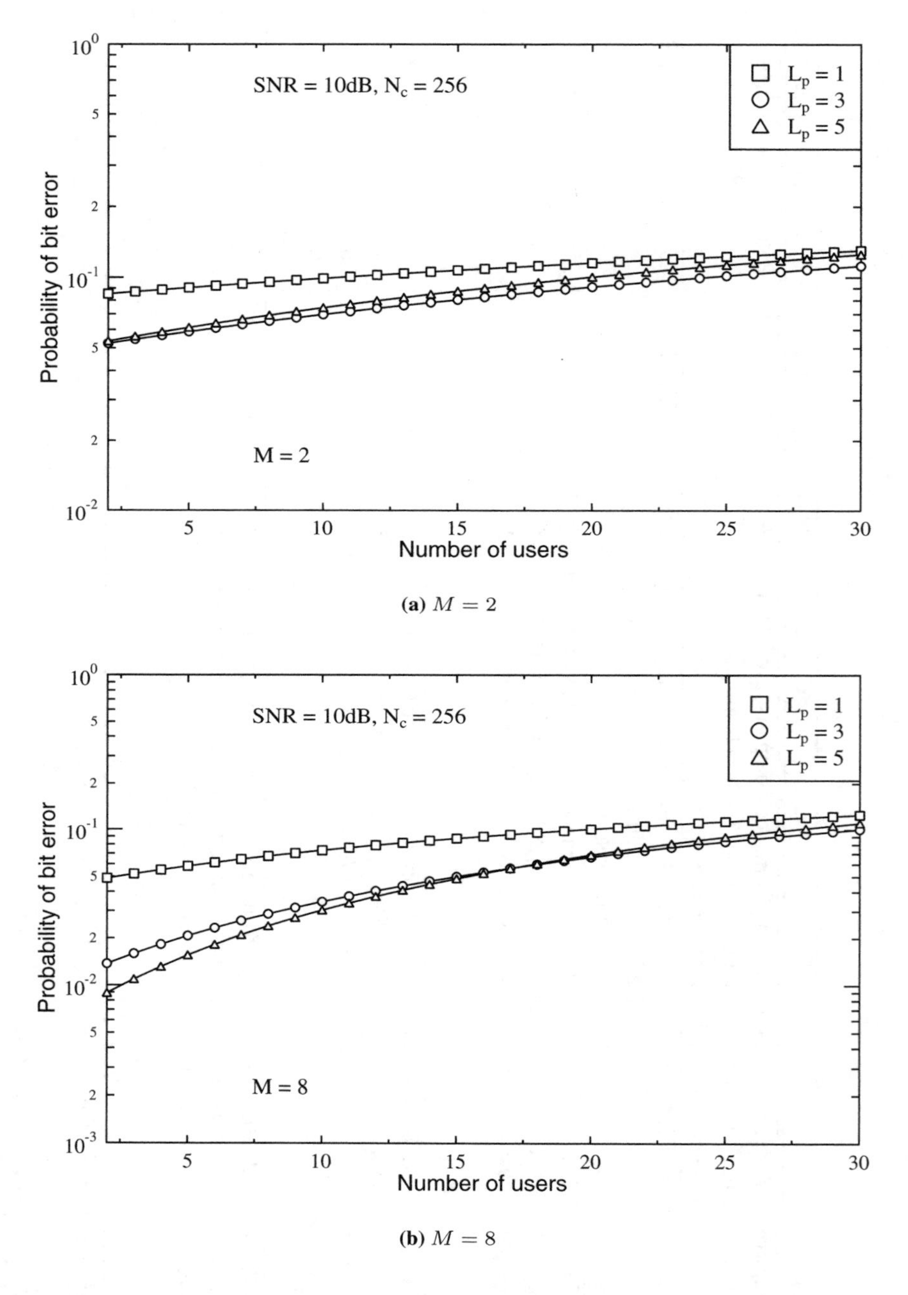

Figure 15.9: BER versus number of users for the non-coherent M-ary orthogonal modulated DS-CDMA system for a frequency-selective Rayleigh fading channel with different orders of diversity and M. Parameters used are $N_c = 256$ with asynchronous transmission, employing Equation 15.50 for the numerical computations using a three-path, symbol-spaced, uniform-magnitude CIR profile, where the tap positions are randomly distributed over the length of the 256-chip spreading code assuming symbol-invariant slow fading.

M	Order of diversity		
	2	4	6
2	0.01172	0.03125	0.03516
4	0.02344	0.05469	0.0625
8	0.03516	0.07031	0.08203
16	0.03125	0.09375	0.09375
32	0.03906	0.09766	0.1172
64	0.04688	0.1172	0.1172

Table 15.3: Relative multi-user bandwidth efficiency according to Equation 15.52 for an average bit error probability of 10^{-3} and SNR = 30dB with a constant chip rate of 256 per symbol and $L_P = 2, 4$ and 6. These results were derived from curves similar to those in Figure 15.9.

the number of users based on Equation 15.50 using a three-path, symbol-spaced, uniform-magnitude channel impulse response (CIR) profile, where the tap positions are randomly distributed over the length of the 256-chip spreading code assuming symbol-invariant slow fading. We see that increasing the order of diversity and M improves the BER performance at lower user capacity, i.e., for a low number of users. In this case, the order of diversity equals the number of multipath components. Increasing the number of multipath components is analogous to increasing the number of users, since both give the same interference magnitude, as seen in Equation 15.37. Hence, the performance for $L_P = 3$ becomes better than for $L_P = 5$ at higher capacity, above about 20 users.

The previous results were based on the fact that the number of chips per symbol N_c, remained constant for all M. Hence, this improves the bandwidth efficiency according to Equation 15.52, since $\eta = K\beta/N_c$ is increased upon increasing $\beta = \log_2 M$, which also means a higher data rate, since the bandwidth is determined by the constant chip rate. Table 15.3 shows the improvement of the relative multi-user bandwidth efficiency with M, which was computed from Equation 15.52 upon finding the BER = 10^{-3} points for the various scenarios.

Now let us consider the BER performance of the system, if the data rate is fixed and N_c is varied along with M, such that the processing gain $N = N_c/\beta$ remains constant. Figures 15.10 and 15.11 show the bit error rate for $K = 1$ and 10 users, respectively, with varying N_c. The processing gain of $N = \frac{N_c}{\beta}$ is kept constant at 256. Now the performance is improved, as M increases for all SNRs. This is due to the fact that, according to Equation 15.37, increasing the number of chips per signalling waveform reduces the interference. Hence, the BER performance improves.

Figure 15.12 shows the bit error rate against the number of users for diversity orders of 3 and 5. It is clear from the figure that increasing M gives a better performance than increasing the order of diversity, assuming that the number of resolvable paths equals the order of diversity. Table 15.4 shows the improvement in the relative multi-user bandwidth efficiency due to the fact that more users can be accommodated in the system at a given BER of 10^{-3} upon increasing M, which was computed from Equation 15.52 upon finding the BER = 10^{-3} points for the various scenarios. The improvement in bandwidth efficiency is comparable to that shown in Table 15.3. Having reviewed the design trade-offs of conventional M-ary modulation-based CDMA, let us now consider a new technique, which is based on signal processing in the so-called Residue Number System.

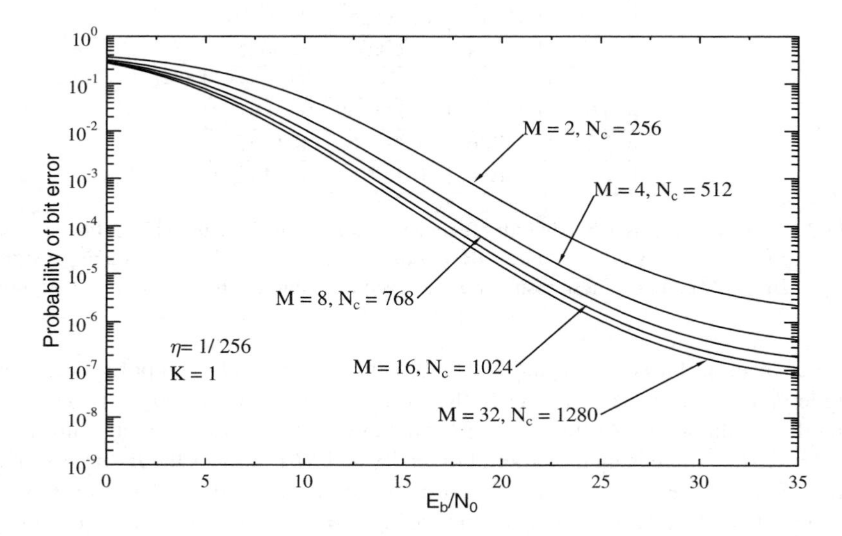

Figure 15.10: BER versus E_b/N_0 for the non-coherent M-ary orthogonal modulated DS-CDMA system for different values of M evaluated from Equation 15.50 in a frequency-selective Rayleigh fading channel having a constant processing gain of $N = \frac{N_c}{\beta} = 256$. Parameters used are $L_P = 3$ and $\boldsymbol{K} = \mathbf{1}$ and a three-path, symbol-spaced, uniform-magnitude CIR profile was invoked, where the tap positions are randomly distributed over the length of the 256-chip spreading code assuming symbol-invariant slow fading.

M	Order of diversity		
	2	4	6
2	0.01172	0.03125	0.03516
4	0.02344	0.05859	0.0625
8	0.03125	0.07422	0.08594
16	0.03516	0.08984	0.1055
32	0.03906	0.1016	0.1211
64	0.03906	0.1133	0.1328

Table 15.4: Relative multi-user bandwidth efficiency for the non-coherent M-ary orthogonal modulated DS-CDMA system according to Equation 15.52 for an average bit error probability of 10^{-3} and SNR = 30dB with a constant processing gain N of 256 and $L_P = 2, 4$ and 6. These results were derived from curves similar to those in Figure 15.12.

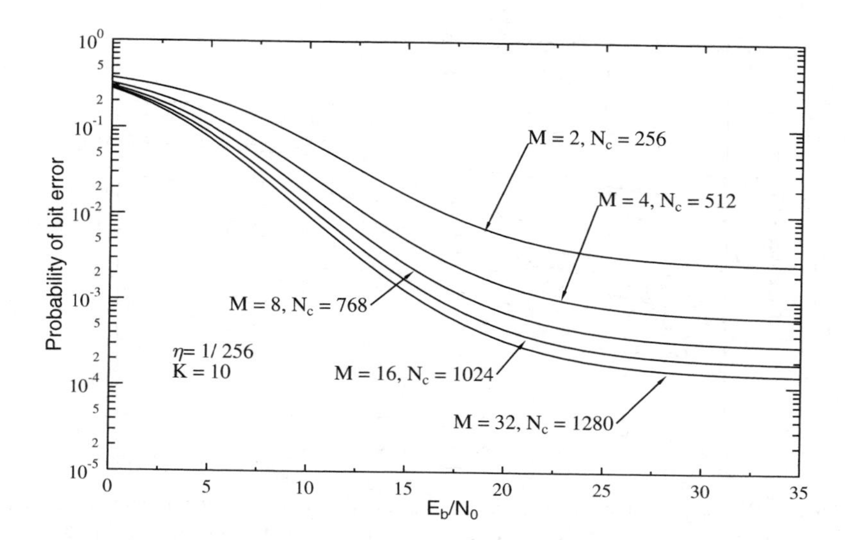

Figure 15.11: BER versus E_b/N_0 for the non-coherent M-ary orthogonal modulated DS-CDMA system for different values of M evaluated from Equation 15.50 in a frequency-selective Rayleigh fading channel having a constant processing gain of $N = N_c/\beta = 256$, where an increased chip-rate and bandwidth are required in order to maintain a constant N upon increasing β. Parameters used are $L_P = 3$ and $\boldsymbol{K} = \mathbf{10}$ and a three-path, symbol-spaced, uniform-magnitude CIR profile was employed, where the tap positions are randomly distributed over the length of the 256-chip spreading code assuming symbol-invariant slow fading.

15.6.1 Summary of Performance Results

We have seen that, regardless of the value of M, the processing gain is still the overiding factor in determining the BER performance of the CDMA system as demonstrated by Figure 15.8. However, given a certain processing gain, increasing the value of M can indeed improve the BER performance as suggested by Figures 15.10 and 15.11. However, increasing M increases the complexity of the receiver exponentially in conventional M-ary signalling in terms of the number of correlators required. Hence, in conventional M-ary signalling large values of M are not feasible.

By contrast, in M-ary orthogonal modulation based on the residue number system, the required number of correlators is determined by the sum of the moduli. The performance can be improved by exploiting the properties of the RNS through error correction and/or erasure. Here we curtail our discussions on RNS-based orthogonal schemes, noting that the proposed techniques are quite attractive in implementational terms, requiring, however, further research, which we intended to stimulate with the aid of the above brief discussions.

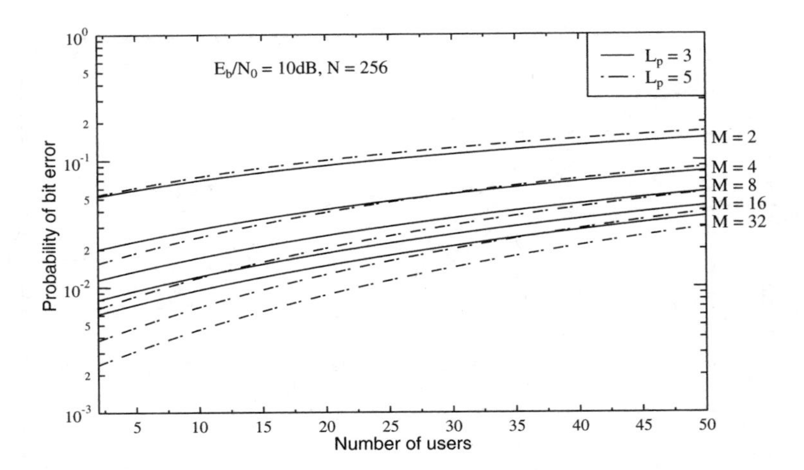

Figure 15.12: BER versus number of users for the non-coherent M-ary orthogonal modulated DS-CDMA system for different values of M evaluated from Equation 15.50 in a 3- and 5-path frequency-selective Rayleigh fading channel having a constant processing gain of $N = N_c/\beta$. Parameters used are $N = 256$, $E_b/N_0 = 10$dB and a three-path, symbol-spaced, uniform-magnitude CIR profile was employed, where the tap positions are randomly distributed over the length of the 256-chip spreading code assuming symbol-invariant slow fading.

15.7 Chapter Summary and Conclusion

M-ary orthogonal modulation techniques applicable to CDMA have been studied. The simulations and numerical results of this chapter demonstrated that for a given number of chips per symbol, increasing M does not significantly improve the BER performance of the system, as illustrated in Figures 15.8 and 15.9, unless the number of users is quite low. By contrast, the bandwidth efficiency of $\eta = \frac{K \log_2 M}{N_c}$ is higher for higher values of M, as shown in Table 15.3. However, for a given processing gain, increasing M gives a better BER performance, regardless of the user capacity, as shown in Figures 15.10 and 15.11. Furthermore, the overall bandwidth efficiency of $\eta = \frac{K \log_2 M}{N_c}$ is also higher for higher values of M due to the fact that the system can accommodate more users and still achieve the required BER, as illustrated in Table 15.4.

Despite its advantages M-ary orthogonal modulation has rarely been employed in practical systems because of the potentially high complexity of the receiver, which requires 2^M number of correlators and hence increases exponentially with M, as becomes explicit in Figure 15.7. However, the employment of a novel modulation technique based on the residue number system (RNS) can be used for overcoming this impediment, as will be shown in Chapters 17 and 18. Furthermore, we will demonstrate that the inherent properties of the RNS can be exploited for providing error correction and/or detection capabilities during the detection of the CDMA signals.

Chapter 16

Reed-Solomon Coded Non-Coherent M-ary Orthogonal Modulation in CDMA for Transmission Over Multipath Fading Channels[1]

16.1 Motivation[2]

In Chapter 15 we studied the performance of M-ary orthogonal signalling-aided CDMA without invoking error correction coding. In this chapter we invoke channel coding for improving the system's performance. Since the system communicates using M-ary orthogonal signalling, it is a natural choice to invoke non-binary Reed-Solomon (RS) codes [5], where each non-binary CDMA symbol is mapped to a non-binary RS coded symbol. Hence both the orthogonal modulation-based CDMA system and the RS channel coded scheme rely on M-ary symbols, resulting in a non-binary system. Therefore in this chapter the performance of Reed-Solomon (RS) coded DS-CDMA systems using non-coherent M-ary orthogonal mod-

[1] *Single- and Multi-Carrier CDMA Multi-User Detection, Space-Time Spreading, Synchronization, Networking and Standards.* L.Hanzo, L-L.Yang, E-L.Kuan and K.Yen,

[2] This chapter is based on ©IEEE L.L. Yang, Yen Kai, L. Hanzo: A Reed-Solomon Coded DS-CDMA System Using Non-coherent M-ary Orthogonal Modulation over Multipath Fading Channels, *IEEE Journal on Selected Areas of Communications, November 2000, Vol. 18, No. 11, pp. 2240-2251.*

ulation will be investigated, when communicating over multipath Rayleigh fading channels. We note at this stage that there exist other non-binary codes, which are also amenable to employment in conjunction with M-ary orthogonal modulation aided CDMA, which will be studied in Chapter 18.

Furthermore, in this chapter diversity reception techniques relying on equal gain combining (EGC) or selection combining (SC) will be invoked and the related performance will be evaluated in the context of both uncoded and RS-coded DS-CDMA systems. 'Errors-and-erasures' decoding will be considered, where the erasures are based on Viterbi's so-called ratio threshold test (RTT). The probability density functions (PDF) of the ratio associated with the RTT conditioned on both the correct detection and erroneous detection of the M-ary signals will be derived. These PDFs will then be used for computing the codeword decoding error probability of the RS coded DS-CDMA system using 'errors-and-erasures' decoding. Furthermore, the performance of the 'errors-and-erasures' decoding technique employing the RTT will be compared to that of 'error-correction-only' decoding refraining from using side-information over multipath Rayleigh fading channels. We will demonstrate with the aid of numerical results that when using 'errors-and-erasures' decoding, RS codes of a given code rate are capable of achieving a higher coding gain than without erasure information.

In recent years there has been significant interest in the application of CDMA various wireless communications systems [106] [10]. CDMA has also been adopted for employment in the third-generation global standard wireless systems, which will be outlined in Chapter 22. On the uplink of a DS-CDMA cellular system, due to the high complexity of coherent modulation/demodulation, which would require a pilot signal for each user, non-coherent M-ary orthogonal modulation using $M = 64$, i.e. 6-bit symbols has been proposed, for example, for the reverse link of IS-95 [11]. Analysis of DS-CDMA systems using M-ary orthogonal signalling has been provided for example in [364] – [386] for both AWGN and multipath fading channels.

In cellular DS-CDMA systems, forward error-correction (FEC) is often used for mitigating the effects of fading and interference. The so-called 'errors-and-erasures' decoding schemes [387] are often preferable to 'error-correction-only' decoding, since typically more erasures than errors can be corrected. Hence, it is beneficial to determine the reliability of the received symbols and to erase the low-reliability symbols prior to the decoding process. There are a number of methods for generating reliability-based side information and their performance has been analyzed, for example, in [386], [388] – [389]. An erasure insertion scheme suitable for M-ary orthogonal modulation is the ratio-threshold test (RTT), which was proposed by Viterbi [94].

In RTT an erasure is declared, whenever the ratio of the maximum to the second maximum of the inputs of the maximum-likelihood detector (MLD) does not exceed a pre-set threshold. The RTT was originally proposed, in order to mitigate the partial-band interference or multitone interference in M-ary frequency shift keying (MFSK) systems, but it was later invoked also in the context of M-ary orthogonal signaling, in order to generate channel-quality related information. Kim and Stark [386] have also employed it for mitigating the effect of Rayleigh-fading and have analyzed some of the performance limits of an RS coded DS-CDMA system by using 'errors-and-erasures' decoding. In Chapter 18 the performance of RS codes and redundant residue number system (RRNS) codes will be evaluated, when both the RTT and the so-called likelihood ratio threshold (LRT) based erasure scheme are considered. It can be shown that the RTT-based erasure scheme's optimum decision threshold is typically in the interval of [1.5,2] for a wide range of signal-to-noise ratio (SNR) per bit values, only slightly depending on the SNR per bit value encountered.

However, in this chapter we will focus our attention on investigating the achievable performance of an RS coded DS-CDMA system, when M-ary orthogonal signaling is employed in conjunction with an RTT-based erasure insertion scheme over multipath Rayleigh-fading channels. Two different diversity combining schemes [12], [13] - namely equal-gain combining (EGC) and selection combining (SC) - are considered and their performance is evaluated in the context of the proposed RTT-based erasure insertion scheme. We derive the probability density function (PDF) of the ratio defined in the RTT conditioned on both the correct detection hypothesis (H_1) and erroneous detection hypothesis (H_0) of the received M-ary signals over multipath Rayleigh fading channels. These PDFs are then used for computing the symbol erasure probability and the random symbol error probability after erasure insertion, and finally to compute the codeword decoding error probability. These analytical results allow us to quantify the performance of the system considered using a numerical approach. Furthermore, with the aid of these analytical results, we gain an insight into the basic characteristics of Viterbi's RTT, and determine the optimum decision threshold for practical systems, since the formulae obtained can be evaluated numerically.

The remainder of this chapter is organized as follows. In the next section, the system and channel model are described. Section 16.3 addresses the non-coherent detection of DS-CDMA signals and derives the required PDFs associated with EGC and SC-based detection. In Section 16.4, the average error probability of the uncoded system is analyzed, while the expressions of the conditional PDFs of the required ratio associated with the RTT and the RS-codeword decoding error probability are derived in Section 16.5. Numerical results are provided in Section 16.6, and finally in Section 16.7 we present our conclusions.

16.2 System Description and Channel Model

16.2.1 The Transmitted Signals

The transmitter block diagram of the coded M-ary orthogonal modulation-based DS-CDMA system is shown in Figure 16.1. The information bits are first grouped into b-bit symbols, where $b = \log_2 M$. Then, $\mathcal{K}$ information symbols are encoded into a $\mathcal{N}$-symbol RS codeword. In order to randomize the effects of bursty symbol errors, the RS codewords are then interleaved. Finally, each RS coded symbol is M-ary modulated, DS spread and carrier-modulated using the approach of [12], in order to form the transmitted signal.

For the sake of notational convenience, we shall adopt notations similar to those used by Holtzman in [12]. In a DS-CDMA system in which there are K active users transmitting their signals simultaneously, the signal transmitted by user i, $i = 1, 2, \ldots, K$, can be expressed as [12]:

$$s_i(t) = \sqrt{P}W^j(t)a_I^i(t)\cos\omega_c t + \sqrt{P}W^j(t-T_0)a_Q^i(t-T_0)\sin\omega_c t, \; 0 \leq t \leq T_s, \quad (16.1)$$

where P is the transmitted power, T_s is the symbol duration, ω_c is the carrier's angular frequency, and T_0 is an offset time. Furthermore, $W^j(t)$ is the jth Walsh-Hadamard orthogonal function, which represents the jth orthogonal signal of the ith user's symbols, while $a_I^i(t)$

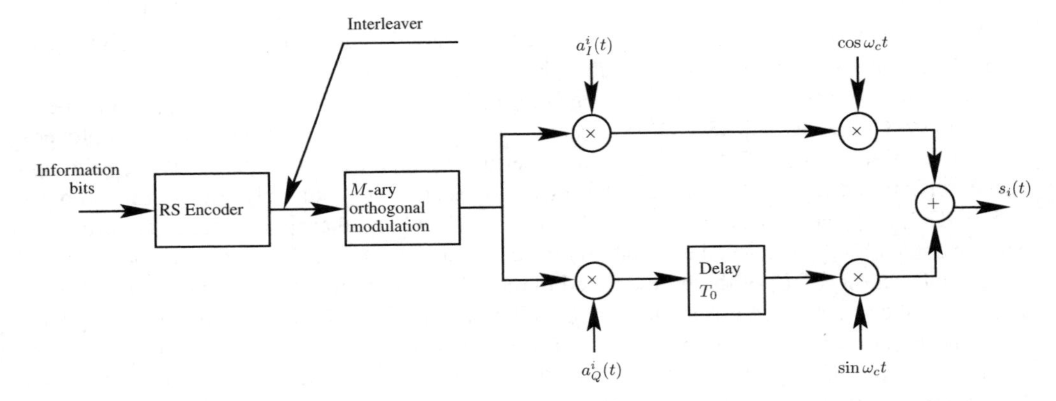

Figure 16.1: Transmitter block diagram of the DS-CDMA system using M-ary orthogonal modulation, RS encoding and symbol interleaving.

and $a_Q^i(t)$ represent the spreading waveforms of the in-phase (I) and quadrature (Q) phase channels, respectively. These quadrature components are expressed as:

$$a_I^i(t) = \sum_{h=-\infty}^{\infty} a_h^{I,i} p(t - hT_c) \tag{16.2}$$

$$a_Q^i(t) = \sum_{h=-\infty}^{\infty} a_h^{Q,i} p(t - hT_c), \tag{16.3}$$

where $a_h^{I,i}$, $a_h^{Q,i}$ are assumed to be independent identically distributed (i.i.d) random variables assuming values of +1 and -1 with equal probability of 1/2. Furthermore, T_c represents the chip duration, and $p(t)$ is assumed to be the rectangular chip waveform, which is defined over the interval $(0, T_c]$. Moreover, we assume that $N_s = T_s/T_c = bT/T_c = bN$, where T is the bit-duration and $N = T/T_c$.

16.2.2 The Channel Model

We assume that the channel between the ith transmitter and the corresponding receiver is a multipath Rayleigh fading channel [90]. The complex lowpass equivalent representation of the impulse response experienced by user i is given by:

$$h_i(t) = \sum_{l_p=1}^{L_p} \alpha_{il_p} \delta(t - \tau_{il_p}) \exp(-j\phi_{il_p}), \tag{16.4}$$

where α_{il_p}, τ_{il_p} and ϕ_{il_p} represent the attenuation factor, delay and phase-shift for the l_pth multipath component of the channel, respectively, while L_p is the total number of diversity paths and $\delta(t)$ is the Delta-function. We assume that the ith user's multipath attenuations $\{\alpha_{il_p}, l_p = 1, 2, \ldots, L_p\}$ in Equation 16.4 are independent Rayleigh-distributed random variables having a PDF given by Equation (14-1-23) of [90]:

$$\begin{aligned} f(\alpha_{il_p}) &= M(\alpha_{il_p}, \Omega), \\ M(R, \Omega) &= \frac{2R}{\Omega} \exp\left(-\frac{R^2}{\Omega}\right), \end{aligned} \tag{16.5}$$

where $\Omega = E\left[(\alpha_{il_p})^2\right]$. The phases $\{\phi_{il_p},\ l_p = 1, 2, \ldots, L_p\}$ of the different paths are assumed to be uniformly distributed random variables in $[0, 2\pi)$, while the ith user's path delays of $\left\{\tau_{il_p},\ l_p = 1, 2, \ldots, L_p\right\}$ are modelled as random variables that are mutually independent of each other and uniformly distributed in $[0, T_s)$. We also assume that ideal power control is employed, in order that the received signal powers are the same for all K users. Then the received signal at the base station generated by the K users can be expressed as:

$$r(t) = \sum_{i=1}^{K} y_i(t) + n(t), \tag{16.6}$$

where

$$\begin{aligned} y_i(t) = \sum_{l_p=1}^{L_p} \sqrt{P}\alpha_{il_p} & \left[W^j(t - \tau_{il_p}) a_I^i(t - \tau_{il_p}) \cos\left(\omega_c(t - \tau_{il_p}) - \phi_{il_p}\right)\right. \\ & \left. + W^j(t - T_0 - \tau_{il_p}) a_Q^i(t - T_0 - \tau_{il_p}) \sin\left(\omega_c(t - \tau_{il_p}) - \phi_{il_p}\right)\right], \end{aligned} \tag{16.7}$$

$n(t)$ represents the AWGN, which is modelled as a random variable with zero mean and double-sided power spectral density of $N_0/2$. After the receiver's bandpass filter, the noise $n(t)$ becomes a narrow-band noise process, which can be expressed as [12]:

$$n(t) = n_c(t) \cos \omega_c t + n_s(t) \sin \omega_c t, \tag{16.8}$$

where $n_c(t)$ and $n_s(t)$ represent low-pass-filtered Gaussian processes.

16.3 The Detection Model

The receiver schematic of the studied DS-CDMA system including non-coherent M-ary demodulation, multipath diversity combining, maximum-likelihood detection (MLD), RTT-based-erasure insertion, deinterleaving and RS 'errors-and-erasures' decoding is shown in Figure 16.2, where the square-law-based non-coherent M-ary demodulation block is the same as that used in [12] and hence interested readers are referred to [12] for further details. Multipath diversity combining invoking the EGC or the SC principle was implemented in the 'EGC or SC' block. The MLD block selects the largest from its input variables and computes the value of λ - the ratio of the largest to the second largest of the MLD's inputs.

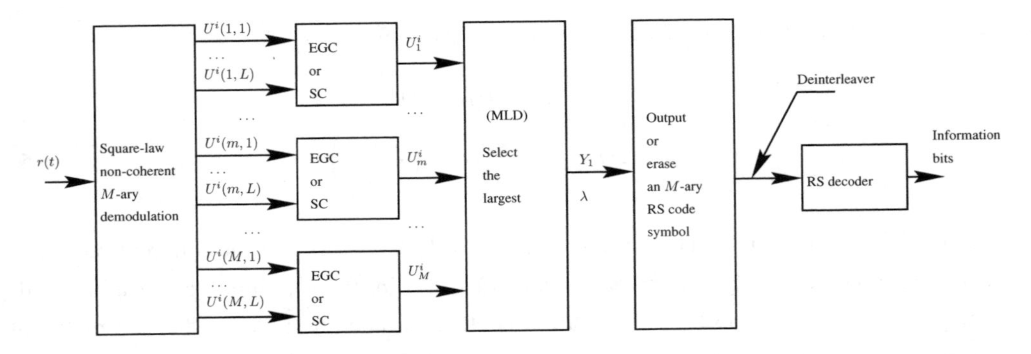

Figure 16.2: Receiver model of the DS-CDMA system investigated using non-coherent M-ary demodulation, multipath diversity combining, maximum-likelihood detection (MLD), erasure insertion based on the RTT, deinterleaving and RS 'errors-and-erasures' decoding.

Following the MLD stage, the next block may output an M-ary RS code symbol or insert an erasure. Finally, after symbol-based deinterleaving, the RS decoder invokes 'errors-and-erasures' decoding and then outputs the received information bits.

As shown in Figure 4 of [12], the outputs of the low-pass filters of the in-phase (I) and quadrature (Q) channels are first multiplied by the spreading sequences, and the resulting signals are correlated with each of the M orthogonal Walsh-Hadamard functions. Let the ith user be the user of interest and assume that the jth symbol was transmitted. According to Jalloul and Holtzman [12], the output variable of the lth branch of the mth correlator in Figure16.2 can be expressed as [12]:

$$U^i(m,l) = \left[\int_{\tau_{il}}^{T_s+\tau_{il}} r(t)a_I^i W^m \cos\omega_c t dt + \int_{\tau_{il}}^{T_s+\tau_{il}} r(t)a_Q^i W^m \sin\omega_c t dt\right]^2 + \left[\int_{\tau_{il}}^{T_s+\tau_{il}} r(t)a_Q^i W^m \cos\omega_c t dt - \int_{\tau_{il}}^{T_s+\tau_{il}} r(t)a_I^i W^m \sin\omega_c t dt\right]^2,$$
$$m = 1, 2, \ldots, M; \; l = 1, 2, \ldots, L. \quad (16.9)$$

Upon assuming that Gaussian approximation of the multipath interference and that of the multiple access interference can be employed, consequently, $U^i(m,l)$ is the sum of the square of two Gaussian random variables, each having a variance of [12] (Equation (28)):

$$\sigma^2 = \frac{N_0}{2} + \frac{E_s\Omega}{3N_s}(KL_p - 1), \quad (16.10)$$

where $\Omega = E\left[(\alpha_{il_p})^2\right]$, as we assumed in Equation 16.5.

Since $U^i(m,l)$ is the sum of the square of two Gaussian random variables, it can be shown that, conditioned on α_{il}, $U^i(m,l)$ becomes Chi-square distributed [90] with two degrees of freedom, and its PDF can be expressed as:

$$f_{U^i(m,l)}(y|\mathcal{R}_{il}) = \frac{1}{2\sigma^2} \exp\left(-\frac{y + \mathcal{R}_{il}}{2\sigma^2}\right) I_0\left(\frac{\sqrt{y\mathcal{R}_{il}}}{\sigma^2}\right), \quad y \geq 0 \tag{16.11}$$

if $m = j$, where $\mathcal{R}_{il} = E_s \alpha_{il}^2$ is the non-centrality parameter of the Chi-square distribution and $I_0(\cdot)$ is the modified Bessel function of the zero-th order. If $m \neq j$, then [90]:

$$f_{U^i(m,l)}(y) = \frac{1}{2\sigma^2} \exp\left(-\frac{y}{2\sigma^2}\right), \quad y \geq 0. \tag{16.12}$$

Using $2\sigma^2$ to normalize $U^i(m,l)$, we obtain the normalized PDFs of $U^i(m,l)$ as:

$$f_{U^i(m,l)}(y|\gamma_{il}) = \exp\left(-[y + \gamma_{il}]\right) I_0\left(\sqrt{4y\gamma_{il}}\right), \quad m = j, \; y \geq 0, \tag{16.13}$$

$$f_{U^i(m,l)}(y) = \exp(-y), \quad m \neq j, \; y \geq 0, \tag{16.14}$$

where

$$\gamma_{il} = \frac{\mathcal{R}_{il}}{2\sigma^2} = \overline{\gamma}_c \cdot \frac{\alpha_{il}^2}{\Omega}, \tag{16.15}$$

and

$$\overline{\gamma}_c = \left[\frac{2}{3N_s}(KL_p - 1) + \left(\frac{E_s\Omega}{N_0}\right)^{-1}\right]^{-1}. \tag{16.16}$$

Since $\{\alpha_{il}\}$ are independent Rayleigh-distributed random variables, $\{\alpha_{il}^2\}$ are exponentially-distributed random variables. Consequently, with the aid of Equation16.5 it can be shown that the PDF of γ_{il} defined in Equation 16.15 can be expressed as:

$$f(\gamma_{il}) = \frac{1}{\overline{\gamma}_c} \exp\left(-\frac{\gamma_{il}}{\overline{\gamma}_c}\right). \tag{16.17}$$

16.3.1 Equal Gain Combining

For a receiver using L-th order ($L \leq L_p$) diversity reception and EGC the L branches are equally weighted and then added in order to form the decision variables, which can be expressed as:

$$U_m^i = \sum_{l=1}^{L} U^i(m,l) \tag{16.18}$$

for $m = 1, 2, \ldots, M$. Assuming that the jth symbol was transmitted, then, since the quantities $\{U^i(m,l)\}$ are Chi-square distributed random variables with two degrees of freedom and have a PDF given by Equation 16.13 and (16.14) for $m = j$ and $m \neq j$, respectively, it can be shown that, for a given set of multipath attenuations $\{\alpha_{i1}, \alpha_{i2}, \ldots, \alpha_{iL}\}$, U_m^i is Chi-square distributed with $2L$ degrees of freedom, and it obeys the normalized PDF of [90]:

$$f_{U_j^i}(y|\gamma) = \left(\frac{y}{\gamma}\right)^{(L-1)/2} \exp\left(-[y+\gamma]\right) I_{L-1}\left(\sqrt{4y\gamma}\right), \quad y \geq 0, \qquad (16.19)$$

$$f_{U_m^i}(y) = \frac{1}{(L-1)!} y^{L-1} \exp(-y), \qquad m \neq j, \; y \geq 0, \qquad (16.20)$$

where $I_L(\cdot)$ is the modified Bessel function of the Lth order [90], and

$$\gamma = \sum_{l=1}^{L} \gamma_{il}, \qquad (16.21)$$

where the PDF of γ_{il} is given by Equation 16.17. We assume throughout that all the path gains are i.i.d random variables, which in turn means that $\{\gamma_{il}\}$ are i.i.d random variables. Consequently, it can be readily shown that γ obeys the central Chi-square distribution with $2L$ degrees of freedom [90], which can be expressed as:

$$f(\gamma) = \frac{1}{(L-1)! \overline{\gamma}_c^L} \gamma^{L-1} \exp\left(-\frac{\gamma}{\overline{\gamma}_c}\right). \qquad (16.22)$$

Since the decision variables $\{U_1^i, U_2^i, \ldots, U_M^i\}$ are independent random variables, the conditioning in Equation 16.19 may be removed[3] by averaging $f_{U_j^i}(y|\gamma)$ over the valid range of γ, which can be expressed as:

$$f_{U_j^i}(y) = \int_0^\infty f_{U_j^i}(y|\gamma) f(\gamma) d\gamma. \qquad (16.23)$$

Upon substituting Equation 16.19 and Equation 16.22 into Equation 16.23, it can be shown that:

$$f_{U_j^i}(y) = \frac{1}{(L-1)!(1+\overline{\gamma}_c)^L} y^{L-1} \exp\left(-\frac{y}{1+\overline{\gamma}_c}\right), \quad y \geq 0. \qquad (16.24)$$

[3] Note that, since we have assumed that $\left\{\alpha_{il_p}\right\}$ are independent random variables and we also assumed that $\{U_1^i, U_2^i, \ldots, U_M^i\}$ are independent random variables, the conditioning due to the fading of the channel can be removed at the branch outputs by averaging the PDF of $U^i(m,l)$ for $m = j$ in Equation 16.13 over the valid range of γ_{il}, which results in $f_{U^i(m,l)}(y) = \frac{1}{1+\overline{\gamma}_c} \exp\left(-\frac{y}{1+\overline{\gamma}_c}\right)$ for $m = j$. The approach of this section can be followed in order to derive Equation 16.24 and Equation 16.31

16.3.2 Selection Combining

For a L-branch diversity receiver with SC - where the branch signal having the largest amplitude is selected for demodulation - the conventional decision variable of a M-ary orthogonal modulation system is defined as:

$$U^i = \max\left\{U^i(1,1), U^i(1,2), \ldots, U^i(1,L), \ldots, U^i(M,1), \ldots, U^i(M,L)\right\}. \tag{16.25}$$

However, this definition is not suitable for generating reliability information by invoking Viterbi's RTT, since there exist L branch signals, which are matched to the transmitted signal. This issue will become explicit during our further discussions. Hence, in this chapter a two-stage maximum selection scheme is proposed, in order to invoke the RTT. The first-stage maximum selection is defined as the maximum selection from the L diversity components for each specific given m, i.e.:

$$U_m^i = \max\left\{U^i(m,1), U^i(m,2), \ldots, U^i(m,L)\right\}, \tag{16.26}$$

for $m = 1, 2, \ldots, M$. Explicitly, after the above selection process, only one output, namely U_j^i is matched to the transmitted signal and the other $(M-1)$ outputs are mismatched with respect to the transmitted signal, and hence the RTT [94] can be invoked, in order to obtain the required reliability measures. Based on the above maximum selection, the symbol is decided correctly, if and only if the output U_j^i, which is matched to the transmitted signal - remembering that the jth symbol was transmitted - is larger than any other outputs mismatched to the transmitted symbol. Hence, the 2nd-stage maximum selection finds the largest one from the set $\left\{U_1^i, U_2^i, \ldots, U_M^i\right\}$, which can be expressed as:

$$U^i = \max\left\{U_1^i, U_2^i, \ldots, U_M^i\right\}. \tag{16.27}$$

Let us now derive the PDFs of U_m^i. Since the normalized PDF of $U^i(m,l)$ for $l = 1, 2, \ldots, L$ was given by Equation 16.14, the PDF of U_m^i for $m \neq j$ can be formulated as:

$$\begin{aligned} f_{U_m^i}(y) &= \frac{d}{dy}\left[1 - \exp(-y)\right]^L \\ &= L\exp(-y)\left[1 - \exp(-y)\right]^{L-1}. \end{aligned} \tag{16.28}$$

Similarly, the conditional PDF of U_j^i can be derived, which can be expressed as:

$$\begin{aligned} f_{U_j^i}(y|\gamma_{i1}, \gamma_{i2}, \ldots, \gamma_{iL}) &= \frac{d}{dy}\left[\prod_{l=1}^{L}\int_0^y f_{U^i(j,l)}(x|\gamma_{il})dx\right] \\ &= \sum_{l=1}^{L} f_{U^i(j,l)}(y|\gamma_{il})\left[\prod_{\substack{u=1 \\ u\neq l}}^{L}\int_0^y f_{U^i(j,u)}(x|\gamma_{iu})dx\right], \end{aligned} \tag{16.29}$$

where $f_{U^i(j,u)}(x|\gamma_{iu})$ for $u = 1, 2, \ldots, L$ was given by Equation 16.13. The conditioning with respect to $\{\gamma_{il}\}$ can be removed[4] at this stage by integrating $f_{U_j^i}(y|\gamma_{i1}, \gamma_{i2}, \cdots, \gamma_{iL})$ over the valid ranges of $\gamma_{i1}, \gamma_{i2}, \cdots, \gamma_{iL}$, which can be expressed as:

$$f_{U_j^i}(y) = \underbrace{\int \cdots \int_0^\infty}_{L} \sum_{l=1}^{L} f_{U^i(j,l)}(y|\gamma_{il}) \left[\prod_{\substack{u=1 \\ u \neq l}}^{L} \int_0^y f_{U^i(j,u)}(x|\gamma_{iu}) dx \right] \cdot f(\gamma_{i1}) f(\gamma_{i2}) \ldots f(\gamma_{iL}) d\gamma_{i1} d\gamma_{i2} \ldots d\gamma_{iL}. \tag{16.30}$$

Substituting $f_{U^i(j,l)}(x|\gamma_{il})$ and $f(\gamma_{il})$ from Equation 16.13 and Equation 16.17 into Equation 16.30, and remembering that $\{\gamma_{il}\}$ are i.i.d random variables, Equation 16.30 can be simplified to:

$$f_{U_j^i}(y) = \frac{L}{1+\overline{\gamma}_c} \exp\left(-\frac{y}{1+\overline{\gamma}_c}\right) \left[1 - \exp\left(-\frac{y}{1+\overline{\gamma}_c}\right)\right]^{L-1}, \quad y \geq 0, \tag{16.31}$$

where $\overline{\gamma}_c$ is given by Equation 16.16.

Above we have derived the PDFs of the decision variables that are input to the MLD block. With the aid of these PDFs the average symbol error probabilities can now be derived for both EGC and SC.

16.4 Average Error Probability without FEC

Let $\{U_1^i, U_2^i, \ldots, U_M^i\}$ represent the decision variables after diversity combining, which are input to the MLD block, and assume that the jth symbol is transmitted. Let H_1 and H_0 represent the hypotheses of correct decision and erroneous decision of the MLD block. Then the average correct symbol probability of $P(H_1)$ and erroneous symbol probability of $P(H_0)$ can be expressed as:

$$P(H_1) = P(U_1^i < U_j^i, \ldots, U_{j-1}^i < U_j^i, U_{j+1}^i < U_j^i, \ldots, U_M^i < U_j^i), \tag{16.32}$$

$$P(H_0) = 1 - P(H_1), \tag{16.33}$$

respectively. Given the symbol error probability, the average bit error rate (BER) can be expressed as [90]:

$$P_b = \frac{2^{b-1}}{M-1} P(H_0). \tag{16.34}$$

[4] See 3.

16.4.1 Equal Gain Combining

For EGC, referring to Equation 16.32, the probability of $P(H_1)$ can be expressed as:

$$P(H_1) = \int_0^{\infty} f_{U_j^i}(y) \left[\int_0^{y} f_{U_m^i}(x) dx \right]^{M-1} dy, \tag{16.35}$$

where $f_{U_m^i}(x)$ and $f_{U_j^i}(y)$ are given by Equation 16.20 and Equation 16.24, respectively. Substituting them into the above equation, it can be shown that:

$$P(H_1) = \int_0^{\infty} \frac{1}{(1+\overline{\gamma}_0)^L (L-1)!} y^{L-1} \exp\left(-\frac{y}{1+\overline{\gamma}_0}\right) \cdot \left[1 - \exp(-y) \sum_{k=0}^{L-1} \frac{y^k}{k!}\right]^{M-1} dy. \tag{16.36}$$

The average symbol error probability and bit error probability can be computed using Equations 16.33 and 16.34, respectively.

16.4.2 Selection Combining

For SC, referring to Equation 16.32, the probability of $P(H_1)$ can also be expressed in the form of Equation 16.35, where $f_{U_m^i}(x)$ and $f_{U_j^i}(y)$ are given by Equation 16.28 and Equation 16.31, respectively. Upon substituting them into Equation 16.35 the correct symbol probability can be simplified to:

$$P(H_1) = \sum_{m=1}^{L} (-1)^{m+1} \binom{L}{m} \prod_{n=1}^{L(M-1)} \left(\frac{n}{n + m/(1+\overline{\gamma}_c)} \right). \tag{16.37}$$

The derivation of Equation 16.37 is provided in Appendix I.[5]

The average symbol error probability and bit error probability can be computed with the aid of Equations 16.33 and 16.34.

So far, we have obtained the expression of the average error probability per bit, while deriving some further expressions for our forthcoming study of the coded DS-CDMA system using M-ary orthogonal modulation. Hence, let us now analyze the performance of the system in conjunction with RS coding using both 'error-correction-only' decoding and 'errors-and-erasures' decoding by invoking Viterbi's RTT techniques [94].

[5]The erroneous symbol probability, $P(H_0)$, of the SC scheme has been given in Equation (11) of [390] without detailed derivation.

16.5 Performance Using RS Forward Error-Correction Codes

Reed-Solomon codes [387] constitute an efficient class of linear codes using multi-bit symbols and having the capability of correcting/detecting symbol errors and symbol erasures. A RS($\mathcal{N}, \mathcal{K}$) code - where $\mathcal{N}$ is the total length of the codeword and $\mathcal{K}$ is the number of information symbols, respectively - can correct up to $t_c = \lfloor(\mathcal{N} - \mathcal{K})/2\rfloor$ random symbol errors, where $\lfloor x \rfloor$ represents the largest integer not exceeding x, or detect up to $(\mathcal{N} - \mathcal{K})$ symbol errors. Alternatively, it is capable of correcting up to $(\mathcal{N} - \mathcal{K})$ symbol erasures. Moreover, it is capable of correcting t or less random symbol errors and e symbol erasures, simultaneously, provided that $2t + e \leq (\mathcal{N} - \mathcal{K})$. Typically, 'errors-and-erasures' decoding is preferable to 'error-correction-only' decoding, since more erasures than errors can be corrected. Hence, it is advantageous to determine the reliability of the received RS-coded symbols and to erase the low-reliability symbols prior to the decoding process. Consequently, in this section, both 'error-correction-only' decoding and 'errors-and-erasures' decoding based on the RTT will be discussed, while their performance comparison will be given in Section 16.6. Let us first analyze the decoding performance of RS codes using 'error-correction-only' decoding.

16.5.1 Error-Correction-Only Decoding

Assuming that sufficiently long channel interleaving was invoked, in order to result in independent symbol errors in a RS codeword, the codeword decoding error probability after 'error-correction-only' decoding can be expressed using Equation (8-1-119) of [90] as follows:

$$P_W = \sum_{n=\lfloor(\mathcal{N}-\mathcal{K})/2\rfloor+1}^{\mathcal{N}} \binom{\mathcal{N}}{n} [P(H_0)]^n [1 - P(H_0)]^{\mathcal{N}-n}, \tag{16.38}$$

where the random symbol error probability $P(H_0)$ before decoding is given by Equation 16.33.

16.5.2 Errors-and-Erasures Decoding

In order to take advantage of the powerful 'errors-and-erasures' correction capability of the RS code concerned, it is essential to design an efficient erasure insertion scheme. In this subsection, erasure insertion using Viterbi's RTT [94] is investigated and the PDFs of the quantities involved in Viterbi's RTT are derived.

Let $\{U_1^i, U_2^i, \ldots, U_M^i\}$ represent the decision variables input to the MLD block of Figure 16.2. The ratio involved in Viterbi's RTT is computed according to the following definition [94]:

$$\lambda = \frac{Y_1 = \max_1 \{U_1^i, U_2^i, \ldots, U_M^i\}}{Y_2 = \max_2 \{U_1^i, U_2^i, \ldots, U_M^i\}}, \tag{16.39}$$

where $Y_1 = \max_1\{\cdot\}$ and $Y_2 = \max_2\{\cdot\}$ represent the maximum and the 'second' maximum of the decision variables of $\{U_1^i, U_2^i, \ldots, U_M^i\}$, respectively. Viterbi [94] pointed out that the demodulated symbols having relatively high ratio of λ were more reliable than those having relatively low values of λ. Consequently, a pre-set threshold λ_T can be invoked, in order to erase these low-reliability symbols associated with a ratio of $\lambda \leq \lambda_T$, which constitutes the so-called RTT. In this subsection we derive the PDFs of λ under the hypotheses H_1 of correct decision and H_0 of erroneous decision, and under the assumption that the decision variables are independent random variables [391]. With the aid of these PDFs, not only the characteristics of the threshold λ_T can be investigated, but also the RS decoding performance can be quantified, while determining the optimum thresholds using a numerical approach.

For reasons of space, the PDFs of the maximum and the 'second maximum' of $Y_1 = \max_1\{\cdot\}$ and $Y_2 = \max_2\{\cdot\}$ under the hypotheses H_1 of correct decision and H_0 of erroneous decision are summarized below by omitting the less informative details, which have been given in the Appendix of [391] for EGC. For SC, the corresponding PDFs can be derived following a similar approach to that applied in the context of the EGC in the Appendix of [391]. The corresponding results may be summarized as follows:

EGC: the PDFs of Y_1 and Y_2 conditioned on the hypothesis of H_1 are given by:

$$f_{Y_1}(y|H_1) = \frac{1}{P(H_1)} \cdot \frac{1}{(1+\overline{\gamma}_c)^L (L-1)!} y^{L-1} \exp\left(-\frac{y}{1+\overline{\gamma}_c}\right) [1-\Psi(y)]^{M-1}, \quad y \geq 0, \tag{16.40}$$

$$f_{Y_2}(y|H_1) = \frac{1}{P(H_1)} \cdot \frac{M-1}{(L-1)!} y^{L-1} \exp(-y) [1-\Psi(y)]^{M-2} \Psi\left(\frac{y}{1+\overline{\gamma}_c}\right), \quad y \geq 0, \tag{16.41}$$

where $P(H_1)$ is the *a-priori* probability of the hypothesis H_1 for EGC, which is given by Equation 16.36, while the short-hand $\Psi(x)$ is defined as:

$$\Psi(x) = \exp(-x) \sum_{k=0}^{L-1} \frac{x^k}{k!}. \tag{16.42}$$

EGC: the PDFs of Y_1 and Y_2 conditioned on the hypothesis of H_0 are as follows:

$$f_{Y_1}(y|H_0) = \frac{1}{P(H_0)} \cdot \frac{M-1}{(L-1)!} y^{L-1} \exp(-y) [1-\Psi(y)]^{M-2} \left[1 - \Psi\left(\frac{y}{1+\overline{\gamma}_c}\right)\right], \quad y \geq 0, \tag{16.43}$$

$$f_{Y_2}(y|H_0) = \frac{1}{P(H_0)} \cdot \frac{M-1}{(L-1)!} y^{L-1} \Psi(y) \left[1 - \Psi(y)\right]^{M-3}$$
$$\cdot \left\{ \frac{1}{(1+\overline{\gamma}_c)^L} \exp\left(-\frac{y}{1+\overline{\gamma}_c}\right) \left[1 - \Psi(y)\right] + (M-2) \exp(-y) \left[1 - \Psi\left(\frac{y}{1+\overline{\gamma}_c}\right)\right] \right\},$$
$$y \geq 0, \tag{16.44}$$

where $P(H_0)$ is the *a-priori* probability of the hypothesis H_0 for EGC, which is given by Equation 16.33 with $P(H_1)$ described by Equation 16.36.

SC: the PDFs of Y_1 and Y_2 conditioned on the hypothesis of H_1 are formulated as:

$$f_{Y_1}(y|H_1) = \frac{1}{P(H_1)} \cdot \frac{L}{1+\overline{\gamma}_c} \exp\left(-\frac{y}{1+\overline{\gamma}_c}\right) \left[1 - \exp\left(-\frac{y}{1+\overline{\gamma}_c}\right)\right]^{L-1} \left[1 - \Omega(y)\right]^{M-1},$$
$$y \geq 0, \tag{16.45}$$

$$f_{Y_2}(y|H_1) = \frac{1}{P(H_1)} \cdot L(M-1) \exp(-y) \left[1 - \exp(-y)\right]^{L-1} \Omega\left(\frac{y}{1+\overline{\gamma}_c}\right) \left[1 - \Omega(y)\right]^{M-2},$$
$$y \geq 0, \tag{16.46}$$

where $P(H_1)$ is the *a-priori* probability of the hypothesis of H_1 for SC, which is given by Equation 16.37, and the short-hand $\Omega(x)$ was defined as:

$$\begin{aligned} \Omega(x) &= \sum_{k=1}^{L} (-1)^{k+1} \binom{L}{k} \exp(-kx) \\ &= 1 - \left[1 - \exp(-x)\right]^L . \end{aligned} \tag{16.47}$$

SC: the PDFs of Y_1 and Y_2 conditioned on the hypothesis of H_0 can be shown to obey:

$$\begin{aligned} f_{Y_1}(y|H_0) = \frac{1}{P(H_0)} &\cdot L(M-1) \exp(-y) \left[1 - \exp(-y)\right]^{L-1} \\ &\cdot \left[1 - \Omega\left(\frac{y}{1+\overline{\gamma}_c}\right)\right] \left[1 - \Omega(y)\right]^{M-2}, \ y \geq 0, \end{aligned} \tag{16.48}$$

$$\begin{aligned} f_{Y_2}(y|H_0) = \frac{1}{P(H_0)} &\cdot L(M-1) \Omega(y) \left[1 - \Omega(y)\right]^{M-3} \\ &\cdot \left\{ \frac{1}{1+\overline{\gamma}_c} \exp\left(-\frac{y}{1+\overline{\gamma}_c}\right) \left[1 - \exp\left(-\frac{y}{1+\overline{\gamma}_c}\right)\right]^{L-1} \left[1 - \Omega(y)\right] \right. \\ &\left. + (M-2) \exp(-x) [1 - \exp(-x)]^{L-1} \left[1 - \Omega\left(\frac{y}{1+\overline{\gamma}_c}\right)\right] \right\}, \ y \geq 0, \end{aligned} \tag{16.49}$$

where $P(H_0)$ is the *a-priori* probability of the hypothesis H_0 for SC, which is given by Equation 16.33 with $P(H_1)$ formulated in Equation 16.37.

It can be shown that the PDF of λ defined in Equation 16.39 can be expressed using Equation 16.50:

$$f_\lambda(y|H_\vartheta) = \frac{1}{P(Y_2 < Y_1|H_\vartheta)} \int_0^\infty y_2 f_{Y_1}(y_2 y|H_\vartheta) f_{Y_2}(y_2|H_\vartheta) dy_2, \; y \geq 1, \tag{16.50}$$

as was shown in the Appendix of [391] for an EGC scheme, where $H_\vartheta \in \{H_1, H_0\}$, $P(Y_2 < Y_1|H_\vartheta)$ represents the probability of $Y_2 < Y_1$ conditioned on the hypothesis of H_ϑ. Substituting the corresponding PDFs into Equation 16.50, the PDFs of λ under the hypotheses of H_1 and H_0 can be expressed as:

$$f_\lambda(y|H_1) = C_{H_1} \times g_\lambda(y|H_1), \; y \geq 1, \tag{16.51}$$

$$f_\lambda(y|H_0) = C_{H_0} \times g_\lambda(y|H_0), \; y \geq 1, \tag{16.52}$$

where $g_\lambda(y|H_1)$ and $g_\lambda(y|H_0)$ can be shown to obey:

EGC:

$$\begin{aligned} g_\lambda(y|H_1) = y^{L-1} \int_0^\infty & x^{2L-1} \exp\left(-x - \frac{xy}{1+\overline{\gamma}_c}\right) \\ & \cdot [1 - \Psi(xy)]^{M-1} [1 - \Psi(x)]^{M-2} \Psi\left(\frac{x}{1+\overline{\gamma}_c}\right) dx, \end{aligned} \tag{16.53}$$

$$\begin{aligned} g_\lambda(y|H_0) = y^{L-1} \int_0^\infty & x^{2L-1} \exp(-xy) [1 - \Psi(xy)]^{M-2} \\ & \cdot \left[1 - \Psi\left(\frac{xy}{1+\overline{\gamma}_c}\right)\right] \Psi(x)[1 - \Psi(x)]^{M-3} \\ & \cdot \left\{ \frac{1}{(1+\overline{\gamma}_c)^L} \exp\left(-\frac{x}{1+\overline{\gamma}_c}\right) [1 - \Psi(x)] \right. \\ & \left. + (M-2)\exp(-x) \left[1 - \Psi\left(\frac{x}{1+\overline{\gamma}_c}\right)\right] \right\} dx. \end{aligned} \tag{16.54}$$

SC:

$$\begin{aligned} g_\lambda(y|H_1) = & \int_0^\infty \left[\left(1 - \exp\left(-\frac{xy}{1+\overline{\gamma}_c}\right)\right)(1 - \exp(-x))\right]^{L-1} \\ & \cdot [1 - \Omega(xy)]^{M-1} \Omega\left(\frac{x}{1+\overline{\gamma}_c}\right) [1 - \Omega(x)]^{M-2} x \exp\left(-x - \frac{xy}{1+\overline{\gamma}_c}\right) dx, \end{aligned} \tag{16.55}$$

$$g_\lambda(y|H_0) = \int_0^\infty [1-\exp(-xy)]^{L-1}\left[1-\Omega\left(\frac{xy}{1+\overline{\gamma}_c}\right)\right][1-\Omega(xy)]^{M-2}\,\Omega(x)$$
$$\cdot[1-\Omega(x)]^{M-3}\left\{\frac{1}{1+\overline{\gamma}_c}\exp\left(-\frac{x}{1+\overline{\gamma}_c}\right)\left[1-\exp\left(-\frac{x}{1+\overline{\gamma}_c}\right)\right]^{L-1}[1-\Omega(x)]\right.$$
$$\left.+(M-2)\exp(-x)[1-\exp(-x)]^{L-1}\left[1-\Omega\left(\frac{x}{1+\overline{\gamma}_c}\right)\right]\right\}x\exp(-xy)\,dx, \tag{16.56}$$

Upon integrating both sides of Equation 16.51 and Equation 16.52 from one to infinity, we obtain:

$$C_{H_1} = \frac{1}{\int_1^\infty g_\lambda(y|H_1)dy}, \tag{16.57}$$
$$C_{H_0} = \frac{1}{\int_1^\infty g_\lambda(y|H_0)dy}. \tag{16.58}$$

In order to erase the low-reliability RS coded symbols, we assume that λ_T is the threshold, which activates an erasure insertion, if $\lambda \leq \lambda_T$. Consequently, the correct symbol probability, P_c, and the random symbol error probability, P_t, after erasure insertion can be formulated as:

$$\begin{aligned} P_c &= P(H_1)\cdot\int_{\lambda_T}^\infty f_\lambda(y|H_1)dy \\ &= P(H_1) - P(H_1)\cdot\int_1^{\lambda_T} f_\lambda(y|H_1)dy, \end{aligned} \tag{16.59}$$
$$\begin{aligned} P_t &= P(H_0)\cdot\int_{\lambda_T}^\infty f_\lambda(y|H_0)dy \\ &= P(H_0) - P(H_0)\cdot\int_1^{\lambda_T} f_\lambda(y|H_0)dy, \end{aligned} \tag{16.60}$$

and the symbol erasure probability P_e can be expressed as:

$$P_e = 1 - P_c - P_t. \tag{16.61}$$

Again, if we assume that the symbol errors and symbol erasures within a codeword are independent due to the ideal interleaving, then the codeword decoding error probability after 'errors-and-erasures' decoding can be expressed in the form of [99]:

$$P_W = \sum_{i=0}^{\mathcal{N}}\sum_{j=j_0(i)}^{\mathcal{N}-i}\binom{\mathcal{N}}{i}\binom{\mathcal{N}-i}{j}P_t^iP_e^j(1-P_t-P_e)^{\mathcal{N}-i-j}, \tag{16.62}$$

where $j_0(i) = \max\{0, \mathcal{N}-\mathcal{K}+1-2i\}$, while P_t, and P_e represent the random symbol error probability and symbol erasure probability, respectively, before decoding, which are given by Equations 16.60 and 16.61.

Above we have analyzed the performance of the uncoded and RS-coded DS-CDMA system using non-coherent M-ary orthogonal modulation, when diversity reception with EGC and SC were considered. Let us now evaluate the system's performance numerically.

16.6 Numerical Results

In this section the average bit or codeword decoding error probability of the DS-CDMA system using non-coherent M-ary orthogonal modulation - with or without RS coding - is evaluated as a function of the average signal-to-noise ratio (SNR) per bit, $\overline{\gamma}_b$, that of the pre-set threshold, λ_T and (or) versus the number of simultaneous users in the system. The average SNR per bit is obtained by $\overline{\gamma}_b = L_p\overline{\gamma}_c/b$, where $b = \log_2 M$.

Figure 16.3 shows the BER performance of the EGC and SC schemes for $L = 1, 2, 3, 4, 5$ diversity branches upon evaluating Equations 16.36, 16.33 and 16.34 for EGC, while Equations 16.37, 16.33 and 16.34 for SC. We assumed that an uncoded DS-CDMA system using 64-ary orthogonal modulation was employed, the receiver was capable of combining all the multipath signals, i.e. we had $L = L_p$, the number of simultaneous users was $K = 10$ and the ratio of bit duration to chip duration was $N = 128$. As expected, both the EGC and SC schemes provide BER improvements for moderate to high SNRs per bit, when the number of combined diversity paths, L, increases. Furthermore, the results show that the EGC scheme has a lower BER than the SC scheme for a given number of combined paths, L and for a given SNR per bit, $\overline{\gamma}_b$. This is because EGC is the optimal diversity combining scheme for a non-coherent demodulation technique. However, BER floors are observed for both the EGC and SC schemes. This implies that in DS-CDMA systems, due to the multiple-access interference, the BER performance cannot be improved simply by increasing the transmission power alone. Multi-user detection [106] constitutes a powerful technique mitigating the multiple-access interference.

As an example, Figure 16.4 shows the codeword decoding error probability of Equation 16.62 over Rayleigh fading channels for both the EGC and SC schemes, when employing the RS (32,20) code over the Galois field GF (32) = GF(2^5) corresponding to 5-bit symbols using 'errors-and-erasures' decoding. In these figures the codeword decoding error probabilities were computed for different values of SNR per bit and for different thresholds, in order to find the optimum thresholds for the RTT-based erasure insertion scheme. From the results we observe that for a constant SNR per bit, $\overline{\gamma}_b$, there exists an optimum threshold, for which the 'errors-and-erasures' decoding achieves the minimum codeword decoding error probability. Hence, an inappropriate threshold leads to higher codeword decoding error probability, than the minimum seen in the figure. Observe furthermore that for both the EGC and SC schemes the optimum threshold assumes values around 1.5 to 2.0, even though the SNR per bit changes over a large dynamic range from about 6 to 15dB.

Equivalently, in Figure 16.5 the codeword decoding error probability of Equation 16.62 was evaluated for different values of the threshold λ_T and for different number of simultaneous users, in order to find the optimum thresholds for the RTT-based erasure insertion scheme versus the number of users and the SNR per bit. Similarly to our previous results in Figure 16.4, we observe that for a constant number of simultaneous users, K, there exists an optimum threshold, for which the 'errors-and-erasures' decoding achieves the minimum

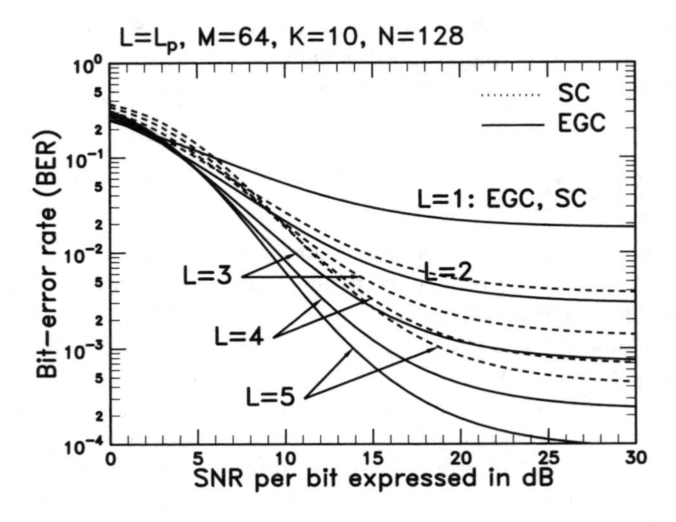

Figure 16.3: EGC, SC: BER versus average SNR per bit for DS-CDMA systems using non-coherent 64-ary modulation without FEC, when the number of simultaneous users is $K = 10$, the ratio of bit duration to chip duration is $N = 128$, the receiver can combine all the multipath signals, i.e., $L = L_p$ and $L = 1, 2, 3, 4, 5$. The results were evaluated from Equations 16.36, 16.33 and 16.34 for EGC, while from Equations 16.37, 16.33 and 16.34 for SC.

codeword decoding error probability and an inappropriate threshold leads to higher codeword decoding error probability than the minimum seen in the figure. Moreover, we observe again that for both the EGC and SC schemes the optimum threshold assumes values around 1.5 to 2.0, when the number of simultaneous users changes over a large dynamic range from about 20 to 100. However, due to the multiple-access interference, the codeword decoding error probability for any given value of threshold increases dramatically, when increasing the number of simultaneous users, K.

In summary, from Figure 16.4 and Figure 16.5 we gain an explicit insight into the characteristics of Viterbi's RTT over the dispersive Rayleigh fading channels studied, suggesting that the optimum threshold was typically around 1.5 to 2. Since the optimum threshold can be derived numerically for a given number of simultaneous users, K, and for a given average SNR per bit, $\overline{\gamma}_b$, in our further investigations we have assumed that the optimum threshold was employed, whenever 'errors-and-erasures' decoding was used. However, the optimum threshold depends on the SNR per bit and slightly increases, when increasing the SNR per bit value.

In Figure 16.6 we evaluated the codeword decoding error probability of a M-ary DS-CDMA system employing either 'error-correction-only' or 'errors-and-erasures' decoding. We assume that there were $L_p = 5$ resolvable paths, and $L = 1,\ 2,\ 3,\ 4$ or 5 paths were actually combined in the receiver using a EGC (Figure 16.6(a)) or SC (Figure 16.6(b)) scheme.

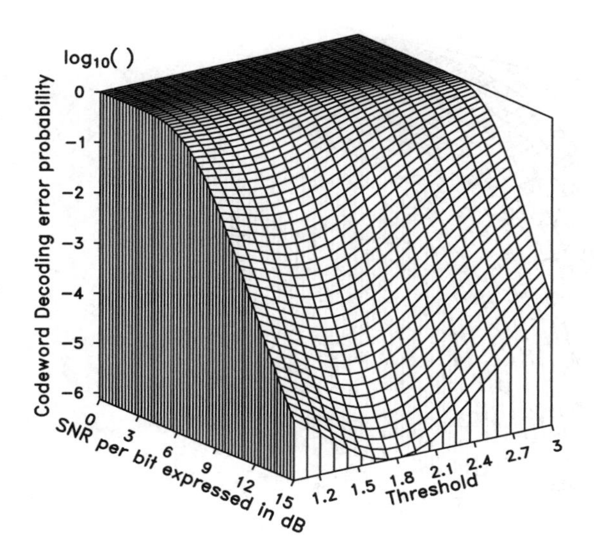

(a) EGC: M=32, $L = L_p = 2$, $K = 10$, $N = 128$, RS(32,20)

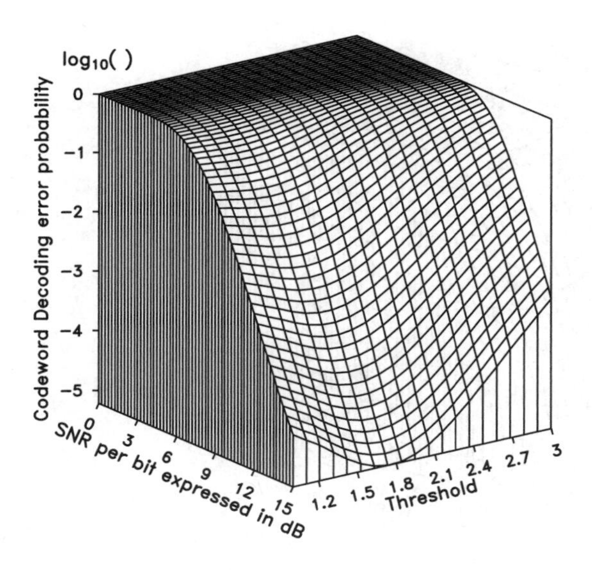

(b) SC: $M = 32$, $L = L_p = 2$, $K = 10$, $N = 128$, RS(32,20)

Figure 16.4: Codeword decoding error probability versus the average SNR per bit, $\overline{\gamma}_b$ and the threshold, λ_T for the RS (32,20) FEC code using 'errors-and-erasures' decoding over Rayleigh fading channels, evaluated from Equations 16.51 – 16.62.

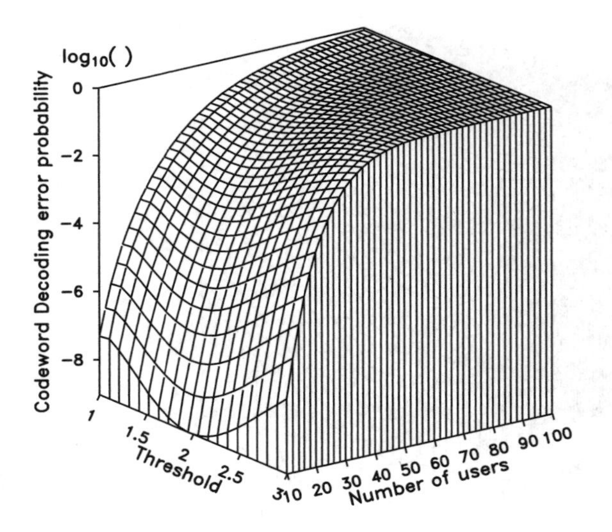

(a) **EGC**: M=32, $L = L_p = 2$, $\overline{\gamma}_b = 20dB$, $N = 128$, RS(32,20)

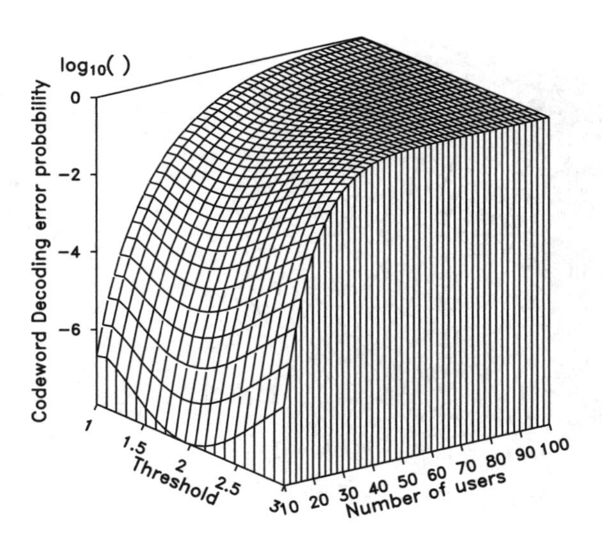

(b) **SC**: $M = 32$, $L = L_p = 2$, $\overline{\gamma}_b = 20$, $N = 128$, RS(32,20)

Figure 16.5: Codeword decoding error probability versus the number of simultaneous users, K and the threshold, λ_T for the RS (32,20) FEC code using 'errors-and-erasures' decoding over Rayleigh fading channels. The results were evaluated from Equations 16.51 – 16.62.

Again, we also assumed that an optimal threshold, λ_T, was employed for any given value of the average received SNR per bit. The remaining parameters were specified in the figures. The results show that under dispersive multipath Rayleigh fading conditions, while assuming a constant average SNR per bit and a constant number of diversity components combined, the decoding algorithms using 'errors-and-erasures' decoding based on the RTT erasure insertion scheme outperform 'error-correction-only' decoding. As shown in Figure 16.6(a) for EGC, for $L = 3, 4, 5$ and at the error probability of 10^{-3}, the 'errors-and-erasures' decoding scheme can achieve a gain of about 1.6dB, 1dB and 0.8dB, respectively, over the 'error-correction-only' decoding scheme. Similarly, for $L = 4, 5$ and at the error probability of 10^{-6}, the 'errors-and-erasures' decoding scheme can achieve a gain of about 1.8dB and 1.2dB, respectively, over the 'error-correction-only' decoding scheme. Similar results can also be observed in Figure 16.6(b) for SC.

Similarly, Figure 16.7 shows the codeword decoding error probability of the RS (32,20) code as a function of the number of simultaneous users, K, for parameters of $\overline{\gamma}_b = 20dB$, $N = 128$, $L = L_p = 1, 2, 3$ over dispersive multipath Rayleigh fading channels. The codeword decoding error probability of the 'error-correction-only' and 'errors-and-erasures' decoding scheme was compared for the EGC and SC considered. We assumed that the receiver used the optimum threshold for any number of users. The results show that for a constant number of users, for a constant number of diversity components combined and upon invoking the optimum threshold, the codeword decoding error probability of the system employing 'errors-and-erasures' decoding based on the RTT erasure insertion scheme was lower than that employing 'error-correction-only' decoding. At the codeword decoding error probability of 10^{-6}, the EGC-assisted 'errors-and-erasures' scheme in Figure 16.7(a) can support 7 or 6 more simultaneous users for $L = 2$ or 3, respectively, than the 'error-correction-only' scheme. By contrast, the SC-aided 'errors-and-erasures' scheme in Figure 16.7(b) can support 6 or 4 more simultaneous users for $L = 2$ or 3, respectively, than the 'error-correction-only' scheme.

Finally, in Figure 16.8 we evaluated the influence of the code rate, R_c, on the codeword decoding error probability of the DS-CDMA systems using both 'error-correction-only' decoding and 'errors-and-erasures' decoding, for both the EGC and the SC combining schemes investigated. The codeword decoding error probability was evaluated as a function of the number of simultaneous users for the code rates of $R_c = 7/8,\ 3/4,\ 5/8$ and $1/2$. We assumed that the 64-symbol-long RS code family of RS (64,$\mathcal{K}$) over the Galois field GF (64) = GF (2^6) using 'error-correction-only' and 'errors-and-erasures' employing RTT was invoked. As expected, the results imply that for all of the code rates considered, the decoding scheme using 'errors-and-erasures' employing RTT outperformed the arrangement using 'error-correction-only' decoding. For a constant codeword decoding error probability and for a given code rate, the DS-CDMA systems employing RS codes using 'errors-and-erasures' decoding can support 4 to 8 more simultaneous users at the codeword decoding error probability of 10^{-6}, than those employing 'error-correction-only' decoding.

16.7 Chapter Summary and Conclusion

In summary, in this chapter the performance of RS-coded DS-CDMA systems using 'errors-and-erasures' decoding has been investigated and compared to that using 'error-correction-

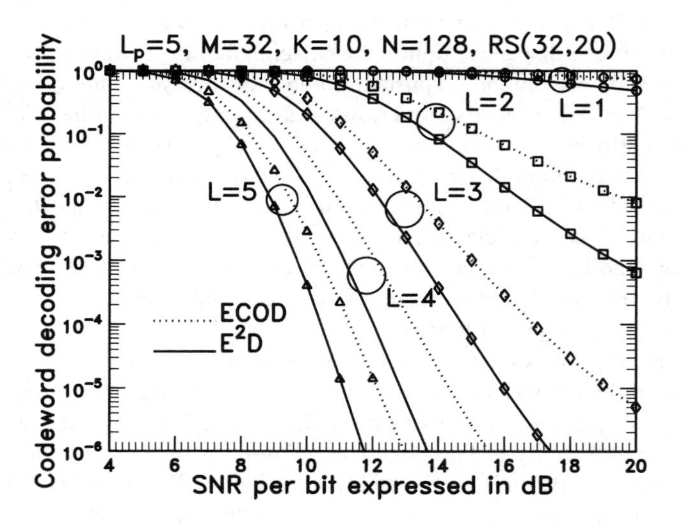

(a) EGC

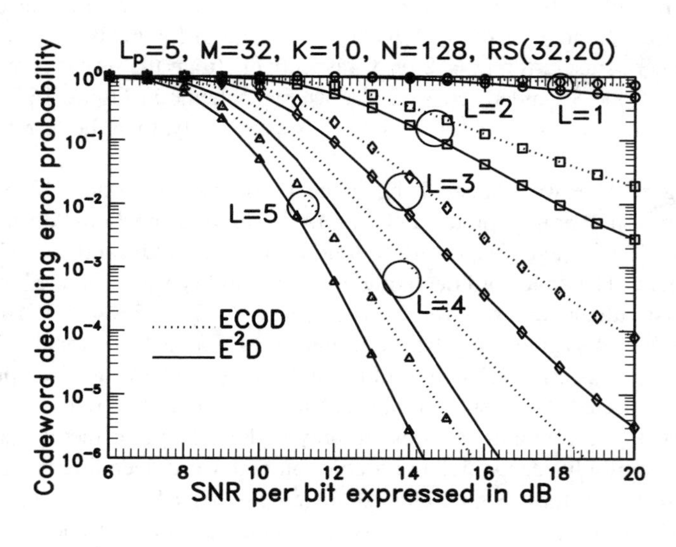

(b) SC

Figure 16.6: Codeword decoding error probability versus the average SNR per bit, $\overline{\gamma}_b$ for the RS (32,20) FEC code using 'error-correction-only' decoding (ECOD) and 'errors-and-erasures' decoding (E^{2}D) with parameters of $M = \mathcal{N} = 32$, $\mathcal{K} = 20$, $K = 10$, $N = 128$, $L_p = 5$ resolvable multipath components and a receiver diversity combining capability of $L = 1, 2, 3, 4, 5$ over dispersive Rayleigh fading channels.

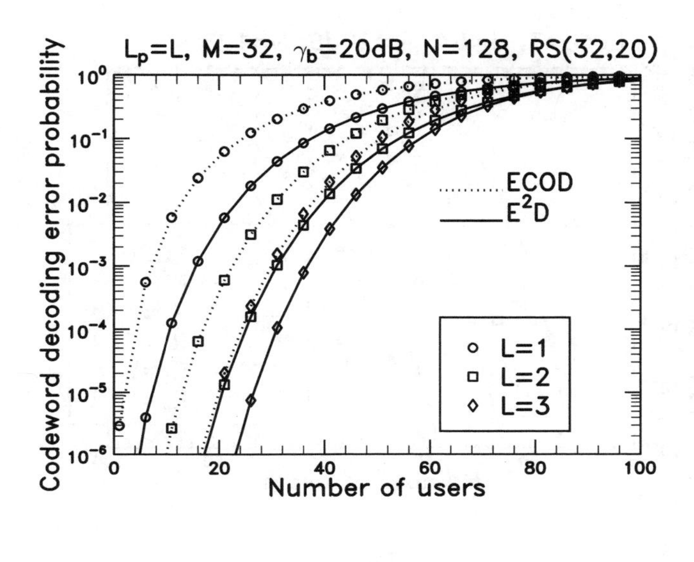

(a) EGC

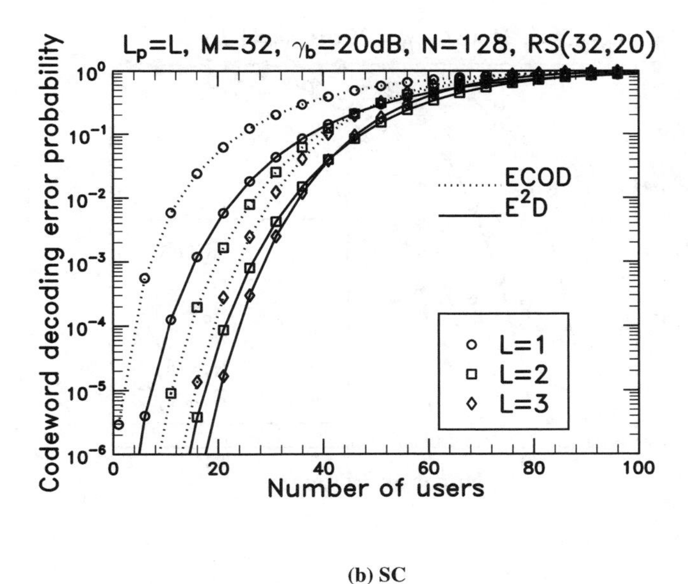

(b) SC

Figure 16.7: Codeword decoding error probability versus the number of simultaneous users, K for the RS (32,20) FEC code using 'error-correction-only' decoding (ECOD) and 'errors-and-erasures' decoding (E^{2}D) with parameters of $M = \mathcal{N} = 32$, $\mathcal{K} = 20$, $\overline{\gamma}_b = 20dB$, $N = 128$, $L = L_p = 1, 2, 3$ over dispersive Rayleigh fading channels.

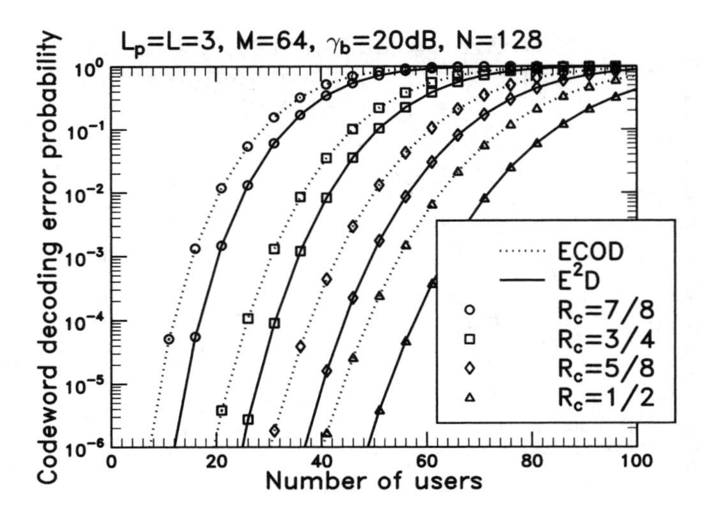

(a) EGC

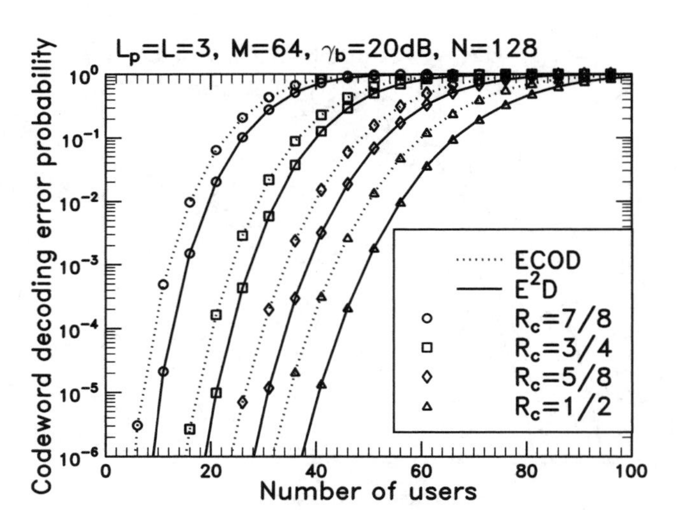

(b) SC

Figure 16.8: Codeword decoding error probability versus the number of simultaneous users, K for various RS FEC codes using 'error-correction-only' decoding (ECOD) and 'errors-and-erasures' decoding (E^{2}D) at different code rates with parameters of $M = \mathcal{N} = 64$, $\mathcal{K} = 32, 40, 48$ and 56, i.e., $R_c = 1/2, 5/8, 3/4$ and 7/8, $\overline{\gamma}_b = 20dB$, $N = 128$, $L = L_p = 3$ over dispersive Rayleigh fading channels.

only' decoding over dispersive multipath Rayleigh fading channels, when non-coherent M-ary orthogonal modulation and diversity reception using EGC or SC schemes were considered. Viterbi's RTT technique has been studied, in order to quantify the reliability of the received RS coded symbols. The symbols having a low-confidence were erased before RS 'errors-and-erasures' decoding. The PDFs associated with the RTT under the hypotheses of correct detection and erroneous detection of the M-ary signals have been derived and the optimum thresholds have been determined. Our numerical results showed that by using 'errors-and-erasures' decoding associated with the RTT, RS codes of a given code rate can achieve a higher coding gain than by using 'error-correction-only' decoding without erasure information. DS-CDMA systems using the proposed 'errors-and-erasures' decoding could also support more simultaneous users than without erasure information. Moreover, the numerical results showed that the optimum threshold for the RTT over multipath Rayleigh fading channels was around [1.5, 2] for the practical range of average SNR per bit values and for a given number of simultaneous users.

Chapter 17

Residue Number System Arithmetic[1]

17.1 Introduction

The theory of residue number systems (RNS) was discovered by the Chinese, dating back as early as the third century. However, it was not widely accepted as an alternative arithmetic approach in digital systems, since complex computations are encountered in the conversion from the so-called residue to decimal or decimal to residue number conversion process. Nevertheless, commencing with the rapid evolution of digital computer and microelectronics technology in the 1950s, RNS arithmetics have attracted considerable attention in the field of designing high-speed special-purpose digital hardware that is suitable for large-scale integration (VLSI) [392–394]. Digital systems that invoke RNS arithmetic units may play an important role in future ultra-speed dedicated real-time systems that support pure parallel processing of integer-valued data.

RNS-based arithmetics exhibit a modular structure that leads naturally to parallelism in digital hardware. The RNS has two inherent features that are attractive in comparison to conventional weighted number systems, such as, for example, the binary weighted number system representation. These two features are [392, 394]: (1) the carry-free arithmetic and (2) the lack of ordered significance among the residue digits. The first property implies that the operations related to the different residue digits are mutually independent and hence the errors occurring during addition, subtraction and multiplication operations, or due to the noise induced by transmission and processing, remain confined to their original residue digits [392, 394]. In other words, these errors do not propagate and hence do not contaminate other residue digits due to the absence of a carry forward. The above second property of the RNS

[1] *Single- and Multi-Carrier CDMA Multi-User Detection, Space-Time Spreading, Synchronization, Networking and Standards.* L.Hanzo, L-L.Yang, E-L.Kuan and K.Yen,

arithmetic implies that redundant residue digits can be discarded without affecting the result, provided that a sufficiently high dynamic range is retained by the resultant reduced RNS system in order to unambiguously describe the non-redundant information symbol. This issue will be developed during our further discourse.

As is well known in VLSI design, usually so-called systolic architectures [392] are invoked to divide a processing task into several simple tasks performed by small, (ideally) identical, easily designed processors. Each processor communicates only with its nearest neighbour, simplifying the interconnection design and test, while reducing signal delays and hence increasing the processing speed. Due to its carry-free property, the RNS arithmetic further simplifies the computations by decomposing a problem into a set of parallel, independent residue computations.

The properties of the RNS arithmetic suggest that a redundant residue number system (RRNS) can be used for self-checking, error-detection and error-correction in digital processors. The RRNS technique provides a useful approach to the design of general-purpose systems, capable of sensing and rectifying their own processing and transmission errors. For example, if a digital filter is implemented using the RRNS having sufficient redundancy, then errors in the processed signals can be detected and corrected by the RRNS-based decoding. Furthermore, the RRNS approach [392] is the only one, where it is possible to use the very same arithmetic module in the very same way for the generation of both the information part and the parity part of an RRNS codeword. Moreover, due to the inherent properties of the RNS, residue arithmetic offers a variety of new approaches to the realization of digital signal processing algorithms, such as digital modulation and demodulation and the fault-tolerant design of arithmetic units. It also offers new approaches to the design of error-detection and error-correction codes.

17.2 Residue Number System Transform and Its Inverse

This section introduces a basic mathematical model for the conversion of operands from the weighted number system to the RNS, or conversely, from the RNS to the weighted number system. For convenience, we define the residue number system transform (RNST) as the algorithm that transforms any conventional weighted number system, such as, for example, the natural binary coded decimal (NBCD) number system to the RNS [392]. The inverse RNST (IRNST) is defined as the algorithm which transforms the RNS to the weighted number system, an issue to be elaborated on below.

Let X be a non-negative integer number in the range of $0 \leq X < \varrho^n$, where ϱ is a base and X is represented in the weighted number system, as $X = \{b_{n-1}\, b_{n-2}\, \ldots\, b_2\, b_1\, b_0\} = \sum_{j=0}^{n-1} b_j \varrho^j$, where $b_j \in \{0, 1, \ldots, \varrho - 1\}$. Specifically, a weighted binary representation is considered, when $\varrho = 2$.

Let $\{m_1, m_2, \ldots, m_u\}$, the set of so-called moduli involved in the RNS, be a set of pairwise relative prime numbers, where, no two moduli have a common integer divisor greater than 1. If $M = \prod_{i=1}^{u} m_i \geq \varrho^n$, then the RNST is realized by uniquely representing X - which was defined in the range of $0 \leq X < \varrho^n$ - as a u-tuple residue digit sequence $(r_1, r_2, \ldots, r_u)$, where $r_i = X \pmod{m_i} = X - \lfloor \frac{X}{m_i} \rfloor m_i$ for $i = 1, 2, \ldots, u$, $\lfloor z \rfloor$ denotes the largest integer not exceeding z, and $0 \leq r_i \leq m_i - 1$. Furthermore, $[0, M-1]$ is defined as the dynamic range of the RNS and any integer X in this range is uniquely represented by

Message X	RNS		
	r_1	r_2	r_3
$X_1 = 3$	3	3	3
$X_2 = 21$	1	1	0
$X_3 = 36$	0	1	1
$X_4 = 39$	3	4	4
$X_5 = 63$	3	3	0
$X_6 = 100$	0	0	2
$X_7 = 139$	3	4	6

Table 17.1: RNS representation of the integer messages X with the aid of the moduli $m_1 = 4,\ m_2 = 5$ and $m_3 = 7$, where $r_i = X\ (mod\ m_i)$.

a u-tuple residue sequence, which is expressed as:

$$X \Leftrightarrow (r_1, r_2, \dots, r_u),\ \ 0 \leq X < M,$$
$$r_i = X\ (\text{mod}\ m_i),\ \ i = 1, 2, \dots, u. \tag{17.1}$$

Example 1 *In Table 17.1 we summarized a few examples for the moduli values of* $m_1 = 4,\ m_2 = 5$ *and* $m_3 = 7$*, where the system's dynamic range is* $[0,\ 139]$*, since* $M = (4{*}5{*}7) = 140$*. Consequently, any integer in this range can be unambiguously represented by the help of the RNS representation, as shown in the table.*

Assuming that the integers X_1 and X_2 have RNS representations of $X_1 = (r_{11}, r_{12}, \dots, r_{1u})$ and $X_2 = (r_{21}, r_{22}, \dots, r_{2u})$, respectively, then $X_1 \bullet X_2$, where $\bullet$ denotes addition, subtraction or multiplication, yields another unique residue sequence X_3, provided that X_3 remains within the dynamic range of $[0,\ M)$. The arithmetic operations over the RNS can be carried out on a residue-by-residue basis, which can be expressed as:

$$X_1 \bullet X_2 \Leftrightarrow \left[(r_{1i} \bullet r_{2i})\ (\text{mod}\ m_i)\right]_{i=1}^{u}, \tag{17.2}$$

where on the left the operation $\bullet$ represents ordinary addition, subtraction or multiplication of two integers, and on the right it represents the same operations performed on the basis of the appropriate residue digits r_{1i} and r_{2i}, with respect to their corresponding modulus m_i. This allows us to convert binary arithmetic operations carried out with large integers to residue arithmetic operations involving smaller residue digits, where the operations can be in parallel executed and there is no carry forward between the residue digits. The lack of carry forward prevents the propagation of errors between residue digits.

Example 2 *Using the moduli values of* $m_1 = 4,\ m_2 = 5,\ m_3 = 7$ *and Table 17.1, a simple addition example is provided in Table 17.2. The corresponding subtraction and multiplication examples are given in Tables 17.3 and 17.4.*

Message X	RNS		
	r_1	r_2	r_3
$X_4 = 39$	3	4	4
$+ X_6 = 100$	0	0	2
$X_7 = 139$	3	4	6

Table 17.2: RNS component-wise addition example using the operands of Table 17.1.

Message X	RNS		
	r_1	r_2	r_3
$X_4 = 39$	3	4	4
$- X_3 = 36$	0	1	1
$X_1 = 3$	3	3	3

Table 17.3: RNS component-wise subtraction example using the operands of Table 17.1.

In contrast to the RNST, the IRNST is defined as the algorithm that transforms the operands from the RNS domain to the conventional weighted number system representation. A well-known implementation of the IRNST is the so-called Chinese remainder theorem (CRT) [395–397]. According to the CRT, for any given u-tuple $(r_1, r_2, \ldots, r_u)$, where $0 \leq r_i < m_i$ for $i = 1, 2, \ldots, u$ there exists one and only one integer X such that $0 \leq X < M$ and $r_i = X \pmod{m_i}$. It can be shown that the numerical value of X can be computed by [397]:

$$X = \sum_{i=1}^{u} r_i T_i M_i \pmod{M}, \tag{17.3}$$

where $M_i = M/m_i$ and the integers T_i are computed *a priori* by solving the so-called congruences $T_i M_i = 1 \pmod{m_i}$. The coefficients T_i are also often referred to as the multiplicative inverses of M_i. According to $T_i M_i = 1 \pmod{m_i}$ it can be readily verified that T_i is the minimum integer number obeying the following condition:

$$T_i M_i - 1 = d_i m_i, \tag{17.4}$$

Message X	RNS		
	r_1	r_2	r_3
$X_1 = 3$	3	3	3
$\times X_2 = 21$	1	1	0
$X_5 = 63$	3	3	0

Table 17.4: RNS component-wise multiplication example using the operands of Table 17.1.

where d_i is also the minimum integer number obeying Equeation 17.4, which takes the integer value of $d_i = \lfloor T_i M_i / m_i \rfloor$. By substituting this into Equation17.4, it can be simplified to:

$$\frac{T_i M_i}{m_i} - \lfloor \frac{T_i M_i}{m_i} \rfloor - \frac{1}{m_i} = 0\,. \tag{17.5}$$

Consequently, after obtaining M_i, T_i can be readily found by solving Equation17.5 for it.

Hence, given the residue digits $r_1, r_2, \ldots, r_u$, M_i and hence also the multiplicative inverses T_i, for $i = 1, 2, \ldots, u$ as well as the equivalent decimal integer can be calculated using Equation 17.3.

Example 3 *Given the residue digits of* $r_1 = 3$*,* $r_2 = 4$ *and* $r_3 = 6$ *for the decimal value of* $X_7 = 139$ *in Table 17.1 let us attempt to recover this decimal value by means of the CRT. Let us commence by calculating* M_i*,* $i = 1, 2, 3$ *and the corresponding multiplicative inverses:*

$$\begin{aligned} M_1 &= \frac{M}{m_1} = \frac{4 \times 5 \times 7}{4} = 5 \times 7 = 35 \\ M_2 &= \frac{M}{m_2} = \frac{4 \times 5 \times 7}{5} = 4 \times 7 = 28 \\ M_3 &= \frac{M}{m_3} = \frac{4 \times 5 \times 7}{7} = 4 \times 5 = 20. \end{aligned}$$

The corresponding multiplicative inverses can be found here by searching Equation 17.5 for the corresponding M_i *values, which are as follows:*

$$\begin{aligned} &\tfrac{T_1 \times 35}{4} - \lfloor \tfrac{T_1 \times 35}{4} \rfloor - \tfrac{1}{4} = 0 \longrightarrow T_1 = 3 \\ &\tfrac{T_1 \times 28}{5} - \lfloor \tfrac{T_1 \times 28}{5} \rfloor - \tfrac{1}{5} = 0 \longrightarrow T_2 = 2 \\ &\tfrac{T_1 \times 20}{7} - \lfloor \tfrac{T_1 \times 20}{7} \rfloor - \tfrac{1}{7} = 0 \longrightarrow T_3 = 6 \end{aligned}$$

Therefore, upon using Equation 17.3 we arrive at:

$$\begin{aligned} X &= [3 \times (3 \times 35) + 4 \times (2 \times 28) + 6 \times (6 \times 20)] \pmod{140} \\ &= [315 + 224 + 720] \pmod{140} \\ &= 1259 \pmod{140} \\ &= 139, \end{aligned}$$

as seen in Table 17.1.

The CRT is a classical algorithm for the implementation of the IRNST. However, the real-time implementation of the CRT is not practical, since it requires modular operations with respect to a large integer M. In order to avoid processing large integers, especially in the context of error control invoking the RRNS, a frequently used approach is the so-called base extension (BEX) operation in conjunction with the mixed radix conversion approach [398]. The mixed radix conversion is an operation, which can be invoked, in order to represent the integer X in the form:

$$X = a_1 + a_2 m_1 + a_3 m_1 m_2 + \cdots + a_l \prod_{i=1}^{l-1} m_i + \cdots + a_u \prod_{i=1}^{u-1} m_i, \qquad (17.6)$$

where $0 \leq a_l < m_l$ for $l = 1, 2, \ldots, u$ are - as implied by the associated terminology - the mixed radix digits, and we define $\prod_{i=1}^{0} m_i = 1$. For a given residue sequence $(r_1, r_2, \ldots, r_u)$ these mixed radix digits are computed by [398]:

$$\begin{aligned} a_1 &= r_1, \\ a_2 &= ((r_2 - a_1)m_1^{-1}) \pmod{m_2}, \\ a_3 &= (((r_3 - a_1)m_1^{-1} - a_2)m_2^{-1}) \pmod{m_3}, \\ &\cdots \\ a_u &= (((\cdots((r_u - a_1)m_1^{-1} - a_2)m_2^{-1} - \cdots) - a_{u-1})m_{u-1}^{-1}) \pmod{m_u}. \end{aligned} \qquad (17.7)$$

It can be shown with the aid of Equation 17.6 and Equation 17.7 that this algorithm is independent of the large integer M and all operations are related to relatively small operands on the order of the moduli values involved in the RNS. Let us now invoke the mixed radix conversion to recover the integer decimal value $X = 139$.

Example 4 *According to Equation 17.7 we have:*

$$\begin{aligned} a_1 &= r_1 = 3, \\ a_2 &= ((r_2 - a_1)m_1^{-1}) \; (mod\; m_2) \\ &= ((4-3)/4) \pmod 5 \Leftrightarrow 4a_2 = 1 \pmod 5 \;\rightarrow\; a_2 = 4, \\ a_3 &= (((r_3 - a_1)m_1^{-1} - a_2)m_2^{-1}) \; (mod\; m_3) \\ &= (((6-3)4^{-1} - 4)5^{-1}) \pmod 7 \Leftrightarrow 20a_3 = -13 \pmod 7 \;\rightarrow\; a_3 = 6. \end{aligned}$$

The integer X can be computed by using Equation 17.6, yielding:

$$\begin{aligned} X &= a_1 + a_2 m_1 + a_3 m_1 m_2 \\ &= 3 + 4 \times 4 + 6 \times 4 \times 5 = 139, \end{aligned}$$

where the associated integers operands are on the order of the moduli involved in the RNS.

We have discussed a number of techniques that can be invoked, in order to convert the operands from the weighted number system to the RNS, or from the RNS to the weighted number system, which we referred to as the RNST and IRNST, respectively. However, the benefits of processing our operands in the RNS become even more attractive, when the system is designed, also for self-checking as well as error-detection and/or error-correction. In such a RNS a certain number of redundant moduli have to be introduced, in order to achieve the required self-checking, error-detection and/or error-correction capability. The RNS involving redundant moduli is referred to as the redundant residue number system (RRNS) [392]. Let us now provide a rudimentary introduction to the RRNS.

17.3 Redundant Residue Number System

As we argued in the previous section, if a residue number system is designed not only for the representation of data, but also for the protection of data, usually we design the RNS using so-called redundant moduli, so that the system has the capability of self-checking, error-detection and error-correction [392]. In this case X is limited to the so-called information dynamic range of $[0, M = \prod_{i=1}^{v} m_i)$, where $v \leq u$, and $m_1, m_2, \ldots, m_v$ are referred to as the information moduli, while $m_{v+1}, m_{v+2}, \ldots, m_u$ are the so-called redundant moduli. We express the product of the redundant moduli as $M_R = \prod_{j=1}^{u-v} m_{v+j}$. The interval $[0, M)$ constitutes the legitimate range of the operand X, and the interval $[M, MM_R)$ is the illegitimate range. Any integer belonging to the legitimate range will be labelled as legitimate and those belonging to the illegitimate range as illegitimate, since it does not represent a number or operand, directly accruing from the non-redundant information-bearing residue digits.

An important property of the RRNS is that an integer represented by the residues of the RRNS can be recovered by any group of v number of moduli and their corresponding residue digits. In order to develop this property, let us consider an example.

Example 5 *Let us consider the moduli* $3, 4, 5, 7$*, where* 7 *is the redundant modulus. In this case the integer message* X *has to be constrained to the range* $[0, M = 3 \times 4 \times 5)$ = *[0,60). Upon considering an integer message of* $X = 21$*, the corresponding RNS values are* X = *(0,1,1,0). According to the above property, we can opt for any three out of the four residue digits and their related moduli in order to convert the RNS representation back to the decimal representation, when using the CRT. Let us retain the first three residue digits using their related moduli, as an example, yielding* $X_v = (0, 1, 1)$ *and* $M = 3 \times 4 \times 5 = 60$*. Following the steps of the CRT in Equation 17.3, we have:*

$$
\begin{aligned}
&M_1 = 4 \times 5 = 20,\ M_2 = 3 \times 5 = 15,\ M_3 = 3 \times 4 = 12 \\
&T_1 = 2,\ T_2 = 3,\ T_3 = 3 \\
&X = [0 \times (2 \times 20) + 1 \times (3 \times 15) + 1 \times (3 \times 12)] \pmod{60} \\
&\quad = 21,
\end{aligned}
$$

demonstrating that upon discarding the residue digit related to the redundant modulus 7, the original decimal message can be recovered correctly.

Similarly, let us retain the last three residue digits using their related moduli yielding

Message X	Non-redundant residue digits			Redundant residue digits			
	r_1	r_2	r_3	r_4	r_5	r_6	r_7
$X_0 = 0$	0	0	0	0	0	0	0
$X_1 = 1$	1	1	1	1	1	1	1
$X_2 = 2$	2	2	2	2	2	2	2
$X_3 = 5$	1	0	5	5	5	5	5
$X_4 = 10$	2	0	3	1	10	10	10
$X_5 = 20$	0	0	6	2	9	7	3
$X_6 = 50$	2	0	1	5	6	11	16
$X_7 = 100$	0	0	2	1	1	9	15

Table 17.5: RRNS codewords of some typical decimal integer messages X in the RRNS with moduli $m_1 = 4,\ m_2 = 5,\ m_3 = 7,\ m_4 = 9,\ m_5 = 11,\ m_6 = 13$ and $m_7 = 17$, where $r_i = X\ (mod\, m_i)$ and $M = 4 \times 5 \times 7 = 140$.

$X_v = (1, 1, 0)$ and $M = 4 \times 5 \times 7 = 140$. Following the steps of the CRT in Equation 17.3, we have:

$$\begin{aligned}
M_2 &= 5 \times 7 = 35,\ M_3 = 4 \times 7 = 28,\ M_4 = 4 \times 5 = 20 \\
T_2 &= 3,\ T_3 = 2,\ T_4 = 6 \\
X &= [1 \times (3 \times 35) + 1 \times (2 \times 28) + 0 \times (6 \times 20)]\ (\text{mod } 140) \\
&= 21,
\end{aligned}$$

which demonstrates that upon discarding the residue digit related to the redundant modulus 3, the original decimal message can also be recovered correctly.

The property that an RRNS-based integer can be recovered from any group of v out of u moduli and their corresponding residue digits constitutes the basis in the design of RRNS-based error-detection and error-correction schemes, an issue we will discuss in more depth in the next section. An RRNS having v number of information moduli and $u - v$ number of redundant moduli is denoted as an RRNS(u, v) code. The operation associated with generating the redundant residue digits of an integer operand X can also be viewed or interpreted, as an encoding operation generating the parity residue digits of a RRNS(u, v) code. The RRNS encoding operation can be implemented based on a group of codewords having typical integer values, by invoking exclusively addition operations. In order to augment this statement, below we provide an example showing the associated groups of codewords for the RRNS(7,3) code having information moduli of $m_1 = 4,\ m_2 = 5,\ m_3 = 7$ and redundant moduli of $m_4 = 9,\ m_5 = 11,\ m_6 = 13,\ m_7 = 17$. The legitimate range of this code is [0,139]. Table 17.5 portrays a range of typical integer messages, which are based on the integers '1, 2, 5, 10, 20, 50, 100' often used in monetary systems. It is well known that on the average any integer in the legitimate range of [0,139] can be expressed by the sum of a subset of the lowest possible number of typical integers above.

Message X	Non-redundant residue digits			Redundant residue digits			
	r_1	r_2	r_3	r_4	r_5	r_6	r_7
$X_0 = 0000000$	0	0	0	0	0	0	0
$X_1 = 0000001$	1	1	1	1	1	1	1
$X_2 = 0000010$	2	2	2	2	2	2	2
$X_3 = 0000100$	0	4	4	4	4	4	4
$X_4 = 0001000$	0	3	1	8	8	8	8
$X_5 = 0010000$	0	1	2	7	5	3	16
$X_6 = 0100000$	0	2	3	5	10	6	15
$X_7 = 1000000$	0	4	1	1	9	12	13

Table 17.6: RRNS codewords of some typical binary integer messages X in the RRNS with moduli $m_1 = 4$, $m_2 = 5$, $m_3 = 7$, $m_4 = 9$, $m_5 = 11$, $m_6 = 13$ and $m_7 = 17$, where $r_i = X \ (mod \ m_i)$ and $M = 4 \times 5 \times 7 = 140$.

Example 6 *In Table 17.5 and Table 17.6 we summarized a range of typical decimal and binary integer numbers and their corresponding RRNS codewords. By using these codewords from Table 17.5 and Table 17.6, any integer in the legitimate range can be readily encoded based on these codewords by simply using the addition operation. For example, for the encoding of the decimal integer* $X = 123$*, since* X *can be expressed as* $123 = 100+20+2+1$*, the codeword of* X *can simply be obtained from the codewords corresponding to* X_7*,* X_5*,* X_2 *and* X_1 *on the basis of modulo addition. It can be readily shown that the codeword of* $X = 123$ *is (3,3,4,6,2,6,4). Similarly, for a binary integer* $X = 1011011$*, the associated codeword is simply constituted by the corresponding modulo addition of* X_7, X_5, X_4, X_2 *and* X_1*, where the subscripts 7, 5, 4, 2, 1 are simply the indeces of the binary 1 positions in* X*. The associated codeword of* $X = 1011011$ *is hence (3,1,0,1,3,0,6).*

17.4 Error-Detection and Error-Correction in RRNS

In this section, we briefly discuss error-detection and error-correction in the context of the RRNS. Error-correction decoding of RRNS(u, v) codes has been studied, for example, in references [392, 394–405]. The concepts of error-correction in RRNS codes were developed by Mandelbaum [401]. Barsi and Maestrini addressed the problems of single residue digit error correction in [399]. Krishna, Lin and Sun have discussed the correction of single or double residue digit errors and simultaneously the detection of multiple residue digit errors in [395–397]. In [406] the work of Barsi and Maestrini [399] in the field of the error-correction techniques applicable to RRNS codes has been developed and unified into the so-called minimum-distance decoding algorithm. Hence, in this section the conditions and error-detection/correction capabilities of the RRNS(u, v) codes are summarized without proof. Readers interested in the intricacies of RRNS coding theory are referred to for example [395–397, 406]. Before outlining the general theory, let us first invoke two simple examples, in order to gain insight into the error-detection and error-correction mechanism in the RRNS.

Example 7 *Let us again consider the moduli* $3, 4, 5, 7$, *where 3, 4 and 5 are the information moduli and* 7 *is the redundant modulus. The information dynamic range is* $[0, M = 3 \times 4 \times 5)$ = *[0,60). Upon considering an integer message of* $X = 21$, *the corresponding residue values are X = (0,1,1,0). If there is an error in the RNS representation due to transmission or processing, for example,* r_3 *is changed from 1 to 3, i.e. the received RNS representation becomes* $(0, 1, \tilde{3}, 0)$. *Then, following the general approach of the CRT in Equation 17.3 and using the first three residue digits and their moduli, we obtain:*

$$\begin{aligned} X &= [20 \times (0 \times 2) + 15 \times (1 \times 3) + 12 \times (3 \times 3)] \bmod 60 \\ &= 33. \end{aligned}$$

However, due to $X = 33$ *(mod 7)* $= 5 \neq r_4 = 0$, *hence we can conclude that there were errors in the RNS representation. Hence, designing the RNS using one redundant modulus, the residue digit error of* r_3 *can be detected.*

Example 8 *Let us now invoke an additional redundant modulus, namely 11 in the above example, which results in a total of two redundant moduli, namely 7 and 11 in the RRNS. Let us also consider the integer message* $X = 21$, *now having corresponding residue digits of X = (0,1,1,0,10) and that* r_3 *is in error and it was changed from 1 to 3, i.e. the received RNS representation becomes* $(0, 1, \tilde{3}, 0, 10)$. *According to the CRT approach in our above examples, the integer X in the range [0, 60) can be recovered by invoking any three moduli and their corresponding residue digits, if no errors happened in the received RNS representation. Let us now consider all possible cases and attempt to recover the integer X represented by* $(0, 1, \tilde{3}, 0, 10)$, *upon retaining all possible combinations of three out of five residue digits,* *which results in:*

$$
\begin{aligned}
(r_1, r_2, r_3) &= (0, 1, 3) &\Leftrightarrow\quad & X_{123} = 33 \ (mod\, 60),\\
(r_1, r_2, r_4) &= (0, 1, 0) &\Leftrightarrow\quad & X_{124} = 21 \ (mod\, 84),\\
(r_1, r_2, r_5) &= (0, 1, 10) &\Leftrightarrow\quad & X_{125} = 21 \ (mod\, 132),\\
(r_1, r_3, r_4) &= (0, 3, 0) &\Leftrightarrow\quad & X_{134} = 63 \ (mod\, 105),\\
(r_1, r_3, r_5) &= (0, 3, 10) &\Leftrightarrow\quad & X_{135} = 153 \ (mod\, 165),\\
(r_1, r_4, r_5) &= (0, 0, 10) &\Leftrightarrow\quad & X_{145} = 21 \ (mod\, 231),\\
(r_2, r_3, r_4) &= (1, 3, 0) &\Leftrightarrow\quad & X_{234} = 133 \ (mod\, 140),\\
(r_2, r_3, r_5) &= (1, 3, 10) &\Leftrightarrow\quad & X_{235} = 153 \ (mod\, 220),\\
(r_2, r_4, r_5) &= (1, 0, 10) &\Leftrightarrow\quad & X_{245} = 21 \ (mod\, 308),\\
(r_3, r_4, r_5) &= (3, 0, 10) &\Leftrightarrow\quad & X_{345} = 98 \ (mod\, 385),
\end{aligned}
$$

where X_{ijk} *represents the recovered result by using moduli* m_i, m_j *and* m_k *as well as their corresponding residue digits* r_i, r_j *and* r_k. *From these results we observe that* $X_{134}, X_{135}, X_{234}$, X_{235} *and* X_{345} *are all illegitimate numbers, since their values are out of the legitimate range [0,60). In the remaining five cases, except for* X_{123}*, all the results are the same and equal to 21. Moreover, all these results were recovered from three moduli without including* m_3*, i.e. from* $X_{124}, X_{125}, X_{145}$ *and* X_{245}*, which are equal to 21. Hence, we might conclude that the correct result is 21 and that there was an error in* r_3*, which can be corrected by computing* $\hat{r}_3 = 21 \pmod 5 = 1$.

Let us now provide a brief summary of our previous elaborations in the context of the general coding theory of RRNSs by first providing some definitions. Let $\boldsymbol{x} = (r_1, r_2, \ldots, r_v, r_{v+1} \ldots, r_u)$, $u > v$ be a codeword of the RRNS(u, v) code.

Definition 1 [406] *The Hamming weight* $Wt(\boldsymbol{x})$ *of a code vector* $\boldsymbol{x}$ *in an RRNS code is defined as the number of non-zero residue components of* $\boldsymbol{x}$.

Definition 2 [406] *The Hamming distance* $d(\boldsymbol{x}_i, \ \boldsymbol{x}_j)$ *between two RRNS code-vectors* $\boldsymbol{x}_i$ *and* $\boldsymbol{x}_j$ *is defined as the number of positions in which* $\boldsymbol{x}_i$ *and* $\boldsymbol{x}_j$ *differ.*

Definition 3 [406] *The minimum Hamming distance* δ *of the RRNS code is defined as:*

$$\delta = \min \left\{ d(\boldsymbol{x}_i, \ \boldsymbol{x}_j) : \ \boldsymbol{x}_i, \ \boldsymbol{x}_j \in \Omega, \ \boldsymbol{x}_i \neq \boldsymbol{x}_j \right\}, \tag{17.8}$$

where Ω *is defined as the code space incorporating all legitimate codewords.*

The necessary and sufficient conditions for an RRNS (u, v) code to achieve the maximum possible minimum distance among legitimate codewords is:

Theorem 17.1 **[406]** *Let $m_1, m_2, \ldots, m_v, m_{v+1}, \ldots, m_u$ be a group of moduli constituting the RRNS (u,v) code and $[0, M)$ - where $M = \prod_{i=1}^{v} m_i$ - be the legitimate range. Then the necessary and sufficient condition for the RRNS (u,v) code to achieve the maximum possible minimum-distance of $\delta = u - v + 1$ is that any of the redundant moduli is larger than the largest non-redundant modulus, which is formulated as:*

$$m_{v+j} > \max\{m_1, m_2, \ldots, m_v\}, \textit{ for } j = 1, 2, \ldots, u-v. \tag{17.9}$$

According to Theorem 17.1, RRNS (u,v) codes constitute a class of codes achieving the maximum possible minimum-distance of $\delta = u-v+1$, provided that the condition of Equation 17.9 is satisfied. Hence, RRNS codes constitute a class of maximum distance separable (MDS) codes [387]. Consequently, in the remainder of our analysis and in the various proposed applications all RRNS (u,v) codes discussed are assumed to satisfy the condition of Equation 17.9, and consequently their minimum distance is $\delta = u - v + 1$. Furthermore, we assume that the u number of moduli of the RRNS(u,v) code satisfy the relationship:

$$m_1 < m_2 < \ldots < m_v < m_{v+1} < \ldots < m_u. \tag{17.10}$$

Since the minimum distance of the RRNS(u,v) code is $\delta = u - v + 1$, when using minimum-distance based decoding according to the minimum distance decoding theory of reference [387] the following Lemma can be stated.

Lemma 1 **[406]** *An RRNS (u,v) code can detect up to $(u-v)$ residue digit errors, or correct up to $\lfloor \frac{u-v}{2} \rfloor$ residue digit errors, where $\lfloor x \rfloor$ represents the largest integer not exceeding x. Alternatively, an RRNS (u,v) code can correct up to t residue digit errors, and simultaneously detect up to ν, $(\nu > t)$ residue digit errors, provided that $t + \nu \leq u - v$.*

Barsi and Maestrini [399] have described a method for error detection in RRNS codes based upon determining whether a given received integer, $X \Leftrightarrow (r_1, r_2, \ldots, r_v, r_{v+1}, \ldots, r_u)$, $u > v$, falls within the legitimate range $[0,\ M)$ or in the illegitimate range $[M,\ MM_R)$. It was shown that any single residue digit error inflicted upon a legitimate integer message X, will map to the illegitimate range of the RRNS. Barsi and Maestrini also developed a method for locating and correcting single residue digit errors $(u \geq v + 2)$ based upon the values of so-called *projections* computed for the RRNS (u,v) code concerned. The so-called *modulus m_i-projection* of X in the context of an RRNS (u,v) code is defined as [399]:

$$X_i \equiv X \left(\bmod \frac{MM_R}{m_i} \right), \tag{17.11}$$

where the m_i-projection, X_i, can be interpreted as $X_i \Leftrightarrow (r_1, r_2, \ldots, r_{i-1}, r_{i+1}, \ldots, r_u)$, which is the residue representation of X in a reduced RRNS with the ith residue digit r_i deleted. This m_i-projection results in the reduced RRNS $(u-1, v)$ code, where the modulus m_i is deleted from the RRNS(u,v) code. Since the RRNS is a data representation system with at least one residue digit of redundancy, the original information is retained, when the redundancy is removed.

Correspondingly, the $M_\wedge$-projection of X, denoted by $X_\wedge$, is defined by [396]:

$$X_\wedge \equiv X \left(\bmod \frac{MM_R}{M_\wedge}\right), \tag{17.12}$$

where $M_\wedge = \prod_{m=1}^{\kappa} m_{i_m}$, $\wedge = \{i_1, i_2, \ldots, i_\kappa\}$, and $\kappa \leq u - v$. In other words, $X_\wedge$ can also be represented as a reduced residue representation of X with the residue digits $r_{i_1}, r_{i_2}, \ldots, r_{i_\kappa}$ deleted. Hence, X can be represented by a reduced RRNS$(u - \kappa, v)$ code with the moduli $m_{i_1}, m_{i_2}, \ldots, m_{i_\kappa}$ deleted from the RRNS(u, v) code. The legitimate and illegitimate ranges of the reduced RRNS code are $[0,\ M')$ and $[M',\ MM_R/M_\wedge)$, respectively, where M' is the product of the retained moduli of the reduced RRNS code. It follows from Equation 17.12 that the $M_\wedge$-projection of any legitimate integer X of the RRNS is still a legitimate integer in $[0,\ M)$, i.e., $X_\wedge = X$, provided that a sufficiently high dynamic range is retained in the reduced RRNS, in order to unambiguously represent X.

The first step of error-correction in RRNS codes (and the only step in the error-detection procedure) is to check whether the received residue vector denoting by $\boldsymbol{y}$ is a legitimate codevector in the RRNS. A simple way of doing this is to compute the corresponding integer Y and to check whether $0 \leq Y < M$. If $0 \leq Y < M$, then the received vector $\boldsymbol{y}$ is a codevector. Otherwise, if $M \leq Y \leq MM_R$, then the received residue vector $\boldsymbol{y}$ included at least one residue digit error. However, this 'range-checking' approach may require processing large-valued integers. A suitable method - as we noted previously - for avoiding this is invoking the so-called base-extension method using mixed radix conversion. According to Equation 17.6, the mixed radix conversion of the integer $Y \Leftrightarrow \boldsymbol{y}$ can now be expressed in the form of:

$$Y = a_1 + a_2 m_1 + a_3 m_1 m_2 + \cdots + a_n \prod_{i=1}^{u-1} m_i, \tag{17.13}$$

where $0 \leq a_l < m_l$, $l = 1, 2, \ldots, u$, and $\prod_{i=1}^{0} m_i = 1$ as we defined previously. For a given received residue digit sequence $\boldsymbol{y} = (r_1, r_2, \ldots, r_u)$ the mixed radix digits $a_1, a_2, \ldots, a_u$ are computed by using Equation 17.7 from the residue digits $r_1, r_2, \ldots, r_u$ as follows:

$$\begin{aligned} a_1 &= r_1, \\ a_2 &= (r_2 - a_1) m_1^{-1} \pmod{m_2}, \\ &\cdots \\ a_u &= ((\cdots((r_u - a_1) m_1^{-1} - a_2) m_2^{-1} - \cdots) - a_{u-1}) m_{(u-1)}^{-1} \pmod{m_u}. \end{aligned} \tag{17.14}$$

In an RRNS, the digits $a_1, a_2, \ldots, a_v$ are referred to as the non-redundant mixed radix digits, and $a_{v+1}, a_{v+2}, \ldots, a_u$ are termed as the redundant mixed radix digits. It can be shown that if $0 \leq Y < M$, then all the redundant mixed radix digits will be zeros.

Furthermore, the integer message Y corresponding to a received residue vector $\boldsymbol{y} = (r_1, r_2, \ldots, r_u)$ can be computed from any v out of the u number of residue digits and their related moduli, according to the projection theory of [399].

Let $r_{i_1}, r_{i_2}, \cdots, r_{i_v}$ be v out of the u number of residue digits from $\boldsymbol{y} = (r_1, r_2, \cdots, r_u)$, where - without loss of any generality - we assume that $i_1 < i_2 < \ldots < i_v$. Then, the integer Y can be computed from these v residue digits and their related moduli according to Equations 17.13 and 17.14, which can be expressed with the aid of the mixed radix conversion as:

$$Y = a_{i_1} + a_{i_2} m_{i_1} + a_{i_3} m_{i_1} m_{i_2} + \cdots + a_{i_v} \prod_{l=1}^{v-1} m_{i_l}. \tag{17.15}$$

Based on Y computed according to Equation 17.15, the remaining $(u - v)$ residue digits can be recomputed using the approach proposed by Claudio, Orlandi and Piazza [394], which is expressed as:

$$\begin{aligned} r'_j &\equiv \left(Y \right)_{m_j} \\ &\equiv (\ldots(((a_{i_1})_{m_j} + (a_{i_2} m_{i_1})_{m_j})_{m_j} + (a_{i_3} m_{i_1} m_{i_2})_{m_j})_{m_j} \\ &\quad + \cdots + (a_{i_v} \prod_{l=1}^{v-1} m_{i_l})_{m_j}))_{m_j} \end{aligned} \tag{17.16}$$

for all $j = 1, 2, \cdots, n;\ j \notin \{i_1, i_2, \cdots, i_v\}$, where $(\ \cdot\ )_{m_j}$ represents the modulo m_j operation for simplicity. The process summarized in Equations 17.15 and 17.16 is the above-mentioned base-extension algorithm. It can be shown - in contrast to the above-mentioned range-checking technique - that according to this approach no large-valued integers have to be processed and that the value of Y did not have to be recovered in order to compute the redundant residue digit r'_j. The difference between the residue digits r'_j, which were computed by the decoder according to Equation 17.16 and the received residue digits r_j in $\boldsymbol{y}$ can be expressed as:

$$\Delta_j = r'_j - r_j,\ j = 1, 2, \cdots, u;\ j \notin \{i_1, i_2, \cdots, i_v\}, \tag{17.17}$$

where the quantities $\{\Delta_j\}$ are referred to as syndromes [395], and all the received residue digits of $\boldsymbol{y}$ corresponding to $\{\Delta_j,\ j = 1, 2, \cdots, u;\ j \notin \{i_1\ ,\ i_2, \cdots, i_v\}\}$ are referred to as parity residue digits.

A fundamental property of the syndrome digits has been given in [395] and [400], which is restated in the following theorem, noting that the proof can be found in [400].

Theorem 17.2 *Under the assumption that no more than $\lfloor \frac{u-v}{2} \rfloor$ residue digit errors occur in an RRNS code having a minimum distance of $\delta = u - v + 1$, based on the $(u - v)$ number of syndrome digits - $\Delta_{v+1}, \Delta_{v+2}, \cdots, \Delta_u$ - computed from Equation 17.17 the following three observations can be stated:*

Observation 1 *A received residue digit vector $\boldsymbol{y}$ is an error-free code-vector if and only if all the $(u - v)$ syndrome digits are zero.*

Observation 2 *λ ($1 \leq \lambda \leq \lfloor \frac{u-v}{2} \rfloor$) out of the $(u - v)$ parity residue digits in $\boldsymbol{y}$ are in error if and only if λ corresponding syndrome digits are non-zero.*

Observation 3 *At least one of the received information residue digits in $\boldsymbol{y}$ - which correspond to the first v residue digits of $r_1, r_2, \ldots, r_v$ - is in error if and only if more than $\lfloor \frac{u-v}{2} \rfloor$ syndrome digits are non-zero.*

Having provided a rudimentary overview of the associated RRNS throry, readers interested in more intricate details of the decoding theory and approaches in RRNS are referred to [395–397, 406] for further details.

A further useful property of the RRNS is that an RRNS codeword does not change its error-detection and error-correction characteristic after arithmetic operations such as addition, subtraction and multiplication [406]. Let us use an example to illustrate this property.

Example 9 *Let us consider the RRNS codes based on an RRNS using non-redundant moduli of $m_1 = 4$, $m_2 = 5$ and $m_3 = 7$ as well as redundant moduli of $m_4 = 9$, $m_5 = 11$, $m_6 = 13$ and $m_7 = 17$. Table 17.5 summarized a range of typical decimal integers and their corresponding RRNS codewords. Let us now consider an operation in the decimal integer field:*

$$Y = 5 \times Y_1 + Y_2 \times Y_3, \tag{17.18}$$

where Y_1, Y_2 and Y_3 are possibly from different sources and may include the transmission and processing errors. However, due to the inherent properties of the RRNS codes, the operations in Equation 17.18 can be carried out as follows. First, before we evaluate Equation 17.18, Y_1, Y_2 and Y_3 are first error-correction decoded. Since four redundant moduli are included, up to two residue digit errors in Y_1, Y_2 or Y_3 can be corrected. Second, Table 17.5 is used in order to evaluate Equation 17.18. Third, since an RRNS codeword does not change its error-detection/correction capability when the codewords are subjected to the arithmetic operations of addition and multiplication, the resultant RRNS codeword corresponding to Y in Equation 17.18 can be further error-correction decoded, in order to remove the errors that may have happened in the evaluation of Equation 17.18. Note that no conventional codes, such as Hamming codes, BCH codes and convolutional codes have this property, since all these codes will change their structure when the associated codewords are subjected to the multiplication operation.

Using the operand values of Table 17.5, let $Y_1 = X_1 = 1$, $Y_2 = X_2 = 2$ and $Y_3 = X_6 = 50$. For example, the RRNS codeword corresponding to 5 is (1,0,5,5,5,5,5) according to Table 17.5. It can readily be shown that the codeword corresponding to the *operation in Equation 17.18 is given by* $Y = 5 \times 1 + 2 \times 50 = 105$, *yielding the residue-sequence of (1,0,0,6,6,1,3) - provided that no more than two residue digit errors occurred in* Y_1, Y_2 *or* Y_3. *Moreover, even if two residue digit errors occurred during the last operation of Equation 17.18, and hence the above codeword changed to* $(1, \tilde{1}, 0, 6, \tilde{3}, 1, 3)$, *i.e.,* r_2 *and* r_5 *are in error, by using RRNS error-correction decoding, the value of* $Y = 105$ *can still be correctly recovered, since four redundant moduli were invoked.*

Let us now demonstrate a potential application of the above RRNS codes in the context of the generic communication system depicted in Figure 17.1. In this framework, redundancy has to be added for the implementation of fault-tolerant computing/processing, protection of information over both wired and wireless channels, in order to achieve reliable processing and communications. Conventionally, the encoding and decoding at different stages have been treated separately. The encoded codewords of the wired channels have been historically decoded in order to remove the related redundancy, before forwarding the information to the encoder of the air interface of Figure17.1. The information at the air interface is then often re-encoded using a different channel code, in order to implement reliable transmission over the associated wireless channels. Further additional encoding/decoding operations may take place at other interfaces of a global communications system, as illustrated in Figure 17.1.

However, with the advent of RRNS codes, the above mentioned encoding/decoding procedures can be substantially simplified, since the required redundancy of the entire global system can be jointly designed. Let us invoke an example to support this argument.

Example 10 *With reference to Table 17.5, let us assume that the fault-tolerant terminals use only one redundant modulus for the self-checking of the associated arithmetic units, employing, for example, the RRNS (4,3) code. Let us assume furthermore that two redundant moduli are required for the protection of information over the wired channel section of Figure 17.1, employing the one-residue digit error-correcting RRNS (5,3) code. Finally, four redundant moduli have to be employed for the protection of information over the low-reliability wireless channels, using the two-residue digit error-correction RRNS (7,3) code. Let* m_1, m_2, m_3, m_4, m_5, m_6 *and* m_7 *be the moduli invoked in the required RRNS, where* m_1, m_2 *and* m_3 *are the non-redundant moduli, while* m_4, m_5, m_6 *and* m_7 *are the redundant moduli. The RRNS codes protecting the entire system of Figure 17.1 can be designed as follows.*

Step 1: *The fault-tolerant terminal No.1 uses moduli* m_1, m_2, m_3 *and* m_4, *in order to form the RRNS (4,3) codewords, which are expressed as* (r_1, r_2, r_3, r_4), *where* r_1, r_2 *and* r_3 *are the non-redundant residue digits, while* r_4 *is the redundant residue digit. Following all the associated signal processing operations to be carried out by the terminal - which may involve additions, subtractions and multiplications on the basis of the RNS - and invoking self-checking, the fault-tolerant terminal forwards the RRNS (4,3) codewords to the encoder of the wired channels. For simplicity the output codewords of the fault-tolerant terminal No.1* *are still expressed as* $(r_1,\ r_2,\ r_3,\ r_4)$*, which will also be used throughout our further discurse.*

Step 2: *With the aid of the Base Extension algorithm described in Equations 17.15 and 17.16, the wired channel's encoder - namely Encoder 1 in Figure 17.1 - computes an additional redundant residue digit* r_5 *based on* $(r_1,\ r_2,\ r_3,\ r_4)$ *and* $(m_1, m_2, m_3, m_4, m_5)$*, in order to form the RRNS (5,3) codewords* $(r_1,\ r_2,\ r_3,\ r_4,\ r_5)$*. Then, these codewords are transmitted over the wired channel in Figure 17.1.*

Step 3: *The wired channel's decoder - namely Decoder 1 in Figure 17.1 - receives the codewords* $(r_1,\ r_2,\ r_3,\ r_4,\ r_5)$*, which are error-correction decoded. The corrected codewords are expressed as* $(r_1,\ r_2,\ r_3,\ r_4,\ r_5)$*, which are forwarded to the encoder of the wireless channel.*

Step 4: *Similarly to Step 2, with the aid of the base extension algorithm, the wireless channel's encoder computes the additional redundant residue digit* r_6 *and* r_7*, based on* $(r_1,\ r_2,\ r_3,\ r_4,\ r_5)$ *and* $(m_1, m_2, m_3, m_4, m_5, m_6, m_7)$*, in order to form the RRNS (7,3) codewords* $(r_1,\ r_2,\ r_3,\ r_4,\ r_5,\ r_6,\ r_7)$*. Then these codewords are transmitted over the wireless channel of Figure 17.1.*

Step 5: *After the wireless channel's decoder has received the codewords* $(r_1,\ r_2,\ r_3,\ r_4,\ r_5,\ r_6,\ r_7)$*, the codewords are error-correction decoded and the redundant residue digits* r_6 *and* r_7 *are removed from the corrected codewords. The wireless channel decoder then passes the RRNS (5,3) codewords to the wired channel's encoder, namely to Encoder 3 in Figure 17.1. These RRNS (5,3) codewords are expressed as* $(r_1,\ r_2,\ r_3,\ r_4,\ r_5)$*.*

Step 6: *The RRNS (5,3) codewords are transmitted over the wired channel to the wired channel's decoder, namely to Decoder 3 in Figure 17.1.*

Step 7: *After the wired channel's decoder has received the RRNS (5,3) codewords* $(r_1,\ r_2,\ r_3,\ r_4,\ r_5)$*, error-correction decoding is invoked, in order to correct the residue digit errors and then the redundant residue digit* r_5 *is removed. The resulting RRNS (4,3) codewords* $(r_1,\ r_2,\ r_3,\ r_4)$ *are then forwarded to the fault-tolerant terminal No. 2.*

Step 8: *The fault-tolerant terminal No. 2 finally carries out its tasks using the RNS representation of its operands in the form of the RRNS (4,3) codewords, invokes self-checking, and*

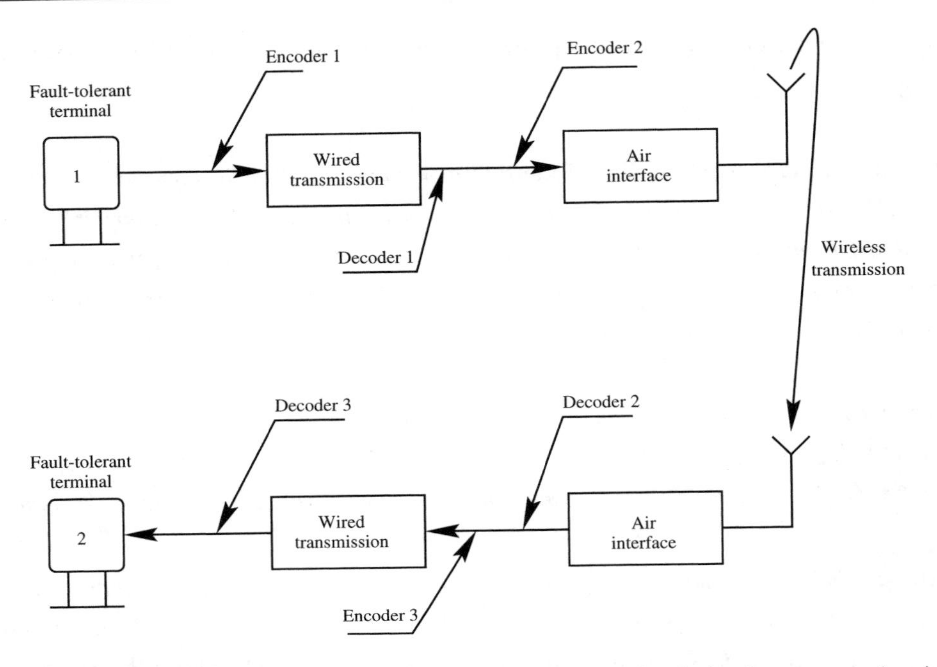

Figure 17.1: A transmission and computing system framework including fault-tolerant terminals, wired transmissions and wireless transmissions.

finally outputs the recovered information.

In the upper branch of Figure 17.1 the wired channel's encoder only had to compute one additional redundant residue digit, in contrast to invoking an independent encoder, as in a conventional system, where each section of the communications links is protected independently. Similarly, the wireless channel's encoder only had to compute two additional redundant residue digits, in contrast to complete re-encoding in independently protected conventional systems. In the bottom branch of Figure 17.1, the encoder of the wired channel, and the encoder of the fault-tolerant terminal No. 2 - which is not shown in the figure - can be totally removed, while in conventional independently protected systems all these encoders would have to be included. Based on these facts, we can conclude that in conjunction with appropriate joint system design using RRNS codes, the complexity of the error-correction/detection sub-systems in global telecommunication systems can be decreased.

17.5 Chapter Summary and Conclusion

In this chapter we have discussed the mathematical models of RNSs and RRNSs as well as the associated principle of error-detection and error-correction in RRNSs by using a range of examples. Moreover, we have summarized the basic coding theory of RRNS codes, in order to support our forthcoming deliberations. A RRNS codeword does not change its error-detection and error-correction characteristics after arithmetic operations such as addition, subtraction and multiplication, while conventional linear codes such as Hamming codes, BCH codes and convolutional codes usually cannot provide their error-control properties after multiplyng codewords with each other. Hence, in conventional communication system design, error protection of signals in signal processing and signal transmission procedures are treated separately. However, in most situations in conjunction with the RRNS representation introduced before signal processing, the RRNS can be used not only for the protection of the signals while they are being processed in the transceivers, but also for enhancing the system's error resilience over the communication channels. From this point of view, we can argue that, to some extent communication systems might be simplified by unifying the whole encoding and decoding procedure across global communication systems.

In digital communications, upon exploiting the inherent properties of the RNS arithmetic, an RRNS can be designed for improving the system's bit error rate (BER) performance. In order to fulfil this potential, an RNST unit, which implements the binary-to-residue transformation and an IRNST unit, which carries out the residue-to-binary transformation have to be incorporated in the transmitter and the receiver of the system, respectively. In the following chapter, as an example of the application of the RNS and RRNS, we will focus our attention on an RNS or RRNS-based parallel direct-sequence code division multiple-access (DS-CDMA) system, and analyze its performance over Additive White Gaussian Noise (AWGN) channels or over multipath Rayleigh fading wireless channels.

Chapter 18

Redundant Residue Number System Coded M-ary DS-CDMA Using Orthogonal Modulation[1]

18.1 State of the Art

Flexible, high-bit-rate, low bit error rate (BER) communication is becoming an issue of increasing importance. Conventionally, communication system design has been based on the well-known weighted number system representation, using, for example, a base of 2, 8, 16, etc. for implementation, ultimately favouring the weighted binary system. In a conventional system – where the operands of the signal processing steps are represented by the conventional weighted number system – due to the carry forward required by the weighted number system a bit error may affect all the bits of the result. By contrast, the so-called residue number system (RNS) [392] portrayed in the previous chapter is a non-weighted, carry-free number system, which has received wide attention due to its robust self-checking, error-detection, error-correction and fault-tolerant signal processing properties [375–378, 392–421].

As we have discussed in the previous chapter, an RNS is defined [392] by the choice of V number of positive integers m_v, $(v = 1, 2, \ldots, V)$ referred to as moduli. If all the moduli are pairwise relative primes, any integer X, describing the information symbols to be transmitted in this chapter, can be uniquely and unambiguously represented by the so-called residue sequence $(r_1, r_2, \ldots, r_V)$ in the range of $0 \leq X < M$, where $r_v = X \pmod{m_v}$ represents the residue digits of X upon division by the moduli m_v, and $M = \prod_{v=1}^{V} m_v$ can be referred to as the information dynamic range, i.e., the legitimate range of the information

[1] *Single- and Multi-Carrier CDMA Multi-User Detection, Space-Time Spreading, Synchronization, Networking and Standards.* L.Hanzo, L-L.Yang, E-L.Kuan and K.Yen,
 ISBN 0-470-86309-9

symbols X. This is true, simply because the V number of residue digits unambiguously describe any integer information symbol X in this range. In the above process the algorithm that transforms any conventional weighted number system to the residue number system was defined as the residue number system transform (RNST) in the previous chapter. According to the so-called Chinese reminder theorem (CRT) [397] – which was also highlighted in the previous chapter, for any given V-tuple of residue digits $(r_1, r_2, \ldots, r_V)$, where $0 \leq r_v < m_v$, there exists one and only one integer X such that $0 \leq X < M$ and $r_v = X \pmod{m_v}$, which allows us to uniquely recover the message X from the received residue digits. The process that transforms the residue number system to the weighted number system was defined as the inverse RNST (IRNST) in the previous chapter.

For incorporating error control [392, 395–397, 406], the RNS has to be designed with redundant moduli, yielding a so-called redundant RNS (RRNS) code. An RRNS code is obtained by appending an additional $(U - V)$ number of moduli $m_{V+1}, m_{V+2}, \ldots, m_U$, to the previously introduced RNS, in order to form an RRNS code of U positive, pair-wise relative prime moduli. The so-called redundant moduli have to obey $m_{V+j} \geq max\{m_1, m_2, \ldots, m_V\}$, in order that the RRNS code has the maximum possible minimum distance of $\delta = U - V + 1$. Now an integer X in the range $[0, M)$ is represented as a U-tuple residue digit sequence, $(r_1, r_2, \ldots, r_U)$ with respect to the U number of moduli and consequently forms an RRNS (U, V) codeword. Again, these issues were discussed in more depth in the previous chapter.

The M-ary orthogonal keyed (MOK) communication scheme [90] – where a set of M mutually orthogonal signals is utilized for the transmission of data – is a widely used arrangement [422]. An example of orthogonal signals suitable for MOK is constituted by sine waves having M number of uniformly spaced frequencies, leading to M-ary Frequency Shift Keying (MFSK). In the field of code division multiple access (CDMA) spread-spectrum communications, typically waveforms using M number of orthogonal pseudo-random spreading codes [364, 365, 385, 423, 424] or Hadamard-Walsh codes [109, 277, 366, 367] are employed for M-ary signalling. A practical manifestation of such a scheme is the 64-ary uplink of the well-known IS-95 system [11], transmitting six bits per 64-ary symbol. In simple terms the corresponding receiver can be viewed as a bank of 64 correlators, where the transmitted 64-ary symbol is deemed to be the one which exhibits the highest correlation with the received one.

By contrast, in this chapter we set out to propose and characterize a novel M-ary CDMA scheme, which exploits the advantages of RRNS-based implementations that were highlighted in the previous chapter. Among a range of attractive properties we will demonstrate that the number of correlators required by an M-ary system can be substantially reduced – well below M. This issue will be developed in more detail during our further discourse.

Our basic approach in this chapter is that the M-ary information symbols are converted to the residue digits of an RNS and all required operations are then carried out in the RNS domain. It is readily seen in this context that the M-ary symbols are effectively mapped to a number of residue digits and since now the product of the corresponding moduli in the RNS has to be M, in order to unambiguously recover the M-ary symbols from the residue digits, the moduli in the RNS are typically significantly smaller than M. Hence, this mapping operation can be interpreted as breaking up the M-ary symbols into U number of m_u-ary symbols, where U is the number of moduli in the RNS, m_u, $u = 1, 2, \ldots, U$ represents the moduli of the RNS.

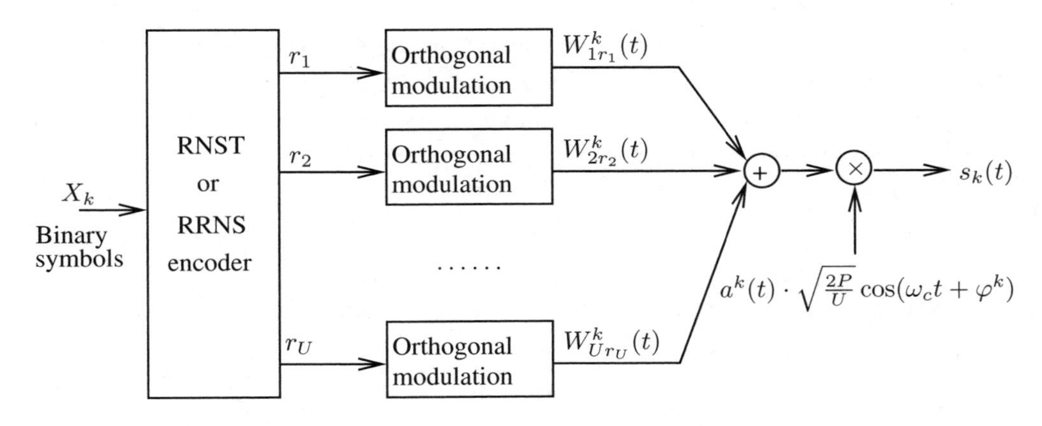

Figure 18.1: The transmitter block diagram of the RNS-based parallel DS-CDMA system.

In a somewhat simplistic, but conceptually feasible manner one could argue then that for the transmission of the U number of reduced-size m_u-ary symbols we now need only $\sum_{u=1}^{U} m_u < M$ correlators, which in bold terms is the motivation of this chapter. Hence, in this chapter, we focus our attention on studying the performance of RNS and RRNS-based parallel direct sequence CDMA (DS-CDMA) systems in the context of M-ary orthogonal modulation, over Additive White Gaussian Noise (AWGN) channels and multipath fading channels. Let us first explore the performance over AWGN channels.

18.2 Performance in AWGN channels

In this section we concentrate on the description of the RNS-based parallel DS-CDMA system using M-ary orthogonal modulation, on its performance analysis and evaluation over AWGN channels, considering both coherent or non-coherent receivers.

18.2.1 Communication Model

The transmitter block diagram of the proposed RNS-based parallel DS-CDMA communication system is shown in Figure 18.1 for the kth user. As mentioned before, the information to be transmitted is transformed (or encoded) by the RNST (or RRNS encoder) block to the residue sequence $(r_1, r_2, \ldots, r_U)$, and the residue digits are then mapped to a set of orthogonal signals $(W^k_{1r_1}(t), W^k_{2r_2}(t), \ldots, W^k_{Ur_U}(t))$ and multiplexed for transmission. More rigorously, let

$$\left\{ \begin{array}{l} W^k_{10}(t), W^k_{11}(t), \ldots, W^k_{1(m_1-1)}(t); \\ W^k_{20}(t), W^k_{21}(t), \ldots, W^k_{2(m_2-1)}(t); \\ \quad \ldots\ldots; \\ W^k_{U0}(t), W^k_{U1}(t), \ldots, W^k_{U(m_U-1)}(t). \end{array} \right\} \tag{18.1}$$

be a set of $\sum_{u=1}^{U} m_u$ real-valued orthogonal signals, such as the orthogonal Hadamard-Walsh codes, which are used for transmission. The subset $\left\{W_{u0}^k(t), W_{u1}^k(t), \ldots, W_{u(m_u-1)}^k(t)\right\}$ of Equation 18.1 for $u = 1, 2, \ldots, U$ is used for the transmission of the residue digit r_u of user k. In Equation 18.1, the orthogonal signals' power is given by:

$$\xi_u = \int_0^{T_s} |W_{uj}^k(t)|^2 dt \tag{18.2}$$

for $u = 1, 2, \ldots, U$ and $j = 0, 1, \ldots, m_u - 1$, which are normalized to T_s, yielding,

$$\frac{1}{T_s}\int_0^{T_s} |W_{uj}^k(t)|^2 dt = 1, \tag{18.3}$$

where T_s represents the signalling interval duration. The orthogonality is expressed as:

$$\frac{1}{T_s}\int_0^{T_s} W_{ij}^k(t)W_{i'j'}^k(t)dt = \begin{cases} 1, & (i = i', j = j') \\ 0, & (\text{otherwise}). \end{cases} \tag{18.4}$$

In order to transmit an information symbol for the kth user, whose symbol is X_k, which is confined to the dynamic range $0 \leq X_k < \prod_{u=1}^{U} m_u$ of the system using the non-redundant RNS, the symbol is first transformed to the RNS representation of $(r_1, r_2, \ldots, r_U)$, which we referred to as the residue sequence in the previous chapter. Then the orthogonal signal set $\{W_{1r_1}^k(t), W_{2r_2}^k(t), \ldots, W_{Ur_U}^k(t)\}$ is obtained from the residue sequence, as seen in Figure 18.1, i.e., by assigning an orthogonal code to each residue digit r_u. The set of U orthogonal signals, $\{W_{1r_1}^k(t), W_{2r_2}^k(t), \ldots, W_{Ur_U}^k(t)\}$, is combined in the transmitter of Figure 18.1 and then it is spread by the kth user's signature spreading sequence $a^k(t)$. Finally, the spread signal modulates the carrier, yielding the transmitted signal expressed as:

$$s_k(t) = \sum_{u=1}^{U} \sqrt{\frac{2P}{U}} W_{ur_u}^k(t) a^k(t) \cos(\omega_c t + \varphi^k), \tag{18.5}$$

for $0 \leq t < T_s$, where $\omega_c = 2\pi f_c$ and f_c is the carrier frequency, φ^k is the kth user's carrier phase, which is assumed to be uniformly distributed in $[0, 2\pi]$. Furthermore, $a^k(t)$ represents the signature spreading waveform of the kth user, which is expressed as:

$$a^k(t) = \sum_{h=-\infty}^{\infty} a_h^k p(t - hT_c), \tag{18.6}$$

where the so-called chip-amplitudes a_h^k are assumed to be independent identically distributed (i.i.d) random variables assuming values of +1 and -1 with equal probability of 1/2. Furthermore, T_c represents the chip duration, and $p(t)$ is assumed to be the rectangular chip waveform, which is defined over the interval $(0,\ T_c]$. Moreover, we assume that $N_s = T_s/T_c = bT/T_c = bN$, where T is the bit duration and $N = T/T_c$ is the number of chips per bit. Since the dynamic range of the RNS system is $[0,\ M)$, then $b = \lfloor \log_2 M \rfloor$, due to b bits being transmitted in one M-ary symbol duration.

In the proposed system, since U number of orthogonal signals are combined linearly, as seen in Equation 18.5, high peak-to-average amplitude ratios can be encountered, potentially resulting in nonlinear distortion, unless 'infinite dynamic-range' linear amplification is assumed. Techniques minimizing the envelope fluctuations after linear combining of multiple signals have been proposed in a number of publications [425,426]. However, these issues are beyond the scope of our discussions here and hence they are not considered in this book.

According to Equations 18.3 and 18.4, the total energy of $s_k(t)$ per symbol period can be directly computed as

$$\begin{aligned} E_s &= E\left[\int_0^{T_s} s_k^2(t)dt\right] \\ &= E\left[\int_0^{T_s}\left(\sum_{i=1}^{U}\sqrt{\frac{2P}{U}}W_{ir_i}^k(t)a^k(t)\cos(\omega_c t+\varphi^k)\right)\right. \\ &\quad \left.\cdot\left(\sum_{j=1}^{U}\sqrt{\frac{2P}{U}}W_{jr_j}^k(t)a^k(t)\cos(\omega_c t+\varphi^k)\right)dt\right] \\ &= E\left[\sum_{i=1}^{U}\int_0^{T_s}\frac{2P}{U}(W_{ir_i}^k(t))^2(a^k(t))^2\cos^2(\omega_c t+\varphi^k)dt\right] \\ &\quad +E\left[\underbrace{\sum_{i=1}^{U}\sum_{j=1}^{U}}_{i\neq j}\int_0^{T_s}\left(\sqrt{\frac{2P}{U}}W_{ir_i}^k(t)a^k(t)\cos(\omega_c t+\varphi^k)\right)\right. \\ &\quad \left.\cdot\left(\sqrt{\frac{2P}{U}}W_{jr_j}^k(t)a^k(t)\cos(\omega_c t+\varphi^k)\right)dt\right]. \end{aligned} \tag{18.7}$$

Since random signature sequences a_h^k are considered, the expectation value of the last term in the Equation 18.7 is zero. Hence, Equation 18.7 can be simplified to:

$$\begin{aligned}
E_s &= E\left[\sum_{i=1}^{U}\int_0^{T_s}\frac{2P}{U}(W_{ir_i}^k(t))^2(a^k(t))^2\cos^2(\omega_c t+\varphi^k)dt\right] \\
&= \frac{P}{U}E\left[\sum_{i=1}^{U}\int_0^{T_s}(W_{ir_i}^k(t))^2(a^k(t))^2dt\right] \\
&= \frac{P}{U}E\left[\sum_{i=1}^{U}\sum_{l=0}^{N_s-1}\int_{lT_c}^{(l+1)T_c}(W_{ir_i}^k(t))^2dt\right] \\
&= \frac{P}{U}E\left[\sum_{i=1}^{U}\int_0^{T_s}(W_{ir_i}^k(t))^2dt\right] \\
&= PT_s. \qquad (18.8)
\end{aligned}$$

We assume that ideal power control is employed, in order that the received signal powers are the same for all K users. Then the received signal at the base station generated by the K asynchronously transmitting users can be expressed as:

$$r(t)=\sum_{k=1}^{K}y_k(t)+n(t), \qquad (18.9)$$

where

$$y_k(t)=\sum_{u=1}^{U}\sqrt{\frac{2P}{U}}W_{ur_u}^k(t-\tau^k)a^k(t-\tau^k)\cos(\omega_c t+\phi^k), \qquad (18.10)$$

and $\phi^k=\varphi^k-\omega_c\tau^k$ is a uniformly distributed random variable over $[0,2\pi]$, while τ^k represents the asynchronous transmission plus channel delay, which is also assumed to be a uniformly distributed random variable in $[0,T_s]$. Furthermore, $n(t)$ represents the AWGN, which is modelled as a random variable with zero mean and double-sided power spectral density of $N_0/2$. Let us now consider the performance of the RNS-based parallel DS-CDMA system using either a coherent or a non-coherent receiver.

18.2.2 Performance of Coherent Receiver

18.2.2.1 Receiver Model

Figure 18.2 portrays the proposed coherent receiver designed for receiving the RNS-based orthogonal modulated M-ary DS-CDMA signals in the form of Equation 18.9. The receiver is constituted by three sections. In Section I the received signals are coherently carrier demodulated and de-spread. Section II consists of U number of banks of correlators, where each bank is dedicated to receiving one m_u-ary residue digit from the set of U residue digits $\{r_1,r_2,\ldots,r_U\}$. According to one of the important inherent properties of the RNS arith-

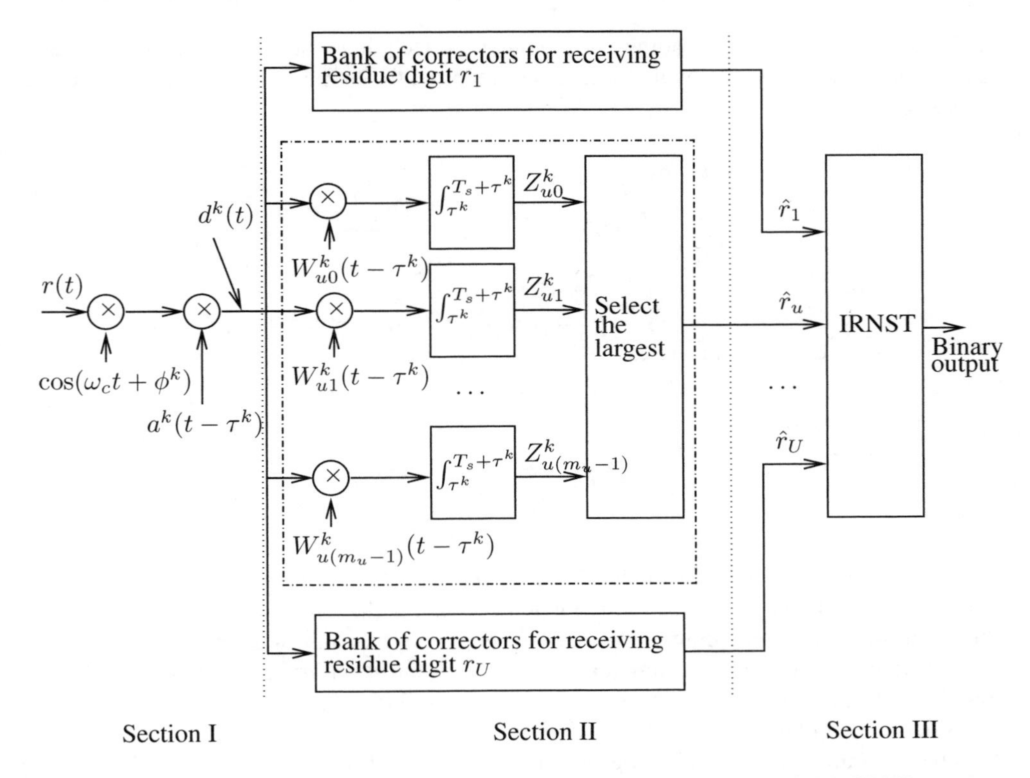

Figure 18.2: The coherent receiver's block diagram for the RNS-based parallel DS-CDMA system using M-ary orthogonal modulation.

metic noted in the previous chapter, the operations based on the residue digits belonging to the different moduli, m_u, $u = 1, 2, \ldots, U$ are mutually independent, hence the receiver banks of different residue digits in Figure 18.2 are independent. Therefore, each bank structure of the receiver in Figure 18.2 is optimum for the AWGN channels considered, when the multiple user interference can be treated as additional additive noise. After each maximum likelihood detector derives an estimate of its corresponding transmitted residue digit r_u, $u = 1, 2, \ldots, U$, the set of U residue digits is input to the IRNST block of Section III in Figure 18.2, and the estimation of the information symbol X_k ensues according to the RNS decoding algorithms discussed in our previous chapter. Having portrayed the basic outline of the RNS-based M-ary CDMA receiver, let us now focus our attention on the performance analysis of the system using a coherent receiver.

18.2.2.2 Coherent RNS-CDMA System's Probability of Error

Let the first user, i.e. user $k = 1$, be the reference user and consider the coherent receiver using maximum likelihood detection (MLD) for each m_u-ary residue digit. Note that the superscript of the reference user's signal is omitted for convenience. We also assume perfect carrier phase tracking and spreading code phase tracking for the reference user, i.e. $\phi = 0$

and $\tau = 0$. A range of spreading code acquisition techniques will be the topic of Chapter 21. Then, after carrier demodulation and despreading, the output $d(t)$ in Figure 18.2 can be expressed as:

$$\begin{aligned} d(t) &= r(t) \cdot a(t) \cos(\omega_c t) \\ &= \sum_{u=1}^{U} \sqrt{\frac{P}{2U}} W_{ur_u}(t) \\ &+ \sum_{k=2}^{K} \sum_{u=1}^{U} \sqrt{\frac{P}{2U}} \cos(\phi^k) W_{ur_u}^k (t - \tau^k) a^k (t - \tau^k) a(t) \\ &+ n(t) a(t) \cos(\omega_c t). \end{aligned} \tag{18.11}$$

Due to the independence of the different m_u-ary residue digits of the RNS, we may compute the system's error probability by first computing the error probabilities for receiving the U number of residue digits separately. Then the system's average error probability can be obtained by exploiting the inherent properties of the RNS arithmetic highlighted in the previous chapter, as will become explicit during our further discourse. Let us specifically consider receiving the m_1-ary residue digit $r_1 = 0$. After correlating it with each of the m_1 number of orthogonal signals in the subset $\{W_{10}(t), W_{11}(t), \ldots, W_{1(m_1-1)}(t)\}$, the set of decision variables can be written as:

$$Z_{1m} = \int_0^{T_s} d(t) W_{1m}(t) dt, \; m = 0, 1, \ldots, m_u - 1, \tag{18.12}$$

which are necessary for deciding upon the m_1-ary residue digit r_1. Since we have assumed that $r_1 = 0$, hence the decision variables can be expressed as:

$$Z_{10} = \sqrt{\frac{P}{2U}} T_s \left[1 + \sum_{k=2}^{K} \sum_{u=1}^{U} I_{u,10}^k + N_{10} \right], \tag{18.13}$$

$$Z_{1m} = \sqrt{\frac{P}{2U}} T_s \left[\sum_{k=2}^{K} \sum_{u=1}^{U} I_{u,1m}^k + N_{1m} \right], \quad m = 1, 2, \ldots, (m_1 - 1), \tag{18.14}$$

where $I_{u,1m}^k$ for $m = 0, 1, \ldots, (m_1 - 1)$ represent the interference from user k associated with the transmission of the m_u-ary residue digit r_u, which can be formulated as:

$$I_{u,1m}^k = \frac{\cos(\phi^k)}{T_s} \int_0^{T_s} W_{ur_u}^k (t - \tau^k) W_{1m}(t) a^k (t - \tau^k) a(t) dt, \tag{18.15}$$

while N_{1m} for $m = 0, 1, \ldots, (m_1 - 1)$ represent the terms due to the presence of AWGN, which can be expressed as:

$$N_{1m} = \left(\sqrt{\frac{P}{2U}}T_s\right)^{-1} \int_0^{T_s} n(t)W_{1m}(t)a(t)\cos(\omega_c t)dt. \quad (18.16)$$

It can be shown that N_{1m} is a Gaussian random variable with zero-mean and variance $(UN_0/2E_s)$, where $E_s = PT_s = bE_b$ is the average energy per symbol. The multiple user interference $I_{u,1m}^k$, due to the kth interfering user can also be approximated as a Gaussian random variable with zero-mean and variance [103]:

$$\text{Var}(I_{u,1m}^k) = \frac{1}{3N_s}. \quad (18.17)$$

Consequently, after the normalization using $\sqrt{\frac{P}{2U}}T_s$ in Equation 18.13 and Equation 18.14, Z_{10} and Z_{1m} for $m = 1, 2, \ldots, (m_1 - 1)$ are Gaussian distributed random variables having a mean value of one and zero, respectively, and a common variance of:

$$\sigma^2 = \frac{(K-1)U}{3N_s} + \frac{UN_0}{2E_s}, \quad (18.18)$$

i.e.,

$$f_{Z_{10}}(x) = \frac{1}{\sqrt{2\pi}\sigma}\exp\left(-\frac{(x-1)^2}{2\sigma^2}\right), \quad (18.19)$$

$$f_{Z_{1m}}(x) = \frac{1}{\sqrt{2\pi}\sigma}\exp\left(-\frac{x^2}{2\sigma^2}\right), \; m = 1, 2, \ldots, (m_1 - 1). \quad (18.20)$$

After further normalization by the mean-square noise power of σ, the above distributions are given by:

$$f_{Z_{10}}(x) = \frac{1}{\sqrt{2\pi}}\exp\left(-\frac{(x-\sqrt{2\gamma})^2}{2}\right), \quad (18.21)$$

$$f_{Z_{1m}}(x) = \frac{1}{\sqrt{2\pi}}\exp\left(-\frac{x^2}{2}\right), \; m = 1, 2, \ldots, (m_1 - 1), \quad (18.22)$$

where γ denotes the output signal-to-noise ratio (SNR) of the demodulator dedicated to receiving residue digit r_1, which is expressed as:

$$\gamma = \frac{1}{2} \cdot \left[\frac{(K-1)U}{3N_s} + \frac{UN_0}{2E_s}\right]^{-1}. \quad (18.23)$$

Let $Z_{1m,max} = \max\{Z_{1m} \text{ for all } m \neq 0\}$, then the PDF of $Z_{1m,max}$ can be expressed as (Appendix C.1):

$$f_{Z_{1m,max}}(x) = \frac{m_1 - 1}{\sqrt{2\pi}} [1 - \mathrm{Q}(x)]^{m_1 - 2} \exp\left(-\frac{x^2}{2}\right), \tag{18.24}$$

where $\mathrm{Q}(x)$ is the Q-function, which is defined as:

$$\mathrm{Q}(x) = \frac{1}{\sqrt{2\pi}} \int_x^{\infty} \exp\left(-\frac{t^2}{2}\right) dt. \tag{18.25}$$

The probability $P_{r_1}(C)$ of correctly receiving the m_1-ary residue digit r_1 is the probability that Z_{10} exceeds all other decision variables $Z_{11}, Z_{12}, \cdots, Z_{1(m_1-1)}$ within its residue bank in Figure 18.2, or the probability that Z_{10} exceeds $Z_{1m,max}$, which can be expressed as:

$$\begin{aligned} P_{r_1}(C) &= P(Z_{1m,max} < Z_{10}) \\ &= P(Z_{1m,max} < y, Z_{10} = y) \\ &= \int_{-\infty}^{\infty} P(Z_{1m,max} < y) f_{Z_{10}}(y) dy. \end{aligned} \tag{18.26}$$

It can be shown that:

$$\begin{aligned} P(Z_{1m,max} < y) &= \int_{-\infty}^{y} f_{Z_{1m,max}}(x) dx \\ &= \int_{-\infty}^{y} \frac{m_1 - 1}{\sqrt{2\pi}} [1 - \mathrm{Q}(x)]^{m_1 - 2} \exp\left(-\frac{x^2}{2}\right) dx \\ &= [1 - \mathrm{Q}(y)]^{m_1 - 1}. \end{aligned} \tag{18.27}$$

Substituting Equation 18.21 and Equation 18.27 into Equation 18.26 we obtain that:

$$P_{r_1}(C) = \frac{1}{\sqrt{2\pi}} \int_{-\infty}^{\infty} [1 - \mathrm{Q}(y)]^{m_1 - 1} \exp\left(-\frac{\left(y - \sqrt{2\gamma}\right)^2}{2}\right) dy. \tag{18.28}$$

Similarly, the probability $P_{r_u}(C)$ for $u = 2, 3, \cdots, U$ of correctly receiving the m_u-ary residue digit r_u can be expressed as:

$$P_{r_u}(C) = \frac{1}{\sqrt{2\pi}} \int_{-\infty}^{\infty} [1 - \mathrm{Q}(y)]^{m_u - 1} \exp\left(-\frac{\left(y - \sqrt{2\gamma}\right)^2}{2}\right) dy. \tag{18.29}$$

Since no redundant moduli are considered in this section, an M-ary information symbol is received correctly if and only if all the U number of m_u-ary residue digits are received correctly. Hence, the probability $P_s(C)$ of correct M-ary symbol recovery can be expressed as the product of receiving all the U number of m_u-ary residue digits constituting the M-ary symbol correctly, which is given by:

$$P_s(C) = \prod_{u=1}^{U} P_{r_u}(C) \tag{18.30}$$

and hence the bit error probability $P_b(\varepsilon)$ can be derived following a similar approach to Proakis' [90][p.262], yielding:

$$P_b(\varepsilon) \approx \frac{1}{2}(1 - P_s(C)), \tag{18.31}$$

when $M = \prod_{u=1}^{U} m_u$ is a large integer, implying a high number of bits/symbol.

When $U = 1$, Equation 18.31 is simply reduced to the bit error probability of the conventional M-ary orthogonal signalling system [90][p.262]. However, if the DS-CDMA system having conventional M-ary orthogonal modulation has the same number of bits per symbol period, as the RNS-based system, then

$$M = \prod_{u=1}^{U} m_u \tag{18.32}$$

must be satisfied. In other words, the product of the moduli determines the legitimate range of the non-binary integer message X_k and a specific combination of the received U number of m_u-ary residue digits uniquely describes the message X_k. This implies that for the conventional M-ary orthogonal signalling scheme $M = \prod_{u=1}^{U} m_u$ number of correlators are required, in order to demodulate $b = \left(\log_2 \prod_{u=1}^{U} m_u\right)$-bit symbols, in contrast to the proposed RNS-based parallel DS-CDMA orthogonal modulation scheme, in which only $\left(\sum_{u=1}^{U} m_u\right)$ number of correlators are necessitated.

As an example, the 64-ary IS-95 system requires 64 correlators. If, however a 64-ary symbol is conveyed with the aid of a number of moduli, the number of correlators depends on the choice of the moduli. Let us, for example, use the relative prime moduli of 3, 4 and 7 for conveying the 6-bit symbols of an RNS-based CDMA system. As seen in the previous chapter, the information symbol's dynamic range becomes $3 \cdot 4 \cdot 7 = 84$, which is sufficient for conveying 6 bits per symbol. The required number of correlators is $3 + 4 + 7 = 14$, instead of the 64 correlators of the IS-95 system.

18.2.2.3 Numerical Results and Analysis

Figure 18.3 shows the bit error rate (BER) performance of both the conventional M-ary and that of the RNS-based systems using orthogonal modulation. In this scenario, $K = 1$ user was supported. In this case there exists no multi-user interference and the system can be viewed as a conventional system using orthogonal modulation without spreading. The number of moduli U was assumed to be three and the associated moduli values were $(4, 5, 7)$, (8,11,13) and (20,21,23), respectively, yielding dynamic ranges of $(4 \cdot 5 \cdot 7 = 140)$,

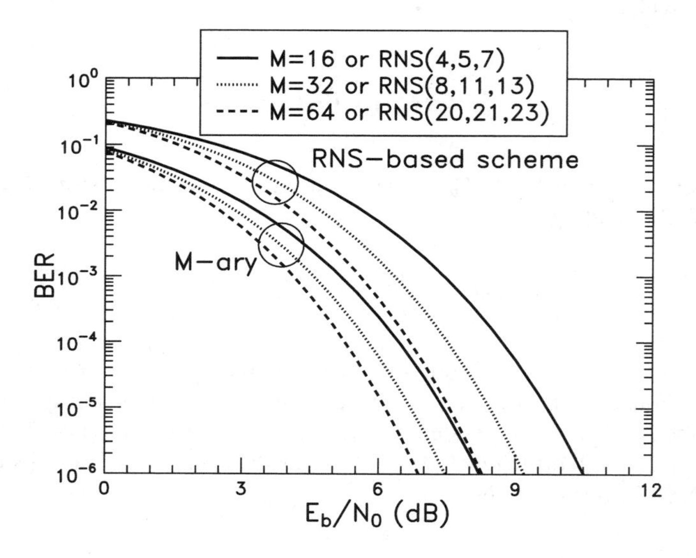

Figure 18.3: Bit error rate (BER) performance of the coherent M-ary orthogonal system ($U = 1$) and the RNS-based orthogonal system of Figure 18.2 using $U = 3$ moduli over AWGN channels. The value of M and the moduli used are listed at the top of the figure.

$(8 \cdot 11 \cdot 13 = 1144)$ and $(20 \cdot 21 \cdot 23 = 9660)$ and hence transmitting $\lfloor \log_2 140 \rfloor = 7$, $\lfloor \log_2 1144 \rfloor = 10$ and $\lfloor \log_2 9660 \rfloor = 13$ bits per symbol. The required number of correlators were $(4+5+7 = 16)$, $(8+11+13 = 32)$ and $(20+21+23 = 64)$, respectively. Hence the conventional M-ary benchmarker scheme was also configured to operate with $M = 16$, 32 and 64 correlators and orthogonal waveforms. From the results of Figure 18.3 it is explicit that the RNS-based system is outperformed by the conventional M-ary based system for each M value in terms of associated the BER. However, the RNS-based system can transmit 7, 10 and 13 bits, respectively, while the M-ary system can only transmit 4, 5 and 6 bits per symbol. This implies that the throughput of the RNS-based system is 7/4, 2, 13/6 times of that of the M-ary system, respectively. Hence, it is only natural that the RNS-based system exhibits a lower error resilience. As stated before, the RNS-based orthogonal system is actually a parallel m_u-ary system using different values of m_u, in contrast to the M-ary system using $M = 2^b$, $b = 1, 2, 3, \ldots$. Consequently, in the RNS-based system we have a larger integer transmission alphabet than the M-ary system,when using the same number of orthogonal waveforms and correlators. Nevertheless, the most important advantage of the RNS-based system is that the system can exploit the inherent advantages of the RNS-based arithmetic and employ RRNS codes for error-correction/detection.

We note furthermore at this stage that the associated E_b/N_0 difference between the RNS-based and the conventional M-ary CDMA system is only about 1-1.5dB at a given BER. This SNR-performance difference is significantly lower than the approximately 6dB extra power requirement encountered, when doubling the throughput of Quadrature Amplitude Modulation (QAM) [76], for example, from 2bit/symbol to 4bit/symbol. A further doubling of the

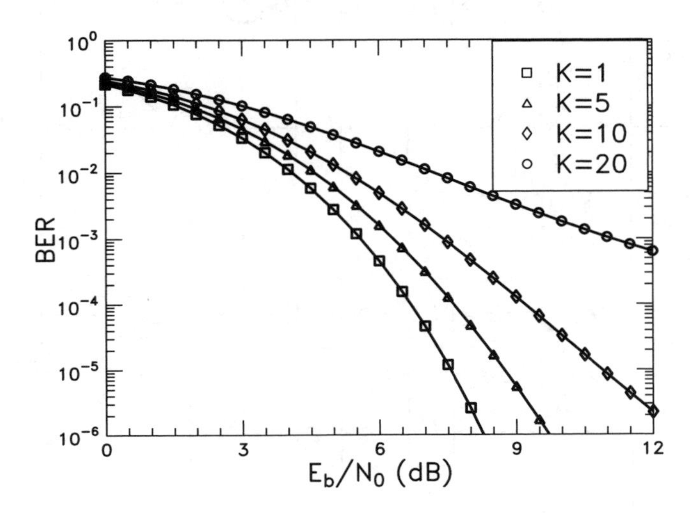

Figure 18.4: Bit error rate (BER) performance of the coherent RNS-based parallel DS-CDMA system of Figure 18.2 using $U = 3$ moduli of $3,\ 4$ and 7 supporting different numbers of users over AWGN channels.

4bit/symbol throughput to 8bit/symbol would require about 12dB extra SNR at a given BER. In the light of these comparisons the 1-1.5dB added SNR requirement of our RNS-based CDMA system is a worthwhile investment for doubling the throughput. Finally, during our further discourse, in the context of Figure 18.10 and Figure 18.11 over multipath Rayleigh fading channels, we will characterize the system's performance upon transmitting the same number of bits per M-ary symbol, where the conventional M-ary system requires 128 correlators, while the RNS-based scheme only 16, implying a factor eight lower complexity. Figure 18.4 portrays the BER versus E_b/N_0 performance of the RNS-based system using $U = 3$ number of moduli, whose values were $m_1 = 20$, $m_2 = 21$, $m_3 = 23$, $N = 64$ chips per bit and the number of CDMA users supported was $K = 1,\ 5,\ 10$ and 20. From these results a graceful degradation of the BER performance can be observed, when the number of simultaneous users, K, increases. The RNS-based system is an interference-limited system and hence multi-user detection techniques [78] can be invoked to mitigate the multi-user interference.

18.2.3 Performance of Non-coherent Receiver

18.2.3.1 Receiver Model

The non-coherent receiver block diagram for the RNS-based parallel DS-CDMA system using orthogonal modulation is similar to that of the coherent receiver block diagram of Figure 18.2, except for the non-coherent carrier demodulation and decorrelation sections, namely

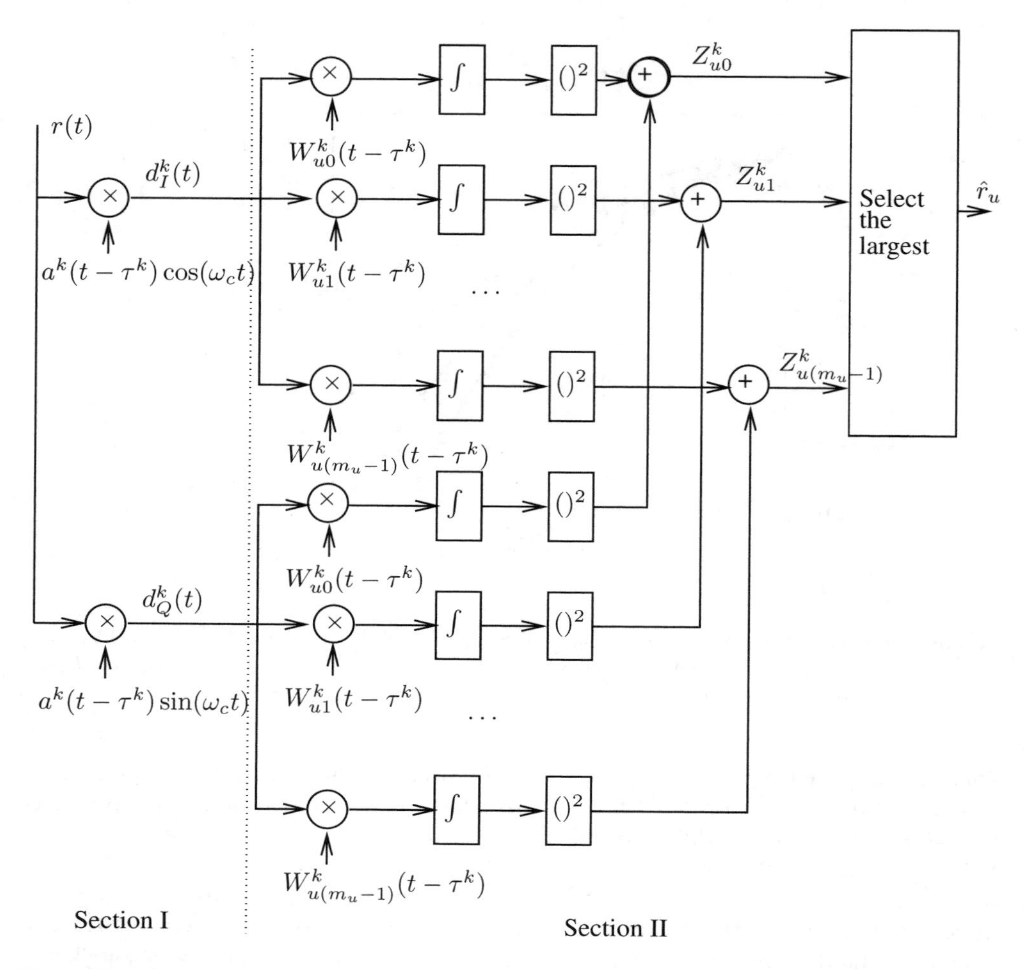

Figure 18.5: The non-coherent receiver block diagram for the RNS-based parallel DS-CDMA system using orthogonal modulation, replacing Section I and Section II of Figure 18.2.

Section I and II, which are now portrayed in Figure 18.5 in the context of receiving residue digit r_u of user k. Let us now derive the error probability expression for the RNS-based parallel DS-CDMA system employing the non-coherent orthogonal demodulation scheme of Figure 18.5.

18.2.3.2 The Non-coherent Receiver's Probability of Error

As we have assumed for the coherent case, let the first user, i.e. user $k = 1$, be the reference user and consider the non-coherent receiver using maximum likelihood detection (MLD) for each residue digit. The superscript of the reference user's signal is also omitted for convenience. We assumed perfect spreading code tracking for the reference user, i.e., $\tau = 0$. However, the carrier phase was set to $\phi \neq 0$, since non-coherent demodulation was employed.

From the Figure 18.5 we infer that the decision variables Z_{1m}, where $m = 0, 1, \ldots, m_1 - 1$ in the context of receiving the first residue digit r_1, can be expressed as:

$$Z_{1m} = (Z_{1m}^I)^2 + (Z_{1m}^Q)^2, \tag{18.33}$$

where the in-phase and quadrature channel outputs Z_{1m}^I and Z_{1m}^Q are expressed as:

$$Z_{1m}^I = \int_0^{T_s} r(t)W_{1m}(t)a(t)\cos(\omega_c t)dt, \tag{18.34}$$

$$Z_{1m}^Q = \int_0^{T_s} r(t)W_{1m}(t)a(t)\sin(\omega_c t)dt. \tag{18.35}$$

Similarly to our approach employed for the analysis of the coherent receiver, Z_{1m}^I can be expressed as:

$$Z_{10}^I = \sqrt{\frac{P}{2U}}T_s\left[\cos\phi + \sum_{k=2}^{K}\sum_{u=1}^{U} I_{u,10}^{k,I} + N_{10}^I\right], \tag{18.36}$$

$$Z_{1m}^I = \sqrt{\frac{P}{2U}}T_s\left[\sum_{k=2}^{K}\sum_{u=1}^{U} I_{u,1m}^{k,I} + N_{1m}^I\right], \quad m = 1, 2, \ldots, (m_1 - 1), \tag{18.37}$$

where $I_{u,1m}^{k,I}$ for $m = 0, 1, \ldots, (m_1 - 1)$ represents the interference inflicted upon the in-phase channel by user k associated with the transmission of residue digit r_u and can be expressed as:

$$I_{u,1m}^{k,I} = \frac{\cos(\theta^k)}{T_s}\int_0^{T_s} W_{ur_u}^k(t-\tau^k)W_{1m}(t)a^k(t-\tau^k)a(t)dt, \tag{18.38}$$

where $\theta^k = \phi^k - \phi$. Furthermore, N_{1m}^I for $m = 0, 1, \ldots, (m_1 - 1)$ represents the terms due to the presence of the AWGN in the in-phase channel, which can be expressed as:

$$N_{1m}^I = \left(\sqrt{\frac{P}{2U}}T_s\right)^{-1}\int_0^{T_s} n(t)W_{1m}(t)a(t)\cos(\omega_c t)dt. \tag{18.39}$$

Similarly, Z_{1m}^Q can be expressed as:

$$Z_{10}^Q = \sqrt{\frac{P}{2U}}T_s\left[-\sin\phi + \sum_{k=2}^{K}\sum_{u=1}^{U} I_{u,10}^{k,Q} + N_{10}^Q\right], \tag{18.40}$$

$$Z_{1m}^Q = \sqrt{\frac{P}{2U}}T_s\left[\sum_{k=2}^{K}\sum_{u=1}^{U} I_{u,1m}^{k,Q} + N_{1m}^Q\right], \quad m = 1, 2, \ldots, (m_1 - 1), \tag{18.41}$$

where $I_{u,1m}^{k,Q}$ for $m = 0, 1, \ldots, (m_1 - 1)$ represents the interference inflicted upon the quadrature phase channel, which can be expressed as:

$$I_{u,1m}^{k,Q} = \frac{-\sin(\theta^k)}{T_s} \int_0^{T_s} W_{ur_u}^k(t - \tau^k) W_{1m}(t) a^k(t - \tau^k) a(t) dt. \quad (18.42)$$

Moreover, N_{1m}^Q for $m = 0, 1, \ldots, (m_1 - 1)$ can be expressed as:

$$N_{1m}^Q = \left(\sqrt{\frac{P}{2U}} T_s\right)^{-1} \int_0^{T_s} n(t) W_{1m}(t) a(t) \sin(\omega_c t) dt. \quad (18.43)$$

Following our discussions in the context of the coherent receiver, it can be shown that N_{1m}^I and N_{1m}^Q are two independent Gaussian random variables with zero-mean and common variance of $(UN_0/2E_s)$. The multiple user interference terms $I_{u,1m}^{k,I}$ and $I_{u,1m}^{k,Q}$ can also be approximated as independent Gaussian random variables with zero-mean and common variance $1/3N_s$. Hence, Z_{1m}^I and Z_{1m}^Q for $m = 0, 1, \ldots, m_1 - 1$ are all independent Gaussian random variables having the common normalized variance expressed in Equation 18.18. Consequently, the decision variable Z_{10} of Equation 18.33 is a non-central chi-square distributed random variable, while the decision variables Z_{1m} of Equation 18.33 for $m \neq 0$ are central chi-square distributed random variables and all of them have two degrees of freedom. Their probability density functions (PDF) can be expressed, respectively, as [90]:

$$f_{Z_{10}}(x) = \frac{1}{2\sigma^2} \exp\left(-\frac{x+1}{2\sigma^2}\right) I_0\left(\frac{\sqrt{x}}{\sigma^2}\right), \quad x \geq 0, \quad (18.44)$$

$$f_{Z_{1m}}(x) = \frac{1}{2\sigma^2} \exp\left(-\frac{x}{2\sigma^2}\right), \quad x \geq 0, \ m = 1, 2, \ldots, m_1 - 1, \quad (18.45)$$

where $I_\alpha(x)$ is the αth-order modified Bessel function of the first kind, which may be represented by the series [90]:

$$I_\alpha(x) = \sum_{k=0}^{\infty} \frac{(x/2)^{\alpha+2k}}{k!\Gamma(\alpha + k + 1)}, \quad x \geq 0, \quad (18.46)$$

where $\Gamma(p)$ is the gamma function, defined as [90]:

$$\begin{aligned} \Gamma(p) &= \int_0^{\infty} t^{p-1} \exp(-t) dt, \quad p > 0, \\ \Gamma(p) &= (p-1)!, \quad \text{if } p \text{ is an integer}, \ p > 0, \\ \Gamma(\tfrac{1}{2}) &= \sqrt{\pi}, \quad \Gamma(\tfrac{3}{2}) = \tfrac{1}{2}\sqrt{\pi}. \end{aligned}$$

Let $Z_{1m,max} = \max\left\{Z_{11}, Z_{12}, \ldots, Z_{1(m_1-1)}\right\}$, where it can be shown that the PDF of

$Z_{1m,max}$ can be expressed as:

$$f_{Z_{1m,max}}(x) = \frac{m_1 - 1}{2\sigma^2}\left[1 - \exp\left(-\frac{x}{2\sigma^2}\right)\right]^{m_1-2} \exp\left(-\frac{x}{2\sigma^2}\right). \tag{18.47}$$

The probability $P_{r_1}(C)$ of correctly receiving the residue digit r_1 is given by the probability that Z_{10} exceeds $Z_{1m,max}$, which can be expressed as:

$$\begin{aligned} P_{r_1}(C) &= P(Z_{1m,max} < Z_{10}) \\ &= \int_0^\infty P(Z_{1m,max} < x) f_{Z_{10}}(x)dx \\ &= \int_0^\infty \left[\int_0^x f_{Z_{1m,max}}(y)dy\right] f_{Z_{10}}(x)dx. \end{aligned} \tag{18.48}$$

Substituting Equation 18.44 and Equation 18.47 into the above equation, it can be simplified to:

$$P_{r_1}(C) = \frac{1}{2\sigma^2}\int_0^\infty \left[1 - \exp\left(-\frac{x}{2\sigma^2}\right)\right]^{m_1-1} \exp\left(-\frac{x+1}{2\sigma^2}\right) I_0\left(\frac{\sqrt{x}}{\sigma^2}\right) \tag{18.49}$$

Let $y = \sqrt{x}/\sigma$ in Equation 18.49 allowing us to express the above distributions as:

$$P_{r_1}(C) = \int_0^\infty y\left[1 - \exp\left(-\frac{y^2}{2}\right)\right]^{m_1-1} \exp\left(-\frac{y^2+2\gamma}{2}\right) I_0\left(y\sqrt{2\gamma}\right), \tag{18.50}$$

where γ is given by Equation 18.23. Since the term $\left[1 - \exp\left(-\frac{y^2}{2}\right)\right]^{m_1-1}$ can be extended as follows:

$$\left[1 - \exp\left(-\frac{y^2}{2}\right)\right]^{m_1-1} = \sum_{n=0}^{m_1-1}(-1)^n \binom{m_1-1}{n} \exp\left(-\frac{ny^2}{2}\right), \tag{18.51}$$

substituting this result into Equation 18.50 and integration over y yields the probability of correct reception for the residue digit r_1 as:

$$P_{r_1}(C) = \sum_{n=0}^{m_1-1}(-1)^n \binom{m_1-1}{n} \frac{1}{n+1} \exp\left(-\frac{n\gamma}{n+1}\right). \tag{18.52}$$

The probability of erroneous reception for the residue digit r_1 can be expressed as:

$$P_{r_1}(\varepsilon) = \sum_{n=1}^{m_1-1}(-1)^{n+1} \binom{m_1-1}{n} \frac{1}{n+1} \exp\left(-\frac{n\gamma}{n+1}\right). \tag{18.53}$$

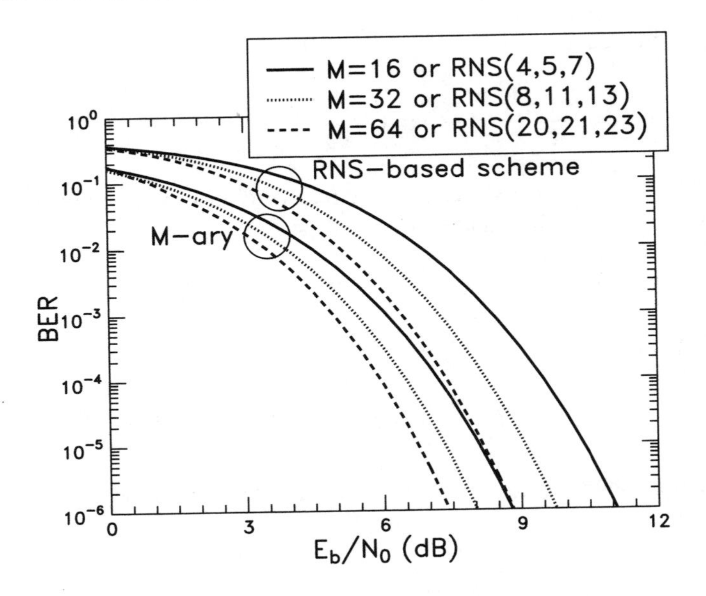

Figure 18.6: Probability of a bit error for non-coherent detection of the RNS-based orthogonal system of Figure 18.5 using $U = 3$ moduli over AWGN channels. The value of M and the moduli are shown at the top of the figure.

Similarly, the probability $P_{r_u}(C)$ for $u = 2, 3, \ldots, U$ of correctly receiving the residue digit r_u can be derived according to the approach we adopted during the derivation of $P_{r_1}(C)$. It can be shown that $P_{r_u}(C)$ takes the same form as Equation 18.52 with m_1 replaced by m_u. Finally, the probability $P_s(C)$ of correct symbol recovery can be expressed as the product of receiving all residue digits constituting the symbol correctly, which is given by Equation 18.30, while the average error probability $P_b(\varepsilon)$ is approximated by Equation 18.31.

18.2.3.3 Numerical Results

Figure 18.6 shows the associated bit error probability of the non-coherent RNS-based CDMA system of Figure 18.5 as a function of E_b/N_0 using $m_1 = 4,\ m_2 = 5,\ m_3 = 7$; $m_1 = 8,\ m_2 = 11,\ m_3 = 13$ and $m_1 = 20,\ m_2 = 21,\ m_3 = 23$ as well as for the conventional M-ary orthogonal system using $M = 16$, 32 and 64, respectively. Here the number of users supported was $K = 1$, i.e. no multi-user interference was inflicted. In this scenario both the RNS-based and the M-ary scheme can be considered conventional orthogonal modulated systems with or without spreading. Since $\sum_{i=1}^{3} m_i = 16$, 32 and 64, both the RNS-based system and the conventional M-ary system require the same bandwidth and the same number of correlators for a given value of M. Similarly to the coherent detection scheme of Figure 18.3, we observe that for any given bit error probability, the required E_b/N_0 decreases as M increases. However, for any given value of M – i.e. for a given bandwidth and a given number of correlators – the BER performance of the M-ary orthogonal system is better than that of the RNS-based orthogonal system. On the other hand, the

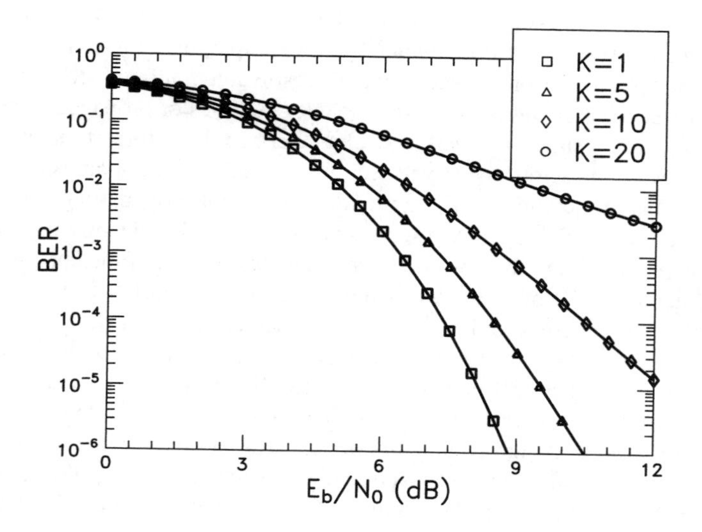

Figure 18.7: Probability of a bit error for non-coherent detection of the RNS-based parallel DS-CDMA orthogonal system using $U = 3$ moduli, whose values are, respectively, $m_1 = 20$, $m_2 = 21$ and $m_3 = 23$ under different numbers of users over AWGN channels.

RNS-based CDMA system exhibits a factor of 7/4, 2 and 13/6 higher throughput within the same bandwidth, which is the reason for its lower error resilience.

As we have seen in Figure 18.3 for the coherently detected RNS-based CDMA BER, only about 1-1.5dB extra E_b/N_0 investment is required, in order to approximately double the throughput of our proposed system. Furthermore, whilst in conventional QAM modems [76] the SNR difference between coherent and non-coherent detection is typically 3dB, this appears to be less than 1dB in our proposed RNS-based CDMA system upon comparing Figure 18.3 and Figure 18.6.

The effect of the number of active users on the bit error rate performance of the RNS-based system is shown in Figure 18.7. The bit error probability was computed as a function of E_b/N_0 for a different number of users with parameters of $m_1 = 20$, $m_2 = 21$, $m_3 = 23$, $U = 3$, and $K = 1$, 5, 10, 20. Since DS-CDMA systems are interference-limited, as expected, we observe that for any given bit error probability, the required E_b/N_0 increases, as the number of users, K, increases. Hence, in the multi-user environment it is desirable to invoke multi-user interference reduction techniques [78].

18.3 Performance in Multipath Fading channels

Wireless communication systems are subject to severe multipath fading, which can seriously degrade their performance. Thus, typically fading compensation is required, in order to mitigate the effects of multipath propagation. Diversity techniques – which essentially exploit receiving the signal over multiple independent fading channels – are used in practice to combat the effects of fading. Diversity reception can be implemented in wideband systems transmitting over frequency-selective fading channels for example, using RAKE reception [90]. The main idea of RAKE reception is to combine the resolvable multipath components, in order to increase the received signal-to-noise ratio (SNR). A series of propagation experiments conducted in typical urban/suburban areas have characterized the nature of the multipath mobile radio channel, which is determined by reflections, refractions and scattering by buildings and other obstructions in the vicinity of the mobile. Multipath propagation can be typically resolved at the receiver of a DS-CDMA system, since usually a significantly higher transmission bandwidth is required than the so-called coherence bandwidth of the channel [78]. Due to the wideband nature of the propagation scenario encountered, in DS-CDMA systems the receiver can exploit the inherent multipath diversity characteristics of the channel, in order to improve the signal's reception. Coherent detection in conjunction with maximum ratio combining (MRC) of the received paths is known to give best performance results, when perfect channel estimation is possible [90]. However, on the uplink of a DS-CDMA cellular system – due to the associated high complexity of coherent modulation/demodulation, which would require a pilot signal for each user – non-coherent schemes using sub-optimum equal gain combining (EGC) or selection combining (SC) are usually employed for the sake of decreasing the complexity of the system. In this section we focus our attention on the effects of the multipath fading channel, each path obeying a Rayleigh probability distribution. The bit error rate performance of the proposed RNS-based parallel DS-CDMA system using orthogonal modulation in conjunction with the above mentioned coherent MRC diversity combining scheme and two non-coherent diversity combining schemes – namely for EGC and SC – is studied and the effects of the RNS arithmetic on the bit error rate performance are characterized. Let us first consider the multipath Rayleigh fading channel model.

18.3.1 Multipath Slow-Fading Channel Model

We consider a multipath channel, where the transmitted signal is received over L independent slowly-varying flat fading channels, as shown in Figure 18.8. The complex low-pass equivalent representation of the impluse response [90] experienced by user k is given by:

$$h_k(t) = \sum_{l=1}^{L} \alpha_l^k \delta\left(t - \tau_l^k\right) \exp(-j\psi_l^k), \tag{18.54}$$

where l is the index of the channel impulse response (CIR) taps, $\delta(\cdot)$ is the Dirac function, $\left\{\alpha_l^k\right\}_{l=1}^L$, $\left\{\psi_l^k\right\}_{l=1}^L$ and $\left\{\tau_l^k\right\}_{l=1}^L$ are the random CIR tap amplitudes, phases and delays, respectively. We assume that $\left\{\alpha_l^k\right\}_{l=1}^L$, $\left\{\psi_l^k\right\}_{l=1}^L$ and $\left\{\tau_l^k\right\}_{l=1}^L$ are mutually independent. Under slowly fading conditions, we can assume that $\left\{\alpha_l^k\right\}_{l=1}^L$, $\left\{\psi_l^k\right\}_{l=1}^L$ and $\left\{\tau_l^k\right\}_{l=1}^L$ are all constant over a symbol interval. The first CIR tap of the reference user – which is assumed to be the first user – is assumed to be the reference channel having a delay of $\tau_1 = 0$. Without loss of generality, we assume that $\tau_1^k < \tau_2^k < \cdots < \tau_L^k$.

For a multipath Rayleigh fading channel the fading amplitudes $\left\{\alpha_l^k\right\}_{l=1}^L$ are assumed to

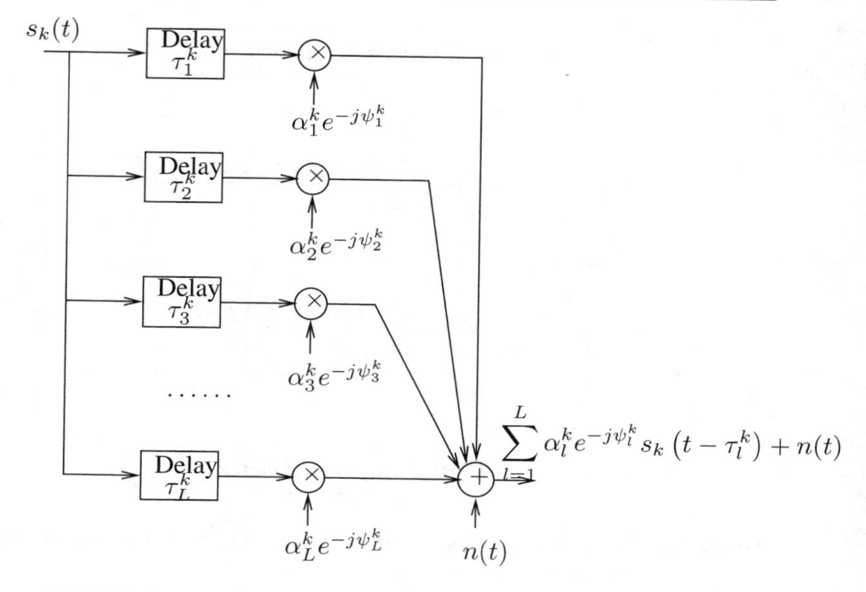

Figure 18.8: Wireless multipath slowly-varying channel model.

be statistically independent random variables having a PDF expressed as:

$$\begin{aligned} f(\alpha_l^k) &= M(\alpha_l^k, \Omega), \\ M(R, \Omega) &= \frac{2R}{\Omega} \exp\left(-\frac{R^2}{\Omega}\right), \end{aligned} \tag{18.55}$$

where $\Omega = E\left[(\alpha_l^k)^2\right]$. The phases $\{\phi_l^k\}_{l=1}^{L}$ of the different paths are assumed to be uniformly distributed random variables in $[0, 2\pi)$, while the kth user's path delays of $\{\tau_l^k\}_{l=1}^{L}$ are uniformly distributed in $[0, T_s)$. We also assume that ideal power control is employed, in order that the received signal powers are the same for all K users. Then for the transmitted signal taking the form of Equation 18.5 the received signal at the base station generated by the K users can be expressed as:

$$r(t) = \sum_{k=1}^{K} y_k(t) + n(t), \tag{18.56}$$

where

$$y_k(t) = \sum_{k=1}^{K} \sum_{l=1}^{L} \sum_{u=1}^{U} \alpha_l^k \sqrt{\frac{2P}{U}} W_{ur_u}^k (t - \tau_l^k) a^k (t - \tau_l^k) \cos\left(\omega_c t - \phi_l^k\right), \tag{18.57}$$

and $\phi_l^k = \varphi^k - \psi^k - \omega_c \tau_l^k$, $n(t)$ represents the AWGN, which is modelled as a Gaussian

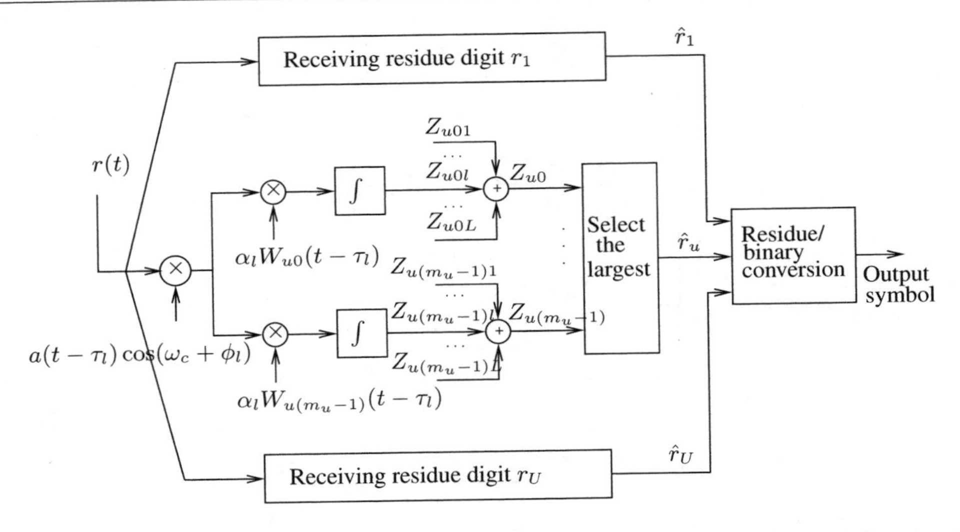

Figure 18.9: The L-branch RAKE receiver for detecting the reference user associated with $k = 1$ and moduli $m_1, m_2, \ldots, m_U$.

random variable with zero mean and double-sided power spectral density of $N_0/2$.

18.3.2 Performance of the Coherent RNS-Based CDMA Receiver Using Maximum Ratio Combining

18.3.2.1 Receiver Model

In order to combat the multipath distortion and achieve an improved performance, in this section diversity combining is employed. For the multipath Rayleigh fading channels, the combiner that achieves the best performance is the one in which each receiver branch output is multiplied by the corresponding complex-valued path gain α_l^k. Let the first user, i.e. user $k = 1$ be the reference user and consider the coherent correlator RAKE receiver invoking MRC [407], as shown in Figure 18.9, where the superscript of the reference user's signal has been omitted for convenience. Moreover, perfect estimations of $\{\phi_l\}$, $\{\tau_1\}$ is assumed, in order to implement coherent detection. Consequently, for the received signal given in the form of Equation 18.56, after demodulation and despreading in Figure 18.9, the baseband signal matched to the nth path, $1 \leq n \leq L$, can be formulated as:

$$\begin{aligned}
& r(t) \cdot a(t-\tau_n)\cos(\omega_c t + \phi_n) \\
&= \sum_{l=1}^{L}\sum_{u=1}^{U} \alpha_l \cos(\theta_l)\sqrt{\frac{P}{2U}} W_{ur_u}(t-\tau_l)a(t-\tau_l)a(t-\tau_n) \\
&+ \sum_{k=2}^{K}\sum_{l=1}^{L}\sum_{u=1}^{U} \alpha_l^k \cos\left(\theta_l^k\right)\sqrt{\frac{P}{2U}} W_{ur_u}^k(t-\tau_l^k)a^k(t-\tau_l^k)a(t-\tau_n) \\
&\quad + n(t)a(t-\tau_n)\cos(\omega_c t + \phi_n),
\end{aligned} \tag{18.58}$$

where the first term on the right-hand side is the signal of interest, the second term is due to the multi-user interference, while the third term is contributed by the AWGN.

Since according to our previous chapter the operations involving the residue digits belonging to the different moduli $m_u, \quad u = 1, 2, \ldots, U$ are mutually independent, let us analyze the bit error rate performance upon receiving the first residue digit r_1. Without loss of generality, we assume that $r_1 = 0$. According to Figure 18.9 the nth branch output of the RAKE receiver matched to the residue digit r_1 can be expressed as:

$$Z_{1mn} = \int_{\tau_n}^{T_s+\tau_n} r(t)a(t-\tau_n)\cos(\omega_c t + \phi_n) \cdot \alpha_n W_{1m}(t-\tau_n)dt, \tag{18.59}$$

where $m = 0, 1, \ldots, m_1 - 1$, $n = 1, 2, \ldots, L$. Upon substituting Equation 18.58 into Equation 18.59 it can be shown that Z_{1ml} assumes the form of:

$$Z_{10n} = \sqrt{\frac{P}{2U}}\left[\alpha_n^2 + \sum_{l=1,l\neq n}^{L}\sum_{u=1}^{U} I_{S,u10} + \sum_{k=2}^{K}\sum_{l=1}^{L}\sum_{u=1}^{U} I_{M,u10}^k + N_{10n}\right], \tag{18.60}$$

$$\begin{aligned}
Z_{1mn} &= \sqrt{\frac{P}{2U}}\left[\sum_{l=1,l\neq n}^{L}\sum_{u=1}^{U} I_{S,u1m} + \sum_{k=2}^{K}\sum_{l=1}^{L}\sum_{u=1}^{U} I_{M,u1m}^k + N_{1mn}\right], \\
& m = 1, 2, \ldots, m_1 - 1,
\end{aligned} \tag{18.61}$$

where α_n^2 is the desired term, $I_{S,u1m}$ is the self-interference inflicted by the user-of-interest, which can be expressed as:

$$I_{S,u1m} = \frac{\alpha_n \alpha_l \cos(\theta_l)}{T_s}\int_{\tau_n}^{T_s+\tau_n} W_{ur_u}(t-\tau_l)W_{1m}(t-\tau_n)a(t-\tau_l)a(t-\tau_n)dt. \tag{18.62}$$

Furthermore, $I_{M,u1m}^k$ in Equation 18.60 and Equation 18.61 for $m = 0, 1, \ldots, m_1 - 1$ represents the interference from the interfering user k, which can be expressed as:

$$I_{M,u1m}^{k} = \frac{\alpha_n \alpha_l^k \cos(\theta_l^k)}{T_s} \int_{\tau_n}^{T_s+\tau_n} W_{ur_u}^{k}(t-\tau_l^k) W_{1m}(t-\tau_n) a(t-\tau_l^k) a(t-\tau_n) dt, \tag{18.63}$$

and finally, the noise term N_{1mn} due to the AWGN is an independent Gaussian random variable, which can be expressed as:

$$N_{1mn} = \left(\sqrt{\frac{P}{2U}} T_s\right)^{-1} \alpha_n \int_{\tau_n}^{T_s+\tau_n} n(t) W_{1m}(t-\tau_n) a(t-\tau_n) \cos(\omega_c t + \phi_n) dt. \tag{18.64}$$

for $m = 0, 1, \ldots, m_1 - 1$ and $n = 1, 2, \ldots, L$.

The output decision variables associated with receiving residue digit r_1 in Figure 18.9 upon using MRC can be expressed as:

$$Z_{10} = \sum_{n=1}^{L} Z_{10n}, \tag{18.65}$$

$$Z_{1m} = \sum_{n=1}^{L} Z_{1mn}, \quad m = 1, 2, \ldots, m_1 - 1. \tag{18.66}$$

The receiver of Figure 18.9 uses the maximum likelihood decision rule for the detection of residue r_1. For an orthogonal signalling scheme this decision rule is reduced to selecting the maximum from the set $\{Z_{1m}, m = 0, 1, \ldots, m_1 - 1\}$, and the index of the largest decision variable in Figure 18.9 denotes the estimation of the transmitted residue digit r_1. This decision rule is optimum for transmission over an AWGN channel using an orthogonal signalling scheme. However, the MAI is not necessarily Gaussian, hence this receiver is actually not optimal. It is commonly used, however, for its simplicity.

Using the same method, the estimates of all other transmitted residue digits of the residue sequence $(r_1, r_2, \ldots, r_U)$ of the user-of-interest can be obtained by selecting the maximum from the set $\{Z_{um}, m = 0, 1, \ldots, m_u - 1\}$ for $u = 1, 2, \ldots, U$. Let the received sequence be expressed by $(\hat{r}_1, \hat{r}_2, \ldots, \hat{r}_u)$, then the transmitted M-ary data symbol can be recovered by transforming this residue sequence into its corresponding binary or integer representation. Let us now derive the expressions of the conditional and the unconditional bit error rate performance by first considering the associated decision statistics. Note that the computation of the exact probability of error for the M-ary orthogonal system using MRC is an arduous task. Hence, in our analysis we derive the upper-bound performance, which has been shown to represent a close approximation of the exact performance [90], when the signal-to-noise ratio is sufficiently high.

18.3.2.2 Decision Statistics

Due to our assumption that $\{\alpha_l^k\}_{l=1}^L$, $\{\phi_l^k\}_{l=1}^L$ and $\{\tau_l^k\}_{l=1}^L$ are modelled as independent random variables for different users k and/or for different diversity paths l, furthermore, since random signature sequences are used, the self-interference term $I_{S,u1m}$ given by Equation 18.62 and the multiple-access interference term $I_{M,u1m}^k$ given by Equation 18.63 are also independent random variables. We infer from Equations 18.62 and 18.63 that the moments of all random variables $I_{S,u1m}$ and $I_{M,u1m}^k$ are finite. Consequently, the Liapounoff version of the central-limit theorem [427] holds. Moreover, N_{1mn} is Gaussian, since $n(t)$ is Gaussian. Thus Z_{1m} is asymptotically Gaussian. Hence, the performance of the RNS-based parallel DS-CDMA system is estimated by employing the often used Gaussian approximations. According to Equations 18.60 to 18.61 we can argue that the approximation becomes tight, as the number of simultaneous users K, the number of moduli U and the number of possible diversity paths L increase.

A. Noise Analysis: The noise term N_{1mn} – which is given by Equation 18.64 – is a Gaussian random variable with zero-mean and variance $U\alpha_n^2 N_0/2E_s$ conditioned on a given fading attenuation α_n, where $E_s = bE_b$ is the transmitted energy per symbol period and E_b is the energy per bit.

B. Multipath Interference Analysis: The multipath-induced self-interference term given by Equation 18.62 is from the reference user, which includes two contributions: the self-interference from the $(L-1)$ number of path signals of residue digit r_1 and the self-interference due to the $(U-1)(L-1)$ number of path signals of the other $(U-1)$ residue digits. The term $I_{S,u1m}$ can be approximated as a Gaussian random variable with zero-mean, while its variance conditioned on a given fading attenuation α_n can be expressed as [428]:

$$\text{Var}\left(I_{S,u1m}|\alpha_n\right) = \frac{\alpha_n^2}{3N_s}\Omega, \tag{18.67}$$

when a rectangular chip waveform is used, where $\Omega = E\left[\left(\alpha_l^k\right)^2\right]$, as we defined previously.

C. Multiple-Access Interference (MAI) Analysis: The multiple-access interference term, due to the kth interfering user is given by Equation 18.63. It can also be approximated as a Gaussian random variable with zero-mean and conditional variance of [428]

$$\text{Var}\left(I_{M,u1m}^k|\alpha_n\right) = \frac{\alpha_n^2}{3N_s}\Omega, \tag{18.68}$$

when a rectangular chip waveform is used.

D. Decision Statistic and Conditional Symbol Error Probability: We have obtained the statistics for the noise, the multipath interference and the multi-user interference. The upper-bound symbol error probability conditioned on the magnitude of the L diversity paths $\{\alpha_n, n = 1, 2, \ldots, L\}$ can be derived by treating all interferences as additional noise. First, we obtain the error probabilities for receiving the residues $(r_1, r_2, \ldots, r_U)$, respectively, and then the symbol error probability is given by the product of the independent residue digit error probabilities according to the properties of the RNS summarized in the previous chapter. After normalization by $\sqrt{\frac{P}{2U}}T_s$, the decision statistics of Equation 18.65 and Equation 18.66 can be approximated as that of Gaussian random variables, having a mean given by:

$$E\left[Z_{10}|\left\{\alpha_n\right\}\right] = \sum_{n=1}^{L} \alpha_n^2, \tag{18.69}$$

for Z_{10}, and zero for Z_{1m} having $m \neq 0$. Their variance can be computed according to Equations (18.60)-(18.68), which is given by:

$$\sigma^2\left(Z_{1m}|\left\{\alpha_n\right\}\right) = \left[\frac{U(KL-1)}{3N_s} + \left(\frac{2\Omega E_s}{UN_0}\right)^{-1}\right] \cdot \Omega \sum_{n=1}^{L} \alpha_n^2, \tag{18.70}$$

where $N_s = N \cdot b = N \cdot \lfloor \log_2 \prod_{u=1}^{U} m_u \rfloor$.

Consequently, the error probability of residue digit r_1 conditioned on the diversity path attenuation $\{\alpha_n, n = 1, 2, \ldots, L\}$ is upper-bounded by [90][p. 263]:

$$P_{r_1}\left(\varepsilon|\left\{\alpha_n\right\}\right) \leq (m_1 - 1)Q\left(\sqrt{\gamma}\right), \tag{18.71}$$

where

$$\gamma = \gamma_c \sum_{n=1}^{L} \frac{\alpha_n^2}{\Omega}, \tag{18.72}$$

$$\gamma_c = \frac{1}{2} \cdot \left[\frac{U(KL-1)}{3N_s} + \left(\frac{2\Omega E_s}{UN_0}\right)^{-1}\right]^{-1}. \tag{18.73}$$

Similarly, the residue digit error probability upper-bound P_{r_u} for $u = 2, 3, \ldots, U$ conditioned on the diversity path attenuation $\{\alpha_n, n = 1, 2, \ldots, L\}$ upon receiving residue digit r_u can be derived using the same approach as in the context of r_1, which takes the same form as Equation 18.71 with r_1 and m_1 replaced by r_u and m_u, respectively.

If we assume that all the moduli are used exclusively for transmitting information and hence there are no redundant moduli in the residue number system, or in other words, if the dynamic range of the binary information data is $\left[0, \prod_{u=1}^{U} m_u\right)$, then the M-ary symbol is received correctly, if and only if, all the received m_u-ary residue digits are correct. Hence, after obtaining the conditional error probabilities of receiving residue digits $(r_1, r_2, \ldots, r_U)$ in Equation 18.71, the total M-ary symbol error probability conditioned on the diversity path attenuations $\{\alpha_n, n = 1, 2, \ldots, L\}$ is upper-bounded by:

$$P(\varepsilon|\{\alpha_n\}) \leq 1 - \prod_{u=1}^{U}[1 - P_{r_u}(\varepsilon|\{\alpha_n\})]. \quad (18.74)$$

The unconditional M-ary symbol error probability will be computed by treating the α_n terms as Rayleigh distributed random variables in the following subsection.

18.3.2.3 Upper-Bound of the Average Probability of Error

The M-ary symbol error probability given in Equation 18.74 is conditioned on the random variables $\{\alpha_n, n = 1, 2, \ldots, L\}$, representing the fading diversity path amplitudes. Hence, the (unconditional) average error probability depends on the probability distribution of $\sum_{n=1}^{L}\alpha_n^2$. Since α_n is Rayleigh-distributed – as shown in Equation 18.55 – $\sum_{n=1}^{L}\alpha_n^2$ is a chi-square-distributed random variable with $2L$ degrees of freedom, and the probability density function of $\gamma = \gamma_c\sum_{n=1}^{L}\alpha_n^2$ in Equation 18.72) is given by [90]:

$$p(\gamma) = \frac{1}{(L-1)!\gamma_c^L}\gamma^{L-1}\exp(-\frac{\gamma}{\gamma_c}), \quad (18.75)$$

where γ_c is given by Equation 18.73.

Due to the individual m_u-ary residue error probabilities $P_{r_u}(\varepsilon|\{\alpha_n\})$ being independent random variables, the upper-bound of the average M-ary symbol error probability can be computed as the expected value of the M-ary symbol error probabilities of Equation 18.74, which is given by:

$$\begin{aligned}\bar{P}_s(\varepsilon) &= E[P(\varepsilon|\{\alpha_n\})] \\ &\leq 1 - \prod_{u=1}^{U}(1 - E[P_{r_u}(\varepsilon|\{\alpha_n\})]),\end{aligned} \quad (18.76)$$

where

$$\begin{aligned}E[P_{r_u}(\varepsilon|\{\alpha_n\})] &\leq \int_0^\infty P_{r_u}(\varepsilon|\{\alpha_n\})p(\gamma)d\gamma \\ &= (m_u - 1)\left(\frac{1-\mu}{2}\right)^L \sum_{k=0}^{L-1}\binom{L-1+k}{k}\left(\frac{1+\mu}{2}\right)^k,\end{aligned} \quad (18.77)$$

and

$$\mu = \sqrt{\frac{\gamma_c}{2+\gamma_c}}. \quad (18.78)$$

Finally, the upper-bound of the bit error probability for the uncoded RNS-based parallel

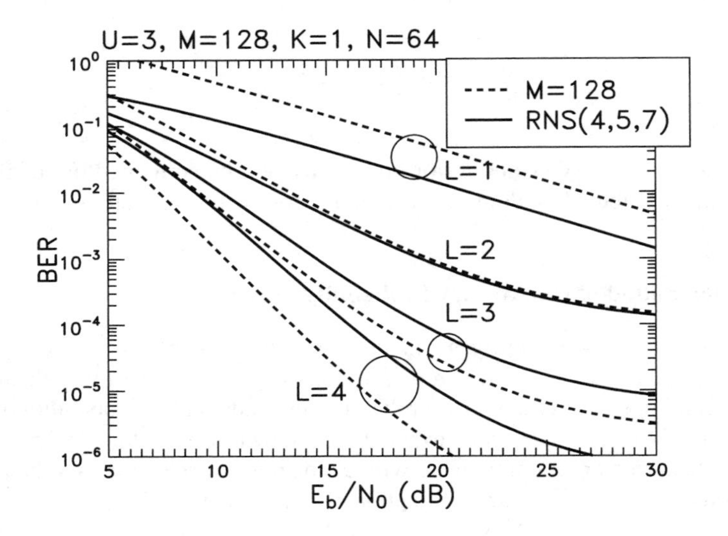

Figure 18.10: Bit error rate (BER) performance of the conventional M-ary DS-CDMA system and the RNS-based parallel DS-CDMA system of Figure 18.9 using coherent MRC and $U = 3$ moduli, namely (4,5,7), for different numbers of resolvable diversity components over multipath fading channels. Both systems are capable of transmitting 7-bits per symbol and the associated number of correlators was 128 and 16, respectively.

DS-CDMA system is computed from the symbol error probability of Equation 18.76, as follows [90]:

$$\bar{P}_b \leq \frac{2^{b-1}}{2^b - 1}\bar{P}_s(\varepsilon) \approx \frac{1}{2}\bar{P}_s(\varepsilon), \tag{18.79}$$

where b is the number of bits per M-ary symbol.

18.3.2.4 Numerical Results

In Figure 18.10 and Figure 18.11, the BER versus the average E_b/N_0 performance of the RNS-based DS-CDMA scheme with $u = 3$ and that of the conventional M-ary orthogonal DS-CDMA (that is $u = 1$,) are evaluated and compared, where the average E_b/N_0 is obtained by $E_b/N_0 = L\gamma_c/b$, and b is the number of bits per M-ary symbol. The number of users was $K = 1$, hence, there is no multiple-access interference inflicted upon the reference signal. In Figure 18.10 we let $m_1 = 4$, $m_2 = 5$, $m_3 = 7$, i.e., employ the RNS (4,5,7), while in Figure 18.11 we used $m_1 = 8$, $m_2 = 11$, $m_3 = 13$, i.e., the RNS (8,11,13). These RNS-based systems are compared with the conventional M-ary orthogonal system under the assumption that both the RNS-based system and the M-ary system have an equal throughput in terms of the number of bits per M-ary symbol. Since the number of bits per symbol

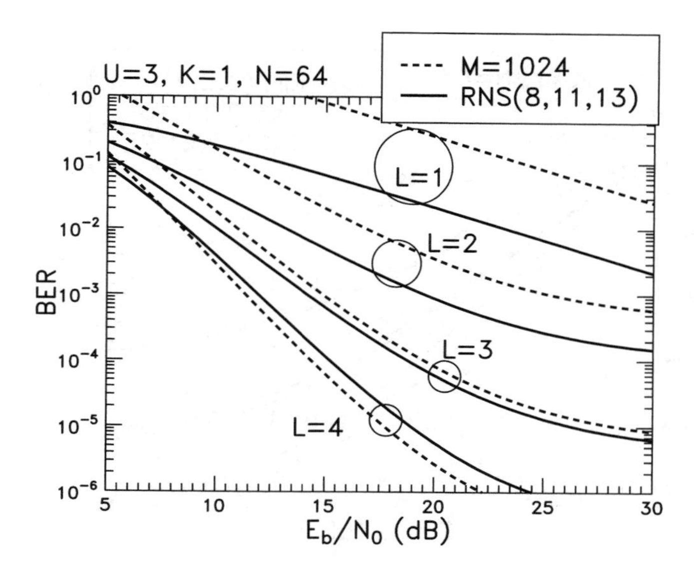

Figure 18.11: Bit error rate (BER) performance of the conventional M-ary DS-CDMA system and the RNS-based parallel DS-CDMA system of Figure 18.9 using coherent MRC and $U = 3$ moduli, namely (8,11,13), for different numbers of resolvable diversity components over multipath fading channels. Both systems are capable of transmitting 10-bits per symbol and the associated number of correlators was 1024 and 32, respectively.

conveyed by the RNS (4,5,7) is 7 and that by the RNS (8,11,13) is 10, consequently, the value of M should be $2^7 = 128$ and $2^{10} = 1024$ corresponding to Figure 18.10 and Figure 18.11.

From the results of Figure 18.10 and Figure 18.11, we observe that all the systems exhibit substantially improved performances, as L increases, although the performance improvement of the conventional M-ary system is relatively higher than that of the RNS-based system upon increasing L. Based on this observation, we found in Figure 18.10 and Figure 18.11 that at relatively low diversity orders of $L = 1, 2$ for RNS (4,5,7) and of $L = 1, 2, 3$ in conjunction with the RNS (8,11,13), the RNS-based system outperformed the conventional M-ary system, while requiring 16 or 32 correlators, instead of 128 or 1024, respectively. By contrast, at the relatively high diversity orders of $L = 3, 4$ for the RNS (4,5,7) and of $L = 4$ for the RNS (8,11,13) the RNS-based system is slightly outperformed by the substantially more complex conventional M-ary system. However, we have to emphasize again that the RNS (4,5,7) and RNS (8,11,13) required only 16 and 32 orthogonal functions, while the $M = 128$ and the $M = 1024$ conventional M-ary systems had to involve 128 and 1024 orthogonal functions, respectively, in order that these two systems exhibited the same throughput of 7bit/symbol and 10bits/symbol, respectively.

Figure 18.12 shows the bit error probability for the RNS-based system using moduli of $m_1 = 4, m_2 = 5, m_3 = 7$; $m_1 = 8, m_2 = 11, m_3 = 13$ and $m_1 = 20, m_2 = 21, m_3 = 23$, respectively, over dispersive Rayleigh fading channels using MRC. The other parameters considered were $N = 64$ chips per bit and diversity orders of $L = 1, 2, 3, 4$. In contrast to

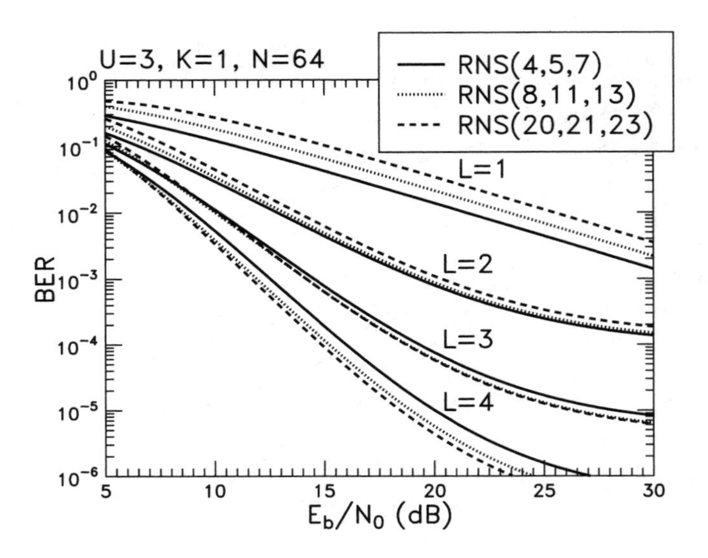

Figure 18.12: Bit error rate (BER) performance of the RNS-based parallel DS-CDMA system of Figure 18.9 using coherent MRC and $U = 3$ moduli for different numbers of resolvable diversity components over multipath fading channels.

the non-fading scenario of Figure 18.3 and Figure 18.6 – where the required E_b/N_0 value decreased for any given bit error probability, when the total number of orthogonal functions increased – the required E_b/N_0 for any given bit error probability increases, when increasing the total number of orthogonal functions in case of $L = 1,\ 2$, while for $L = 3, 4$ it decreases, upon increasing the total number of orthogonal functions. If a practical system has a diversity order of three and employs a MRC scheme, according to the results of Figure 18.12, the bit error probability is comparatively robust to the change of the total number of orthogonal functions. However, if the diversity order is low, for example, $L = 1$, or high, such as $L = 4$, the bit error probability is much more sensitive to the change of the total number of orthogonal functions.

Finally, Figure 18.13 portrays the bit error probability versus E_b/N_0 performance in conjunction with the parameters of $L = 2,\ 5$; $N = 64$; $K = 1,\ 5,\ 10,\ 20$; $u = 3$ and moduli of $m_1 = 20,\ m_2 = 21,\ m_3 = 23$. Notice the graceful degradation of the performance, as the number of active users, K, increases.

18.3.3 Performance of Non-coherent RNS-Based DS-CDMA Receiver Using Equal Gain Combining and Selection Combining

Orthogonal signalling is well suited for non-coherent reception and hence M-ary orthogonal signalling schemes in conjunction with non-coherent demodulation have been used for practical systems [366]. For the non-coherent demodulation of M-ary orthogonal signalling

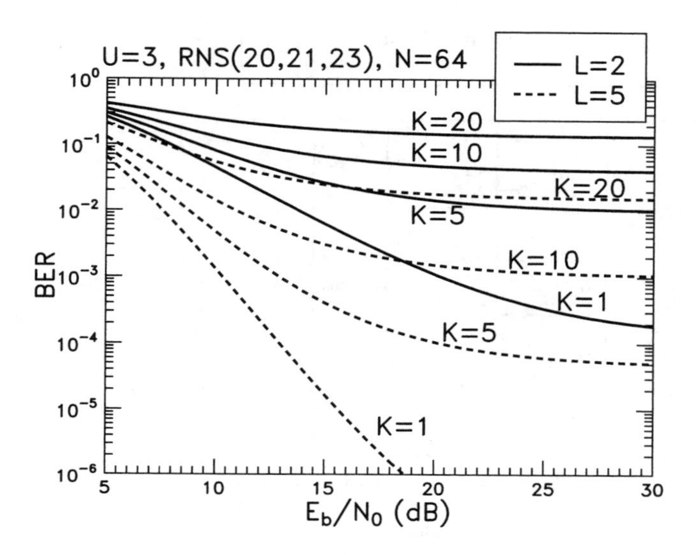

Figure 18.13: Bit error rate (BER) performance of the RNS-based parallel DS-CDMA system of Figure 18.9 using coherent MRC and $U = 3$ moduli, namely (20,21,23), for $L = 2, 5$ number of resolvable diversity components over multipath fading channels. The number of active users was $K = 1, 5, 10, 20$, respectively.

schemes equal gain combining (EGC) has been commonly used, whereby several multipath components are demodulated, equally weighted and then added non0coherently. This technique is considered the optimal combining technique for non-coherent demodulation [78]. However, from an implementation point of view, having a receiver complexity dependent on the number of resolvable paths is undesirable. Alternatively, a simple suboptimal selection combining (SC) scheme can be invoked [390], in which several multipath components are demodulated and the one having the highest amplitude (or signal-to-noise ratio) is selected for decision.

In this section we focus our attention on the multipath fading channel, each path obeying a Rayleigh probability distribution. The error probability of the proposed RNS-based parallel DS-CDMA system using orthogonal modulation for the above mentioned two non-coherent diversity combining schemes – namely for EGC and SC – is studied and the effects of the RNS arithmetic on the BER are characterized.

18.3.3.1 Receiver Model

The receiver schematic of the studied RNS-based parallel DS-CDMA system including square-law-based non-coherent M-ary demodulation, multipath diversity combining, maximum-likelihood detection (MLD), as well as residue-to-binary conversion is shown in Figure 18.14, where the detailed schematic of the square-law-based non-coherent M-ary demodulation block is shown in Figure 18.15, as an example, which is matched to the nth,

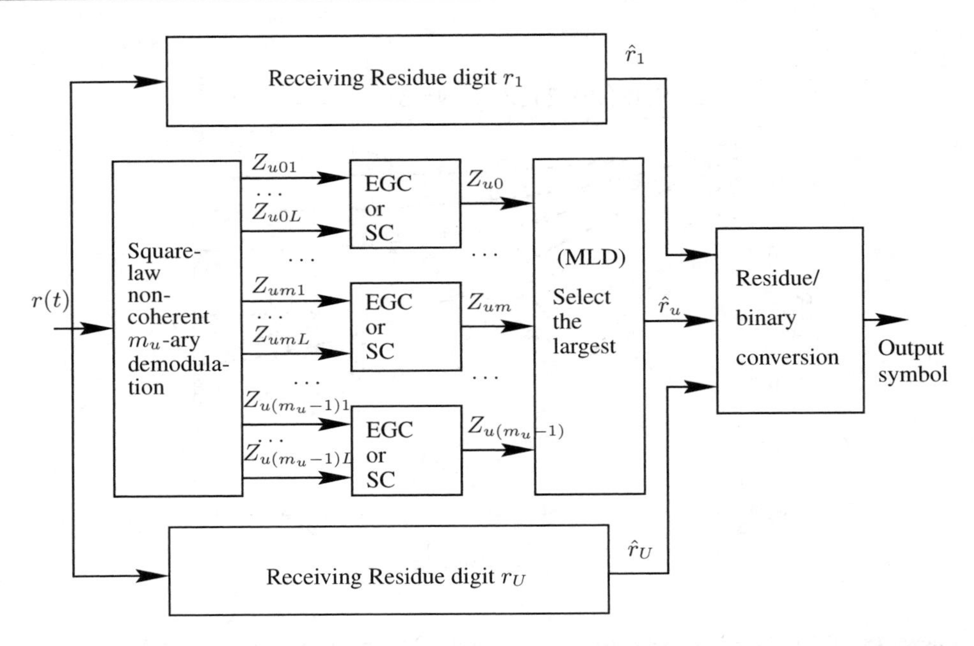

Figure 18.14: Receiver model of the RNS-based parallel DS-CDMA system investigated using non-coherent M-ary demodulation, multipath diversity combining, maximum-likelihood detection (MLD) and residue-to-binary conversion. The unit of square-law non-coherent m_u-ary demodulation is portrayed in Figure 18.15.

$(1 \leq n \leq L)$, path signal and to the uth m_u-ary residue digit r_u of the user-of-interest. Multipath diversity combining invoking the EGC or the SC principle was implemented in the 'EGC or SC' block. The MLD block of Figure 18.14 selects the largest from its input variables and maps it to the estimated m_u-ary residue digit $\hat{r}_u$. Following the MLD stage, the next block in Figure 18.14 implements the residue-to-binary conversion and finally outputs the binary information bits.

As shown in Figure 18.15, after down-conversion and despreading, the outputs of the in-phase and quadrature-phase channels signals are correlated with each of the m_u orthogonal functions. Let the first user be the user-of-interest and assume that the first residue digit $r_1 = 0$ is being received. Referring to Figure 18.15, the output variable of the nth branch of the mth correlator can be expressed as:

$$\begin{aligned} Z_{1mn} = & \left[\int_{\tau_n}^{T_s+\tau_n} r(t) W_{1m}(t-\tau_n) a(t-\tau_n) \cos(\omega_c t) dt \right]^2 \\ & + \left[\int_{\tau_n}^{T_s+\tau_n} r(t) W_{1m}(t-\tau_n) a(t-\tau_n) (\sin \omega_c t) dt \right]^2 \\ & m = 1, 2, \ldots, m_1 - 1; \;\; n = 1, 2, \ldots, L. \end{aligned} \tag{18.80}$$

Pursuing a similar approach to that in the coherent receiver of Figure 18.9 employing

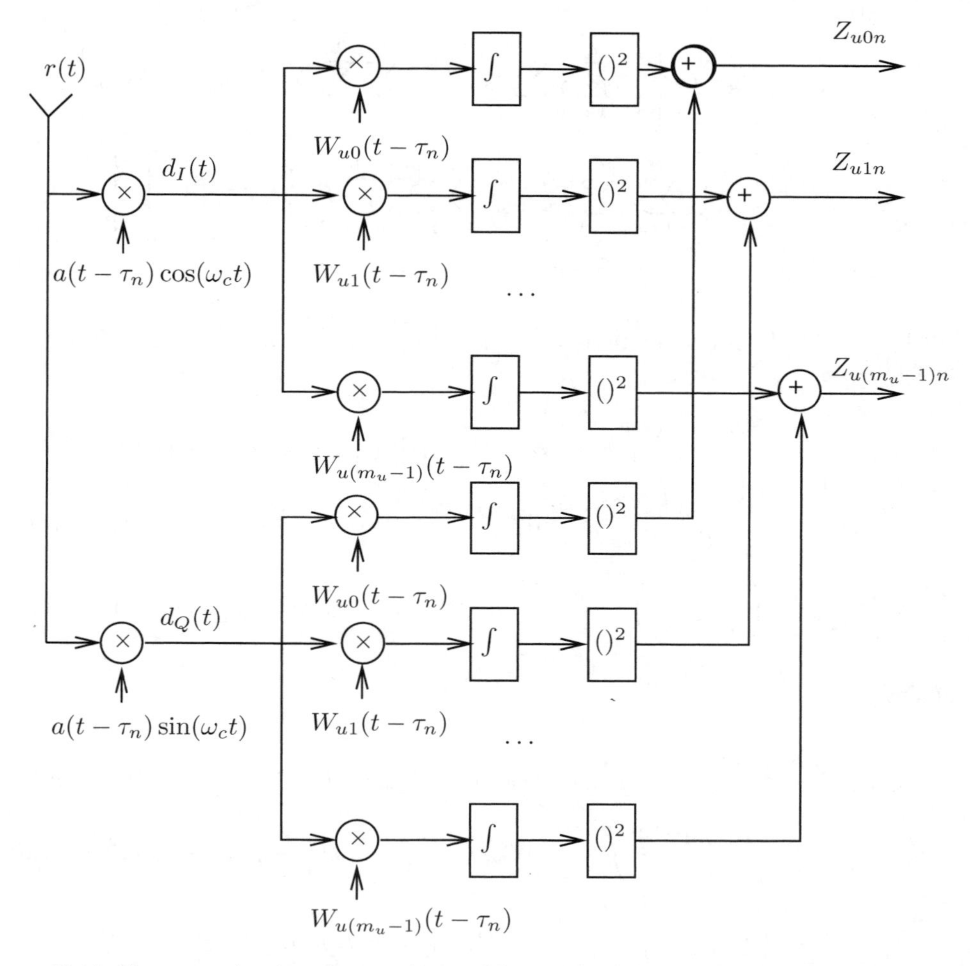

Figure 18.15: The square-law-based non-coherent M-ary demodulation block extracted from Figure 18.14 matched to the nth path signal and the uth residue digit r_u for the user-of-interest.

MRC, it can be shown that Equation 18.80 can be formulated as:

$$Z_{10n} = \frac{E_s}{2U} T_s \left\{ \left[\alpha_n \cos(\phi_n) + \sum_{l=1,l\neq n}^{L} \sum_{u=1}^{U} I_{S,u10}^{I} + \sum_{k=2}^{K} \sum_{l=1}^{L} \sum_{u=1}^{U} I_{M,u10}^{k,I} + N_{10n}^{I} \right]^2 + \left[-\alpha_n \sin(\phi_n) + \sum_{l=1,l\neq n}^{L} \sum_{u=1}^{U} I_{S,u10}^{Q} + \sum_{k=2}^{K} \sum_{l=1}^{L} \sum_{u=1}^{U} I_{M,u10}^{k,Q} + N_{10n}^{Q} \right]^2 \right\}, \tag{18.81}$$

$$\begin{aligned} Z_{1mn} &= \frac{E_s}{2U}T_s\left\{\left[\sum_{l=1,l\neq n}^{L}\sum_{u=1}^{U}I_{S,u1m}^{I}+\sum_{k=2}^{K}\sum_{l=1}^{L}\sum_{u=1}^{U}I_{M,u1m}^{k,I}+N_{1mn}^{I}\right]^2\right. \\ &\quad \left.+\left[\sum_{l=1,l\neq n}^{L}\sum_{u=1}^{U}I_{S,u1m}^{Q}+\sum_{k=2}^{K}\sum_{l=1}^{L}\sum_{u=1}^{U}I_{M,u1m}^{k,Q}+N_{1mn}^{Q}\right]^2\right\}, \\ &\qquad m=1,2,\ldots,m_1-1, \end{aligned} \tag{18.82}$$

where $E_s = PT_s$, $I_{S,u1m}^{I}$, $I_{M,u1m}^{k,I}$, N_{1mn}^{I} and $I_{S,u1m}^{Q}$, $I_{M,u1m}^{k,Q}$, N_{1mn}^{Q} are, respectively, given by:

$$\begin{aligned} I_{S,u1m}^{I} &= \frac{\alpha_l\cos(\phi_l)}{T_s}\int_{\tau_n}^{T_s+\tau_n}W_{ur_u}(t-\tau_l)W_{1m}(t-\tau_n)a(t-\tau_l)a(t-\tau_n)dt, \\ I_{M,u1m}^{k,I} &= \frac{\alpha_l^k\cos(\theta_l^k)}{T_s}\int_{\tau_n}^{T_s+\tau_n}W_{ur_u}^{k}(t-\tau_l^k)W_{1m}(t-\tau_n)a(t-\tau_l^k)a(t-\tau_n)dt, \\ N_{1mn}^{I} &= \left(\sqrt{\frac{P}{2U}}T_s\right)^{-1}\int_{\tau_n}^{T_s+\tau_n}n(t)W_{1m}(t-\tau_n)a(t-\tau_n)\cos(\omega_c t)dt. \end{aligned} \tag{18.83}$$

and by:

$$\begin{aligned} I_{S,u1m}^{Q} &= \frac{-\alpha_l\sin(\phi_l)}{T_s}\int_{\tau_n}^{T_s+\tau_n}W_{ur_u}(t-\tau_l)W_{1m}(t-\tau_n)a(t-\tau_l)a(t-\tau_n)dt, \\ I_{M,u1m}^{k,Q} &= \frac{-\alpha_l^k\sin(\theta_l^k)}{T_s}\int_{\tau_n}^{T_s+\tau_n}W_{ur_u}^{k}(t-\tau_l^k)W_{1m}(t-\tau_n)a(t-\tau_l^k)a(t-\tau_n)dt, \\ N_{1mn}^{Q} &= \left(\sqrt{\frac{P}{2U}}T_s\right)^{-1}\int_{\tau_n}^{T_s+\tau_n}n(t)W_{1m}(t-\tau_n)a(t-\tau_n)\sin(\omega_c t)dt. \end{aligned} \tag{18.84}$$

for $m = 0, 1, \ldots, m_1 - 1$ and $n = 1, 2, \ldots, L$, where $\theta_l^k = \phi_l^k - \phi_n$.

18.3.3.2 Decision Statistics

We have argued in the context of Section 18.3.2 that $I_{S,u1m}^{I}$, $I_{M,u1m}^{k,I}$ and $I_{S,u1m}^{Q}$, $I_{M,u1m}^{k,Q}$ can be approximated as Gaussian variables with zero-mean and variance $\Omega/3N_s$, while N_{1mn} is a zero-mean Gaussian random variable with variance $UN_0/2E_s$. Consequently, Z_{1mn} is the sum of the square of two Gaussian random variables, each having a variance of:

$$\sigma^2 = \frac{\Omega E_s T_s}{2U}\left[\frac{U(KL-1)}{3N_s}+\left(\frac{2\Omega E_s}{UN_0}\right)^{-1}\right].$$

Since Z_{1mn} is the sum of the square of two Gaussian random variables, it can be shown that, conditioned on α_n, Z_{1mn} becomes chi-square distributed [90] with two degrees of freedom, and its PDF can be expressed as:

$$f_{Z_{10n}}(y|\mathcal{R}_{il}) = \frac{1}{2\sigma^2}\exp\left(-\frac{y+\mathcal{R}_{il}}{2\sigma^2}\right) I_0\left(\frac{\sqrt{y\mathcal{R}_{il}}}{\sigma^2}\right), \quad y \geq 0 \tag{18.85}$$

if $m = 0$, where $\mathcal{R}_{il} = (E_s T_s \alpha_n^2)/2U$ is the non-centrality parameter of the chi-square distribution and $I_0(\cdot)$ is the modified Bessel function of the zero-th order. If $m \neq 0$, then [90]:

$$f_{Z_{1mn}}(y) = \frac{1}{2\sigma^2}\exp\left(-\frac{y}{2\sigma^2}\right), \quad y \geq 0. \tag{18.86}$$

Using $2\sigma^2$ to normalize Z_{1mn}, we obtain the normalized PDFs of Z_{1mn} as:

$$f_{Z_{10n}}(y|\gamma_{1n}) = \exp\left(-[y+\gamma_{1n}]\right) I_0\left(\sqrt{4y\gamma_{1n}}\right), \quad y \geq 0, \tag{18.87}$$

$$f_{Z_{1mn}}(y) = \exp(-y), \quad m \neq 0, \ y \geq 0, \tag{18.88}$$

where

$$\gamma_{1n} = \frac{\mathcal{R}_{il}}{2\sigma^2} = \gamma_c \cdot \frac{\alpha_n^2}{\Omega}, \tag{18.89}$$

and

$$\gamma_c = \frac{1}{2} \cdot \left[\frac{U(KL-1)}{3N_s} + \left(\frac{2\Omega E_s}{UN_0}\right)^{-1}\right]^{-1}. \tag{18.90}$$

Since $\{\alpha_n\}$ are independent Rayleigh-distributed random variables, $\{\alpha_n^2\}$ are exponentially distributed random variables. Consequently, with the aid of Equation 18.55 it can be shown that the PDF of γ_{1n} defined in Equation 18.89 can be expressed as:

$$f(\gamma_{1n}) = \frac{1}{\gamma_c}\exp\left(-\frac{\gamma_{1n}}{\gamma_c}\right). \tag{18.91}$$

A. Equal Gain Combining For a receiver using L-th order diversity reception and EGC, the L branches are equally weighted and then added in order to form the decision variables in Figure 18.15, which can be expressed as:

$$Z_{1m} = \sum_{n=1}^{L} Z_{1mn}, \tag{18.92}$$

for $m = 0, 1, \ldots, m_1 - 1$. Assuming that $r_1 = 0$ was transmitted, then, since the quantities $\{Z_{1mn}\}$ are chi-square distributed random variables with two degrees of freedom and have a PDF given by Equations 18.87 and 18.88 for $m = 0$ and $m \neq 0$, respectively, it can be shown that, for a given set of multipath attenuations $\{\alpha_1, \alpha_2, \ldots, \alpha_L\}$, Z_{1m} is chi-square distributed with $2L$ degrees of freedom, and it obeys the normalized PDF of:

$$f_{Z_{10}}(y|\gamma) = \left(\frac{y}{\gamma}\right)^{(L-1)/2} \exp\left(-[y+\gamma]\right) I_{L-1}\left(\sqrt{4y\gamma}\right), \quad y \geq 0, \tag{18.93}$$

$$f_{Z_{1m}}(y) = \frac{1}{(L-1)!} y^{L-1} \exp(-y), \qquad m > 0, \ y \geq 0, \tag{18.94}$$

where $I_L(\cdot)$ is the modified Bessel function of the Lth order, and

$$\gamma = \sum_{n=1}^{L} \gamma_{1n}, \tag{18.95}$$

where the PDF of γ_{1n} is given by Equation 18.91. We assume throughout that all the path gains are i.i.d random variables, which in turn means that $\{\gamma_{1n}\}$ are i.i.d random variables. Consequently, it can readily be shown that γ obeys the central Chi-square distribution with $2L$ degrees of freedom [90], which can be expressed as:

$$f(\gamma) = \frac{1}{(L-1)!\gamma_c^L} \gamma^{L-1} \exp\left(-\frac{\gamma}{\gamma_c}\right). \tag{18.96}$$

Since the decision variables $\{Z_{10}, Z_{11}, \ldots, Z_{1(m_1-1)}\}$ are independent random variables, the conditioning in Equation 18.93 may be removed by averaging $f_{Z_{10}}(y|\gamma)$ over the valid range of γ, which can be expressed as:

$$f_{Z_{10}}(y) = \int_0^\infty f_{Z_{10}}(y|\gamma) f(\gamma) d\gamma. \tag{18.97}$$

Upon substituting Equation 18.93 and Equation 18.96 into Equation 18.97, it can be shown that:

$$f_{Z_{10}}(y) = \frac{1}{(L-1)!(1+\gamma_c)^L} y^{L-1} \exp\left(-\frac{y}{1+\gamma_c}\right), \quad y \geq 0. \tag{18.98}$$

Similarly, the PDFs of $f_{Z_{um}}(y)$, $m = 0, 1, \ldots, m_u - 1$ upon receiving the m_u-ary residue digit r_u can be derived, which take the same form as Equation 18.98 for $m = 0$ and Equation 18.94 for $m > 0$, with the subscript '$1m$' replaced by 'um'.

B. Selection Combining For an L-branch diversity receiver using SC – where the branch signal of Figure 18.15 having the largest amplitude is selected for demodulation – the conventional decision variable of an m_1-ary orthogonal modulation system is defined as:

$$Z_1 = \max\left\{Z_{101}, \ldots, Z_{10L}, \ldots, Z_{1(m_1-1)1}, \ldots, Z_{1(m_1-1)L}\right\}. \quad (18.99)$$

Alternatively, a two-stage maximum selection scheme can be proposed, in order to invoke side-information in the orthogonal demodulation process, an issue which will become explicit during our further discussions in the next section. The first-stage maximum selection is defined as the maximum selection from the L diversity components for each specific given m, which can be formulated as:

$$Z_{1m} = \max\left\{Z_{1m1}, Z_{1m2}, \ldots, Z_{1mL}\right\}, \quad (18.100)$$

for $m = 0, 1, \ldots, m_1 - 1$. Based on the above maximum selection, the symbol is decided correctly, if and only if the output Z_{10}, which is matched to the transmitted signal – remembering that the m_1-ary residue digit $r_1 = 0$ was transmitted – is larger than any other outputs mismatched to the transmitted symbol. Hence, the 2nd-stage maximum selection finds the largest one from the set $\left\{Z_{10}, Z_{11}, \ldots, Z_{1(m_1-1)}\right\}$, which can be expressed as:

$$Z_1 = \max\left\{Z_{10}, Z_{11}, \ldots, Z_{1(m_1-1)}\right\}. \quad (18.101)$$

Let us now derive the PDFs of Z_{1m}. Since the normalized PDF of Z_{1mn} for $n = 1, 2, \ldots, L$ was given by Equation 18.88, the PDF of Z_{1m} for $m > 0$ can be expressed as:

$$\begin{aligned} f_{Z_{1m}}(y) &= \frac{d}{dy}\left[1 - \exp(-y)\right]^L \\ &= L\exp(-y)\left[1 - \exp(-y)\right]^{L-1}. \end{aligned} \quad (18.102)$$

Similarly, the conditional PDF of Z_{10} can be derived, which can be expressed as:

$$\begin{aligned} f_{Z_{10}}(y|\gamma_{11}, \gamma_{12}, \ldots, \gamma_{1L}) &= \frac{d}{dy}\left[\prod_{n=1}^{L}\int_0^y f_{Z_{10n}}(x|\gamma_{1n})dx\right] \\ &= \sum_{n=1}^{L} f_{Z_{10n}}(y|\gamma_{1n})\left[\prod_{\substack{l=1 \\ l\neq n}}^{L}\int_0^y f_{Z_{10l}}(x|\gamma_{1l})dx\right], \end{aligned} \quad (18.103)$$

where $f_{Z_{10l}}(x|\gamma_{1l})$ for $l = 1, 2, \ldots, L$ was given by Equation 18.87. The conditioning with respect to $\{\gamma_{1l}\}$ can be removed at this stage by integrating $f_{Z_{10}}(y|\gamma_{11}, \gamma_{12}, \ldots, \gamma_{1L})$ over the valid ranges of $\gamma_{11}, \gamma_{12}, \ldots, \gamma_{1L}$, which can be expressed as:

$$f_{Z_{10}}(y) = \underbrace{\int \cdots \int_0^\infty}_{L} \sum_{n=1}^{L} f_{Z_{10n}}(y|\gamma_{1n}) \left[\prod_{\substack{l=1 \\ l \neq n}}^{L} \int_0^y f_{Z_{10l}}(x|\gamma_{1l}) dx \right] \cdot f(\gamma_{11}) f(\gamma_{12}) \ldots f(\gamma_{1L}) d\gamma_{11} d\gamma_{12} \ldots d\gamma_{1L}. \quad (18.104)$$

Substituting $f_{Z_{10l}}(x|\gamma_{1l})$ and $f(\gamma_{1l})$ from Equation 18.87 and Equation 18.91 into Equation 18.104, and remembering that $\{\gamma_{1l}, l = 1, 2, \ldots, L\}$ are i.i.d random variables, Equation 18.104 can be simplified to:

$$f_{Z_{10}}(y) = \frac{L}{1+\gamma_c} \exp\left(-\frac{y}{1+\gamma_c}\right) \left[1 - \exp\left(-\frac{y}{1+\gamma_c}\right)\right]^{L-1}, \quad y \geq 0, \quad (18.105)$$

where γ_c is given by Equation 18.90.

Above we have derived the PDFs of the decision variables that are input to the MLD block of Figure 18.14. With the aid of these PDFs the average symbol error probabilities can now be derived for both EGC and SC.

18.3.3.3 Average Probability of Error

Let $\{Z_{10}, Z_{11}, \ldots, Z_{1(m_1-1)}\}$ represent the decision variables after diversity combining, which are input to the MLD block of Figure 18.14 dedicated to receiving residue digit $r_1 = 0$. Then the average correct residue digit detection probability of $P_{r_1}(C)$ and erroneous residue digit detection probability of $P_{r_1}(\varepsilon)$ can be expressed as:

$$P_{r_1}(C) = P(Z_{11} < Z_{10}, Z_{12} < Z_{10}, \ldots, Z_{1(m_1-1)} < Z_{10}), \quad (18.106)$$

$$P_{r_1}(\varepsilon) = 1 - P_{r_1}(C), \quad (18.107)$$

respectively. Since no redundant moduli are invoked, an M-ary symbol is received correctly if and only if all m_u-ary residue digits are received correctly. Hence, the probability $\overline{P}_s(C)$ of correct M-ary symbol recovery can be expressed as the product of receiving all m_u-ary residue digits constituting the M-ary symbol correctly, which is given by:

$$\overline{P}_s(C) = \prod_{u=1}^{U} P_{r_u}(C). \quad (18.108)$$

Lastly, the bit error probability $\overline{P}_b(\epsilon)$ can be approximated as [90][p.150, Equation (4.2.55)]:

$$\overline{P}_b(\epsilon) \approx \frac{1}{2}(1 - \overline{P}_s(C)), \quad (18.109)$$

when the number of bits per symbol is high. Let us now derive the probability $P_{r_u}(C)$ of correct residue digit detection for both EGC and SC by deriving $P_{r_1}(C)$, as an example.

A. Equal Gain Combining For EGC the correct residue digit detection probability of $P_{r_1}(C)$ can be expressed upon referring to Equation 18.106, as:

$$P_{r_1}(C) = \int_0^\infty f_{Z_{10}}(y) \left[\int_0^y f_{Z_{1m}}(x) dx \right]^{m_1 - 1} dy, \tag{18.110}$$

where $f_{Z_{1m}}(x)$ and $f_{Z_{10}}(y)$ are given by Equation 18.94 and Equation 18.98, respectively. Substituting them into the above equation, it can be shown that:

$$\begin{aligned} P_{r_1}(C) = \int_0^\infty & \frac{1}{(1+\gamma_0)^L (L-1)!} y^{L-1} \exp\left(-\frac{y}{1+\gamma_0}\right) \\ & \cdot \left[1 - \exp(-y) \sum_{k=0}^{L-1} \frac{y^k}{k!}\right]^{m_1 - 1} dy. \end{aligned} \tag{18.111}$$

Similarly, the other correct residue digit detection probabilities $P_{r_u}(C)$ for $u = 2, \ldots, U$ can be derived, which have the same form as Equation 18.111 with the index 'r_1' replaced by 'r_u' and m_1 replaced by m_u. The average symbol error probability and bit error probability can be computed using Equations 18.108 and 18.109, respectively.

B. Selection Combining For SC the correct residue digit detection probability of $P_{r_1}(C)$ can also be expressed upon referring to Equation 18.106 in the form of Equation 18.110, where $f_{Z_{1m}}(x)$ and $f_{Z_{10}}(y)$ are given by Equation 18.102 and Equation 18.105, respectively. Upon substituting them into Equation 18.110, then the correct residue digit detection probability can be simplified to:

$$P_{r_1}(C) = \sum_{w=1}^{L} (-1)^{w+1} \binom{L}{w} \prod_{n=1}^{L(m_1-1)} \left(\frac{n}{n + w/(1+\gamma_c)} \right). \tag{18.112}$$

The average symbol error probability and bit error probability can be determined with the aid of Equations 18.108 and (18.109). Let us now evaluate the performance of our EGC and SC-assisted non-coherent RNS-based DS-CDMA of Figure 18.14.

18.3.3.4 Numerical Results

Figure 18.16 shows the BER of our RNS-based DS-CDMA system for the EGC and SC schemes upon evaluating Equations 18.109 and 18.111 for EGC and Equations 18.109 and 18.112 for SC. We assumed that the diversity order was $L = 1, 2, 3, 4$ and the RNS-based parallel DS-CDMA system of Figure 18.14 with its moduli taking values of $m_1 = 4$, $m_2 = 5$, $m_3 = 7$ was considered. As expected, both the EGC and SC schemes provide dramatic BER improvements for moderate to high SNRs per bit, when the number of diversity paths, L, increases. Furthermore, the results suggest that the EGC scheme has a lower BER, than

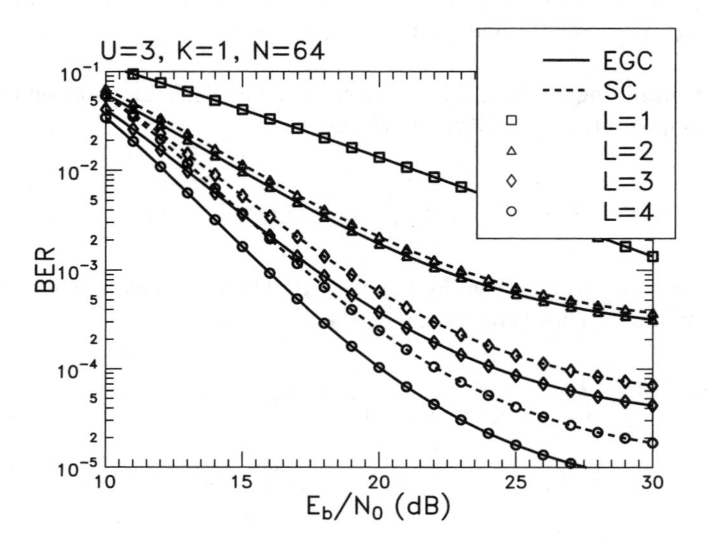

Figure 18.16: EGC, SC: Bit error probability versus E_b/N_0 for the non-coherent RNS-based parallel DS-CDMA system of Figure 18.14 with 3 moduli, namely, $m_1 = 4$, $m_2 = 5$, $m_3 = 7$, $L = 1, 2, 3, 4$ diversity paths, $N = 64$ chips per bit and $K = 1$ active user, i.e., without multi-user interference over dispersive multipath Rayleigh fading channels.

the SC scheme for any given E_b/N_0. Viewing their performance from a different perspective an additional 1-1.5dB E_b/N_0 is required by the SC scheme for maintaining an identical BER to that of the EGC scheme. This is because EGC is the optimal diversity combining scheme for a non-coherent demodulation technique over multipath fading channels.

In Figure 18.17 and Figure 18.18 we evaluated the bit error rate performance of the RNS-based parallel DS-CDMA system using different sets of moduli. Specifically, three sets of moduli were considered and the moduli values were (4,5,7), (8,11,13) and (20,21,23), respectively. We assumed for both EGC and SC that the number of chips per bit was $N = 64$, the number of diversity components was $L = 1, 2, 3$, and the number of active users was $K = 1$, i.e., no multi-user interference was inflicted. We observe from the results that both EGC and SC have similar BER performance curves. However, for the given set of moduli, for the given number of diversity components and any given bit error probability, the EGC scheme required a slightly lower E_b/N_0 than the SC scheme, which was typically 1-1.5dB. Furthermore, it can be observed that, for any given number of diversity components L and for any given E_b/N_0 value, the bit error probability of the system using moduli (4,5,7) was the highest and that of using moduli (20,21,23) was the lowest. This observation became more explicit, when the number of diversity components increased from $L = 1$ to $L = 3$.

Finally, Figure 18.19 portrays the BER of the EGC and SC schemes as a function of E_b/N_0 for $K = 1, 5, 10, 20$ upon evaluating Equations (18.109) and (18.111) for EGC and upon involving Equations 18.109 and 18.112 for SC. We assumed that the diversity order was $L = 3$, and the non-coherent RNS-based parallel DS-CDMA system of Figure 18.14 with its

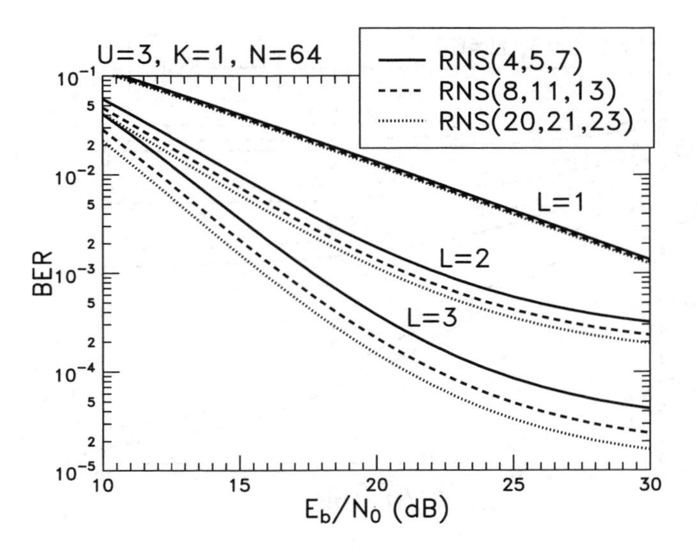

Figure 18.17: **EGC**: Bit error probability versus E_b/N_0 performance for the non-coherent RNS-based parallel DS-CDMA system of Figure 18.14 with $L = 1, 2, 3$ diversity paths, $N = 64$ chips per bit and $K = 1$ active user over dispersive multipath Rayleigh fading channels.

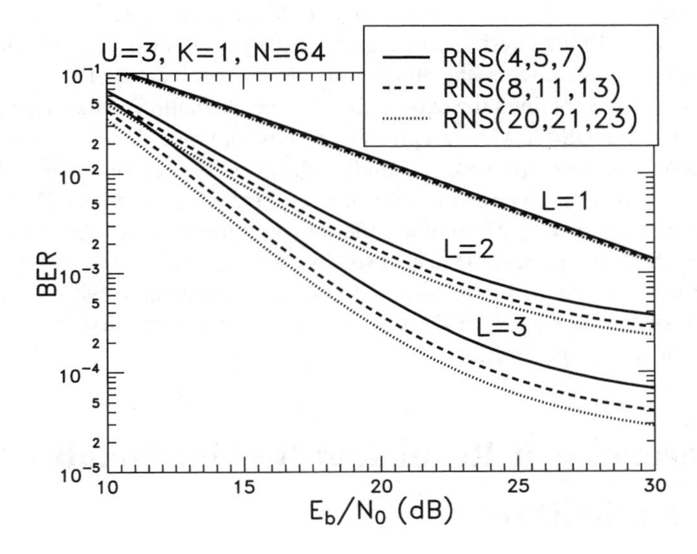

Figure 18.18: **SC**: Bit error probability versus E_b/N_0 performance for the non-coherent RNS-based parallel DS-CDMA system of Figure 18.14 with $L = 1, 2, 3$ diversity paths, $N = 64$ chips per bit and $K = 1$ active user over dispersive multipath Rayleigh fading channels.

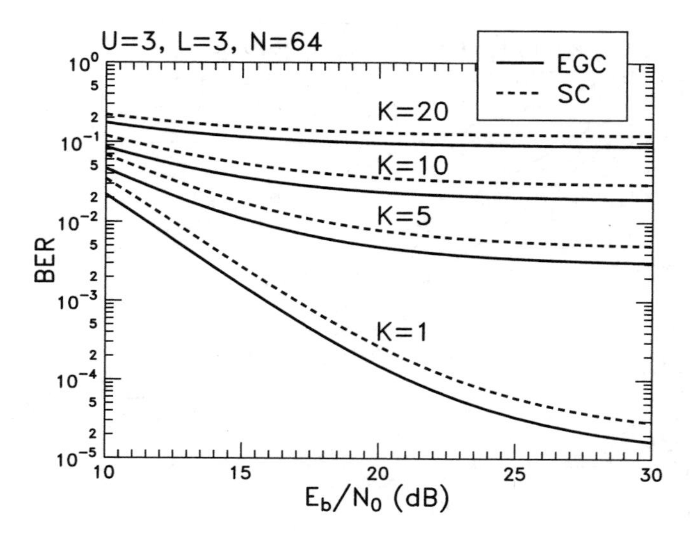

Figure 18.19: EGC, SC: Bit error probability of the non-coherent RNS-based parallel DS-CDMA system of Figure 18.14 using $L = 2, 5$ diversity paths, $N = 64$ chips per bit and supporting $K = 1, 5, 10, 20$ active users over multipath dispersive Rayleigh fading channels.

moduli taking values of $m_1 = 20$, $m_2 = 21$, $m_3 = 23$ was considered. Since DS-CDMA systems are interference-limited systems, as expected, the performance of both the EGC and the SC schemes was degraded, when the number of active users increased.

In this section we have obtained the expressions for the evaluation of the error probability of the RNS-based parallel DS-CDMA system using orthogonal modulation, when coherent or non-coherent receivers were employed, and multipath diversity receivers using MRC, EGC or SC scheme were considered. The error performance of the RNS-based parallel DS-CDMA system can be further improved by exploiting the inherent properties of the RNS arithmetic with the aid of RRNS codes. Hence, in the next section, we will analyze the error performance of the system in conjunction with RRNS-based coding. Specifically, we will analyze the error performance of the RRNS-based DS-CDMA system over dispersive multipath fading channels, when EGC or SC are considered.

18.4 Performance of Redundant Residue Number System Coded DS-CDMA

18.4.1 Introduction

RRNS codes exhibit similar coding properties to the well-known Reed-Solomon (RS) codes [395–397]. RRNS codes constitute a class of maximum minimum-distance separable codes. An RRNS (U, V) code, where the information dynamic range represented by the RRNS (U, V) is $[0, \prod_{u=1}^{V} m_u)$ and the RRNS code's dynamic range is $[0, \prod_{u=1}^{U} m_u)$, has a minimum distance of $(U - V + 1)$ and hence it is capable of detecting $(U - V)$ or less residue digit errors and correct up to $\lfloor (U - V)/2 \rfloor$ random residue digit errors, where $\lfloor x \rfloor$ represents the largest integer not exceeding x. Moreover, an RRNS (U, V) code is capable of correcting t random residue digit errors and simultaneously correcting β residue erasures, if and only if $2t + \beta \leq (U - V)$ [397].

According to the carry-free property and due to the lack of weighted significance of the residue digits in the RNS arithmetic, in RRNS codes some of the channel-impaired received residue digits can be discarded as an error correction measure, provided that a sufficiently high dynamic range is retained by the reduced-range system, in order to unambiguously decode the result. The above statement can be developed as follows. Let $\{m_1, m_2, \ldots, m_U\}$ be a set of U moduli of an RRNS (U, V) code, where $m_1 < m_2 < \ldots < m_U$. Furthermore, let X_k be the integer message of user k associated with an M-ary information symbol, which is expressed with the aid of the residue digits $(r_1, r_2, \ldots, r_U)$ with respect to the above moduli. If the dynamic range of the M-ary integer message X_k is $[0, \prod_{u=1}^{V} m_u)$, where $V \leq U$, then X_k can be recovered from any V out of the U number of residue digits and their relevant moduli. This property implies that – after the maximum likelihood detection (MLD) stage of the U receiver banks in Figure 18.14 – d $(d \leq U - V)$ number of MLD outputs can be dropped before the IRNST stage, while still recovering the transmitted symbol X_k using the retained MLD outputs, provided that the retained MLD outputs are those matched to the related residue digits transmitted. Alternatively, the residue digit errors in the retained MLD outputs can be corrected by using the RRNS (U, V) decoder.

Moreover, if we let d be the number of discarded residue digits, where $d \leq U - V$, then, a RRNS (U, V) code is converted to a RRNS $(U - d, V)$ after d out of the U residue digits are discarded. Hence, the reduced RRNS $(U - d, V)$ code can detect up to $[U - V - d]$ residue digit errors and correct up to $\lfloor (U - V - d)/2 \rfloor$ residue digit errors. This property suggests that the RRNS (U, V) decoding can be designed by first discarding d $(d \leq U - V)$ out of the U outputs of the MLDs in Figure 18.14, which is followed by RRNS $(U - d, V)$ decoding. Since the discarded outputs do not have to be considered in the RRNS $(U - d, V)$ decoding, the decoding procedure is therefore simplified.

RS codes and RRNS codes exhibit similar distance properties and hence they are capable of achieving a similar coding performance for identical-length RS and RRNS codes. However, the RS code symbols possess a constant base of 2^b for b-bit symbols, while RRNS code symbols are related to a group of bases from the set of moduli $\{m_1, m_2, \ldots, m_U\}$. The length U of RS codes is usually related to the legitimate range of code symbols given by $U = 2^b - 1$ and short RS codes having a high symbol dynamic range can only be obtained by shortening long RS codes. However, the decoding complexity of the shortened RS codes is typically similar to that of the original unshortened codes [64, 387]. In closing we note that all equations derived in this section are suitable for the average error probability evaluation of RS coded systems [387] using similar design criteria. In order to characterize the error performance of RS codes, we simply replace the corresponding moduli m_u of the RRNS (U, V) code in the corresponding equation by a constant modulus of $m = 2^b$. In the next subsection let us invoke the so-called Ratio Statistic Test (RST), which can be advantageously used in our RRNS-based DS-CDMA receiver.

18.4.2 Ratio Statistic Test

In cellular DS-CDMA systems, forward error-correction (FEC) is often used for mitigating the effects of fading and interference. The so-called 'errors-and-erasures' decoding schemes [387] are often preferable to 'error-correction-only' decoding, since typically more erasures than errors can be corrected. Hence, it is beneficial to determine the reliability of the received symbols and to erase the low-reliability symbols prior to the decoding process. There are a number of methods that can be invoked for generating reliability-based side information and their performance has been analyzed, for example, in [386, 389, 429–435]. An erasure insertion scheme suitable for M-ary orthogonal modulation is the ratio-threshold test (RTT), which was originally proposed by Viterbi [430].

In RTT an erasure is declared, whenever the ratio of the maximum to the second maximum of the inputs of the maximum-likelihood detector (MLD) involved in an orthogonal modulation system does not exceed a pre-set threshold. AS mentioned above, the RTT was originally proposed by Viterbi [429], in order to mitigate the so-called partial-band interference or multitone interference in M-ary frequency shift keying (MFSK) systems, but it was later invoked also in the context of M-ary orthogonal signalling, in order to generate channel-quality related information. Kim and Stark [386] have employed it also for mitigating the effects of Rayleigh-fading and have analyzed some of the performance limits of an RS coded DS-CDMA system using 'errors-and-erasures' decoding. Since RRNS codes have similar properties to RS codes in terms of using non-binary symbols and they are amenable to transmission using RNS-based M-ary orthogonal modulation, 'errors-and-erasures' decoding using RTT is also suitable for the decoding of RRNS codes. The RRNSs 'errors-and-erasures' decoding procedure can be implemented by first dropping d number of residue digits, where $(d \leq U - V)$ and then decoding the reduced RRNS $(U - d, V)$ code with the aid of the procedures highlighted in the previous chapter. Due to the unique decoding method discussed above, it does not require a pre-set threshold in this kind of RRNS decoding. Hence, a more appropriate term is used, which we refer to as 'Ratio Statistic Test (RST)'. In the rest of this section, we investigate the performance of an RRNS (U, V) coded DS-CDMA system, when RNS-based orthogonal signalling is employed in conjunction with the above-mentioned RST technique over multipath Rayleigh-fading channels. Two different diversity combining schemes [13, 366] – namely equal-gain combining (EGC) and selection combining (SC) – are considered and their performance is evaluated in the context of the RST-based dropping scheme.

Figure 18.20 portrays the receiver model of the RRNS (U, V) coded parallel DS-CDMA system. This figure is similar to Figure 18.14, except that in this receiver decision-reliability information is produced from the outputs of the EGC or SC diversity block by the maximum-likelihood detection unit, and an RRNS decoder is invoked before the IRNST.

Let $\left\{Z_{u0}, Z_{u1}, \ldots, Z_{u(m_u-1)}\right\}$ represent the decision variables input to the MLD block of Figure 18.20, where $\{Z_{ui}\}$ are the output variables after the EGC or SC combining stage. The ratio involved in the RST is computed according to the following definition [430]:

$$\lambda_u = \frac{Y_1 = \max_1\left\{Z_{u0}, Z_{u1}, \ldots, Z_{u(m_u-1)}\right\}}{Y_2 = \max_2\left\{Z_{u0}, Z_{u1}, \ldots, Z_{u(m_u-1)}\right\}}, \tag{18.113}$$

where $Y_1 = \max_1\{\cdot\}$ and $Y_2 = \max_2\{\cdot\}$ represent the maximum and the 'second' max-

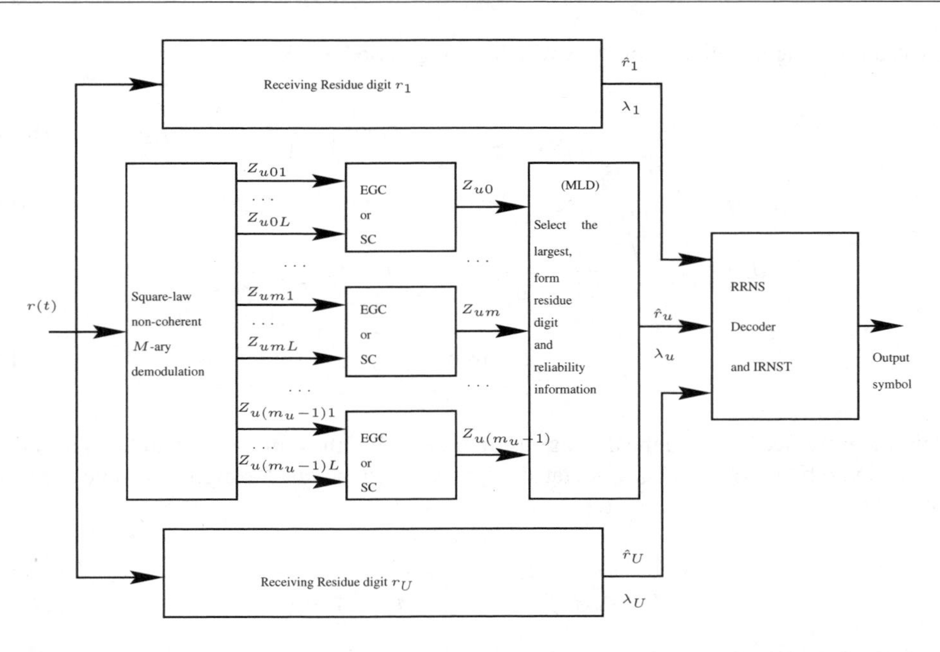

Figure 18.20: Receiver model of the RRNS coded parallel DS-CDMA system investigated using non-coherent M-ary demodulation, multipath diversity combining, maximum-likelihood detection, side-information, RRNS decoder and IRNST.

imum of the decision variables of $\{Z_{u0}, Z_{u1}, \ldots, Z_{u(m_u-1)}\}$, respectively. The RST is based on the fact that an unreliable received signal in Figure 18.20 is likely to have nearly equal energy in both the correlation branch matched to the transmitted signal and the correlation branches mismatched to the transmitted signal, in particular, as far as the second largest one in concerned. Viterbi [430] pointed out that the demodulated symbols having relatively high ratio of λ_u were more reliable than those having relatively low values of λ_j. Consequently, we can assume that the demodulator outputs having the lowest value of λ_u for the RST are the lowest-reliability outputs, hence the corresponding residue digits are likely to be corrupted and can be dropped before the RRNS decoding stage of Figure 18.20, provided that the number of dropped residue digits is less than $(U - V)$.

Let H_1 and H_0 represent the hypotheses that a received residue digit is demodulated correctly and erroneously, respectively, according to the maximum likelihood detection stage of Figure 18.20. Let us now derive the PDFs of the maximum and the 'second maximum' of Y_1 and Y_2 under the correct residue digit detection hypothesis of H_1 and under the erroneous residue digit detection hypothesis of H_0, as well as the corresponding PDFs of the RST defined in Equation 18.113.

18.4.2.1 Probability Density Function of the Ratio Using EGC

The PDFs of the decision variables Z_{10}, Z_{1m} seen in the schematic of Figure 18.20 as well as correct residue detection probability $P_{r_1}(C)$ were determined in Section 18.3, which were given by Equations 18.98, 18.94 and 18.111. The PDFs of $\{Z_{um}\}$ and $P_{r_u}(C)$ can then be derived pursuing a similar approach, which are expressed as:

$$f_{Z_{u0}}(y) = \frac{1}{(L-1)!(1+\gamma_c)^L} y^{L-1} \exp\left(-\frac{y}{1+\gamma_c}\right), \quad y \geq 0. \tag{18.114}$$

$$f_{Z_{um}}(y) = \frac{1}{(L-1)!} y^{L-1} \exp(-y), \quad m > 0, \quad y \geq 0, \tag{18.115}$$

$$P_{r_u}(C) = \int_0^{\infty} \frac{1}{(1+\gamma_0)^L (L-1)!} y^{L-1} \exp\left(-\frac{y}{1+\gamma_0}\right) \cdot \left[1 - \exp(-y) \sum_{k=0}^{L-1} \frac{y^k}{k!}\right]^{m_u - 1} dy. \tag{18.116}$$

Consequently, according to the definitions of H_1 and H_0, the *a priori* probabilities of satisfying the hypotheses H_1 and H_0 associated with receiving the residue digit r_u can be expressed as:

$$P_{r_u}(H_1) = P_{r_u}(C), \tag{18.117}$$

$$P_{r_u}(H_0) = 1 - P_{r_u}(H_1). \tag{18.118}$$

The PDFs of the maximum and the 'second maximum' – namely that of $Y_1 = \max_1\{\cdot\}$ and $Y_2 = \max_2\{\cdot\}$ – under the hypotheses H_1 of correct residue digit decision and H_0 of erroneous residue digit decision have been given in Appendix C.2. Using the PDF $f_{Z_{u0}}(y)$ of Equation 18.114 for $f_{Z_1}(y)$ and the PDFs $f_{Z_{um}}(y)$, $m > 0$ of Equation 18.115 for the $f_{Z_m}(y), m > 1$ and substituting them into the relevant equations of Equation C.9 and Equation C.10 from Appendix C.2 for the hypothesis H_1, the PDFs of Y_1 and Y_2 conditioned on the hypothesis of H_1 are given by:

$$f_{Y_1}(y|H_1) = \frac{1}{P_{r_u}(H_1)} \cdot \frac{1}{(1+\gamma_c)^L (L-1)!} \times y^{L-1} \exp\left(-\frac{y}{1+\gamma_c}\right) [1 - \Psi(y)]^{m_u - 1}, \quad y \geq 0, \tag{18.119}$$

$$f_{Y_2}(y|H_1) = \frac{1}{P_{r_u}(H_1)} \cdot \frac{m_u - 1}{(L-1)!} \times y^{L-1} \exp(-y) [1 - \Psi(y)]^{m_u - 2} \Psi\left(\frac{y}{1+\gamma_c}\right), \quad y \geq 0, \tag{18.120}$$

where the short-hand $\Psi(x)$ is defined as:

$$\Psi(x) = \exp(-x) \sum_{k=0}^{L-1} \frac{x^k}{k!}. \tag{18.121}$$

Similarly, the PDFs of Y_1 and Y_2 – namely that of the maximum and 'second maximum' in Figure 18.20 – conditioned on the hypothesis of H_0 can be derived with the aid of Equation C.13 and Equation C.14 from Appendix C.2, which can be expressed as follows:

$$\begin{aligned} f_{Y_1}(y|H_0) &= \frac{1}{P_{r_u}(H_0)} \cdot \frac{m_u - 1}{(L-1)!} y^{L-1} \exp(-y) \left[1 - \Psi(y)\right]^{m_u - 2} \\ &\quad \cdot \left[1 - \Psi\left(\frac{y}{1+\gamma_c}\right)\right], \; y \geq 0, \quad (18.122) \\ f_{Y_2}(y|H_0) &= \frac{1}{P_{r_u}(H_0)} \cdot \frac{m_u - 1}{(L-1)!} y^{L-1} \Psi(y) \left[1 - \Psi(y)\right]^{m_u - 3} \\ &\quad \cdot \left\{ \frac{1}{(1+\gamma_c)^L} \exp\left(-\frac{y}{1+\gamma_c}\right) \left[1 - \Psi(y)\right] \right. \\ &\quad \left. +(m_u - 2) \exp(-y) \left[1 - \Psi\left(\frac{y}{1+\gamma_c}\right)\right] \right\}, \; y \geq 0. \quad (18.123) \end{aligned}$$

The PDF of the ratio λ_u defined in Equation 18.113 under the hypothesis of correct or erroneous residue digit decision hypothesis, namely under H_1 or H_0 can now be derived by using Equation C.11 of Appendix C.2. Substituting the corresponding PDFs into Equation C.11, the PDFs of λ_u under the hypotheses of H_1 and H_0 can be expressed as:

$$\begin{aligned} f_{\lambda_u}(y|H_1) &= C_{H_1}^u \times g_{\lambda_u}(y|H_1), \; y \geq 1, \quad (18.124) \\ f_{\lambda_u}(y|H_0) &= C_{H_0}^u \times g_{\lambda_u}(y|H_0), \; y \geq 1, \quad (18.125) \end{aligned}$$

where $g_{\lambda_u}(y|H_1)$ and $g_{\lambda}(y|H_0)$ can be shown to obey:

$$\begin{aligned} g_{\lambda_u}(y|H_1) &= y^{L-1} \int_0^\infty x^{2L-1} \exp\left(-x - \frac{xy}{1+\gamma_c}\right) \\ &\quad \cdot \left[1 - \Psi(xy)\right]^{m_u - 1} \left[1 - \Psi(x)\right]^{m_u - 2} \Psi\left(\frac{x}{1+\gamma_c}\right) dx, \quad (18.126) \\ g_{\lambda_u}(y|H_0) &= y^{L-1} \int_0^\infty x^{2L-1} \exp(-xy) \left[1 - \Psi(xy)\right]^{m_u - 2} \\ &\quad \cdot \left[1 - \Psi\left(\frac{xy}{1+\gamma_c}\right)\right] \Psi(x) \left[1 - \Psi(x)\right]^{m_u - 3} \\ &\quad \cdot \left\{ \frac{1}{(1+\gamma_c)^L} \exp\left(-\frac{x}{1+\gamma_c}\right) \left[1 - \Psi(x)\right] \right. \\ &\quad \left. +(m_u - 2) \exp(-x) \left[1 - \Psi\left(\frac{x}{1+\gamma_c}\right)\right] \right\} dx. \quad (18.127) \end{aligned}$$

Upon integrating both sides of Equation 18.124 and Equation 18.125 from one to infinity, we obtain:

$$C_{H_1}^u = \frac{1}{\int_1^\infty g_{\lambda_u}(y|H_1)dy}, \quad (18.128)$$

$$C_{H_0}^u = \frac{1}{\int_1^\infty g_{\lambda_u}(y|H_0)dy}. \quad (18.129)$$

Having derived the relevant PDFs in the context of EGC, let us now focus our attention on SC.

18.4.2.2 Probability Density Function of the Ratio Using SC

The PDFs of the decision variables Z_{10}, Z_{1m} seen in Figure 18.20 as well as that of the correct residue digit decision probability $P_{r_1}(C)$ for the SC scheme have been derived in Section 18.3, which are given by Equations 18.105, 18.102 and 18.112. The PDFs of $\{Z_{um}\}$ and $P_{r_u}(C)$ can then be expressed as:

$$f_{Z_{u0}}(y) = \frac{L}{1+\gamma_c}\exp\left(-\frac{y}{1+\gamma_c}\right)\left[1-\exp\left(-\frac{y}{1+\gamma_c}\right)\right]^{L-1}, \quad y \geq 0, \quad (18.130)$$

$$f_{Z_{um}}(y) = L\exp(-y)\left[1-\exp(-y)\right]^{L-1}, \quad m \neq 0, \quad y \geq 0 \quad (18.131)$$

$$P_{r_u}(C) = \sum_{w=1}^{L}(-1)^{w+1}\binom{L}{w}\prod_{n=1}^{L(m_u-1)}\left(\frac{n}{n+w/(1+\gamma_c)}\right). \quad (18.132)$$

The *a priori* probabilities of the hypotheses H_1 and H_0 associated with receiving the residue digit r_u correctly or erroneously can also be expressed with the help of Equation 18.117 and Equation 18.118 with $P_{r_u}(C)$ replaced by Equation 18.132. The PDF of the ratio λ_u defined in Equation 18.113 under the hypothesis of H_1 or H_0 can be derived by using a similar approach to that used in the context of EGC. The PDFs of the maximum and 'second maximum' used in Equation 18.113, namely Y_1 and Y_2 conditioned on the correct residue digit reception hypothesis of H_1 are formulated as:

$$f_{Y_1}(y|H_1) = \frac{1}{P_{r_u}(H_1)}\cdot\frac{L}{1+\gamma_c}\exp\left(-\frac{y}{1+\gamma_c}\right) \cdot\left[1-\exp\left(-\frac{y}{1+\gamma_c}\right)\right]^{L-1}\left[1-\Omega(y)\right]^{m_u-1}, \quad y \geq 0, \quad (18.133)$$

$$f_{Y_2}(y|H_1) = \frac{1}{P(H_1)}\cdot L(m_u-1)\exp(-y)\left[1-\exp(-y)\right]^{L-1} \cdot\Omega\left(\frac{y}{1+\gamma_c}\right)\left[1-\Omega(y)\right]^{m_u-2}, \quad y \geq 0, \quad (18.134)$$

where the short-hand $\Omega(x)$ was defined as:

$$\Omega(x) = \sum_{k=1}^{L}(-1)^{k+1}\binom{L}{k}\exp(-kx). \tag{18.135}$$

Similarly, the PDFs of Y_1 and Y_2 conditioned on the erroneous residue digit reception hypothesis of H_0 can be shown to obey:

$$\begin{aligned} f_{Y_1}(y|H_0) &= \frac{1}{P_{r_u}(H_0)} \cdot L(m_u-1)\exp(-y)\left[1-\exp(-y)\right]^{L-1} \\ &\cdot \left[1-\Omega\left(\frac{y}{1+\gamma_c}\right)\right]\left[1-\Omega(y)\right]^{m_u-2},\ y \geq 0, \end{aligned} \tag{18.136}$$

$$\begin{aligned} f_{Y_2}(y|H_0) &= \frac{1}{P_{r_u}(H_0)} \cdot L(m_u-1)\Omega(y)\left[1-\Omega(y)\right]^{m_u-3} \\ &\cdot \left\{\frac{1}{1+\gamma_c}\exp\left(-\frac{y}{1+\gamma_c}\right)\left[1-\exp\left(-\frac{y}{1+\gamma_c}\right)\right]^{L-1}\left[1-\Omega(y)\right]\right. \\ &\left.+(m_u-2)\exp(-x)[1-\exp(-x)]^{L-1}\left[1-\Omega\left(\frac{y}{1+\gamma_c}\right)\right]\right\},\ y \geq 0. \end{aligned} \tag{18.137}$$

Finally, the PDF of the ratio λ_u defined in Equation 18.113 under the hypothesis of H_1 or H_0 can be expressed in Equation 18.124 and Equation 18.125 with $g_{\lambda_u}(y|H_1)$ and $g_\lambda(y|H_0)$ given by:

$$\begin{aligned} g_{\lambda_u}(y|H_1) &= \int_0^\infty \left[\left(1-\exp\left(-\frac{xy}{1+\gamma_c}\right)\right)(1-\exp(-x))\right]^{L-1} \\ &\cdot\left[1-\Omega(xy)\right]^{m_u-1}\Omega\left(\frac{x}{1+\gamma_c}\right)\left[1-\Omega(x)\right]^{m_u-2} \\ &\cdot x\exp\left(-x-\frac{xy}{1+\gamma_c}\right)dx, \end{aligned} \tag{18.138}$$

$$\begin{aligned} g_{\lambda_u}(y|H_0) &= \int_0^\infty \left[1-\exp(-xy)\right]^{L-1}\left[1-\Omega\left(\frac{xy}{1+\gamma_c}\right)\right]\left[1-\Omega(xy)\right]^{m_u-2}\Omega(x) \\ &\cdot\left[1-\Omega(x)\right]^{m_u-3}\left\{\frac{1}{1+\gamma_c}\exp\left(-\frac{x}{1+\gamma_c}\right)\right. \\ &\cdot\left[1-\exp\left(-\frac{x}{1+\gamma_c}\right)\right]^{L-1}\left[1-\Omega(x)\right] \\ &\left.+(m_u-2)\exp(-x)[1-\exp(-x)]^{L-1}\left[1-\Omega\left(\frac{x}{1+\gamma_c}\right)\right]\right\} \\ &\cdot x\exp\left(-xy\right)dx. \end{aligned} \tag{18.139}$$

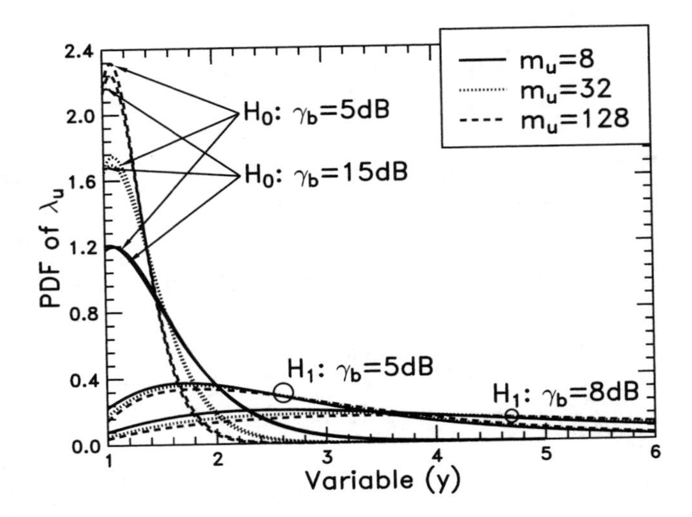

Figure 18.21: EGC: The probability density function (PDF) of the ratio $\lambda_u = \frac{\max_1\{\cdot\}}{\max_2\{\cdot\}}$ under the correct and erroneous residue digit decision hypothesis of H_1 and H_0 using the moduli of $m_u = 8, 32, 128$, $L = 3$ over dispersive Rayleigh fading channels at SNR/bit of $E_b/N_0 = 5dB$, 8dB and 15dB.

The efficiency of the RST in Equation 18.113 depends on the distributions of the conditional ratio PDFs $f_{\lambda_u}(y|H_1)$ and $f_{\lambda_u}(y|H_0)$, specifically, on the overlapping area between these distributions. Conventionally, in conjunction with the ratio threshold test (RTT) in RS or RRNS decoding an erasure is inserted, if the ratio of λ_u is lower than a pre-set threshold [386]. Otherwise, the demodulator passes a code symbol to the channel decoder. 'Errors-and-erasures' decoding [386] can than be invoked to correct the random symbol errors and fill the erasures. In the context of RRNS decoding, alternatively, $d < (U - V)$ out of the U number of residue digits having the lowest ratio values λ_u can be dropped. Then, a reduced RRNS $(U - d, V)$ decoding can be employed to correct the retained random residue digit errors.

Figure 18.21 and Figure 18.22 show the PDFs of the ratio of the maximum to the 'second' maximum under the correct and erroneous residue digit decision hypothesis of H_1 and H_0 for EGC (Figure 18.21) and SC (Figure 18.22) for different values of m_u and for the SNR per bit values of $E_b/N_0 = $ 5dB, 8dB and 15dB. We assumed that there were $L = 3$ resolvable multipath components at the receiver, which were combined with the aid of EGC or SC. Our results were derived from Equation 18.124 and Equation 18.125, when the modulus m_u took values of 8, 32, 128. From the results we observe that for a given value of m_u the peak of the distribution of $f_{\lambda_u}(y|H_1)$ will shift to the right for both EGC and SC, while $f_{\lambda_u}(y|H_0)$ is distributed essentially around $y \approx 1$, but at a point slightly higher than $y = 1$ and the peak of the distribution of $f_{\lambda_u}(y|H_0)$ becomes lower, as the SNR per bit increases from 5 to 15dB. However, for a given value of E_b/N_0 the peaks of the distribution of $f_{\lambda_u}(y|H_0)$ become

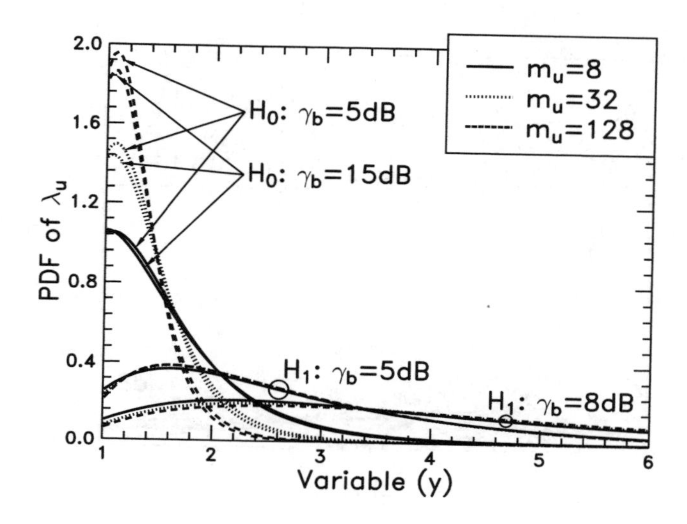

Figure 18.22: SC: The probability density function (PDF) of the ratio $\lambda_u = \frac{\max_1\{\cdot\}}{\max_2\{\cdot\}}$ under the correct and erroneous residue digit decision hypothesis of H_1 and H_0 using the moduli of $m_u = 8, 32, 128$, $L = 3$ over dispersive Rayleigh-fading channels at SNR/bit of $E_b/N_0 = 5dB$, 8dB and 15dB.

higher, when increasing the value of m_u, while the peak of the distribution of $f_{\lambda_u}(y|H_1)$ shifts slightly to the right. Observe furthermore that the PDFs of the RST decision variable λ_u are similar for both EGC and SC, as shown in Figure 18.21 and Figure 18.22.

In Figure 18.23 we used SC as an example, in order to evaluate the effect of the number of multipath components at different SNRs per bit on the distribution of $f_{\lambda_u}(y|H_1)$ and $f_{\lambda_u}(y|H_0)$. We used a modulus of $m_u = 32$ and the multipath diversity order was $L = 1, 2, 3$. The results show that for a given SNR per bit, when increasing the number of the multipath components L combined by the receiver, the peaks of the distributions $f_{\lambda_u}(y|H_1)$ and $f_{\lambda_u}(y|H_0)$ are increased. The above results suggest that when the SNR per bit is sufficiently high – for example $E_b/N_0 = 10dB$ in Figure 18.23 – then the area of overlap is reduced, allowing us to make more reliable erasure decisions. Hence a low-reliability input to the RRNS decoder using a high-order multipath diversity will be dropped with a higher probability than that of a receiver using low-order multipath diversity, since the dropping failure depends on the area of the overlapping regions of $f_{\lambda_u}(y|H_1)$ and $f_{\lambda_u}(y|H_0)$ at a given SNR per bit.

Finally in Figure 18.24 and 18.25, we offer a deeper insight into the influence of the SNR per bit on the PDFs of $f_{\lambda_u}(y|H_1)$ and $f_{\lambda_u}(y|H_0)$, respectively. These results were derived using $m_u = 32$ and $L = 3$. From the results we observe that the peak value of the distribution $f_{\lambda_u}(y|H_0)$ in Figure 18.25 remains approximately constant over a wide range of SNR per bit value, only slightly decreasing, when increasing the SNR per bit value from 0dB to 15dB. However, the peak value of the distribution $f_{\lambda_u}(y|H_1)$ in Figure 18.24 is sig-

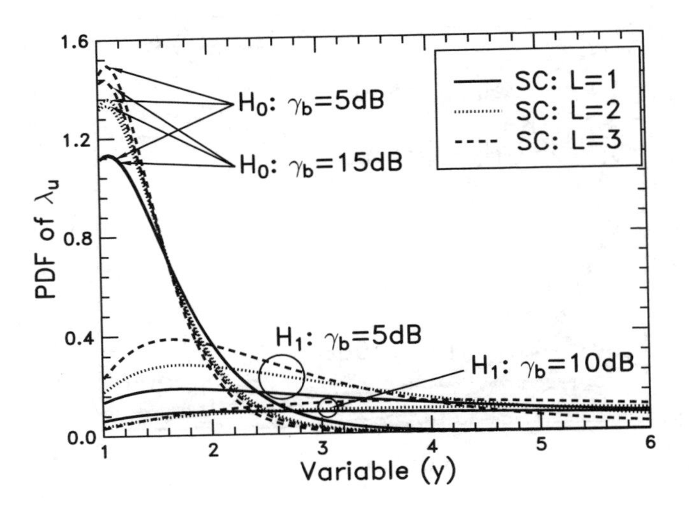

Figure 18.23: SC: The probability density function (PDF) of the ratio $\lambda_u = \frac{\max_1\{\cdot\}}{\max_2\{\cdot\}}$ under the hypothesis of H_1 and H_0 using the modulus of $m_u = 32$ and $L = 1, 2, 3$ over dispersive Rayleigh-fading channels at SNR/bit of $E_b/N_0 = 5dB$, 10dB and 15dB.

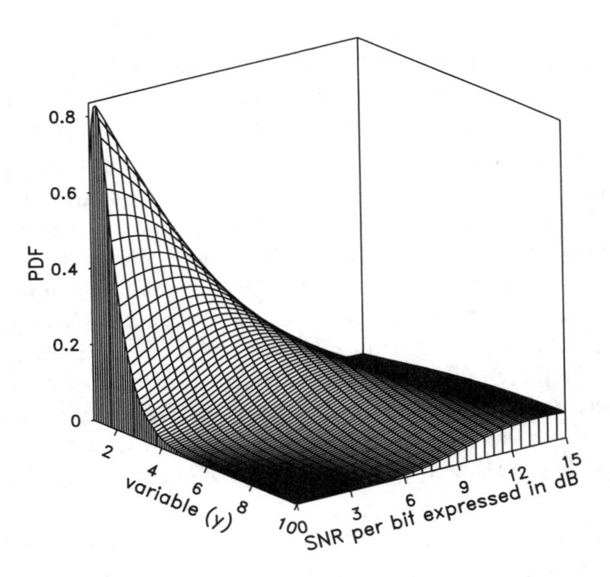

Figure 18.24: SC: The probability density function (PDF) of the ratio $\lambda_u = \frac{\max_1\{\cdot\}}{\max_2\{\cdot\}}$ under the hypothesis of H_1 using the modulus of $m_u = 32$ and $L = 3$ over dispersive multipath Rayleigh-fading channels.

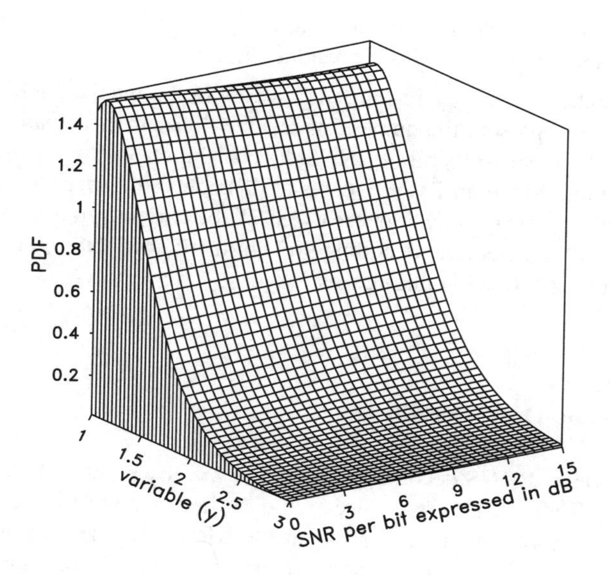

Figure 18.25: SC: The probability density function (PDF) of the ratio $\lambda_u = \frac{\max_1\{\cdot\}}{\max_2\{\cdot\}}$ under the hypothesis of H_0 using the modulus of $m_u = 32$ and $L = 3$ over dispersive multipath Rayleigh-fading channels.

nificantly changed, and its peak shifts to the right, when increasing the SNR per bit value. These results imply that, when the SNR per bit is sufficiently high, the erroneous inputs to the RRNS decoder of Figure 18.20 will be dropped with a significantly higher probability, than the correct inputs. This inherent property of the RST justifies that in a coded system using M-ary orthogonal modulation the relatively low-complexity RST technique can be successfully employed, in order to provide reliable erasure information for the RRNS channel coder and hence enhance the system's performance.

Having determined the PDFs of the ratio λ_u under the correct and erroneous residue digit decision hypotheses H_1 and H_0, we now estimate the system's error probability by first estimating the correct M-ary symbol or codeword probability after the IRNST. After obtaining the correct symbol probability, the average symbol error probability and bit error probability can be estimated using Equation 18.109.

18.4.3 Decoding Error Probability

Let us now assume that a RRNS (U, V) code is introduced for residue digit protection, as seen in the schematic of Figure 18.20. In the RRNS decoding stage of Figure 18.20, d ($d \leq U - V$) number of lowest-reliability inputs of the decoding unit – i.e. the d number of inputs associated with the lowest values of the ratio λ_u among the U inputs – are first dropped. Then the remaining residue digit errors may be corrected by the reduced RRNS $(U-d, V)$ decoding stage of Figure 18.20, as discussed previously. According to the properties of RRNS codes discussed in the previous chapter, we know that a RRNS $(U - d, V)$ code has a minimum distance of $\delta = U - V - d + 1$ and can correct up to $t_{max} = \lfloor (U - V - d)/2 \rfloor$ m_u-ary residue digit errors and detect up to $(U - V - d)$ residue digit errors.

Let t represent the number of m_u-ary residue digit errors in the received RRNS (U, V) codeword, and let s $(s \leq d)$ represent the number of residue digit errors actually discarded by dropping d number of the lowest-reliability RRNS decoder inputs in Figure 18.20 before RRNS $(U - d, V)$ decoding. Since an RRNS $(U - d, V)$ code can correct up to $t_{max} = \lfloor (U - V - d)/2 \rfloor$ number of m_u-ary residue digit errors [397], the correct symbol probabilities of EGC and SC after IRNST, or the correct decoding probability after RRNS (U, V) decoding can be treated according to the following three cases.

18.4.3.1 'Error-Correction-Only' Decoding

When 'Error-Correction Only' RRNS decoding is used, the m_u-ary residue digit errors can only be corrected by the RRNS (U, V) error correction decoding stage of Figure 18.20, i.e., $d = 0$, but in this case no lowest-reliability inputs at the RRNS decoder are dropped before decoding. A codeword – which represents an M-ary symbol or a non-binary symbol expressed by the RNS – is decoded successfully, if and only if all the m_u-ary residue digit errors are corrected by the RRNS (U, V) decoding stage. Consequently, the correct codeword probability after IRNST – remembering that a RRNS codeword expresses a single M-ary symbol – can be expressed as:

$$P_s(C) = \sum_{t=0}^{t_{max}} \left\{ \sum_{Q\binom{U}{t}} \left[\underbrace{\prod_{m=1}^{t} (1 - P_{r_{jm}}(C))}_{t \text{ errors}} \underbrace{\prod_{n=1, n \neq m}^{U} P_{r_{jn}}(C)}_{(U-t) \text{ error-free}} \right] \right\}, \tag{18.140}$$

where $t_{max} = \lfloor \frac{U-V}{2} \rfloor$, since $d = 0$, $\prod_{()=1}^{0}(\cdot) = 1$, while the correct residue digit detection probabilities $P_{r_{jm}}(C)$ and $P_{r_{jn}}(C)$ are given by Equation 18.116 or Equation 18.132 for EGC or SC, respectively. Furthermore, $\{j_1, j_2, \cdots, j_U\}$ is a possible mapping of $\{1, 2, \cdots, U\}$ and $Q\binom{U}{t}$ represents that t out of U of the decoder inputs $\{\hat{r}_1, \hat{r}_2, \cdots, \hat{r}_U\}$ in Figure 18.20 were decided wrongly, i.e. t out of U the m_u-ary residue digits are received in error before the RRNS decoding stage, but the other $(U - t)$ residue digits are error-free, while $\sum_{Q\binom{U}{t}}$ represents all possible selections of t elements from $\{1, 2, \cdots, U\}$. For example, for a four residue digit RRNS codeword (r_1, r_2, r_3, r_4), the possible two-error patterns are $(\hat{r}_1, \hat{r}_2, r_3, r_4)$, $(\hat{r}_1, r_2, \hat{r}_3, r_4)$, $(\hat{r}_1, r_2, r_3, \hat{r}_4)$, $(r_1, \hat{r}_2, \hat{r}_3, r_4)$, $(r_1, \hat{r}_2, r_3, \hat{r}_4)$, $(r_1, r_2, \hat{r}_3, \hat{r}_4)$, where $\hat{r}_i$ represents that the corresponding m_u-ary residue digit was received erroneously. For this RRNS codeword, we have $\sum_{Q\binom{4}{2}} = 6$. Note that this equation is also suitable for the estimation of the well-known RS code performance, since RS codes and RRNS codes exhibit the same minimum distance property. In order to do this, we simply replace $P_{r_{jm}}(C)$, $P_{r_{jn}}(C)$ by the probability $P_j(C)$ – which is the same for all RS-coded symbols – and $\sum_{Q\binom{U}{t}}$ by $\binom{U}{t}$, since RS codes use a constant b-bit symbol size. Consequently, Equation 18.140 is reduced to:

$$P_s(C) = \sum_{t=0}^{t_{max}} \binom{U}{t} [1 - P_j(C)]^t [P_j(C)]^{U-t}. \tag{18.141}$$

Having characterized the performance of 'error-correction-only' decoding, let us now consider 'error-dropping-only' decoding.

18.4.3.2 'Error-Dropping-Only' Decoding

For a RRNS decoder designed for 'Error-Dropping Only' decoding, we have $d = (U - V)$ and $t_{max} = 0$. In this scenario no decoding procedure is invoked, since the remaining V number of residue digits – after $(U - V)$ out of the U residue digits were dropped – all have to be used for the IRNST. Nevertheless, here the residue digit dropping and IRNST processes are referred to as 'Error-Dropping-Only' decoding. In this case, a RNS represented symbol can be recovered correctly by the IRNST, even if there were residue digit errors, provided that the erroneous inputs of the decoder are the lowest-reliability inputs and hence they are dropped during the RRNS decoding by dropping $d = (U - V)$ number of its inputs, namely those having the lowest ratios of λ_u. Hence, the correct M-ary symbol probability after IRNST can be expressed as:

$$P_s(C) = \sum_{t=0}^{d} \left\{ \sum_{Q\binom{U}{t}} \left[\underbrace{\prod_{m=1}^{t} (1 - P_{r_{jm}}(C))}_{t \text{ errors}} \underbrace{\prod_{n=1, n \neq m}^{U} P_{r_{jn}}(C)}_{(U-t) \text{ error-free}} \right] \right\} \cdot P(d, t), \tag{18.142}$$

where the first term on the right-hand side represents the probabilities that there happened at most d number of residue digit errors, while $P(d, t)$ for $t = 0, 1, \ldots, d$ is the probability that t number of m_u-ary residue digit errors are successfully discarded by dropping d number of the lowest-reliability inputs of the RRNS decoding unit of Figure 18.20. Accordingly, $P(d, 0) = 1$.

Let $\mathbf{A} = \{\lambda_{m_1}, \lambda_{m_2}, \ldots, \lambda_{m_{(U-t)}}\}$ represent the ratio set, for which the m_u-ary residue digits are received correctly, while $\mathbf{B} = \{\overline{\lambda}_{n_1}, \overline{\lambda}_{n_2}, \ldots, \overline{\lambda}_{n_t}\}$ represent the ratio set, for which the residue digits are received in error, where the over-bar of λ_{n_j} is used to indicate the erroneous reception hypothesis of H_0. Moreover, let $\overline{\lambda}_{n_\mu} = \max\{\overline{\lambda}_{n_1}, \overline{\lambda}_{n_2}, \ldots, \overline{\lambda}_{n_t}\}$. Consequently, $P(d, t)$ is the probability that, among the $(U - t)$ number of error-free residue digits, there are no more than $(U - V - t) = (d - t)$ residue digits, whose corresponding ratio values are lower than that of $\overline{\lambda}_{n_\mu}$. This means that after residue digit dropping, all the retained V number of residue digits are the error-free residue digits. Hence, the probability

$P(d,t)$ in Equation 18.142 can be expressed as:

$$P(d,t) = \sum_{h=0}^{d-t} \left\{ \sum_{Q\binom{t}{1}} \sum_{Q\binom{U-t}{h}} P\left(\underbrace{\lambda_{m_{i_1}} < \overline{\lambda}_{n_\mu}, \lambda_{m_{i_2}} < \overline{\lambda}_{n_\mu}, \ldots, \lambda_{m_{i_h}} < \overline{\lambda}_{n_\mu};}_{h} \right. \right.$$
$$\underbrace{\lambda_{m_{j_1}} > \overline{\lambda}_{n_\mu}, \lambda_{m_{j_2}} > \overline{\lambda}_{n_\mu}, \ldots, \lambda_{m_{j_{(U-t-h)}}} > \overline{\lambda}_{n_\mu};}_{(U-t-h)}$$
$$\left. \left. \underbrace{\overline{\lambda}_{n_1} < \overline{\lambda}_{n_\mu}, \ldots, \overline{\lambda}_{n_{\mu-1}} < \overline{\lambda}_{n_\mu}, \overline{\lambda}_{n_{\mu+1}} < \overline{\lambda}_{n_\mu}, \ldots, \overline{\lambda}_{n_t} < \overline{\lambda}_{n_\mu}}_{(t-1)} \right) \right\}. \tag{18.143}$$

More explicitly, we have:

$$P(d,t) = \sum_{h=0}^{d-t} \left\{ \sum_{Q\binom{t}{1}} \sum_{Q\binom{U-t}{h}} \int_0^\infty \left[\prod_{\eta=1}^{h} \int_0^x f_{\lambda_{m_{i\eta}}}(y|H_1)dy \right] \right.$$
$$\cdot \left[\prod_{\nu=1}^{U-t-h} \int_x^\infty f_{\lambda_{m_{j\nu}}}(y|H_1)dy \right]$$
$$\left. \cdot \left[\prod_{\zeta=1,\zeta\neq\mu}^{t} \int_0^x f_{\overline{\lambda}_{n_\zeta}}(y|H_0)dy \right] f_{\overline{\lambda}_{n_\mu}}(x|H_0)dx \right\}, \tag{18.144}$$

where the PDFs conditioned on the error-free and erroneous residue digit decision hypotheses, namely $f_{\lambda_{m_{\{\cdot\}}}}(y|H_1)$ and $f_{\overline{\lambda}_{n_{\{\cdot\}}}}(x|H_0)$ are given by Equations 18.124 and 18.125 in conjunction with the receiver using EGC or SC, respectively.

If a RS coded system is considered, Equation 18.144 can be simplified to:

$$P(d,t) = \sum_{h=0}^{d-t} \binom{t}{1} \binom{U-t}{h} \int_0^\infty \left[\int_0^x f_{\lambda_M}(y|H_1)dy \right]^h \left[\int_x^\infty f_{\lambda_M}(y|H_1)dy \right]^{U-t-h}$$
$$\cdot \left[\int_0^x f_{\overline{\lambda}_M}(y|H_0)dy \right]^{t-1} f_{\overline{\lambda}_M}(x|H_0)dx, \tag{18.145}$$

where the conditional PDFs $f_{\lambda_M}(y|H_1)$ and $f_{\overline{\lambda}_M}(y|H_0)$ are given by Equation 18.124 and Equation 18.125 with m_u for $u = 1, 2, \ldots, U$ replaced by a constant value of M.

Having characterized the low-complexity 'error-dropping-only' RRNS decoding technique, let us now consider finally the combined 'error-dropping-and-correction' RRNS decoder.

18.4.3.3 'Error-Dropping-and-Correction' Decoding

If the RRNS decoder of Figure 18.20 is designed using 'error-dropping-and-correction' decoding, i.e., $t_{max} > 0$ and $d > 0$, where $t_{max} = \lfloor (U-d-V)/2 \rfloor$, then the transmitted M-ary symbol can be recovered correctly, if the m_u-ary residue digit errors encountered due to the channel effects are dropped by dropping d number of the RRNS decoder inputs having the lowest values of λ_u or corrected by the RRNS $(U-d, V)$ decoding stage. Hence, the correct symbol probability after the IRNST block of Figure 18.20 can be expressed as:

$$P_s(C) = \sum_{t=0}^{d+t_{max}} \left\{ \sum_{Q\binom{U}{t}} \left[\underbrace{\prod_{m=1}^{t} (1 - P_{r_{jm}}(C))}_{t \text{ errors}} \underbrace{\prod_{n=1, n \neq m}^{U} P_{r_{jn}}(C)}_{(U-t) \text{ error-free}} \right] \right\} \cdot P(s \geq t - t_{max}), \tag{18.146}$$

where the term $P_{r_{jm}}(C)$ is the correct reception probability of r_{jm} given by Equation 18.116 or Equation 18.132 for EGC or SC, respectively. The term $P(s \geq t - t_{max})$ in Equation 18.146 represents the probability that the number of discarded residue digit errors is not less than $t - t_{max}$, i.e., after the low-confidence residue digit dropping operation the number of retained residue digit errors is less than t_{max}, and hence can be corrected by the reduced RRNS $(U-d, V)$ decoding stage. Accordingly, $P(s \geq t - t_{max}) = 1$, if $t \leq t_{max}$. By contrast, when $t > t_{max}$, we have:

$$\begin{aligned} &P(s \geq t - t_{max}) = \\ &\sum_{h=0}^{d+t_{max}-t} \left\{ \sum_{Q\binom{t}{1}} \sum_{Q\binom{U-t}{h}} \sum_{Q\binom{t}{t_{max}}} P \left(\underbrace{\lambda_{m_{i_1}} < \overline{\lambda}_{n_\mu}, \lambda_{m_{i_2}} < \overline{\lambda}_{n_\mu}, \ldots, \lambda_{m_{i_h}} < \overline{\lambda}_{n_\mu};}_{h} \right. \right. \\ &\underbrace{\lambda_{m_{j_1}} > \overline{\lambda}_{n_\mu}, \lambda_{m_{j_2}} > \overline{\lambda}_{n_\mu}, \ldots, \lambda_{m_{j_{(U-t-h)}}} > \overline{\lambda}_{n_\mu};}_{(U-t-h)} \\ &\underbrace{\overline{\lambda}_{n_{k_1}} > \overline{\lambda}_{n_\mu}, \overline{\lambda}_{n_{k_2}} > \overline{\lambda}_{n_\mu}, \ldots, \overline{\lambda}_{n_{k_{t_{max}}}} > \overline{\lambda}_{n_\mu};}_{t_{max}} \\ &\left. \left. \underbrace{\overline{\lambda}_{n_{l_1}} < \overline{\lambda}_{n_\mu}, \overline{\lambda}_{n_{l_2}} < \overline{\lambda}_{n_\mu}, \ldots, \overline{\lambda}_{n_{l_{(t-t_{max}-1)}}} < \overline{\lambda}_{n_\mu};}_{(t-t_{max}-1)} \right) \right\} \end{aligned} \tag{18.147}$$

which results in:

$$\begin{aligned}
&P(s \geq t - t_{max}) = \\
&= \sum_{h=0}^{d+t_{max}-t} \left\{ \sum_{Q\binom{t}{1}} \sum_{Q\binom{U-t}{h}} \sum_{Q\binom{t}{t_{max}}} \int_0^{\infty} \left[\prod_{\eta=1}^{h} \int_0^x f_{\lambda_{m_{i_\eta}}}(y|H_1)dy \right] \right. \\
&\cdot \left[\prod_{\nu=1}^{U-t-h} \int_x^{\infty} f_{\lambda_{m_{j_\nu}}}(y|H_1)dy \right] \left[\prod_{\kappa=1}^{t_{max}} \int_x^{\infty} f_{\overline{\lambda}_{n_{k_\kappa}}}(y|H_0)dy \right] \\
&\cdot \left. \left[\prod_{\zeta=1}^{t-t_{max}-1} \int_0^x f_{\overline{\lambda}_{n_{l_\zeta}}}(y|H_0)dy \right] f_{\overline{\lambda}_{n_\mu}}(x|H_0)dx \right\},
\end{aligned} \tag{18.148}$$

where $\lambda_{m\{\cdot\}}$ and $\overline{\lambda}_{n\{\cdot\}}$ are from the ratio sets **A** and **B**, which were defined previously. Note that, if $t_{max} = 0$, then $P(s \geq t - t_{max})$ of Equation 18.148 becomes equivalent to $P(d,t)$ of Equation 18.145. Furthermore, if an RS coded system is considered, Equation 18.148 can be simplified to:

$$\begin{aligned}
P(s \geq t - t_{max}) &= \sum_{h=0}^{d+t_{max}-t} \binom{t}{1} \binom{U-t}{h} \binom{t}{t_{max}} \\
&\cdot \int_0^{\infty} \left[\int_0^x f_{\lambda_M}(y|H_1)dy \right]^h \left[\int_x^{\infty} f_{\lambda_M}(y|H_1)dy \right]^{U-t-h} \\
&\cdot \left[\int_x^{\infty} f_{\overline{\lambda}_M}(y|H_0)dy \right]^{t_{max}} \left[\int_0^x f_{\overline{\lambda}_M}(y|H_0)dy \right]^{t-t_{max}-1} f_{\overline{\lambda}_M}(x|H_0)dx.
\end{aligned} \tag{18.149}$$

Let us now consider the BER performance of the coded RRNS-based DS-CDMA system.

18.4.4 Numerical Results

In Figure 18.26, which is related to the EGC scheme and in Figure 18.27 characterizing the SC scheme, we evaluated the BER performance of a RNS-based parallel DS-CDMA system without redundancy, as well as upon using one redundant modulus in conjunction with one lowest-reliability input of the RRNS decoder being discarded, which we denote as $d = 0$ and $d = 1$, respectively. In the related probability expressions of Equations 18.142 and 18.144 moduli values of 29, 31, 32, 33, 35, 37, 41 were used, the number of diversity multipath components combined in the receiver was $L = 1, 2$ or 3 using an EGC or SC scheme. The results show that, when the RNS is designed with redundant moduli, the BER performance of both the EGC and that of the SC scheme is substantially improved. Taking $L = 2$, 3 diversity components as examples, the EGC scheme can achieve a BER of 10^{-3} at SNRs per bit of 14.5dB or 12.5dB, when one lowest-reliability input of the RRNS decoder is dropped during

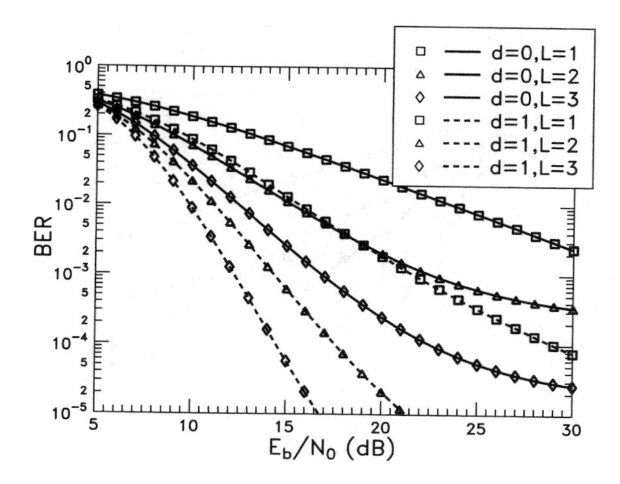

Figure 18.26: EGC: BER versus E_b/N_0 performance for the coded RRNS-based parallel DS-CDMA system of Figure 18.20 using orthogonal modulation with 7 moduli, $m_1 = 29, m_2 = 31, m_3 = 32, m_4 = 33, m_5 = 35, m_6 = 37, m_7 = 41$, $K = 1$, $N = 64$ and $L = 1, 2, 3$, where d is the number of lowest-reliability inputs of the RRNS decoder dropped, the results were evaluated from Equation18.109 for the non-redundant system, while Equations 18.142 and 18.144 for the redundant system over dispersive Rayleigh channels.

RRNS decoding, as demonstrated by the $d = 1,\ L = 2$ and $d = 1,\ L = 3$ curves. However, if the RNS-based system is designed without redundancy, an average of 23dB or 17dB SNR per bit is required for the EGC scheme using $L = 2$ or $L = 3$, in order to achieve the BER of 10^{-3}. This implies that we can obtain about 8.5dB or 4.5dB gain at $P_b(\varepsilon) = 10^{-3}$ by using one lowest-reliability residue digit dropping for $L = 2$ or 3, respectively. Similarly, the SC scheme using $L = 2$ or 3 can achieve a BER of 10^{-3} at SNRs per bit of 15dB or 13.5dB, and obtain about 8.5dB or 5dB gain at $P_b(\epsilon) = 10^{-3}$ by using one lowest-reliability residue digit dropping for $L = 2$ or 3, respectively. By comparing Figures 18.26 and 18.27 we also note that the required E_b/N_0 difference between EGC and SC is typically about 1dB.

In Figure 18.28 and Figure 18.29 the BER performance of the systems considered in Figure 18.26 and Figure 18.27 were evaluated when exposed to multi-user interference using $L = 3$ diversity components, for EGC and SC, respectively. In contrast to Figure 18.27 and Figure 18.27, where we assumed $K = 1$, here we assumed $K = 1, 5, 10$, i.e., the number of interfering users was 0, 4 and 9, respectively. From the results a graceful degradation is observed for both EGC and SC, and also for both the non-redundant system and for redundant system using one lowest-reliability residue digit dropping based RRNS decoding, when the number of active users, K, increases. However, for a given number of active users and a given bit error probability, the redundant system employing one lowest-reliability input dropping based RRNS decoding requires a significantly lower SNR per bit, than the non-redundant RNS-based system.

Similarly to Figure 18.28 and Figure 18.29, in Figure 18.30 and Figure 18.31 we eval-

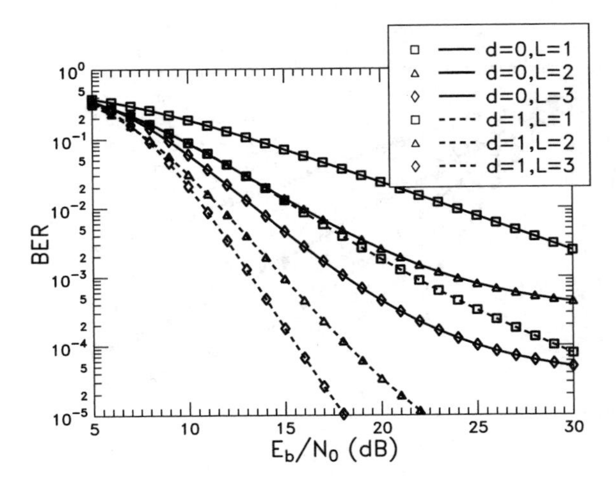

Figure 18.27: **SC**: BER versus E_b/N_0 performance for the coded RRNS-based parallel DS-CDMA system of Figure 18.20 using orthogonal modulation with 7 moduli, $m_1 = 29, m_2 = 31, m_3 = 32, m_4 = 33, m_5 = 35, m_6 = 37, m_7 = 41$, $K = 1$, $N = 64$ and $L = 1, 2, 3$, where d is the number of lowest-reliability inputs of the RRNS decoder dropped, the results were evaluated from Equation 18.109 for the non-redundant system, while Equations 18.142 and 18.144 for the redundant system over dispersive Rayleigh channels.

uated the BER performance of a 10-moduli RNS-based parallel DS-CDMA orthogonal signalling system using EGC or SC schemes, when $d = 2$ lowest-reliability residue digits were dropped in Figure 18.20 or $t_{max} = 1$ residue digit error correction RRNS decoding was considered. These two RNS-based DS-CDMA systems used the same number of redundant moduli and had the same information rate. The parameters related to these investigations were $m_1 = 29$, $m_2 = 31$, $m_3 = 35$, $m_4 = 36$, $m_5 = 37$, $m_6 = 41$, $m_7 = 43$, $m_8 = 47$, $m_9 = 53$, $m_{10} = 59$, and $L = 3$, $N = 64$ and $K = 1$, 10. We notice that the BER of the RNS-based parallel DS-CDMA system discarding the two lowest-reliability inputs of the RRNS decoder was lower than that of the system with one residue digit error correction-based RRNS decoding, when the same number of multipath diversity components was considered. Specifically, both the EGC and SC schemes dropping two lowest-reliability residue digits require 2dB less bit-SNR than that of the systems using one residue digit error correction for achieving the BER of 10^{-5}, when $K = 1$, i.e., without multi-user interference. The results imply that for a ten-moduli RNS-based system, using lowest-reliability residue digit dropping is a highly effective method of improving the system's BER performance. Furthermore, the complexity of the RRNS $(10, 8)$ decoding based on dropping two of the lowest-reliability inputs of the RRNS decoder is lower, than that of the RRNS $(10, 8)$ one residue digit error correction decoding. However, the BER performance of both the EGC and that of SC scheme is adversely affected by the multi-user interference and hence the BER of the system supporting ten active users is significantly higher than that of the system without multi-user interference.

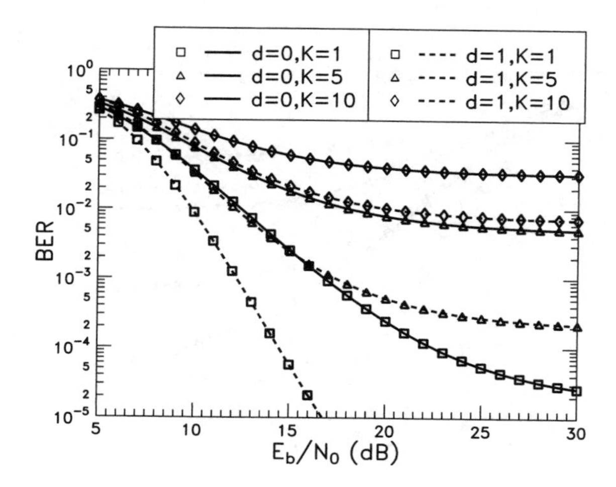

Figure 18.28: EGC: BER versus E_b/N_0 performance for the coded RRNS-based parallel DS-CDMA system of Figure 18.20 using orthogonal modulation with 7 moduli, $m_1 = 29, m_2 = 31, m_3 = 32, m_4 = 33, m_5 = 35, m_6 = 37, m_7 = 41$, $L = 3$ diversity components, $N = 64$ chips per bit duration and $K = 1, 5, 10$ **active users**, where d is the number of lowest-reliability inputs of the RRNS decoder dropped. The results were evaluated from Equation 18.109 for the non-redundant system, while Equations 18.142 and 18.144 for the redundant system over dispersive Rayleigh fading channels.

In summary, above we have analyzed the performance of the proposed RNS-based parallel DS-CDMA system using orthogonal modulation, when non-coherent demodulation, EGC- or SC-assisted diversity reception and RRNS decoding were considered. However, in wireless communications powerful forward error correction (FEC) is required in order to maintain high-reliability communications, while supporting a multiplicity of users. Hence, in the next section let us now discuss the structure and the performance of the proposed RRNS-based DS-CDMA system in conjunction with a concatenated RS-RRNS coding scheme.

18.5 Performance of Concatenated RS-RRNS Coded Systems

18.5.1 Concatenated RS-RRNS codes

Concatenated coding [436–438] is a technique of combining relatively simple channel codes, in order to form a powerful coding system for achieving a high performance and large coding gain with reduced decoding complexity. In practical applications, the inner code is traditionally a relatively short binary block code or a binary convolutional code with relatively short constraint length, and the outer code is usually a RS code with symbols from a Galois field GF(2^b).

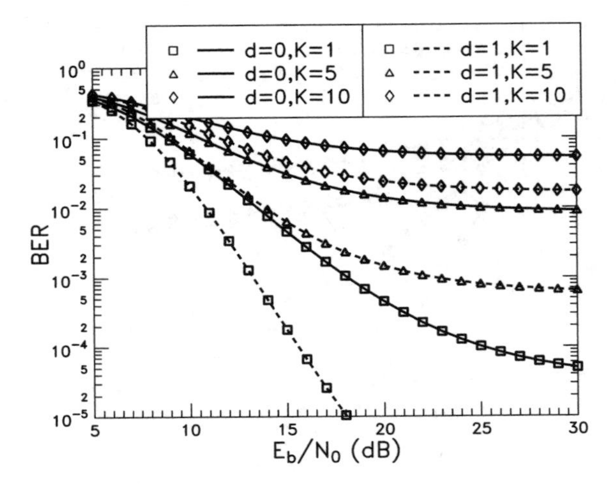

Figure 18.29: **SC**: BER versus E_b/N_0 performance for the coded RRNS-based parallel DS-CDMA system of Figure 18.20 using orthogonal modulation with 7 moduli, $m_1 = 29, m_2 = 31, m_3 = 32, m_4 = 33, m_5 = 35, m_6 = 37, m_7 = 41$, $L = 3$ diversity components, $N = 64$ chips per bit duration and $K = 1, 5, 10$ **active users**, where d is the number of lowest-reliability inputs of RRNS decoder dropped. The results were evaluated from Equation 18.109 for the non-redundant system, while Equations 18.142 and 18.144 for the redundant system over dispersive Rayleigh channels.

As noted in the previous chapter in more detail, the family of RRNS codes constitutes a class of maximum-minimum-distance codes [397, 406], akin to RS codes. RRNS codes can provide a powerful error correction and error detection capability, which is similar to that of RS codes, but the inherent parallel structure of the RNS arithmetic, the associated independent residue processing and the availability of convenient decoding algorithms render RRNS codes an attractive alternative for using as inner codes. Short RRNS codes can be combined with RS codes, where the former is used as the inner code, in order to form a concatenated RS-RRNS code. Furthermore, it is possible to realize high-bit-rate communication using an RNS arithmetic by combining the RNS with highly efficient modulation and demodulation schemes, such as the RNS-based M-ary orthogonal modulation scheme discussed in this chapter.

Using an RS-RRNS concatenated code, after RRNS inner decoding, the symbol error probability is typically decreased to a degree, which may maximize the external RS coding gain, and hence the average symbol error probability is further decreased to the required degree using RS decoding. In addition, the error-detection capability of the RRNS code can provide symbol error information or erasure information for the RS outer decoding. Consequently, the efficiency of the RS code utilizing 'errors-and-erasures' decoding can be enhanced upon exploiting the explicit error detection capability of the inner code, since in this case the error positions are known by the outer RS decoder. Hence all the RS syndrome-equations [64] can be used to determine a doubled number of RS symbol error magnitudes in comparison to the scenario, where no erasure – i.e., no error position information is available

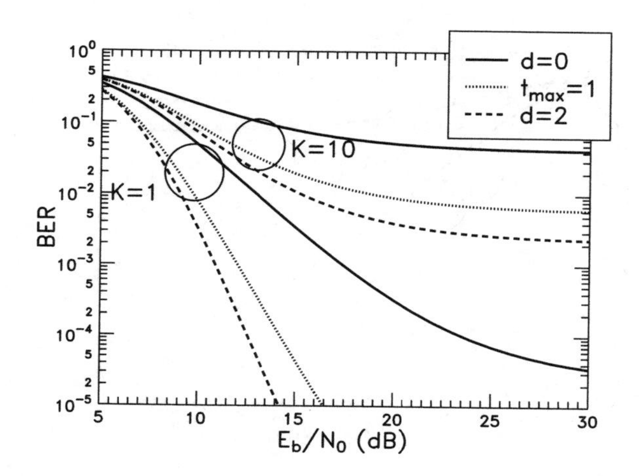

Figure 18.30: **EGC**: BER versus E_b/N_0 performance for the coded RRNS-based parallel DS-CDMA system of Figure 18.20 with 10 moduli, $m_1 = 29, m_2 = 31, m_3 = 35, m_4 = 36, m_5 = 37, m_6 = 41, m_7 = 43, m_8 = 47, m_9 = 53, m_{10} = 59$, and $L = 3$, $K = 1,\ 10$, where t_{max} is the number of errors corrected by the RRNS $(10, 8)$ code, and d is the number of lowest-reliability inputs of the RRNS decoder dropped. The results were evaluated from Equation 18.109 for $d = 0$, Equations 18.142 and 18.144 for $d = 2$, and Equation 18.140 for $t_{max} = 1$ over dispersive Rayleigh fading channels.

– requiring the determination of both the error positions and magnitudes.

Let N_c and K_c be the length of the outer RS code and the number of original information symbols, respectively. We assume that a RRNS code is constructed using the moduli $(m_1, m_2, \ldots, m_U)$, where $(m_1, m_2, \ldots, m_V)$, $(V < U)$ are the information moduli, $(m_{V+1}, m_{V+2}, \ldots, m_U)$ the redundant moduli, and $m_{V+j} \geq \max\{m_1, m_2, \ldots, m_V\}$ for $j = 1, 2, \ldots, U - V$. The interval $\left[0, M = \prod_{u=1}^{V} m_u\right)$ is the information symbols' dynamic range for the RRNS code, and $[0, MM_R)$ is the RRNS (U, V) code's dynamic range, where $M_R = \prod_{u=V+1}^{U} m_u$. Notice that since the information symbols' dynamic range of RRNS is $[0, M)$, not $[0, MM_R)$ as in the RRNS code, and since the RS code rate is $R_c = K_c/N_c$, the energy for transmitting a parallel signal must be scaled according to $E_s = R_c E_b[\log_2 \prod_{u=1}^{V} m_u]$. The encoding procedure of the concatenated code is performed as follows:

- Encode K_c information symbols using an RS (N_c, K_c) code, assuming that a symbol is constituted by b bits, and $2^b \leq M$ but $2^{b+1} > M$.
- Each symbol of the RS (N_c, K_c) code is encoded into a RRNS (U, V) code by computing the residue digits with respect to the moduli $\{m_1, m_2, \ldots, m_U\}$, and expressing the encoded result as $\{r_1, r_2, \ldots, r_U\}$.
- The U residue digits are transmitted by the proposed RRNS-based parallel DS-CDMA system using orthogonal modulation, as discussed previously in this chapter.

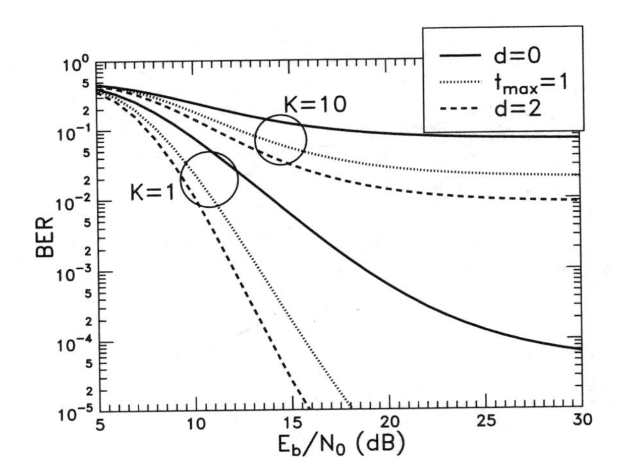

Figure 18.31: **SC**: BER versus E_b/N_0 performance for the coded RRNS-based parallel DS-CDMA system of Figure 18.20 with 10 moduli, $m_1 = 29, m_2 = 31, m_3 = 35, m_4 = 36, m_5 = 37, m_6 = 41, m_7 = 43, m_8 = 47, m_9 = 53, m_{10} = 59$, and $L = 3$, $K = 1,\ 10$, where t_{max} is the number of errors corrected by the RRNS $(10, 8)$ code, and d is the number of lowest-reliability inputs of RRNS decoder dropped. The results were evaluated from Equation 18.109 for $d = 0$, Equations 18.142 and 18.144 for $d = 2$, and Equation 18.140 for $t_{max} = 1$ over dispersive Rayleigh fading channels.

After the receiver provides the estimates of the U residues, the code can be decoded as follows:

- The residue errors are detected and/or corrected using the inner RRNS (U, V) decoding, where RRNS (U, V) decoding might be 'error-detection-only' decoding, 'error-correction-only' decoding, 'error-dropping-only' decoding, 'error-dropping-and-detection' decoding or 'error-dropping-and-correction' decoding. If no residue digit errors are detected, because there are no residue digit errors or there are undetectable residue digit errors in the RRNS (U, V) code, or since the residue digit errors are corrected by the RRNS (U, V) decoding, an estimate of an outer RS code symbol is obtained. Otherwise, if uncorrectable residue errors are detected in the RRNS (U, V) code, the RS code symbol is marked as an erasure.

- Following the above estimation of all symbols of a full RS (N_c, K_c) codeword, the symbol errors of the outer RS codeword are corrected and the erasures are filled by the RS (N_c, K_c) decoder. Otherwise, a 'decoding failure' is declared.

18.5.2 Decoding Error Probability of RS-RRNS codes

Upon using the above decoding procedure, and assuming ideal interleaving of the RS code symbols, then, after 'error-correction-only' decoding the resulting symbol error probability of RS hard decision decoding is approximately given by [439]:

$$P_{s,d}(\varepsilon) \approx \frac{1}{N_c} \sum_{i=\lfloor \frac{N_c-K_c}{2} \rfloor+1}^{N_c} i \binom{N_c}{i} P_t^i (1-P_t)^{N_c-i}, \tag{18.150}$$

where P_t represents the probability of a random RS symbol error after RRNS decoding, which is given, for example, by $(1-P_s(C))$, depending on the RRNS structure and on the decoding algorithm used. If RS symbol erasure information can be provided by the inner RRNS decoding, 'errors-and-erasures' outer RS decoding can be applied, consequently, the resulting symbol error probability after outer RS decoding is approximately given by [440]:

$$P_{s,d}(\varepsilon) \approx \frac{1}{N_c} \sum_{i=0}^{N_c} \sum_{j=j_0(i)}^{N_c-i} (i+j) \binom{N_c}{i} \binom{N_c-i}{j} P_t^i P_e^j (1-P_t-P_e)^{N_c-i-j}, \tag{18.151}$$

where $j_0(i) = \max\{0, N_c - K_c + 1 - 2i\}$, while P_t and P_e are the probability of a random RS symbol error and a RS symbol erasure after RRNS decoding. Note that, $P_{s,d}(\varepsilon)$ in Equation 18.151 is the average joint probability of uncorrectable random RS symbol errors and RS symbol erasures, in other words the average probability of outer RS decoding failure.

The probability of a random RS symbol error and symbol erasure in Equation 18.150 and Equation 18.150, namely P_t and P_e, can be computed according to the RRNS decoding algorithm discussed previously in previous sections of this chapter, except for the 'error-detection-only' and the 'error-dropping-and-detection' decoding. Let us first consider P_t and P_e encountered in the context of 'error-detection-only' decoding, recalling that one RS symbol was encoded into an RRNS (U, V) codeword.

The inner RRNS (U, V) code can detect up to $(U - V)$ residue digit errors. However, if the number of residue digit errors exceeds $(U - V)$, RRNS decoding can still detect the presence of RRNS codeword errors and deliver RS-symbol erasures to the outer RS code, provided that the erroneous RRNS vector is not a legitimate codeword, i.e., its value is not in the legitimate range of $[0, M)$. Consequently, the probability P_t of a random RS-symbol error – which is equivalent to the undetectable probability of an RRNS code – can be approximated as:

$$\begin{aligned} P_t &\approx \frac{M-1}{MM_R - M_{(U-V)}} \times P_s(\varepsilon|U-V) \\ &\approx \frac{1}{M_R} \times P_s(\varepsilon|U-V), \end{aligned} \tag{18.152}$$

where $M_{(U-V)}$ represents the total number of vectors having distance from the transmitted RRNS codeword not exceeding $(U - V)$ and usually $M_{(U-V)} << MM_R$. Obviously, any error vector of the $M_{(U-V)}$ can be detected by the RRNS (U, V) 'error-detection-only' decoding. $P_s(\varepsilon|U - V)$ in Equation 18.152 is the average probability that there were at least

$(U - V + 1)$ RRNS residue digit errors, which can be expressed as:

$$P_s(\varepsilon|U-V) = 1 - \sum_{t=0}^{U-V} \left\{ \sum_{Q\binom{U}{t}} \left[\underbrace{\prod_{m=1}^{t} (1 - P_{r_{jm}}(C))}_{t \text{ errors}} \underbrace{\prod_{n=1, n\neq m}^{U} P_{r_{jn}}(C)}_{(U-t) \text{ error-free}} \right] \right\}, \quad (18.153)$$

where $P_{r_{jm}}(C)$ and $P_{r_{jn}}(C)$ are the correct detection probabilities for residue digits r_{j_m} and r_{j_n}, which, for example, are given by Equation 18.116 or Equation 18.132 for the multipath fading channel using EGC or SC, respectively.

The probability of a RS-symbol erasure after RRNS decoding can then be expressed as:

$$\begin{aligned} P_e &\approx \left(1 - \frac{1}{M_R}\right) P_s(\varepsilon|U-V) \\ &\approx \frac{M_R - 1}{M_R} \times P_s(\varepsilon|U-V). \end{aligned} \quad (18.154)$$

Comparing Equation 18.154 with Equation 18.152 it can be shown that $P_e >> P_t$, provided that $M_R >> 1$, which means that after RRNS decoding most RS symbol errors will be marked as erasures.

When using inner RRNS 'error-dropping-and-detection' decoding, we assume that the number of dropped residue digits is $d \leq U - V$. Then, after low-reliability residue digit dropping we obtain a reduced RRNS $(U - d, V)$ code. We have shown in Section 18.4.3.2 that after the 'error-dropping-only' RRNS decoding, the correct M-ary symbol probability is given by $P_s(C)$ of Equation 18.142 over dispersive multipath fading channels using EGC or SC, and the probability of M-ary symbol error before the RS decoding can then be computed by $P_s(\varepsilon) = 1 - P_s(C)$.

Let us assume that the residue digits and their corresponding moduli are $(r_{g1}, \ldots, r_{g(U-d)})$ and $(m_{g1}, m_{g2}, \ldots, m_{g(U-d)})$, respectively, after the inner 'error-dropping-only' RRNS decoding. Moreover, we assume that $m_{g1} < m_{g2} < \ldots < m_{g(U-d)}$, and that the product of the redundant moduli in the reduced RRNS $(U - d, V)$ code is $M'_R = \prod_{j=1}^{U-d-V} m_{g(V+j)}$.

P_t and P_e can now be approximated as:[2]

$$P_t \approx \frac{1}{M'_R} \times P_s(\varepsilon), \quad (18.155)$$

$$P_e \approx \frac{M'_R - 1}{M'_R} \times P_s(\varepsilon), \quad (18.156)$$

where $P_s(\varepsilon) = 1 - P_s(C)$, while $P_s(C)$ was given by Equation 18.142, specifically, for the case of dispersive multipath fading channels using EGC or SC diversity reception. Let us now characterize the proposed system in quantitative performance terms.

18.5.3 Numerical Results

In Figure 18.32 to Figure 18.33 we evaluated the BER performance of the RNS-based parallel DS-CDMA orthogonal signalling system, when additional concatenated RS-coding was introduced. Specifically, an outer RS $(255, 223)$ code over the Galois Field $GF(2^8)$ using 8-bit symbols was invoked. This RS $(255, 223)$ scheme has been proposed in the CCSDS (the Consultative Committee for Space Data System) standard as an outer code combined with a half rate, constraint-length seven inner convolutional code for data protection [441]. Since 8 bits per symbol period were transmitted, hence for a non-redundant RNS-based system the moduli of $m_1 = 5$, $m_2 = 7$, $m_3 = 8$ were appropriate for transmitting the 8-bit symbol, since $5 \cdot 7 \cdot 8 = 280 > 256 = 2^8$. Specifically, in Figure 18.32 we evaluated the influence of RRNS coding on the approximations to the average bit error probability after RS $(255, 223)$ decoding. When $t = 0$ and $\beta = 0$, the three moduli of $m_1 = 5$, $m_2 = 7$, $m_3 = 8$ are all information moduli, corresponding to a non-redundant or to a 'no residue error detection' RNS-based parallel DS-CDMA system. The RS symbol errors are corrected by RS $(255, 223)$ 'errors-correction-only' decoding. When $t = 0$ and $\beta = 1$, moduli $m_1 = 5$, $m_2 = 7$, $m_3 = 8$ are the information moduli, and $m_4 = 9$ is the redundant modulus invoked for residue error detection. When using RRNS inner decoding, RS symbol erasure information is generated, which can be filled in by the outer RS decoding. Specifically, the RS symbol errors and erasures are corrected and filled by RS $(255, 223)$ 'errors-and-erasures' decoding. Finally, when $t = 1$ and $\beta = 0$, moduli $m_1 = 5$, $m_2 = 7$, $m_3 = 8$ are the information moduli, and $m_4 = 9, m_5 = 11$ are the redundant moduli used for residue error correction. In this case, the RRNS inner decoding corrects one residue digit error, but will not generate erasure information. The random RS symbol errors of the outer RS code are corrected by RS $(255, 223)$ 'error-correction-only' decoding. The RNS-based parallel DS-CDMA system using one or two redundant moduli can achieve a lower BER than the RNS-based parallel DS-CDMA system without redundant moduli, if the SNR per bit given by E_b/N_0 is suffi-

[2]Note that, following our discussions on inner 'error-detection-only' RRNS decoding, after d out of the U number of inputs of the RRNS decoder are dropped, the reduced RRNS $(U - d, V)$ code can detect up to $(U - V - d)$ residue digit errors and can also detect some of the residue digit errors beyond $(U - d - V)$ residue digits. However, in this case the exact computation of P_t and P_e is complex, since the dropping alters the statistical properties of the retained RRNS decoder inputs. Hence, here we only give approximations of P_t and P_e, as shown in Equation 18.155 and Equation 18.156. The actual value of P_t is lower than that given by Equation 18.155, and P_e is higher than that given by Equation 18.156. Consequently, the decoding error probability resulting from Equation 18.155 and Equation 18.156 is higher than the actual decoding error probability.

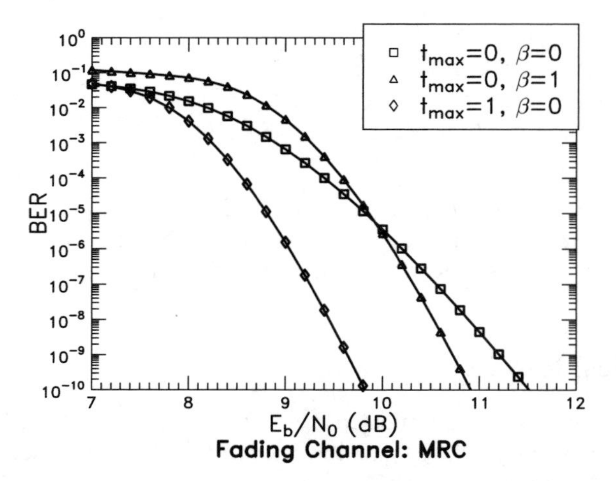

Figure 18.32: BER versus average SNR per bit for the concatenated RS-RRNS coded parallel DS-CDMA orthogonal system over dispersive multipath fading channels using MRC combining and concatenated coding, where moduli $m_1 = 5, m_2 = 7, m_3 = 8$ are employed for $(t_{max} = 0, \beta = 0)$ and RS (255,233) 'error-correction-only' decoding. Moduli $m_1 = 5, m_2 = 7, m_3 = 8, m_4 = 9$ are employed for $(t_{max} = 0, \beta = 1)$ and RS (255,233) 'errors-and-erasures' decoding, while moduli $m_1 = 5, m_2 = 7, m_3 = 8, m_4 = 9, m_5 = 11$ are employed for $(t_{max} = 1, \beta = 0)$ and RS (255,233) 'error-correction-only' decoding, where t_{max} and β are, respectively, the error-correction capability and error-detection capability of the RRNS inner code. The other parameters are $N = 64, K = 1, L = 3$.

ciently high. Specifically, it can be seen from Figure 18.32 that approximately 1dB higher E_b/N_0 is required by the system characterized by $t = 0$ and $\beta = 1$, than for the scheme associated with $t = 1$ and $\beta = 0$ for maintaining the same BER.

In Figure 18.33 two decoding techniques, namely 'error-correction-only' decoding and 'errors-and-erasures' decoding were assumed, depending on the inner RRNS decoding. An 8-bit symbol was transmitted per symbol period, and the symbol erasure information - when required – was provided by the RRNS error-detection decoding. The moduli of $m_1 = 5$, $m_2 = 7$, $m_3 = 8$ were used for the nonredundant RNS-based system. However, for this nonredundant RNS-based system, the symbol errors were corrected solely by the outer RS (255,223) code. The modulus values for the concatenated RRNS-based system were $m_1 = 5$, $m_2 = 7$, $m_3 = 8$, $m_4 = 9$ for the RRNS (4,3) code with $d = 1$ lowest-reliability input of the RRNS decoder being dropped. Similarly, moduli $m_1 = 5$, $m_2 = 7$, $m_3 = 8$, $m_4 = 9$ were used for the RRNS (4,3) scheme in conjunction with RRNS error-detection decoding, i.e., when providing erasure information for the outer RS decoding by the RRNS decoding. Moreover, moduli $m_1 = 5$, $m_2 = 7$, $m_3 = 8$, $m_4 = 9$, $m_5 = 11$ were selected for the RRNS (5,3) code, when $t_{max} = 1$ residue digit error was corrected. The other parameters related to our proposed systems were here $L = 3$, $K = 1$ and $N = 64$.

From the results of Figure 18.33, we observe that the BER performance of the proposed

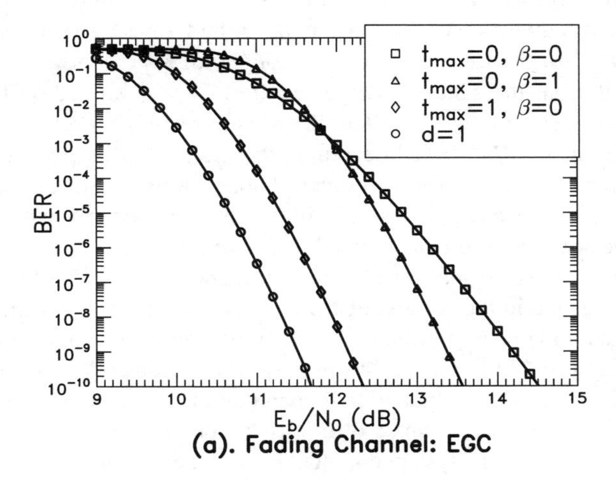

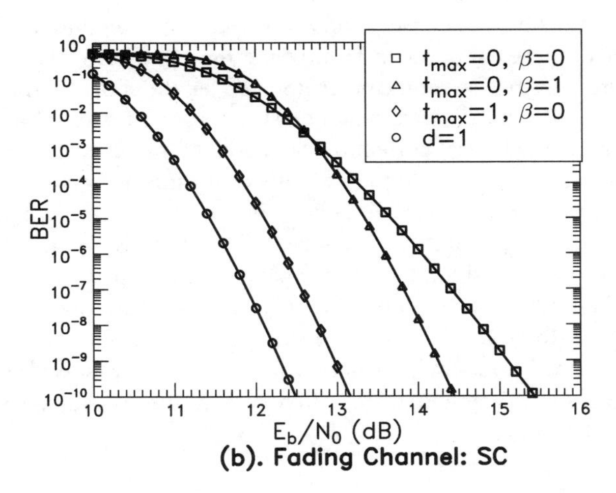

Figure 18.33: BER versus average SNR per bit for the concatenated RS-RRNS coded parallel DS-CDMA orthogonal system over dispersive multipath fading channels using EGC (a) and SC (b), where moduli $m_1 = 5, m_2 = 7, m_3 = 8$ are employed for $(t_{max} = 0, \beta = 0)$ and RS (255,233) 'error-correction-only' decoding, moduli $m_1 = 5, m_2 = 7, m_3 = 8, m_4 = 9$ are employed for both $(t_{max} = 0, \beta = 1)$ and RS (255,233) 'errors-and-erasures' decoding and $d = 0$ and 'error-correction-only' decoding , while moduli $m_1 = 5, m_2 = 7, m_3 = 8, m_4 = 9, m_5 = 11$ are employed for $(t_{max} = 1, \beta = 0)$ and RS (255,233) 'error-correction-only' decoding, where t_{max}, β and d are, respectively, the error-correction capability, error-detection capability and the number of lowest-reliability residue digits dropped by the RRNS inner code. The other parameters are $N = 64, K = 1, L = 3$.

system was dramatically improved upon using RS-RRNS concatenated coding in conjunction with both the EGC and SC combining schemes. For example, for the receiver invoking EGC at a bit error probability of 10^{-10}, the average SNR per bit required for the RRNS (4,3) code in conjunction with the lowest-reliability input of the RRNS decoder dropped $(d = 1)$ is 11.7dB. By contrast, for the RRNS (4,3) code along with $\beta = 1$ in the context of RRNS error-detection decoding at the same BER the required E_b/N_0 was 13.6dB, while for the RRNS (5,3) code along with $t_{max} = 1$ in RRNS error-correction decoding was 12.3dB. By contrast, without the inner RRNS-based decoding, the required SNR per bit was 14.5dB at $BER = 10^{-10}$. This implies that we can obtain about 2.8dB, 0.9dB and 2.2dB gain, respectively, for $d = 1$, $\beta = 1$ and $t_{max} = 1$, with the $d = 1$ scenario constituting the best scheme, although the scheme used is associated with two redundant moduli. Similar results were obtained for the receiver using the SC scheme, demonstrating that the SC combining technique is not as effective in the context of the non-coherent receiver as the EGC scheme, a fact which was noted previously. Furthermore, from the figures we notice that using RRNS error-detection and by providing erasure information for the outer RS decoding stage, the BER performance is improved less dramatically than that of the scheme using RRNS decoding discarding the lowest-reliability inputs of the RRNS decoder and additionally using RS 'error-correction-only' decoding. This phenomenon can be explained as follows. Upon using lowest-reliability input dropping, most erroneously received RS symbols are successfully recovered after inner decoding and consequently this decreased the number of erroneous RS symbols to be corrected by the RS decoder. However, if the inner RNS decoding stage is only used to provide erasure information for the outer RS decoder, the sum of erroneous and erased symbols may exceed the RS code's 'errors-and-erasures' correction capability, although RS decoding using 'errors-and-erasures' correction can correct more erroneous and erased symbols, than RS 'error-correction-only' decoding without erasure information. For example, if the error probability of each residue digit concerning the above computations is less than 0.25, then the average erroneous number of residue digits in the RRNS (4,3) codeword is less than 1. Hence, the erroneous residue digit can typically be dropped by the RRNS decoder using one lowest-reliability input dropping and consequently there are very few erroneous RS symbols in the outer RS code. However, when using 'error-detection-only' inner decoding, the previously dropped erroneous residues will now result in erased RS symbols and consequently the number of RS symbol errors often exceeds the error-and-erasure correction capability of the outer RS code.

18.6 Chapter Summary and Conclusion

In this chapter, we have presented a novel parallel DS-CDMA system that uses a set of orthogonal signals and associated correlators. The output of the correlators processed according to the principles of the RNS arithmetic algorithms were summarized in the previous chapter. This technique is amenable to the transmission of signals conveying a high number of bits per symbol, while using a relatively low number of orthogonal signals, and therefore it is attractive in the context of high-bit-rate communications. We have discussed the characteristics of the RNS arithmetic and demonstrated how these characteristics affect the communications performance in AWGN using coherent or non-coherent receivers and in dispersive multipath Rayleigh fading channels using MRC, EGC or SC schemes. Based on the independence and on the lack of weighted significance of the residue digits of a RNS, novel receivers were proposed for receiving RNS arithmetic-based DS-CDMA signals. In this novel receiver a RRNS code can be decoded by first discarding a number of the lowest-reliability residue digits according to the so-called ratio statistic test, which is based on the ratio of the (first) maximum to the 'second maximum' of the inputs of the RRNS decoder, and then decoding the reduced and simultaneously simplified RRNS code. Concatenated coding techniques have been discussed and evaluated in a dispersive multipath fading scenario, and the concatenated coding performance in conjunction with an RS code as the outer code and the RRNS code as the inner code has been evaluated both analytically and numerically. The results throughout the sections associated with RRNS decoding and concatenated RS-RRNS decoding demonstrated that using RRNS decoding with the lowest-reliability inputs of the RRNS decoder being dropped is a high-efficiency, low-complexity decoding technique.

Part IV

Multi-Carrier CDMA

Chapter 19

Overview of Multicarrier CDMA [1]

19.1 Introduction

Orthogonal frequency division multiplexing (OFDM) [76, 442–444] is a parallel data transmission scheme in which high data rates can be achieved by transmitting U orthogonal subcarriers. The inter-symbol interference (ISI) and inter-channel interference (ICI) in OFDM systems are reduced by the insertion of guard intervals and code synchronization is made easier by the extended symbol period engendered by the associated serial-to-parallel conversion preceding the parallel transmission of low-rate subchannels. Recently, code-division multiple-access (CDMA) systems-based on the combination of CDMA schemes and OFDM signalling, which are referred to as multicarrier CDMA systems, have attracted attention in the field of wireless communications [20, 65–67, 74, 75, 357, 444–481]. This is mainly due to the need to support high data rate services in a wireless environment characterized by highly hostile radio channels. These signals can be efficiently modulated and demodulated using Fast Fourier Transform (FFT) devices without substantially increasing the receiver's complexity. These systems also exhibit the attractive feature of high spectral efficiency, since they can operate using a low Nyquist roll-off factor [76]. Hence, OFDM systems can approach the 2 Baud/Hz maximum bandwidth efficiency associated with Nyquist sampling. The combination of code division and OFDM can combat the effects of fading channels by spreading signals over several carriers, in order to achieve frequency diversity.

Depending on whether all the subcarriers are activated on each transmission, multicarrier CDMA arrangements can be classified as the non-frequency hopping multicarrier CDMA and the frequency-hopping assisted multicarrier CDMA schemes. The family of non-frequency hopping multicarrier CDMA schemes includes multicarrier CDMA using frequency-domain spreading [466,473,476,481], subchannel band-limited multicarrier direct-sequence CDMA (DS-CDMA) [467,472], orthogonal multicarrier DS-CDMA [65,451,468,479] and multitone

[1] *Single- and Multi-Carrier CDMA Multi-User Detection, Space-Time Spreading, Synchronization, Networking and Standards.* L.Hanzo, L-L.Yang, E-L.Kuan and K.Yen,
 ISBN 0-470-86309-9

DS-CDMA [469,482]. The class of frequency-hopping assisted multicarrier CDMA schemes belongs to the extended family of the above multicarrier schemes, which include multicarrier DS-CDMA using adaptive frequency-hopping [480], as well as multicarrier DS-CDMA using adaptive subchannel allocation [459] and the subclass of constant-weight code assisted multicarrier DS-CDMA using slow frequency-hopping [66, 74, 445].

Based on their signal spreading model, multicarrier schemes can also be categorized mainly into two types. In the first class of schemes, the serial data stream is first spread by a spreading code and then converted into N_p parallel chip sequences with each chip modulating a different subcarrier. The number of subcarriers is N_p, which equals the number of chips per data symbol. The spreading operation in this type of multicarrier CDMA arrangements occurs in the frequency-domain. This type of systems combine the robustness of orthogonal modulation with the flexibility of CDMA schemes [75, 357]. In the second type of multicarrier CDMA systems, the original data stream is first serial-to-parallel converted into U substreams. Then, each substream is spread using a given spreading code in the time-domain, and finally, modulates a different subcarrier with each of the data stream. In the second type of multicarrier CDMA each subcarrier's signal is similar to that of a conventional normal single carrier DS-CDMA scheme [65]. The first class of multicarrier CDMA only includes one particular scheme, namely multicarrier CDMA using frequency-domain spreading [466, 473, 476, 481], while other known multicarrier CDMA schemes belong to the second family.

One of the main implementation disadvantages on the transmitter side of OFDM-based multicarrier CDMA systems is the high peak-to-average power ratio [443, 456] of the transmitted signal. Whenever the peak transmitted power is limited by regulatory or implementational constraints – such as the minimum required transmit power or the power efficiency of the amplifier – this has the effect of reducing the average power of the transmitter and limiting the range of transmissions. Moreover, since the multicarrier signal exhibits a high amplitude variation, it is subject to nonlinear distortions inflicted by the power amplifier. This distortion inevitably results in out-of-band emissions and co-channel interference, potentially causing a significant degradation in the system's performance.

In this chapter the performance of different multicarrier CDMA systems is investigated over frequency-selective Rayleigh fading channels. Section 19.2 reviews the family of the multicarrier CDMA schemes, analyzes their characteristics and discusses their advantages as well as disadvantages in terms of their transmitter and receiver structures. We will also consider their spectral efficiency and spreading gain. In Sections 19.4 to 19.8 we analyze a range of typical multicarrier CDMA schemes in depth and derive their corresponding bit error probability. Let us now commence our overview of multicarrier CDMA systems.

19.2 Overview of Multicarrier CDMA Systems

In this section we review the class of multicarrier CDMA schemes, which have been discussed in the literature. Specifically, we discuss their design parameters, spectral characteristics, advantages and disadvantages in terms of the transmitter and receiver structures as well as their spectral efficiency. The bit error rate (BER) performance analysis of these multicarrier CDMA schemes is provided in the following sections. Before considering these schemes in more detail, we first list some their common parameters.

- W_s (Hz): Overall spectral null-to-null system bandwidth;
- W_{ds} (Hz): Null-to-null transmission bandwidth of each subcarrier signal;
- W_d (Hz): null-to-null bandwidth of the binary baseband data signal;
- T_b: Binary bit duration;
- T_s: Transmitted symbol's duration;
- T_c: Chip duration of the spreading codes in the multicarrier CDMA system;
- T_{c1}: Chip duration of the spreading codes in the single-carrier CDMA system;
- f_0: RF carrier frequency;
- $N = W_s/W_d = T_b/T_{c1}$: Spreading gain of BPSK modulated single-carrier DS-CDMA signals;
- $N_e = T_s/T_c$: Spreading gain of BPSK modulated subcarrier DS-CDMA signal, or number of chips per symbol;
- $N_b = T_b/T_c$: Number of chips per bit;
- L_1: Number of resolvable propagation paths experienced by the single-carrier DS-CDMA signal;
- Δ: Spacing between two adjacent subcarriers;
- SG: Spectral gain, which is defined as the ratio between the bandwidth required by the multicarrier scheme without overlapping and the actual bandwidth of the corresponding multicarrier scheme.

Having defined the common multicarrier CDMA parameters, let us first highlight the concepts associated with frequency-domain spreading.

19.2.1 Frequency-Domain Spreading Assisted Multicarrier CDMA Scheme

Figure 19.1 shows the transmitter diagram of the multicarrier CDMA (MC-CDMA) scheme associated with frequency-domain spreading [466, 473, 476, 481]. The MC-CDMA transmitter spreads the original data stream over N_p subcarriers using a given spreading code of $\{c_k[0], c_k[1], \ldots, c_k[N_p-1]\}$ in the frequency domain. Observe in Figure 19.1 that this scheme does not include serial-to-parallel data conversion and there exists no spreading modulation on each subcarrier. Therefore, the data rate on each of the N_p subcarriers is the same as the input data rate. However, by spreading each data bit across all of the N_p subcarriers the fading effects of multipath channels is mitigated. With reference to Figure 19.1 the kth MC-CDMA user's transmitted signal can be expressed as:

$$s_k(t) = \sqrt{\frac{2P}{N_p}} \sum_{n=0}^{N_p-1} b_k(t) c_k[n] \cos(2\pi f_n t), \tag{19.1}$$

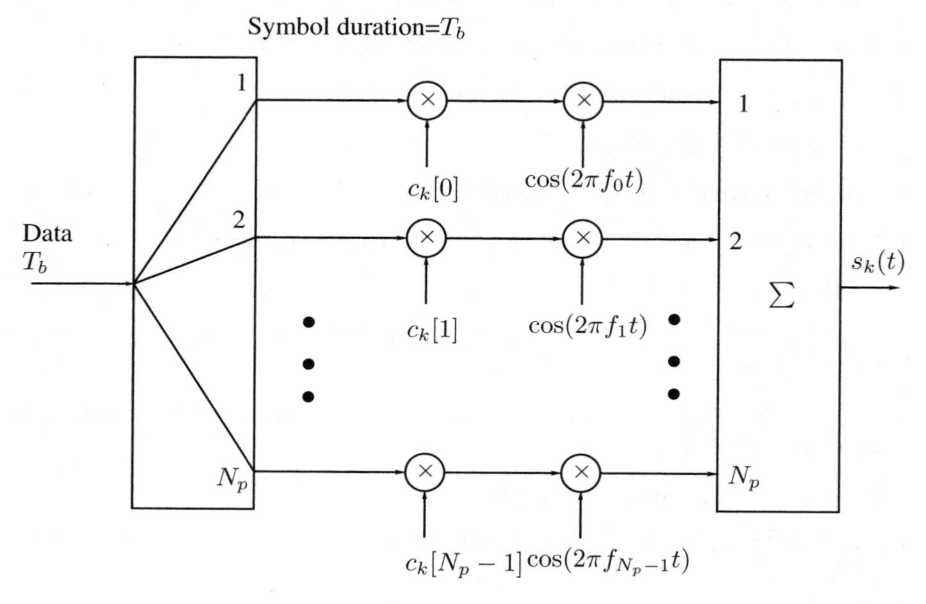

Figure 19.1: The transmitter diagram of frequency-domain spreading assisted MC-CDMA systems.

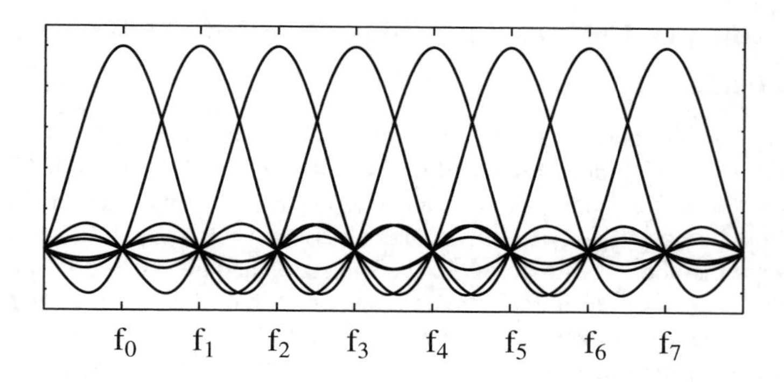

Figure 19.2: Spectrum of the frequency-domain spreading assisted MC-CDMA signal.

where P represents the transmitted power of the MC-CDMA signal, N_p is the number of subcarriers as well as the spreading gain, $\{c_k[0], c_k[1], \ldots, c_k[N_p-1]\}$ is the spreading code, $\{f_n,\ n = 0, 1, \ldots, N_p - 1\}$ are the subcarrier frequencies, and finally, $b_k(t)$ represents the binary data sequence.

The spectrum of the transmitted MC-CDMA signal is shown in Figure 19.2, where we assume that the MC-CDMA system has eight subcarriers and that all the chip values of the spreading code of $\{c_k[0], c_k[1], \ldots, c_k[N_p-1]\}$ are +1. In this MC-CDMA system the subcarrier frequencies are chosen to be orthogonal to each other, i.e. the subcarrier frequencies satisfy the following condition:

$$\int_0^{T_b} \cos(2\pi f_i t + \phi_i) \cdot \cos(2\pi f_j t + \phi_j) dt = 0, \text{ for } i \neq j, \tag{19.2}$$

where T_b is the bit duration. Therefore, the minimum spacing Δ between two adjacent subcarriers satisfies $\Delta = \frac{1}{T_b}$, which is a widely used assumption [466, 473, 476, 481] and is also the case employed in Figure 19.2, where $f_n = f_0 + \frac{n}{T_b}$ for $n = 0, 1, \ldots, N_p - 1$. If $\Delta = \frac{1}{T_b}$ associated with 50% overlap is assumed, then the bandwidth required by the MC-CDMA system is $(N_p + 1) \cdot \frac{1}{T_b}$. However, the MC-CDMA system having N_p non-overlapping subcarriers requires a total bandwidth of $N_p \cdot \frac{2}{T_b}$. Hence, the spectral gain of this MC-CDMA system is given by:

$$SG = \frac{N_p(2/T_b)}{(N_p + 1)(1/T_b)}, \tag{19.3}$$

which approaches two, as N_p increases. Thus, this MC-CDMA system exhibits an increased processing gain given by $N_p \approx 2N$, as explained by the 50% overlap of the main lobes of the adjacent MC-CDMA subcarrier spectra.

Yee [466] *et al.*. have considered a MC-CDMA system, in which the subcarriers' frequency separation is higher than the coherence bandwidth of the channel, and therefore the individual subcarriers experience independent fading. As a result, the frequency diversity is maximized. However, this system requires a considerable transmission bandwidth. Nevertheless, the addition of an interleaver after the chip spreading would lessen this bandwidth requirement.

This scheme can be applied in conjunction with the same set of subcarriers, to a multiple access system by allocating each user a different spreading code. The separation of different users is provided by the spreading codes. In the receiver the fading impaired signals of the subcarriers are first equalized and then the different user's signals are separated by exploiting their different spreading codes. Due to the orthogonality of the subcarrier signals, in a downlink mobile radio communication channel, we can use the Hadamard-Walsh codes as an optimum spreading code set [75], since we do not have to pay attention to the auto-correlation characteristic of the spreading codes.

The receiver structure of the MC-CDMA scheme is shown in Figure 19.3. In this MC-CDMA receiver the received signal is combined, in a sense, in the frequency domain. Therefore the receiver can always make use of all the received signal energy scattered in the frequency domain. This is the main advantage of the MC-CDMA scheme of Figures 19.1 and

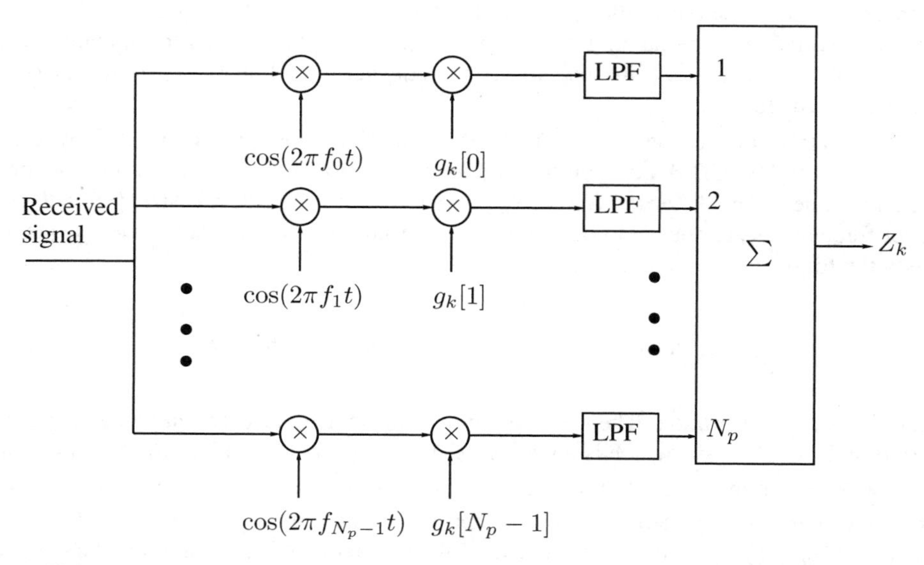

Figure 19.3: The receiver block diagram of frequency-domain spreading assisted MC-CDMA systems.

19.3 over other multicarrier CDMA schemes [75]. However, in a frequency selective fading channel different subcarriers may encounter different amplitude attenuations and phase shifts, which can consequently result in destroying the orthogonality of the subcarriers. There are many combining and detection techniques that can be employed by the MC-CDMA receiver, in order to efficiently exploit the received signal energy [75]. One of the most common approaches is Maximum Ratio Combining (MRC), where the subcarrier signals' weighting factors, $g_k[i]$, $i = 0, 1, \ldots, N-1$ in Figure 19.3, are computed as the complex conjugate of the received subcarrier signal's envelope. This approach can minimize the BER, as long as a single-user system is considered.

Note that the number of subcarriers does not have to be the same as the processing gain. If the original symbol rate is high, the signal experiences frequency selective fading. Then the input data has to be serial-to-parallel converted, mapping the data to a number of reduced-rate streams before spreading over the frequency domain. This is because it is crucial for MC-CDMA signal transmission to have frequency non-selective fading over each subcarrier [75]. This arrangement has been studied in [357,455].

19.2.2 Orthogonal Multicarrier DS-CDMA Scheme – Type I

In [467,472] a multicarrier DS-CDMA system was proposed (we refer to this scheme as multicarrier DS-CDMA-I system), in which a data sequence multiplied by a spreading sequence modulates U subcarriers. The transmitter diagram of the multicarrier DS-CDMA-I system used in [472] is similar to Figure 19.4, except that band-limited subcarrier signals are employed in [472]. Similarly to Figure 19.1, this scheme does not include serial-to-parallel data conversion either. However, each subcarrier signal is direct sequence (DS) spread using a common spreading sequence $c_k(t)$, as shown in Figure 19.4. Therefore, the symbol duration

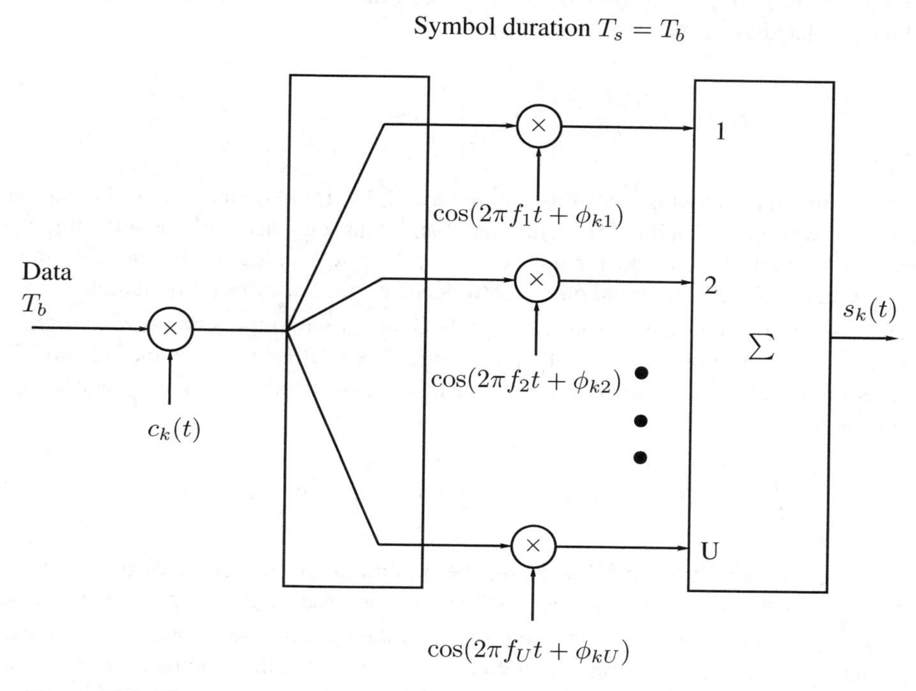

Figure 19.4: The transmitter diagram of the orthogonal multicarrier DS-CDMA-I system.

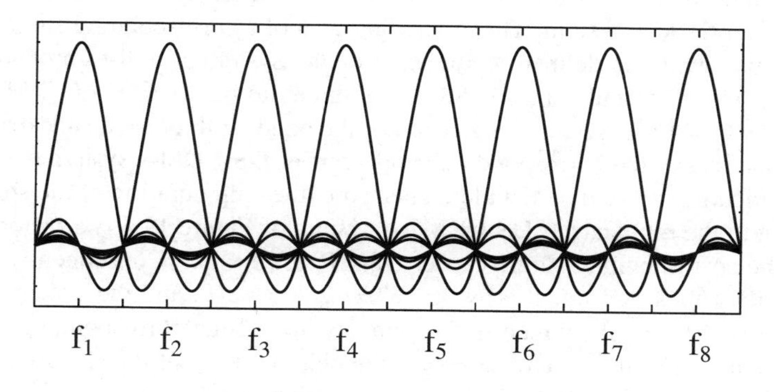

Figure 19.5: Spectrum of time-domain spreading assisted MC-CDMA signals.

of the multicarrier DS-CDMA signal is the same as that of the input data bit duration. With the aid of Figure 19.4, the transmitted signal of user k in the multicarrier DS-CDMA-I system can be formulated as [472]:

$$s_k(t) = \sqrt{\frac{2P}{U}} \sum_{u=1}^{U} b_k(t)c_k(t)\cos(2\pi f_u t + \phi_{ku}), \tag{19.4}$$

where P is the transmitted power of the multicarrier DS-CDMA-I signal, U is the number of subcarriers, while $b_k(t)$ and $c_k(t)$ are the baseband data sequence and the spreading waveforms, respectively. Finally, f_u for $u = 1, 2, \ldots, U$ are the subcarrier frequencies and ϕ_{ku} for $u = 1, 2, \ldots, U$ are the initial phases introduced by the subcarrier modulation.

The spectrum of the multicarrier DS-CDMA-I signal without overlap having eight subcarriers is shown in Figure 19.5. In the multicarrier DS-CDMA-I system the subcarrier frequencies are usually chosen to be orthogonal to each other after spreading, which can be formulated as:

$$\int_0^{T_c} \cos(2\pi f_i t + \phi_i) \cdot \cos(2\pi f_j t + \phi_j)dt = 0, \text{ for } i \neq j, \tag{19.5}$$

where T_c is the chip duration. Therefore, the minimum spacing Δ between two adjacent subcarriers satisfies $\Delta = \frac{1}{T_c}$. Figure 19.5 shows the case of $\Delta = \frac{2}{T_c}$. For the case of $\Delta = \frac{2}{T_c}$, there exists no spectral overlap between the spectral main-lobes of two adjacent subcarriers and hence the spectral gain is $SG = 1$. However, if the spacing between adjacent subcarriers is assumed to be $\Delta = \frac{1}{T_c}$, the spectral gain is then given by:

$$SG = \frac{U(2/T_c)}{(U+1)(1/T_c)}, \tag{19.6}$$

that approaches two, when the number of subcarriers is high.

Let us assume that T_{c1} is the chip duration of the spreading code corresponding to a single-carrier DS-CDMA system. The processing gain of a corresponding single-carrier DS-CDMA system – which is defined as the ratio of the system's 'null-to-null' bandwidth to the binary data's 'null-to-null' bandwidth – is assumed to be $N = T_b/T_{c1}$. Hence, for the multicarrier DS-CDMA-I system using $\Delta = \frac{2}{T_c}$, the bandwidth of each subcarrier is a factor of U lower than that of the corresponding single-carrier DS-CDMA system having identical data rate and identical system bandwidth. Therefore, the chip duration of the spreading code corresponding to the multicarrier DS-CDMA-I system of Figure 19.4 is a factor of U higher than that of the corresponding single-carrier DS-CDMA system. Consequently, the processing gain of each subcarrier signal is $N_e = T_b/UT_{c1} = N/U$ and the system's processing gain is $N_p = U \cdot N_e = N$, which is the same as that of the corresponding single-carrier DS-CDMA system. For the case of $\Delta = \frac{1}{T_c}$, the chip duration of the spreading code corresponding to the multicarrier DS-CDMA-I system is given by $T_c = (U+1)T_{c1}/2$ and hence, the processing gain of each subcarrier signal is $N_e = T_b/T_c = 2N/(U+1)$. By contrast, the multicarrier DS-CDMA-I system's processing gain is $N_p = U \cdot N_e = 2UN/(U+1)$. The multicarrier DS-CDMA-I system increased the processing gain by about a factor of two, if

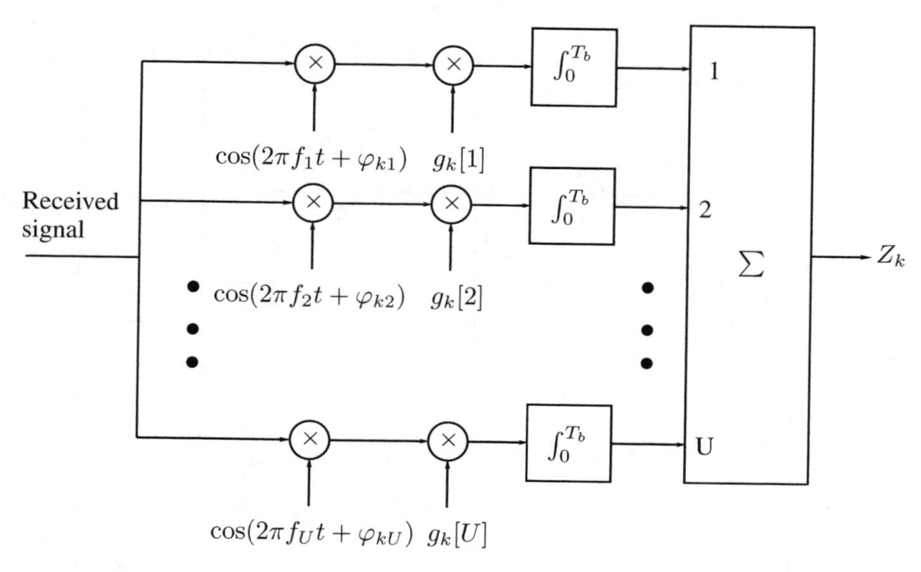

Figure 19.6: The receiver block diagram of the time-domain spreading assisted multicarrier DS-CDMA-I systems.

50% overlap of the main lobes of the adjacent subcarrier spectra in Figure 19.5 is assumed.

The receiver block diagram of the multicarrier DS-CDMA-I system is shown in Figure 19.6. In the multicarrier DS-CDMA-I system frequency diversity is achieved by combining the U correlator's outputs associated with the U subcarriers. The receiver provides a correlator for each of the U subcarriers, and the outputs of the U correlators are combined to yield a processing gain comparable to that of a single-carrier DS system, provided that the spacing between two adjacent subcarriers is $\Delta = \frac{2}{T_c}$. This system has a range of advantages [472]. First, the multicarrier DS-CDMA-I system is robust to multipath fading due to the frequency diversity achieved over the subcarriers. Second, the multicarrier DS-CDMA-I system exhibits narrowband interference suppression effect due to the DS spreading. Third, a lower chip rate is required – which has the advantage of reduced-complexity parallel implementation – since, in a multicarrier DS-CDMA-I system having U subcarriers the entire bandwidth of the system is divided into U equi-width frequency bands. Thus each subcarrier frequency is modulated by a spreading sequence having a chip duration, which is U times as long as that of the corresponding single-carrier DS-CDMA system. In other words, a multicarrier system requires a lower speed, parallel-type of signal processing, in contrast to a fast, serial-type of signal processing in a single carrier RAKE [90] receiver. This, in turn, might be helpful for implementing a low power consumption device. Finally, the multicarrier DS0-CDMA-I system does not require a contiguous frequency band, hence available spectral gaps can efficiently be exploited.

In [472] the performance of the multicarrier DS-CDMA-I system has been investigated by assuming that each subcarrier signal encounters independent non-frequency selective fading. If the DS spreading code's chip rate is high, the signal is subject to frequency selective fading. Then RAKE receivers associated with each of the subcarriers can be implemented, in order

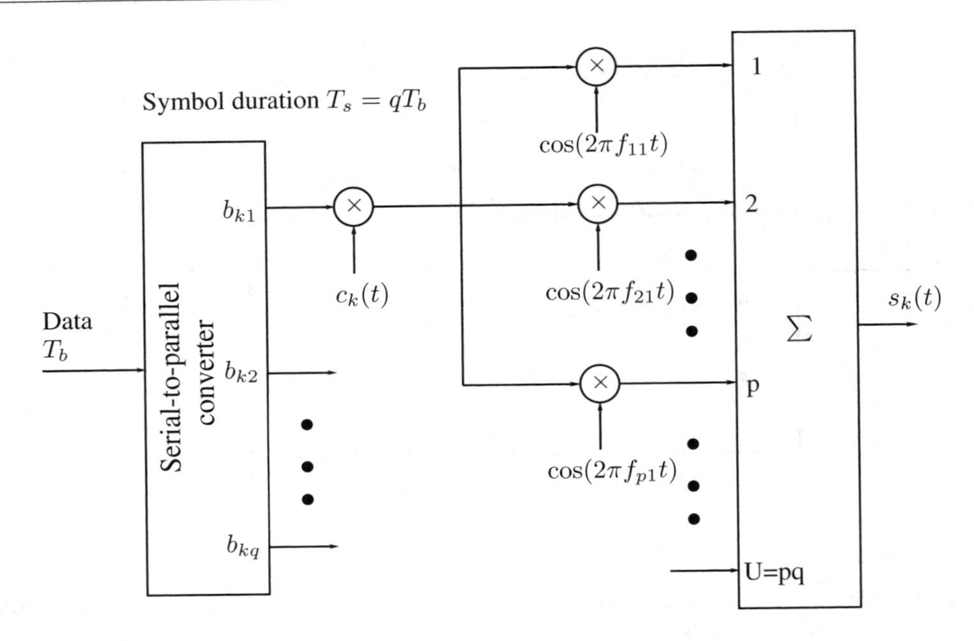

Figure 19.7: The transmitter diagram of the orthogonal multicarrier DS-CDMA-II system.

to combine the energy scattered over different paths. However, if the subcarriers are subject to correlated fading, then interleaving over time can be employed, in order to guarantee the independence of the subcarrier signals. For example [472], suppose the first U symbols emitted by the source modulate the U subcarriers in the first time interval. In the second time interval, the same U symbols modulate the U subcarriers after a cyclic shift by one symbol. In the ith interval, $1 \leq i \leq U$, the $(i-1)$th shifted version of the data modulates the U subcarriers.

19.2.3 Orthogonal Multicarrier DS-CDMA Scheme – Type II

Orthogonal multicarrier DS-CDMA systems – which we refer to here as multicarrier DS-CDMA-II systems – have been studied in [65, 451, 468, 479]. The orthogonal multicarrier DS-CDMA transmitter spreads the serial-to-parallel converted data streams using a given spreading code in the time domain so that the resulting spectrum of each subcarrier can satisfy the orthogonality condition with the minimum frequency separation [65, 468]. This scheme was originally proposed for an uplink communication system, because this characteristic is effective in establishing a quasi-synchronous channel. The transmitter diagram shown in Figure 19.7 for the multicarrier DS-CDMA-II system is the same as that used in [65], which can be interpreted as the extension of the multicarrier DS-CDMA-I system shown in Figure 19.4. In this scheme the initial data stream is serial-to-parallel converted to a number of lower-rate streams. Each of these lower-rate streams is spread by the spreading code $c_k(t)$ of Figure 19.7, feeding a number of parallel streams having the same rate. Each of the latter lower-rate parallel streams is bit-interleaved and these streams modulate the orthogonal

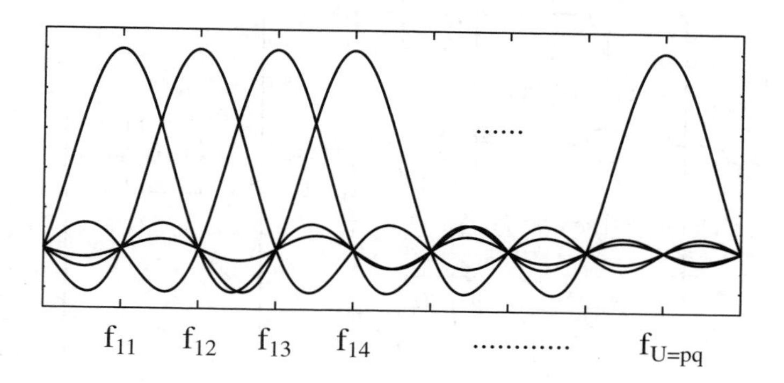

Figure 19.8: Spectrum of the orthogonal multicarrier DS-CDMA signal.

subcarriers with a successively overlapping bandwidth, as shown for example in Figure 19.8.

The spectrum of the multicarrier DS-CDMA-II signal is shown in Figure 19.8. In the multicarrier DS-CDMA-II system subcarrier frequencies are chosen to be orthogonal to each other with the minimum frequency separation. The orthogonality is ensured by Equation 19.5, but the spacing between two adjacent subcarriers is chosen to be $\Delta = 1/T_c$.

Let $\{f_{11}, f_{12}, \ldots, f_{U=pq}\}$ be the subcarrier frequencies, which are arranged according to Figure 19.8. These subcarrier frequencies can be written in the form of a matrix as:

$$\{f_i\} = \begin{pmatrix} f_{11} & f_{12} & \cdots & f_{1q} \\ f_{21} & f_{22} & \cdots & f_{2q} \\ \vdots & \vdots & \ddots & \vdots \\ f_{p1} & f_{p2} & \cdots & f_{pq} \end{pmatrix}. \tag{19.7}$$

Then, according to the transmitter diagram of Figure 19.7, the bit stream having a bit duration of T_b is first serial-to-parallel converted into q parallel streams. The new bit duration on each substream, which is referred to here as the symbol duration, is $T_s = qT_b$. Each substream, b_{ki}, $k = 1, 2, \ldots, K$; $i = 1, 2, \ldots, q$ feeds p parallel streams and modulates p subcarriers from the same column of Equation 19.7. It can be shown from Figure 19.8 that these subcarrier frequencies modulated by the same data bit have maximum frequency separation, which ensures the independence of the fading endured by the subcarriers modulated by the same data bit. The transmitted signal of user k is given by:

$$s_k(t) = \sum_{i=1}^{q} \sum_{j=1}^{p} \sqrt{\frac{2P}{p}} b_{ki}(t) c_k(t) \cos(2\pi f_{ji} t), \tag{19.8}$$

where P is the transmitted power of each stream, $b_{ki}(t)$ represents the data sequence of the ith stream, while $c_k(t)$ is the common spreading code in Figure 19.7.

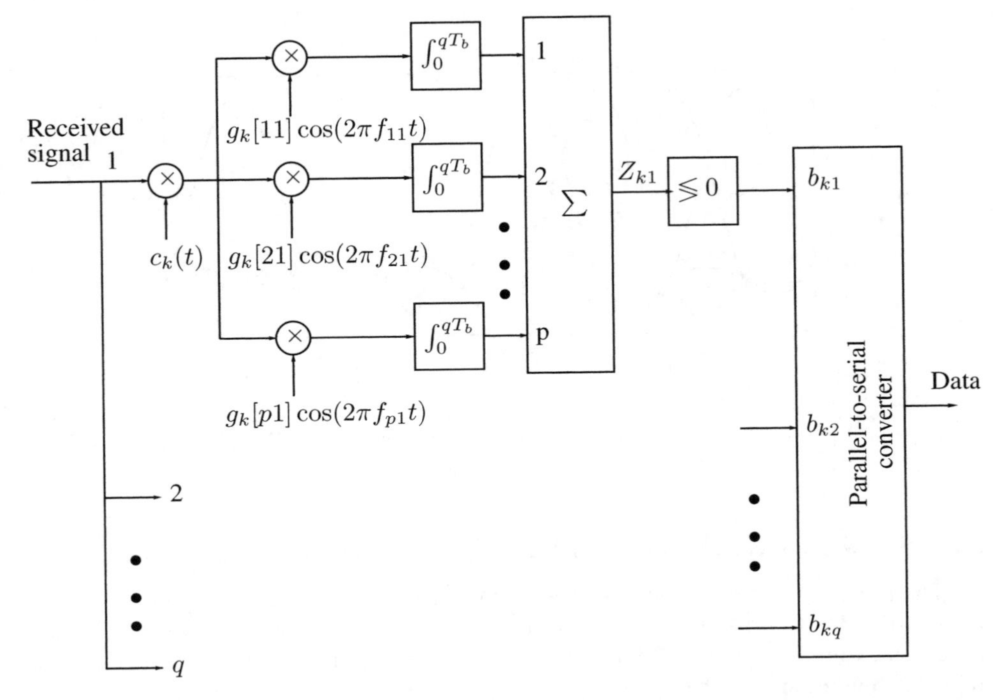

Figure 19.9: The receiver block diagram of the orthogonal multicarrier DS-CDMA systems.

The spectral gain of the multicarrier DS-CDMA-II system is given by:

$$SG = \frac{pq(2/T_c)}{(pq+1)(1/T_c)} = \frac{2U}{U+1}, \tag{19.9}$$

which approaches two as $U = pq$ increases. The chip duration of the spreading code $c_k(t)$ employed by the orthogonal multicarrier DS-CDMA system is $T_c = (pq+1)T_{c1}/2 = (U+1)T_{c1}/2$, where T_{c1} represents the chip duration of a corresponding single-carrier DS-CDMA scheme ($p = q = 1$). The spreading gain of each subcarrier signal is given by:

$$N_e = qT_b/T_c = \frac{qT_b}{(U+1)T_{c1}/2} = \frac{2qN}{U+1}. \tag{19.10}$$

By contrast, the multicarrier DS-CDMA-II system's spreading gain is $N_p = p \cdot N_e = 2UN/(U+1)$. Thus, the multicarrier DS-CDMA-II system exhibits an increased processing gain, which amounts to approximately a factor of two, as explained by the 50% overlap of the main lobes of the adjacent subcarrier spectra.

The receiver block diagram of the multicarrier DS-CDMA-II system is shown in Figure 19.9. The receiver provides a correlator for each subcarrier and the correlator outputs associated with the same data bit are combined to form a decision variable. Finally, a parallel-to-serial converter is employed to recover the serial data stream.

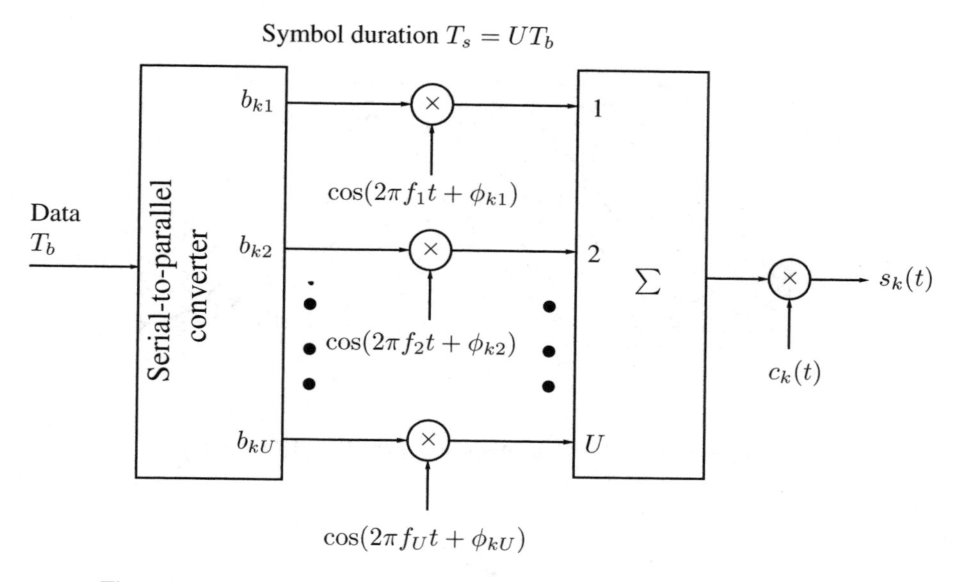

Figure 19.10: The transmitter diagram of the multitone DS-CDMA system.

In the multicarrier DS-CDMA-II system, provided that each subcarrier signal is subject to frequency selective fading, a further RAKE receiver can be employed, in order to match each subcarrier signal and combine the energy scattered by the multipath channel.

The multicarrier DS-CDMA-II scheme can provide the following advantages [65]. First, the spreading processing gain is increased compared to the corresponding single-carrier DS-CDMA scheme. Second, the effect of multipath interference is mitigated because of DS spreading. Third, frequency/time diversity can be achieved. Finally, a longer chip duration may lead to more relaxed synchronization schemes. However, a high complexity receiver has to be implemented and forward error control (FEC) techniques must be used, in order to enhance its associated performance.

19.2.4 Multitone DS-CDMA Scheme

Multitone DS-CDMA scheme was proposed by Vandendorpe in [469, 482]. The multitone DS-CDMA transmitter spreads the serial-to-parallel converted data streams using a given spreading code in the time domain, so that the spectrum of each subcarrier prior to the spreading operation can satisfy the orthogonality condition with the minimum frequency separation [482]. Therefore, there exists strong spectral overlap among the different subcarrier signals after DS spreading. The transmitter diagram and the spectrum associated with eight subcarriers for the multitone DS-CDMA signal are shown in Figure 19.10 and Figure 19.11, respectively. At the transmitter side of Figure 19.10, the binary data stream having a bit duration of T_b is serial-to-parallel converted to U parallel substreams. The new bit duration on each subcarrier, which defines the modulated symbol duration is $T_s = UT_b$. The ith substream modulates the subcarrier frequency f_i, $i = 1, 2, \ldots, U$. The multitone signal is obtained by the addition of the different subcarriers' signals. Then, spectrum spreading is

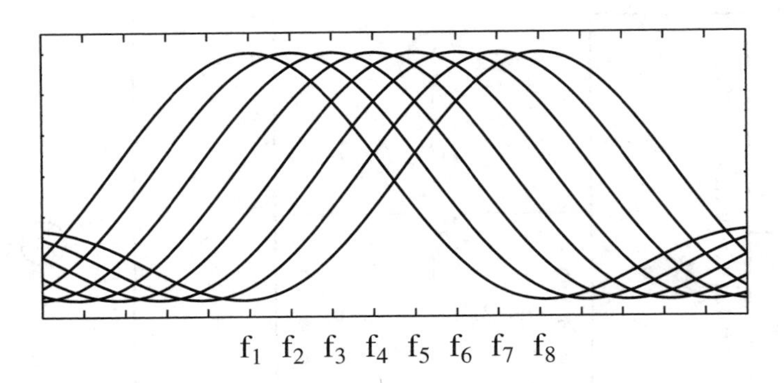

Figure 19.11: Spectrum of the multitone DS-CDMA signal.

imposed on the multitone signal by multiplying it with a spreading code, as shown in Figure 19.10. The transmitted signal of user k can be expressed as:

$$s_k(t) = \sum_{u=1}^{U} \sqrt{2P} b_{ku}(t) c_k(t) \cos(2\pi f_u t + \phi_{ku}), \tag{19.11}$$

where P represents the transmitted power of each subcarrier, $b_{ku}(t)$ represents the data sequence modulating the uth subcarrier, $c_k(t)$ is the spreading code of user k, while f_u and ϕ_{ku} are the uth subcarrier frequency and modulation phase.

In the multitone DS-CDMA system the subcarrier frequencies are chosen to be orthogonal to each other with the minimum frequency separation before spreading, which can be formulated as:

$$\int_0^{T_s} \cos(2\pi f_i t + \phi_i) \cdot \cos(2\pi f_j t + \phi_j) dt = 0, \text{ for } i \neq j. \tag{19.12}$$

It can be shown that the minimum spacing of the subcarrier frequencies is $1/T_s = 1/UT_b$. Let T_{c1} represent the chip duration of the corresponding single-carrier DS-CDMA system. Referring to Figure 19.11, it can be shown that the chip duration of the spreading code for the multitone DS-CDMA system is given by:

$$T_c = \frac{2UN}{2UN - U + 1} T_{c1}, \tag{19.13}$$

where N is the spreading gain corresponding to a single-carrier DS-CDMA system. Observe that T_c approaches T_{c1} as N increases. The spreading gain of the subcarrier signal is given

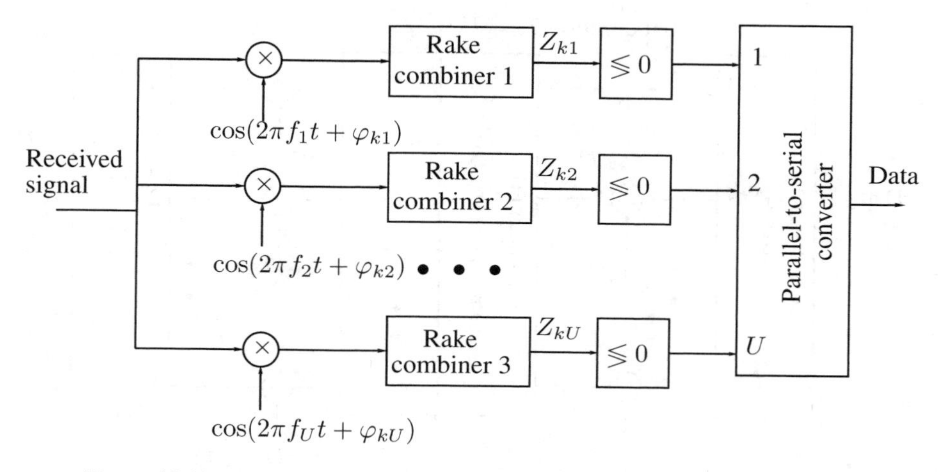

Figure 19.12: The receiver block diagram of the multitone DS-CDMA system.

by:

$$\begin{aligned} N_e &= T_s/T_c \\ &= UT_b/\frac{2UN}{2UN-U+1}T_{c1} \\ &= \frac{2UN-U+1}{2}, \end{aligned} \tag{19.14}$$

which can be approximated by UN, when N is sufficiently high. Since different data bits are transmitted on different subcarriers, the overall system's processing gain is also given by Equation 19.14.

The spectral gain of the multitone DS-CDMA system is given by:

$$\begin{aligned} SG &= \frac{U(2/T_c)}{2/T_c + (U-1)/T_s} \\ &= \frac{2N_eU}{2N_e + U - 1}, \end{aligned} \tag{19.15}$$

that approaches U, when N_e is sufficiently high, which is the highest SG among the multi-carrier schemes considered.

The receiver block diagram of the multitone DS-CDMA system is shown in Figure 19.12. The receiver is composed of U RAKE combiners, each of which has the same structure as the single-carrier DS-CDMA RAKE receiver. This is an optimum receiver for an AWGN channel [482]. Unfortunately, the multitone DS-CDMA scheme suffers from inter-subcarrier interference and requires a high-complexity RAKE-based receiver. However, the capability to use longer spreading codes results in the reduction of self-interference and multiple access interference, as compared to the spreading codes assigned to a corresponding single-carrier DS-CDMA scheme. The multitone DS-CDMA scheme uses longer spreading codes than

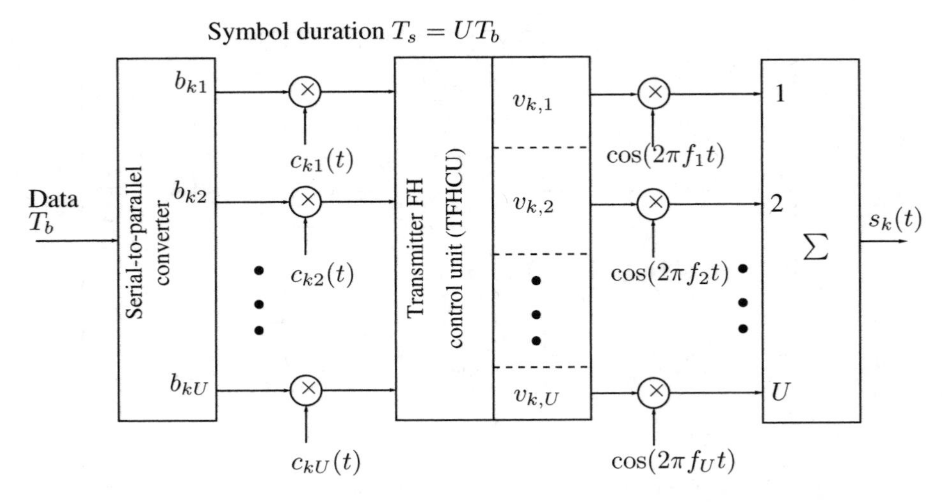

Figure 19.13: The transmitter schematic of the adaptive frequency-hopping assisted multicarrier DS-CDMA system.

the corresponding single-carrier DS-CDMA scheme [75], where the relative code-length extension is in proportion to the number of subcarriers. Therefore, the multitone DS-CDMA system can accommodate more users.

We have reviewed the family of various multicarrier-based CDMA schemes such as MC-CDMA, multicarrier DS-CDMA-I, multicarrier DS-CDMA-II and multitone DS-CDMA in the previous four subsections. A common property of these multicarrier-based schemes is that all the subcarriers in the system are activated for each transmission. By contrast, in the following three subsections, multicarrier-based CDMA schemes are reviewed, where each transmission only exploits a fraction of the subcarriers provided by the system. In these schemes frequency-hopping techniques are employed, in order to efficiently utilize the system bandwidth abailable. Hence, these schemes are referred to as frequency-hopping multicarrier DS-CDMA (FH/MC DS-CDMA) schemes.

19.2.5 Adaptive Frequency-Hopping Assisted Multicarrier DS-CDMA scheme

Multicarrier DS-CDMA systems using adaptive frequency-hopping (AFH/MC DS-CDMA) were proposed by Chen *et al.* [480] for uplink (mobile-to-base) transmissions. This scheme bears close resemblance to the orthogonal multicarrier DS-CDMA-I or multicarrier DS-CDMA-II scheme of figures 19.4 and 19.7, respectively. In the AFH/MC DS-CDMA scheme the total system bandwidth is divided into U equi-width subchannels, and the serial data stream of each user is serial-to-parallel converted to U parallel substreams, where each substream is transmitted as a narrow-band DS signal over one of the subchannels. In contrast to the transmission schemes in [65, 472], where each subcarrier was used to carry part of the transmitted information, in the AFH/MC DS-CDMA system, data substreams can be transmitted on any of the subcarriers, depending on the severity of the fading in a subchannel.

However, in the above transmission scheme it is possible for more than one data substream belonging to the same user to be modulated on to the same subcarrier. Hence, U number of spreading codes have to be assigned to each user, so that every data substream can be uniquely identified. The number of spreading codes required for a system supporting K users is given by KU.

The transmitter diagram of the AFH/MC DS-CDMA system using adaptive frequency-hopping is shown in Figure 19.13. The spectrum of the transmitted signal in the AFH/MC DS-CDMA system is similar to that seen in Figure 19.5 ($\Delta = 2/T_c$) or Figure 19.8 ($\Delta = 1/T_c$). Hence, the spectral gain, the spreading gain of each subcarrier signal as well as the spreading gain of the overall system are the same as that of the orthogonal multicarrier DS-CDMA-I or multicarrier DS-CDMA-II scheme. On the transmitter side the binary data stream is first converted to U parallel substreams $D_k(t) = \{b_{k1}(t), b_{k2}(t), \ldots, b_{kU}(t)\}$. These substreams are then spread by the corresponding spreading codes $C_k(t) = \{c_{k1}(t), c_{k2}(t), \ldots, c_{kU}(t)\}$. Let $F_k = \{f_{k1}(t), f_{k2}(t), \ldots, f_{kU}(t)\}$ – where $f_{k1}(t) \leq f_{k2}(t) \leq \ldots \leq f_{kU}(t)$ – be a set of subcarrier frequencies, which can be activated by the transmitter's FH control unit (TFHCU) in Figure 19.13 based on the subchannel states $V_k = \{v_{k,1}, v_{k,2}, \ldots, v_{k,U}\}$ fed back from the base station receiver. Finally, after modulating each subcarrier by a spread substream according to the activated frequency set, the transmitted signal can be expressed as:

$$s_k(t) = \sum_{u=1}^{U} \sqrt{2P} b_{ku}(t) c_{ku}(t) \cos(2\pi f_{ku} t), \tag{19.16}$$

where P is the transmitted power per substream.

The receiver's block diagram for the AFH/MC DS-CDMA system is shown in Figure 19.14. The receiver is constituted by a conventional correlator detector. The receiver's 'channel quality estimator' of Figure 19.14 estimates the subchannel fading parameters periodically and passes the estimates to the receiver's FH control unit (RFHCU). Based on the subchannel parameters, the RFHCU generates a $K \times U$ matrix F given by [480]:

$$\mathbf{V} = \begin{pmatrix} v_{1,1} & v_{1,2} & \cdots & v_{1,U} \\ v_{2,1} & v_{2,2} & \cdots & v_{2,U} \\ \vdots & \vdots & \ddots & \vdots \\ v_{K,1} & v_{K,2} & \cdots & v_{K,U} \end{pmatrix}, \tag{19.17}$$

where $v_{k,u} \in [0, U]$ denotes the number of substreams hopping to the uth subcarrier of the kth user. Hence, the sum of each row of the matrix $\mathbf{V}$ is U. The kth row of $\mathbf{V}$ is sent to the TFHCU of user k through a feedback control channel [480], in order to control the following activated subcarriers. At the same time these parameters are also used for recovering the order of the transmitted substreams, in order to complete the despreading process, as shown in Figure 19.14.

The success of the AFH/MC DS-CDMA scheme using adaptive frequency-hopping depends on the availability of a reliable channel quality estimator on providing reasonably accurate and timely estimation of the channel, on the availability of a reliable feedback control channel and on the generation of optimal frequency-hopping patterns. In [480] the water-filling algorithm has been proposed for deriving the optimal frequency hopping patterns. The

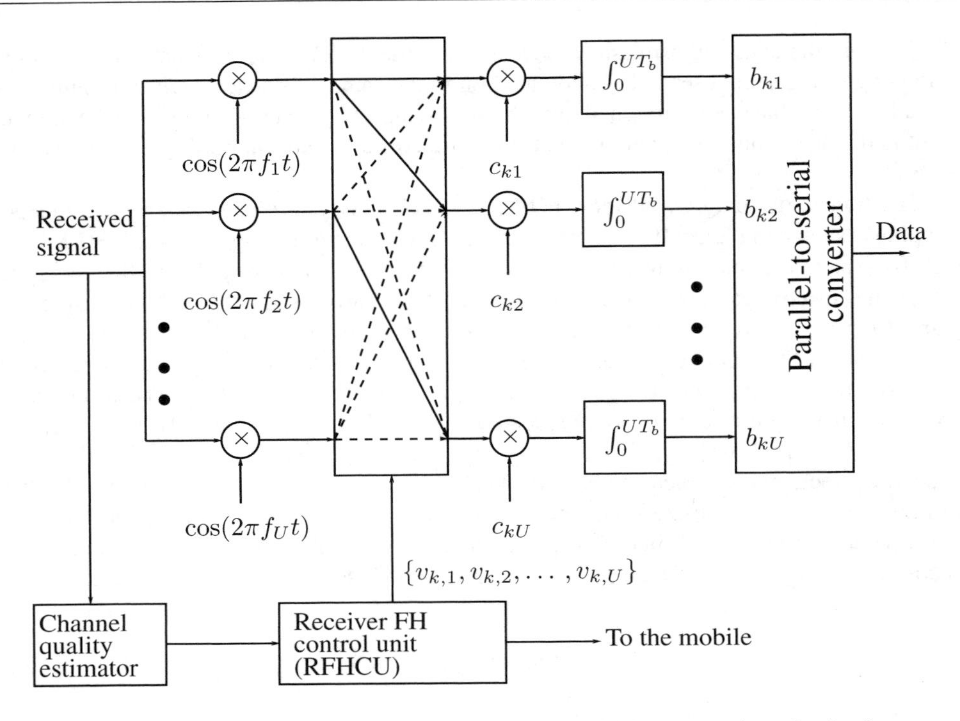

Figure 19.14: The receiver block diagram of the FH/MC DS-CDMA system using adaptive frequency-hopping.

AFH/MC DS-CDMA scheme outperforms the conventional corresponding single-carrier DS-CDMA systems using a RAKE receiver in terms of its average bit error rate (BER). However, the AFH/MC DS-CDMA scheme can only operate in a channel exhibiting slow variations versus time. Furthermore, it needs a channel quality estimator and a complex algorithm for deriving the optimal frequency-hopping patterns.

19.2.6 Adaptive Subchannel Allocation Assisted Multicarrier DS-CDMA

In contrast to the AFH/MC DS-CDMA scheme of reference [480], which was proposed for the uplink (mobile-to-base), the multicarrier DS-CDMA scheme using adaptive subchannel allocation, which is referred to here as adaptive AMC DS-CDMA, was suggested in the downlink (base-to-mobile) transmission [459]. The adaptive AMC DS-CDMA scheme is based on the multicarrier DS-CDMA-I scheme of Figure 19.4 proposed in [472], where a data sequence spread by a narrowband signature sequence modulates U subcarriers. In the adaptive AMC DS-CDMA scheme of Figure 19.15, instead of transmitting the same DS waveforms over all subchannels according to Figure 19.4, each user's DS waveform is transmitted over a specific favourite subchannel for the user. This favourite subchannel is determined as follows [459]. The mobile estimates the fading amplitudes of all subchannels by exploiting its received pilot

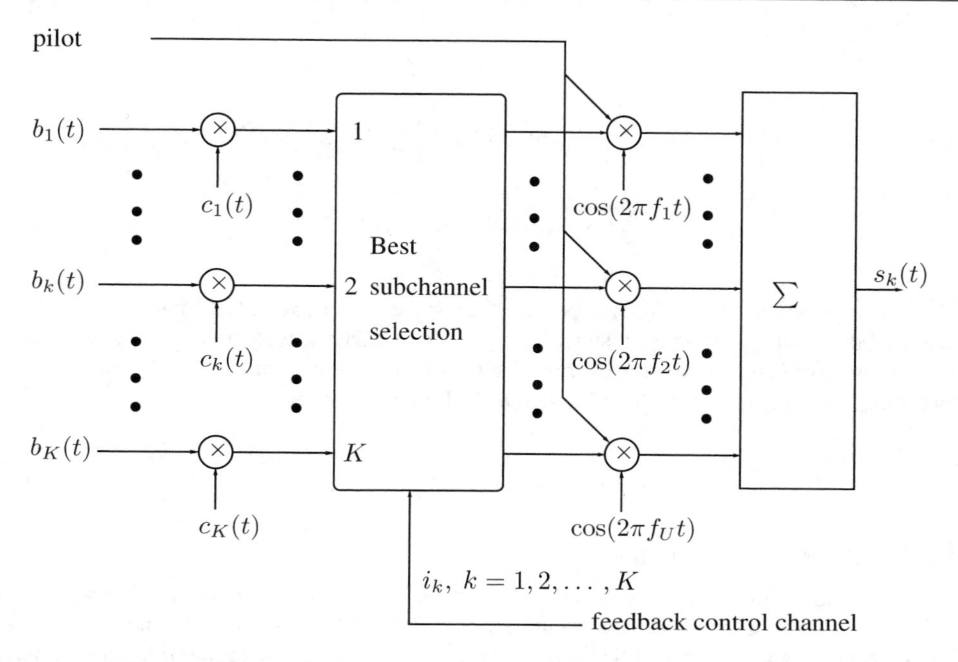

Figure 19.15: The downlink transmitter's schematic for the multicarrier DS-CDMA system using adaptive subchannel allocation.

signal and feeds back the index of the subchannel exhibiting the highest amplitude to the base station. With the aid of this index information, the base station allocates each user's DS waveform to the best subchannel for the user considered. This allocation scheme is optimum, provided that orthogonal spreading codes are employed.

The downlink schematic of the adaptive AMC DS-CDMA system supporting K users is shown in Figure 19.15. The system model proposed in [459] is based on the multicarrier DS-CDMA scheme proposed in [472], where the total bandwidth of the system was divided into U equi-width subchannels. At the transmitter of the base station, first, the favourite subchannel is selected for each user according to the channel state, and a narrowband DS waveform is transmitted through the favourite subchannel. As shown in Figure 19.15, at the base station the binary data waveform $b_k(t)$ of the kth user is first spread by the user's spreading code waveform $c_k(t)$, yielding $b_k(t)c_k(t)$. Then, the 'best subchannel selection' unit of Figure 19.15 chooses the best subchannel index denoted as i_k for the kth user, $k = 1, 2, \ldots, K$. The subcarrier of the best subchannel is then modulated by $b_k(t)c_k(t)$. Finally, the K number of user signals as well as the downlink pilot signal – which modulates all subcarriers – is multiplexed for transmission to the mobile receivers. The base station's transmitted signal can be written as:

$$\begin{aligned} s_k(t) &= \sum_{k=1}^{K} \sqrt{2P} b_k(t) c_k(t) \sum_{u=1}^{U} \Delta_u(i_k) \cos(2\pi f_u t) \\ &+ \sum_{u=1}^{U} \sqrt{2P_0} c_0(t) \cos(2\pi f_u t), \end{aligned} \quad (19.18)$$

where P represents the transmitted power of each user, P_0 denotes the identical transmitted power of the pilot signal on each subcarrier, $b_k(t)$ is the data waveform of the kth user, while $c_k(t)$ is the spreading code waveform of the kth user. Furthermore, $c_0(t)$ represents the spreading code waveform of the pilot signal. In Equation 19.18,

$$\Delta_u(i_k) = \begin{cases} 1, & \text{for } u = i_k \\ 0, & \text{for } u \neq i_k \end{cases} \quad (19.19)$$

indicates the subcarrier activation function.

The spectrum of the adaptive AMC DS-CDMA signal using rectangular chip waveforms is the same as that of the orthogonal multicarrier DS-CDMA-I, which is shown in Figure 19.5. The spectral gain and the spreading gain corresponding to a subcarrier are the same as that of the orthogonal multicarrier DS-CDMA-I scheme using the minimum spacing of $\Delta = 2/T_c$, i.e., we have $SG = 1$ and $N_e = N/U$, respectively. The overall system's spreading gain is also N.

The receiver schematic of the multicarrier DS-CDMA mobile receiver using adaptive subchannel allocation is shown in Figure 19.16, where both the coherent demodulation phases with respect to the ith subcarrier and the despreading code are incorporated in the term $g_0[i]$, $i = 1, 2, \ldots, U$ corresponding to the pilot detection. By contrast, $g_k[i_k]$ corresponds to the kth user's detection. At the mobile's receiver, the index of the best subchannel is determined by estimating the received amplitudes of the subchannels based on the mobile's received pilot signal. As shown in Figure 19.16, the received signal is first coherently demodulated using each subcarrier and then correlated with the despreading code of the pilot signal, in order to obtain the decision variables characterizing the fading amplitude. The best subchannel estimation block estimates the fading amplitudes of all the subchannels using the outputs of the pilot correlators. The index of the best subchannel corresponding to the subchannel having highest amplitude is determined, which is denoted as i_k in Figure 19.16. This favorite subchannel index information on the one hand is sent to the base station by using a feedback control channel and is also preserved in the mobile over the subchannel quality update period, in order to provide the subchannel index for the kth user's data demodulation. According to the index i_k, the i_kth subchannel of user k is selected and correlated with the kth user's despreading code $c_k(t)$ – which is included in the term of $g_k[i_k]$ in Figure 19.16 – in order to form the decision variable for the transmitted data.

In [459] the performance of the adaptive AMC DS-CDMA system was studied over slow Rayleigh fading channels. It was shown that the adaptive AMC DS-CDMA scheme outperforms the multicarrier DS-CDMA scheme proposed in [472], when both orthogonal PN spreading codes and random spreading codes were considered. Since a common pilot signal

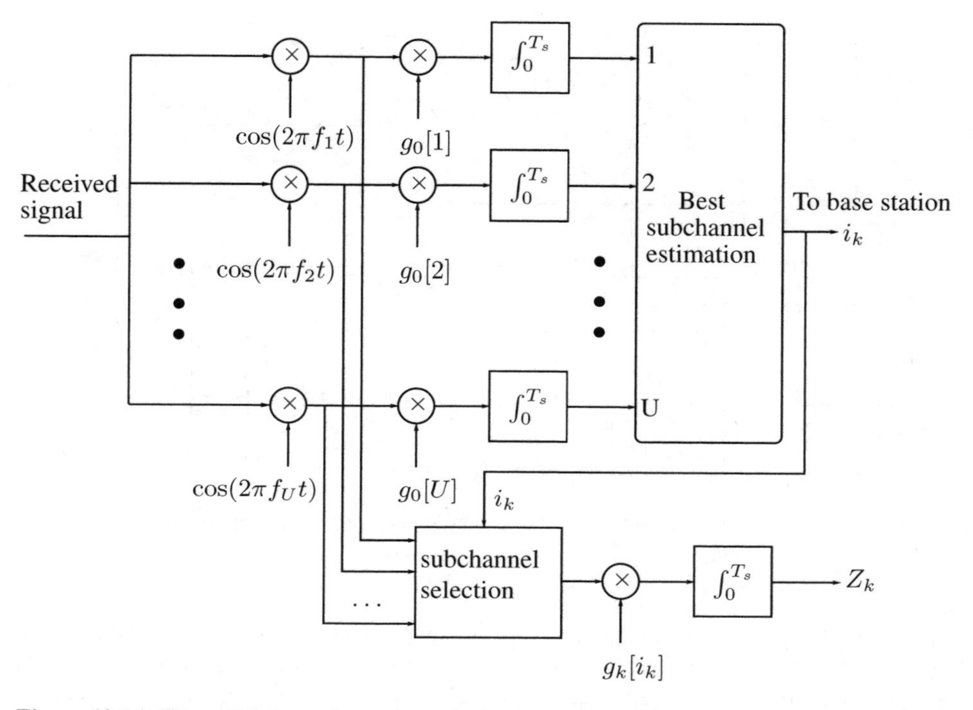

Figure 19.16: The mobile's receiver schematic for the multicarrier DS-CDMA system using adaptive subchannel allocation.

is transmitted to all users in the downlink, the complexity associated with the pilot signal's transmission is low. However, a reliable, low-delay feedback control channel has to be assigned to each user, in order to send the index of the best subchannel provided by the mobile station's receiver to the base station, in order to control the subcarrier activation.

The adaptive AMC DS-CDMA scheme can also be extended to allocate a number of subchannels instead of one for data transmission, in order to support variable-rate or multi-rate services.

19.2.7 Slow Frequency-Hopping Multicarrier DS-CDMA

The slow FH multicarrier DS-CDMA (SFH/MC DS-CDMA) scheme using constant-weight codes-based FH patterns has been proposed in [66, 74, 445]. This scheme can efficiently amalgamate the techniques of slow FH, OFDM and DS-CDMA. In SFH/MC DS-CDMA nonlinear constant-weight codes have been introduced, in order to control the associated FH patterns and hence to activate a number of subcarriers, in order to support multi-rate services. At the same time, a set of constant-weight codes satisfying the required minimum distance conditions is employed, in order to assist in the determination and initial synchronization of the FH patterns employed.

The model of the transmitter and the multiple access channel used in the analysis of SFH/MC DS-CDMA is depicted in Figure 19.17. Each of the K users in the system is

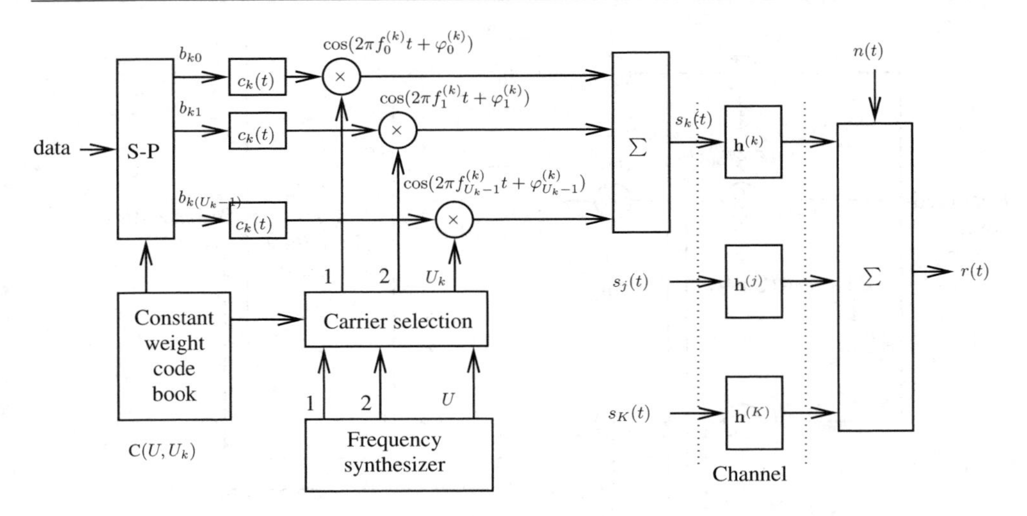

Figure 19.17: Transmitter and channel block diagram of the frequency-hopping multicarrier DS-CDMA system.

assigned a randomly generated spreading code, which produce spread signals. In the figure, $C(U, U_k)$ represents a constant-weight code (CWC) of user k with U_k number of '1's and $(U - U_k)$ number of '0's assigned. Hence, the weight of $C(U, U_k)$ is U_k. This code is read from a so-called CWC book, which represents the frequency-hopping patterns. Theoretically, the size of the CWC book is $\binom{U}{U_k} = U!/U_k!(U-U_k)!$ The CWC $C(U, U_k)$ plays two different roles. Its first role is that its weight – namely U_k – determines the number of activated subcarriers involved, while its second function is that the positions of the U_k number of binary '1's determines the selection of a set of U_k number of activated subcarrier frequencies from the U number of outputs of the frequency synthesizer.

At the transmitter of the kth user in Figure 19.17 the bit stream having a bit duration of T_b is first serial-to-parallel (S-P) converted, yielding U_k parallel streams, which is controlled by the CWC $C(U, U_k)$. The symbol duration, T_s, of the SFH/MC DS-CDMA signal is $T_s = U_k T_b$. Multi-rate transmission can be supported by controlling the weight of the code $C(U, U_k)$. As seen in Figure 19.17, after serial-to-parallel conversion each stream is DS spread, in order to form the spread signal and this spread signal then modulates one of the selected subcarriers. Finally, the transmitted signal of the kth user is yielded in the form of:

$$s_k(t) = \sum_{u=0}^{U_k-1} \sqrt{2P} b_{ku}(t) c_k(t) \cos\left(2\pi f_u^{(k)} t + \varphi_u^{(k)}\right), \tag{19.20}$$

where P represents the transmitted power per carrier, while U_k indicates the weight of the CWC currently employed by the kth user. Furthermore, $b_{ku}(t)$ represents the current data stream's waveforms, $c_k(t)$ denotes the kth user's DS spreading waveforms, while $\left\{f_u^{(k)}\right\}$ and $\left\{\varphi_u^{(k)}\right\}$ represent the current subcarrier frequency set and modulation phase set, respectively.

The spectrum of the SFH/MC DS-CDMA signal is similar to that of the orthogonal multicarrier DS-CDMA-I system of Figure 19.5 or orthogonal multicarrier DS-CDMA-II of Figure 19.8, depending on the spacing, Δ, between two adjacent subcarriers, which may assume $\Delta = 2/T_c$ or $\Delta = 1/T_c$, respectively.

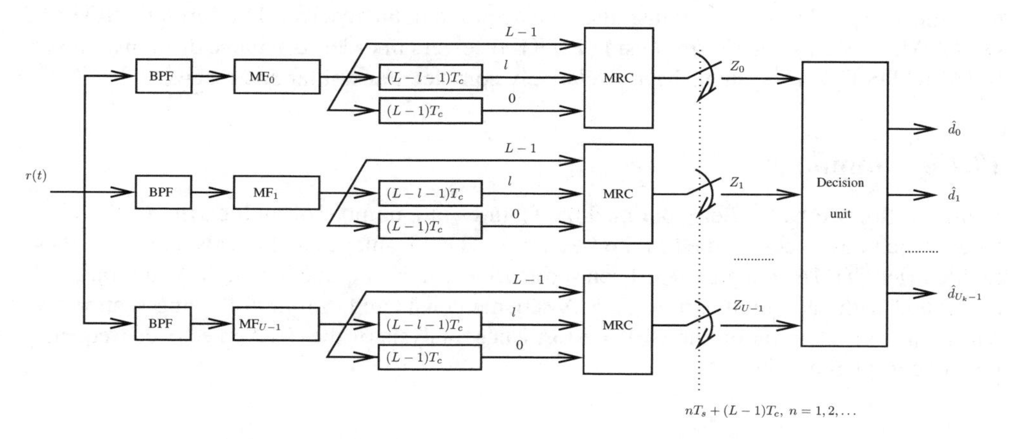

Figure 19.18: Receiver block diagram of the frequency-hopping multicarrier DS-CDMA system.

The conventional matched filter-based RAKE receiver using MRC can be invoked for its detection, as shown in Figure 19.18, where we assume that L number of diversity paths is available. In contrast to the transmitter side, where only U_k out of U activated subcarriers are transmitted by user k, at the receiver all U subcarriers are always tentatively demodulated. The information bits transmitted over the different subcarriers are detected in the decision block of Figure 19.18. Specifically, if the receiver has the explicit knowledge of the FH patterns, the information bits conveyed by different subcarriers can be detected, as in conventional FH systems [445]. However, if no explicit *a-priori* knowledge of the FH patterns is available, the information bits can be blindly detected with the aid of the CWC codes [66]. Blind joint soft-detection based on maximum-likelihood sequence detection (MLSD) has been proposed and studied in [66] over multipath Nakagami-m fading channels. It was shown that the proposed blind joint soft-detection techniques were capable of detecting the transmitted information and simultaneously acquiring the FH patterns without explicit signalling.

In SFH/MC DS-CDMA systems the entire bandwidth of future wireless communication systems can be divided into a number of sub-bands and each sub-band can be assigned a subcarrier. According to the prevalent service requirements, the set of legitimate subcarriers can be allocated in line with the instantaneous information rate requirements. A range of FH techniques can be employed for each user, in order to occupy the whole system bandwidth and to efficiently utilize the system's frequency resources. Specifically, slow FH, fast FH or adaptive FH techniques can be introduced, in order to further improve the system's performance. In SFH/MC DS-CDMA systems the sub-bands are not required to be of equal bandwidth. Hence existing second and third generation CDMA systems can be supported using one or more subcarriers, consequently simplifying the frequency resource management and efficiently utilizing the entire bandwidth available. This regime can also remove the spectrum segmentation of existing 'legacy' systems of the past, while ensuring compatibility with future Broadband Access Networks (BRAN) and unlicensed systems. Furthermore, a number of sub-channels associated with variable processing gains can be employed, in order to support various services requiring low- to very high-rate transmissions, for example, for wireless Internet access. However, in SFH/MC DS-CDMA systems the CWC-based FH patterns must be produced in the transmitter and acquired in the receiver. Furthermore, SFH/MC DS-CDMA systems benefit from using RAKE receivers in order to achieve diversity. Hence, SFH/MC DS-CDMA systems have a relatively complex transmitter and receiver.

19.2.8 Summary

In this section we have briefly outlined the features of a number of multicarrier CDMA systems, which have been studied in the literature. The advantages and disadvantages of these multicarrier CDMA schemes have been summarized. It can be shown that there are trade-offs associated with each multicarrier CDMA scheme considered. In the following sections we will focus our attention on the BER performance analysis of these schemes over frequency selective Rayleigh fading channels.

19.3 Channel Model

The channel model considered in this chapter is a frequency selective channel, where the transmitted signal is received over L independent slowly-varying flat fading channels, as shown in Figure 18.8. The complex low-pass equivalent representation of the impulse response experienced by subcarrier u of user k is given by:

$$h_{ku}(t) = \sum_{l=1}^{L} \alpha_{ul}^{(k)} \delta\left(t - \tau_{kl}\right) \exp(j\psi_{ul}^{(k)}), \tag{19.21}$$

where l is the index of the channel impulse response (CIR) taps, $\delta(\cdot)$ is the Dirac function, $\left\{\alpha_{ul}^{(k)}\right\}_{l=1}^{L}$, $\left\{\psi_{ul}^{(k)}\right\}_{l=1}^{L}$ and $\{\tau_{kl}\}_{l=1}^{L}$ are the random CIR tap amplitudes, phases and delays, respectively. However, we assume that all subcarrier signals of the same user encounter the same delay, i.e., τ_{kl} is independent of u. Note that, if $L = 1$, these terms will be represented by $\left\{\alpha_{u}^{(k)}\right\}$, $\left\{\psi_{u}^{(k)}\right\}$ and $\{\tau_{kl}\}$, respectively. We assume that $\left\{\alpha_{ul}^{(k)}\right\}_{l=1}^{L}$, $\left\{\psi_{ul}^{(k)}\right\}_{l=1}^{L}$ and $\{\tau_{kl}\}_{l=1}^{L}$ are mutually independent. Without loss of generality, we also assume that $\tau_{k1} < \tau_{k2} < \cdots < \tau_{kL}$.

If we let $\tau_{kl} = (l-1)T_c + \tau_k$, then Equation 19.21 can be expressed as:

$$h_{ku}(t) = \sum_{l=1}^{L} \alpha_{ul}^{(k)} \delta\left(t - (l-1)T_c - \tau_k\right) \exp(j\psi_{ul}^{(k)}), \tag{19.22}$$

which represents the widely used tap-delay line mode of the frequency-selective channel [90].

For a multipath Rayleigh fading channel the fading amplitudes $\left\{\alpha_{ul}^{(k)}\right\}_{l=1}^{L}$ are assumed to be statistically independent random variables having a probability density function (PDF) expressed as [90]:

$$f(\alpha_{ul}^{(k)}) = M(\alpha_{ul}^{(k)}, \Omega),$$
$$M(R, \Omega) = \frac{2R}{\Omega} \exp\left(-\frac{R^2}{\Omega}\right), \tag{19.23}$$

where $\Omega = E\left[\left(\alpha_{ul}^{(k)}\right)^2\right]$. We assume that $\sum_{l=1}^{L}\left\{E\left[\left(\alpha_{ul}^{(k)}\right)^2\right]\right\} = 1$, hence, we have $\Omega = 1/L$. The phases $\left\{\psi_{ul}^{(k)}\right\}_{l=1}^{L}$ of the different paths and of different subcarriers are assumed to be uniformly distributed random variables in $[0, 2\pi)$, while the path delays of $\{\tau_{kl}\}_{l=1}^{L}$ of user k are uniformly distributed in $[0, T_s)$.

19.4 Performance of Multicarrier CDMA Systems Using Frequency-Domain Spreading

In the subsection of 19.2.1 the features of the MC-CDMA scheme using frequency-domain spreading were summarized. As shown in Figure 19.1, in this MC-CDMA scheme the transmitter spreads the original signal using a given spreading code in the frequency domain. In other words, a fraction of the symbol corresponding to a chip of the spreading code is transmitted through a different subcarrier. For efficient multicarrier transmission it is crucial to have frequency non-selective fading over each subcarrier [75]. Therefore, if the original symbol rate is high, resulting in frequency selective fading, then serial-to-parallel conversion of the original data stream to a number of reduced-rate substreams may be needed, in order to increase the chip duration and hence to avoid frequency selective fading.

In this section we derive the BER for the MC-CDMA system seen in Figure 19.19 employing serial-to-parallel conversion. Specifically, we derive the BER expressions for the asynchronous MC-CDMA system [357] over frequency selective Rayleigh fading channels. The expressions derived in this section are suitable for the analysis of uplink transmissions. Let us first consider the system model.

19.4.1 System Description

19.4.1.1 The Transmitter

The transmitter schematic of the MC-CDMA system using frequency-domain spreading, which is based on Figure 19.1 but additionally involves serial-to-parallel data conversion is shown in Figure 19.19. The MC-CDMA system transmits N_p chips of a data symbol in parallel on N_p different subcarriers, one chip per subcarrier, where N_p is the total number of chips per data bit or the processing gain. Hence, the chip duration is the same as the symbol duration of the transmitted MC-CDMA signal, i.e., T_s, which is given by $T_s = qT_b$, since q bits are transmitted in a symbol duration. Throughout our discussions we assume that random

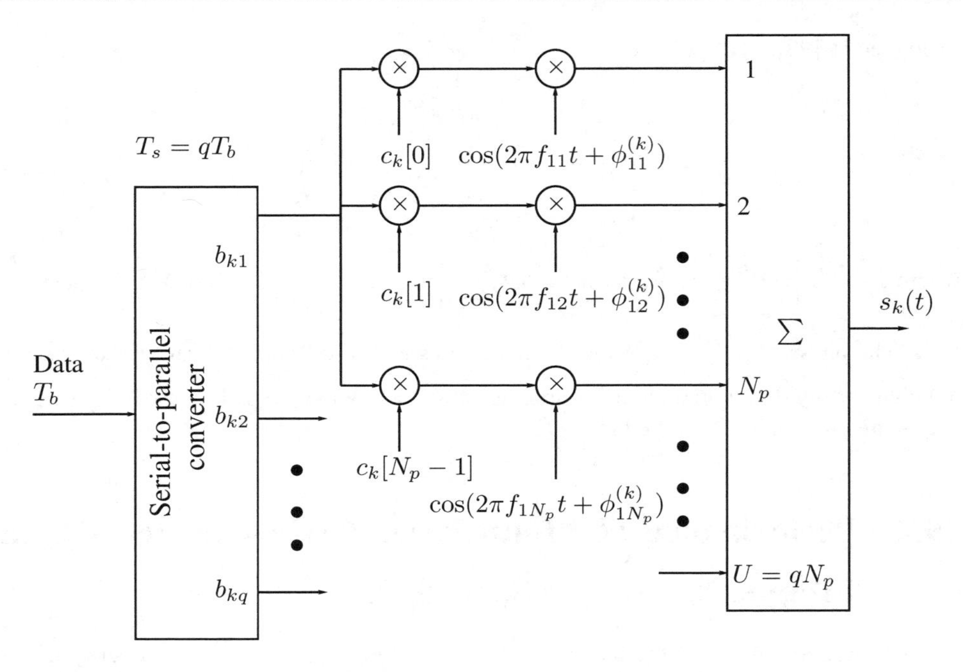

Figure 19.19: The transmitter schematic of the MC-CDMA system using frequency-domain spreading and serial-to-parallel data conversion.

spreading codes are employed and the frequency separation between two adjacent subcarriers is $1/T_s$. Hence, the subcarriers are orthogonal over chip duration T_s. Let

$$\{f_{ij}\} = \begin{pmatrix} f_{11} & f_{12} & \cdots & f_{1N_p} \\ f_{21} & f_{22} & \cdots & f_{2N_p} \\ \vdots & \vdots & \ddots & \vdots \\ f_{q1} & f_{q2} & \cdots & f_{qN_p} \end{pmatrix} \tag{19.24}$$

be the subcarrier frequency set, which are arranged according to:

$$f_{ij} = f_0 + \frac{1}{T_s}\left[(i-1) + (j-1)q\right]. \tag{19.25}$$

Hence, according to Equation 19.25 each column represents q number of adjacent subcarriers, while the minimum spacing between two subcarriers from the same row is q/T_s. In Figure 19.19 after the serial-to-parallel conversion stage, q data bits of user k are spread by the spreading code of user k in the frequency domain. Note that the data bits mapped to different substreams use the same spreading code. The spread signal then modulates the subcarriers mapping each chip to a subcarrier. The subcarrier frequencies modulated by the same data bit are from the same row of the matrix in Equation 19.25, and these subcarriers are sufficiently separated in frequency to avoid correlated fading. Consequently, referring to

Figure 19.19 the transmitted signal can be formulated as:

$$s_k(t) = \sum_{n=-\infty}^{\infty} \sum_{i=1}^{q} \sum_{j=1}^{N_p} \sqrt{\frac{2P}{N_p}} b_{ki}[n] c_k[j-1] P_{T_s}(t - nT_s) \cos\left(2\pi f_{ij} t + \phi_{ij}^{(k)}\right), \quad (19.26)$$

where $b_{ki}[n]$ represents the ith data substream of user k and $b_{ki}[n]$ is assumed to be a random variable, assuming values of +1 or -1 with equal probability. Furthermore, $c_k[j]$ is the jth chip of the spreading code $\{c_k[1]\ c_k[2]\ \ldots\ c_k[N_p - 1]\}$ assigned to user k, which is also assumed to be a random variable taking values +1 or -1 with equal probability of 1/2. The variable $P_\tau(t)$ represents the rectangular modulation waveform defined as $P_\tau(t) = 1$, if $0 \leq t \leq \tau$, and zero otherwise. Furthermore, f_{ij} is the subcarrier frequency associated with the ith data substream and the $(j-1)$th chip of the spreading code. Finally, $\phi_{ij}^{(k)}$ is the random phase introduced by the carrier modulation, which is assumed to be uniformly distributed in $[0, 2\pi)$.

We assume that frequency non-selective fading is encountered by each subcarrier signal. Hence, the following condition is satisfied:

$$1/T_s << (\Delta B)_c, \quad (19.27)$$

where $(\Delta B)_c$ represents the coherence bandwidth of the channel [90]. The reciprocal of $(\Delta B)_c$ is a measure of the multipath delay spread of the channel, which is denoted by T_m, $T_m \approx 1/(\Delta B)_c$. In order to guarantee independent fading of the subcarrier signals carrying chips associated with the same data bit, we assume that the following condition is satisfied:

$$(\Delta B)_c << q/T_s. \quad (19.28)$$

The conditions of Equation 19.27 and Equation 19.28, i.e., $1/T_s << (\Delta B)_c << q/T_s$ imply that the modulated subcarriers having transmission bandwidth of $1/T_s$ do not experience significant dispersion.

However, if the conditions of $1/T_s << (\Delta B)_c << q/T_s$ cannot be satisfied, the system model of Figure 19.19 can be modified, in order to satisfy this conditions. Specifically, if $1/T_s << (\Delta B)_c$ cannot be satisfied, we can decrease the term $1/T_s$ by transmitting more bits per symbol, i.e, by increasing the value of q in Figure 19.19. By contrast, if the condition of $(\Delta B)_c << q/T_s$ cannot be satisfied, the independence between the adjacent subcarriers conveying the same data bit can be further guaranteed by incorporating sufficient interleaving at the cost of an increased delay. For example, assuming that the first N_p symbols, i.e., $= N_p q\ bits$, emitted by the source modulate the $N_p q$ subcarriers in the first time interval, where each symbol is modulated by the subcarrier frequencies from the same column of Equation 19.24. In the second time interval, the same N_p symbols modulate the $N_p q$ subcarriers, but cyclically shifted by one symbol. Similarly, in the ith interval, $1 \leq i \leq N_p$, the N_p symbols after the $(i-1)$th cyclic shift modulate the $N_p q$ subcarriers. At the receiver, a deinterleaver is employed to recover the original ordering of the symbols. However, interleaving techniques can only be used for the transmission of delay-insensitive data.

Here, we assume that there are K asynchronous CDMA users in the system and that all of them use the same q and N_p values. The average power received from each user at the base

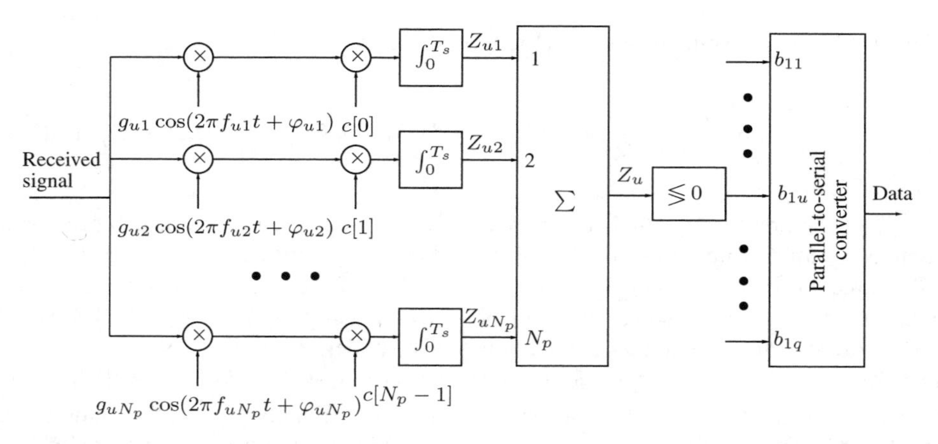

Figure 19.20: The receiver block diagram of the MC-CDMA system using frequency-domain spreading for the reference user.

station is also assumed to be the same, i.e. perfect power control is assumed. Consequently, the received signal at the base station can be expressed as:

$$r(t)=\sum_{k=1}^{K}\sum_{n=-\infty}^{\infty}\sum_{i=1}^{q}\sum_{j=1}^{N_p}\sqrt{\frac{2P}{N_p}}\alpha_{ij}^{(k)}b_{ki}[n]c_k[j-1]P_{T_s}(t-nT_s-\tau_k)$$

$$\cdot\cos\left(2\pi f_{ij}t+\varphi_{ij}^{(k)}\right)+n(t), \qquad (19.29)$$

where $\varphi_{ij}^{(k)}=\phi_{ij}^{(k)}+\psi_{ij}^{(k)}-2\pi f_{ij}\tau_k$ and the random phase $\psi_{ij}^{(k)}$ corresponds to the subcarrier frequency f_{ij} of user k, which was introduced by the channel. The phase $\varphi_{ij}^{(k)}$ is a random variable uniformly distributed in $[0,2\pi)$. In Equation 19.29, τ_k is the misalignment of user k with respect to the reference user – we assumed that $k=1$ and $\tau_1=0$ – at the receiver, which is i.i.d for different k values and uniformly distributed in $[0,T_s)$. Furthermore, $\alpha_{ij}^{(k)}$ represents the amplitude attenuation due to the fading channel, which is assumed to be a Rayleigh distributed random variable obeying the PDF given by Equation 19.23 and $\Omega=E\left[\left(\alpha_{ij}^{(k)}\right)^2\right]=1$. Finally, $n(t)$ is the Additive White Gaussian Noise (AWGN) with zero mean and double-sided power spectral density of $N_0/2$.

19.4.1.2 Receiver

The receiver of the first user ($k=1$) is shown in Figure 19.20, where the superscript denoting the reference user is omitted. The MC-CDMA receiver considered is based on coherent correlator detector. As shown in Figure 19.20, the decision variable Z_u of the 0th data bit in the uth data substream of the reference user is given by:

$$Z_u = \sum_{v=1}^{N_p} Z_{uv}, \tag{19.30}$$

$$Z_{uv} = \int_0^{T_s} r(t)c[v-1]g_{uv}\cos(2\pi f_{uv}t + \varphi_{uv})dt, \tag{19.31}$$

where g_{uv} is a parameter determining which type of combining model of the chips belonging to the same data bit is used. Depending on the choice of $\{g_{uv}, v = 1, 2, \ldots, N_p\}$, in our analysis there are two ways of combining the chips of the same data bit, namely MRC and equal gain combining (EGC). In this section the associated BER using the MRC scheme will be investigated.

Upon substituting Equation 19.29 into Equation 19.30, the variable Z_{uv} can be simplified to:

$$Z_{uv} = \sqrt{\frac{P}{2N_p}}T_s \left[D_{uv} + \sum_{k=2}^{K} I_1^{(k)} + \sum_{k=2}^{K} \underbrace{\sum_{i=1}^{q}\sum_{j=1}^{N_p}}_{j\neq v \text{ for } i=u} I_2^{(k)} + \eta_{uv} \right], \tag{19.32}$$

where D_{uv} is the desired term given by:

$$D_{uv} = \alpha_{uv} g_{uv} b_u[0], \tag{19.33}$$

and $b_u[0]$ is the 0th data bit of the reference user transmitted by the uvth subcarrier. Due to the orthogonality of the subcarrier signals associated with the same user, no self-interference is inflicted by the reference signal to the vth subcarrier of the uth data stream. The multi-user interference imposed by user k can be divided into two terms. The first term is constituted by the subcarrier signal having the same subcarrier frequency, f_{uv}, as the branch considerd. This term is $I_1^{(k)}$ in Equation 19.32, which can be written as:

$$\begin{aligned} I_1^{(k)} &= \left(\sqrt{\frac{P}{2N_p}}T_s\right)^{-1} \int_0^{T_s} \sum_{n=-\infty}^{\infty} \sqrt{\frac{2P}{N_p}} \alpha_{uv}^{(k)} b_{ku}[n] c_k[v-1] \\ &\quad \cdot P_{T_s}(t - nT_s - \tau_k)\cos\left(2\pi f_{uv}t + \varphi_{uv}^{(k)}\right) \\ &\quad \cdot c[v-1] g_{uv}\cos(2\pi f_{uv}t + \varphi_{uv})dt \\ &= \frac{\alpha_{uv}^{(k)} g_{uv}\cos\theta_{uv}^{(k)}}{T_s} \\ &\quad \cdot \left[b_{ku}[-1]c_k[v-1]c[v-1]\tau_k + b_{ku}[0]c_k[v-1]c[v-1](T_s - \tau_k)\right], \end{aligned} \tag{19.34}$$

where $\theta_{uv}^{(k)} = \varphi_{uv}^{(k)} - \varphi_{uv}$. The second term of the multi-user interference imposed by user k is contributed by the other subcarrier signals associated with $\{f_{ij}, i = 1, 2, \ldots, q; j = 1, 2, \ldots, N_P; j \neq v \text{ for } i = u\}$. The interference, $I_2^{(k)}$, arising from the subcarrier signal associated with the frequency f_{ij} of user k can be formulated as:

$$\begin{aligned} I_2^{(k)} &= \left(\sqrt{\frac{P}{2N_p}}T_s\right)^{-1} \int_0^{T_s} \sum_{n=-\infty}^{\infty} \sqrt{\frac{2P}{N_p}} \alpha_{ij}^{(k)} b_{ki}[n] c_k[j-1] \\ &\cdot P_{T_s}(t - nT_s - \tau_k) \cos\left(2\pi f_{ij} t + \varphi_{ij}^{(k)}\right) \\ &\cdot c[v-1] g_{uv} \cos(2\pi f_{uv} t + \varphi_{uv}) dt \\ &= \frac{\alpha_{uv}^{(k)} g_{uv}}{T_s} \left[R_{ij}^{(k)}(\tau_k) + \hat{R}_{ij}^{(k)}(\tau_k)\right], \end{aligned} \tag{19.35}$$

where $R_{ij}^{(k)}(\tau_k)$ and $\hat{R}_{ij}^{(k)}(\tau_k)$ are chip-partial cross-correlation functions defined as:

$$\begin{aligned} R_{ij}^{(k)}\left(\tau_k, \theta_{ij}^{(k)}\right) &= \int_0^{\tau_k} b_{ki}[-1] c_k[j-1] c[v-1] \\ &\cdot \cos\left[2\pi(f_{ij} - f_{uv})t + \theta_{ij}^{(k)}\right] dt, \end{aligned} \tag{19.36}$$

$$\begin{aligned} \hat{R}_{ij}^{(k)}\left(\tau_k, \theta_{ij}^{(k)}\right) &= \int_{\tau_k}^{T_s} b_{ki}[0] c_k[j-1] c[v-1] \\ &\cdot \cos\left[2\pi(f_{ij} - f_{uv})t + \theta_{ij}^{(k)}\right] dt, \end{aligned} \tag{19.37}$$

and $\theta_{ij}^{(k)} = \varphi_{ij}^{(k)} - \varphi_{uv}$. Upon using Equation 19.25, Equation 19.36 and Equation 19.37 can be written as:

$$\begin{aligned} R_{ij}^{(k)}\left(\tau_k, \theta_{ij}^{(k)}\right) &= \int_0^{\tau_k} b_{ki}[-1] c_k[j-1] c[v-1] \\ &\cdot \cos\left[\frac{2\pi\left[(i-u) + (j-v)q\right]t}{T_s} + \theta_{ij}^{(k)}\right] dt \\ &= \frac{T_s b_{ki}[-1] c_k[j-1] c[v-1]}{2\pi\left[(i-u) + (j-v)q\right]} \\ &\cdot \left\{\sin\left[\frac{2\pi\left[(i-u) + (j-v)q\right]\tau_k}{T_s} + \theta_{ij}^{(k)}\right] - \sin\theta_{ij}^{(k)}\right\} \end{aligned} \tag{19.38}$$

and as:

$$\begin{aligned}\hat{R}_{ij}^{(k)}\left(\tau_k,\theta_{ij}^{(k)}\right) &= \int_{\tau_k}^{T_s} b_{ki}[0]c_k[j-1]c[v-1] \\ &\quad \cdot \cos\left[\frac{2\pi\left[(i-u)+(j-v)q\right]t}{T_s}+\theta_{ij}^{(k)}\right]dt \\ &= \frac{T_s b_{ki}[0]c_k[j-1]c[v-1]}{2\pi\left[(i-u)+(j-v)q\right]} \\ &\quad \cdot\left\{\sin\left[\frac{2\pi\left[(i-u)+(j-v)q\right]T_s}{T_s}+\theta_{ij}^{(k)}\right]\right. \\ &\quad \left. -\sin\left[\frac{2\pi\left[(i-u)+(j-v)q\right]\tau_k}{T_s}+\theta_{ij}^{(k)}\right]\right\}, \end{aligned} \tag{19.39}$$

where $(i-u)+(j-v)q \neq 0$. Finally, η_{uv} is the noise term engendered by the AWGN $n(t)$, which can be written as:

$$\eta_{uv} = \left(\sqrt{\frac{P}{2N_p}}T_s\right)^{-1}\int_0^{T_s} n(t)c[v-1]g_{uv}\cos(2\pi f_{uv}t+\varphi_{uv})dt. \tag{19.40}$$

19.4.2 Performance Analysis

In order to derive the system's bit error probability, we have to derive the PDF of the decision variable Z_u in Equation 19.30. First, we examine the terms in the Z_{uv} expression of Equation 19.31.

19.4.2.1 Noise and Interference Analysis

A. Noise Analysis The noise term η_{uv} is given by Equation 19.40, which is a Gaussian random variable with zero mean and variance given by:

$$\text{Var}[\eta_{uv}] = \frac{N_p N_0}{2E_b}g_{uv}^2, \tag{19.41}$$

where $E_b = PT_s$ is the energy per bit.

B. Analysis of the Multi-user Interference Term $I_1^{(k)}$ The first multi-user interference term $I_1^{(k)}$ expressed by Equation 19.34 can be modelled as a Gaussian random variable with zero mean. Its variance is given by:

$$\text{Var}[I_1^{(k)}] = E_{\tau_k,\theta_{uv}^{(k)}}\left[(I_1^{(k)})^2\right], \tag{19.42}$$

where $E_{\tau_k, \theta_{uv}^{(k)}}[\ ]$ represents the average associated with the random variables $\tau_k \in [0, T_s)$, $\theta_{uv}^{(k)} \in [0, 2\pi)$. Upon substituting $I_1^{(k)}$ from Equation 19.34 into Equation 19.42 it can be shown that the variance of $I_1^{(k)}$ can be expressed as:

$$\text{Var}[I_1^{(k)}] = \frac{1}{3} g_{uv}^2, \tag{19.43}$$

where we used $E\left[\left(\alpha_{uv}^{(k)}\right)^2\right] = 1$, since frequency non-selective fading is assumed for each subcarrier signal, i.e., for each subcarrier signal there exists only one resolvable path.

C. Analysis of the Multi-user Interference Term $I_2^{(k)}$ The second multi-user interference term $I_2^{(k)}$ expressed by Equation 19.35 can also be modelled as a Gaussian random variable with zero mean. Its variance is derived as follows. According to Equation 19.35, the variance of $I_2^{(k)}$ is given by:

$$\begin{aligned} \text{Var}[I_2^{(k)}] &= E_{\tau_k, \theta_{ij}^{(k)}}\left[(I_2^{(k)})^2\right] \\ &= \frac{g_{uv}^2}{T_s^2}\left\{E_{\tau_k, \theta_{ij}^{(k)}}\left[\left(R_{ij}^{(k)}\left(\tau_k, \theta_{ij}^{(k)}\right)\right)^2\right] + E_{\tau_k, \theta_{ij}^{(k)}}\left[\left(\hat{R}_{ij}^{(k)}\left(\tau_k, \theta_{ij}^{(k)}\right)\right)^2\right]\right\}, \end{aligned} \tag{19.44}$$

where $E_{\tau_k, \theta_{ij}^{(k)}}\left[\left(R_{ij}^{(k)}\left(\tau_k, \theta_{ij}^{(k)}\right)\right)^2\right]$ and $E_{\tau_k, \theta_{ij}^{(k)}}\left[\left(\hat{R}_{ij}^{(k)}\left(\tau_k, \theta_{ij}^{(k)}\right)\right)^2\right]$ are the variances of the chip-partial cross-correlation functions defined in Equation 19.38 and Equation 19.39, respectively. Let us first derive the first term. According to Equation 19.38, it can be shown that:

$$\begin{aligned} E_{\tau_k, \theta_{ij}^{(k)}}\left[\left(R_{ij}^{(k)}\left(\tau_k, \theta_{ij}^{(k)}\right)\right)^2\right] &= \frac{T_s^2}{4\pi^2\left[(i-u)+(j-v)q\right]^2} \\ &\cdot E_{\tau_k, \theta_{ij}^{(k)}}\left\{\sin^2\left[\frac{2\pi\left[(i-u)+(j-v)q\right]\tau_k}{T_s} + \theta_{ij}^{(k)}\right]\right. \\ &+ \sin^2\theta_{ij}^{(k)} \\ &\left.-2\sin\left[\frac{2\pi\left[(i-u)+(j-v)q\right]\tau_k}{T_s} + \theta_{ij}^{(k)}\right]\sin\theta_{ij}^{(k)}\right\}. \end{aligned} \tag{19.45}$$

It is not difficult to shown that $E_{\tau_k, \theta_{ij}^{(k)}}\left\{\sin^2\left[\frac{2\pi[(i-u)+(j-v)q]\tau_k}{T_s} + \theta_{ij}^{(k)}\right]\right\} = 1/2$, $E_{\theta_{ij}^{(k)}}\left\{\sin^2\theta_{ij}^{(k)}\right\} = 1/2$ and $E_{\tau_k, \theta_{ij}^{(k)}}\left\{2\sin\left[\frac{2\pi[(i-u)+(j-v)q]\tau_k}{T_s} + \theta_{ij}^{(k)}\right]\sin\theta_{ij}^{(k)}\right\} = 0$. By substituting these terms into the above equation, finally, we obtain that:

$$E_{\tau_k,\theta_{ij}^{(k)}}\left[\left(R_{ij}^{(k)}\left(\tau_k,\theta_{ij}^{(k)}\right)\right)^2\right]=\frac{T_s^2}{4\pi^2\left[(i-u)+(j-v)q\right]^2}. \quad (19.46)$$

Similarly, the second term of Equation 19.44 can be derived and it can be shown that it is the same as Equation 19.46

$$E_{\tau_k,\theta_{ij}^{(k)}}\left[\left(\hat{R}_{ij}^{(k)}\left(\tau_k,\theta_{ij}^{(k)}\right)\right)^2\right]=\frac{T_s^2}{4\pi^2\left[(i-u)+(j-v)q\right]^2}. \quad (19.47)$$

Upon substituting Equation 19.46 and Equation 19.47 into Equation 19.44, it can be readily shown that the variance of $I_2^{(k)}$ is given by:

$$\text{Var}[I_2^{(k)}]=\frac{g_{uv}^2}{2\pi^2\left[(i-u)+(j-v)q\right]^2}. \quad (19.48)$$

19.4.2.2 Decision Statistics and Error Probability

Now we derive the PDF of the decision variable Z_u of Equation 19.30 associated with the MRC scheme. We approximate the multi-user interference by Gaussian noise, since it is the sum of many independent random variables. Consequently, it can be shown that Z_u of Equation 19.32 can be approximated as a Gaussian random variable with normalized[2] mean given by:

$$E[Z_u]=\sum_{v=1}^{N_p}E[Z_{uv}]=b_u[0]\sum_{v=1}^{N_p}\alpha_{uv}g_{uv}, \quad (19.49)$$

and its normalized variance can be expressed as:

$$\sigma^2=\left[\frac{N_pN_0}{2E_b}+\frac{(K-1)}{3}+(K-1)q(N_p-1)\bar{I}_M\right]\cdot\sum_{v=1}^{N_P}g_{uv}^2, \quad (19.50)$$

[2]The normalization is obtained from Z_u divided by $\left(\sqrt{\frac{P}{2N_p}}\right)$.

where $\overline{I}_M$ represents the average of Var$[I_2^{(k)}]$ associated with the variables of u and v as well as i and j. For the given values of u and v, let $I(u,v)$ be expressed as:

$$\begin{aligned} I(u,v) &= E_{i,j}\left[\text{Var}[I_2^{(k)}]\right] \\ &= \frac{1}{q(N_p-1)} \underbrace{\sum_{i=1}^{q}\sum_{j=1}^{N_p}}_{j\neq v \text{ for } i=u} \left[\frac{1}{2\pi^2\left[(i-u)+(j-v)q\right]^2}\right], \end{aligned} \tag{19.51}$$

then, $\overline{I}_M$ can be formulated as:

$$\overline{I}_M = \frac{1}{qN_p}\sum_{u=1}^{q}\sum_{v=1}^{N_p} I(u,v). \tag{19.52}$$

Consequently, for a given set of channel attenuations $\alpha_{uv},\ v=1,2,\ldots,N_p$, the conditional BER is given by [90]:

$$P_b(\gamma) = Q\left(\sqrt{2\gamma}\right), \tag{19.53}$$

where $Q(\ )$ represents the Gaussian Q-function, which is defined as [90]:

$$Q(x) = \frac{1}{\sqrt{2\pi}}\int_x^{\infty} e^{-t^2/2}dt. \tag{19.54}$$

For the use of MRC, the channel gain associated with each subcarrier is continuously estimated and multiplied by the correlator outputs. If perfect channel gain estimation is assumed, then, $g_{uv} = \alpha_{uv}$. Consequently, the variable γ in Equation 19.53 using MRC can be expressed as:

$$\gamma = \overline{\gamma}_c \cdot \sum_{v=1}^{N_p} \alpha_{uv}^2, \tag{19.55}$$

$$\overline{\gamma}_c = \left[\frac{2(K-1)}{3} + 2(K-1)q(N_p-1)\overline{I}_M + \left(\frac{E_b}{N_0 N_p}\right)^{-1}\right]^{-1}. \tag{19.56}$$

The unconditional BER P_b can be derived by weighting $P_b(\gamma)$ of Equation 19.53 with its probability of occurrence expressed in terms of its PDF and then averaging, i.e., integrating it over the valid range of γ, which can be written as:

$$P_b = \int_0^{\infty} Q\left(\sqrt{2\gamma}\right) f(\gamma)dr, \tag{19.57}$$

where $f(\gamma)$ is the PDF of γ. Since α_{uv} is a Rayleigh distributed random variable having PDF given by Equation 19.23 and $\Omega = 1$ for the MC-CDMA scheme, $\sum_{v=1}^{N_p} \alpha_{uv}^2$ obeys the central chi-square distribution with $2N_p$ degrees of freedom [90]. Finally, γ given by Equation 19.55 is also a random variable obeying the central chi-square distribution with $2N_p$ degrees of freedom, which can be formulated as [90]:

$$f(\gamma) = \frac{1}{\overline{\gamma}_c^{N_p}(N_p - 1)!} \gamma^{N_p - 1} \exp\left(-\frac{\gamma}{\overline{\gamma}_c}\right), \ \gamma \geq 0. \tag{19.58}$$

Upon substituting Equation 19.58 into Equation 19.57 and upon simplifying the integral, it can be shown that the BER of the MC-CDMA system using MRC can be written as:

$$P_b = \left[\frac{(1-\mu)}{2}\right]^{N_p} \sum_{n=0}^{N_p-1} \binom{N_p - 1 + n}{n} \left[\frac{(1+\mu)}{2}\right]^n, \tag{19.59}$$

where, by definition:

$$\mu = \sqrt{\frac{\overline{\gamma}_c}{1 + \overline{\gamma}_c}}. \tag{19.60}$$

When the average SNR of $\overline{\gamma}_c$ satisfies the condition $\overline{\gamma}_c >> 1$, we have $\frac{(1+\mu)}{2} \approx 1$ and $\frac{(1-\mu)}{2} \approx \frac{1}{4\overline{\gamma}_c}$ [90]. Furthermore, we have:

$$\sum_{n=0}^{N_p-1} \binom{N_p - 1 + n}{n} = \binom{2N_p - 1}{N_p}. \tag{19.61}$$

Therefore, when $\overline{\gamma}_c$ is sufficiently high, the BER of Equation 19.59 can be formulated as:

$$P_b \approx \left(\frac{1}{4\overline{\gamma}_c}\right)^{N_p} \binom{2N_p - 1}{N_p}. \tag{19.62}$$

According to Equation 19.62 the probability of error for the MC-CDMA system shown in Figure 19.19 varies as $1/\overline{\gamma}_c$ raised to the N_pth power. Thus, when using frequency-domain spreading in the MC-CDMA system employing a spreading code of length N_p, the bit error rate decreases inversely proportionately with the N_pth power of the SNR.

Figure 19.21 shows the BER performance of the MC-CDMA system despicted in Figure 19.19 using frequency-domain spreading. The BER results were computed as a function of the SNR per bit, namely E_b/N_0, using parameters of $q = 4$, $N_p = 32$ and the number of users of $K = 1, 5, 10, 20$. From the results we observe that the BER performance degrades, as the number of users increases. Furthermore, as expected, the BER decreases as the SNR per bit increases. However, when the number of users supported is sufficiently high, the BER decreases slowly, as shown by the curve associated with $K = 20$ marked by the 'triangles'.

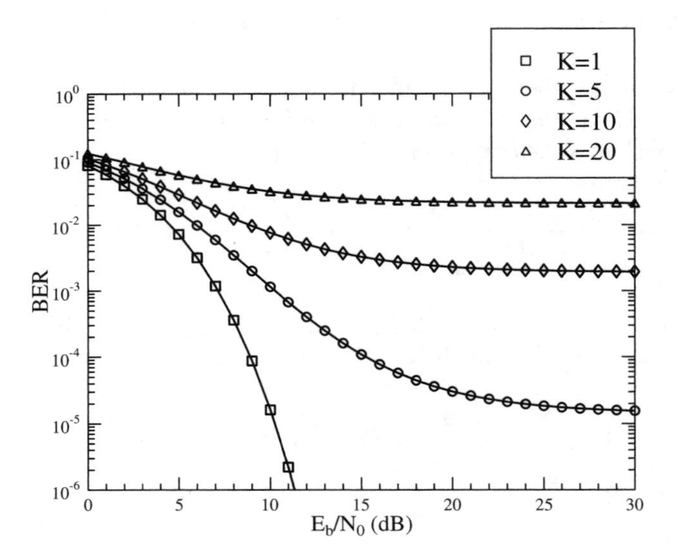

Figure 19.21: BER versus SNR per bit performance for the MC-CDMA system using frequency-domain spreading over Rayleigh fading channels with parameters $q = 4$ number of bits per symbol, processing gain of $N_p = 32$ and number of users $K = 1, 5, 10, 20$. The results were computed from Equation 19.59.

This can be also seen in Equation 19.59 and Equation 19.60. According to Equation 19.60, if the number of interfering users and the SNR per bit are sufficiently high, in Equation 19.59 we have $\mu \approx 1/\sqrt{2}$. Hence, for a given N_p, the BER remains approximately constant according to Equation 19.59 for $\mu \approx 1/\sqrt{2}$.

19.5 Performance of Overlapping Multicarrier DS-CDMA Systems

19.5.1 Preliminaries

In Sections 19.2.3 and 19.2.4 we have reviewed the basic characteristics of two orthogonal multicarrier DS-CDMA systems, namely those of the multicarrier DS-CDMA-II, and the multitone DS-CDMA systems seen in Figures 19.7 and 19.10. In the multitone DS-CDMA system of Figure 19.10, the subcarrier frequencies are chosen to be orthogonal to each other with minimum frequency separation before the associated DS spreading, which is formulated as:

$$\int_0^{T_s} \cos(2\pi f_i t + \phi_i) \cdot \cos(2\pi f_j t + \phi_j) dt = 0, \text{ for } i \neq j. \tag{19.63}$$

Hence, the spacing Δ between two adjacent subcarrier frequencies is $\Delta = 1/T_s$, where T_s is the multitone DS-CDMA signal's symbol duration. The subcarrier frequencies, therefore, take the values of $f_0 + i/T_s$ for $i = 0, 1, \ldots, U-1$, where f_0 is the main carrier frequency. In contrast to the multitone DS-CDMA system of Figure 19.10, in the orthogonal multicarrier DS-CDMA-II system of Figure 19.7, the subcarrier frequencies are chosen to be orthogonal to each other with a minimum frequency separation after the DS spreading, which can be written as:

$$\int_0^{T_c} \cos(2\pi f_i t + \phi_i) \cdot \cos(2\pi f_j t + \phi_j) dt = 0, \ \text{for } i \neq j, \tag{19.64}$$

implying that the spacing Δ between two adjacent subcarrier frequencies is $\Delta = 1/T_c$, as shown in Figure 19.8, where T_c is the chip duration of the DS spreading codes and the subcarrier frequencies assume the value of $f_0 + i/T_c$ for $i = 0, 1, \ldots, U-1$.

Let $N_e = T_s/T_c$ be the spreading gain of the DS spread subcarrier signals and we assume that each subcarrier signal has the same 'null-to-null' bandwidth of $2/T_c$. Then, it can be shown that the condition of Equation 19.63, which was stipulated for the multitone DS-CDMA system of Figure 19.10, in fact, includes the orthogonality condition of Equation 19.64 stated in the context of the multicarrier DS-CDMA-II systems of Figure 19.7. This observation can be proven by setting $f_i = f_j + N_e/T_s$ in Equation 19.63, yielding:

$$\int_0^{T_s} \cos(\frac{2\pi}{T_c} t + \psi_i) dt = 0, \tag{19.65}$$

after a few steps, where $\psi_i = \phi_i - \phi_j$. Note that, in Equation 19.65 we used $N_e/T_s = 1/T_c$. To proceed further, Equation 19.65 can be extended to:

$$\sum_{l=0}^{N_e-1} \int_{lT_c}^{(l+1)T_c} \cos(\frac{2\pi}{T_c} t + \psi_i) dt = 0. \tag{19.66}$$

According to Equation 19.66 it can be shown that each term of the sum is zero and hence we have:

$$\int_0^{T_c} \cos(\frac{2\pi}{T_c} t + \psi_i) dt = 0, \tag{19.67}$$

which reflects the orthogonality between subcarrier frequencies having minimum frequency separation after DS spreading. In other words, Equation 19.63 constitutes the orthogonality condition of the subcarrier frequencies in the multicarrier DS-CDMA-II system of Figure 19.8.

Furthermore, it can readily be shown that the orthogonality condition of (19.63 is obeyed, whenever the spacing Δ takes the form of $\Delta = \lambda/T_s$, $\lambda = 1, 2, \ldots$, where λ is referred to as the normalized spacing between two adjacent subcarriers. The multicarrier DS-CDMA scheme belongs to the family of multitone DS-CDMA arrangements having the spectrum of Figure 19.11, if $\lambda = 1$, while to the class of orthogonal multicarrier DS-CDMA systems with spectrum shown in Figure 19.8, if $\lambda = N_e$. Furthermore, there exists no overlap between the mainlobes of the modulated subcarrier signals after DS spreading, when $\lambda = 2N_e$, which is the bandwidth requirement of Figure 19.11 for the multicarrier DS-CDMA system proposed in [472].

Based on the above observations both the multitone DS-CDMA system and the orthogonal multicarrier DS-CDMA system can be viewed as a member of the class of generalized multicarrier DS-CDMA systems having arbitrary subcarrier spacing of $\lambda \in \{1, 2, \ldots, \}$. Hence, the above generalized multicarrier DS-CDMA system model includes a number of specific multicarrier DS-CDMA schemes. Furthermore, based on the analysis of this general model, the results generated can be extended to different multicarrier DS-CDMA systems by simply varying single parameter, namely λ. Finally, the subcarrier spacing λ can be optimized according to specific design requirements tailored to the communication environments encountered, in order to achieve the optimum performance in term of λ. For example, for a given total system bandwidth, λ can be optimized, in order to minimize the multi-user interference, since λ has an influence on both the overlap of the modulated signals of the subcarriers and on the processing gain. In this context a clear trade-off exists between the overlap and the processing gain. On the one hand, if λ is low – for example $\lambda = 1$ – in the context of multitone DS-CDMA, then a subcarrier signal will overlap with a high number of subcarrier signals of both the same user and with those of the interfering users. On the other hand, given a total bandwidth and a low value of λ, a high spreading gain can be maintained, which leads to the reduction of the multi-user interference. By contrast, if λ is high – for example $\lambda = 2N_e$ – which means that there exists no spectral overlap between the mainlobes of the subcarrier signals, then the modulated subcarrier signals benefit from a low interference inflicted by the other subcarrier signals of both the reference and the interfering users. However, in this case the spreading gain of each subcarrier signal is low, which leads to the increase of the multi-user interference. The influence of the subcarrier spacing λ on both the spreading gain and on the spectral overlap of the subcarrier signals highlights that there exists an optimum spacing λ_{opt}, that may minimize the multi-user interference inflicted upon each of the subcarrier signals. Another example in the context of optimizing λ is that the subcarrier spacing λ can be set appropriately, in order to match the receiver requirements. For example, assuming that the receiver employs a three-finger RAKE receiver. Then a specific spacing λ can be selected such that the number of resolvable paths associated with each subcarrier becomes three in the propagation environment typically encountered, since in this case, in addition to achieving the diversity gain, the receiver can combine all the energy scattered over the multipath components.

Hence, in the remainder of section we investigate the performance of the generalized overlapping multicarrier DS-CDMA system having an arbitrary spacing of Δ, when transmitting over frequency selective Rayleigh fading channels. We commence by first considering the system model and the parameters involved in our analysis.

19.5.2 System Description

19.5.2.1 Transmitted Signal

The transmitter schematic of the overlapping multicarrier DS-CDMA system considered is the same as that seen in Figure 19.10. Hence, the transmitted signal can be expressed as:

$$s_k(t) = \sum_{u=1}^{U} \sqrt{2P} b_{ku}(t) c_k(t) \cos(2\pi f_u t + \phi_{ku}), \tag{19.68}$$

where P is the transmitted power of each subcarrier signal, $b_{ku}(t)$ represents the uth data substream of user k after the serial-to-parallel conversion, $b_{ku}(t) = \sum_{i=-\infty}^{\infty} b_{ui}^{(k)} P_{T_s}(t - iT_s)$ consists of a sequence of mutually independent rectangular signalling pulses of duration T_s and of amplitude +1 or -1 with equal probability. Furthermore, $c_k(t) = \sum_{j=-\infty}^{\infty} c_j^{(k)} P_{T_c}(t - jT_c)$ denotes the random spreading code waveform of the kth user, where $P_\tau(t) = 1$ for $0 \leq t \leq \tau$ and equals zero otherwise, where $c_j^{(k)}$ takes values of +1 and -1 with equal probability. The variable ϕ_{ku} represents the modulation phase imposed on the uth subcarrier of user k. Finally, $\{f_u, u = 1, 2, \ldots, U\}$ are the subcarrier frequencies, which are arranged according to:

$$f_u = f_0 + \frac{\lambda(u-1)}{T_s}, \quad u = 1, 2, \ldots, U, \tag{19.69}$$

where $\lambda = 1, 2, \ldots, 2N_e$ corresponding to $\Delta = 1/T_s, 2/T_s, \ldots, 2N_e/T_s$, if we assume that the maximum spacing between two adjacent subcarriers is $2N_e$. The spectrum of the overlapping multicarrier DS-CDMA signal is shown in Figure 19.22, where W_s is the bandwidth of the system considered, T_{c1} is the chip-duration corresponding to a single-carrier DS-CDMA system, while W_{ds} represents the null-to-null bandwidth of the subcarrier signals.

According to Figure 19.22 the system's total transmission bandwidth, the subcarrier spacing Δ and the bandwidth of the subcarrier signal obey the following relationship:

$$W_s = (U-1)\Delta + W_{ds}. \tag{19.70}$$

This in turn means that:

$$\frac{2}{T_{c1}} = (U-1)\frac{\lambda}{T_s} + \frac{2}{T_c}. \tag{19.71}$$

Since $T_s = UT_b = UNT_{c1}$ and $T_s = N_e T_c$, after substituting these relationships into Equation 19.71, the spreading gain associated with each subcarrier signal can be expressed as:

$$N_e = UN - \frac{(U-1)\lambda}{2}. \tag{19.72}$$

According to Equation 19.72, it can be shown that the spreading gain of the multicarrier

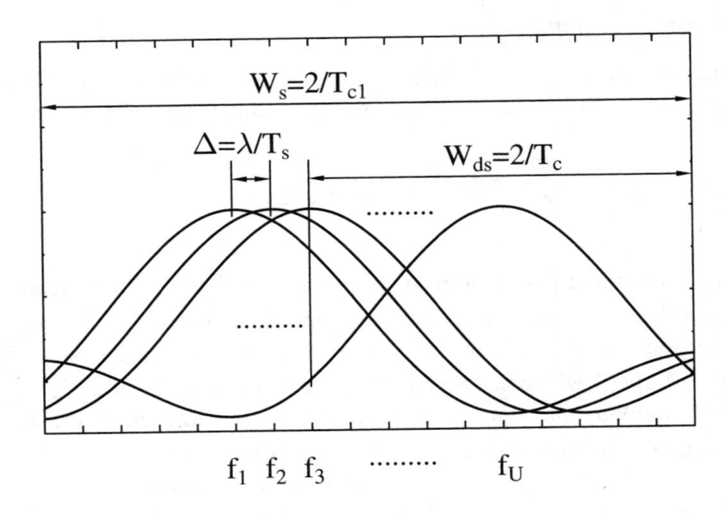

Figure 19.22: Spectrum of the overlapping multicarrier DS-CDMA signals.

signal corresponding to the normalized spacing of $\lambda = 1$, i.e. that of the multitone DS-CDMA system [482], is $N_e = \left(UN - \frac{U-1}{2}\right)$, as shown in Equation 19.14. The spreading gain of the system associated with $\lambda = N_e$, i.e. that of the orthogonal multicarrier DS-CDMA system of Figure 19.7 [65], is $N_e = 2UN/(U+1)$, as shown in Equation 19.10.

Let L_1 be the number of resolvable paths of a corresponding single-carrier DS-CDMA system. Then the number of resolvable paths associated with the subcarrier signals in the overlapping multicarrier DS-CDMA system having spectrum shown in Figure 19.22 can be approximated by:

$$L \approx \lfloor \frac{2N_e L_1}{2N_e + (U-1)\lambda} \rfloor. \tag{19.73}$$

Let us assume that we support K asynchronous CDMA users in the system and all of them use the same parameters U and N_e. The average power received from each user at the base station is also assumed to be the same, corresponding to the perfect power control assumption. Consequently, when K signals described by Equation 19.68 are transmitted over the frequency selective fading channels characterized by Equation 19.21, the received signal at the base station can be formulated as:

$$r(t) = \sum_{k=1}^{K} \sum_{u=1}^{U} \sum_{l=1}^{L} \sqrt{2P} \alpha_{ul}^{(k)} b_{ku}(t - \tau_{kl}) c_k(t - \tau_{kl}) \cos\left(2\pi f_u t + \varphi_{ul}^{(k)}\right) + n(t), \tag{19.74}$$

where $\varphi_{ul}^{(k)} = \phi_{ku} + \psi_{ul}^{(k)} - 2\pi f_u \tau_{kl}$ and $\psi_{ul}^{(k)}$ is contributed by the channel.

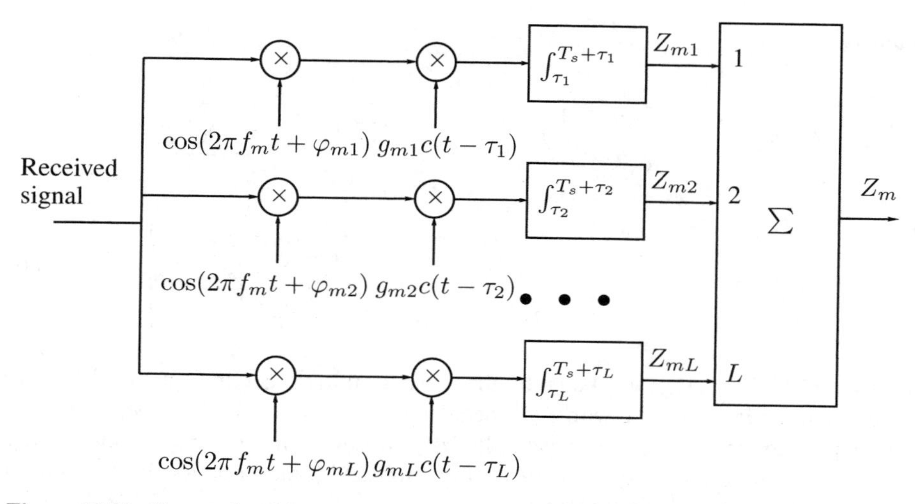

Figure 19.23: The receiver block diagram of the overlapping multicarrier DS-CDMA system.

19.5.2.2 Receiver Model

Let us assume that we want to receive the signal transmitted by the first user, which is treated as the reference user or reference signal in our analysis, and that the receiver is capable of acquiring perfect time-domain synchronization with each subcarrier signal of the reference user. The receiver considered is a conventional correlator-based RAKE receiver, as shown in Figure 19.12. The mth branch of Figure 19.12 corresponding to the mth subcarrier is shown in Figure 19.23. Referring to Figure 19.23, the decision variable Z_m associated with the 0th data bit corresponding to the mth substream of the reference user can be written as:

$$Z_m = \sum_{v=1}^{L} Z_{mv}, \; m = 1, 2, \ldots, U, \tag{19.75}$$

$$Z_{mv} = \int_{\tau_v}^{T_s+\tau_v} r(t) \cdot g_{mv} c(t-\tau_v) \cos(2\pi f_m t + \varphi_{mv})\, dt, \tag{19.76}$$

where g_{mv} is a parameter controlling which type of combining scheme is used, such as MRC, EGC or SC, respectively. However, in this section only the performance of the system using MRC is investigated, hence, $g_{mv} = \alpha_{mv}$ is assumed. Moreover, in Equation 19.75 the subscripts and superscripts associated with the reference user are omitted for notational simplicity. Furthermore, without loss of any generality, we let $\tau_v = 0$ in the following analysis. Upon substituting Equation 19.74 into Equation 19.76, it can be shown that Z_{mv} can be written as:

$$Z_{mv} = \sqrt{\frac{P}{2}}T_s \left\{ D_{mv} + \eta_{mv} + \sum_{\substack{l=1\\l\neq v}}^{L} I_1^{(s)} + \sum_{\substack{u=1\\u\neq m}}^{U}\sum_{\substack{l=1\\l\neq v}}^{L} I_2^{(s)} + \sum_{k=2}^{K}\sum_{l=1}^{L} I_1^{(k)} + \sum_{k=2}^{K}\sum_{\substack{u=1\\u\neq m}}^{U}\sum_{l=1}^{L} I_2^{(k)} \right\}. \tag{19.77}$$

In Equation 19.77, η_{mv} is contributed by the noise term $n(t)$ seen in Equation 19.74, which is a Gaussian random variable with zero mean and variance $g_{mv}^2 N_0/2E_b$, where $E_b = PT_s$ represents the energy per bit. The variable D_{mv} in Equation 19.77 represents the desired output given by combining Equation 19.74, Equation 19.76 and Equation 19.77 upon setting $k = 1, l = v$ and $u = m$, which yields:

$$D_{mv} = b_m[0]\alpha_{mv}g_{mv}. \tag{19.78}$$

The output of the correlator matched to subcarrier m, corresponding to the vth path of the reference user contains four types of interference in Equation 19.77. The interference of $I_1^{(s)}$ is contributed by the path $l,\ l = 1, 2, \ldots, L,\ l \neq v$, on the same subcarrier m received from the reference user, which can be written as:

$$I_1^{(s)} = \frac{\alpha_{ml}g_{mv}\cos\theta_{ml}}{T_s}\int_0^{T_s} b_m(t-\tau_l)c(t-\tau_l)c(t)dt, \tag{19.79}$$

where $\theta_{ml} = \varphi_{ml} - \varphi_{mv}$, which is a random variable uniformly distributed in $[0, 2\pi)$. With the aid of the partial auto-correlation functions given below, Equation 19.79 can be expressed as:

$$I_1^{(s)} = \frac{\alpha_{ml}g_{mv}\cos\theta_{ml}}{T_s}\left[b_m[-1]R_1(\tau_l) + b_m[0]\hat{R}_1(\tau_l)\right], \tag{19.80}$$

where $R_1(\tau_l)$ and $\hat{R}_1(\tau_l)$ are the partial auto-correlation functions defined as:

$$R_1(\tau_l) = \int_0^{\tau_l} c(t-\tau_l)c(t)dt, \tag{19.81}$$

$$\hat{R}_1(\tau_l) = \int_{\tau_l}^{T_s} c(t-\tau_l)c(t)dt. \tag{19.82}$$

The interference term of $I_2^{(s)}$ in Equation 19.77 is contributed by the path $l,\ l = 1, 2, \ldots, L,\ l \neq v$, on the subcarrier $u,\ u = 1, 2, \ldots, U,\ u \neq m$ by the reference user,[3]

[3] Due to the orthogonality between the subcarrier signals from the same path and the same user, i.e.,

which can be formulated as:

$$I_2^{(s)} = \frac{\alpha_{ul} g_{mv}}{T_s} \left[b_u[-1] R_1(\tau_l, \theta_{ul}) + b_u[0] \hat{R}_1(\tau_l, \theta_{ul}) \right], \tag{19.83}$$

where $\theta_{ul} = \varphi_{ul} - \varphi_{mv}$. Due to the difference of the frequencies associated with the subcarriers u and m, the corresponding partial auto-correlation functions are now defined as:

$$R_1(\tau_l, \theta_{ul}) = \int_0^{\tau_l} c(t - \tau_l) c(t) \cos(2\pi (f_u - f_m) t + \theta_{ul}) dt, \tag{19.84}$$

$$\hat{R}_1(\tau_l, \theta_{ul}) = \int_{\tau_l}^{T_s} c(t - \tau_l) c(t) \cos(2\pi (f_u - f_m) t + \theta_{ul}) dt. \tag{19.85}$$

With the aid of Equation 19.69, Equation 19.84 and Equation 19.85 can then be written as:

$$R_1(\tau_l, \theta_{ul}) = \int_0^{\tau_l} c(t - \tau_l) c(t) \cos\left(\frac{2\pi\lambda(u - m)t}{T_s} + \theta_{ul} \right) dt, \tag{19.86}$$

$$\hat{R}_1(\tau_l, \theta_{ul}) = \int_{\tau_l}^{T_s} c(t - \tau_l) c(t) \cos\left(\frac{2\pi\lambda(u - m)t}{T_s} + \theta_{ul} \right) dt. \tag{19.87}$$

The multi-user interference term $I_1^{(k)}$ in Equation 19.77 is due to path l, $l = 1, 2, \ldots, L$ on the same subcarrier m inflicted by the interfering users for which we have $k \neq 1$. Then, $I_1^{(k)}$ in Equation 19.77 can be written as:

$$I_1^{(k)} = \frac{\alpha_{ml}^{(k)} g_{mv} \cos \theta_{ml}^{(k)}}{T_s} \left[b_{km}[-1] R_k(\tau_{kl}) + b_{km}[0] \hat{R}_k(\tau_{kl}) \right], \tag{19.88}$$

where $\theta_{ml}^{(k)} = \varphi_{ml}^{(k)} - \varphi_{mv}$ is a random variable uniformly distributed in $[0, 2\pi)$, $R_k(\tau_{kl})$ and $\hat{R}_k(\tau_{kl})$ are the partial cross-correlation functions defined as:

$$R_k(\tau_{kl}) = \int_0^{\tau_{kl}} c_k(t - \tau_{kl}) c(t) dt, \tag{19.89}$$

$$\hat{R}_k(\tau_{kl}) = \int_{\tau_{kl}}^{T_s} c_k(t - \tau_{kl}) c(t) dt. \tag{19.90}$$

Finally, the multi-user interference term $I_2^{(k)}$ in Equation 19.77 is due to path l, $l = 1, 2, \ldots, L$ on the subcarrier u, $u = 1, 2, \ldots, U$; $u \neq m$ inposed by the interfering users

$\int_0^{T_s} \cos(2\pi f_i + \varphi_{il}) \cos(2\pi f_j + \varphi_{jl}) dt = 0$, the interference of $I_2^{(s)}$ due to the path v on the subcarrier u, $u = 1, 2, \ldots, U$, $u \neq m$ from the reference user is zero. Hence, the associated terms are excluded by letting $l \neq v$ in Equation 19.77.

associated with $k \neq 1$. Hence, $I_2^{(k)}$ in Equation 19.77 can be expressed as:

$$I_2^{(k)} = \frac{\alpha_{ul}^{(k)} g_{mv}}{T_s} \left[b_{ku}[-1] R_k(\tau_{kl}, \theta_{ul}^{(k)}) + b_{ku}[0] \hat{R}_k(\tau_{kl}, \theta_{ul}^{(k)}) \right], \quad (19.91)$$

where $\theta_{ul}^{(k)} = \varphi_{ul}^{(k)} - \varphi_{mv}$ is a random variable uniformly distributed in $[0, 2\pi)$. The associated partial cross-correlation functions are defined as:

$$R_k(\tau_{kl}, \theta_{ul}^{(k)}) = \int_0^{\tau_{kl}} c_k(t - \tau_{kl}) c(t) \cos\left(\frac{2\pi\lambda(u-m)t}{T_s} + \theta_{ul}^{(k)} \right) dt, \quad (19.92)$$

$$\hat{R}_k(\tau_{kl}, \theta_{ul}^{(k)}) = \int_{\tau_{kl}}^{T_s} c_k(t - \tau_{kl}) c(t) \cos\left(\frac{2\pi\lambda(u-m)t}{T_s} + \theta_{ul}^{(k)} \right) dt. \quad (19.93)$$

Having analyzed the receiver model and the decision variable, in the next subsection the multipath interference imposed by the reference user and the multi-user interference inflicted by the interfering users will be analyzed.

19.5.3 Interference Analysis

In this subsection the multipath interference engendered by the reference user itself and the multi-user interference will be evaluated by assuming that all these interference sources are Gaussian distributed and hence can be treated as independent additional noise. According to Equation 19.80, Equation 19.83, Equation 19.88 and Equation 19.91, the interference terms $I_1^{(s)}$, $I_2^{(s)}$ and $I_1^{(k)}$ constitute the special cases of the term $I_2^{(k)}$ in Equation 19.91. Specifically, if we set $u = m$ in Equation 19.91, we obtain Equation 19.88. If we let $k = 1$ in Equation 19.91, then we get Equation 19.83. Finally, if we let $k = 1$ and $u = m$, then we obtain Equation 19.80. Hence, we can analyze the multipath interference engendered by the reference user as well as the multi-user interference by first analyzing the multi-user interference term of Equation 19.91. Based on the standard Gaussian approximation [103, 483], the interference term $I_2^{(k)}$ of Equation 19.91 can be modelled as a Gaussian random variable with zero mean and a variance expressed as:

$$\text{Var}\left[I_2^{(k)}\right] = \frac{\Omega g_{mv}^2}{T_s^2} \left\{ E_{\tau_{kl}, \theta_{ul}^{(k)}} \left[R_k^2\left(\tau_{kl}, \theta_{ul}^{(k)}\right) \right] + E_{\tau_{kl}, \theta_{ul}^{(k)}} \left[\hat{R}_k^2\left(\tau_{kl}, \theta_{ul}^{(k)}\right) \right] \right\}, \quad (19.94)$$

where $\Omega = E\left[\left(\alpha_{ul}^{(k)}\right)^2\right]$, $E_{\tau_{kl}, \theta_{ul}^{(k)}}[\,]$ represents taking the expected value with respect to the random variables τ_{kl} and $\theta_{ul}^{(k)}$. In our following discourse, we analyze the partial cross-correlation functions of Equation 19.92 and Equation 19.93. Specifically, we will evaluate their expectations by assuming that the phase angles and time delays are modelled as mutually independent random variables, each of which is uniformly distributed over the appropriate interval. Below we only analyze the expectation of $R_k^2\left(\tau_{kl}, \theta_{ul}^{(k)}\right)$, the expectation of $\hat{R}_k^2\left(\tau_{kl}, \theta_{ul}^{(k)}\right)$ can be obtained following the same approach.

Equation 19.92 can be written as [483]:

$$
\begin{aligned}
R_k(\tau,\theta) &= \sum_{j=0}^{h} a_i(j)a_k(N_e-h-1+j)\int_{jT_c}^{jT_c+\tau-hT_c}\cos\left(\frac{2\pi\lambda(u-m)t}{T_s}+\theta\right)dt \\
&\quad+\sum_{j=0}^{h-1} a_i(j)a_k(N_e-h+j)\int_{jT_c+\tau-hT_c}^{(j+1)T_c}\cos\left(\frac{2\pi\lambda(u-m)t}{T_s}+\theta\right)dt,
\end{aligned}
\tag{19.95}
$$

where $hT_c \leq \tau_{kl} < (l+1)T_c$ and N_e represents the number of chips per symbol period or the spreading factor of the subcarrier signal. Furthermore, in Equation 19.95 we set $\tau_{kl} = \tau$ and $\theta_{ul}^{(k)} = \theta$ for simplicity. Upon evaluating the integrals in Equation 19.95, we find that:

$$
\begin{aligned}
R_k(\tau,\theta) &= (\tau-hT_c)\mathrm{sinc}\left[\frac{\pi(u-m)\lambda(\tau-hT_c)}{T_s}\right] \\
&\quad\cdot\sum_{j=0}^{h} a_i(j)a_k(N_e-h-1+j)\cos\beta_{ka}(j,\tau) \\
&\quad+[(h+1)T_c-\tau]\mathrm{sinc}\left[\frac{\pi(u-m)\lambda((h+1)T_c-\tau)}{T_s}\right] \\
&\quad\cdot\sum_{j=0}^{h-1} a_i(j)a_k(N_e-h+j)\cos\beta_{kb}(j,\tau),
\end{aligned}
\tag{19.96}
$$

where $\mathrm{sinc}(x) = \sin(x)/x$, and

$$
\begin{aligned}
\beta_{ka}(j,\tau) &= \frac{\pi(u-m)\lambda}{T_s}(2jT_c+\tau-hT_c)+\theta, \\
\beta_{kb}(j,\tau) &= \frac{\pi(u-m)\lambda}{T_s}((2j+1)T_c+\tau-hT_c)+\theta.
\end{aligned}
$$

The expression of $R^2(\tau,\varphi)$ is computed from Equation 19.96 as:

$$\begin{aligned}
R_k^2(\tau,\theta) = {} & (\tau - hT_c)^2 \mathrm{sinc}^2\left[\frac{\pi(u-m)\lambda(\tau-hT_c)}{T_s}\right] \\
& \cdot\left[\sum_{j=0}^{h} a_i^2(j)a_k^2(N_e-h-1+j)\cos^2\beta_{ka}(j,\tau)\right. \\
& +\sum_{r=0}^{h}\sum_{\substack{s=0\\ s\neq r}}^{h} a_i(r)a_k(N_e-h-1+r)a_i(s)a_k(N_e-h-1+s) \\
& \left.\cdot\cos\beta_{ka}(r,\tau)\cos\beta_{ka}(s,\tau)\right] \\
& +(\tau-hT_c)[(h+1)T_c-\tau]\mathrm{sinc}\left[\frac{\pi(u-m)\lambda(\tau-hT_c)}{T_s}\right] \\
& \cdot\mathrm{sinc}\left[\frac{\pi(u-m)\lambda((h+1)T_c-\tau)}{T_s}\right] \\
& \cdot\left[\sum_{r=0}^{h}\sum_{s=0}^{h-1} a_i(r)a_k(N_e-h-1+r)a_i(s)a_k(N_e-h+s)\right. \\
& \cdot\cos\beta_{ka}(r,\tau)\cos\beta_{kb}(s,\tau) \\
& +\sum_{r=0}^{h-1}\sum_{s=0}^{h} a_i(r)a_k(N_e-h+r)a_i(s)a_k(N_e-h-1+s) \\
& \left.\cdot\cos\beta_{kb}(r,\tau)\cos\beta_{ka}(s,\tau)\right] \\
& +[(h+1)T_c-\tau]^2\mathrm{sinc}^2\left[\frac{\pi(u-m)\lambda((h+1)T_c-\tau)}{T_s}\right] \\
& \cdot\left[\sum_{r=0}^{h-1} a_i^2(r)a_k^2(N_e-h+r)\cos^2\beta_{kb}(r,\tau)\right. \\
& +\sum_{r=0}^{h-1}\sum_{\substack{s=0\\ s\neq r}}^{h-1} a_i(r)a_k(N_e-h+r)a_i(s)a_k(N_e-h+s) \\
& \left.\cdot\cos\beta_{kb}(r,\tau)\cos\beta_{kb}(s,\tau)\right]. \qquad (19.97)
\end{aligned}$$

Since we assumed that random spreading codes are employed in our system, $a_i(r)$ is statistically independent of $a_k(s)$ when $i \neq k$ or $r \neq s$, and takes values of +1 or -1 with equal probability. Hence, upon taking the expectation of $R_k^2(\tau,\theta)$ with respect to $a_i(r)$,

$a_k(s)$, θ and τ we arrive at:

$$\begin{aligned}
\mathrm{E}_{\tau,\theta}\left[R_k^2(\tau,\theta)\right] &= \frac{1}{2T_s}\sum_{h=0}^{N_e-1} \\
&\cdot \int_{hT_c}^{(h+1)T_c}\left\{(\tau - hT_c)^2\mathrm{sinc}^2\left[\frac{\pi(u-m)\lambda(\tau-hT_c)}{T_s}\right]\cdot(h+1)\right. \\
&\left.+[(h+1)T_c-\tau]^2\mathrm{sinc}^2\left[\frac{\pi(u-m)\lambda((h+1)T_c-\tau)}{T_s}\right]\cdot h\right\}d\tau.
\end{aligned} \tag{19.98}$$

Upon evaluating the resulting integral we find that:

$$\mathrm{E}_{\tau,\theta}\left[R_k^2(\tau,\theta)\right] = \frac{N_e T_s^2}{4\pi^2(u-m)^2\lambda^2}\left[1-\mathrm{sinc}\left(\frac{2\pi(u-m)\lambda}{N_e}\right)\right]. \tag{19.99}$$

The corresponding expression for $\mathrm{E}_{\tau,\theta}\left[\hat{R}_k^2(\tau,\theta)\right]$ can be obtained by evaluating the expectation of $\hat{R}_k^2(\tau,\theta)$ in the same way, as for $R_k^2(\tau,\theta)$. We found that the result is the same as Equation 19.99, i.e.:

$$\mathrm{E}_{\tau,\theta}\left[\hat{R}_k^2(\tau,\theta)\right] = \frac{N_e T_s^2}{4\pi^2(u-m)^2\lambda^2}\left[1-\mathrm{sinc}\left(\frac{2\pi(u-m)\lambda}{N_e}\right)\right]. \tag{19.100}$$

Substituting Equation 19.99 and Equation 19.100 into Equation 19.94, we finally found that the variance of the multi-user interference $I_2^{(k)}$ can be formulated as:

$$\mathrm{Var}\left[I_2^{(k)}\right] = \frac{\Omega g_{mv}^2 N_e}{2\pi^2(u-m)^2\lambda^2}\left[1-\mathrm{sinc}\left(\frac{2\pi(u-m)\lambda}{N_e}\right)\right]. \tag{19.101}$$

Let $u - m = x$ and compute the limit $\lim_{x\to 0}\mathrm{Var}\left[I_2^{(k)}\right]$. Then we find that the variance of the interference $I_2^{(k)}$ equals to $\frac{\Omega g_{mv}^2}{3N_e}$. This is the interference variance, when a subcarrier signal of the reference user is totally overlapped by the subcarrier signal of the kth interfering user, i.e. the variance of $I_1^{(k)}$ in Equation 19.88, which can be expressed as:

$$\mathrm{Var}\left[I_1^{(k)}\right] = \frac{\Omega g_{mv}^2}{3N_e}. \tag{19.102}$$

By using a similar approach determining the term $I_2^{(k)}$, we can derive the variance of $I_2^{(s)}$, which was found to be the same as Equation 19.101. This result in fact is predictable, since we have assumed that random spreading codes are used, when the signal associated with each chip is assumed to be an i.i.d random variable. Moreover, it can be seen that Equation 19.101

is independent of the user index k. Hence, the variance of $I_2^{(s)}$ can be written as:

$$\text{Var}\left[I_2^{(s)}\right] = \frac{\Omega g_{mv}^2 N_e}{2\pi^2(u-m)^2\lambda^2}\left[1 - \text{sinc}\left(\frac{2\pi(u-m)\lambda}{N_e}\right)\right]. \quad (19.103)$$

Finally, the variance of $I_1^{(s)}$ can be obtained from Equation 19.103 by computing the limit expressed as $\lim_{u-m=x\to 0}\left(\text{Var}\left[I_2^{(s)}\right]\right)$, which results in:

$$\text{Var}\left[I_1^{(s)}\right] = \frac{\Omega g_{mv}^2}{3N_e}. \quad (19.104)$$

Before concluding this subsection, we note that the variance of the multi-user interference in the multitone DS-CDMA system of Figure 19.10 [482] can be obtained by setting $\lambda = 1$, while that in the orthogonal multicarrier DS-CDMA system of Figure 19.7 [65] can be obtained by setting $\lambda = N_e$. It can be shown that for the case of $\lambda = N_e$, the variance of $I_2^{(k)}$ can be expressed as:

$$\text{Var}\left[I_2^{(k)}\right] = \frac{\Omega g_{mv}^2}{2\pi^2(u-m)^2 N_e}, \quad (19.105)$$

if $u - m \neq 0$. Otherwise, if $u - m = 0$, it can be shown that:

$$\text{Var}\left[I_2^{(k)}\right] = \frac{\Omega g_{mv}^2}{3N_e}. \quad (19.106)$$

In summary, in this subsection we have analyzed the multipath interference engendered by the reference user as well as the multi-user interference inflicted by the interfering users. With the aid of the analysis in this subsection, the bit error rate performance of the overlapping multicarrier DS-CDMA system can now be derived.

19.5.4 Performance Analysis

In order to derive the BER expressions of the generalized multicarrier DS-CDMA system using overlapping subcarriers, we first have to derive the PDF of the associated decision variables. We have assumed that all the interference terms in Equation 19.77 are independent additive Gaussian variables, hence, the variable Z_{mv} of Equation 19.77 is a Gaussian variable with normalized mean given by Equation 19.78, and normalized variance given by:

$$\sigma^2 = \left[\frac{KL-1}{3N_e} + (U-1)(L-1)\overline{I}_s + (K-1)(U-1)L\overline{I}_M + \left(\frac{2\Omega E_b}{N_0}\right)^{-1}\right] \cdot \Omega g_{mv}^2, \quad (19.107)$$

where $\overline{I}_s$ and $\overline{I}_M$ are the average associated with the value of u and m, respectively, which can be expressed as:

$$\begin{aligned} \overline{I}_s &= \overline{I}_M \\ &= \frac{1}{U(U-1)} \sum_{m=1}^{U} \sum_{\substack{u=1 \\ u \neq m}}^{U} \frac{N_e}{2\pi^2 (u-m)^2 \lambda^2} \left[1 - \text{sinc}\left(\frac{2\pi(u-m)\lambda}{N_e} \right) \right] \end{aligned} \quad (19.108)$$

Upon substituting Equation 19.108 into Equation 19.107, we obtain that:

$$\sigma^2 = \left[\frac{KL-1}{3N_e} + (U-1)(KL-1)\overline{I}_M + \left(\frac{2\Omega E_b}{N_0} \right)^{-1} \right] \cdot \Omega g_{mv}^2. \quad (19.109)$$

Since Z_m of Equation 19.75 is the sum of L independent Gaussian variables, Z_m is also a Gaussian variable. Let $g_{mv} = \alpha_{mv}$, i.e. we assume that the correlator outputs of the same data bit are combined according to the MRC scheme. Then, the normalized mean of Z_m can be formulated as:

$$E[Z_m] = b_m[0] \sum_{v=1}^{L} \alpha_{mv}^2, \quad (19.110)$$

and the normalized variance of Z_m is given by:

$$\sigma^2 = \left[\frac{KL-1}{3N_e} + (U-1)(KL-1)\overline{I}_M + \left(\frac{2\Omega E_b}{N_0} \right)^{-1} \right] \cdot \Omega \sum_{v=1}^{L} \alpha_{mv}^2. \quad (19.111)$$

Consequently, the BER conditioned on encountering the subcarrier fading attenuation $\{\alpha_{mv}\}$ can be expressed as:

$$P_b(\gamma) = Q\left(\sqrt{2\gamma} \right), \quad (19.112)$$

where

$$\gamma = \overline{\gamma}_c \cdot \frac{1}{\Omega} \sum_{v=1}^{L} \alpha_{mv}^2, \quad (19.113)$$

and

$$\overline{\gamma}_c = \left[\frac{2(KL-1)}{3N_e} + 2(U-1)(KL-1)\overline{I}_M + \left(\frac{\Omega E_b}{N_0} \right)^{-1} \right]^{-1}. \quad (19.114)$$

As we have shown in Section 19.4, the average BER, P_b, can be derived by the weighted

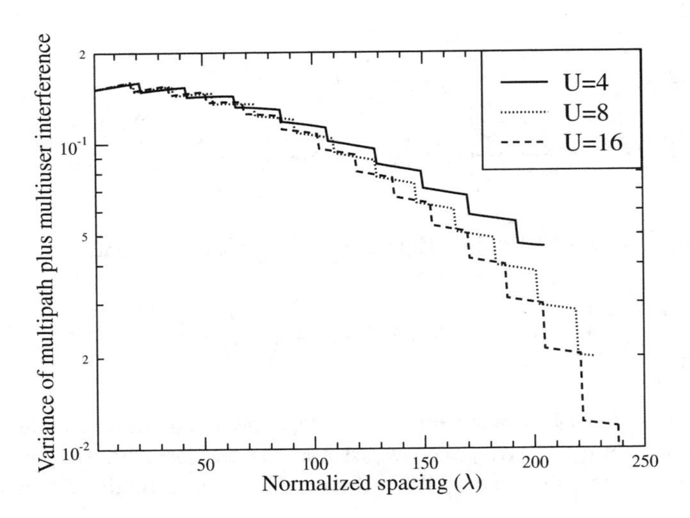

Figure 19.24: Variance of the multipath plus multi-user interference versus normalized subcarrier spacing λ performance for the overlapping multicarrier DS-CDMA system over the dispersive Rayleigh fading channels with parameters $N = 128, L_1 = 16$ and $K = 2$, where N and L_1 are the spreading gain and the number of resolvable paths of a corresponding single-carrier DS-CDMA system having the same data rate as well as the same system bandwidth. The results were generated be evaluating Equation 19.111.

averaging $P_b(\gamma)$ of Equation 19.113 over the valid range of γ, which can be expressed according to Equation 19.57. The PDF of γ can be derived from Equation 19.58 with N_p replaced by L. With the aid of these equations, finally, the average BER of the overlapping multicarrier DS-CDMA system using the MRC scheme can be written as:

$$P_b = \left[\frac{(1-\mu)}{2}\right]^L \sum_{n=0}^{L-1} \binom{L-1+n}{n} \left[\frac{(1+\mu)}{2}\right]^n, \qquad (19.115)$$

where μ is given by:

$$\mu = \sqrt{\frac{\overline{\gamma}_c}{1+\overline{\gamma}_c}}, \qquad (19.116)$$

and $\overline{\gamma}_c$ is given by Equation 19.114.

In Figure 19.24 the variance of the multipath interference plus the multi-user interference was estimated as a function of the normalized spacing between two adjacent subcarrier frequencies. The parameters employed were $N = 128, L_1 = 16, K = 2$ and $U = 4, 8, 16$, where N and L_1 were the spreading gain and the number of resolvable paths corresponding

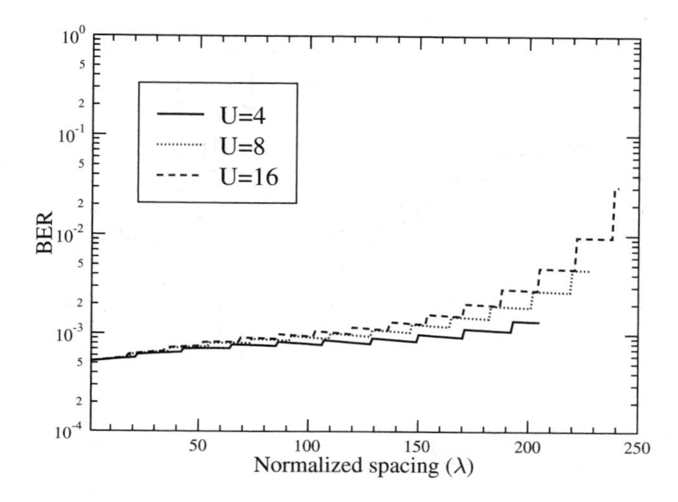

Figure 19.25: BER versus normalized subcarrier spacing λ performance for the overlapping multicarrier DS-CDMA system over the dispersive Rayleigh fading channels with parameters $N = 128, L_1 = 16, E_b/N_0 = 10dB$ and $K = 10$, where N and L_1 are the spreading gain and the number of resolvable paths of a corresponding single-carrier DS-CDMA system having the same data rate as well as the same system bandwidth.. The results were computed from Equation 19.115 by assuming that the receiver can combine all the resolvable paths.

to the single-carrier DS-CDMA system having the same data rate as well as the same system bandwidth. For a corresponding overlapping multicarrier DS-CDMA system having the same system bandwidth as the single-carrier DS-CDMA system, the spreading gain N_e, associated with each subcarrier signal and the number of resolvable paths, L of each subcarrier signal were given by Equation 19.72 and Equation 19.73, respectively. Note that, the system associated with $\lambda = 1$ corresponds to the multitone DS-CDMA scheme of Figure 19.10 [482], while the system using $\lambda = N_e$ corresponds to the orthogonal multicarrier DS-CDMA-II scheme of Figure 19.7 [65]. From the results seen in Figure 19.24 we find that the interference power engendered by the multipath signals and the multi-user signals decreases, as the spacing between two adjacent subcarrier frequencies becomes wider. This also implies that the multipath plus multi-user interference encountered by a given subcarrier signal in the multitone DS-CDMA scheme is higher than that experienced by a given subcarrier signal in the orthogonal multicarrier DS-CDMA-II scheme. Furthermore, observe in Figure 19.24 for a given value of λ, the variance of the multipath interference plus the multi-user interference decreases, as the number of subcarriers increases. Note that the variance curves in Figure 19.24 appear in the shape of steps. This is because the spreading gain and the number of resolvable paths associated with each subcarrier signal only assume discrete values according to Equation 19.72 and Equation 19.73.

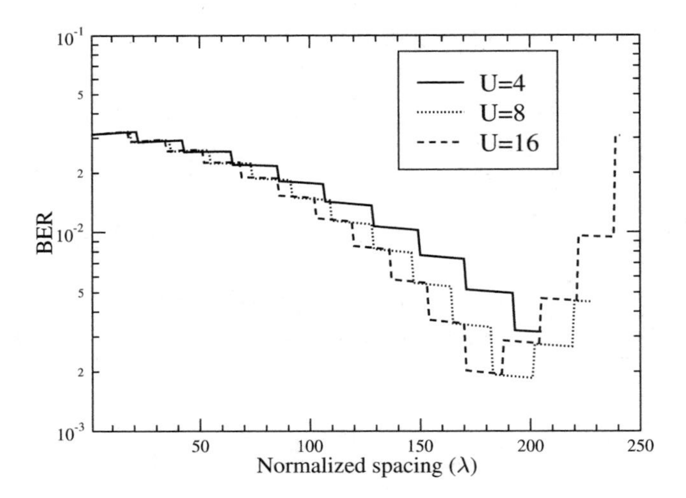

Figure 19.26: BER versus normalized subcarrier spacing λ performance for the overlapping multicarrier DS-CDMA system over the dispersive Rayleigh fading channels with parameters $N = 128, L_1 = 16, E_b/N_0 = 10dB$ and $K = 10$, where N and L_1 are the spreading gain and the number of resolvable paths of a corresponding single-carrier DS-CDMA system having the same data rate as well as the same system bandwidth. The results were computed from Equation 19.115 by assuming that the receiver can only combine at most five paths.

According to Equation 19.72 and Equation 19.73 both the spreading gain and the number of resolvable paths decreases, as the spacing between two adjacent subcarrier frequencies increases. Furthermore, according to Equation 19.115, for a given value of $\overline{\gamma}_c$, the BER will increase, as the number of diversity branches, L that can be resolved for each subcarrier decreases. This in turn implies that the BER will increase, as the spacing between two adjacent subcarrier frequencies increases. However, according to Figure 19.24, the interference engendered by both multipath and multi-user signals becomes less severe when increasing the normalized subcarrier spacing, λ. The resulting BER performance is determined by these two tendencies.

In Figure 19.25 the BER performance of the overlapping multicarrier DS-CDMA system was evaluated according to Equation 19.115 as a function of the normalized subcarrier spacing λ over the dispersive Rayleigh fading channels of Section 19.3. The associated parameters were $N = 128$, $L_1 = 16$, $K = 10$, $E_b/N_0 = 10dB$ and $U = 4,\ 8,\ 16$. The subcarrier spreading gain and the number of resolvable paths of each subcarrier signal were computed with the aid of Equation 19.72 and Equation 19.73. With the aid of Equation 19.73, it can be shown that the number of resolvable paths of each subcarrier decreases, as the normalized spacing, λ increases. We assumed that the receiver has the ability to combine all the resolvable paths associated with each subcarrier. From the results of Figure 19.25 we observe that, for any given number of subcarriers U, the BER increases, as the normalized spacing, λ, increases. However, the BER increase is less severe, when $\lambda < 200$. Therefore, over dispersive multipath Rayleigh fading channels diversity reception appears more important than the reduction of multipath and multi-user interference. According to the results of Figure 19.25, if the receiver can combine all the resolvable paths, the multitone DS-CDMA scheme of Figure 19.10 ($\lambda = 1$) outperforms the orthogonal multicarrier DS-CDMA scheme of Figure 19.7 ($\lambda = N_e$). Moreover, for any given value of λ, the overlapping multicarrier DS-CDMA scheme using a low number of subcarriers performs better than that using a high number of subcarriers.

If the complexity of the receiver is considered in terms of the diversity fingers of the RAKE receiver, it can be shown that the associated complexity of the receiver decreases, as the normalized subcarrier spacing, λ, increases. This is because, according to Equation 19.73, the number of resolvable paths decreases, as the normalized subcarrier spacing, λ, increases. However, due to the complexity limitation of the receiver, usually the receiver cannot combine an arbitrary number of resolvable paths. Hence, in Figure 19.26 the BER of the overlapping multicarrier DS-CDMA system was estimated by assuming that the receiver can combine at most five resolvable paths for each subcarrier. If there were more than five resolvable paths, we assumed that the receiver was capable of combining five paths. However, again, if there were less than five resolvable paths for each subcarrier, we assumed that the receiver was capable of combining all the resolvable paths. The other system parameters were the same as those used in Figure 19.25 and were listed in the caption of Figure 19.26. The corresponding results of Figure 19.26 were computed from Equation 19.115 with L replaced by the number of paths that the receiver was capable of combining. From the results of Figure 19.26 we find that, for the overlapping multicarrier DS-CDMA system having a sufficiently high number of subcarriers, both the multitone and orthogonal multicarrier DS-CDMA schemes are the suboptimum schemes. From Figure 19.26 we found that for $U = 8$ and $U = 16$, there exists an optimum normalized subcarrier spacing region in the range of $1 < \lambda < N_e$, where N_e is about 195 for $U = 8$ and 185 for $U = 16$. In this range of Figure 19.26 the overlapping multicarrier DS-CDMA scheme can achieve the minimum BER value. However, for the scheme using $U = 4$ the orthogonal multicarrier DS-CDMA system associated with $\lambda = N_e = 204$ has the best BER performance, since the number of resolvable paths at $\lambda = N_e = 204$ is still $L = 6$ according to Equation 19.73, which is higher than the diversity combining capability of the receiver.

19.6 Performance of Multicarrier DS-CDMA-I Systems

19.6.1 Decision Variable Statistics

In this section we analyze the performance of the multicarrier DS-CDMA system of Figure 19.4 described in Subsection 19.2.2, which we referred to as the multicarrier DS-CDMA-I system. In this system a data sequence is multiplied by a spreading code and then modulates U subcarriers. Observe in Figure 19.4 that this scheme does not include serial-to-parallel data conversion. Each subcarrier signal is direct sequence (DS) spread using a common spreading sequence, as shown in Figure 19.4. Therefore, the symbol duration of the multicarrier DS-CDMA signal is the same as that of the input data bits. The transmitted signal of user k in the multicarrier DS-CDMA-I system can be formulated as:

$$s_k(t) = \sqrt{\frac{2P}{U}} \sum_{u=1}^{U} b_k(t)c_k(t)\cos(2\pi f_u t + \phi_{ku}), \qquad (19.117)$$

where P is the transmitted power of the multicarrier DS-CDMA-I signal, U is the number of subcarriers, while $b_k(t)$ and $c_k(t)$ are the baseband data sequence and the spreading waveforms, respectively, as we discussed previously in Section 19.5. Finally, f_u and ϕ_{ku} for $u = 1, 2, \ldots, U$ are the subcarrier frequencies and modulation phases.

The users transmit over a slowly varying frequency selective Rayleigh fading channel having a delay spread of T_m and coherence bandwidth $(\Delta B)_c$, which are related by $(\Delta B)_c \approx 1/T_m$. Assuming that the spacing between two adjacent subcarrier frequencies is $\Delta = \frac{2}{T_c}$, where T_c is the chip duration of the spreading codes, no overlap exists between the mainlobes of the subcarrier signals. In this kind of multicarrier DS-CDMA systems the number of subcarriers U is chosen so as to meet the following conditions [472]:

- The subcarrier signals of the multicarrier DS-CDMA-I system experience no frequency selectivity or no significant dispersion. Hence $T_m/T_c \leq 1$.
- All subcarrier signals are subject to independent fading, which implies that $2/T_c \geq (\Delta B)_c$.

These conditions can be expressed as:

$$\frac{1}{2} \leq \frac{T_m}{T_c} \leq 1. \qquad (19.118)$$

Let T_{c1} be the chip duration of a corresponding single-carrier DS-CDMA system having the same data rate and the same bandwidth as the multicarrier DS-CDMA system considered. Then, the chip duration of the multicarrier DS-CDMA-I system considered is $T_c = UT_{c1}$. Upon substituting this into the above equation, we can see that the number of subcarriers should satisfy the condition of:

$$\frac{T_m}{T_{c1}} \leq U \leq \frac{2T_m}{T_{c1}}. \qquad (19.119)$$

Since $L_1 = \lfloor T_m/T_{c1} \rfloor + 1$ represents the number of resolvable paths in the corresponding single-carrier DS-CDMA system, hence, we choose $U = L_1$.

Since in the multicarrier DS-CDMA-I system of Figure 19.4 satisfying the condition of Equation 19.118 each subcarrier signal is subject to independent frequency non-selective fading, the received signal associated with K number of asynchronous users is given by:

$$r(t) = \sum_{k=1}^{K} \sum_{u=1}^{U} \sqrt{\frac{2P}{U}} \alpha_{ku} b_k(t-\tau_k)c_k(t-\tau_k)\cos(2\pi f_u t + \varphi_{ku}) + n(t), \qquad (19.120)$$

where $\varphi_{ku} = \phi_{ku} + \psi_{ku} - 2\pi f_u \tau_k$ is a random variable uniformly distributed in $[0, 2\pi)$, while

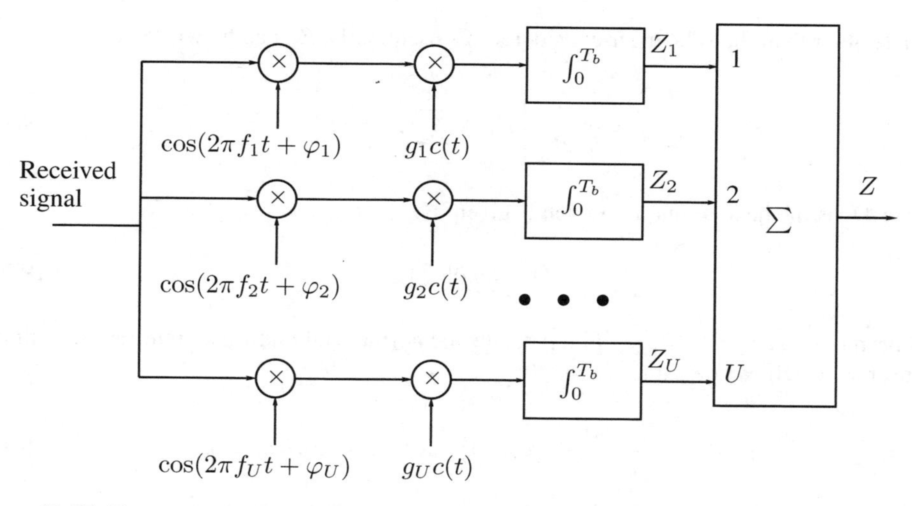

Figure 19.27: The receiver block diagram matched to the reference user $k = 1$ in the multicarrier DS-CDMA-I system.

α_{ku}, τ_k and ψ_{ku} are the channel amplitude, delay and phase, respectively, corresponding to the subcarrier u of user k. Finally, $n(t)$ represents the AWGN with double-sided power spectral density of $N_0/2$.

The receiver block diagram for the reference user is shown in Figure 19.27, where the index of $k = 1$ is omitted and $\tau_1 = 0$ is assumed for notational convenience. In the multicarrier DS-CDMA-I system, frequency diversity is achieved by combining the correlator's outputs associated with the subcarriers. The receiver provides a correlator for each subcarrier and the outputs of the correlators are combined in order to yield a processing gain comparable to that of a single-carrier DS system, provided that the spacing between two adjacent subcarriers is $\Delta = \frac{2}{T_c}$. Referring to Figure 19.27, the decision variable corresponding to the 0th data bit can be formulated as:

$$Z = \sum_{v=1}^{U} Z_v, \tag{19.121}$$

$$Z_v = \int_0^{T_b} r(t) g_v c(t) \cos(2\pi f_v t + \varphi_v) dt, \tag{19.122}$$

where $g_v = \alpha_v$ is assumed, associated with perfect channel estimation and a MRC diversity combining scheme. According to Equation 19.101, in the multicarrier DS-CDMA-I system of Figure 19.4, the normalized variance of the multi-user interference engendered by an adjacent subcarrier signal is $\frac{g_v^2}{8\pi^2 N_e}$, which is derived from Equation 19.101 by setting $u - m = 1$, $\lambda = 2N_e$, $g_{mv} = g_v$ and $\Omega = 1$. It can be shown that $\frac{g_v^2}{8\pi^2 N_e}$ is significantly lower than $g_v^2/3N_e$. Therefore, in our following analysis we ignore the interference imposed by the subcarrier signals other than the subcarrier considered. Consequently, Z_v can be written as:

$$Z_v = \sqrt{\frac{P}{2U}}\left[D_v + \sum_{k=2}^{K} I_k + \eta_v\right], \tag{19.123}$$

where D_v is the desired output, which is given by:

$$D_v = b[0]\alpha_v^2. \tag{19.124}$$

Furthermore, $I_k,\ k = 2, 3, \ldots, K$ represents the normalized multi-user interference inflicted by user k, which can be expressed as:

$$I_k = \frac{\alpha_{kv}\alpha_v \cos\theta_{kv}}{T_b}\left[b_k[-1]R_k(\tau_k) + b_k[0]\hat{R}_k(\tau_k)\right], \tag{19.125}$$

where $\theta_{kv} = \varphi_{kv} - \varphi_v$. The variables $R_k(\tau_k)$ and $\hat{R}_k(\tau_k)$ are the partial cross-correlation functions defined by Equations 19.89 and (19.90, while I_k can be approximated as a Gaussian random variable with zero mean and variance of $\alpha_v^2/3N_e$. Finally, η_v in Equation 19.123 is a Gaussian random variable with zero mean and variance of $U\alpha_v^2 N_0/2E_b$, where $E_b = PT_b$ is the energy per bit.

Therefore, Z_v can be approximated as a Gaussian random variable having a normalized mean given by D_v and a normalized variance of $[1/3N_e + UN_0/2E_b]\, g_v^2$. Since Z is the sum of U independent Gaussian random variables, Z is also approximately Gaussian distributed with a mean given by:

$$E[Z] = b[0]\sum_{v=1}^{U}\alpha_v^2, \tag{19.126}$$

and variance of:

$$\text{Var}[Z] = \left[\frac{K-1}{3N_e} + \left(\frac{2E_b}{UN_0}\right)^{-1}\right]\sum_{v=1}^{U}\alpha_v^2. \tag{19.127}$$

Having obtained the statistics of the decision variable, let us now derive the BER expression for the multicarrier DS-CDMA-I system in the following subsection.

19.6.2 Performance Analysis

Since the decision variable Z can be modelled as a Gaussian variable having a normalized mean given by Equation 19.126 and a normalized variance given by Equation 19.127, the BER conditioned on a set of channel fading amplitudes $\{\alpha_v,\ v = 1, 2, \ldots, U\}$ can be expressed as:

$$P_b(\gamma) = Q\left(\sqrt{2\gamma}\right), \tag{19.128}$$

where

$$\gamma = \overline{\gamma}_c \sum_{v=1}^{U} \alpha_v^2, \tag{19.129}$$

and

$$\overline{\gamma}_c = \left[\frac{2(K-1)}{3N_e} + \left(\frac{E_b}{UN_0}\right)^{-1}\right]^{-1}. \tag{19.130}$$

The unconditional BER can be derived by averaging Equation 19.128 – after weighting it with the PDF of γ – over the valid range of γ using a similar approach to that in Section 19.5. Finally, the unconditional BER can be formulated as:

$$P_b = \left[\frac{(1-\mu)}{2}\right]^U \sum_{n=0}^{U-1} \binom{U-1+n}{n} \left[\frac{(1+\mu)}{2}\right]^n, \tag{19.131}$$

where, μ is given by:

$$\mu = \sqrt{\frac{\overline{\gamma}_c}{1+\overline{\gamma}_c}}, \tag{19.132}$$

and $\overline{\gamma}_c$ is given by Equation 19.130.

19.7 Performance of Multicarrier DS-CDMA Systems Using Adaptive Subchannel Allocation

The multicarrier DS-CDMA scheme using adaptive subchannel allocation – which was portrayed in Figure 19.15 and referred to as the adaptive AMC DS-CDMA scheme – has been proposed and studied in [459]. The adaptive AMC DS-CDMA scheme represents the extension of the multicarrier DS-CDMA scheme [472] of Figure 19.4 for the forward links. In contrast to the multicarrier DS-CDMA scheme of [472], where identical DS waveforms are transmitted over each of the subchannels, in adaptive AMC DS-CDMA each user's DS waveform is transmitted over the user's favourite subchannel exhibiting the highest amplitude [459]. The channel's fading amplitudes are estimated at the mobile for all subchannels and the index of the best subchannel is fed back to the base station.

In this section, we investigate the performance of the adaptive AMC DS-CDMA system of

Figure 19.15 discussed in Subsection 19.2.6. Again, this scheme can be viewed as the extension of the multicarrier DS-CDMA scheme investigated in Section 19.6. In our investigations we assume that random spreading codes are employed and the channel is frequency-selective Rayleigh fading, but each subchannel's fading amplitude is an i.i.d random variable obeying the Rayleigh distribution. We assume, for simplicity that the channel's fading amplitudes are perfectly estimated and fed back from the mobile station to the base station without delay and feedback transmission errors. This assumption has often been used in the literature in order to derive the best-case performance estimate of systems with the aid of perfect channel quality side-information [71]. Furthermore, we assume that the multi-user interference can be approximated by an additive Gaussian random variable having zero mean, as we assumed in the previous sections.

In the adaptive AMC DS-CDMA system of Figure 19.15 the signal transmitted on the downlink by the base station can be expressed as in Equation 19.18, which can be rewritten as:

$$\begin{aligned} s_k(t) &= \sum_{k=1}^{K} \sqrt{2P} b_k(t) c_k(t) \sum_{u=1}^{U} \Delta_u(i_k) \cos(2\pi f_u t + \phi_{ku}) \\ &+ \sum_{u=1}^{U} \sqrt{2P_0} c_0(t) \cos(2\pi f_u t + \phi_u), \end{aligned} \tag{19.133}$$

where P represents the power transmitted by the base station to each user, P_0 represents the power of the pilot signal that is transmitted by the base station on each subcarrier, in order to facilitate the selection of the best subchannel for the transmission of the useful data to the mobiles, $b_k(t)$ and $c_k(t)$ represent the data waveform and spreading code waveform of the kth user, while $c_0(t)$ represents the spreading code waveform of the pilot signal. Finally,

$$\Delta_u(i_k) = \begin{cases} 1, & \text{for } u = i_k \\ 0, & \text{for } u \neq i_k, \end{cases} \tag{19.134}$$

indicating that only one subcarrier is activated by each user.

Assuming that frequency non-selective fading is experienced by the U subcarrier signals and that the K number of downlink user signals plus the associated pilot signal used for estimating the downlink subchannel quality are transmitted synchronously, the signal received by the mobile can be expressed as:

$$\begin{aligned} r(t) &= \sum_{k=1}^{K} \sqrt{2P} \alpha_{i_k} b_k(t) c_k(t) \cos(2\pi f_{i_k} t + \varphi_{ki_k}) \\ &+ \sum_{u=1}^{U} \sqrt{2P_0} \alpha_u c_0(t) \cos(2\pi f_u t + \varphi_u) \\ &+ n(t), \end{aligned} \tag{19.135}$$

where $\varphi_{ki_k} = \phi_{ki_k} + \psi_{ki_k}$, $\varphi_u = \phi_u + \psi_u$, $n(t)$ is the AWGN.

The mobile's receiver block diagram for the AMC DS-CDMA scheme was shown in Figure 19.16, where the coherent demodulation phases associated with the ith subchannel and the despreading code are incorporated in the term of $g_0[i]$, $i = 1, 2, \ldots, U$ corresponding to the detection of the pilot signal used for the estimation of the subchannel quality. Similarly, the term of $g_k[i_k]$ corresponds to the kth user's data detection. Explicitly, at the mobile, the index of the best subchannel is determined by estimating the fading amplitudes of the subchannels with the aid of the known pilot signal. We assume that the first user is the reference user and ignore the user index related to the reference user associated with $k = 1$. Let i_1 be the best subchannel, which is being used for the 0th data bit's transmission. We assume that there are K_h out of $K - 1$ active interfering users who share the i_1th subchannel with the reference user. Therefore, assuming perfect estimation of the sub-channels' amplitudes and phases, the decision variable, Z seen at the output of the mobile's receiver schematic, can be expressed in the form of:

$$\begin{aligned} Z &= \int_0^{T_b} r(t)\alpha_{i1}c_1(t)\cos\left(2\pi f_{i_1}t + \varphi_{1i_1}\right)dt \\ &= \sqrt{\frac{P}{2}}T_b\left[D_{i_1} + \sum_{k_h=1}^{K_h} I_{k_h} + I_0 + \eta_{i_1}\right], \end{aligned} \tag{19.136}$$

where D_{i1} represents the term associated with the desired user, which can be written as:

$$D_{i_1} = b[0]\alpha_{i_1}^2, \tag{19.137}$$

and I_{k_h} is the multi-user interference imposed by the kth user, which can be expressed as:

$$\begin{aligned} I_{k_h} &= \frac{\alpha_{i_k}\alpha_{i_1}\cos(\theta_{i_k})}{T_b}\int_0^{T_b} b_{k_h}(t)c_k(t)c_1(t)dt \\ &= \frac{\alpha_{i_k}\alpha_{i_1}\cos(\theta_{i_k})b_{k_h}[0]}{N_e}\sum_{n=0}^{N_e-1} c_n^{(k)}c_n^{(1)}. \end{aligned} \tag{19.138}$$

The multi-user interference term I_{k_h} can be approximated by a Gaussian random variable with mean zero and variance of:

$$\text{Var}\left[I_{k_h}\right] = \frac{\Omega_M\alpha_{i_1}^2}{2N_e}, \tag{19.139}$$

where $\Omega_M = E\left[\alpha_M^2\right]$, while α_M represents the fading amplitude corresponding to the best subchannel of user k. Since $\{\alpha_{ki},\ i = 1, 2, \ldots, U\}$ are Rayleigh distributed random variables obeying a PDF given by Equation 19.23, $\left\{\alpha_{ki}^2,\ i = 1, 2, \ldots, U\right\}$ obey exponential distributions having a PDF given by:

$$f_{\alpha_{ki}^2}(R) = \frac{1}{\Omega}\exp\left(-\frac{R}{\Omega}\right),\ R \geq 0. \tag{19.140}$$

The PDF of $\alpha_M^2 = \max\{\alpha_{k1}^2, \alpha_{k2}^2, \ldots, \alpha_{kU}^2\}$ can be derived using the following formula:

$$f_{\alpha_M^2}(r) = \frac{d}{dr}\left[\int_0^r f_{\alpha_{ki}^2}(R)dR\right]^U. \tag{19.141}$$

Upon substituting Equation 19.140 into Equation 19.141, we obtain:

$$f_{\alpha_M^2}(r) = \frac{U}{\Omega}\exp\left(-\frac{r}{\Omega}\right)\left[1-\exp\left(-\frac{r}{\Omega}\right)\right]^{U-1}, \; r \geq 0. \tag{19.142}$$

With the aid of Equation 19.142 Ω_M is given by:

$$\begin{aligned} \Omega_M &= E\left[\alpha_M^2\right] \\ &= \int_0^\infty r \cdot f_{\alpha_M^2}(r)dr \\ &= \int_0^\infty r\frac{U}{\Omega}\exp\left(-\frac{r}{\Omega}\right)\left[1-\exp\left(-\frac{r}{\Omega}\right)\right]^{U-1}dr \\ &= U\Omega\sum_{n=0}^{U-1}\binom{U-1}{n}\frac{(-1)^n}{(n+1)^2}. \end{aligned} \tag{19.143}$$

In Equation 19.136 I_0 represents the interference engendered by the pilot signal. Assuming that the interference due to the subcarriers other than subcarrier i_1 of interest can be ignored, I_0 can be formulated as:

$$\begin{aligned} I_0 &= \left(\sqrt{\frac{P}{2}}T_b\right)^{-1}\int_0^{T_b}\sqrt{2P_0}\alpha'_{i_1}c_0(t)\cos(2\pi f_{i_1}t+\varphi_{i_1}) \\ &\quad \alpha_{i_1}c_1(t)\cos(2\pi f_{i_1}t+\varphi_{1i_1})dt \\ &= \frac{\sqrt{\rho}\alpha'_{i_1}\alpha_{i_1}\cos\theta_{i_1}}{N_e}\sum_{n=0}^{N_e-1}c_n^{(0)}c_n^{(1)}, \end{aligned} \tag{19.144}$$

where $\rho = P_0/P$ represents the ratio between the power of the pilot signal transmitted by the base station on each subcarrier and that of the reference signal. The fading amplitude of the pilot signal on subcarrier i_1 is denoted by α'_{i_1}. For a given value of α_{i_1}, I_0 can also be approximated by a Gaussian random variable with zero mean and variance of:

$$\text{Var}[I_0] = \frac{\rho\Omega_M\alpha_{i_1}^2}{2N_e}. \tag{19.145}$$

Finally, η_{i_1} is a Gaussian random variable with mean zero and variance given by $N_0/2E_b$.

Consequently, for a given fading amplitude of α_{i_1}, the decision variable Z at the output of the mobile's receiver schematic seen in Figure 19.16 can be approximated by a Gaussian random variable having a normalized mean given by Equation 19.137, and a normalized variance given by:

$$\text{Var}[Z] = \left[\frac{K_h + \rho}{2N_e} + \left(\frac{2\Omega_M E_b}{N_0}\right)^{-1}\right] \Omega_M \alpha_{i_1}^2. \tag{19.146}$$

The bit error probability for a given fading amplitude of α_{i_1} and for K_h number of users activating the i_1th subcarrier can be written as:

$$P_b(\alpha_{i_1}^2, K_h) = Q\left(\sqrt{\overline{\gamma}_c \cdot \frac{2\alpha_{i_1}^2}{\Omega_M}}\right), \tag{19.147}$$

where

$$\overline{\gamma}_c = \left[\frac{K_h + \rho}{N_e} + \left(\frac{\Omega_M E_b}{N_0}\right)^{-1}\right]^{-1}. \tag{19.148}$$

Since the i_1th subchannel was assumed to be the best subchannel during the 0th bit's transmission, i.e., $\alpha_{i_1}^2 = \max\left\{\alpha_1^2, \alpha_2^2, \ldots, \alpha_U^2\right\}$, $\alpha_{i_1}^2$ has the same PDF as α_M^2 which was given by Equation 19.142. Therefore, the conditioning of $P_b(\alpha_{i_1}^2, K_h)$ on $\alpha_{i_1}^2$ can be removed by the weighted averaging of Equation 19.147 according to its PDF $f_{\alpha_{i_1}^2}(r)$ of Equation 19.142 over the valid range of $\alpha_{i_1}^2$, which can be expressed as:

$$\begin{aligned}
P_b(K_h) &= \int_0^\infty P_b(r, K_h) f_{\alpha_{i_1}^2}(r) dr \\
&= \int_0^\infty Q\left(\sqrt{\overline{\gamma}_c \cdot \frac{2r}{\Omega_M}}\right) \times \frac{U}{\Omega} \exp\left(-\frac{r}{\Omega}\right)\left[1 - \exp\left(-\frac{r}{\Omega}\right)\right]^{U-1} dr \\
&= \sum_{n=0}^{U-1} \frac{(-1)^n}{\sqrt{2\pi}} \binom{U-1}{n} \frac{U}{\Omega} \int_0^\infty \int_{\sqrt{\overline{\gamma}_c \frac{2r}{\Omega_M}}}^\infty \exp\left(-\frac{t^2}{2}\right) \exp\left(-\frac{(n+1)r}{\Omega}\right) dt dr.
\end{aligned} \tag{19.149}$$

Interchanging the order of integration associated with t and r, Equation 19.149 can be written as:

$$P_b(K_h) = \sum_{n=0}^{U-1} \frac{(-1)^n}{\sqrt{2\pi}} \binom{U-1}{n} \frac{U}{\Omega} \int_0^\infty \int_0^{\frac{\Omega_M t^2}{2\overline{\gamma}_c}} \exp\left(-\frac{t^2}{2}\right) \exp\left(-\frac{(n+1)r}{\Omega}\right) dt dr. \tag{19.150}$$

Upon evaluating out the above integrals, Equation 19.150 can be simplified to:

$$P_b(K_h) = \frac{U}{2}\sum_{n=0}^{U-1}\frac{(-1)^n}{(n+1)}\binom{U-1}{n}\left[1-\sqrt{\frac{\overline{\gamma}_c\Omega}{\overline{\gamma}_c\Omega+(n+1)\Omega_M}}\right]. \quad (19.151)$$

Since

$$U\sum_{n=0}^{U-1}\frac{(-1)^n}{(n+1)}\binom{U-1}{n} = 1, \quad (19.152)$$

Equation 19.151 can be rewritten as:

$$\begin{aligned} P_b(K_h) &= \frac{1}{2}+\frac{U}{2}\sum_{n=0}^{U-1}\frac{(-1)^{n+1}}{(n+1)}\binom{U-1}{n}\sqrt{\frac{\overline{\gamma}_c\Omega}{\overline{\gamma}_c\Omega+(n+1)\Omega_M}} \\ &= \frac{1}{2}+\frac{1}{2}\sum_{n=1}^{U}(-1)^n\binom{U}{n}\sqrt{\frac{\overline{\gamma}_c\Omega}{\overline{\gamma}_c\Omega+n\Omega_M}}. \end{aligned} \quad (19.153)$$

If we assume that the channel attenuation for a given mobile is independent of the channel quality for any other mobile and that the probability of choosing any specific subchannel as the best subchannel from the set of U subchannels is equiprobable, the probability that there are K_h out of $K-1$ number of interfering users activating the i_1th subchannel can be formulated as:

$$P(K_h) = \binom{K-1}{K_h}\left(\frac{1}{U}\right)^{K_h}\left(1-\frac{1}{U}\right)^{K-1-K_h}. \quad (19.154)$$

The unconditional BER, consequently, can be expressed as [459]:

$$\begin{aligned} P_b &= \sum_{K_h=0}^{K-1} P(K_h)P_b(K_h) \\ &= \sum_{K_h=0}^{K-1}\binom{K-1}{K_h}\left(\frac{1}{U}\right)^{K_h}\left(1-\frac{1}{U}\right)^{K-1-K_h} P_b(K_h), \end{aligned} \quad (19.155)$$

where $P_b(K_h)$ is given by Equation 19.153.

In Figure 19.28 and Figure 19.29 we compared the downlink BER performance of the multicarrier DS-CDMA-I system of Figure 19.4, which was analyzed in Section 19.6, and that of the adaptive AMC DS-CDMA system of Figure 19.15 and Figure 19.16 investigated in this section. In Figure 19.28 the BER was evaluated as a function of the SNR per bit E_b/N_0 by assuming that the spreading gain of a corresponding single-carrier DS-CDMA system was $N = N_eU = 256$ and the number of active users was $K = 20$. By contrast, in Figure 19.29 the BER was evaluated as a function of the number of active users K by

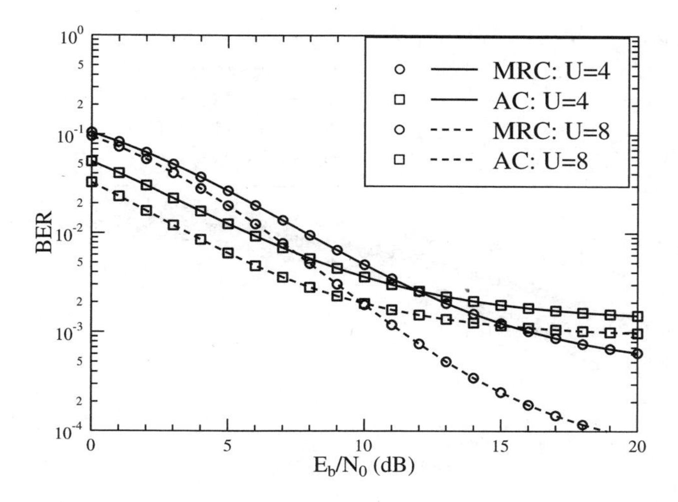

Figure 19.28: Downlink BER versus SNR per bit, E_b/N_0, performance comparison for the multicarrier DS-CDMA-I system of Figure 19.4 and the adaptive AMC DS-CDMA system of Figure 19.15 over dispersive Rayleigh fading channels with parameters of $N = 256, K = 20$ and $U = 4,\ 8$, where N is the spreading gain of a corresponding single-carrier DS-CDMA system. The results were computed from Equation 19.131 for the multicarrier DS-CDMA-I scheme, while from Equation 19.155 for the AMC DS-CDMA scheme.

assuming that the spreading gain of the corresponding single-carrier DS-CDMA system was also $N = N_eU = 256$ and the SNR per bit was $E_b/N_0 = 8$dB. Therefore, the spreading gain of each subcarrier signal was $N_e = 64$ for $U = 4$ and $N_e = 32$ for $U = 8$. The results for the adaptive AMC DS-CDMA scheme were computed according to Equation 19.155 with $\overline{\gamma}_c$ given by Equation 19.148, while the results for the multicarrier DS-CDMA-I scheme were computed from Equation 19.131 with $\overline{\gamma}_c$ given by Equation 19.130. However, since synchronous transmissions are usually employed in the downlink, the term $3N_e$ in Equation 19.130 should be replaced by $2N_e$, according to the derivation of Equation 19.148 in this section. From the results of Figure 19.28 we observe that for low to moderate SNR per bit value, the adaptive AMC DS-CDMA scheme outperforms the multicarrier DS-CDMA-I scheme, as long as the receiver can maintain near-perfect channel estimation with the aid of the pilot signals and provided that there are no feedback errors concerning the choice of the best subchannel. However, if the SNR per bit in excess of $E_b/N_0 > 14dB$ for $U = 4$ and $E_b/N_0 > 12dB$ for $U = 8$, the above-mentioned trend is reversed, in other words, the adaptive AMC DS-CDMA scheme is outperformed by the multicarrier DS-CDMA-I scheme. In Figure 19.29 the adaptive AMC DS-CDMA scheme outperforms the multicarrier DS-CDMA-I scheme over the entire range of the number of active users investigated, namely from $K = 1$ to $K = 50$, when SNR per bit of $E_b/N_0 = 8$dB was assumed. Furthermore, since multicarrier systems using a high number of subcarriers can achieve a high diversity or-

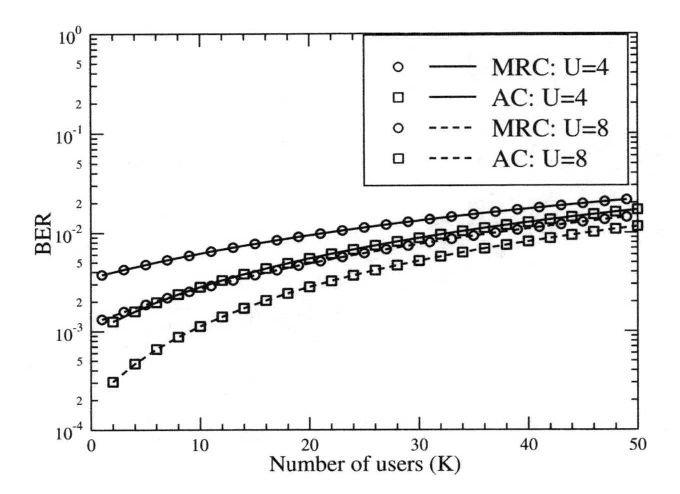

Figure 19.29: Downlink BER versus number of users, K, performance comparison for the multicarrier DS-CDMA-I of Figure 19.4 system and the adaptive AMC DS-CDMA system of Figure 19.15 over dispersive Rayleigh fading channels with parameters of $N = 256, E_b/N_0 = 8dB$ and $U = 4,\ 8$, where N is the spreading gain of a corresponding single-carrier DS-CDMA system. The results were computed from Equation 19.131 for the multicarrier DS-CDMA-I scheme, while from Equation 19.155 for the AMC DS-CDMA scheme.

der, we infer from the results of both Figure 19.28 and Figure 19.29 that for any given value of E_b/N_0 or any given number of active users, K, the BER performance of the system using a high number of subcarriers, such as $U = 8$, is better than that of the system employing a low number of subcarriers, such as $U = 4$.

19.8 Performance of Slow Frequency-Hopping Multicarrier DS-CDMA Systems

In this section we investigate the performance of the SFH/MC DS-CDMA system seen in Figure 19.17, when using BPSK data modulation. Its performance is evaluated over a range of multipath Rayleigh fading channels. Two detection schemes are investigated in conjunction with the receiver having perfect knowledge or no knowledge of the FH patterns employed. When the receiver invokes the explicit knowledge of the FH patterns used, then conventional hard-detection – which is often applied in direct-sequence, slow frequency-hopping CDMA (DS/SFH CDAM) [428, 484] systems – is employed for the sake of simplifying the receiver. By contrast, when the receiver does not have any knowledge concerning the FH pattern used, then the joint soft-detection techniques are used, in order to simultaneously accomplish both demodulation and FH pattern acquisition.

19.8.1 System Description

19.8.1.1 The Transmitted Signal

The SFH/MC DS-CDMA system has been reviewed in Subsection 19.2.7, where the model of the transmitter and the multiple access channel was depicted in Figure 19.17. Referring to Figure 19.17, the transmitted signal of user k can be expressed as:

$$s_k(t) = \sum_{u_k=0}^{U_k-1} \sqrt{2P} b_{ku_k}(t) c_k(t) \cos\left(2\pi f_{u_k}^{(k)} t + \varphi_{u_k}^{(k)}\right), \tag{19.156}$$

where P represents the transmitted power per subcarrier, while U_k indicates the weight of the constant-weight code currently employed by the kth user, and $c_k(t)$ represents the spreading waveform. Furthermore, $\{b_{ku_k}(t)\}$, $\left\{f_{u_k}^{(k)}\right\}$ and $\left\{\varphi_{u_k}^{(k)}\right\}$ represent the current data stream's waveforms, the subcarrier frequency set and the phase angles introduced in the carrier modulation process. The data stream's waveform $b_{ku_k}(t) = \sum_{i=-\infty}^{\infty} b_{ku_k}^{(i)} P_{T_s}(t - iT_s)$ consists of a sequence of mutually independent rectangular pulses of duration T_s and of amplitude +1 or -1 with equal probability. The spreading sequence $c_k(t) = \sum_{j=-\infty}^{\infty} c_k^{(j)} P_{T_c}(t - jT_c)$ denotes the signature sequence waveform of the kth user, where $c_k^{(j)}$ assumes values of +1 or -1 with equal probability, while $P_{T_c}(t)$ is the common chip waveform for all signals. In our analysis for the sake of simplicity we assume that there exists no spectral overlap between the spectral mainlobes of two adjacent subcarriers, and also assume that there exists no interference between subcarriers. More explicitly, interference is inflicted only when an interfering user activates the same subcarrier, as the reference user. Let $N_b = T_b/T_c$ be the number of chips per bit, where T_b is the bit duration before the serial-to-parallel conversion stage of Figure 19.17. Let the processing gain of each subcarrier signal be N_e, which can be expressed as $N_e = T_s/T_c$. Furthermore, we assume that the frequency hopping duration is T_h, and that the number of data bits $N_h = T_h/T$ transmitted per hop is a positive integer, which is strictly larger than 1, i.e. we assume using slow frequency hopping.

19.8.1.2 Channel Description

The channel model considered in this section is the finite-length tapped delay line model of a frequency-selective multipath channel [90], whose complex low-pass impulse response for subcarrier u_k of user k is given by:

$$h_{u_k}^{(k)}(t) = \sum_{l_p=0}^{L_p-1} \alpha_{u_k,l_p}^{(k)} e^{j\phi_{u_k,l_p}^{(k)}} \delta(t - l_p T_c). \tag{19.157}$$

In Equation 19.157 $l_p T_c$ is the relative delay of the l_p-th path of user k with respect to the main path, the phases $\left\{\phi_{u_k,l_p}^{(k)}\right\}$ are i.i.d random variables uniformly distributed in the interval $[0, 2\pi)$, whilst the L_p tap weights $\left\{\alpha_{u_k,l_p}^{(k)}\right\}$ are independent Rayleigh distributed random variables with a PDF given by Equation 19.23. Consequently, for an asynchronous CDMA system supporting K users, the received signal takes the form of:

$$\begin{aligned} r(t) &= n(t) \\ &+ \sqrt{2P} \cdot \sum_{k=1}^{K} \sum_{u_k=0}^{U_k-1} \sum_{l_p=0}^{L_p-1} \alpha_{u_k,l_p}^{(k)} b_{ku}^{(k)}(t - l_p T_c - \tau_k) \\ &\cdot c_k(t - l_p T_c - \tau_k) \cos\left(2\pi f_{u_k}^{(k)} t + \psi_{u_k,l_p}^{(k)}\right), \end{aligned} \tag{19.158}$$

where $n(t)$ represents the AWGN with zero mean and double-sided power spectral density of $N_0/2$, $\psi_{u_k,l_p}^{(k)} = \left[\varphi_{u_k}^{(k)} + \phi_{u_k,l_p}^{(k)} - 2\pi f_{u_k}^{(k)}(\tau_k + l_p T_c)\right]$ $(mod\ 2\pi)$, which is assumed to be an i.i.d random variable having a uniform distribution in $[0, 2\pi)$, and τ_k represents the propagation delay of user k.

19.8.1.3 Receiver Model

Let the first user be the user-of-interest and consider the conventional matched filter-based RAKE receiver using MRC, as shown in Figure 19.18, where the superscript and subscript of the reference user's signal have been omitted for notational convenience. In Figure 19.18 L – the number of diversity branches used by the receiver – $1 \leq L \leq L_p$ is a variable, allowing us to study the effect of different diversity orders. In contrast to the transmitter side of Figure 19.17, where only U_k out of U subcarriers are transmitted by the user k, at the receiver of Figure 19.18 all U subcarriers are always tentatively demodulated. Furthermore, the information bits transmitted over the different subcarriers might be detected using hard-detection separately, or using blind soft-detection jointly for all subcarriers. The choice of soft-detection depends on whether the receiver is capable of acquiring the FH patterns blindly, i.e., without their explicit signalling. This important issue will be discussed in the forthcoming sections. Consequently, from the point of view of the receiver, each subcarrier can be viewed as an On-Off type signalling scheme. When a subcarrier is actively used for signalling and hence it is in the On-state, the MRC output samples give +1 or -1 information. Otherwise, while passive and hence in the Off-state, the MRC stage outputs noise.

The U number of matched filters in Figure 19.18 are matched to the reference user's spreading code, and are assumed to have achieved time synchronization. Let us assume that perfect estimates of the channel tap weights and phases are available. Then – after appropriately delaying the individual matched filter outputs, in order to coherently combine the L number of path signals used by the RAKE combiner – the qth MRC output sampled at $t = T + (L-1)T_c$, in order to detect the first symbol can be expressed as:

$$Z_q = D_q[0] + I_q, \tag{19.159}$$

where $D_q[0]$ represents the desired direct component, which can be expressed as:

$$D_q[0] = \sqrt{\frac{P}{2}} T_s b_q[0] \sum_{l=0}^{L-1} \alpha_{q,l}^2. \tag{19.160}$$

In Equation 19.160 $b_q[0]$ is the first bit transmitted on subcarrier q by the reference user and $b_q[0] \in \{+1, -1, 0\}$ with '0' representing the Off-state. Since we have assumed that there exists no interference among the subcarriers, interference is inflicted only when an interfering user activates the same subcarrier, as the reference user. Hence, the interference plus noise term I_q in Equation 19.159 can be written as:

$$I_q = I_q[S] + I_q[M] + N_q, \tag{19.161}$$

where $I_q[S]$ represents the multipath interference imposed by the q-th subcarrier of the user-of-interest, which can be expressed as:

$$\begin{aligned} I_q[S] &= \left(\sqrt{\frac{P}{2}} T_s\right) \sum_{l=0}^{L-1} \alpha_{q,l} \sum_{\substack{l_p=0 \\ l_p \neq l}}^{L_p-1} \frac{\alpha_{q,l_p} \cos\theta_{q,l_p}}{T_s} \\ &\cdot \int_{(L-1)T_c}^{T_s+(L-1)T_c} b_q[t - (L + l_p - l - 1)T_c] \\ &\cdot c[t - (L + l_p - l - 1)T_c] c[t - (L-1)T_c] dt, \end{aligned} \tag{19.162}$$

where $I_q[M]$ represents the multi-user interference inflicted on the q-th subcarriers of the interfering users. Let us assume that there exists K_h, $(0 \leq K_h \leq K-1)$ number of interfering signals, all of which activate the q-th subcarrier during the first symbol's transmission of the reference signal. The event, when an interferer activates the same subcarrier as the reference user, is often referred to in the literature as a so-called *hit* – an event that will be discussed in detail in the next subsection. Then, $I_q[M]$ can be expressed as:

$$\begin{aligned} I_q[M] &= \left(\sqrt{\frac{P}{2}} T_s\right) \sum_{l=0}^{L-1} \alpha_{q,l} \sum_{h=1}^{K_h} \sum_{l_p=0}^{L_p-1} \frac{\alpha_{q,l_p}^{(h)} \cos\theta_{q,l_p}^{(h)}}{T} \\ &\cdot \int_{(L-1)T_c}^{T_s+(L-1)T_c} b_q^{(h)}[t - (L + l_p - l - 1)T_c - \tau_h] \\ &\cdot c_h[t - (L + l_p - l - 1)T_c - \tau_h] c[t - (L-1)T_c] dt. \end{aligned} \tag{19.163}$$

In Equation 19.162 and (19.163 the $\cos(\cdot)$ terms are contributed by the phase differences between the incoming subcarriers and the locally generated subcarrier used in the demodulation. Finally, the noise term of Equation 19.161 can be formulated as:

$$N_q = \sum_{l=0}^{L-1} \alpha_{q,l} \int_{(L-1)T_c}^{T_s+(L-1)T_c} n(t)c[t-(L-1)T_c] \cos(2\pi f_q t + \psi_{q,l}) dt, \quad (19.164)$$

which is a Gaussian random variable with zero mean and variance of $\frac{N_0 T_s}{4} \sum_{l=0}^{L-1} \alpha_{q,l}^2$, where $\{\alpha_{q,l}\}$ represents the path attenuations.

We have obtained the decision variables of the MRC output samples. Let us now analyze the BER performance of the SFH/MC DS-CDMA system using hard-detection by invoking the often-used Gaussian approximation.

19.8.2 Performance of the SFH/MC DS-CDMA Receiver with Explicit Knowledge of the FH Patterns: Hard-Detection

19.8.2.1 Probability of Error

In the analysis of this section, we employ the Gaussian approximation and hence model the multi-user interference and the self-interference terms of Equation 19.161 as an AWGN process having zero mean and a variance equal to the corresponding variances. Consequently, for a set of given channel amplitudes $\{\alpha_{q,l}\}$ – according to the analysis of the previous sections – the q-th MRC output sample can be approximated as an AWGN variable having a mean value given by Equation 19.160 and a variance of:

$$\sigma^2 = \frac{PT_s^2}{2}\left[\frac{K_h L_p + L_p - 1}{3N_e} + \left(\frac{2\Omega E_b}{N_0}\right)^{-1}\right] \cdot \Omega \sum_{l=0}^{L-1} \alpha_{q,l}^2, \quad (19.165)$$

where $E_b = PT_s$ is the energy per bit and $\Omega = E\left[(\alpha_{q,l})^2\right]$.

Since the receiver has the explicit knowledge of the FH pattern employed by the transmitter, the information transmitted on the U_k number of activated subcarriers can be detected without taking into account the Off-state carriers. Hence, the average bit error probability can be expressed as [428, 484]:

$$P_b = \sum_{K_h=0}^{K-1} \binom{K-1}{K_h} P_h^{K_h} (1-P_h)^{K-1-K_h} P_b(K_h, \overline{\gamma}_c), \quad (19.166)$$

where $0 \le K_h \le K-1$ and P_h is the probability of a hit – as defined above – imposed by an interfering signal. In an asynchronous system, if we assume that the FH pattern is determinated randomly by a constant-weight code chosen from the set of $\binom{U}{U_k}$ codes, then the probability of a hit engendered by the interfering user k can be approximated by:

$$P_h(k) = \frac{\binom{U-1}{U_k-1}}{\binom{U}{U_k}} = \frac{U_k}{U}. \quad (19.167)$$

The average probability of a hit, P_h, can be computed by averaging Equation 19.167, taking into account the weights of the constant-weight codes used and the number of users, K, which yields:

$$P_h = \frac{\overline{U}}{Q}, \quad (19.168)$$

where $\overline{U}$ represents the average weight of the constant-weight codes used. The probability of $P_b(K_h, \overline{\gamma}_c)$ in Equation 19.166 denotes the conditional bit error probability of the hard-detections, given that K_h hits were inflicted by the other $K-1$ interfering users, i.e., K_h out of $K-1$ users in the system activated the same subcarrier as the reference user.

Before proceeding to the evaluation of the average probability of $P_b(K_h, \overline{\gamma}_c)$ for a given K_h, we first have to determine the error probability conditioned on the multipath component attenuations $\{\alpha_{q,l}\}$. Following from our previous discussions, for the receiver having an explicit knowledge of the FH pattern, only the case, when $b_q[0] \in \{+1, -1\}$ has to be considered in deriving the error probability. The associated conditional bit error probability of the BPSK modulated bits may be written as:

$$P_b(K_h, \gamma) = Q(\sqrt{2\gamma}), \quad (19.169)$$

where

$$\gamma = \overline{\gamma}_c \cdot \frac{1}{\Omega} \sum_{l=0}^{L-1} \alpha_{q,l}^2 \quad (19.170)$$

$$\overline{\gamma}_c = \left[\frac{2(K_h L_p + L_p - 1)}{3N_e} + \left(\frac{\Omega E_b}{N_0} \right)^{-1} \right]^{-1}, \quad (19.171)$$

where $\overline{\gamma}_c$ represents the average SNR per path.

The average bit error probability for K_h number of hits is calculated from the conditional error probability upon weighting $P_b(K_h, \gamma)$ by the PDF of γ, $f(\gamma)$, and then averaging or integrating the weighted product over its legitimate range, as shown in Equation 19.57. According to our analysis in Section 19.4, the average BER for a given number of interfering users can be expressed with the aid of Equation 19.57 – Equation 19.59 as:

$$P_b(K_h, \overline{\gamma}_c) = \left[\frac{(1-\mu)}{2} \right]^L \sum_{n=0}^{L-1} \binom{L-1+n}{n} \left[\frac{(1+\mu)}{2} \right]^n, \quad (19.172)$$

where, by definition:

$$\mu = \sqrt{\frac{\overline{\gamma}_c}{1+\overline{\gamma}_c}}, \tag{19.173}$$

with $\overline{\gamma}_c$ given by Equation 19.171.

Consequently, the average BER of the receiver using hard-detection can be computed by substituting Equation 19.168 and Equation 19.172 into Equation 19.166.

19.8.2.2 Numerical Results

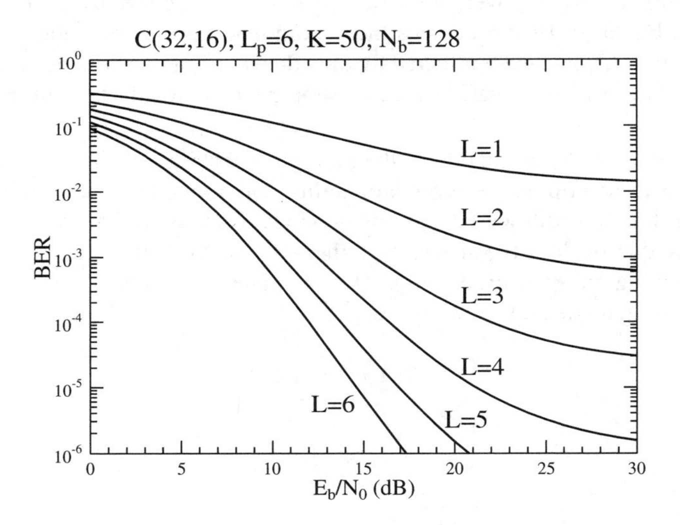

Figure 19.30: BER versus SNR per bit performance for constant-weight code based SFH/MC DS-CDMA systems using hard-detection over multipath Rayleigh fading channels upon varying the diversity order L, using the constant-weight code C(32,16) for FH patterns, $L_p = 6$ resolvable paths, $K = 50$ users and a bit-duration to chip-duration ratio $N_b = 128$. For the receiver using maximum ratio combining (MRC), the optimum diversity order L is its maximum possible value of $L = L_p$, corresponding to combining all the resolvable multipath components.

In order to quantify the system's performance improvements due to diversity, Figure 19.30 depicts the BER as a function of the SNR per bit, namely E_b/N_0. The individual curves in each figure are parametrized by the diversity order $L = 1, 2, \ldots, 6$. We assumed that the FH patterns were designed from the constant-weight code C(32,16), there were $L_p = 6$ resolvable paths, the number of active users was $K = 50$ and the spreading gain of each subcarrier signal was $N_e = 128U$, i.e., the bit-duration to chip-duration ratio was $N =$

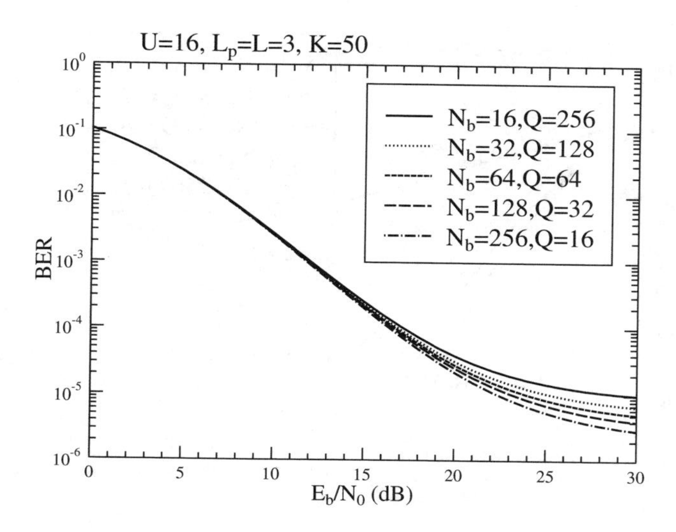

Figure 19.31: BER versus SNR per bit performance for constant-weight code based SFH/MC DS-CDMA systems using hard-detection and $L_p = L = 3$ upon varying the value of N and Q. For a constant bandwidth associated with $N_b Q = 2^{12}$, the combination of $N_b = 256$ and $Q = 16$ provides the best performance while also imposing the highest complexity, requiring a high bit/chip duration ratio N_b.

128. From the results it is seen that the system's BER performance is dramatically improved by increasing the diversity order L. For MRC, the value of L has to be maximized. The corresponding choice of $L = L_p$ implies combining all the resolvable multipath components, irrespective of the receiver's associated complexity.

In Figure 19.31 SFH/MC DS-CDMA systems having a constant system bandwidth associated with the product $N_b Q = 2^{12}$, but using various combinations of the number of subcarriers Q and bit-duration to chip-duration ratio of N_b were considered. In this system, increasing the number of subcarriers implies decreasing the 'hit' probability inflicted by the interfering users and, simultaneously, decreasing the direct-sequence spread bandwidth of each subcarrier. The parameters used are shown in the figure. For a constant system bandwidth and for $L_p = L = 3$, we observe that although Q and N_b change over a wide range, the BER performance remains indistinguishable for relatively low SNR per bit values, namely below 15dB. However, for higher SNR per bit values, in excess of 21dB, the BER performance improves upon increasing N_b.

Since increasing the value of N_b implies increasing the DS spread bandwidth and results in decreasing the chip-duration, consequently, for a given multipath fading environment having an average delay spread of T_m, the number of resolvable paths $L_p = T_m/T_c$, increases upon decreasing T_c. Hence, the assumption of $L_p = L = 3$ in Figure 19.31 is impractical. Therefore, in Figure 19.32 the performance of a SFH/MC DS-CDMA system using

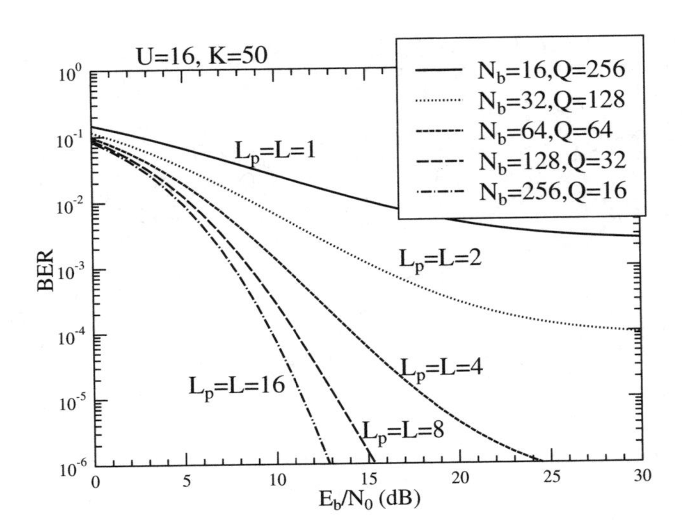

Figure 19.32: BER versus SNR per bit performance for constant-weight code based SFH/MC DS-CDMA systems using hard-detection and $L_p = L = 1, 2, 4, 8, 16$ upon varying the value of N_b and Q. Best BER performance is achieved for a high diversity order of $L_p = L = 16$, combining all associated multipath components, regardless of the associated receiver complexity.

hard-detection over a given multipath fading environment having an average delay spread of T_m was considered, as a function of $L_p = L = 1, 2, 4, 8$ and 16. More explicitly, here we assumed that there was one resolvable path at the receiver for the system associated with $N_b = 16$. Consequently, the number of resolvable paths was $L_p = 2, 4, 8, 16$ for the systems using $N_b = 32, 64, 128, 256$, respectively. Furthermore, we assumed that the receiver was capable of combining all the resolvable paths regardless of the associated complexity. Other parameters related to the computations were the same as in Figure 19.31, which are shown in the figure. The results indicate that the BER performance is significantly improved, when increasing N_b. Hence, for signals undergoing severe fading, a high number of independent subchannels are required, in order to enhance the system's performance.

For the systems considered in Figure 19.32 to achieve the best BER performance the receiver has to have a high diversity order. For example, for the system with $N_b = 256$ to achieve the best BER performance, the receiver has to combine all the 16 multipath signals. However, at the time of writing the implementation of such a complex receiver is impractical. Hence, in Figure 19.33 we considered a receiver with a maximum diversity order of $L = 3$, although a higher number of resolvable paths, L_p, were available at the receiver. From the results we infer that the curve associated with $Q = 64,\ N_b = 64$, or $L_p = 4,\ L = 3$ achieves the best BER performance.

From the results of Figure 19.31 to Figure 19.33 we conclude that, for a SFH/MC DS-

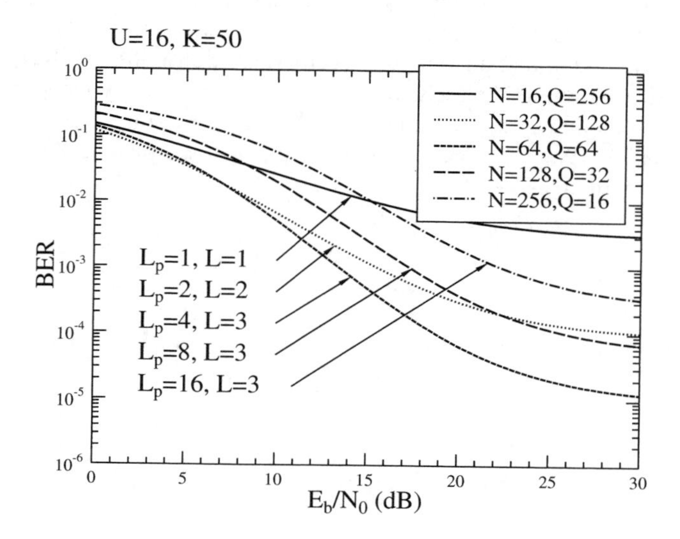

Figure 19.33: BER versus SNR per bit performance for constant-weight code based SFH/MC DS-CDMA systems using hard-detection upon varying the value of N_b and Q for a maximum diversity order of $L = 3$ and $L_p = 1, 2, 4, 8, 16$. Given the complexity constraint of $L = 3$, the system associated with $N_b = 64,\ Q = 64,\ L_p = 4,\ L = 3$ achieved the best performance.

CDMA system having a constant system bandwidth and using hard-detection, from a system optimization point of view, the DS spread bandwidth of each subchannel has to be adjusted, so that the resulting number of resolvable paths, L_p, is as close to L as possible. Then the receiver can efficiently utilize the energy dispersed over the multipath components. The required number of subchannels, consequently, can be obtained by dividing the system bandwidth by the required DS spread bandwidth.

19.8.3 Detection of SFH/MC DS-CDMA Signals Without Knowledge of the FH Patterns: Blind Joint Soft-Detection

In the previous subsection, we studied the detection of the SFH/MC DS-CDMA signals, when the receiver employed the explicit knowledge of the FH pattern of the transmitter used, whilst the perfect estimation of the channel impulse response (CIR) was also assumed. Often, however, the receiver is not supplied with the explicit knowledge of the FH patterns. This happens, for example, at the commencement of communications before the receiver detected this information, or in case of soft hand-overs to other base stations. However, the loss of information under these circumstances is unacceptable and hence more powerful detection algorithms than the hard-detection techniques of the last subsection are required.

If we assume that the transmitted signals are equally probable, symbol-by-symbol

Maximum-Likelihood Sequence Detection (MLSD) is considered to be the optimum receiver scheme [78, 90, 485]. In this subsection we investigate the MLSD of SFH/MC DS-CDMA signals under the assumption that all symbols (vectors) are transmitted with equal probability. We assume furthermore that the MRC output samples are mutually independent random variables having mean value given by Equation 19.160 and, for simplicity, a common variance given by Equation 19.165. Hence the PDF of the q-th sampled subchannel output of Figure 19.18 can be written as:

$$p_q(z) = \frac{1}{\sqrt{2\pi}\sigma} e^{-\frac{(z - D_q[0])^2}{2\sigma^2}}, \tag{19.174}$$

where $D_q[0]$ is given by Equation 19.160 with $b_q[0] \in \{+1, -1, 0\}$, and σ^2 is given by Equation 19.165.

19.8.3.1 Maximum Likelihood Detection

Since the receiver is only aware of the weight of the transmitted constant-weight code, but not the positions of the binary '1's, the receiver now has to detect not only the positions of the '1's, which indicate the subcarriers used, but also the antipodal binary modulated information conveyed by the activated subcarriers. Let us express the input data of the scheme seen in Figure 19.18 in the vectorial form of $\mathbf{D_i} = \{d_{i,0}, d_{i,1}, \ldots, d_{i,U-1}\}$, where $0 \leq i \leq M-1$ represents the constant-weight code set of weight U_1 and M is the number of constant-weight codes included in this set. Then a set of U MRC samples $\mathbf{Z} = \{Z_1, Z_2, \ldots, Z_{U-1}\}$ in Figure 19.18 – which we refer to here as a received symbol or vector – are input to the 'decision unit' invoking a joint ML detection rule, which is based on the determination of the probabilities defined as:

$$p(\mathbf{Z}|\mathbf{D_i}) = \frac{1}{(2\pi\sigma^2)^{U/2}} \exp\left(-\frac{\sum_{q=0}^{U-1} \left(Z_q - \sqrt{\frac{P}{2}} T_s d_{i,q} \sum_{l=0}^{L-1} \alpha_l^2 \right)^2}{2\sigma^2} \right),$$

$$i = 0, 1, \ldots, M-1. \tag{19.175}$$

The decision criterion is based on selecting the signal corresponding to the maximum of the set of probabilities $\{p(\mathbf{Z}|\mathbf{D_i})\}$.

Upon taking the logarithm of both sides of the above equation we arrive at:

$$\ln(p(\mathbf{Z}|\mathbf{D_i}) = -\frac{U}{2}\ln(2\pi\sigma^2) - \frac{1}{2\sigma^2} D(\mathbf{Z}, \mathbf{D_i}), \tag{19.176}$$

where

$$D(\mathbf{Z}, \mathbf{D_i}) = \sum_{q=0}^{U-1} |Z_q - \sqrt{\frac{P}{2}} T_s d_{i,q} \sum_{l=1}^{L-1} \alpha_l^2 |^2 \tag{19.177}$$

represents the Euclidean distance between the decision variable vector of $\mathbf{Z}$ and the transmitted data vector of $\mathbf{D_i}$. Finding the maximum of $\ln(p(\mathbf{Z}|\mathbf{D_i})$ over the whole set of legitimate transmitted data vectors $\mathbf{D_i}$ is equivalent to finding the vector $\mathbf{D_i}$ that minimizes the Euclidean distance of $D(\mathbf{Z}, \mathbf{D_i})$, $i = 0, 1, \ldots, M-1$. Furthermore, by extending Equation 19.177 we arrive at:

$$\begin{aligned} D(\mathbf{Z}, \mathbf{D_i}) &= \sum_{q=0}^{U-1} Z_q^2 - \left(\sqrt{\frac{P}{2}} T_s \sum_{l=0}^{L-1} \alpha_l^2\right) \sum_{q=0}^{U-1} Z_q d_{i,q} \\ &+ \left(\sqrt{\frac{P}{2}} T_s \sum_{l=0}^{L-1} \alpha_l^2\right)^2 \sum_{q=0}^{U-1} d_{i,q}^2, \end{aligned} \tag{19.178}$$

where the first term on the right-hand side is a constant for all i values, i.e., for all transmitted data vectors. Furthermore, since we assumed that the constant-weight code is from the code set $\{\mathbf{D_i}\}$, the third term is also constant for $i = 0, 1, \ldots, M-1$. Consequently, the minimization of $D(\mathbf{Z}, \mathbf{D_i})$ is equivalent to the maximization of the correlation metrics of:

$$C(\mathbf{Z}, \mathbf{D_i}) = \sum_{q=0}^{U-1} Z_q d_{i,q}, \; i = 0, 1, \ldots, M-1, \tag{19.179}$$

which defines the corresponding decision rule based on selecting the signal corresponding to the maximum of the set of correlation metrics $C(\mathbf{Z}, \mathbf{D_i})$ over the set of M legitimate data vectors may be input to the transmitter schematic of Figure 19.17.

The detection complexity of a received symbol or vector is determined by both the length and the weight of the constant-weight code. For a constant-weight code $\mathrm{C}(U, U_k)$, the detection complexity is proportional to $O\left(\binom{U}{U_k} 2^{U_k}\right)$ – where O indicates the 'order' of complexity – since each activated subcarrier associated a binary '1' in the constant-weight code $\mathrm{C}(U, U_k)$ may convey a +1 or -1 data bit. However, if U number of different-rate transmission schemes are supported, and each rate is invoked with an equal *a-priori* probability, then the average detection complexity can be expressed as $O\left(3^U/U\right)$. This detection complexity becomes excessive, when evaluating Equation 19.179 for all possible code words, if the value of U is high. In the next two sections we impose some limitations on the transmission scheme by limiting the minimum distance of the constant-weight codes representing the FH patterns, in order to simplify the detection process.

19.8.3.2 Approach I

To this effect, let us assume that the synthesizer of the reference user in Fig 19.17 generates a total of $U = 2^m$ subcarriers. This design choice conveniently coincides with the practical implementational constraints of invoking the Fast Fourier Transform (FFT) for modulation. Assume furthermore that $U_1 = 2^n$, $n = 0, 1, \ldots, m$, number of subcarriers are activated. Then the U number of subcarriers can be divided into U_1 number of groups, each group having $W = U/U_1 = 2^{m-n}$ subcarriers. A U_1-bit symbol or vector now can be transmitted by U_1 number of subcarriers randomly selected from the U_1 groups, where each group contributes one activated subcarrier. Under the above constraints, each of the U_1 groups of W subcarriers can be independently detected. This reduces the number of combinations to be considered by the detector and hence the correlation metrics of Equation 19.179 can be reformulated as:

$$C(\mathbf{Z}, \mathbf{D_i}) = \sum_{q=0}^{W-1} Z_q d_{i,q} + \sum_{q=W}^{2W-1} Z_q d_{i,q} + \ldots + \sum_{q=(U_1-1)W}^{U_1 W-1} Z_q d_{i,q}, \tag{19.180}$$

where $C_j = \{d_{i,jW}, d_{i,jW+1}, \ldots, d_{i,(j+1)W-1}\}$ for $j = 0, 1, \ldots, U_1 - 1$ represents a constant-weight code $\text{C}(W, 1)$ having a weight of one. Since the MRC outputs in Figure 19.18 are mutually independent random variables, the U_1 terms in Equation 19.180 can be computed in a parallel fashion, which implies that each of the U_1 bits can be detected separately by simply considering the W number of MRC output samples. Consequently, the detection complexity of the U_1-bit symbol is now proportional to $O(2U_1W) \equiv O(2U)$, which is linearly dependent on the total number of subcarriers, but independent of the information rate.

Although the complexity of the above detection approach is low, it results in a reduced detection performance. In order to enhance the detection performance and at the same time simplify the computations, the FH patterns can be designed by selecting a subset of codes having a minimum distance of d from the constant-weight codes $\text{C}(U, U_1)$, which is discussed in the next subsection.

19.8.3.3 Approach II

Let $\text{C}(U, d, U_1)$ represent a constant-weight code set having a code length of U and weight of U_1, as discussed previously. Furthermore, let the minimum distance between any pair of codes from $\text{C}(U, d, U_1)$ be d. Then this code constitutes a specific subset of the constant-weight code $\text{C}(U, U_1)$, where the number of codewords was $M = \binom{U}{U_1}$. By contrast, let $\text{A}(U, d, U_1)$ represent the number of codewords of the constant-weight code $\text{C}(U, d, U_1)$. Then, if the frequency hopping patterns are determined now by all the $\text{A}(U, d, U_1)$ codewords, the detection complexity will be reduced from $O\left[\binom{U}{U_1}2^{U_1}\right]$ to $O\left[\text{A}(U, d, U_1)2^{U_1}\right]$. In fact, from our numerical results in Subsection 19.8.4.3, we will see that, if the value of the minimum distance d is sufficiently high and the SNR is also sufficiently high, then the BER performance will be very close to that of the receiver having an explicit knowledge of the FH pattern. This allows the receiver to blindly acquire a restricted set of FH patterns exhibiting a minimum distance of d, during the call-initiation process. This limited set of FH codes can be used by the transmitter to signal the actual FH code used for the transmission of 'payload' information to the receiver. According to the above philosophy a set of constant-weight codewords having a minimum distance of d is used to convey the side-information constituted by the index of the FH code in the fixed weight codebook to be used by the receiver. During the consecutive information transmission phase then the whole set of $\binom{U}{U_1}$ number of FH codes or patterns can be used without imposing any minimum distance limitations.

The nonlinear constant-weight code $C(U, d, U_1)$, where $d = 2v$ with v being a positive integer, has some well-known properties. Specifically, the constant-weight codes constitute a class of efficient codes suitable for error-correction or error-detection over both binary symmetrical and asymmetrical channels [486–488]. For completeness, some of them, which are related to the forthcoming analysis are characterized below.

Proposition 1 *Johnson bound [487]:*

$$A(U, 2v, U_1) \leq \frac{Uv}{Uv - U_1(U - U_1)}, \tag{19.181}$$

provided that $Uv > U_1(U - U_1)$.

Proposition 2 *[488][pp.528]: Given that q is the integer power of a prime number, we have:*

$$\begin{aligned} A(q^2 + 1, 2q - 2, q + 1) &= q(q^2 + 1), \\ A(q^3 + 1, 2q, q + 1) &= q^2(q^2 - q + 1). \end{aligned} \tag{19.182}$$

Proposition 3 *[488][pp.528]: Assuming that a* $4r \times 4r$ *Hadamard matrix exists, we have:*

$$\begin{aligned} A(4r - 2, 2r, 2r - 1) &= 2r, \\ A(4r - 1, 2r, 2r - 1) &= 4r - 1, \\ A(4r, 2r, 2r) &= 8r - 2. \end{aligned} \tag{19.183}$$

Based on the above limitations in the context of Approach I and II, we can exploit a range of further properties of the code $C(U, 2v, U_1)$. Let us now derive the expressions of the error probability for the SFH/MC DS-CDMA system using blind joint soft-detection.

19.8.4 Performance of the SFH/MC DS-CDMA Receiver Without Knowledge of the FH Patterns: Blind Joint Soft-Detection

Figure 19.34 portrays an example of the previously discussed constant-weight codes. The SFH/MC DS-CDMA system of Figure 19.17 uses a constant-weight code $C(U, 2v, U_1)$ for activating the subcarriers. In this section we employ a receiver relying on no knowledge of the FH pattern itself. However, the receiver is aware of the number of the active subcarriers. The receiver's task is then to evaluate the previously derived correlation metrics seen in Equation 19.179. These metrics must be evaluated for two different scenarios in order to demodulate a parallel U_1-bit symbol or vector.

Specifically, we have to consider those codes, which have identical '1' positions in the FH code, corresponding to identical activated subcarriers, but potentially conveying different data symbols on these active subcarriers. We refer to these as intra-set codes or intra-codes, for short, which are exemplified in Figure 19.34. Explicitly, the above intra-codes are derived from the same constant-weight code and activate the same subcarriers. By contrast, the so-called inter-codes of Figure 19.34 are derived from different constant-weight codes, which have the same number of '1' FH code positions, but activate different subcarriers.

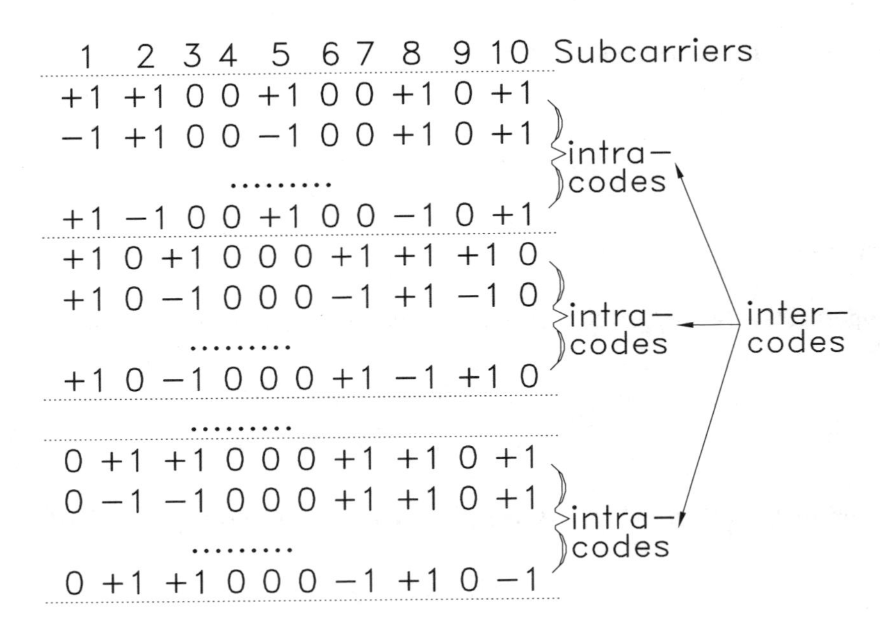

Figure 19.34: Example of intra-codes and inter-codes of a constant-weight code $C(10, 4, 5)$ for SFH/MC DS-CDMA system using 10 subcarriers.

Consequently, in our proposed blind joint soft-detection scheme two different types of errors exist, the intra-code errors and the inter-code errors. The probability of these errors is denoted by $P_{intra}(\cdot)$ and $P_{inter}(\cdot)$, respectively. The intra-code errors do not result in opting for a code other than the transmitter's code and hence do not inflict a constant-weight FH code decision error, they simply result in some bits being demodulated erroneously. By contrast, the inter-code errors may lead not only to erroneous bit decisions, but also to opting for another constant-weight code and hence to potentially more severe bit decision error events. The derivation of the exact expression for the probability of correct vector or symbol decoding is arduous, since the error probabilities depend on all the $A(U, 2v, U_1)2^{U_1}$ number of correlation metrics evaluated according to Equation 19.179. Hence below we derive the union upper-bound of the error probability.

Let us divide the entire set of $A(U, 2v, U_1)2^{U_1}$ number of information vectors or symbols into $A(U, 2v, U_1)$ number of sets, where the vectors from the same set activate the same subcarriers. Hence the vectors of the different sets constitute 'inter-codes', which were exemplified in Figure 19.34. Hence each set has 2^{U_1} vectors, since the information transmitted by an active subcarrier might be +1 or -1 and there are U_1 active subcarriers. Let the first vector in the first set, which is denoted by X_{11}, be transmitted. Then the union upper-bound probability of an intra-code error can be expressed as [90, 485]:

$$P_{intra}(Up) = \sum_{j=2}^{2^{U_1}} Q\left(\sqrt{\frac{|X_{11} - X_{1j}|^2}{4\sigma^2}}\right), \tag{19.184}$$

where X_{1j} is the jth vector of the first set, $|X_{11} - X_{1j}|^2$ represents the distance between X_{11} and X_{1j}, and σ^2 is the common variance of the MRC output samples, which is given by Equation 19.165. By contrast, the union upper-bound probability of an inter-code error can be written as:

$$P_{inter}(Up) = \sum_{i=2}^{A(U,2v,U_1)} \sum_{j=1}^{2^{U_1}} Q\left(\sqrt{\frac{|X_{11} - X_{ij}|^2}{4\sigma^2}}\right), \tag{19.185}$$

where again, $|X_{11} - X_{ij}|^2$ represents the distance between X_{11} and X_{ij}.

Consequently, the union upper-bound probability of error is the sum of Equation 19.184 and Equation 19.185, yielding:

$$P(Up) = P_{intra}(Up) + P_{inter}(Up). \tag{19.186}$$

Furthermore, the joint probability of a correct detection plus that of an intra-code decision error represents the probability of the event that the receiver selected the correct constant-weight code matched to the transmitted one. This event corresponds to the correct FH pattern detection probability, which can be interpreted as the correct acquisition - or acquisition success - probability expressed as:

$$\lambda = 1 - P_{inter}(Up). \tag{19.187}$$

With the aid of Equations 19.184 to (19.187, we can now approximate the union upper-bound bit error probability performance of the SFH/MC DS-CDMA scheme. For the previously introduced Approach II of subsection 19.8.3.3 we will also consider a more accurate error probability expression.

19.8.4.1 Approach I

Recall from our previous discussions that according to our Approach I in subsection 19.8.3.2 each group of $W = U/U_1$ subcarriers of the U_1 groups can be separately detected, since each hosts one active subcarrier. The codes utilized by the W subcarriers of a group are equivalent to a class of $C(W,2,1)$ code with $A(W,2,1) = W$. This class of constant-weight codes includes W number of orthogonal inter-codes, each activating a single carrier in one of the W possible positions. In other words, this scenario can be viewed as considering the independent detection of U_1 number of length W codes, which have a weight of 1 and a minimum distance of 2. Reduced detection complexity is attained due to decomposing the original detection problem into U_1 number of low-complexity detection steps.

Since the cross-correlation between two intra-codes is -1, hence, the intra-code error probability conditioned on the CIR tap attenuations can be written using Equation 19.184 as:

$$P_{intra}(Up,\gamma) = Q\left(\sqrt{2\gamma}\right), \tag{19.188}$$

where γ is given by Equation 19.171. By contrast, the union upper-bound of an inter-code FH sequence error conditioned on the CIR tap attenuations can be written as:

$$P_{inter}(Up,\gamma) = 2(W-1)Q\left(\sqrt{\gamma}\right). \tag{19.189}$$

Hence the union upper-bound probability of intra- or inter-code FH pattern errors conditioned on the CIR tap attenuations is the sum of Equation 19.188 and Equation 19.189, as we have seen in Equation 19.186. The average union upper-bound error probability per bit for a given number of hits, K_h, is calculated from the conditional union upper-bound probability by weighting it with the aid of its probability of occurrence expressed in terms of the PDF of γ, $f(\gamma)$, which is given by Equation 19.58 with N_p replaced by L and $\overline{\gamma}_c$ being given by Equation 19.171. It can be shown that:

$$P^I(K_h,\overline{\gamma}_c) = P(K_h,\overline{\gamma}_c) + 2(W-1)P\left(K_h,\frac{\overline{\gamma}_c}{2}\right), \tag{19.190}$$

where $P(K_h,x)$ is given by Equation 19.172.

The average union upper-bound bit error probability for the receiver using the previously introduced blind soft-detection Approach I of subsection 19.8.3.2 can be computed by substituting Equation 19.168 and Equation 19.190 into Equation 19.166 with $P(K_h,\overline{\gamma}_c)$ replaced by $P^I(K_h,\overline{\gamma}_c)$.

The acquisition success probability for the case of potential hits inflicted by the $K-1$ interfering users is expressed as:

$$\lambda(K_h) = 1 - 2(W-1)P_b(K_h,\overline{\gamma}_c). \tag{19.191}$$

Then the *average* acquisition probability for the proposed blind joint soft-detection Approach I of subsection 19.8.3.2 can be computed by averaging Equation 19.191 over the distribution of the number of hits, yielding:

$$\lambda = \sum_{K_h=0}^{K-1} \binom{K-1}{K_h} P_h^{K_h}(1-P_h)^{K-1-K_h}\lambda(K_h). \tag{19.192}$$

19.8.4.2 Approach II

For the previously introduced blind joint soft-decision Approach II of subsection 19.8.3.3, the knowledge of the distance between any pairs of constant-weight codewords is required. Explicitly, this distance information is necessary for the computation of the union upper-bounds for the probability of intra-code and inter-code FH pattern detection errors in Equation 19.184 and Equation 19.185, as well as for the determination of the union upper-bound of the FH pattern detection error probability in Equation 19.186. For specific sets of constant-weight codes exhibiting an equal distance in the context of arbitrary pairs, the exact union upper-bound can be determined from the above-mentioned equations.

Note that for intra-codes and for sufficiently high SNRs, the probability of a single bit error in a U_1-bit symbol is significantly higher than that of two bit errors, assuming that the MRC outputs are i.i.d random variables. Hence the union upper-bound probability of an intra-code error in Equation 19.184 can be approximated by:

$$P_{intra}(Up) \approx P_{intra}(II) = \sum_{j=2}^{U_1} Q\left(\sqrt{\frac{|X_{11} - X_{1j}|^2}{4\sigma^2}}\right), \tag{19.193}$$

where we replaced Up by II, in order to indicate that this probability is now not the union-upper bound of an intra-code error according to Approach II. Furthermore, X_{11} and X_{1j} are two intra-codes activating the same subcarrier, but having a distance of ‘one’ between them, representing a one-bit error. Hence, the cross-correlation of X_{11} and X_{1j} is $(1 - \frac{2}{U_1})$, and consequently the error probability of Equation 19.193 can be written as [90]:

$$P_{intra}(II, \gamma) = (U_1 - 1)Q\left(\sqrt{2\gamma}\right). \tag{19.194}$$

By contrast, for an inter-code error under sufficiently high SNRs, according to Figure 19.34, the metrics computed from Equation 19.179 for the specific data vectors $\mathbf{D_i}$ other than the transmitted vector are maximized, if the bits of $\mathbf{D_i}$ in the positions corresponding to the activated subcarriers of the transmitter were identical to the transmitted bits. These cases constitute the most probable events of inter-code errors and hence Equation 19.185 can be approximated by:

$$P_{inter}(Up) \approx P_{inter}(II) = 2^v \left[A(U, 2v, U_1) - 1\right] Q\left(\sqrt{\frac{|X_{11} - X_{ij}|^2}{4\sigma^2}}\right), \tag{19.195}$$

where again, Up was replaced by II, in order to avoid confusion with the union upper-bound of the inter-code error probability. Furthermore, X_{11} and X_{ij} are two inter-codes having a minimum distance of $d = 2v$. The conditional probability of error in Equation 19.195 can be hence simplified to:

$$P_{inter}(II, \gamma) = 2^v \left[A(U, 2v, U_1) - 1\right] Q\left(\sqrt{v\gamma}\right). \tag{19.196}$$

Note that in deriving Equation 19.196, according to Figure 19.34, the cross-correlation between X_{11} and X_{ij} is $1 - \frac{v}{U_1}$, since the '0' elements (Off-state) of the constant-weight code are included in the correlation computations.

The total bit error probability conditioned on the CIR tap attenuations is constituted by both intra- and inter-code errors, which can be approximated as:

$$P(II, \gamma) = \frac{1}{U_1} P_{intra}(II, \gamma) + \frac{1}{2} P_{inter}(II, \gamma) \tag{19.197}$$

according to the above analysis. The average bit error probability for a given number of hits, K_h, is calculated from the conditional bit error probability of Equation 19.197 by averaging it with respect to the PDF of γ, $f(\gamma)$, which is given by Equation 19.58 with N_p replaced by L and $\overline{\gamma}_c$ being given by Equation 19.171. Following this averaging or integration we have:

$$P^{II}(K_h, \overline{\gamma}_c) = \frac{U_1 - 1}{U_1} P(K_h, \overline{\gamma}_c) + \frac{2^v \left[A(U, 2v, U_1) - 1\right]}{2} P\left(K_h, \frac{v\overline{\gamma}_c}{2}\right), \tag{19.198}$$

where $P(K_h, x)$ is given by Equation 19.172.

The average bit error probability of the receiver using the blind soft-detection Approach II of subsection 19.8.3.3 can be computed by substituting Equation 19.168 and Equation 19.198 into Equation 19.166 with $P(K_h, \overline{\gamma}_c)$ replaced by $P^{II}(K_h, \overline{\gamma}_c)$.

The acquisition success probability for the given number of hits inflicted by the $K - 1$ interfering users can be expressed as:

$$\lambda(K_h) = 1 - 2^v \left[A(U, 2v, U_1) - 1\right] P(K_h, \frac{v\overline{\gamma}_c}{2}). \tag{19.199}$$

Finally, the average acquisition success probability of the blind joint soft-detection Approach II can be computed by averaging Equation 19.199 over the distribution of the number of hits, which was given in Equation 19.192 with $\lambda(K_h)$ replaced by Equation 19.199. Let us now characterized the performance of the blind soft detection Approach I and Approach II.

19.8.4.3 Numerical Results

In Figure 19.35 we estimated the upper-bound BER of Approach I upon combining $L = 1, 3, 5$ paths in the receiver. The BER of hard-detection based on the approach of subsection 19.8.2 was also plotted as a benchmark, assuming that the receiver exploited the explicit knowledge of the FH patterns. The parameters related to the computations were shown in the figures. The results demonstrate that the system provides dramatic BER improvements, when the number of combined diversity paths, L, increases. However, the results also demonstrate that opting for the blind joint soft-detection Approach I of subsection 19.8.3.2 increased the BER with respect to hard-detection.

In Figure 19.36 we evaluated the intra-code Word Error Rate (WER) and the inter-code word error rate, as well as their sum for the SFH/MC DS-CDMA system using the blind joint soft-detection Approach II of subsection 19.8.3.3. We employed the FH description code of C(16,12,8), having a minimum distance of $d = 12$ between the constant-weight codes,

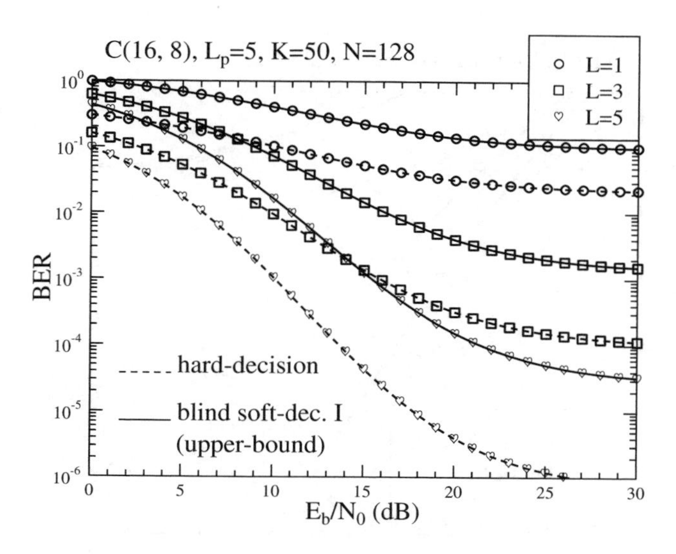

Figure 19.35: Comparison of the BER versus bit-SNR performance between the hard-detection technique of subsection 19.8.2 and blind soft-detection using Approach I evaluated from Equations 19.190, (19.166 and (19.168 for the constant-weight code C(16,8), $L_p = 5$ resolvable paths, diversity combining orders $L = 1, 3, 5$, $K = 50$ users and a bit-duration to chip-duration ratio of $N_b = 128$. The remaining parameters are explicitly stated at the top of the illustration. The BER performance of hard-detection is superior to that of the blind joint soft-detection using Approach I.

which corresponded to $v = 6$. Since the intra-code word error probability is the codeword error probability of the soft-detection technique proposed, when the receiver has the *a-priori* knowledge of the FH patterns, the results shown in this figure explicitly illustrate the comparison of the word error rate performance between the SFH/MC DS-CDMA system using blind joint soft-detection and that using soft-detection with *a-priori* knowledge of the FH patterns. From the results we concluded that under various SNR conditions the total WER was dominated by one of its contributing factors. Namely, for very low SNR per bit values or a very low number of diversity fingers, such as $L = L_p = 1$, the performance was dominated by the word error rate of the inter-code decisions. By contrast, for moderate to high SNRs per bit and for high diversity orders $L = L_p = 5$, the word error rate of the intra-code errors was more dominant. However, for the case of $L = L_p = 3$ and for sufficiently high SNRs per bit, the resultant word error rate was dependent on both the intra-code and the inter-code error events. Based on the fact that the word error rate of blind joint soft-detection is the sum of the intra-code and inter-code WER, the blind joint soft-detection approach is outperformed by the soft-detection technique using the *a-priori* knowledge of the FH patterns. This becomes explicit for $L = 1$, $L = 3$ and for $L = 5$ at relatively low SNR per bit values in Figure 19.36. However, for $L = 5$ at sufficiently high SNR per bit values there is effec-

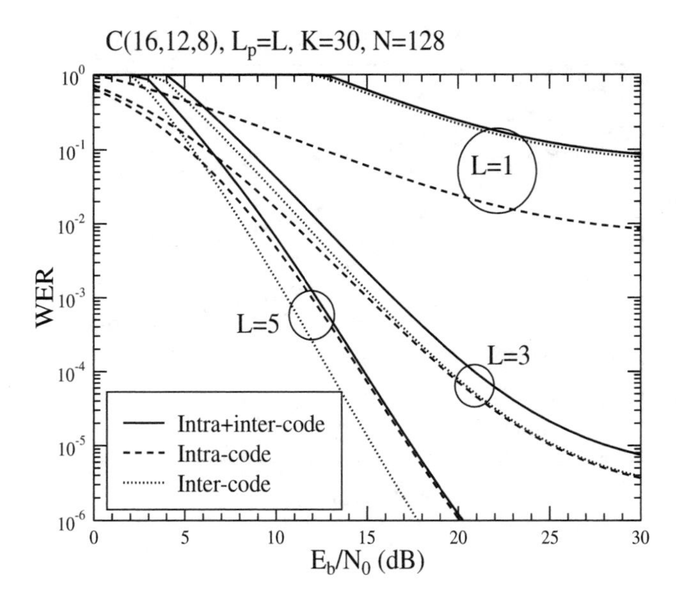

Figure 19.36: Word error rate (WER) versus SNR per bit performance of the constant-weight code based SFH/MC DS-CDMA system using the blind soft-detection Approach II. For $v > 2$, sufficiently high SNR per bit values and sufficiently high diversity orders, the average error probability of blind joint soft-detection using Approach II is dominated by the intra-code errors.

tively very little difference between the word error rate performance curves of the blind joint soft-detection scheme (Intra+Inter-code curve) and the soft-detection arrangement using the *a-priori* knowledge of the FH patterns (Intra-code curve). Since the inter-code word error rate is a function of v, we infer from Equation 19.194 and Equation 19.196 that if $v > 2$, the total word error rate will be dominated by Equation 19.194, provided that the SNR per bit is sufficiently high and the channel fading is sufficiently benign or the diversity order is sufficiently high. This property suggests that for $v > 2$ intra-code error-control techniques can be introduced, in order to correct the intra-code errors by increasing the minimum distance between the intra-codes and hence decreasing the intra-code word error rate of Equation 19.194. This would decrease the total word error rate of the blind soft-detection using Approach II.

Finally, in Figure 19.37 we characterized the probability of the successful FH code acquisition for the blind soft-detection Approach I and that of the blind soft-detection Approach II for the multi-rate transmission scenario discussed previously, under the assumption that all the interfering users employed the same constant-weight codes, namely C(32,16). From the results we observe that for Approach I the probability of successful FH code acquisition becomes higher, when the constant-weight codes associated with higher information rates are used. By contrast, for Approach II and for the specific set of constant-weight codes we used,

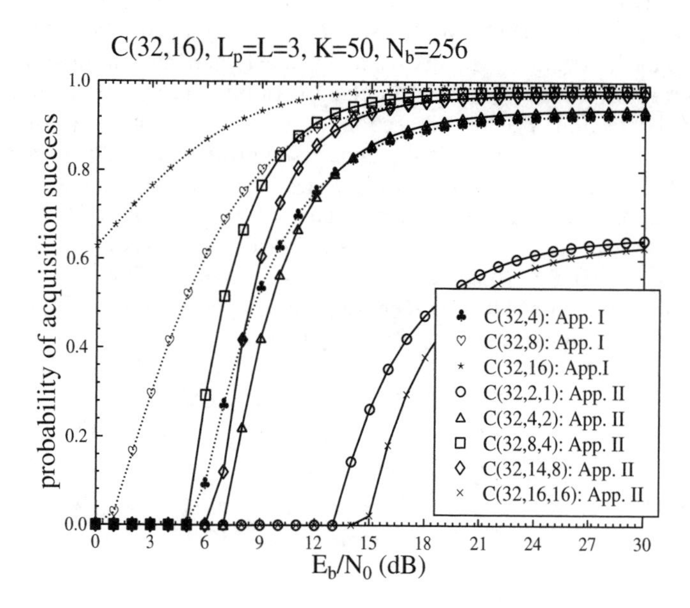

Figure 19.37: Acquisition success probability of the constant-weight code based SFH/MC DS-CDMA system using the blind soft-detection Approach I and II, under the assumption of constant spreading gain for multi-rate based systems. If the SNR per bit is sufficiently high, blind joint soft-detections can acquire the FH patterns used with a high probability, while also detecting the transmitted information bits.

C(32,8,4) achieved the best performance. Nevertheless, if the SNR per bit is sufficiently high, the curves corresponding to the constant-weight codes of C(32,4,2), C(32,8,4) and C(32,14,8) will tend to a successful acquisition probability of unity.

This allows the receiver to blindly acquire a restricted set of FH patterns exhibiting a minimum distance of d. This restricted set of FH patterns can then be used by the transmitter to signal the index of the actual FH codes used for the transmission of 'payload' information to the receiver. During the information transmission phase then a randomly selected set of FH patterns from the $\binom{Q}{U}$ number of FH codes can be used without imposing any minimum distance limitations. In other words, following the blind detection of the 'side-information' constituted by the FH codes used by the transmitter, successive communications can be based on the explicit knowledge of the FH patterns.

19.9 Chapter Summary and Conclusion

Following a brief classification of the most popular MC-CDMA systems in this chapter we introduced slow frequency hopping for improving the system's performance and flexibility. The proposed SFH/MC DS-CDMA system is capable of efficiently amalgamating the techniques of slow FH, OFDM and DS-CDMA. Nonlinear constant-weight codes have been introduced, in order to control the associated FH patterns, and hence to activate a number of subcarriers, in order to support multi-rate services. Hard-detection or blind joint soft-detection have been proposed for signal detection. The hard-detection exploits the explicit knowledge of of the FH pattern employed, while the soft-detection assumes no *a priori* knowledge of the FH patterns used. The properties of the constant-weight FH codes have been investigated in conjunction with specific system parameters, and the performance of the proposed SFH/MC DS-CDMA systems using either hard-detection or blind joint soft-detection has been evaluated, under the conditions of supporting constant-rate or multi-rate services. From the results we concluded that the blind joint soft-detection techniques are capable of detecting the transmitted information and simultaneously acquiring the FH patterns.

Chapter 20

Space-Time Spreading-Assisted Broadband Multicarrier DS-CDMA: System Design and Performance Analysis[1]

20.1 Motivation and Outline

Following the philosophy of the generic communications system framework proposed in the introduction, in Chapter 8 an adaptive STS-aided W-CDMA system was proposed, which was capable of adapting the mode of operation of its STS scheme and its corresponding data rate according to the near-instantaneous frequency-selectivity information fed back from the mobile receiver to the base station's transmitter. The numerical results of Chapter 8 showed that this adaptive STS scheme is capable of efficiently exploiting the diversity potential provided by the channel's frequency-selectivity, hence significantly improving the effective throughput of W-CDMA systems.

In this chapter our discussions evolve further by initially considering the advantages and disadvantages of single-carrier CDMA, MC-CDMA and MC-DS-CDMA in Section 20.2, with particular attention devoted to communicating in diverse propagation environments exhibiting a time-variant amount of dispersion. More specifically, the benefits and deficiencies of these three systems are analyzed, when aiming for supporting ubiquitous communications

[1] *Single- and Multi-Carrier CDMA Multi-User Detection, Space-Time Spreading, Synchronization, Networking and Standards.* L.Hanzo, L-L.Yang, E-L.Kuan and K.Yen,
 ISBN 0-470-86309-9

over a variety of propagation channels encountered in indoor, open rural, suburban and urban environments. We will demonstrate that, when communicating in such diverse environments, both SC DS-CDMA and MC-CDMA exhibit certain limitations that are hard to avoid. By contrast, when appropriately selecting the system parameters and using transmit diversity, MC DS-CDMA becomes capable of adapting to such diverse propagation environments at a reasonable detection complexity.

Hence again, following the philosophy of the generic communications system framework proposed in the introduction, in Section 20.3 we focus our attention on MC DS-CDMA schemes using STS. Specifically, MC DS-CDMA using STS is investigated in the context of broadband communications over frequency-selective Rayleigh fading channels. We also consider the attainable capacity extension of broadband MC DS-CDMA with the advent of using Time-Frequency-domain (TF-domain) spreading. Furthermore, the detection issues of broadband MC DS-CDMA using TF-domain spreading are investigated. Our study shows that by appropriately selecting the system parameters, STS-assisted broadband MC DS-CDMA is capable of supporting ubiquitous communications over various communication environments including indoor, open rural, suburban and urban areas without BER performance degradation. Furthermore, we demonstrate that STS-based transmit diversity schemes can be designed according to the required diversity gain, while maintaining a near-constant BER in various communication environments, as long as frequency-selective Rayleigh fading channels are encountered.

20.2 Comparison of CDMA, MC-CDMA and MC DS-CDMA Communicating in Diverse Propagation Environments

In this section we identify some of the key problems that may be encountered when designing a broadband multiple-access systems having bandwidth on the order of tens or even hundreds of MHz. We commence with a comparative discussion in terms of the characteristics of three typical code-division multiple-access (CDMA) schemes, namely single-carrier direct-sequence CDMA (SC DS-CDMA), multicarrier CDMA (MC-CDMA) and multicarrier DS-CDMA (MC DS-CDMA). Specifically, their benefits and deficiencies are analyzed, when aiming for supporting ubiquitous communications over a variety of channels encountered in indoor, open rural, suburban and urban environments. We will show that, when communicating in such diverse environments, both SC DS-CDMA and MC-CDMA exhibit certain limitations that are hard to avoid. By contrast, we will show that when appropriately selecting the system parameters and using transmit diversity, MC DS-CDMA becomes capable of adapting to such diverse propagation environments at a reasonable detection complexity. Specifically, in the second part of this section we consider space-time spreading (STS) assisted broadband MC DS-CDMA and discuss its system architecture, its achievable capacity improvement as well as its signal detection techniques and the system performance attained.

20.2.1 Introduction

The future generations of broadband wireless mobile systems will aim for supporting a wide range of services and bit rates by employing a variety of techniques capable of achieving the highest possible spectrum efficiency. The list of desirable characteristics is long, including high user capacity, the ability of mitigating the effects of interference, convenience of frequency resource management, support of soft hand-over, etc. Code-division multiple-access (CDMA) schemes have been considered attractive multiple-access schemes in both the second generation (2G) and third generation (3G) wireless systems. Recently the interest in wireless communications has been shifting in the direction of broadband systems [59]. This is mainly due to the explosive growth of the Internet and the continued dramatic increase in demand for all types of advanced wireless multimedia services, including conventional voice and data transmissions. Furthermore, the future generations of broadband wireless systems are expected to support ubiquitous communications, regardless of the propagation environment encountered, while maintaining the required Quality of Service (QoS).

In the context of broadband wireless communications using CDMA without the assistance of frequency-hopping, the main multiple-access options include single-carrier direct-sequence CDMA (SC DS-CDMA) using Time-domain (T-domain) DS spreading [78], multicarrier CDMA (MC-CDMA) using Frequency-domain (F-domain) spreading [75], as well as multicarrier DS-CDMA (MC DS-CDMA) using T-domain DS spreading of the individual sub-carrier signals [75]. The power spectrum as well as the spreading model of the above three CDMA arrangements are shown in Figure refsignal-waveform, all of which have approximately the same bandwidth. This figure will be discussed in more depth in the next subsection.

It is well recognized that the achievable capacity of broadband wireless communication systems is limited by the time-varying characteristics of the broadband channels encountered. Hence, in the first part of this section we investigate the behaviour of the above three CDMA schemes, when communicating over broadband wireless channels. We will show that regardless of the communication environments encountered, both SC DS-CDMA and MC-CDMA exhibit certain limitations that are hard to avoid. We will also demonstrate that by appropriately selecting the system parameters, broadband MC DS-CDMA augmented by space-time spreading (STS) based transmit diversity is capable of mitigating the problems encountered by both SC DS-CDMA and MC-CDMA. Hence, in the second part of this section we investigate STS-enhanced broadband MC DS-CDMA, highlighting the system architecture, the achievable capacity improvement and the appropriate signal detection techniques. Let us first embark on a rudimentary overview of the above three CDMA schemes.

20.2.2 Overview of CDMA Schemes Using No Frequency-Hopping

Prasad and Hara have provided an excellent overview of both SC DS-CDMA and the different multicarrier CDMA schemes in [75]. In this subsection we provide a different digest of SC DS-CDMA, MC-CDMA and MC DS-CDMA, with specific emphasis on the transmitted signals' structures. As discussed in [75], multicarrier CDMA using DS-spread sub-carrier signals can be further divided into multitone DS-CDMA [482], orthogonal MC DS-CDMA [65] and MC DS-CDMA using no sub-carrier overlapping [472]. The authors of this section have shown in [489] that the above three types of MC DS-CDMA schemes using

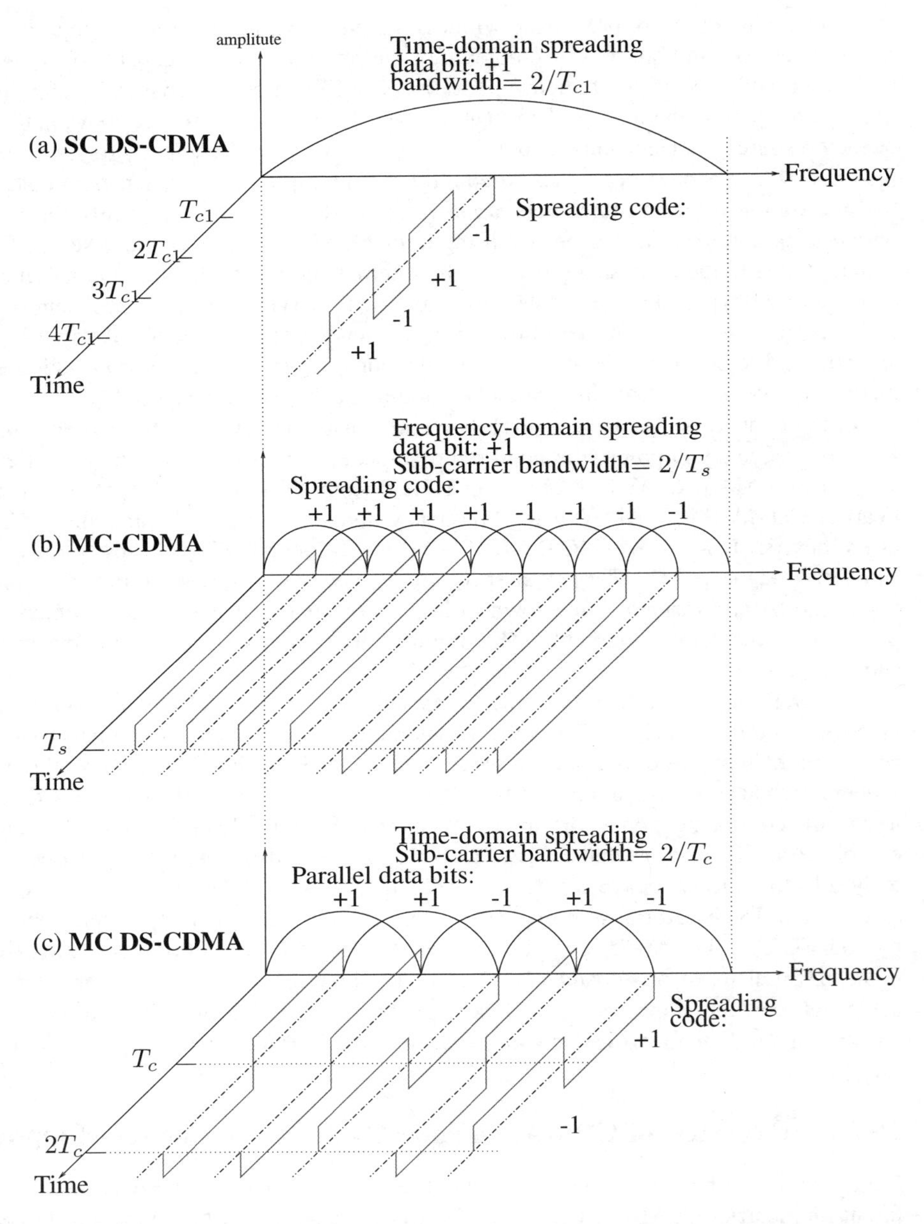

Figure 20.1: Power spectra and time-domain signal waveforms associated with single-carrier DS-CDMA (SC DS-CDMA), MC-CDMA as well as MC DS-CDMA assuming the same total system bandwidth.

a specific frequency spacing between two adjacent sub-carriers can be unified in the family of generalized MC DS-CDMA schemes using an arbitrary frequency spacing, Δ, between two adjacent sub-carriers. Furthermore, it was shown in [489] that the orthogonal MC DS-CDMA scheme, which uses the specific frequency spacing of $\Delta = 1/T_c$, where T_c represents the chip-duration of the DS spreading codes invoked, is capable of achieving the optimum Bit Error Ratio (BER) of MC DS-CDMA using an arbitrary frequency spacing, Δ, provided that the channel-induced fading is not too severe. Therefore, in this section we typically refer to orthogonal MC DS-CDMA simply as MC DS-CDMA, unless it would cause confusion. However, our analysis can be readily extended to MC DS-CDMA schemes using different frequency spacing values of Δ.

20.2.2.1 SC DS-CDMA

By definition, a SC DS-CDMA scheme transmits DS-spread signals using a single carrier. In SC DS-CDMA the original data stream is spread using a given spreading code in the T-domain, as shown in Figure 20.1(a). Hence, the transmitted signal in SC DS-CDMA using Binary Phase Shift Keying (BPSK) modulation can be expressed as:

$$s_{DS}(t) = \sqrt{2P} \sum_{i=-M}^{M} \sum_{j=0}^{N-1} b[i]c[j]p_{T_{c1}}(t - iT_b - jT_{c1}) \cos(2\pi f_c t), \tag{20.1}$$

where P and f_c represent the transmitted power and transmitted carrier frequency, respectively, T_b and T_{c1} represent the bit duration and chip-duration, respectively, and the processing gain of $N = T_b/T_{c1}$ represents the number of chips per bit. Furthermore, in (20.1) $2M + 1$ represents the number of bits conveyed by a transmitted data burst, $b[i] \in \{+1, -1\}$ is the ith transmitted bit, while $c[j] \in \{+1, -1\}$ is the jth chip of the spreading code, and finally, $p_\tau(t)$ represents the chip waveform defined over the interval $[0, \tau)$.

The number of users supported by SC DS-CDMA depends on the achievable processing gain and on the cross-correlation characteristics of the spreading codes. When a frequency-selective fading channel is characterized by the superposition of several signals having different delays in the T-domain, the number of users supported by SC DS-CDMA is also determined by the auto-correlation characteristics of the spreading codes.

20.2.2.2 MC-CDMA

MC-CDMA conveys the transmitted signals using a number of sub-carriers. In MC-CDMA the transmitter spreads the original data stream across the N_p number of sub-carriers using a given N_p-chip spreading code of $\{c[0], c[1], \ldots, c[N_p - 1]\}$ [75]. As seen in Figure 20.1(b), the transmitted signal of MC-CDMA using BPSK modulation can be expressed as:

$$s_{MC}(t) = \sqrt{\frac{2P}{N}} \sum_{i=-M}^{M} \sum_{j=0}^{N_p-1} b[i]c[j]p_{T_s}(t - iT_s) \cos\left[2\pi(f_c + f_j)t\right], \tag{20.2}$$

where P, $b[i]$, $c[j]$, f_c and $p_T(t)$ have the same meaning as in (20.1). In (20.2) N_p represents the number of sub-carriers having corresponding frequencies of $\{f_j\}_{j=0}^{N_p-1}$, which are invoked for conveying the N_p chips of the data stream $b[i]$, $i = -M, \ldots, M$. Hence T_s in (20.2) represents both the symbol duration and the chip-duration, i.e. in MC-CDMA, the symbol duration and the chip-duration assume the same value.

The number of users supported by MC-CDMA depends on the processing gain and on the cross-correlation characteristics of the different spreading codes. However, in MC-CDMA each sub-carrier signal is assumed to be experiencing flat-fading, and no multipath-induced intersymbol interference (ISI) is imposed on the sub-carrier signals, which would impair their auto-correlation. Hence the number of users supported by MC-CDMA is independent of the auto-correlation characteristics of the spreading codes.

If, however, the MC-CDMA transmission bit rate is high [75], the signal would experience frequency selective fading. In order to prevent the sub-carriers from experiencing frequency-selective fading in case of high-rate transmission, the input data has to be serial-to-parallel (S-P) converted, mapping the original high-rate data to U number of reduced-rate streams – which have an extended symbol duration of $T_s = UT_b$ – before mapping the N_p chips of the reduced-rate substreams to the frequency domain sub-carriers. This is because it is crucial for the MC-CDMA signal to experience frequency non-selective fading for each sub-carrier [75]. The number of sub-carriers in MC-CDMA invoking U-bit S-P conversion is about a factor of U times the number of sub-carriers in MC-CDMA without invoking S-P conversion but having the same system bandwidth. However, the spreading gain associated with each data bit is independent of U and remains a constant, namely N_p.

20.2.2.3 MC DS-CDMA

Generally, MC DS-CDMA transmits T-domain DS-spread signals using multiple sub-carriers, as shown in Figure 20.1(c). The transmitter of MC DS-CDMA usually includes a S-P converter, which reduces the sub-carrier data rate and increases the processing gain associated with each sub-carrier signal. As shown in Figure 20.1(c), the transmitted signal in MC DS-CDMA using BPSK modulation associated with each sub-carrier can be expressed as:

$$s_{MDS}(t) = \sqrt{2P} \sum_{i=-M}^{M} \sum_{u=1}^{U} \sum_{j=0}^{N-1} b_u[i]c[j]p_{T_c}(t - iT_s - jT_c) \cos\left[2\pi(f_c + f_u)t\right], \quad (20.3)$$

where P is each sub-carrier's transmitted power, $\{f_u\}_{u=1}^{U}$ are the frequencies corresponding to the U number of sub-carriers, $b_u[i]$ represents the ith bit on the uth sub-carrier, while $(2M + 1)$ is the total number of bits transmitted by the uth sub-carrier during a transmission burst. Furthermore, the MC DS-CDMA signal has a symbol duration of T_s and a chip-duration of T_c.

Similarly to SC DS-CDMA but in contrast to MC-CDMA, the DS spread sub-carrier signals in MC DS-CDMA may experience frequency-selective fading. Hence, the number of users supported by MC DS-CDMA is determined by the processing gain, the auto-correlation as well as the cross-correlation characteristics of the spreading codes employed.

In MC DS-CDMA the chips of a DS spread sub-carrier signal can be further interleaved across different sub-carriers, i.e., using F-domain interleaving, in order to achieve a higher frequency diversity [65]. For a given number of sub-carriers, each having a constant chip rate, the number of bits per symbol and the processing gain constituted by the number of chips per symbol decrease upon increasing the interleaving depth, S. Hence, the number of users supported by MC DS-CDMA also decreases upon increasing the interleaving depth. This is because the number of DS spreading codes having good correlation characteristics is determined by the number of chips per symbol.

Instead of the F-domain interleaving, in MC DS-CDMA the transmitted data stream can be spread in both the T-domain and the F-domain, i.e. in the TF-domain, in order to mitigate the problem that the number of users supported by MC DS-CDMA decreases upon increasing the interleaving depth. However, this is a topic rarely investigated so far in the literature [490]. Referring to Figure 20.1(c) as an example, in order to elaborate further on this philosophy, let us assume that only a single data bit will be transmitted, but let the parallel data bits of the $S = 5$ sub-carriers be replaced by the $S = 5$ chip values of $\{+1, +1, -1, +1, -1\}$ of a spreading code invoked for spreading the data in the F-domain across $S = 5$ number of different sub-carriers. The resultant bandwidth is again approximately the same as that of the other two schemes due to the overlapping of the sub-carriers. Then, the transmitted MC DS-CDMA signal benefits from both T-domain spreading and F-domain spreading. At the receiver, the MC DS-CDMA signal is despread using both the T-domain spreading code – having a length of $N = 2$ in Figure 20.1(c) – and the F-domain spreading code associated with a length of $S = 5$ in Figure 20.1(c). The total processing gain will be the product of the T-domain spreading code's processing gain and the F-domain spreading code's processing gain, namely $N \cdot S = 10$. Furthermore, the number of users supported by the MC DS-CDMA system is also determined by the above product of $N \cdot S = 10$, which is determined by the system bandwidth.

20.2.2.4 Flexibility Comparison

Above we have reviewed three typical CDMA schemes, namely SC DS-CDMA, MC-CDMA and MC DS-CDMA. In the context of broadband wireless communications the design and reconfiguration flexibility constitute important considerations. The flexibility of a multiple-access scheme depends on its degree of freedom defined as the number of independent parameters that can be controlled and adapted during the system design phase or that can be reconfigured near instantaneously during the communications session with the aid of advanced techniques facilitated by the concept of software radios. Let us assume that the above three typical CDMA schemes employ a given zero-to-zero system bandwidth of $2/T_{c1}$. Furthermore, they use a common chip waveform and employ the same data modulation scheme, namely BPSK. We also assume that these CDMA systems support a common data rate of $R_b = 1/T_b$. Then, in addition to the aforementioned degrees of freedom, the range of other parameters that can be reconfigured by the CDMA schemes considered are as follows:

- In the context of SC DS-CDMA no other degrees of freedom are available. In other words, the characteristics of the transmitted signal in SC DS-CDMA are fully determined by the above four degrees of freedom.

- In the context of MC-CDMA, another degree of freedom is the number of bits, U, involved in the S-P conversion. This parameter determines both the symbol duration and the chip-duration, which are expressed as $T_s = T_c = UT_b$. It also determines the total number of sub-carriers, which can be expressed as $Q = (2UT_b/T_{c1} - 1)$.

- In MC DS-CDMA systems there exist another three degrees of freedom. The first is the chip-duration, T_c, which determines the total number of sub-carriers, yielding, for example, $Q = (2T_c/T_{c1} - 1)$ for orthogonal MC DS-CDMA. The second is the number of bits, U, involved in the S-P conversion, which determines the symbol duration ($T_s = UT_b$). Furthermore, the above two degrees of freedom determine the spreading gain of each sub-carrier signal, which can be expressed as $N = UT_b/T_c$. They also determine the F-domain interleaving depth, S, across the sub-carriers, which can be expressed as $S = Q/U$. Finally, the third degree of freedom associated with MC DS-CDMA is the frequency spacing Δ between two adjacent sub-carriers [489]. The spacing between two adjacent sub-carriers can assume a value spanning the range from $1/T_s$ to $2/T_c$. MC DS-CDMA using an arbitrary spacing value of Δ includes the subclasses of multitone DS-CDMA [482] using $\Delta = 1/T_s$ and orthogonal MC DS-CDMA [65] using $\Delta = 1/T_c$ as special cases. The results of [489] show that the parameter Δ can be optimized for enhancing the MC DS-CDMA system's BER performance. Furthermore, the findings of [489] also demonstrated that the orthogonal MC DS-CDMA scheme can asymptotically achieve the optimum BER performance of the MC DS-CDMA using an arbitrary spacing value, provided that the channel fading is not too severe.

20.2.3 Broadband Wireless Communications Based on CDMA

The future generations of broadband wireless systems will aim at supporting a wide range of services and bit rates in a bandwidth on the order of tens or even hundreds MHz. These broadband wireless signals hence may have a bandwidth significantly higher than the coherence bandwidth of the channels encountered, therefore they will inevitably experience severe frequency-selective fading. Furthermore, broadband wireless systems using, for example, multicarrier transmissions may encounter a different Doppler frequency shift for the lowest and highest sub-carriers, due to the high frequency difference between them.

20.2.3.1 Deficiencies of Broadband SC DS-CDMA and Broadband MC-CDMA

In Table 20.1 we summarize some of the typical signal characteristics and the corresponding receiver models in the context of the SC DS-CDMA, MC-CDMA and MC DS-CDMA systems communicating over time-varying wireless communication channels. When aiming at supporting transmissions in diverse propagation environments, both SC DS-CDMA and MC-CDMA exhibit certain deficiencies. Specifically, in the context of broadband wireless communications, the system may have a 20 MHz bandwidth and may be required to support a bit rate of 1 Mbits/s. In this context both the broadband SC DS-CDMA and MC-CDMA will experience the following problems.

- **Communications** in diverse propagation environments cannot readily be supported by SC DS-CDMA or MC-CDMA. Assuming binary transmissions, such as BPSK modulation as an example, the transmitted symbol's duration and the data bit's duration

Alamo Regional Academy of Science and Engineering

A Non-Profit Organization Dedicated To Encouraging Youth Achievement in Science
P.O Box 701175 • San Antonio, Texas 78270
(210) 736-2716

____/____/____

OFDM
OFDMA
SC-FDMA
MC-CDMA
MC-DS CDMA

3 4 5 → 18
8 7 6 / 8 7 6 → 45
8 7 6 → 82
8 7 6 → 67
8 7 6 → 40

Table 20.1: Typical signal and receiver characteristics associated with SC DS-CDMA, MC-CDMA and MC DS-CDMA communicating over wireless communications channels

Multiple-access scheme	Number of sub-carriers	Spreading gain	Number of resolvable paths	Diversity combining
SC DS-CDMA	1	$\frac{T_b}{T_{c1}}$	$\left[\lfloor\frac{T_m}{T_{c1}}\rfloor+1, \lfloor\frac{T_M}{T_{c1}}\rfloor+1\right]$	RAKE
MC-CDMA	$\frac{2UT_b}{T_{c1}}-1$	$\frac{2T_b}{T_{c1}}$	1	F-domain
MC DS-CDMA	$\frac{2T_c}{T_{c1}}-1$	$\frac{T_s}{T_c}$	$\left[\lfloor\frac{T_m}{T_c}\rfloor+1, \lfloor\frac{T_M}{T_c}\rfloor+1\right]$	RAKE and/or F-domain
	Achievable diversity order	**Slow fading**	**Strong ISI**	**Correl. between combined comp.**
SC DS-CDMA	$\left[\lfloor\frac{T_m}{T_{c1}}\rfloor+1, \lfloor\frac{T_M}{T_{c1}}\rfloor+1\right]$	Yes, if $T_b < \frac{1}{\Delta F}$	Yes, if $T_b < T_M$	No
MC-CDMA	$\left(\frac{2UT_b}{T_{c1}}-1\right)/U$	Yes, if $T_s < \frac{1}{\Delta F}$	Yes, if $UT_b < T_M$	Yes, if $\frac{U}{T_s} < \frac{1}{T_m}$, or if $(T_b > T_m)$
MC DS-CDMA	$\left[\lfloor\frac{T_m}{T_c}\rfloor+1, \lfloor\frac{T_M}{T_c}\rfloor+1\right] \times \left(\frac{2T_c}{T_{c1}}-1\right)/U$	Yes, if $T_s < \frac{1}{\Delta F}$	Yes, if $UT_b < T_M$	Yes, if $\frac{U}{T_c} < \frac{1}{T_m}$

T_{c1}: Chip-duration of spreading codes in SC DS-CDMA
T_c: Chip-duration of spreading codes in MC DS-CDMA
T_b: Bit-duration of input information
T_s: Symbol duration
U: Number of bits involved in S-P conversion
T_m: Delay-spread of the environment having the lowest delay-spread;
T_M: Delay-spread of the environment having the highest delay-spread;
ΔF: Maximum Doppler frequency shift.

are the same in both of the above schemes. Propagation measurements conducted in typical wireless environments including indoor, open rural, suburban and urban areas show that the delay-spread is typically distributed over the range of $[0.1\mu s, 3\mu s]$ [273]. Hence, when communicating at 1 Mbits/s, these two schemes cannot perform well in environments having a delay-spread higher than $1\mu s$. Otherwise, severe inter-symbol interference (ISI) will be imposed on the adjacent symbols due to the delayed and unresolvable paths having relative delays higher than $1\mu s$. We might argue that in broadband MC-CDMA S-P conversion can be employed for rendering the symbol duration higher than the highest delay-spread of, for example, $> 4\mu s$, which would mitigate the ISI. However, employing MC-CDMA will result in an increased peak-to-average power fluctuation [456], due to the increased symbol duration and hence the increased number of sub-carriers. Furthermore, using S-P conversion cannot mitigate the following problem.

- **Frequency-diversity** may not be efficiently exploited in broadband SC DS-CDMA. By the same token, achieving frequency-diversity in MC-CDMA may be hampered, since significant correlation may exist between the fading envelopes of adjacent sub-carriers of a broadband MC-CDMA system. A broadband SC DS-CDMA scheme designed using a high number of RAKE fingers for propagation environments having a high delay-spread will combine noise, if the number of resolvable paths is low, since the environment encountered exhibits a low delay-spread. By contrast, a broadband SC DS-CDMA scheme designed using a low number of RAKE fingers is suitable for environments having a low delay-spread, but it will waste some of the effective received signal energy delivered by the paths that cannot be combined due to the low number of RAKE fingers in environments having a high delay-spread. As a remedy, hybrid schemes based on Selection Combining and Maximal Ratio Combining (SC/MRC) schemes may be employed, depending on the specific environment, which select only a fraction of the resolvable paths for combining, namely the dominant ones and then combine these ones using MRC. However, the high complexity imposed by the estimation of all resolvable paths and the sub-optimum performance due to the received signal energy loss associated with the uncombined paths are unavoidable. Consequently, a highly efficient diversity combining arrangement has to invoke an adaptive MRC scheme, which is capable of combining a time-variant number of resolvable paths encountered in various propagation environments. However, the cost of such a combining scheme is the associated increase of complexity. Again, in the context of broadband MC-CDMA the adjacent sub-carriers may be exposed to correlated fading, especially, if the delay-spread of the channel is relatively low, resulting in the relatively high coherence bandwidth. Consequently, combining the adjacent sub-carrier signals may not achieve the expected BER performance, when transmitting over such low-dispersion fading channels. As indicated in Table 20.1, the correlation of the fading envelopes of the adjacent combined sub-carriers cannot be removed by S-P conversion.

- **Multiuser detection (MUD)** is a highly efficient detection technique, which is capable of attaining near-single-user performance, while supporting a multiplicity of users and hence increasing the user capacity of wireless systems. The downlink (base to mobile) is expected to become the bottleneck in the forthcoming wireless Internet, which is likely to convey more down-loading type traffic than uplink (mobile to base) traffic. Moreover, the mobile handset has to be light and has to have a relatively low signal processing complexity. Hence at the Mobile Station (MS) relatively low-complexity detectors have to be employed. However, in order for broadband SC DS-CDMA or for broadband MC-CDMA to achieve the best possible BER performance, both require high-complexity MUDs. The detection algorithm's complexity typically increases at least linearly with the total number of users detected by the system. The complexity of using MUD cannot be sufficiently relaxed, even if orthogonal spreading codes and synchronous downlink transmissions are used. The orthogonality of the spreading codes used in broadband SC DS-CDMA will be destroyed by the delay-spread of the channel, which results in multipath interference imposed both by the desired user and the other users. By contrast, the orthogonality of the spreading codes used in broadband MC-CDMA employing F-domain spreading will be destroyed by the frequency-selective fading experienced by the different sub-carriers. Furthermore, the receiver of SC DS-CDMA requires signal processing at a rate comparable to the chip rate, which is extremely high in broadband SC DS-CDMA systems.

- **Transmit diversity** using multiple Base Station (BS) antennas has been proposed for boosting the capacity and data rate of CDMA systems [274]. In the context of SC DS-CDMA communicating over various propagation channels including indoor, open rural, suburban and urban areas, transmit diversity schemes, such as space-time spreading (STS), designed on the basis of a low number of resolvable paths or based on the premise of encountering a constant number of resolvable paths may not achieve the maximum effective throughput. In the context of MC-CDMA, usually each sub-carrier signal is assumed to be experiencing flat-fading. However, the actual diversity gain achieved by using multiple BS antennas and by combining the sub-carrier signals is time-variant, since the coherence bandwidth associated with the above-mentioned various communication environments is different and since the coherence bandwidth is time-varying even in the same propagation environment. A multiple-access scheme designed for diverse propagation environments is typically expected to be able to achieve a constant diversity order and to maintain a similar BER performance, regardless of what communication channels are encountered.

20.2.3.2 Using Broadband MC DS-CDMA for Supporting Ubiquitous Wireless Communications

First of all, to a certain extent MC DS-CDMA constitutes a trade-off between SC DS-CDMA and MC-CDMA in the context of the system's architecture and performance. MC DS-CDMA typically requires lower chip rate spreading codes than SC DS-CDMA due to employing multiple sub-carriers, while it necessitates a lower number of sub-carriers than MC-CDMA due to imposing DS spreading on each sub-carrier's signal. Consequently, MC DS-CDMA typically requires lower-rate signal processing than SC DS-CDMA and has a lower worst-case peak-to-average power fluctuation than MC-CDMA. However, MC DS-CDMA is more attractive than an arbitrary *ad hoc* scheme that constitutes a trade-off multiple-access scheme, positioned between SC DS-CDMA and MC-CDMA, since it exhibits a number of advantageous properties, which can be exploited in the context of broadband MC DS-CDMA for supporting ubiquitous wireless communications in diverse propagation environments. In Subsection 20.2.2 we have shown that MC DS-CDMA has the highest degree of freedom in the family of CDMA schemes that can be beneficially exploited during the system design procedure. Below we investigate how the specific parameters of MC DS-CDMA, which determine the degree of design freedom, can be adjusted for satisfying the requirements of ubiquitous communications in diverse propagation environments.

The channels are assumed to be slowly varying frequency-selective fading channels and the delay-spreads are assumed to be limited to the range of $[T_m, T_M]$, where T_m corresponds to the environment having the shortest delay-spread considered, experienced, for example, in an indoor environment. By contrast, T_M is associated with an environment having the highest possible delay-spread, as in an urban area. First, in order to ensure that the MC DS-CDMA system considered maintains the required frequency-diversity order in different communication environments, the simplest approach is to configure the system such that each sub-carrier signal is guaranteed to experience flat-fading. Then the required frequency-diversity gain is attained by combining the independently faded sub-carrier signals, which is achieved with the aid of F-domain interleaving or F-domain spreading. Let the delay-spread be limited to the range of $[T_m, T_M]$. The flat-fading condition of each sub-carrier signal is satisfied, if the chip-duration, T_c, is higher than the highest delay-spread, T_M, i.e. when $T_c > T_M$.

Second, in order to achieve the highest possible grade of frequency-diversity, as mentioned above, the sub-carrier signals combined must experience independent F-domain fading. This implies that the F-domain spacing between the combined sub-carriers must be higher than the maximum coherence bandwidth of $(\Delta f)_{cM} \approx 1/T_m$. Let U be the number of data streams after the S-P conversion stage [65]. Then, the above condition is satisfied if $\frac{U}{T_c} \geq \frac{1}{T_m}$, i.e., $U \geq \frac{T_c}{T_m}$, where U/T_c is the minimum frequency spacing between two adjacent sub-carriers conveying the same data bit.

The above philosophy might be developed with the aid of an example. Let us assume that the total bandwidth of the broadband MC DS-CDMA system is about 20 MHz. The delay-spread is assumed to be limited to the range of $[T_m = 0.1\mu s, T_M = 3\mu s]$, which includes the typical delay-spread values experienced in indoor, open rural, suburban and urban areas. Based on the philosophy discussed earlier in this subsection, we can set $T_c = 4\mu s > T_M = 3\mu s$ and $U = T_c/T_m = 4\mu s/0.1\mu s = 40$. When, for example, a total of 80 sub-carriers occupying the spectrum of Figure 20.1(c) and having a total system bandwidth of about 20 MHz ($= (80 + 1)/T_c$) are employed, then the interleaving depth is two, i.e. each data stream can be transmitted on two sub-carriers. Consequently, this MC DS-CDMA system will operate efficiently over a wide range of communication environments and will achieve a total diversity order of two, again provided that the delay-spread of the specific environment encountered is in the range of $[0.1\mu s, 3\mu s]$.

Advantages – Based on the above rules of selecting the system parameters, broadband MC DS-CDMA is capable of mitigating the problems encountered by both SC DS-CDMA and MC-CDMA. Specifically, broadband MC DS-CDMA has the following advantages:

- MC DS-CDMA is capable of supporting ubiquitous communication in environments as diverse as indoor, open rural, suburban and urban areas. This is achieved by avoiding or at least mitigating the problems imposed by the different-dispersion fading channels associated with the above communication environments.

- Broadband MC DS-CDMA guarantees that the combined sub-carrier signals experience independent fading.

- Broadband MC DS-CDMA is capable of mitigating the requirements of high chip rate based signal processing, as in broadband SC DS-CDMA. This is achieved by introducing computationally efficient Fast Fourier Transform (FFT) based parallel processing, carrying out modulation for all sub-carriers in a single FFT-step. Broadband MC DS-CDMA is also capable of mitigating the worst-case peak-to-average power fluctuation experienced, since due to using DS spreading of the sub-carriers we typically have a decreased number of sub-carriers in comparison to MC-CDMA.

- In broadband MC DS-CDMA the orthogonality of the T-domain spreading codes, which are assigned to different users, is unimpaired by fading-induced dispersion, since each sub-carrier signal experiences flat fading. Therefore, we may be able to dispense with using MUD and the desired signal can be detected using conventional low-complexity single-user detectors, if no F-domain spreading is employed. This is because, if in addition to T-domain spreading, also F-domain spreading across the different sub-carriers is employed, the orthogonality of the F-domain spreading codes cannot be retained due to the independent frequency-selective fading experienced by the sub-carrier signals, and hence requiring MUD for achieving the best possible BER performance. However, the MUD complexity of broadband MC DS-CDMA using TF-domain spreading can be substantially mitigated, since only a fraction of the total number of users supported by the system has to be detected by the MUD, as it will become explicit during our further discourse in sub-section 20.2.5.

- The above example shows that the achievable frequency diversity order is a constant value, namely two in the context of the above example, when communicating over a variety of fading channels. Consequently, when multiple BS antennas are employed, the STS-based transmit diversity scheme employed can be designed under the assumption of a constant frequency diversity order. Hence, the proposed STS-assisted broadband MC DS-CDMA scheme is capable of achieving a similar BER performance over a variety of fading channels.

Disadvantages and Their Counter-measures – With the aid of Table 20.1 we conclude that the two main deficiencies associated with broadband MC DS-CDMA are as follows.

- The Doppler frequency shift of the lowest and highest sub-carriers may be substantially different. This is because broadband MC DS-CDMA may occupy a system bandwidth on the order of tens or even hundreds of MHz. The different Doppler frequency shifts corresponding to different sub-carriers will destroy the orthogonality of the sub-carriers and a given sub-carrier signal will experience inter-sub-carrier interference (ICI) imposed by the adjacent sub-carrier signals. However, in most cases the ICI imposed by the other sub-carrier signals becomes neglectably low, provided that the mobile terminal's travelling speed is not too high. This is because the orthogonality between the desired sub-carrier and its adjacent sub-carriers remains relatively intact due to their similar frequencies. By contrast, the distant sub-carriers impose a relatively low cross-talk on the desired sub-carrier, since the cross-talk between the sub-carriers decays inverse-proportionally with their F-domain separation [489].

- In broadband MC DS-CDMA each sub-carrier signal experiences flat fading. According to the second column of Table 20.1 the achievable frequency-diversity order also depends inverse-proportionally on the number of parallel sub-streams. Hence, the frequency-diversity order alone in broadband MC DS-CDMA may be insufficient for maintaining the required BER performance. However, the diversity order achieved can be developed by using transmit diversity, as has been discussed in the context of the 3G CDMA systems in reference [272].

20.2.4 MC DS-CDMA Using Two-Antenna-Assisted STS

In this subsection, as an example, the MC DS-CDMA using the STS-based on two transmission antennas is discussed. The generalized broadband MC DS-CDMA using the STS-based on more than two transmission antennas will be investigated in Section 20.3. The proposed STS-assisted broadband MC DS-CDMA transmitter's schematic is shown in Figure 20.2. As

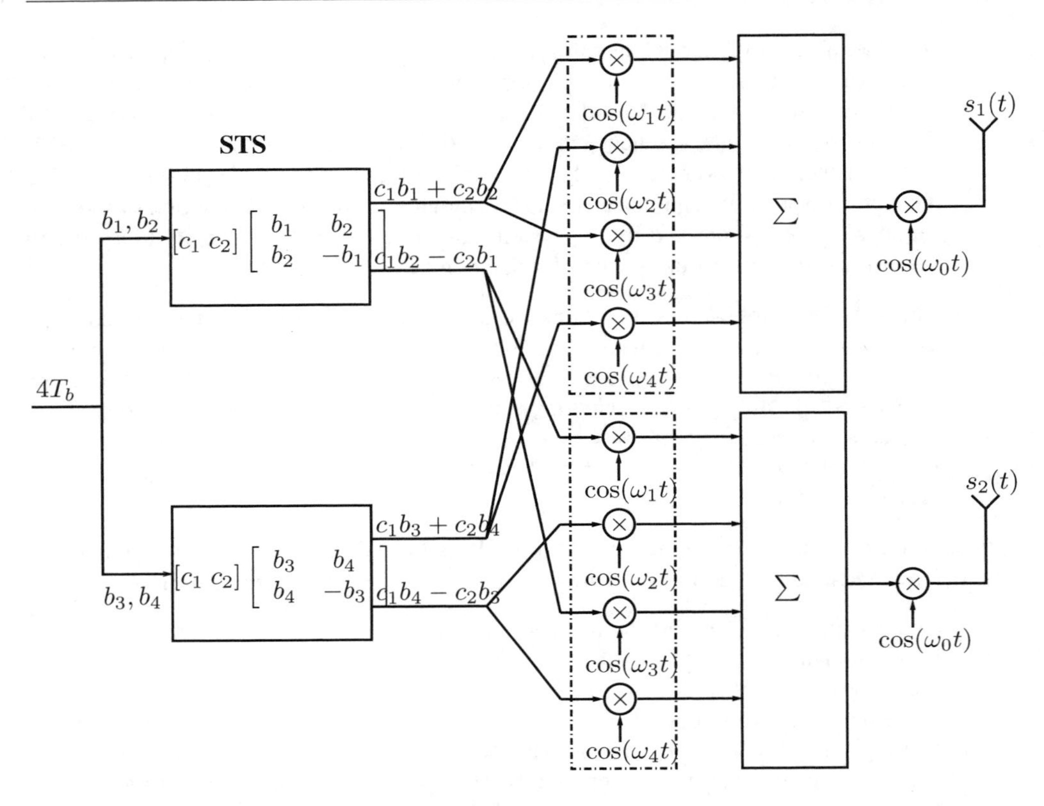

Figure 20.2: Transmitter schematic of the MC DS-CDMA system using space-time spreading.

an example we assumed that the transmitter uses $T_x = 2$ transmit antennas, $Q = 4$ sub-carriers, $U = 2$ parallel sub-streams and an interleaving depth of $S = 2$. The corresponding receiver schematic is shown in Figure 20.3. As shown in Figure 20.2, at the transmitter side $U \cdot T_x = 4$ data bits, each having a bit duration of T_b are S-P converted to $U = 2$ parallel sub-streams. Each parallel sub-stream has two data bits, which are space-time (ST) spread using the schemes of [274, 491] with the aid of two orthogonal spreading codes $\{c_1, c_2\}$, for example, Walsh codes. As seen in Figure 20.2, the resultant STS signals of $\{c_1b_1 + c_2b_2\}$, $\{c_1b_2 - c_2b_1\}$, $\{c_1b_3 + c_2b_4\}$ and $\{c_1b_4 - c_2b_3\}$ modulate the corresponding sub-carriers of $\{\omega_1, \omega_3\}$, $\{\omega_1, \omega_3\}$, $\{\omega_2, \omega_4\}$ and $\{\omega_2, \omega_4\}$, respectively and they are then mapped to $T_x = 2$ transmitter antennas. Again, as seen in Figure 20.2, following STS, each STS block generates two parallel signals, each of which will be mapped to one of the two transmitter antennas. The specific STS signals that will be transmitted using the same antenna are then interleaved using a F-domain interleaving depth of $S = 2$, so that each STS signal is transmitted on $S = 2$ sub-carriers. The interleavers guarantee that the same STS signal is transmitted by the two specific sub-carriers having the maximum possible frequency spacing, so that they experience as much independent fading as possible and hence achieve the maximum possible frequency diversity.

The receiver designed for the demodulation of STS-based MC DS-CDMA is shown in

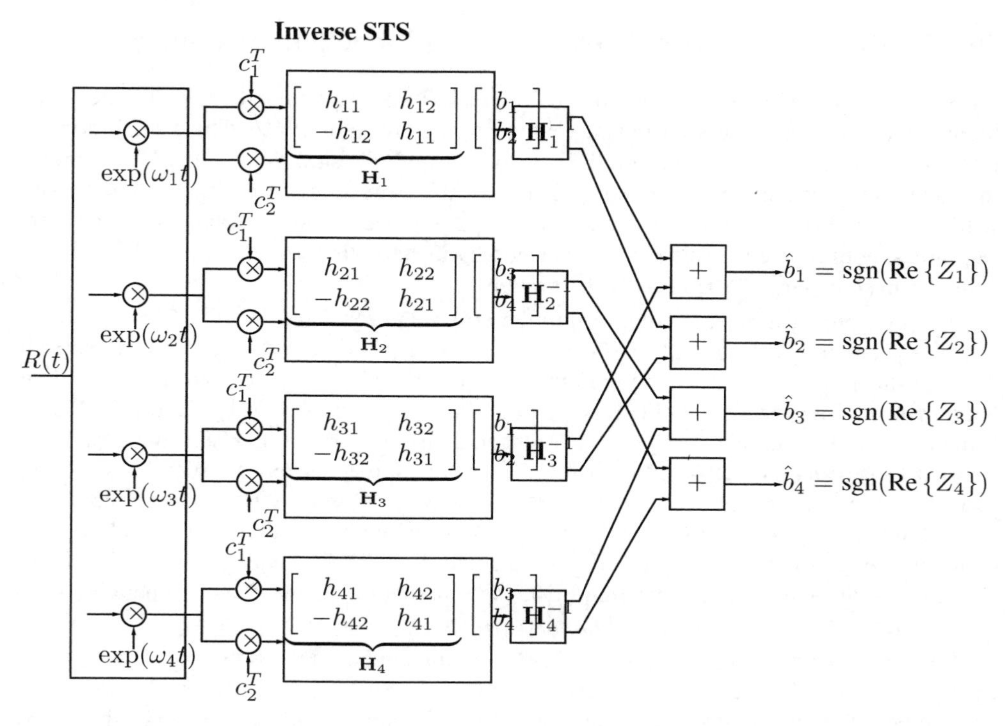

Figure 20.3: Receiver schematic of the MC DS-CDMA system using space-time spreading.

Figure 20.3, which essentially carries out the inverse operations of those seen in Figure 20.2. In Figure 20.3 the down-converted received signal is demodulated using FFT-based multicarrier demodulation. After FFT-based multicarrier demodulation we obtain $Q = 4$ parallel streams, corresponding to the signals transmitted on the $Q = 4$ sub-carriers. Each stream is space-time despread using the approach of [274], in order to obtain two separate variables, corresponding to the two data bits transmitted in the given sub-stream. Following space-time despreading, a decision variable is formed for each transmitted data bit by combining the corresponding variables associated with the $S = 2$ interleaved sub-carriers.

The above STS-assisted MC DS-CDMA scheme is capable of achieving a total diversity order of $T_x \cdot S = 4$, which was contributed by the transmit diversity order of two achieved on both of the F-domain interleaved sub-carriers. Furthermore, as shown in [490], the above STS-based MC DS-CDMA scheme can be generalized to the case of using T_x number of transmitter antennas, to employing U number of sub-streams after the S-P converter and to invoking an F-domain interleaving depth of S, respectively. Specifically, if four transmitter antennas are employed, a total diversity order of eight can be achieved by the above STS-assisted broadband scheme, which is the result of the transmit diversity order of four on both of the F-domain interleaved sub-carriers.

20.2.5 Capacity Improvement Based on TF-domain Spreading

It has been argued before in sub-section 20.2.2, for a given total system bandwidth, the maximum number of users supported and the frequency diversity gain achieved by using F-domain interleaving have to obey a trade-off. This is not a desirable result. We would like to achieve the maximum possible transmit diversity gain as well as the required frequency diversity gain without having to accept any other trade-off, i.e. without decreasing the total number of users supported by the system. This can be achieved by using both T-domain and F-domain, i.e. TF-domain spreading in MC DS-CDMA systems.

The TF-domain spreading employed in the proposed broadband MC DS-CDMA system can be understood with the aid of Figures 20.1(c) and 20.2. As mentioned in sub-section 20.2.2, if the parallel data bits seen in Figure 20.1(c) are replaced by the chips of a spreading code invoked now for the F-domain spreading, the signal represented by Figure 20.1(c) is actually a TF-domain spread signal. Correspondingly, the transmitter schematic of STS-assisted broadband MC DS-CDMA using TF-domain spreading has a structure similar to that of Figure 20.2, except that the S-depth interleaver of Figure 20.2 is now replaced by the F-domain spreading across S sub-carriers associated with an orthogonal spreading code of length S. Furthermore, the receiver of the MC DS-CDMA scheme using TF-domain spreading is also similar to that of Figure 20.3, except that the de-interleaving operation seen in Figure 20.3 is now correspondingly replaced by the F-domain de-spreading.

Spreading Code Assignment – Let the maximum number of users that can be supported by the T-domain STS codes in Figure 20.1(c) and Figure 20.2 be $\mathcal{K}_{max}$. It is well known that the number of length S orthogonal codes used for F-domain spreading is S. Since the above two sets of orthogonal codes, namely the T-domain and F-domain codes, can be assigned to users independently, we can see that even if S number of users share the same sub-set of T-domain spreading codes, these S user signals might be distinguishable with the aid of the associated S number of F-domain spreading codes. Hence, the maximum number of users supported by the proposed broadband MC DS-CDMA scheme using TF-spreading is $\mathcal{K}_{max} \times S$. The orthogonal spreading codes used in the broadband MC DS-CDMA system can be assigned as follows.

If the number of users is in the range of $0 \leq K \leq \mathcal{K}_{max}$, these users will be assigned the required T-domain STS codes and the same F-domain spreading code. The resultant scheme is the same as the one we studied in sub-section 20.2.4, and we achieve the BER performance seen in Figure 20.7 for transmissions over frequency-selective Rayleigh fading channels. However, when the number of users is in the range of $s\mathcal{K}_{max} \leq K \leq (s+1)\mathcal{K}_{max}$, $s = 1, 2 \ldots, S-1$, then the same sub-set of STS codes must be assigned to s or $(s+1)$ users, and hence these s or $(s+1)$ users have to be assigned different F-domain spreading codes. These s or $(s+1)$ users employing the same sub-set of T-domain STS codes are identified by their corresponding F-domain spreading codes.

Signal Detection – The sub-carrier signals conveying the different chips of the F-domain spreading code encounter independent fading, therefore the orthogonality of the orthogonal codes used for F-domain spreading cannot be retained. Hence, multiuser interference is inevitably introduced, which degrades the BER performance, when increasing the number of users sharing the same sub-set of T-domain STS orthogonal codes. However, in synchronous downlink broadband MC DS-CDMA using TF-domain spreading, the number of users sharing the same sub-set of T-domain STS codes is only a fraction of the total number of users supported by the system, which assumes the maximum value of S. Hence, for this fraction of the users advanced multiuser detection algorithms [78] can be invoked, in order to achieve an enhanced BER performance while maintaining an acceptable complexity. Accordingly, the receiver of broadband MC DS-CDMA will have a substantially lower complexity than that of broadband SC DS-CDMA or that of broadband MC-CDMA. Specifically, in the proposed system the MUD complexity depends on $\lceil K/\mathcal{K}_{max} \rceil$, where $\lceil x \rceil$ represents the minimum integer larger than x, while in broadband SC DS-CDMA or in broadband MC-CDMA the receiver's complexity depends on the total number of users supported, namely on K.

20.2.6 Summary

The systems of interest we studied in this section were broadband CDMA schemes supporting ubiquitous communications, regardless of the specific propagation environments encountered. Specifically, a broadband multicarrier DS-CDMA (MC DS-CDMA) was proposed for satisfying the requirements of broadband systems and for supporting ubiquitous communications, with the aid of STS-based transmit diversity. We have shown how the system parameters can be designed, in order that the STS-assisted broadband MC DS-CDMA scheme becomes capable of supporting ubiquitous communications without compromising the achievable BER. We have also considered the user capacity extension achievable by STS-assisted broadband MC DS-CDMA with the aid of TF-domain spreading. The studies suggest that broadband MC DS-CDMA using STS constitutes a promising multiple-access scheme, which is capable of avoiding the various design limitations that are unavoidable, when using single-carrier DS-CDMA or MC-CDMA.

20.3 Performance of Broadband Multicarrier DS-CDMA Using Space-Time Spreading

20.3.1 Introduction

In the above section we have shown that the STS-assisted broadband MC DS-CDMA scheme consititutes one of the promising broadband multiple-access systems, which are capable of supporting ubiquitous communications without compromising the achievable BER. In this section we investigate the performance of the broadband MC DS-CDMA systems using STS-assisted transmit diversity. Specifically, synchronous downlink (base-to-mobile) transmission of the user signals is considered and the BER performance is evaluated for a range of parameter values. We also investigate the performance of the extended broadband MC DS-CDMA, which uses Time-Frequency-domain (TF-domain) spreading. Both the correlation-based single-user detector and the de-correlating multiuser detector [78] are investigated.

The remainder of this section is organized as follows. In the next sub-section, the STS-assisted broadband MC DS-CDMA scheme is described and the required parameter values are investigated. In sub-section 20.3.3 we derive the achievable BER and characterize the BER performance of the broadband MC DS-CDMA scheme using STS. Sub-section 20.3.4 considers the achievable system capacity improvement of broadband MC DS-CDMA using STS and TF-domain spreading. Finally, our conclusions are offered in sub-section 20.3.5.

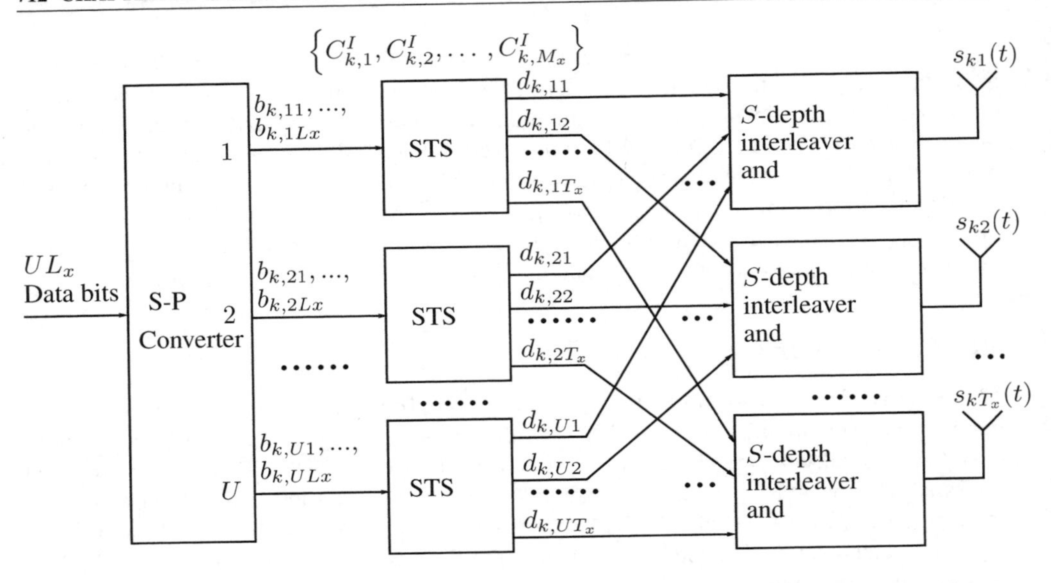

Figure 20.4: The transmitter schematic of the MC DS-CDMA system using space-time spreading.

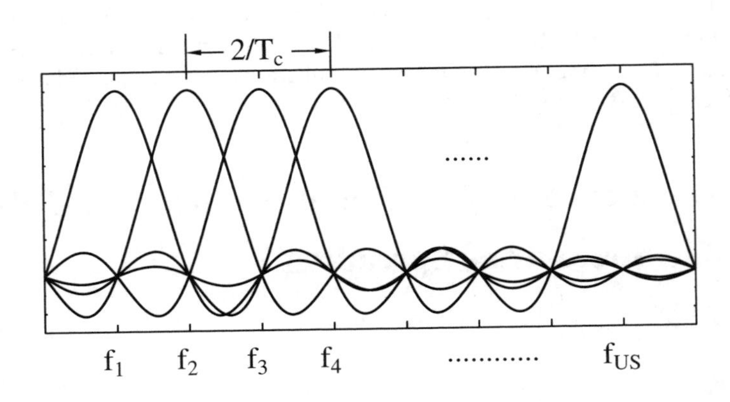

Figure 20.5: Spectrum of orthogonal MC DS-CDMA signals having a minimum sub-carrier spacing of $1/T_c$, where the zero-to-zero bandwidth of each DS spread signal is $2/T_c$.

20.3.2 System Description

20.3.2.1 Transmitter Model

The system considered in this section is an orthogonal MC DS-CDMA scheme [65] using $U \cdot S$ number of sub-carriers, T_x number of transmitter antennas and one receiver antenna. Furthermore, in this section a synchronous MC DS-CDMA scheme is investigated, where the K user signals are transmitted synchronously. The transmitter schematic of the kth user is shown in Figure 20.4, where real-valued data symbols using BPSK modulation and real-valued spreading [274] were considered. Note, however, that the analysis provided in this section may be extended to MC DS-CDMA systems using both complex-valued data symbols as well as complex-valued spreading. Figure 20.5 shows the frequency arrangement of the $U \cdot S$ sub-carriers. As shown in Figure 20.4, at the transmitter side a block of $U \cdot L_x$ data bits each having a bit duration of T_b is S-P converted to U parallel sub-blocks. Each parallel sub-block has L_x data bits, which are space-time spread using the schemes of [274, 491] – which will be further discussed in this sub-section – with the aid of M_x orthogonal spreading codes – for example, Walsh codes – $\left\{c_{k,1}^I(t), c_{k,2}^I(t), \ldots, c_{k,M_x}^I(t)\right\}$, $k = 1, 2, \ldots, K$ and mapped to T_x transmitter antennas. The symbol duration of the STS signals is UL_xT_b, and the discrete period of the orthogonal codes is $UL_xT_b/T_c = UL_xN$, where $N = T_b/T_c$ and T_c represents the chip-duration of the orthogonal spreading codes. The orthogonal codes take the form of $c_{k,i}^I(t) = \sum_{j=0}^{UL_xN-1} c_{k,i}^I[j]P_{T_c}(t - jT_c)$, where $c_{k,i}^I[j] \in \{+1, -1\}$ and they obey the relationship of $\sum_{l=0}^{UL_xN} c_{i,m}[l]c_{j,n}[l] = 0$, whenever $i \neq j$ or $m \neq n$. Furthermore, $P_{T_c}(t)$ represents the chip impulse waveform defined over the interval of $[0, T_c)$. Since the total number of orthogonal codes having a discrete period of UL_xN is UL_xN and since each user requires M_x orthogonal codes for STS, the maximum number of users supported by these orthogonal codes is UL_xN/M_x. As seen in Figure 20.4, following STS, each STS block generates T_x parallel signals to be mapped to the T_x transmitter antennas. The specific U STS signals of Figure 20.4, which are output by the U STS blocks and which will be transmitted using the same antenna from the set $1, 2, \ldots, T_x$ are then interleaved by an S-depth interleaver, so that each STS signal is transmitted on S sub-carriers. The interleavers guarantee that the same STS signal is transmitted by the specific S sub-carriers having the maximum possible frequency spacings, so that they experience independent fading and hence achieve maximum frequency diversity. Specifically, let $\{f_1, f_2, \ldots, f_{US}\}$ be the sub-carrier frequencies, which are arranged according to Figure 20.5. These sub-carrier frequencies can be written in the form of a matrix as

$$\{f_i\} = \begin{pmatrix} f_1 & f_{U+1} & \cdots & f_{(S-1)U+1} \\ f_2 & f_{U+2} & \cdots & f_{(S-1)U+2} \\ \vdots & \vdots & \ddots & \vdots \\ f_U & f_{2U} & \cdots & f_{SU} \end{pmatrix}. \tag{20.4}$$

Then, a STS signal will be transmitted using the sub-carrier frequencies from the same row of (20.4). Finally, as shown in Figure 20.4, the Inverse Fast Fourier Transform (IFFT) is invoked for carrying out multicarrier modulation, and the IFFT block's output signal is transmitted using one of the transmitter antennas.

The general form of the kth user's transmitted signal corresponding to the T_x transmitter antennas can be expressed as

$$\mathbf{s}_k(t) = \text{Re}\left\{\sqrt{\frac{2E_b}{UT_b}\frac{1}{SM_xT_x}}\left[\mathbf{C}_k\mathbf{B}_k\right]^T\mathbf{P}\mathbf{w} \times \exp(j2\pi f_c t)\right\}, \tag{20.5}$$

where E_b/UT_b represents the transmitted power per sub-carrier expressed as $L_xE_b/UL_xT_b = E_b/UT_b$, the factor S in the denominator is due to the S-depth interleaving, while the factor of M_xT_x represents STS using M_x orthogonal codes and T_x transmitter antennas. In

(20.5) $\mathbf{s}_k(t) = [s_{k1}(t)\ s_{k2}(t)\ ...\ s_{kT_x}(t)]^T$ - where the superscript T denotes the vector or matrix transpose – represents the transmitted signal vector of the T_x transmitter antennas, $\mathbf{P}$ represents the S-depth interleaving operation, which is a $U \times US$ matrix expressed as $\mathbf{P} = [\mathbf{I}_U\ \mathbf{I}_U\ \ldots\ \mathbf{I}_U]$ with $\mathbf{I}_U$ being a unity matrix of rank U. Furthermore, $\mathbf{C}_k$ is a $U \times UM_x$ dimensional matrix constituted by the orthogonal codes, which can be expressed as

$$\mathbf{C}_k^T = \begin{pmatrix} c_{k,1}^I(t) & 0 & \cdots & 0 \\ c_{k,2}^I(t) & 0 & \cdots & 0 \\ \vdots & \vdots & \ddots & \vdots \\ c_{k,M_x}^I(t) & 0 & \cdots & 0 \\ 0 & c_{k,1}^I(t) & \cdots & 0 \\ 0 & c_{k,2}^I(t) & \cdots & 0 \\ \vdots & \vdots & \ddots & \vdots \\ 0 & c_{k,M_x}^I(t) & \cdots & 0 \\ \vdots & \vdots & \ddots & \vdots \\ 0 & 0 & \vdots & c_{k,1}^I(t) \\ 0 & 0 & \vdots & c_{k,2}^I(t) \\ \vdots & \vdots & \ddots & \vdots \\ 0 & 0 & \vdots & c_{k,M_x}^I(t) \end{pmatrix}. \tag{20.6}$$

In (20.5) $\mathbf{B}_k$ is a $UM_x \times T_x$ matrix mapped from the U sub-block data bits, according to the requirements of the STS [274]. Specifically, the matrix $\mathbf{B}_k$ can be expressed as

$$\mathbf{B}_k = \left[\mathbf{B}_{k1}^T\ \mathbf{B}_{k2}^T\ \ldots\ \mathbf{B}_{kU}^T\right]^T, \tag{20.7}$$

where $\mathbf{B}_{ku}$ for $u = 1, 2, \ldots, U$ are $M_x \times T_x$ dimensional matrices, which obey the structure of

$$\mathbf{B}_{ku} = \begin{pmatrix} a_{11}b'_{k,11} & a_{12}b'_{k,12} & \cdots & a_{1L_x}b'_{k,1T_x} \\ a_{21}b'_{k,21} & a_{22}b'_{k,22} & \cdots & a_{2L_x}b'_{k,2T_x} \\ \vdots & \vdots & \ddots & \vdots \\ a_{M_x1}b'_{k,M_x1} & a_{U2}b'_{k,M_x2} & \cdots & a_{M_xL_x}b'_{k,M_xT_x} \end{pmatrix}, \quad u = 1, 2, \ldots, U, \tag{20.8}$$

where a_{ij} represents the sign of the element at the ith row and the jth column, which is determined by the STS design rule, while $b'_{k,ij}$ in $\mathbf{B}_{ku}$ is the data bit assigned to the (i,j)th element, which is one of the L_x input data bits $\{b_{k,u1}, b_{k,u2}, \cdots, b_{kL_x}\}$ of user k. For $L_x =$

$M_x = T_x = 2$ and 4, the corresponding $\mathbf{B}_{ku}$ matrices are given by [274]

$$\begin{pmatrix} b_{k,u1} & b_{k,u2} \\ b_{k,u2} & -b_{k,u1} \end{pmatrix}, \text{ and } \begin{pmatrix} b_{k,u1} & b_{k,u2} & b_{k,u3} & b_{k,u4} \\ b_{k,u2} & -b_{k,u1} & b_{k,u4} & -b_{k,u3} \\ b_{k,u3} & -b_{k,u4} & -b_{k,u1} & b_{k,u2} \\ b_{k,u4} & b_{k,u3} & -b_{k,u2} & -b_{k,u1} \end{pmatrix}, \quad u = 1, 2, \ldots, U \tag{20.9}$$

Finally, in (20.5) $\mathbf{w}$ represents the multicarrier modulated vector of length US, which can be expressed as

$$\mathbf{w} = \left[\exp(j2\pi f_1 t) \ \exp(j2\pi f_2 t) \ \ldots \ \exp(j2\pi f_{SU} t)\right]^T. \tag{20.10}$$

Equation (20.5) represents the general form of the transmitted signals using STS, regardless of the values of L_x, M_x and T_x. However, the study conducted in [274] has shown that STS schemes using $L_x = M_x = T_x$, i.e. those having an equal number of data bits, orthogonal STS-related spreading sequences as well as transmission antennas constitute attractive schemes, since they are capable of providing maximal transmit diversity without requiring extra STS codes. Furthermore, [274] indicates that such attractive STS schemes exist for $L_x = M_x = T_x = 2, 4, 8$, etc. Note that for the specific values of $U = 2, 4$ the above mentioned attractive STS schemes have been unambiguously specified with the aid of (20.9). In this section, we investigate only these attractive STS schemes, and our results are mainly based on MC DS-CDMA systems using two or four transmitter antennas.

For the case of $L_x = M_x = T_x = 2$, the MC DS-CDMA signals transmitted by antenna 1 and 2 can be simply expressed as

$$\begin{aligned} \mathbf{s}_k(t) &= \begin{pmatrix} s_{k1}(t) \\ s_{k2}(t) \end{pmatrix} \\ &= \sqrt{\frac{2E_b}{4UST_b}} \begin{pmatrix} \sum_{u=1}^{U}\sum_{s=1}^{S}[c_{k,1}^I b_{k,u1} + c_{k,2}^I b_{k,u2}] \\ \sum_{u=1}^{U}\sum_{s=1}^{S}[c_{k,1}^I b_{k,u2} - c_{k,2}^I b_{k,u1}] \end{pmatrix} \\ &\quad \times \cos\left[2\pi(f_c + f_{(s-1)U+u})t\right]. \end{aligned} \tag{20.11}$$

By contrast, for the case of $L_x = M_x = T_x = 4$, the MC DS-CDMA signals transmitted by antenna 1, 2, 3 and 4 can be expressed as

$$\mathbf{s}_k(t) = \begin{pmatrix} s_{k1}(t) \\ s_{k2}(t) \\ s_{k3}(t) \\ s_{k4}(t) \end{pmatrix} = \sqrt{\frac{2E_b}{16UST_b}}$$

$$\times \begin{pmatrix} \sum_{u=1}^{U}\sum_{s=1}^{S}[c_{k,1}^I b_{k,u1} + c_{k,2}^I b_{k,u2} + c_{k,3}^I b_{k,u3} + c_{k,4}^I b_{k,u4}] \\ \sum_{u=1}^{U}\sum_{s=1}^{S}[c_{k,1}^I b_{k,u2} - c_{k,2}^I b_{k,u1} - c_{k,3}^I b_{k,u4} + c_{k,4}^I b_{k,u3}] \\ \sum_{u=1}^{U}\sum_{s=1}^{S}[c_{k,1}^I b_{k,u3} + c_{k,2}^I b_{k,u4} - c_{k,3}^I b_{k,u1} - c_{k,4}^I b_{k,u2}] \\ \sum_{u=1}^{U}\sum_{s=1}^{S}[c_{k,1}^I b_{k,u4} - c_{k,2}^I b_{k,u3} + c_{k,3}^I b_{k,u2} - c_{k,4}^I b_{k,u1}] \end{pmatrix}$$

$$\times \quad \cos\left[2\pi(f_c + f_{(s-1)U+u})t\right], \tag{20.12}$$

Note that the explicit notation indicating the time dependence of $c_{k,i}^I(t)$ has been omitted for notational convenience in (20.11) and (20.12), since in this section only synchronous transmissions are considered.

20.3.2.2 Receiver Model

When the broadband MC DS-CDMA system is designed based on the limitations of subsection 20.2.3.2, each of the sub-carrier signals will experience flat fading and the sub-carrier signals conveying the same symbol will experience independent fading. Hence, assuming that K user signals in the form of (20.5) are transmitted synchronously over Rayleigh fading channels, the received complex low-pass equivalent signal can be expressed as

$$R(t) = \sum_{k=1}^{K}\sum_{g=1}^{T_x}\sqrt{\frac{2E_b}{UT_b}\frac{1}{SM_xT_x}}\left(\left[\mathbf{C}_k\mathbf{B}_k\right]^T\mathbf{P}\right)_g \mathbf{H}\mathbf{w} + N(t), \tag{20.13}$$

where $(\mathbf{X})_g$ represents the gth row of the matrix $\mathbf{X}$, $N(t)$ is the complex valued low-pass-equivalent Additive White Gaussian Noise (AWGN) having a double-sided spectral density of N_0, while

$$\mathbf{H} = \text{diag}\left\{h_{1g}\exp(j\psi_{1g}), h_{2g}\exp(j\psi_{2g}), \ldots, h_{(US)g}\exp(j\psi_{(US)g})\right\}, \quad g = 1, 2, \ldots, T_x, \tag{20.14}$$

is a diagonal matrix of rank US, which represents the channel's complex impulse response in the context of the gth antenna. The coefficients h_{ig}, $i = 1, 2, \ldots, US$; $g = 1, 2, \ldots, T_x$ in $\mathbf{H}$ are independent identically distributed (i.i.d) random variables obeying the Rayleigh

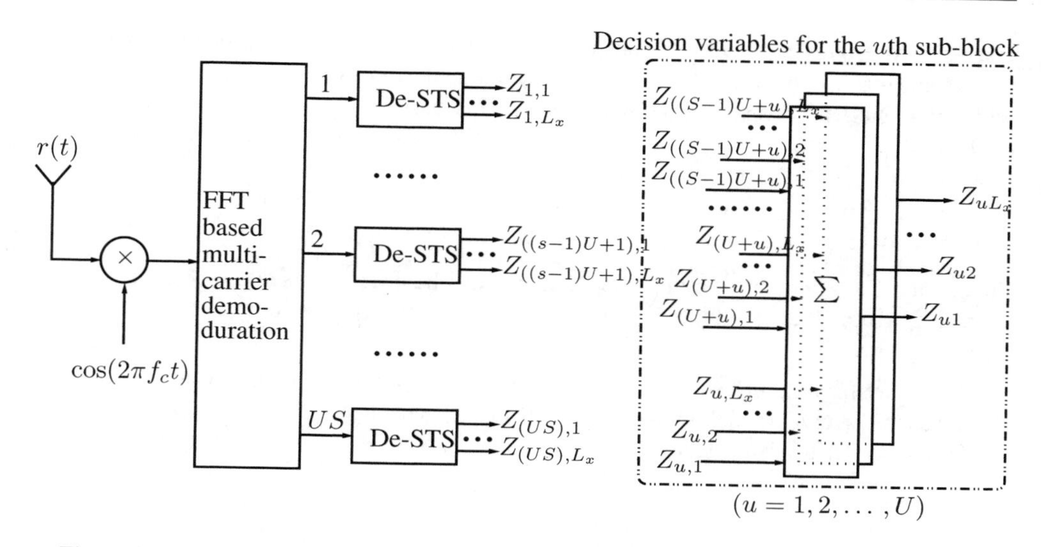

Figure 20.6: The receiver schematic of the MC DS-CDMA system using space-time spreading.

distribution, which can be expressed as

$$f_{h_{ig}}(y) = \frac{2y}{\Omega}\exp\left(-\frac{y^2}{\Omega}\right),\ y \geq 0, \tag{20.15}$$

where $\Omega = \text{E}[(h_{ig})^2]$. Furthermore, the phases ψ_{ig}, $i = 1, 2, \ldots, US$; $g = 1, 2, \ldots, T_x$ are introduced by the fading channels and are uniformly distributed in the interval $(0, 2\pi)$.

Specifically, for the case of $L_x = M_x = T_x = 2$, when the transmitted signal obeys (20.11), the received complex low-pass equivalent signal can be expressed as

$$\begin{aligned} R(t) &= \sum_{k=1}^{K}\sum_{u=1}^{U}\sum_{s=1}^{S}\sqrt{\frac{2E_b}{4UST_b}}\Big(h_{(us)1}\exp(j\psi_{(us)1})[c_{k,1}^{I}(t)b_{k,u1} + c_{k,2}^{I}(t)b_{k,u2}] \\ &\quad + h_{(us)2}\exp(j\psi_{(us)2})[c_{k,1}^{I}(t)b_{k,u2} - c_{k,2}^{I}(t)b_{k,u1}]\Big) \\ &\quad \times \exp\left(2\pi f_{(s-1)U+u}t\right) + N(t). \end{aligned} \tag{20.16}$$

Let the first user be the user-of-interest and consider a receiver employing FFT-based multicarrier demodulation, space-time despreading as well as diversity combining, as shown in Figure 20.6. The receiver of Figure 20.6 essentially carries out the inverse operations of those seen in Figure 20.4. In Figure 20.6 the received signal is first down-converted using the carrier frequency f_c, and then demodulated using FFT-based multicarrier demodulation. After FFT-based multicarrier demodulation we obtain US number of parallel streams corresponding to the signals transmitted on US sub-carriers, and each stream is space-time despread using the approach of [274], in order to obtain L_x separate variables, $\{Z_{u,l}, Z_{u,2}, \ldots, Z_{u,L_x}\}_{u=1}^{US}$, corresponding to the L_x data bits transmitted on the uth stream, where $u = 1, 2, \ldots, US$, respectively. Following space-time despreading, a decision variable is formed for each transmitted data bit $\{b_{u1}, b_{u2}, \ldots, b_{uL_x}\}_{u=1}^{U}$ by combining the corresponding variables associated with the S interleaved sub-carriers, which can be expressed as

$$Z_{ui} = \sum_{s=1}^{S} Z_{((s-1)U+u),i}, \; u = 1, 2, \ldots, U; \; i = 1, 2, \ldots, L_x. \tag{20.17}$$

Finally, the UL_x number of transmitted data bits can be decided based on the decision variables $\{Z_{ui}, \; u = 1, 2, \ldots, U; \; i = 1, 2, \ldots, L_x\}$ using the conventional decision rule of a BPSK scheme.

Above we have described the transmitter model and characterized the system parameters as well as the receiver model of broadband MC DS-CDMA using STS. Let us now investigate the achievable BER performance.

20.3.3 Bit Error Rate Analysis

In this sub-section we derive the BER expression of the broadband MC DS-CDMA system using STS, which was described in Subsection 20.3.2. As an example, we derive the BER expression in detail for STS-based MC DS-CDMA using the parameters of $L_x = M_x = T_x = 2$. The generalized BER expression of MC DS-CDMA using the set of attractive STS schemes, i.e. using $L_x = M_x = T_x = 2, 4, 8$, etc. is then derived from the case of $L_x = M_x = T_x = 2$ without providing the detailed derivations, since the extension is relatively straightforward.

For the case of $L_x = M_x = T_x = 2$, the analysis can be commenced from (20.16). Let $d_{u,1}, \; d_{u,2}$ – where $u = 1, 2, \ldots, US$ – represent the correlator's output variables corresponding to the first two data bits transmitted on the uth sub-carrier, where

$$d_{u,1} = \int_0^{2UT_b} R(t) c_{1,1}^I(t) \exp\left(-j2\pi f_u t\right) dt, \tag{20.18}$$

$$d_{u,2} = \int_0^{2UT_b} R(t) c_{1,2}^I(t) \exp\left(-j2\pi f_u t\right) dt. \tag{20.19}$$

Since orthogonal multicarrier signals, orthogonal STS codes, synchronous transmission of the K user signals as well as slowly flat-fading of each sub-carrier are assumed, there is no interference between the different users and the different sub-carrier signals. Therefore, when substituting (20.16) into (20.18), it can be shown that

$$d_{u,1} = \sqrt{\frac{2UE_bT_b}{S}} \left[h_{u1} \exp(j\psi_{u1}) b_{1,u1} + h_{u2} \exp(j\psi_{u2}) b_{1,u2}\right] + N_{u,1}, \tag{20.20}$$

$$d_{u,2} = \sqrt{\frac{2UE_bT_b}{S}} \left[h_{u1} \exp(j\psi_{u1}) b_{1,u2} - h_{u2} \exp(j\psi_{u2}) b_{1,u1}\right] + N_{u,2}, \tag{20.21}$$

where $N_{u,i}, i = 1, 2$ is due to the AWGN expressed as

$$N_{u,i} = \int_0^{2UT_b} N(t)c^I_{1,i}(t)\exp(-j2\pi f_u t)\,dt, \tag{20.22}$$

which is a complex Gaussian distributed variable having zero mean and a variance of $2UN_0T_b$.

Assuming that the receiver has perfect knowledge of the fading parameters of $h_{ui}\exp(j\psi_{ui})$, $i = 1, 2$, the decision variables corresponding to the data bits $b_{1,ui}, i = 1, 2$ associated with the uth sub-carrier can be expressed as

$$\begin{aligned} Z_{u,1} &= \text{Re}\{d_{u,1}h_{u1}\exp(-j\psi_{u1}) - d_{u,2}h_{u2}\exp(-j\psi_{u2})\} \\ &= \sqrt{\frac{2UE_bT_b}{S}}[h^2_{u1} + h^2_{u2}]b_{1,u1} \\ &\quad +\text{Re}\{N_{u,1}h_{u1}\exp(-j\psi_{u1}) - N_{u,2}h_{u2}\exp(-j\psi_{u2})\}, \end{aligned} \tag{20.23}$$

$$\begin{aligned} Z_{u,2} &= \text{Re}\{d_{u,1}h_{u2}\exp(-j\psi_{u2}) + d_{u,1}h_{u1}\exp(-j\psi_{u1})\} \\ &= \sqrt{\frac{2UE_bT_b}{S}}[h^2_{u1} + h^2_{u2}]b_{1,u2} \\ &\quad +\text{Re}\{N_{u,1}h_{u2}\exp(-j\psi_{u2}) + N_{u,2}h_{u1}\exp(-j\psi_{u1})\}, \end{aligned} \tag{20.24}$$

for $u = 1, 2, \ldots, US$.

Finally, after combining the replicas of the same signal transmitted on the S sub-carriers, the decision variables corresponding to the two bits in the uth sub-block can be expressed as

$$Z_{u1} = \sum_{s=1}^{S} Z_{((s-1)U+u),1}, \tag{20.25}$$

$$Z_{u2} = \sum_{s=1}^{S} Z_{((s-1)U+u),2}, \tag{20.26}$$

for $u = 1, 2, \ldots, U$.

Since $Z_{((s-1)U+u),1}$ of (20.23) and $Z_{((s-1)U+u),2}$ of (20.24) are i.i.d random variables for different s values associated with a mean of $\sqrt{\frac{2UE_bT_b}{S}}[h^2_{((s-1)U+u)1} + h^2_{((s-1)U+u)2}]$ and variance of $UN_0T_b[h^2_{((s-1)U+u)1} + h^2_{((s-1)U+u)2}]$, conditioned on $h_{((s-1)U+u)1}$ and $h_{((s-1)U+u)2}$, it is well known [274] that the conditional BER can be expressed as

$$P_b\left(E|\left\{h_{((s-1)U+u)1}, h_{((s-1)U+u)2}\right\}\right) = Q\left(\sqrt{2\cdot\sum_{s=1}^{S}\sum_{l=1}^{2}\gamma_{sl}}\right), \tag{20.27}$$

where $Q(x)$ represents the Gaussian Q-function, $\gamma_{sl} = \overline{\gamma} \times \frac{(h_{((s-1)U+u),l})^2}{\Omega}$ and $\overline{\gamma} = E_b\Omega/SN_0$. Finally, with the aid of Equation 20.15 and [90],[p. 781], it can readily be shown that the average BER can be expressed as

$$P_b = \left[\frac{1-\mu}{2}\right]^{2S} \sum_{k=0}^{2S-1} \binom{2S-1+k}{k} \left[\frac{1+\mu}{2}\right]^k, \tag{20.28}$$

where $\mu = \sqrt{\bar{\gamma}/(1+\bar{\gamma})}$.

For the general case of $L_x = M_x = T_x = 2, 4, 8$, etc. the decision variable Z_{u1} corresponding to the first bit in the sub-block u can also be expressed as in (20.25), but with $Z_{((s-1)U+u),1}$ given by

$$Z_{((s-1)U+u),1} = \sqrt{\frac{2UE_bT_b}{S}} \sum_{l=1}^{T_x} h^2_{((s-1)U+u)l} \times b_{1,u1} + \text{Re}\left\{N'_{((s-1)U+u),1}\right\}, \quad s = 1, 2, \ldots, S, \tag{20.29}$$

where $\text{Re}\left\{N'_{((s-1)U+u),1}\right\}$ is an AWGN process having a zero mean and a variance of $UN_0T_b \sum_{l=1}^{T_x} h^2_{((s-1)U+u)l}$. The average BER for the case of $L_x = M_x = T_x = 2, 4, 8$, etc. can be expressed as

$$P_b = \left[\frac{1-\mu}{2}\right]^{T_xS} \sum_{k=0}^{T_xS-1} \binom{T_xS-1+k}{k} \left[\frac{1+\mu}{2}\right]^k. \tag{20.30}$$

Before proceeding to presenting our numerical and simulation results, we note that according to (20.25) and (20.29), by using STS and F-domain sub-carrier interleaving, the diversity order achieved is T_xS, provided that T_x transmitter antennas and S-depth F-domain interleaving schemes were used. Furthermore, the diversity gains achieved by the T_x transmitter antennas and by the process of interleaving over S sub-carriers are independent and hence their product determines the total diversity order.

Figure 20.7 shows both the numerical and simulation-based BER results, which are drawn using lines and markers, respectively, for $S = 1$, $T_x = 1, 2, 4$ as well as for $S = 3$, $T_x = 1, 2, 4$. From the results we observe that at a BER of 0.01, using two transmitter antennas rather than one yields a gain of approximately 5.0 dB. Furthermore, when $T_x = 4$ transmitter antennas and an interleaving depth of $S = 3$ are considered instead of $T_x = 1, S = 1$, the diversity gain achieved is approximately 9.0 dB.

The BER performance of Figure 20.7 can be achieved, provided that the number of orthogonal STS codes is sufficiently high for supporting the K number of users without reuse. Based on our arguments in Subsections 20.3.2 and 20.3.3, the number of orthogonal STS codes is given by $UL_xT_b/T_c = UL_xN$. Since each user requires M_x orthogonal STS codes, the maximum number of users supported by the UL_xN orthogonal STS codes is given by

$$\mathcal{K}_{max} = UL_xN/M_x. \tag{20.31}$$

In other words, if the number of users supported by the synchronous broadband MC DS-

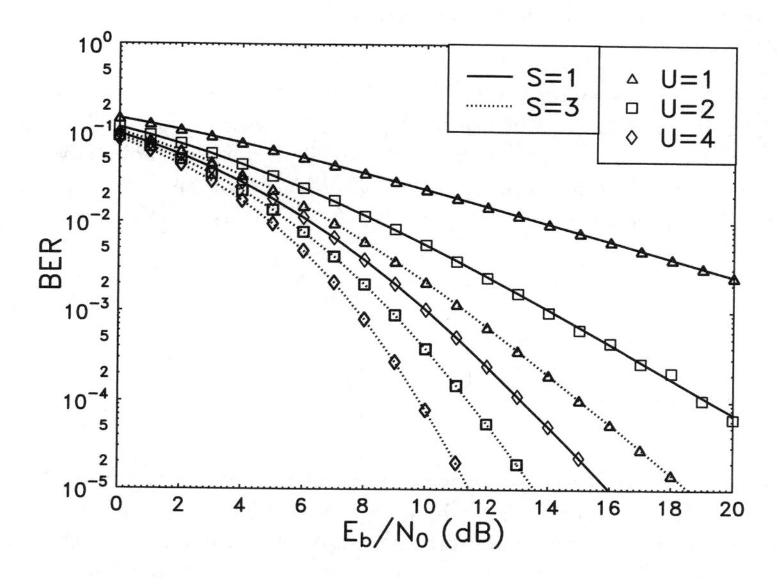

Figure 20.7: Numerical (lines) and simulated (markers) BER versus the SNR per bit, E_b/N_0, performance for the broadband MC DS-CDMA using STS-based transmit diversity, when communicating over frequency-selective Rayleigh fading channels evaluated from (20.30).

CDMA system designed according to the philosophy of this section obeys $K \leq \mathcal{K}_{max}$, the BER performance of Figure 20.7 can be achieved for transmissions over frequency-selective Rayleigh fading channels. For the attractive STS schemes using $L_x = M_x = T_x = 2, 4, 8$, etc. the maximum number of users supported by the orthogonal STS codes is $\mathcal{K}_{max} = UN$. It can be seen that in this case the number of users supported is independent of the number of transmitter antennas, which emphasizes the advantages of the STS schemes using the specific values of $L_x = M_x = T_x = 2, 4, 8$, etc. [274].

Let $N = T_b/T_c$ and the total number of sub-carriers US be constants. Then, the maximum number of users supported by the broadband MC DS-CDMA system depends only on the interleaving depth S. Specifically, the maximum number of users supported decreases, when increasing the interleaving depth, S, i.e. when increasing the frequency diversity order. Consequently, the maximum number of users supported and the frequency diversity gain achieved have to obey a trade-off. This is not a desirable result. We would like to achieve the maximum transmit diversity gain as well as the required frequency diversity gain without having to accept any other trade-off, i.e. without decreasing the total number of users supported by the system. Let us now consider this issue in more detail in the next subsection.

20.3.4 Performance of STS-Assisted Broadband MC DS-CDMA Using Time-Frequency-Domain Spreading

So far, the STS or the DS spreading used in MC DS-CDMA systems was carried out in the time-domain (T-domain) based on orthogonal (Walsh) DS-spreading codes. However, in the context of multicarrier modulation another degree of freedom can be exploited by employing spreading in the F-domain, as was proposed for MC-CDMA schemes in [75,357]. Let us now investigate the STS-assisted broadband MC DS-CDMA scheme using TF-domain spreading.

20.3.4.1 System Description

The transmitter schematic of the STS-assisted broadband MC DS-CDMA using TF-domain spreading is similar to that seen in Figure 20.4, except that the S-depth interleaver of Figure 20.4 is now replaced by the F-domain spreading associated with an orthogonal spreading code of length S. Specifically, let $\{c_k^{II}[0], c_k^{II}[1], \ldots, c_k^{II}[S-1]\}$ be the kth user's orthogonal code in discrete form, which will be used for the F-domain spreading. By contrast, the kth user's T-domain orthogonal codes used for STS have been expressed in Subsection 20.3.2 as $c_{k,i}^{I}(t),\ i = 1, 2, \ldots, M_x$ in continuous form. In the broadband MC DS-CDMA system using TF-domain spreading, instead of employing interleaving over S sub-carriers, the U sub-block signals of user k after STS are now further spread over the F-domain using the above F-domain spreading codes. The signals transmitted from the T_x transmitter antennas can be expressed as

$$\mathbf{s}_k(t) = \text{Re}\left\{\sqrt{\frac{2E_b}{UT_b}\frac{1}{SM_xT_x}}\left[\mathbf{C}_k\mathbf{B}_k\right]^T\mathbf{Q}\mathbf{w} \times \exp(j2\pi f_c t)\right\}, \tag{20.32}$$

where all variables have the same interpretations as those in (20.5), except that the interleaving matrix $\mathbf{P}$ of (20.5) is now replaced by $\mathbf{Q}$ in (20.32), which represents the F-domain spreading and is a $U \times US$ dimensional matrix expressed as

$$\mathbf{Q} = \left[\mathbf{C}_k^{II}[0]\ \mathbf{C}_k^{II}[1]\ \ldots\ \mathbf{C}_k^{II}[S-1]\right], \tag{20.33}$$

where $\mathbf{C}_k^{II}[s],\ s = 0, 1, \ldots, S-1$ are diagonal matrices of rank U, which can be expressed as

$$\mathbf{C}_k^{II}[s] = \text{diag}\left\{c_k^{II}[s], c_k^{II}[s], \ldots, c_k^{II}[s]\right\}. \tag{20.34}$$

Specifically, for $L_x = M_x = T_x = 2$, i.e. for MC DS-CDMA using two transmitter antennas, the MC DS-CDMA signals transmitted by antenna 1 and 2 can simply be expressed as

$$\begin{aligned}\mathbf{s}_k(t) &= \begin{pmatrix} s_{k1}(t) \\ s_{k2}(t) \end{pmatrix} \\ &= \sqrt{\frac{2E_b}{4USТ_b}} \begin{pmatrix} \sum_{u=1}^{U}\sum_{s=1}^{S}[c_{k,1}^{I}b_{k,u1} + c_{k,2}^{I}b_{k,u2}]c_k^{II}[s-1] \\ \sum_{u=1}^{U}\sum_{s=1}^{S}[c_{k,1}^{I}b_{k,u2} - c_{k,2}^{I}b_{k,u1}]c_k^{II}[s-1] \end{pmatrix} \\ &\quad \times \cos\left[2\pi(f_c + f_{(s-1)U+u})t\right]. \end{aligned} \tag{20.35}$$

The total number of users supported by the broadband MC DS-CDMA system using TF-domain spreading and the assignment of orthogonal codes to the users is analyzed as follows. According to our analysis in Subsections 20.3.2 and 20.3.3, the total number of orthogonal codes that can be used for STS is UL_xN and the maximum number of users supported by these orthogonal codes is $\mathcal{K}_{max} = UL_xN/M_x$. By contrast, the total number of orthogonal codes that can be used for F-domain spreading is S. This implies that even if S number of users share the same set of STS codes, these S user signals might be distinguishable with the aid of the associated S number of F-domain spreading codes. Explicitly, the total number of users supported is $S\mathcal{K}_{max} = USL_xN/M_x$. Therefore, the orthogonal spreading codes can be assigned as follows. If the number of users is in the range of $0 \leq K \leq \mathcal{K}_{max}$, these users will be assigned the required orthogonal STS codes and the same F-domain orthogonal spreading code. The resultant scheme is the same as the one we studied in Subsections 20.3.2 as well as 20.3.3, and we achieve the BER performance of Figure 20.7 for transmissions over frequency-selective Rayleigh fading channels. However, when the number of users is in the range of $s\mathcal{K}_{max} \leq K \leq (s+1)\mathcal{K}_{max}$, $s = 1, 2 \ldots, S-1$, then the same set of M_x STS orthogonal codes must be assigned to s or $(s+1)$ users, but these s or $(s+1)$ users are assigned different F-domain spreading codes. These s or $(s+1)$ users employing the same set of M_x STS codes are identified by their corresponding F-domain spreading codes. Since the sub-carrier signals across which F-domain spreading takes place encounter independent fading, the orthogonality of the F-domain spreading codes cannot be retained. Hence, multiuser interference is inevitably introduced, which degrades the BER performance, when increasing the number of users sharing the same set of M_x STS orthogonal codes.

Let $1 \leq K' \leq S$ be the number of users sharing the same set of M_x STS orthogonal codes. We also assume that any set of M_x STS orthogonal codes is shared by the same K' number of users. Then, when the $K'\mathcal{K}_{max}$ signals expressed in the form of (20.32) are transmitted over frequency-selective fading channels, the received complex low-pass equivalent signal can be expressed as

$$R(t) = \sum_{k=1}^{K'\mathcal{K}_{max}} \sum_{g=1}^{T_x} \sqrt{\frac{2E_b}{UT_b}\frac{1}{SM_xT_x}} \left(\left[\mathbf{C}_k\mathbf{B}_k\right]^T \mathbf{Q}\right)_g \mathbf{Hw} + N(t), \tag{20.36}$$

where $(X)_g$ and $N(t)$ have the same interpretation as in (20.13). Specifically, for the case of

$L_x = M_x = T_x = 2$, $R(t)$ can be expressed as

$$\begin{aligned} R(t) &= \sum_{k=1}^{K'\mathcal{K}_{max}} \sum_{u=1}^{U} \sum_{s=1}^{S} \sqrt{\frac{2E_b}{4UST_b}} \\ &\quad \Big(h_{(us)1} \exp(j\psi_{(us)1})[c_{k,1}^{I}(t)b_{k,u1} + c_{k,2}^{I}(t)b_{k,u2}]c_k^{II}[s-1] \\ &\quad + h_{(us)2} \exp(j\psi_{(us)2})[c_{k,1}^{I}(t)b_{k,u2} - c_{k,2}^{I}(t)b_{k,u1}]c_k^{II}[s-1] \Big) \\ &\quad \times \exp\left(2\pi f_{(s-1)U+u}t\right) + N(t). \end{aligned} \tag{20.37}$$

The receiver schematic of the MC DS-CDMA scheme using TF-domain spreading is similar to that of Figure 20.6, except that the de-interleaving operation seen in Figure 20.6 is now replaced by the F-domain despreading. The signals at the output of the De-STS' block of Figure 20.6 can be detected by invoking a range of single- or multi-user detection schemes.

20.3.4.2 Signal Detection

Following the derivations in Subsection 20.3.3, for the general case of $L_x = M_x = T_x = 2, 4, 8$, etc. the decision variable $Z_{((s-1)U+u),1}$ in terms of the first data bit in the sub-block u and the sub-carrier $(s-1)U+u$ can now be expressed as

$$\begin{aligned} Z_{((s-1)U+u),1} = \sqrt{\frac{2UE_bT_b}{S}} \sum_{k=1}^{K'} \left(\sum_{l=1}^{T_x} h^2_{((s-1)U+u)l} \right) \times c_k^{II}[s-1]b_{k,u1} \\ + \text{Re}\left\{ N'_{((s-1)U+u),1} \right\}, \quad s = 1, 2, \ldots, S, \end{aligned} \tag{20.38}$$

where Re$\left\{ N'_{((s-1)U+u),1} \right\}$ is an AWGN process having zero mean and a variance of $UN_0T_b \sum_{l=1}^{T_x} h^2_{((s-1)U+u)l}$.

Let

$$\mathbf{z}_{u1} = \left[Z_{u,1}\ Z_{(U+u),1}\ \ldots\ Z_{((S-1)U+u),1}\right]^T, \tag{20.39}$$

$$\mathbf{A} = \text{diag}\left\{\sum_{l=1}^{T_x} h_{ul}^2, \sum_{l=1}^{T_x} h_{(U+u)l}^2, \ldots, \sum_{l=1}^{T_x} h_{((S-1)U+u)l}^2\right\}, \tag{20.40}$$

$$\mathbf{C} = \begin{pmatrix} c_1^{II}[0] & c_2^{II}[0] & \ldots & c_{K'}^{II}[0] \\ c_1^{II}[1] & c_2^{II}[1] & \ldots & c_{K'}^{II}[1] \\ \vdots & \vdots & \ddots & \vdots \\ c_1^{II}[S-1] & c_2^{II}[S-1] & \ldots & c_{K'}^{II}[S-1] \end{pmatrix}, \tag{20.41}$$

$$\mathbf{b} = \left[b_{1,u1}\ b_{2,u1}\ \ldots\ b_{K',u1}\right]^T, \tag{20.42}$$

$$\mathbf{n} = \text{Re}\left[N'_{u,1}\ N'_{(U+u),1}\ \ldots\ N'_{((S-1)U+u),1}\right]^T. \tag{20.43}$$

Then (20.38) can be written in a matrix form as

$$\mathbf{z}_{u1} = \sqrt{\frac{2UE_bT_b}{S}}\mathbf{ACb} + \mathbf{n}. \tag{20.44}$$

Based on (20.44) the multiuser MC DS-CDMA signals can be detected by invoking different detection algorithms [78]. In this section, as examples, we investigate two detection algorithms, namely the single-user correlation-based detector and the multiuser decorrelating detector [78]. In the context of the single-user correlation-based detector, let $\mathbf{z} = [Z_{u1}\ Z_{u2}\ \ldots\ Z_{uK'}]^T$ represent the decision variables. Then, these decision variables are obtained by multiplying both sides of (20.44) with $\mathbf{C}^T$, which can be expressed as

$$\mathbf{z} = \sqrt{\frac{2UE_bT_b}{S}}\left(\sum_{s=1}^{S}\sum_{l=1}^{T_x} h_{((s-1)U+u)l}^2\right)\mathbf{Rb} + \mathbf{C}^T\mathbf{n}, \tag{20.45}$$

where

$$\mathbf{R} = \begin{pmatrix} 1 & \rho_{12} & \ldots & \rho_{1K'} \\ \rho_{21} & 1 & \ldots & \rho_{2K'} \\ \vdots & \vdots & \ddots & \vdots \\ \rho_{K'1} & \rho_{K'2} & \ldots & 1 \end{pmatrix} \tag{20.46}$$

is the correlation matrix of the K' user signals, while ρ_{ij} represents the correlation factor between user i and user j, which can be expressed as

$$\rho_{ij} = \frac{\sum_{s=1}^{S} \left(c_i^{II}[s-1] c_j^{II}[s-1] \sum_{l=1}^{T_x} h^2_{((s-1)U+u)l} \right)}{\sum_{s=1}^{S} \sum_{l=1}^{T_x} h^2_{((s-1)U+u)l}}. \tag{20.47}$$

Equation (20.45) suggests that the diversity gain contributed both by the transmit diversity and frequency diversity can be retained, since we have a double sum of the components h^2 corresponding to the transmit and frequency diversity orders of S and T_x, respectively. However, multiuser interference is introduced by the channel's time-varying characteristics. Finally, the corresponding data bits, $b_{k,u1}$, $k = 1, 2, \ldots, K'$, are decided according to $\hat{b}_{k,u1} = \text{sgn}((\mathbf{z})_k)$ for $k = 1, 2, \ldots, K'$, where $(\mathbf{z})_k$ represents the kth row of $\mathbf{z}$, while $\text{sgn}(\cdot)$ is the sign function [78].

Note that the correlation factors $\{\rho_{ij}\}$ in (20.47) are time-variant due to the time-varying nature of the channel's fading envelope. However, since $\sum_{s=1}^{S} c_i^{II}[s-1] c_j^{II}[s-1] = 0$, it can be shown that $\rho_{ij} = 0$, provided that the sum of $\sum_{l=1}^{T_x} h^2_{((s-1)U+u)l}$ is identical for different values of s. Moreover, it can be shown that the correlation factors $\{\rho_{ij}\}$ are contributed by the differences of the sums $\sum_{l=1}^{T_x} h^2_{((s-1)U+u)l}$ experienced according to the different values of s, while the common part of $\sum_{l=1}^{T_x} h^2_{((s-1)U+u)l}$ in terms of the different values of s can successfully be removed due to the orthogonality of the F-domain spreading codes. Specifically, let $\sum_{l=1}^{T_x} h^2_{((s-1)U+u)l} = A_h + \Delta_s$, where A_h represents the average value of $\sum_{l=1}^{T_x} h^2_{((s-1)U+u)l}$ in terms of s, while $\Delta_s = \sum_{l=1}^{T_x} h^2_{((s-1)U+u)l} - A_h$. Then, (20.47) can be written as

$$\rho_{ij} = \frac{\sum_{s=1}^{S} \left(c_i^{II}[s-1] c_j^{II}[s-1] \Delta_s \right)}{\sum_{s=1}^{S} (A_h + \Delta_s)}. \tag{20.48}$$

Figure 20.8 shows the probability density function (PDF) of the correlation factor ρ_{12} between the signals of user 1 and user 2 for a two-user MC DS-CDMA system employing STS-based transmit diversity, when communicating over frequency-selective Rayleigh fading channels. The curves show that the correlation factor using $T_x = 1, 2$ and 4 transmitter antennas is symmetrically distributed around $\rho_{12} = 0$. An important observation is that the correlation factor's value is predominantly distributed in the vicinity of $\rho_{12} = 0$ and becomes similar to a truncated Gaussian random variable distributed within [-1,1] having a relatively low variance, when increasing the number of transmitter antennas. *This observation implies that STS using several transmitter antennas is capable of suppressing the multiuser interference without decreasing the achievable transmit diversity gain.*

In the context of the decorrelating multiuser detector, the decision variables associated with $b_{k,u1}$, $k = 1, 2, \ldots, K'$ are obtained by multiplying both sides of (20.45) with the

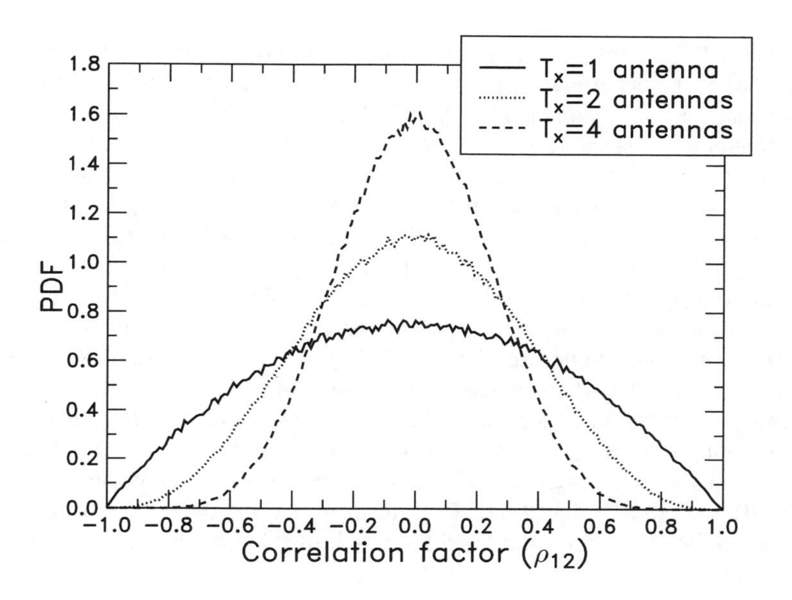

Figure 20.8: Simulated PDF of the correlation factor ρ_{12} for a two-user MC DS-CDMA using $T_x = 1, 2$ or 4 transmitter antennas and STS, when communicating over frequency-selective Rayleigh fading channels.

inverse of $\mathbf{R}$, i.e. with $\mathbf{R}^{-1}$, which can be expressed as

$$\mathbf{R}^{-1}\mathbf{z} = \sqrt{\frac{2UE_bT_b}{S}} \left(\sum_{s=1}^{S} \sum_{l=1}^{T_x} h^2_{((s-1)U+u)l} \right) \mathbf{b} + \mathbf{R}^{-1}\mathbf{C}^T\mathbf{n}, \tag{20.49}$$

and the corresponding data bits $b_{k,u1}$, $k = 1, 2, \ldots, K'$ are decided according to $\hat{b}_{k,u1} = \text{sgn}((\mathbf{R}^{-1}\mathbf{z})_k)$, $k = 1, 2, \ldots, K'$. Equation (20.49) shows that each user's data can be decided independently of the other users' data and the diversity order achieved is T_xS.

Unfortunately, deriving the closed-form BER formulae for both the correlation-based single-user detector and for the decorrelating multiuser detector is an arduous task, since – as we noted previously in this subsection – the correlation factors between the signals of two users are random variables. The closed-form BER formulae can be obtained, when the Gaussian approximation of the correlation factors is invoked and when the F-domain spreading codes are sufficiently long. However, if S, i.e. the length of the F-domain spreading codes assumes a low value, the approximate BER might be far from the actual BER. Hence, in the next subsection the BER performance of the broadband MC DS-CDMA system using TF-domain spreading is investigated with the aid of simulations.

20.3.4.3 BER Performance

The BER versus SNR per bit, E_b/N_0, performance of both the correlation-based single-user detector and that of the decorrelating multiuser detector is shown in Figures 20.9 and 20.10 for the TF-domain spread broadband MC DS-CDMA systems. In both figures we considered $T_x = 2$ transmitter antennas and supporting $K = \mathcal{K}_{max}$, $2\mathcal{K}_{max}$, $3\mathcal{K}_{max}$ and $4\mathcal{K}_{max}$ users corresponding to $K' = 1, 2, 3, 4$, where $\mathcal{K}_{max}$ represented the maximum number of users supported by the T-domain orthogonal spreading codes without imposing multiuser interference. The difference between Figure 20.9 and Figure 20.10 is that in Figure 20.9 the length of the F-domain spreading codes was $S = 4$, while in Figure 20.10 it was $S = 8$. As expected, we observe in both figures that the BER performance is significantly improved, when the correlation-based single-user detector is replaced by the decorrelating multiuser detector. For both the correlation-based single-user detector and the decorrelating multiuser detector the BER performance degrades, when increasing the number of users sharing the same T-domain spreading code, i.e. when increasing the value of K'. However, the BER increase due to increasing the value of K' is significantly lower for the decorrelating multiuser detector, than that of the correlation-based single-user detector. Furthermore, upon comparing Figure 20.9 to Figure 20.10, we observe that the BER performance of the decorrelating multiuser detector is closer to the BER performance without multiuser interference, when using $S = 8$ (Figure 20.10) instead of $S = 4$ (Figure 20.9).

In Figure 20.11 we investigated the BER performance of both the correlation-based single-user detector and that of the decorrelating multiuser detector for various numbers of transmitter antennas, namely for $T_x = 1, 2, 4$ and for $S = 8, 4, 2$ F-domain spreading codes, while maintaining a constant T_xS value of eight. In our experiments we assumed that $K' = 2$, i.e. that each set of T-domain STS codes was shared by two users. Let the maximum number of users supported by the T-domain codes be $\mathcal{K}_{max}$, while using the parameters of $(T_x = 4,\ S = 2)$. Then, for a broadband MC DS-CDMA system having a constant system bandwidth, the maximum number of users, $\mathcal{K}_{max}$ supported by the T-domain codes and using the parameters of $(T_x = 1,\ S = 8)$ or $(T_x = 2,\ S = 4)$ is $\mathcal{K}_{max}/4$ or $\mathcal{K}_{max}/2$, respectively. In other words, there is a maximum of two, four or eight users sharing the same set of orthogonal STS codes, corresponding to the cases of $(T_x = 4,\ S = 2)$, $(T_x = 2,\ S = 4)$ or $(T_x = 1,\ S = 8)$, respectively. From the results we infer the following observations: (a) all schemes achieve the same total diversity gain; (b) the number of transmitter antennas has the same effect on the BER performance as the length of the F-domain spreading codes, i.e. the same BER can be maintained, regardless of what values T_x and S take, provided that the product T_xS remains a constant; (c) the decorrelating multiuser detector significantly outperforms the correlation-based single-user detector. The gain achieved at the BER of 10^{-3} by using multiuser detection instead of single-user detection is about 5-6 dB; (d) the maximum number of users, K'_{max}, sharing the same set of STS orthogonal codes is two, four or eight, when we use the parameters $(T_x = 4,\ S = 2)$, $(T_x = 2,\ S = 4)$ or $(T_x = 1,\ S = 8)$. Furthermore, since, according to Figures 20.9, 20.10 the BER performance degrades upon increasing the number of users sharing the same set of STS orthogonal codes, consequently, for a fully loaded system using the maximum values of K'_{max}, we can surmise that MC DS-CDMA employing $(T_x = 4,\ S = 2)$ outperforms the scheme using the parameter combinations of both $(T_x = 2,\ S = 4)$ and $(T_x = 1,\ S = 8)$. Furthermore, the MC DS-CDMA system using the parameters $(T_x = 2,\ S = 4)$ outperforms that employing $(T_x = 1,\ S = 8)$.

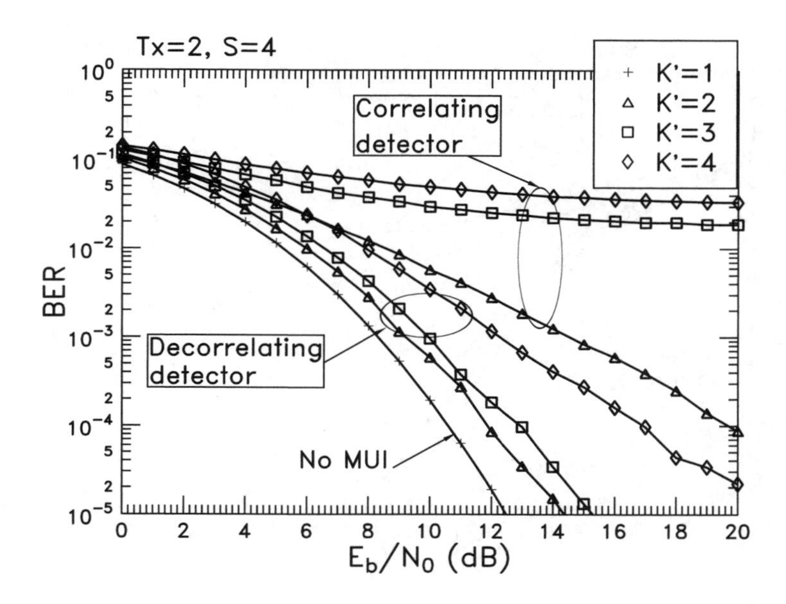

Figure 20.9: Simulation-based BER versus the SNR per bit, E_b/N_0, performance of both the single-user correlator and decorrelating multiuser detectors for STS-based broadband MC DS-CDMA using TF-domain spreading, when communicating over frequency-selective Rayleigh fading channels.

The above arguments suggest that *the best broadband MC DS-CDMA system will only use transmit diversity and no frequency diversity at all, i.e. use the parameters* ($T_x = 8,\ S = 1$), *which simultaneously suggests that no multiuser detection is required.*

20.3.5 Summary

In summary, in this section we have investigated the performance of a broadband MC DS-CDMA using STS, when frequency-selective Rayleigh fading channels are considered. The BER performance of the broadband MC DS-CDMA using STS-based transmit diversity has been estimated by using simulations. Furthermore, we have also investigated the performance of the extended STS-assisted broadband MC DS-CDMA using TF-domain spreading. Specifically, the corresponding BER performance of the TF-domain spread broadband MC DS-CDMA has been investigated in the context of the correlation-based single-user detector and the decorrelating multiuser detector for transmissions over frequency-selective Rayleigh fading channels.

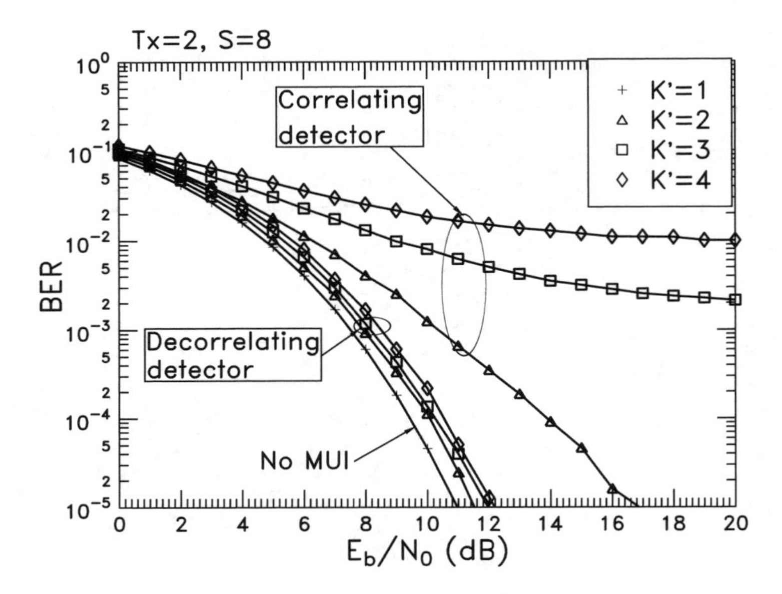

Figure 20.10: Simulation-based BER versus the SNR per bit, E_b/N_0, performance of both the single-user correlator and decorrelating multiuser detector for STS-based broadband MC DS-CDMA using TF-domain spreading, when communicating over frequency-selective Rayleigh fading channels.

20.4 Time- and Frequency-Domain Spreading Aided Quasi-Synchronous MC-CDMA Using Generalized Orthogonal Codes and Loosely Synchronized Codes

Hua Wei and L. Hanzo

20.4.1 Introduction

As argued in the context of Section 20.3.4, in code-division multiple-access (CDMA) communication there are two fundamental types of spread-spectrum schemes. The first scheme spreads the original data stream in the time (T)-domain [83, 492], while the second in the frequency (F)-domain, resulting in a scheme known as MC-CDMA [75, 354]. In Section 20.3.4 an amalgam of the above spreading schemes has been developed and analyzed mathematically. Explicitly, the philosophy of TF-domain spreading is that the original data stream is spread not only in the T-Domain, but also in the F-domain. Hence each user is assigned two

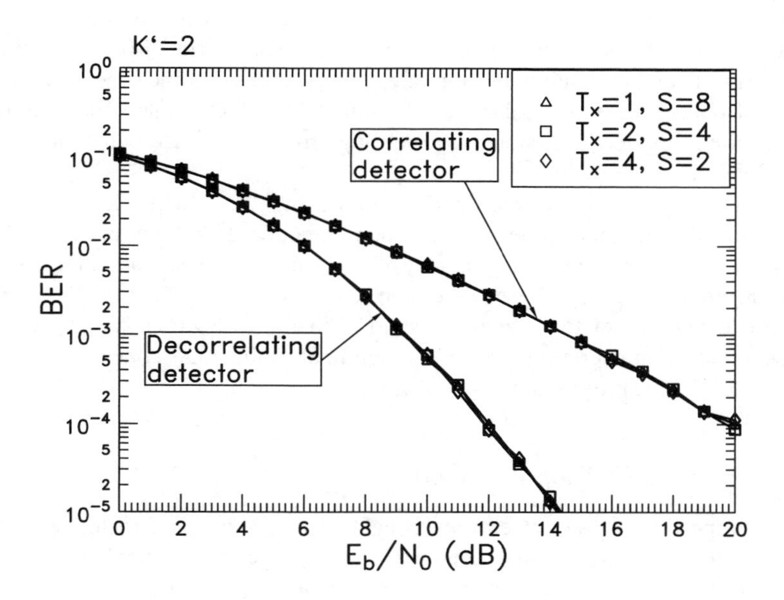

Figure 20.11: Simulation-based BER versus the SNR per bit, E_b/N_0, performance of both the single-user correlator and decorrelating multiuser detector for STS-based broadband MC DS-CDMA using TF-domain spreading, when communicating over frequency-selective Rayleigh fading channels.

spreading sequences for this operation, namely a T-domain and a F-domain sequence. As we have shown in Section 20.3.4, this system exhibits a high flexibility as well as a reduced Multiuser Detection (MUD) complexity. In Section 20.3.4 and in [493] several TF-domain spreading-assisted MC DS-CDMA MUD schemes have been considered. However, the complexity imposed may become excessive, when the number of users supported is high. Hence in this section we will consider the employment of two specific families of T-domain spreading codes, which have the potential of reducing the MUD complexity. These codes are the set of generalized orthogonal codes [494] and Loosely Synchronized (LS) codes [495] [496], which exhibit a so-called Interference Free Window (IFW). We will show that since over the duration of the IFW both the cross-correlation and the auto-correlation of the spreading codes are zero, no T-domain MUD is necessitated. More explicitly, the benefit of employing these specific codes as the T-domain code is that we are capable of eliminating the T-domain MUD, while achieving a frequency diversity gain. Interpreting the benefits of the IFW further, the MUD's complexity is reduced because only a small fraction of the total number of users has to be separated and detected by the MUD, which belong to a given MUD group. By contrast, the set of users which are differentiated with the aid of unique user-specific T-domain spreading codes having an IFW does not interfere with each other, as a benefit of the IFW provided by the T-Domain codes used. Viewing the benefits of TF-domain spreading from a different perspective, the advantage of the proposed scheme is that we can significantly extend the width of the IFW in comparison to a single-carrier DS-CDMA system, because as a benefit of distributing the bits to be transmitted to several sub-carriers MC DS-CDMA has the potential of significantly reducing the chip rate, thereby extending the time-duration T_c of the chips. This also allows us to extend the width of the IFW, which renders the system more insensitive to timing imperfections, since larger timing errors can be accommodated without imposing interference.

This section is organized as follows. Sections 20.4.2 and 20.4.3 will briefly describe the family of generalized orthogonal codes and LS codes. Section 20.4.4 characterizes the philosophy of TF-domain spreading in the context of MC DS-CDMA signals. Section 20.4.5 considers two different correlation-based detection schemes, while in Section 20.4.6 we discuss the beneficial features of this specific system. Finally, in Section 20.4.7 we provide simulation results for characterizing different generalized orthogonal codes and LS codes, while in Section 20.4.8 we present our conclusions.

20.4.2 Generalized Orthogonal Codes

Since the basic properties of generalized orthogonal codes have been characterized in [494], we will concentrate our attention on procedures used for creating generalized orthogonal codes. First, we define a sequence set $\mathbf{c}_1, \ldots, \mathbf{c}_N$, where $\mathbf{c}_n = [c_{n,0}, \ldots, c_{n,L-1}]$ is a spreading sequence having a length of L. The spreading codes result in an IFW width of τ_{IFW}, if the cross-correlation of the spreading codes satisfies:

$$R_{jk}(\tau) = \sum_{l=0}^{L-1} c_{j,l} c_{k,(l+\tau) \bmod L} = \begin{cases} L, & for \;\; \tau = 0, j = k \\ 0, & for \;\; \tau = 0, j \neq k \\ 0, & for \;\; 0 < \mid \tau \mid \leq \tau_{\text{IFW}}. \end{cases} \tag{20.50}$$

The family of generalized orthogonal binary codes is generated from a pair of so-called complementary sequences also referred to as mates [497, 498], which can be recursively generated as follows:

$$[X_0, Y_0] = [1, 1]. \tag{20.51}$$

$$[X_m, Y_m] = [X_{m-1}Y_{m-1}, (-X_{m-1})Y_{m-1}]. \tag{20.52}$$

Hence, the length of X_m, Y_m is $L_m = 2^m$, and for a given complementary sequence pair $[X_m, Y_m]$, we can construct the 0th order generalized orthogonal code's mother matrix $F^{(0)}$, which can be expressed as [494]:

$$F^{(0)} = \left[\begin{array}{cc} -X_m & Y_m \\ -\tilde{Y}_m & -\tilde{X}_m \end{array} \right], \tag{20.53}$$

where $\tilde{Y}_m$ is generated by reversing the order of the sequence Y_m, while $-X_m$ is the negated version of X_m. Each row of the mother matrix $F^{(0)}$ constitutes a spreading sequence, hence two spreading sequences are hosted by the matrix $F^{(0)}$. Once we obtained the mother matrix $F^{(0)}$, the so-called $(n+1)$th-order generalized orthogonal code's matrix can be recursively generated according to:

$$F^{(n+1)} = \begin{bmatrix} F^{(n)} \otimes F^{(n)} & (-F^{(n)}) \otimes F^{(n)} \\ (-F^{(n)}) \otimes F^{(n)} & F^{(n)} \otimes F^{(n)} \end{bmatrix}, \tag{20.54}$$

where $\otimes$ denotes an operation referred to as interleaving, and the interleaving interval is 2^{m+n-1}. For the example of two vectors, namely for $A = [a_1, a_2, a_3, a_4, \ldots]$, $B = [b_1, b_2, b_3, b_4, \ldots]$ and for the interleaving interval of 2 we have:

$$A \otimes B = [a_1, a_2, b_1, b_2, a_3, a_4, b_3, b_4, \ldots]. \tag{20.55}$$

For simplicity, we denote the nth-order generalized orthogonal codes as $F(L, M, Z)$, where L is the length of generalized orthogonal code, M is the number of the codes generated and Z is the width of the IFW. Explicitly, we have $L = 2^{2n+m+1}$, $M = 2^{n+1}$, $Z = 2^{n+m-1}$. As an example, we consider $m = 2$ complementary pairs, which can be obtained with the aid of Equation 20.51 and 20.52 as follows:

$X_m = \{- - + - - - - +\}$

$Y_m = \{- + + + - + - -\}$.

Once we obtained the complementary pair X_m, Y_m, we can construct the 0th-order mother matrix $F^{(0)}$ according to Equation 20.53, and the matrix $F(32, 4, 4)$ of the codes can be constructed according to Equation 20.54, while the interleaving interval is $2^{m+n-1} = 2^{2+1-1} = 4$. Therefore, following the interleaving operation, the spreading code $F(32, 4, 4)$ can be constructed as:

$\{- - + - - - + - - - - + - - - + + + - + - - + - + + + - - - - +\}$

$\{- + + + - + + + - + - - - + - - + - - - - + + + + - + + - + - -\}$

$\{+ + - + - - + - + + + - - - - + - - + - - - + - - - - + - - - +\}$

$\{+ - - - - + + + + - + + - + - - - + + + - + + + - + - - - + - -\}$

All four different codes of the $F(32, 4, 4)$ family exhibited the same auto-correlation magnitudes, namely that seen in Figure 20.12(a). It can be observed in Figure 20.12(a) that the off-peak auto-correlation $R_p[\tau]$ $p = 1, \ldots, 4$ becomes zero for $|\tau| \leq 4$. The cross-correlation magnitudes $|R_{j,k}(\tau)|$ depicted in Figure 20.12(b) are also zero for $|\tau| \leq 4$. From the observations made as regards to the aperiodic correlations we may conclude that the $F(32, 4, 4)$ codes exhibit an IFW within an offset duration of ± 4 chip intervals.

Furthermore, we can shorten the generalized orthogonal code set $F(L, M, Z)$ for creating a new generalized orthogonal code set $F'(L', M, Z')$, where we have $L' = 2^{2n+m+1-t}$, $M = 2^{n+1}$, $Z' = 2^{n+m-1-t}$, provided that we have $t < min\{m, n\}$. This shortening operation will reduce the code length L, however, it also reduces the width τ_{IFW} of the interference-free window. Our general objective is to maximize the relative duration of the IFW in comparison to the code length, while generating the highest possible number of codes. Broadly spreading this allows us to support the highest possible number of users without imposing multiuser interference. Having described the process of creating generalized orthogonal codes, in the next section we will consider another family of codes, which also exhibits an IFW, namely the family of LS codes.

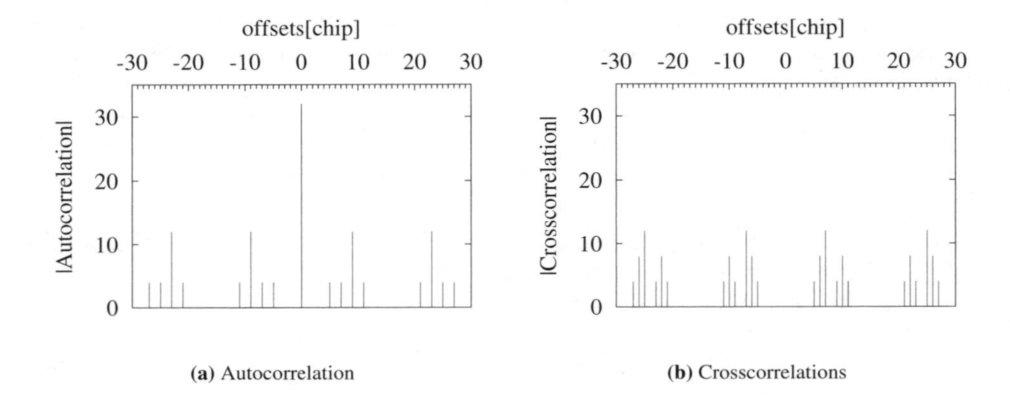

(a) Autocorrelation

(b) Crosscorrelations

Figure 20.12: The auto- and cross-correlation magnitudes of the $F(32, 4, 4)$ codes. (a) All four codes of the family exhibit the same auto-correlation magnitude. (b) The cross-correlation magnitudes of the four codes are also identical.

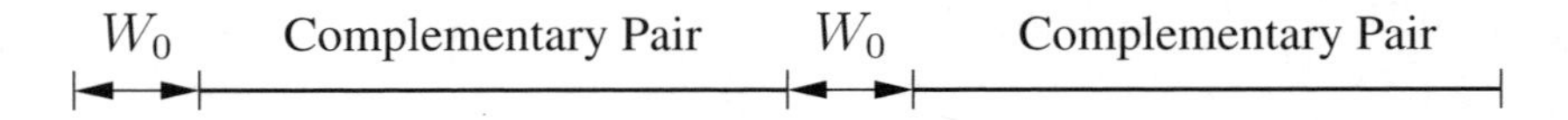

Figure 20.13: The LS code structure.

20.4.3 Loosely Synchronized Codes

There exists a specific family of LS codes [495], which also exhibits an IFW. Specifically, LS codes exploit the properties of the so-called orthogonal complementary sets [495, 497]. To expound further, let us introduce the notation of LS(N, P, W_0) for denoting the family of LS codes generated by applying a $P \times P$ dimensional Walsh-Hadamard matrix to an orthogonal complementary code set of length N, while inserting W_0 number of zeros both in the centre and at the beginning of the LS code, as shown in Figure 20.13, using the procedure described in [495]. Then, the total length of the LS(N, P, W_0) code is given by $L = 2NP+2W_0$ and the number of codes available is given by $2P$. Since the number of codes having a certain IFW limits the number of users that can be supported without imposing multiuser interference, this number has to be maximized for a given code length.

Since the construction method of binary LS codes was described in [495], here we refrain from providing an in-depth discourse and we will focus our attention on the employment of orthogonal complementary sets [498, 499] for the generation of LS codes. The aperiodic correlation $R_{j,k}(\tau)$ of two sequences $\mathbf{g}_j$ and $\mathbf{g}_k$ has to satisfy Equation 20.50 for the sake of maintaining an IFW of τ_{IFW} chip intervals.

For a given complementary pair $\{\mathbf{c}_0, \mathbf{s}_0\}$ of length N, one of the corresponding mate pairs can be written as $\{\mathbf{c}_1, \mathbf{s}_1\}$, where we have:

$$\mathbf{c}_1 = \tilde{\mathbf{s}}_0^*, \tag{20.56}$$

$$\mathbf{s}_1 = -\tilde{\mathbf{c}}_0^*, \tag{20.57}$$

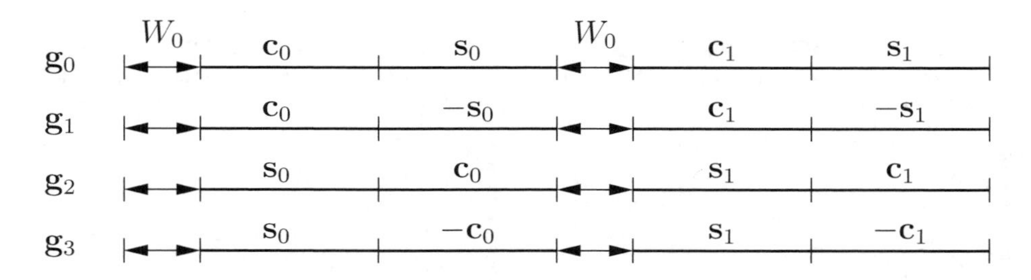

Figure 20.14: Generating four LS codes

and where $\tilde{\mathbf{s}}_0$ denotes the reverse-ordered sequence and $-\mathbf{s}_0$ is the negated version of $\mathbf{s}_0$, respectively. Note that in (20.56) and (20.57) additional complex conjugation of the polyphase complementary sequences is required for deriving the corresponding mate pair in comparison to binary complementary sequences [495]. Having obtained a complementary pair and its corresponding mate pair, we may employ the construction method of [495] for generating a family of LS codes. The LS codes generated exhibit an IFW, where we have $R_{jk}(\tau)=0$ for $|\tau| \leq min\{N-1, W_0\}$. Hence, we may adopt the choice of $W_0 = N-1$ in order to minimize the total length of the LS codes generated, while providing as long an IFW as possible.

For example, the LS(4,2,3) codes can be generated based on the complementary pair of [499]:

$$\mathbf{c}_0 = + + + - \tag{20.58}$$

$$\mathbf{s}_0 = + + - - . \tag{20.59}$$

Upon substituting (20.56) and (20.57) into (20.58) and (20.59), the corresponding mate pair can be obtained as:

$$\mathbf{c}_1 = \tilde{\mathbf{s}}_0^* = + - + + \tag{20.60}$$

$$\mathbf{s}_1 = -\tilde{\mathbf{c}}_0^* = + - - - . \tag{20.61}$$

The generation of this set of the four LS codes can be viewed in Fig 20.14. Upon invoking the 2×2-dimensional Hadamard expansion of [495] in the context of the above orthogonal complementary pairs, we can generate a family of four LS(4,2,3) codes, $\mathbf{g}_p$, $p = 0, \cdots, 3$. All four different codes of the LS(4,2,3) code family exhibited the same auto-correlation magnitudes, namely that seen in Figure 20.15(a). It can be observed in Figure 20.15(a) that the off-peak auto-correlation $R_p[\tau]$ becomes zero for $|\tau| \leq W_0 = 3$. The cross-correlation magnitudes $|R_{j,k}(\tau)|$ depicted in Figure 20.15(b) are also zero for $|\tau| \leq W_0 = 3$. Based on the observations made as regards to the aperiodic correlations we may conclude that the LS(4,2,3) codes exhibit an IFW of ± 3 chip durations.

In this section we demonstrated that the family of LS(N, P, W_0) codes can be constructed for almost any arbitrary code-length related parameter N by employing binary sequences. Since LS codes exhibit an IFW, they also constitute an attractive alternative for employment in the joint TF-domain spreading assisted MC-DS CDMA system of Section 20.4.4.

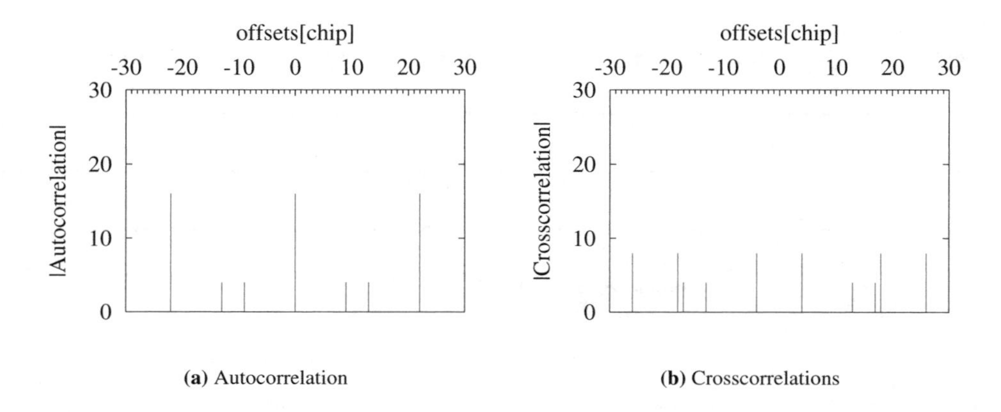

(a) Autocorrelation

(b) Crosscorrelations

Figure 20.15: Correlation magnitudes of the LS$(4, 2, 3)$ codes. (a) All four codes exhibit the same auto-correlation magnitude. (b) The cross-correlation magnitudes of $\mathbf{g}_0$ and $\mathbf{g}_2$.

20.4.4 System Model

The transmitter schematic of the MC DS-CDMA scheme using both T-domain and F-domain, i.e. TF-domain spreading is shown in Figure 20.16 in the context of the kth user. At the transmitter side, the binary data stream $b_k(t)$ is first direct-sequence (DS) spread using the T-domain signature sequence $a_k(t)$. Following T-domain spreading, the T-domain DS spread signal is divided into M parallel branches, where each branch-signal is multiplied by a corresponding chip value of the F-domain spreading sequence $\mathbf{c}_k = [c_k[1], c_k[2], \ldots, c_k[M]]^T$ of length M. Following F-domain spreading, each of the M branch signals modulates a sub-carrier frequency using binary phase shift keying (BPSK). Then, the M number of sub-carrier-modulated substreams are added in order to form the transmitted signal. Hence, the transmitted signal of user k can be expressed as

$$s_k(t) = \sqrt{\frac{2P}{M}} \sum_{m=1}^{M} b_k(t) a_k(t) c_k[m] \cos(\omega_m t), \; k = 1, 2, \ldots, K, \tag{20.62}$$

where P represents the transmitted power of each user and $\{\omega_m\}, \quad m = 1, \ldots, M$ represents the sub-carrier frequency set. The binary data stream's waveform $b_k(t) = \sum_{i=0}^{\infty} b_k P_{T_b}(t - iT_b), \quad k = 1, \ldots, K$ consists of a sequence of mutually independent rectangular pulses P_{T_b} of duration T_b and of amplitude +1 or -1 with equal probability. The spreading sequence $a_k(t) = \sum_{j=0}^{\infty} a_{kj} P_{T_c}(t - jT_c), \quad k = 1, \ldots, K$ denotes the T-domain spreading sequence waveform of the kth user, where $P_{T_c}(t)$ is the rectangular chip waveform, which is defined over the interval $[0, T_c)$. We assume that the T-domain spreading factor is $N = T_b/T_c$, which represents the number of chips per bit-duration. Furthermore, we assume that the sub-carrier signals are orthogonal and the spectral main-lobes of the sub-

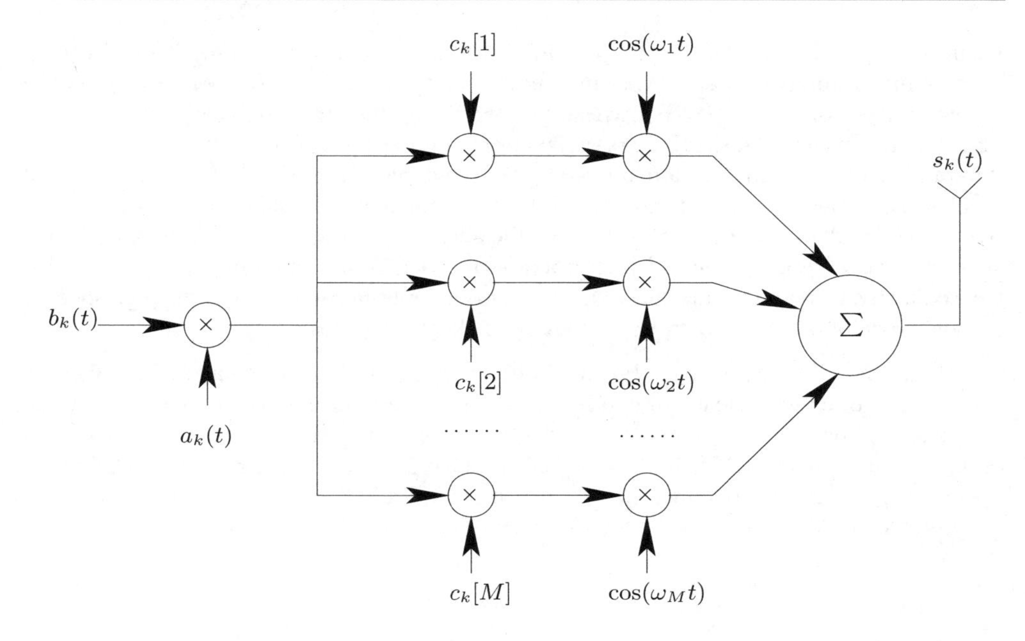

Figure 20.16: Transmitter model of MC DS-CDMA using both time-domain and frequency-domain spreading. T-domain spreading is achieved using the code $a_k(t)$ and each chip of $a_k(t)$ is spread in the frequency-domain by mapping it to M sub-carriers, each carrying one of the M F-domain chips c_k.

carrier signals are not overlapping with each other.

We assume that K quasi-synchronous[2] TF-domain spread MC DS-CDMA signals obeying the form of (20.62) are transmitted over the uplink frequency selective channel, but each sub-carrier of each user experiences statistically independent single-path flat Rayleigh fading. Hence, the mth sub-carrier's Channel Impulse Response (CIR) can be written as $\gamma_{m,k}e^{j\phi_{m,k}}, \quad m = 1, \ldots, M, \quad k = 1, \ldots, K$, where the amplitude $\gamma_{m,k}$ is a Rayleigh distributed random variable and the phase $\phi_{m,k}$ is uniformly distributed between $[0, 2\pi)$. We also assume furthermore that each user experiences a different delay of τ_k, which obeys $\tau_k \leq \tau_{\text{IFW}}$. This can be achieved by invoking a Global Positioning System (GPS) assisted synchronization protocol. Then the received signal may be expressed as:

$$r(t) = \sum_{k=1}^{K} \sqrt{\frac{2P}{M}} \sum_{m=1}^{M} b_k(t)a_k(t - \tau_k)c_k[m]\gamma_{m,k} \cos(\omega_m t + \phi_{m,k}) + n(t), \qquad (20.63)$$

where $n(t)$ represents the AWGN having zero mean and double-sided power spectral density of $N_0/2$.

Moreover, let $\{a_1(t), a_2(t), \ldots, a_N(t)\}$ and $\{\mathbf{c}_1, \mathbf{c}_2, \ldots, \mathbf{c}_M\}$ be the N number of T-domain spreading sequences and M number of F-domain spreading sequences, respectively,

[2]Quasi-synchronous in this context implies that the delay-differences of the individual users are within the IFW.

where $\mathbf{c}_u = [c_u[1]\ c_u[2]\ \ldots\ c_u[M]]^T$, $u = 1, \ldots, M$ represents a F-domain spreading code. Furthermore, we assume that the number of active users is K. We also introduce a new variable of $\mathcal{K} = \lfloor K/N \rfloor$, where $\lfloor x \rfloor$ represents the smallest integer not less than x, for denoting the number of users associated with a specific T-domain code, which are differentiated by a unique F-domain code. Then, we have $1 \leq \mathcal{K} \leq M$, since the total number of users is less than the product of the T-domain and F-domain spreading factor, i.e. we have $K \leq NM$ and $\mathcal{K} \leq M$. Based on the above assumptions, the K number of users supported can be grouped into N user groups, with each group supporting at most $\mathcal{K}$ users. Consequently, it can be readily shown that each of the N user groups can be distinguished by assigning one of the N number of T-domain spreading sequences.

As shown in Figure 20.16 and Equation (20.62), each TF-domain spread MC DS-CDMA signal is identified with the aid of two spreading sequences, one applied in the context of the T-domain and one in the F-domain. In the following sections we will analyze the detection of TF-domain spread MC DS-CDMA signals by invoking two different detection schemes. Specifically, in Section 20.4.5 we used both Maximum Ratio Combining (MRC) as well as a low-complexity Maximum Likelihood (ML) decision-based MUD.

20.4.5 Detection Schemes

In this section we will discuss the receiver model of the MC DS-CDMA schemes employed, which is shown in Fig 20.17. We consider a correlation-based receiver, which essentially carries out the inverse operation seen in Fig 20.16. In Fig 20.17 the output variable $Z_{m,k}$ corresponding to the mth sub-carrier of the kth user can be expressed as:

$$Z_{m,k} = \int_0^{T_b} r(t) a_k(t - \tau_k) \cos(\omega_m t) dt, \ k = 1, \ldots, K, \ m = 1, \ldots, M. \qquad (20.64)$$

According to [78], the received signal vector $\mathbf{Z}_m$ at the output of the bank of matched filters related to the mth sub-carrier can be expressed as:

$$\begin{aligned} \mathbf{Z}_m^{(i)} &= [z_{m,1}, \ldots, z_{m,K}] \\ &= \mathbf{R}[1]\mathbf{W}_m\mathbf{C}_m\mathbf{b}^{(i-1)} + \mathbf{R}[0]\mathbf{W}_m\mathbf{C}_m\mathbf{b}^{(i)} + \mathbf{R}^T[1]\mathbf{W}_m\mathbf{C}_m\mathbf{b}^{(i+1)} + \mathbf{n_m}, \end{aligned} \qquad (20.65)$$

where we have:

$$\begin{aligned} \mathbf{C}_m &= \mathrm{diag}[c_1[m], \ldots, c_K[m]] \\ \mathbf{W}_m &= \mathrm{diag}[\gamma_{m,1} e^{j\phi_{m,1}}, \ldots, \gamma_{m,K} e^{j\phi_{m,K}}] \\ \mathbf{b} &= [b_1, \ldots, b_K]^T \\ \mathbf{n} &= [n_1, \ldots, n_K]^T \end{aligned} \qquad (20.66)$$

i is the time index,

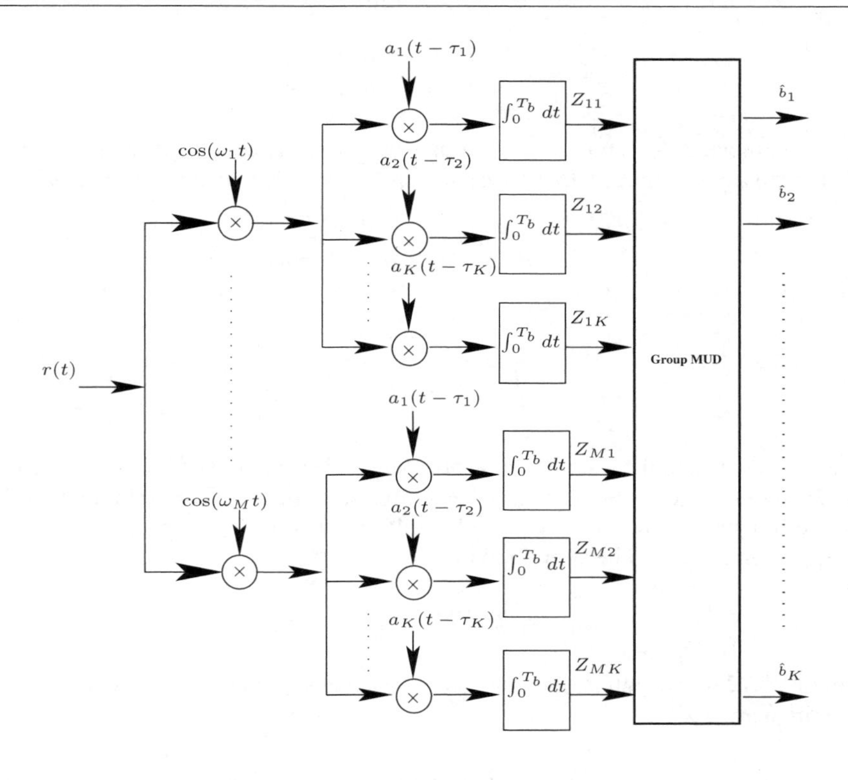

Figure 20.17: Receiver model of MC DS-CDMA using both time-domain and frequency-domain spreading.

and the zero-mean Gaussian noise vector $\mathbf{n_m}$ has the cross-correlation matrix of:

$$E[\mathbf{n}[i]\mathbf{n}^T[j]] = \begin{cases} \sigma^2\mathbf{R}^T[1], & \text{if } j = i+1; \\ \sigma^2\mathbf{R}[0], & \text{if } j = i; \\ \sigma^2\mathbf{R}[1], & \text{if } j = i-1; \\ \mathbf{0}, & otherwise. \end{cases} \tag{20.67}$$

The partial cross-correlation matrix $\mathbf{R}[1]$ and the cross-correlation matrix $\mathbf{R}[0]$ of the T-domain spreading codes, when communicating over an asynchronous channel, are defined as [78]:

$$\mathbf{R}_{jk}[0] = \begin{cases} 1, & \text{if } j = k; \\ \rho_{jk}, & \text{if } j < k; \\ \rho_{kj}, & \text{if } j > k, \end{cases} \tag{20.68}$$

$$\mathbf{R}_{jk}[1] = \begin{cases} 0, & \text{if } j \geq k; \\ \rho_{kj}, & \text{if } j < k, \end{cases} \tag{20.69}$$

where the coefficients ρ_{kj} and ρ_{jk} are the pair of cross-correlations of the spreading codes recorded in the asynchronous CDMA system considered, which can be written as [78]:

$$\rho_{jk} = \int_{\tau}^{T_b} a_j(t) a_k(t - \tau) dt \tag{20.70}$$

$$\rho_{kj} = \int_{0}^{\tau} a_j(t) a_k(t - \tau + T_b) dt. \tag{20.71}$$

However, in our system the time-domain spreading codes are either LS codes or generalized orthogonal codes, which exhibit an IFW. Therefore, the partial cross-correlation matrix $\mathbf{R}[1]$ is an all-zero matrix, provided that the delay differences obey $\tau_k \leq \tau_{\text{IFW}}, \;\; k = 1, \ldots, K$. Hence, Equation 20.65 can be simplified to:

$$\mathbf{Z}_m^{(i)} = \mathbf{R}[0] \mathbf{W}_m \mathbf{C}_m \mathbf{b}^{(i)} + \mathbf{n}_m. \tag{20.72}$$

In Equation 20.72 we can ignore the time index (i) in the superscripts. Hence, Equation 20.72 can be written as:

$$\mathbf{Z}_m = \mathbf{R}[0] \mathbf{W}_m \mathbf{C}_m \mathbf{b} + \mathbf{n}_m. \tag{20.73}$$

20.4.5.1 Maximum Ratio Combining

Following the philosophy of MRC, the output variable of the kth user can be written as:

$$\mathbf{Z}_{\text{MRC},k} = \sum_{m=1}^{M} Z_{m,k} \gamma_{m,k} e^{-j\phi_{m,k}}, \tag{20.74}$$

which is the superposition of the correlator outputs matched to the individual sub-carriers $m = 1, \ldots, M$, weighted by the complex conjugate of the corresponding complex-valued CIR coefficient, which is expressed as $\gamma_{m,k} \cdot e^{-j\phi_{m,k}}$. Finally, the corresponding data bits b_k are decided according to the decision rule of $\hat{b}_k = \text{sgn}(\mathbf{Z}_{\text{MRC},k})$ for $k = 1, 2, \ldots, K$.

20.4.5.2 Low-Complexity Multiuser Detection

Let us now interpret Equation 20.73 in more detail. In Section 20.4.4 we have divided the K number of users supported into $\mathcal{K}$ user groups, each group having N users, which are distinguished by their time-domain spreading code $a_k(t)$. In this scenario, the system activates N time-domain spreading codes and $\mathcal{K}$ frequency-domain spreading codes. Therefore, provided that both the ith and jth user's delay are within the range of the IFW and the ith user and jth user are in the same group, the element ρ_{ij} of the correlation-matrix $\mathbf{R}[0]$ of Equation 20.73

will satisfy $\rho_{ij} = 0$. More specially, in this system each user will encounter Multiple Access Interference (MAI) imposed by a reduced number of $\mathcal{K} - 1$ users, rather than K users, since all these $\mathcal{K}$ users of each of the N user groups employ the same T-domain spreading code but a different F-domain spreading code. Therefore, we have to detect a reduced number of $\mathcal{K}$ rather than K users, which facilitates the employment of low-complexity Multiuser Detection (MUD). For example, let us consider the first user in the context of supporting a total of $K = \mathcal{K}N$ users. Then the first user encounters interference inflicted by the Nth,...,$\mathcal{K}N$th user, because they share the same T-domain spreading code, but they are identified by the different F-domain spreading codes. Hence, Equation 20.73 can be simplified as:

$$\tilde{\mathbf{Z}}_m = \tilde{\mathbf{R}}[0]\tilde{\mathbf{W}}_m\tilde{\mathbf{C}}_m\tilde{\mathbf{b}} + \tilde{\mathbf{n}}_m, \tag{20.75}$$

where we have:

$$\begin{aligned}
\tilde{\mathbf{Z}}_m &= [Z_1, Z_N, \ldots, Z_{\mathcal{K}N}]^T, \\
\tilde{\mathbf{C}}_m &= \text{diag}[c_1[m], c_N[m] \ldots, c_{\mathcal{K}N}[m]], \\
\tilde{\mathbf{W}}_m &= \text{diag}[\gamma_{1,m}e^{j\phi_{1,m}}, \gamma_{N,m}e^{j\phi_{N,m}} \ldots, \gamma_{\mathcal{K}N,m}e^{j\phi_{\mathcal{K}N,m}}], \\
\tilde{\mathbf{b}} &= [b_1, b_N, \ldots, b_{\mathcal{K}N}]^T, \\
\tilde{\mathbf{n}}_m &= [n_1, n_N, \ldots, n_{\mathcal{K}N}]^T,
\end{aligned}$$

$$\tilde{\mathbf{R}}[0] = \begin{bmatrix} \rho_{1,1} & \rho_{1,N} & \cdots & \rho_{1,\mathcal{K}N} \\ \rho_{N,1} & \rho_{N,N} & \cdots & \rho_{N,\mathcal{K}N} \\ \vdots & \vdots & \vdots & \vdots \\ \rho_{\mathcal{K}N,1} & \rho_{\mathcal{K}N,N} & \cdots & \rho_{\mathcal{K}N,\mathcal{K}N} \end{bmatrix}. \tag{20.76}$$

The Maximum Likelihood (ML) decision-based MUD of the mth sub-carrier has to evaluate:

$$\hat{\mathbf{b}} = \arg\{\min_{\mathbf{b}}\{\| \tilde{\mathbf{Z}}_m - \tilde{\mathbf{R}}[0]\tilde{\mathbf{W}}_m\tilde{\mathbf{C}}_m\tilde{\mathbf{b}} \|^2\}\}, \tag{20.77}$$

where $\| \cdot \|^2$ denotes the Euclidean norm. Upon combining all the M sub-carriers, the ML-based MUD's decision function can be written as:

$$\hat{\mathbf{b}} = \arg\{\min_{\mathbf{b}}\{\sum_{m=1}^{M} \| \tilde{\mathbf{Z}}_m - \tilde{\mathbf{R}}[0]\tilde{\mathbf{W}}_m\tilde{\mathbf{C}}_m\tilde{\mathbf{b}} \|^2\}\}. \tag{20.78}$$

According to Equation 20.78, the complexity of the ML decision-based MUD invoked in the context of the TF-domain spreading assisted MC DS-CDMA system is on the order of $2^{\mathcal{K}}$, rather than on $2^{\mathcal{K}N}$. Let us now briefly summarize the basic features of the system considered, before we characterize the achievable system performance.

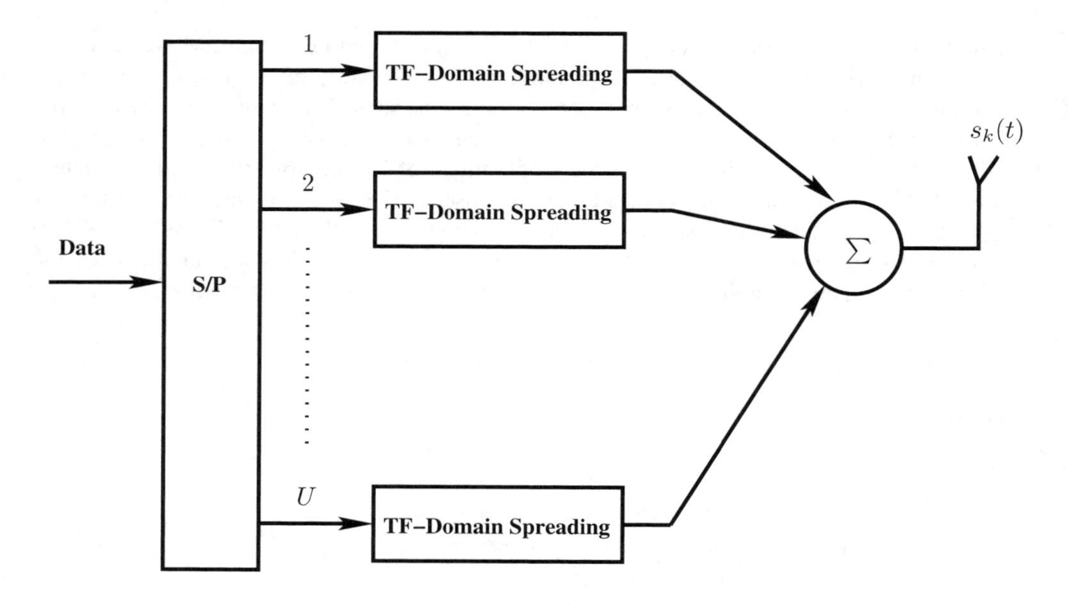

Figure 20.18: Transmitter model of MC DS-CDMA using serial to parallel conversion.

20.4.6 Characteristics of TF-domain Spreading Assisted MC DS-CDMA Employing Generalized Orthogonal Codes and Loosely Synchronized Codes

The system considered in Figure 20.16 of the previous section can be extended by using $U \cdot M$ number of sub-carriers. Specifically, as shown in Fig 20.18, the original serial data stream is first subjected to serial to parallel conversion, resulting in U independent parallel data streams. Moreover, each parallel data stream is further spread to M sub-carriers using the TF-domain spreading philosophy of Fig 20.16. The M sub-carriers are arranged for maintaining the maximum possible frequency spacing, so that they experience independent fading and hence achieve the maximum possible frequency diversity gain. Specifically, let $\{f_1, f_2, \cdots, f_{US}\}$ be the $U \cdot M$ number of sub-carrier frequencies, where each of the U parallel bits is spread according to the following $U \times M$-dimensional matrix:

$$\mathbf{F} = \begin{pmatrix} f_1 & f_{U+1} & \cdots & f_{(M-1)U+1} \\ f_2 & f_{U+2} & \cdots & f_{(M-1)U+2} \\ \vdots & \vdots & \ddots & \vdots \\ f_U & f_{2U} & \cdots & f_{MU} \end{pmatrix}. \quad (20.79)$$

Therefore, the chip rate of the MC DS-CDMA signal can be reduced by a factor of $U \cdot M$ and hence the width of the IFW can be extended by a factor of $U \cdot M$ in comparison to a

DS-CDMA system, which can be expressed as:

$$\tau_{IFW,MCDS-CDMA} = U \cdot M \cdot \tau_{IFW,DS-CDMA}. \tag{20.80}$$

Based on Equation 20.80, we argue that the width of the interference-free window of MC DS-CDMA systems may be significantly higher than that of the DS-CDMA. This beneficial feature allows us to have significantly larger cells, which result in higher propagation delay differences and/or allows the system to reliably operate even in case of higher absolute code synchronization errors, as long as they do not exceed the IFW width.

Another advantage of this system is that it is capable of achieving a high frequency diversity gain, because the chips of each bit are transmitted on independently fading sub-carriers. Hence when the transmitted signal experiences frequency selective fading, the chances are that only some of the chips of a F-domain spreading code are corrupted and therefore the corresponding bit conveyed by the specific spreading code concerned may still be recovered. Consequently this system may be capable of achieving Mth-order F-domain diversity and benefit from a factor of $2^{\mathcal{K}N}/2^{\mathcal{K}} = 2^N$ lower multiuser detection complexity.

20.4.7 Simulation Results

Let us consider a system having $M = 4$ sub-carriers and using different T-domain spreading codes. The F-domain spreading codes are generated by the 4×4-dimensional orthogonal Walsh code generator matrix given by:

$$\mathbf{F} = \begin{pmatrix} + & + & + & + \\ + & - & + & - \\ + & + & - & - \\ + & - & - & + \end{pmatrix}. \tag{20.81}$$

Furthermore, we considered four different T-domain spreading codes, which are the $F(16, 8, 1)$, $F(16, 4, 2)$, LS$(2, 4, 1)$ and the LS$(4, 2, 3)$ codes, where again, generalized orthogonal codes are defined with the aid of the parameters $F(L, M, Z)$, with L being the length of the code, M being the number of codes generated and Z being the IFW width. Furthermore, the LS codes are defined with the aid of the parameters LS(N, P, W), where the length of the LS code is $2NP + 2W$, the number of codes generated is $2P$, and the width of the IFW is W. Moreover, we assume that each sub-carrier of each user experienced independent flat Rayleigh fading.

From Figures 20.19, 20.20, 20.21 and 20.22 we can observe that the low-complexity MUD is capable of approaching the single-user bound. By contrast, it becomes explicit from Figures 20.23, 20.24 20.25 and 20.26 that – as expected – the performance of MRC based detection is substantially worse than that of the MUD.

20.4.8 Summary

In this section we employed a specific family of spreading codes which exhibit an interference-free window in the context of the TF-domain spreading-assisted MC DS-CDMA

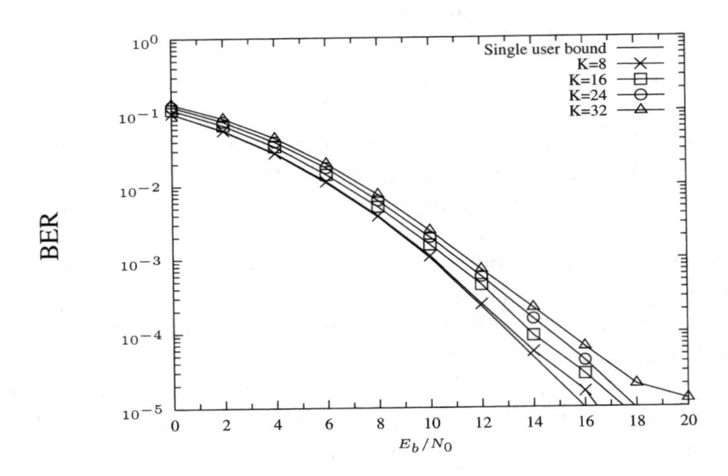

Figure 20.19: BER performance of the TF-domain spreading-aided MC DS-CDMA system in conjunction with low-complexity MUD, while employing the $F(16, 8, 1)$ generalized orthogonal code as the T-domain spreading code. The F-domain spreading code was a 4×4-dimensional Walsh code, and each of the $M = 4$ sub-carriers experienced independent narrowband Rayleigh fading. The MUD complexity reduction factor was $2^N = 2^8 = 512$.

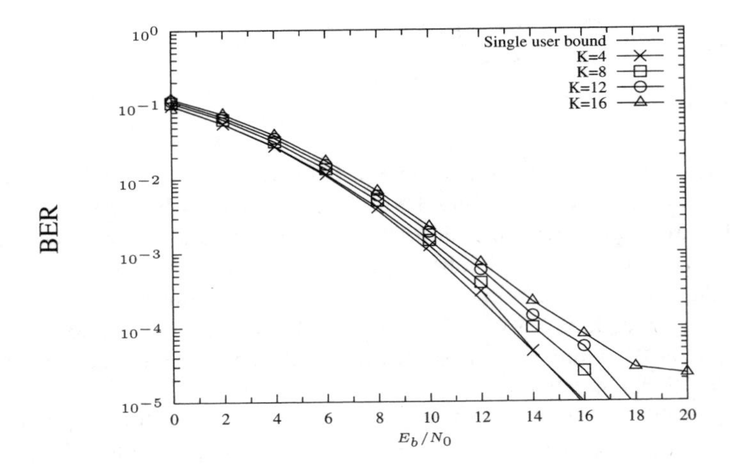

Figure 20.20: BER performance of the TF-domain spreading-aided MC DS-CDMA system in conjunction with low-complexity MUD, while employing the $F(16, 4, 2)$ generalized orthogonal code as the T-domain spreading code. The F-domain spreading code was a 4×4-dimensional Walsh code, and each of the $M = 4$ sub-carriers experienced independent narrowband Rayleigh fading. The MUD complexity reduction factor was $2^N = 2^4 = 16$.

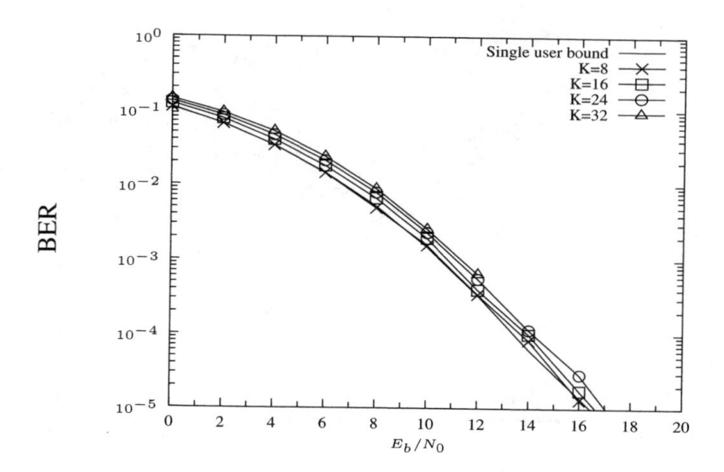

Figure 20.21: BER performance of the TF-domain spreading-aided MC DS-CDMA system in conjunction with low-complexity MUD, while employing the $\mathrm{LS}(2, 4, 1)$ loosely synchronized code as the T-domain spreading code. The F-domain spreading code was a 4×4-dimensional Walsh code, and each of the $M = 4$ sub-carriers experienced independent narrowband Rayleigh fading. The MUD complexity reduction factor was $2^N = 2^8 = 512$.

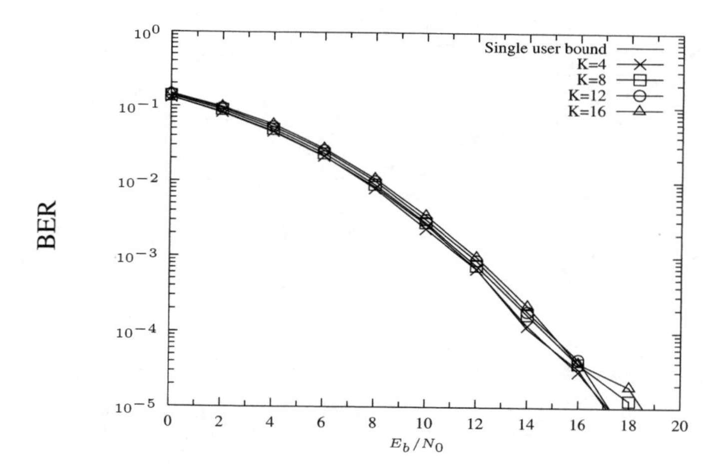

Figure 20.22: BER performance of the TF-domain spreading-aided MC DS-CDMA system in conjunction with low-complexity MUD, while employing the $\mathrm{LS}(4, 2, 3)$ loosely synchronized code as the T-domain spreading code. The F-domain spreading code was a 4×4-dimensional Walsh code, and each of the $M = 4$ sub-carriers experienced independent narrowband Rayleigh fading. The MUD complexity reduction factor was $2^N = 2^4 = 16$.

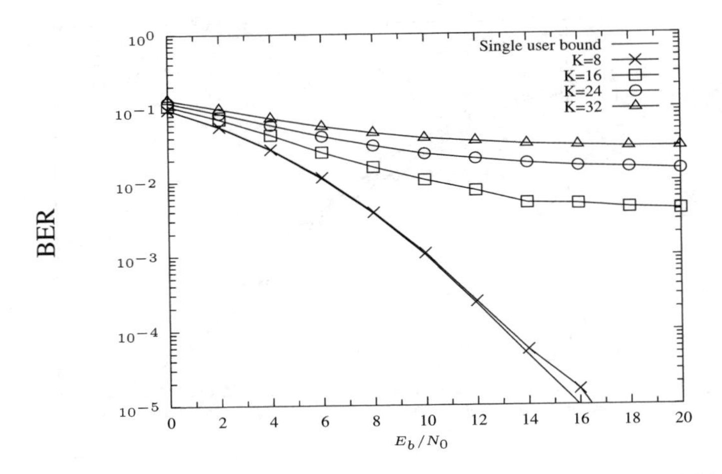

Figure 20.23: BER performance of the TF-domain spreading-aided MC DS-CDMA system when MRC detection was employed, while employing the $F(16, 8, 1)$ generalized orthogonal code as the T-domain spreading code. The F-domain spreading code was a 4×4-dimensional Walsh code, and each of the $M = 4$ sub-carriers experienced independent narrowband Rayleigh fading.

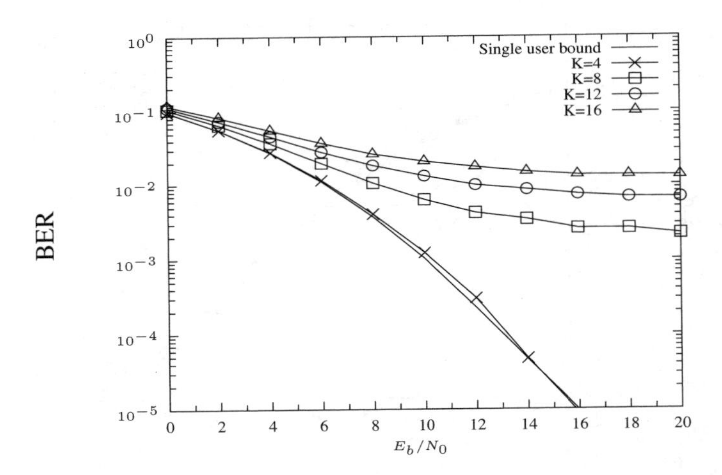

Figure 20.24: BER performance of the TF-domain spreading-aided MC DS-CDMA system when MRC detection was employed, while employing the $F(16, 4, 2)$ generalized orthogonal code as the T-domain spreading code. The F-domain spreading code was a 4×4-dimensional Walsh code, and each of the $M = 4$ sub-carriers experienced independent narrowband Rayleigh fading.

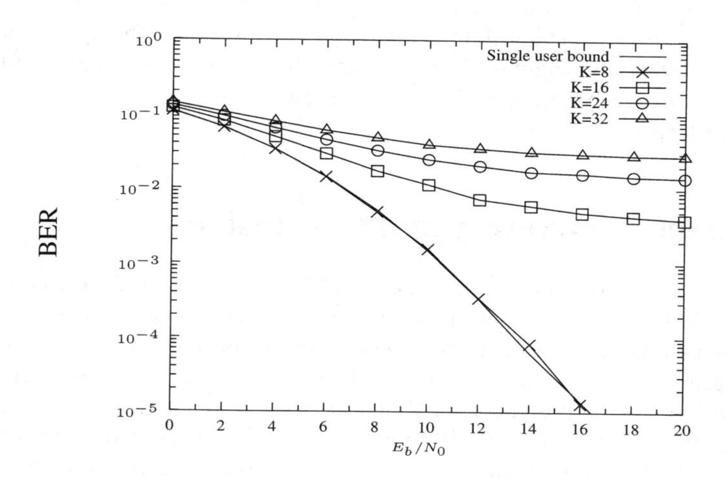

Figure 20.25: BER performance of the TF-domain spreading-aided MC DS-CDMA system when MRC detection was employed, while employing the LS$(2, 4, 1)$ loosely synchronized code as the T-domain spreading code. The F-domain spreading code was a 4×4-dimensional Walsh code, and each of the $M = 4$ sub-carriers experienced independent narrowband Rayleigh fading.

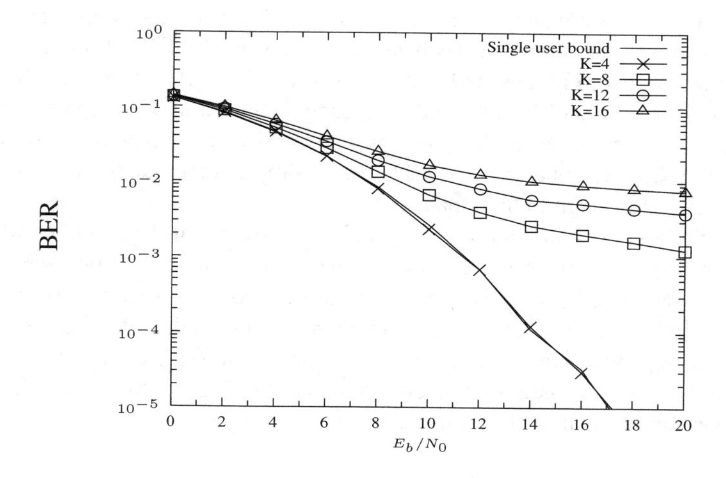

Figure 20.26: BER performance of the TF-domain spreading-aided MC DS-CDMA system when MRC detection was employed, while employing the LS$(4, 2, 3)$ loosely synchronized code as the T-domain spreading code. The F-domain spreading code was a 4×4-dimensional Walsh code, and each of the $M = 4$ sub-carriers experienced independent narrowband Rayleigh fading.

system considered. In this system we reduced the complexity of the MUD by a factor of $2^{\mathcal{K}N}/2^{\mathcal{K}} = 2^N$, while achieving Mth-order frequency diversity. The system is capable of significantly extending the width of the interference-free window as a benefit of the serial to parallel conversion invoked in Figure 20.18, which additionally renders the system insensitive to timing imperfections.

20.5 Chapter Summary and Conclusion

Following the brief motivational notes of Section 20.1 in Section 20.2 we compared various broadband CDMA schemes supporting ubiquitous communications in diverse propagation environments and maintaining the target integrity aimed for, regardless of the dispersion of the specific propagation environments encountered. Specifically, multicarrier DS-CDMA (MC DS-CDMA) was proposed for satisfying the various requirements of broadband systems and for supporting error-resilient communications, with the aid of STS-based transmit diversity. We have discussed the design of the various system parameters rendering the STS-assisted broadband MC DS-CDMA scheme capable of supporting ubiquitous communications without compromising the achievable BER. We have also considered the user capacity extension achievable by STS-assisted broadband MC DS-CDMA with the aid of TF-domain spreading. Our studies suggest that broadband MC DS-CDMA using STS constitutes a promising multiple-access scheme, which is capable of avoiding the various design limitations that are unavoidable, when using single-carrier DS-CDMA or MC-CDMA.

In Section 20.3 we investigated the performance of the broadband MC DS-CDMA system and that of the extended broadband MC DS-CDMA system, which employs both time-domain spreading and frequency-domain spreading. Both of the above broadband MC DS-CDMA systems are assumed to be designed based on the limitations of Section 20.2 and are assumed to use STS-assisted transmit diversity. Furthermore, in the context of the STS-assisted broadband MC DS-CDMA arrangement using TF-domain spreading, the corresponding BER performance has been investigated in the context of the correlation-based single-user detector and the decorrelating multiuser detector for transmissions over frequency-selective Rayleigh fading channels.

Finally, in Section 20.4 we employed a specific family of spreading codes which exhibit an interference-free window in the context of the TF-domain spreading-assisted MC DS-CDMA system considered. With the advent of the LS codes we reduced the complexity of the MUD by a factor of $2^{\mathcal{K}N}/2^{\mathcal{K}} = 2^N$, while achieving Mth-order frequency diversity. The system was capable of significantly extending the width of the interference-free window as a benefit of the serial to parallel conversion invoked in Figure 20.18, which additionally rendered the system insensitive to timing imperfections.

Chapter 21

Initial Synchronization in Single- and Multi-Carrier Direct-Sequence Code Division Multiple-Access Systems[1]

21.1 Introduction

Accurate and fast synchronization plays a cardinal role in the efficient utilization of any spread-spectrum system. Typically, the first step in the process of synchronization between the received spread pseudo noise (PN) code (sequence) and the locally generated despread code (sequence) is code acquisition [14–16, 500], which is also referred to as initial synchronization. More explicitly, initial synchronization is constituted by a process of successive decisions, wherein the ultimate goal is to bring the two codes into coarse time alignment, namely within one code-chip interval. Once initial code acquisition has been accomplished, usually a code tracking loop [14] is employed for achieving fine alignment of the two codes and for maintaining their alignment during the whole data transmission process. The aim of this chapter, however, is to focus on the task of initial synchronization in the context of direct-sequence code division multiple-access (DS-CDMA) systems.

Initial synchronization has been lavishly treated in the literature [15–53] in recent years. Initial code acquisition in DS-CDMA systems is usually achieved with the aid of non-

[1] *Single- and Multi-Carrier CDMA Multi-User Detection, Space-Time Spreading, Synchronization, Networking and Standards.* L.Hanzo, L-L.Yang, E-L.Kuan and K.Yen,
 ISBN 0-470-86309-9

coherent correlation or matched filtering, since, prior to despreading, the signal-to-noise ratio (SNR) is usually insufficiently high for attaining a satisfactory performance with the aid of coherent carrier phase estimators based on carrier-phase tracking loops. Initial code synchronization methods can broadly be classified as serial search acquisition [15,22,28,30,39,44] and parallel search acquisition [19,20,23–25,31,46,49] techniques. In serial search-based initial code synchronization all potentially possible code phases are searched serially, until synchronization is achieved. More explicitly, in serial search-based code acquisition each reference phase is evaluated by attempting to despread the received signal. If the estimated code phase is correct, successful despreading will take place, which can be detected. If the code phase is incorrect, the received signal will not be successfully despread, and the local reference code replica will be shifted to a new tentative phase for evaluation. By contrast, in parallel search-based code acquisition potentially all of the possible code phases are tested simultaneously. In this chapter both serial and parallel search-based acquisition schemes will be investigated.

It is well known that the near–far interference inflicted upon low-power signals by high-power interfering signals may substantially degrade the system's performance. In the context of DS-CDMA the above-discussed serial and parallel search-based code timing acquisition are as interference-limited and as vulnerable to the near–far problem, as conventional matched-filter-based single-user detection. Recently, a range of timing acquisition algorithms having a high near–far resistance in multi-user environments has also been proposed [17,18,21,37,54–56]. These timing-acquisition algorithms can be classified as maximum-likelihood synchronization scheme [21,55], minimum-mean-square-error (MMSE) timing estimation arrangements [17,37,54], per-survivor processing (PSP) based blind acquisition techniques [18] and MUltiple Signal Classification Algorithm (MUSIC) based timing estimation scheme [54–56]. Note that in this chapter only the conventional serial and parallel search-based initial synchronization schemes are considered, while the advanced timing acquisition schemes having near–far resistance will not be discussed in more detail in this chapter.

Following the above rudimentary introduction, in this chapter we will investigate both serial and parallel searched code acquisition schemes in more depth. In Section 21.2 we commence our discourse by describing the transmitted signal as well as the channel models. The serial and parallel searched based code acquisition schemes are discussed in Section 21.3. In Section 21.4 we derive the mean acquisition time expressions for both serial and parallel search-based acquisition schemes. Finally, from Section 21.5 to Section 21.9 we investigate the performance of a number of acquisition schemes associated with single-carrier or multi-carrier DS-CDMA systems. We note at this early stage that typically there exist multiple correct timing states (so-called H_1 cells) in almost any practical acquisition system, since the typical search step size usually takes value of 1/2 code chip duration and since the delay spread caused by the channel fading is on the order of several chip durations. Hence, our analysis in Sections 21.5, 21.6, 21.7 and 21.9 is based on the assumption of having multiple correct timing states (multiple H_1 cell), although most previous work was based on the assumption of having a single correct timing state or single H_1 cell [15,23–25,28,31,39, 49].

In Section 21.8 two novel sequential estimation methods are proposed for the initial synchronization of CDMA systems communicating using PN signals derived from m-sequences. These sequential estimation methods are designed based on the novel principle of iterative soft-in-soft-out (SISO) decoding. Specifically, two types of sequential estimation methods are proposed and investigated, namely Coherent Recursive Soft Sequential Estimation (CRSSE) based acquisition and Differential Recursive Soft Sequential Estimation (DRSSE) acquisition. The study in Section 21.8 shows that both the CRSSE and DRSSE acquisition schemes exhibit a complexity similar to that of a conventional m-sequence generator, which increases only linearly with the number of stages used in the m-sequence generator. Furthermore, the initial acquisition time of the CRSSE and the DRSSE acquisition schemes is also linearly dependent on the number of stages used in the m-sequence generator. Let us now characterize the DS-CDMA signal mode suitable for the analysis of the initial synchronization performance in the next section.

21.2 System Model

21.2.1 The Transmitted Signal

The communication model under consideration consists of K simultaneous transmitters, which includes $K-1$ actively communicating data transmission users, who have completed their code acquisition and an additional user attempting to establish initial synchronization. For the sake of brevity, we will often refer to the user whose PN sequence is being acquired by the receiver as the initial synchronization user. We assume that the first user is the initial synchronization user, whose performance is to be evaluated in conjunction with different acquisition schemes. Each user is assigned a unique CDMA spreading sequence, which spreads the data sequence. In this chapter random signature sequences having a common chip rate of $1/T_c$ are considered, unless otherwise noted. The processing gain is given by $N = T_b/T_c$, and $1/T_b$ is the information bit rate. Let $\{c_j^{(k)}\}$ denote a binary $\{+1,-1\}$ sequence with $c_j^{(k)}$ taking values of +1 and -1 with equal probability, and let $c_k(t) = \sum_{j=-\infty}^{\infty} c_j^{(k)} P_{T_c}(t - jT_c)$ denote the signature sequence waveform of the kth user, where $P_\tau(t) = 1$ over the time interval of $0 \le t \le \tau$ and equals zero otherwise. The data waveform $b_k(t) = \sum_{i=-\infty}^{\infty} b_i^{(k)} P_{T_b}(t - iT_b)$ consists of a sequence of mutually independent rectangular pulses of duration T_b and of amplitude +1 or -1 with equal probability.

Using the data sequence for modulating the signature sequence waveform and the carrier, the transmitted signal of the kth user is expressed as:

$$s_k(t) = \sqrt{2P_k}b_k(t)c_k(t)\cos(2\pi f_c t + \varphi_k), \qquad (21.1)$$

where P_k represents the average transmitted power of the kth signal, f_c is the common carrier frequency and φ_k is the phase introduced by the kth modulator, which is modelled as a random variable uniformly distributed over $[0, 2\pi)$.

However, the presence of data modulation in the initial synchronization signal complicates the code synchronization process at the receiver in various ways [24]. Therefore, in practical DS-CDMA systems [14–16, 28, 30, 39, 44, 49] the transmitter often aids the initial synchronization process by transmitting the phase-coded carrier signal without data modulation at the beginning of each transmission. Hence, in this chapter we shall assume for the sake of simplicity that no data modulation is imposed on the signals transmitted during initial synchronization process.

21.2.2 The Channel Model and the Received Signal

The various acquisition systems will be analyzed for transmissions over typical Additive White Gaussian Noise (AWGN) channels as well as while communicating over typical frequency-selective slow fading channels. When encountering an AWGN channel and asynchronous transmissions are assumed, the received signal can be viewed as the sum of the phase-coded carrier signal of interest, for which initial synchronization is considered, the $K-1$ data transmission signals (multi-user interference) and the AWGN. This composite signal can be expressed in the form of:

$$\begin{aligned} r(t) &= \sqrt{2P_r}c_1(t-\tau_1)\cos(2\pi f_c t+\theta_1) \\ &\quad +\sum_{k=2}^{K}\sqrt{2P_k}b_k(t-\tau_k)c_k(t-\tau_k)\cos(2\pi f_c t+\theta_k) \\ &\quad +n(t), \end{aligned} \tag{21.2}$$

where the subscript 'r' explicitly identifies the reference signal, and $\tau_k,\ k=r,2,3,\ldots,K$ denotes the relative time delay associated with the propagation delay in the context of an asynchronous transmission scheme. Furthermore, $\theta_k=\varphi_k-2\pi f_c\tau_k$ for $k=r,2,3,\ldots,K$ denotes the initial carrier phase, which are modelled as independent identical distributed (iid) random variables uniformly distributed in $[0,2\pi)$, while $n(t)$ represents the AWGN having a double-sided power spectral density of $N_0/2$. Note that since the $K-1$ interfering users are in the active data transmission phase, we assume that their signals are ideally power-controlled and the average received power of each interfering signal is expressed as $P_k=P_I$ for $k=2,3,\ldots,K$. However, for the initial synchronization user it is unrealistic to invoke near-ideal power-control, before successful synchronization is achieved. Hence this user can only rely on open-loop power control operating on the basis of estimating the channel's attenuation. Therefore, the average received power arriving from the initial synchronization user at the base station is usually different from that of the active data transmission users. The power of the reference signal is denoted in Equation 21.2 as P_r. These assumptions are practical in conventional DS-CDMA systems, such as, for example, the IS-95 system [11], employing a correlation or a matched-filtering-based single user receiver. However, when advanced DS-CDMA systems are considered, which employing multi-user detection techniques [78] for eliminating or mitigating the *near-far* problem, the above assumptions can be relaxed.

When frequency-selective multipath fading channels are assumed, a widely accepted channel model is the finite-length tapped delay line having a tap spacing of one chip duration [90]. This scenario is shown in Figure 21.1 for the kth signal, where the L channel tap weights $\{\alpha_{kl}\}$ are assumed to be iid Rayleigh random variables having a probability density function (pdf) given by [90]:

$$f_{\alpha_{kl}}(x)=\frac{2x}{\Omega}\exp\left(-\frac{x^2}{\Omega}\right),\ x\geq 0, \tag{21.3}$$

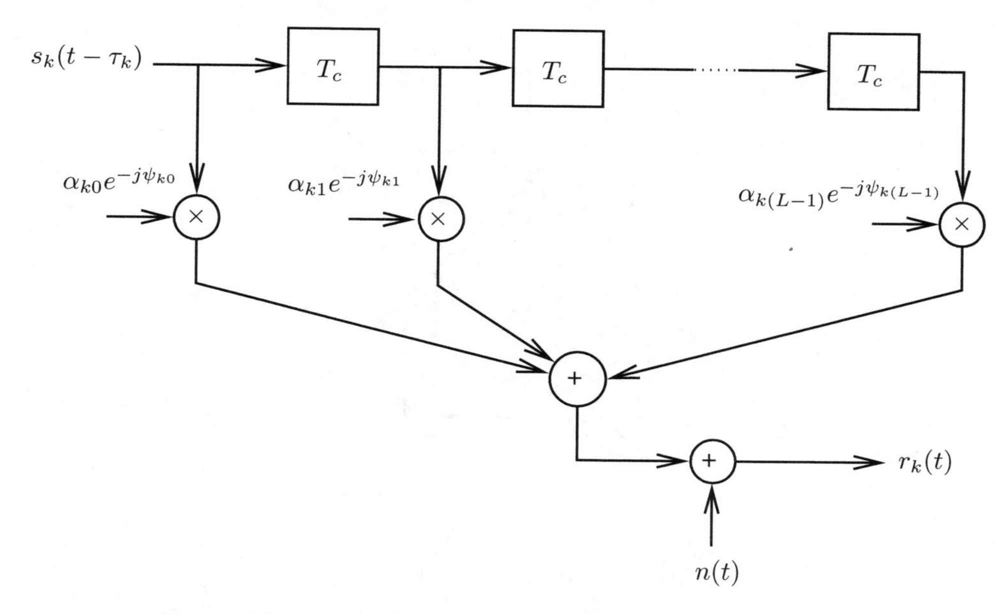

Figure 21.1: Tapped delay line model of frequency-selective channels.

and $\Omega = E[\alpha_{kl}^2]$, while the phases $\{\psi_{kl}\}$ in Figure 21.1 are assumed to be iid random variables uniformly distributed in the interval of $[0, 2\pi]$. Additionally, the phases are independent of $\{\alpha_{kl}\}$. Furthermore, since during the acquisition process, the integration dwell time[2] is usually shorter than the information bit duration, we assume that the fading is sufficiently slow in order to guarantee that the amplitudes of the chips spanning the integral dwell time are faded identically. Then, the received signal at the base station can be written as:

$$\begin{aligned} r(t) &= \sum_{l=0}^{L-1} \sqrt{2P_r}\alpha_{1l}c_1(t-\tau_1-lT_c)\cos(2\pi f_c t+\theta_{1l}) \\ &\quad +\sum_{k=2}^{K}\sum_{l=0}^{L-1} \sqrt{2P_k}\alpha_{kl}b_k(t-\tau_k-lT_c)c_k(t-\tau_k-lT_c)\cos(2\pi f_c t+\theta_{kl}) \\ &\quad +n(t), \end{aligned} \tag{21.4}$$

where τ_k is the relative time delay associated with an asynchronous transmission scheme and $\theta_{kl} = \varphi_k - \psi_{kl} - 2\pi f_c(\tau_k + lT_c)$. The phases are modelled as iid random variables uniformly distributed in $[0, 2\pi]$.

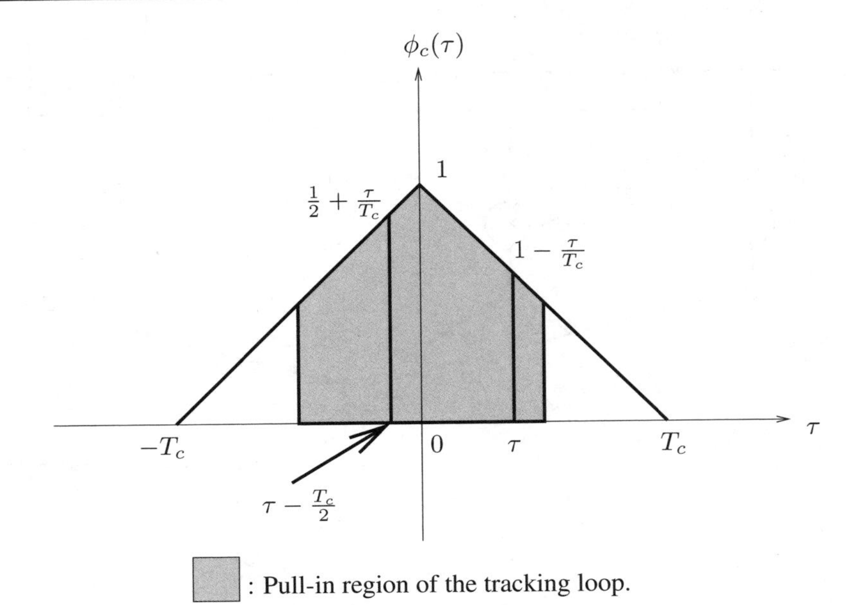

Figure 21.2: The auto-correlation function of a random binary signal.

21.3 Acquisition Procedure Description

The auto-correlation function of a random binary sequence $c_k(t)=\sum_{j=-\infty}^{\infty} c_j^{(k)} P_{T_c}(t-jT_c)$ can be expressed as:

$$\phi_c(\tau)=\lim_{A\to\infty}\frac{1}{2A}\int_{-A}^{A} c_k(t)c_k(t-\tau)dt. \tag{21.5}$$

For a signal displacement of $\tau=0$, this integral is equal to one indicating that the power of the signal $c_k(t)$ is unity, since $c_k^2(t)=1$. By contrast, for a time-displacement of $\tau\geq T_c$ the integral is zero since the spreading code has been modelled as a sequence of an infinite number of independent random binary chips. Finally, for $0<\tau<T_c$ the integral is equal to the fraction of the chip duration, for which we have $c_k(t+\tau)=c_k(t)$. Therefore, the auto-correlation function can be expressed as:

$$\phi_c(\tau)=\begin{cases} 1-\frac{|\tau|}{T_c}, & |\tau|<T_c \\ 0, & |\tau|\geq T_c \end{cases}, \tag{21.6}$$

which takes the shape illustrated in Figure 21.2. Explicitly, the auto-correlation function of a random binary sequence is of an ideal triangular shape, where the shaded area represents the pull-in range of a code-tracking loop, which will be defined during our forthcoming discus-

[2]The integration dwell time is defined as the time-interval of the integration associated with the correlators in the receiver. This definition will become more clear in the context of Figure 21.8 during our forthcoming discourse.

sion in this section.

In the context of the spreading sequence acquisition process, the so-called cell H_1 is usually defined as a phase interval – or synonymously, as a delay interval – corresponding to a small timing offset between the noise-freely received PN code and the locally generated PN code replica, where their auto-correlation value exceeds a pre-set threshold. In other words, the cell H_1 can be interpreted as the range, where the received and locally generated spreading code replica can be considered to be aligned within a delay that is less than the required fraction of the chip duration. By contrast, any phase or delay interval, where the auto-correlation value does not exceed the pre-set threshold is defined as cell H_0. In short, cell H_1 represents the synchronized, while cell H_0 the unsynchronized conditions.

For a discrete-time PN code acquisition system the uncertainty region associated with a PN sequence is defined in terms of the number of cells to be searched. In a practical PN sequence acquisition system, usually there exist multiple H_1 cells within the uncertainty region defined as the time-interval, where the incoming PN sequence and the locally generated PN sequence reside within the required small timing offset facilitating the successive fine-alignment by the code tracking loop [14]. More specifically, the number of H_1 cells is determined by the pull-in range of the code tracking loop, the search step size and the channel characteristics. For example, for a baseband early-late non-coherent tracking loop [14], where the total normalized time difference between the early and late discriminator channels is Δ, the search step size expressed in terms of time should be less than $T_c\Delta$, where T_c is the chip-duration. This is necessary for ensuring that there is always at least one H_1 cell, and hence the time offset between the received PN sequence and the locally generated PN sequence remains within the code tracking loop's pull-in region. However, in practice, there are always at least two adjacent H_1 cells in the uncertainty region and their related offsets are within the pull-in range of the tracking loop, provided that the search step size does not exceed $T_c\Delta/2$, when a non-fading channel is considered [15, 501]. Furthermore, if a dispersive multipath fading channel is considered [44], and the multipath-induced delay spread of the channel is higher than the chip duration of the spreading code, then more than two H_1 cells may be encountered in the uncertainty region associated with the search process due to receiving several strong echoes of the transmitted signal.

As argued above, in practice there are multiple H_1 cells in real mobile communication environments, which is a consequence of having a search step size smaller than the chip interval (typically 1/2 PN chip duration) and that of the delay spread caused by channel fading. Moreover, the majority of published results assumed that there was a single H_1 cell [15, 23–25, 28, 31, 39, 49]. Hence, the code acquisition performance documented in these seminal references may be significantly different from the actual performance. Therefore, in this chapter a range of acquisition schemes are investigated under the hypothesis that there are multiple H_1 cells in the uncertainty region of the PN sequence. Specifically, we assume that the pull-in region of the code tracking loop is $(-T_c/2, T_c/2)$, as seen in Figure 21.2, i.e the normalized time difference between the early and late sampling position is $\Delta = 1$ and the search step size is $T_c/2$. We note that Δ, which was defined as the total normalized time difference between the early and late discriminator channels of the tracking loop, must not be set to an excessive value. For example, $\Delta = 2$ is never used in the non-coherent delay-lock tracking loop, because the slope of the corresponding 'S' curve near the optimum timing position would become zero [14] and hence the tracking performance would become unstable. As expected, the choice of the search step-size has to be based on a compromise.

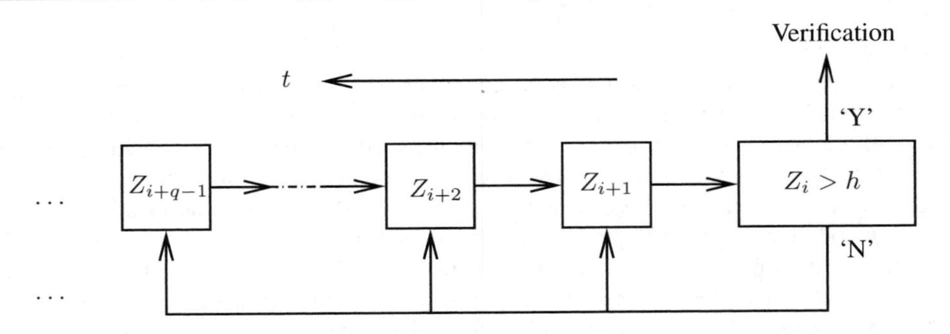

Figure 21.3: The block diagram of serial search schemes.

On the one hand, if the search step size is set to an excessively low value, which implies increasing the number of cells in the uncertainty region, the search time will be excessive, when serial search is used. By contrast, in the context of a pure parallel search scheme the complexity of the acquisition system becomes excessively high, if a very low search step size is used. Furthermore, there will be an excessively high number of so-called false alarms in the search process designed for a given false alarm probability, since the average number of false alarms is given by the false-alarm probability multiplied by the number of cells in the uncertainty region. On the other hand, if the step size is set to an excessively high value, for example, higher than ΔT_c, there may be no H_1 cells registered during the search process.

So far, the general concepts of various acquisition schemes have been considered. As highlighted above, the family of conventional acquisition methods applicable to DS-CDMA can be broadly classified as serial search acquisition and parallel search acquisition schemes. Following these introductory notes, let us now describe these acquisition schemes in more depth, under the hypothesis of encountering multiple H_1 cells in the uncertainty region of the PN code.

21.3.1 Serial Search Acquisition Scheme

Let the decision variable sequence $\{Z_0, Z_1, \dots, Z_{q-1}\}$ denote a sequence of period q, where each element Z_i corresponds to a phase to be searched, or time-domain raster position to be tested in order to find the most accurately aligned received and locally generated codes. Hence, q represents the number of raster positions to be tested in the uncertainty region – or simply the uncertainty region – of the PN sequence. Then, the operation of the serial search scheme can be characterized with the aid of Figure 21.3, where $Z_i = Z_{i \pmod q}$. In Figure 21.3, h represents a threshold set for making a decision, whether the currently tested phase of the received PN signature sequence and the locally generated PN sequence is within the required small timing offset, so as to terminate the search mode and proceed to the so-called verification mode. The threshold is selected according to a trade-off among a number of contradicting design parameters, such as the detection probability, missing probability, false-alarm probability and mean acquisition time. Note that the verification mode is usually used for confirming a stronger decision concerning the correct delay assumed by the search mode [15,30], an issue not analyzed in greater detail in this chapter. The serial search scheme can be described as follows. Whenever the decision variable Z_i constituted, for example, by

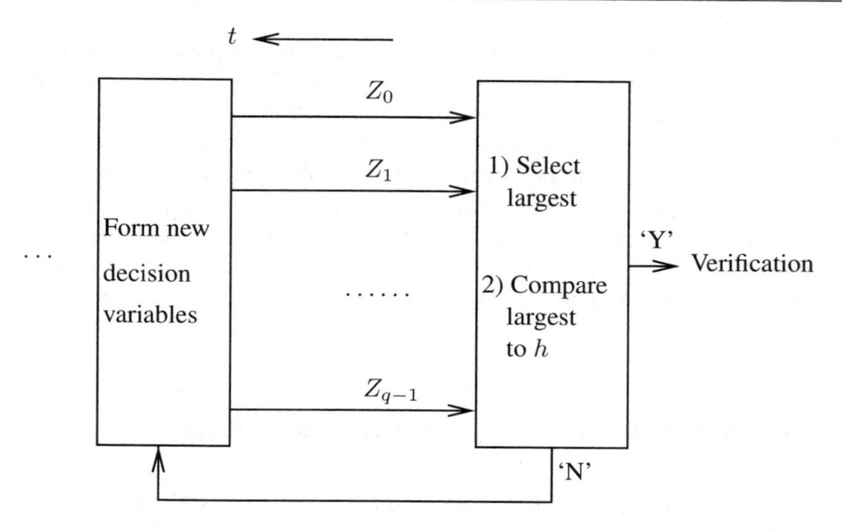

Figure 21.4: The block diagram of a parallel search-based acquisition schemes.

the correlation value between the received and the appropriately delayed local code replica exceeds the threshold h, the system assumes that the corresponding delay of the locally generated PN sequence is the correct delay and proceeds to the verification mode. If this delay is also determined by the verification mode, then acquisition is declared and the successive fine tracking loop is enabled. Otherwise, if Z_i does not exceed h or the delay is not confined by the verification mode, then the relative phase of the locally generated PN sequence is readjusted and a new decision variable Z_{i+1} is obtained, in order to replace the old decision variable Z_i in the thresholding operation. The above process is repeated, until either a delay is accepted by the verification mode, or the maximum acceptable acquisition time is exceeded.

21.3.2 Parallel Acquisition Scheme

In contrast to the serial search acquisition scheme of Section 21.3.1, where each possible code phase of the uncertainty range is tested one at a time, in a pure parallel search scheme, all of the possible code phases are tested simultaneously, as shown in Figure 21.4. Let $\{Z_0, Z_1, \ldots, Z_{q-1}\}$ denote the decision variables corresponding to all the q number of possible code phases, where each of the q number of elements Z_i corresponds to one of the possible q phases to be searched. The philosophy of the parallel search scheme can be described as follows. If the largest decision variable exceeds a threshold h, the corresponding phase is assumed, tentatively, to be the correct phase of the received signal and accordingly the acquisition system proceeds to the verification mode. If this phase is also confirmed by the verification mode, acquisition is declared and the fine code tracking system is enabled. Otherwise, if the largest decision variable does not exceed the threshold h, or the corresponding phase is not confirmed by the verification mode, even when the largest decision variable does exceed the threshold h, then another q number of decision variables are collected and the above process is repeated.

Clearly, the trade-off between the serial search scheme and the pure parallel search scheme is acquisition speed versus hardware cost and complexity. Simple modifications of these two schemes results in various hybrid schemes. For instance, we might partition the uncertainty code phase region into subsets. One subset is considered at each step. The decision variables associated with all code-phase offsets in the current subset are compared. Acquisition is confirmed if the largest output exceeds a preselected threshold. Otherwise, the next subset is examined. This strategy potentially offers a rich set of compromise schemes, which provide different trade-offs between acquisition speed versus hardware cost. However, in this chapter we only investigate the two extremes – serial search schemes and pure parallel search schemes. Let us now investigate the mean acquisition time achieved by both serial and pure parallel schemes.

21.4 Asymptotic Mean Acquisition Time

21.4.1 Mean Acquisition Time of Serial Acquisition

Holmes and Chen [28], as well as Polydoros and Weber [15] have provided the equations for computing the mean acquisition time of a serial search-based code acquisition system in both exact and asymptotic forms. They assumed having either a single H_1-cell or a double H_1-cell in the uncertainty region, i.e. having one or two time-displacements often referred to as states in the uncertainty region, in which the received sequence and the locally generated PN sequence can be considered to be in-phase, i.e. within a synchronization error less than a single chip duration. However, due to the finite search step size and due to receiving multiple delayed replicas of the transmitted PN sequences in a dispersive multipath fading environment, the actual situation is that more than two synchronized states may exist, in which the locally generated and the received PN sequence may become synchronous, hence satisfying the hypothesis of having multiple H_1 cells. Hence, in this section we extend Viterbi's result [501] in order to obtain a unified asymptotic equation for the evaluation of the mean acquisition time of such systems, which includes Holmes' and Polydoros' asymptotic results [15, 28] as special cases. the results are valid for both the one H_1-cell and two H_1-cell hypotheses, respectively, in the uncertainty region.

Viterbi's equivalent circular state diagram of [501] can be adapted for employment in a serial search-based code acquisition system having multiple H_1 cells in the uncertainty region, which is redrawn in Figure 21.5. However, in Viterbi's corresponding illustration P_{D_i} was replaced by $1 - \beta_i$ for notational convenience in our derivation, where P_{D_i} and β_i represent the detection and missing probabilities corresponding to the i-th H_1 cell, leading to the relationship of $\beta_i = 1 - P_{D_i}$. Explicitly, the missing probability of a H_1 cell indicates the probability of the event that the decision variable corresponding to the H_1 cell is lower than the detection threshold of h. In Figure Figure 21.5 the nodes represent states corresponding to different displacements, while the labelled branches between two nodes represent state transitions. Furthermore, z indicates the unit-delay operator, while the power of z represents the time-delay expressed in terms of the number of time-domain raster positions. Let us assume that there are a total of q number of states that have to be searched one by one serially in the uncertainty region, and that λ out of the q number of states are H_1 cells. Let α represent the false-alarm probability associated with a H_0 cell, τ_D represent the integral dwell time, while $W\tau_D$ represent the 'penalty time' associated with determining or realizing that there

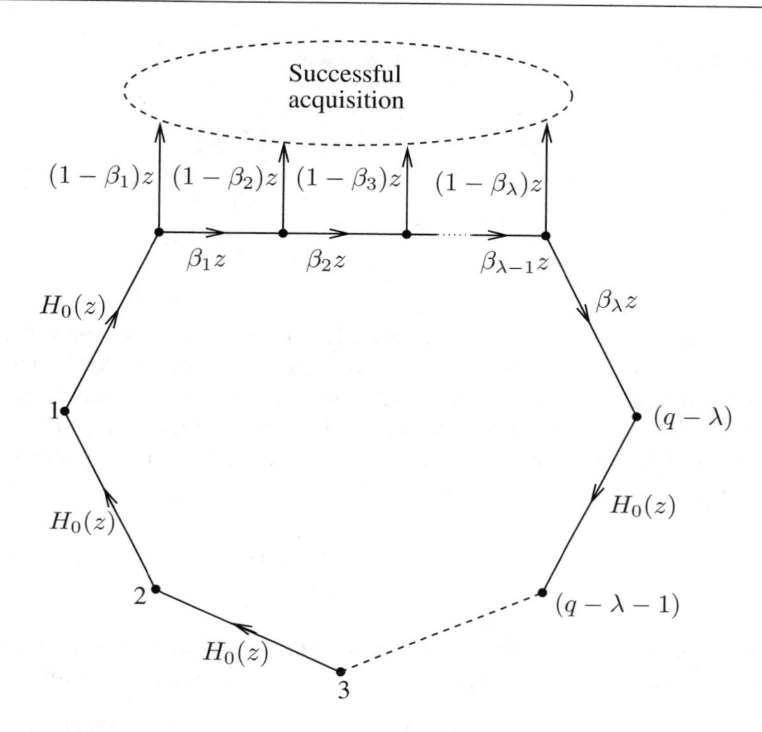

Figure 21.5: Equivalent circular state diagram for a serial search-based code acquisition system having λ number of synchronous states associated with H_1 cells and $(q-\lambda)$ number of unsynchronous states associated with H_0 cells. Furthermore, β_i is the missing probability associated with the ith H_1 state.

is a false alarm and with re-entering the search mode. Then, following Viterbi's philosophy [501] and assuming that we commence the search anywhere at an arbitrary displacement outside the correct states, the transfer function associated with the transitions leading to the state corresponding to successful acquisition for the general serial search-based acquisition scheme, while assuming equal prior probability of each state can be expressed as:

$$U(z) = \frac{H_D(z)}{(q-\lambda)\left[1 - H_M(z)H_0^{q-\lambda}(z)\right]} \cdot \sum_{i=1}^{q-\lambda} H_0^i(z), \tag{21.7}$$

where $H_D(z)$ represents the transition functions including all paths leading to a successful acquisition detection (D), which can be expressed as:

$$\begin{aligned} H_D(z) &= \sum_{j=1}^{\lambda}(1-\beta_j)z\prod_{i=1}^{j-1}\beta_i z, \\ &= \sum_{j=1}^{\lambda}\left[(1-\beta_j)\prod_{i=1}^{j-1}\beta_i\right]Z^j, \end{aligned} \quad (21.8)$$

where we define $\prod_{i=1}^{0}\beta_i z = 1$. In Equation 21.7 $H_0(z)$ is the transition function from one H_0 cell to its successive H_0 cell. This transition function can be divided into two events: the first event is no false-alarm at the current H_0 test corresponding to the transition function of $(1-\alpha)z$, while the other event is false-alarm taking place at the current H_0 test, and hence having the transition function of αz^{W+1} from the current H_0 cell to its successive H_0 cell. Therefore, the transition function $H_0(z)$ can be expressed as:

$$H_0(z) = (1-\alpha)z + \alpha z^{W+1}, \quad (21.9)$$

Finally, $H_M(z)$ in Equation 21.7 represents the transition function including all paths leading the H_1 cells to a miss (M), i.e., all H_1 cells are missed due to their corresponding correlation values being lower than the threshold h. Let $P_M(\lambda) = \prod_{i=1}^{\lambda}\beta_i$, which represents the overall missing probability of a search over the full uncertainty region. Then $H_M(z)$ in Equation 21.7 is given by $H_M(z) = P_M(\lambda)z^{\lambda}$. The mean acquisition time can be expressed following Holmes and Chen [28] as:

$$\overline{T}_{acq} = \left[\frac{dU(z)}{dz}\right]_{z=1} \times \tau_D. \quad (21.10)$$

After differentiating Equation 21.7 with respect to the variable z, and then setting $z = 1$, we arrive at:

$$\overline{T}_{acq} = \left(\frac{2\cdot\sum_{j=0}^{\lambda-1}\beta_j\prod_{i=0}^{j-1}\beta_i + (q-\lambda+1)\left[1-P_M(\lambda)\right](1+W\alpha)}{2[1-P_M(\lambda)]} + \frac{2(q-\lambda)P_M(\lambda)(1+W\alpha)}{2[1-P_M(\lambda)]}\right) \times \tau_D, \quad (21.11)$$

where we define $\beta_0 = 1$.

In order to obtain the asymptotic form of the mean acquisition time for the scenario of having a high number of states in the uncertainty region, i.e., for $q >> 1$, we divide both sides of Equation 21.11 by the total number of states of q, and then compute its limit with respect to $q \to \infty$, i.e for an infinite number of states in the uncertainty region of Figure 21.5, leading to:

$$\lim_{q\to\infty} \frac{\overline{T}_{acq}}{q} = \frac{[1 + P_M(\lambda)](1 + W\alpha)}{2[1 - P_M(\lambda)]} \times \tau_D. \tag{21.12}$$

Consequently, if the total number of states q in the uncertainty region is significantly higher than the number of correct states, i.e. $q >> \lambda$, then the mean acquisition time can be approximated as:

$$\overline{T}_{acq} \approx \frac{[1 + P_M(\lambda)](1 + W\alpha)}{2[1 - P_M(\lambda)]} \times (q\tau_D). \tag{21.13}$$

Equation 21.13 is a unified equation that can be used for computing the asymptotic mean acquisition time of the serial search-based acquisition system having multiple H_1 states in the uncertainty region, which includes Holmes' [28] and Polydoros' [15] results as special cases. If there is only one H_1 cell in the uncertainty region of q states and the detection probability of this H_1 cell as well as the false-alarm probability associated with a H_0 cell are P_D and α, respectively, then the overall missing probability is given by $P_M(1) = 1 - P_D$. Upon substituting these probabilities into Equation 21.13, we found that the approximate mean acquisition time associated with the single H_1 cell scenario can be expressed as:

$$\overline{T}_{acq} \approx \frac{[2 - P_D](1 + W\alpha)}{2P_D} \times (q\tau_D). \tag{21.14}$$

This is identical to the equation derived by Holmes and Chen in [28]. In order to elaborate a little further, when there are two H_1 cells in the uncertainty region spanning q number of cells, then we have $P_M(2) = \beta_1\beta_2 = (1 - P_{D_1})(1 - P_{D_2})$, where P_{D_1} and P_{D_2} represent the detection probabilities of the first and the second H_1 cells for the serial acquisition system having two H_1 cells. Upon substituting these conditions in Equation 21.13, the asymptotic mean acquisition time for a serial search acquisition system having two correct timing states in the uncertainty region can be expressed as:

$$\overline{T}_{acq} \approx \frac{[2 - P_{D_1} - (1 - P_{D_1})P_{D_2}](1 + W\alpha)}{2[P_{D_1} + (1 - P_{D_1})P_{D_2}]} \times (q\tau_D), \tag{21.15}$$

which is the asymptotic result given in Equation (20) of [15], when uniform distribution of the two H_1 cells over the entire uncertainty region spanning a total of q number of cells is assumed.

Note that in practical acquisition systems the search process can commence at any phase state of the uncertainty region. Viterbi's method [501] ignores the case of starting searching at the correct phase state, hence it is in fact an asymptotic analysis. However, in any practical serial acquisition system the total number of states in the uncertainty region is usually significantly higher than the number of correct phase states. Consequently, sufficiently accurate mean acquisition time results can be obtained by using the results of Equation 21.13. Furthermore, according to Equation 21.13 the mean acquisition time is effectively constituted by the delay introduced by false alarm events and that due to the events of missing the detection of synchronized states. This may be the consequence of having no high code correlation peaks because of the impairment and by the effects of co-channel interference.

21.4.2 Parallel Acquisition Scheme

Sourour and Gupta [31] have derived an expression for computing the mean acquisition time of a parallel search-based code acquisition system complemented also by a verification mode. The expression is sufficiently general for computing the mean acquisition time of parallel search-based code acquisition systems having an arbitrary number of H_1 cells. Following the approach used in [31], the mean acquisition time of the parallel acquisition system described in sub-section 21.3.2 can be derived as follows.

The state transition diagram of a parallel code acquisition system is shown in Figure 21.6(a), where $H_D(z)$, $H_M(z)$ and $H_F(z)$ represent the state transitions to the successful detection (D), missing (M) and false-alarm (F) states during the search, respectively, while $H_R(z)$ represents the transition from the false-alarm state to the next search state. In the state transition diagram of Figure 21.6 there is one absorbing state, which is the detection state symbolized by the outer-most ring, and one false-alarm state indicated by the inner-most ring. If an arbitrarily long acquisition process can be tolerated, then Figure 21.6(a) can be equivalently simplified to Figure 21.6(b).

Let P_D, P_M and P_F, respectively represent the overall detection probability, the overall missing probability and the overall false-alarm probability of a search-based on the decision variables $(Z_0, Z_1, \dots, Z_{q-1})$, which correspond to λ number of H_1 cells and $(q-\lambda)$ number of H_0 cells. More specifically, the test process can be described as follows. During the first τ_D seconds, q number of decision variables are loaded into the parallel acquisition scheme, and during the following τ_D seconds q decision variables are tested, as was described in sub-section 21.3.2. The probability that none of the q decision variables exceeds the threshold h is given by P_M. In this case, a new set of q decision variables is loaded into the parallel acquisition scheme during the next τ_D seconds. If one or more samples exceed the threshold h, the largest decision variable is assumed tentatively to correspond to the correct displacement of the received and locally stored code replicas. This assumption is correct with probability of P_D or false with a probability P_F. This tentative phase is then forwarded to the verification mode. If the correct phase was forwarded to the verification mode, the acquisition process is completed. If, however, the tentative phase is not confirmed by the verification mode, the system reverts back to the search mode after a penalty time of $W\tau_D$.

According to the above acquisition process, it can be shown that we have:

$$
\begin{aligned}
P_D + P_M + P_F &= 1, & (21.16)\\
H_D(z) &= P_D z, & (21.17)\\
H_M(z) &= P_M z, & (21.18)\\
H_F(z) &= P_F z, & (21.19)\\
H_R(z) &= z^W. & (21.20)
\end{aligned}
$$

Then, according to Figure 21.6, the transfer function associated with the transitions leading

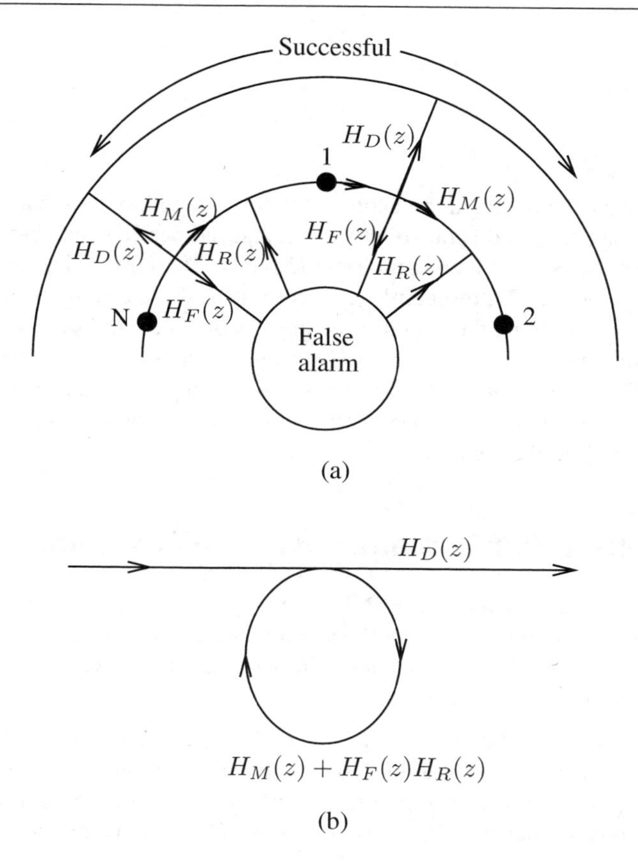

Figure 21.6: Equivalent circular state diagram for a parallel search-based code acquisition system having multiple correct timing states, where $H_D(z)$, $H_M(z)$ and $H_F(z)$ represent the transition functions associated with state transitions to the successful detection (D), unsuccessful or missing (M), and false-alarm (F) states, respectively, while $H_R(z)$ represents the transition function from the false-alarm state to the next search state.

to successful acquisition can be expressed as:

$$U(z) = \frac{H_D(z)}{1 - H_M(z) - H_F(z)H_R(z)} = \frac{P_D z}{1 - P_M z - P_F Z^{W+1}}. \quad (21.21)$$

The resultant mean acquisition time of the parallel acquisition scheme is given by:

$$\overline{T}_{acq} = \left[\frac{dU(z)}{dz}\right]_{z=1} \times \tau_D. \quad (21.22)$$

After differentiating Equation 21.22 with respect to the variable z and then setting $z = 1$, we arrive at:

$$\overline{T}_{acq} = \frac{1 + WP_F}{P_D} \times \tau_D. \tag{21.23}$$

In summary, we have obtained the expressions for computing the mean acquisition time of both the serial search-based acquisition system and the parallel search-based acquisition system under the hypothesis of having multiple H_1 cells. These formulae are given by Equation 21.13 and Equation 21.23, respectively. According to these equations, in order to evaluate the mean acquisition time, the overall probabilities of successful detection, unsuccessful detection or missing and false-alarm have to be evaluated first. Hence, in the following sections we will analyze these probabilities, when considering transmissions over AWGN or multipath fading channels in conjunction with different detection schemes. Let us first consider AWGN channels in the next section.

21.5 Detection Performance over AWGN Channels

In this section we derive the overall probabilities of successful detection, unsuccessful detection or missing and false-alarm for both the serial and the parallel non-coherent acquisition schemes in the context of AWGN channels. In deriving the associated expressions of the statistics, we adopt the following assumptions:

- The pull-in range of the tracking loop is $[-T_c/2, T_c/2]$, and the search step size is $T_c/2$; consequently, there are two possible correct synchronization points, namely one at each side of the perfect synchronization instant in the uncertainty region corresponding to the correct phase. Hence, there are two adjacent H_1 cells, as shown in Figure 21.7.[3]
- All correlator output samples are independent.
- $M >> 1$, where M is the number of chips per integration dwell-time such that the mean value of the correlation between the received signal and the locally generated PN sequence is zero, when they are not in-phase, i.e., in the context of H_0 cells.
- The multiple-access interference inflicted by the $K-1$ data transmission users can be modelled by additive white Gaussian random processes.

21.5.1 Statistics of the Decision Variables

The detection schematic of generating the decision variable Z_i is shown in Figure 21.8. The received signal is first down-converted to its baseband in-phase (I) and quadrature (Q) components. Two I-Q correlators[4] evaluate the correlation between the locally generated

[3] According to Figure 21.2 and Figure 21.7, there is one cell before the first H_1 cell and another cell following the second H_1 cell, which will result in correlator output samples having a non-zero mean, due to the partial synchronization of the incoming PN sequence with the local PN sequence. However, since the phases of these two cells are out of the pull-in range of the tracking loop of $[-T_c/2, T_c/2]$, we will treat them as H_0 cells and approximate their mean by zero (the corresponding non-zero mean is actually $\frac{1}{4}MT_c$), as we assume that $q >> 1$.

[4] These correlators can be replaced by matched filters. The small difference between PN code acquisition using matched filters and correlators has been described, for example, in [30].

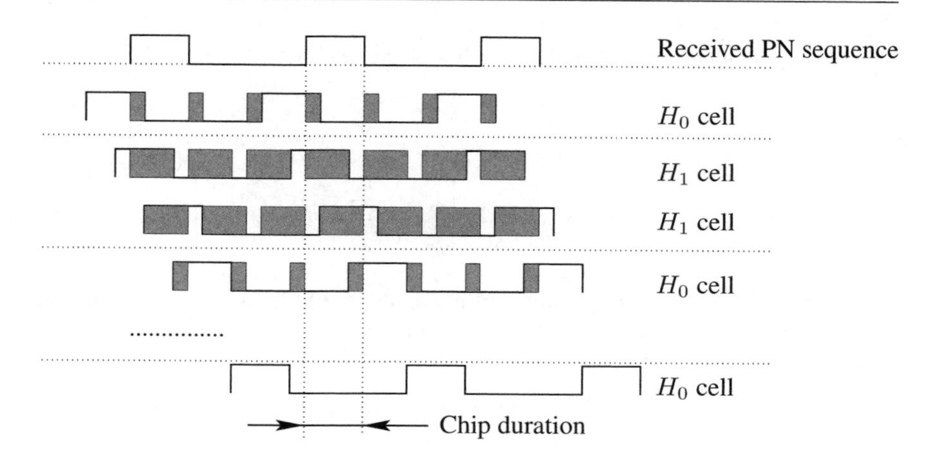

Figure 21.7: Illustration of H_1 cells and H_0 cells for transmission over AWGN channels. The shaded areas indicate the synchronized parts between the local PN sequence replica and the received PN sequence.

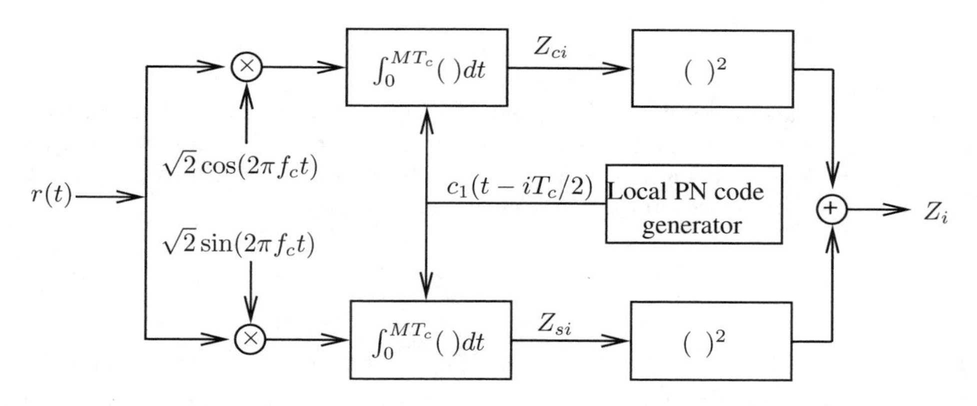

Figure 21.8: The block diagram of generating the decision variable Z_i.

PN code and the baseband in-phase as well as quadrature-phase signals. The I-Q correlator performs the multiplication of the I or Q component of the received signal with its locally stored code replica $c(t - iT_c/2)$ in Figure 21.8 and integrates this product over the interval of $\tau_D = MT_c$ seconds, where τ_D represents the integral dwell time and M is an integer. The outputs of the I-Q correlators are then squared and summed, in order to generate an output variable, which is denoted by Z_i in Figure 21.8.

Referring to Equation 21.2 and Figure 21.8, the correlator output of the in-phase branch, Z_{ci}, gives the wanted signal component, the interference-induced noise term due to the non-zero cross correlation of the locally generated code sequence with the signals of the other users plus the Gaussian thermal noise term, which can be expressed as:

$$\begin{aligned} Z_{ci} &= \int_0^{MT_c} r(t)c_1(t - iT_c/2)\sqrt{2}\cos(2\pi f_c t)dt \\ &= \sqrt{P_r}MT_c\left[S_I + \sum_{k=2}^{K} M_I(k) + N_I\right]. \end{aligned} \tag{21.24}$$

In Equation 21.24 S_I is the component associated with the user-of-interest, which includes the wanted signal component that can be expressed as:

$$S_I = \frac{\cos\theta_1}{MT_c}\left[\tau R_M(i, g+1) + (T_c - \tau)R_M(i, g)\right], \tag{21.25}$$

if $\tau_1 - iT_c/2 = gT_c + \tau$, where g is an integer, τ_1 was defined in Equation 21.2 and τ is a random variable uniformly distributed in $[0, T_c]$. In Equation 21.25, $R_M(i, g)$ represents the partial auto-correlation function of the spreading sequences, which is defined as:

$$R_M(i, g) = \sum_{v=0}^{M-1} c_{v+i}^{(1)} \cdot c_{v+i+g}^{(1)}. \tag{21.26}$$

Since random spreading codes are assigned for all users of the system, random data modulation does not change the statistics of the random spreading codes. Consequently, assuming $\tau_k - iT_c/2 = gT_c + \tau$, where τ is a uniformly distributed random variable over $[0, T_c]$, the multiple-access interference term, $M_I(k)$ can be expressed as:

$$M_I(k) = \frac{\sqrt{\rho}\cos\theta_k}{MT_c}\left[\tau R_M^{(k)}(i, g+1) + (T_c - \tau)R_M^{(k)}(i, g)\right], \tag{21.27}$$

where we have $\rho = P_k/P_r = P_I/P_r$, since ideal power control was assumed for all data transmission users. Furthermore, $R_M^{(k)}(i, g)$ represents the partial cross-correlation between the sequence of the kth user and the locally generated sequence, which is defined as:

$$R_M^{(k)}(i, g) = \sum_{v=0}^{M-1} c_{v+i}^{(1)} \cdot c_{v+i+g}^{(k)}. \tag{21.28}$$

Finally, the noise term due to the presence of the AWGN $n(t)$ can be expressed as:

$$N_I = \frac{1}{\sqrt{P_r}MT_c}\int_0^{MT_c} n(t)c_1(t - iT_c/2)\sqrt{2}\cos(2\pi f_c t)dt. \tag{21.29}$$

The correlator output, Z_{si} of the quadrature branch can be obtained from that of the in-phase component, if we replace cos( ) by $-$sin( ) in Equation 21.25 as well as in Equation 21.27, and replace cos( ) by sin( ) in Equation 21.29.

Recall that the spreading sequences of the system are modelled as random binary sequences of equi-probable ± 1s. Therefore, it can be shown that the terms of Z_{ci} and Z_{si} are asymptotically Gaussian, as the number of users increases due to the central limit theorem [29], [427]. In our analysis, we employ the Gaussian approximation and model the multiple-access interference inflicted by the $K-1$ data transmission users and the thermal noise as AWGN. Consequently, provided that in Equation 21.25 we have $g=0$ and τ is uniformly distributed in $[0, T_c/2]$ as well as if $g=-1$ and τ is uniformly distributed in $[T_c/2, T_c]$, which correspond to the two adjacent H_1 cells in Figure 21.2, then the mean value of the wanted component, S_I, contributed by the user-of-interest becomes:

$$E[S_I]_{g=0} = E[S_I]_{g=-1} = \frac{3}{4}\cos\theta_1. \tag{21.30}$$

Otherwise, if $g \neq 0, -1$, then S_I can be approximated as a Gaussian random variable having zero-mean, while its variance can be readily computed and expressed as:

$$\text{Var}[S_I] = \frac{1}{3M}. \tag{21.31}$$

The multiple-access interference term, due to the kth interfering user given by Equation 21.27 can be approximated as a Gaussian random variable having zero-mean and variance of:

$$\text{Var}[M_I(k)] = \frac{\rho}{3M}. \tag{21.32}$$

The thermal noise term N_I, which is given by Equation 21.29 is a Gaussian random variable having zero-mean and variance of:

$$\text{Var}[N_I] = \frac{1}{2M\gamma_c}, \tag{21.33}$$

where $\gamma_c = P_r T_c / N_0$ represents the signal-to-noise ratio per chip, i.e., the SNR/chip.

Consequently, if we assume that the hypothesis of residing in a H_1 cell is being tested, according to Equation 21.24 the variance of Z_{ci} normalized by $(\sqrt{P_r}MT_c)$ can be expressed as:

$$\text{Var}[Z_{ci}] = \frac{(K-1)\rho}{3M} + \frac{1}{2M\gamma_c}. \tag{21.34}$$

If we assume that the hypothesis of encountering a H_0 cell is being tested, the variance of Z_{ci} will be slightly higher than that given by Equation 21.34, since the reference signal also contributes interference having a variance given by Equation 21.31. However, since our aim is to analyze the acquisition performance in a multiple-access communication environment, we assume for simplicity that the variance of Z_{ci} under the H_0 hypothesis is the same, as that given by Equation 21.34.

The statistics of the quadrature-phase component Z_{si} can also be obtained according to the above method, resulting in:

$$E[S_s]_{g=0} = E[S_s]_{g=-1} = \frac{3}{4}(-\sin\theta_1), \tag{21.35}$$

and the normalized variance of Z_{si} becomes $\text{Var}[Z_{si}] = \text{Var}[Z_{ci}]$.

Since the decision variable is constituted by $Z_i = Z_{ci}^2 + Z_{si}^2$, consequently, given the Gaussian nature of Z_{ci} and Z_{si}, and assuming that Z_i constitutes a H_1 sample, the PDF of Z_i is the chi-square distribution with two degrees of freedom, which can be expressed as Equation (2-1-121) of [90]:

$$f_{Z_i}(y|H_1) = \frac{1}{2\sigma_o^2}\exp\left(-\frac{m^2+y}{2\sigma_o^2}\right) I_0\left(\frac{m\sqrt{y}}{\sigma_o^2}\right), \; y \geq 0, \tag{21.36}$$

where

$$\sigma_o^2 = \text{Var}[V_{ci}] = \text{Var}[V_{si}] = \frac{(K-1)\rho}{3M} + \frac{1}{2M\gamma_c}, \tag{21.37}$$

and $I_0(\;)$ is the modified Bessel function of the first kind with zero order, while m^2 is the normalized non-central metric, given by:[5]

$$m^2 = \frac{9}{16}. \tag{21.38}$$

When Z_i constitutes a H_0 sample, Z_i is the central chi-square distribution of two degrees of freedom and its PDF can be expressed as Equation (2-1-110) of [90]:

$$f_{Z_i}(y|H_0) = \frac{1}{2\sigma_o^2}\exp\left(-\frac{y}{2\sigma_o^2}\right), \; y \geq 0. \tag{21.39}$$

Having derived the statistics of the decision variables, we can now proceed with deriving the probabilities required for the computation of the mean acquisition time.

21.5.2 Probabilities of False-Alarm and Overall Missing for the Serial Search Acquisition Scheme

In this subsection, the probability of false-alarm P_F and the overall unsuccessful detection or missing probability of P_M are derived for the serial search-based acquisition scheme, under the hypothesis of having $\lambda = 2$ number of H_1 cells. The mean acquisition time $\overline{T}_{acq}$ expressed as a function of P_F and P_M was given in Equation 21.13. According to the derivation of Equation 21.13 P_M is a function of the successful detection probabilities P_{Di} of all the H_1 cells, expressed as $P_M = \prod_{i=1}^{\lambda}(1-P_{Di})$, where P_{Di} is the successful detection probability of the ith H_1 cell. Hence, in order to derive the overall missing probability of P_M, we

[5] This m^2 value is only suitable for the step size of $T_c/2$ when transmitting over AWGN channels.

first have to derive $\{P_{Di}\}$. Given the PDF of the decision variable Z_i corresponding to a H_1 cell and remembering that all decision variables corresponding to the H_1 cells obey the same distribution given by Equation 21.36, the detection probability P_{Di} can be expressed as:

$$\begin{aligned} P_{Di} &= \int_h^{\infty} f_{Z_i}(y|H_1)dy \\ &= \int_h^{\infty} \frac{1}{2\sigma_o^2} \exp\left(-\frac{m^2+y}{2\sigma_o^2}\right) I_0\left(\frac{m\sqrt{y}}{\sigma_o^2}\right) dy \\ &= Q(R, \sqrt{h'}), \end{aligned} \tag{21.40}$$

where $Q(a, b)$ is the generalized Marcum Q function, which is defined as [90]:

$$\begin{aligned} Q_m(a,b) &= \int_b^{\infty} x \left(\frac{x}{a}\right)^{m-1} \exp\left(-\frac{x^2+a^2}{2}\right) I_{m-1}(ax)dx && (21.41) \\ &= Q_1(a,b) + \exp\left(\frac{a^2+b^2}{2}\right) \sum_{k=1}^{m-1} \left(\frac{b}{a}\right)^k I_k(ab), && (21.42) \end{aligned}$$

where

$$\begin{aligned} Q_1(a,b) &= Q(a,b) \\ &= \exp\left(-\frac{a^2+b^2}{2}\right) \sum_{k=0}^{\infty} \left(\frac{a}{b}\right)^k I_k(ab), \;\; b > a > 0. \end{aligned} \tag{21.43}$$

In Equation 21.40 $h' = h/\sigma_o^2$ is the normalized threshold, while R is expressed as:

$$R = m/\sigma_o = \frac{3}{4}\left[\frac{(K-1)\rho}{3M} + \frac{1}{2M\gamma_c}\right]^{-1/2}, \tag{21.44}$$

where $\gamma_c = P_r T_c / N_0$ is the SNR/chip.

Since all the decision variables corresponding to the H_1 cells have the same distribution, the overall unsuccessful detection or missing probability can be expressed as:

$$\begin{aligned} P_M &= \prod_{i=1}^{\lambda} (1 - P_{Di}) \\ &= \left[1 - Q(R, \sqrt{h'})\right]^{\lambda}. \end{aligned} \tag{21.45}$$

Given the PDF of Equation 21.39 for the decision variable corresponding to a H_0 cell,

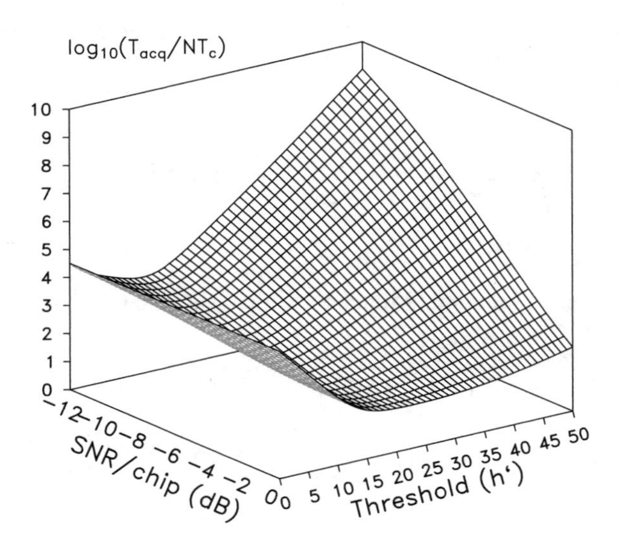

Figure 21.9: Mean acquisition time versus SNR/chip and normalized threshold h' performance for the serial search acquisition scheme over AWGN channels using parameters $M = 32,\ N = 64,\ W = 1000,\ q = 128,\ K = 1$ computed from Equation 21.13, Equation 21.45 and Equation 21.46.

the false-alarm probability P_F of the serial acquisition system can be expressed as:

$$\begin{aligned} P_F &= \int_h^{\infty} f_{Z_i}(y|H_0)dy \\ &= \int_h^{\infty} \frac{1}{2\sigma_o^2} \exp\left(-\frac{y}{2\sigma_o^2}\right) dy \\ &= \exp\left(-\frac{h'}{2}\right), \end{aligned} \tag{21.46}$$

where again $h' = h/\sigma_o^2$ is the normalized threshold, as we defined previously.

Based on the overall unsuccessful detection or missing probability and the false-alarm probability given by Equation 21.45 and Equation 21.46, respectively, finally the mean acquisition time can be computed from Equation 21.13.

Figure 21.9 shows the normalized mean acquisition time performance versus the normalized threshold h' and the SNR/chip for the serial search-based acquisition scheme, when transmitting over AWGN channels. In the computations the number of potential synchronization instants in the uncertainty region was set to $q = 128$, the processing gain was $N = 64$, the integral dwell time was $\tau_D = 32T_c$, i.e. we had $M = 32$, and no multi-user interference was imposed, i.e. we had $K = 1$ user. For convenience, the mean acquisition time

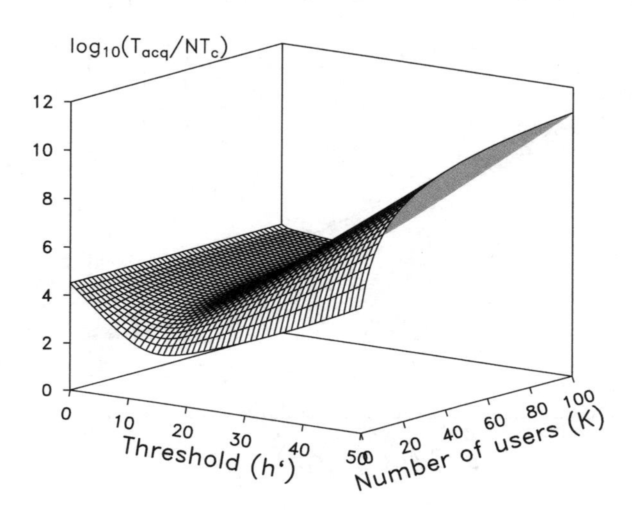

Figure 21.10: Mean acquisition time versus normalized threshold h' and versus the number of active users K performance for the serial search acquisition scheme over AWGN channels using parameters $M = 32$, $N = 64$, $W = 1000$, $q = 128$, $\gamma_c = -5dB$, $\rho = 1$ computed from Equation 21.13, Equation 21.45 and Equation 21.46.

was normalized by the data bit duration. More explicitly, the mean acquisition time given by Equation 21.13 was divided by the bit duration of $T_b = NT_c$. It is clear from Figure 21.9 that an inappropriate choice of the detection threshold h' can lead to a severe increase of the mean acquisition time, although the sensitivity of the mean acquisition time to the threshold decreases, as the SNR/chip increases. For any given SNR/chip value there exists an optimal choice of the threshold h', which minimizes the value of the mean acquisition time. Additionally, for any given normalized threshold h', the mean acquisition time decreases, as the SNR/chip increases, and it is expected to reach a residual value, which is essentially due to the non-zero false-alarm probability experienced. This value can be computed from Equation 21.13 by setting $P_M(\lambda) = 0$, resulting in $\overline{T}_{acq} \approx \frac{(1+W\alpha)}{2} \cdot (q\tau_D)$.

The effect of the number of active users K and the threshold h' on the mean acquisition time performance of the serial search-based acquisition scheme is shown in Figure 21.10. The curves were plotted versus the number of simultaneously transmitting users, K and versus the normalized threshold h', while the other parameters were given by $M = 32$, $N = 64$, $W = 1000$, $q = 128$, $\rho = 1$ and for SNR/chip values of $\gamma_c = -5$dB. As expected, the mean acquisition time of the serial search-based schemes increases, when the number K of active users increases. From Equation 21.44 it can be seen that the average output energy-to-noise ratio, R^2 over the integral dwell time MT_c is a function of both the SNR/chip value, γ_c, and that of the number of active users, K, for a fixed number of chips, M, per integral dwell-time. The average output energy-to-noise ratio, R^2 and hence the ac-

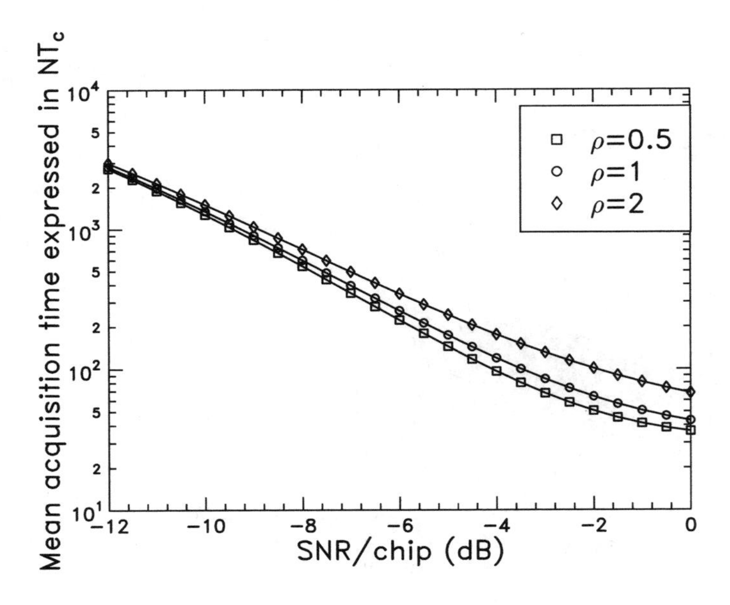

Figure 21.11: Mean acquisition time versus SNR/chip performance for the serial search acquisition scheme over AWGN channels using parameters $M = 32,\ N = 64,\ W = 1000,\ q = 128,\ K = 2$ and $\rho = 0.5,\ 1,\ 2$ computed from Equation 21.13, Equation 21.45 and Equation 21.46.

quisition time are predominantly determined by the noise contribution expressed in terms of SNR/chip, when the number K of active users is low, and hence a low multi-user interference is imposed. This trend becomes explicit in Figure 21.10. Otherwise, it is mainly dictated by the number of active users. Specifically, when the number of active user is high, we have $R \approx \frac{3}{4}\left(\frac{(K-1)\rho}{3M}\right)^{-1/2}$.

In Figure 21.11 we evaluated the effect of varying the transmitted power of the user-of-interest on the mean acquisition time performance of the serial acquisition scheme. The curves of Figure 21.11 were evaluated as a function the SNR/chip. The ratio ρ of the interfering users' power to that of the user-of-interest was taking values of $\rho = 0.5, 1, 2$, respectively. The other parameters were $M = 32,\ N = 64,\ W = 1000,\ q = 128$ and $K = 2$. From the results we observe that, in the SNR/chip range of -12dB ... -0dB the mean acquisition time performance was substantially improved when we increased the power of the user-of-interest. This argument is supported by comparing the curve associated with $\rho = P_I/P_r = 0.5$ and that with $\rho = P_I/P_r = 2$. However, we should note that from a total system capacity point of view, increasing the power of the initial synchronization users implies increasing the interference imposed on the active data transmission users, which consequently decreases the total user capacity of the system. Hence, in mobile DS-CDMA systems increasing the initial synchronization user's transmitting power to aid acquisition should consider the underlying trade-off between the mean acquisition time and the total system capacity.

21.5.3 Overall Detection, Missing and False-Alarm Probabilities of the Parallel Acquisition Scheme

As we discussed previously in Subsection 21.3.2, the parallel search-based scheme computes the test statistics corresponding to all of the possible code phases or displacements and opts for using the displacement associated with the most reliable correlator output. These test statistics were given by $\{Z_0, Z_1, \ldots, Z_{q-1}\}$, where we assume that there are λ number of H_1 cells, whose corresponding decision variables obey the PDF expressed by Equation 21.36. Correspondingly there are $(q-\lambda)$ number of H_0 cells, whose corresponding decision variables obey the PDF expressed by Equation 21.39. The overall detection probability is defined as the probability that the largest of the set $\{Z_0, Z_1, \ldots, Z_{q-1}\}$ is that corresponding to one of the λ number of H_1 cells, and its amplitude is higher than the threshold h. Hence, the overall successful detection probability can be expressed as:

$$P_D = \lambda \int_h^\infty f_{Z_i}(y|H_1)\left[\int_0^y f_{Z_j}(x|H_1)dx\right]^{\lambda-1}\left[\int_0^y f_{Z_k}(x|H_0)dx\right]^{q-\lambda} dy. \quad (21.47)$$

Based on Equation 21.36 and Equation 21.39, it can be shown that we have

$$\begin{aligned}\int_0^y f_{Z_j}(x|H_1)dx &= 1-\int_y^\infty f_{Z_j}(x|H_1)dx \\ &= 1-Q\left(R, \sqrt{y/\sigma_o^2}\right), \quad (21.48)\\ \int_0^y f_{Z_k}(x|H_0)dx &= 1-\int_y^\infty f_{Z_k}(x|H_0)dx \\ &= 1-\exp\left(-\frac{y}{2\sigma_o^2}\right). \quad (21.49)\end{aligned}$$

When substituting these results into Equation 21.47, P_D can be expressed as:

$$\begin{aligned}P_D &= \lambda \int_h^\infty \frac{1}{2\sigma_o^2}\exp\left(-\frac{m^2+y}{2\sigma_o^2}\right) I_0\left(\frac{m\sqrt{y}}{\sigma_o^2}\right) \\ &\cdot\left[1-Q\left(R, \sqrt{y/\sigma_o^2}\right)\right]^{\lambda-1}\left[1-\exp\left(-\frac{y}{2\sigma_o^2}\right)\right]^{q-\lambda} dy. \quad (21.50)\end{aligned}$$

Let $x^2 = y/\sigma_o^2$, then Equation 21.50 can be simplified to:

$$\begin{aligned}P_D &= \lambda \int_{\sqrt{h'}}^\infty x\exp\left(-\frac{R^2+x^2}{2}\right) I_0(Rx) \\ &\cdot\left[1-Q(R,x)\right]^{\lambda-1}\left[1-\exp\left(-\frac{x^2}{2}\right)\right]^{q-\lambda} dx. \quad (21.51)\end{aligned}$$

The overall missing probability, P_M is defined as the probability that no decision variables of the set $\{Z_0, Z_1, \ldots, Z_{q-1}\}$ exceed the threshold, which can be expressed as:

$$P_M = \left[\int_0^h f_{Z_i}(x|H_1)dx\right]^{\lambda} \left[\int_0^h f_{Z_j}(x|H_0)dx\right]^{q-\lambda}. \tag{21.52}$$

Upon substituting Equation 21.48 and Equation 21.49 in the above equation and replacing the variable y by h, it can be shown that:

$$P_M = \left[1 - Q\left(R, \sqrt{h'}\right)\right]^{\lambda} \left[1 - \exp\left(-\frac{h'}{2}\right)\right]^{q-\lambda}. \tag{21.53}$$

Finally, the overall false-alarm probability – which is defined as the probability that the largest of the set $\{Z_0, Z_1, \ldots, Z_{q-1}\}$ is that corresponding to one of the $(q - \lambda)$ number of H_0 cells, and of simultaneously, its amplitude is higher than the threshold h – can be computed by using the relationship:

$$P_F = 1 - P_D - P_M. \tag{21.54}$$

Finally, the mean acquisition time of the parallel search-based acquisition scheme can be obtained by Equation 21.23 with P_D and P_F given by Equation 21.51 and Equation 21.54, respectively.

Figure 21.12 shows the mean acquisition time relative to the bit duration of $T_b = NT_c$ versus the SNR/chip, γ_c and versus the normalized threshold, h', when communicating over an AWGN channel. In our computations the other parameters were assumed to be $M = 32, N = 64, W = 1000, q = 128$ and no multi-user interference was inflicted in the single-user scenario of $K = 1$. Similarly to the serial search-based acquisition scheme of Figure 21.9, it can be seen that, for any given SNR/chip value, there exists an optimum threshold, which minimizes the mean acquisition time. However, an inappropriate choice of the thresholds can severely increase the mean acquisition time. This sensitivity of the mean acquisition time to the threshold h' is found to decrease, when the SNR/chip increases. Hence, the curve in Figure 21.12 flattens, when increasing the SNR/chip value. Upon comparing Figure 21.9 to Figure 21.12 we can observe that for any given SNR/chip value, the mean acquisition time of the parallel search-based acquisition scheme is significantly lower than that of the serial search-based acquisition scheme, provided that both schemes are working at the optimum thresholds. However, the cost of the parallel search-based acquisition scheme's lower mean acquisition time is its higher hardware complexity.

In Figure 21.13 the mean acquisition time performance is evaluated versus the number of simultaneously transmitting users, K and versus the normalized threshold h' with the other parameters given by $M = 32, N = 64, W = 1000, q = 128, \gamma_c = -5dB, \rho = 1.0$. Similarly to Figure 21.10, for any given threshold, the mean acquisition time of the parallel acquisition scheme increases, when the number of active users increases. However, for any given number of active users, there exists an optimum threshold value, which results in the minimum mean acquisition time. The sensitivity of the mean acquisition time to the number of simultaneously transmitting users decreases, when the number of active users decreases. Upon comparing Figure 21.10 with Figure 21.13, we can observe that for a given number of

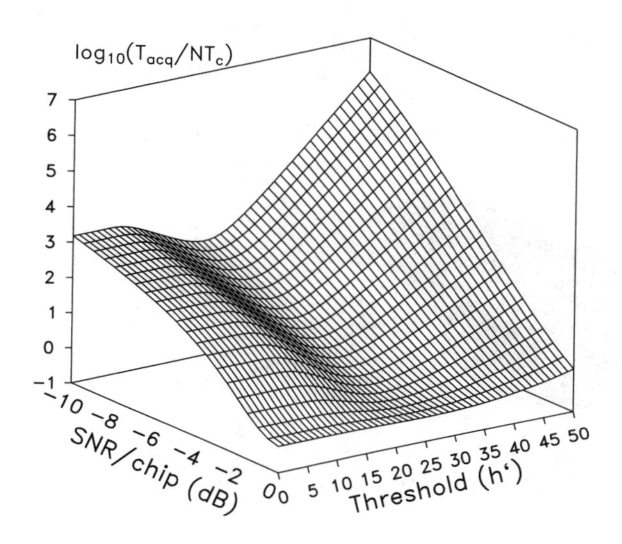

Figure 21.12: Mean acquisition time versus SNR/chip, γ_c and normalized threshold, h' performance for the parallel search-based acquisition scheme for transmission over AWGN channels using the parameters of $M = 32, N = 64, W = 1000, q = 128, K = 1$. The results were computed from Equation 21.23, Equation 21.51 and Equation 21.54.

active users the mean acquisition time of the parallel acquisition scheme is lower than that of the serial acquisition scheme, provided that both schemes are operated at the optimum thresholds.

In the parallel search-based acquisition schemes, the the overall successful detection probability, the overall missing probability and the overall false-alarm probability obey the relationship of $P_D+P_M+P_F = 1$. By contrast, in serial search-based acquisition systems, where the successful detection probability is not directly related to the false-alarm probability, this relationship does not hold. In serial search-based acquisition schemes, the mean acquisition time is the only important metric. However, in parallel search-based acquisition schemes, due to the relationship of $P_D + P_M + P_F = 1$ and as a consequence of the dependence of these probabilities on the mean acquisition time of Equation 21.23, the successful detection probability of P_D or the false-alarm probability of P_F also constitutes important performance metrics. For example, when the normalized threshold is set to $h' = 0$, which in turn means that we have $P_M = 0$ and hence $P_D + P_F = 1$, with reference to Equation 21.23 we find that the knowledge of P_D or P_M is sufficient for characterizing the scheme's performance, since under the above conditions and with the remaining parameters fixed, the mean acquisition time of Equation 21.23 is a simple function of P_D or P_M. Hence, in Figure 21.14 the overall detection probability, P_D was plotted against the normalized threshold of h' and the SNR/chip of γ_c by assuming the parameters of $M = 32, N = 64, q = 128, K = 5, \rho = 1.0$. As expected, the overall detection probability increases for a given threshold, when increas-

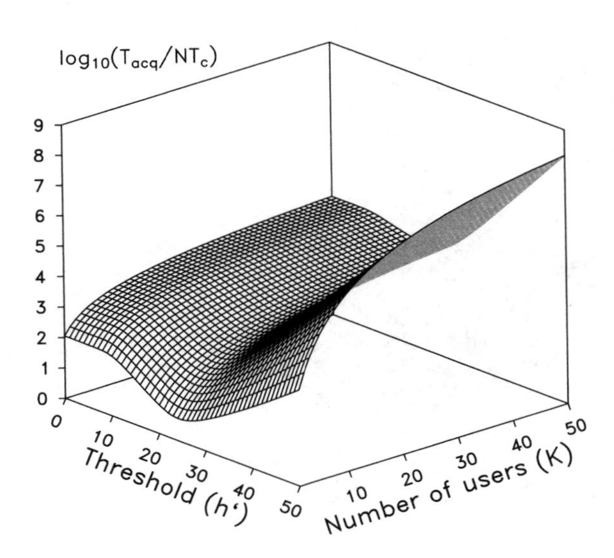

Figure 21.13: Mean acquisition time versus normalized threshold, h' and versus the number of active users, K performance for the parallel search-based acquisition scheme for transmitting over AWGN channels using the parameters of $M = 32,\ N = 64,\ W = 1000,\ q = 128,\ \gamma_c = -5dB,\ \rho = 1.0$. The results were computed from Equation 21.23, Equation 21.51 and Equation 21.54.

ing the SNR/chip, and decreases for any given SNR/chip, when increasing the threshold.

Finally, in Figure 21.15 we evaluated the effect of varying the relative transmitted power of the user-of-interest on the false-alarm probability of the parallel search-based acquisition schemes. The ratio ρ of the interfering users' power to that of the user-of-interest was taking values of $\rho = 0.5, 1, 2$, respectively. The remaining parameters were $M = 32,\ N = 64,\ q = 128,\ W = 1000,\ K = 2,\ h' = 0$. Since the threshold was set to $h' = 0$, the knowledge of the false-alarm probability is sufficient for characterizing the mean acquisition time as we argued above. From the results we observe that in the SNR/chip range of -12dB ... -0dB the false-alarm probability decreased, when increasing the transmitted power of the user-of-interest.

Having characterized the serial and parallel search-based acquisition schemes in terms of our standard performance metrics when communicating over AWGN channels, in the next section their performance will be studied for transmissions over Rayleigh fading channels.

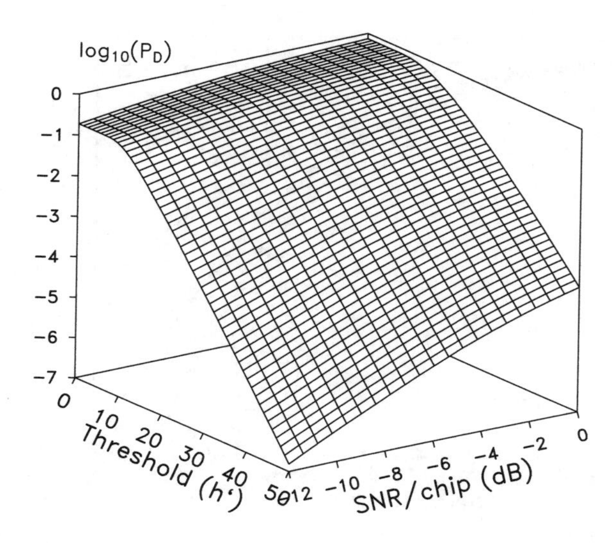

Figure 21.14: Overall detection probability, P_D versus normalized threshold, h' and SNR/chip, γ_c performance for the parallel search-based acquisition scheme for transmitting over AWGN channels using the parameters of $M = 32,\ N = 64,\ q = 128,\ K = 5,\ \rho = 1.0$. The results were computed from Equation 21.51.

21.6 Detection Performance over Multipath Fading Channels

21.6.1 Statistics of Decision Variables

In this section we derive the statistics of the decision variables and the overall successful detection probability, missing probability and false-alarm probability for the serial search and parallel search-based acquisition schemes in the context of a non-coherent acquisition system, when transmitting over frequency-selective multipath fading channels, as we discussed in Subsection 21.2.2. All derivations are subject to the same assumptions as in the case of AWGN channels. Additionally, we assumed that the self-noise of the desired signal due to multipath transmission and due to the multiple-access interference inflicted by the $K-1$ data transmission users can be modelled by additive white Gaussian random processes.

According to the received signal expression of Equation 21.4, which is valid for transmission over dispersive multipath fading channels, and referring to Figure 21.8, the correlator output of the in-phase branch, namely Z_{ci}, gives the composite signal constituted by the wanted signal component, the self-interference induced noise term due to multipath transmission, the multiple-access interference term generated by the $K - 1$ data transmission users

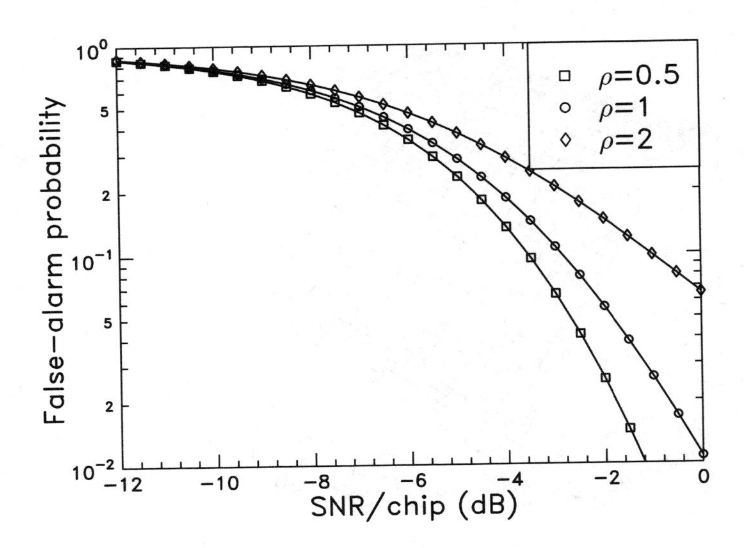

Figure 21.15: False-alarm probability versus SNR/chip performance for the parallel search-based acquisition scheme over AWGN channels using the parameters of $M = 32$, $N = 64$, $q = 128$, $K = 2$, $h' = 0$ and $\rho = 0.5, 1, 2$. the results were computed from Equation 21.54.

plus the Gaussian thermal noise term, which can be expressed as:

$$\begin{aligned} Z_{ci} &= \int_0^{MT_c} r(t)c_1(t - iT_c/2)\sqrt{2}\cos(2\pi f_c t)dt \\ &= \sqrt{P_r}MT_c\left[\sum_{l=0}^{L-1} S_I(l) + \sum_{k=2}^{K}\sum_{l=0}^{L-1} M_I(k,l) + N_I\right]. \end{aligned} \quad (21.55)$$

In Equation 21.55 $S_I(l)$ is the signal component generated by the user-of-interest, which includes the wanted signal component as well as the self-noise that can be expressed as:

$$S_I(l) = \frac{\alpha_{1l}\cos\theta_{1l}}{MT_c}\left[\tau R_M(i, g+1) + (T_c - \tau)R_M(i, g)\right], \quad (21.56)$$

if $\tau_1 + (l - i/2)T_c = gT_c + \tau$, where τ_1 was defined in Equation 21.4 and τ is a random variable uniformly distributed in the interval of $[0, T_c]$. In Equation 21.56, $R_M(i, g)$ represents the partial auto-correlation function of the spreading sequences, which was defined in Equation 21.26.

Note that since a non-coherent acquisition system is capable of acquiring only one path of the multipath signal at time, the other $L - 1$ path signals contribute interference to the in-phase path. Moreover, if the locally generated sequence is not in-phase with any of the L multipath signals generated by the desired signal, all L multipath signals of the user-of-interest constitute interference.

Assuming that we have $\tau_k + (l - i/2)T_c = gT_c + \tau$, where τ is a uniformly distributed random variable over the interval of $[0, T_c]$, the multiple-access interference term of Equation 21.55, namely $M_I(k,l)$, can be expressed as:

$$M_I(k,l) = \frac{\sqrt{\rho}\alpha_{kl}\cos\theta_{kl}}{MT_c}\left[\tau R_M^{(k)}(i, g+1) + (T_c - \tau)R_M^{(k)}(i,g)\right], \tag{21.57}$$

where $\rho = P_I/P_r$ and $R_M^{(k)}(i,g)$ represents the partial cross-correlation between the spreading sequence of the kth user and the locally generated spreading sequence, which was defined in Equation 21.28.

Finally, the noise term N_I in Equation 21.55, which is due to the presence of the AWGN $n(t)$ can be expressed as:

$$N_I = \frac{1}{\sqrt{P_r}MT_c}\int_0^{MT_c} n(t)c_1(t - iT_c/2)\sqrt{2}\cos(2\pi f_c t)dt. \tag{21.58}$$

The correlator output of the quadrature branch, namely Z_{si}, can be obtained from the expression of the in-phase component in Equation 21.55, if we replace $\cos(\,)$ by $-\sin(\,)$ in Equation 21.56 as well as in Equation 21.57, and replace $\cos(\,)$ by $\sin(\,)$ in Equation 21.58.

Figure 21.16 shows the locations of H_1 cells and H_0 cells, when transmitting over a dispersive three-path fading channel is assumed. If a search step size of $T_c/2$ is assumed, then six H_1 cells can be observed in the figure, where each path contributes two H_1 cells. The task of initial synchronization in dispersive multipath fading channels is to ensure that the code tracking loop locks on one of the correct phases.

Let us invoke the Gaussian approximation for characterizing the self-noise and the multiple-access interference. Then, following the approach used in Section 21.5 for AWGN channels and taking into account that if $g = 0$ and τ is uniformly distributed in $[0, T_c/2]$ as well as if $g = -1$ and τ is uniformly distributed in $[T_c/2, T_c]$, which are the scenarios corresponding to the two adjacent H_1 cells of the lth path in Figure 21.16, the mean value of the wanted component, $S_I(l)$ in Equation 21.56, contributed by the user-of-interest can be expressed as:

$$E[S_I(l)]_{g=0} = E[S_I(l)]_{g=-1} = \frac{3}{4}\alpha_{1l}\cos\theta_{1l}. \tag{21.59}$$

Furthermore, the normalized variance of Z_{ci} can be expressed as:

$$\text{Var}[Z_{ci}] = \frac{(L-1)\Omega}{3M} + \frac{(K-1)L\rho\Omega}{3M} + \frac{1}{2M\gamma_c}, \tag{21.60}$$

where $\Omega = E[\alpha_{kl}^2]$.

If we assume that the hypothesis of encountering a H_0 cell is being tested, all L paths

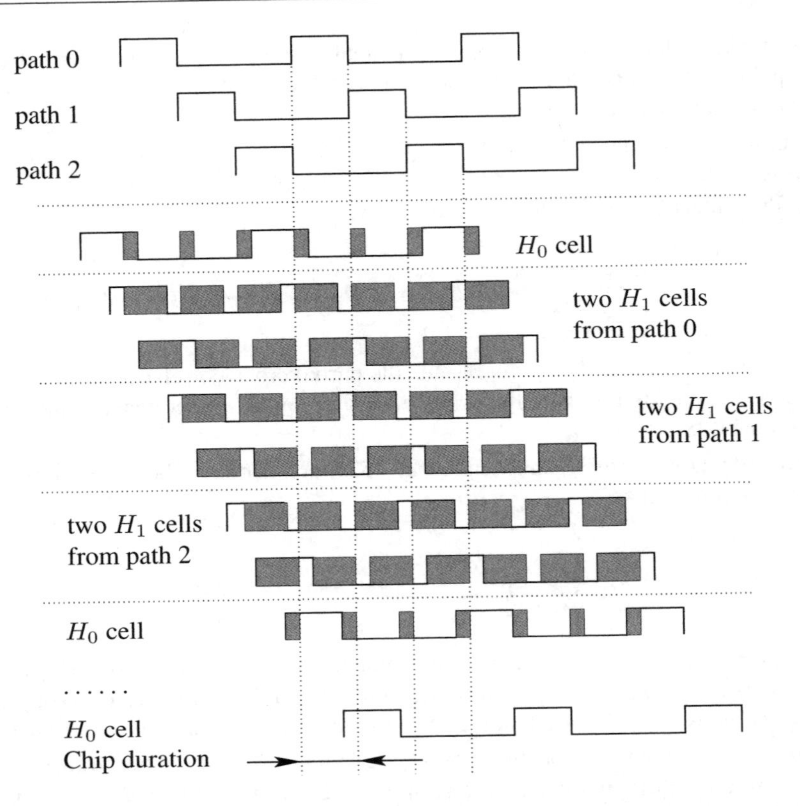

Figure 21.16: Illustration of H_1 cells and H_0 cells for transmission over a dispersive fading channel. The shaded areas indicate the synchronized parts between the local PN sequence replica and one of the three paths of the received PN sequence.

arriving from the user-of-interest constitute interference. Consequently, the variance of Z_{ci} can be computed from Equation 21.60, if we replace the '$L-1$' term by 'L'. However, since our aim is to analyze the acquisition performance of a serial search-based acquisition scheme in a multiple-access communication environment, we assume for simplicity that the variance of Z_{ci} under the H_0 hypothesis is the same as that given by Equation 21.60.

The statistics of the quadrature-phase component Z_{si} can also be obtained according to the above method, resulting in:

$$E[S_s(l)]_{g=0} = E[S_s(l)]_{g=-1} = \frac{3}{4}\alpha_{1l}(-\sin\theta_{1l}), \tag{21.61}$$

and the normalized variance of Z_{si} is given by $\text{Var}[Z_{si}] = \text{Var}[Z_{ci}]$.

Above we have derived the statistics of the correlator's outputs, Z_{ci} and Z_{si} for the in-phase branch and the quadrature branch, respectively, which are approximately Gaussian random processes. Since the decision variable $Z_i = Z_{ci}^2 + Z_{si}^2$ represents either an H_1 state or an H_0 state, hence, assuming that Z_i constitutes an H_1 sample, the PDF of Z_i for a given

α_{1l} is the chi-square distribution with two degrees of freedom, which can be expressed with the aid of Equation (2-1-121) of [90] as:

$$f_{Z_i}(y|H_1) = \frac{1}{2\sigma_o^2} \exp\left(-\frac{m^2+y}{2\sigma_o^2}\right) I_0\left(\frac{m\sqrt{y}}{\sigma_o^2}\right), \; y \geq 0, \tag{21.62}$$

where

$$\begin{aligned}\sigma_o^2 &= \text{Var}[Z_{ci}] = \text{Var}[Z_{si}] \\ &= \frac{(L-1)\Omega}{3M} + \frac{(K-1)L\rho\Omega}{3M} + \frac{1}{2M\gamma_c},\end{aligned} \tag{21.63}$$

and m^2 is the normalized non-central metric, which is given by:[6]

$$m^2 = \frac{9}{16}\alpha_{1l}^2. \tag{21.64}$$

When Z_i constitutes an H_0 sample, Z_i is the central chi-square distribution and its PDF can be expressed with the aid of Equation (2-1-110) of [90] as:

$$f_{Z_i}(y|H_0) = \frac{1}{2\sigma_o^2} \exp\left(-\frac{y}{2\sigma_o^2}\right), \; y \geq 0. \tag{21.65}$$

We have derived the conditional PDFs of the decision variables under different hypotheses suitable for both the serial search mode and parallel search mode. Let us now derive the probabilities of the overall missing, and of the false-alarm for the serial search-based acquisition scheme and the probabilities of the overall successful detection, overall false-alarm and overall missing for the parallel search-based acquisition scheme. Note that, due to the multipath nature of the fading channels, according to Figure 21.16 a PN code having q number of phases in the uncertainty range for transmission over nonfading channels will have $[q + 2(L-1)]$ number of phases in the uncertainty range, if we assume that L resolvable paths are encountered. Moreover, assuming L resolvable paths and a search step size of $T_c/2$, the number of H_1 cells becomes $\lambda = 2L$, where each path contributes two of them.

21.6.2 Probabilities of Overall Missing and False-Alarm for the Serial Acquisition Scheme

At this stage we recall that we have assumed that the L number of resolvable multipath signals are independent and obey an identical distribution. Furthermore, we have also assumed that the decision variables corresponding to the two H_1 cells due to the same propagation path are also independent, since random PN sequences are used. Hence, all the decision variables $\{Z_0, Z_1, \ldots, Z_q, \ldots, Z_{q+2L-3}\}$ – which include $\lambda = 2L$ number of H_1 cells and $(q-2)$ number of H_0 cells – are independent random variables. The decision variables correspond-

[6]This m^2 value is only suitable for the step size of $T_c/2$, when transmitting over multipath fading channels.

ing to H_1 cells are distributed according to Equation 21.62, while those corresponding to H_0 cells are distributed according to Equation 21.65.

Let P_{Di} represent the successful detection probability of the ith H_1 cell. Then the overall missing probability for a serial search-based scheme can be expressed as:

$$P_M = \prod_{i=1}^{2L}(1 - P_{Di}) = [1 - P_{Di}]^{2L}. \tag{21.66}$$

In order to derive the unconditional successful detection probability, namely of P_{Di}, first we must derive the conditional probability of $P_{Di}(R)$, which can be expressed as:

$$\begin{aligned} P_{Di}(R) &= \int_h^\infty f_{Z_i}(y|H_1)dy \\ &= \int_h^\infty \frac{1}{2\sigma_o^2}\exp\left(-\frac{m^2+y}{2\sigma_o^2}\right) I_0\left(\frac{m\sqrt{y}}{\sigma_o^2}\right) dy \\ &= Q\left(R, \sqrt{h'}\right), \end{aligned} \tag{21.67}$$

where $h' = h/\sigma_o^2$ with σ_o^2 given by Equation 21.63, R is a function of α_{1l}, which is expressed as:

$$R = m/\sigma_o = \overline{\gamma}_c \cdot \frac{\alpha_{1l}}{\sqrt{\Omega}}, \tag{21.68}$$

where

$$\overline{\gamma}_c = \frac{3}{4}\left[\frac{(L-1)}{3M} + \frac{(K-1)L\rho}{3M} + \frac{1}{2M\Omega\gamma_c}\right]^{-1/2}. \tag{21.69}$$

The unconditional successful detection probability is calculated from the conditional detection probability upon weighting $P_{Di}(R)$ by the PDF of R, and then averaging or integrating the weighted product over its legitimate range, as will be made explicit below. However, since α_{1l} is Rayleigh distributed according to Equation 21.3, it can be shown that R is also Rayleigh distributed having a PDF given by:

$$p_R(r) = \frac{2r}{\overline{\gamma}_c^2}\exp\left(-\frac{r^2}{\overline{\gamma}_c^2}\right). \tag{21.70}$$

Consequently, upon using the above equation the unconditional successful detection probability P_{Di} can be written as:

$$\begin{aligned} P_{Di} &= \int_0^\infty P_{Di}(r)p_R(r)dr \\ &= \int_0^\infty Q\left(R, \sqrt{h'}\right) p_R(r)dr. \end{aligned} \tag{21.71}$$

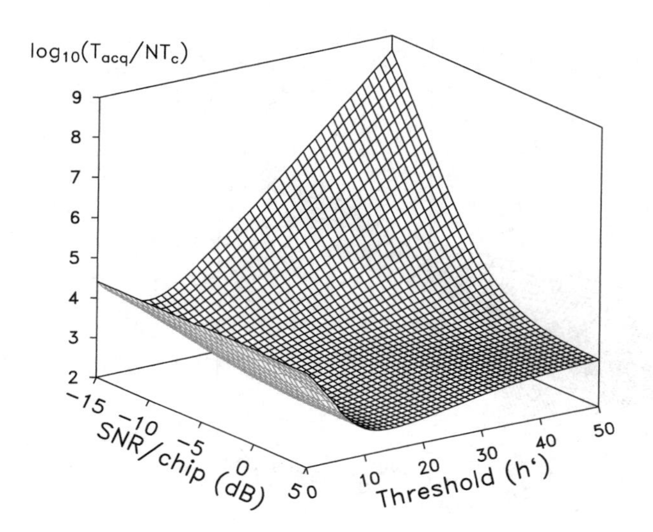

Figure 21.17: Mean acquisition time versus SNR/chip, $\overline{\gamma}_c$ and normalized threshold, h' performance for the serial search-based acquisition scheme for transmission over multipath fading channels using the parameters $M = 64,\ N = 128,\ q = 1024,\ K = 5,\ W = 100,\ \rho = 1.0,\ L = 1$. The results were computed from Equation 21.13, Equation 21.66 and Equation 21.73.

Finally, using the result of Appendix D, we can obtain the unconditional detection probability P_{Di} by setting $\mu = 1,\ p = 1,\ \Omega = \overline{\gamma}_c^2$ and $b^2 = h'$ in Equation D.7, resulting in:

$$P_{Di} = \sum_{i=0}^{\infty}\sum_{j=0}^{i} \frac{(h')^{i-j}}{(i-j)!2^{i-j-1}} \cdot \frac{\exp(-h'/2)}{\overline{\gamma}_c^2(1+2/\overline{\gamma}_c^2)^{i+1}}. \tag{21.72}$$

The false-alarm probability is the probability that the output decision variable Z_i under the hypothesis of H_0 will exceed the threshold h, which can be expressed as:

$$P_F = \int_h^{\infty} f_{Z_i}(y|H_0)dy = \exp\left(-\frac{h'}{2}\right). \tag{21.73}$$

With the aid of our results generated so far, we can now evaluate the mean acquisition time performance, $\overline{T}_{acq}$, of the serial search mode by substituting the overall missing probability of Equation 21.66 as well as the false-alarm probability of Equation 21.73 and the other related parameters into Equation 21.13.

Figure 21.17 presents the sensitivity of the mean acquisition time with respect to the nor-

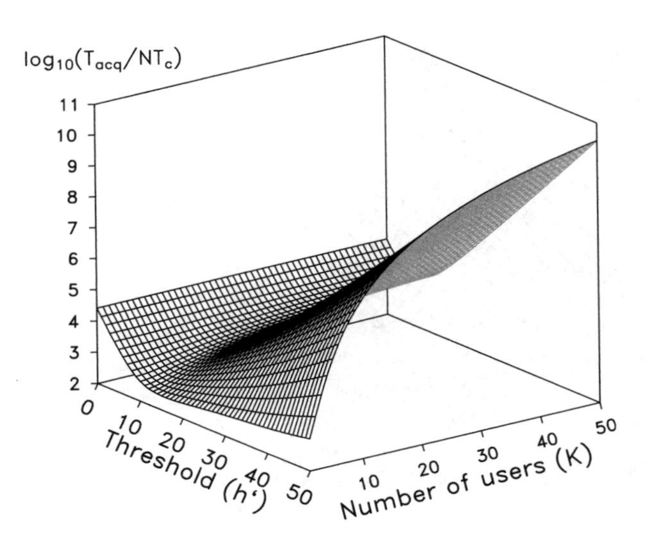

Figure 21.18: Mean acquisition time versus normalized threshold, h' and versus the number of active users, K performance for the serial search-based acquisition scheme for transmission over multipath fading channels using the parameters of $M = 64$, $N = 128$, $q = 1024$, $\gamma_c = 0dB$, $W = 100$, $\rho = 1.0$, $L = 3$. The results were computed from Equation 21.13, Equation 21.66 and Equation 21.73.

malized threshold h' and the SNR/chip for the serial search-based acquisition scheme for transmissions over multipath Rayleigh fading channels. In the computations, the number of cells in the uncertainty region was set to $q = 1024$, the processing gain was $N = 128$, the integral dwell time was $\tau_D = 64T_c$, i.e. we had $M = 64$, there were $L = 1$ resolvable paths at the receiver, the system supported $K = 5$ users and the base station received the same power from all users, i.e we had $\rho = 1$. The results show that an inappropriate choice of the detection threshold h' may lead to a severe degradation of the mean acquisition time. However, the sensitivity of the mean acquisition time to the threshold decreases, as the SNR/chip value increases. For any given SNR/chip value there exists an optimal choice of the threshold h', which minimizes the mean acquisition time. In addition, for any given normalized threshold h', the mean acquisition time decreases, as the SNR/chip value increases, and finally reaches a residual value. This residual acquisition time is essentially due to the non-zero false-alarm probability, which is associated with a certain penalty time required by the system for retaining to its search mode. This value is $\overline{T}_{acq} \approx \frac{(1+W\alpha)}{2} \cdot (q\tau_D)$, which is computed according to Equation 21.13 by setting $P_M(\lambda) = 0$.

The effect of the number of active users on the mean acquisition time performance is shown in Figure 21.18, where the mean acquisition time was plotted versus the normalized threshold, h' and versus the number of active users, K. The remaining parameters used were $M = 64$, $N = 128$, $q = 1024$, $\gamma_c = 0dB$, $W = 100$, $\rho = 1.0$, and $L = 3$. The results

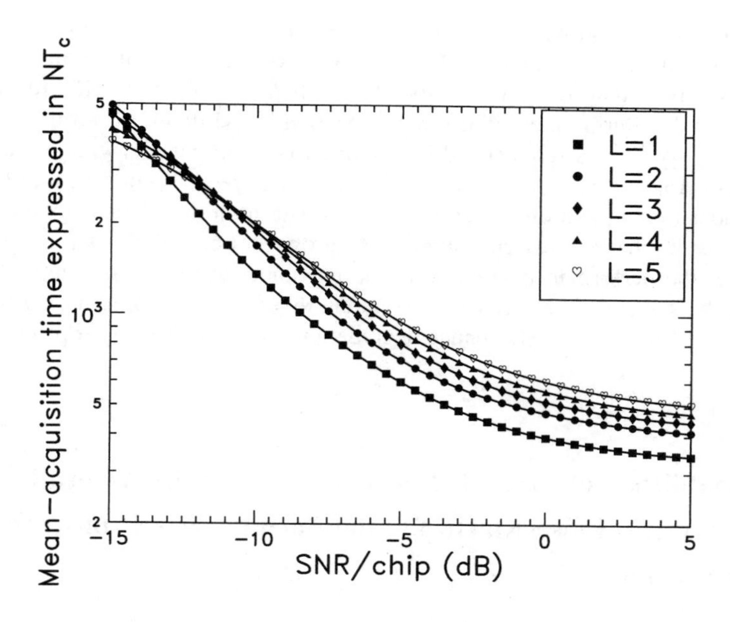

Figure 21.19: Mean acquisition time versus SNR/chip performance for the serial search-based acquisition scheme for transmission over multipath fading channels using the parameters of $M = 64,\ N = 128,\ q = 1024,\ W = 100,\ \rho = 1.0,\ L = 1, 2, 3, 4, 5,\ K = 5$. The results were computed from Equation 21.13, Equation 21.66 and Equation 21.73, under assumption that the acquisition system always operates at the optimum threshold.

demonstrated that the mean acquisition time performance degrades when the number of simultaneously transmitting users increases. However, for any given number of active users, there is an optimum threshold value resulting in the minimum mean acquisition time. According to Figure 21.18, the optimum threshold happens to be around $h' = 10$ for any given number of active users. Since $h' = h/\sigma_o^2$ in both AWGN channels and fading channels, and σ_o^2 is a function of the self-interference imposed by the user-of-interest, that of the multi-user interference and that of the background AWGN, the optimum threshold inevitably changes with the changing of the communication environment. Hence, any initial synchronization system using a constant threshold for initial correlation-based code acquisition scheme becomes a sub-optimum scheme. Optimum initial synchronization schemes must adaptively set their detection threshold, in order to match the requirements of the continuously varying wireless communication environment.

In Figure 21.19 we show the effect of multipath transmission on the mean acquisition time performance of the serial search-based acquisition scheme. The results were plotted against the SNR/chip, γ_c for different number of resolvable paths, L. The other parameters were $M = 64,\ N = 128,\ q = 1024,\ W = 100,\ \rho = 1.0,\ K = 5$. In the computations, we assumed that the acquisition scheme was capable of invoking the optimum threshold for any number of resolvable paths. From the results we observe that the mean acquisition time performance degrades, when the number of paths resolved by the receiver increases, provided that the SNR/chip value is sufficiently high ($\geq$ 11dB). However, the associated increase is not very significant. It can be observed that the normalized mean acquisition time associated with $L = 5$ is no more than twice that associated with $L = 1$ for the SNR/chip range of [-10dB,5dB]. Note that during the initial synchronization it is difficult, if not impossible, to combine all the energy dispersed over the different multipath components with the aid of such techniques, as coherent diversity combining. Hence, in this respect, multipath propagation inevitably degrades the acquisition performance. However, multipath propagation is capable of producing more in-phase states than single-path propagation, which has the potential of compensating for the performance loss due to the inevitable energy dispersion, which is a consequence of the system's inability to use coherent diversity combining during the initial acquisition. Consequently, the performance degradation due to multipath propagation is not significant.

21.6.3 Probabilities of Overall Successful Detection, Overall Missing and Overall False-Alarm for the Parallel search-based Acquisition Scheme

In the context of the parallel search-based acquisition scheme communicating over multipath fading channels, the largest decision variable in the set of $\{Z_0, Z_1, \ldots, Z_q, \ldots, Z_{q+2L-3}\}$ is found and compared to the pre-set threshold h. Without loss generality, let $\{Z_0, Z_1, \ldots, Z_{2L-1}\}$ represent the decision variables corresponding to the $2L$ number of H_1 cells, while $\{Z_{2L}, Z_{2L+1}, \ldots, Z_{q+2L-3}\}$ the decision variables corresponding to the $(q-2)$ number of H_0 cells. Hence, the decision variables from the set $\{Z_0, Z_1, \ldots, Z_{2L-1}\}$ obey the distribution of Equation 21.62, while variables from the set $\{Z_{2L}, Z_{2L+1}, \ldots, Z_{q+2L-3}\}$ obey the distribution of Equation 21.65. Using σ_o^2 for normalizing Z_i, we obtain the normalized PDFs of Z_i as:

$$f_{Z_i}(y|R^2, H_1) = \frac{1}{2}\exp\left(-\frac{R^2+y}{2}\right) I_0\left(\sqrt{R^2 y}\right), \quad y \geq 0, \tag{21.74}$$

if Z_i is the decision variable corresponding to an H_1 cell. Otherwise, if Z_i is the decision variable corresponding to an H_0, cell, the normalized PDFs of Z_i can be expressed as:

$$f_{Z_i}(y|H_0) = \frac{1}{2}\exp\left(-\frac{y}{2}\right). \tag{21.75}$$

In Equation 21.74, R^2 is defined as:

$$R^2 = \frac{m^2}{\sigma_o^2} = \overline{\gamma}_c^2 \cdot \frac{\alpha_{1l}^2}{\Omega}. \tag{21.76}$$

Since $\{\alpha_{1l}\}_{l=0}^{L-1}$ are independent Rayleigh-distributed random variables, $\{\alpha_{1l}^2\}$ are exponentially-distributed random variables. Consequently, with the aid of Equation 21.3 it can be shown that the PDF of R^2 defined in Equation 21.76 can be expressed as:

$$f_{R^2}(r) = \frac{1}{\overline{\gamma}_c^2} \exp\left(-\frac{r}{\overline{\gamma}_c^2}\right). \tag{21.77}$$

Since the decision variables $\{Z_0, Z_1, \ldots, Z_{q+2L-3}\}$ are independent random variables, the conditioning in Equation 21.74 may be removed by averaging $f_{Z_i}(y|R^2, H_1)$ over the valid range of R^2, which can be expressed as:

$$f_{Z_i}(y|H_1) = \int_0^\infty f_{Z_i}(y|R^2, H_1) f_{R^2}(r) dr. \tag{21.78}$$

Upon substituting Equation 21.74 and Equation 21.77 into Equation 21.78, it can be shown that:

$$f_{Z_i}(y|H_1) = \frac{1}{2+\overline{\gamma}_c^2} \exp\left(-\frac{y}{2+\overline{\gamma}_c^2}\right). \tag{21.79}$$

Consequently, the overall detection probability of the parallel search-based acquisition scheme for transmission over the dispersive multipath Rayleigh fading channel can be formulated as:

$$P_D = 2L \int_{h'}^\infty f_{Z_i}(y|H_1) \left[\int_0^y f_{Z_j}(x|H_1)dx\right]^{2L-1} \left[\int_0^y f_{Z_k}(x|H_0)dx\right]^{q-2} dy, \tag{21.80}$$

where $h' = h/\sigma_o^2$, since the variable Z_i has been normalized by σ_o^2. Substituting the related PDFs from Equation 21.75 and Equation 21.79 into the above equation, P_D can be expressed as:

$$\begin{aligned}
P_D &= 2L \int_{h'}^\infty \frac{1}{2+\overline{\gamma}_c^2} \exp\left(-\frac{y}{2+\overline{\gamma}_c^2}\right) \\
&\quad \cdot \left[1 - \exp\left(-\frac{y}{2+\overline{\gamma}_c^2}\right)\right]^{2L-1} \left[1 - \exp\left(-\frac{y}{2}\right)\right]^{q-2} dy \\
&= 2L \sum_{i=0}^{2L-1} \sum_{j=0}^{q-2} \binom{2L-1}{i} \binom{q-2}{j} \frac{(-1)^{i+j}}{2+\overline{\gamma}_c^2} \int_{h'}^\infty \exp\left(-\frac{2+2(i+j)+j\overline{\gamma}_c^2}{2(2+\overline{\gamma}_c^2)} y\right) dy \\
&= 2L \sum_{i=0}^{2L-1} \sum_{j=0}^{q-2} \binom{2L-1}{i} \binom{q-2}{j} \frac{(-1)^{i+j}}{1+(i+j)+j\overline{\gamma}_c^2/2} \\
&\qquad \cdot \exp\left(-\frac{1+(i+j)+j\overline{\gamma}_c^2/2}{1+\overline{\gamma}_c^2/2} \cdot \frac{h'}{2}\right).
\end{aligned} \tag{21.81}$$

The overall missing probability can be expressed as:

$$P_M = \left[\int_0^{h'} f_{Z_i}(x|H_1)dx\right]^{2L} \left[\int_0^{h'} f_{Z_j}(x|H_0)dx\right]^{q-2} dy. \quad (21.82)$$

Upon substituting the related PDFs from Equation 21.75 and Equation 21.79 into the above equation, P_M can be formulated as:

$$P_M = \left[1 - \exp\left(-\frac{h'}{2+\overline{\gamma}_c^2}\right)\right]^{2L} \left[1 - \exp\left(-\frac{h'}{2}\right)\right]^{q-2}. \quad (21.83)$$

By expanding the terms on the right-hand side of Equation 21.83, we obtain

$$P_M = \sum_{i=0}^{2L}\sum_{j=0}^{q-2}(-1)^{i+j}\binom{2L}{i}\binom{q-2}{j}\exp\left(-\frac{i+j+j\overline{\gamma}_c^2/2}{1+\overline{\gamma}_c^2/2}\cdot\frac{h'}{2}\right). \quad (21.84)$$

Finally, the overall false-alarm probability can be computed by using the relationship of:

$$P_F = 1 - P_D - P_M, \quad (21.85)$$

and the mean acquisition time of the parallel search-based acquisition scheme communicating over dispersive multipath Rayleigh fading channels can be obtained by Equation 21.23 with P_D and P_F given by Equation 21.81 and Equation 21.85, respectively.

Figure 21.20 and Figure 21.21 show the normalized mean acquisition time performance for the parallel search-based acquisition scheme communicating over dispersive multipath Rayleigh fading channels. In Figure 21.20 the normalized mean acquisition time was plotted against the SNR/chip and the normalized threshold, while in Figure 21.21 the normalized mean acquisition time was plotted against the normalized threshold and the number of active users. The other related parameters were shown in the figures. From the results we observe that, for any given SNR/chip value or for any given number of active users, there exists an optimal choice of the threshold h', which minimizes the value of the mean acquisition time. However, an inappropriate choice of the detection threshold h' may lead to a severe degradation of the mean acquisition time, although the sensitivity of the mean acquisition time to the threshold decreases, as the SNR/chip value increases or as the number of active users decreases.

In Figure 21.22 and Figure 21.23 the effect of multipath transmission on the false-alarm probability was evaluated, where we set the normalized threshold to $h' = 0$. Hence, the false-alarm probability represents an unambiguous performance characterization of the parallel search-based acquisition scheme. In Figure 21.22 the false-alarm probability was plotted versus the SNR/chip for different number of resolvable paths, while in Figure 21.23 it was plotted versus the number of active users for different number of resolvable paths. The other parameters related to Figure 21.22 and Figure 21.23 were given in the figures. The results of both figures show that the false-alarm probability increases, when increasing the number of resolvable paths. However, the increase of the false-alarm probability due to the increasing

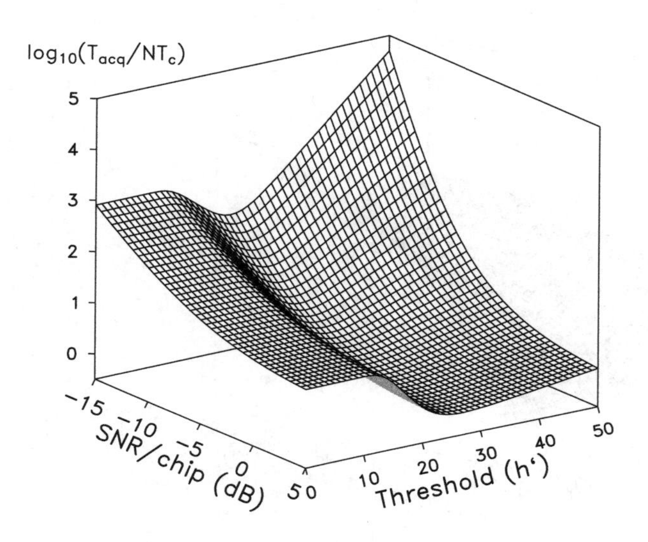

Figure 21.20: Mean acquisition time versus SNR/chip, γ_c and versus normalized threshold, h' performance for the parallel search-based acquisition scheme for transmission over multipath fading channels using the parameters $M = 64,\ N = 128,\ q = 1024,\ W = 100,\ \rho = 1.0,\ L = 1,\ K = 5$. The results were computed from Equation 21.23, Equation 21.81 and Equation 21.85.

of the number of resolvable paths is insignificant.

So far we have characterized the performance of serial as well as parallel acquisition schemes communicating over both AWGN and dispersive Rayleigh channels. In the next section we will consider more sophisticated arrangements.

21.7 Multicell Joint Detection-Based Serial Search Acquisition Scheme

So far, we have investigated the performance of both serial and parallel acquisition schemes, under the hypothesis of multiple H_1 cells in the uncertainty region of the PN code being acquired. According to Figure 21.16, we can assume that all H_1 cells are adjacent. Moreover, we can assume that, if the search step size is set to $T_c/2$, there exist at least two H_1 cells in the uncertainty region, where the received PN sequence and the locally generated PN sequence are within the desired small timing offset falling in the pull-in region of the code tracking loop. This is true, regardless whether non-fading or dispersive multipath fading channels are encountered. This *a-priori* information concerning the position of the H_1 cells might be efficiently exploited for enhancing the acquisition performance. We note that

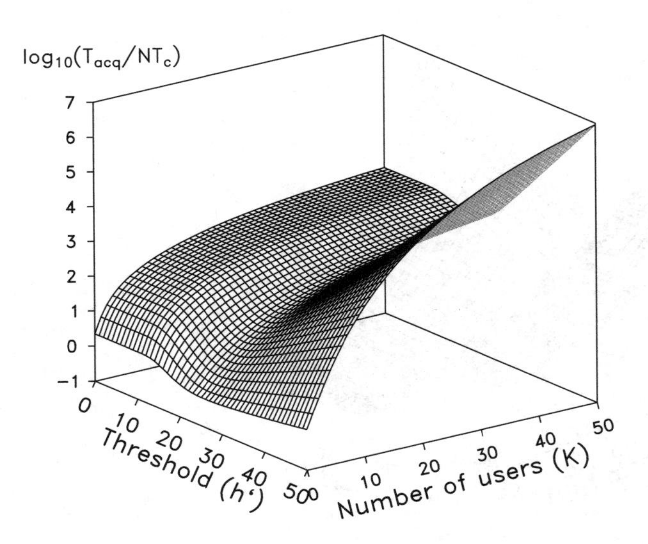

Figure 21.21: Mean acquisition time versus normalized threshold, h' and versus the number of active users, K performance for the parallel search-based acquisition scheme for transmission over multipath fading channels using the parameters $M = 64$, $N = 128$, $q = 1024$, $W = 100$, $\rho = 1.0$, $L = 3$, $\gamma_c = 0dB$. The results were computed from Equation 21.23, Equation 21.81 and Equation 21.85.

from an information theoretic point of view, any prior knowledge concerning the received sequence in an acquisition scheme facilitates the employment of efficient detection schemes, consequently potentially decreasing the average acquisition time. Our goal in this section is to quantify the performance of serial search-based acquisition systems by using a novel joint two-cell detection arrangement [50, 51], in which two adjacent cells are combined, as will be detailed below. Note that the analysis can be readily extended to the case of using parallel search-based acquisition schemes. However, if joint detection combining more than two cells is invoked, the analysis becomes intractable, despite the fact that indeed there exist more than two H_1 cells in the uncertainty region, for example, in the dispersive multipath fading environment of Figure 21.16. This can be exploited in practical acquisition schemes and the potential performance gains can be quantified with the aid of simulation. Nevertheless, there is always at least two H_1 cells in the uncertainty region, regardless, whether non-fading or dispersive multipath fading channels are encountered. Hence, in our following analysis, only joint detection combining two adjacent cells is considered. Again, we refer to this detection scheme as joint two-cell detection. Let us first discuss the joint two-cell detection model in the next section.

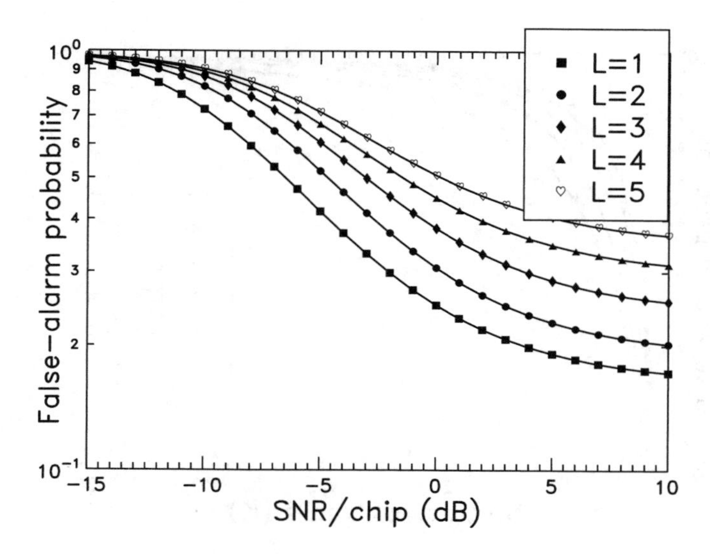

Figure 21.22: False-alarm probability versus SNR/chip, γ_c performance for the parallel search-based acquisition scheme communicating over multipath fading channels using the parameters $M = 64$, $N = 128$, $q = 1024$, $\rho = 1.0$, $K = 5$ and $h' = 0$. The results were computed from Equation 21.85.

21.7.1 Joint Two-Cell Detection Model

As we discussed in Section 21.3, the set $\{Z_0, Z_1, \ldots, Z_{q-1}\}$ represents the decision variables, where each decision variable corresponds to a phase to be tested. For the so-called joint two-cell detection scheme, the statistics of the final decision variables $U_i, i = 1, 2, \ldots$ are defined now by those of the sum of two adjacent correlator's output variables, which is expressed as:

$$U_i = Z_{i-1} + Z_i, \quad i = 0, 1, \ldots, q, \tag{21.86}$$

where q still represents the number of states in the uncertainty region, and $Z_{q+j} = Z_j$. However, if a false-alarm was encountered in the search mode during the detection of U_{i-1}, the term Z_{i-1} in the decision variable U_i would be set to zero. Let h be a threshold provided for making a decision, as to whether the phase of the received PN signature sequence and the locally generated PN sequence is within the required small timing offset in terms of the pull-in region of the code tracking loop, so as to conclude the search mode and to proceed to the so-called verification mode. The search mode during joint two-cell detection is described as follows. Whenever the decision variable U_i exceeds the threshold h, the system assumes that the corresponding current delay of the locally generated PN sequence is the correct delay and proceeds to the verification mode. Otherwise, if U_i does not exceed h, the relative phase

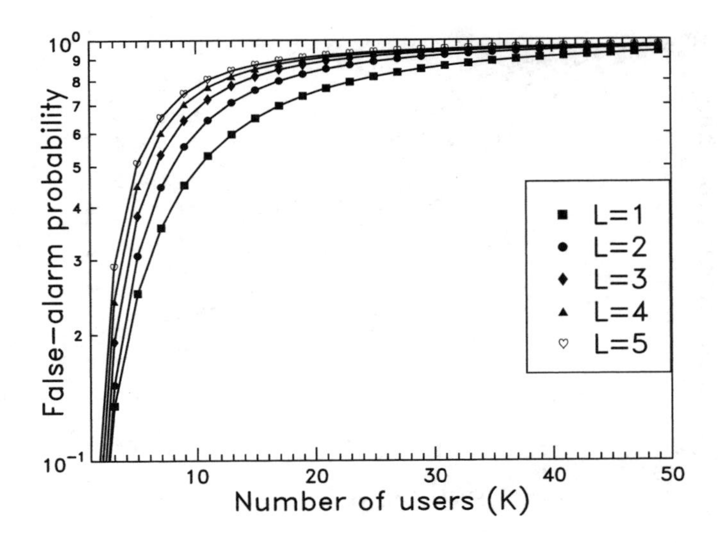

Figure 21.23: False-alarm probability versus the number of active users, K performance for the parallel search-based acquisition scheme for transmission over multipath fading channels using the parameters $M = 64$, $N = 128$, $q = 1024$, $\rho = 1.0$, $\gamma_c = 0dB$ and $h' = 0$. The results were computed from Equation 21.85.

of the locally generated PN sequence is readjusted, in order to obtain the decision variable U_{i+1}, and the above process is repeated.

Note that if a dispersive multipath fading channel having L resolvable paths is considered, then the number of cells in the uncertainty region of the joint two-cell detection scheme equals to $q + 2(L-1)$, since $\{Z_i\}$ has $q + 2(L-1)$ states, as we discussed previously in section 21.6.

Since the variable Z_i corresponds to either an H_1 or an H_0 cell, the adjacent cells associated with the indexes $\{i-1, i\}$ consequently correspond to four possible states, i.e. to $S_{01} = \{H_0, H_1\}$, $S_{11} = \{H_1, H_1\}$, $S_{10} = \{H_1, H_0\}$ and $S_{00} = \{H_0, H_0\}$. The acquisition procedure can be described as a four-state random Markov process [50] and the sequence of states corresponding to the decision variables $\{U_i\}$ takes, for example, the form of $\{\ldots, S_{00}, S_{00}, S_{01}, S_{11}, S_{10}, S_{00}, S_{00}, \ldots\}$, when transmitting over an AWGN channel and a search step size of $T_c/2$ are assumed. By contrast, it may take the form of $\{\ldots, S_{00}, S_{00}, S_{01}, S_{11}, \ldots, S_{11}, S_{10}, S_{00}, S_{00}, \ldots\}$, when a dispersive multipath fading channel and a search step size of $T_c/2$ are assumed. Moreover, if we assume that L resolvable paths are possible, then, $(2L-1)$ number of S_{11} states will be included in the uncertainty range.

Let the states 'S_{01}', 'S_{11}' and 'S_{10}' represent the synchronized states, while 'S_{00}' corresponds to an unsynchronized state. Then, the acquisition process is concluded and the process successfully proceeds to the fine tracking process, if the two adjacent samples belong to 'S_{01}', 'S_{11}' or 'S_{10}' and the combined correlation-related output decision variable $U_i = Z_{i-1} + Z_i$ exceeds the threshold h. Similarly, a false-alarm is encountered, if the two adjacent correlator's output samples belong to 'S_{00}', i.e., they were $\{H_0, H_0\}$ and yet and the combined correlation-related output decision variable $U_i = Z_{i-1}+Z_i$ exceeded the detection threshold h.

Since in the proposed detection scheme the combined decision variable of $U_i = Z_{i-1}+Z_i$ is used, provided that the search process terminates in the state S_{11}, the tracking-loop will be able to successfully track the correct spreading code phase, since both phases corresponding to Z_{i-1} and Z_i of the state S_{11} are correctly synchronized phases corresponding to H_1 cells. However, if the search process terminates in the state S_{01} or in the state S_{10}, which are constituted by one H_1 cell and one H_0 cell, then the tracking-loop may or may not be able to successfully track the correct phase of the spreading code. In this case usually an extra decision has to be invoked for determining, which of the two associated phases will be forwarded to the verification mode. This extra decision implies that if we have $Z_{i-1} > Z_i$, then the phase corresponding to Z_{i-1} will be forwarded to the verification mode. Otherwise, if $Z_{i-1} < Z_i$, then the phase corresponding to Z_i will be forwarded.

Assuming, for example that state S_{01} is being tested and the decision variable U_i corresponding to the state S_{01} exceeds the threshold h. Then, conditioned on $U_i = Z_{i-1}+Z_i > h$, an incorrect phase corresponding to Z_{i-1} will be erroneously forwarded to the verification with a probability of:

$$\begin{aligned} P(Z_{i-1} > Z_i | U_i = Z_{i-1} + Z_i > h) \\ &= \frac{P(Z_{i-1} > Z_i;\ U_i = Z_{i-1} + Z_i > h)}{P(U_i = Z_{i-1} + Z_i > h)} \\ &= \frac{P(Z_i < \frac{h}{2}, Z_{i-1} > h - Z_i) + P(Z_i > \frac{h}{2}, Z_{i-1} > Z_i)}{P(U_i = Z_{i-1} + Z_i > h)}. \end{aligned} \tag{21.87}$$

Again, conditioned on $U_i = Z_{i-1} + Z_i > h$, the probability that the correct phase corresponding to Z_i will be forwarded to the verification is expressed as:

$$P(Z_{i-1} < Z_i | U_i = Z_{i-1} + Z_i > h) = 1 - P(Z_{i-1} > Z_i | U_i = Z_{i-1} + Z_i > h). \tag{21.88}$$

Although an extra decision has to be invoked in the joint two-cell detection scheme, for simplicity, its effect on the acquisition performance will not be considered in our forthcoming analysis. In the following subsections, we simply assume that the acquisition process will hand over the correct phase with a unity probability to the fine tracking-loop, if the search process terminates in any of the states S_{01}, S_{11} or S_{10}. This assumption is true only, if $P(Z_{i-1} < Z_i | U_i = Z_{i-1} + Z_i > h) \approx 1$, i.e. if $P(Z_{i-1} > Z_i | U_i = Z_{i-1} + Z_i > h) >> (Z_{i-1} < Z_i | U_i = Z_{i-1} + Z_i > h)$, when S_{01} is considered. This is practically the case, since the amplitude of Z_i corresponding to an H_1 cell is usually higher, than that of Z_j corresponding to an H_0 cell, and the optimum threshold associated with the joint two-cell detection is relatively high. According to Equation 21.87 all these conditions will result in the reduction of the probability of forwarding an incorrect phase to the verification mode.

21.7.2 Detection Performance over AWGN Channels

In this subsection, we derive the probabilities of the overall missing and false-alarm for the serial search-based acquisition scheme using joint two-cell detection, when communicating over AWGN channels. We have obtained the statistics of the decision variables $\{Z_i\}$ in Section 21.5 for transmission over AWGN channels. The corresponding formulae of Equation 21.36 and Equation 21.39 are re-written as:

$$f_{Z_i}(y|H_1) = \frac{1}{2\sigma_o^2}\exp\left(-\frac{m^2+y}{2\sigma_o^2}\right) I_0\left(\frac{m\sqrt{y}}{\sigma_o^2}\right), \; y \geq 0, \qquad (21.89)$$

$$f_{Z_i}(y|H_0) = \frac{1}{2\sigma_o^2}\exp\left(-\frac{y}{2\sigma_o^2}\right), \; y \geq 0. \qquad (21.90)$$

In addition to the above PDFs, we may also need the PDFs of U_i conditioned on encountering the states S_{00}, S_{01}, S_{11} or S_{10}. At this stage we note that $U_i = Z_{i-1} + Z_i = Z^2_{c(i-1)} + Z^2_{s(i-1)} + Z^2_{ci} + Z^2_{si}$ and that $Z_{c(i-1)}, Z_{s(i-1)}$ as well as Z_{ci}, Z_{si} are the correlator's outputs in the context of the $(i-1)$th and the ith tests of the serial acquisition scheme. Furthermore, we note that these variables are independent Gaussian random variables. Hence, assuming that Z_{i-1} corresponds to an H_0 cell and Z_i corresponds to an H_1 cell, which corresponds the state S_{01}, or assuming that Z_{i-1} constitutes an H_1 cell and Z_i constitutes an H_0 cell, which corresponds the state S_{10}, the PDF of U_i is a non-central chi-square distribution with four degrees of freedom, which can be expressed as:

$$\begin{aligned} f_{U_i}(y|S_{01}) &= f_{U_i}(y|S_{10}) \\ &= \frac{1}{2\sigma_o^2}\left(\frac{y}{m^2}\right)^{1/2}\exp\left(-\frac{m^2+y}{2\sigma_o^2}\right) I_1\left(\frac{m\sqrt{y}}{\sigma_o^2}\right), \; y \geq 0. \end{aligned} \qquad (21.91)$$

In Equation 21.91 $I_1(\,)$ is the modified Bessel function of the first kind with one order, and σ_o^2, is given by Equation 21.37, while the non-centrality parameter m^2 is contributed by the H_1 cell, where m^2 is given by Equation 21.38.

If both of the decision variables Z_{i-1} and Z_i correspond to H_1 cells, which in turn correspond to a state of S_{11}, then U_i obeys the non-central chi-square distribution with four degrees of freedom. This PDF can be expressed as:

$$f_{U_i}(y|S_{11}) = \frac{1}{2\sigma_o^2}\left(\frac{y}{2m^2}\right)^{1/2}\exp\left(-\frac{2m^2+y}{2\sigma_o^2}\right) I_1\left(\frac{m\sqrt{2y}}{\sigma_o^2}\right), \; y \geq 0, \qquad (21.92)$$

where the non-centrality parameter $2m^2$ is contributed by two H_1 samples.

Finally, if both of the decision variables Z_{i-1} and Z_i correspond to H_0 cells, which in turn correspond to the state of S_{00}, then the PDF of U_i is a central chi-square distribution with four degrees of freedom, which can be expressed as:

$$f_{U_i}(y|S_{00}) = \frac{y}{4\sigma_o^4}\exp\left(-\frac{y}{2\sigma_o^2}\right), \quad y \geq 0. \qquad (21.93)$$

Having determined the PDFs of Z_i given that Z_i corresponds to either an H_1 cell or an H_0 cell which are seen in Equations 21.89 and 21.90, and with the PDFs of U_i formulae from Equation 21.91 to Equation 21.93 given that U_i corresponds to state S_{00}, S_{10}, S_{01} or S_{11}, let us now estimate the overall missing probability and the false-alarm probability P_F. We commence by considering the false-alarm probability. Due to the existence of correlation between two adjacent output decision variables, the false-alarm probability at the nth decision is the probability that $U_n = Z_{n-1} + Z_n$ exceeds the threshold h, which can be written as:

$$P_F = P(U_n > h). \tag{21.94}$$

Repeatedly expanding Equation 21.94 we arrive at:

$$\begin{aligned}
P_F &= P(U_{n-1} > h)P(U_n > h|U_{n-1} > h) &(21.95)\\
&\quad +P(U_{n-1} < h)P(U_n > h|U_{n-1} < h) &(21.96)\\
&= P(U_{n-1} > h)P(Z_n > h) &(21.97)\\
&\quad +P(U_{n-1} < h)P(Z_n > h, U_n > h|U_{n-1} < h) &(21.98)\\
&\quad +P(U_{n-1} < h)P(Z_n < h, U_n > h|U_{n-1} < h) &(21.99)\\
&= P(Z_n > h) &(21.100)\\
&\quad +P(U_{n-1} < h, Z_n < h, U_n > h) &(21.101)\\
&= P(Z_n > h) &(21.102)\\
&\quad +P(U_{n-2} > h)P(U_{n-1} < h, Z_n < h, U_n > h|U_{n-2} > h) &(21.103)\\
&\quad +P(U_{n-2} < h)P(U_{n-1} < h, Z_n < h, U_n > h|U_{n-2} < h) &(21.104)\\
&= P(Z_n > h) &(21.105)\\
&\quad +P(U_{n-2} > h)P(Z_{n-1} < h, Z_n < h, U_n > h) &(21.106)\\
&\quad +P(U_{n-2} < h, U_{n-1} < h, Z_n < h, U_n > h), &(21.107)
\end{aligned}$$

To elaborate a little further, Equation 21.95 and Equation 21.96 were derived by expanding Equation 21.94 conditioned on the fact that a false-alarm did or did not occur during the search of the $(n - 1)$st H_0 state. Furthermore, Equation 21.97 is due to Equation 21.95, since we assumed that $Z_{n-1} = 0$, if a false-alarm occurred during the $(n - 1)$st H_0 state, in order to eliminate the influence of the forthcoming search, while Equations 21.98 and 21.99 are derived from Equation 21.96 conditioned on the fact that the current decision variable obeys $Z_n > h$ or $Z_n < h$. Combining Equation 21.97 and Equation 21.98, we arrive at Equation 21.100, while Equation 21.99 can be written in the form of Equation 21.101. Alternatively, by extending Equation 21.101 under the condition of $U_{n-2} > h$ or $U_{n-2} < h$, we obtain Equation 21.103 and Equation 21.104. Since $Z_{n-2} = 0$ if $U_{n-2} > h$, Equation 21.103 can be simplified to Equation 21.106. Finally, Equation 21.104 can be expressed in the form of Equation 21.107. In Equation 21.106 $P(U_{n-2} > h)$ represents the false-alarm probability of the $(n - 2)$nd state, which is independent of $P(Z_{n-1} < h, Z_n < h, U_n > h)$, since we have $U_{n-2} = Z_{n-3} + Z_{n-2}$. If we assume that $q >> 1$, then $P(U_{n-2} > h) = P_F$. Equation 21.107 is still dependent on the previous search. However, this probability can be expressed as $P(U_{n-2} < h, U_{n-1} < h, Z_n < h, U_n > h) \lessapprox P(Z_{n-2} < h, U_{n-1} < h, Z_n < h, U_n > h)$ – where $\lessapprox$ means 'less than' and 'approximately equal to', since the event of

$Z_{n-2} < h$ includes that of $U_{n-2} < h$ and the probability $P(U_n > h)$ depends on Z_{n-3} through $U_{n-1} < h$ and $U_{n-2} < h$. Hence, the influence of Z_{n-3} on $P(U_n > h)$ can be ignored. Consequently, using these relationships in the corresponding equations, we finally arrive at:

$$P_F \lessapprox \frac{P(Z_n > h) + P(Z_{n-2} < h, U_{n-1} < h, Z_n < h, U_n > h)}{1 - P(Z_{n-1} < h, Z_n < h, U_n > h)}. \quad (21.108)$$

In Equation 21.108 $P(Z_n > h)$ can be computed as follows:

$$\begin{aligned} P(Z_n > h) &= \int_h^\infty f_{Z_n}(y|H_0)dy \\ &= \int_h^\infty \frac{1}{2\sigma_o^2} \exp\left(-\frac{1}{2\sigma_o^2}\right) \\ &= \exp\left(-\frac{h'}{2}\right). \end{aligned} \quad (21.109)$$

The second term in the numerator of Equation 21.108 can be expressed as:

$$\begin{aligned} &P(Z_{n-2} < h, U_{n-1} < h, Z_n < h, U_n > h) \\ &= \int_0^h f_{Z_{n-1}}(y|H_0) \left[\int_0^{h-y} f_{Z_{n-2}}(x|H_0)dx\right] \left[\int_{h-y}^h f_{Z_n}(x|H_0)dx\right] dy. \end{aligned} \quad (21.110)$$

By substituting the corresponding PDFs of $\{Z_i\}$ conditioned on either H_1 or H_0 in the above equation, we can simplify Equation 21.110 to:

$$\begin{aligned} &P(Z_{n-2} < h, U_{n-1} < h, Z_n < h, U_n > h) \\ &= \left(\frac{h'}{2} - 2\right) \exp\left(-\frac{h'}{2}\right) + \left(\frac{h'}{2} + 2\right) \exp\left(-h'\right). \end{aligned} \quad (21.111)$$

Finally, the second term in the denominator of Equation 21.108 can be expressed as:

$$\begin{aligned} P(Z_{n-1} < h, Z_n < h, U_n > h) &= \int_0^h f_{Z_{n-1}}(y|H_0) \left[\int_{h-y}^h f_{Z_n}(x|H_0)dx\right] dy \\ &= \left(\frac{h'}{2} - 1\right) \exp\left(-\frac{h'}{2}\right) + \exp(-h'). \end{aligned} \quad (21.112)$$

Finally, upon substituting Equations 21.109, 21.111 and 21.112 into Equation 21.108, the

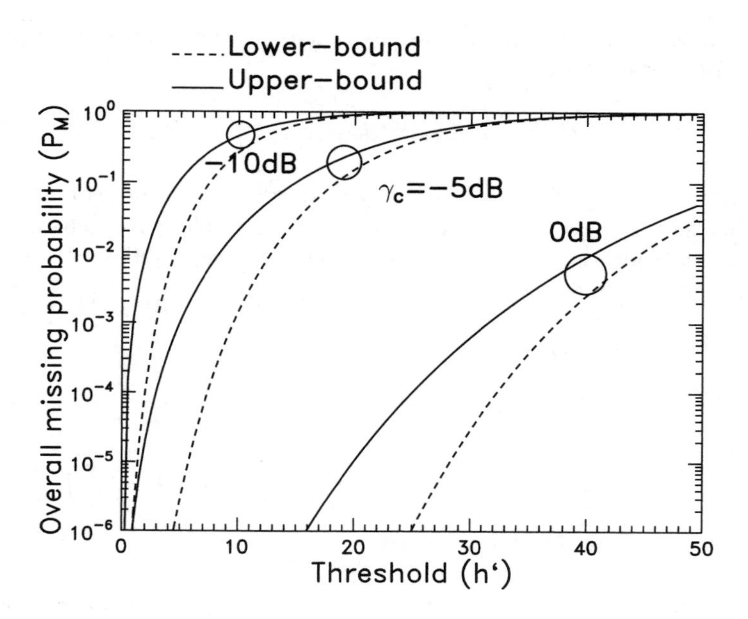

Figure 21.24: Lower-bound and upper-bound of the overall missing probability versus normalized threshold, h' performance for the serial search-based acquisition scheme using joint two-cell detection for transmission over AWGN channels using the parameters $M = 32$, $N = 64$, $q = 128$, $W = 100$, $K = 1$, $\rho = 1.0$, $\gamma_c = 10dB, -5dB, 0dB$. The results were computed from Equation 21.118 and Equation 21.120.

false-alarm probability can be expressed as:

$$P_F \lessapprox \frac{\left(\frac{h'}{2} - 1\right)\exp\left(-\frac{h'}{2}\right) + \left(\frac{h'}{2} + 2\right)\exp\left(-h'\right)}{1 - \left(\frac{h'}{2} - 1\right)\exp\left(-\frac{h'}{2}\right) - \exp(-h')}. \tag{21.113}$$

Let us now derive the overall missing probability. Let us assume that two H_1 cells are located at the raster positions i and $i+1$. Then the search process may be terminated at the raster positions i, $i+1$ or $i+2$ associated with the states 'S_{01}', 'S_{11}' or 'S_{10}', as we assumed previously. The overall missing probability of the search mode is the probability that none of the composite output decision variables U_i, U_{i+1} and U_{i+2} exceeds the threshold, h, which can be expressed as:

$$P_M = P(U_i < h, U_{i+1} < h, U_{i+2} < h). \tag{21.114}$$

However, due to the correlation of the successive decision variables, it is difficult to derive a closed form expression for the overall missing probability. Hence, here we aim for deriving its lower-bound and upper-bound. The lower-bound can be derived by ignoring the dependence or correlation between the adjacent decisions variables, which allows the lower-bound of Equation 21.114 to be expressed as:

$$P_M \geq P(U_i < h)P(U_{i+1} < h)P(U_{i+2} < h), \tag{21.115}$$

where $P(U_i < h)$, $P(U_{i+1} < h)$ or $P(U_{i+2} < h)$ represents the probability of the event that the decision variable U_i, U_{i+1} or U_{i+2} does not exceed the threshold h, respectively, because we considered the decision variables to be independent. The probabilities $P(U_i < h)$ and $P(U_{i+2} < h)$ can be expressed as:

$$\begin{aligned} P(U_i < h) &= P(U_{i+2} < h) \\ &= \int_0^h f_{U_i}(y|S_{01})dy \\ &= \int_0^h \frac{1}{2\sigma_o^2}\left(\frac{y}{m^2}\right)^{1/2} \exp\left(-\frac{m^2+y}{2\sigma_o^2}\right) I_1\left(\frac{m\sqrt{y}}{\sigma_o^2}\right) dy \\ &= 1 - Q_2(R, \sqrt{h'}), \end{aligned} \tag{21.116}$$

where $R = m/\sigma_o$ was given by Equation 21.44. Similarly, $P(U_{i+1} < h)$ in Equation 21.115 can be expressed as:

$$\begin{aligned} P(U_{i+1} < h) &= \int_0^h f_{U_{i+1}}(y|S_{11})dy \\ &= \int_0^h \frac{1}{2\sigma_o^2}\left(\frac{y}{2m^2}\right)^{1/2} \exp\left(-\frac{2m^2+y}{2\sigma_o^2}\right) I_1\left(\frac{m\sqrt{2y}}{\sigma_o^2}\right) dy \\ &= 1 - Q_2(\sqrt{2}R, \sqrt{h'}). \end{aligned} \tag{21.117}$$

Substituting Equation 21.116 and Equation 21.117 into Equation 21.115, the lower-bound of the overall missing probability for the joint two-cell detection scheme communicating over AWGN channels can be expressed as:

$$P_M \geq \left[1 - Q_2(R, \sqrt{h'})\right]^2 \left[1 - Q_2(\sqrt{2}R, \sqrt{h'})\right]. \tag{21.118}$$

Let us now derive the upper-bound of the overall missing probability. According to Equation 21.114, the upper-bound of the overall missing probability can be derived, if we set $Z_{i-1} = 0$, since the event of $Z_i < h$ includes the event of $U_i = Z_{i-1} + Z_i < h$. Hence, the upper-bound of the overall missing probability can be expressed as:

$$\begin{aligned} P_M &\leq P(Z_i < h, U_{i+1} < h, U_{i+2} < h) \\ &= P(U_{i+1} < h, U_{i+2} < h) \\ &= \int_0^h f_{Z_{i+1}}(y|H_1) \left[\int_0^{h-y} f_{Z_i}(x|H_1) dx \right] \left[\int_0^{h-y} f_{Z_{i+2}}(x|H_0) dx \right]. \end{aligned} \tag{21.119}$$

Substituting the corresponding PDFs from Equation 21.89 and Equation 21.90 into the above equations, finally, the upper-bound of the overall missing probability of the serial search-based acquisition scheme using joint two-cell detection can be expressed as:

$$\begin{aligned} P_M \leq & \int_0^{\sqrt{h'}} x \exp\left(-\frac{R^2 + x^2}{2}\right) I_0(Rx) \\ & \left[1 - Q\left(R, \sqrt{h' - x^2}\right)\right] \left[1 - \exp\left(-\frac{h' - x^2}{2}\right)\right] dx. \end{aligned} \tag{21.120}$$

Figure 21.24 shows the lower-bound and upper-bound of the overall missing probability for the serial search-based acquisition scheme using joint two-cell detection. The numerical results were plotted against the normalized threshold and these results were computed from Equation 21.118 and Equation 21.120 using the parameters of $M = 32,\ N = 64,\ q = 128,\ W = 100,\ K = 1,\ \rho = 1.0$ and $\gamma_c = 10dB,\ -5dB,\ 0dB$. The results show that the overall missing probability increases, when increasing the threshold, but decreases upon increasing the SNR/chip value. Moreover, as expected, both the lower-bound and the upper-bound of the overall missing probability will tend to zero, when the normalized threshold is close to zero, while it will asymptotically reach unity, when the normalized threshold is close to infinity.

Figure 21.25 and Figure 21.26 show the normalized mean acquisition time performance both versus the normalized threshold h' and the SNR/chip for the serial acquisition scheme using joint two-cell detection, respectively. In Figure 21.25 the results were computed from Equation 21.13, Equation 21.113 and Equation 21.118, where Equation 21.118 represents the lower-bound of the overall missing probability, while in Figure 21.26 the results were generated from Equation 21.13, Equation 21.113 and Equation 21.120, where Equation 21.120 represents the upper-bound of the overall missing probability. A range of other parameters were shown in the figures. By using joint two-cell detection, according to Figure 21.25 and Figure 21.26, we arrive at similar conclusions to those drawn from Figure 21.9, which was based on single-cell detection. It is shown that an inappropriate choice of the detection threshold h' degrades the mean acquisition time, especially when the SNR/chip value is low. In other words the sensitivity of the mean acquisition time to the detection threshold decreases, as the SNR/chip value increases. For any given SNR/chip value there exists an optimal choice of the threshold h', which minimizes the value of the mean acquisition time. Additionally, for any given normalized threshold h', the mean acquisition time decreases, as the SNR/chip value increases.

In Figure 21.27 the mean acquisition time versus the normalized threshold, h' and

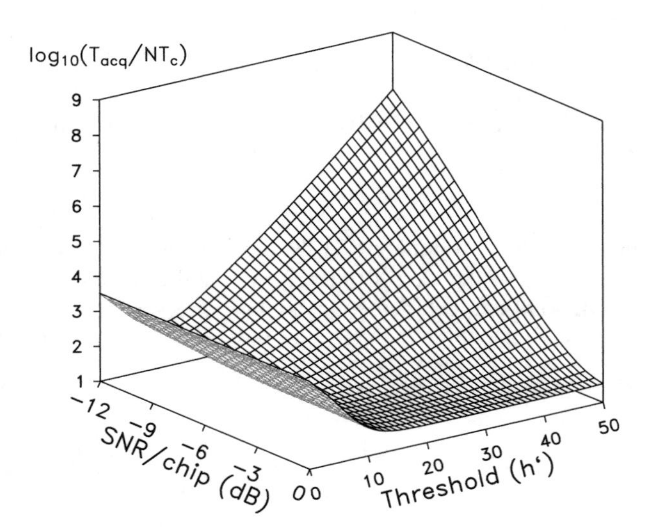

Figure 21.25: Mean acquisition time versus SNR/chip, γ_c and versus normalized threshold, h' performance for the serial search-based acquisition scheme using joint two-cell detection for transmission over AWGN channels using the parameters $M = 32,\ N = 64,\ q = 128,\ W = 100,\ K = 1,\ \rho = 1.0$. The results were computed from Equation 21.13, Equation 21.113 and Equation 21.118.

the number of active users, K performance of the serial acquisition scheme using joint two-cell detection was evaluated. The parameters employed for the computations were $M = 32,\ N = 64,\ q = 128,\ W = 100,\ \gamma_c = -5dB$, and $\rho = 1.0$. Note that since both the false-alarm probability evaluated from Equation 21.113 and the overall missing probability of Equation 21.120 represent their corresponding upper-bound, consequently the mean acquisition time computed from Equation 21.13 with the aid of the false-alarm probability expression of Equation 21.113 and the overall missing probability formula of Equation 21.120 also represents the upper-bound. According to the results, we arrive at similar conclusions to those inferred from Figure 21.10 for the serial acquisition scheme using single-cell detection. From the results we observe that for any given number of active users, there exists an optimal choice of the threshold h', which minimizes the value of the mean acquisition time. However, an inappropriate choice of the detection threshold h' may lead to severe degradation of the mean acquisition time, although again the sensitivity of the mean acquisition time to the threshold value decreases, as the SNR/chip value increases or as the number of active users decreases. Moreover, for any given threshold, the mean acquisition time increases, when the number of active users increases.

In Figure 21.28 we compared the mean acquisition time performance of the serial search-based acquisition system using joint two-cell detection and that using single-cell-assisted detection when communicating over AWGN channels. The normalized mean acquisition

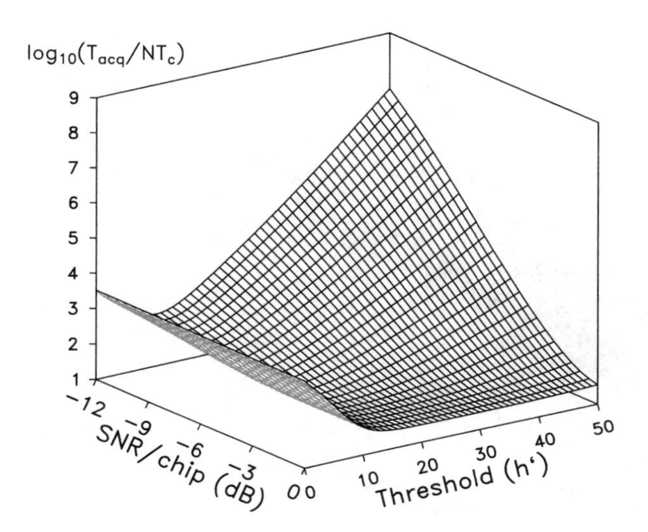

Figure 21.26: Mean acquisition time versus SNR/chip, γ_c and versus normalized threshold, h' performance for the serial search-based acquisition scheme using joint two-cell detection for transmission over AWGN channels using the parameters $M = 32$, $N = 64$, $q = 128$, $W = 100$, $K = 1$, $\rho = 1.0$. The results were computed from Equation 21.13, Equation 21.113 and Equation 21.120.

time was computed against the value of the normalized threshold, h' for different SNR/chip values. The other parameters and conditions were shown in the figures. From the results it can be observed that the acquisition time performance of the serial acquisition system using joint two-cell detection is better than that using single-cell-based detection, when the threshold is set to a sufficiently high value. This may be explained as follows. In the joint two-cell detection scheme, due to the presence of two adjacent cells being used jointly for detection, the average signal energy used for detection is higher than that in case of single-cell-based detection, since the single-cell-based detection scheme benefits only from a single state's energy during the acquisition process. Hence, if two H_1 cells are combined, the detection probability will be significantly higher than that using a single H_1 cell. However, if two H_0 cells are combined, potentially, the false-alarm probability will also be inevitably higher than that when using a single H_0 cell. This is why we cannot observe a more significant difference between the mean acquisition time curves recorded for the optimum threshold value, which is argued by the fact that for the joint two-cell detection the upper bound was considered. However, from the figures we observe that the mean acquisition performance of the joint two-cell detection scheme is more robust to the mismatch of the threshold than that of the single-cell-based detection scheme.

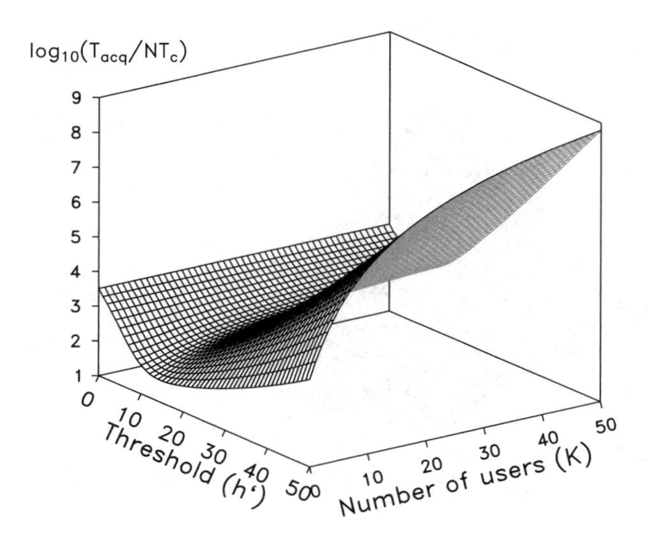

Figure 21.27: Mean acquisition time versus normalized threshold, h' and number of active users, K performance for the serial search-based acquisition scheme using joint two-cell detection for transmission over AWGN channels using the parameters $M = 32$, $N = 64$, $q = 128$, $W = 100$, $\gamma_c = -5dB$, $\rho = 1.0$. The results were computed from Equation 21.13, Equation 21.113 and Equation 21.120.

21.7.3 Detection Performance over Dispersive Multipath Rayleigh Fading Channels

In this subsection we investigate the acquisition performance of the joint two-cell detection scheme for transmissions over the multipath Rayleigh fading channels. The analysis is based on the discussions of Section 21.6.

21.7.3.1 Conditional Probability Density Functions

In Section 21.6 we derived the statistics of the decision variable $Z_i = Z_{ci}^2 + Z_{si}^2$, which represents either an H_1 state or an H_0 state for transmissions over multipath fading channels, which are repeated here as:

$$f_{Z_i}(y|H_1) = \frac{1}{2\sigma_o^2} \exp\left(-\frac{m^2+y}{2\sigma_o^2}\right) I_0\left(\frac{m\sqrt{y}}{\sigma_o^2}\right), \; y \geq 0, \tag{21.121}$$

$$f_{Z_i}(y|H_0) = \frac{1}{2\sigma_o^2} \exp\left(-\frac{y}{2\sigma_o^2}\right), \; y \geq 0. \tag{21.122}$$

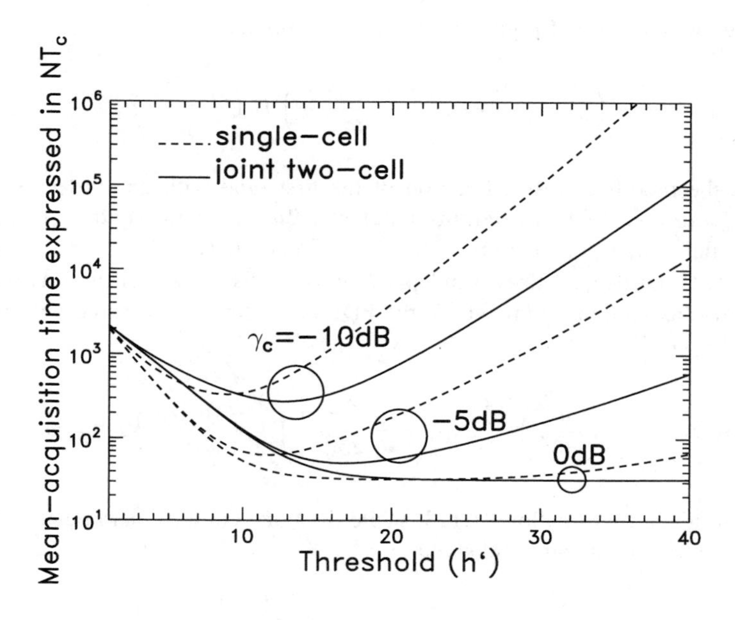

Figure 21.28: Comparison of mean acquisition time performance between the single-cell detection and the joint two-cell detection schemes for transmission over AWGN channels using the parameters of $M = 32$, $N = 64$, $q = 128$, $W = 100$, $K = 1$, $\rho = 1.0$. The results for the single-cell-based detection were computed from Equation 21.13, Equation 21.45 and Equation 21.46, while the results for the joint two-cell detection were generated from Equation 21.13, Equation 21.113 and Equation 21.120.

where

$$\sigma_o^2 = \frac{(L-1)\Omega}{3M} + \frac{(K-1)L\rho\Omega}{3M} + \frac{1}{2M\gamma_c}, \tag{21.123}$$

and m^2 is the normalized non-central metric, given by:

$$m^2 = \frac{9}{16}\alpha_{1l}^2. \tag{21.124}$$

As it becomes clear from Equations 21.121 and 21.122 $f_{Z_i}(y|H_1)$ and $f_{Z_i}(y|H_0)$ are conditioned on the fading parameters $\{\alpha_{1l}\}$.

For the search mode using joint two-cell detection we have $U_i = Z_{i-1} + Z_i = Z_{c(i-1)}^2 + Z_{s(i-1)}^2 + Z_{ci}^2 + Z_{si}^2$. Hence, U_i has four states, namely S_{00}, S_{01}, S_{11} and S_{10}. Consequently, when Z_{i-1} and Z_i represent two independent H_1 cells associated with the same multipath component, as seen in Figure 21.16, the PDF of U_i is the non-central chi-square distribution with four degrees of freedom due to the Gaussian nature of the terms included in U_i, which can be expressed with the aid of Equation (2-1-121) of [90] as:

$$f_{U_i}^1(y|S_{11}) = \frac{1}{2\sigma_o^2}\left(\frac{y}{2m^2}\right)^{1/2} \exp\left(-\frac{2m^2+y}{2\sigma_o^2}\right) I_1\left(\frac{m\sqrt{2y}}{\sigma_o^2}\right), \; y \geq 0, \quad (21.125)$$

where $I_1(\,)$ is the modified Bessel function of the first kind with one order, σ_o^2 is given by Equation 21.123 and the non-centrality parameter $2m^2$ is contributed by two H_1 cells associated with the same path, where m^2 is given by Equation 21.124.

If Z_{i-1} and Z_i correspond to two independent H_1 cells associated with two different propagation paths, as seen in Figure 21.16, the PDF of U_i can be expressed with the aid of Equation (2-1-121) of [90] as:

$$f_{U_i}^2(y|S_{11}) = \frac{1}{2\sigma_o^2}\left(\frac{y}{m'^2}\right)^{1/2} \exp\left(-\frac{m'^2+y}{2\sigma_o^2}\right) I_1\left(\frac{m'\sqrt{y}}{\sigma_o^2}\right), \; y \geq 0, \quad (21.126)$$

where σ_o^2 is also given by Equation 21.123, while the non-centrality parameter m'^2 is contributed by two independent paths and given by:

$$m'^2 = \frac{9}{16}(\alpha_{1l_1}^2 + \alpha_{1l_2}^2), \quad (21.127)$$

where m'^2 is conditioned on the amplitudes of both α_{1l_1} and α_{1l_2}.

When Z_{i-1} represents a H_0 cell and Z_i a H_1 cell, or conversely, Z_{i-1} a H_1 sample and Z_i a H_0 sample, the PDF of U_i can be computed by straightforward convolution of the PDFs of Z_{i-1} and Z_i, which results in:

$$\begin{aligned} f_{U_i}(y|S_{01}) &= f_{U_i}(y|S_{10}) \\ &= \frac{1}{2\sigma_o^2}\left(\frac{y}{m^2}\right)^{1/2} \exp\left(-\frac{m^2+y}{2\sigma_o^2}\right) I_1\left(\frac{m\sqrt{y}}{\sigma_o^2}\right), \; y \geq 0, \quad (21.128) \end{aligned}$$

where m^2 is given by Equation 21.124, which is conditioned on the distribution of $\{\alpha_{1l}\}$.

Above we have derived the conditional PDFs of the decision variable $\{U_i\}$ under different hypotheses for the search mode using joint two-cell detection. In order to compute the mean acquisition time, let us now derive the expressions of the overall missing probability and the false-alarm probability.

21.7.3.2 Probabilities of Overall Missing and False-Alarm for Joint Two-Cell Detection

In this subsection we derive the overall missing probability and the false-alarm probability for the serial search-based acquisition scheme using joint two-cell detection under the condition of a given instantaneous fading attenuation α_{1l}, $l = 1, 2, \ldots, L$ of the dispersive paths. The average probabilities can be obtained by averaging or integrating the related conditional probabilities over the dynamic range of α_{1l}, $l = 1, 2, \ldots, L$. The average overall missing probability of the search mode will be computed according to the detection mode and the channel characteristics. Since the ith decision Z_i corresponds to either an H_1 or an H_0 cell, for the joint two-cell detection the decision variable U_i, which is derived from $\{Z_{i-1}, Z_i\}$ corresponds to four possible phase states, namely to $S_{01} = \{H_0, H_1\}$, $S_{11} = \{H_1, H_1\}$, $S_{10} = \{H_1, H_0\}$ and $S_{00} = \{H_0, H_0\}$. Moreover, the two adjacent $T_c/2$-spaced decision variables Z_{i-1} and Z_i in Figure 21.16 can be generated by the same propagation path or by two different paths.

The false-alarm probability of the joint two-cell detection scheme at the nth decision is the probability that $U_n = Z_n + Z_{n-1}$ exceeds the threshold h, conditioned on two H_0 cells. This false-alarm probability can be computed using a similar method to that used in Section 21.7.2 for transmission over AWGN channels. This is because under the hypothesis of H_0 cells the PDF of Z_i when communicating over multipath fading channels – which is given by Equation 21.122 – takes the same form as the PDF of Z_i for transmissions over AWGN channels – which is given by Equation 21.90. Hence, the false-alarm probability can be expressed with the aid of Equation 21.113, which is repeated here for convenience:

$$P_F \lessapprox \frac{\left(\frac{h'}{2} - 1\right) \exp\left(-\frac{h'}{2}\right) + \left(\frac{h'}{2} + 2\right) \exp\left(-h'\right)}{1 - \left(\frac{h'}{2} - 1\right) \exp\left(-\frac{h'}{2}\right) - \exp(-h')}. \tag{21.129}$$

where $h' = h/\sigma_o^2$ with σ_o^2 now given by Equation 21.123.

Let us assume that there are L number of resolvable paths. As we discussed in Section 21.6, for the receiver the frequency-selective faded sequence of the user-of-interest appears to be a sequence with total number of $q + 2(L-1)$ cells in the uncertainty region. For the joint two-cell detection scheme the frequency-selective faded sequence of the user-of-interest can be approximated as a sequence having a total of $q+2(L-1)$ number of states in the uncertainty region and $2L+1$ number of correctly synchronized states associated with H_1 cells, where the received PN sequence and the local PN sequence are within a displacement falling in the pull-in range of the PN code tracking loop. The $2L+1$ number of correctly synchronized states include one S_{01}, one S_{10}, L number of S_{11} states associated with two H_1 samples due to the same propagation path and $L-1$ number of S_{11} states associated with two H_1 samples due to two different paths. Consequently, the overall missing probability of the serial search mode can be expressed as:

$$\begin{aligned} P_M = \mathrm{E}\big[P\,(U_1 = Z_0 + Z_1^1 < h, U_2 = Z_1^1 + Z_2^1 < h, U_3 = Z_2^1 + Z_3^2 < h,\\ U_4 = Z_3^2 + Z_4^2 < h, \ldots, U_{2L-1} = Z_{2L-2}^{L-1} + Z_{2L-1}^{L} < h,\\ U_{2L} = Z_{2L-1}^{L} + Z_{2L}^{L} < h, U_{2L+1} = Z_{2L}^{L} + Z_0 < h)\big], \end{aligned} \tag{21.130}$$

where Z_0 represents an H_0 cell, while the other decision variables $\{Z_i,\ i \neq 0\}$ in the above equation represent H_1 cells. Furthermore, $U_{2i-1} = Z_{2i-2}^{i-1} + Z_{2i-1}^{i}$ represents the sum of two decision variables associated with H_1 cells from the $(i-1)$-st and the i-th paths, $U_{2i} = Z_{2i-1}^{i} + Z_{2i}^{i}$ represents the sum of two decision variables associated with H_1 cells due to the i-th path, while $\mathrm{E}[\cdot]$ represents averaging over the fading attenuations.

However, for the search mode using joint two-cell detection, due to the inherent correlation between two adjacent decision variables, i.e., for example, between U_{2i-1} and U_{2i}, it is difficult to derive a closed form expression for the overall missing probability. Hence, in our following discourse we aim for deriving its lower-bound and upper-bound. Let us first derive the lower-bound of the overall missing probability. The lower-bound of Equation 21.130 is achieved, if we ignore the dependence between two adjacent decision variables, which can be expressed as:

$$\begin{aligned} P_M \geq\ & \mathrm{E}\left[P\left(U_1 = Z_0 + Z_1^1 < h\right)\right] \prod_{i=1}^{L} \mathrm{E}\left[P\left(U_{2i} = Z_{2i-1}^i + Z_{2i}^i < h\right)\right] \\ & \prod_{j=1}^{L-1} \mathrm{E}\left[P\left(U_{2j+1} = Z_{2j}^j + Z_{2j+1}^{j+1} < h\right)\right] \mathrm{E}\left[P\left(U_{2L+1} = Z_{2L}^L + Z_0 < h\right)\right], \end{aligned} \quad (21.131)$$

where each term taking the form of $\mathrm{E}\left[P\left(U_i < h\right)\right]$ represents the related average missing probability of the i-th correct state without considering the correlation with the other decision variables. The relevant terms in Equation 21.131 can be obtained by computing the corresponding detection probabilities first as follows. If Z_0 represents an H_0 cell and Z_1^1 an H_1 cell, which corresponds to the first term of Equation 21.131, or conversely, Z_{2L}^L represents a H_1 cell and Z_0 an H_0 cell, which corresponds to the last term of Equation 21.131, then the successful detection probability conditioned on the fading amplitude of path l, α_{1l}, is the probability that the combined decision variable U_i will exceed the threshold h, which can be expressed as:

$$\begin{aligned} P_{D_{01}}(R) &= P_{D_{10}}(R) \\ &= \int_h^{\infty} f_{U_i}(y|S_{10})dy = Q_2(R, \sqrt{h'}). \end{aligned} \quad (21.132)$$

In Equation 21.132 $f_{Y_i}(y|S_{10})$ was given by Equation 21.128, $Q_2(a, b)$ represents Marcum's Q function, which is defined by Equation 21.41, while R was defined in Equation 21.68 and its PDF was given in Equation 21.70. Furthermore, the subscript D_{01} indicates that Equation 21.132 is contributed by an H_0 cell as well as an H_1 cell.

The unconditional detection probability of the first synchronized state 'D_1 (S_{01})' or the last synchronized state 'D_{2L+1} (S_{10})' can be obtained by averaging Equation 21.132 over the range of R, which can be obtained from Equation D.7 of the Appendix D by setting $\mu = 2,\ p = 1,\ \Omega = \overline{\gamma_c}^2$ and $b^2 = h'$, yielding:

$$P_{D_{01}} = P_{D_{10}} = \sum_{m=0}^{\infty} \sum_{n=0}^{m+1} \frac{(h')^{m-n+1}}{(m-n+1)!2^{m-n}} \cdot \frac{\exp(-h'/2)}{\overline{\gamma}_c^2(1 + 2/\overline{\gamma}_c^2)^{m+1}}. \quad (21.133)$$

Similarly, if Z_{2i-1}^i and Z_{2i}^i represent two H_1 cells due to the same path, which is a situation that corresponds to the second term of Equation 21.131, then the conditional detection probability for $i = 1, 2, \ldots, L$ can be expressed as:

$$P_{D_{11}}^{(1)}(R) = \int_h^\infty f_{U_{2i}^1}(y|S_{11})dy = Q_2(\sqrt{2}R, \sqrt{h'}), \tag{21.134}$$

where R is given by Equation 21.68. The unconditional detection probability can be similarly computed from Equation D.7 of Appendix D by setting $\mu = 2,\ p = 1,\ \Omega = 2\overline{\gamma}_c^2$ and $b^2 = h'$, which results in:

$$P_{D_{11}}^{(1)} = \sum_{m=0}^{\infty} \sum_{n=0}^{m+1} \frac{(h')^{m-n+1}}{(m-n+1)!2^{m-n+1}} \cdot \frac{\exp(-h'/2)}{\overline{\gamma}_c^2(1+1/\overline{\gamma}_c^2)^{m+1}}, \tag{21.135}$$

for $i = 1, 2, \ldots, L$.

If Z_{2j}^j and Z_{2j+1}^{j+1} represent two H_1 cells due to two different paths, which is a scenario that corresponds to the third term of Equation 21.131, then the detection probability conditioned on α_{1l_1} and α_{1l_2} can be expressed as:

$$P_{D_{11}}^{(2)}(R') = \int_h^\infty f_{U_{2j+1}^2}(y|S_{11})dy = Q_2(R', \sqrt{h'}), \tag{21.136}$$

where

$$R' = \overline{\gamma}_c \cdot \frac{\sqrt{\alpha_{1l_1}^2 + \alpha_{1l_2}^2}}{\sqrt{\Omega}}. \tag{21.137}$$

Since α_{1l_1} and α_{1l_2} are two iid Rayleigh distributed random variables, the PDF of R' can be derived using Equation D.5 of Appendix D with $p = 2$, which can be expressed as:

$$f_R'(r) = \frac{2r^3}{\overline{\gamma}_c^4} \exp\left(-\frac{r^2}{\overline{\gamma}_c^2}\right),\ r \geq 0. \tag{21.138}$$

The unconditional detection probability can be obtained by averaging Equation 21.136 using Equation 21.138 in the valid range of R', which can also be obtained from Equation D.7 of Appendix D by setting $\mu = 2,\ p = 2,\ \Omega = \overline{\gamma}_c^2$ and $b^2 = h'$, yielding:

$$P_{D_{11}}^{(2)} = \sum_{m=0}^{\infty} \sum_{n=0}^{m+1} \frac{(m+1)(h')^{m-n+1}}{(m-n+1)!2^{m-n-1}} \cdot \frac{\exp(-h'/2)}{\overline{\gamma}_c^4(1+2/\overline{\gamma}_c^2)^{m+2}}, \tag{21.139}$$

for $j = 1, 2, \ldots, L-1$.

Above we have obtained the unconditional successful detection probabilities for the states from 'D_1' to 'D_{2L+1}'. The related missing probabilities of the states 'D_1' to 'D_{2L+1}' can be computed using $1 - P_{D_{()}}$. The lower-bound of the overall missing probability of the serial search mode using joint two-cell detection can be expressed as:

$$P_M \geq (1 - P_{D_{01}})(1 - P_{D_{10}})\left(1 - P_{D_{11}}^{(1)}\right)^L \left(1 - P_{D_{11}}^{(2)}\right)^{L-1}. \quad (21.140)$$

Let us now derive the upper-bound of the overall missing probability. The upper-bound of the overall missing probability can be derived, if we set Z_0 and Z_{2i-2}^{i-1} in U_{2i-1} for $i = 2, 3, \ldots, L$ equal to zero, yielding,

$$\begin{aligned} P_M \leq \; & \mathrm{E}\Big[P\left(Z_1^1 < h, U_2 = Z_1^1 + Z_2^1 < h; Z_3^2 < h,\right. \\ & \left. U_4 = Z_3^2 + Z_4^2 < h; \ldots; Z_{2L-1}^L < h, U_{2L} = Z_{2L-1}^L + Z_{2L}^L < h\right)\Big]. \end{aligned} \quad (21.141)$$

Note that Equation 21.141 corresponds to the overall missing probability of the search mode, where we assume that the delayed decision variable in U_i is set to zero, when the 'present' decision variable is from a different path compared to the 'last' decision variable, namely, the delayed decision. Since on the right-hand side of Equation 21.141 the terms $(Z_{2i-1}^i < h, U_{i+1} = Z_{2i-1}^i + Z_{2i}^i < h)$ for $i = 1, 2, \ldots, L$ are mutually independent, hence Equation 21.141 can be written as:

$$\begin{aligned} P_M & \leq \prod_{l=1}^{L} \mathrm{E}\left[P\left(Z_{2i-1}^i < h, U_{2i} = Z_{2i-1}^i + Z_{2i}^i < h\right)\right] \\ & = \prod_{l=1}^{L} \mathrm{E}\left[P\left(U_{2i} = Z_{2i-1}^i + Z_{2i}^i < h\right)\right]. \end{aligned} \quad (21.142)$$

The probability of $P\left(U_{2i} = Z_{2i-1}^i + Z_{2i}^i < h\right)$ is the probability that the decision variable U_{2i} does not exceed the threshold, h, where Z_{2i-1}^i and Z_{2i}^i represent two decision variables associated with H_1 cells due to the same propagation path. Hence we have:

$$\begin{aligned} P\left(U_{2i} = Z_{2i-1}^i + Z_{2i}^i < h\right) & = \int_0^h f_{U_{2i}}^1(y|S_{11})dy \\ & = \int_0^h \frac{1}{2\sigma_o^2}\left(\frac{y}{2m^2}\right)^{1/2} \exp\left(-\frac{2m^2 + y}{2\sigma_o^2}\right) I_1\left(\frac{m\sqrt{2y}}{\sigma_o^2}\right) dy \\ & = 1 - Q_2(\sqrt{2}R, \sqrt{h'}), \end{aligned} \quad (21.143)$$

where R is given by Equation 21.68 and its PDF is given by Equation 21.70. Taking into account the factor $\sqrt{2}$ before R, the unconditional probability can be computed from Equation D.7 of Appendix D by setting $\mu = 2$, $p = 1$, $\Omega = 2\overline{\gamma}_c^2$ and $b^2 = h'$, which results

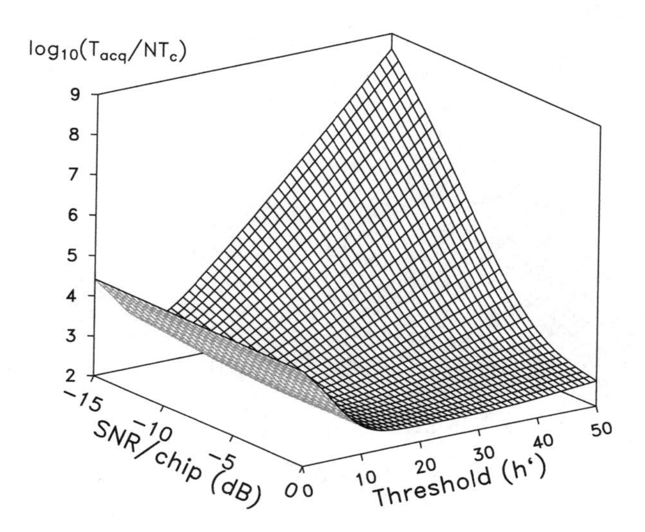

Figure 21.29: Mean acquisition time versus SNR/chip, γ_c and normalized threshold, h' performance for the serial search-based acquisition scheme using joint two-cell detection for transmission over dispersive multipath fading channels using the parameters of $M = 64,\ N = 128,\ q = 1024,\ W = 100,\ K = 1,\ L = 3,\ \rho = 1.0$. The results were computed from Equation 21.13, Equation 21.129 and Equation 21.140.

in:

$$E\left[P\left(U_{2i} = Z_{2i-1}^i + Z_{2i}^i < h\right)\right]$$
$$= 1 - \sum_{m=0}^{\infty} \sum_{n=0}^{m+1} \frac{(h')^{m-n+1}}{(m-n+1)!2^{m-n+1}} \cdot \frac{\exp(-h'/2)}{\overline{\gamma}_c^2(1 + 1/\overline{\gamma}_c^2)^{m+1}}, \quad (21.144)$$

for $i = 1, 2, \ldots, L$.

Finally, upon substituting Equation 21.144 into Equation 21.142, the upper-bound of the overall missing probability of the serial search-based acquisition scheme using joint two-cell detection for transmissions over dispersive multipath fading channels can be expressed as:

$$P_M \leq \left[1 - \sum_{m=0}^{\infty} \sum_{n=0}^{m+1} \frac{(h')^{m-n+1}}{(m-n+1)!2^{m-n+1}} \cdot \frac{\exp(-h'/2)}{\overline{\gamma}_c^2(1 + 1/\overline{\gamma}_c^2)^{m+1}}\right]^L. \quad (21.145)$$

Figure 21.29 and Figure 21.30 show the normalized mean acquisition time performance both versus the normalized threshold h' and the SNR/chip for the serial search modes using joint two-cell detection, when communicating over dispersive multipath Rayleigh fading

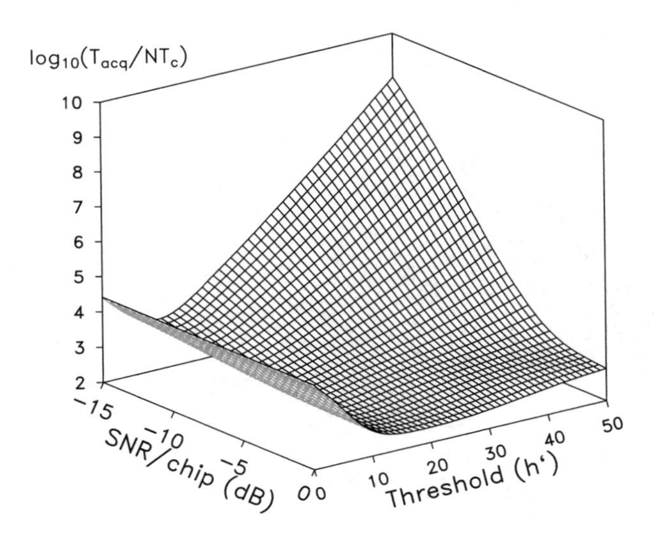

Figure 21.30: Mean acquisition time versus SNR/chip, γ_c and normalized threshold, h' performance for the serial search-based acquisition scheme using joint two-cell detection for transmission over dispersive multipath fading channels using the parameters of $M = 64$, $N = 128$, $q = 1024$, $W = 100$, $K = 1$, $L = 3$, $\rho = 1.0$. The results were computed from Equation 21.13, Equation 21.129 and Equation 21.145.

channels. In Figure 21.29 the results were computed from Equation 21.13, Equation 21.129 and Equation 21.140, where Equation 21.140 represents the lower-bound of the overall missing probability, while in Figure 21.30 the results were generated from Equation 21.13, Equation 21.129 and Equation 21.145, where Equation 21.145 represents the upper-bound of the overall missing probability. Here we emphasize again that the normalized mean acquisition time computed from Equation 21.13, Equation 21.129 and Equation 21.140 is not the achievable lower-bound, since Equation 21.129 represents the upper-bound. However, the mean acquisition time computed from Equation 21.13, Equation 21.129 and Equation 21.145 definitely represents the upper-bound of the mean acquisition time, since both Equation 21.129 and Equation 21.145 represent the upper-bound of the corresponding false-alarm and that of the overall missing probability. In our numerical computations the number of cells in the uncertainty region was set to $q = 1024$, the processing gain was $N = 128$, the integral dwell time was $\tau_D = 64T_c$, i.e. we had $M = 64$, there were $L = 3$ resolvable paths at the receiver, the system supported $K = 1$ users and the base station received the same power from all users, i.e. we had $\rho = 1$. The results suggest similar conclusions to those inferred from Figure 21.17 for the serial search-based acquisition scheme using single-cell-based detection, which indicate that an inappropriate choice of the detection threshold h' can lead to a severe increase of the mean acquisition time. Moreover, as in the context of Figure 21.17, the sensitivity of the mean acquisition time to the threshold decreases, as the SNR/chip value

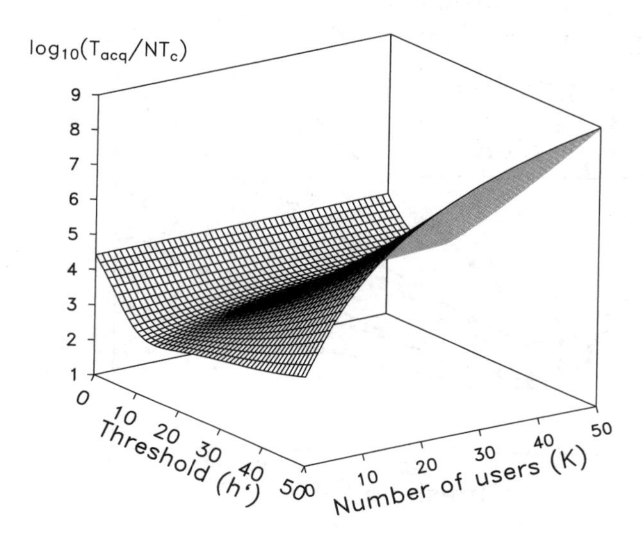

Figure 21.31: Upper-bound mean acquisition time versus normalized threshold, h' and number of active users performance for the serial search-based acquisition scheme using joint two-cell detection for transmission over dispersive multipath fading channels using the parameters of $M = 64,\ N = 128,\ q = 1024,\ W = 100,,\ L = 3,\ \rho = 1.0,\ \gamma_c = -5dB$. The results were computed from Equation 21.13, Equation 21.129 and Equation 21.145.

increases. For any given SNR/chip value there exists an optimal choice of the threshold h', which minimizes the value of the mean acquisition time.

In Figure 21.31, the mean acquisition time versus the normalized threshold, h' and versus the number of active users, K performance of the serial search-based acquisition scheme using joint two-cell detection was evaluated for transmissions over dispersive multipath Rayleigh fading channels. The parameters employed for the computations were $M = 64,\ N = 128,\ q = 1024,\ W = 100,\ L = 3,\ \gamma_c = -5dB$, and $\rho = 1.0$. From the results we arrive at similar observation to those noted in the context of Figure 21.18 for the case using single-cell-based detection, when communicating over dispersive multipath Rayleigh fading channels. The results show that for any given number of active users, there exists an optimal choice of the threshold h', which minimizes the value of the mean acquisition time. However, an inappropriate choice of the detection threshold h' leads to the degradation of the mean acquisition time. We also observed that the sensitivity of the mean acquisition time to the threshold decreases, as the number of active users decreases. Moreover, for any given threshold the mean acquisition time increases, when the number of active users increases.

Figure 21.32 presents the mean acquisition time performance of the joint two-cell detection and single-cell-based detection schemes against the normalized threshold h' for $M = 64,\ N = 128,\ q = 1024,\ W = 100,\ L = 3,\ K = 5,\ \rho = 1.0$ and $\gamma_c = -10dB,\ -5dB,\ 0dB$, respectively. For any given SNR/chip value, there is an op-

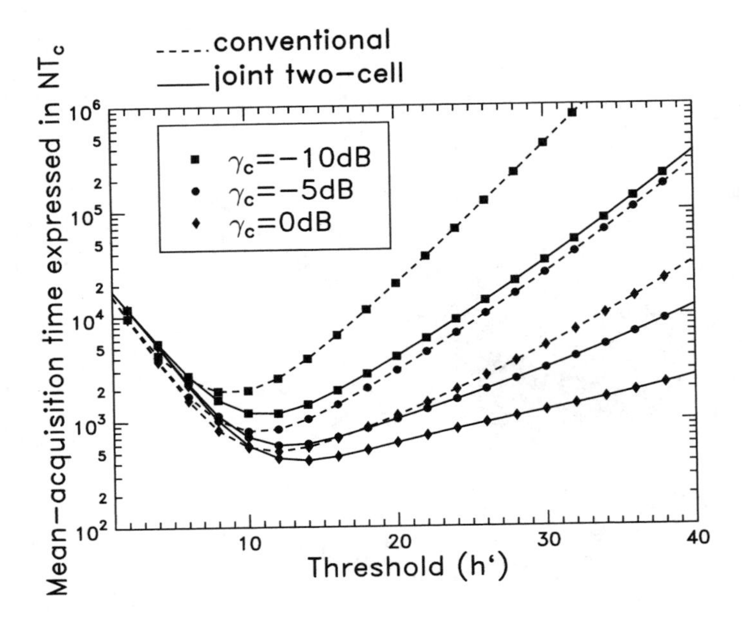

Figure 21.32: Comparison of mean acquisition time performance between the single-cell detection and the joint two-cell detections schemes for transmissions over AWGN channels using the parameters of $M = 64$, $N = 128$, $q = 1024$, $W = 100$, $L = 3$, $K = 5$, $\rho = 1.0$. The results for the single-cell-based detection scheme were computed from Equation 21.13, Equation 21.66 and Equation 21.73, while the results for the joint two-cell detection scheme were evaluated from Equation 21.13, Equation 21.129 and Equation 21.145.

timal choice of the threshold h', which leads to the minimum mean acquisition time. At the optimal value of h', we notice that for the single-cell-based detection or for the joint two-cell detection based scheme the mean acquisition time performance is improved, when the SNR/chip increases. The mean acquisition time performance of the joint two-cell detection scheme improved more significantly and became also more robust to the change of the threshold in a wide range of the normalized threshold h', namely from about 20 to 40, than that of the single-cell-based detection scheme, when assuming a given SNR/chip value. Note furthermore that in a practical acquisition system the threshold must not be set to an excessively low value, because too low a threshold usually leads to too high a number of false alarms, and hence, due to the associated penalty time delay, to an increased mean acquisition time.

Finally, in Figure 21.33 we show the influence of the number of resolvable paths on the mean acquisition time performance. The parameters used were shown in the figure. From these results we observe that the mean acquisition time increases, when the number of resolvable paths increases. However, in the range considered, the mean acquisition time for $L = 5$ does not exceed more than twice of that for $L = 1$.

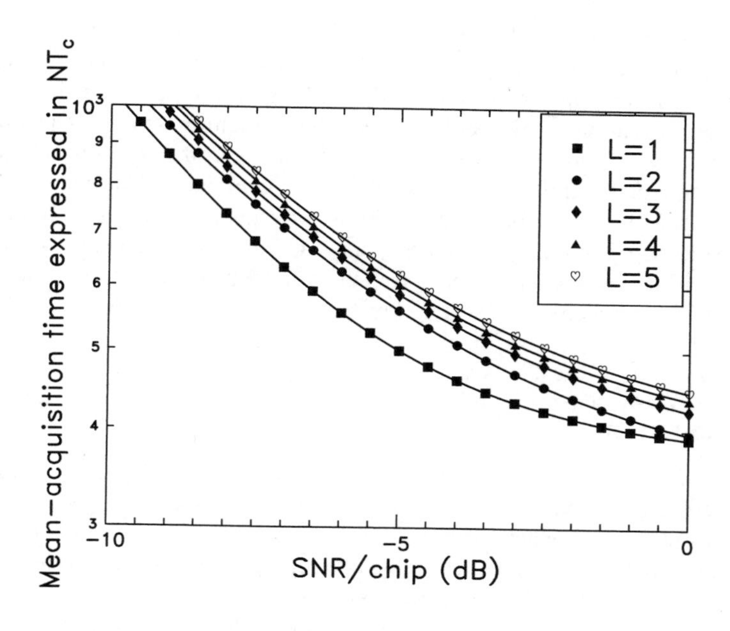

Figure 21.33: Upper-bound mean acquisition time versus SNR/chip performance for the serial search acquisition scheme using joint two-cell detection for transmission over dispersive multipath fading channels using the parameters of $M = 64,\ N = 128,\ q = 1024,\ W = 100,\ K = 5,\ \rho = 1.0$ and $L = 1, 2, 3, 4, 5$. The results were computed from Equation 21.13, Equation 21.129 and Equation 21.145.

21.8 Acquisition of m-Sequences Using Soft Sequential Estimation

In this section two novel sequential estimation methods are proposed for the initial synchronization of CDMA systems communicating using pseudonoise (PN) signals derived from m-sequences. These sequential estimation methods are designed based on the novel principle of iterative soft-in-soft-out (SISO) decoding. Specifically, two types of sequential estimation methods are proposed and investigated, namely Coherent Recursive Soft Sequential Estimation (CRSSE) based acquisition and Differential Recursive Soft Sequential Estimation (DRSSE) acquisition. Both the CRSSE and DRSSE acquisition schemes exhibit a complexity similar to that of a conventional m-sequence generator, which increases only linearly with the number of stages used in the m-sequence generator. Furthermore, the initial acquisition time of the CRSSE and the DRSSE acquisition schemes is also linearly dependent on the number of stages used in the m-sequence generator.

An important feature of the proposed acquisition schemes is that the acquisition device is capable of determining the real-time reliabilities associated with the decision concerning a set of, say S, consecutive chips. These S number of correctly detected consecutive chips uniquely describe the m-sequence of period $N = 2^S - 1$ and hence are sufficient for the local

m-sequence generator to produce a locally synchronized despreading m-sequence replica. In terms of these characteristics the proposed CRSSE and DRSSE acquisition scheme is more attractive than the conventional serial search and parallel search-based acquisition schemes. This is because the conventional serial search-based scheme requires a mean-acquisition time, which is exponentially increasing with the number of stages in the m-sequence generator. Similarly, the conventional parallel search-based scheme exhibits a complexity, which also grows exponentially with the number of stages in the m-sequence generator. Hence, owing to the linear dependence of both the complexity and the initial synchronization time on the number of stages in the m-sequence generator, the employment of the proposed CRSSE and DRSSE acquisition schemes is particularly beneficial for the acquisition of long m-sequences.

21.8.1 Introduction

The sequential estimation acquisition arrangement proposed by Ward [502] constitutes one of the simplest acquisition schemes. The philosophy behind the sequential estimation acquisition scheme is as follows. The acquisition of an m-sequence of length $(2^S - 1)$ is deemed successful, provided that S consecutive chips are correctly received by the acquisition device and are loaded into the local m-sequence generator, where successive shifts of the chips in the generator will generate chips that exactly match the forthcoming received chips of the transmitted m-sequence. However, in the presence of noise, one or more of the S consecutive chips might be in error, potentially resulting in erroneous loading of the m-sequence generator. In this case a new set of S chips can be processed similarly.

Clearly, the most critical requirement for attaining the successful acquisition of PN sequences based on sequential estimation [502] is that S consecutive chips of the received and noise-contaminated PN sequence have to be correctly estimated. Ward [502] has shown that for moderate SNR per chip values this acquisition scheme is capable of providing a shorter expected acquisition time than the conventional sliding correlator-based acquisition scheme [28]. However, in spread-spectrum communications the transmitted signal energy is spread over dozens or even hundreds of chips and hence the SNR per chip value is typically low. Therefore, the estimation of S consecutive chips using chip-by-chip-based hard-decisions is typically unreliable. In order to improve the reliabilities associated with deciding upon the value of S consecutive initial chips, Kilgus [503] proposed a majority logic decoding aided scheme for estimating the required S number of consecutive chips. By contrast, in [504] Ward and Yiu have proposed a recursive sequential estimation-assisted acquisition scheme. It was shown that both the above schemes are capable of significantly improving the acquisition performance by exploiting the inherent properties of the m-sequences for aiding the estimation of the S consecutive chips. However, all existing sequential estimation-based acquisition schemes operate on the basis of hard-decision chips. Furthermore, in the existing sequential estimation-based acquisition schemes coherent detection was assumed, which is often unrealistic to achieve, since prior to despreading, the SNR is usually insufficiently high for attaining a satisfactory carrier-phase tracking.

In this section we invoke the iterative soft-in-soft-out (SISO) decoding principle – which was originally developed for turbo channel decoding [79, 505, 506] – for improving the reliabilities associated with deciding upon the S consecutive chips to be loaded into the local m-sequence generator. Upon exploiting the inherent properties of m-sequences, a Coherent

Recursive Soft Sequential Estimation (CRSSE) acquisition scheme and a Differential Recursive Soft Sequential Estimation (DRSSE) arrangement are proposed, both of which estimate S consecutive chips using a recursive SISO decoder. In the CRSSE scheme the recursive SISO decoder receives soft information from the channel's output (*intrinsic* information) and soft *extrinsic* information [79,505,506] from the soft channel outputs associated with the previous chips. The soft output of the recursive SISO decoder is then shifted into a so-called soft-chip-register, which provides *extrinsic* information for the following decoding steps. By contrast, in the DRSSE acquisition scheme the transmitted m-sequence is first mapped to another m-sequence, where the time-varying carrier-phase is removed with the aid of differential pre-processing. Then, S consecutive chips of the resultant m-sequence are estimated using a recursive SISO decoder. The recursive SISO decoder receives soft information from the differential processor's output (*intrinsic* information) and, similarly to the CRSSE scheme, soft *extrinsic* information from the soft channel outputs associated with the previous chips. Then, in a similar fashion to the CRSSE arrangement, the soft output of the recursive SISO decoder is shifted into a soft-chip-register, which provides *extrinsic* information for the forthcoming decoding steps.

An important feature of the proposed CRSSE and DRSSE acquisition schemes is that it exploits the real-time knowledge of the reliabilities associated with the S consecutive chips. By exploiting this real-time knowledge of the chip-reliabilities the acquisition device becomes capable of astutely managing the loading of S consecutive chips into the local m-sequence generator. Both the CRSSE and the DRSSE acquisition schemes have an algorithmic complexity similar to that of an m-sequence generator. Our simulation results show that the acquisition time of both the CRSSE and DRSSE acquisition schemes is a linear function of the number of stages in the m-sequence generator. Therefore, the proposed CRSSE and DRSSE acquisition schemes constitute promising techniques, especially for the acquisition of long m-sequences. Furthermore, the DRSSE acquisition scheme is suitable for m-sequence acquisition, when communicating over various environments including additive white Gaussian noise (AWGN), slow fading and fast fading channels. This is because the differential processing takes place at the chip-level of the m-sequences and during a short chip-interval the fading-induced phase-changes and amplitude-changes are low, even when the Doppler frequency shift is relatively high.

Finally, the maximum-likelihood (ML) soft sequential estimation-based scheme is also investigated, where S consecutive chips are estimated using the maximum-likelihood sequential estimation (MLSE) method [90]. It is shown that the ML soft sequential estimation-based acquisition scheme is equivalent to the conventional parallel search-based acquisition arrangement [507] – [23], which has an implementational complexity that is exponentially related to the number of stages in the m-sequence generator.

The remainder of this section is organized as follows. Sub-section 21.8.2 describes the principle of the sequential estimation acquisition. Sub-section 21.8.3 investigates the ML soft sequential estimation-based acquisition scheme. In sub-section 21.8.4 the proposed CRSSE acquisition scheme is described and investigated. In sub-section 21.8.5 we provide simulation results for the CRSSE acquisition scheme. Subsection 21.8.6 describes and investigates the proposed DRSSE acquisition scheme, while simulation results in the context of the DRSSE are provided in Subection 21.8.7. Finally, in Section 21.8.8 we present our conclusions.

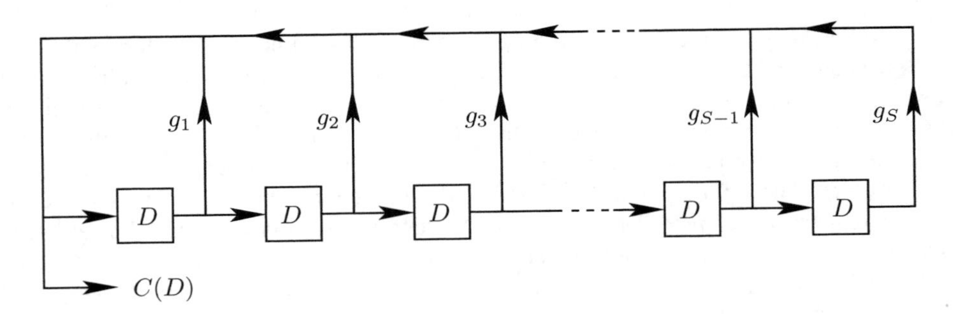

Figure 21.34: Maximal-length sequence (m-sequences) generator that outputs binary sequences.

21.8.2 Principle of Sequential Estimation Acquisition

The well-established maximum-length sequences, which are also known as m-sequences are generated using feedback shift registers of the form shown in Figure 21.34. In Figure 21.34 D represents a unity time-delay operation, while each of the coefficients, $g_1, g_2, \cdots, g_S$, represents the presence of a connection if it is a 1 or the absence of a connection if it is a 0. Since spread-spectrum communication systems usually employ binary spreading sequences having chip values of $\{+1, -1\}$, in this section we assume that the m-sequence generator outputs duo-binary $\{+1, -1\}$ symbols, representing a logical zero with +1. Consequently, the conventional modulo-2 addition defined over the field of $\{1, 0\}$ is now replaced by the modulo-2 multiplication operation defined in the field of $\{+1, -1\}$, as shown in Figure 21.34. More specifically, let the output binary sequence be $\ldots, c_{-2}, c_{-1}, c_0, c_1, c_2, \ldots$, where $c_i \in \{+1, -1\}$. Furthermore, we assume without loss any generality that in Figure 21.34, the coefficients $g_{s_1} = g_{s_2} = \ldots = g_{s_l} = \ldots = g_{s_M=S} = 1$, where s_i is an integer in the range of $[1, S]$, while the remaining coefficients are 0s. The above configuration corresponds to the generator polynomial of

$$g(D) = 1 + D^{s_1} + D^{s_2} + \ldots + D^{s_l} + \ldots + D^{s_M=S}, \qquad (21.146)$$

where $g(D)$ must be a primitive polynomial [14], i.e., a polynomial that cannot be factorized, in order to generate an m-sequence. Based on the above assumptions, it can be shown that the output symbols of Figure 21.34 obey the recursive relationship of

$$\begin{aligned} c_i &= c_{i-s_1} c_{i-s_2} \cdots c_{i-s_l} \cdots c_{i-(s_M=S)} \\ &= \prod_{m=1}^{M} c_{i-s_m} \text{ for } i = \ldots, -1, 0, 1, \ldots, \end{aligned} \qquad (21.147)$$

where $\prod$ represents the product of the coefficients.

In spread-spectrum communications using m-sequences as spreading sequences, Equation 21.147 implies that if the receiver has the knowledge of the chip values $c_{i-1}, c_{i-2}, \cdots, c_{i-S}$ before the transmitter generates the ith chip c_i, and if the receiver uses the same m-sequence generator as the transmitter, then the chip values $c_{i-1}, c_{i-2}, \cdots, c_{i-S}$ can be loaded into the corresponding registers of the m-sequence generator at the receiver for generating the forthcoming chips. Consequently, the corresponding replicas of the ith as well as the forthcoming chips, namely $c_i, c_{i+1}, \ldots$, can be obtained at the receiver, which exactly match the received chips as a result of the transmitted m-sequence. Hence, the despreading of the spread transmitted signal can be successfully carried out by correlating it with the m-sequence replica generated at the receiver. What we have described above constitutes the principle behind the sequential estimation acquisition scheme proposed by Ward [502].

As we discussed in Section 21.8.1, so far all the known sequential estimation acquisition schemes have been based on the hard chip-decisions. Below we propose and investigate the philosophy of acquisition of m-sequences using soft sequential estimation.

21.8.3 Maximum Likelihood Soft Sequential Estimation-Based Acquisition

Let $Z_i = \alpha_i c_i + n_i$ represent the received channel output corresponding to chip c_i, where $i = \ldots, -1, 0, 1, \ldots$. When communicating over a fading channel, α_i denotes the fading amplitude, whereas for an Additive White Gaussian Noise (AWGN) channel we set $\alpha_i = 1$. Furthermore, n_i denotes the AWGN having zero mean and a normalized variance of $N_0/2E_c$, where N_0 represents the single-sided power spectral density of the AWGN, E_c represents the transmitted chip energy and E_c/N_0 represents the signal-to-noise ratio (SNR) per chip. Let S be the number of stages in the m-sequence generator. Then, having received $L \geq S$ number of channel outputs – where the channel outputs are simply expressed as $Z_0, Z_1, \ldots, Z_{L-1}$ – corresponding to the L successive transmitted chips, the S consecutive chips can be estimated using the ML scheme as follows. It is well known that the m-sequence generator employing S stages has a total of $(2^S - 1)$ states, which are determined by the contents of the S number of register stages in the generator.[7] Each of the $(2^S - 1)$ states corresponds to a specific m-sequence phase or delay. The objective of the PN code acquisition process is to find a specific spreading code phase that matches the received sequence, in order to carry out the despreading operation. In other words, the acquisition device must load the local m-sequence generator with a correct initial state constituted by S consecutive chips, so that the locally generated m-sequence replica based on this initial state exactly matches the received m-sequence. This correct initial state can be obtained with the aid of ML soft sequential estimation. More specifically, let $\mathbf{C}_k = \left\{c_{k0}, c_{k1}, \ldots, c_{k(L-1)}\right\}$, $k = 1, 2, \ldots, 2^S - 1$ represent $(2^S - 1)$ number of sequences of length L, which are extracted from the $(2^S - 1)$ number of possible states of the m-sequence generator, where $c_{kl} \in \{+1, -1\}$ denotes the chip value in the lth position of the kth sequence. From the knowledge of the $(2^S - 1)$ possible transmitted phases and upon the reception of the L-chip sequence of $\mathbf{Z} = \{Z_i\}_{i=0}^{L-1}$, the ML soft sequence estimator computes the following $(2^S - 1)$ number of correlation-like metrics

$$CM_k = C(\mathbf{Z}, \mathbf{C}_k) = \sum_{l=0}^{L-1} c_{kl} Z_l, \quad k = 1, 2, \ldots, 2^S - 1. \tag{21.148}$$

[7] Note that the all-one state has to be excluded, since it generates an all-one sequence that is not a valid PN sequence.

The index $\hat{k}$ associated with the maximum of the metrics $\{CM_k\}$ is then selected as the correct code phase estimate and its last S number of hard-decision based chips can then be loaded into the m-sequence generator, in order to generate the correctly aligned local m-sequence replica for despreading the received spread-spectrum signals.

According to Equation 21.148 and [90],[pp. 439-440] the acquisition performance of the ML soft sequential estimation-based acquisition scheme depends on the sequence length L, in addition to the propagation conditions encountered. Since the minimum distance among the $(2^S - 1)$ number of sequences $\mathbf{C}_k = \left\{c_{k0}, c_{k1}, \ldots, c_{k(L-1)}\right\}$, for $k = 1, 2, \ldots, 2^S - 1$ increases, as the length L increases, and finally reaches 2^{S-1}, when $L = 2^S - 1$ [503], the acquisition performance is expected to improve, when we increase the length L.

The above ML soft sequential estimation-based acquisition scheme can be implemented as a parallel bank of $(2^S - 1)$ cross-correlators matched to the $(2^S - 1)$ possible received phases. The outputs of the $(2^S - 1)$ cross-correlators are compared at the end of estimation interval, which encompasses the transmission of L chips in the transmitted sequence and the phase – which represents a specific state of the m-sequence generator – corresponding to the highest cross-correlation output is selected. The corresponding state consisting of S consecutive chips is then loaded into the local m-sequence generator and the forthcoming chips of the local despreading sequence can be generated with the aid of this initial state. Based on these arguments, the ML soft sequential estimation-based acquisition scheme is equivalent to the conventional parallel search-based acquisition scheme [507] – [23]. Hence, both acquisition schemes are capable of achieving a similar acquisition performance, provided that both the schemes use the same correlation interval and communicate over the same propagation environment.

Although the computations involved in forming the correlation metrics according to Equation 21.148 are relatively simple, it may still be impractical to compute Equation 21.148 for all possible spreading code phases, when the number of the generator's stages is high, e.g. $S > 10$. In such a case the ML soft sequential estimation-aided acquisition scheme can still be invoked for estimating the initial state of the m-sequence generator. It may be implemented using complexity-reduction techniques, such as, for example, the Chase algorithm [508], for discarding the low-probability phases without computing their correlation metrics given by Equation 21.148. However, if the number of stages, S, of the m-sequence generator is excessive, the employment of the ML soft sequential estimation-based acquisition scheme may become impractical, even in conjunction with complexity-reduction techniques. In case of the acquisition of long PN sequences generated by m-sequence generators having high stages, we have to find more efficient acquisition techniques. In the following sections, we propose and investigate low-complexity acquisition schemes, which are based on the philosophy of the Recursive Soft Sequential Estimation (CRSSE) concept.

21.8.4 Coherent Recursive Soft Sequential Estimation Acquisition

Since the invention of turbo codes [79, 505], iterative decoding techniques have attracted wide attention [506]. The employment of iterative decoding is facilitated by using multiple encoders generating either parallel or serial concatenated codes, which are decoded with the aid of soft-in-soft-out (SISO) decoding algorithms. In iterative decoding each code is decoded separately, but the soft output arising from one of the decoding stages is used as the soft input of the next decoding stage. In the context of the sequential estimation-based acquisition used for short m-sequences, the generator's state can principally be estimated using iterative decoding techniques, since each m-sequence produced by an S-stage generator has a period of $(2^S - 1)$ chips and can be considered to be a cyclic BCH codeword of length $(2^S - 1)$ having minimum distance of 2^{S-1} [503]. Consequently, after the receiver obtained two sets of $2 \times (2^S - 1)$ number of consecutive samples of the transmitted m-sequence, the m-sequence generator's initial state of S chips can be estimated by iteratively decoding the received m-sequence with the aid of its $2 \times (2^S - 1)$ number of samples. However, in practical spread-spectrum systems typically long PN sequences are employed for ensuring secure communications and for the sake of efficiently randomizing and spreading the transmitted signals. Hence, the iterative decoding of $2 \times (2^S - 1)$ number of chips may be beyond the practical complexity limitations imposed. Therefore, in this section we propose a coherent recursive soft sequential estimation (CRSSE) acquisition scheme, which has a complexity that is only linearly dependent on the number of stages in the m-sequence generator used.

21.8.4.1 Description of the CRSSE

The schematic diagram of the proposed CRSSE acquisition scheme is shown in Figure 21.35, which includes four fundamental building blocks, namely an m-sequence generator, a soft-chip-register, a SISO decoder and a code phase tracking loop. The soft-chip-register has the same S number of delay-units – which we refer to as soft-chip-delay-units (SCDUs) – as the m-sequence generator. The SCDUs store the instantaneous log-likelihood ratio (LLR) values of S consecutive chips. With the aid of these S number of LLR values, S number of consecutive chips can be determined and are loaded into the corresponding delay-units of the m-sequence generator of Figure 21.35. The SISO decoder estimates the corresponding LLR soft output after receiving a soft channel output sample associated with a given chip of the m-sequence. In addition to the so-called intrinsic information of this chip, which was received from the channel, we also exploit the so-called *a-priori* (*extrinsic*) information related to the chip considered, which is provided by the previous decoded LLR values stored in the SCDUs of Figure 21.35. The soft output of the SISO decoder is then shifted to the left-most position of the SCDUs in the soft-chip-register, while the soft value in the right-most SCDUs is shifted out and discarded. Note that both the m-sequence generator and the soft-chip-register use the same feedback branches. However, in the m-sequence generator the feedback elements are duo-binary values and the product of these feedback elements is used for generating a binary feedback quantity. By contrast, the feedback elements from the soft-chip-register to the SISO decoder consist of the LLR values and the specific operations must be employed in the soft-value domain to provide *extrinsic* information for the SISO decoding.

21.8.4.2 CRSSE Acquisition Algorithm

As shown in Figure 21.35, the SISO decoder requires both soft channel output information and *extrinsic* information provided by the previous estimates of the SISO decoder, in order to compute the soft output for updating the contents of the soft-chip-register. In the context of the iterative decoding principle [506], the soft input information is derived from the channel outputs. Let $Z_0, Z_1, Z_2, \ldots$ be the channel outputs, which correspond to the transmitted chips $c_0, c_1, c_2, \ldots$. The soft input information in terms of c_i entered into the SISO decoder is the LLR of c_i conditioned on the channel output Z_i, $i = 0, 1, 2, \ldots$, which is given

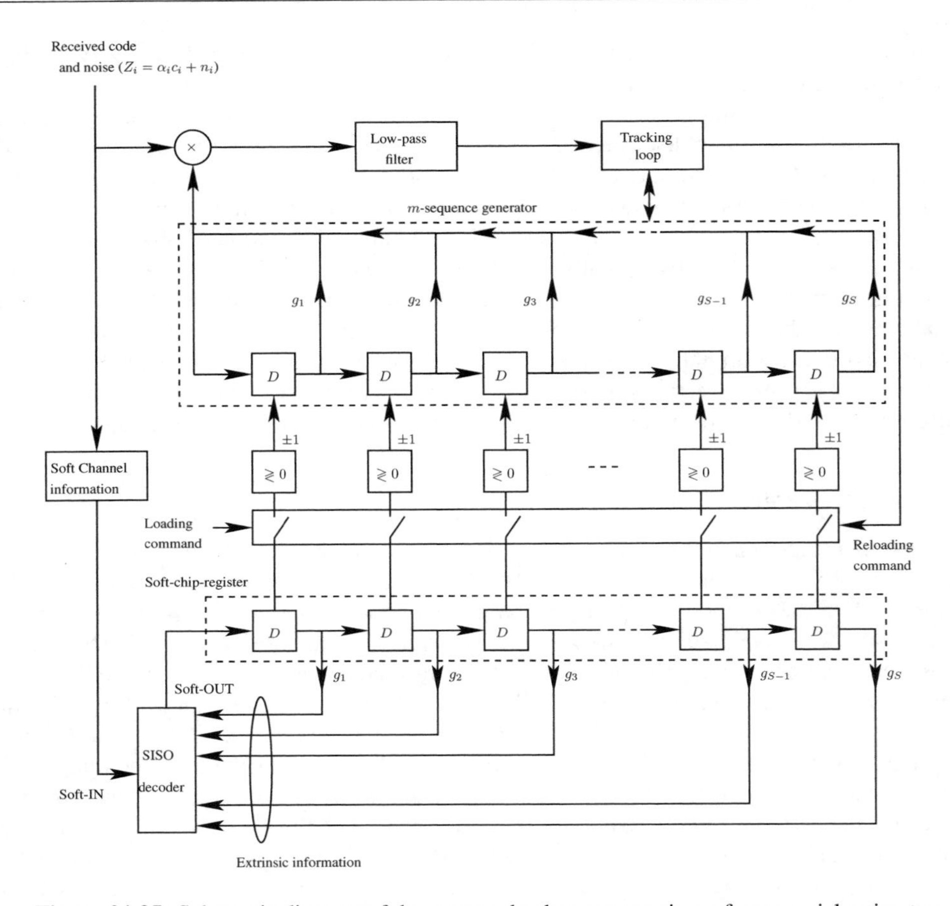

Figure 21.35: Schematic diagram of the proposed coherent recursive soft sequential estimator.

by [506],[Equation (15)]

$$L(c_i|Z_i) = L_c \cdot Z_i + L(c_i), \;\; i = 0, 1, 2, \dots , \tag{21.149}$$

where $L_c = 4\alpha_i \cdot E_c/N_0$, α_i denotes the fading amplitude associated with the transmission of chip c_i over a fading channel, while we have $\alpha_i = 1$ for an AWGN channel. Explicitly, L_c is referred to as the reliability value of the channel. In Equation 21.149 $L(c_i)$ is the LLR of a random variable c_i, which is defined as [506]

$$L(c_i) = \log \frac{P(c_i = +1)}{P(c_i = -1)}, \;\; i = 0, 1, 2, \dots , \tag{21.150}$$

where $L(c_i) = 0$, if we have no *a-priori* information related to c_i, which hence assume the chip value of +1 or -1 with equal probability, i.e. if $P(c_i = +1) = P(c_i = -1)$.

As in Section 21.8.3 we assume that in Figure 21.35 the generator coefficients are given by $g_{s_1} = g_{s_2} = \ldots = g_{s_l} = \ldots = g_{s_M=S} = 1$, while the other coefficients are 0s, i.e. the m-sequence generator obeys a recursive equation described by Equation 21.147. Consequently, the previous soft outputs of the SISO decoder of Figure 21.35 obtained at the time-indices of $(i - s_1), (i - s_2), \ldots, (i - (s_M = S))$ are fed back to the SISO decoder, in order to provide *extrinsic* information for enhancing the correct decoding probability of chip c_i. Let the previous S number of soft outputs of the SISO decoder be $L(y_{i-1}), L(y_{i-2}), \ldots, L(y_{i-S})$. According to Equation 21.147 the *extrinsic* information used for enhancing the correct decoding probability of c_i can be approximately expressed as [506],[Equation (12)]

$$\begin{aligned} L_e(c_i) &\approx \left[\prod_{m=1}^{M} \text{sign}(L(y_{i-s_m}))\right] \\ &\quad \times \min\left\{|L(y_{i-s_1})|, |L(y_{i-s_2})|, \ldots, |L(y_{i-(s_M=S)})|\right\}, \; i = 0, 1, 2, \ldots, \end{aligned} \tag{21.151}$$

where we assumed that we have $L_e(c_{-\infty}) = \ldots = L_e(c_{-2}) = L_e(c_{-1}) = 0$.

Finally, with the aid of the channel output information $L_c \cdot Z_i + L(c_i)$ of Equation 21.149 and the *extrinsic* information $L_e(c_i)$ of Equation 21.151, the soft output of the SISO decoder associated with chip c_i can be expressed as

$$\begin{aligned} L(y_i) &= L(c_i|Z_i) + L_e(c_i) \\ &= L_c \cdot Z_i + L(c_i) + \left[\prod_{m=1}^{M} \text{sign}(L(y_{i-s_m})\right] \\ &\quad \cdot \min\left\{|L(y_{i-s_1})|, |L(y_{i-s_2})|, \ldots, |L(y_{i-(s_M=S)})|\right\}, \; i = 0, 1, 2, \ldots, \end{aligned} \tag{21.152}$$

where, again, we assumed that we have $L_e(c_{-\infty}) = \ldots = L_e(c_{-2}) = L_e(c_{-1}) = 0$. Note that Equation 21.152 represents a recursive equation that can be used for estimating the S number of consecutive chips required by the receiver's m-sequence generator for producing the full m-sequence. Provided that the channel output SNR per chip value of E_c/N_0 is sufficiently high, the LLR values of the S consecutive chips will increase upon increasing the depth of this recursion. In other words, the reliabilities associated with the S consecutive chips increase, while the erroneous loading probability of the generator – which is defined as the probability of the event that the m-sequence generator is loaded with one or more erroneous chips – decreases upon increasing the number of recursions or iterations. Therefore, the acquisition device is capable of observing the reliabilities of the S consecutive chips through observing the amplitudes of the LLR values stored in the SCDUs. If the amplitudes of the LLR values in the SCDUs become sufficiently high after a number of update operation using Equation 21.152 and they result in a sufficiently low erroneous loading probability, then, as shown in Figure 21.35, a 'loading command' can be activated to load the corresponding hard-decision based binary +1 or -1 chip values into the delay-units of the m-sequence generator according to the signs of the corresponding LLR values stored in the SCDUs.

The operation of the CRSSE acquisition scheme can be summarized in the following steps:

1. All the soft-chip-delay-units (SCDUs) are initialized to zeros.

2. Whenever the SISO decoder receives a channel output sample Z_i corresponding to the chip c_i, the SISO decoder computes the LLR of c_i conditioned on Z_i using Equation 21.149 and computes the *extrinsic* information of $L_e(c_i)$ using Equation 21.151. Finally, the soft output $L(y_i)$ of the SISO decoder related to chip c_i is computed according to Equation 21.152.

3. The soft output $L(y_i)$ is then shifted into the left-most SCDU of Figure 21.35 after all the other soft outputs $L(y_{i-1}), L(y_{i-2}), L(y_{i-S+1})$ have been shifted to the right by one position, while $L(y_{i-S})$ is removed from the soft-chip-register. In other words, the soft-chip-register always stores the most recent S number of soft outputs of the SISO decoder, which correspond to S consecutive chips of the transmitted m-sequence.

4. Following a number of recursions according to Equation 21.152, when the amplitudes of the most recent S number of soft outputs of the SISO decoder become sufficiently high for guaranteeing a sufficiently low erroneous loading probability, a 'loading command' is activated. Then, S consecutive chips are determined using hard-decisions based on the most recent S LLR values stored in the soft-chip-register of Figure 21.35. Then the S consecutive binary chips are loaded into the corresponding delay-units of the local m-sequence generator.

5. Once the m-sequence generator is loaded with the initial binary chip values, the received spread-spectrum signal can be despread using the locally generated m-sequence replica, provided that the initial chip values of the m-sequence generator have been correctly loaded. The despread signal is then low-pass filtered and is then sent to the code tracking loop. If the code tracking loop is capable of tracking the phase, the code acquisition process is completed. By contrast, if the tracking loop is incapable of tracking the phase, the code tracking loop activates a 'reloading command', in order to load another group of S consecutive chips into the delay-units of the m-sequence generator. The above process can be repeated until successful code tracking is accomplished.

Note that since the acquisition device is capable of observing the reliabilities of the most recent S consecutive chips through observing the amplitudes of the corresponding soft outputs stored in the soft-chip-register of Figure 21.35, the acquisition scheme can decide, when it should activate the 'loading command' for loading the initial chips into the m-sequence generator. When the erroneous loading probability is typically deemed sufficiently low in the region of for example 10^{-4}, the PN sequence can be acquired with a high probability after the first loading of the initial chips. Therefore, the total acquisition time of the CRSSE acquisition scheme can be approximated by the time duration required by the CRSSE for carrying out the recursive SISO decoding, in order to reach a sufficiently low erroneous loading probability.

21.8.4.3 Complexity

In this subsection we compare the complexity of the conventional as well as the proposed CRSSE acquisition schemes, in order to demonstrate the advantages of the CRSSE acquisition scheme. It is widely recognized that the conventional serial search-based acquisition scheme [15, 28] – [509] is one of the simplest PN code acquisition schemes. In a serial search scheme each possible code phase associated with a given raster position in the uncertainty region is tested one at a time. The mean-acquisition time of a serial search-based acquisition scheme is a function of the period of the PN sequences [15, 28, 509]. In the context of m-sequences the mean acquisition increases exponentially with the number of stages in the m-sequence generator. By contrast, conventional parallel search-based acquisition schemes [507] – [23] test all the possible code phases simultaneously by using a bank of parallel correlators or matched filters having the same number of possible code phases in the uncertainty region of the PN code. Hence, conventional parallel search-based acquisition schemes are capable of rapidly establishing initial synchronization. However, in the context of m-sequences conventional parallel search-based acquisition schemes impose an implementational complexity that is linearly (or exponentially) dependent on the period (or the number of stages) of the m-sequences. There is also a third type of acquisition scheme referred to as hybrid acquisition scheme [25, 31], which relies on a trade-off between the conventional serial search and parallel search-based acquisition schemes. It was suggested in [25, 31] that these hybrid acquisition schemes may achieve initial synchronization within a shorter time than serial search-based acquisition schemes, although at the cost of imposing a higher complexity than the family of serial search-based acquisition schemes. By contrast, hybrid schemes are expected to impose a lower hardware complexity than the class of parallel search-based acquisition schemes.

If we consider the complexity of initial synchronization and its relationship with the related mean-acquisition time, then all the above-mentioned three types of initial synchronization schemes have a complexity dependent on the period of the related PN sequences. In the context of m-sequences, the complexity of these initial synchronization schemes increases exponentially with the number of stages in the m-sequence generators, since the period of an m-sequence is an exponential function of the number of stages in its generator. Therefore, conventional initial acquisition schemes are unsuitable for the acquisition of long PN sequences. These long PN sequences have so far been acquired with the aid of short constituent sequences of the long PN sequence [510]. Once all short constituent sequences have been acquired, the long sequence's phase can be estimated by exploiting the concatenated relationship of its short constituent sequences.

As shown in Figure 21.35, the implementation of the proposed CRSSE acquisition scheme requires a soft-chip-register having the same length as the m-sequence generator and a simple SISO decoder. We have argued in the previous subsection that the acquisition time of the CRSSE acquisition scheme is determined by the time that the CRSSE requires for carrying out the recursive SISO decoding, so that the erroneous loading probability reaches a sufficiently low value. According to the iterative decoding principle [79, 506] and to our results to be presented in Section 21.8.5, it can be shown that for a reasonable SNR per chip value the time duration required by the CRSSE for carrying out recursive SISO decoding is linearly dependent on the number of stages in the related m-sequence generator. Therefore, the proposed CRSSE acquisition scheme is suitable for the acquisition of long m-sequences. Furthermore, in the context of the initial synchronization of long m-sequences, the proposed CRSSE acquisition scheme has a significantly lower complexity than the family of conventional acquisition schemes. Let us now turn our attention to providing performance.

21.8.5 Performance Results of CRSSE Acquisition Scheme

In this section we provide a range of simulation results in the context of the proposed CRSSE acquisition scheme. Our simulation results were mainly based on the acquisition of two different-length m-sequences having $S = 5$ and $S = 13$ stages, which correspond to having periods of $N = 2^5 - 1 = 31$ and $N = 2^{13} - 1 = 8191$, respectively. In the context of the m-sequence associated with $S = 5$ and $N = 31$, the generator polynomial used was $g(D) = 1 + D^2 + D^5$ or $g(D) = 1 + D + D^2 + D^4 + D^5$. By contrast, when the m-sequence associated with $S = 13$ and $N = 8191$ was considered, the generator polynomial used was $g(D) = 1 + D + D^3 + D^4 + D^{13}$. Note that the curves were drawn either versus the signal-to-noise ratio (SNR) per chip, namely E_c/N_0 or versus the normalized number of chips received, which also represents the normalized number of chips that the SISO decoder processed. The normalized number of chips L/S was defined as the total number of received chips, L, divided by the number of generator stages, S, of the corresponding m-sequence generator.

Figure 21.36 shows the reliability associated with correctly deciding the polarity of a specific chip of an m-sequence, when the m-sequence was transmitted over AWGN channels. The m-sequence was generated by the generator polynomial of $g(D) = 1 + D^2 + D^5$ using an $S = 5$-stage generator, or by a generator polynomial of $g(D) = 1 + D + D^3 + D^4 + D^{13}$ employing an $S = 13$-stage generator. The decision reliability associated with using SISO decoding was defined as the absolute value of the SISO decoder's output evaluated according to Equation 21.152 in the context of each chip received. By contrast, for conventional hard-decision the decision reliability associated with the decision concerning the polarity of a specific chip was defined as the absolute value of the channel output corresponding to that chip. From the simulation results of Figure 21.36 we observe that for the conventional hard-decision based scheme the decision reliability associated with a chip is mainly distributed within the range spanning from 0 to 10 and does not increase as more chips received, since the polarity of each chip is decided separately. By contrast, for the proposed CRSSE acquisition scheme the correct decision reliability increases, when receiving more chips from the channel. Furthermore, we observe that the average correct decision reliability increases nearly linearly upon increasing the normalized number of chips received. This is because the proposed recursive SISO decoder is capable of efficiently exploiting the *a-priori* information provided by the previous chips received. Therefore, according to Figure 21.36 we expect that the CRSSE acquisition scheme will outperform the conventional sequential estimation acquisition scheme [502] without exploiting the *a-priori* information provided by the previous chips. The following results will support this argument.

In Figure 21.37 we show the erroneous loading probability, P_e, versus the SNR per chip, E_c/N_0 performance for an m-sequence generated using the generator polynomial of $g(D) = 1 + D^2 + D^5$ and transmitted over AWGN channels. Note that in Figure 21.37 the curve corresponding to the parameter of $L = 1S$ represents the erroneous loading probability of the conventional sequential estimation acquisition scheme [502]. From the results we can see that when more channel output chips are involved in the recursive SISO decoding process, a higher correct detection reliability and hence a lower erroneous loading probability can be achieved. Let us assume that the transmitted m-sequence can reliably be acquired, once the erroneous loading probability is lower than 10^{-4}. Then, from the results of Figure 21.37 we can observe that the m-sequence can reliably be acquired at an SNR per chip

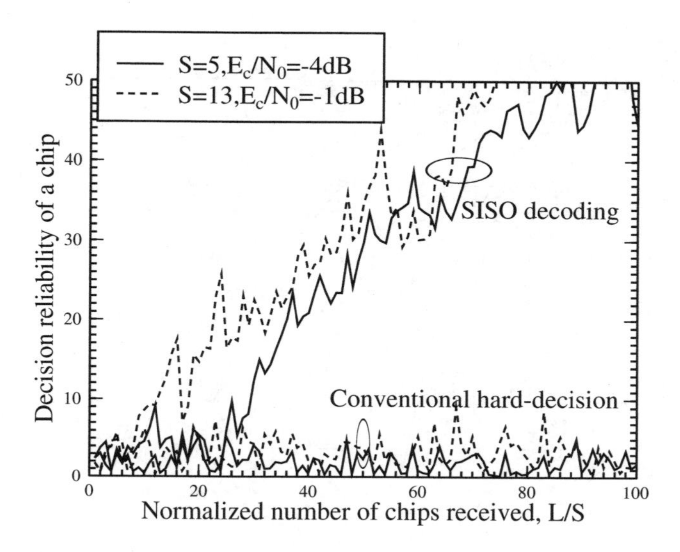

Figure 21.36: Decision reliability versus the normalized number of received chips invoked in the recursive SISO decoding, when communicating over an AWGN channel and using an SNR per chip value of $E_c/N_0 = -4$dB for $S = 5$ and $E_c/N_0 = -1$dB for $S = 13$.

value of $E_c/N_0 = -2.8$dB by invoking about $L = 40S = 40 \times 5 = 200$ chips into the recursive SISO decoding scheme. By contrast, without exploiting any *a-priori* information the conventional sequential estimation acquisition scheme [502] has to operate at an SNR per chip value of $E_c/N_0 = 9.2$dB, in order to achieve the same erroneous loading probability of 10^{-4}. At the erroneous loading probability of 10^{-4} the SNR per chip gain of using the proposed CRSSE acquisition scheme instead of the conventional sequential estimation acquisition scheme relying on hard-decisions is about 12dB, when $L = 200$ chips are invoked into the recursive SISO decoding scheme. In general, it is expected that when more chips can be used during the recursive SISO decoding process, an increasing gain can be achieved by the proposed CRSSE acquisition scheme in comparison to the conventional sequential estimation acquisition scheme.

It is widely recognized that different m-sequences having the same period can be generated using various generator polynomials. Figure 21.38 demonstrates the effect of using various generator polynomials on the erroneous loading probability of the m-sequences having the period of $N = 31$, when communicating over AWGN channels. The generator polynomials used for generating the m-sequences of period $N = 31$ were $g(D) = 1 + D^2 + D^5$ and $g(D) = 1 + D + D^2 + D^4 + D^5$, respectively. Explicitly, the generator using $g(D) = 1 + D + D^2 + D^4 + D^5$ has more feedback branches than that using $g(D) = 1 + D^2 + D^5$. According to the results of Figure 21.38, it can be observed that given a fixed number of chips invoked during the recursive SISO decoding process, the generator using $g(D) = 1+D^2+D^5$ outperforms that using $g(D) = 1 + D + D^2 + D^4 + D^5$. The reason for these results is that

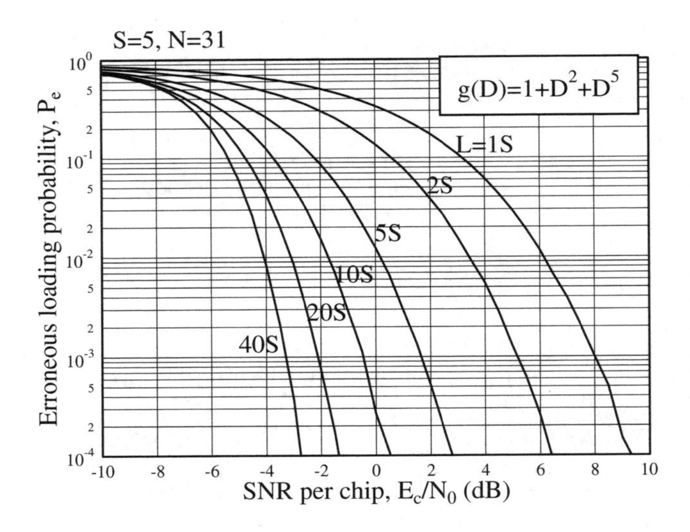

Figure 21.37: Erroneous loading probability, P_e, versus the SNR/chip, E_c/N_0, performance for various numbers of chips invoked into the proposed recursive SISO decoder, when transmitting the m-sequence generated using the generator polynomial of $g(D) = 1 + D^2 + D^5$ over AWGN channels.

in the context of the generator using $g(D) = 1 + D^2 + D^5$, the *extrinsic* information is only based on two feedback branches. By contrast, in the context of the generator using $g(D) = 1 + D + D^2 + D^4 + D^5$, the *extrinsic* information is based on four feedback branches. However, according to Equation 21.151 the *extrinsic* information is determined by the minimum of the LLR values of all the feedback branches. Hence, the generator using a low number of feedback branches has a higher probability of providing high-valued, i.e. reliable *extrinsic* information to the SISO decoder than that using a high number of feedback branches. Nevertheless, the results of Figure 21.38 show that regardless of which generator polynomial is used, the proposed CRSSE acquisition scheme has a significantly lower erroneous loading probability than the conventional sequential estimation acquisition scheme, provided that a sufficiently high number of received chips are invoked in the recursive SISO decoding process.

Above the m-sequences considered were generated by five-stage m-sequence generators and their period was 31 chips. By contrast, Figure 21.39 shows the acquisition performance of an m-sequence having a period of $N = 8191$ chips, which was generated by a thirteen-stage ($S = 13$) generator using the generator polynomial of $g(D) = 1 + D + D^3 + D^4 + D^{13}$. As shown in Figure 21.39, the m-sequence can be reliably acquired at an SNR per chip value of $E_c/N_0 = -0.5$dB by invoking about $L = 40S = 40 \times 13 = 520$ chips into the recursive SISO decoder. By contrast, without using any *a-priori* information, i.e., using the conventional sequential estimation acquisition scheme [502], the PN code acquisition scheme

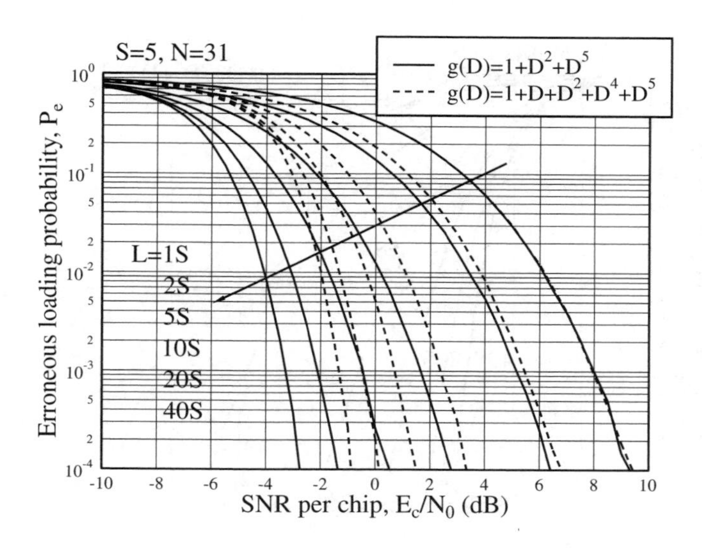

Figure 21.38: Comparison of erroneous loading probabilities for the m-sequences having the same length of $N = 31$, but generated by the generating functions of $g(D) = 1 + D^2 + D^5$ and $g(D) = 1 + D + D^2 + D^4 + D^5$, respectively, when communicating over AWGN channels.

has to operate at the SNR per chip value of $E_c/N_0 = 9.5$dB, in order to achieve the erroneous loading probability of 10^{-4}. Hence, the SNR per chip gain of using the proposed CRSSE acquisition scheme instead of using the conventional sequential estimation acquisition scheme at the erroneous loading probability of 10^{-4} is about 10dB, when $L = 520$ chips are invoked in the recursive SISO decoder. Let τ_D represent the integration dwell time in the context of a conventional serial search-based acquisition scheme [28], which typically assumes values from tens to hundreds of chip intervals. In [28] it is shown that the mean-acquisition time of a conventional serial search-based acquisition scheme cannot be lower than $N\tau_D/2 = 4095\tau_D >> 520$ chips for a sequence of length 8191, even if the SNR per chip value is sufficiently high, when the corresponding detection probability approaches one. Hence, the proposed CRSSE acquisition scheme significantly outperforms the conventional serial search-based acquisition scheme in terms of the achievable acquisition time performance.

The results in the above figures were evaluated when communicating over AWGN channels. In Figures 21.40 and 21.41 we investigated the acquisition performance of the CRSSE acquisition scheme for transmission over Rayleigh fading channels. In Figure 21.40 the m-sequence was generated by a five-stage m-sequence generator using the generator polynomial $g(D) = 1 + D^2 + D^5$. By contrast, the m-sequence associated with Figure 21.41 was generated by a thirteen-stage m-sequence generator using the generator polynomial of $g(D) = 1 + D + D^3 + D^4 + D^{13}$. From the results of Figures 21.40 and 21.41, we ob-

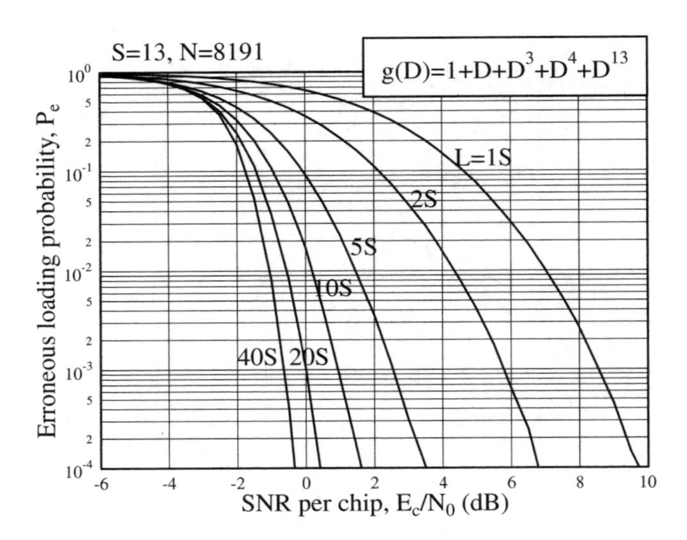

Figure 21.39: Erroneous loading probability, P_e, versus the SNR/chip, E_c/N_0, performance for various numbers of chips invoked in the proposed recursive SISO decoder, when transmitting the m-sequence generated using the generator polynomial of $g(D) = 1 + D + D^3 + D^4 + D^{13}$ over AWGN channels.

serve that at a given SNR per chip value of E_c/N_0, similarly to the results of Figures 21.37 and 21.39, the erroneous loading probability decreases, when increasing the number of received chips invoked in the recursive SISO decoder. Furthermore, by comparing the results of Figures 21.40 and 21.41 to those of Figures 21.37 and 21.39 valid for communicating over AWGN channels, it can be shown that for a given number of received chips invoked in the recursive SISO decoder, the achievable SNR gain over Rayleigh fading channels is significantly higher than that over AWGN channels, when using the proposed CRSSE acquisition scheme instead of the conventional sequential estimation acquisition scheme. However, for a given number of received chips used by the recursive SISO decoder and for a given SNR per chip value, the acquisition scheme communicating over AWGN channels achieves a lower erroneous loading probability than that over Rayleigh fading channels.

Finally, in Figure 21.42 we show the erroneous loading probability performance versus the normalized number of received chips used by the recursive SISO decoder for an m-sequence generated by an $S = 13$-stage m-sequence generator and hence having a period of 8191 chips. The results were generated for transmissions over both AWGN and Rayleigh fading channels at the SNR per chip value of $E_c/N_0 = -1$dB. According to the results of Figure 21.42 it can be shown that for AWGN channels the CRSSE acquisition scheme is capable of achieving an erroneous loading probability of 10^{-3} after receiving approximately $80 \times 13 = 1040$ chips. By contrast, when communicating over Rayleigh fading channels, the CRSSE acquisition scheme is capable of achieving the erroneous loading probability of

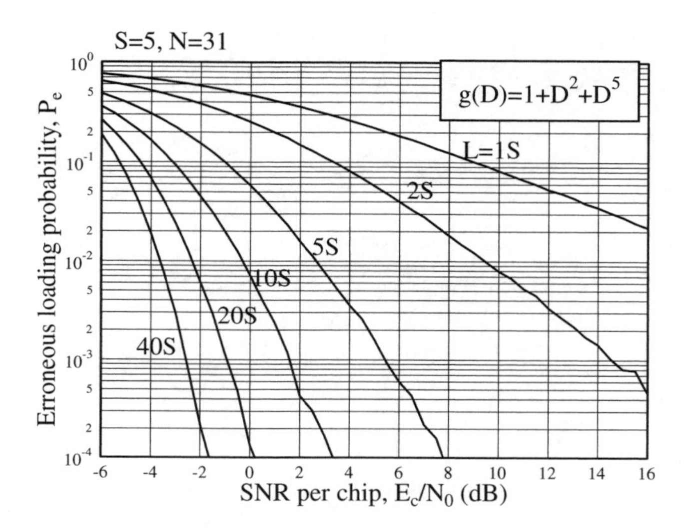

Figure 21.40: Erroneous loading probability, P_e, versus the SNR/chip, E_c/N_0, performance for various numbers of chips invoked in the proposed recursive SISO decoder, when transmitting the m-sequence generated using the generator polynomial of $g(D) = 1 + D^2 + D^5$ over Rayleigh fading channels.

10^{-3} after receiving approximately $500 \times 13 = 6500$ chips. These results imply that with the aid of the proposed recursive SISO decoder the required target performance can be achieved, regardless of the specific communication environment encountered by invoking a corresponding number of chips in the SISO decoder. The results of Figure 21.42 indicate that successful acquisition can be achieved by the proposed CRSSE acquisition scheme, when communicating over Rayleigh fading channels at $E_c/N_0 = -1$dB within about 6500 chips, a value which is still significantly lower than the required $4095\tau_D$ chip duration of the best mean-acquisition time associated with the convention serial search-based acquisition scheme [28].

21.8.6 Differential Recursive Soft Sequential Estimation Acquisition

The motivation of employing a chip-based differential pre-processing operation is two-fold. Firstly, prior to despreading the signal-to-noise ratio (SNR) is usually insufficiently high for attaining a satisfactory performance with the aid of coherent carrier phase estimators based on carrier-phase tracking loops. By contrast, the chip-based differential operation is capable of reducing the effects of the time-varying phase fluctuations imposed by fading and frequency offset. Consequently, chip-based differential pre-processing is capable of significantly enhancing the performance of the following sequential estimation process. Secondly, from the shift-and-add property of m-sequences [42], the product of the spreading waveform and its chip-time delayed phase yields another phase. This indicates that the differential processing

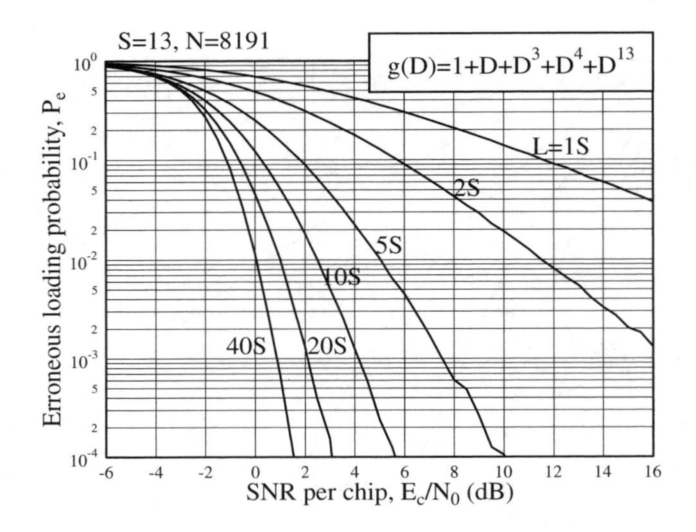

Figure 21.41: Erroneous loading probability, P_e, versus the SNR/chip, E_c/N_0, performance for various numbers of chips invoked in the proposed recursive SISO decoder, when transmitting the m-sequence generated using the generator polynomial of $g(D) = 1 + D + D^3 + D^4 + D^{13}$ over Rayleigh fading channels.

of an m-sequence preserves the characteristics of the m-sequences. Specifically, this property can be demonstrated as follows.

The recursive equation of the m-sequence generator using the generator polynomial of Equation 21.146 is given by Equation 21.147. We multiply both sides of Equation 21.147 by c_{i-1} and upon substituting c_{i-1} at the right-hand side by $c_{i-1} = \prod_{n=1}^{M} c_{i-1-s_n}$, we obtain

$$\begin{aligned} c_i \cdot c_{i-1} &= \prod_{m=1}^{M} c_{i-s_m} \cdot \prod_{n=1}^{M} c_{i-1-s_n} \\ &= \prod_{m=1}^{M} (c_{i-s_m} \cdot c_{i-1-s_m}) \text{ for } i = \ldots, -1, 0, 1, \ldots. \end{aligned} \tag{21.153}$$

Let $b_j = c_j \cdot c_{j-1}$ in Equation 21.153, where j is an integer. Then Equation 21.153 can be expressed as

$$b_i = \prod_{m=1}^{M} b_{i-s_m} \text{ for } i = \ldots, -1, 0, 1, \ldots. \tag{21.154}$$

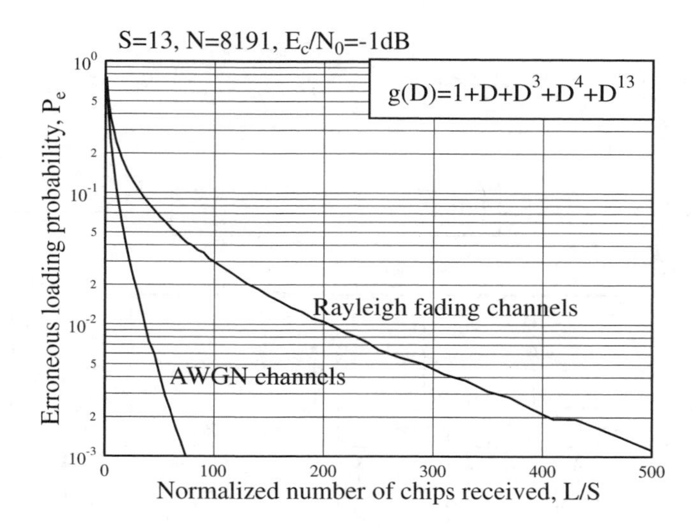

Figure 21.42: Erroneous loading probability, P_e versus the normalized number of received chips invoked in the proposed recursive SISO decoder for an m-sequence having a period of 8191, when communicating over AWGN or Rayleigh fading channels and using an SNR per chip value of $E_c/N_0 = -1$dB for $S = 13$.

Explicitly, both recursive equations, namely Equations 21.147 and 21.154 describe the same m-sequence generator. The m-sequence generated by Equation 21.154 represents a specifically delayed or phase-shifted version of the m-sequence generated by Equation 21.147. Consequently, once the m-sequence of Equation 21.154 has been acquired, the acquisition of the m-sequence generated by Equation 21.147 can also be achieved. The objective of the DRSSE is to achieve the acquisition of the m-sequence generated by Equation 21.154 using the proposed soft recursive sequential estimation scheme.

21.8.6.1 Description of the DRSSE Scheme

The schematic diagram of the proposed DRSSE acquisition arrangement is shown in Figure 21.43, which includes five fundamental building blocks, namely a chip-based differential processor, an m-sequence generator, a soft-chip-register, a SISO decoder and a code phase tracking loop. The chip-based differential processor executes differential processings at the chip level, which requires that the carrier phases between two adjacent chips remain similar. Hence, the DRSSE acquisition scheme is suitable for communicating over fading channels exhibiting a high fading rate. The soft-chip-register of Figure 21.43 has the same S number of delay-units – which we refer to as soft-chip-delay-units (SCDUs) – as the m-sequence generator. The SCDUs store the instantaneous log-likelihood ratio (LLR) values of S consecutive chips of the m-sequence in the form of Equation 21.154. With the aid of these S number

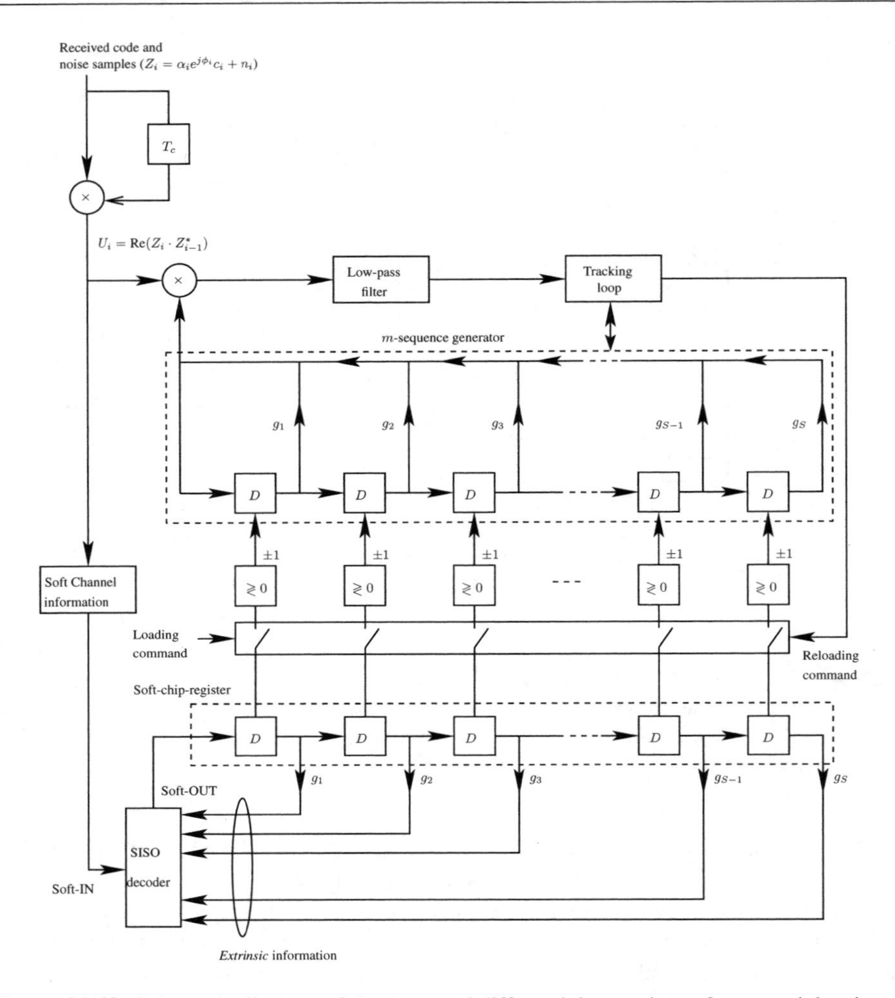

Figure 21.43: Schematic diagram of the proposed differential recursive soft sequential estimator.

of LLR values, S number of consecutive chips can be determined, which are loaded into the corresponding delay-units of the m-sequence generator of Figure 21.43. The SISO decoder estimates the corresponding LLR soft output after receiving a soft output sample from the differential processor associated with a given chip of the m-sequence of Equation 21.154. In addition to the so-called *intrinsic* information of this chip, which characterizes its the reliability at the output of the channel, we also exploit the so-called *a-priori* (*extrinsic*) information related to the chip considered. This *extrinsic* information is provided by the previously decoded LLR values stored in the SCDUs of Figure 21.43. The soft output of the SISO decoder is then shifted to the left-most position of the SCDUs in the soft-chip-register, while the soft value in the right-most position of the SCDUs is shifted out and discarded. Note that both the m-sequence generator and the soft-chip-register use the same feedback branches. However, in the m-sequence generator the feedback elements are duo-binary values and the product of these feedback elements is used for generating a binary feedback quantity. By contrast, the feedback elements of the soft-chip-register feeding information back to the input of the SISO decoder consist of the LLR values. Hence, the specific operations must be carried out in the soft-value domain for the sake of providing *extrinsic* information for the SISO decoding.

21.8.6.2 DRSSE Acquisition Algorithm

As shown in Figure 21.43, the SISO decoder requires both *intrinsic* information conveyed to the input of the SISO decoder by the differential processor and *extrinsic* information provided by the previous estimates of the SISO decoder, in order to generate the soft output required for updating the contents of the soft-chip-register. Let us now consider these sources of information. Let $Z_i = \alpha_i e^{j\phi_i} c_i + n_i$ represent the received channel output sample corresponding to chip c_i, where $i = 0, 1, 2, \ldots$. The parameter α_i denotes the fading amplitude, while ϕ_i represents the phase shift due to carrier modulation and channel fading. Note that when communicating over Additive White Gaussian Noise (AWGN) channels, the term of $\alpha_i e^{j\phi_i}$ is set to one. Furthermore, n_i denotes the complex AWGN having zero mean and a normalized variance of N_0/E_c, where N_0 represents the single-sided power spectral density of the AWGN, E_c represents the transmitted chip energy and E_c/N_0 represents the signal-to-noise ratio (SNR) per chip. Assuming that the channel-induced fading and phase-rotation remain constant over two adjacent chip time-durations, i.e., that we have $\alpha_i = \alpha_{i-1}$ and $\phi_i = \phi_{i-1}$, the differential processor's output can be expressed as:

$$\begin{aligned} U_i &= \mathrm{Re}(Z_i \cdot Z_{i-1}^*) \\ &= \alpha_i^2 c_i \cdot c_{i-1} + \mathrm{Re}(\alpha_i e^{j\phi_i} c_i n_{i-1}^* + \alpha_i e^{-j\phi_i} c_{i-1} n_i + n_i n_{i-1}^*), \end{aligned} \tag{21.155}$$

where * represents the complex conjugate operation. Using $b_i = c_i \cdot c_{i-1}$ and, without changing their statistical properties, absorbing the terms of $e^{j\phi_i} c_i$ and $e^{-j\phi_i} c_{i-1}$ into the Gaussian noise components n_{i-1}^* and n_i, the above equation can be written as

$$U_i = \alpha_i^2 b_i + \alpha_i \cdot \mathrm{Re}(n_{i-1}^* + n_i) + \mathrm{Re}(n_i n_{i-1}^*). \tag{21.156}$$

It was shown in [90] that at the SNRs of practical interest the term $n_i n_{i-1}^*$ is small relative to the dominant noise term of $\alpha_i(n_{i-1}^* + n_i)$. Consequently, if we neglect the term $n_i n_{i-1}^*$, then U_i can be described as a Gaussian variable having mean given by $\alpha_i^2 b_i$ and variance given by $\Omega N_0/E_c$, where $\Omega = \overline{\alpha_i^2}$ represents the ensemble average of α_i^2. Hence, the probability density function (PDF) of U_i can be approximated as

$$p_{U_i}(y|b_i) \approx \frac{1}{\sqrt{2\pi\Omega N_0/E_c}} \exp\left(-\frac{E_c}{2\Omega N_0}(y - \alpha_i^2 b_i)^2\right). \tag{21.157}$$

The *intrinsic* information is derived from the LLR of b_i conditioned on the channel-related variable U_i, which can be expressed as

$$\begin{aligned} L(b_i|U_i) &= \log \frac{p_{U_i}(b_i = +1|y = U_i)}{p_{U_i}(b_i = -1|y = U_i)} \\ &= \log \left(\frac{p_{U_i}(y = U_i|b_i = +1)}{p_{U_i}(y = U_i|b_i = -1)} \cdot \frac{P(b_i = +1)}{P(b_i = -1)} \right). \end{aligned} \quad (21.158)$$

Upon substituting $p_{U_i}(y = U_i|b_i)$ of Equation 21.157 associated with $b_i = +1$ and $b_i = -1$ into Equation 21.158, finally, we obtain

$$\begin{aligned} L(b_i|U_i) &= \log \frac{\exp\left(-\frac{E_c}{2\Omega N_0}(U_i - \alpha_i^2)^2\right)}{\exp\left(-\frac{E_c}{2\Omega N_0}(U_i + \alpha_i^2)^2\right)} + \log \frac{P(b_i = +1)}{P(b_i = -1)} \\ &= L_c \cdot U_i + L(b_i), \quad i = 0, 1, 2, \ldots, \end{aligned} \quad (21.159)$$

where we have $L_c = 2\alpha_i^2 E_c/\Omega N_0$, which is referred to as the reliability value of the channel, while $L(b_i)$ is the LLR of a random variable b_i, which is defined as [506] $L(b_i) = \log \frac{P(b_i = +1)}{P(b_i = -1)}$. Furthermore, we set $L(b_i) = 0$, if we have no *a-priori* information related to b_i, which corresponds to assuming a chip value of +1 or -1 with equal probability, i.e., what we have if $P(b_i = +1) = P(b_i = -1)$.

Note that at the initial synchronization stage the estimation of the channel parameters might be unreliable. If we have no prior knowledge concerning the channel parameters, we can simply set $L_c = 2E_c/N_0$. However, our simulation results in Section 21.8.7 will show that the acquisition performance degrades in the absence of perfect knowledge of the channel. Furthermore, for transmission over AWGN channels the reliability value of the channel is set to $L_c = 2E_c/N_0$.

In the context of the *extrinsic* information, as in Section 21.8.2, we assume that in Figure 21.43 the generator coefficients are given by $g_{s_1} = g_{s_2} = \ldots = g_{s_l} = \ldots = g_{s_M=S} = 1$, while the other coefficients are 0s, i.e., the m-sequence generator obeys the recursive equation Equation 21.147 or Equation 21.154. Consequently, the previous soft outputs of the SISO decoder of Figure 21.43 obtained at the time instants of $(i - s_1), (i - s_2), \ldots, (i - (s_M = S))$ are fed back to the input of the SISO decoder, in order to provide *extrinsic* information for enhancing the correct decoding probability of chip b_i. Let the previous S number of soft outputs of the SISO decoder be $L(y_{i-1}), L(y_{i-2}), \ldots, L(y_{i-S})$. According to Equation 21.147 the *extrinsic* information used for enhancing the correct decoding probability of b_i can approximately be expressed as [506][Equation (12)]:

$$\begin{aligned} L_e(b_i) \approx & \left[\prod_{m=1}^{M} \text{sign}(L(y_{i-s_m})) \right] \\ & \times \min \left\{ |L(y_{i-s_1})|, |L(y_{i-s_2})|, \ldots, |L(y_{i-(s_M=S)})| \right\}, \ i = 0, 1, 2, \ldots, \end{aligned} \quad (21.160)$$

where we assumed that we have $L_e(b_{-\infty}) = \ldots = L_e(b_{-2}) = L_e(b_{-1}) = 0$.

Finally, with the aid of the *intrinsic* information $L_c \cdot U_i + L(b_i)$ of Equation 21.159 and the *extrinsic* information $L_e(b_i)$ of Equation 21.160, the soft output of the SISO decoder associated with chip b_i can be expressed as

$$\begin{aligned} L(y_i) &= L(b_i|Z_i) + L_e(b_i) \\ &\approx L_c \cdot U_i + L(b_i) + \left[\prod_{m=1}^{M} \text{sign}(L(y_{i-s_m})\right] \\ &\quad \cdot \min\left\{|L(y_{i-s_1})|, |L(y_{i-s_2})|, \ldots, |L(y_{i-(s_M=S)})|\right\}, \; i = 0, 1, 2, \ldots, \end{aligned} \tag{21.161}$$

where, again, we assumed that we have $L_e(b_{-\infty}) = \ldots = L_e(b_{-2}) = L_e(b_{-1}) = 0$. Note that Equation 21.161 represents a recursive equation that can be used for estimating the S number of consecutive chips required by the receiver's m-sequence generator for producing the full m-sequence of Equation 21.154. Provided that the channel's output SNR per chip value of E_c/N_0 is sufficiently high, the LLR values of the S consecutive chips will increase upon increasing the depth of this recursion. In other words, the reliabilities associated with the S consecutive chips increase, while the erroneous loading probability of the m-sequence generator – which is defined as the probability of the event that the m-sequence generator is loaded with one or more erroneous chips – decreases upon increasing the number of recursions or iterations. Therefore, the acquisition device is capable of observing the reliabilities of the S consecutive chips through observing the amplitudes of the LLR values stored in the SCDUs. If the amplitudes of the LLR values in the SCDUs become sufficiently high after a number of update operations using Equation 21.161 and they result in a sufficiently low erroneous loading probability, then, as shown in Figure 21.43, a 'loading command' can be activated for loading the corresponding hard-decision based binary +1 or -1 chip values into the delay-units of the m-sequence generator according to the signs of the corresponding LLR values stored in the SCDUs.

The operation of the DRSSE acquisition scheme can be summarized in the following steps:

1. All the soft-chip-delay-units (SCDUs) are initialized to zero. The delay-unit in the differential processor is initialized to one.

2. Whenever the SISO decoder receives a sample U_i corresponding to the chip b_i from the differential processor, the SISO decoder computes the LLR of b_i conditioned on U_i using Equation 21.159 and computes the *extrinsic* information of $L_e(b_i)$ using Equation 21.160. Finally, the soft output $L(y_i)$ of the SISO decoder, which is related to chip b_i is computed according to Equation 21.161.

3. The soft output $L(y_i)$ is then shifted into the left-most SCDU of Figure 21.43 after all the other soft outputs $L(y_{i-1}), L(y_{i-2}), L(y_{i-S+1})$ have been shifted to the right by one position, while $L(y_{i-S})$ is removed from the soft-chip-register. In other words, the soft-chip-register always stores the most recent S number of soft outputs of the SISO decoder, which correspond to S consecutive chips of the m-sequence at the output of the differential processor.

4. Following a number of recursions according to Equation 21.161, when the amplitudes of the most recent S number of soft outputs of the SISO decoder become sufficiently high for guaranteeing a sufficiently low erroneous loading probability, a 'loading command' is activated. Then, S consecutive chips are determined using hard-decisions based on the most recent S LLR values stored in the soft-chip-register of Figure 21.43. Next, the S consecutive binary chips are loaded into the corresponding delay-units of the local m-sequence generator.

5. Once the m-sequence generator is loaded with the initial binary chip values, the received differentially processed spread-spectrum signal can be despread using the locally generated m-sequence replica, provided that the initial chip values of the m-sequence generator have been correctly loaded. The despread signal is then low-pass filtered and forwarded to the code tracking loop. If the code tracking loop is capable of tracking the phase, the code acquisition process is completed. By contrast, if the tracking loop is incapable of tracking the phase, the code tracking loop activates a 'reloading command', in order to load another group of S consecutive chips into the delay-units of the m-sequence generator. The above process can be repeated, until successful code tracking is accomplished.

Finally, for the family of spread-spectrum communication systems using coherent demodulation, we have to derive the S number of consecutive chips of the originally transmitted m-sequence, in order to generate a local m-sequence replica for directly despreading the transmitted m-sequence, instead of despreading the differentially processed m-sequence. These S number of consecutive chips of the originally transmitted m-sequence, namely $\{c_i\}$, can be determined using the relationship between the m-sequences $\{c_i\}$ and $\{b_i\}$. Below we specify three methods for achieving this objective.

1. Since each specific delay in the m-sequence of $\{c_i\}$ corresponds to a specific delay of the m-sequence $\{b_i\}$, we can design a look-up table of size $(2^S - 1) \times 2S$, where each row contains S consecutive chips of $\{b_i\}$ and the corresponding S consecutive chips of $\{c_i\}$. Once the S consecutive chips of $\{b_i\}$ have been obtained, the S number of consecutive chips of $\{c_i\}$ can be obtained by referring to this table. However, since this table has a dimension of $(2^S - 1) \times 2S$, it may becomes impractical to use this method, if S is large.

2. The second method is based on the recursive equation Equation 21.154 and the differential processing operation of $b_j = c_j \cdot c_{j-1}$. Let $b_G, b_{G+1}, \ldots, b_{G+S-1}$ be the S number of consecutive chips of $\{b_i\}$. From Equation 21.154 we obtain that

$$b_j = b_{j+S} \cdot \prod_{m=1}^{M-1} b_{j+S-s_m}, \quad j = G-1, G-2, \ldots, 0. \tag{21.162}$$

With the aid of Equation 21.162 we can generate all the chip values of $b_0, b_1, \ldots, b_{G+S-1}$. Then, from $b_j = c_j \cdot c_{j-1}$ we arrive at $c_j = b_j \cdot c_{j-1}$. Furthermore, we assumed that the delay-unit in the differential processor was initialized to one, i.e., we have $c_0 = 1$. Explicitly, based on these quantities, S number of consecutive chips $c_G, c_{G+1}, \ldots, c_{G+S-1}$ associated with the originally transmitted m-sequence can be obtained by solving the recursive equation $c_j = b_j \cdot c_{j-1}$ for $j = 1, 2, \dots, G+S-1$. Note that the complexity of this scheme is linearly dependent on the SISO recursive decoding depth. Hence this procedure is efficient, provided that the SISO recursive decoding depth is sufficiently low.

3. The third proposed scheme is independent of both the length $(2^S - 1)$ of the m-sequence and of the SISO recursive decoding depth. It has a complexity, which is linearly dependent on the number of stages, S, in the soft-chip-register. However, in order to determine the S number of consecutive chips, say $c_G, c_{G+1}, \dots, c_{G+S-1}$, this scheme requires a further stage of decision. More explicitly, let $b_G = c_G c_{G-1}, b_{G+1} = c_{G+1} c_G, \dots, b_{G+S-1} = c_{G+S-1} c_{G+S-2}$ be the S number of consecutive chips of $\{b_i\}$, which have been reliably determined. By setting $c_{G-1} = +1$ or $c_{G-1} = -1$ we obtain two different S-chip sequences, which are expressed as $\{c_G, c_{G+1}, \dots, c_{G+S-1}\}_{c_{G-1}=+1}$ and $\{c_G, c_{G+1}, \dots, c_{G+S-1}\}_{c_{G-1}=-1}$. It can be readily shown that these two S-chip sequences have a Hamming distance of S. However, only one of them is the desirable one, which can be further determined using the following approach. By loading them into two local m-sequence generators, two despread PN sequences having different phases can be generated, although only one of them has the same phase as the originally transmitted m-sequence. Consequently, by correlating the received m-sequence with both of the above-mentioned locally generated PN sequence replicas, we can select the specific S number of consecutive chips corresponding to the higher correlation output as the desirable initial chips. This is because the correlator using the desirable S consecutive chips has higher correlation outputs associated with a significantly higher probability, than that using the undesirable S consecutive chips.

21.8.6.3 Acquisition Time and Complexity

According to our analysis in Section 21.8.6.2, the DRSSE acquisition scheme is capable of observing the reliabilities of the most recent S consecutive chips through observing the amplitudes of the corresponding soft outputs stored in the soft-chip-register of Figure 21.43. The DRSSE acquisition scheme then decides as to when it should activate the 'loading command' for loading the initial chips into the m-sequence generator. When the erroneous loading probability is deemed sufficiently low, i.e., it is in the region of, for example, 10^{-4}, successful PN-sequence acquisition can typically be declared with a high probability right after the first loading of the initial chips. Therefore, the total acquisition time of the DRSSE acquisition scheme can be approximated by the time duration required by the DRSSE for carrying out the recursive SISO decoding, in order to reach a sufficiently low erroneous loading probability. According to the iterative decoding principle [79, 506] and to our results to be presented in Section 21.8.7, it can be shown that for a reasonable SNR per chip value the time duration required by the DRSSE for carrying out recursive SISO decoding is linearly dependent on the number of stages, S, in the related m-sequence generator.

As shown in Figure 21.43, the implementation of the proposed DRSSE acquisition scheme requires a soft-chip-register having the same length S, as the m-sequence generator, plus a simple differential processor and a low-complexity SISO decoder. The complexity of the DRSSE acquisition scheme is dominated by that of the soft-chip-register, which is linearly dependent on the number of stages, S, in the soft-chip-register. Therefore, due to the fact that both the complexity and the acquisition time are linearly dependent on the number of stages S in the m-sequence generator, the proposed DRSSE acquisition scheme is particularly attractive for the acquisition of long m-sequences.

21.8.7 Performance Results of DRSSE Acquisition Scheme

In this section we provide a range of simulation results for characterizing the proposed DRSSE acquisition scheme. Our simulation results were mainly based on the acquisition of two different-length m-sequences having $S = 5$ and $S = 13$ stages, which correspond to having PN-sequence periods of $N = 2^5 - 1 = 31$ and $N = 2^{13} - 1 = 8191$, respectively. The generator polynomial associated with $S = 5$ and $N = 31$ was $g(D) = 1 + D^2 + D^5$, while that associated with $S = 13$ and $N = 8191$ was $g(D) = 1 + D + D^3 + D^4 + D^{13}$. Note that the curves were drawn either versus the signal-to-noise ratio (SNR) per chip, namely E_c/N_0 or versus the normalized number of chips received, which also represents the normalized number of chips that the SISO decoder processed. The normalized number of chips L/S was defined as the total number of received chips, L, divided by the number of generator stages, S, of the corresponding m-sequence generator.

In Figure 21.44 we show the erroneous loading probability, P_e, versus the SNR per chip, E_c/N_0 performance for an m-sequence generated using the generator polynomial of $g(D) = 1 + D^2 + D^5$ and transmitted over AWGN channels. Note that in Figure 21.44 the curve corresponding to the parameter of $L = 1 \times S$ represents the erroneous loading probability of the DRSSE acquisition scheme using no recursive SISO decoding. From the results we can see that when more channel output chips are involved in the recursive SISO decoding process, a higher correct detection reliability and hence a lower erroneous loading probability can be achieved. For the sake of illustration, let us assume that the transmitted m-sequence can be reliably acquired, once the erroneous loading probability is lower than 10^{-4}. Then, from the results of Figure 21.44 we can observe that the m-sequence can be reliably acquired at an SNR per chip value of $E_c/N_0 = 0$dB by invoking about $L = 40 \times S = 40 \times 5 = 200$ chips in the recursive SISO decoding scheme. By contrast, without exploiting the power of recursive SISO decoding, the DRSSE acquisition scheme has to operate at an SNR per chip value of $E_c/N_0 = 9.5$dB, in order to achieve the same erroneous loading probability of 10^{-4}. Explicitly, at the erroneous loading probability of 10^{-4} the SNR per chip gain is about 9.5dB, when $L = 200$ chips are invoked by the recursive SISO decoding scheme. In general, it is expected that when more chips can be used during the recursive SISO decoding process, an increasing gain can be achieved by the proposed DRSSE acquisition scheme. For example, when invoking $L = 200 \times S$ instead of $L = 40 \times S$ chips in the recursive SISO decoder, another 1.7dB SNR per chip gain can be achieved at the erroneous loading probability of 10^{-4}.

Figure 21.45 shows the acquisition performance for an m-sequence having a period of $N = 8191$ chips, which was generated by a thirteen-stage ($S = 13$) generator using the generator polynomial of $g(D) = 1 + D + D^3 + D^4 + D^{13}$. As shown in Figure 21.45, the m-sequence can be reliably acquired at an SNR per chip value of $E_c/N_0 = 1.7$dB by invoking about $L = 40 \times S = 40 \times 13 = 520$ chips by the recursive SISO decoder. This m-sequence can also be reliably acquired at a reduced SNR per chip value of $E_c/N_0 = 1$dB by invoking about $L = 200 \times S = 200 \times 13 = 2600$ chips in the recursive SISO decoder. By

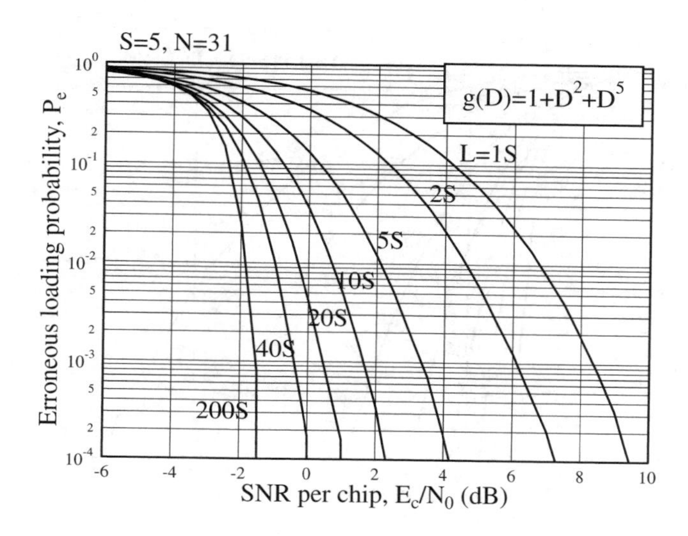

Figure 21.44: Erroneous loading probability, P_e, versus the SNR/chip, E_c/N_0, performance for various numbers of chips invoked by the proposed recursive SISO decoder, when transmitting the m-sequence generated using the generator polynomial of $g(D) = 1 + D^2 + D^5$ over AWGN channels.

contrast, without the recursive SISO decoding, the PN code acquisition scheme has to operate at the SNR per chip value of $E_c/N_0 = 10$dB, in order to achieve the erroneous loading probability of 10^{-4}. Hence, the SNR per chip gain at the erroneous loading probability of 10^{-4} is about 8.3dB or 9dB, respectively, when $L = 520$ or $L = 2600$ chips are invoked by the recursive SISO decoder. Note that even though $L = 2600$ chips are invoked by the recursive SISO decoder for achieving reliable acquisition at the SNR per chip value of $E_c/N_0 = 1$dB, the acquisition time of $L = 2600$ chip durations is still significantly lower than that of any conventional serial search acquisition schemes [15, 28], which demand a mean acquisition time on the order of the period of the m-sequences considered, i.e., 8191 chips in this example.

The results portrayed in the above figures were evaluated, when communicating over AWGN channels. In Figure 21.46 we investigated the acquisition performance of the DRSSE acquisition scheme for transmission over Rayleigh fading channels. As we analyzed in Section 21.8.6, when communicating over fading channels, depending on whether the recursive SISO decoder employs perfect knowledge or no knowledge of the fading channel, the term L_c in the *intrinsic* information of Equation 21.159 can be set to $L_c = 2\alpha_i^2 E_c/\Omega N_0$ or to $L_c = 2E_c/N_0$. In Figure 21.46 we also compared the acquisition performance recorded in the context of these two cases. In Figure 21.46 the m-sequence was generated by a thirteen-stage m-sequence generator using the generator polynomial of $g(D) = 1 + D + D^3 + D^4 + D^{13}$. From the results of Figure 21.46 we observe that the acquisition performance improves in

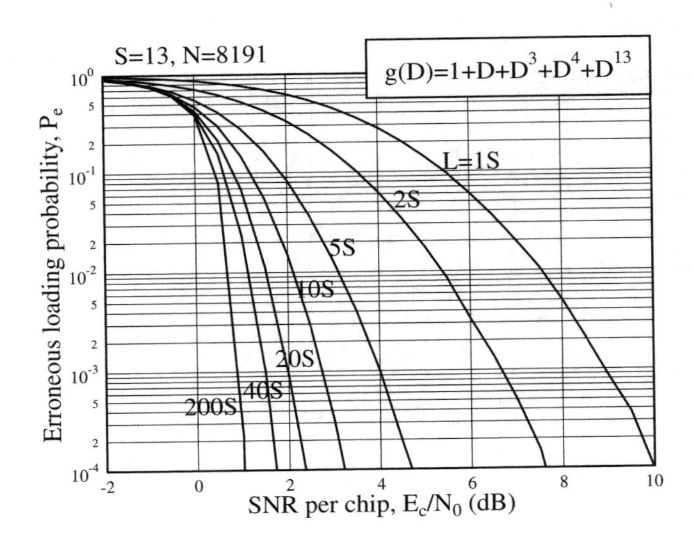

Figure 21.45: Erroneous loading probability, P_e, versus the SNR/chip, E_c/N_0, performance for various numbers of chips invoked by the proposed recursive SISO decoder, when transmitting the m-sequence generated using the generator polynomial of $g(D) = 1+D+D^3+D^4+D^{13}$ over AWGN channels.

both cases, when increasing the number of received chips invoked by the recursive SISO decoder. However, compared to the acquisition performance using no recursive SISO decoding (the curve marked with circles), for a given number of received chips, the recursive SISO decoder using perfect channel knowledge provides a higher SNR per chip gain than that attained without using any channel knowledge. This observation can be explained with the aid of Equation 21.161. When perfect knowledge of the channel is employed, Equation Equation 21.161, in fact, is processed based on the 'maximal-ratio combining' principle. By contrast, without the knowledge of the channel, Equation Equation 21.161 is processed based on the 'equal-gain combining' principle. It is well known that the maximal-ratio combining (MRC) scheme outperforms the equal gain combining (EGC) scheme in multipath fading channels.

Finally, in Figure 21.47 we show the erroneous loading probability performance versus the normalized number of received chips used by the recursive SISO decoder for an m-sequence generated by an $S = 13$-stage m-sequence generator and hence having a period of 8191 chips. The results were generated for transmissions over both AWGN and Rayleigh fading channels at the SNR per chip value of $E_c/N_0 = 2$dB. According to the results of Figure 21.47 it can be shown that for AWGN channels the DRSSE acquisition scheme is capable of achieving an erroneous loading probability of 10^{-3} after receiving approximately $20 \times 13 = 260$ chips. By contrast, when communicating over Rayleigh fading channels, the DRSSE acquisition scheme is capable of reaching the erroneous loading probability of

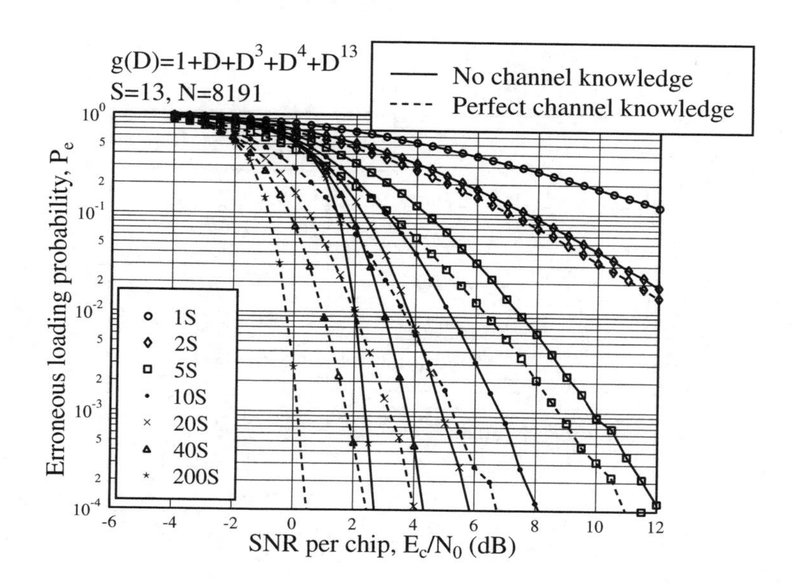

Figure 21.46: Erroneous loading probability, P_e, versus the SNR/chip, E_c/N_0, performance for various numbers of chips invoked by the proposed recursive SISO decoder, when transmitting the m-sequence generated using the generator polynomial of $g(D) = 1+D+D^3+D^4+D^{13}$ over Rayleigh fading channels.

10^{-3} after receiving approximately $500 \times 13 = 6500$ chips. These results imply that with the aid of the proposed recursive SISO decoder the required target performance can be achieved, regardless of the specific communication environment encountered by invoking a sufficiently high number of chips in the SISO decoder. The results of Figure 21.47 indicate that successful acquisition can be achieved by the proposed DRSSE acquisition scheme, when communicating over Rayleigh fading channels at $E_c/N_0 = 2$dB within about 6500 chip durations, which is lower than the 8191-chip period of the transmitted m-sequence.

21.8.8 Summary and Conclusion

In summary, in this section the acquisition of m-sequences using the proposed soft-decision based sequential estimation schemes has been investigated. We have shown that the ML soft sequential estimation-based acquisition scheme is equivalent to a conventional parallel search-based acquisition scheme. Using the principles of iterative SISO decoding, we have proposed two recursive soft sequential estimation schemes, namely the CRSSE acquisition scheme and the DRSSE acquisition scheme. The acquisition performance of both the CRSSE scheme and the DRSSE scheme has been investigated. It has been shown that both the CRSSE and DRSSE acquisition schemes have an implementational complexity as well as an initial synchronization time, which are linearly dependent on the number of stages in the m-sequence generator. In terms of these characteristics, the CRSSE and DRSSE acqui-

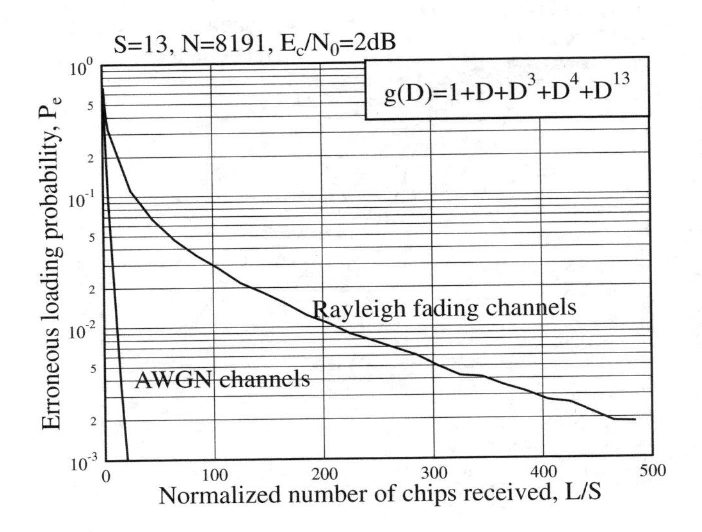

Figure 21.47: Erroneous loading probability, P_e versus the normalized number of received chips invoked by the proposed recursive SISO decoder for an m-sequence having a period of 8191, when communicating over AWGN or Rayleigh fading channels and using an SNR per chip value of $E_c/N_0 = 2\text{dB}$ for $S = 13$.

sition schemes outperform both the family of conventional serial search-based acquisition schemes and the class of conventional parallel search-based acquisition schemes, which either result in a mean-acquisition time or impose an implementational complexity, that are exponentially dependent on the number of stages in the m-sequence generator. Owing to these attractive characteristics, the proposed CRSSE and DRSSE acquisition schemes constitute promising initial synchronization schemes for the acquisition of long m-sequences. Furthermore, the DRSSE acquisition scheme is suitable for m-sequence acquisition, when communicating over various environments including AWGN, slow fading and fast fading channels. Again, this is because the differential processing employed is carried out at the chip-level of the m-sequence and the channel-effects do not change dramatically during a short chip-duration.

21.9 Code-Acquisition in Multi-carrier DS-CDMA Systems

So far, we have studied a range of acquisition schemes proposed for the family of classic single-carrier DS-CDMA systems and investigated their acquisition performance when communicating over both AWGN channels and dispersive multipath fading channels. The performance of multi-carrier DS-CDMA systems has been widely investigated, which have been proposed as candidates for high data rate wireless communications [65, 66, 450]. However,

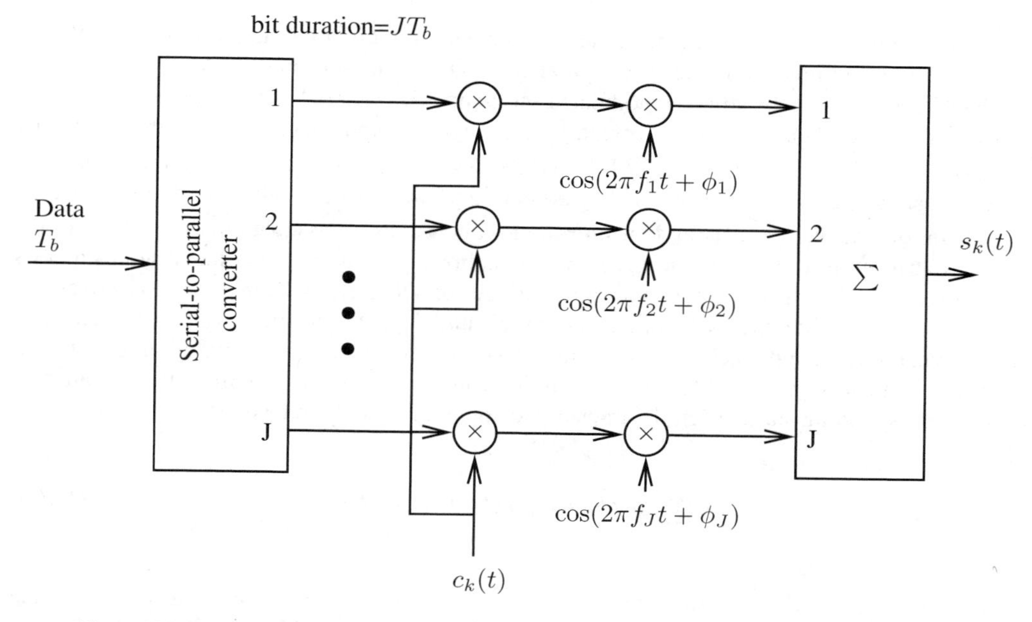

Figure 21.48: The transmitter schematic of the multi-carrier DS-CDMA system studied.

in the analysis of the multi-carrier CDMA systems' performance it was assumed that there is perfect synchronization between the spreading codes of the received signal and the locally generated despreading code of the receiver. Hence, the purpose of this section is to investigate the issues of PN code acquisition in the context of multi-carrier DS-CDMA systems. Specifically, we will investigate the performance of a pure parallel acquisition scheme in the context of a multi-carrier DS-CDMA system communicating over dispersive multipath fading channels, under the hypothesis of encountering multiple H_1 cells. We will also investigate the advantages and disadvantages of multi-carrier and single-carrier acquisition systems studied in the context of dispersive multipath fading channels from the perspective of initial synchronization and provide performance comparisons.

21.9.1 System Model

21.9.1.1 The transmitter

The kth user's transmitter is shown in Figure 21.48 in the framework of the multi-carrier DS-CDMA system considered, which is similar to that considered in [65], except that no interleaving over the sub-carriers is assumed in our analysis. However, the analysis of this section can readily be extended to the PN code acquisition study of multi-carrier DS-CDMA systems using interleaving over the associated sub-carriers. We note at this stage that frequency-domain interleaving is often used in multi-carrier CDMA systems, in order to map a given information symbol to be transmitted to several sub-carriers that are sufficiently far apart from each other for avoiding the simultaneous corruption of them due to frequency-selective fading. This measure is capable of providing substantial diversity gain.

As shown in Figure 21.48, the bit stream having a bit duration of T_b is serial-to-parallel converted into J number of reduced-rate parallel streams at the transmitter. Specifically, the bit duration of each of the streams modulated on to the sub-carriers hence becomes $T_{bs} = JT_b$. All sub-carrier data streams are spread by the same signature sequence, which is given by $c_k(t) = \sum_{n=-\infty}^{\infty} c_n^{(k)} P_{T_c}(t - nJT_c)$. This fact does not impose any problems, since we will assume that the sub-carrier signals do not interfere with each other due to their spectral seperation. Again, $c_n^{(k)}$ is a random spreading sequence, as we assumed in the analysis of the single-carrier system studied in the previous sections. We assume that the system supports K users and that all users have the same number of sub-carriers, namely J. Moreover, we assume that the first user is the one that requires initial synchronization, while the other users are synchronized and reached the data transmission stage. Furthermore, we assume that ideal power control is employed for all the communicating users. Consequently, after modulating the corresponding sub-carrier, the transmitted signal of user k can be expressed as:

$$s_k(t) = \sum_{j=1}^{J} \sqrt{2P_k} b_k(t) c_k(t) \cos(2\pi f_j t + \phi_{jk}), \tag{21.163}$$

where the sub-carrier data stream $b_k(t) = \sum_{i=-\infty}^{\infty} b_i^{(k)} P_{T_b}(t - iJT_b)$ consists of a sequence of mutually independent rectangular pulses of duration $T_{bs} = JT_b$ and of amplitude +1 or -1 having an equal probability. Furthermore, the associated user powers are $P_1 = P_r$, while $P_k = P_I$ for $k \neq 1$. finally, f_j is the jth sub-carrier frequency and ϕ_{jk} is the random phase of each sub-carrier. As already alluded to earlier, in the following analysis – for the sake of simplicity – we assume that there exists no spectral overlap between the spectral main-lobes of two adjacent sub-carriers [472]. In fact, we stimulate an even more stringent constraint, namely that there exists no interference between sub-carriers. More explicitly, interference is inflicted only, when an interfering user activates the same sub-carrier, as the reference user. Readers interested in the quantitative effects of overlapping sub-carriers are referred to, for example, [65] and [482]. Let $N = JT_b/JT_c = T_b/T_c$, where T_b and T_c are the corresponding bit duration and chip duration, respectively, for the single-carrier DS-CDMA benchmark system considered. Consequently, both the single-carrier and multi-carrier DS-CDMA systems considered have the same overall information rate and the same system bandwidth, which allows their direct comparison in the forthcoming discourse.

The channel model considered here is the frequency selective Rayleigh fading channel, which was characterized by Figure 21.1. Hence, when signals obeying the form of Equation 21.163 are transmitted and no data modulation is imposed on the signal of the user attempting initial synchronization, then the received signal can be written as:

$$\begin{aligned} r(t) &= \sum_{j=1}^{J} \sum_{l=0}^{L-1} \sqrt{2P_r} \alpha_{1jl} c_1(t - \tau_1 - lT_c) \cos(2\pi f_j t + \theta_{1jl}) \\ &+ \sum_{k=2}^{K} \sum_{j=1}^{J} \sum_{l=0}^{L-1} \sqrt{2P_k} \alpha_{kjl} b_k(t - \tau_k - lT_c) c_k(t - \tau_k - lT_c) \cos(2\pi f_j t + \theta_{kjl}) \\ &+ n(t), \end{aligned} \tag{21.164}$$

where τ_k is the relative time delay among the users' transmitted signals associated with an asynchronous transmission scheme, $\theta_{kjl} = \phi_{jk} - \psi_{kjl} - 2\pi f_j(\tau_k + lT_c)$ is a phase term, which is modelled with the aid of iid random variables uniformly distributed in the interval $[0, 2\pi]$ for different values of j, k and l, while ψ_{kjl} is the channel-induced phase rotation. All the other quantities in Equation 21.164 were defined before in the context of Equation 21.4 in Section 21.2.2.

21.9.1.2 Code-Acquisition Description

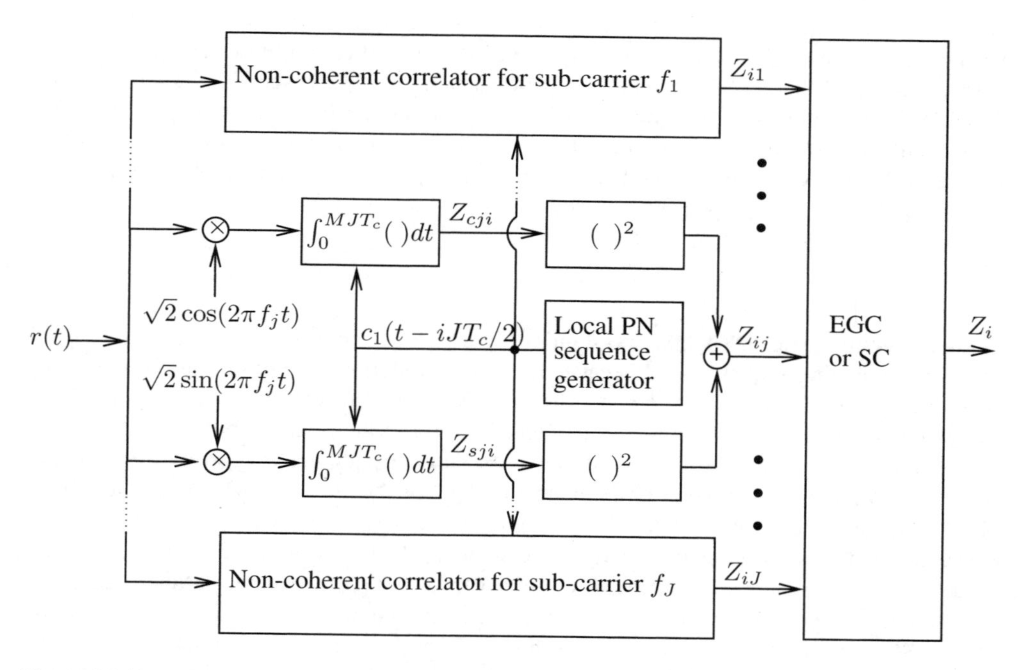

Figure 21.49: Block diagram of generating the decision variable Z_i in the multi-carrier DS-CDMA code-acquisition system studied.

We consider a pure parallel implementation of the maximum-likelihood acquisition scheme, as in Subsection 21.3.2, where all possible PN code phases of the received signal are tested simultaneously, as was shown in Figure 21.4. We assume that the acquisition search step size of the multi-carrier DS-CDMA system concerned is $JT_c/2$, which is J times higher than that of the single-carrier CDMA benchmarker scheme. However, since the sub-carrier symbol rate of the multi-carrier CDMA scheme is also J times lower, the same relative time-domain synchronization resolution is maintained. We also assume that there are L resolvable paths, where $L = \lfloor T_m/JT_c \rfloor + 1$ and T_m represents the maximum delay-spread of the fading channel, as we assumed in Subsection 21.2.2. Hence, according to the analysis of Section 21.6, for a PN sequence having q number of cells in the uncertainty region, which communicating over a non-fading channel, the total number of cells to be searched in the context of receiving over multipath fading channels will be $q + 2(L-1)$. The number of H_1 cells is $2L$. Let $\{Z_0, Z_1, \ldots, Z_{q+2L-3}\}$ denote the decision variables corresponding to all possible PN code phases, where each element Z_i of the $q + 2(L - 1)$ number of decision variables corresponds to one of the possible $q+2(L-1)$ phase differences between the received and locally generated PN code. The block diagram of the proposed scheme used for generating the decision variable Z_i is shown in Figure 21.49. Here we refrain from detailing the operation of the scheme seen in Figure 21.49, since its philosophy is identical to that of Figure 21.8, which was discussed in the context of single-carrier DS-CDMA. It is important to note, however, that the scheme of Figure 21.8 is repeated in Figure 21.49 for each sub-carrier. The approach of the parallel search-based acquisition scheme is that, if the largest decision variable Z_i of Figure 21.49 exceeds a threshold h, the corresponding phase is tentatively assumed to be the correct phase of the received signal and the acquisition system proceeds to the associated verification mode. If this phase is confirmed during the verification mode, acquisition is declared and the PN code phase tracking system is enabled. Otherwise, if the largest decision variable Z_i of Figure 21.49 does not exceed the threshold h, or the tentative acquisition phase is not confirmed by the verification mode – even though the largest decision variable Z_i of Figure 21.49 did exceed the threshold h – then a set of q new decision variables are collected, and the above process is repeated.

In the multi-carrier DS-CDMA system of Figure 21.48, since all sub-carriers employ the same PN spreading sequence and because no data modulation is imposed during the acquisition stage, the outputs of the J non-coherent sub-carrier correlators seen in Figure 21.49 and associated with the same phase of the local PN code replica can be combined. This is expected to be advantageous, since even if one of the sub-carrier signals is corrupted by channel effects, it is unlikely that several of them are simultaneously corrupted. Hence, the reliability of the acquisition scheme is expected to be better than in a single-carrier DS-CDMA scheme. However, as we will see during our further discourse, there is a range of other facts, which influence the multi-carrier DS-CDMA system's acquisition performance. In this section, two non-coherent combining schemes – namely equal gain combining (EGC) and selection combining (SC) – will be investigated. Since the PN code chip rate of the multi-carrier system using J sub-carriers is a factor of $1/J$ lower than that of the corresponding single-carrier system, the number of chips collected during the integration period of the integrator in Figure 21.49 is also reduced by a factor of $1/J$. Let $\tau_D = MJT_c$ be the integration dwell-time for both the multi-carrier and the single-carrier DS-CDMA systems considered. Then, it can be shown that for the multi-carrier DS-CDMA system the average energy captured by the correlator of Figure 21.49 during the integration dwell-time is only $1/J$ times of that for the corresponding single-carrier DS-CDMA system. However, since the chip rate of the single-carrier DS-CDMA system is J times of that of the multi-carrier DS-CDMA system, the number of resolvable paths will be at least JL, where L represents the number of resolvable paths of the corresponding multi-carrier DS-CDMA system. Moreover, since the acquisition search step size is $T_c/2$, which is $1/J$ times of $JT_c/2$ of the multi-carrier DS-CDMA system, in a single-carrier DS-CDMA system the number of cells in the uncertainty region is $q + 2(JL - 1)$ and the number of H_1 cells is $2JL$. Bearing these relations between the multi-carrier and single-carrier DS-CDMA systems in mind, let us now commence our analysis of the overall successful detection, overall missing and overall false-alarm probabilities. We proceed by deriving the relevant statistics for the decision variables.

21.9.2 Decision Variable Statistics

Since we have assumed that there exists no spectral overlap between the spectral main-lobes of two adjacent sub-carriers and that there exists no interference between the sub-carriers, hence the statistics associated with sub-carrier f_j are identical to those of a single-carrier system communicating over the dispersive multipath fading channel. Referring to Section 21.6, the output variable of Figure 21.49 matched to the sub-carrier f_j can be expressed as:

$$\begin{aligned} Z_{ij} &= Z_{cji}^2 + Z_{sji}^2 \\ &= \left[\int_0^{MJT_c} r(t)c_1(t - iJT_c/2)\sqrt{2}\cos(2\pi f_j t)dt\right]^2 \\ &+ \left[\int_0^{MJT_c} r(t)c_1(t - iJT_c/2)\sqrt{2}\sin(2\pi f_j t)dt\right]^2, \end{aligned} \quad (21.165)$$

where both Z_{cji} and Z_{sji} can be approximated by Gaussian random variables having normalized means given by $\frac{3}{4}\alpha_{1jl}\cos\theta_{ijl}$ and $\frac{3}{4}\alpha_{1jl}(-\sin\theta_{ijl})$, provided that an H_1 cell is being tested. By contrast the normalized means are zero, if an H_0 cell is being tested. We have also shown in Section 21.6 that the normalized variance of both Z_{cji} and Z_{sji} is $\sigma_o^2 = \frac{(L-1)\Omega}{3M} + \frac{(K-1)L\rho\Omega}{3M} + \frac{1}{2M\gamma_c}$, where we have $\gamma_c = E_c/N_0$ and E_c is the chip energy. Hence, assuming that Z_{ij} in Figure 21.49 constitutes an H_1 cell, the PDF of Z_{ij}, for a given α_{1jl} is the chi-square distribution with two degrees of freedom, which can be expressed as [90]:

$$f_{Z_{ij}}(y|\alpha_{1jl}, H_1) = \frac{1}{2\sigma_o^2}\exp\left(-\frac{m^2 + y}{2\sigma_o^2}\right) I_0\left(\frac{m\sqrt{y}}{\sigma_o^2}\right), \ y \geq 0, \quad (21.166)$$

where

$$m^2 = \frac{9}{16}\alpha_{1jl}^2. \quad (21.167)$$

By contrast, when Z_{ij} constitutes an H_0 cell, Z_{ij} in Figure 21.49 is the central chi-square distribution and its PDF can be expressed as [90]:

$$f_{Z_{ij}}(y|H_0) = \frac{1}{2\sigma_o^2}\exp\left(-\frac{y}{2\sigma_o^2}\right), \ y \geq 0. \quad (21.168)$$

Using the above expression of σ_o^2 for normalizing Z_{ij}, from Equations 21.166 and 21.168 we obtain the normalized PDFs of Z_{ij} as:

$$f_{Z_{ij}}(y|R^2, H_1) = \frac{1}{2}\exp\left(-\frac{R^2 + y}{2}\right) I_0\left(\sqrt{R^2 y}\right), \ y \geq 0, \quad (21.169)$$

$$f_{Z_{ij}}(y|H_0) = \frac{1}{2}\exp\left(-\frac{y}{2}\right), \quad (21.170)$$

where R^2 is defined as:

$$R^2 = \frac{m^2}{\sigma_o^2} = \overline{\gamma}_c^2 \cdot \frac{\alpha_{1jl}^2}{\Omega}, \tag{21.171}$$

and

$$\overline{\gamma}_c = \frac{3}{4}\left[\frac{(L-1)}{3M} + \frac{(K-1)L\rho}{3M} + \frac{1}{2M\Omega\gamma_c}\right]^{-1/2}, \tag{21.172}$$

where the variables K, L, M, ρ and γ_c were defined earlier in the context of Equation 21.69.

Assuming that the decision variables $\{Z_{i1}, Z_{i2}, \ldots, Z_{iJ}\}$ are iid random variables, the conditioning upon α_{1jl} in Equation 21.169 can be removed at this stage, resulting in:

$$f_{Z_{ij}}(y|H_1) = \frac{1}{2+\overline{\gamma}_c^2}\exp\left(-\frac{y}{2+\overline{\gamma}_c^2}\right), \quad y \geq 0, \tag{21.173}$$

which is the central chi-square distribution with two degrees of freedom.

We have obtained the conditional PDF of the output Z_{ij} in Figure 21.49 matched to the sub-carrier $\{f_j\}$. The specific Z_{ij} outputs of Figure 21.49, which are associated with those from the set of J sub-carriers that exhibit the same phase difference with respect to the local code replica can be combined using EGC or SC, before inputting them to the detection unit of Figure 21.49. Let us hence derive the corresponding PDFs for both EGC and SC.

21.9.2.1 Equal Gain Combining

The output of the EGC scheme is given by the sum of the synchronization outputs $\{Z_{ij}\}$ for $j = 1, 2, \ldots, J$ seen in Figure 21.49:

$$Z_i = \sum_{j=1}^{J} Z_{ij}, \tag{21.174}$$

where $\{Z_{ij}\}$ are iid random variables having PDFs given by Equation 21.173 or Equation 21.170 under the hypothesis of encountering H_1 or H_0 cells, respectively. Both Equation 21.170 and Equation 21.174 are central chi-square distributed with two-degrees of freedom and variance of unity or $(2+\overline{\gamma}_c^2)/2$ for the H_0 or H_1 cell hypothesis, respectively. Hence, Z_i of Figure 21.49 obeys the central chi-square distribution with $2L$ degrees of freedom. Assuming that an H_1 cell is being tested, the PDF of Z_i can be formulated as [90]:

$$f_{Z_i}(y|H_1) = \frac{1}{(2+\overline{\gamma}_c^2)^J (J-1)!} y^{J-1} \exp\left(-\frac{y}{2+\overline{\gamma}_c^2}\right), \quad y \geq 0. \tag{21.175}$$

Assuming that a H_0 cell is being tested, the PDF of Z_i can then be expressed as [90]:

$$f_{Z_i}(y|H_0) = \frac{1}{2^J(J-1)!}y^{J-1}\exp\left(-\frac{y}{2}\right), \quad y \geq 0. \tag{21.176}$$

21.9.2.2 Selection Combining

The output of the SC scheme is given by the maximum output of Figure 21.49, which is expressed as:

$$Z_i = \max\left\{Z_{i1}, Z_{i2}, \ldots, Z_{iJ}\right\}. \tag{21.177}$$

The conditioned PDFs of Z_i can be formulated as follows:

$$f_{Z_i}(y|H_i) = \frac{\partial Pr\left\{Z_{i1} \leq y, Z_{i2} \leq y, \ldots, Z_{iJ} \leq y|H_i\right\}}{\partial y}, \tag{21.178}$$

where H_i represents H_1 or H_0.

Upon invoking the conditional PDFs of Equation 21.173 and Equation 21.170, it can be shown that the PDFs of the SC-assisted scheme conditioned on testing the H_1 and H_0 cells can be expressed, respectively, as:

$$f_{Z_i}(y|H_1) = \frac{J}{2+\overline{\gamma}_c^2}\exp\left(-\frac{y}{2+\overline{\gamma}_c^2}\right)\left[1-\exp\left(-\frac{y}{2+\overline{\gamma}_c^2}\right)\right]^{J-1}, \quad y \geq 0, \tag{21.179}$$

$$f_{Z_i}(y|H_0) = \frac{J}{2}\exp\left(-\frac{y}{2}\right)\left[1-\exp\left(-\frac{y}{2}\right)\right]^{J-1}, \quad y \geq 0. \tag{21.180}$$

21.9.3 Probability of Detection, Missing and False-Alarm

21.9.3.1 Equal Gain Combining

The overall successful detection probability is the probability that the largest of the input decision variables $\{Z_0, Z_1, \ldots, Z_{q+2L-3}\}$, which correspond to $q + 2(L-1)$ number of cells in the uncertainty region, seen in Figure 21.4, corresponds to one of the $2L$ number of H_1 cells, and additionally, this largest decision variable exceeds the pre-set threshold h. Hence, the overall detection probability can be expressed as:

$$P_D = 2L\int_{h'}^{\infty} f_{Z_i}(y|H_1)\left[\int_0^y f_{Z_j}(x|H_1)dx\right]^{2L-1}\left[\int_0^y f_{Z_k}(x|H_0)dx\right]^{q-2} dy, \tag{21.181}$$

where $h' = h/\sigma_o^2$, since the variable Z_i of Figure 21.49 has been normalized by σ_o^2 in Equation 21.169 and Equation 21.170. In Equation 21.181, $f_{Z_j}(x|H_1)$ and $f_{Z_k}(x|H_0)$ are given by Equation 21.175 and Equation 21.176, respectively. It can be shown that:

$$\int_0^y f_{Z_j}(x|H_1)dx = 1 - \exp\left(-\frac{y}{2+\overline{\gamma}_c^2}\right)\sum_{k=0}^{J-1}\frac{1}{k!}\left(\frac{y}{2+\overline{\gamma}_c^2}\right)^k, \qquad (21.182)$$

$$\int_0^y f_{Z_k}(x|H_0)dx = 1 - \exp\left(-\frac{y}{2}\right)\sum_{k=0}^{J-1}\frac{1}{k!}\left(\frac{y}{2}\right)^k. \qquad (21.183)$$

Note that in Equation 21.181 the variable L represents the number of resolvable paths, while the variable J in Equation 21.182 and Equation 21.183 represents the number of sub-carriers in the multi-carrier DS-CDMA system of Figure 21.49. Substituting $f_{Z_i}(y|H_1)$ from Equation 21.175 associated Equation 21.182 and Equation 21.183 into Equation 21.181, the overall successful detection probability can be expressed as:

$$\begin{aligned} P_D &= 2L\int_{h'}^{\infty}\frac{1}{(2+\overline{\gamma}_c^2)^J(J-1)!}y^{J-1}\exp\left(-\frac{y}{2+\overline{\gamma}_c^2}\right) \\ &\times\left[1-\exp\left(-\frac{y}{2+\overline{\gamma}_c^2}\right)\sum_{k=0}^{J-1}\frac{1}{k!}\left(\frac{y}{2+\overline{\gamma}_c^2}\right)^k\right]^{2L-1} \\ &\times\left[1-\exp\left(-\frac{y}{2}\right)\sum_{k=0}^{J-1}\frac{1}{k!}\left(\frac{y}{2}\right)^k\right]^{q-2} dy. \qquad (21.184)\end{aligned}$$

The overall missing probability of the EGC assisted acquisition scheme is the probability that none of the decision variables in the set $\{Z_0, Z_1, \ldots, Z_{q+2L-3}\}$ – which includes $2L$ number of H_1 cells and $(q-2)$ number of H_0 cells – exceeds the pre-set threshold h, which can be expressed as:

$$P_M = \left[\int_0^{h'} f_{Z_j}(x|H_1)dx\right]^{2L}\left[\int_0^{h'} f_{Z_k}(x|H_0)dx\right]^{q-2} dy. \qquad (21.185)$$

Upon substituting Equation 21.182 and Equation 21.183 into the above equation, we obtain:

$$\begin{aligned} P_M &= \left[1-\exp\left(-\frac{h'}{2+\overline{\gamma}_c^2}\right)\sum_{k=0}^{J-1}\frac{1}{k!}\left(\frac{h'}{2+\overline{\gamma}_c^2}\right)^k\right]^{2L} \\ &\times\left[1-\exp\left(-\frac{h'}{2}\right)\sum_{k=0}^{J-1}\frac{1}{k!}\left(\frac{h'}{2}\right)^k\right]^{q-2}. \qquad (21.186)\end{aligned}$$

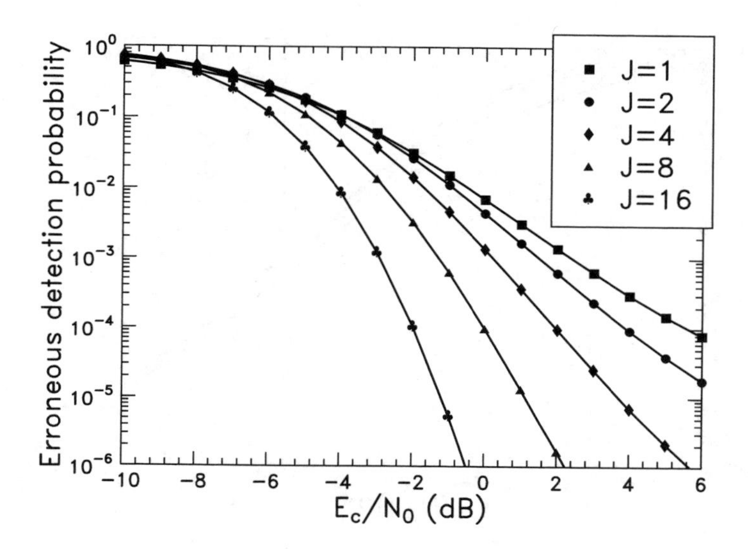

Figure 21.50: EGC: Erroneous detection probability versus SNR/chip performance for the parallel acquisition scheme employed in multi-carrier systems communicating over dispersive multipath Rayleigh fading channels using the parameters of $N = 128, q = 256, K = 1, \rho = 1.0, L_p = 16, M_p = 128, L = L_p/J, M = M_p/J$. The results were computed from Equation 21.187 with P_D given by Equation 21.184 and P_M given by Equation 21.186.

Finally, the false-alarm probability can be expressed as:

$$P_F = 1 - P_D - P_M. \tag{21.187}$$

Having considered the various probabilities associated with the EGC-assisted synchronization scheme of Figure 21.49, in the next section let us now consider SC-aided synchronization arrangements.

21.9.3.2 Selection Combining

For an SC assisted acquisition scheme the overall successful detection probability can also be expressed by Equation 21.181 with $f_{Z_j}(x|H_1)$ and $f_{Z_k}(x|H_0)$ given by Equation 21.179 and Equation 21.180, respectively. It can be shown that in Equation 21.181:

$$\int_0^y f_{Z_i}(x|H_1)dx = \left[1 - \exp\left(-\frac{y}{2+\overline{\gamma}_c^2}\right)\right]^J, \quad y \geq 0, \tag{21.188}$$

$$\int_0^y f_{Z_i}(x|H_0)dx = \left[1 - \exp\left(-\frac{y}{2}\right)\right]^J, \quad y \geq 0. \tag{21.189}$$

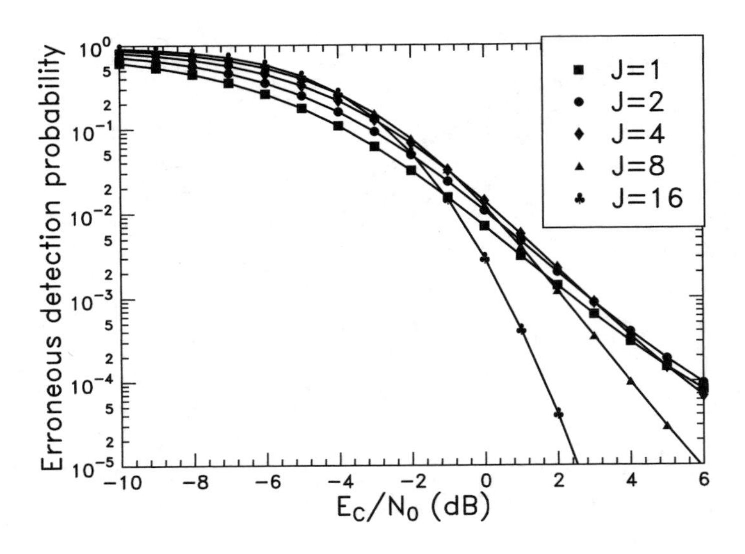

Figure 21.51: **SC**: Erroneous detection probability versus SNR/chip performance for the parallel acquisition scheme employed in multi-carrier systems communicating over dispersive multipath fading channels using the parameters of $N = 128, q = 256, K = 1, \rho = 1.0, L_p = 16, M_p = 128, L = L_p/J, M = M_p/J$. The results were computed from Equation 21.187 with P_D given by Equation 21.191 and P_M given by Equation 21.192.

Substituting these equations and the PDF of $f_{Z_i}(y|H_1)$ from Equation 21.179 into Equation 21.181, the overall successful detection probability for SC can be expressed as:

$$
\begin{aligned}
P_D &= \frac{2LJ}{2+\overline{\gamma}_c^2} \int_{h'}^{\infty} \exp\left(-\frac{y}{2+\overline{\gamma}_c^2}\right) \left[1-\exp\left(-\frac{y}{2+\overline{\gamma}_c^2}\right)\right]^{J-1} \\
&\quad \times \left[1-\exp\left(-\frac{y}{2+\overline{\gamma}_c^2}\right)\right]^{(2L-1)J} \cdot \left[1-\exp\left(-\frac{y}{2}\right)\right]^{(q-2)J} dy \\
&= \frac{2LJ}{2+\overline{\gamma}_c^2} \int_{h'}^{\infty} \exp\left(-\frac{y}{2+\overline{\gamma}_c^2}\right) \left[1-\exp\left(-\frac{y}{2+\overline{\gamma}_c^2}\right)\right]^{2LJ-1} \\
&\quad \times \left[1-\exp\left(-\frac{y}{2}\right)\right]^{(q-2)J} dy.
\end{aligned}
\tag{21.190}
$$

Upon extending the related terms in the above equation, finally we obtain:

$$P_D = 2LJ \sum_{i=0}^{2LJ-1} \sum_{j=0}^{(q-2)J} \frac{(-1)^{i+j}}{1+(i+j)+j\overline{\gamma}_c^2/2} \binom{2LJ-1}{i} \binom{(q-2)J}{J} \cdot \exp\left(-\frac{1+(i+j)+j\overline{\gamma}_c^2/2}{1+\overline{\gamma}_c^2/2} \cdot \frac{h'}{2}\right). \quad (21.191)$$

In Equation 21.184 and Equation 21.191, if we let $J = 1$, then the expressions characterize the successful detection probability of the corresponding single-carrier DS-CDMA system, which was given earlier in Equation 21.81.

The missing probability using SC can be expressed employing Equation 21.185 with $f_{Z_j}(x|H_1)$ and $f_{Z_k}(x|H_0)$ given by Equation 21.179 and Equation 21.180, respectively. Upon substituting these PDFs into Equation 21.185, we obtain:

$$\begin{aligned} P_M &= \left[1-\exp\left(-\frac{h'}{2+\overline{\gamma}_c^2}\right)\right]^{2LJ} \left[1-\exp\left(-\frac{h'}{2}\right)\right]^{(q-2)J} \\ &= \sum_{i=0}^{2LJ} \sum_{j=0}^{(q-2)J} (-1)^{i+j} \binom{2LJ}{i} \binom{(q-2)J}{j} \exp\left(-\frac{i+j+j\overline{\gamma}_c^2/2}{1+\overline{\gamma}_c^2/2} \cdot \frac{h'}{2}\right). \end{aligned} \quad (21.192)$$

Similarly to the EGC- and SC-aided successful detection probabilities expressed in Equations 21.184 and 21.191, if we set $J = 1$ in Equation 21.186 and Equation 21.192, the associated equations characterize the missing probability of the corresponding single-carrier DS-CDMA system, which was given in Equation 21.84.

Finally, the false-alarm probability of the SC-aided acquisition scheme can be evaluated from Equation 21.187 with P_D and P_M given by Equation 21.191 and Equation 21.192, respectively.

The probability of erroneous detection, which is defined as $(1 - P_D) = P_M + P_F$ for EGC- and SC-assisted acquisition, when communicating over dispersive multipath fading channels was shown in Figure 21.50 and Figure 21.51, respectively. Here, the erroneous detection probability is essentially synonymous to the false-alarm probability, since we have set $h' = 0$ and hence $P_M = 0$ according to Equation 21.185. In Figure 21.50 and Figure 21.51, L_p and M_p represent the number of resolvable paths and the number of chips in the integral dwell-time corresponding to a single-carrier ($J = 1$) DS-CDMA system, respectively. From the results we observe that when transmitting over the dispersive multipath fading channels considered, the initial synchronization performance of the multi-carrier system using equal gain combining is better than that of the single-carrier system. By contrast, the multi-carrier system using selection combining and the single-carrier system exhibit a similar performance, when communicating over dispersive multipath fading channels in the SNR/chip range considered. Specifically, the single-carrier system outperforms the multi-carrier system, if the SNR/chip is low. This situation is reversed, when the SNR/chip is high. This issue can be readily explained upon invoking Equation 21.190 combined with Equation 21.172. In Equa-

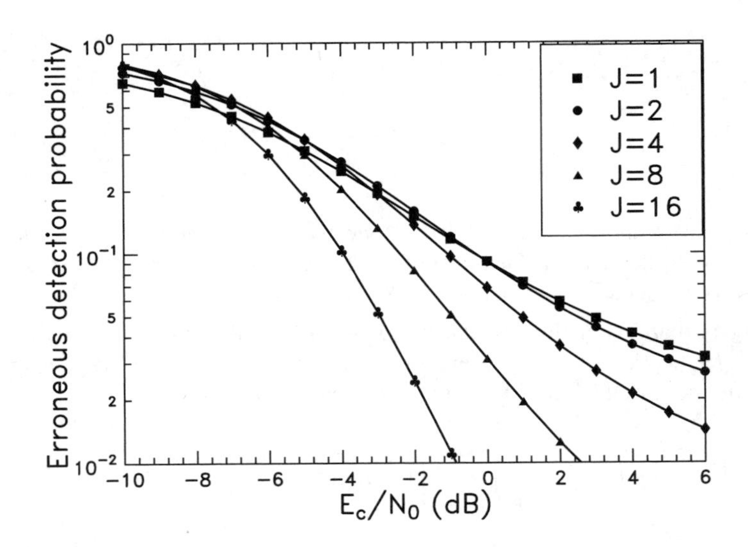

Figure 21.52: EGC: Erroneous detection probability versus SNR/chip performance for the parallel acquisition scheme employed in multi-carrier systems communicating over dispersive multipath Rayleigh fading channels using the parameters of $N = 128, q = 256, K = 3, \rho = 1.0, L_p = 16, M_p = 128, L = L_p/J, M = M_p/J$. The results were computed from Equation 21.187 with P_D given by Equation 21.184 and P_M given by Equation 21.186.

tion 21.190, if we let $L_p = JL$, then we have:

$$\begin{aligned} P_D &= \frac{2L_p}{2+\overline{\gamma}_c^2} \int_{h'}^{\infty} \exp\left(-\frac{y}{2+\overline{\gamma}_c^2}\right) \left[1 - \exp\left(-\frac{y}{2+\overline{\gamma}_c^2}\right)\right]^{2L_p-1} \\ & \quad \times \left[1 - \exp\left(-\frac{y}{2}\right)\right]^{(q-2)J} dy. \end{aligned} \tag{21.193}$$

It can be shown that P_D in Equation 21.190 will decrease, as the number of sub-carriers, J increases. However, the average received SNR/chip, $\overline{\gamma}_c$ increases, when J increases, since the number of resolvable paths $L = L_p/J$ for each sub-carrier decreases, when J increases. Consequently, the two-fold influence of J results in the observation stated in the context of Figure 21.51.

In contrast to Figure 21.50 and Figure 21.51, where the number of active users was $K = 1$, i.e., no multi-user interference was inflicted, in Figure 21.52 and Figure 21.53 we let $K = 3$, i.e., there were two interfering users. The other parameters were the same, as in Figure 21.50 and Figure 21.51, which were listed in the caption of the figures. From the results we observe that for the equal gain combining scheme the performance is similar to that of $K = 1$, provided that the SNR/chip is sufficiently high. However, for the selection combining scheme the single-carrier system outperforms the multi-carrier system over the

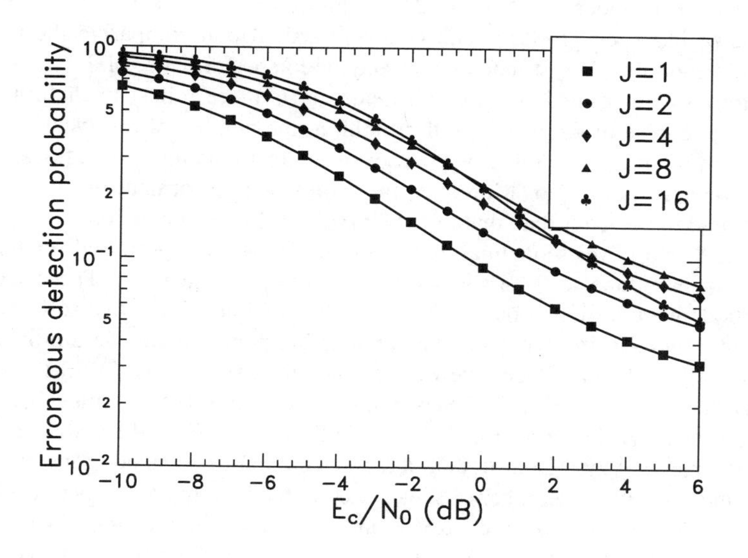

Figure 21.53: SC: Erroneous detection probability versus SNR/chip performance for the parallel acquisition scheme employed in multi-carrier systems communicating over multipath Rayleigh fading channels using the parameters of $N = 128, q = 256, K = 3, \rho = 1.0, L_p = 16, M_p = 128, L = L_p/J, M = M_p/J$. The results were computed from Equation 21.187 with P_D given by Equation 21.191 and P_M given by Equation 21.192.

entire SNR/chip range considered.

21.10 Chapter Summary and Conclusion

In this chapter a range of initial synchronization or PN code acquisition schemes applicable to DS-CDMA systems has been considered. Specifically, following an introductory review of the system and channel models in Section 21.2, a variety of classic serial and parallel PN code acquisition principles were discussed in Section 21.3. Since practical PN code acquisition systems tend to encounter multiple correct timing states (or H_1 cells), in Section 21.4 the mean acquisition time of both serial and parallel acquisition schemes has been derived under the assumption of encountering more than one H_1 cells in the uncertainty range associated with the search process.

In Sections 21.5, 21.6, 21.7 and 21.9 we have considered the acquisition performance of the classic serial and parallel search-based acquisition schemes. The analysis and results provided in these sections have been based on the assumption that multiple correct timing states (H_1 cells) exist in the uncertainty range of the PN sequence. In Sections 21.5 and 21.6 the acquisition performance of the classic serial and parallel acquisition schemes has been investigated, when communicating over AWGN and multipath fading channels. The results of Sections 21.5 and 21.6 show that the acquisition performance degrades, when the transmission channel degrades. In Section 21.7 a joint two-cell detection-based PN code acquisition scheme has been proposed and characterized. Upon comparing the results of Section 21.6 and Section 21.7, we found that the acquisition performance of the serial search-based acquisition system using joint two-cell detection is more robust to multiple access interference, to the choice of the detection threshold, to the number of resolvable multipath components and to the received power of the user-of-interest than the conventional cell-by-cell detection arrangement of 21.6. Moreover, the acquisition performance of joint two-cell detection was found to be better than that of the classic cell-by-cell detection.

In Section 21.8 two novel sequential estimation methods were proposed for the initial synchronization of pseudonoise (PN) signals derived from m-sequences. These sequential estimation methods were designed based on the novel principle of iterative soft-in-soft-out (SISO) decoding. Two specific types of sequential estimation methods were proposed and investigated, namely Coherent Recursive Soft Sequential Estimation (CRSSE) based acquisition and Differential Recursive Soft Sequential Estimation (DRSSE) aided acquisition. The analysis and results of Section 21.8 show that both the CRSSE and DRSSE acquisition schemes exhibit a complexity similar to that of a conventional m-sequence generator, which increases only linearly with the number of stages used in the m-sequence generator. Furthermore, the initial acquisition time of the CRSSE and the DRSSE acquisition schemes is also linearly dependent on the number of stages used in the m-sequence generator. Hence, the proposed CRSSE and DRSSE acquisition schemes constitute promising initial synchronization schemes, in particular in the context of the acquisition of long m-sequences.

Finally, in Section 21.9 the code acquisition performance of multi-carrier DS-CDMA systems has been investigated and compared, when communicating over multipath Rayleigh fading channels under the hypothesis of multiple synchronous states (H_1 cells) in the uncertainty region of the PN code. The acquisition performance of the multi-carrier DS-CDMA system has been evaluated, when the correlator outputs of the sub-carriers associated with the same phase of the local PN code replica are non-coherently combined using both EGC and SC schemes. From the results we conclude that the code acquisition performance of a multi-carrier DS-CDMA system using EGC improves significantly, as the number of sub-carrier signals combined increases. By contrast, the code acquisition performance of the equivalent multi-carrier DS-CDMA system using SC improves only marginally and becomes similar to that of the corresponding single-carrier DS-CDMA system, when increasing the number of sub-carriers. Furthermore, the code acquisition performance of a multi-carrier DS-CDMA system using EGC is better than that of a multi-carrier DS-CDMA system using SC.

Part V

Standards and Networking

Chapter 22

Third-Generation CDMA Systems[1]

22.1 Introduction

The number of mobile subscribers is expected to further increase in the new millenium and it is anticipated that the tele-traffic will be dominated by multimedia applications [58]. Hence, the popular speech-, data- and e-mail-oriented services are expected to be enriched by a whole host of more advanced services in the near future, such as for example videoconferencing, multimedia document transfer and teleshopping, just to name a few. Thus the next generation of mobile communication systems must be capable of handling not only a multitude of services having different data rates and matching Quality of Service (QoS) constraints, but their performance is also expected to become comparable to, if not better than, that of their wired counterparts.

These ambitious objectives are beyond the capabilities of the currently operational Second-Generation (2G) mobile systems such as the Global System for Mobile Communications known as GSM [64], the Interim Standard-95 (IS-95) based Pan-American system, or the Personal Digital Cellular (PDC) system [511] in Japan. Thus, in recent years, a range of new system concepts and objectives were defined and these are incorporated in the Third-Generation (3G) mobile systems. Specifically, the evolution of 3G wireless systems began in the late 1980s when the International Telecommunication Union's Radiocommunication Sector (ITU-R) Task Group 8/1 defined the requirements for the 3G mobile radio systems. This initiative was then termed as the Future Public Land Mobile Telecommunication System (FPLMTS) initiative [512, 513]. Later the terminology was changed to IMT-2000, an acronym used for the International Mobile Telecommunications system in the year 2000.

[1] *Single- and Multi-Carrier CDMA Multi-User Detection, Space-Time Spreading, Synchronization, Networking and Standards.* L.Hanzo, L-L.Yang, E-L.Kuan and K.Yen,
©2003 John Wiley & Sons, Ltd. ISBN 0-470-86309-9

Besides possessing the ability to support services from rates of a few kbps to as high as 2 Mbps in a spectrally efficient way, IMT-2000 aimed to provide a seamless global radio coverage for global roaming. This implied the ambitious goal of aiming to connect virtually any two mobile terminals worldwide. The IMT-2000 system was designed to be sufficiently flexible in order to operate in any propagation environment, such as indoor, outdoor to indoor, and vehicular scenarios. It is also aiming to be sufficiently flexible to handle circuit as well as packet mode services and to handle services of variable data rates. In addition, as mentioned previously, these requirements must be fulfilled with a QoS comparable to that of the current wired network at an affordable cost.

During the World Administrative Radio Conference (WARC) in 1992, the frequency spectrum for IMT-2000 was identified on a worldwide basis as the bands 1885–2025 MHz and 2110–2200 MHz [513]. This allocated frequency spectrum was originally based on the assumption that speech and low data rate transmission would constitute the dominant services offered by IMT-2000. However, this assumption has been superseded, as the trend has shifted toward services that require high-speed data transmission, such as Internet access and multimedia services. A study conducted by the UMTS Forum [514] forecast that these frequency bands allocated for IMT-2000 are only sufficient for the initial deployment of the system. According to the current estimates, it was foreseen that an additional frequency bandwidth of 187 MHz is required for IMT-2000 in high-traffic demand areas by the year 2010. Consequently, three additional bands were identified for IMT-2000 during the World Radio Conference (WRC)-2000 and these are 806-960 MHz, 1710-1885 MHz and 2500-2690 MHz.

Several regional standard organizations — led by the European Telecommunications Standards Institute (ETSI) in Europe, the Association of Radio Industries and Businesses (ARIB) in Japan, and the Telecommunications Industry Association (TIA) in the United States — have been dedicating their efforts to specifying the standards for IMT-2000. A total of 15 Radio Transmission Technology (RTT) IMT-2000 proposals were submitted to ITU-R in June 1998, five of which are satellite-based solutions, while the rest are terrestrial solutions. Table 22.1 shows a list of the terrestrial-based proposals submitted by the various organizations and their chosen radio access technology.

As shown in Table 22.1, Code Division Multiple Access (CDMA) is the favoured multiple access technique proposed for the 3G wireless communications systems worldwide, partly motivated by the success of the IS-95 system. However, there is still a marked disparity in the type of radio interface being proposed by the different standardization bodies. Despite the ultimate goal of harmonizing the standards for 3G, an agreement as to which particular radio access technology to adopt globally has not been reached. As a result, a compromise was made such that IMT-2000 is now comprised of five different radio interfaces, namely W-CDMA (UTRA FDD), cdma2000, TD-SCDMA/TD-CDMA (UTRA TDD), UWC 136 and DECT. Among these, only W-CDMA, cdma2000 and TD-CDMA can be considered to be genuine 3G standards. UWC-136 is actually an upgraded version of GSM and is usually considered as a 2.5G standard. The inclusion of DECT as a 3G technology, a standard, which was originally designed for cordless phones, is because it enables these phones to operate in hot-spot areas within the 3G network. Hence in this chapter, we will focus our attention only on the UTRA FDD and TDD proposals as well as the cdma2000 proposal. A rudimentary discourse on both the UWC-136 and the DECT RTT proposals can be found for example at http://www.tiaonline.org and http://www.etsi.org, respectively.

Several of the regional standard organizations have agreed to cooperate and jointly prepare the Technical Specifications (TS) for the 3G mobile systems in order to assist as well as to accelerate the ITU process for standardization of IMT-2000. This led to the formation of two Partnership Projects (PPs), which are known as 3GPP1 [515] and 3GPP2 [516]. 3GPP1 was officially launched in December 1998 with the aim of establishing the TS for IMT-2000 based on the evolved GSM [64] core networks and the UMTS[2] Terrestrial Radio Access (UTRA) RTT proposal. There are six organizational partners in 3GPP1: ETSI, ARIB, the China Wireless Telecommunication Standard (CWTS) group, the Standards Committee T1 Telecommunications (T1, USA), the Telecommunications Technology Association (TTA, Korea), and the Telecommunication Technology Committee (TTC, Japan). The first set of specifications for UTRA was released in December 1999, which contained detailed information on not just the physical layer aspects for UTRA, but also on the protocols and services provided by the higher layers. Here we will concentrate on the UTRA physical layer specifications, and a basic familiarity with CDMA principles is assumed.

In contrast to 3GPP1, the objective of 3GPP2 is to produce the TS for IMT-2000 based on the evolved ANSI-41 core networks, the cdma2000 RTT. 3GPP2 is spearheaded by TIA, and its members include ARIB, CWTS, TTA, and TTC. Despite evolving from completely diversified core networks, members from the two PPs have agreed to cooperate closely in order to produce a globally applicable TS for the 3G mobile systems.

This chapter serves as an overview of the UTRA specifications, which is based on the evolved GSM core network and cdma2000 specifications, which is based on the IS-95 core network. However, the information provided here is by no means based on the final specifications for UTRA or indeed for IMT-2000. It is expected that the parameters and technologies presented in this chapter will evolve further. Readers may also want to refer to the following books on related topics [517–521], which focus exclusively on the 3G systems.

22.2 UMTS Terrestrial Radio Access (UTRA) [242, 515, 517, 522–528]

Research activities for UMTS [511, 512, 522, 523, 529–531] within ETSI have been spearheaded by the European Union's (EU) sponsored programmes, such as the Research in Advanced Communication Equipment (RACE) [96, 532] and the Advanced Communications Technologies and Services (ACTS) [522, 529, 532] initiative. The RACE programme, which is comprised of two phases, commenced in 1988 and ended in 1995. The objective of this programme was to investigate and develop testbeds for the air interface technology candidates. The ACTS programme succeeded the RACE programme in 1995. Within the ACTS Future Radio Wideband Multiple Access System (FRAMES) project, two multiple access modes have been chosen for intensive study, as the candidates for UMTS terrestrial radio access (UTRA). They are based on Time Division Multiple Access (TDMA) with and without spreading, and on W-CDMA [242, 533, 534].

As early as January 1997, ARIB decided to adopt W-CDMA as the terrestrial radio access technology for its IMT-2000 proposal and proceeded to focus its activities on the detailed specifications of this technology [531]. Driven by a strong support behind W-CDMA

[2]UMTS, an abbreviation for Universal Mobile Telecommunications System, is a term introduced by ETSI for the 3G wireless mobile communication system in Europe.

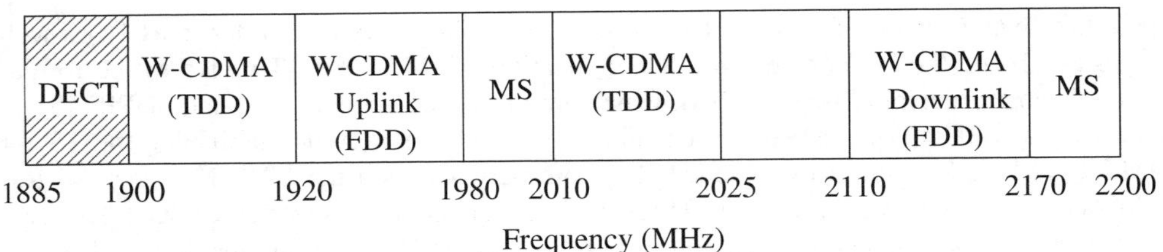

Figure 22.1: The proposed spectrum allocation in UTRA.

worldwide and this early decision from ARIB, ETSI reached a consensus agreement in January 1998 to adopt W-CDMA as the terrestrial radio access technology for UMTS. In this section, we highlight the key features of the physical layer aspects of UTRA that have been developed since then. Most of the material in this section is based on an amalgam of [242, 515, 517, 522–528].

22.2.1 Characteristics of UTRA

The proposed spectrum allocation for UTRA is shown in Figure 22.1. As can be seen, UTRA is unable to utilize the full frequency spectrum allocated for the 3G mobile radio systems during the WARC'92, since those frequency bands have also been partially allocated to the Digital Enhanced Cordless Telecommunications (DECT) systems. The radio access supports both *Frequency Division Duplex* (FDD) and *Time Division Duplex (TDD)* operations. The operating principles of these two schemes are discussed here in the context of Figure 22.2.

Specifically, the uplink (UL) and downlink (DL) signals are transmitted using different carrier frequencies f_1 and f_2, respectively, separated by a frequency guard band in FDD mode. On the other hand, the UL and DL messages in the TDD mode are transmitted using the same carrier frequency f_c, but in different time-slots, separated by a guard period. As seen from the spectrum allocation in Figure 22.1, the paired bands of 1920–1980 MHz and 2110–2170 MHz are allocated for FDD operation in the UL and DL, respectively, whereas the TDD mode is operated in the remaining unpaired bands [522]. The parameters designed for FDD and TDD operations are mutually compatible so as to ease the implementation of a dual-mode terminal capable of accessing the services offered by both FDD and TDD operators.

We note furthermore that recent research advocates the TDD mode quite strongly in the context of burst-by-burst adaptive CDMA modems [535], in order to adjust the modem parameters, such as the spreading factor or the number of bits per symbol on a burst-by-burst basis. This allows the system to more efficiently exploit the time-variant wireless channel capacity, hence maintaining a higher bits/s/Hz bandwidth efficiency. Furthermore, there have been proposals in the literature for allowing TDD operation in certain segments of the FDD spectrum as well, since FDD is incapable of surrendering the UL or DL frequency band of the

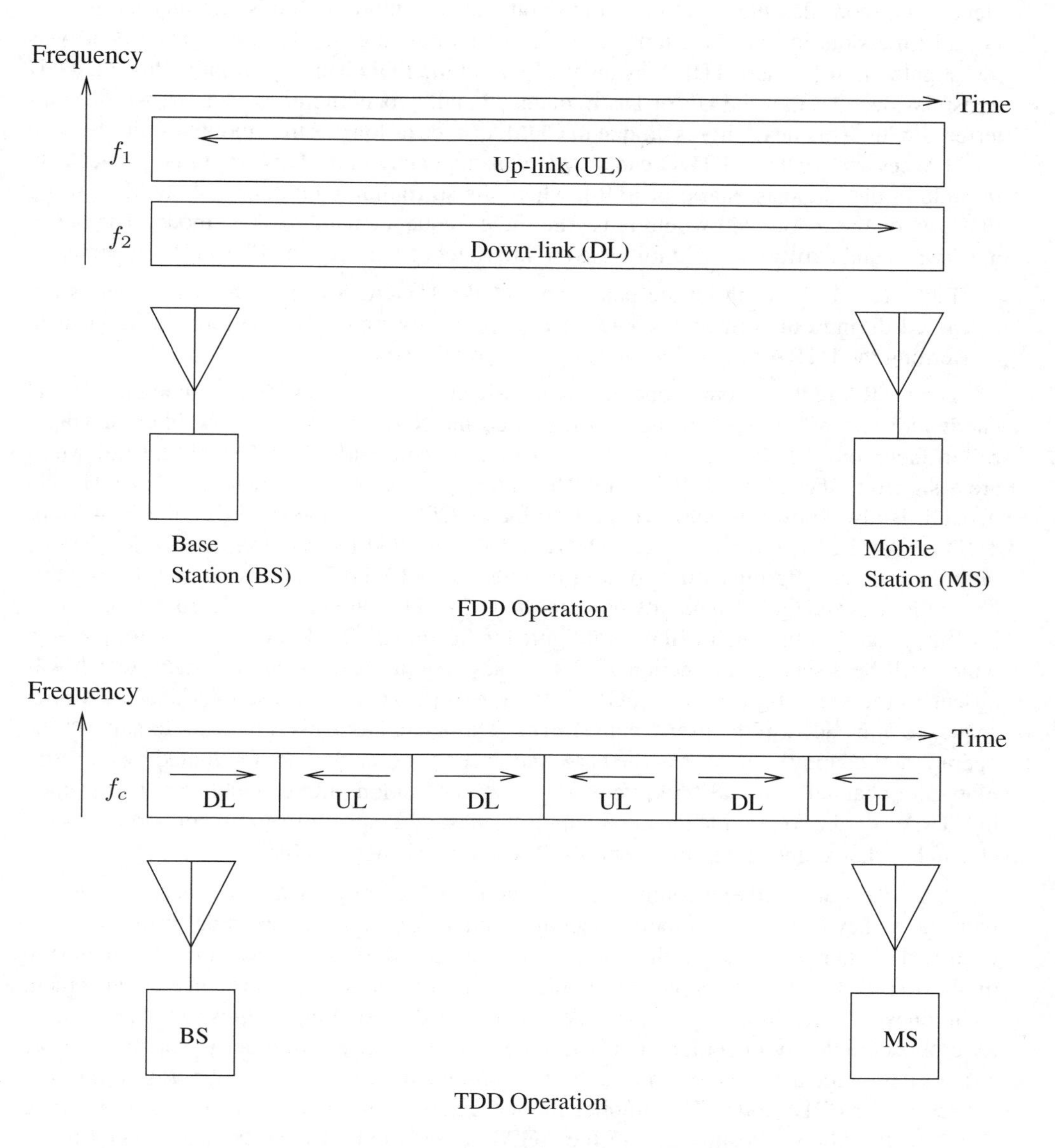

Figure 22.2: Principle of FDD and TDD operation.

duplex link, when the traffic demand is basically simplex. In fact, segmenting the spectrum in FDD/TDD bands inevitably results in some inefficiency in bandwidth utilization terms, especially in case of asymmetric or simplex traffic, when only one of the FDD bands is required. Hence, the more flexible TDD link could potentially double the link's capacity by allocating all time-slots in one direction. The idea of eliminating the dedicated TDD band was investigated [536], where TDD was invoked within the FDD band by simply allowing TDD transmissions in either the UL or DL frequency band, depending on which one was less interfered with. This flexibility is unique to CDMA, since as long as the amount of interference is not excessive, FDD and TDD can share the same bandwidth. This would be particularly feasible in the indoor scenario of [536], where the surrounding outdoor cell could be using FDD, while the indoor cell would reuse the same frequency band in TDD mode. The buildings' walls and partitions could mitigate the interference between the FDD/TDD schemes.

Table 22.2.1 shows the basic parameters of the UTRA. Some of these parameters are discussed during our further discourse, but significantly more information can be gleaned concerning the UTRA system by carefully studying the table.

The UTRA FDD system is operated at a basic chip rate of 3.84 Mcps,[3] giving a nominal bandwidth of 5 MHz, when using root-raised cosine Nyquist pulse-shaping filters having a rolloff factor of 0.22. On the other hand, two different chip rates of 3.84 Mcps and 1.28 Mcps are assigned to the UTRA TDD mode. The latter option gives a bandwidth of 1.6 MHz and hence it is also known as Narrowband CDMA (N-CDMA), while the FDD and 3.84 Mcps TDD modes are also known as W-CDMA. UTRA fulfilled the requirements of 3G mobile radio systems by offering a range of user bit rates up to 2 Mbps. Various services having different bit rates and QoS can readily be supported using Orthogonal Variable Spreading Factor (OVSF) codes [240], which will be highlighted in Section 22.2.6.1, and service multiplexing, which will be discussed in Section 22.2.4. A key feature of the UTRA system, which was absent in the second-generation (2G) IS-95 system [537] was the use of a dedicated pilot sequence embedded in the users' data stream. These can be invoked in order to support the operation of adaptive antennas at the base station (BS), which was not facilitated by the common pilot channel of the IS-95 system. However, a common pilot channel was still retained in UTRA in order to provide the demodulator's phase reference for certain common physical channels, when embedding pilot symbols for each user is not feasible.

Regardless of whether a common pilot channel is used or dedicated pilots are embedded in the data, they facilitate the employment of *coherent detection*. Coherent detection is known to provide better performance than noncoherent detection [90]. Furthermore, the inclusion of short spreading codes enables the implementation of various performance enhancement techniques, such as interference cancellers and joint-detection algorithms, which results in excessive complexity in conjunction with long spreading codes. In order to support flexible system deployment in indoor and outdoor environments, *intercell-asynchronous operation* is used in the FDD mode. This implies that no external timing source, such as a reference signal or the Global Positioning System (GPS) is required. However, in the TDD mode intercell synchronization is required in order to be able to seamlessly access the time-slots offered by adjacent BSs during handovers. This is achieved by maintaining synchronization between the BSs.

[3]In the UTRA RTT proposal submitted by ETSI to ITU, the chip rate was actually set to 4.096 Mcps.

22.2.2 Transport Channels

Transport channels are offered by the physical layer to the higher Open Systems Interconnection (OSI) layers, and they can be classified into two main groups, as shown in Table 22.3 [242, 522]. The Dedicated transport CHannel (DCH) is related to a specific Mobile Station (MS)-BS link, and it is used to carry user and control information between the network and an MS. Hence, the DCHs are bidirectional channels. There are eight transport channels within the common transport channel group, as shown in Table 22.3. The Broadcast CHannel (BCH) is used to carry system- and cell-specific information on the DL to all MSs in the entire cell. This channel conveys information, such as the initial UL transmit power of the MS during a random access transmission and the cell-specific scrambling code, as we shall see in Section 22.2.7. The Forward Access CHannel (FACH) of Table 22.3 is a DL common channel used for carrying control information and short user data packets to MSs, if the system knows the serving BS of the MS. On the other hand, the Paging CHannel (PCH) of Table 22.3 is used to carry control information to an MS if the serving BS of the MS is unknown, in order to page the MS, when there is a call for the MS. The Random Access CHannel (RACH) of Table 22.3 is an UL channel used by the MS to carry control information and short user data packets to the BS, in order to support the MS's access to the system, when it wishes to set up a call. As the terminology implies, the Downlink Shared CHannel (DSCH) and the High Speed DSCH (HS-DSCH) are DL transport channels that are shared by several users for bursty DL packet data traffic. Similarly, the Uplink Shared CHannel (USCH) is the uplink counterpart of DSCH, which is used for carrying packet-based traffic on the UL and it is shared by several users. Note that the USCH is only present in the TDD mode. Lastly, the Common Packet CHannel (CPCH) is an UL channel used for transmitting bursty data traffic in a contention-based random access manner and it only appears in the FDD mode.

The philosophy of these channels is fairly plausible, and it is informative as well as enlightening to explore the differences between the somewhat less flexible control regime of the 2G GSM [64] system and the more advanced 3G proposals, which we leave for the motivated reader due to lack of space. Unfortunately it is not feasible to design the control regime of a sophisticated mobile radio system by 'direct synthesis' and so some of the solutions reviewed throughout this section in the context of the 3G proposals may appear somewhat heuristic and quite ingenious. These solutions constitute an amalgam of the wireless research community's experience in the design of the existing 2G systems and of the lessons learned from their operation. Further contributing factors in the design of the 3G systems were based on solving the signalling problems specific to the favoured physical layer traffic channel solutions, namely, CDMA. In order to mention only one of them, the TDMA-based GSM system [64] was quite robust against power control inaccuracies, while the Pan-American IS-95 CDMA system [537] required an accurate power control. As we will see in Section 22.2.8, the power control problem was solved quite elegantly in the 3G proposals. We will also see that statistical multiplexing schemes — such as ALOHA, the original root of the recently more familiar Packet Reservation Multiple Access (PRMA) procedure — found their way into public mobile radio systems. A variety of further interesting solutions have also found applications in these 3G proposals, which are the results of the past decade of wireless system research. Let us now review the range of physical channels in the next section.

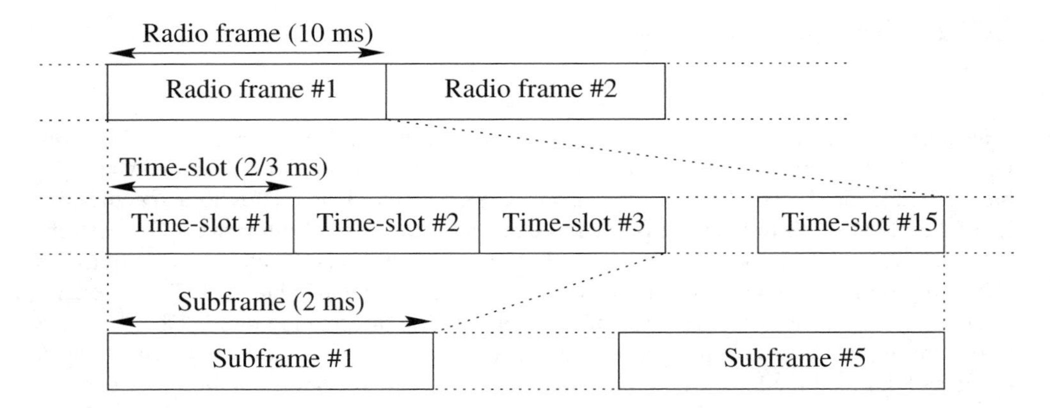

Figure 22.3: UTRA FDD and 3.84 Mcps TDD physical channel structure. The subframes are only used in high speed (HS) channels.

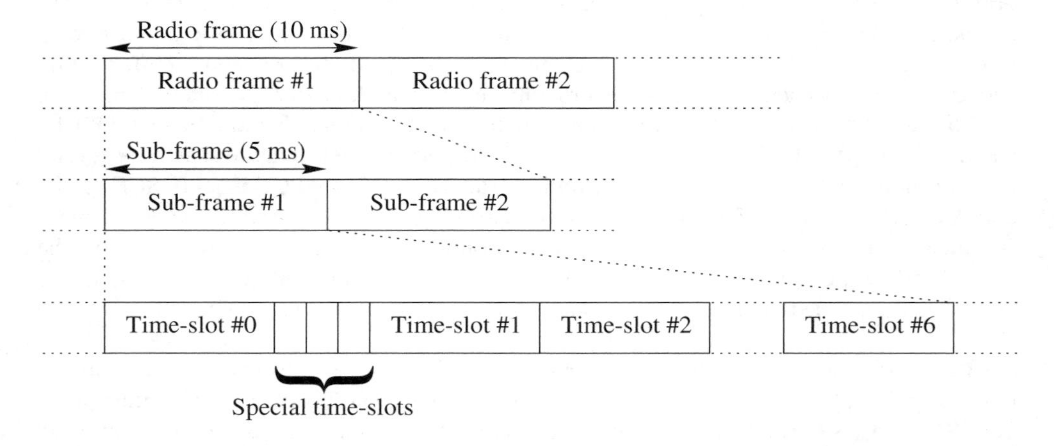

Figure 22.4: UTRA 1.28 Mcps TDD physical channel structure.

22.2.3 Physical Channels

The transport channels are transmitted using the physical channels. The physical channels are typically organized in terms of radio frames and time-slots, as shown in Figures 22.3 and 22.4 for W-CDMA and N-CDMA, respectively. For high speed (HS) channels supporting fast packet access, the radio frames are further subdivided into subframes, as shown in Figure 22.3. The philosophy of this hierarchical frame structure is also reminiscent to a certain degree of the GSM TDMA frame hierarchy of [64]. However, while in GSM each TDMA user had an exclusive slot allocation, in UTRA the number of simultaneous users supported is dependent on the users' required bit rate and their associated spreading factors. The MSs can transmit continuously in all slots or discontinuously, for example, when invoking a voice activity detector (VAD). Some of these issues will be addressed in Section 22.2.4. Under the regime of both the FDD mode and the 3.84 Mcps TDD mode, there are 15 time-slots within each radio frame, as depicted in Figure 22.3. The duration of each time-slot is $\frac{2}{3}$ ms, which gives a duration of 10 ms for the radio frame. As we shall see later in this section, the configuration of the information in the time-slots of the physical channels differs from one another in the UL and DL, as well as in the FDD and TDD modes. The 10 ms frame duration also conveniently coincides, for example, with the frame length of the ITU's G729 speech codec for speech communications, while it is a 'submultiple' of the GSM system's various full- and half-rate speech codecs' frame durations [64]. We also note that a convenient mapping of the video stream of the H.263 videophone codec can be arranged on the 10 ms-duration radio frames for supporting interactive video services, while on the move. Furthermore, the spreading factor (SF) can be varied on a 10 ms burst-by-burst (BbB) basis, in order to adapt the transmission mode in harmony with channel quality fluctuations, while maintaining a given target bit error rate. Although it is not part of the standard proposal, we found that it was more beneficial to adapt the number of bits per symbol on a BbB basis than varying the SF [535]. As for the 1.28 Mcps TDD mode, the 10 ms frame is divided into two 5 ms sub-frames, each having 7 normal time-slots and 3 special time-slots, as illustrated in Figure 22.4.

In the FDD mode, a DL physical channel is defined by its spreading code and frequency. Furthermore, in the UL, the modem's orthogonal in-phase (I) and quadrature-phase (Q) branches are used to deliver the data and control information simultaneously in parallel (as will be discussed in Figure 22.26). Thus, knowledge of the relative carrier phase, namely whether the I or Q branch is involved, constitutes part of the physical channel's identifier. On the other hand, in the TDD mode, a physical channel is defined by its spreading code, frequency, and time-slot.

Similarly to the transport channels of Table 22.3, the physical channels of UTRA may also be classified as dedicated and common channels. Table 22.4 shows the type of physical channels and the corresponding mapping of transport channels on the physical channels in FDD UTRA. The mapping used in TDD UTRA will be shown later in the context of Figure 22.5.

22.2.3.1 Dedicated Physical Channels of the FDD Mode

The dedicated physical channels of UTRA FDD shown in Table 22.4 consist of the Dedicated Physical Data CHannel (DPDCH) and the Dedicated Physical Control CHannel (DPCCH), both of which are bidirectional, as well as the UL High Speed-Dedicated Physical Control CHannel (HS-DPCCH), which is used for carrying UL feedback signalling information associated with the HS-DSCH transmission. The DPDCH is used for transmitting the DCH information between the BS and MS, while the DPCCH is used for conveying the Layer 1 information. The time-slot structures of the UL and DL DPDCH and its associated DPCCH are shown in Figures 22.5 and 22.6, respectively. Notice that on the DL, as illustrated by Figure 22.6, the DPDCH and DPCCH are interspersed by time-multiplexing to form a single Dedicated Physical CHannel (DPCH), as will be discussed in the context of Figure 22.27. On the other hand, the DPDCH and DPCCH on the UL are transmitted in parallel on the I and Q branches of the modem, as will become more explicit in the context of Figure 22.26 [242]. The reason for the parallel transmission on the UL is to avoid Electromagnetic Compatibility (EMC) problems due to Discontinuous Transmission (DTX) of the DPDCH of Table 22.4 [531]. DTX occurs when temporarily there are no data to transmit, but the link is

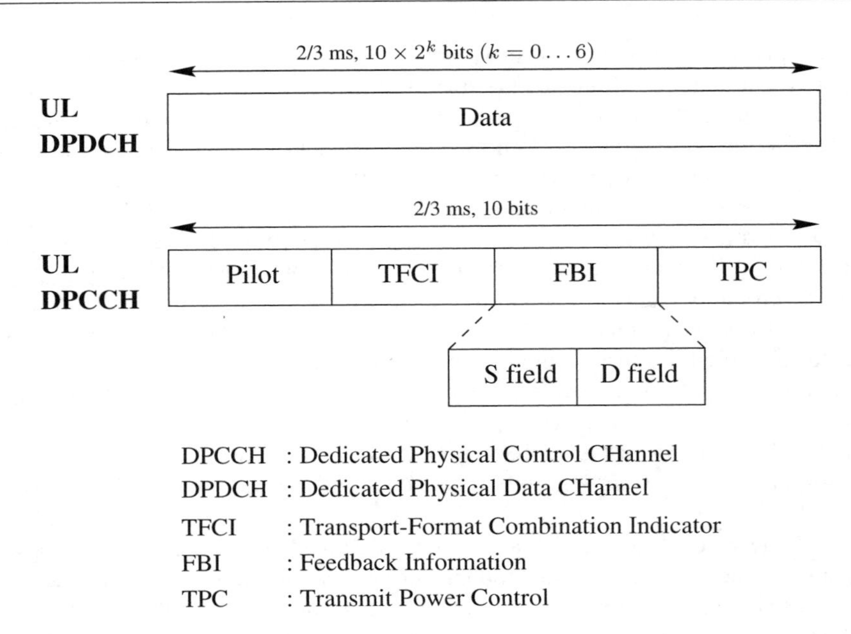

Figure 22.5: UTRA UL FDD DPDCH and DPCCH time-slot configuration, which is mapped to the time-slots of Figure 22.3. The UL DPDCH and DPCCH messages are transmitted in parallel on the I and Q branches of the modem of Figure 22.26. By contrast, the DPDCH and DPCCH bursts are time-multiplexed on the DL as shown in Figure 22.6.

still maintained by the DPCCH. If the UL DPCCH is time-multiplexed with the DPDCH, as in the DL of Figure 22.6, this can create short, sharp energy spikes. Since the MS may be located near sensitive electrical equipment, these spikes may affect this equipment.

As seen in Figures 22.5 and 22.6, the Layer 1 control information carried by the DPCCH corresponding to the DPDCH includes the pilot bits, Transmit Power Control (TPC) commands, and an optional Transport-Format Combination Indicator (TFCI). In addition, on the UL the Feedback Information (FBI) is also mapped to the DPCH in Figure 22.5. The pilot bits are used to facilitate coherent detection on both the UL and DL as well as to enable the implementation of performance enhancement techniques, such as adaptive antennas and interference cancellation. Since the pilot sequences are known, they can also be used as frame synchronization words in order to maintain transmission frame synchronization between the BS and MS. The TPC commands support an agile and efficient power control scheme, which is essential in DS-CDMA using the techniques to be highlighted in Section 22.2.8. The TFCI carries information concerning the instantaneous parameters of each transport channel multiplexed on the physical channel in the associated radio frame. The FBI is used to provide the capability to support certain transmit diversity techniques. The FBI field is further divided into two smaller fields as shown in Figure 22.5, which are referred to as the S field and D field. The S field is used to support the *Site Selection DiversiTy* (SSDT), which can reduce the amount of interference caused by multiple transmissions during a soft handover operation, while assisting in fast cell selection. On the other hand, the D field is used to provide attenuation and phase information in order to facilitate *closed-loop transmit diversity*, a

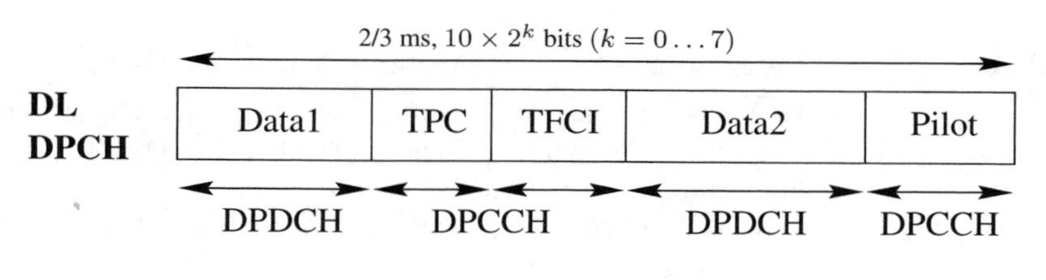

DPCCH : Dedicated Physical Control CHannel
DPDCH : Dedicated Physical Data CHannel
TFCI : Transport-Format Combination Indicator
TPC : Transmit Power Control

Figure 22.6: UTRA DL FDD dedicated physical channels time-slot configuration, which is mapped to the time-slots of Figure 22.3. The DPDCH and DPCCH messages are time-multiplexed on the DL, as it will be discussed in Figure 22.27. By contrast, the UL DPDCH and DPCCH bursts are transmitted in parallel on the I and Q branches of the modem as shown in Figure 22.5.

technique highlighted in Section 22.4.1.3. Given that the TPC and TFCI segments render the transmission packets 'self-descriptive', the system becomes very flexible, supporting burst-by-burst adaptivity, which substantially improves the system's performance [535], although this side-information is vulnerable to transmission errors.

The parameter k in Figures 22.5 and 22.6 determines the number of bits in each time-slot, which in turn corresponds to the bit rate of the physical channel. Therefore, the channel bit rates available for the UL DPDCH are 15/30/60/120/240/480/960 kbps, due to the associated 'payload' of 10×2^k bits per $\frac{2}{3}$ ms burst in Figure 22.5, where $k = 0 \ldots 6$. Note that the UL DPCCH has a constant channel bit rate of 15 kbps. Similarly, the channel bit rates available for the DL DPCH are 15/30/60/120/240/480/960 and 1920 kbps. However, since the user data are time-multiplexed with the Layer 1 control information, the actual user data rates on the DL will be slightly lower than those mentioned above. Even higher channel bit rates can be achieved using a technique known as multicode transmission [218], which will be highlighted in more detail in the context of Figure 22.24 in Section 22.2.5.

Apart from the DPDCH, a DL DPCCH may also correspond to a CPCH transmission. The time-slot structure of this DPCCH is shown in Figure 22.7, which consists of pilot bits, TFCI, TPC and CPCH Control Commands (CCC). Examples of the CCC include the Start of Message Indicator and the Emergency Stop. The spreading factor and the channel bit rate of the DPCCH used for CPCH are 512 and 15 kbps, respectively.

The subframe structure of the HS-DPCCH is illustrated in Figure 22.8, which consists of the Hybrid ARQ-Acknowledgement (HARQ-ACK) field and the Channel Quality Indication (CQI) field. According to Figure 22.8, the spreading factor of the HS-DPCCH is 256 and it has a channel bit rate of 15 kbps. Only one HS-DPCCH can exist with each UL DPCCH. Let us now consider the common physical channels summarized in Table 22.4.

Proposal	Description	Multiple Access	Source
DECT	Digital Enhanced Cordless Telecommunications	Multicarrier TDMA (TDD)	ETSI Project (EP) DECT
UWC-136	Universal Wireless Communications	TDMA (FDD and TDD)	USA TIA TR45.3
WIMS W-CDMA	Wireless Multimedia and Messaging Services Wideband CDMA	Wideband CDMA (FDD)	USA TIA TR46.1
TD-CDMA	Time Division Synchronous CDMA	Hybrid with TDMA/CDMA/ SDMA (TDD)	Chinese Academy of Telecommunication Technology (CATT)
W-CDMA	Wideband CDMA	Wideband DS-CDMA (FDD and TDD)	Japan ARIB
CDMA II	Asynchronous DS-CDMA	DS-CDMA (FDD)	South Korean TTA
UTRA	UMTS Terrestrial Radio Access	Wideband DS-CDMA (FDD and TDD)	ETSI SMG2
NA: W-CDMA	North America Wideband CDMA	Wideband DS-CDMA (FDD and TDD)	USA T1P1-ATIS
cdma2000	Wideband CDMA (IS-95)	DS-CDMA (FDD and TDD)	USA TIA TR45.5
CDMA I	Multiband synchronous DS-CDMA	Multiband DS-CDMA	South Korean TTA

Table 22.1: Proposals for the radio transmission technology of terrestrial IMT-2000 (obtained from ITU's web site: http://www.itu.int/imt).

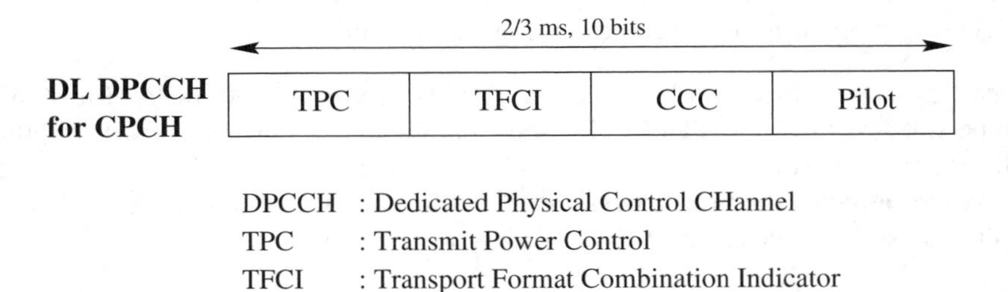

Figure 22.7: UTRA UL FDD DL-DPCCH for CPCH time-slot configuration.

Radio Access Technology	FDD : DS-CDMA TDD : TDMA/CDMA
Operating environments	Indoor/Outdoor to indoor/Vehicular
Chip rate (Mcps)	3.84 (FDD & TDD), 1.28 (TDD)
Channel bandwidth (MHz)	5 for 3.84 Mcps, 1.6 for 1.28 Mcps
Nyquist rolloff factor	0.22
Duplex modes	FDD and TDD
Channel bit rates (kbps)	FDD (UL) : 15/30/60/120/240/480/960 FDD (DL) : 15/30/60/120/240/480/960/1920 TDD (UL)† : variable, from 366 to 6624 TDD (DL)† : 366/414/5856/6624
Frame length	10 ms
Spreading factor	FDD (UL) : variable, 4 to 256 FDD (DL) : variable, 4 to 512 TDD (UL) : variable, 1 to 16 TDD (DL) : 1, 16
Detection scheme	Coherent with time-multiplexed pilot symbols Coherent with common pilot channel
Intercell operation	FDD : Asynchronous TDD : Synchronous
Power control	Inner-loop Open loop (TDD UL)
Transmit power dynamic range	80 dB (UL), 30 dB (DL)
Handover	Soft handover Inter-frequency handover

† Channel bit rate per time-slot.

Table 22.2: Basic parameters of the UTRA system

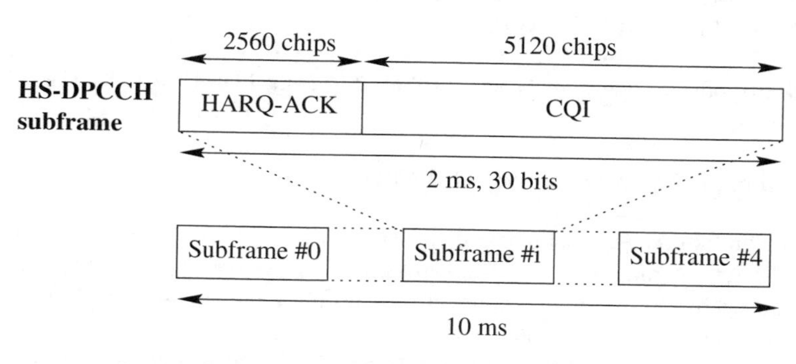

Figure 22.8: UTRA UL FDD HS-DPCCH subframe configuration.

Dedicated Transport Channel	Common Transport Channel
Dedicated CHannel (DCH) (UL/DL)	Broadcast CHannel (BCH) (DL)
	Forward Access CHannel (FACH) (DL)
	Paging CHannel (PCH) (DL)
	Random Access CHannel (RACH) (UL)
	Downlink Shared CHannel (DSCH) (DL)
	High Speed Downlink Shared CHannel (HS-DSCH) (DL)
	Common Packet CHannel† (CPCH) (UL)
	Uplink Shared CHannel‡ (USCH) (UL)

†Only for FDD mode.
‡Only for TDD mode.

Table 22.3: UTRA transport channels.

Dedicated Physical Channels	Transport Channels
Dedicated Physical Data CHannel (DPDCH) (UL/DL)†	DCH
Dedicated Physical Control CHannel (DPCCH) (UL/DL)	
High Speed-Dedicated Physical Control CHannel (HS-DPCCH) (UL)	
Common Physical Channels	**Transport Channels**
Physical Random Access CHannel (PRACH) (UL)	RACH
Physical Common Packet CHannel (PCPCH) (UL)	CPCH
Common Pilot CHannel (CPICH) (DL)	
Primary Common Control Physical CHannel (P-CCPCH) (DL)	BCH
Secondary Common Control Physical CHannel (S-CCPCH) (DL)	FACH, PCH
Synchronisation CHannel (SCH) (DL)	
Physical Downlink Shared CHannel (PDSCH) (DL)	DSCH
Acquisition Indicator CHannel (AICH) (DL)	
Access Preamble Acquisition Indicator CHannel (AP-AICH) (DL)	
Paging Indicator CHannel (PICH) (DL)	
CPCH Status Indicator CHannel (CSICH) (DL)	
Collision Detection/Channel Assignment Indicator CHannel (CD/CA-ICH) (DL)	
High Speed-Physical Downlink Shared CHannel (HS-PDSCH) (DL)	HS-DSCH
High Speed-Shared Control CHannel (HS-SCCH) (DL)	

†On the DL, the DPDCH and DPCCH are time-multiplexed in each time-slot to form a single Dedicated Physical CHannel (DPCH)

Table 22.4: Mapping the transport channels of Table 22.3 to the FDD UTRA physical channels.

22.2.3.2 Common Physical Channels of the FDD Mode

The Physical Random Access CHannel (PRACH) of Table 22.4 is used to carry the RACH message on the UL. A random access transmission is activated whenever the MS has data to transmit and wishes to establish a connection with the local BS. Although the procedure of this transmission will be elaborated on in Section 22.2.7, here we will briefly highlight the structure of a random access transmission burst. Typically, a random access burst consists of one or several so-called preambles and a message. Each preamble contains a signature that is constructed of 256 repetitions of a 16-chip Hadamard code, which yields a $256 \times 16 = 4096$-chip-long signature. Similarly to the UL dedicated physical channels of Figure 22.5, the message part of the random access transmission consists of data information and control information that are transmitted in parallel on the I/Q channels of the modulator,

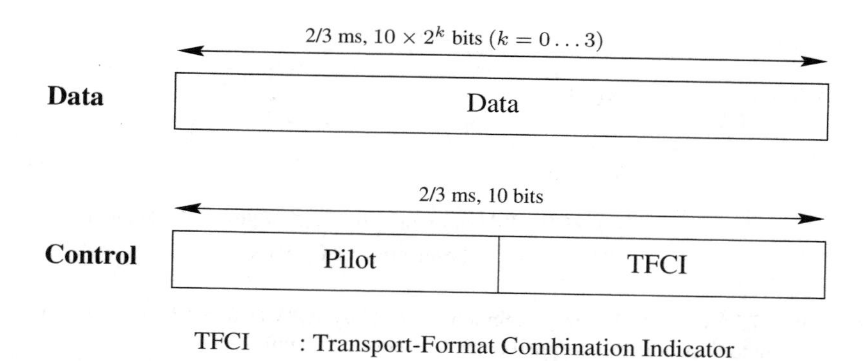

Figure 22.9: The time-slot configuration of the message part during a random access transmission in UTRA, which are mapped to the frame structure of Figure 22.3. The data and control information are multiplexed on the I/Q channels of the modulator and the frame is transmitted at the beginning of an access slot, as it will be discussed in Section 22.2.7.1.

as shown in Figure 22.9. The channel bit rates available for the data part of the message are 15/30/60/120 kbps. By contrast, the control information, which contains an 8-bit pilot and a 2-bit TFCI, is transmitted at a fixed rate of 15 kbps. Obviously in this case, no FBI and TPC commands are required, since transmission is initiated by the MS.

The Physical Common Packet CHannel (PCPCH) of Table 22.4 is used to carry the CPCH message on the UL, based on a Digital Sense Multiple Access-Collision Detection (DSMA-CD) random access technique. A CPCH random access burst consists of one or several Access Preambles (A-P), one Collision Detection Preamble (CD-P), a DPCCH Power Control Preamble (PC-P), and a message. The length of both the A-P and CD-P spans a total of 4096 chips, while the duration of the PC-P can be equivalent to either 0 or 8 time-slots. Each time-slot of the PC-P contains the pilot, the FBI, and the TPC bits, which is similar to the DPCCH shown in Figure 22.5, where the TFCI field is filled with '1' bits. The message part of the CPCH burst consists of a data part and a control part, which is identical to the UL dedicated physical channel shown in Figure 22.5 in terms of its structure and available channel bit rates. A 15 kbps DL DPCCH is always associated with an UL PCPCH, as highlighted in Section 22.2.3. Hence, both the FBI and TPC information are included in the message conveyed by a CPCH burst in order to facilitate the employment of DL transmit diversity and power control, unlike in a RACH burst. The procedure of a CPCH transmission will be further elaborated on in Section 22.2.7.

The DL Primary Common Control Physical CHannel (P-CCPCH) of Table 22.4 is used by the BS in order to broadcast the BCH information at a fixed rate of 30 kbps to all MSs in the cell. The P-CCPCH is transmitted only after the first 256 chips of each slot, as shown in Figure 22.10. During the first 256 chips of each slot, the Synchronization CHannel (SCH) message is transmitted instead, as will be discussed in Section 22.2.10. The P-CCPCH is used as a timing reference directly for all the DL physical channels and indirectly for all the UL physical channels. Hence, as long as the MS is synchronized to the DL P-CCPCH of a specific cell, it is capable of detecting any DL messages transmitted from that BS by listening at the predefined times. For example, the DL DPCH will commence transmission at an offset, which is a multiple of 256 chips from the start of the P-CCPCH radio frame seen in

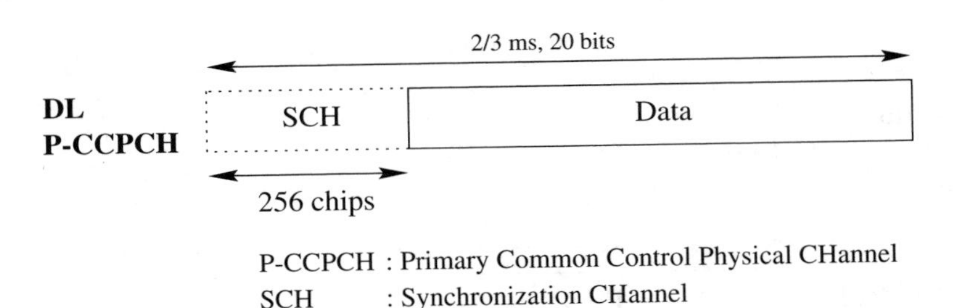

Figure 22.10: UTRA DL FDD Primary Common Control Physical CHannel (P-CCPCH) time-slot configuration, which is mapped to the time-slots of Figure 22.3.

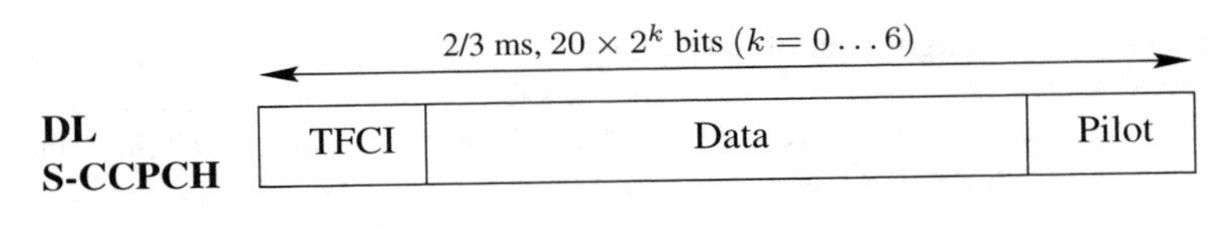

Figure 22.11: UTRA DL FDD Secondary Common Control Physical CHannel (S-CCPCH) time-slot configuration, which is mapped to the time-slots of Figure 22.3.

Figure 22.10. Upon synchronization with the P-CCPCH, the MS will know precisely when to begin receiving the DL DPCH. The UL DPDCH/DPCCH is transmitted 1024 chips after the reception of the corresponding DL DPCH.

The Secondary Common Control Physical CHannel (S-CCPCH) of Table 22.4 carries the FACH and PCH information of Table 22.3 on the DL, and they are transmitted only when data are available for transmission. The S-CCPCH will be transmitted at an offset, which is a multiple of 256 chips from the start of the P-CCPCH message seen in Figure 22.10. This will allow the MS to know exactly when to detect the S-CCPCH, as long as the MS is synchronized to the P-CCPCH. The time-slot configuration of the S-CCPCH is shown in Figure 22.11. Notice that the S-CCPCH message can be transmitted at a variable bit rate, namely, at 30/60/120/240/480/960/1920 kbps.

At this stage it is worth mentioning that the available control channel rates are significantly higher in the 3G systems than in their 2G counterparts. For example, the maximum BCH signalling rate in GSM [64] is more than an order of magnitude lower than the above-mentioned 30 kbps UTRA BCH rate. In general, this increased control channel rate will support a significantly more flexible system control than the 2G systems.

The Physical Downlink Shared CHannel (PDSCH) of Table 22.4 is used for carrying the DSCH message at rates of 30/60/120/240/480/960/1920 kbps. Similarly, the High Speed-PDSCH (HS-PDSCH) is used for conveying the HS-DSCH message at rates of 480/960 kbps. Both these channels are shared among several users based on code multiplexing. Each PDSCH or HS-PDSCH is associated with one or more DL DPCH. The Layer 1 control information of the PDSCH is transmitted on the DPCCH of the associated DL DPCH. However, since the structure of the HS-PDSCH is configured in terms of subframes rather than radio

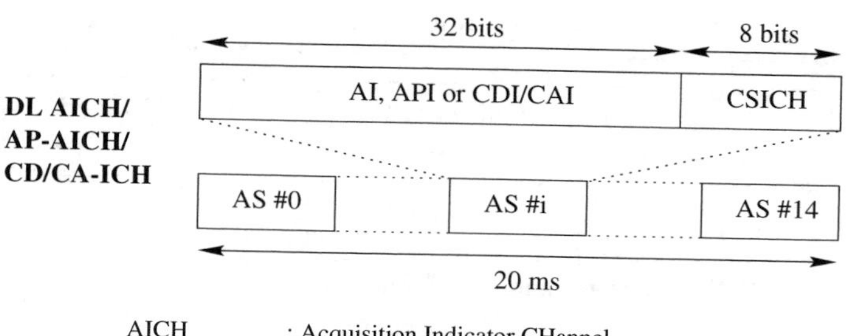

AICH : Acquisition Indicator CHannel
AP-AICH : Access Preamble Acquisition Indicator CHannel
CD/CA-ICH : Collision Detection/Channel Assignment Indicator CHannel
AI : Acquisition Indicator
API : Access Preamble Indicator
CDI/CAI : Collision Detection/Channel Assignment Indicator
CSICH : CPCH Status Indicator CHannel
AS : Access Slot

Figure 22.12: UTRA DL Acquisition Indicator CHannel (AICH), the CPCH Access Preamble Acquisition Indicator CHannel (AP-AICH), and the CPCH Collision Detection/Channel Assignment Indicator Channel (CD/CA-ICH) Access Slot (AS) configuration, which is mapped to the AS of the corresponding channel. Since it has a duration of 20 ms, it is mapped to every other 10 ms frame in Figure 22.3.

frames, the Layer 1 information of the HS-PDSCH will be transmitted using the HS-Shared Control CHannel (HS-SCCH) at a rate of 60 kbps.

The Acquisition Indicator CHannel (AICH), the CPCH Access Preamble Acquisition Indicator CHannel (AP-AICH) and the CPCH Collision Detection/Channel Assignment Indicator Channel (CD/CA-ICH) of Table 22.4 are used for carrying Acquisition Indicators (AI), AP acquisition Indicators (API) of CPCH and CD and/or CA Indicator (CDI/CAI) messages, respectively. More specifically, the AI is a response to a PRACH transmission, while the API and the CDI are responses to the CPCH's A-P and CD-P, respectively and they correspond to the signatures used by the associated PRACH preamble, a PCPCH A-P or a PCPCH CD-P, which were defined above. The structure of the AICH, AP-AICH and the CD/CA-ICH consists of a repeated sequence of 15 consecutive Access Slots (AS). Each AS consists of a 32-symbol AI, API or CDI/CAI part and an eight-symbol part, reserved for the CPCH Status Indicator CHannel (CSICH), as shown in Figure 22.12. The CSICH carries the status information about the PCPCH, which is based on the DSMA-CD protocol. The AS#0 will commence at the start of every other 10 ms P-CCPCH radio frame seen in Figure 22.3, since its duration is 20 ms. The phase reference required for the AICH is obtained from the Primary CPICH.

The Page Indicator CHannel (PICH) is used to carry Page Indicator (PI) messages. A PI message is used to signal to the MS on the associated S-CCPCH that there are data addressed to it, in order to facilitate a power-efficient sleep-mode operation. A PICH, illustrated in Figure 22.13, is a 10 ms frame consisting of 300 bits, out of which 288 bits are used to carry PIs, while the remaining 12 bits are unused. Each PICH frame can carry a total of N PIs, where $N = 18, 36, 72$, and 144. The PICH is also transmitted at an offset with respect to the start of the P-CCPCH, which is a multiple of 256 chips. The associated S-CCPCH will be

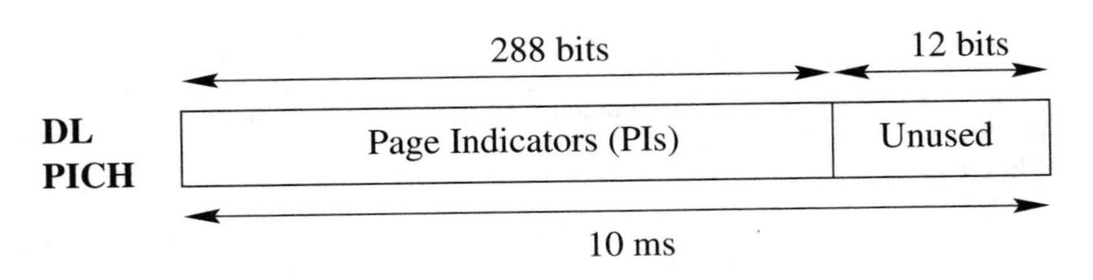

Figure 22.13: UTRA DL Page Indicator CHannel (PICH) configuration. Each PICH frame can carry a total of N PIs, where $N = 18, 36, 72$, and 144.

transmitted 7680 chips later.

Finally, the Common PIlot CHannel (CPICH) of Table 22.4 is a 30 kbps DL physical channel that carries a predefined bit sequence. It provides a phase reference for the SCH, P-CCPCH, AICH, AP-AICH, CD/CA-ICH, CSICH and the PICH, since these channels do not carry pilot bits, as shown in Figures 22.10, 22.12, and 22.13, respectively. The CPICH is transmitted synchronously with the P-CCPCH.

22.2.3.3 Physical Channels of the TDD Mode

In contrast to the previously discussed FDD structures of Figures 22.5–22.13, in TDD operation the burst structures of Figure 22.14 and Figure 22.15, which are proposed for the 3.84 Mcps and 1.28 Mcps modes, respectively, are used for all the physical channels. In the 3.84 Mcps TDD mode, each time-slot's transmitted information can be arbitrarily allocated to the DL or UL, as shown in the three possible TDD allocations of Figure 22.16. On the other hand, in the 1.28 Mcps TDD mode, the first slot is always allocated for the DL, while the second slot is always dedicated to the UL, as illustrated in Figure 22.17. The remaining slots can be arbitrarily allocated to the DL or UL. Hence, this flexible allocation of the UL and DL burst in the TDD mode enables the use of an adaptive modem [4, 535, 538] whereby the modem parameters, such as the spreading factor or the number of bits per symbol can be adjusted on a burst-by-burst basis to optimise the link quality. The terminology of symmetric UL/DL allocation refers to a scenario, in which an approximately equal number[4] of DL and UL bursts are allocated within a TDD frame, while in asymmetric UL/DL allocation, there is an unequal number of UL and DL bursts, such as, for example, in 'near-simplex' file download from the Internet or in video-on-demand type applications.

In UTRA, three different TDD burst structures, known as *Burst Type 1*, *Burst Type 2* and *Burst Type 3*, are defined, as shown in Figure 22.14. Type 1 and 2 bursts can be used for both the UL and DL, while the Type 3 burst is only used on the UL. Notice that the Type 3 burst has a longer guard period and this makes it more suitable for initial access or access to a new cell after a handover. The mid-amble sequences that are allocated to the different TDD bursts in each time-slot belong to a so-called *mid-amble code set*. The codes in each mid-amble code set are derived from a unique cell-specific *Basic Mid-amble Code*. Adjacent cells are

[4]Since there are 15 time-slots per frame, there will always be one more additional DL or UL burst per frame in a symmetric allocation.

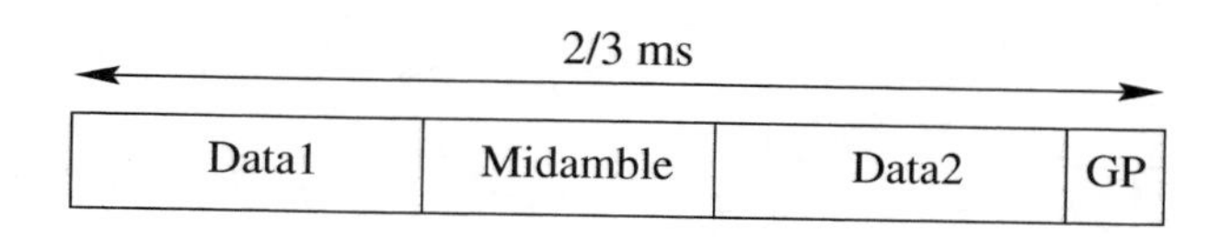

	Data1	Midamble	Data2	Guard Period (GP)
Burst Type 1	976	512	976	96
Burst Type 2	1104	256	1104	96
Burst Type 3	976	512	880	192
Length in chips				

Figure 22.14: Burst configuration mapped on the TDD burst structure of Figure 22.16 in the UTRA 3.84 Mcps TDD mode. Three different types of TDD bursts are defined in UTRA, namely, Burst Type 1, Burst Type 2 and Burst Type 3.

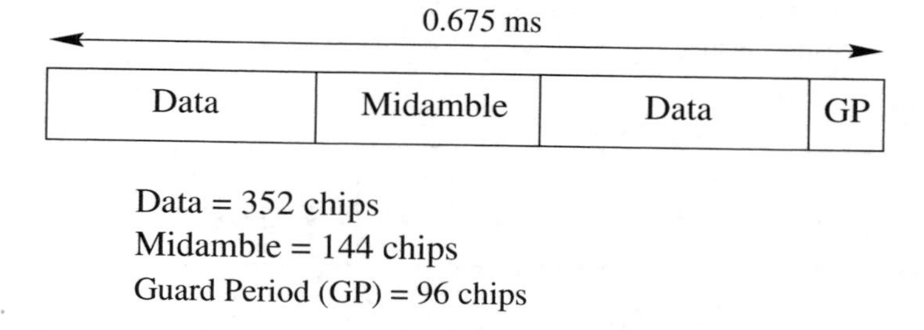

Figure 22.15: Burst configuration mapped on the TDD burst structure of Figure 22.17 in the UTRA 1.28 Mcps TDD mode.

allocated different mid-amble code sets, i.e. a different basic mid-amble code. This can be exploited for assisting in cell identification.

The mapping of the transport channels and the physical channels in TDD UTRA is shown in Table 22.5. As we can see, unlike in the FDD mode, there is only one type of Dedicated Physical CHannel (DPCH) in the TDD mode. Hence, the Layer 1 control information — such as the TPC command and the TFCI information — will be transmitted in the data field of Figure 22.14, if required. The TDD burst structures that incorporate the TFCI information as well as the TFCI+TPC information are shown in Figure 22.18a and Figure 22.18b, respectively, while the transmission of the Synchronization Shift (SS) symbols with TPC is shown in Figure 22.18c. This should be contrasted with their corresponding FDD allocations in Figures 22.5 and 22.6. The TFCI field is divided into two parts, which reside immediately before and after the mid-amble (or after the TPC command, if power control is invoked) in the data field. The TPC command is always transmitted immediately after the mid-amble, or immediately after the SS field, as portrayed in Figure 22.18. As a result of these control information segments, the amount of user data is reduced in each time-slot. Note that the TPC command is only transmitted on the UL in the 3.84 Mcps TDD mode. Also the SS symbols are only transmitted in the 1.28 Mcps TDD UL mode, in order to facilitate synchronization

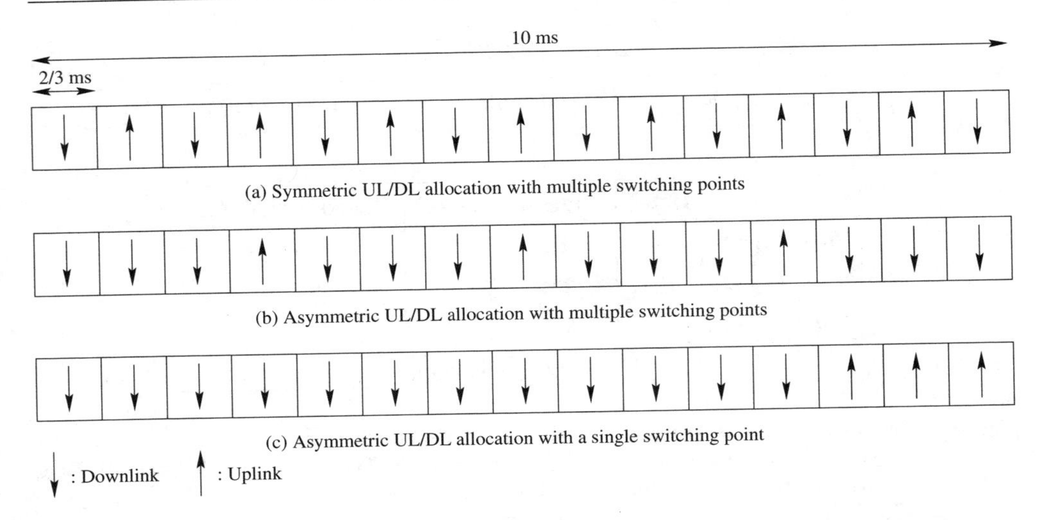

Figure 22.16: Uplink/downlink allocation examples for the 15 slots in UTRA 3.84 Mcps TDD operation using the time-slot configurations of Figure 22.14.

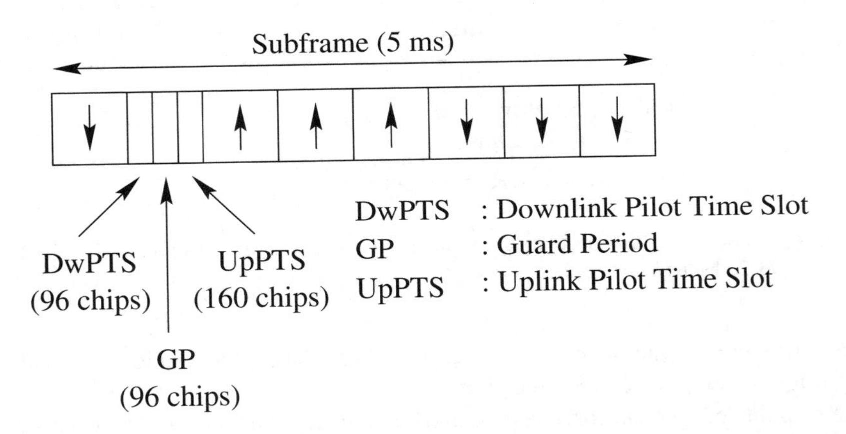

Figure 22.17: uplink/downlink allocation examples for the 7 slots in UTRA 1.28 Mcps TDD operation using the time-slot configurations of Figure 22.15.

control and these symbols are transmitted immediately after the mid-amble.

Most of the common physical channels in the TDD mode shown in Table 22.5 are similar to those in the FDD mode, as given in Table 22.4. Hence in this section, we will only highlight those channels that are unique to TDD.

The Physical Uplink Shared CHannel (PUSCH) carries the USCH messages. Since the Layer 1 information can be embedded into a TDD burst as illustrated in Figure 22.14, the PUSCH as well as the PDSCH used in the TDD mode do not have to be associated with any DPCH. The Physical Node B Synchronization CHannel (PNBSCH) carries the cell sync bursts, in order to maintain intercell synchronization via the air interface in the TDD mode. The High Speed-Shared Information CHannel (HS-SICH) carries the higher layer control in-

Dedicated Physical Channels	Transport Channels
Dedicated Physical CHannel (DPCH) (UL/DL)	DCH
Common Physical Channels	**Transport Channels**
Physical Random Access CHannel (PRACH) (UL)	RACH
Physical Uplink Shared CHannel (PUSCH) (UL)	USCH
Primary Common Control Physical CHannel (P-CCPCH) (DL)	BCH
Secondary Common Control Physical CHannel (S-CCPCH) (DL)	FACH, PCH
Synchronisation CHannel (SCH) (DL)	
Physical Node B Synchronisation CHannel (PNBSCH)‡	
Physical Downlink Shared CHannel (PDSCH) (DL)	DSCH
Paging Indicator CHannel (PICH) (DL)	
Downlink Pilot CHannel (DwPCH) †	
Uplink Pilot CHannel (UpPCH) †	
Fast Physical Access CHannel (FPACH) †	
High Speed-Physical Downlink Shared CHannel (HS-PDSCH) (DL)	HS-DSCH
High Speed-Shared Control CHannel (HS-SCCH) (DL)	
High Speed-Shared Information CHannel (HS-SICH)	

†Only in 1.28 Mcps TDD mode ‡ Only in 3.84 Mcps TDD mode

Table 22.5: Mapping the transport channels of Table 22.3 to the TDD UTRA physical channels.

Data	TFCI	Midamble	TFCI	Data	GP

a) Burst structure with TFCI information only

Data	TFCI	Midamble	TPC	TFCI	Data	GP

b) Burst structure with TFCI and TPC information only

Data	Midamble	SS	TPC	Data	GP

c) Burst structure with SS and TPC information only

TFCI : Transport Format Combination Indicator
TPC : Transmit Power Control
GP : Guard Period
SS : Synchronization Shift

Note : The SS symbols are only transmitted in the 1.28 Mcps TDD mode

Figure 22.18: Burst configuration mapped on the TDD burst configuration of Figure 22.16 in the UTRA TDD mode incorporating TFCI, TPC and SS information.

formation and the CQI for the HS-DSCH in TDD, similar to the HS-DPCCH used in the FDD mode. The Downlink Pilot CHannel (DwPCH) is used by the mobile in order to establish the DL synchronization with the cell during a cell search procedure. Once the DL synchronization is achieved, an UL synchronization procedure will commence, which will involve the Uplink Pilot CHannel (UpPCH). Note that synchronization using DwPCH and UpPCH is only invoked in the 1.28 Mcps TDD mode. Synchronization in the 3.84 Mcps TDD mode is still established by the SCH. Finally, the Fast Physical Access CHannel (FPACH) is used by the Node B to carry the acknowledgement of a detected signature with timing and power level adjustment indication to a mobile.

Having highlighted the basic features of the various UTRA channels, let us now consider how the various services are error protected, interleaved, and multiplexed on to the physical channels. This issue is discussed with reference to Figures 22.19 and 22.20 in the context of UTRA.

22.2.4 Service Multiplexing and Channel Coding in UTRA

Service multiplexing is employed when multiple services of identical or different bit rates requiring different QoS belonging to the same user's connection are transmitted. An example would be the simultaneous transmission of a voice and video service for a multimedia application. Each service is represented by its corresponding transport channels, as described in Section 22.2.2. The coding and multiplexing of the transport channels are performed in sets of transport blocks that arrived from the higher layers at fixed intervals of 10, 20, 40 or 80 ms. These intervals are known as the *Transmission Time Interval* (TTI). Note that the number of bits on each transport channel can vary between different TTIs, as well as between different transport channels. A possible method of transmitting multiple services is by using code-multiplexing with the aid of orthogonal codes. Every service could have its own DPDCH and DPCCH, each assigned to a different orthogonal code. This method is not very efficient, however, since a number of orthogonal codes would be reserved by a single user, while on the UL it would also inflict self-interference when the multiple DPDCH and DPCCH codes' orthogonality is impaired by the fading channel. Alternatively, these services can be time-multiplexed into one or several DPDCHs, as shown in Figures 22.19 and 22.20 for the FDD UL and DL, respectively. The algorithm of Figure 22.19 also applies to the TDD mode for both the UL and DL. Note that here we only highlight the multiplexing process for the DPCH transport channel. The multiplexing algorithms used for other transport channels may differ and interested readers might like to refer to the specifications found in [515] for further details. The function of the individual processing steps is detailed below.

22.2.4.1 CRC Attachment

A Cyclic Redundancy Checksum (CRC) is first calculated for each incoming transport block within a TTI. The CRC consists of either 24, 16, 12, 8, or 0 parity bits, which is decided by the higher layers. The CRC is then attached to the end of the corresponding transport block in order to facilitate reliable error detection at the receiver. This facility is very important, for example, for generating the video packet acknowledgement flag in wireless video telephony using standard video codecs, such as H.263 [539].

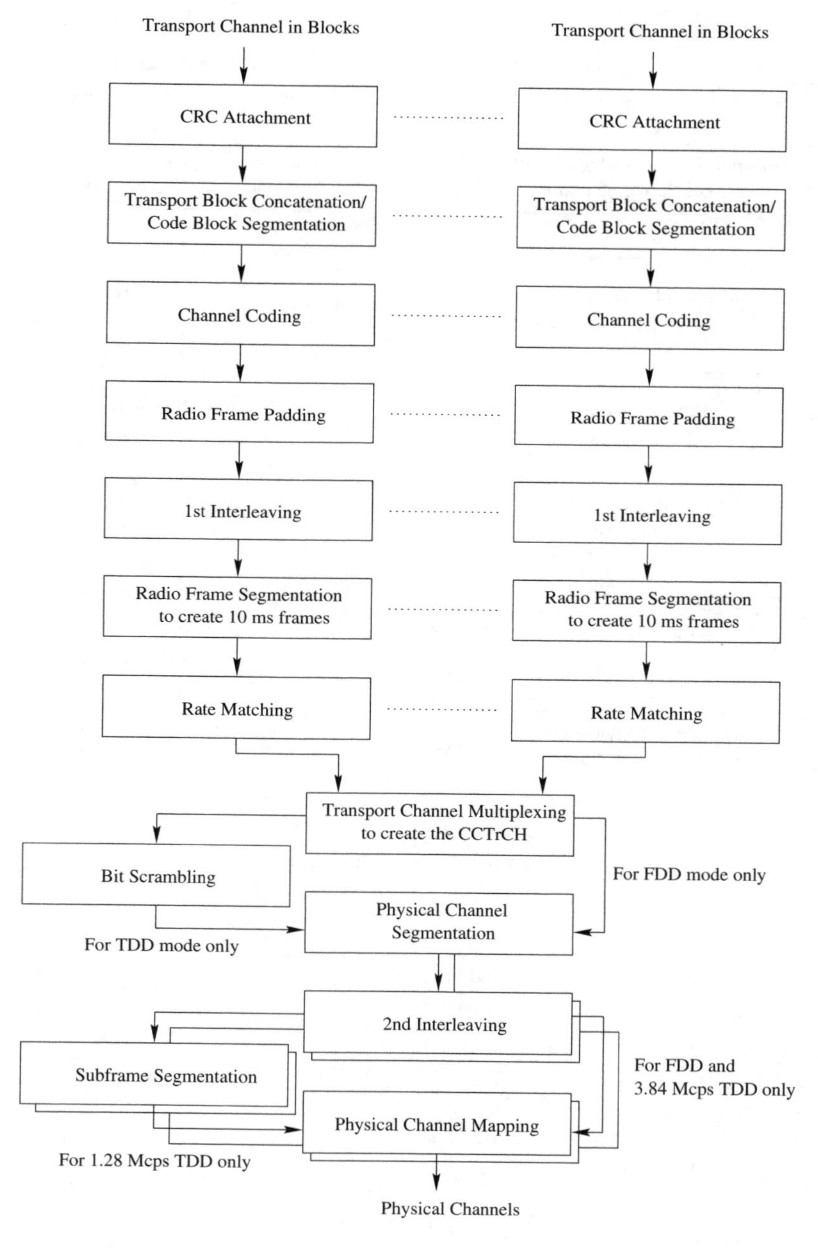

Figure 22.19: Transport channel-coding/multiplexing flowchart for the FDD UL as well as for both the TDD UL and DL in UTRA.

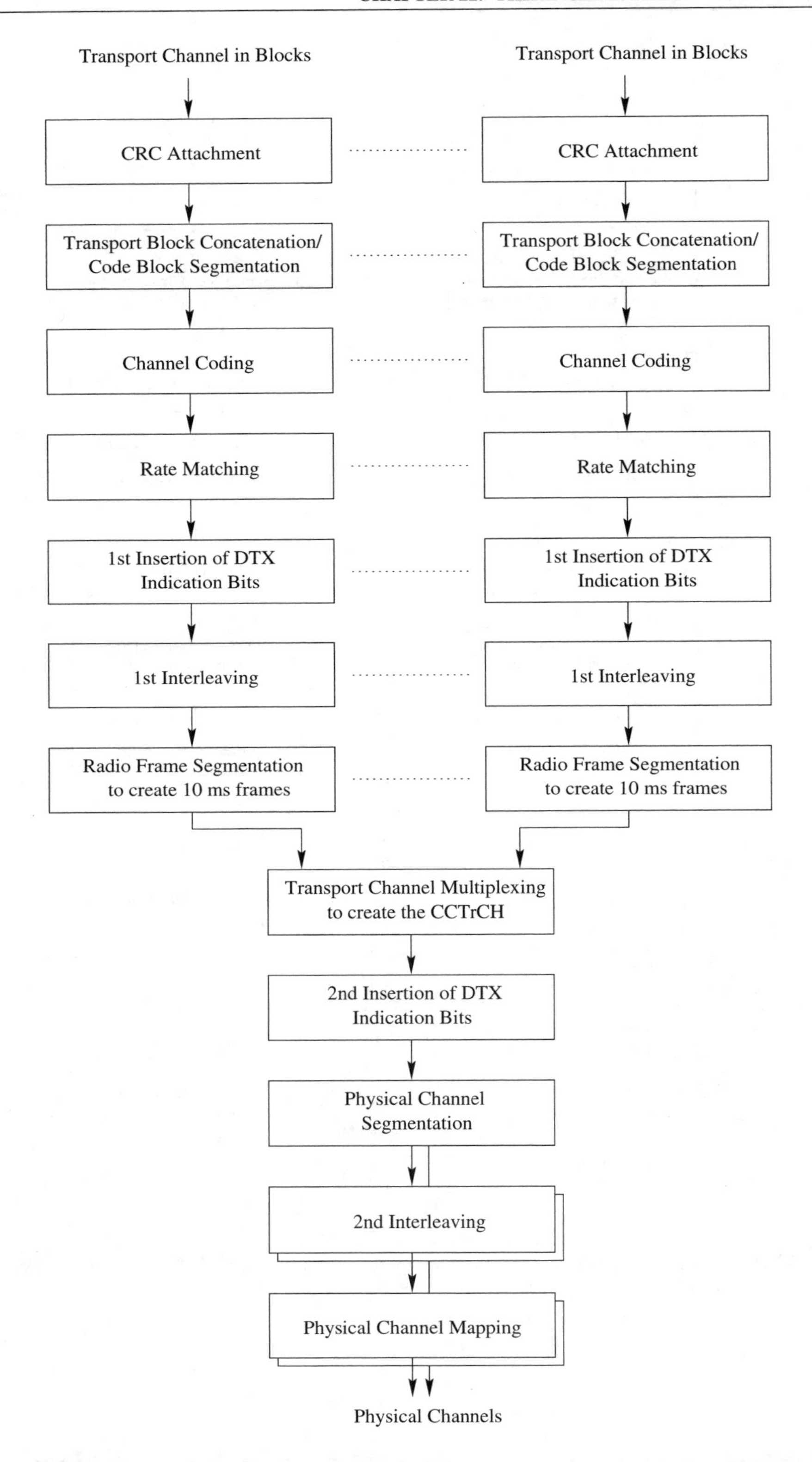
Transport Channel in Blocks
Transport Channel in Blocks
CRC Attachment
CRC Attachment
Transport Block Concatenation/ Code Block Segmentation
Transport Block Concatenation/ Code Block Segmentation
Channel Coding
Channel Coding
Rate Matching
Rate Matching
1st Insertion of DTX Indication Bits
1st Insertion of DTX Indication Bits
1st Interleaving
1st Interleaving
Radio Frame Segmentation to create 10 ms frames
Radio Frame Segmentation to create 10 ms frames
Transport Channel Multiplexing to create the CCTrCH
2nd Insertion of DTX Indication Bits
Physical Channel Segmentation
2nd Interleaving
Physical Channel Mapping
Physical Channels

Transport Channels	Channel-Coding Schemes	Coding Rate
BCH, PCH, RACH	Convolutional code	1/2
CPCH,DCH,DSCH,FACH USCH	Convolutional code Turbo code	1/3, 1/2 1/3
HS-DSCH	Turbo code	1/3
DCH,DSCH,FACH,USCH	No coding (TDD mode only)	

Table 22.6: UTRA channel-coding parameters for the channels of Table 22.3.

22.2.4.2 Transport Block Concatenation

Following the CRC attachment, the incoming transport blocks within a TTI are serially concatenated in order to form a code block. If the number of bits exceeds the maximum code block length, denoted as Z, then the code block is segmented into shorter ones. Furthermore, filler bits (zeros) are added to the beginning of the first block, if neccessary, in order to generate code blocks of the same length. The maximum code block length Z is dependent on the type of channel-coding invoked. For convolutional coding we have $Z = 504$, while for turbo coding $Z = 5114$ is used, since turbo codes require a long coded block length [79] for achieving a high performance. If no channel-coding is invoked, then the code block can be of unlimited length.

22.2.4.3 Channel-Coding

Each of the code blocks is then delivered to the channel-coding unit. Several Forward Error Correction (FEC) techniques are proposed for channel-coding. The FEC technique used is dependent on the QoS requirement of that specific transport channel. Table 22.6 shows the various types of channel-coding techniques invoked for different transport channels. Typically, *convolutional coding* is used for services having a bit error rate requirement on the order of 10^{-3}. For services requiring a lower BER, namely, on the order of 10^{-6}, *turbo coding* with two 8-state constituent encoders is applied. Turbo coding is known to guarantee a high performance [5, 540] over AWGN channels at the cost of increased interleaving-induced latency or delay. The implementational complexity of the turbo codec (TC) does not necessarily have to be higher than that of the convolutional codes (CC), since a constraint-length $K = 9$ CC is invoked, while the constraint-length of the turbo codes employed may be as low as $K = 3$. In somewhat simplistic but plausible terms, one could argue that a $K = 3$ TC using two decoders per iteration and employing four iterations has a similar complexity to a $K = 6$ CC, since they are associated with the same number of trellis states. The encoded code blocks within a TTI are then serially concatenated after the channel-coding unit, as seen in Figures 22.19 and 22.20.

22.2.4.4 Radio Frame Padding

Radio frame padding is only performed on the FDD UL. By contrast, radio frame padding is utilized on both the TDD UL and DL, whereby the input bit sequence (the concatenated encoded code blocks emanating from the channel-coding unit) is padded in order to ensure that the output can be segmented into (TTI/10 ms) number of equal-length, 10 ms-duration radio frames. Note that radio frame padding is not required on the FDD DL, since DTX is invoked, as seen in Figure 22.20. This process was termed Radio Frame Equalization in the standard. However, in order to avoid confusion with channel equalization, we used the terminology 'padding'.

22.2.4.5 First Interleaving

The depth of this first interleaver seen in Figures 22.19 and 22.20 may range from one radio frame (10 ms) to as high as 80 ms, depending on the TTI.

22.2.4.6 Radio Frame Segmentation

The input bit sequence after the first interleaving is then segmented into consecutive radio frames of 10 ms duration, as highlighted in Section 22.2.3. The number of radio frames required is equivalent to (TTI/10). Because of the Radio Frame Padding step performed prior to the segmentation on the UL in Figure 22.19 and also because of the Rate Matching step on the DL in Figure 22.20, the input bit sequence can be conveniently divided into the required number of radio frames.

22.2.4.7 Rate Matching

The rate matching process of Figures 22.19 and 22.20 implies that bits on a transport channel are either repeated or punctured in order to ensure that the total bit rate after multiplexing all the associated transport channels will be identical to the channel bit rate of the corresponding physical channel, as highlighted in Section 22.2.3. Thus, rate matching must be coordinated among the different coded transport channels, so that the bit rate of each channel is adjusted to a level that fulfils its minimum QoS requirements [522]. On the FDD DL as well as on both the TDD DL and UL, the bit rate is also adjusted so that the total instantaneous transport channel bit rate approximately matches the defined bit rate of the physical channel, as listed in Table 22.2.1.

22.2.4.8 Discontinuous Transmission Indication

On the FDD DL, the transmission is interrupted if the bit rate is less than the allocated channel bit rate. This is known as discontinuous transmission (DTX). DTX indication bits are inserted into the bit sequence in order to indicate when the transmission should be turned off. The first insertion of the DTX indication bits shown in Figure 22.20 is performed only if the position of the transport channel in the radio frame is fixed. In this case, a fixed number of bits is reserved for each transport channel in the radio frame. For the second insertion step shown in Figure 22.20, the DTX indication bits are inserted at the end of the radio frame.

22.2.4.9 Transport Channel Multiplexing

One radio frame from each transport channel that can be mapped to the same type of physical channel is delivered to the transport channel multiplexing unit of Figures 22.19 and 22.20, where they are serially multiplexed to form a *Coded Composite Transport CHannel* (CC-TrCH). At this point, it should be noted that the bit rate of the multiplexed radio frames may be different for the various transport channels. In order to successfully de-multiplex each transport channel at the receiver, the TFCI — which contained information about the bit rate of each multiplexed transport channel — can be transmitted together with the CCTrCH information (which will be mapped to a physical channel), as highlighted in Section 22.2.3. Alternatively, *blind transport format detection* can be performed at the receiver without the explicit knowledge of the TFCI, where the receiver acquires the transport format combination through some other means, such as, for example, the received power ratio of the DPDCH to the DPCCH.

22.2.4.10 Bit Scrambling

Bit scrambling is only performed in the TDD mode. The bits h_k, $k = 1, \ldots, S$ emanating from the output of the transport channel multiplexer, where S is the total number of bits, are scrambled according to the following equation :

$$s_k = h_k \oplus p_k, \quad k = 1, \ldots, S \tag{22.1}$$

where

$$p_k = \begin{cases} 0; & k < 1 \\ 1; & k = 1 \\ \left(\sum_{i=1}^{16} g_i \cdot p_{k-i}\right) \bmod 2; & k > 1 \end{cases} \tag{22.2}$$

and g_i is the ith element in $\boldsymbol{G} = [0, 0, 0, 0, 0, 0, 0, 0, 0, 0, 1, 0, 1, 1, 0, 1]$.

22.2.4.11 Physical Channel Segmentation

If more than one physical channel is required in order to accommodate the bits of a CCTrCH, then the bit sequence is segmented equally into different physical channels, as seen in Figures 22.19 and 22.20. A typical example of this scenario would be, where the bit rate of the CCTrCH exceeds the maximum allocated bit rate of the particular physical channel. Thus, multiple physical channels are required for its transmission. Furthermore, restrictions are imposed on the number of transport channels that can be multiplexed onto a CCTrCH. Hence, several physical channels are required to carry any additional CCTrCHs.

22.2.4.12 Second Interleaving

The depth of the second interleaving stage shown in Figures 22.19 and 22.20 is equivalent to one radio frame. Hence, this process does not increase the system's delay.

22.2.4.13 Physical Channel Mapping

Finally, the bits are mapped to their respective physical channels summarized in Table 22.4, as portrayed in Figures 22.19 and 22.20. Note that for the 1.28 Mcps TDD mode, the bits generated by the second interleaver must be further segmented into the 5 ms subframes, as discussed in Figure 22.4.

Having highlighted the various channel-coding and multiplexing techniques as well as the structures of the physical channels illustrated by Figures 22.5–22.14, let us now consider how the services of different bit rates are mapped to the UL and DL dedicated physical data channels (DPDCH) of Figures 22.5 and 22.6, respectively. In order to further augment the process, we will present three examples. Specifically, we consider the mapping of two multirate services on a UL DPDCH and an example of the mapping of a 4.1 kbps data service on a DL DPDCH in the FDD mode. We will then use the same parameters as employed in the first example and show how multirate services can be mapped to the corresponding UL DPCH in TDD mode.

22.2.4.14 Mapping Several Multirate Services to the UL Physical Channels in FDD Mode [515]

In this example, we assume that a 4.1 kbps speech service and a 64 kbps video service are to be transmitted simultaneously on the UL. The parameters used for this example are shown in Table 22.7. As illustrated in Figure 22.21, a 16-bit CRC checksum is first attached to

	Service 1, DCH#1	**Service 2, DCH#2**
Transport Block Size	640 bits	164 bits
Transport Block Set Size	4 * 640 bits	1 * 164 bits
TTI	40 ms	40 ms
Bit Rate	64 kbps	4.1 kbps
CRC	16 bits	16 bits
Coding	Turbo Rate: 1/3	Convolutional Rate: 1/3

Table 22.7: Parameters for the multimedia communication example of Section 22.2.4.14.

each transport block of DCH#1, that is, #1a,...,#1d, as well as the transport block of DCH#2 for the purpose of error detection. As a result, the number of bits in the transport block of Service 1 and Service 2 is increased to 640 + 16 = 656 bits and 164 + 16 = 180 bits, respectively. The four transport blocks of Service 1 are then concatenated, as illustrated in Figure 22.21. Notice that no code block segmentation is invoked, since the total number of bits in the concatenated transport block is less than $Z = 5114$ for turbo coding, as highlighted in Section 22.2.4.2. Since the video service typically requires a low BER — unless specific measures are invoked for mitigating the video effects of transmission errors [538] — turbo coding is invoked, using a coding rate of $\frac{1}{3}$. Hence, after turbo coding and the attachment of tailing bits, the resulting 40 ms segment would contain $(656 \times 4) \times 3 + 12 = 7884$ bits, as shown in Figure 22.21. By contrast, the speech service can tolerate a higher BER. Hence, convolutional coding is invoked. First, a block of 8 tail bits is concatenated to the

	Service 1, DCH#1
Transport Block Size	164 bits
TTI	40 ms
Bit Rate	4.1 kbps
CRC	16 bits
Coding	Convolutional Rate: 1/3

Table 22.8: Parameters for the example of Section 22.2.4.15.

transport block in order to flush the assumed constraint-length $K = 9$ shift registers of the convolutional encoder. Thus, a total of 180 + 8 = 188 bits are conveyed to the convolutional encoder of DCH#2, as shown in Figure 22.21. Again, no code block segmentation is invoked, since the total number of bits in the transport channel is less than $Z = 504$ for convolutional coding, as highlighted in Section 22.2.4.2. A coding rate of $\frac{1}{3}$ is used for the convolutional encoding of DCH#2, as exemplified in Table 22.7. The output of the convolutional encoder of DCH#2 will have a total of $188 \times 3 = 564$ bits per 40 ms segment. Since the TTI of these transport channels is 40 ms, four radio frames are required to transmit the associated data. At this stage, notice that there are a total of 7884 bits and 564 bits for DCH#1 and DCH#2, respectively. Since these numbers are divisible by four, they can be divided equally into four radio frames. Thus, no padding is required as illustrated in the Radio Frame Padding step of Figure 22.21. Interleaving is then performed across the 40 ms segment for each transport channel before being segmented into four 10 ms radio frames.

At this point, we note that these two transport channels can be mapped to the same DPDCH, since they belong to the same MS. Hence, the 10 ms radio frames, marked 'A' in Figure 22.21 will be multiplexed, in order to form a CCTrCH. Similarly, the frames marked 'B', 'C' (not shown in Figure 22.7 due to lack of space), and 'D' will be multiplexed, in order to form another three CCTrCHs. The rate of these CCTrCHs must be matched to the allocated channel bit rate of the physical channel. Without rate matching, the bit rate of these CCTrCHs is (1971 + 141)/10 ms = 211.2 kbps, which does not fit any of the available channel bit rates of the UL DPDCH, as listed in Table 22.2.1. Hence, the Rate Matching step of Figures 22.19, 22.20, and 22.21 must be invoked in order to adapt the multiplexed bit rate to one of the available UL DPDCH bit rates of Table 22.2.1. Let us assume that the allocated channel bit rate is 240 kbps. Thus, a number of bits must be punctured or repeated for each service, in order to increase the total number of bits per 10 ms segment after multiplexing from 2171 to 2400. This would require coordination among the different services, as was highlighted in Section 22.2.4.7. After multiplexing the transport channels, a second interleaving is performed across the 10 ms radio frame before finally mapping the bits to the UL DPDCH.

22.2.4.15 Mapping of a 4.1 Kbps Data Service to the DL DPDCH in FDD Mode

The parameters for this example are shown in Table 22.8. In this context, we assume that a single DCH consisting of one transport block within a TTI duration of 40 ms is to be transmitted on the DL. As illustrated in Figure 22.22, a 16-bit CRC sum segment is appended

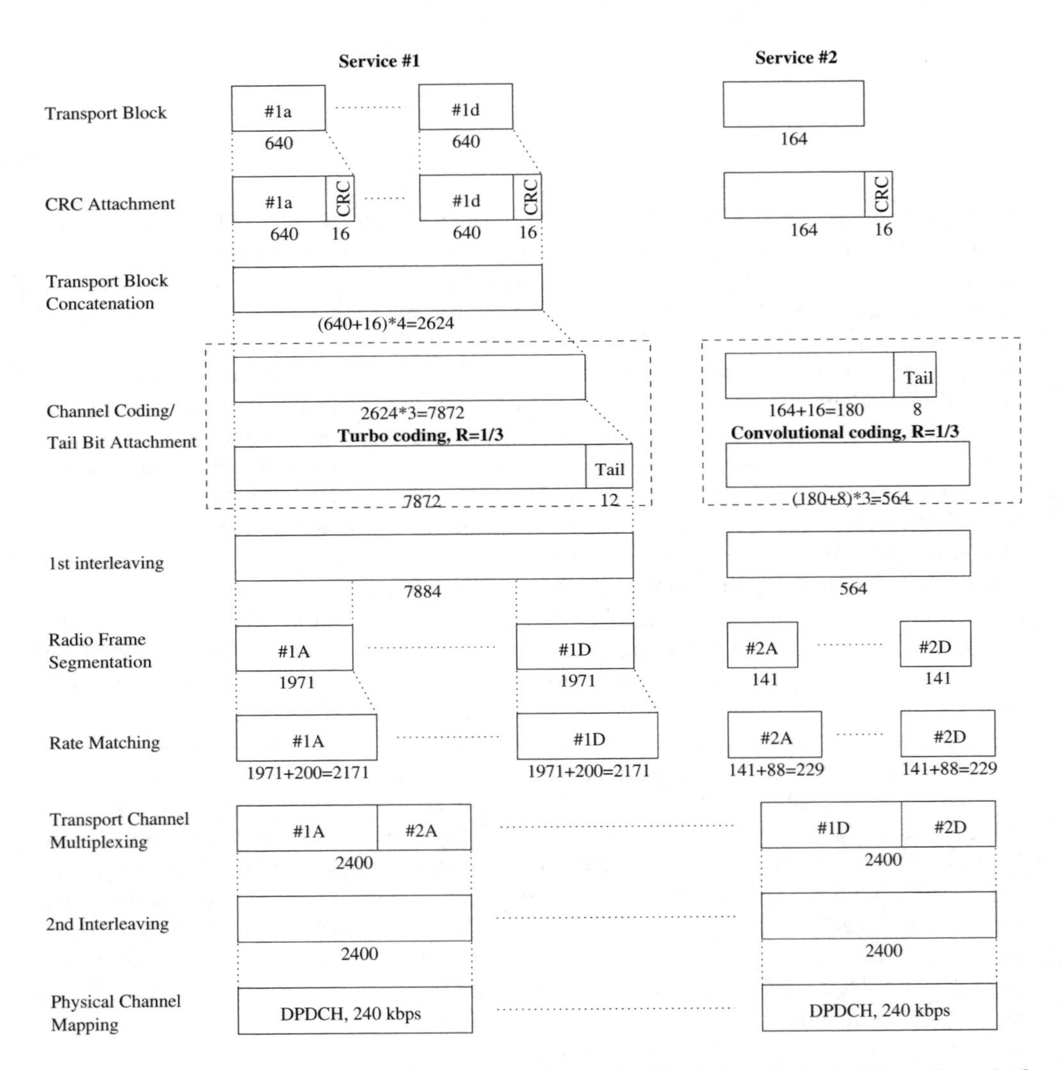

Figure 22.21: Mapping of several multimedia services to the UL dedicated physical data channel of Figure 22.5 in FDD mode. The corresponding schematic diagram is seen in Figure 22.19.

to the transport block. A 8-bit tailing block is then attached to the end of the segment in order to form a 188-bit code block. Similarly to the previous example, the length of the code block is less than $Z = 504$, since CC is used. Hence, no segmentation is invoked. The 188-bit data block is convolutional coded at a rate of $\frac{1}{3}$, which results in a $3 \times 188 = 564$-bit segment. According to Figure 22.20, rate matching is invoked for the encoded block. Since the TTI duration is 40 ms, four radio frames are required to transmit the data. Without rate matching, the bit rate per radio frame is 564/40 ms = 14.1 kbps, which does not fit any of the available bit rates listed in Table 22.3 for the DL. Note that for the case of the DL dedicated physical channels, the channel bit rate will include the additional bits required for the pilot and TPC, as shown explicitly in Figure 22.6. Since there is only one transport channel in this case, no TFCI bits are required. We assume that an 8-bit pilot and a 2-bit TPC per slot are assigned to this tranmission, which yields a total rate of 15 kbps for the DPCCH. Hence all the bits in the encoded block will be repeated in order to increase its bit rate of 15 kbps to 30 kbps for the DL DPCH. In this case the number of padding bits appended becomes $N = 36$. After the second interleaving stage of Figure 22.20, the segmented radio frames are mapped to the corresponding DPDCH, which are then multiplexed with the DPCCH, as shown in Figure 22.22.

22.2.4.16 Mapping Several Multirate Services to the UL Physical Channels in 3.84 Mcps TDD Mode [515]

In this example, we will demonstrate how the multirate multimedia services, considered previously in the example of Section 22.2.4.14 in an FDD context, are mapped to the corresponding dedicated physical channels (DPCH) in the 3.84 Mcps TDD mode. The channel-coding/multiplexing process is identical in the FDD and TDD mode, and so both are based on Figures 22.19 and 22.20. The only difference is in the mapping of the transport channels to the corresponding physical channels seen at the bottom of Figures 22.19 and 22.20, since the FDD and TDD modes have a different frame structure, as shown previously in Figures 22.5–22.13 and Figure 22.14, respectively. In this example, we are only interested in the process of service mapping to the physical channel, which follows the second interleaving stage of Figure 22.23. Here we assumed that for the TDD UL scenario of Table 22.7 the total number of bits per segment after DCH multiplexing is 2186 as a result of rate matching. In the FDD example of Section 22.2.4.14, this was 2400. Each segment is divided into two bursts, which can be transmitted either by orthogonal code multiplexing onto a single time-slot, or using two time-slots within a 10 ms radio frame. Note that only one burst in each segment is required to carry the TFCI and the TPC information. The multiplexing process associated with the 1.28 Mcps TDD mode will be similar to that of the 3.84 Mcps TDD mode, except that the bits are segmented into 5 ms subframes after the 2nd interleaver before they are mapped onto the physical channels.

Following these brief discussions on service multiplexing, channel coding, and interleaving, let us now concentrate on the aspects of variable-rate and multicode transmission in UTRA in the next section.

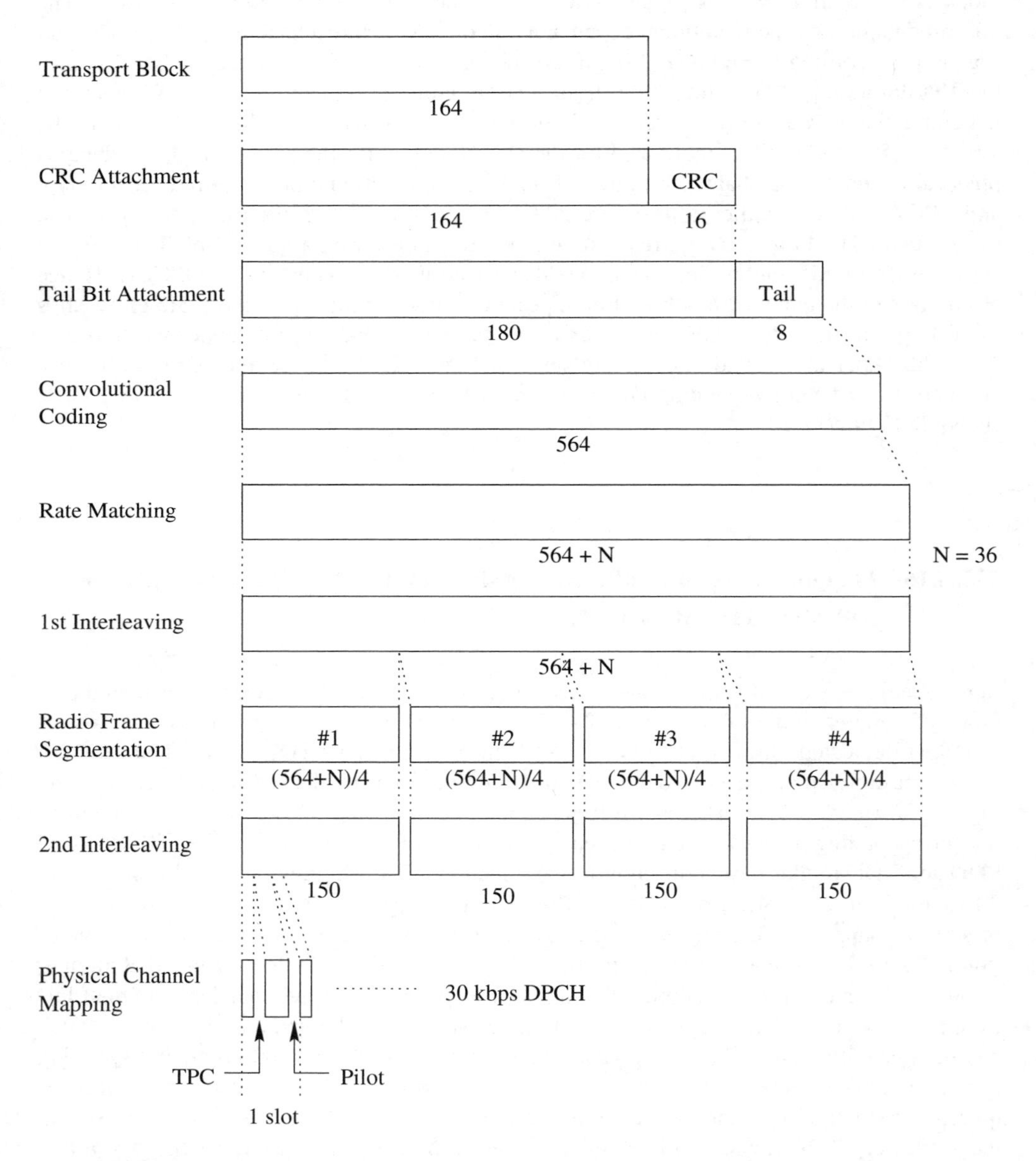

Figure 22.22: Mapping of a 4.1 kbps data service to the DL dedicated physical channel of Figure 22.6 in FDD mode. The corresponding schematic diagram is seen in Figure 22.20.

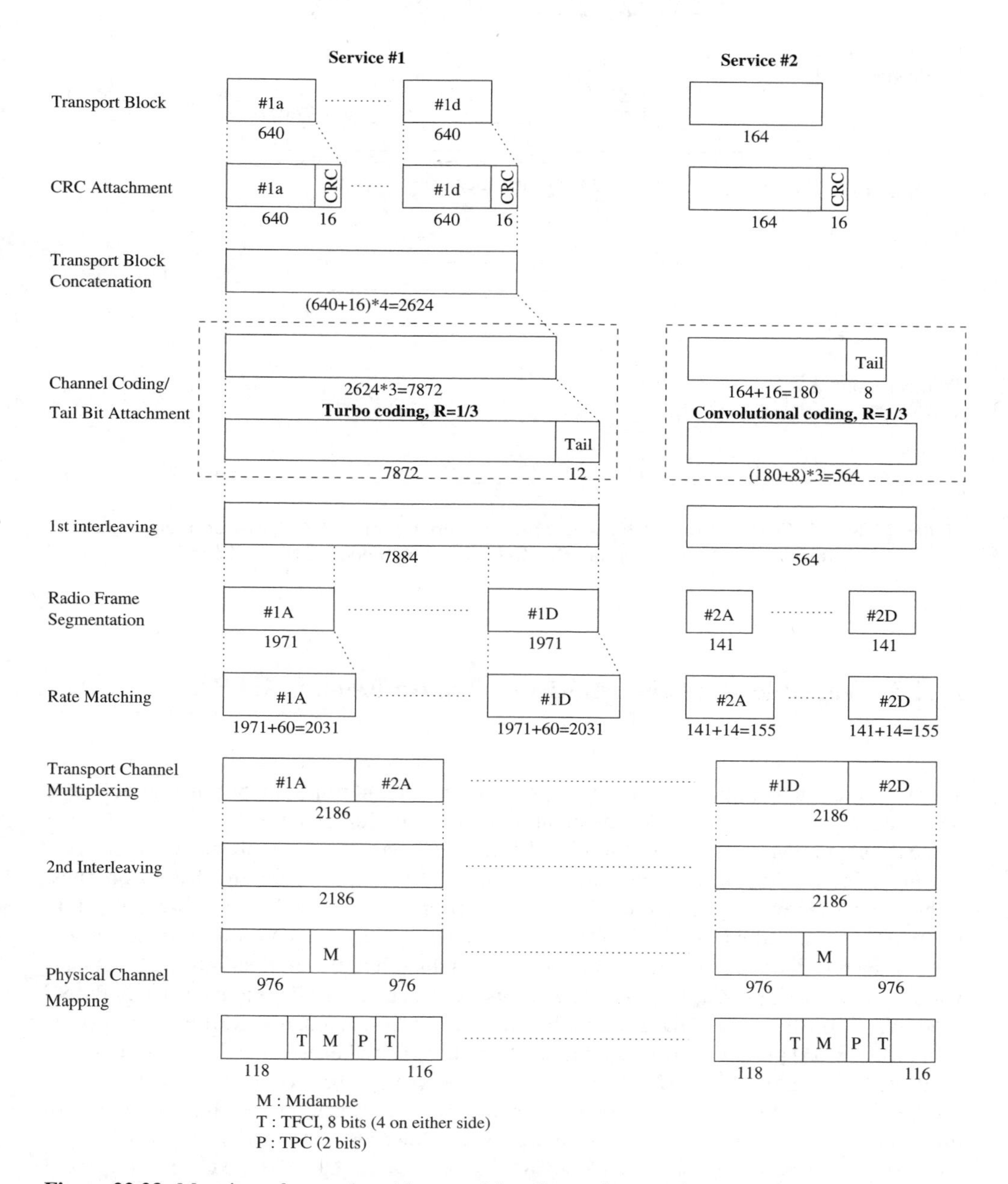

Figure 22.23: Mapping of several multirate multimedia services to the UL dedicated physical data channel of Figure 22.5 in TDD mode. The corresponding schematic diagram is seen in Figure 22.19.

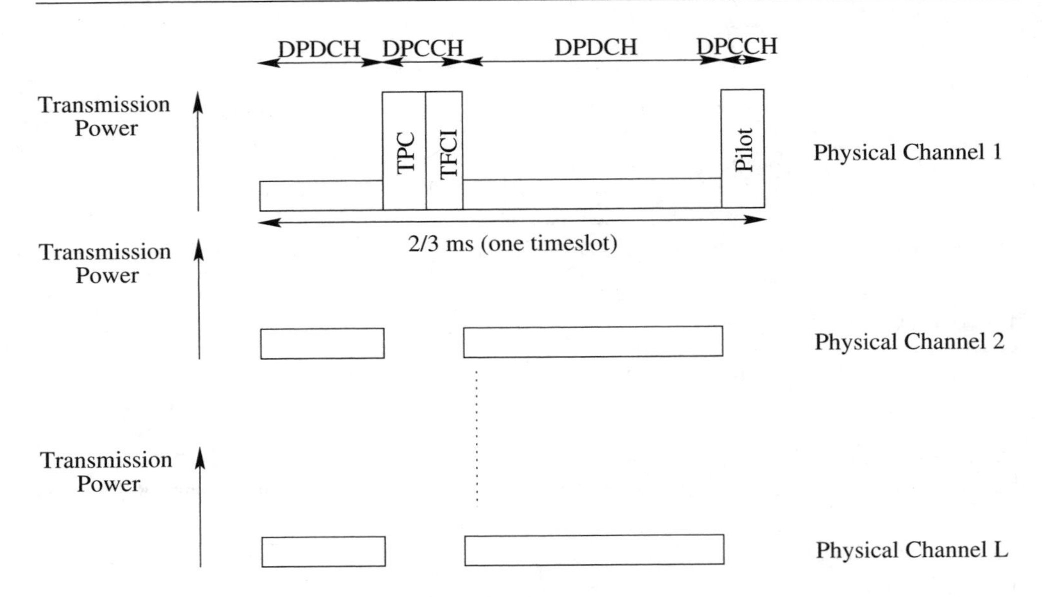

Figure 22.24: DL FDD slot format for multicode transmission in UTRA, based on Figure 22.6, but dispensing with transmitting DPCCH over all multicode physical channels.

22.2.5 Variable-Rate and Multicode Transmission in UTRA

Three different techniques have been proposed in the literature for supporting variable-rate transmission, namely, multicode-, modulation-division multiplexing- (MDM), and multiple processing gain (MPG)-based techniques [541]. UTRA employs a number of different processing gains, or variable spreading factors, in order to transmit at different channel bit rates, as highlighted previously in Section 22.2.3. The spreading factor (SF) has a direct effect on the performance and capacity of a DS-CDMA system. Since the chip rate is constant, the SF — which is defined as the ratio of the spread bandwidth to the original information bandwidth — becomes lower, as the bit rate increases. Hence, there is a limit to the value of the SF used, which is SF = 4 in FDD mode in the proposed UTRA standards. Multicode transmission [98, 218, 351] is used if the total bit rate to be transmitted exceeds the maximum bit rate supported by a single DPDCH, which was stipulated as 960 kbps for the UL and 1920 kbps for the DL. When this happens, the bit rate is split among a number of spreading codes and the information is transmitted using two or more codes. However, only one DPCCH is transmitted during this time. Thus, on the UL one DPCCH and several DPDCH are code-multiplexed and transmitted in parallel, as will be discussed in the context of Figure 22.26. On the DL, the DPDCH and DPCCH are time-multiplexed on the first physical channel associated with the first spreading code as seen in Figure 22.24. If more physical channels are required, the DPCCH part in the slot will be left blank again, as shown in Figure 22.24. The transmit power of the DPDCH is also reduced.

22.2.6 Spreading and Modulation

It is well known that the performance of DS-CDMA is interference limited [103]. The majority of the interference originates from the transmitted signals of other users within the same cell, as well as from neighbouring cells. This interference is commonly known as *Multiple Access Interference* (MAI). Another source of interference, albeit less dramatic, is a result of the wideband nature of CDMA, yielding several delayed replicas of the transmitted signal, which reach the receiver at different time instants, thereby inflicting what is known as *interpath interference*. However, the advantages gained from wideband transmissions, such as multipath diversity and the noise-like properties of the interference, outweigh the drawbacks.

The choice of the spreading codes [82, 542] used in DS-CDMA will have serious implications for the amount of interference generated. Suffice to say that the traditional measures used in comparing different codes are their *cross-correlations* (CCL) and *autocorrelation* (ACL). If the CCL of the spreading codes of different users is nonzero, this will increase their interference, as perceived by the receiver. Thus a low CCL reduces the MAI. The so-called out-of-phase ACL of the codes, on the other hand, plays an important role during the initial synchronization between the BS and MS, which has to be sufficiently low to minimise the probability of synchronizing to the wrong ACL peak.

In order to reduce the MAI and thereby improve the system's performance and capacity, the UTRA physical channels are spread using two different codes, namely, the *channelization code* and a typically longer so-called *scrambling code*. In general, the channelization codes are used to maintain orthogonality between the different physical channels originating from the same source. On the other hand, the scrambling codes are used to distinguish between different cells, as well as between different MSs. All the scrambling codes in UTRA are in complex format. Complex-valued scrambling balances the power on the I and Q branches. This can be shown by letting c_s^I and c_s^Q be the I and Q branch scrambling codes, respectively. Let $d(t)$ be the complex-valued data of the transmitter, which can be written as:

$$d(t) = d_I + jd_Q, \tag{22.3}$$

where d_I and d_Q represent the data on the I and Q branches, respectively. Let us assume for the sake of argument that the power level in the I and Q branches is unbalanced due to, for instance, their different bit rates or different QoS requirements. If only real-valued scrambling is used, then the output becomes:

$$s(t) = c_s^I \left(d_I + jd_Q\right), \tag{22.4}$$

which is also associated with an unbalanced power level on the I and Q branches. By contrast, if complex-valued scrambling is used, then the output would become:

$$\begin{aligned} s(t) &= (d_I + jd_Q) \cdot \left(c_s^I + jc_s^Q\right) && (22.5) \\ &= c_s^I \cdot d_I - c_s^Q \cdot d_Q + j\left(c_s^Q \cdot d_I + c_s^I \cdot d_Q\right). && (22.6) \end{aligned}$$

As can be seen, the power on the I and Q branches after complex scrambling is the same, regardless of the power level of the unscrambled data on the I and Q branches. Hence,

	Channelization Codes	Scrambling Codes
Type of codes	OVSF (Section 22.2.6.1)	FDD UL : Gold codes (long codes) (Section 22.2.6.2) FDD UL : S(2) codes (short codes) (Section 22.2.6.2) FDD DL : Gold codes (Section 22.2.6.3) TDD UL & DL : 128 16-chip codes
Code length	Variable	FDD UL : 10 ms of $(2^{25}-1)$-chip Gold code FDD UL : 256-chip S(2) code FDD DL : 10 ms of $(2^{18}-1)$-chip Gold code TDD UL & DL : 16-chip code
Type of spreading	BPSK (UL/DL)	QPSK (UL/DL)

Table 22.9: UL/DL spreading and modulation parameters in UTRA.

complex scrambling potentially improves the power amplifier's efficiency by reducing the peak-to-average power fluctuation. This also relaxes the linearity requirements of the UL power amplifier used.

Table 22.9 shows the parameters and techniques used for spreading and modulation in UTRA, which will be discussed in depth in the following sections.

22.2.6.1 Orthogonal Variable Spreading Factor Codes

The channelization codes used in the UTRA systems are derived from a set of orthogonal codes known as *Orthogonal Variable Spreading Factor* (OVSF) codes [240]. OVSF codes are generated from a tree-structured set of orthogonal codes, such as the Walsh-Hadamard codes, using the procedure shown in Figure 22.25. Specifically, each channelization code is denoted by $c_{N,n}$, where $n = 1, 2, \ldots, N$ and $N = 2^x, x = 2, 3, \ldots 8$. Each code $c_{N,n}$ is derived from the previous code $c_{(N/2),n}$ as follows [240]:

$$\begin{bmatrix} c_{N,1} \\ c_{N,2} \\ c_{N,3} \\ \vdots \\ c_{N,N} \end{bmatrix} = \begin{bmatrix} c_{(N/2),1}|c_{(N/2),1} \\ c_{(N/2),1}|\bar{c}_{(N/2),1} \\ c_{(N/2),2}|c_{(N/2),2} \\ \vdots \\ c_{(N/2),(N/2)}|\bar{c}_{(N/2),(N/2)} \end{bmatrix}, \qquad (22.7)$$

where [|] at the right-hand side of Equation 22.7 denotes an augmented matrix and $\bar{c}_{(N/2),n}$ is the binary complement of $c_{(N/2),n}$. For example, according to Equation 22.7 and Figure 22.25, $c_{N,1} = c_{8,1}$ is created by simply concatenating $c_{(N/2),1}$ and $c_{(N/2),1}$, which doubles the number of chips per bit. By contrast, $c_{N,2} = c_{8,2}$ is generated by attaching $\bar{c}_{(N/2),1}$ to $c_{(N/2),1}$. From Equation 22.7, we see that, for example, $c_{N,1}$ and $c_{N,2}$ at the left-hand side of Equation 22.7 are not orthogonal to $c_{(N/2),1}$, since the first half of both was derived from $c_{(N/2),1}$ in Figure 22.25, but they are orthogonal to $c_{(N/2),n}, n = 2, 3, \ldots, (N/2)$. The code $c_{(N/2),1}$ in Figure 22.25 is known as the mother code of the codes $c_{N,1}$ and $c_{N,2}$, since these two codes are derived from $c_{(N/2),1}$. The codes on the 'highest'-order branches ($k = 6$) of the tree at the left of Figure 22.25 have a spreading factor of 4, and they are used for transmission at the highest possible bit rate for a single channel, which is 960 kbps. On the other hand, the codes on the 'lowest'-order branches ($k = 0$) of the tree at the right of Figure 22.25 result in a spreading factor of 256, and these are used for transmission at the lowest bit rate, which is

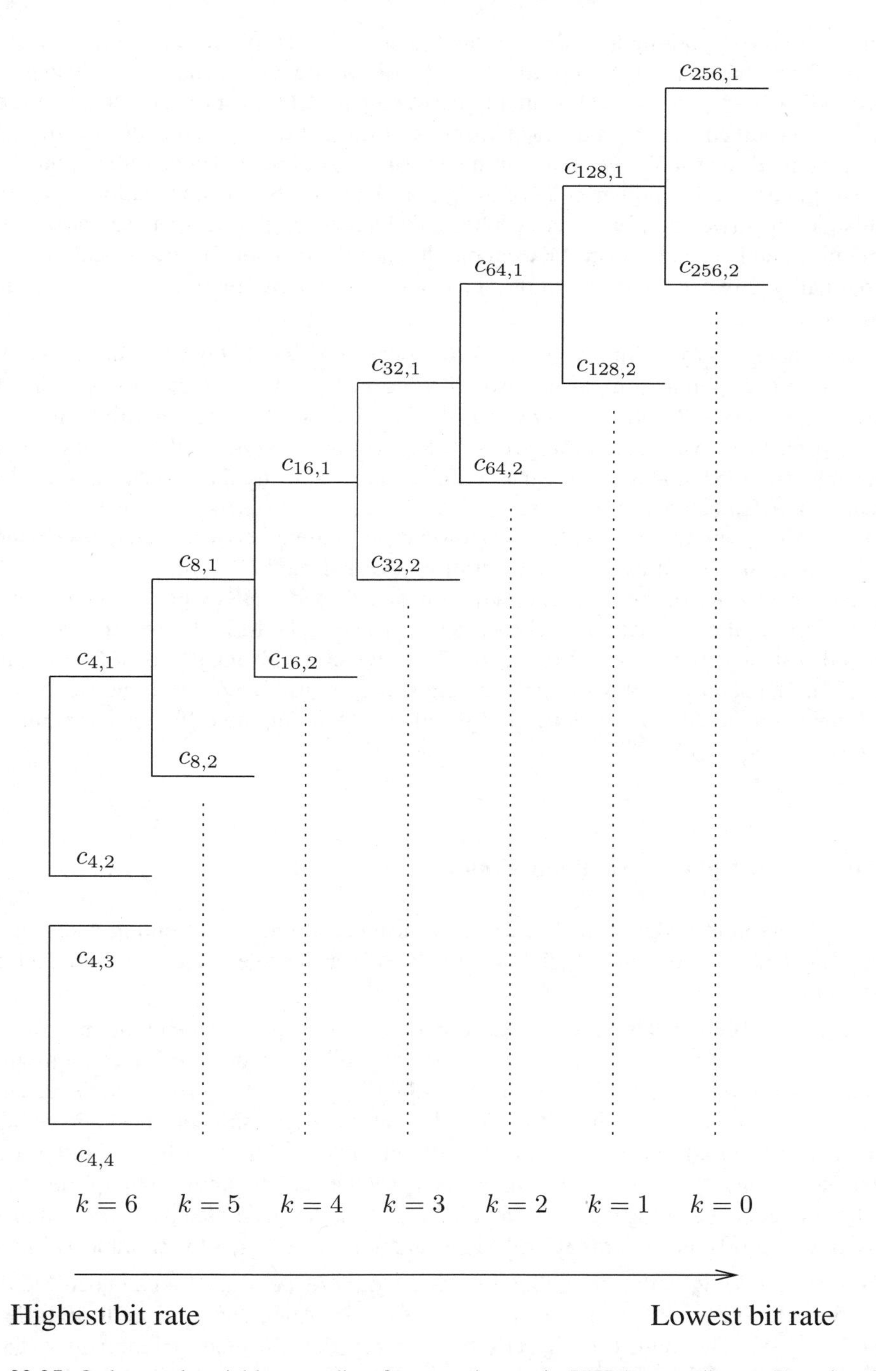

Figure 22.25: Orthogonal variable-spreading factor code tree in UTRA according to Equation 22.7. The parameter k in the figure is directly related to that found in Figures 22.5–22.11.

15 kbps. It is worth noting here that an intelligent BbB adaptive scheme may vary its SF on a 10 ms frame basis in an attempt to adjust the SF on a near-instantaneous channel-quality motivated basis [535, 538]. Orthogonality between parallel transmitted channels of the same bit rate is preserved by assigning each channel a different orthogonal code accordingly. For channels with different bit rates transmitting in parallel, orthogonal codes are assigned, ensuring that no code is the mother-code of the other. Thus, OVSF channelization codes provide total isolation between different users' physical channels on the DL that are transmitted synchronously and hence eliminate MAI among them. OVSF channelization codes also provide orthogonality between the different physical channels seen in Figure 22.24 during multicode transmission.

Since there is only a limited set of OVSF codes, which is likely to be insufficent to support a large user-population, while also allowing identification of the BSs by the MSs on the DL, *each cell will reuse the same set of OVSF codes.* Statistical multiplexing schemes such as packet reservation multiple access (PRMA) [543] can be used for allocating and deallocating the OVSF codes on a near-instantaneous basis using the DSCH transport channel instead of the dedicated DCH, for example, depending on the users' voice activity in the case of DTX-based communications [544]. However, orthogonal codes, such as the orthogonal OVSF codes, in general exhibit poor out-of-phase ACL and CCL properties [545]. Therefore, the correlations of the OVSFs of adjacent asynchronous BSs will become unacceptably high, degrading the correlation receiver's performance at the MS. On the other hand, certain long codes such as Gold codes exhibit low CCL, which is advantageous in CDMA applications [83]. Hence in UTRA, cell-specific long codes are used in order to reduce the intercell interference on the DL. On the UL, MAI is reduced by assigning different scrambling codes to different users.

22.2.6.2 FDD Uplink Scrambling Codes

The UL scrambling codes in UTRA can be classified into long scrambling codes and short scrambling codes. A total of 2^{24} UL scrambling codes can be generated for both the long and short codes.

Long scrambling codes are constructed from two m-sequences using the polynomials of $1 + X^3 + X^{25}$ and $1 + X + X^2 + X^3 + X^{25}$, following the procedure highlighted by Proakis [90] in order to produce a set of *Gold codes* for the I branch. The Q-branch Gold code is a shifted version of the I-branch Gold code, where a shift of 16,777,232 chips was recommended. Gold codes are rendered different from each other by assigning a unique initial state to one of the shift registers of the m-sequence. The initial state of the other shift register is a sequence of logical 1. Although the Gold codes generated have a length of $2^{25} - 1$ chips, only 38,400 chips (10 ms at 3.84 Mcps) are required in order to scramble a radio frame.

Short scrambling codes are defined from a family of periodically extended S(2) codes. This 256-chip S(2) code was introduced to ease the implementation of multi-user detection at the BS [531]. The multi-user detector has to invert the so-called system matrix [546], the dimension of which is proportional to the sum of the channel impulse response duration and the spreading code duration. Thus, using a relatively short scrambling code is an important practical consideration in reducing the size of the system-matrix to be inverted.

22.2.6.3 FDD Downlink Scrambling Codes

Unlike the case for the UL, only Gold codes are used on the DL. The DL Gold codes on the I branch are constructed from two m-sequences using the polynomials of $1 + X^7 + X^{18}$ and $1 + X^5 + X^7 + X^{10} + X^{18}$. These Gold codes are shifted by 131,072 chips in order to produce a set of Gold codes for the Q branch.

Although a total of $2^{18} - 1 = 262,143$ Gold codes can be generated, only 8192 of them will be used as the DL scrambling code. These codes are divided into 512 groups, each of which contains a *primary scrambling code* and 15 *secondary scrambling codes*. Altogether there are 512 primary scrambling codes and $8192 - 512 = 7680$ secondary scrambling codes. Each cell is allocated one primary scrambling code, which is used on the CPICH, P-CCPCH AICH, AP-ICH, CD/CA-ICH, CSICH and the S-CCPCH channels of Table 22.4. This primary scrambling code will be used to identify the BS for the MS. All the other physical channels belonging to this cell can use either the primary scrambling code or any of the 15 secondary scrambling codes that belong to the same group, as the primary scrambling code. In order to facilitate fast cell or BS identification, the set of 512 primary scrambling codes is further divided into 64 subsets, each consisting of eight primary scrambling codes, as will be shown in Section 22.2.10.

22.2.6.4 FDD Uplink Spreading and Modulation

A model of the UL transmitter for a single DPDCH is shown in Figure 22.26 [242]. We have seen in Figure 22.5 that the DPDCH and DPCCH are transmitted in parallel on the I and Q branches of the UL, respectively. Hence, to avoid I/Q channel interference in case of I/Q inbalance of the quadrature carriers, different orthogonal spreading codes are assigned to the DPDCH and DPCCH on the I and Q branch, respectively. These two channelization codes for DPDCH and DPCCH, denoted by $c_{D,1}$ and c_C in Figure 22.26, respectively, are allocated in a predefined order. From Figure 22.5, we know that the SF of the DPCCH is 256. Hence, $c_C = c_{256,1}$ in the context of Figure 22.25. This indicates that the high SF of the DPCCH protects the vulnerable control channel message against channel impairments. On the other hand, we have $c_{D,1} = c_{SF,(SF/4)+1}$, depending on the SF of the DPDCH. In the event of multicode transmission portrayed by the dashed lines in Figure 22.26, different additional orthogonal channelization codes, namely, $c_{D,2}$ and $c_{D,3}$, are assigned to each DPDCH for the sake of maintaining orthogonality, and they can be transmitted on either the I or Q branch. In this case, the BS and MS have to agree on the number of channelization codes to be used. After spreading, the BPSK modulated I and Q branch signals are summed in order to produce a complex Quadrature Phase Shift Keying (QPSK) signal. The signal is then scrambled by the complex scrambling code, c_{scramb}. The pulse-shaping filters, $p(t)$, are root-raised cosine Nyquist filters using a roll-off factor of 0.22.

The transmitter of the UL PRACH and PCPCH message part is also identical to that shown in Figure 22.26. As we have mentioned in Section 22.2.3.2 in the context of Figure 22.9, the PRACH and the CPCH message consist of a data part and a control part. In this case, the data part will be transmitted on the I branch, and the control part on the Q branch. The choice of the channelization codes for the data and control part for the PRACH message depends on the signature of the preambles transmitted beforehand. As highlighted in Section 22.2.3.2, the preamble signature is a 256-chip sequence generated by the repetition of a

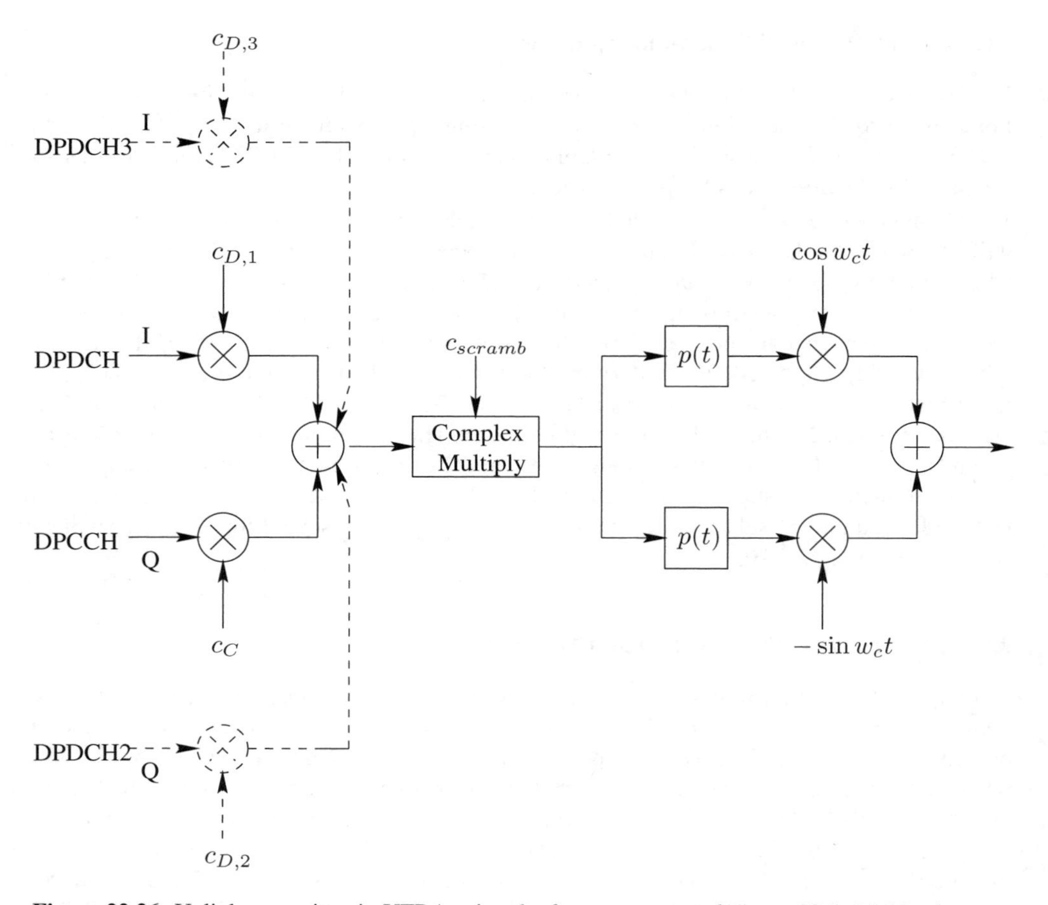

Figure 22.26: Uplink transmitter in UTRA using the frame structure of Figure 22.5. Multicode transmissions are indicated by the dashed lines.

16-chip Hadamard code. This 16-chip code actually corresponds to one of the OVSF codes, namely, to $c_{16,n}$, where $n = 1, \ldots, 16$. The codes in the subtree of Figure 22.25 below this specific 16-chip code n will be used as the channelization codes for the data part and control part. Specifically, the channelization codes assigned to the control part and the message part of the PRACH will be $c_{256,16\times(n+1)}$ and $c_{SF,m+1}$, where $m = SF \times n/16$, respecitvely. On the other hand, the channelization codes assigned to the control part and the message part of the PCPCH will be always $c_{256,0}$ and $c_{SF,(SF/4)+1}$, respectively.

22.2.6.5 FDD Downlink Spreading and Modulation

The schematic diagram of the DL transmitter is shown in Figure 22.27. All the DL physical channel bursts (except for the SCH) are first QPSK modulated in order to form the I and Q branches, before spreading to the chip rate. The HS-PDSCH can also use 16-level Quadrature Amplitude Modulation (QAM) [4]. In contrast to the UL of Figure 22.26, the same OVSF channelization code c_{ch} is used on the I and Q branches. Different physical channels are assigned different channelization codes in order to maintain their orthogonality. For instance, the channelization codes used for the CPICH and P-CCPCH of Table 22.4 are fixed to the codes $c_{256,1}$ and $c_{256,2}$ of Figure 22.25, respectively. The channelization codes invoked for all the other physical channels are assigned by the network.

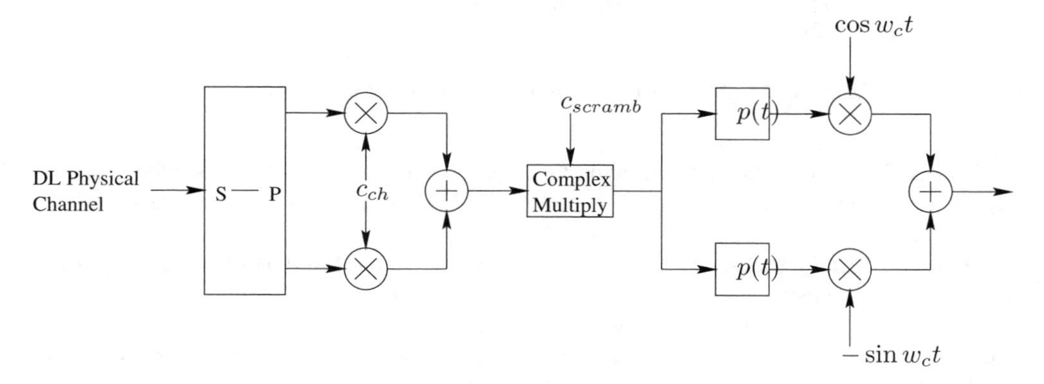

Figure 22.27: Downlink transmitter in UTRA using the frame structure of Figure 22.6.

The resultant signal seen in Figure 22.27 is then scrambled by a cell-specific scrambling code c_{scramb}. Similarly to the DL, the pulse-shaping filters are root-raised cosine Nyquist filters using a rolloff factor of 0.22.

22.2.6.6 TDD Spreading and Modulation

In the TDD mode the transmitter structure used for both the UL and DL is similar to that of the FDD DL transmitter of Figure 22.27. QPSK and 16-QAM [4] schemes are used for modulating the incoming bit sequences. Additionally, 8PSK modulation can be used for the 1.28 Mcps TDD mode. There are a total of 128 16-chip complex valued scrambling codes in the TDD mode. Hence, if a spreading factor of 8 is invoked, then the scrambling code will encompass two symbols. Since each time-slot can be used for transmitting several TDD bursts from the same source or from different sources, the OVSF codes are invoked in order to maintain orthogonality between the transmission bursts of different TDD/CDMA users/messages. An advantage of the TDD/CDMA mode is that the user population is separated in both the time and the code domain. In other words, only a small number of CDMA users/services will be supported within a TDD time-slot, which dramatically reduces the complexity of the multi-user detector that can be used in both the UL and DL for mitigating the MAI or multi-code interference.

22.2.7 Random Access

22.2.7.1 Mobile-Initiated Physical Random Access Procedures

If data transmission is initiated by an MS, it is required to send a random access request to the BS. Since such requests can occur at any time, collisions may result when two or more

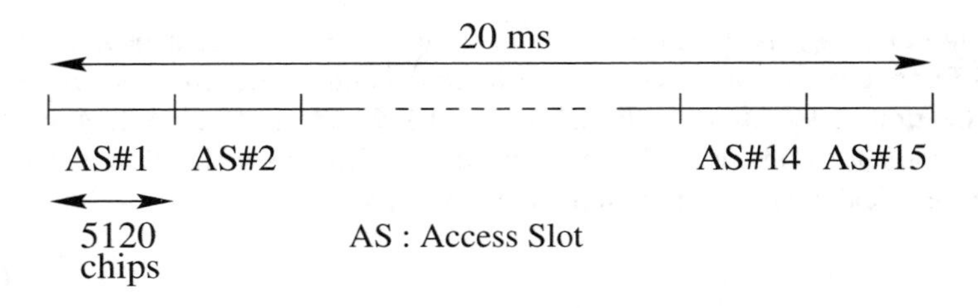

Figure 22.28: ALOHA-based physical UL random access slots in UTRA.

MSs attempt to access the network simultaneously. Hence, in order to reduce the probability of a collision, the random access procedure in UTRA is based on the slotted ALOHA technique [522].

Random access requests are transmitted to the BS via the PRACH of Table 22.4. Each random access transmission request may consist of one or several preambles and a message part, whose time-slot configuration was shown in Figure 22.9. According to the regime of Figure 22.28, the preambles and the message part can only be transmitted at the beginning of one of those 15 so-called *access slots*, which span two radio frames (i.e., 20 ms). Thus, each access slot has a length equivalent to 5120 chips or $\frac{4}{3}$ ms.

Before any random access request can be transmitted, the MS has to obtain certain information via the DL BCH transmitted on the P-CCPCH of Table 22.4 according to the format of Figure 22.10. This DL BCH/PCCPCH information includes the identifier of the cell-specific scrambling code for the preamble and message part of Figure 22.9, the available preamble signatures, the available access slots of Figure 22.28, which can be contended for in ALOHA mode, the initial preamble transmit power, the preamble power ramping factor, and the maximum number of preamble retransmissions necessitated by their decoding failure due to collisions at the BS. All this information may become available once synchronization is achieved, as will be discussed in Section 22.2.10. After acquiring all the necessary information, the MS will randomly select a preamble signature from the available signatures and transmit a preamble at the specific power level specified by the BS on a randomly selected access slot chosen from the set of available access slots seen in Figure 22.28. Note that the preamble is formed by multiplying the selected signature with the preamble scrambling code.

After the preamble is transmitted, the MS will listen for the acknowledgement of reception transmitted from the BS on the AICH of Table 22.4. Note that the AICH is also transmitted at the beginning of the access slot that corresponds to the access slot selected for the RACH. Furthermore, the phase reference used for coherent detection is obtained from the DL CPICH of Table 22.4. The acknowledgement is represented by a positive AI conveyed by the AICH of Table 22.4 that corresponds to the selected preamble signature. The MS will then proceed to transmit the message part at the beginning of a predefined access slot. If, however, a negative acknowledgement is received, i.e. a negative AI, a 'NACK on AICH received' message will be relayed to the higher layers and the random access procedure will be terminated. On the other hand, if the MS fails to receive any acknowledgement after a predefined time-out, it will retransmit the preamble in another randomly selected access slot of Figure 22.28 with a newly selected signature, provided that the maximum number of preamble retransmissions was not exceeded. The transmit power of the preamble is also increased, as specified by the above-mentioned preamble power ramping factor. This procedure is repeated until either an acknowledgement is received from the BS or the maximum number of preamble retransmissions is reached. For the latter case, a 'No ACK on AICH' will be relayed to the higher layers.

22.2.7.2 Common Packet Channel Access Procedures

The transmission of the CPCH of Table 22.4 is somewhat similar to that of the RACH transmission regime highlighted in Figure 22.28. Before commencing any CPCH transmission, the MS must acquire vital information from the BCH message transmitted on the P-CCPCH. This information includes the scrambling codes, the available signatures and the access slots for both the A-P and CD-P messages introduced in Section 22.2.3.2, the scrambling code of the message part, the DL AICH and the associated DL DPCCH channelization code, the initial transmit power of the preambles, the preamble power ramping factor, and the maximum allowable number of retransmissions.

At the commencement of any CPCH transmission, the MS will acquire the up-to-date status information from the CSICH regarding the availability of the PCPCH and/or the available data rates. If there is no PCPCH available or if the requested data rate is higher than the available data rate, the access procedure will be terminated. Otherwise, a PCPCH is randomly selected for the A-P transmission using a procedure which is identical to that of the random access transmission process highlighted in Section 22.2.7.1. We will accordingly omit the details here.

Once a positive acknowledgement is received from the BS on the DL AP-AICH, the MS will transmit the CD-P on a randomly selected access slot of Figure 22.28 using a randomly selected CD-P signature. However, if a negative acknowledgement is received, the MS will terminate the access. On the other hand, if no acknowledgement was received, the MS will monitor the status information on the CSICH again, in order to ensure that at least one PCPCH is available, and then commence another access procedure. Upon receiving a positive acknowledgement from the BS on the CD/CA-ICH, the MS will begin transmitting the PC-P followed immediately by the message part shown in Figure 22.5 at a predefined access slot of Figure 22.28. Following this process it will monitor the DL DPCCH associated with the CPCH, as discussed in Section 22.2.3.1, for the sake of detecting the beginning of the message indicator. The MS is enabled to continue transmitting the data if this indicator is received. Otherwise, the transmission will cease, if an emergency stop indicator is received or if the MS fails to detect the DL DPCCH.

22.2.8 Power Control

Accurate power control is essential in CDMA in order to mitigate the so-called *near-far problem* [547, 548]. Furthermore, power control has a dramatic effect on the coverage and capacity of the system: we will therefore consider the UTRA power control issues in detail.

22.2.8.1 Closed-Loop Power Control in UTRA

Closed-loop power control is employed on both the UL and DL of the FDD mode through the TPC commands that are conveyed in the UL and DL according to the format of Figures 22.5

and 22.6, respectively. Since the power control procedure is the same on both links, we will only elaborate further on the UL procedure.

UL closed-loop power control is invoked in order to adjust the MS's transmit power such that the received Signal-to-Interference Ratio (SIR) at the BS is maintained at a given target SIR. The value of the target SIR depends on the required quality of the connection. The BS measures the received power of the desired UL transmitted signal for both the DPDCH and the DPCCH messages shown in Figure 22.5 after RAKE combining, and it also estimates the total received interference power in order to obtain the estimated received SIR. This SIR estimation process is performed every $\frac{2}{3}$ ms, or a time-slot duration, in which the SIR estimate is compared to the target SIR. According to the values of the estimated and required SIRs, the BS will generate a TPC command, which is conveyed to the MS using the burst of Figure 22.6. If the estimated SIR is higher than the target SIR, the TPC command will instruct the MS to lower the transmit power of the DPDCH and DPCCH of Figure 22.5 by a step size of Δ_{TPC} dB. Otherwise, the TPC command will instruct the MS to increase the transmit power by the same step size. The step size Δ_{TPC} is typically 1 dB or 2 dB. Transmitting at an unnecessarily high power reduces the battery life, while degrading other users' reception quality, who — as a consequence — may request a power increment, ultimately resulting in an unstable overall system operation.

In some cases, BS-diversity combining may take place, whereby two or more BSs transmit the same information to the MS in order to enhance its reception quality. These BSs are known as the active BS set of the MS. The received SIR at each BS will be different and so the MS may receive different TPC commands from its active set of BSs. In this case, the MS will adjust its transmit power according to a simple algorithm, increasing the transmit power only if the TPC commands from all the BSs indicate an 'increase power' instruction. Similarly, the MS will decrease its transmit power if all the BSs issue a 'decrease power' TPC command. Otherwise, the transmit power remains the same. In this way, the multi-user interference will be kept to a minimum without significant deterioration of the performance, since at least one BS has a good reception. Again, the UL and DL procedures are identical, obeying the TPC transmission formats of Figures 22.5 and 22.6, respectively.

22.2.9 Open-Loop Power Control in TDD Mode

As mentioned previously in Section 22.2.3, in contrast to the closed-loop power control regime of the FDD mode, no TPC commands are transmitted on the DL in TDD mode. Instead, open-loop power control is used to adjust the transmit power of the MS. Prior to any data burst transmission, the MS would have acquired information about the interference level measured at the BS and also about the BS's P-CCPCH transmitted signal level, which are conveyed to the MS via the BCH according to the format of Figure 22.14. At the same time, the MS would also measure the power of the received P-CCPCH. Hence, with knowledge of the transmitted and received power of the P-CCPCH, the DL path-loss can be found. Since the interference level and the estimated path-loss are now known, the required transmitted power of the TDD burst can be readily calculated based on the required SIR level. Let us now consider how the MS identifies the different cells or BSs with which it is communicating.

22.2.10 Cell Identification

22.2.10.1 Cell Identification in the FDD Mode

System- and cell-specific information is conveyed via the BCH transmitted by the P-CCPCH of Table 22.4 in the context of Figure 22.10 in UTRA. This information has to be obtained before the MS can access the network. The P-CCPCH information broadcast from each cell is spread by the system-specific OVSF channelization code $c_{256,2}$ of Figure 22.25. However, each P-CCPCH message is scrambled by a cell-specific primary scrambling code as highlighted in Section 22.2.6.3 in order to minimise the intercell interference as well as to assist in identifying the corresponding cell. Hence, the first step for the MS is to recognise this primary scrambling code and to synchronise with the corresponding BS.

As specified in Section 22.2.6.3, there are a total of 512 DL primary scrambling codes available in the network. Theoretically, it is possible to achieve scrambling code identification by cross-correlating the P-CCPCH broadcast signal with all the possible 512 primary scrambling codes. However, this would be an extremely tedious and slow process, unduly delaying the MS's access to the network. In order to achieve a fast cell identification by the MS, UTRA adopted a three-step approach [549], which invoked the SCH broadcast from all the BSs in the network. The SCH message is transmitted during the first 256 chips of the P-CCPCH, as illustrated in Figure 22.10. The concept behind this three-step approach is to divide the set of 512 possible primary scrambling codes into 64 subsets, each containing a smaller set of primary scrambling codes, namely, eight codes. Once knowledge of which subset the primary scrambling code of the selected BS belongs to is acquired, the MS can proceed to search for the correct primary scrambling code from a smaller subset of the possible codes.

The frame structure of the DL SCH message seen in Figure 22.10 is shown in more detail in Figure 22.29. It consists of two subchannels, the *Primary SCH* and *Secondary SCH*, transmitted in parallel using code multiplexing. As seen in Figure 22.29, in the Primary SCH a so-called *Primary Synchronization Code* (PSC), based on a generalized hierarchical Golay sequence [550] of length 256 chips, is transmitted periodically at the beginning of each slot, which is denoted by c_p in Figure 22.29. The same PSC is used by all the BSs in the network. This allows the MS to establish slot-synchronization and to proceed to the frame-synchronization phase with the aid of the secondary SCH. On the secondary SCH, a sequence of 15 *Secondary Synchronization Codes* (SSCs), each of length 256 chips, is transmitted with a period of one 10 ms radio frame duration, that is, 10 ms, as seen in Figure 22.29. An example of this 15-SSC sequence would be:

$$c_1^1\ c_1^2\ c_2^3\ c_8^4\ c_9^5\ c_{10}^6\ c_{15}^7\ c_8^8\ c_{10}^9\ c_{16}^{10}\ c_2^{11}\ c_7^{12}\ c_{15}^{13}\ c_7^{14}\ c_{16}^{15}, \tag{22.8}$$

where each of these 15 SSCs is selected from a set of 16 legitimate SSCs. The specific sequence of 15 SSCs denoted by $c_i^1, \ldots, c_i^{15}$ — where $i = 1, \ldots, 16$ in Figure 22.29 — is used as a code in order to identify and signal to the MS which of the 64 subsets the primary scrambling code used by the particular BS concerned belongs to. The parameter a in Figure 22.29 is a binary flag used to indicate the presence ($a = +1$) or absence ($a = -1$) of a Space Time Block Coding Transmit Diversity (STTD) encoding scheme [551] in the P-CCPCH, as will be discussed in Section 22.4.1.1. Specifically, when each of the 16 legitimate 256-chip SSCs can be picked for any of the 15 positions in Figure 22.29 and assuming no other further

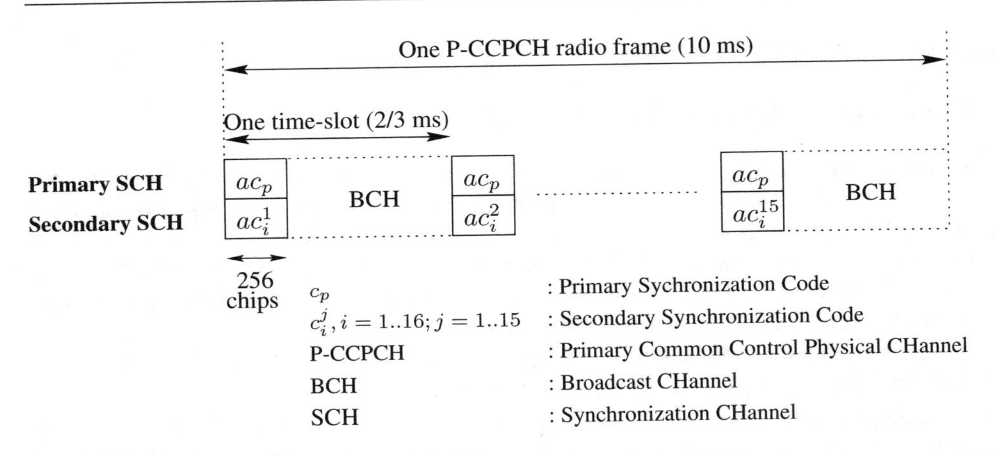

Figure 22.29: Frame structure of the UTRA DL synchronization channel (SCH), which is mapped to the first 256 chips of the P-CCPCH of Figure 22.10. The primary and secondary SCH are transmitted in parallel using code multiplexing. The parameter a is a gain factor used to indicate the presence ($a = +1$) or absence ($a = -1$) of STTD encoding in the P-CCPCH.

constraints, one could construct

$$\begin{aligned} c_{i,j}^{\text{repeated}} &= \binom{i+j-1}{j} \\ &= \frac{(i+j-1)!}{j!(i-1)!} \\ &= \frac{30!}{15! \cdot 15!} \\ &= 155,117,520 \end{aligned} \tag{22.9}$$

different such sequences, where $i = 16$ and $j = 15$. However, the 15 different 256-chip SSCs of Figure 22.29 must be constructed so that their cyclic shifts are also unique, since these sequences have to be uniquely recognized before synchronization. In other words, none of the cyclic shifts of the 64 required $15 \times 256 = 3840$-chip sequences can be identical to any of the other sequences' cyclic shifts. Provided that these conditions are satisfied, the 15 specific 256-chip secondary SCH sequences can be recognized within one 10 ms-radio frame-duration of 15 slots. Thus, both slot and frame synchronization can be established within the particular 10 ms frame received. Using this technique, initial cell identification and synchronization can be carried out in the following three basic steps.

Step 1: The MS uses the 256-chip PSC of Figure 22.29 to perform cross-correlation with all the received Primary SCHs of the BSs in its vicinity. The BS with the highest correlator output is then chosen, which constitutes the best cell site associated with the lowest path-loss. Several periodic correlator output peaks have to be identified in order to achieve a high BS detection reliability, despite the presence of high-level interference. *Slot synchronization is also achieved* in this step by recognizing the 15 consecutive c_p sequences, providing 15 periodic correlation peaks.

Step 2: Once the best cell site is identified, the primary scrambling code subset of that cell site is found by cross-correlating the Secondary SCH with the 16 possible SSCs in each of the 15 time-slots of Figure 22.29. This can be easily implemented using 16 correlators, since the timing of the SSCs is known from Step 1. Hence, there are a total of $15 \times 16 = 240$ correlator outputs. From these outputs, a total of $64 \times 15 = 960$ decision variables corresponding to the 64 possible secondary SCH sequences and 15 cyclic shifts of each $15 \times 256 = 3840$-chip sequence are obtained. The highest decision variable determines the primary scrambling code subset. Consequently, *frame synchronization is also achieved.*

Step 3: With the primary scrambling code subset identified and frame synchronization achieved, the primary scrambling code itself is acquired in UTRA by cross-correlating the received CPICH signal — which is transmitted synchronously with the P-CCPCH — on a symbol-by-symbol basis with the eight possible primary scrambling codes belonging to the identified primary scrambling code subset. Note that the CPICH is used in this case, because it is scrambled by the same primary scrambling code as the P-CCPCH and also uses a predefined pilot sequence and so it can be detected more reliably. By contrast, the P-CCPCH carries the unknown BCH information. Once the exact primary scrambling code is identified, the BCH information of Table 22.4, which is conveyed by the P-CCPCH of Figure 22.10, can be detected.

22.2.10.2 Cell Identification in the 3.84 Mcps TDD Mode

The procedure of cell identification in the TDD mode is somewhat different from that in FDD mode. In the TDD mode, a combination of three 256-chip SSCs out of 12 unique SSCs are used to identify one of 32 SSC code groups allocated to that cell. Each code group contains four different scrambling codes and four corresponding long (for Type 1 and 3 bursts) and short (for Type 2 burst) basic mid-amble codes, which were introduced in the context of Figure 22.14. Each code group is also associated with a specific time offset, t_{offset}. The three SSCs, c_i^1, c_i^2, and c_i^3, are transmitted in parallel with the PSC, c_p, at a time offset t_{offset} measured from the start of a time-slot, as shown in Figure 22.30. Similarly to the FDD mode, the PSC is based on a so-called generalized hierarchical Golay sequence [550], which is common to all the cells in the system. Initial cell identification and synchronization in the TDD mode can also be carried out in three basic steps.

Step 1: The MS uses the 256-chip PSC of Figure 22.30 to perform cross-correlation with all the received PSC of the BSs in its vicinity. The BS associated with the highest correlator output is then chosen, which constitutes the best cell site exhibiting the lowest path-loss. Slot synchronization is also achieved in this step. If only one time-slot per frame is used to transmit the SCH as outlined in the context of Figure 22.14, then frame synchronization is also achieved.

Step 2: Once the PSC of the best cell site is identified, the three SSCs transmitted in parallel with the PSC in Figure 22.30 can be identified by cross-correlating the received signal with the 12 possible prestored SSCs. The specific combination of the three SSCs will identify the code group used by the corresponding cell. The specific frame timing of that cell also becomes known from the time offset t_{offset} associated with that code group.

If two time-slots per frame are used to transmit the SCH as outlined in the context of Figure 22.14, then the second PSC must be detected at an offset of seven or eight time-slots

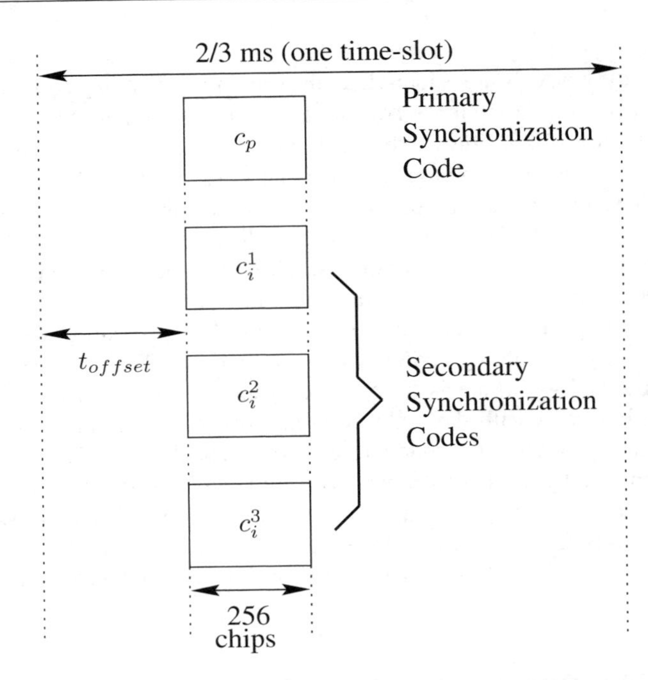

Figure 22.30: Time-slot structure of the UTRA TDD DL synchronization channel (SCH), which obeys the format of Figure 22.3. The primary and three secondary synchronization codes are transmitted in parallel at a time offset t_{offset} from the start of a time-slot.

with respect to the first one in order to achieve frame synchronization.

Step 3: As mentioned in Section 22.2.3, each cell-specific basic mid-amble code defined in the context of Figure 22.14 is associated with a mid-amble code set. The P-CCPCH of Table 22.4 is always associated with the first mid-amble of that set. Hence, with the code group identified and frame synchronization achieved, the cell-specific scrambling code and the associated basic mid-amble code are acquired in the TDD mode of UTRA by cross-correlating the four possible mid-amble codes with the P-CCPCH. Once the exact basic mid-amble code is identified, the associated scrambling code will be known, and the BCH information of Table 22.4, which is conveyed by the P-CCPCH of Figure 22.10, can be detected. Having highlighted the FDD and TDD UTRA cell-selection and synchronization solutions, let us now consider some of the associated handover issues.

In contrast to the FDD mode, the SCH in the TDD mode is not time-multiplexed onto the P-CCPCH of Table 22.4. Instead, the SCH messages are transmitted on one or two time-slots per frame.[5] The P-CCPCH will be code-multiplexed with the first SCH time-slot in each frame.

[5]If two time-slots are allocated to the SCH per frame, they will be spaced seven slots apart.

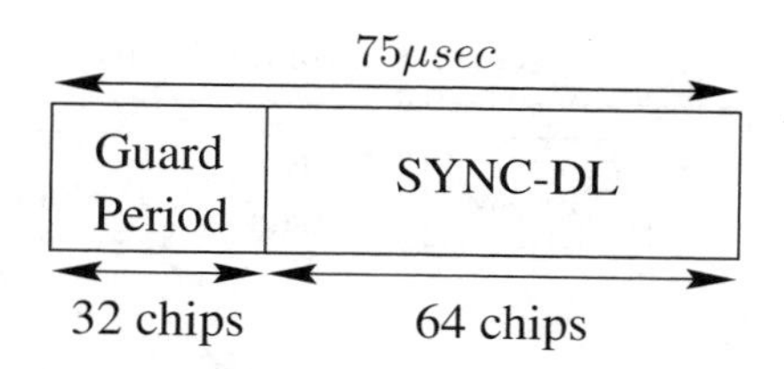

Figure 22.31: Structure of the UTRA 1.28 Mcps TDD DL Pilot Time-Slot (DwPTS), which obeys the format of Figure 22.17.

22.2.10.3 Cell Identification in the 1.28 Mcps TDD Mode

Recall that there is no SCH in the 1.28 Mcps TDD mode. Instead the DwPTS, as discussed in Section 22.2.3.3 and illustrated in Figure 22.17, will be used for assisting in the the cell identification procedure. The structure of the DwPTS is shown in Figure 22.31. There are 32 different basic SYNC-DL 64-chip codes and they are not scrambled, since the MS has no knowledge concerning the scrambling code used by the cell. Each QPSK modulated SYNC-DL code corresponds to four different basic mid-amble codes. One of these mid-amble codes is used by the P-CCPCH, which was transmitted before the DwPTS. Each mid-amble code is also associated with a scrambling code. Four consecutive phases of the SYNC-DL are used for indicating the presence of the P-CCPCH in the next four subframes, from which the BCH information is transmitted.

Step 1: The MS first identifies, which of the SYNC-DL codes is used by cross-correlating the received signal with the 32 possible locally stored SYNC-DL codes. Again, the code that exhibits the highest correlator output will be identified as the SYNC-DL.

Step 2: The P-CCPCH used in the 1.28 Mcps TDD mode is always transmitted on time-slot 0 or immediately before the DwPTS, as shown in Figure 22.4 and Figure 22.17, respectively. Hence, once the position of the DwPTS is known, the MS can determine the basic mid-amble code used by the P-CCPCH by simply correlating it with the four possible codes associated with the SYNC-DL code. Acquiring the mid-amble code will lead to the knowledge of the scrambling code used by the cell.

Step 3: By now subframe synchronization has been achieved. In order to accomplish frame synchronization, the MS must acquire four consecutive phases of the SYNC-DL. Information about these four phases informs the MS as regards to where the BCH data is being conveyed by the P-CCPCH.

22.2.11 Handover

In this section, we consider the handover issues in the context of the FDD mode, since the associated procedures become simpler in the TDD mode, where the operations can be carried out during the unused time-slots. Theoretically, DS-CDMA has a frequency reuse factor of one [552]. This implies that neighbouring cells can use the same carrier frequency without interfering with each other, unlike in TDMA or FDMA. Hence, seamless uninterrupted handover can be achieved when mobile users move between cells, since no switching of carrier frequency and synthesiser retuning is required. However, in *hierarchical cell structures*

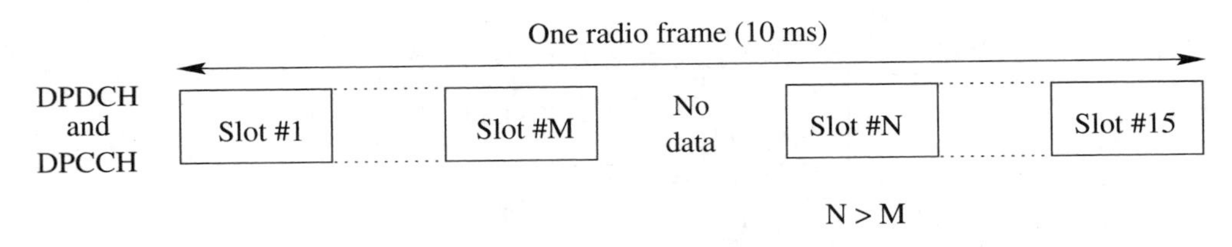

Figure 22.32: Uplink frame structure in compressed mode operation during UTRA handovers.

(HCS)[6] catering, for example, for high-speed mobiles with the aid of a macrocell oversailing a number of microcells, using a different carrier frequency is necessary in order to reduce the intercell interference. In this case, inter-frequency handover is required. Furthermore, because the various operational GSM systems used different carrier frequencies, handover from UTRA systems to GSM systems will have to be supported during the transitory migration phase, while these systems will coexist. Thus, handovers in terrestrial UMTSs can be classified into inter-frequency and intra-frequency handovers.

22.2.11.1 Intra-Frequency Handover or Soft Handover

Soft handover [553, 554] involves no frequency switching because the new and old cell use the same carrier frequency. The MS will continuously monitor the received signal levels from the neighbouring cells and compares them against a set of thresholds. This information is fed back to the network. Based on this information, if a weak or strong cell is detected, the network will instruct the MS to drop or add the cell from/to its active BS set. In order to ensure a seamless handover, a new link will be established before relinquishing the old link, using the *make before break* approach.

22.2.11.2 Inter-Frequency Handover or Hard Handover

In order to achieve handovers between different carrier frequencies without affecting the data flow, a technique known as *compressed mode* can be used [555]. With this technique, the UL data, which normally occupies the entire 10 ms frame of Figure 22.3 is time-compressed, so that it only occupies a portion of the frame, that is, slot#1-slot#M and slot#N-slot#15, while no data is transmitted during the remaining portion, that is, slot#(M+1)-slot#(N-1). The latter interval is known as the idle period, as shown in Figure 22.32. There are two types of frame structures for the DL compressed mode, as shown in Figure 22.33. In the Type A structure, shown at the top of Figure 22.33, no data is transmitted after the pilot field of slot#M until the start of the pilot field of slot#(N-1) in order to maximise the transmission gap length. By contrast, in the Type B structure shown at the bottom of Figure 22.33, a TPC command is transmitted in slot#(M+1) during the idle period in order to optimise the power control.

The idle period has a variable duration, but the maximum period allowable within a 15-slot, 10 ms radio frame is seven slots. The idle period can occur either at the centre of a 10 ms frame or at the end and the beginning of two consecutive 10 ms frames, such that the idle period spans over two frames. However, in order to maintain the seamless operation of

[6] Microcells overlaid by a macrocell.

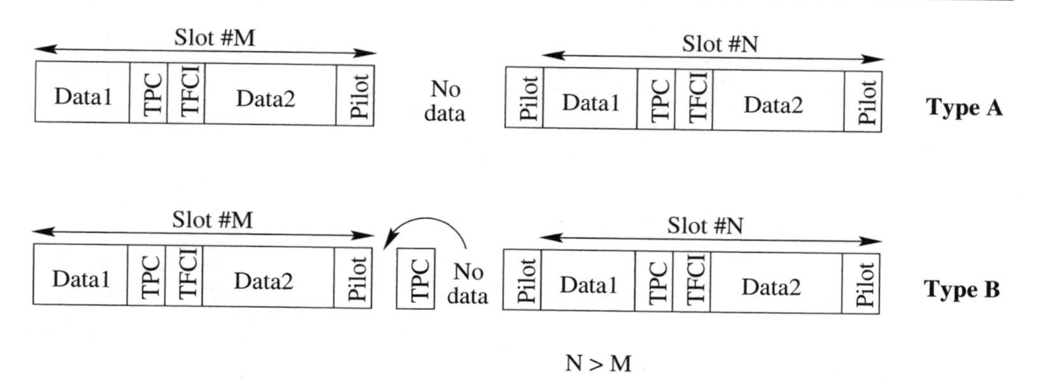

Figure 22.33: Downlink frame structure in compressed mode operation during UTRA handovers using the transmission formats of Figure 22.3.

all MSs occupying the uncompressed 15-slot, 10 ms frame, the duration of all time-slots has to be shortened by 'compressing' their data. The compression of data can be achieved by channel-code puncturing, a procedure that obliterates some of the coded parity bits, thereby slightly reducing the code's error correcting power, or by adjusting the spreading factor. In order to maintain the quality of the link, the instantaneous power is also increased during the compressed mode operation. After receiving the data, the MS can use this idle period in the 10 ms frame, to switch to other carrier frequencies of other cells and to perform the necessary link-quality measurements for handover.

Alternatively, a twin-receiver can be used in order to perform inter-frequency handovers. One receiver can be tuned to the desired carrier frequency for reception, while the other receiver can be used to perform handover link-quality measurements at other carrier frequencies. This method, however, results in a higher hardware complexity at the MS.

The 10 ms frame length of UTRA was chosen so that it is compatible with the multiframe length of 120 ms in GSM. Hence, the MS is capable of receiving the Frequency Correction Channel (FCCH) and Synchronization Channel (SCH) messages in the GSM [64] frame using compressed mode transmission and to perform the necessary handover link-quality measurements [517].

22.2.12 Intercell Time Synchronization in the UTRA TDD Mode

Time synchronization between BSs is required when operating in the TDD mode in order to support seamless handovers. A simple method of maintaining intercell synchronization is by periodically broadcasting a reference signal from a source to all the BSs. The propagation delay can be easily calculated, and hence compensated, from the fixed distance between the source and the receiving BSs. There are three possible ways of transmitting this reference signal, namely, via the terrestrial radio link, via the physical wired network, or via the Global Positioning System (GPS).

Global time synchronization in 3G mobile radio systems is achieved by dividing the synchronous coverage region into three areas, namely, the so-called sub-area, main area and coverage area, as shown in Figure 22.34. Intercell synchronization within a sub-area is pro-

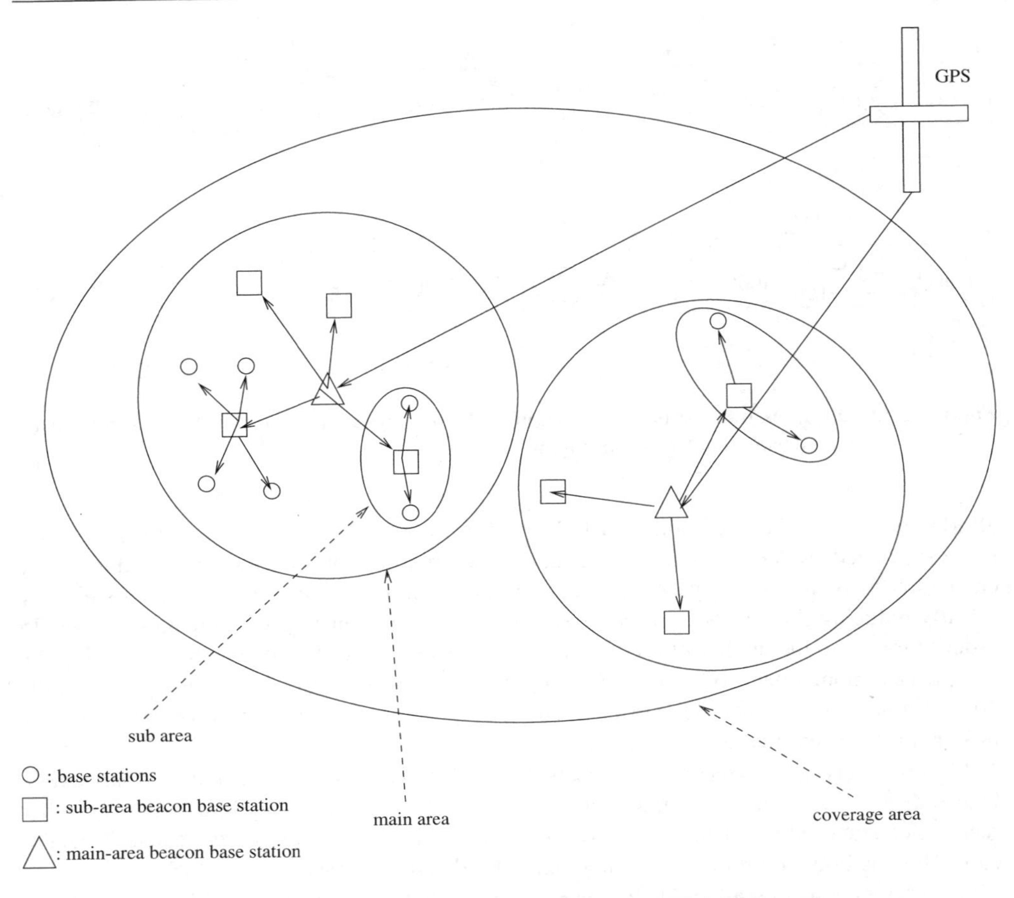

Figure 22.34: Intercell time synchronization in UTRA TDD mode.

vided by a sub-area reference BS. Since the sub-area of Figure 22.34 is smaller than the main area, transmitting the reference signal via the terrestrial radio link or the physical wired network is more feasible. All the sub-area reference BSs in a main area are in turn synchronized by a main-area reference BS. Similarly, the reference signal can be transmitted via the terrestrial radio link or the physical wired network. Finally, all the main-area reference BSs are synchronized using the GPS. The main advantage of dividing the coverage regions into smaller areas is that each lower hierarchical area can still operate on its own, even if the synchronization link with the higher hierarchical areas is lost.

22.3 The cdma2000 Terrestrial Radio Access [556–558]

The current 2G mobile radio systems standardized by TIA in the United States are IS-95-A and IS-95-B [556]. The radio access technology of both systems is based on narrowband DS-CDMA with a chip rate of 1.2288 Mcps, which gives a bandwidth of 1.25 MHz. IS-

Radio Access Technology	DS-CDMA, Multicarrier CDMA
Operating environments	Indoor/Outdoor to indoor/Vehicular
Chip rate (Mcps)	1.2288/3.6864
Channel bandwidth (MHz)	1.25/3.75
Duplex modes	FDD
Frame length	Typically 5 and 20 ms
Spreading factor	Variable, 4 to 256
Detection scheme	Coherent with common pilot channel
Intercell operation	FDD : Synchronous
Power control	Open and closed loop
Handover	Soft-handover Inter-frequency handover

Table 22.10: The cdma2000 system's basic parameters.

95-A was commercially launched in 1995, supporting circuit and packet mode transmissions at a maximum bit rate of only 14.4 kbps [556]. An enhancement to the IS-95-A standards, known as IS-95-B, was developed and introduced in 1998 in order to provide higher data rates, on the order of 115.2 kbps [531]. This was feasible without changing the physical layer of IS-95-A. However, this still falls short of the 3G mobile radio system requirements. Hence, the technical committee TR45.5 within TIA has proposed cdma2000, a 3G mobile radio system that is capable of meeting all the requirements laid down by ITU. One of the problems faced by TIA is that the frequency bands allocated for the 3G mobile radio system, identified during WARC'92 to be 1885–2025 MHz and 2110–2200 MHz, have already been allocated for Personal Communications Services (PCS) in the United States from 1.8 GHz to 2.2 GHz. In particular, the CDMA PCS based on the IS-95 standards has been allocated the frequency bands of 1850–1910 MHz and 1930–1990 GHz. Hence, the 3G mobile radio systems have to fit into the allocated bandwidth without imposing significant interference on the existing applications. Thus, the framework for cdma2000 was designed so that it can be overlaid on IS-95 and it is backwards compatible with IS-95. Most of this section is based on [556–558].

22.3.1 Characteristics of cdma2000

The basic parameters of cdma2000 are shown in Table 22.10. The cdma2000 system has a basic chip rate of 1.2288 Mcps, which is identical to that used in the IS-95 standard. Hence the existing IS-95 network can also be used for supporting the operation of cdma2000. In addition, a chip rate of 3.6864 Mcps was also introduced in cdma2000, which is accommodated in a bandwidth of 3.75 MHz. This chip rate is in fact three times the chip rate of 1.2288 Mcps. Two different modulation techniques are employed in cdma2000, namely *direct-spread (DS)* modulation and *multicarrier (MC) modulation*. In DS modulation, the symbols are spread according to the chip rate and transmitted using a single carrier, giving a bandwidth of 1.25 MHz or 3.75 MHz for the chip rate of 1.2288 Mcps and 3.6864 Mcps, respectively. The 1.2288 Mcps chip rate is used on both the UL and DL, which is referred to in cdma2000 as Spreading Rate 1 (SR1) or 1x, as shown in Figure 22.35. The 3.6864 Mcps

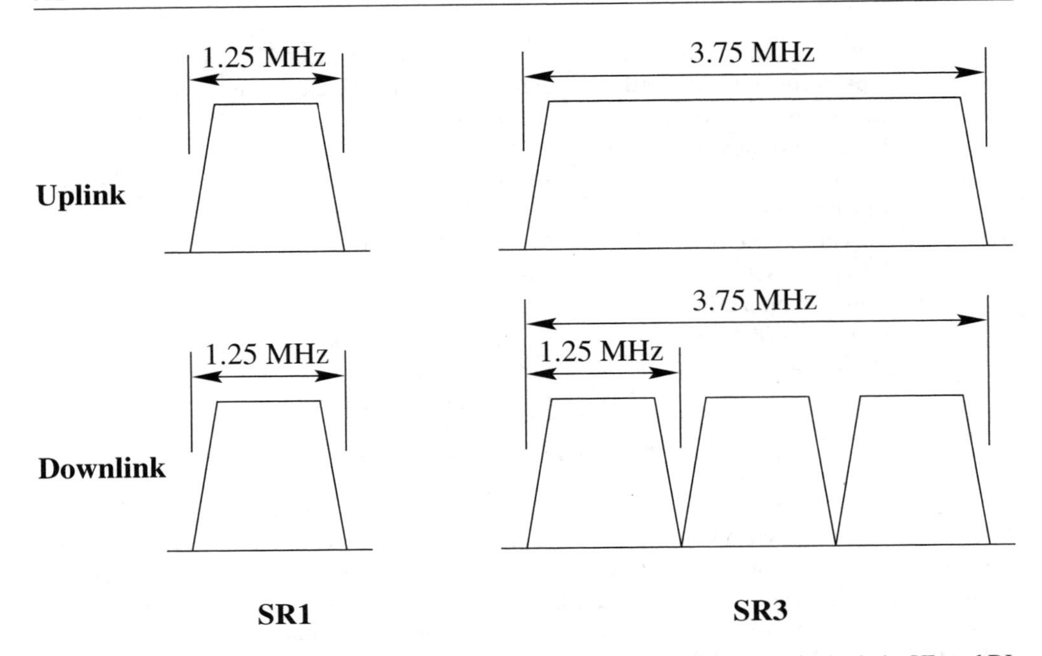

Figure 22.35: The Spreading Rate 1 (SR1) and Spreading Rate 3 (SR3) modes for both the UL and DL in cdma2000.

chip rate DS modulation is only used on the UL. In MC modulation, the symbols to be transmitted are de-multiplexed into separate signals, each of which is then spread at a chip rate of 1.2288 Mcps. Three different carrier frequencies are used for transmitting these spread signals, each of which has a bandwidth of 1.25 MHz. This method is used for the DL only, because in this case, transmit diversity can be achieved by transmitting the different carrier frequencies over spatially separated antennas. The DS method operating at 3.6864 Mcps and the MC method are known in cdma2000 as Spreading Rate 3 (SR3) or 3x, as illustrated in Figure 22.35. Note that transmit diversity can also be achieved in SR1 by using either an orthogonal transmit diversity scheme or a space time spreading scheme.

In contrast to UTRA and IMT-2000, where the pilot symbols of Figure 22.6 are time-multiplexed with the dedicated data channel on the DL, cdma2000 employs a common code multiplexed continuous pilot channel on the DL, as in the IS-95 system. The advantage of a common DL pilot channel is that no additional pilot overhead is incurred for the individual users. However, if adaptive antennas are used for the sake of supporting spatially selective beamforming, then additional pilot or reference signals have to be transmitted from each antenna, which are used by the beamformer as the reference signal during the adjustment of its array weights [268]. This process is analogous to the training of a channel equalizer [4].

Another difference with respect to UTRA and IMT-2000 is that the BSs are operated in a synchronous mode in cdma2000. As a result, the same PN code having different phase offsets can be used for distinguishing the BSs. Using a single common PN sequence can expedite cell acquisition as compared to using a set of PN sequences, as we have seen in Section 22.2.10 for IMT-2000/UTRA. Let us now consider the cdma2000 physical channels.

Dedicated PHysical CHannels (DPHCH)
Fundamental CHannel (FCH) (UL/DL)
Supplemental CHannel (SCH) (UL/DL)
Dedicated Control CHannel (DCCH) (UL/DL)
Reverse PIlot CHannel (R-PICH) (UL)
Common PHysical CHannels (CPHCH)
Forward PIlot CHannel (F-PICH) (DL)
Auxiliary PIlot CHannel (APICH) (DL)
Transmit Diversity PIlot CHannel (TDPICH) (DL)
Auxiliary Transmit Diversity PIlot CHannel (ATDPICH) (DL)
Quick Paging CHannel (QPCH) (DL)
Paging CHannel (PCH) (DL)
Common Control CHannel (CCCH) (UL/DL)
SYNc Channel (SYNC) (DL)
Broadcast Control CHannel (BCCH) (DL)
Access CHannel (ACH) (UL)
Enhanced Access CHannel (EACH) (UL)
Common Assignment CHannel (CACH) (DL)

Table 22.11: The cdma2000 system's physical channels.

22.3.2 Physical Channels in cdma2000

The PHysical CHannels (PHCH) in cdma2000 can be classified into two groups, namely Dedicated PHysical CHannels (DPHCH) and Common PHysical CHannels (CPHCH). DPHCHs carry information between the BS and a single MS, while CPHCHs carry information between the BS and several MSs. Table 22.11 shows the collection of physical channels in each group. These channels will be elaborated on during our further discourse. Typically, the basic frame length in cdma2000 is 20 ms. However, the control information mapped to the so-called Fundamental CHannel (FCH) and Dedicated Control CHannel (DCCH) can also be transmitted in 5 ms duration frames. In addition, further frame lengths as long as 40 ms and 80 ms are also introduced in cdma2000, in order to increase the interleaving depth and hence the achievable coding and diversity gains. Furthermore, frame lengths as short as 1.25, 2.5 and 5 ms are also included in cdma2000 in order to allow for faster exchanges of messages. An exception with respect to these frame durations is constituted by the Sync CHannel, which has a frame length of 80/3 ms - a length, which is the same as that used in IS-95.

As shown in Table 22.11, there are 4 types of DL common pilot channels transmitted continuously from the BS, which is shared by all the MSs within the coverage area of the BS, namely the Forward PIlot CHannel (F-PICH), Auxiliary PIlot CHannel (APICH), Transmit Diversity PIlot CHannel (TDPICH) and the Auxiliary Transmit Diversity PIlot CHannel (ATDPICH). Typically these pilot channels are unmodulated DS signals and the MSs can use these pilot channels in order to perform channel estimation for coherent detection, soft handover, and fast acquisition of strong multipath rays for RAKE combining. The transmission of the F-PICH is mandatory in all cells and is transmitted along with all the other DL physical channels from the BS by using a unique orthogonal code (Walsh code 0) as in the IS-95 system. On the other hand, the transmission of the TDPICH, APICH and ATDPICH is optional,

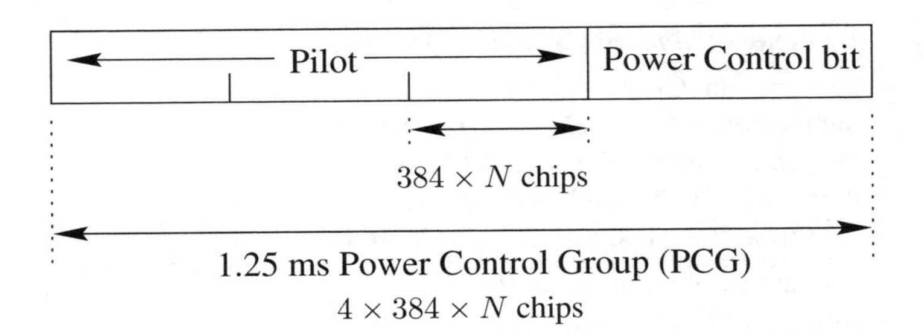

Figure 22.36: Uplink pilot channel structure in cdma2000 for a 1.25 ms duration PCG, where $N = 1$ and 3 corresponds to SR1 and SR3, respectively.

which are used for supporting the implementation of transmit diversity, antenna arrays for beamforming or both, respectively. Every MS also transmits an orthogonal code-multiplexed Reverse link PIlot CHannel (R-PICH), which enables the BS to perform coherent detection in the UL as well as to detect strong multipaths and to invoke power control measurements. This differs from IS-95, which supports only noncoherent detection in the UL due to the absence of a coherent UL reference. In addition to the pilot symbols, the R-PICH also contains time-multiplexed power control bits assisting in DL power control. A power control bit is multiplexed onto the 20 ms frame every 1.25 ms, giving a total of 16 power control bits per 20 ms frame or 800 power updates per second, implying an agile, prompt-response power control regime. Each 1.25 ms duration is referred to as a Power Control Group, as shown in Figure 22.36. The power control bits are inserted only in the FCH and the DCCH on the DL.

In order to conserve battery power, the MSs are usually operated in idling mode when they are not transmitting or receiving. Nevertheless, they still monitor the channel in case the network has messages intended for them. Hence, if a BS has messages for a particular MS, it will wake up the mobile by transmitting its request on the Quick Paging CHannel (QPCH), and conveys information such as paging indicators, configuration change indicators or broadcast indicators. In any case, these indicators are dedicated to informing the MS that there are messages for it in either the Paging CHannel (PCH) or the Common Control CHannel (CCCH), that are transmitted at a predetermined time offset after the indicators. The data rate of the QPCH is either 2.4 or 4.8 kbps. Paging functions and MS-specific message transmissions are handled by the PCH and the CCCH. The uncoded data rate of the PCH can be either 4.8 kbps or 9.6 kbps transmitted in 20 ms duration frames. The PCH only operates in the SR1 mode and its specification is exactly the same as that in IS-95, which was included in cdma2000 for the sake of maintaining backward compatibility. The CCCH is an improved version of the PCH, which can support additional higher data rates, such as 19.2 and 38.4 kbps on both the SR1 and SR3 modes. In this case, in addition to the 20 ms duration frames, 5 ms or 10 ms duration frames will also be used.

The SYNc Channel (SYNC) — note the different acronym in comparison to the SCH abbreviation in UTRA/IMT-2000 — is used for aiding the initial synchronization of a MS with the BS and for providing the MS with system-related information. This system-related information includes the Pseudo Noise (PN) sequence offset, which is used for identifying the BSs and the long code mask, which will be defined explicitly in Section 22.3.4. The SYNC has an unprotected data rate of 1.2 kbps and a channel coded data rate of 4.8 kbps. In the SR3

RC	SR	Characteristics
1	1	1.2, 2.4, 4.8 and 9.6 kbps data rate at coding rate $R = 1/3$, 64-ary orthogonal modulation
2	1	1.8, 3.6, 7.2 and 14.4 kbps data rate at $R = 1/2$, 64-ary orthogonal modulation
3	1	1.2, 1.35, 1.5, 2.4, 2.7, 4.8, 9.6, 19.2, 38.4, 76.8 and 153.6 kbps data rate at $R = 1/4$, 307.2 kbps data rate at $R = 1/2$, BPSK modulation with a pilot
4	1	1.8, 3.6, 7.2, 14.4, 28.8, 57.6, 115.2 and 230.4 kbps data rate at $R = 1/4$, BPSK modulation with a pilot
5	3	1.2, 1.35, 1.5, 2.4, 2.7, 4.8, 9.6, 19.2, 38.4, 76.8 and 153.6 kbps data rate at $R = 1/4$, 307.2 and 614.4 kbps data rate at $R = 1/3$, BPSK modulation with a pilot
6	3	1.8, 3.6, 7.2, 14.4, 28.8, 57.6, 115.2, 230.4 and 460.8 kbps data rate at $R = 1/4$, 1036.8 kbps data rate at $R = 1/2$, BPSK modulation with a pilot

Table 22.12: Radio Configuration (RC) characteristics for the UL traffic channels.

mode, the SYNC is demultiplexed into three streams, one for each carrier. The operation of the SYNC is exactly the same as that used in the IS-95 system. Hence, it shall not be further elaborated on here.

The Broadcast Control CHannel (BCCH) is used for the transmission of control information from the BS to a MS at a data rate of 4.8, 9.6 or 19.2 kbps. The BCCH is divided into slots that are 40, 80 or 160 ms in duration, each containing 1, 2 and 4 frames, respectively. The Access CHannel (ACH), which is exactly the same as that in IS-95, is based on a slotted random access procedure, carries short signalling messages such as call initialization and paging responses conveyed from the MS to the BS. It is transmitted at a fixed data rate of 4.8 kbps in the SR1 mode. A higher speed version of the ACH is introduced in cdma2000, known as the Enhanced Access CHannel (EACH), offering data rates of 9.6, 19.2 and 38.4 kbps. The Common Assignment CHannel (CACH) is associated with the EACH during a reservation access mode, namely one of the random access modes, which will be further discussed in Section 22.3.5.

The employment of three dedicated physical data channels, namely, the so-called Fundamental CHannel (FCH), Supplemental CHannel(SCH) and Supplemental Code CHannel (SCCH), optimizes the system's operation during multiple simultaneous service transmissions. Each channel carries a different type of service and is coded as well as interleaved independently. However, in any connection, there can be only one FCH, while up to 7 SCCHs and up to 2 SCHs can be supported. The transmission format of these channels is specified by a set of so-called Radio Configurations (RC). Each RC is characterized by the physical layer parameters such as the SR, data rate and the FEC coding rate. A list of the RCs is shown in Table 22.12 and Table 22.13 for the UL and the DL, respectively, where there are 6 RCs for the UL and 9 RCs for the DL [559]. For the UL FCH, fixed data rates specified in RC1 and RC2 given in Table 22.12 are supported. Additionally, fixed and variable data rates at or between 1.5, 2.7, 4.8 and 9. kbps in RC3 and RC5 as well as 1.8, 3.6, 7.2, 9.6 and 14.4 kbps in RC4 and RC6 are also supported. Similarly, for the DL FCH fixed data rates specified in RC1 and RC2 given in Table 22.13 are supported. In addition, fixed and variable data rates at or between 1.5, 2.7, 4.8 and 9.6 kbps in RC3, RC4, RC6 and RC7 as well as 1.8, 3.6, 7.2, 9.6 and 14.4 kbps in RC5, RC8 and RC9 are also supported. Both the SCH and the SCCH are

RC	SR	Characteristics
1	1	1.2, 2.4, 4.8 and 9.6 kbps data rate at coding rate $R = 1/2$, BPSK pre-spreading symbols
2	1	1.8, 3.6, 7.2 and 14.4 kbps data rate at $R = 1/2$, BPSK pre-spreading symbols
3	1	1.5, 2.7, 4.8, 9.6, 19.2, 38.4, 76.8 and 153.6 kbps data rate at $R = 1/4$, QPSK pre-spreading symbols, transmit diversity allowed
4	1	1.5, 2.7, 4.8, 9.6, 19.2, 38.4, 76.8, 153.6 and 307.2 kbps data rate at $R = 1/2$, QPSK pre-spreading symbols, transmit diversity allowed
5	1	1.8, 3.6, 7.2, 14.4, 28.8, 57.6, 115.2 and 230.4 kbps data rate at $R = 1/4$, QPSK pre-spreading symbols, transmit diversity allowed
6	3	1.5, 2.7, 4.8, 9.6, 19.2, 38.4, 76.8, 153.6 and 307.2 kbps data rate at $R = 1/6$, QPSK pre-spreading symbols
7	3	1.5, 2.7, 4.8, 9.6, 19.2, 38.4, 76.8, 153.6, 307.2 and 614.4 kbps data rate at $R = 1/3$, QPSK pre-spreading symbols
8	3	1.8, 3.6, 7.2, 14.4, 28.8, 57.6, 115.2, 230.4 and 460.8 kbps data rate at $R = 1/4$ for 20 ms frame or $R = 1/3$ for 5 ms frame, QPSK pre-spreading symbols
9	3	1.8, 3.6, 7.2, 14.4, 28.8, 57.6, 115.2, 230.4, 460.8 and 1036.8 kbps data rate at $R = 1/2$ for 20 ms frame or $R = 1/3$ for 5 ms frame, QPSK pre-spreading symbols

Table 22.13: Radio Configuration (RC) characteristics for the DL traffic channels.

capable of transmitting higher data rates than the FCH. Hence for the UL SCCH and SCH, fixed and variable data rates specified in RC1-RC2 and RC3-RC6, respectively, as given in Table 22.12 are supported. Similarly for the DL SCH, fixed and variable data rates specified in RC3-RC9 given in Table 22.13 will be supported. Blind rate detection [560] is used for SCHs not exceeding 14.4 kbps, while the rate information is explicitly provided for higher data rates. As we can see, the parameters of RC1 and RC2 employed in both the UL and DL correspond to the values specified in IS-95 and hence we will not be highlighting their features here. Thus only the channels that utilized RC3-RC6 and RC3-RC9 for the UL and DL, respectively, will be discussed subsequently. The Dedicated Control CHannel (DCCH) has a fixed uncoded data rate of 9.6 kbps in both the 5 ms and 20 ms duration frames or 14.4 kbps in the 20 ms duration frames. This control channel rate is more than an order of magnitude higher than that of the IS-95 system, hence it supports a substantially enhanced system control. Having described the cdma2000 physical channels of Table 22.11, let us now consider the service multiplexing and channel-coding aspects.

22.3.3 Service Multiplexing and Channel Coding

Services of different data rates and different QoS requirements are carried by different physical channels, namely, by the FCH and SCH of Table 22.11. This differs from UTRA and IMT-2000, whereby different services were time-multiplexed onto one or more physical channels, as highlighted in Section 22.2.4. These channels in cdma2000 are code-multiplexed using Walsh codes. Two types of coding schemes are used in cdma2000, as shown in Table 22.14. Basically, all channels use convolutional codes for forward error correction. However, for SCHs at rates higher than 14.4 kbps, turbo coding [540] is preferable. The rate of the input data stream is matched to the given channel rate either by adjusting the coding rate or

	Convolutional	**Turbo**
Rate	1/2 or 1/3 or 1/4	1/2 or 1/3 or 1/4 or 1/5
Constraint length	9	4

Table 22.14: The cdma2000 system's channel-coding parameters.

using symbol repetition with and without symbol puncturing, or alternatively, by sequence repetition. Tables 22.15 and 22.16 show the coding rate and the associated rate matching procedures for the various DL and UL physical channels, respectively. As we can see, the achievable channel rates of the FCH and SCCH based on RC1 and RC2 are rather low. Higher channel rates can be obtained using RC3-RC9, which was not explicitly shown here for reasons of space economy.

Following the above brief notes on the cdma2000 channel coding and service multiplexing issues, let us now turn our attention to the spreading and modulation processes.

22.3.4 Spreading and Modulation

There are generally three layers of spreading in cdma2000, as shown in Table 22.17. Each user's UL signal is identified by different offsets of a long code, a procedure that is similar to that of the IS-95 system briefly portrayed in [64]. As seen in Table 22.17, this long code is an m-sequence having a period of $2^{42} - 1$ chips. The construction of m-sequences was highlighted by Proakis [90]. Different user offsets are obtained using a long code mask.

Orthogonality between the different physical channels of the same user belonging to the same connection in the UL is maintained by spreading using Walsh codes. However, unlike in IS-95 whereby the Walsh codes are selected according to the transmitted symbols based on the 64-ary orthogonal modulation, in cdma2000, each physical channel is associated with a predefined Walsh codes, as shown in Table 22.18. The notation of a Walsh code W_n^N is given such that N refers to the code length, i.e. number of chips per symbol, and n is the row number of a Walsh-Hadamard matrix. However, backward compatible channels such as the ACH, the FCH utilizing RC1 and RC2 and the SCCH still use the 64-ary orthogonal modulation in cdma2000.

In contrast to the IS-95 DL of Figure 1.42 of [64], whereby Walsh code spreading is performed prior to QPSK modulation, the data in cdma2000 is first QPSK modulated before spreading the resultant I and Q branches with the aid of the same Walsh code. In this way, the number of Walsh codes available is increased twofold due to the orthogonality of the I and Q carriers. Furthermore, quasi-orthogonal codes are also employed for the sake of increasing the number of channelization codes on the DL. Since there may be numerous DL physical channels corresponding to the different users, the assignment of Walsh codes will not be predetermined like those on the UL. All the BSs in the system are distinguished by different offsets of the same complex DL m-sequence, as indicated by Table 22.17. This DL m-sequence code is the same as that used in IS-95, which has a period of $2^{15} = 32768$, and it is derived from m-sequences. The feedback polynomials of the shift registers used for the I and Q sequences are $X^{15} + X^{13} + X^9 + X^8 + X^7 + X^5 + 1$ and $X^{15} + X^{12} + X^{11} + X^{10} + X^6 + X^5 + X^4 + X^3 + 1$, respectively, for SR1 and for the DL of SR3 in the context of each carrier. For the UL of SR3 based on DS modulation, the polynomial invoked for both

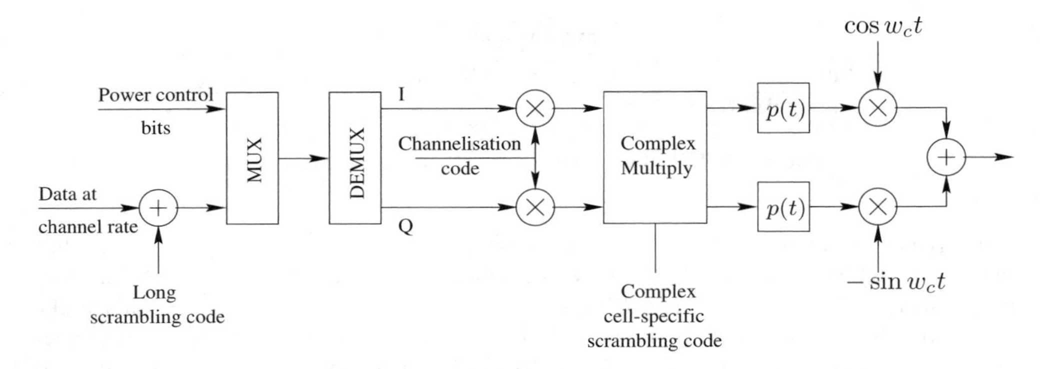

Figure 22.37: The cdma2000 downlink transmitter. The long scrambling code is used for the purpose of improving user privacy. Hence, only the paging channels and the traffic channels are scrambled with the long code. The common pilot channel and the SYNC channel are not scrambled by this long code (the terminology of Table 22.17 is used). Power control bits are multiplexed into the FCH and the DCCH only.

the I and Q sequences is $X^{20} + X^9 + X^5 + X^3 + 1$, giving an m-sequence of $2^{20} - 1$ in length. Both the I and Q PN sequences differ in position by 2^{19} chips. These sequences are then truncated after 3×2^{15} chips. Let us now focus our attention on the various DL spreading issues more closely.

22.3.4.1 Downlink Spreading and Modulation

Figure 22.37 shows the structure of a DL transmitter for a physical channel. The user data are first scrambled by the long scrambling code by assigning a different offset to different users for the purpose of improving user privacy, which is then mapped to the I and Q channels. This long scrambling code is identical to the user-specific scrambling code of the UL, as given in Table 22.17, which is also the same as that used in IS-95. The DL pilot channels of Table 22.11 (F-PICH, APICH, TDPICH and ATDPICH) and the SYNC channel are not scrambled with a long code since there is no need for user-specificity. The DL power control symbols are inserted into the FCH or the DCCH at a rate of 800 Hz, as shown in Figure 22.37. The I and Q channels are then spread using a Walsh code or a quasi-orthogonal code and multiplied with the complex-valued cell-specific PN sequence of Table 22.17, as portrayed in Figure 22.37. Each BS's DL channel is assigned a different Walsh code or quasi-orthogonal code in order to eliminate or reduce any intracell interference, since all Walsh codes transmitted by the serving BS are received synchronously. The length of the DL channelization Walsh code of Table 22.17 is determined by the type of physical channel and its data rate. Typically, DL FCHs having data rates belonging to RC4, RC6 and RC8 use a 128-chip Walsh code, while those in RC3 and RC5 employ a 64-chip Walsh code. The length of the Walsh codes used for the DL SCHs may range from 4 chips to 256 chips, depending on the type of RC. The F-PICH is an unmodulated sequence (all 0s) spread by the Walsh code W_0^{64}. Finally, the complex-valued spread data signal seen in Figure 22.37 is filtered in the baseband using the Nyquist filter impulse response $p(t)$ of Figure 22.37 and modulated onto a carrier frequency.

In case of MC modulation, the data is split into three branches immediately after the long

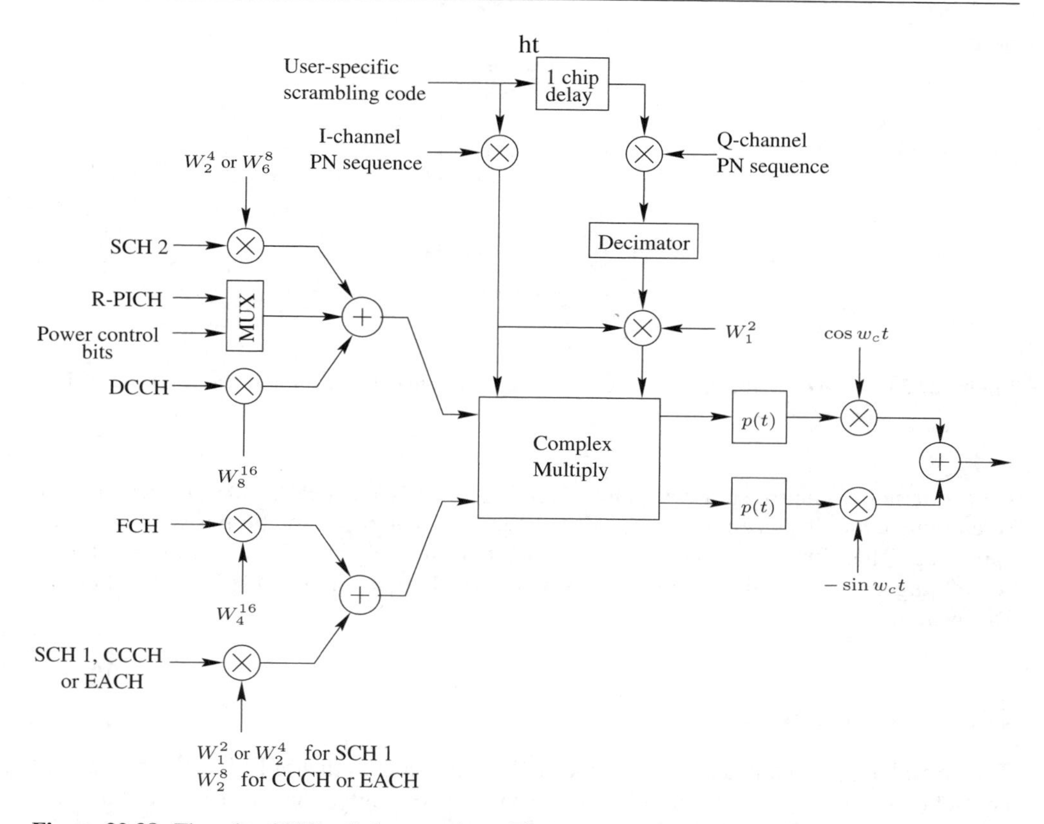

Figure 22.38: The cdma2000 uplink transmitter. The complex scrambling code is identical to the DL cell-specific complex scrambling code of Table 22.17 used by all the base stations in the system (the terminology of Table 22.17 is used).

code based scrambling operation of Figure 22.37, which was omitted in the figure for the sake of simplicity. Each of the three branches is then treated as a separate transmitter and modulated using different carrier frequencies.

22.3.4.2 Uplink Spreading and Modulation

The UL cdma2000 transmitter is shown in Figure 22.38. Notice how the different physical channels are summed and mapped to the I and Q channel. Each of these UL physical channels belonging to the same user is assigned different Walsh channelization codes in order to maintain orthogonality. The higher-rate channels use shorter Walsh codes. The I and Q data channels are then multiplied by a complex-valued spreading sequence, where the in-phase spreading sequence is formed by the modulo-2 addition of an I-channel PN sequence and a long code. On the other hand, the quadrature spreading sequence is formed by the modulo-2 addition of the W_1^2 Walsh code, the in-phase spreading sequence and the decimated, one-chip-delayed long code. The generation of the user-specific long code of Table 22.17 in cdma2000 is exactly the same as that in IS-95. The I-channel and the Q-channel PN se-

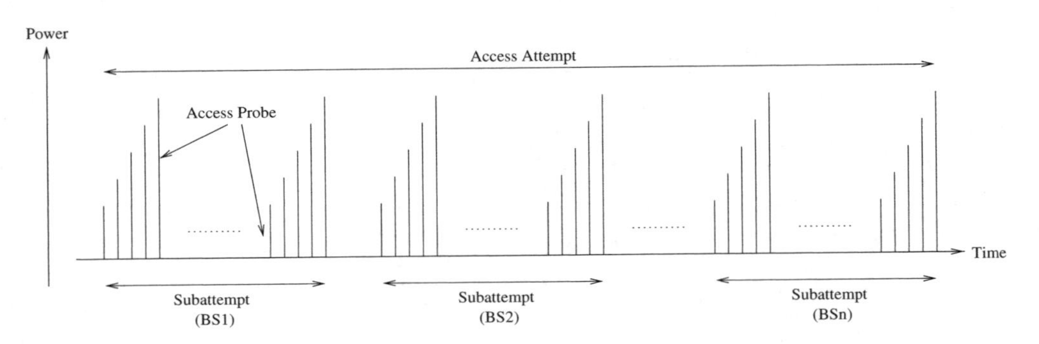

Figure 22.39: An access attempt by a mobile station in cdma2000 using the access probe of Figure 22.40.

quence is not explicitly shown in Table 22.17, since it is identical to the DL cell-specific scrambling code. This complex scrambling code is only used for the purpose of quadrature spreading. Thus, in order to reduce the complexity of the base station receiver, this complex scrambling code is identical to the cell-specific scrambling code of Table 22.17 used on the DL by all the BSs.

22.3.5 Random Access

The procedure of random access using the ACH is the same as that in IS-95. Hence we shall not elaborate on its operation here. Instead we will focus our attention on the access procedure using the EACH of Table 22.11. There are two possible modes of access in conjunction with the EACH, namely the Basic Access Mode and the Reservation Access Mode. In the basic access mode, the access probe consists of the access preamble followed by the access data. On the other hand, in the reservation access mode, the access probe will consist of the access preamble followed by the access header. The access data in the reservation access mode will be transmitted on the DL CCCH upon receiving permission from the BS. In either case, the MS submits an access request to the network by repeatedly transmitting a so-called access probe, until a request acknowledgement is received. This entire process of sending a request is known as an access attempt, as shown in Figure 22.39. Within a single access attempt, the request may be sent to several base stations. An access attempt addressed to a specific base station is known as a sub-attempt. Within a sub-attempt of Figure 22.39, a sequence of access probes is transmitted until an acknowledgement is received from the BS. Each successive access probe is transmitted at a higher power compared to the previous access probe, as shown in Figure 22.41.The initial power (IP) of the first probe is determined by the open-loop power control plus a nominal offset power that corrects for the open-loop power control imbalance between UL and DL. Subsequent probes are transmitted at a power level higher than the previous probe. This increased level is indicated by the Power Increment (PI). In the basic access mode, the acknowledgement will be sent via the DL CCCH for the sake of informing the MS that its access request was received. In the reservation access mode, the acknowledgement is sent via the CACH, which will assign an unused channel on the UL CCCH for the MS to transmit the access data. The access probe transmission follows the

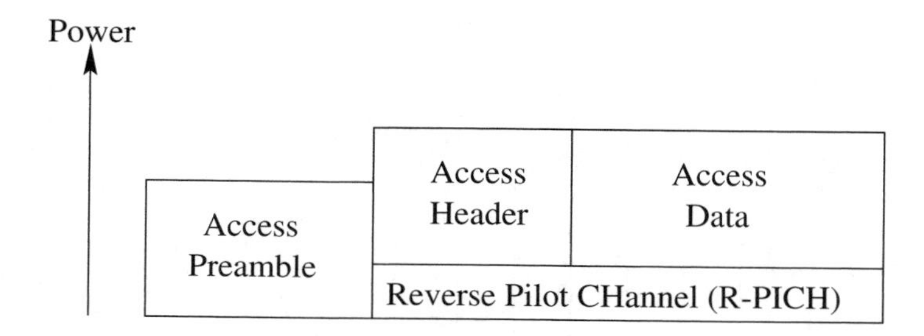

Figure 22.40: A cdma2000 access probe transmitted using the regime of Figure 22.39. The access preamble is always transmitted regardless of the access mode while the access header is transmitted only during the reservation access mode. The access data will be transmitted on the access probe during a basic access mode. However, it will be transmitted on the DL CCCH during a reservation access mode.

slotted ALOHA algorithm, which is a close relative of PRMA [543]. The transmission must begin at the commencement of each 1.25 ms slot. If an acknowledgement of the most recently transmitted probe is not received by the mobile station after a time-out period, another probe is transmitted in another randomly chosen slot, obeying the regime of Figure 22.39. An access probe can be divided into three parts, as shown in Figure 22.40. The access preamble carries a nondata-bearing R-PICH at an increased power level, in order to aid the BS in acquiring access to the network. The R-PICH will also be transmitted along with the access header. The structure of the pilot channel assoicated with the EACH is similar to that of the UL R-PICH of Figure 22.36 except that in this case there are no time-multiplexed power control bits. The preamble length in Figure 22.40 is an integer multiple of the 1.25 ms slot intervals. The specific access preamble length is indicated by the BS, which depends on how fast it can search the PN code space in order to recognise an access attempt. The length of the access header is 5 ms in duration and is transmitted at a data rate of 9.6 kbps. On the other hand, the length of the access data transmitted on the EACH during a basic access mode or on the UL CCCH during a reservation access mode can be 5, 10 or 20 ms in duration, having data rates of either 9.6, 19.2 or 38.4 kbps. Let us now highlight some of the cdma2000 handover issues.

22.3.6 Handover

Intra-frequency or soft-handover is initiated by the MS. While communicating, the MS may receive the same signal from several BSs. These BSs constitute the Active Set of the MS. The MS will continuously monitor the power level of the received pilot channels (F-PICH) transmitted from neighbouring BSs, including those from the MS's active set. The power levels of these BSs are then compared to a set of thresholds according to an algorithm, which will be highlighted later in this chapter. The set of thresholds consists of the static thresholds, which are maintained at a fixed level, and the dynamic thresholds, which are dynamically adjusted based on the total received power. Subsequently, the MS will inform the network when any of the monitored power levels exceed the thresholds.

Whenever the MS detects a F-PICH, whose power level exceeds a given static threshold,

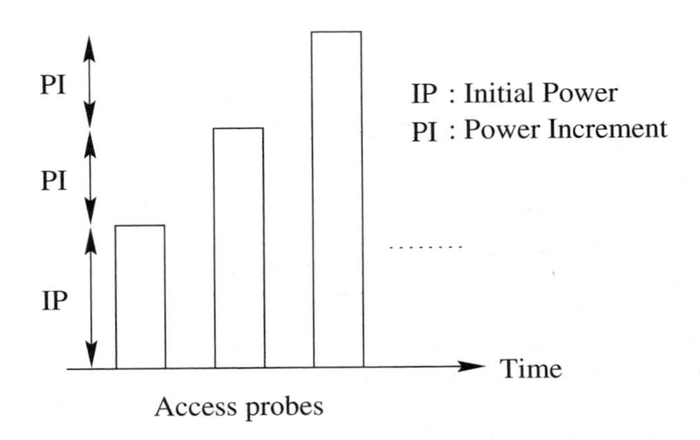

Figure 22.41: Access probes within a sub-attempt of Figure 22.39.

denoted as T_1, this F-PICH will be moved to a candidate set and will be searched and compared more frequently against a dynamically adjusted threshold denoted as T_2. This value of T_2 is a function of the received power levels of the F-PICHs of the BSs in the active set. This process will determine whether the candidate BS is worth adding to the active set. If the overall power level in the active set is weak, then adding a BS of higher power will improve the reception. By contrast, if the overall power level in the active set is relatively high, then adding another high-powered BS may not only be unnecessary, but may actually utilize more network resources.

For the BSs that are already in the active set, the power level of their corresponding F-PICH is compared to a dynamically adjusted threshold, denoted as T_3, which is also a function of the total power of the F-PICH in the active set, similar to T_2. This is to ensure that each BS in the active set is contributing sufficiently to the overall power level. If any of the F-PICH's power level dropped below T_3 after a specified period of time allowed in order to eliminate any uncertainties due to fading which may have caused fluctuations in the power level, the BS will again be moved to the candidate set where it will be compared with a static threshold T_4. At the same time, the MS will report to the network the identity of the low-powered BS in order to allow the corresponding BS to increase its transmit power. If the power level decreases further below a static threshold, denoted as T_4, then the MS will again report this to the network and the BS will subsequently be dropped from the candidate set.

Inter-frequency or hard-handovers can be supported between cells having different carrier frequencies. Here we conclude our discussions on the cdma2000 features and provide some rudimentary notes on a number of advanced techniques, which can be invoked in order to improve the performance of the 3G W-CDMA systems.

22.4 Performance-Enhancement Features

The treatment of adaptive antennas, multi-user detection, interference cancellation, or the portrayal of transmit diversity techniques is beyond the scope of this chapter. Here we simply provide a few pointers to the associated literature.

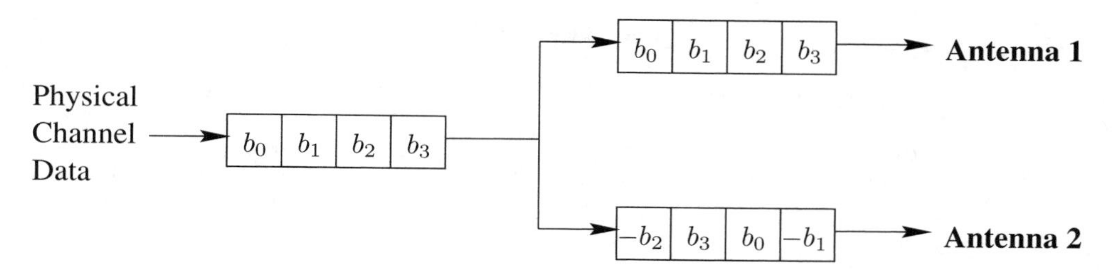

Figure 22.42: Transmission of a physical channel using Space Time block coding Transmit Diversity (STTD).

22.4.1 Downlink Transmit Diversity Techniques

22.4.1.1 Space Time Block Coding-Based Transmit Diversity

Further diversity gain can be provided for the MSs by upgrading the BS with the aid of Space Time block coding assisted Transmit Diversity (STTD) [551], which can be applied to all the DL physical channels with the exception of the SCH. Typically the data of physical channels are encoded and transmitted using two antennas, as shown in Figure 22.42.

22.4.1.2 Time-Switched Transmit Diversity

Time-Switched Transmit Diversity (TSTD) [561] is only applicable to the SCH, and its operation becomes explicit in Figure 22.43.

22.4.1.3 Closed-Loop Transmit Diversity

Closed-loop transmit diversity is only applicable to the DPCH and PDSCH messages of Table 22.4 on the DL, which is illustrated in Figure 22.44. The weights w_1 and w_2 are related to the DL channel's estimated phase and attenuation information, which are determined and transmitted by the MS to the BS using the FBI D field, as portrayed in Figure 22.5. The weights for each antenna are independently measured by the MS using the corresponding pilot channels CPICH1 and CPICH2.

22.4.2 Adaptive Antennas

The transmission of time-multiplexed user-specific pilot symbols on both the UL and DL as seen for UTRA in Figures 22.5–22.11 facilitates the employment of adaptive antennas. Adaptive antennas are known to enhance the capacity and coverage of the system [271, 562].

22.4.3 Multi-User Detection/Interference Cancellation

Following Verdú's seminal paper [106], extensive research has shown that Multi-user Detection (MUD) [184, 546, 563–568] and Interference Cancellation techniques [138, 147, 154,

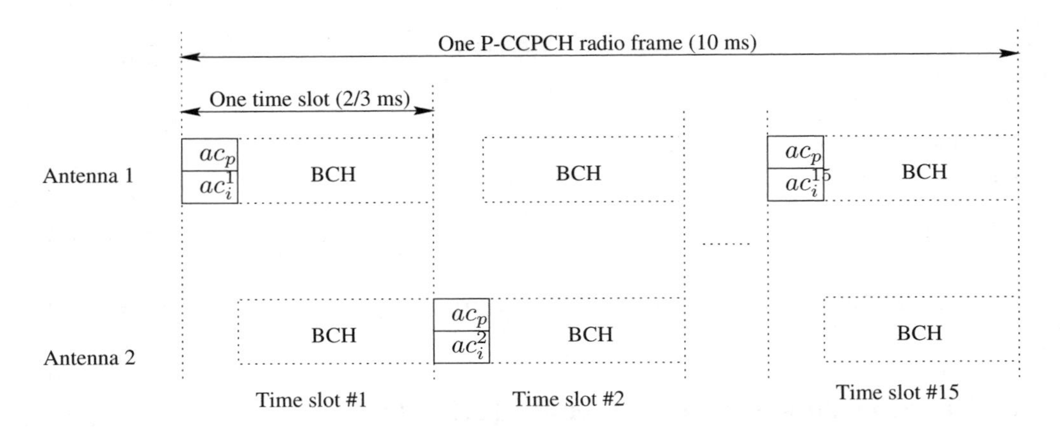

Figure 22.43: Frame structure of the UTRA DL synchronization channel (SCH), transmitted by a TSTD scheme. The primary and secondary SCH are transmitted alternatively from Antennas 1 and 2. The parameter a is a binary flag used to indicate the presence ($a = +1$) or absence ($a = -1$) of STTD encoding in the P-CCPCH.

569–577] can substantially improve the performance of the CDMA link in comparison to conventional RAKE receivers. However, using long scrambling codes increases the complexity of the MUD [531]. As a result, UTRA introduced an optional short scrambling code, namely, the S(2) code of Table 22.9, as mentioned in Section 22.2.6.4, in order to reduce the complexity of MUD [522]. Another powerful technique is invoking burst-by-burst adaptive CDMA [535, 538] in conjunction with MUD.

However, interference cancellation and MUD schemes require accurate channel estimation, in order to reproduce and deduct or cancel the interference. Several stages of cancellation are required in order to achieve a good performance, which in turn increases the canceller's complexity. It was shown that recursive channel estimation in a multistage interference canceller improved the accuracy of the channel estimation and hence gave improved BER performance [153].

Because of the complexity of the multi-user or interference canceller detectors, they were originally proposed for the UL. However, recently reduced-complexity DL MUD techniques have also been proposed [578].

22.5 Chapter Summary and Conclusions

We have presented an overview of the terrestrial radio transmission technology of 3G mobile radio systems proposed by ETSI, ARIB, and TIA. All three proposed systems are based on Wideband-CDMA. Despite the call for a common global standard, there are some differences

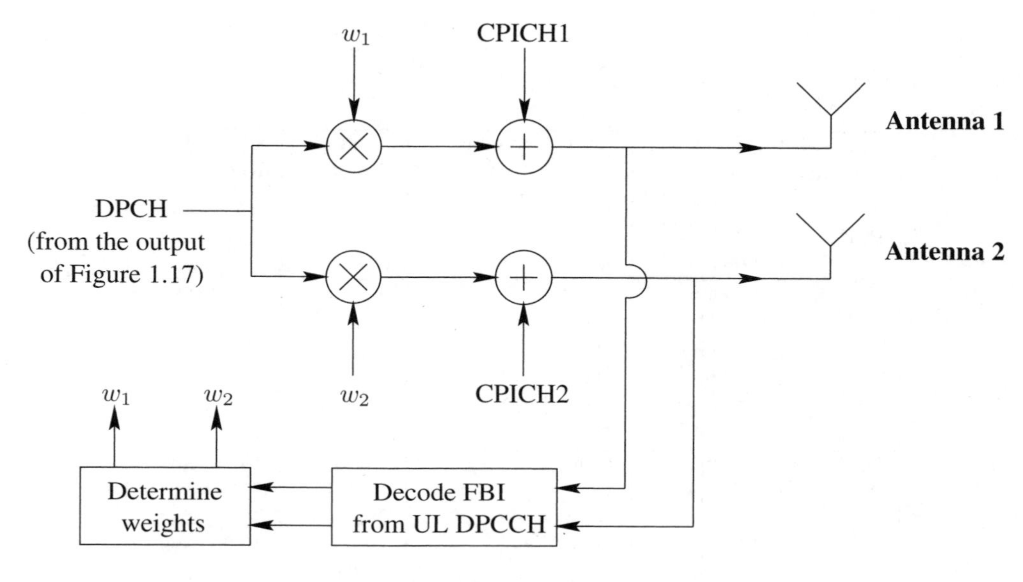

DPCH : Dedicated Physical CHannel
FBI : FeedBack Information
CPICH : Common PIlot CHannel

Figure 22.44: Transmission of the DL DPCH using a closed-loop transmit diversity technique.

in the proposed technologies, notably, the chip rates and intercell operation. These differences are partly due to the existing 2G infrastructure already in use all over the world, and are specifically due to the heritage of the GSM and the IS-95 systems. Huge capital has been invested in these current 2G mobile radio systems. Therefore, the respective regional standard bodies have endeavoured to ensure that the 3G systems are compatible with the 2G systems. Because of the diversified nature of these 2G mobile radio systems, it is not an easy task to reach a common 3G standard that can maintain perfect backwards compatibility.

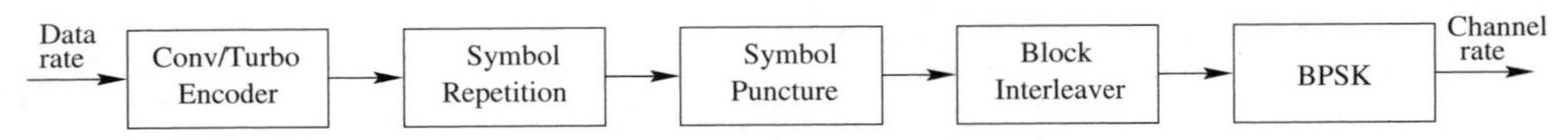

Physical Channel	Data Rate	Code Rate	Symbol Repetition	Symbol Puncture	Channel Rate
SYNC	1.2 kbps	1/2	×2	0	4.8 ksps
PCH	4.8 kbps	1/2	×2	0	19.2 ksps
	9.6 kbps	1/2	×1	0	19.2 ksps
CCCH	9.6 kbps	1/4	×1	0	38.4 ksps
	19.2 kbps	1/4	×1	0	76.8 ksps
	38.4 kbps	1/4	×1	0	153.6 ksps
	9.6 kbps	1/2	×1	0	19.2 ksps
	19.2 kbps	1/2	×1	0	38.4 ksps
	38.4 kbps	1/2	×1	0	76.8 ksps
FCH/ SCH	1.5 kbps	1/2n	×8	1 of 5	19.2n ksps
	1.8 kbps	1/4	×8	4 of 12	38.4 ksps
	1.8 kbps	1/2w	×8	0	28.8w ksps
	2.7 kbps	1/2n	×4	1 of 9	19.2n ksps
	3.6 kbps	1/4	×4	4 of 12	38.4 ksps
	3.6 kbps	1/2w	×4	0	28.8w ksps
	4.8 kbps	1/2n	×2	0	19.2n ksps
	7.2 kbps	1/4	×2	4 of 12	38.4 ksps
	7.2 kbps	1/2w	×2	0	28.8w ksps
	9.6 kbps	1/2n	×1	0	19.2n ksps
	14.4 kbps	1/4	×1	4 of 12	38.4 ksps
	14.4 kbps	1/2w	×1	0	28.8w ksps
	19.2 kbps	1/2n	×1	0	38.4n ksps
	28.8 kbps	1/4	×1	4 of 12	76.8 ksps
	28.8 kbps	1/2w	×1	0	57.6w ksps
	38.4 kbps	1/2n	×1	0	76.8n ksps
	57.6 kbps	1/4	×1	4 of 12	153.6 ksps
	57.6 kbps	1/2w	×1	0	115.2w ksps
	76.8 kbps	1/2n	×1	0	153.6n ksps
	115.2 kbps	1/4	×1	4 of 12	307.2 ksps
	115.2 kbps	1/2w	×1	0	230.4w ksps
	153.6 kbps	1/2n	×1	0	307.2n ksps
	230.4 kbps	1/4	×1	4 of 12	614.4 ksps
	230.4 kbps	1/2w	×1	0	460.8w ksps
	307.2 kbps	1/2m	×1	0	614.4m ksps
	460.8 kbps	1/2w	×1	0	921.6w ksps
	614.4 kbps	1/3	×1	0	1843.2 ksps
	1036.8 kbps	1/2	×1	2 of 18	1843.2 ksps
DCCH	9.6 kbps	1/4	×1	0	38.4 ksps
	9.6 kbps	1/2	×1	0	19.2 ksps
	14.4 kbps	1/4	×1	4 of 12	38.4 ksps

Table 22.15: The cdma2000 downlink physical channel (see Table 22.11) coding parameters, where repetition $\times$ 2 implies transmitting a total of two copies. The parameters $n = \times 1, \times 1.5, \times 2, \times 3, m = \times 1, \times 1.5, \times 3, w = \times 2$.

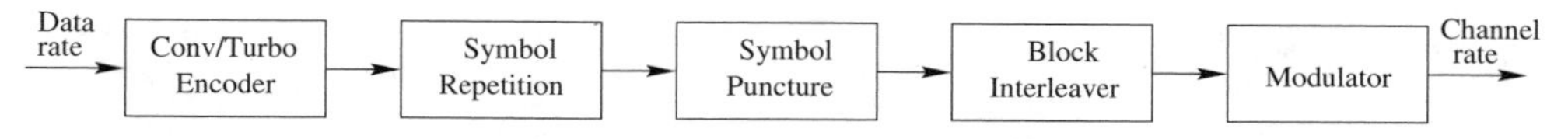

Physical Channel	Data Rate	Code Rate	Symbol Repetition	Symbol Puncture	Channel Rate
CCCH	9.6 kbps	1/4	×4	0	153.6 ksps
	19.2 kbps	1/4	×2	0	153.6 ksps
	38.4 kbps	1/4	×1	0	153.6 ksps
FCH/ SCH	1.5 kbps	1/4	×16	1 of 5	76.8 ksps
	1.8 kbps	1/4	×16	8 of 24	76.8 ksps
	2.7 kbps	1/4	×8	1 of 9	76.8 ksps
	3.6 kbps	1/4	×8	8 of 24	76.8 ksps
	4.8 kbps	1/4	×4	0	76.8 ksps
	7.2 kbps	1/4	×4	8 of 24	76.8 ksps
	9.6 kbps	1/4	×2	0	76.8 ksps
	14.4 kbps	1/4	×2	8 of 24	76.8 ksps
	19.2 kbps	1/4	×1	0	76.8 ksps
	28.8 kbps	1/4	×1	4 of 12	76.8 ksps
SCH	38.4 kbps	1/4	×1	0	153.6 ksps
	57.6 kbps	1/4	×1	4 of 12	153.6 ksps
	76.8 kbps	1/4	×1	0	307.2 ksps
	115.2 kbps	1/4	×1	4 of 12	307.2 ksps
	153.6 kbps	1/4	×1	0	614.4 ksps
	230.4 kbps	1/4	×1	4 of 12	614.4 ksps
	307.2 kbps	1/2	×1	0	614.4 ksps
	460.8 kbps	1/4	×1	0	1843.2 ksps
	614.4 kbps	1/3	×1	0	1843.2 ksps
	1036.8 kbps	1/2	×1	2 of 18	1843.2 ksps
EACH	9.6 kbps	1/4	×4	0	153.6 ksps
	19.2 kbps	1/4	×2	0	153.6 ksps
	38.4 kbps	1/4	×1	0	153.6 ksps
DCCH	9.6 kbps	1/4	×2	0	76.8 ksps
	14.4 kbps	1/4	×2	8 of 24	76.8 ksps

Table 22.16: The cdma2000 uplink physical channel (see Table 22.11) coding parameters, where repetition $\times$ 2 implies transmitting a total of two copies.

	Channelization Codes	User-specific Scrambling Codes (UL)	Cell-specific Scrambling Codes (DL)
Type of codes	- Walsh codes (UL/DL) - Quasi-orthogonal codes (DL)	Different offsets of a real m-sequence	Different offsets of a complex m-sequence
Code length	Variable	$2^{42} - 1$ chips	- 2^{15} chips for SR1(UL/DL) and SR3(DL) - 3×2^{15} chips for SR3(UL)

Table 22.17: Spreading parameters in cdma2000.

ht

Channel Type	Walsh Code
R-PICH	W_0^{32}
EACH	W_2^{8}
CCCH	W_2^{8}
DCCH	W_8^{16}
FCH	W_4^{16}
SCH 1	W_1^{2} or W_2^{4}
SCH 2	W_2^{4} or W_6^{8}

Table 22.18: Walsh codes associated with the UL physical channels

Chapter 23

Adaptive DS-CDMA Networking Using Variable Spreading Factors and Adaptive Beamforming

23.1 Motivation

In this chapter adaptive rate transmissions are investigated in the context of direct-sequence code-division multiple-access (DS-CDMA) systems using variable spreading factors (VSF). In the context of the recently established family of adaptive rate transmission schemes [4] the transmission rate is typically adapted in response to the perceived near-instantaneous channel quality, as we have seen, for example, in Chapter 4. The perceived channel quality is influenced by numerous factors, such as the variation of the number of users supported, which imposes a time-variant MUI load or the fading-induced channel quality fluctuation of the user considered. In Chapter 4 the impact of both of these channel quality factors was considered in the context of multi-user detection aided adaptive modulation as well as variable spreading factor assisted transmissions. It was found that the system performed best, when both the spreading factors and the modulation modes were controlled in a near-instantaneous fashion.

However, the number of users supported was varied in Chapter 4 on a long-term basis. Hence the average MUI level was constant throughout the simulations. By contrast, in this chapter the individual users' instantaneous transmission rate will be adapted in response to the near-instantaneous MUI fluctuations encountered. We present the philosophy of the proposed adaptive rate transmission scheme and analyze the achievable effective throughput as well as the achievable BER performance, when communicating over Additive White Gaussian Noise (AWGN) channels. We will demonstrate in Section 23.2 that by employing the proposed VSF-assisted adaptive rate transmission scheme, the effective throughput may be increased by up to 40%, when compared to that of DS-CDMA systems using constant spreading factors. This increased throughput is achieved without wasting power, without imposing extra interference upon other users and without increasing the BER.

In Section 23.3 our discussions evolve further by considering an entire IMT2000-like network constituted by 49 cells. Recall that in Chapter 22 we provided an overview of the standardized IMT2000 wireless communications system, while in Chapter 4 we considered a detailed comparative study of various near-instantaneously adaptive MUD-, PIC- and SIC-aided CDMA transceivers. In Section 23.3 the user capacity and network performance of an IMT2000-like MUD-aided Frequency Division Duplex (FDD) and Time Division Duplex (TDD) network will be investigated. The new call blocking and call dropping probabilities, the probability of low-quality access as well as the required average transmit power are quantified both with and without adaptive antenna arrays as well as with and without shadow fading. We will demonstrate that in some of the scenarios investigated the system's user capacity is doubled with the advent of adaptive antennas. The employment of adaptive modulation techniques in conjunction with adaptive antenna arrays resulted in further significant network capacity gains. This is particularly so in the context of TDD CDMA, where the system's capacity becomes poor without adaptive antennas and adaptive modulation owing to the high base station to base station interference inflicted as a consequence of potentially using all time-slots in both the up-link and down-link.

23.2 Variable-Spreading Factor Aided Adaptive CDMA Networks

23.2.1 Introduction

DS-CDMA is the prevalent technique in the third-generation (3G) wireless communications systems of Chapter 22, because it is capable of providing numerous advantages compared to other solutions. The capacity of DS-CDMA systems is limited by both the time-varying characteristics of the wireless channel and that of the MUI. The family of efficient techniques designed for compensating the time-varying nature of the wireless channels include the popular RAKE receiver [579] contrived for achieving frequency diversity. Alternatively, multiple transmit and/or receiver antennas can be employed for achieving spatial diversity [5,274,580]. The most efficient technique of combating the MAI is multi-user detection (MUD) [78], which was discussed throughout Part I of the book. The above techniques have attracted world-wide attention in recent years.

Another efficient technique of increasing the capacity of time-varying wireless channels is the employment of adaptive rate transmissions [4, 71, 72], where the transmission rate can be adaptively adjusted according to the instantaneous channel conditions. The main philosophy behind adaptive rate transmissions is the real-time balancing of the link budget through adaptive variation of the symbol rate, modulation constellation size and format, spreading factor, coding rate/scheme, etc, or in fact any combination of these parameters. However, the results of [72] and [4, 5] have shown that when a sufficiently high diversity order is available, regardless, whether this is due to transmitter or receiver diversity achieved in the time or frequency domain, the advantages of adaptive rate transmissions erode. Hence, in the context of the 3G DS-CDMA systems using power control, the channel fading can be efficiently mitigated by employing both the RAKE receiver and multiple transmitter/receiver antennas. In order to combat the MUI in DS-CDMA systems, as we have mentioned above, the most efficient approach is to use multi-user detection receivers [78]. The main obstacle of employing DS-CDMA MUD receivers is, however, the high complexity of the multi-user detection algorithms. Therefore, the conventional matched filter based receiver remains popular because of its simplicity, despite its suboptimal performance.

In this chapter we consider the problem of how the effective throughput of DS-CDMA systems can be increased, when the conventional matched filter based receiver is employed. Specifically, we consider the up-link transmission of a single cell DS-CDMA system, where the number of active mobile users obeys the Poisson distribution [581] and all the signals transmitted by the mobile users are power controlled. Hence, the multi-user interference level can be modelled as a discrete Markov process [581], which describes the number of active mobile users. In order to exploit the time-varying nature of the multi-user interference level, an adaptive rate transmission scheme using variable spreading factors (VSF) [526] is proposed for increasing the effective throughput. In contrast to the conventional VSF-assisted adaptive rate transmission scheme, where the transmission rate is adapted in response to the channel quality fluctuation recorded at the output of the MUD [582–584], the transmission rate in the proposed scheme is adapted in response to the time-varying interference level due to the MUI, while maintaining the required target BER value. More explicitly, the mobile users increase their transmission rate, when the number of active interfering users decreases, while decreases its transmission rate in response to an increased number of active interfering users. The number of active interfering users is broadcast to the mobile users by the central base station. In this chapter the performance of the DS-CDMA systems using the proposed VSF-assisted adaptive rate transmission scheme is evaluated when communicating over AWGN channels. The reasons for us to consider only AWGN channels are as follows. Firstly, as we have mentioned above, the fading effects encountered in power-controlled DS-CDMA systems can be efficiently mitigated by using RAKE receivers and multiple transmitter/receiver antennas. Secondly, our aim is to study the effects of the MUI in isolation, without the obfuscating effects of the channel's fading and then gain insight into the effects of the MUI on the system's effective throughput, when the conventional matched filter receiver is considered. Note that our study can be readily extended for considering various fading channels as well as to multi-cell CDMA systems. *Our results show that by employing VSF-assisted adaptive rate transmissions, the effective throughput of a DS-CDMA system may be increased by* 40% *upon exploiting the Markovian distributed number of active users in the system. The increased effective throughput is achieved without wasting power and without increasing the bit error ratio (BER).*

The remainder of this chapter is organized as follows. In the next section we give a rudimentary overview in the context of DS-CDMA systems and introduce the Markov model describing the number of active users, while providing simulation results. In Section 23.2.3 we describe the VSF-assisted adaptive rate transmission scheme. Section 23.2.4 derives the effective throughput of a DS-CDMA system, when the proposed adaptive rate transmission scheme is employed, and provides the corresponding BER expression. Our numerical results are provided in Section 23.2.5, and finally in Section 23.4 we present our conclusions.

23.2.2 System Overview

We consider a single cell DS-CDMA system, where a single base station (BS) is located at the centre of the cell, while the mobile users are uniformly distributed in the area covered by this base station. The base station is capable of simultaneously processing a maximum number of K calls, i.e. the maximum number of active users supported by the cell is K. We assume that each active user's data is BPSK modulated and it is transmitted to the base station asynchronously over AWGN channels. Furthermore, we assume ideal power control, i.e. the received power of each active user is the same at the BS. Based on the above assumptions and assuming furthermore that there are K_l+1 active users (the reference user plus K_l interfering users), then the received signal at the base station can be expressed as [103]

$$r(t) = n(t) + \sum_{k=1}^{K_l+1} \sqrt{2P}a_k(t-\tau_k)b_k(t-\tau_k)\cos(\omega_c t + \phi_k), \tag{23.1}$$

where $n(t)$ is the AWGN having a two-sided spectral density of $N_0/2$, P represents the power received from each active user, $a_k(t)$ is the spreading code, $b_k(t)$ is the data signal, τ_k is the time delay parameter that accounts for the propagation delay as well as for the lack of synchronism between the transmitters, while ϕ_k is the phase angle due to carrier modulation and channel delay.

According to the analysis of [103] the bit errors in DS-CDMA systems communicating over AWGN channels are caused by the effect of multiple access interference and the AWGN. The BER of an asynchronous DS-CDMA system having received signals given by (23.1) can be closely approximated by [318]:

$$P_e(K_l) = \frac{2}{3}Q\left[\left(\frac{K_l}{3N} + \frac{1}{2N\gamma_c}\right)^{-1/2}\right] + \frac{1}{6}Q\left[\left(\frac{K_lN/3+\sqrt{3}\sigma}{N^2} + \frac{1}{2N\gamma_c}\right)^{-1/2}\right] + \frac{1}{6}Q\left[\left(\frac{K_lN/3-\sqrt{3}\sigma}{N^2} + \frac{1}{2N\gamma_c}\right)^{-1/2}\right], \tag{23.2}$$

where $\gamma_c = PT_c/N_0 = E_c/N_0$ represents the signal to noise ratio (SNR) per chip, T_c is the chip-duration, while the variable N represents the spreading factor (number of chips per bit). In our further discourse N will be controlled as a function of the number of active users, and the parameter σ can be derived from the following equation [318]:

$$\sigma^2 = (K_l+1)\left[N^2\frac{23}{360} + N\left(\frac{1}{20} + \frac{K_l}{36}\right) - \frac{1}{20} - \frac{K_l}{36}\right]. \tag{23.3}$$

Furthermore, in (23.2) the Gaussian Q-function is given by $Q(x) = \frac{1}{\sqrt{2\pi}}\int_x^\infty \exp(-t^2/2)dt$.

The number of interfering users, K_l can be modelled with the aid of a Markov chain having K states [581]. The state transition diagram modelling the number of users determining the interference level is shown in Figure 23.1, which represents a $M/M/m/m$ queueing system [581][1]. The arrival rate of new calls or users λ corresponds to the probability of the event that a new interfering user is activated within a unit-length time-duration, while the average service time of $1/\mu$ is the average duration of an active interfering connection. For the $M/M/m/m$ queueing system, the probability that there are K_l active interfering users (customers) can be expressed as [581]:

$$p_{K_l} = \frac{\left(\frac{\lambda}{\mu}\right)^{K_l} \frac{1}{K_l!}}{\sum_{m=0}^{K-1} \left(\frac{\lambda}{\mu}\right)^m \frac{1}{m!}}, \quad K_l = 0, 1, \dots, K-1, \tag{23.4}$$

where the probability of simultaneously supporting $K_l = K - 1$ users is known as the *Erlang B formula* [581], which determines the call *blocking probability* of the system considered.

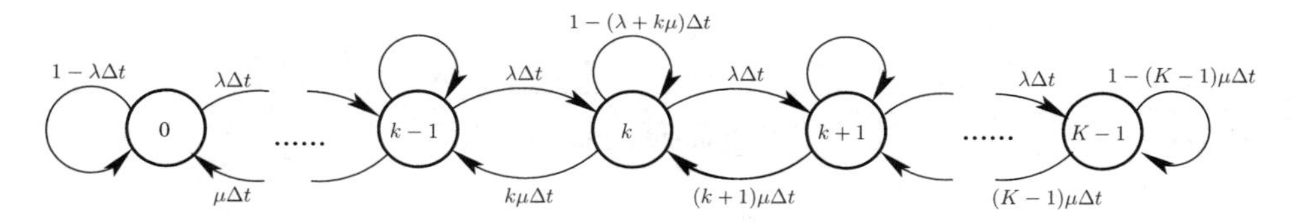

Figure 23.1: State transition diagram modelling the number of active interfering users with the aid of a Markov chain having K states.

Figure 23.2 shows number of active users generated by the above-mentioned Markov chain for the first 3000 normalized time slots (Figure 23.2(a)) and for the normalized time slots spanning the index-range of 1000 to 2000 (Figure 23.2(b)). The parameters used for the simulations were $\lambda = 0.35$, $\mu = 0.009$ and the maximum number of users supported was $K = 64$. From Figure 23.2(a) and Figure 23.2(b) we can observe that the number of active interfering users is a slowly time-variant variable, fluctuating as a function of the normalized time slot index.

It is widely recognized that DS-CDMA systems are interference-limited systems and the systems' BER performance is highly sensitive to the number of interfering users. Figure 23.3 shows the achievable BER performance with respect to the number of active users for a DS-CDMA system communicating over AWGN channels, when the spreading factors of N =8,16,24,40,56,80,112,120 are employed. From the results of Figure 23.3 we can infer following observations:

- For a given spreading factor, the BER increases when supporting an increased number of active users;
- For a given number of users, the BER decreases upon increasing the value of the spreading factor;

[1]In the $M/M/m/m$ queueing system [581], the first parameter M indicates that the arrival process is a Poisson process, the second parameter M indicates that the service time obeys a negative exponential distribution, the third parameter m quantifies the number of servers, while the last letter m indicates the limit of the number of customers in the system.

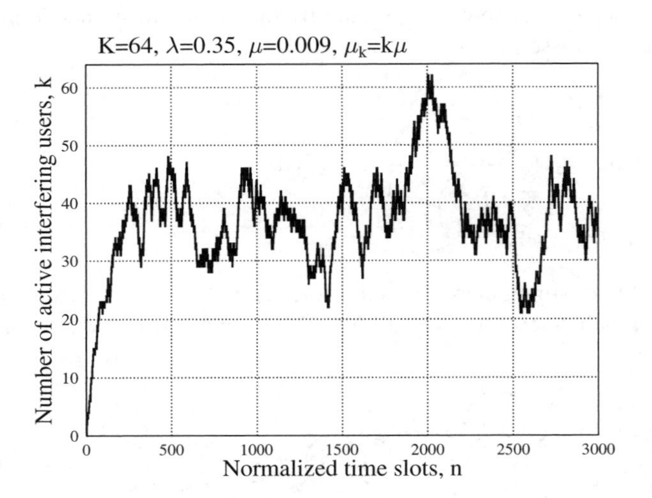

(a) Time slot interval, [0, 3000]

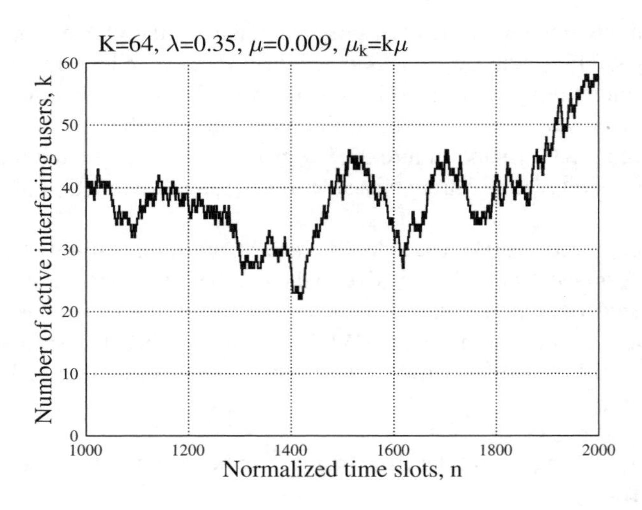

(b) Time slot interval, [1000, 2000]

Figure 23.2: Markov characteristics of the number of active interfering users.

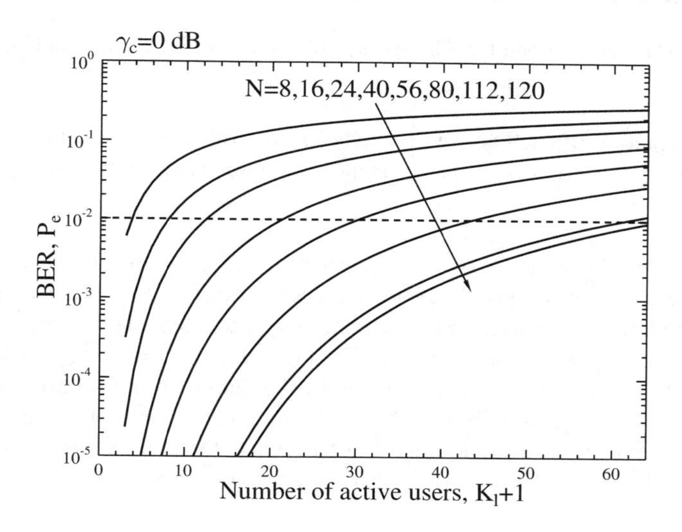

Figure 23.3: BER performance versus the number of active users for the parameters of $\gamma_c = 0dB$ and spreading factors of $N = 8, 16, 24, 40, 56, 80, 112, 120$ computed from (23.2).

- For a given target BER - for example for $P_E = 10^{-2}$ - and for a given number of active users, there exists a spreading factor, which results in a specific BER for the DS-CDMA system matching the target BER requirement. A DS-CDMA system using a higher spreading factor is capable of supporting a higher number of active users, than that using a lower spreading factor, while maintaining the target BER.

Therefore, based on the Markov chain characterized in Figure 23.2 and on the BER performance of the DS-CDMA system as a function of the number of active users shown in Figure 23.3, we argue that an appropriate spreading factor can be employed within a specific time slot for maximizing the number of bits transmitted in this specific time slot, while maintaining the required BER performance. Furthermore, when the number of active users dynamically fluctuates, variable spreading factors can be employed by the DS-CDMA system for achieving the maximum throughput, while guaranteeing the required BER performance. In other words, VSF-assisted adaptive DS-CDMA system is capable of increasing the effective throughput of the system, while maintaining a given target BER.

In this section we have reviewed the behaviour of interference-limited DS-CDMA systems and highlighted the philosophy of an adaptive DS-CDMA system, in order to increase the effective throughput of the system, when the number of active interfering users is time-varying. Let us now investigate the behaviour of adaptive VSF-assisted schemes in more detail.

23.2.3 Adaptive Transmission Scheme

The requirements for adaptive rate DS-CDMA systems may be listed - without completeness - as follows.

- The rate adaptation of each active user may be controlled independently, i.e., without co-operation with other active users. Hence the associated complexity is reasonably low.

- Since DS-CDMA systems are typically interference-limited, an adaptive rate transmission scheme must not impose extra interference on the system. Hence an attractive adaptive rate DS-CDMA scheme is expected to maintain the interference state, regardless of the active users' transmission rates. In other words, an active user's interference environment is expected to be affected only by the number of active users corresponding to a certain level of MUI, but not by their individual transmission rates. The transmission rate and the achievable Quality of Service (QoS) of a user have to obey a trade-off for this particular user, regardless of the other users of the system.

- For DS-CDMA systems, where some active users may communicate at constant rates, while the remaining active users communicate at a variable rate, the BER and throughput of the active users communicating at constant rates are expected to be unaffected by those communicating using adaptive rate transmissions.

- For adaptive DS-CDMA systems using VSF, the set of legitimate spreading factors must be appropriately designed, so that the effective throughput can be maximized with the aid of readily realizable spreading codes.

Below we propose and investigate a specific adaptive rate transmission scheme, which is capable of meeting the above requirements. Let us assume that each user transmits a block of data as shown in Figure 23.4. As shown in Figure 23.4, the data block is divided into L frames, where L is assumed to be a random variable distributed over a certain range potentially extending to ∞. We assume that at the beginning of the lth frame there are K_l interfering users, which is the *a priori* knowledge for determining the required spreading factor and the corresponding transmission rate during the lth frame. Let us assume that each frame consists of a constant number of chips, which is expressed as N_f, i.e. each frame has a constant duration of $N_f T_c$ seconds. Let us also assume that there are m spreading factors having values expressed as $N_1 < N_2 < \ldots < N_m$, where each of them is a multiple of N_f. Furthermore, let us assume that the target BER is P_E. Then the required spreading factor of the lth frame and the corresponding number of bits conveyed by the lth frame can be determined as follows:

- A specific spreading factor is selected from the set $\{N_1, N_2, \ldots, N_m\}$ according to Equation (23.2) based on the number of interfering users, K_l, and on the target BER, P_E, such that the selected spreading factor's value is as low as possible, while guaranteeing the required BER performance. We denote the selected spreading factor by $N_l(K_l, P_E)$.

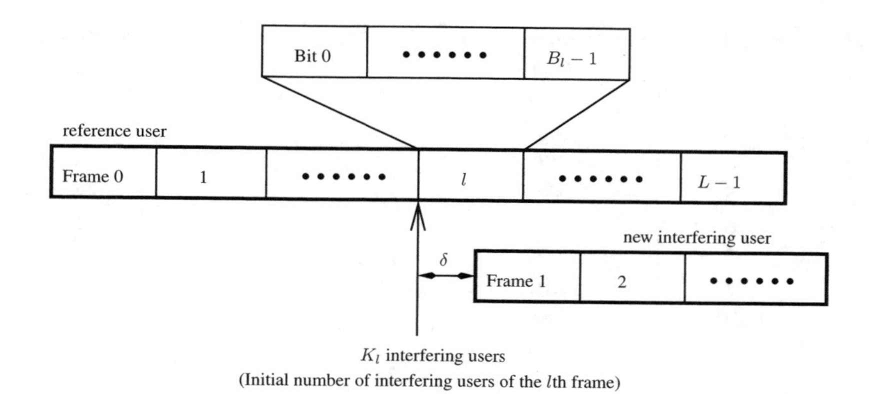

Figure 23.4: Data structure of the transmitted signal in adaptive rate DS-CDMA systems using VSF-assisted adaptive rate transmissions.

- Once the spreading factor, $N_l(K_l, P_E)$ of the lth frame was selected, the corresponding number of bits, $B_l(K_l, P_E)$, conveyed by the lth frame can be determined as

$$B_l(K_l, P_E) = \frac{N_f}{N_l(K_l, P_E)}. \tag{23.5}$$

Based on (23.5) and on our previous arguments it is readily seen that the maximum or minimum throughput of the lth frame is obtained, when using the lowest or the highest spreading factor from the set $\{N_1, N_2, \dots, N_m\}$, respectively. Specifically, the maximum throughput of the lth frame is given by $B_{lM}(K_l, P_E)/N_f T_c = 1/(N_1 T_c)(bits/second) = 1/N_1(bits/chip)$, while the minimum throughput is given by $B_{lm}(K_l, P_E)/N_f T_c = 1/(N_m T_c)(bits/second) = 1/N_m(bits/chip)$.

The proposed adaptive transmission scheme is capable of meeting the requirements listed at the beginning of this section. Specifically, each user's rate adaptation procedure is independent of that of the other users. The knowledge required by each mobile user for adjusting his/her transmission rate is the number of actively interfering users at the beginning of each transmitted frame. This knowledge can be broadcast to each mobile user by the base station. Secondly, since each user's received power is constant when different spreading factors as well as different bit-durations are used, the interference level is only a function of the number of active interfering users. However, the SNR per bit expressed as $N_l(K_l, P_E)\gamma_c$ exhibits different values, when various spreading factors or various date rates are employed. Finally, the communication environment of the mobile users employing constant rate transmissions is not affected by the rate adaptation operations of the mobile users invoking adaptive rate transmissions.

In [71] Goldsmith and Chua have shown that the highest effective throughput can be achieved by using continuous rate adaptation. Hence, the results of [71] suggest that we should use as many different spreading factors as possible and, simultaneously, have a near-continuous spreading factors value set. Table 23.1 and Table 23.2 show two design examples for the appropriate choice of the spreading factors. In Table 23.1 we assumed that random

Index	Bits/frame	Spreading factors	Index	Bits/frame	Spreading factors
1	1	1680	21	42	40
2	2	840	22	48	35
3	3	560	23	56	30
4	4	420	24	60	28
5	5	336	25	70	24
6	6	280	26	80	21
7	7	240	27	84	20
8	8	210	28	105	16
9	10	168	29	112	15
10	12	140	30	120	14
11	14	120	31	140	12
12	15	112	32	168	10
13	16	105	33	210	8
14	20	84	34	240	7
15	21	80	35	280	6
16	24	70	36	336	5
17	28	60	37	420	4
18	30	56	38	560	3
19	35	48	39	840	2
20	40	42	40	1680	1

Table 23.1: Number of bits transmitted in a frame by assuming that the total number of chips per frame is $N_f = 1680 = 2^4 \times 3 \times 5 \times 7$ when the spreading factors seen in the right column are employed. *Random spreading codes were assumed.*

spreading sequences having various lengths were employed, and that the frame's length is $N_f = 2^4 \times 3 \times 5 \times 7 = 1680$ chips. As shown in Table 23.1, we can obtain 40 different spreading factors and hence can support 40 different transmission rates. In the context of Table 23.2, we assumed that orthogonal spreading sequences were employed, and that the frame's length is also $N_f = 4 \times 2^2 \times 3 \times 5 \times 7 = 1680$ chips. These orthogonal spreading sequences having variable lengths were derived based on the fact that there exist $n \times n$ $(n > 2)$ Hadamard matrices, provided that $n = 4t$, where t is a positive integer. Table 23.2 shows that there exist 26 different orthogonal spreading codes having different VSF values in conjunction with $N_f = 1680$. Hence, the DS-CDMA systems using the spreading factors of Table 23.2 are capable of supporting 26 different transmission rates.

23.2.4 Throughput and BER Analysis

In this section we analyze the effective throughput as well as the resultant average BER, when achieving this effective throughput. The effective throughput is defined as the total number of bits successfully transmitted within a unity-duration time-interval by all users supported by the system. Our analysis is based on the following assumptions. (a) All active users communicate using adaptive rate transmissions based on the same set of spreading factors, as described in Section 23.2.3. The transmitted data block-length of each active user obeys an

Index	Bits/frame	Spreading factors	Index	Bits/frame	Spreading factors
1	1	1680	14	28	60
2	2	840	15	30	56
3	3	560	16	35	48
4	4	420	17	42	40
5	5	336	18	60	28
6	6	280	19	70	24
7	7	240	20	84	20
8	10	168	21	105	16
9	12	140	22	140	12
10	14	120	23	210	8
11	15	112	24	420	4
12	20	84	25	840	2
13	21	80	26	1680	1

Table 23.2: Number of bits transmitted in a frame by assuming that the total number of chips per frame is $N_f = 1680 = 4 \times 2^2 \times 3 \times 5 \times 7$ when the various spreading factors seen in the right column are employed. *Orthogonal Walsh-Hadamard spreading codes were assumed.*

independent identical distribution (i.i.d). (b) Assuming that the number of interfering users at the beginning of the lth frame is K_l, the probability of increasing or decreasing this number by one within a frame's time duration is given by $P_{K_l+} = \lambda$ or $P_{K_l-} = K_l\mu$, respectively. The probability of increasing or decreasing the number of interfering users within a frame's time duration by more than one is zero. Therefore, the probability that the number of interfering users remains unchanged, i.e. K_l, within a frame's time duration can be written as:

$$P_{K_l} = \begin{cases} 1-\lambda & \text{if } K_l = 0 \\ 1-\lambda-K_l\mu & \text{if } 1 \le K_l \le K-2 \\ 1-K_l\mu & \text{if } K_l = K-1 \end{cases} \tag{23.6}$$

(c) When the number of active interfering users increases by one or decreases by one within the lth frame, we assume that this happens at the moment having a time difference of δ from the beginning of the lth frame (see Figure 23.4), where δ is assumed to be uniformly distributed over the interval $[0, N_fT_c)$.

The effective throughput can be derived as follows. According to our analysis in Sections 23.2.2 and 23.2.3 we know that the spreading factor, $N_l(K_l, P_E)$, as well as the number of bits, $B_l(K_l, P_E) = N_f/N_l(K_l, P_E)$ transmitted during the lth frame are determined by the number of active interfering users at the beginning of the lth frame as well as by the target

BER to be maintained. The effective throughput can be expressed as:

$$\begin{aligned} B &= E_\delta \Bigg[p_{K_l=0} \left(\frac{B_l(0, P_E)}{N_f} + P_{K_l+} \times \frac{(N_f T_c - \delta)/T_b}{N_f} \right) \\ &+ \sum_{K_l=1}^{K-2} p_{K_l} \left(\frac{(K_l+1) B_l(K_l, P_E)}{N_f} + P_{K_l+} \times \frac{(N_f T_c - \delta)/T_b}{N_f} \right. \\ &\qquad \left. - P_{K_l-} \times \frac{(N_f T_c - \delta)/T_b}{N_f} \right) \\ &+ p_{K_l=K-1} \left(\frac{K B_l(K-1, P_E)}{N_f} - P_{K_l-} \times \frac{(N_f T_c - \delta)/T_b}{N_f} \right) \Bigg], \end{aligned} \quad (23.7)$$

where the first term is contributed by the event that there exist no active interfering users, the second term is by the event that there are K_l, $K_l = 1, 2, \ldots, K-2$ number of active interfering users, while the last term is contributed by the scenario that there are $K-1$ active interfering users in the system. Since $P_{K_l+} = \lambda$ and $P_{K_l-} = K_l \mu$, Equation (23.7) can be written as

$$\begin{aligned} B &= E_\delta \Bigg[\sum_{K_l=0}^{K-1} p_{K_l} \left(\frac{(K_l+1) B_l(K_l, P_E)}{N_f} + P_{K_l+} \times \frac{(N_f T_c - \delta)/T_b}{N_f} \right. \\ &\left. - P_{K_l-} \times \frac{(N_f T_c - \delta)/T_b}{N_f} \right) - p_{K_L=K-1} P_{0+} \times \frac{(N_f T_c - \delta)/T_b}{N_f} \Bigg] \\ &= E_\delta \Bigg[\sum_{K_l=0}^{K-1} p_{K_l} \left(\frac{(K_l+1) B_l(K_l, P_E)}{N_f} + \lambda \times \frac{(N_f T_c - \delta)/T_b}{N_f} \right. \\ &\left. - K_l \mu \times \frac{(N_f T_c - \delta)/T_b}{N_f} \right) - p_{K_L=K-1} \lambda \times \frac{(N_f T_c - \delta)/T_b}{N_f} \Bigg]. \end{aligned} \quad (23.8)$$

Upon taking the expectation with respect to δ, considering that $T_b = N_l(K_l, P_E) T_c$ and $B_l(K_l, P_E) = N_f / N_l(K_l, P_E)$, and substituting p_{K_l} from (23.4) into (23.7), finally we obtain the effective throughput as follows:

$$\begin{aligned} B &= \sum_{K_l=0}^{K-1} \frac{\left(\frac{\lambda}{\mu}\right)^{K_l} \frac{1}{K_l!}}{\sum_{m=0}^{K-1} \left(\frac{\lambda}{\mu}\right)^m \frac{1}{m!}} \left[\frac{K_l+1}{N_l(K_l, P_E)} + (\lambda - K_l \mu) \times \frac{1}{2 N_l(K_l, P_E)} \right] \\ &- \frac{\left(\frac{\lambda}{\mu}\right)^{K-1} \frac{1}{(K-1)!}}{\sum_{m=0}^{K-1} \left(\frac{\lambda}{\mu}\right)^m \frac{1}{m!}} \times \frac{\lambda}{2 N_l(K-1, P_E)} \quad \text{[bits/chip]}. \end{aligned} \quad (23.9)$$

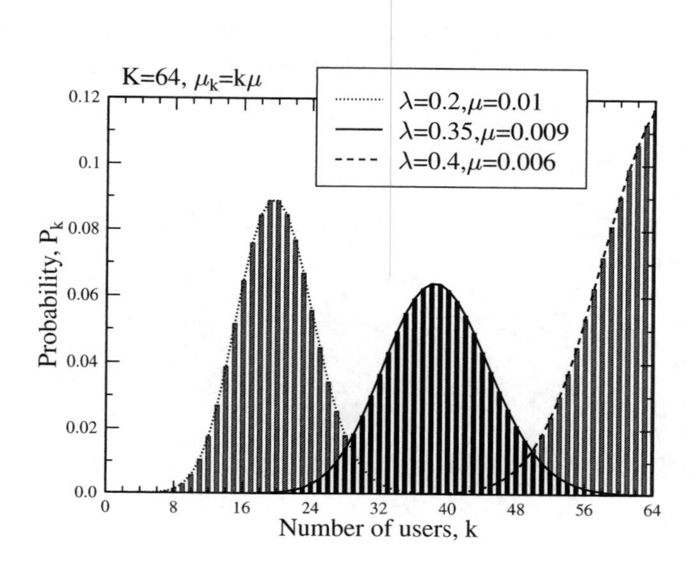

Figure 23.5: Probability density functions of the number of active users associated with various call generation rates of λ and for various average call duration of $1/\mu$ computed from (23.4).

In order to derive the average BER, we have to consider the following three 'events':

Event 1: The number of interfering users during the lth frame remains a constant, namely K_l. The probability of this event is given by (23.6), and the average BER is given by $P_e(K_l)$ of (23.2).

Event 2: The number of interfering users increases to $K_l + 1$ within the lth frame, where the new interfering user commences its transmission δ seconds after the beginning of the lth frame of the reference user, as shown in Figure 23.4. The probability of this event is $P_{K_l+} = \lambda$ for $K_l = 0, 1, \ldots, K-2$ and $P_{(K-1)+} = 0$ for $K_l = K-1$. Let $h = \lfloor \delta/T_b \rfloor$, where $\lfloor x \rfloor$ represents the largest integer not exceeding x. Then, δ can be expressed as $\delta = hT_b + g$, where h is a discrete random variable uniformly distributed in the interval $\{0, 1, \ldots, B_l(K_l, P_E)\}$, while g is continuous random variable uniformly distributed over $[0, T_b)$. Hence, in the context of this event, there exist three types of interference patterns within the lth frame. *The first interference pattern* is encountered in the scenario, that the number of active interfering users associated with the first h number of bits of the lth frame is K_l. Therefore, the corresponding BER is given by $P_e(K_l)$ of (23.2). *The second interference pattern* corresponds to the scenario, that the number of active interfering users associated with the last $[B_l(K_l, P_E) - h - 1]$ number of bits of the lth frame is $K_l + 1$, and hence the corresponding BER is given by $P_e(K_l + 1)$ of (23.2). *The third interference pattern* is characteristic of the scenario, that there exists a bit in the lth frame, where the leading section of $g = \delta - hT_b$ time-duration is associated with K_l interfering users, while the remaining $T_b - g = (h+1)T_b - \delta$ time-duration is associated with $K_l + 1$ interfering users. Since g is uniformly distributed in $[0, T_b)$, the average BER associated with this bit can be approximated as $[P_e(K_l) + P_e(K_l+1)]/2$. Consequently, upon considering all the above-mentioned three interfering patterns, the average BER can be formulated as:

$$\begin{aligned} P_e^+(K_l) &= E_h\left[\frac{1}{B_l(K_l,P_E)}\left(hP_e(K_l)\right.\right. \\ &\quad \left.\left. +[B_l(K_l,P_E)-h-1]P_e(K_l+1)+\frac{P_e(K_l)+P_e(K_l+1)}{2}\right)\right] \\ &= \frac{1}{[B_l(K_l,P_E)]^2}\sum_{h=0}^{B_l(K_l,P_E)-1}\left(hP_e(K_l)\right. \\ &\quad \left. +[B_l(K_l,P_E)-h-1]P_e(K_l+1)+\frac{P_e(K_l)+P_e(K_l+1)}{2}\right) \\ &= \frac{1}{2}\left[P_e(K_l)+P_e(K_l+1)\right]. \end{aligned} \tag{23.10}$$

Event 3: The number of interfering users decreases to $K_l - 1$ within the lth frame, where one of the active interfering users terminates his/her transmission δ seconds after the beginning of the lth frame. The probability of this event is $P_{K_l-} = K_l\mu$ for $K_l = 0, 1, \ldots, K-1$. The average BER of Event 3 can be analyzed following the same steps as in the context of Event 2, except that $K_l + 1$ is now replaced by $K_l - 1$. The average BER of Event 3 can be written as:

$$P_e^-(K_l) = \frac{1}{2}\left[P_e(K_l) + P_e(K_l-1)\right]. \tag{23.11}$$

Finally, by considering the above three possible events, the average BER of the bits in the lth frame can be written as:

$$\begin{aligned} \hat{P}_e(K_l) &= P_{K_l+}P_e^+(K_l) + P_{K_l-}P_e^-(K_l) + P_{K_l}P_e(K_l) \\ &= \begin{cases} \frac{\lambda}{2}\left[P_e(K_l)+P_e(K_l+1)\right]+(1-\lambda)P_e(K_l), & \text{if } K_l = 0 \\ \frac{\lambda}{2}\left[P_e(K_l)+P_e(K_l+1)\right]+\frac{K_l\mu}{2}\left[P_e(K_l)+P_e(K_l-1)\right] & \\ \quad +(1-\lambda-K_l\mu)P_e(K_l), & \text{if } 1 \le K_l \le K-2 \\ \frac{K_l\mu}{2}\left[P_e(K_l)+P_e(K_l-1)\right]+[1-K_l\mu]P_e(K_l), & \text{if } K_l = K-1. \end{cases} \end{aligned} \tag{23.12}$$

The overall average BER can be computed by averaging $\hat{P}_e(K_l)$ with respect to the distribution of the number of active interfering users, K_l, which can be expressed as:

$$\begin{aligned} P_e &= \sum_{K_l=0}^{K-1} p_{K_l} \hat{P}_e(K_l) \\ &= \sum_{K_l=0}^{K-1} \frac{\left(\frac{\lambda}{\mu}\right)^{K_l} \frac{1}{K_l!}}{\sum_{m=0}^{K-1} \left(\frac{\lambda}{\mu}\right)^m \frac{1}{m!}} \times \left(\frac{\lambda}{2} [P_e(K_l) + P_e(K_l+1)] \right. \\ &\quad \left. + \frac{K_l \mu}{2} [P_e(K_l) + P_e(K_l - 1)] + (1 - \lambda - K_l \mu) P_e(K_l) \right) \\ &\quad - \frac{\left(\frac{\lambda}{\mu}\right)^{K-1} \frac{1}{(K-1)!}}{\sum_{m=0}^{K-1} \left(\frac{\lambda}{\mu}\right)^m \frac{1}{m!}} \times \left(\frac{\lambda}{2} [P_e(K-1) + P_e(K)] - \lambda P_e(K-1) \right). \end{aligned} \tag{23.13}$$

Note that in (23.13) $P_e(K_l)$, $P_e(K_l+1)$ and $P_e(K_l-1)$ are functions of $N_l(K_l, P_E)$, since the spreading factor assumes different values in response to the time-variant number of active interfering users in the context of the VSF-assisted adaptive rate transmission scheme.

Above we have derived both the effective throughput formula of a single cell DS-CDMA system and the resultant BER expression, when all the mobile users in the cell transmit using the proposed adaptive rate transmission scheme. In the following section, we characterize the performance of the DS-CDMA system using VSF-assisted adaptive rate transmissions.

23.2.5 Numerical Results and Discussions

In this section we provide some performance results, in order to demonstrate the advantages of VSF-assisted adaptive DS-CDMA systems. We will also compare the throughput performance of the proposed adaptive rate and that of the conventional constant rate DS-CDMA transmission scheme. The spreading factors employed for adaptive rate transmissions are shown in Table 23.1. Furthermore, in the context of the DS-CDMA system using constant rate transmissions we assumed that at any given SNR/chip value the specific spreading factor was used, which was capable of maximizing the effective throughput, while guaranteeing the target BER performance, for the given distribution of the number of interfering users.

Figure 23.5 shows the probability density function (PDF) of the number of active users computed from (23.4) associated with various values of the call generation rate, λ and for various average call durations, $1/\mu$, as shown in the figure. In our computations we assumed that the system was capable of supporting a maximum of $K = 64$ active users. The results show that when the call generation rate and the average call duration are relatively low, such as $\lambda = 0.2$, $1/\mu = 1/0.01$, the number of active users is distributed over bottom end of the range seen in Figure 23.5. This observation implies that the *blocking probability* derived according to the *Erlang B formula* [581] is low, resulting in a high probability of successful call establishment. By contrast, when the values of the call generation rate and the average

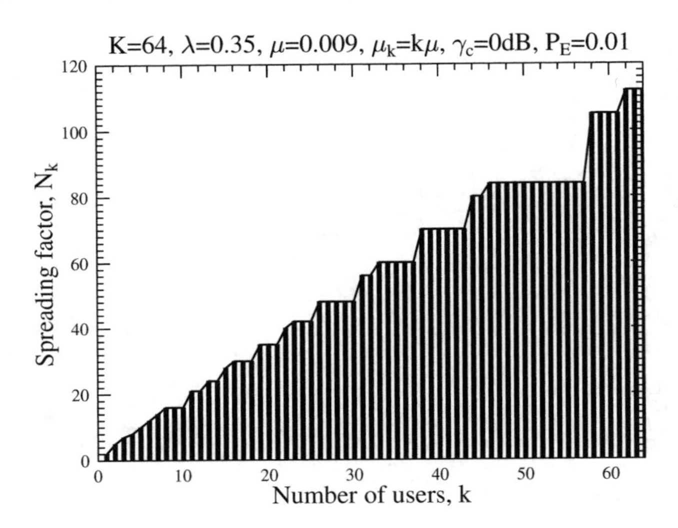

Figure 23.6: Spreading factor versus the number of users required for achieving the target BER of $P_e = 0.01$.

call duration are relatively high, such as $\lambda = 0.4$, $1/\mu = 1/0.006$, the number of active users is distributed over the higher end of the range seen in Figure 23.5. In this case the *blocking probability* derived by the *Erlang B formula* is high and therefore new calls will be blocked with a high probability. Finally, the corresponding PDF has also been portrayed in Figure 23.5 using the solid line and black bars for the parameters of $\lambda = 0.35$, $\mu = 0.009$ and $K = 64$.

Figure 23.6 shows the spreading factors required by the DS-CDMA system for achieving the target BER of $P_E = 0.01$, when supporting different number of active users ranging from one to 64. The spreading factors were chosen from Table 23.1, each of which guarantee maintaining the target BER for a given number of active users, while using the lowest possible spreading factor value. As expected, in order to achieve the required BER, the spreading factor's value has to be increased appropriately, as the number of active users increases.

Figure 23.7 shows the resultant BER performance against the number of active users for the DS-CDMA system using the specific variable spreading factors of Figure 23.6. The results of Figure 23.7 demonstrate that the BER achieved always lies below the target BER of $P_E = 0.01$ and it is predominantly distributed within the range of $[0.005, 0.01)$. A substantial advantage of the associated near-constant BER versus the number of active users is that this results in a significantly improved channel coding performance in comparison to a fixed spreading factor assisted scenario, when the BER fluctuates over a wider range.

Figure 23.8 shows the throughput density as well as the throughput cumulative function achieved by the DS-CDMA system, when the number of active users obeyed the distribution portrayed in Figure 23.5 for $\lambda = 0.35$, $\mu = 0.009$, when the VSFs used assumed the value

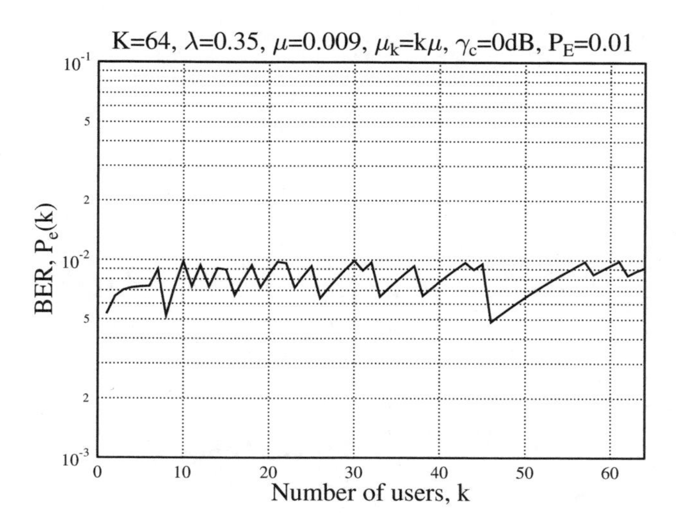

Figure 23.7: BER performance versus the number of interfering users, when the number of users obeys the distribution of Figure 23.5, while employing the spreading factors according to Figure 23.6.

shown in Figure 23.6 and when the DS-CDMA system achieved the BER performance of Figure 23.7. As shown in Figure 23.8, a large fraction of the effective throughput achieved is contributed, when the system supports about 38 active users, which corresponds to the peak of the PDF characterizing the distribution of the number of active users, as shown in Figure 23.5. In accordance with the throughput density, the throughput cumulative function increases sharply, when the number of active users increases from about 30 to about 50. Above this range, it increases slowly and finally reaches the effective throughput of about 0.58 bits/chip. This is the effective throughput provided by the DS-CDMA system at the SNR/chip value of $\gamma_c = 0dB$, under the experimental conditions used in Figure 23.5, Figure 23.6 and Figure 23.7.

In Figure 23.9 we compared the throughput versus SNR/chip performance of the DS-CDMA system using VSF-assisted adaptive transmissions to that of the DS-CDMA system using constant rate transmissions, when various call generation rates and average call durations are considered. The results related to adaptive rate transmissions were computed from (23.9) based on the spreading factors of Table 23.1 conditioned on the target BER of $P_E = 0.01$. For the constant rate transmission scheme the effective throughput of Figure 23.9 recorded at a given SNR/chip value represents the maximum throughput that the DS-CDMA system is capable of achieving. More explicitly, these results were computed as follows. For a given SNR/chip value, we computed the corresponding effective throughput results, when all possible spreading factors seen in Table 23.1 were considered, while taking into account the given distribution of the number of active users as well as the target BER. Then the maximum

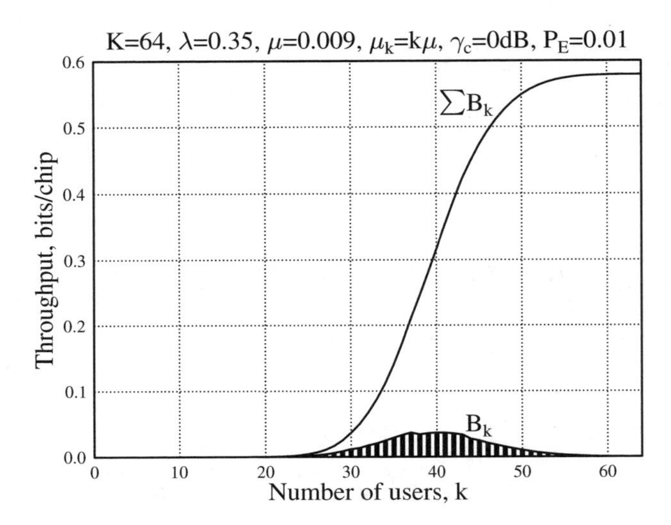

Figure 23.8: Throughput density and throughput cumulative functions versus the number of users, when the number of interfering users obeys the distribution of Figure 23.5, while employing the variable spreading factors according to Figure 23.6 for the target BER of $P_e = 0.01$.

of these throughput values was selected as the achievable effective throughput of the constant transmission scheme. The results of Figure 23.9 show that the adaptive rate transmission scheme significantly outperforms the constant rate transmission scheme. Specifically, the adaptive rate transmission scheme is capable of providing an approximately 40% higher effective throughput than the constant rate transmission scheme. The plausible justification for these results is that when adaptive rate transmission is employed, the system is capable of accommodating the interference level experienced by activating an appropriate spreading factor according to the number of active interfering users at a given SNR/chip value. By contrast, for the constant rate transmission scheme, only single spreading factor is employed at a given SNR/chip value, regardless of the number of users supported. Consequently, when the number of active interfering users is low, the BER performance will be better than the target BER and hence the effective throughput is correspondingly lower than necessary. However, when the number of active interfering users is excessive, the BER of the received data is higher than the target BER. Hence the received data may have to be discarded without contributing to the effective throughput.

Finally, in Figure 23.10 we show the resultant BER performance of the DS-CDMA system using both constant rate transmissions and the proposed adaptive rate transmission scheme, when they achieve the effective throughput values shown in Figure 23.9. From the results of Figure 23.10 we observe that for each group of $(\lambda,\ \mu)$ values, the constant rate scheme has a lower BER, than the adaptive rate scheme, while the BER of the adaptive rate

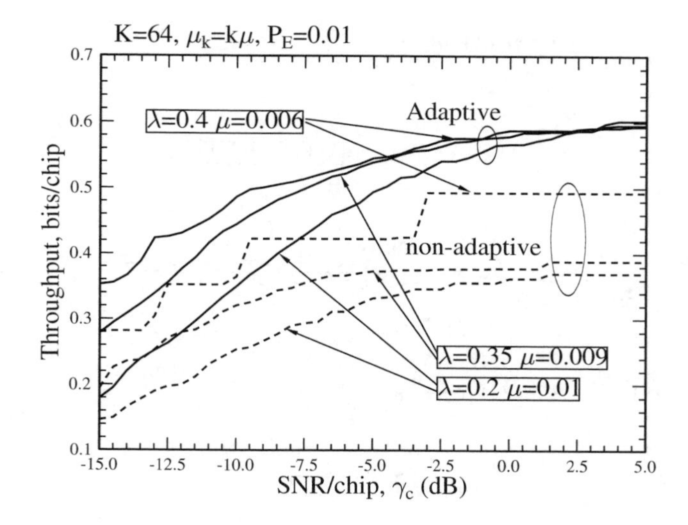

Figure 23.9: Throughput performance comparison of the constant spreading factor assisted non-adaptive DS-CDMA scheme and the VSF-assisted adaptive DS-CDMA arrangement.

transmission scheme is closer to the target BER, than that of the constant rate transmission scheme. The above observation explains why the adaptive rate transmission scheme is capable of providing a higher effective throughput, than the constant rate transmission scheme. The results of Figure 23.10 demonstrate that neither the BER of the constant rate nor that of the adaptive rate transmission scheme fluctuates dramatically. Indeed, they both remain in the BER range of $(0.001, 0.01)$ for various SNR/chip values. The reason for observing a near-constant BER even for the constant rate transmission scheme is because we deliberately adjusted the data rate in response to the SNR/chip value experienced for the sake of fair comparison, although a practical constant-rate system is incapable of doing so. By contrast, the adaptive rate transmission scheme was capable of adapting the data rate in response to both the SNR/chip value as well as the number of active interfering user supported at each specific SNR/chip value.

23.3 Adaptive Antennas, Adaptive Modulation and Multi-User Detection in IMT2000-like FDD and TDD Networks

J.S. Blogh, S. Ni, L. Hanzo

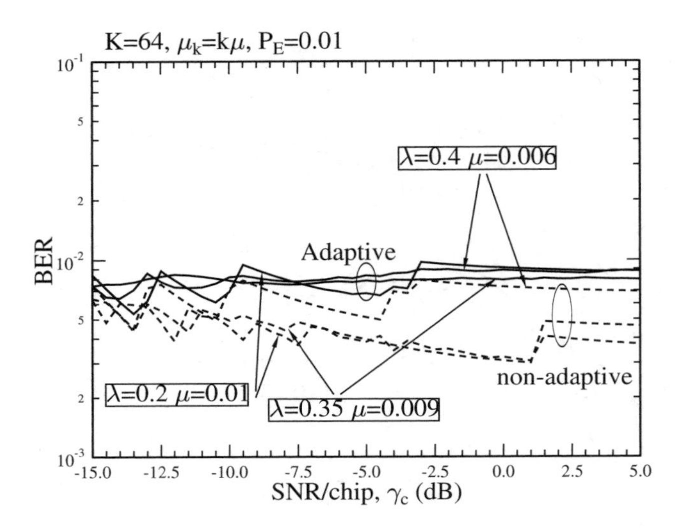

Figure 23.10: BER performance comparison between the constant spreading factor assisted non-adaptive DS-CDMA and the VSF-assisted adaptive DS-CDMA schemes, when they achieve the effective throughputs of Figure 23.9.

23.3.1 State-of-the-art

As we reported in Chapter 22, at the time of writing there is a range of activities in various parts of the globe concerning the standardization, research and development of the third-generation (3G) mobile systems termed as IMT-2000 by the International Telecommunications Union (ITU) [57, 58, 268, 543]. The wealth of activities reported in these references also indicates that due to the rapid evolution of the various proposals under consideration no single winning candidate is emerging. As a compromise, at the time of writing the IMT-2000 standard framework is in fact constituted by a loose conglomerate of five different standards. These are the UTRA Frequency Division Duplex (FDD) Wideband Code Division Multiple Access (W-CDMA) mode [64, 242, 268, 543], the UTRA Time Division Duplex (TDD) CDMA mode [64, 242, 268, 543], the Pan-American multi-carrier CDMA configuration mode known as cdma2000 [64,268,543], the Pan-American Time Division Multiple Access (TDMA) mode known as UWT-136 and finally, the Digital European Cordless Telecommunications (DECT) [64] mode. In addition, a wealth of proposals are under consideration in the 3GPP and 3GPP2 standardization fora.

The next few years will witness the emergence of the UMTS Terrestrial Radio Access (UTRA) mode of IMT2000, which consists of two modes, a Frequency Division Duplex (FDD) mode, where the up-link and down-link are transmitted on different frequencies, and a Time Division Duplex (TDD) mode, where the up-link and the down-link are transmitted on the same carrier frequency, but multiplexed in time. The standard recommends the employment of W-CDMA for UTRA FDD and that of Time Division - Code Division Multiple Access (TD-CDMA) for UTRA TDD. TD-CDMA is based on a combination of Time Division Multiple Access (TDMA) and CDMA, whereas W-CDMA is a pure CDMA-based system. The UTRA scheme can be used for operation within a minimum spectrum of 2 x 5 MHz for UTRA FDD and 5 MHz for UTRA TDD. Both duplex or paired and simplex or unpaired frequency bands have been identified in the region of 2 GHz to be used for the UTRA third generation mobile radio system. Both modes of UTRA have been harmonized with respect to the basic system parameters, such as carrier spacing, chip rate and frame length. Thereby, FDD/TDD dual mode operation is facilitated, which provides a basis for the development of low cost terminals. Furthermore, the interworking of UTRA with GSM [64, 268, 543] is ensured.

In UTRA, the different service needs are supported in a spectrally efficient way by a combination of FDD and TDD. The FDD mode is intended for applications in both macro- and micro-cellular environments, supporting data rates of up to 384kbps and high mobility. The TDD mode, on the other hand, is more suited to micro and pico-cellular environments, as well as for licensed and unlicensed cordless and wireless local loop applications. It makes efficient use of the unpaired spectrum - for example, in wireless Internet applications, where much of the teletraffic is in the down-link - and supports data rates of up to 2 Mbps. Therefore, the TDD mode is particularly well suited for indoor coverage, where the applications tend to have highly asymmetric traffic, again, as in Internet access.

Against this backdrop, a number of studies have been conducted, in order to characterize the network capacity of WCDMA-assisted 3G networks [585–587], which is also the basic objective of this contribution, with emphasis on a UTRA-like FDD and TDD system.

Although the various 3G system parameters undergo perpetual evolution, it is beneficial to study the network performance of a typical advanced UTRA-like FDD CDMA system. Albeit the initial 3G systems are expected to refrain from employing the most powerful performance enhancement techniques available at the current state-of-the-art, in this contribution we embark on quantifying their potential joint performance benefits. Specifically, in contrast to the previous literature, the novelty of this paper is that it jointly investigates the performance benefits of adaptive antennas [268], adaptive modulation [4] and multi-user detection [4, 106], while jointly optimizing the physical and network layer. We will show that these advanced techniques have the potential of doubling the achievable network capacity of both FDD and TDD CDMA. Furthermore, we will demonstrate that the capacity of TDD CDMA is rather poor without adaptive antennas and adaptive modulation owing to the high bases tation to base station interference inflicted as a consequence of potentially using all time-slots in both the up-link and down-link.

The outline of this section is as follows. In Section 23.3.2 we briefly introduce soft handovers, while in Section 23.3.3 power control issues are discussed. The subject of Section 23.3.4 is the allocation of spreading codes in UTRA, while Section 23.3.5 introduces the network architecture, leading to a discussion of the performance metrics and system parameters used in Sections 23.3.6 and 23.3.7. The adaptive antenna [268] assisted network performance is quantified in Section 23.3.8, while the additional performance benefits of using adaptive modulation [4] in a pedestrian scenario are quantified in Section 23.3.9.

23.3.2 Soft Handover

As mentioned above, **our aim is to study a UTRA-like adaptive antenna [268] assisted, adaptive modulation aided [4], MUD-based 3G system, while some system parameters still continue evolving further in the standardization bodies.** The process of soft handovers is based on a *make-before-break* approach, where a new communications link is established before the existing link is relinquished due to the associated link quality degradation. The mobile station (MS) continuously monitors the power level of the received PIlot CHannels (PICH) transmitted from the neighbouring base stations (BSs). The power levels of these base stations are compared against two thresholds, T_{acc} and T_{drop}. If the power level is above the bases tation's acceptance threshold, T_{acc}, then assuming the base station is not already in the Active Base station Set (ABS), it is added to the ABS. If, however, the PICH of a base-station in the ABS is found to be below the dropping threshold, T_{drop}, then the base station is removed from the ABS. If the threshold T_{acc} is set to too low a value, then base stations are added unnecessarily to the ABS, which results in extraneous network resource utilization. Conversely, if T_{acc} is excessively high, then it is possible that no base stations may exist within the ABS at the cell extremities. A mobile station is in simultaneous communication with two or more base stations during the soft handover, hence optimal combining of the down-link signals of several BSs is performed at the MS.

By contrast, the network invokes selective combining of the MSs' signals decoded at each base station. Since a dropped call is less desirable from the user's point of view, than a blocked call, two resource allocation queues were invoked, one for new calls and the other - higher priority - queue, for handovers. By forming a queue of the handover requests, which have a higher priority during contention for network resources than new calls, it is possible to reduce the number of dropped calls at the expense of an increased blocked call probability. A further advantage of the Handover Queueing System (HQS) is that during the time, while a handover is in the queue, previously allocated resources may become available, hence increasing the probability of a successful handover.

A disadvantage of using fixed handover thresholds is that in some locations all the pilot signals may be weak, whereas in other locations they may all be strong. Hence, dynamic thresholds are advantageous. An additional benefit of using dynamic thresholds is experienced in a fading environment, where the received pilot strength may drop momentarily below a fixed threshold and thus may cause an ABS removal and addition. However, this base station may be the only base station in the ABS, which would result in a dropped call. Using dynamic thresholds this scenario would not have occurred, since the pilot strength would not have dropped below that of any of the other pilot signals.

23.3.3 Power Control

Accurate power control is essential in CDMA in order to mitigate the near-far problem, which affects the network capacity and coverage [588]. Closed-loop power control is employed on both the up-link and down-link. The mobiles and base stations estimate the Signal-to-Interference Ratio (SIR) every 0.667ms, or in each time slot, and compare this estimated SIR to a target SIR. If the estimated SIR is higher than the target SIR, then the relevant transmitter is instructed to reduce its transmit power. Likewise, if the estimated SIR is lower than the target SIR, the associated transmitter is instructed to increase its transmit power.

Transmitting at an unnecessarily high power increases the power consumption and degrades the other users' signal quality by inflicting excessive co-channel interference. Hence, the other users may request a power increase in an effort to maintain their target link quality, potentially leading to an unstable system.

If the mobile is in soft handover, and therefore base station diversity combining is performed, then the base stations' transmit powers are controlled independently. Hence, the mobile station may receive different power control commands from the BSs in its ABS. Thus, the mobile only increases its transmit power, if all of the BSs in the ABS instruct it to do so. However, if any one of the base stations in the ABS instructs the mobile to decrease its power, then the mobile will reduce its transmit power. This method ensures that the multi-user interference is kept to a minimum, since at least one base station has a sufficiently high quality link.

23.3.4 Code Allocation

Again, an overview of the UTRA standard can be found for example in [64,215,242,268,543]. The down-link is assumed to be synchronous under the control of a base station, however, the base stations are asynchronous with respect to the other base stations. The UMTS channelization codes are known as Orthogonal Variable Spreading Factor (OVSF) codes [268, 543], which provide total isolation between different users on the synchronous down-link under perfect channel conditions, thus perfectly eliminating intra-cell Multiple Access Interference (MAI). However, the OVSF codes exhibit poor asynchronous cross-correlation properties and hence the inter-cell MAI may be high, unless the same code is only allocated to BSs exhibiting a sufficiently high geographic separation. By contrast, other codes such as Gold codes, exhibit a low asynchronous cross-correlation. Therefore, cell-specific long codes are used for reducing the inter-cell interference on the down-link. These so-called scrambling codes are Gold codes of $2^{18}-1$ chip-duration and each user served by a given base station has the same down-link scrambling code. There are a total of 512 scrambling codes, potentially allowing the system to assign a different cell-specific scrambling code to 512 cell sites, which eases the task of code planning and allocation.

The down-link OVSF codes are allocated by the base station, again, facilitating perfectly interference-free isolation between different users on the synchronous down-link, if the channel coditions are perfect. Thus each user supported by a given base station has a different down-link OVSF code, while MSs served by a different base station may be using the same OVSF code.

On the asynchronous up-link the MAI is reduced by assigning different scrambling codes to different users, emphasizing again that the employment of scrambling codes exhibiting low asynchronous cross-correlation is important [268]. The primary scrambling code is constructed from the so-called extended Kasami code set [64, 268, 543] of length 256, where the short length enables low complexity multi-user detection [106] to be implemented. A range of further detection techniques has been reviewed in [589].

For single-user detector assisted base stations, a long secondary scrambling code is used [64, 215, 242, 268, 543], which is a Gold-code having a length of $2^{41}-1$ chips. Since the up-link transmissions are asynchronous and hence each user has a unique Gold code exhibiting a high asynchronous cross-correlation, every user can employ the same set of channelization codes.

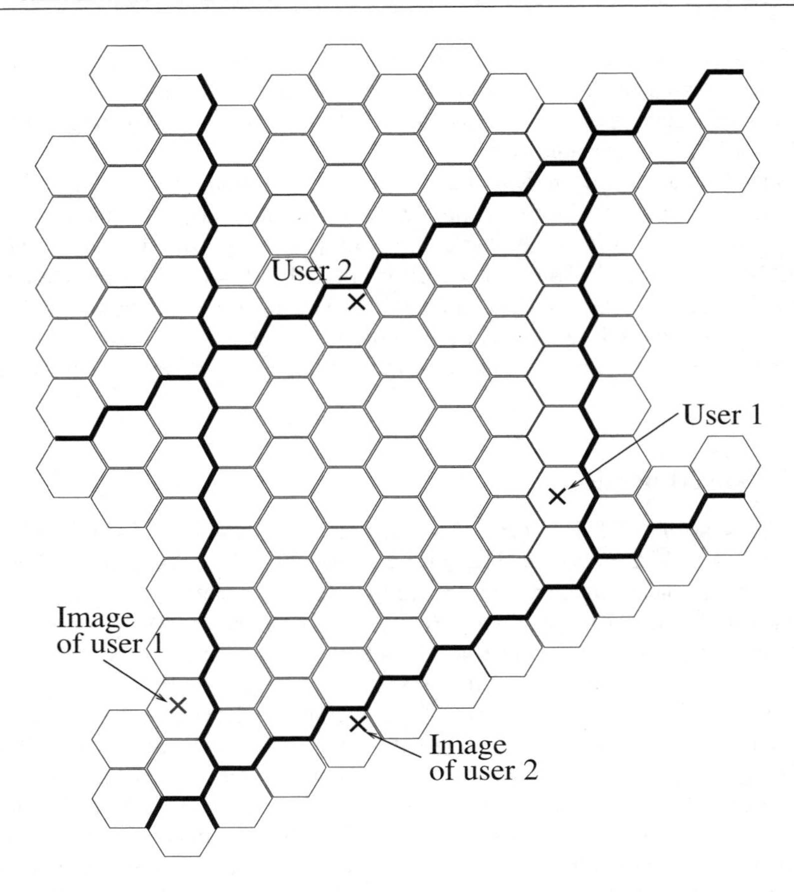

Figure 23.11: The 7x7 rhombic simulation area showing a user and its "wrapped" image.

23.3.5 Simulating an Infinite-Plane Network

A tessellating rhombic simulation area of 7 cells by 7 cells was used, as shown in Figure 23.11, thus allowing the simulation area to be replicated around itself. This approach avoids the border or edge effects associated with simulating a "desert-island-like" cellular network, which would result in the central cells experiencing significantly higher interference, than the edge cells. A further advantage is that all cells encounter a somewhat higher level of interference, than that experienced by an "island-type" system. Thus a higher call dropping rate occurs across the network, allowing statistically valid results to be obtained within a reduced-duration simulation time frame, hence expediting our simulations. The infinite plane network replicates itself, such that when a mobile leaves the central network, it enters an adjacent network which is mapped onto the central network. More explicitly, as a mobile leaves the network, it is "wrapped around" and enters the network at the opposite edge, whilst inflicting co-channel interference to *all* users of the network, who may be located at either edge of the network. In order to enable a call to be maintained under these conditions, both the signals and the interference are "wrapped around" the edges of the network.

Parameter	Value	Parameter	Value
Noisefloor	-100 dBm	Pilot power	-5 dBm
Frame length	10 ms	Cell radius	150 m
Multiple access	FDD/CDMA	Number of base stations	49
Modulation scheme	4QAM/QPSK	Spreading factor	16
Minimum BS transmit power	-44 dBm	Minimum MS transmit power	-44 dBm
Maximum BS transmit power	21 dBm	Maximum MS transmit power	21 dBm
Power control stepsize	1 dB	Power control hysteresis	1 dB
Low quality access:		Outage (BER $\geq$ 1%) SINR	6.6 dB
SINR, where BER $\geq$ 0.5 %	7.0 dB		
Pathloss exponent	-3.5	Size of Active Base station Set	2
Average inter-call-time	300 sec	Max. new-call queue-time	5 sec
Average call length	60 sec	MS speed	3 mph
Maximum consecutive outages	5	Signal bandwidth	5 MHz
Target SINR (at BER=0.1%)	8.0 dB		

Table 23.3: Simulation parameters.

23.3.6 System Parameters

New call channel allocation requests were placed in a resource allocation queue for up to 5s. If during this period a call was not serviced, it was classed as blocked. The mobiles moved freely, in random directions, at a speed of 3 miles/hour (mph) within the simulation area, which consisted of 49 cells, as described in Section 23.3.5. The cell-radius was 150 m. The call duration and inter-call periods were Poisson distributed with the mean values shown in Table 23.3. For our initial investigations we have assumed that the base stations and mobiles form a synchronous network, both in the up- and the down-link.

Furthermore, the base stations are assumed to be equipped with the Minimum Mean Squared Error Block Decision Feedback Equalizer (MMSE-BDFE) based Multi-User Detector (MUD) [106]. The post-despreading SINRs required by this MUD for obtaining the target BERs were determined with the aid of physical-layer simulations using a 4-QAM modulation scheme, in conjunction with $1/2$ rate turbo coding [5] and MUD over a COST 207 seven-path Bad Urban channel [4]. Using this turbo-coded MUD-assisted transceiver and a spreading factor of 16, the post-despreading SINR required for maintaining the target BER of 1×10^{-3} was 8.0 dB. The BER corresponding to low-quality access was stipulated at 5×10^{-3}. This BER was exceeded for SINRs below 7.0dB. Furthermore, a low-quality outage was declared, when the BER of 1×10^{-2} was exceeded, namely for SINRs below 6.6 dB. These values can be seen along with the other system parameters in Table 23.3.

23.3.7 Performance Metrics

There are several performance metrics that can be used for quantifying the performance or quality of service provided by a mobile cellular network. The following performance metrics have been widely used in the literature and were also advocated by Chuang [590]:

- New call blocking probability, P_B.
- Call dropping or forced termination probability, P_{FT}. A call is dropped when the lower of the up-link and down-link SINRs dips consecutively below the outage SINR, where the BER exceeds 1% a given number of times.
- Probability of a low quality access, P_{low}, quantifies the chances of either the up-link or down-link signal quality being sufficiently poor, resulting in a low quality access, where the BER exceeds 0.5%.
- Probability of outage, P_{out}, is defined as the probability that the SINR is below the value at which the call is deemed to be in outage.
- Grade-Of-Service (GOS) was defined by Cheng and Chuang [590] as :

$$\begin{aligned} GOS &= P\{\text{unsuccessful or low-quality call accesses}\} \\ &= P\{\text{call is blocked}\} + P\{\text{call is admitted}\} \times \\ &\quad P\{\text{low signal quality and call is admitted}\} \\ &= P_B + (1 - P_B)P_{low}. \end{aligned} \quad (23.14)$$

In order to determine the number of users that may be supported with adequate call quality by the network, we have defined a conservative and a lenient scenario which are formed from a combination of the performance metrics, as follows [268]:

- *Conservative scenario* :
  $P_B \leq 3\%$, $P_{FT} \leq 1\%$, $P_{low} \leq 1\%$ and $GOS \leq 4\%$.
- *Lenient scenario* :
  $P_B \leq 5\%$, $P_{FT} \leq 1\%$, $P_{low} \leq 2\%$ and $GOS \leq 6\%$.

As argued above, employing *relative received pilot power based handover thresholds* is important in realistic propagation environments exposed to shadow fading. More explicitly, in contrast to using absolute handover thresholds, which were expressed in terms of dBm, i.e. with respect to 1 mW in [268], we also investigated the employment of a pair of relative handover thresholds. Accordingly, both the call acceptance threshold T_{acc} and the call dropping threshold T_{drop} were expressed in terms of dB relative to the received pilot strength of the base stations in the ABS. The employment of these relative thresholds also caters for situations, where the absolute pilot power may be too low for use in conjunction with fixed thresholds, but nonetheless sufficiently high for reliable communications.

Another soft handover activation metric that we used in [268] for determining "cell ownership" was the *pilot to down-link interference power ratio of a cell*, which we denoted by E_c/I_o. This handover metric was proposed for employment in the third generation systems [242]. The pilot to down-link interference ratio, or E_c/I_o, may be calculated thus as [591]:

$$\frac{E_c}{I_o} = \frac{P_{pilot}}{P_{pilot} + N_0 + \sum_{k=1}^{N_{cells}} P_k T_k}, \quad (23.15)$$

where P_k is the total transmit power of cell k, T_k is the transmission gain which includes antenna gain and pathloss as well as shadowing, N_0 is the thermal noise and N_{cells} is the number of cells in the network. The advantage of using such a scheme is that it is not an absolute measurement that is used, but the ratio of the pilot power to the interference power. Thus, if fixed thresholds were used, a form of admission control may be employed for new calls, if the interference level became too high. A further advantage of this technique is that it takes into account the time-varying nature of the interference level in a shadowed environment.

In [268] we concluded that it was beneficial to combine the employment of the received E_c/I_o ratio and the relative soft handover thresholds, thus ensuring that variations in both the received pilot signal strength and interference levels were monitored during the soft handover process.

Since in [268] we identified an attractive handover algorithm, in this contribution *we focus our attention on the impact of adaptive modulation [4] and Adaptive Antenna Arrays [268] (AAAs)* on a UTRA-like network's performance in a pedestrian scenario. Specifically, our investigations were conducted using the relative E_c/I_o based soft handover algorithm in conjunction with T_{acc}=-10 dB and T_{drop}=-18 dB, using a spreading factor of 16. Given that the chip rate of UTRA is 3.84 Mchips/sec, this spreading factor corresponds to a channel data rate of $3.84\times10^6/16 = 240$ kbps. Applying $1/2$ rate error correction coding would result in an effective data throughput of 120 kbps, whereas utilizing a $2/3$ rate error correction code would provide a useful throughput of 160 kbps. Again, a cell radius of 150 m was assumed and a pedestrian walking velocity of 3 miles/hour was used, while the remaining system characteristics - including the power control scheme, the OVSF code allocation algorithm [64] and the multi-user detector [106] - were identical to those used in [268], which are also summarized in Table 23.3.

23.3.8 Adaptive Antenna Assisted Network Performance

In our previous investigations employing AAAs at the base station [268] we observed quite significant performance gains as a direct result of the interference rejection capabilities of the AAAs invoked. Since the CDMA-based network considered here has a frequency reuse of 1, the levels of co-channel interference are significantly higher than in [268], and hence the adaptive antennas may be able to null the interference more effectively. On the other hand, the high number of interference sources may limit the achievable interference rejection.

In order to render the simulations realistic, we used two multipath rays, in addition to the line-of-sight ray, each having a third of the direct-path's power. The angle-of-arrival of each multipath ray was determined using the so-called Geometrically Based Single-Bounce Elliptical Model (GBSBEM) of [592, 593] with parameters chosen such that the multipath rays had one-third of the received power of the direct ray. The Probability Density Function (PDF) of the angle-of-arrival distribution used in the simulations generated using the GBSBEM is shown in Figure 23.12. It was assumed that the multipath rays arrived with no time delay relative to the LOS path. However, in a practical system a space-time equalizer [594, 595] would be required to prevent the nulling of the delayed paths.

Network performance results were obtained using two and four element adaptive antenna arrays, both in the absence of shadow fading, and in the presence of 0.5 Hz and 1.0 Hz frequency shadow fading exhibiting a standard deviation of 3 dB. *The adaptive beamforming*

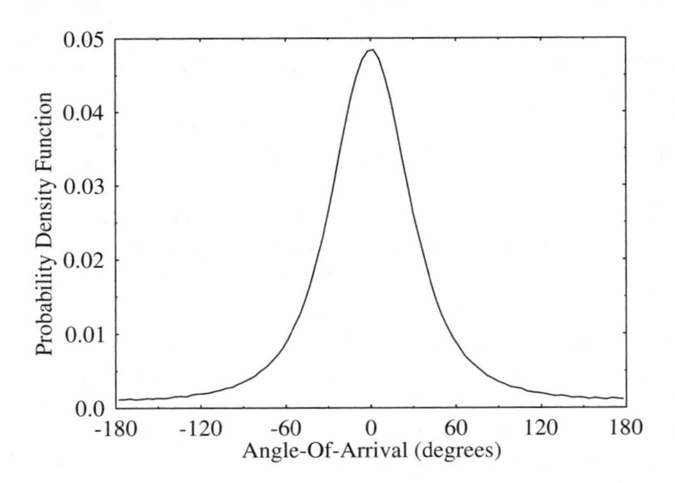

Figure 23.12: Probability density function of angle-of-arrival of the multipath rays, centred about the angle-of-arrival of the line-of-sight path.

algorithm used was the Sample Matrix Inversion (SMI) algorithm [268]. Below the specific adaptive beamforming implementation used for calculating the AAA weights in the CDMA-based network studied here is briefly highlighted as follows [268].

Specifically, one of the eight possible 8-bit BPSK reference signals was used for identifying the desired user, and the remaining interfering users were assigned the other seven 8-bit reference signals. The received signal's autocorrelation matrix was then calculated, and from the knowledge of the desired user's reference signal, the receiver's optimal antenna array weights were determined with the aid of the SMI algorithm [268]. Since this implementation of the algorithm only calculated the base station receiver's antenna array weights, i.e. the antenna arrays weights used by the base station in the up-link, these weights may not be suitable for use in the down-link, when independent up-/down-link shadow fading is experienced. Hence, investigations were conducted in two specific scenarios, namely where the up-link and down-link AAA weights were identical, as well as when they were separately determined for the up-link and down-link. The corresponding up-link and down-link beamforming scenarios are portrayed for the sake of illustration in Figure 23.13.

The two separate up-link and down-link AAA weight calculation scenarios allowed us to determine the potential extra performance gain that may be achieved by separately calculating the AAA weights to be used in the down-link. The AAA weights were re-calculated for every power control step, i.e. 15 times per UTRA data frame, due to the potential significant changes in terms of the desired signal and interference powers that may occur during one UTRA frame as a result of the maximum possible 15 dB change in power transmitted by each user. The performance of both of these scenarios will be characterized in Table 23.4 during our further discourse.

Figure 23.14 shows the significant reduction in the probability of a dropped call achieved

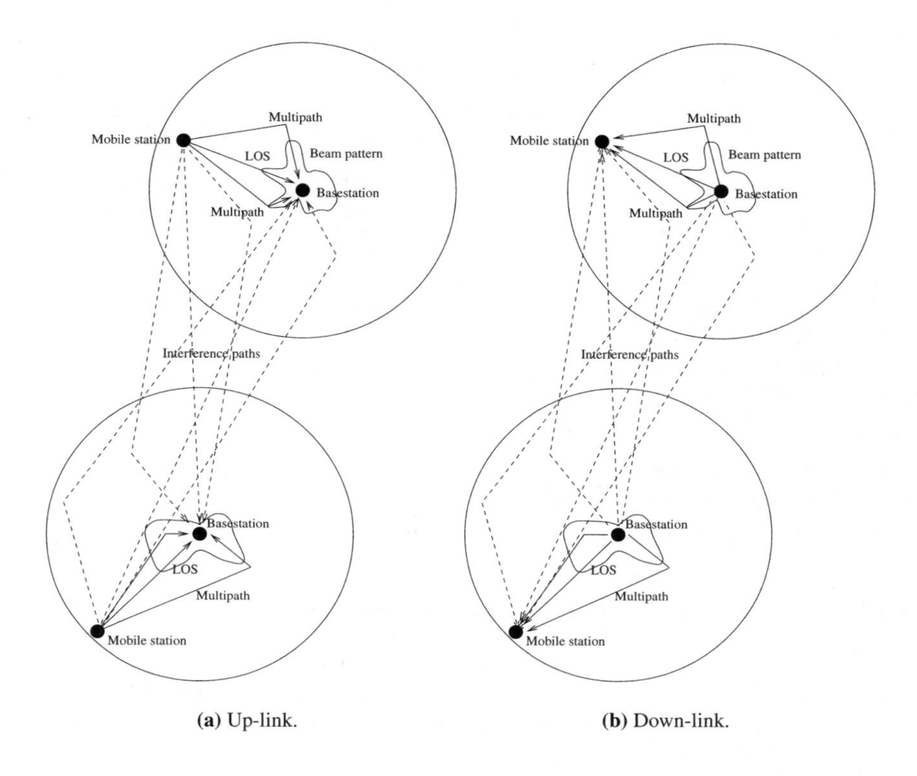

Figure 23.13: The multipath environments of both the up-link and down-link, showing the multipath components of the desired signals, the line-of-sight interference and the associated base station antenna array beam patterns.

by employing AAAs in a non-shadowed propagation environment. The figure has demonstrated that, even with only two antenna elements, the AAAs have considerably reduced the levels of co-channel interference, leading to a reduced call dropping probability. This has been achieved in spite of the numerous sources of co-channel interference resulting from the frequency reuse factor of one, which was remarkable in the light of the limited grade of freedom attained by the two element array. Without employing antenna arrays at the base stations the network capacity was limited to 256 users, or to a teletraffic load of approximately 1.4 Erlangs/km^2/MHz. However, with the advent of two element adaptive antenna arrays at the base stations the number of users supported by the network was increased by 27% to 325 users, or almost 1.9 Erlangs/km^2/MHz. Replacing the two element AAAs with four element arrays led to a further rise of 48%, or 88% with respect to the capacity of the network using no AAAs. This is associated with a network capacity of 480 users, or 2.75 Erlangs/km^2/MHz. A summary of the network capacities achieved under different conditions are summarized in Table 23.4.

The probability of low quality outage, presented in Figure 23.15 also exhibited a substantial improvement with the advent of two element AAAs. However, the performance gains obtained when invoking four element AAAs were less pronounced. It can be seen from the

Shadowing	Beamforming:	independent up/down-link	Conservative scenario, P_{FT}=1%, P_{low}=1% Users	Erlangs /km²/MHz	Power (dBm) MS	BS
No	No	-	256	1.42	3.1	2.7
No	2 elements	-	325	1.87	3.75	0.55
No	4 elements	-	480	2.75	4.55	1.85
0.5 Hz, 3 dB	No	-	≈150	0.87	-1.2	-1.7
0.5 Hz, 3 dB	2 elements	No	203	1.16	0.1	-1.1
0.5 Hz, 3 dB	4 elements	No	349	2.0	2.0	0.65
0.5 Hz, 3 dB	2 elements	Yes	233	1.35	0.2	-0.8
0.5 Hz, 3 dB	4 elements	Yes	≈375	2.2	2.15	0.85
1.0 Hz, 3 dB	No	-	144	0.82	-1.1	-1.6
1.0 Hz, 3 dB	2 elements	No	201	1.12	-0.3	-1.1
1.0 Hz, 3 dB	4 elements	No	333	1.88	1.6	0.5
1.0 Hz, 3 dB	2 elements	Yes	225	1.31	0.1	-0.9
1.0 Hz, 3 dB	4 elements	Yes	365	2.05	1.65	0.6

Table 23.4: Maximum mean carried traffic and maximum number of mobile users that can be supported by the network, whilst meeting the conservative quality constraints. The carried traffic is expressed in terms of normalized Erlangs (Erlang/km^2/MHz) for the network described in Table 23.3 both **with and without beamforming (as well as with and without independent up/down-link beamforming), and also with and without shadow fading having a standard deviation of 3 dB** for SF = 16.

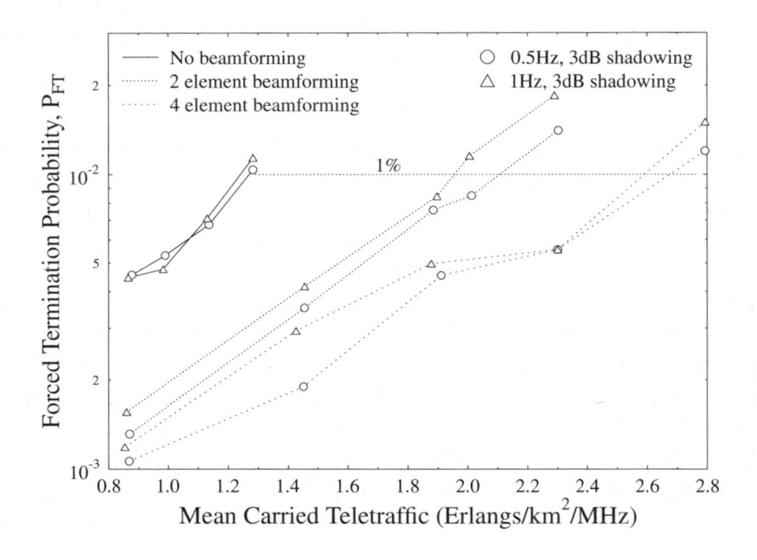

Figure 23.14: Call dropping probability versus mean carried traffic of a CDMA-based cellular network using **relative received** E_c/I_o based soft handover thresholds **with and without beamforming and without shadowing** for SF = 16.

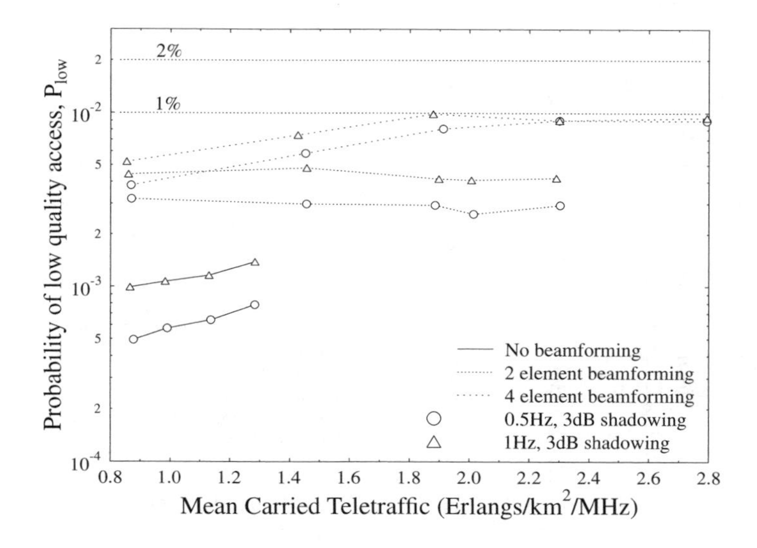

Figure 23.15: Probability of low quality access versus mean carried traffic of a CDMA-based cellular network using **relative received** E_c/I_o based soft handover thresholds **with and without beamforming and without shadowing** for SF = 16.

figure that higher traffic loads were carried with a sufficiently low probability of a low quality occurring, and at higher traffic loads the probability of a low quality access was lower than that achieved using a two element array. However, at lower traffic loads the performance was worse than that obtained when using two element arrays, and the gradient of the performance curve was significantly lower. Further in-depth analysis of the results suggested that the majority of the low quality outages were occurring when new calls started. When a user decided to commence communications with the base station, the current interference level was measured, and the target transmission power was determined in order to reach the target SINR necessary for reliable communications. However, in order to avoid dropping existing calls, the transmission power was ramped up slowly, until the target SINR was reached. A network using no AAAs, i.e. employing omni-directional antennas, can be viewed as offering equal gain to all users of the network, which we assumed to be 1.0, or 0 dB. Thus, when a new call is initiated, the level of interference rises gradually, and the power control algorithm ensures that the existing users compensate for the increased level of co-channel interference by increasing their transmission power. By contrast, in our network the AAAs are used for nulling the sources of interference, and in doing so the array may reduce the antenna gain in the direction of the desired user, in order to maximise the SINR. Hence a user starting a new call, even if it has a low transmission power, may alter the AAA's response, and thus the antenna gain experienced by the existing users. This phenomenon is more pronounced, when using four element arrays due to their sharper directivity and as a result of their increased sensitivity to interfering signals.

Even though the employment of AAAs may result in the attenuation of the desired signal, this is performed in order to maximise the received SINR, and thus the levels of interference

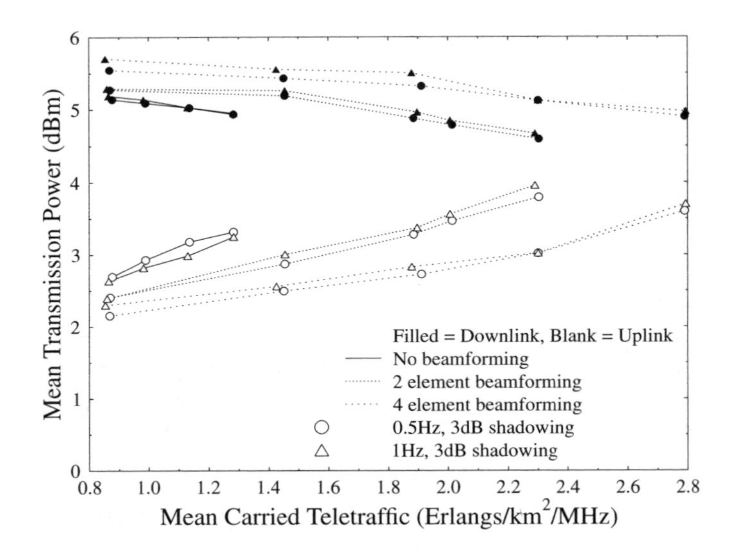

Figure 23.16: Mean transmission power versus mean carried traffic of a CDMA-based cellular network using **relative received** E_c/I_o based soft handover thresholds **with and without beamforming and without shadowing** for SF = 16.

are attenuated more strongly, ultimately leading to the reduction of the mean transmission power, as emphasized by Figure 23.16. This figure clearly shows the lower levels of transmission power required for maintaining an acceptable performance, whilst using AAAs at the base stations. A mean mobile transmission power reduction of 3 dB was achieved by invoking two element AAAs, and a further reduction of 1.5 dB resulted from using four element arrays. These power budget savings were obtained in conjunction with reduced levels of co-channel interference, leading to superior call quality, as illustrated in Figures 23.14 and 23.15. A higher performance advantage was evident in the up-link scenario, suggesting that the selective base station diversity techniques employed in the up-link are amenable to amalgamation with adaptive antenna arrays. By contrast, the maximum ratio combining performed at the mobile inherently reduces the impact of co-channel interference, and hence benefits to a lesser extent from the employment of adaptive antenna arrays.

The impact of AAAs in a propagation environment subjected to shadow fading was then investigated. The associated call dropping performance is shown in Figure 23.17. This figure illustrates the substantial network capacity gains achieved with the aid of both two and four element AAAs under shadow fading propagation conditions. Simulations were conducted in conjunction with log-normal shadow fading having a standard deviation of 3 dB, and maximum shadowing frequencies of both 0.5 Hz and 1.0 Hz. As expected, the network capacity was reduced at the higher shadow fading frequency. The effect of performing independent up- and down-link beamforming, as opposed to using the base station's receive antenna array weights in the down-link was also studied, and a small, but not insignificant call dropping probability reduction can be seen in the Figure 23.17. The network supported just over 150 users, and 144 users, when subjected to 0.5 Hz and 1.0 Hz frequency shadow fading, re-

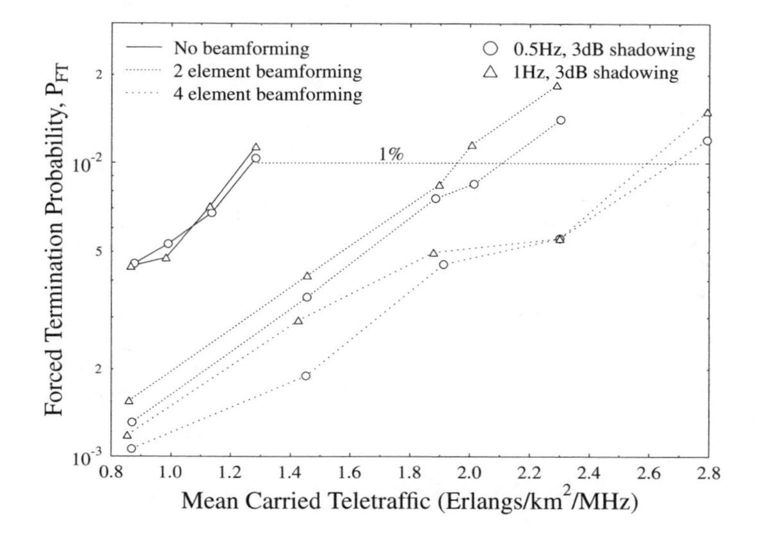

Figure 23.17: Call dropping probability versus mean carried traffic of a CDMA-based cellular network using **relative received** E_c/I_o based soft handover thresholds **with and without beamforming and with shadowing having a standard deviation of 3 dB** for SF = 16.

spectively. With the application of two element AAAs these capacities were increased by 35% and 40%, to 203 users and 201 users, when re-using the base station's up-link receiver weights on the down-link. Performing independent up- and down-link beamforming resulted in a further 13% increase of the network's capacity. The implementation of four element AAAs led to a network capacity of 349 users at a 0.5 Hz shadowing frequency, and 333 users at a 1.0 Hz shadowing frequency. This corresponded to relative gains of 133% and 131% over the capacity provided without beamforming. Invoking independent up- and down-link beamforming provided another 7% and 10% boost of the network capacity for the 0.5 Hz and 1.0 Hz frequency shadowing environments, respectively, giving final network capacities of just over 375 users and 365 users.

Similar trends were observed regarding the probability of low quality outage to those found in the non-shadowing scenarios. However, the trend was much more prevalent under shadowing, due to the higher variation of the received signal strengths, as a result of the shadow fading, as shown in Figure 23.18. The figure indicates that the trend is also evident, when using two element AAAs in conjunction with shadow fading. As expected, the performance deteriorated, as the number of antenna elements increased, and when the maximum shadow fading frequency was increased from 0.5 Hz to 1.0 Hz. It should be noted, however, that the probability of low quality access always remained below the 1% constraint of the conservative scenario, and the call dropping probability was considerably reduced by the AAAs.

The mean transmission power performance is depicted in Figure 23.19, suggesting that as for the non-shadowing scenario of Figure 23.16, the number of antenna elements had only a limited impact on the base stations' transmission power, although there was some reduction

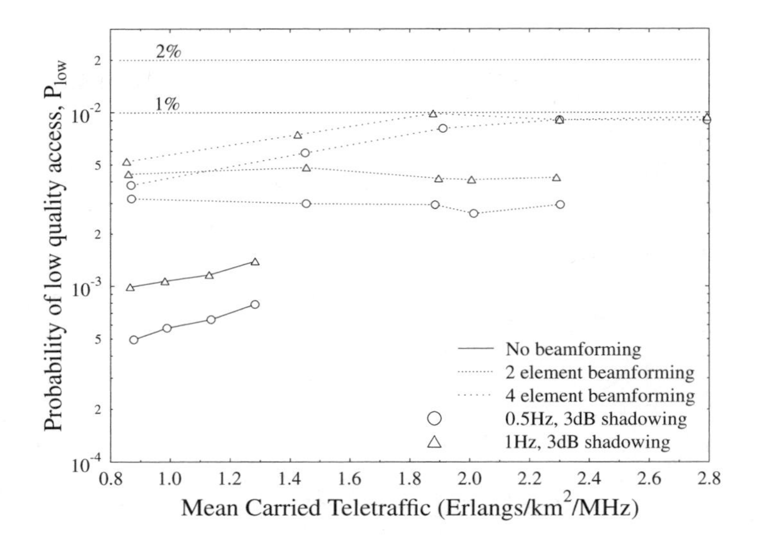

Figure 23.18: Probability of low quality access versus mean carried traffic of a CDMA-based cellular network using **relative received** E_c/I_o based soft handover thresholds **with and without beamforming and with shadowing having a standard deviation of 3 dB** for SF = 16.

in the mobile stations' mean transmission power. The mean transmission powers required, when using independent up- and down-link beamforming are not explicitly shown, but were slightly less than those presented here, with a mean reduction of about 0.4 dB.

A summary of the maximum achievable network capacities both with and without shadowing, employing beamforming using two and four element arrays is given in Table 23.4, along with the teletraffic carried and the mean mobile and base station transmission powers required. In the next section we will show the benefits of employing adaptive modulation [4].

23.3.9 Performance of Adaptive Arrays and Adaptive Modulation in a High Data Rate Pedestrian Environment

In this section we build upon the results presented in the previous section by applying Adaptive Quadrature Amplitude Modulation (AQAM) techniques [4]. *There are two main objectives, when employing AQAM, namely counter-acting the effects of time-variant channel quality fluctuations as well as the effects of the time-variant interference load imposed by the time-variant number of variable-rate users supported.* [2] The various scenarios and channel conditions investigated were identical to those of the previous section, except for the application of AQAM [4]. Since in the previous section an increased network capacity was achieved due to using independent up- and down-link beamforming, this procedure was invoked in

[2]Unless otherwise stated, for the sake of simplicity we will refer to time-variant channel quality fluctuations, regardless, whether these were imposed by fading effects or by co-channel interference fluctuations.

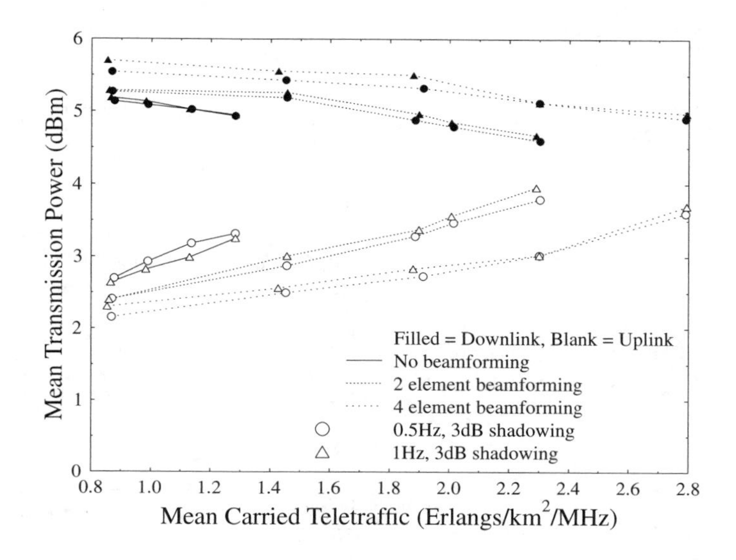

Figure 23.19: Mean transmission power versus mean carried traffic of a CDMA-based cellular network using **relative received** E_c/I_o based soft handover thresholds **with and without beamforming and shadowing having a standard deviation of 3 dB** for SF = 16.

these simulations. AQAM involves the selection of the appropriate modulation mode in order to maximise the achievable data throughput over a channel, whilst minimizing the Bit Error Ratio (BER). More explicitly, the philosophy behind AQAM is the most appropriate selection of a modulation mode according to the instantaneous radio channel quality experienced [4, 92]. Therefore, if the SINR of the channel is high, then a high-order modulation mode may be employed, thus exploiting the temporal fluctuation of the radio channel's quality. Similarly, if the channel is of low quality, exhibiting a low SINR, a high-order modulation mode would result in an unacceptably high BER or FER, and hence a more robust, but lower throughput modulation mode would be employed. Therefore, AQAM combats the effects of time-variant channel quality, while also attempting to maximise the achieved data throughput, and maintaining a given BER or FER. In the investigations conducted, the modulation modes of the up and down-link were determined independently, thus taking advantage of the lower levels of co-channel interference on the up-link, or of the potentially higher transmit power of the base stations.

The particular implementation of AQAM used in these investigations is illustrated in Figure 23.20. This figure describes the algorithm in the context of the down-link, but the same implementation was used also in the up-link. The first step in the process was to establish the current modulation mode. If the user was invoking 16-QAM and the SINR was found to be below the Low Quality (LQ) outage SINR threshold after the completion of the power control iterations, then the modulation mode for the next data frame was 4-QAM. Alternatively, if the SINR was above the LQ outage SINR threshold, but any of the base stations in the ABS were using a transmit power within 15 dB of the maximum transmit power, then the 4-QAM modulation mode was selected. This transmit power "headroom" was introduced in order to

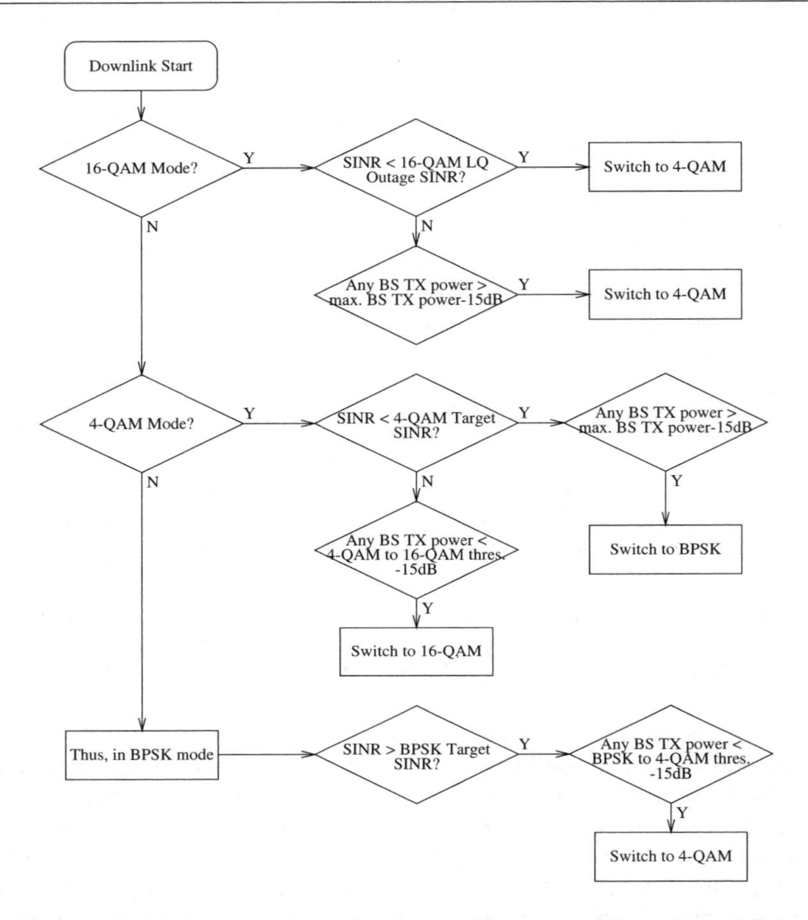

Figure 23.20: The AQAM mode switching algorithm used in the down-link of the CDMA-based cellular network.

provide a measure of protection, since if the interference conditions degraded, then at least 15 dB of increased transmit power would be available in order to mitigate the consequences of the SINR reduction experienced.

A similar procedure was invoked, when switching to other legitimate AQAM modes from the 4-QAM mode. If the SINR was below the 4-QAM target SINR and any one of the base stations in the ABS was within 15 dB (the maximum possible power change during a 15-slot UTRA data frame) of the maximum transmit power, then the BPSK modulation mode was employed for the next data frame. However, if the SINR exceeded the 4-QAM target SINR and there would be 15 dB of headroom in the transmit power budget in excess of the extra transmit power required for switching from 4-QAM to 16-QAM, then the 16-QAM modulation mode was invoked.

And finally, when in the BPSK mode, the 4-QAM modulation mode was selected if the SINR exceeded the BPSK target SINR, and the transmit power of any of the base stations in the ABS was less than the power required for transmitting reliably using 4-QAM, while

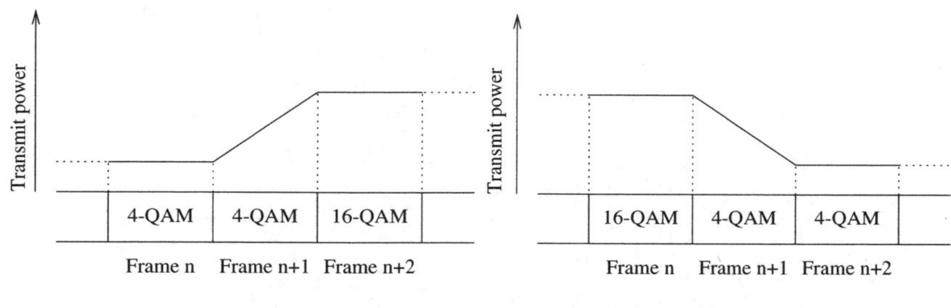

(a) Ramping up the transmit power whilst remaining in the lower order modulation mode.

(b) Ramping down the transmit power whilst switching to the lower order modulation mode.

Figure 23.21: Power ramping requirements whilst switching modulation modes.

SINR Threshold	BPSK	4-QAM	16-QAM
Outage SINR	2.6 dB	6.6 dB	12.1 dB
Low Quality Outage SINR	3.0 dB	7.0 dB	12.5 dB
Target SINR	4.0 dB	8.0 dB	13.5 dB

Table 23.5: The target SINR, low quality outage SINR and outage SINR thresholds used for the BPSK, 4-QAM and 16-QAM modulation modes of the adaptive modem.

being at least 15 dB below the maximum transmit power. The algorithm was activated at the end of each 15-slot UTRA data frame, after the power control algorithm had performed its 15 iterations per data frame, and thus the AQAM mode selection was performed on a UTRA transmission frame-by-frame basis. Similarly, when changing from a lower-order modulation mode to a higher-order mode, the lower-order mode was retained for an extra frame duration, in order to ramp up the transmit power to the required level, as shown in Figure 23.21(a). Conversely, when changing from a higher-order modulation mode to a lower-order modulation mode, the lower-order modulation mode was employed, whilst ramping the power down, in order to avoid excessive outages in the higher-order modulation mode due to the reduction of the transmit power, as illustrated in Figure 23.21(b).

Table 23.5 shows the BPSK, 4-QAM and 16-QAM reconfiguration SINR thresholds used in the simulations. The BPSK SINR thresholds were 4 dB lower than those necessary when using 4-QAM, while the 16-QAM SINR thresholds were 5.5 dB higher, as shown in Chapter 4. In other words, in moving from the BPSK modulation mode to the 4-QAM modulation mode, the target SINR, low quality outage SINR and outage SINR all increased by 4 dB. When switching to the 16-QAM mode from the 4-QAM mode, the SINR thresholds were increased by 5.5 dB. However, it was necessary to set the BPSK to 4-QAM and the 4-QAM to 16-QAM mode switching thresholds to a value 7 dB higher than the SINR required for maintaining the target BER/FER, in order to prevent excessive outages due to sudden dramatic channel-induced variations in the SINR levels.

Performance results were obtained both with and without beamforming in a log-normal shadow fading environment, at maximum fading frequencies of 0.5 Hz and 1.0 Hz, and a

Shadowing	Beamforming	Conservative scenario Users	Erlangs /km^2/MHz
0.5 Hz, 3 dB	No	3568	20.3
0.5 Hz, 3 dB	2 elements	5856	33.8
0.5 Hz, 3 dB	4 elements	7616	42.9
1.0 Hz, 3 dB	No	3488	19.8
1.0 Hz, 3 dB	2 elements	5456	31.7
1.0 Hz, 3 dB	4 elements	7360	41.4

Table 23.6: A lower bound estimate of the maximum mean carried traffic and maximum number of mobile **speech-rate** users that can be supported by the network, whilst meeting the conservative quality constraints. The carried traffic is expressed in terms of normalized Erlangs (Erlang/km^2/MHz), for the network described in Table 23.3 both **with and without beamforming (using independent up/down-link beamforming), in conjunction with shadow fading having a standard deviation of 3 dB, whilst employing adaptive modulation techniques** for SF = 256. The number of users supported in conjunction with a spreading factor of 256 was calculated by multiplying the capacities obtained in Table 23.7 by 256/16=16.

standard deviation of 3 dB. Again, a pedestrian velocity of 3 mph, a cell radius of 150 m and a spreading factor of 16 were used, as in our previous investigations.

Figure 23.22 shows the significant reduction in the probability of a dropped call, achieved by employing AAAs in conjunction with AQAM in a log-normal shadow faded environment. The figure demonstrates that, even with the aid of a two element AAA and its limited degrees of freedom, a substantial call dropping probability reduction was achieved. The performance benefit of increasing the array's degrees of freedom, achieved by increasing the number of antenna elements, becomes explicit from the figure, resulting in a further call dropping probability reduction. Simulations were conducted in conjunction with log-normal shadow fading having a standard deviation of 3 dB, and maximum shadowing frequencies of 0.5 Hz and 1.0 Hz. As expected, the call dropping probability was generally higher at the higher shadow fading frequency, as demonstrated by Figure 23.22. The network was found to support 223 users, corresponding to a traffic load of 1.27 Erlang/km^2/MHz, when subjected to 0.5 Hz frequency shadow fading. The capacity of the network was reduced to 218 users, or 1.24 Erlang/km^2/MHz, upon increasing the maximum shadow fading frequency to 1.0 Hz. On employing two element AAAs, the network capacity increased by 64% to 366 users, or to an equivalent traffic load of 2.11 Erlang/km^2/MHz, when subjected to 0.5 Hz frequency shadow fading. When the maximum shadow fading frequency was raised to 1.0 Hz, the number of users supported by the network was 341, carrying a traffic load of 1.98 Erlang/km^2/MHz, representing an increase of 56% in comparison to the network without AAAs. Increasing the number of antenna elements to four, whilst imposing shadow fading with a maximum frequency of 0.5 Hz, resulted in a network capacity of 2.68 Erlang/km^2/MHz or 476 users, corresponding to a gain of an extra 30% with respect to the network employing two element arrays, and of 113% in comparison to the network employing no AAAs. In conjunction with a maximum shadow fading frequency of 1.0 Hz the network capacity was 460 users or 2.59

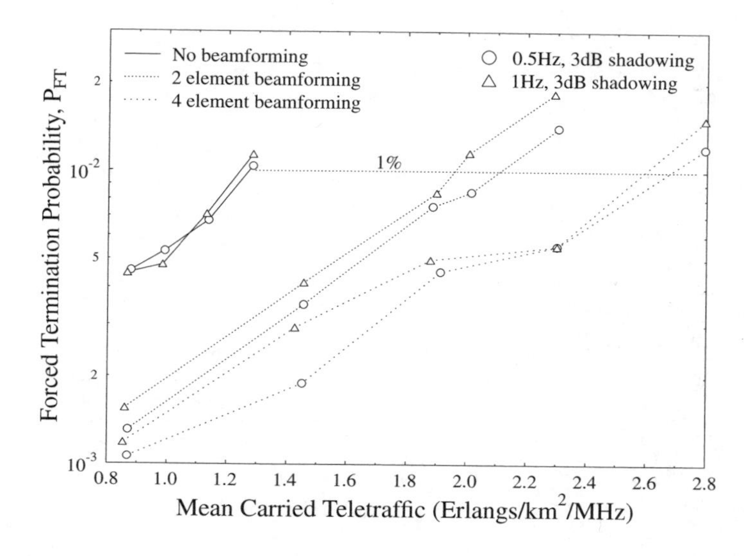

Figure 23.22: Call dropping probability versus mean carried traffic of a CDMA-based cellular network using **relative received** E_c/I_o based soft handover thresholds both **with and without beamforming in conjunction with AQAM as well as with shadowing having a standard deviation of 3 dB** for SF = 16. See Figure 23.17 for corresponding results without adaptive modulation.

Erlang/km^2/MHz, which represented an increase of 35% with respect to the network invoking two element antenna arrays, or 111% relative to the identical network without AAAs.

The probability of low quality outage, presented in Figure 23.23, did not benefit from the application of AAAs, or from the employment of AQAM. Figure 23.18 depicts the probability of low quality outage without AQAM, and upon comparing these results to those obtained in conjunction with AQAM in Figure 23.23, the performance degradation due to AQAM can be explicitly seen. However, the increase in the probability of low quality access can be attributed to the employment of less robust, but higher throughput, higher-order modulation modes invoked by the AQAM scheme. Hence, under given propagation conditions and using the fixed 4-QAM modulation mode a low quality outage may not occur, yet when using AQAM and a higher order modulation mode, the same propagation conditions may inflict a low quality outage. This phenomenon is further exacerbated by the AAAs, where the addition of a new source of interference, constituted by a user initiating a new call, results in an abrupt change in the gain of the antenna in the direction of the desired user. This in turn leads to low quality outages, which are more likely to occur for prolonged periods of time, when using a higher order modulation mode. Again, increasing the number of antenna elements from two to four results in an increased probability of a low quality outage due to the sharper antenna directivity. This results in a higher sensitivity to changes in the interference incident upon it.

The mean transmission power versus teletraffic performance is depicted in Figure 23.24, suggesting that the required mean up-link transmission power was always significantly below the mean down-link transmission power, which can be attributed to the pilot power interfer-

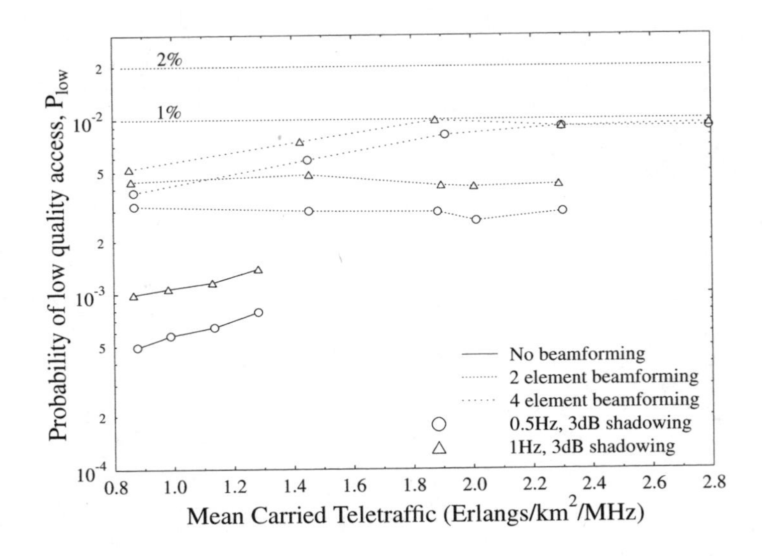

Figure 23.23: Probability of low quality access versus mean carried traffic of a CDMA-based cellular network using **relative received** E_c/I_o based soft handover thresholds both **with and without beamforming in conjunction with AQAM as well as with shadowing having a standard deviation of 3 dB** for SF = 16. See Figure 23.18 for corresponding results without adaptive modulation.

ence encountered by the mobiles in the down-link. This explanation can be confirmed by examining Figure 23.25, which demonstrates that the mean modem throughput in the down-link, without AAAs, was lower than that in the up-link even in conjunction with increased down-link transmission power. Invoking AAAs at the base stations reduced the mean up-link transmission power required in order to meet the service quality targets of the network. The attainable down-link power reduction increased as the number of antenna array elements increased, as a result of the superior interference rejection achieved with the aid of a higher number of array elements. A further advantage of employing a larger number of antenna array elements was the associated increase in the mean up-link modem throughput, which became more significant at higher traffic loads. In the down-link scenario, however, increasing the number of AAA elements led to an increased mean down-link transmission power, with the benefit of a substantially improved mean down-link modem throughput. This suggests that there was some interaction between the AAAs, the AQAM mode switching algorithm and the maximal ratio combining performed at the mobiles. By contrast, simple switched diversity was performed by the base stations on the up-link, thus avoiding such a situation. However, the increase in the mean down-link transmission power resulted in a much more substantial increase in the mean down-link modem throughput, especially with the advent of the four element antenna arrays, which exhibited an approximately 0.5 BPS throughput gain over the two element arrays for similarly high traffic loads which can be seen in Figure 23.25.

A summary of the maximum user capacities of the networks considered in this section in conjunction with log-normal shadowing having a standard deviation of 3 dB, both with and

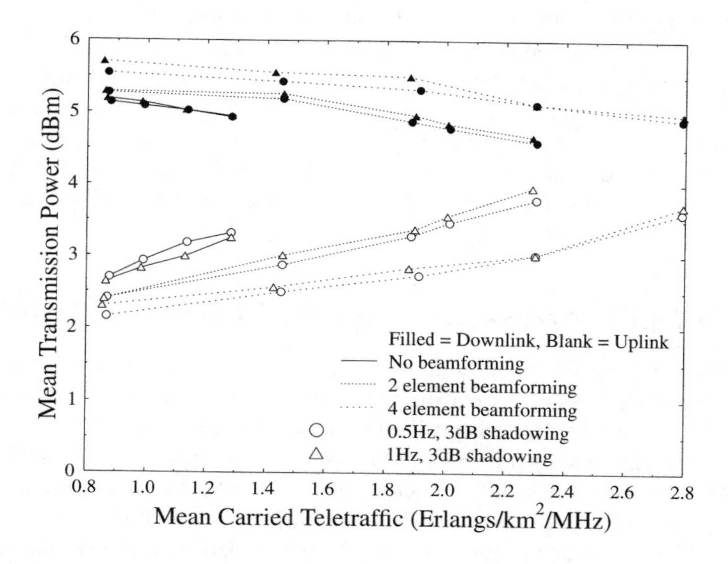

Figure 23.24: Mean transmission power versus mean carried traffic of a CDMA-based cellular network using **relative received** E_c/I_o based soft handover thresholds both **with and without beamforming in conjunction with AQAM as well as with shadowing having a standard deviation of 3 dB** for SF = 16. See Figure 23.19 for corresponding results without adaptive modulation.

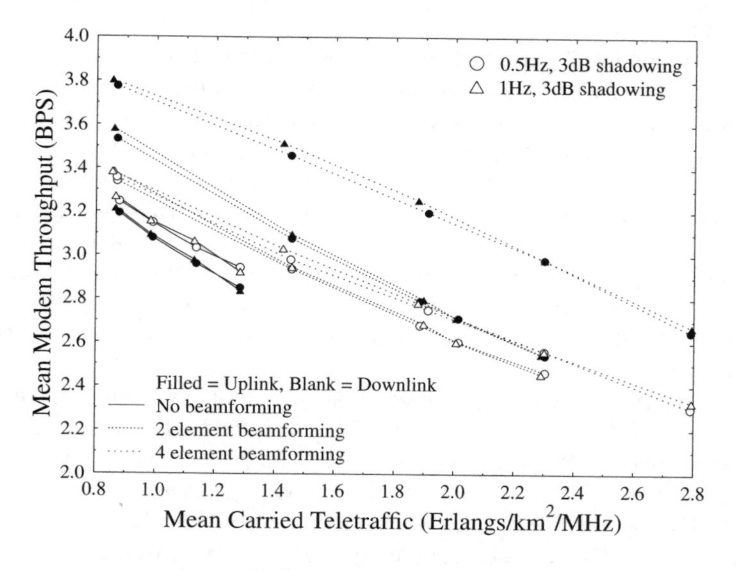

Figure 23.25: Mean modem throughput versus mean carried traffic of a CDMA-based cellular network using **relative received** E_c/I_o based soft handover thresholds both **with and without beamforming in conjunction with AQAM as well as with shadowing having a standard deviation of 3 dB** for SF = 16.

without employing beamforming using two and four element arrays is given in Table 23.7. The teletraffic carried, the mean mobile and base station transmission powers required, and the mean up- and down-link modem data throughputs achieved are also shown in Table 23.7. Similarly, the lower bounds of the maximum network capacities obtained under identical scenarios in conjunction with a spreading factor of 256, leading to a bit rate of 15 kbps, suitable for speech-rate users are presented in Table 23.6. The network capacity calculations were performed by scaling the number of users supported, as presented in Table 23.7, by the ratio of their spreading factors, i.e. by 256/16=16.

23.3.10 Adaptive Arrays and Adaptive Modulation in TDD/CDMA

TDD is attractive in terms of facilitating the allocation of asymmetric or uneven resources to the up-link and down-link, which supports a more efficient exploitation of the frequency bands available. However, the associated interference scenario is markedly different from that experienced in FDD, as shown in Figure 23.26. Mobile to mobile (MS-to-MS) interference occurs in the situation displayed in Figure 23.26, if MS_1 is transmitting, while MS_0 is receiving in a specific time slot mapped to the same carrier frequency in an adjacent cell. The MS-to-MS interference cannot be completely avoided by network planning, since the geographic location of mobiles cannot be controlled. TDD/CDMA is also prone to base station (BS-to-BS) interference. In fact, as will be shown later, it is the most serious source of intercell interference in a TDD/CDMA cellular scenario. As seen in Figure 23.26, if BS_1 is transmitting and BS_0 is receiving at the same time in a given time slot, BS-to-BS interference takes place, provided that these base stations are in adjacent cells. The severity of the BS-to-BS interference depends heavily on the path loss between the two base stations, hence it can be reduced with the aid of careful network planning.

In [596] we analytically studied the achievable network performance of UTRA-like TDD/CDMA systems, where conventional fixed-mode modulation was assumed. By contrast, in [268] the performance of a UTRA-like FDD/CDMA system was quantified, when supported by adaptive beam-steering and adaptive modulation [4]. These performance improvements have approximately doubled the network capacity of the system.

Our current research is building on our previous findings recorded in the context of a UTRA-like FDD system [268], where we found that invoking adaptive modulation as well as beam-steering proved to be a powerful means of enhancing the capacity of FDD/CDMA. Hence they are expected to be even more powerful in the context of TDD/CDMA, where the capacity of TDD/CDMA is poor as a consequence of the excessive base station to base station interference experienced.

The advanced UTRA FDD system level simulator [268] employing adaptive antenna arrays at the base station was extended to the UTRA TDD mode for evaluating the system achievable performance. We observed quite significant performance gains as a direct result of the interference rejection capabilities of the adaptive antenna arrays invoked. Network performance results were obtained using two and four element adaptive antenna arrays, both in the absence of shadow fading, and in the presence of 0.5 Hz and 1.0 Hz frequency shadow fading exhibiting a standard deviation of 3 dB. The adaptive beamforming algorithm used was the Sample Matrix Inversion (SMI) algorithm. The specific adaptive beamforming implementation used in our TDD/CDMA based network was identical to that used in the network simulations of [268]. Briefly, one of the eight possible 8-bit BPSK reference signals was used

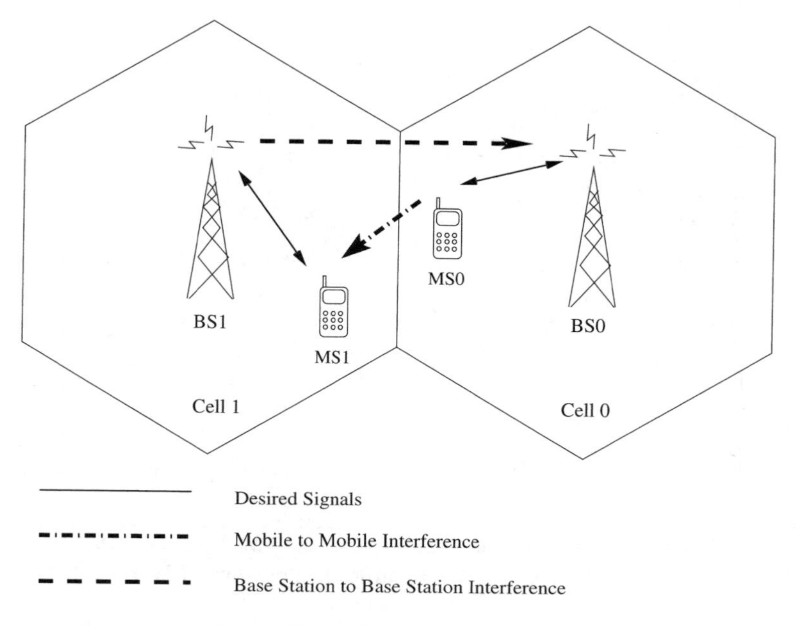

Figure 23.26: Inter-cell interference.

for identifying the desired user, while the remaining interfering users were assigned the other seven 8-bit reference signals. The received signal's autocorrelation matrix was then calculated, and from the knowledge of the desired user's reference signal the receiver's optimal antenna array weights were determined with the aid of the SMI algorithm. This implementation of the algorithm only calculated the receiver's antenna array weights, i.e. the antenna array weights used by the base station for receiving the mobiles' up-link transmissions. However, it was demonstrated in [268] that further performance gains are attainable, if the UL and DL array pattern are optimized individually. The antenna array weights were re-calculated for every power control step, i.e. 15 times per UTRA data frame, owing to the potential significant changes in terms of the desired signal and interference powers that may occur during one UTRA frame as a result of the maximum possible 15 dB change in the power transmitted by each user.

Figure 23.27 shows the call dropping probability associated with a variety of traffic loads without shadowing, measured in terms of the mean normalized carried traffic expressed in Erlangs/km^2/MHz. The figure suggests that the network's performance was poor without employing antenna arrays at the base stations. The "No beamforming" scenario suffered from the highest call dropping probability of the three traffic scenarios at a given load. When using "2 element beamforming", the adaptive antenna arrays have considerably reduced the levels of interference, leading to a reduced call dropping probability. Without employing antenna arrays at the base stations the network capacity was limited to 142 users, or to a teletraffic load of approximately 0.81 Erlangs/km^2/MHz. However, with the advent of employing two-element adaptive antenna arrays at the base stations the number of users supported by the network increased by 45% to 206 users, or almost to 1.18 Erlangs/km^2/MHz. Replacing

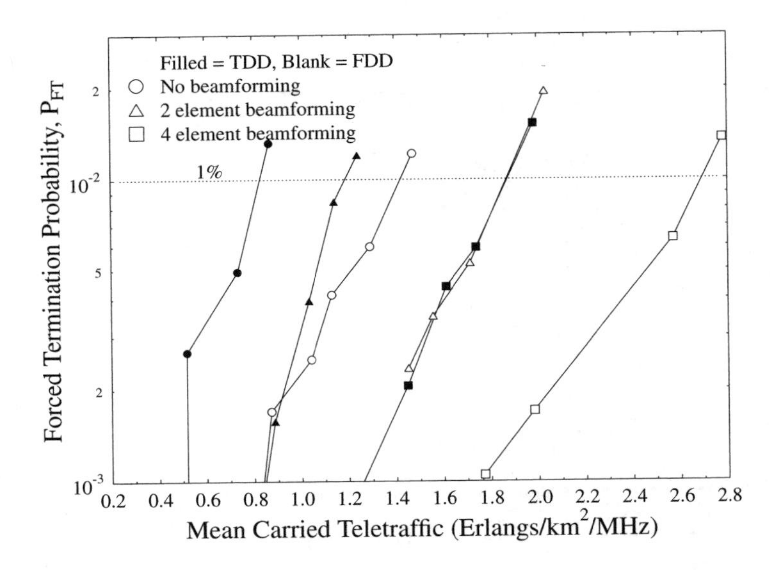

Figure 23.27: Call dropping probability versus mean carried traffic of the UTRA-like TDD/CDMA-based cellular network of Table 23.3 both **with as well as without beamforming and without shadowing** for SF = 16.

the two-element adaptive antenna arrays with four-element arrays led to a further capacity increase of 56%, or 127% with respect to the capacity of the network using no antenna arrays. This is associated with a network capacity of 322 users, or 1.85 Erlangs/km^2/MHz. We can also see in Figure 23.27 that the capacity of the UTRA-like TDD/CDMA celluar system is poorer than that of the UTRA-like FDD/CDMA system under the same propagation conditions. The "TDD 4 element beamforming" scenario has a similar performance to the "FDD 2 element beamforming" scenario. This is because the TDD system suffers from the effects of the extra inter-cell interference, which we alluded to in Section 23.3.10.

Figure 23.28 portrays the probability of low quality access versus various traffic loads. Again, it can be seen from the figure that higher traffic loads were carried with the aid of the four-element array at a sufficiently low probability of a low quality, than that achieved using a two-element array.

Figure 23.29 shows the achievable Grade-Of-Service (GOS) for a range of teletraffic loads. Similar trends were observed regarding the probability of call blocking to those shown in Figure 23.27. The grade of service is better (i.e. lower) when the traffic load is low, and vice versa for high traffic loads. This is mainly attributable to the higher call blocking probability of the "No beamforming" scenario, particularly in the region of the highest traffic loads.

The impact of adaptive antenna arrays recorded in a propagation environment subjected to shadow fading was then investigated. The associated call dropping performance is shown in Figure 23.30. This figure illustrates the substantial network capacity gains achieved with the aid of both two- and four-element adaptive antenna arrays under shadow fading propaga-

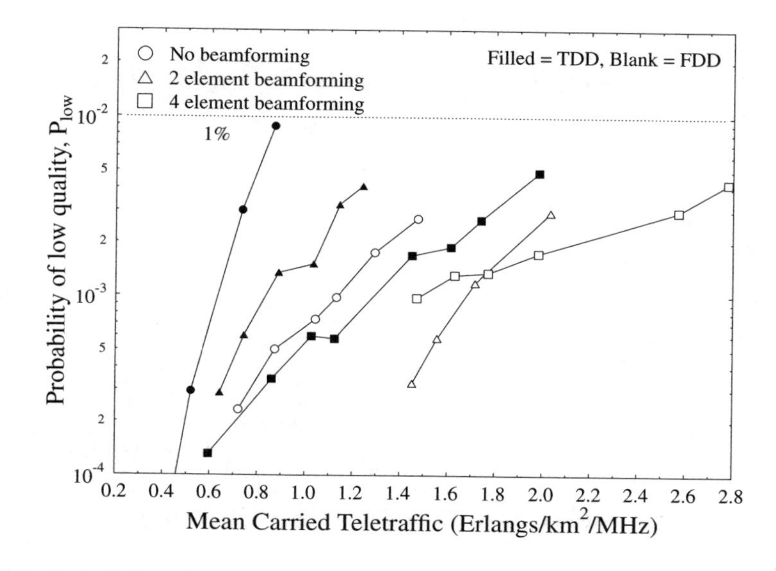

Figure 23.28: Probability of low quality access versus mean carried traffic of the UTRA-like TDD/CDMA-based cellular network both **with as well as without beamforming and without shadowing** for SF = 16.

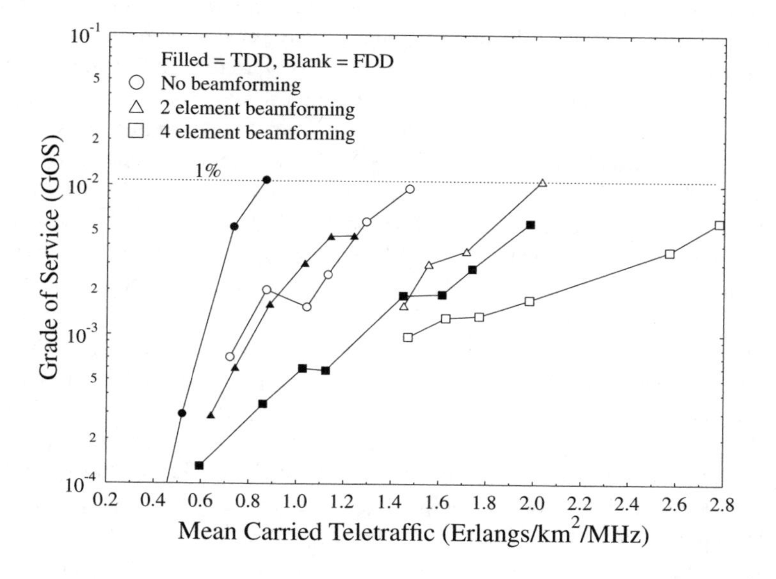

Figure 23.29: Grade-Of-Service (GOS) versus mean carried traffic of the UTRA-like TDD/CDMA-based cellular network both **with as well as without beamforming and without shadowing** for SF = 16.

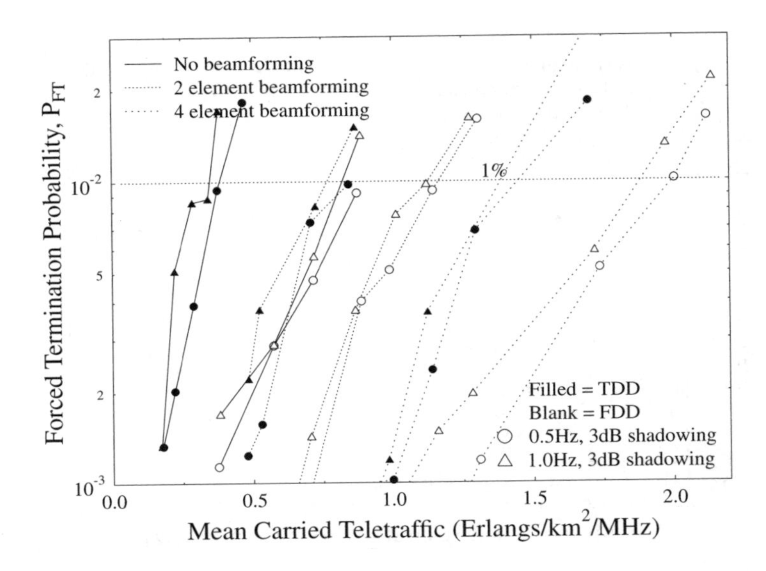

Figure 23.30: Call dropping probability versus mean carried traffic of the UTRA-like TDD/CDMA-based cellular network both **with as well as without beamforming and with shadowing** for SF = 16.

tion conditions. Simulations were conducted in conjunction with log-normal shadow fading having a standard deviation of 3 dB, experiencing maximum shadowing frequencies of both 0.5 Hz and 1.0 Hz. As expected, the network capacity was reduced at the higher shadow fading frequency. Without employing adaptive antenna arrays, the network supported just over 71 users and 62 users, when subjected to 0.5 Hz and 1.0 Hz frequency shadow fading, respectively. With the application of two-element adaptive antenna arrays, these capacities increased by 111% and 113%, to 151 users and 131 users, respectively. The employment of four-element adaptive antenna arrays led to a network capacity of 245 users at a 0.5 Hz shadowing frequency, and 234 users at a 1.0 Hz shadowing frequency. This corresponded to relative gains of 62% and 78% over the capacity provided with the aid of two-element adaptive antenna arrays.

The probability of low quality access performance is depicted in Figure 23.31. As expected, a given P_{low} value was associated with a higher traffic load, as the number of antenna elements increased. When the maximum shadow fading frequency was increased from 0.5 Hz to 1.0 Hz, P_{low} also increased. The probability of low quality seen in Figure 23.31 is similar in the scenarios employing adaptive antenna arrays in the UTRA TDD and FDD CDMA systems. It should be noted, however that the probability of low quality access always remained below the 1% constraint of the conservative scenario under the scenarios studied, and the call dropping probability was considerably reduced by the adaptive antenna arrays, as seen in Figure 23.32.

More explicitly, Figure 23.32 shows the significant reduction in the probability of a dropped call, achieved by employing adaptive antenna arrays in conjunction with adaptive

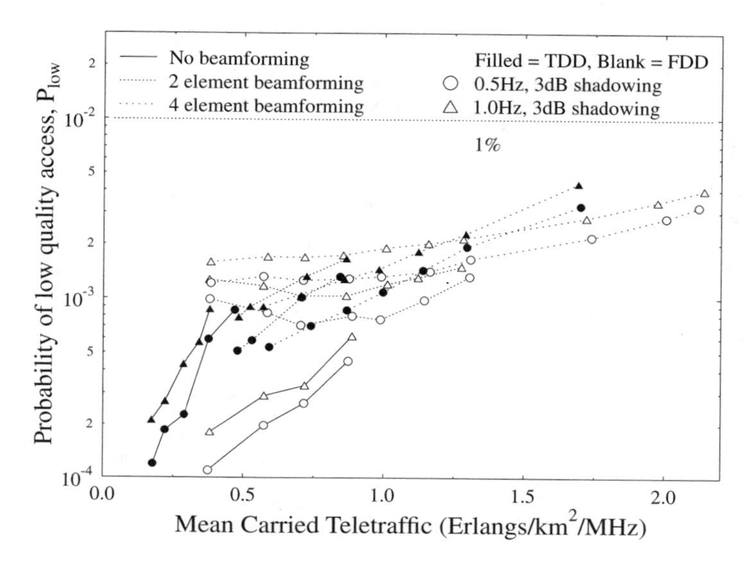

Figure 23.31: Probability of low quality access versus mean carried traffic of the UTRA-like TDD/CDMA-based cellular network both **with as well as without beamforming and with shadowing** for SF = 16.

modulation [4, 268] in a log-normal shadow faded environment. The figure demonstrates that, even with the aid of a two-element adaptive antenna array, a substantial call dropping probability reduction was achieved. The single-antenna based network was found to support 153 users, corresponding to a traffic load of 0.875 Erlang/km^2/MHz, when subjected to 0.5 Hz frequency shadow fading. The capacity of the single-antenna aided network was slightly reduced to 152 users, or 0.874 Erlang/km^2/MHz, upon increasing the maximum shadow fading frequency to 1.0 Hz. Upon employing two-element adaptive antenna arrays, the network capacity increased by 109% to 320 users, or to an equivalent traffic load of 1.834 Erlang/km^2/MHz, when subjected to 0.5 Hz frequency shadow fading. When the maximum shadow fading frequency was increased to 1.0 Hz, the number of users supported by the network was 307, or 1.82 Erlang/km^2/MHz, representing an increase of 102% in comparison to the network refraining from using adaptive antenna arrays. It is seen in Figure 23.32 that the forced termination probability of the UTRA-like TDD/CDMA scenarios is close to that of the FDD/CDMA scenarios, when employing adaptive antenna arrays in conjunction with adaptive modulation.

The probability of low quality outage, presented in Figure 23.33, did not benefit from the application of adaptive antenna arrays, in fact on the contrary. Furthermore, recall that Figure 23.31 depicted the probability of low quality outage without adaptive modulation, i.e. using fixed modulation, and upon comparing these results to those obtained in conjunction with adaptive modulation shown in Figure 23.33, the performance degradation owing to the employment of adaptive modulation can be explicitly seen. This is because the increase in the probability of low quality access can be attributed to the employment of less robust,

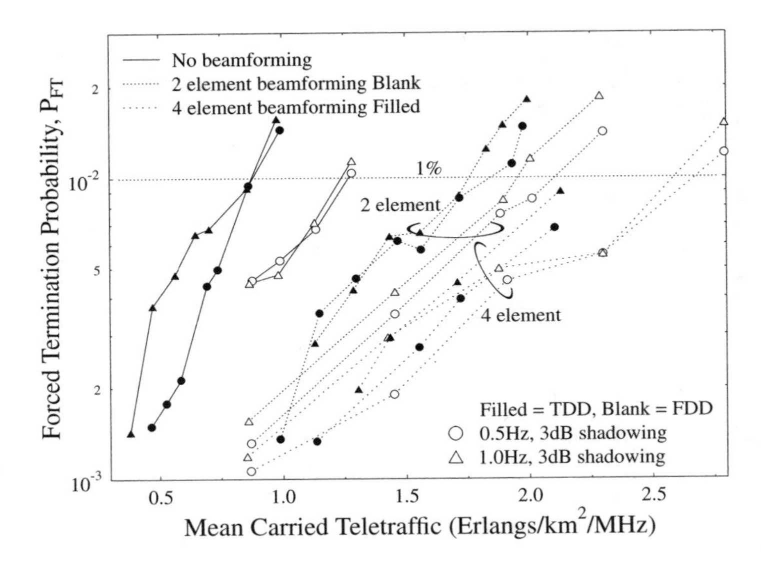

Figure 23.32: Call dropping probability versus mean carried traffic of the UTRA-like TDD/CDMA-based cellular network both **with and without beamforming in conjunction with AQAM as well as with shadowing having a standard deviation of 3 dB** for SF = 16.

but higher throughput, higher-order modulation modes invoked by the adaptive modulation scheme. Hence, under given propagation conditions and using the interference-resilient fixed 4-QAM modulation mode, as in Figure 23.31, a low quality outage may not occur. By contrast, when using adaptive modulation invoking a less resilient, but higher-throughput and higher-order modulation mode, the same propagation conditions may inflict a low quality outage.

23.4 Chapter Summary and Conclusion

In summary, in Section 23.2 we have shown that, when the number of active users in a DS-CDMA system is a time-varying random variable and when the conventional matched filter based receiver is employed, an adaptive rate transmission scheme using variable spreading factors (VSF) can be employed for compensating the effects of the time-varying MUI level experienced. The VSF-assisted adaptive rate transmission scheme is capable of significantly increasing the system's effective throughput. Specifically, our results show that the effective throughput may be increased by up to 40%, when compared to that of DS-CDMA systems using constant spreading factors. This increased throughput is achieved without wasting power, without imposing extra interference upon other users and without increasing the BER.

In Section 23.3 our discussions evolved further to consider an entire 49-cell IMT2000-like FDD and TDD network. The impact of AAAs upon the IMT2000 / UTRA network

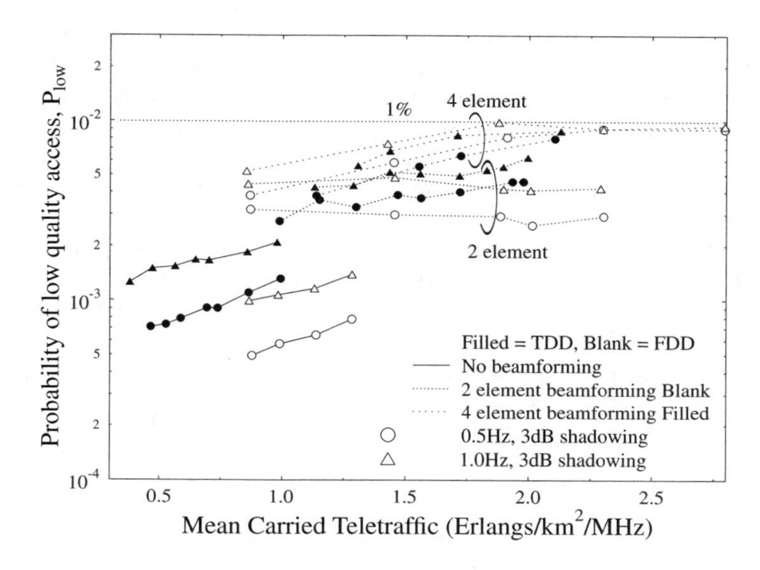

Figure 23.33: Probability of low quality access versus mean carried traffic of the UTRA TDD/CDMA-based cellular network both **with and without beamforming in conjunction with AQAM as well as with shadowing having a standard deviation of 3dB** for SF = 16.

Shadowing	Beamforming	Conservative scenario Users	Erlangs /km^2/MHz	Power (dBm) MS	Power (dBm) BS	Throughput (BPS) Up-link	Throughput (BPS) Down-link
0.5 Hz, 3 dB	No	223	1.27	3.25	4.95	2.86	2.95
0.5 Hz, 3 dB	2 elements	366	2.11	3.55	4.7	2.56	2.66
0.5 Hz, 3 dB	4 elements	476	2.68	3.4	5.0	2.35	2.72
1.0 Hz, 3 dB	No	218	1.24	3.3	4.95	2.87	2.96
1.0 Hz, 3 dB	2 elements	341	1.98	3.5	4.9	2.62	2.73
1.0 Hz, 3 dB	4 elements	460	2.59	3.5	4.95	2.4	2.8

Table 23.7: Maximum mean carried traffic and maximum number of mobile 0.users that can be supported by the network, whilst meeting the 0.conservative quality constraints. The carried traffic is 0.expressed in terms of normalized Erlangs (Erlang/km^2/MHz), 0.for the network described in Table 23.3 both 0.**with and without beamforming (using independent 0.up/down-link beamforming), in conjunction with shadow fading 0.having a standard deviation of 3 dB, whilst employing adaptive 0.modulation techniques** for SF = 16.

capacity was considered in both non-shadowed and log-normal shadow faded propagation environments. Considerable network capacity gains were achieved, employing both two and four element AAAs. This work was then extended by the application of AQAM techniques in conjunction with the previously studied AAAs in a log-normal shadow faded propagation environment, which elicited further significant network capacity gains. Explicitly, the employment of AAAs facilitated the doubling of the network's capacity in some of the investigated scenarios. Furthermore, invoking AQAM increased both the average throughput and the robustness of the network, since a sudden channel quality reduction did not result in dropping the call supported, it rather activated a lower-throughput modulation mode.

The network performance of the UTRA FDD and TDD systems was also compared. It was shown that the employment of adaptive arrays in conjunction with AQAM limited the detrimental effects of co-channel interference on the TDD system and resulted in performance improvements both in terms of the achievable call quality and the system's capacity. Our further research will quantify the design trade-offs associated with the joint employment of AQAM, space-time coding and MC-CDMA, as well as their impact on the network's capacity.

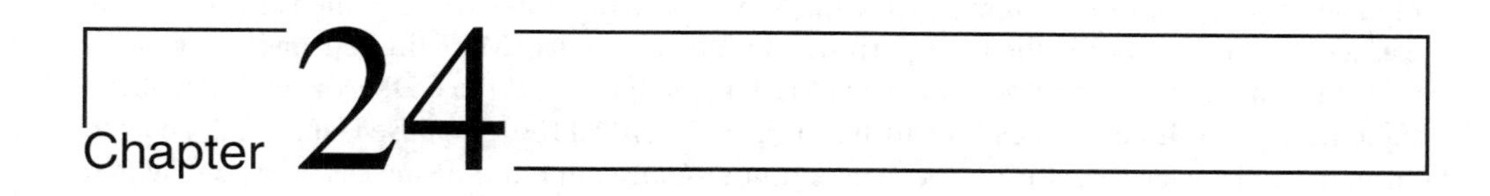

Book Summary and Conclusions

This concluding chapter summarizes the results that were presented in this monograph, and our suggestions for further work are outlined.

24.1 Part I: Multi-User Detection for Adaptive Single-Carrier CDMA

Part I of the monograph was concerned with the performance of multi-user receivers [127] in the context of adaptive rate CDMA transmission. Joint detection (JD) receivers constitute a class of sub-optimum multi-user receivers that were developed in order to mitigate the MAI-limited nature of CDMA as a multiple access scheme.

As an introduction, a basic exposure to CDMA transmission schemes was presented. The recovery of the desired user's signal in the presence of MAI was highlighted and it was shown that the matched filter or correlator catered for adequate signal recovery in a Gaussian channel. This was followed by a brief discussion on some examples of binary spreading sequences and the relevant properties of these sequences, such as their auto-correlation (ACL) and cross-correlation (CCL) values were examined. The theoretical BER performance of the CDMA user over the asynchronous Gaussian channel was presented, following the Gaussian approximation approach proposed by Pursley [103]. Over the asynchronous Gaussian channel where the received signals of the various users arrived at the receiver with different delays at the receiver, it was shown that the BER performance was dependent on the number of users, K, and on the spreading sequence length, Q. In the theoretical BER expression of Equation 2.72, the extra term of $(K-1)/3Q$ represented the MAI that resulted in a degraded BER performance. The theoretical BER performance was also derived over the synchronous Gaussian channel and it was shown that the MAI was directly related to the CCL between the spreading sequences. The theoretical BER performance of a synchronous CDMA system was compared to our simulation results, which exhibited a good match, as can be observed in Figure 2.14.

The performance of CDMA over multipath fading channels was investigated next and it was shown that conventional CDMA receivers, such as the matched filter and the RAKE receiver, were unable to provide good BER performance results. This was because these receivers employed only the knowledge of the desired user's spreading sequence and channel impulse response (CIR) estimates, while the MAI due to other users was treated as noise. This resulted in an error floor in the BER performance and created an MAI-limited environment. In order to remove this error floor, the available knowledge on *all* the CDMA users' spreading sequences and CIR estimates had to be utilized. Verdú [106] proposed and analyzed the optimum multi-user detector for the asynchronous Gaussian multiple access channels, which was the maximum likelihood sequence estimator. Following his seminal work, numerous sub-optimum CDMA receivers were proposed. A survey of the various CDMA receivers was conducted and loosely classified into multi-user and single-user types. The multi-user detector types included the adaptive linear receivers, the JD receivers and the interference cancellation (IC) receivers, along with the tree-search algorithms and the various iterative schemes. The single-user receivers could be divided into two types, the conventional matched filter or RAKE receiver and the single-user blind receivers.

In order to concentrate our investigations on a specific type of multi-user receiver rather than on a variety of receivers, the joint detector was chosen because of its attractive performance reported, for example, by Klein, Baier *et al.* [126–128] and since its block-detection oriented structure was suitable for combination with burst-by-burst adaptive rate schemes. Since the MAI exhibited similar characteristics to ISI, the single-user equalizers designed for wideband channels were extended to sub-optimum multi-user receivers that equalized both MAI and ISI. The single-user zero-forcing and minimum mean square error (MMSE) equalizers were briefly discussed in Section 3.1 and this led to the modifications and extensions required for the joint detection of CDMA users. Four different joint detection algorithms were introduced in Section 3.3, namely the ZF-BLE, the MMSE-BLE, ZF-BDFE and the MMSE-BDFE. For all four algorithms, a CCL matrix was constructed, which consisted of the CCL values between each received symbol and all of the other symbols transmitted by the same user and the other users. In the context of the zero-forcing algorithms this CCL matrix was inverted, in order to obtain the data estimates of all the users. This had the effect of removing all the interference from the data estimates, but unfortunately resulted in noise enhancement at the output of the joint detector. For the MMSE algorithms, this CCL matrix was biased with the covariance matrix of the transmitted symbols in Equation 3.123 before inversion was carried out. Here, not all of the interference was removed but the mean square error at the output of the joint detector was minimized which in most cases, resulted in a better BER performance. For the feedback algorithms of ZF-BDFE and MMSE-BDFE, the inversion algorithm was decomposed such that the data symbols that had already been estimated were fed back into the detector and the interference due to these symbols was cancelled from the remaining signal. This helped the detector to produce more reliable estimates of the forthcoming symbols.

The complexity of all the four JD algorithms was derived in terms of the addition and multiplication operations required to detect each data symbol, as presented in Section 3.4. The complexity expression was dependent on the number of symbols per detection block, KN, the spreading factor, Q, the delay spread, W, of the CIR, and the number of multipaths, L, in the wideband channel. In these complexity derivations, it was noted that the highest complexity was incurred by the construction of the CCL matrix. In order to reduce the complexity of the JD algorithms, some assumptions were stipulated, where the noise was assumed to be uncorrelated with a variance of σ^2 and the transmitted data symbols were also assumed to be uncorrelated and were normalized to a variance of unity. It was also observed that the CCL matrix was a sparse matrix and the sparsity was dependent on the delay spread of the wideband channel, where the lower the delay spread, the sparser the matrix. These properties were exploited in the matrix multiplication routines in order to reduce the complexity. For the matrix inversion, the reduced complexity Cholesky decomposition [122] was invoked. With the implementation of these complexity reduction routines, a drop in complexity was observed and the step that incurred the highest complexity was now the Cholesky decomposition routine. Taking an arbitrary CDMA system supporting $K = 10$ users, which transmitted $N = 20$ symbols per burst each with a spreading factor of $Q = 31$ over a COST 207 [105] Typical Urban channel of $L = 4$ paths and channel spread of $W = 10$ chips, it was found that the complexity of the simplified JD algorithms was reduced by two orders of magnitude. Further complexity reduction schemes for JD were proposed [208–210], but these remained beyond the scope of this monograph. Finally, it was observed that all four JD algorithms resulted in a similar complexity per detected symbol. The full and reduced complexities of all the four algorithms were tabulated in Table 3.1.

Experimental investigations were carried out, in order to ascertain the performance of the four JD algorithms in Section 3.6. Simulations were conducted over three types of COST 207 [105] channels, namely the Rural Area, Typical Urban and Bad Urban channels, which were shown in Figures 3.15, 3.18 and 3.20, respectively. It was found that as the number of paths increased and correspondingly the multipath diversity increased, the performance difference between the four algorithms became more apparent, as was observed in Figures 3.16, 3.19 and 3.21. The best performance was observed for the MMSE-BDFE, followed by either the MMSE-BLE or the ZF-BDFE and finally the ZF-BLE. Since the MMSE-BDFE provided the best BER performance and the complexity of all four algorithms was similar, the MMSE-BDFE was chosen as the multi-user receiver in subsequent investigations. In Section 3.6.3, the effect of multipath diversity on the BER performance was investigated, as can be observed in Figure 3.22. At the BER of 1%, the performance gain due to the Bad Urban channel was approximately 2.5 dB over the Typical Urban channel, and approximately 6 dB over the Rural Area channel. In conjunction with a higher order of multipath diversity, the MMSE-BDFE was capable of compensating for the fading of the mobile channel and hence it improved the BER performance.

In Section 3.6.4, the BER performance of the MMSE-BDFE versus the number of CDMA users was studied, as shown in Figure 3.23. We observe from the figure that the BER performance slowly degraded as the number of users, K, was increased and this characterized the "soft capacity" [127] property of CDMA systems. The effect of two different types of channel coding schemes in combination with joint detection was evaluated in Section 3.6.5. Binary BCH coding [64] and turbo convolutional coding [197] were combined with RAKE receivers and with the MMSE-BDFE, in order to compare their performances. As can be observed from Figure 3.24, the employment of channel coding and interleaving assisted in improving the BER performance, but did not eliminate the existing error floor. In Figures 3.25 and 3.26, the MMSE-BDFE was capable of providing more reliable data estimates to the channel decoder and thus helped improve the BER performance. The performance gain of channel coding over the uncoded case was higher in conjunction with the MMSE-BDFE than for the RAKE receiver. Finally, since the implementation of the JD algorithms required the knowledge of all the users' CIR estimates, the effect of channel estimation errors on the BER performance was investigated in Section 3.6.6. Following the approach adopted by Yoon, Kohno and Imai [141], the channel estimation errors for both the phase and amplitude were modelled as independent and identically distributed zero mean Gaussian random variables and these were used to perturb the perfect CIR estimates, which were then used in the construction of the CCL matrix. In Figure 3.27, it was observed that in order to maintain a loss of no more than 3 dB in terms of SNR performance at a target BER of 0.1%, the path amplitude and phase estimation errors must not exceed $\pm 10\%$ per path.

Due to the fading nature of the mobile channel, the instantaneous received signal quality also fluctuates. When the channel quality is low, the performance is degraded and if the channel quality remains inferior for a significant amount of time, burst errors will occur. In order to accommodate and in fact exploit the fluctuating quality of the channel, adaptive-rate transmission was introduced. When the instantaneous channel quality was high, the transmission rate was increased in order to increase the throughput. Conversely, when the channel conditions were hostile, the transmission rate was reduced and a more robust transmission mode was employed in order to maintain the required BER performance target. Two methods of adaptively varying the rate were considered, namely the AQAM/JD-CDMA system and the VSF/JD-CDMA scheme. The AQAM scheme was based on the seminal work by Steele and Webb [216] as well as on the work carried out by Sampei *et al.* [234], Goldsmith *et al.* [237], Torrance *et al.* [597] and Wong *et al.* [238]. AQAM is an adaptive-rate technique, whereby the data modulation mode is chosen according to some criterion related to the instantaneous channel quality. When the instantaneous channel quality is high, a higher-order modulation mode that transmits a higher number of bits per symbol is employed and when the channel quality is low, a more robust modulation mode is employed that transmits a lower number of bits per symbol. On the other hand, in VSF transmissions, the information rate is varied by adapting the spreading factor of the CDMA codes used, while keeping the chip rate constant. For both adaptive-rate schemes, a channel quality metric is required. In our investigations, the SINR at the output of the joint detector was chosen as this metric and an expression was derived for this SINR value in Section 4.3.2 for the JD algorithm in general and specifically for the MMSE-BDFE in Section 4.3.3. Following these derivations, a semi-analytical method of quantifying the BER performance of the MMSE-BDFE was presented, where the output SINR was utilized. The semi-analytical and simulation performance results were compared in Figure 4.7 and it was observed that the semi-analytical expression provided a reasonably accurate performance measure of the BER. This semi-analytical expression was extended to include the BER and the BPS throughput performance of the AQAM/JD-CDMA system in Section 4.3.4.

In Section 4.3.5, the performance of various AQAM/JD-CDMA systems was evaluated, where perfect channel estimation and perfect SINR estimation were assumed. Firstly, a two-user, twin-mode AQAM/JD-CDMA system was analyzed, where the modulation mode was switched between BPSK and 4-QAM. In Figure 4.11, the effect of different modem mode switching thresholds on the BER and BPS throughput performances was observed. Comparing the BER curves, it was revealed that as the switching threshold, t_1, was reduced from 10 dB to 4 dB, the average BER increased, since the 4-QAM mode was chosen for data modulation more often, thus increasing the number of errors. This was followed by an increase in the BPS throughput and the $t_1 = 4$ dB–system achieved the highest throughput, followed by the $t_1 = 8$ dB–system and the $t_1 = 10$ dB–system. Comparing all three performance curves, the

$t_1 = 8$ dB–system appeared to achieve the optimum performance having a BER below the target rate of 1% in the range of 6 to 18 dB, while the throughput approached the maximum of 2 BPS offered by a 4-QAM system at $E_s/N_0 = 14$ dB. The effect of increasing the number of users, K, on the same AQAM JD-CDMA system was also investigated. From Figure 4.13 it was concluded that increasing K did not significantly degrade the performance of the AQAM JD-CDMA system and the BER and BPS performances degraded only slightly when $K = 8$ users were supported compared to the $K = 2$–user system. The semi-analytical expression derived in Section 4.3.4 for the AQAM/JD-CDMA performance was compared with our simulation results in Figure 4.14. The semi-analytical and simulation results showed a good match and hence the semi-analytical expressions proved to provide a reasonable approximation of the AQAM JD-CDMA performance. The twin-mode AQAM/JD-CDMA scheme was then extended to a triple-mode scheme, where the modulation mode was chosen to be BPSK, 4-QAM or 16-QAM. The performance results for $K = 2$ and $K = 8$ users are depicted in Figures 4.15 and 4.17. Here again, similar performance characteristics were observed. As the switching thresholds were decreased, the BPS throughput improved at the expense of a degraded BER performance. This is because the higher throughput modulation modes were chosen with increasing probability, thus allowing more bits to be transmitted. By the same token, this caused the degradation in the BER performance due to the higher BERs inflicted by the use of higher modulation modes. However, it should be noted that the BER kept to the level of 1%, as targeted. Figure 4.18 shows our comparison between the numerical and simulation performance for the triple-mode AQAM/JD-CDMA system, where it can be seen that the numerical expressions provided a reasonably accurate estimation of the upper-bound performance of the AQAM/JD-CDMA scheme. The AQAM/JD-CDMA system was further extended to a five-mode scheme, which included BPSK, 4-QAM, 16-QAM and 64-QAM as the modulation modes, as well as the NOTX or blocking mode, where no information bits were transmitted. Instead, known bits that were required for SINR estimation were transmitted and discarded from the received composite signal at the receiver. The performance of this five-mode AQAM/JD-CDMA scheme for three CIRs of varying multipath diversity orders are presented in Figures 4.19, 4.20 and 4.21, namely for a seven-path Bad Urban channel, four-path Typical Urban channel and a two-path Rural Area channel, respectively. It was observed that for the target BER of 0.01%, the performance gain of the AQAM/JD-CDMA scheme over fixed-mode schemes was in the range of 1–2 dB for the Bad Urban channel, 4–6 dB for the Typical Urban channel and 11–14 dB for the least dispersive, low diversity Rural Area channel. Figure 4.23 shows the performance gain of the five-mode AQAM scheme over BPSK, 4-QAM and 16-QAM versus channel diversity order. It was revealed that as the diversity order decreased, the performance gain of the AQAM scheme increased. This was because in conjunction with a higher diversity order, the MMSE-BDFE was capable of utilizing this diversity in order to produce more reliable data estimates and hence to reduce the variation in output SINR. However, when the multipath diversity was low, the output SINR and hence the channel quality exhibited larger fluctuations, resulting in higher performance gains when the AQAM scheme was capable of exploiting this fluctuation. Since the performance of the AQAM/JD-CDMA system was dependent on the estimation of the SINR, the effect of latency in the SINR estimation was evaluated. The latency was quantified in terms of the number of TDMA slots that elapsed between the SINR estimation and the actual transmission of the burst. Therefore, a zero-slot lag indicated perfect zero-latency SINR estimation and as the SINR estimation lag increased, the accuracy of the SINR estimation also decreased. In Figures 4.27 and 4.30, it can be observed that as the latency increased, the BER performance degraded. This was because the decreasing accuracy in SINR estimation often caused an inappropriate modulation mode to be chosen. Thus a less robust modulation mode may have been opted for when the instantaneous channel quality was low, and while the channel quality was high, this was not always properly exploited in BPS terms. However, this performance degradation was shown to be a function of the fading rate, as revealed in Figures 4.29 and 4.32. The latency values were kept the same, but the Doppler frequency was lowered from 80 Hz to 26.7 Hz. In these figures, a latency of up to 8 slots, which is equivalent to one FMA1 frame, did not have a significant effect on the AQAM/JD-CDMA performance. This was because the fading rate was sufficiently slow, such that the SINR estimation was still reasonably accurate at the time of burst transmission despite the latency.

Our other adaptive-rate method, namely VSF/JD-CDMA, was investigated in Section 4.4. In this technique, the spreading factor, Q, was varied adaptively according to the quality of the channel. In general, when RAKE receivers or matched filters are used as CDMA receivers, the BER performance degrades as the spreading factor is decreased. Therefore, when the channel conditions were favourable, a lower value of Q was employed, thus enabling more bits to be transmitted within the same time period. When the instantaneous channel quality was low, a higher value of Q was chosen in order to maintain the required target BER performance. However, when JD was employed in the receiver, the BER performance remained the same for various spreading factor values, as shown in Figure 4.33(a). Therefore, in order to combine VSF transmission with JD-CDMA, the E_s/N_0 value was varied when Q was varied. This implied that a constant transmitted power, P, was maintained, since $P = E_s/Q$. When the channel conditions allowed for a low value of Q to be employed, the E_s/N_0 value was also lowered by the same factor, in order to maintain a constant transmitted power. Conversely, when Q was increased, the E_s/N_0 value was also increased by the same factor. Figures 4.34, 4.35 and 4.36 showed the BER and normalized throughput performance of the VSF/JD-CDMA system for the seven-path Bad Urban channel, four-path Typical Urban channel and a two-path Rural Area channel, respectively. In this VSF/JD-CDMA scheme, the spreading factor was switched between $Q_1 = 64$, $Q_2 = 32$ and $Q_3 = 16$. The throughput values of the VSF/JD-CDMA scheme were normalized to the minimum throughput of the fixed-mode $Q = 64$ scheme. In all three figures, it can be observed that the BER performance of the VSF/JD-CDMA scheme always matched the performance of the fixed-mode $Q = 64$ scheme and as the E_s/N_0 value increased, the performance gain of the VSF/JD-CDMA scheme also increased, especially for the less dispersive, lower diversity Typical Urban and Rural Area channels. As the performance gain increased, the normalized throughput of the VSF/JD-CDMA scheme also increased, eventually approaching the maximum of 4 at high E_s/N_0, while still maintaining a high SNR performance gain over the fixed-mode scheme exhibiting the lowest throughput. In Figure 4.37, the effect of increasing the number of users, K, on the VSF/JD-CDMA scheme was investigated, and it was concluded that there was only a minimal performance degradation, when the number of users was increased from $K = 1$ to $K = 8$.

In order to amalgamate the benefits of both AQAM and VSF, a combined AQAM and VSF JD-CDMA scheme was investigated in Section 4.6. The performance of the system is shown in Figure 4.39. Also shown for comparison are the triple-mode AQAM/JD-CDMA scheme and the triple-mode VSF/JD-CDMA scheme that provided the same minimum and maximum throughput values as the combined AQAM-VSF/JD-CDMA system. Considering the BER curves, we can see that the combined scheme outperformed both the triple-mode AQAM/JD-CDMA scheme and the VSF/JD-CDMA scheme. The target BER of 1% was achieved and as E_s/N_0 increased, the BER also gradually decreased. In throughput terms, the combined scheme also outperformed the other two individual schemes. Table 4.12 summarized our performance comparisons for the triple-mode AQAM, triple-mode VSF and the triple-mode combined AQAM-VSF/JD-CDMA schemes, where all three schemes were seen to achieve the target BER of 1% at $E_s/N_0 \leqslant 8$ dB. As the average E_s/N_0 increased, the combined AQAM-VSF/JD-CDMA scheme outperformed the purely VSF-based scheme in terms of throughput, while still achieving a BER of 0.001%, which was better than the 0.1% BER offered by the AQAM/JD-CDMA scheme.

Another class of multi-user receivers is constituted by the IC receivers, where the most common schemes are the (SIC) receiver [147] and the PIC receiver [138]. These IC receivers employ RAKE receivers in order to produce initial data estimates, which are subsequently used to cancel the MAI from the received composite signal so that more reliable data estimates can be obtained at the output of the cancellation stages. In comparison to JD receivers, IC algorithms have a complexity that increases only linearly with the number of users. SIC [147] receivers rank the users' signals according to a certain criterion, by estimating the received signal quality, for example, or the auto-correlation value for each user and the user receiving the highest quality signal is designated user 1, the next is user 2 and so on. This signal quality ranking is carried out, in order to determine the reliability of the initial data estimates for each user. After ranking, the received composite signal is processed by a matched filter or RAKE receiver, in order to obtain the initial data estimates of user 1. The received signal of this user is then reconstructed using the initial data estimates, the CIR and the spreading sequence of this user and subtracted from the composite received signal. The remaining signal is then processed by the matched filter bank or RAKE receiver of user 2, in order to obtain the data estimates for this user. Employing these data estimates, as well as the CIR and spreading sequence of this second user, the transmitted signal is reconstructed and subtracted again from the remaining composite multi-user signal that has already had the strongest user's signal cancelled from it. This is repeated, until the data estimates of all the users are obtained. The SIC receiver was discussed in Section 5.1, where its complexity was derived and its performance was evaluated. From Figures 5.2 and 5.3, it was observed that the BER performance improved as the spreading sequence length, Q, and hence the symbol period, QT_c, increased relative to the delay spread of the wideband channel. When the channel delay spread was negligible compared to the symbol period of QT_c, increasing the spreading sequence length further ceased to result in BER performance improvement.

Adaptive-rate SIC-CDMA transmission was investigated in Section 5.1.3. Similarly to the adaptive-rate JD-CDMA systems investigated earlier in Chapter 4, a measure of reception quality was required, in order to determine the information rate for the next transmission. This quality metric was chosen to be the SINR at the output of the SIC receiver and this SINR estimate was derived by modifying the approach presented by Patel and Holtzmann [147]. The BER and BPS throughput performance of AQAM SIC-CDMA was investigated in Section 5.1.4. Figure 5.7 showed our performance comparisons for a twin-mode AQAM scheme for different values of spreading factor, Q. It was concluded that the use of $Q = 64$ resulted in the best BER and BPS throughput performance of the three values of Q investigated and hence $Q = 64$ was employed in our subsequent investigations. Figures 5.8 and 5.9 depicted our performance results for a twin-mode AQAM/JD-CDMA system and triple-mode

AQAM/JD-CDMA system, respectively, for supporting $K = 2$ users. In the twin-mode AQAM scheme, the SIC receiver was capable of attaining performance gains by maintaining a lower BER performance than the fixed-mode 4-QAM scheme, while achieving a throughput of close to the maximum of 4 BPS offered by the fixed-mode 4-QAM scheme. However, in the triple-mode AQAM/JD-CDMA system, the inclusion of 16-QAM did not result in improved throughput performance, even though the target BER of 1% was maintained. This was because the BER of 16-QAM had an irreducible error floor which was higher than the target BER of 1%. Thus the 16-QAM mode could not be employed most of the time and there was no visible gain in the throughput over the twin-mode scheme. The performance of the twin-mode AQAM/JD-CDMA scheme was then evaluated for the four-path Typical Urban channel and two-path Rural Area channel in Figures 5.10 and 5.11, respectively. Similarly to the AQAM/JD-CDMA performance, as the number of paths decreased, the performance gain improved. As the dispersive channel spread decreased relative to the symbol period, the RAKE receiver was capable of producing reasonably reliable estimates for interference cancellation, thus improving the BER performance. The effect of increasing the number of users, K, from 2 to 8 on the performance of the AQAM SIC-CDMA system is depicted in Figure 5.12. Although both systems were capable of achieving the same BER performance, the throughput of the $K = 8$–user system was almost constant at 1 BPS. This is due to the increase in MAI, resulting in the output SINR values being lower than the required switching threshold and hence the 4-QAM mode was rarely used in the $K = 8$–user system.

The triple-mode VSF/SIC-CDMA system was analyzed and the performance results were presented in Figures 5.13 and 5.14 for $K = 2$ and $K = 8$ users. The spreading factor was switched between $Q_1 = 64$, $Q_2 = 32$ and $Q_3 = 16$. Previously, for the VSF JD-CDMA schemes, the transmission power was kept constant regardless of the spreading factor chosen for transmission, since the BER performance of the joint detector was not affected by the spreading factor. However, for the VSF SIC-CDMA schemes, the transmitter power was increased when a lower spreading factor was used, and decreased when a higher spreading factor was employed, in order to keep the E_s/N_0 value constant. This was because the BER performance of the SIC receiver was dependent on the spreading factor and the E_s/N_0 value had to be kept constant in order to exploit the potential available from the VSF scheme. From Figures 5.13 and 5.14, we observed that although the $K = 2$–user system exhibited performance gains over the fixed-rate schemes, this performance gain was eroded when K was increased to eight due to the increase in MAI.

PIC algorithms usually consist of multiple stages of interference cancellation. In each cancellation stage, the signal of each user is reconstructed using the data estimates from the previous cancellation stage. Then, for each user, the reconstructed signals of all the other users are subtracted from the received composite signal, and the resulting signal is processed through a matched filter or RAKE receiver, in order to obtain a new set of data estimates for this user. This reconstruction, cancellation and re-estimation stage is repeated as many times, as required by the system. The PIC receiver was discussed in Section 5.2, where its complexity was derived and its performance was evaluated in the context of fixed-rate schemes. From Figure 5.16, the performance improvement due to an increase in the number of cancellation stages was observed for a $K = 8$–user system transmitting over the COST 207 seven-path Bad Urban channel [105]. The BER performance improved, as the number of stages was increased from one to two, but there was no further improvement as the number of stages was increased to three and four. In subsequent investigations the number of cancellation stages was fixed to two in order to reduce the complexity.

In order to combine the employment of the PIC receiver with adaptive-rate schemes, the output SINR of the PIC receiver was derived and used as the criterion for adapting the transmission rate in Section 5.2.3. The performance of our twin-mode AQAM/PIC-CDMA scheme was compared for two different thresholds in Figure 5.18. Also shown for comparison are the BER curves for fixed-mode BPSK and 4-QAM. Again, it was observed that increasing the switching threshold resulted in an improvement in BER performance, but this was accompanied by a drop in BER throughput. For $t_1 = 20$ dB, the BER achieved at $E_s/N_0 = 14$ dB was similar to the BER of fixed-mode BPSK, but the BPS throughput was approximately 1.8 BPS, which was higher than the 1 BPS offered by fixed-mode BPSK. When the AQAM scheme was extended to the triple-mode AQAM/PIC-CDMA scheme by including 16-QAM as one of the modulation mode choices, the BER performance suffered from an error floor, as was shown in Figure 5.20. This was because the reduced distance between constellation points in the 16-QAM mode led to inaccuracy in data estimation, thus leading to error propagation in the cancellation stages.

The performance of the three types of multi-user receivers discussed previously, namely the MMSE-BDFE, the SIC receiver and the PIC receiver, in conjunction with AQAM schemes was compared in Section 5.3. Figure 5.22 compared the performance of the JD, the SIC receiver and the PIC receiver for our twin-mode, two-user AQAM-CDMA system. Here, the BER performance of all three receivers was kept as similar as possible and the performance comparison was evaluated on the basis of the BPS throughput. From Figure 5.22 it was observed that the SIC and PIC receivers were capable of achieving a higher BPS throughput than the JD at low E_s/N_0 values and matched the JD performance at high E_s/N_0 values. The performance comparison for twin-mode AQAM transmission was extended to systems supporting $K = 8$ users and the results were presented in Figure 5.23. In this case, the BPS throughput of the JD was the highest, where approximately 1.9 BPS was achieved at $E_s/N_0 = 14$ dB. The PIC receiver outperformed the SIC receiver in BPS throughput terms, where the BPS throughput of the PIC receiver was approximately 1.55 BPS at $E_s/N_0 = 14$ dB compared to the approximately 1.02 BPS achieved by the SIC receiver for the same E_s/N_0 value. The two IC receivers suffered from MAI and were unable to match the performance of the JD receiver. The PIC receiver outperformed the SIC receiver, since two parallel cancellation stages were implemented and the PIC receiver was capable of cancelling out more interference than the SIC receiver. The performance comparisons between the three multi-user receiver types were extended to triple-mode AQAM systems that had a system load of $K = 8$ users, as portrayed in Figure 5.24. For these systems, we observed from the results that both the PIC and SIC receivers were unable to match the BPS performance of the JD, when both multi-user receivers achieved similar BERs. This was because the increase in the number of users increased the MAI, thus degrading the ability of the RAKE receivers to provide reliable data estimates for interference cancellation. The twin-mode AQAM scheme comparisons between the three receivers were also extended to the Typical Urban and Rural Area channels, as presented in Figures 5.25 and 5.26. The BPS throughput of the JD was the highest of all three receivers, followed by the PIC receiver and lastly, the SIC receiver. Figure 5.27 showed the BER and throughput comparisons for the adaptive VSF schemes using 4-QAM, where the throughput performance values were normalized to the minimum throughput, thus giving the minimum and maximum normalized throughput values of 1 and 4, respectively. At low E_s/N_0 values, the PIC and SIC receivers outperformed the JD receiver in throughput terms. However, as E_s/N_0 increased, the JD gradually outperformed the two cancellation receivers in BER and BPS throughput performance terms, with a normalized throughput of 3.5 compared to 2.7 and 2 for the PIC and SIC receivers, respectively, at $E_s/N_0 = 16$ dB. Tables 5.13, 5.14 and 5.15 summarized our performance comparisons for all three multi-user detectors in terms of E_s/N_0 required to achieve a target BER of 1% or less; the normalized throughput performance; and the complexity in terms of the number of operations required per detected symbol.

In Chapter 6 a powerful PIC system was investigated further, which benefited from the employment of various channel codecs, such as convolutional coding, as well as Trellis Coded Modulation (TCM) and Turbo Trellis Coded Modulation (TTCM). The latter two schemes achieved an enhanced system performance without any bandwidth expansion. Additionally, the principle of turbo equalization was invoked, where the receiver jointly carried out soft interference cancellation and channel decoding. The attraction of this combined operation is that the channel decoder refrains from making premature "hard" decisions concerning the transmitted data. Instead, it provides soft information for the PIC scheme and after a few iterations a substantially improved performance may be achieved, which may approach the single-user bound at the cost of a moderate implementational complexity.

The PIC-based iterative multi-user detector's complexity per decoded user bit is independent of the number of users, which is an attractive property. Figures 6.29 and 6.28 showed the coding gain versus complexity of all these channel coded schemes. From these figures we observed that the iterative MUD is capable of approaching the single-user bound after two iterations when QPSK modulation was employed for transmission over an AWGN channel. Figure 6.30 shows the coding gain versus complexity of all these schemes when communicating over a two-path uncorrelated Rayleigh fading channel. From this figure we can observe that the performance of the iterative PIC scheme converges slower when communicating over a Rayleigh fading channel than over an AWGN channel. More specifically, in this scenario the system required about three or four iterations for approaching the single user bound at a BER of 10^{-3}. Figures 6.31 and 6.32 showed the coding gain versus complexity of TCM and TTCM when 8PSK and 16QAM are employed. Form these figures we infered that the iterative PIC scheme converges slower in the asynchronous environment than in the synchronous environment, since it requires four iterations for approaching the single-user bound instead of two or three iterations.

In Section 3.6.6, the impact of channel estimation errors on the performance of the JD-CDMA receiver was analyzed and it was observed that, in order to maintain a loss of no more than 3 dB in terms of SNR performance at a target BER of 0.1%, the path amplitude and phase estimation errors must not exceed ±10% per path. Following this discovery, a joint data and CIR estimation receiver for CDMA was proposed in Chapter 7, which combined PSP [172, 173] algorithms with adaptive RLS CIR estimators [125]. The approach of Xie *et al.* [171, 175] for asynchronous, static, narrowband channels, was modified for Rayleigh-faded, multipath channels in Sections 7.2, 7.3 and 7.4. The performance of PSP-CDMA receiver supporting $K = 2$ users was investigated for chip-spaced, two-path channels, and the four-path Typical Urban channel of Figure 3.18. It was concluded that the PSP receiver performed reliably for static channels but resulted in an error floor when fading was introduced into the channels, as shown in Figure 7.7. In order to remove the error floor, turbo convolutional coding [79, 197] was employed and the sub-optimum but reduced complexity Max-Log-MAP decoder [266, 267] was utilized for turbo decoding. The blind PSP receiver was modified to generate soft reliability measurements for individual turbo decoders of each of the users. It was shown that the error floor in the BER curve for the PSP-CDMA receiver was eliminated when turbo coding was employed in conjunction with large turbo and channel interleaver sizes, as can be observed in Figures 7.9, 7.10, 7.11 and 7.12.

In Chapter 8 we investigated the performance of STS-assisted W-CDMA systems, when multipath Nakagami-m fading, multi-user interference and background noise induced impairments were considered. Our analysis and numerical results demonstrated that the achievable transmit diversity gain formulated in Equation (8.59) was independent of the frequency-selective diversity gain and that both the transmit diversity and the frequency-selective diversity have the same order of importance. Furthermore, both the transmit diversity and the frequency-selective diversity have a similar influence on the BER performance of the W-CDMA systems considered. Since W-CDMA signals typically experience high dispersion and hence high-dynamic frequency-selective fading in both urban and suburban areas, determining the number of transmission antennas on the premise of assuming a low number of resolvable paths would result in wasteful over-engineering of the system. Based on the above scenarios, we proposed an adaptive STS transmission scheme, which adapts its STS configuration using Equations (8.61) - (8.63) according to the frequency-selectivity information fed back from the mobile receivers. The numerical results show that by efficiently exploiting the channel's frequency-selectivity, the proposed adaptive STS scheme is capable of significantly improving the throughput of W-CDMA systems. For W-CDMA systems transmitting at a data rate of $3R_b$ instead of R_b, only an extra of 0.4dB and 1.2dB transmitted power is required in the urban and suburban areas considered, respectively. Our future work will endeavor to contrive similar MC DS-CDMA schemes.

24.2 Part II: Genetic Algorithm Assisted Multi-user Detection

Part II of the book was concerned with the application of GAs in CDMA multi-user detection, in order to mitigate the complexity problem that arises when employing an optimum multi-user detector.

A basic overview of GAs employed as optimization tools was presented in Chapter 9. The terms and definitions that are associated with GAs were introduced. In order to justify the employment of GAs, an example was provided, which showed how a GA performed a search for the optimum solution. From the example, we observed that the GA identifies high-fitness individuals and exploits similarities among the individuals. Based on this concept, the notion of schema was introduced, which can be used in predicting the behaviour of GAs over the course of their evolution. This led to the derivation of the schema theorem, as formulated by Holland [6]. Since its introduction in the 1970s, numerous variants of GAs were developed, each of which was found to be suited for solving specific problems. Hence the general consensus is that GAs should not be treated as off-the-shelf optimization algorithms that are expected to solve every problem efficiently. Some of the more popular operations and strategies that constituted a GA were highlighted, which were subsequently investigated, in order to determine their usefulness in the context of our multi-user detection problem.

We commenced our study of GA-assisted CDMA multi-user detectors by first considering a symbol-synchronous CDMA system supporting K users over an AWGN channel as well as a non-dispersive Rayleigh fading channel. Having established the mathematical notations that were adopted in Part II, we proceeded to derive the correlation metric of Equation (10.23) as formulated by Verdu [106] for the optimum multi-user detector. It can be seen that for attaining the optimum BEP performance, every possible K-bit combination must be tested, in order to find the combination that maximizes the correlation metric. Hence the optimum multi-user detector has a computational complexity that is exponentially proportional to the number of users to be detected. Thus its implementation becomes impractical, when there is a high number of users. Hence the GA-assisted multi-user detector was proposed, in order to circumvent this complexity problem. According to our results obtained in Section 10.5, we demonstrated that the type of GA operations and strategies invoked can have a significant impact on the convergence rate and hence also on the resulting BEP and on the complexity of the detector. Upon comparing our proposed schemes to other similar GA-assisted multi-user detectors proposed for example by Juntti *et al.* [311], Wang *et al.* [312] and Ergün *et al.* [313, 314], we can see the reasons why they concluded that GAs are not particularly promising for implementation in CDMA multi-user detectors. However, we made no claims concerning the optimality of our adopted GA configuration, which was invoked in our GA-assisted multi-user detector, since we only investigated a fraction of the potentially suitable GA implementations. Based on our adopted GA configuration characterized in Table 10.9, we can see that the GA-assisted multi-user detector is capable of achieving a near-optimum BEP performance over both AWGN channels as well as over non-dispersive Rayleigh fading channels with the assistance of perfect CIR coefficient estimation. This can be achieved at a significantly lower computational complexity than that required by the optimum multi-user detector.

We demonstrated that the GA-assisted multi-user detector is capable of achieving the near-optimum BEP performance up to a certain channel SNR value with the aid of the perfect knowledge of the users' CIR coefficients. In reality, these CIR coefficients must be estimated at the receiver. Conventional CIR estimation schemes require the transmission of known pilot symbols from each user, in order to assist the CIR estimator at the receiver in acquiring the users' CIR coefficients. However, pilot-symbol-based estimation will reduce the throughput and bandwidth efficiency of the system. Furthermore, a training period is required for adaptive filter-based CIR estimation, such as the Kalman filter [7], in order to acquire and update the filter tap weights. We noticed that the correlation metric of Equation (10.23) requires that the users' CIR coefficients be known. However, by treating these CIR coefficients as unknown variables, we can employ GAs also for jointly estimating the CIR coefficients as well as the transmitted bits simultaneously for all users by optimizing the correlation metric given by Equation (11.6). In the context of joint CIR estimation and symbol detection solely by the GAs, the search space became continuous, having an infinite number of possible points, simply because the fading attenuation and phase trajectories are continuous. Hence, the GA configuration that was adopted previously had to be modified. Specifically, the mutation operation and the mating pool size had to be adapted since the number of quantities to be determined was doubled. Through simulation, we determined the value of the maximum mutation size λ_{max} as well as the mating pool size T that were capable of offering an acceptable CIR estimation performance, having an MSE as low as 0.001 in conjunction with known data bits. Having determined the GA parameters, the BEP performance of the joint GA-assisted multi-user CIR estimator and symbol detector was evaluated by simulations. It was shown that the BEP performance was limited due to the imperfect CIR estimation. However, the same phenomenon is also experienced in a single-user scenario employing a matched filter having the same channel estimation MSE. Since the CIR estimation can be conducted without explicit training sequences or decision feedback, our proposed detector is capable of offering a potentially higher throughput and a shorter detection delay, than that of explicitly trained CDMA multi-user detectors.

It is well known that diversity techniques can be used for mitigating the hostile effects of the transmission channel, in order to attain a BEP performance improvement. A symbol-synchronous CDMA system in conjunction with multiple receiving antennas was investigated in Chapter 12. By assuming that the antennas are placed at a distance higher than half the signals' wavelength, every user's signal received at each antenna is uncorrelated, which gives rise to an independent correlation metric at each antenna. From an optimization point of view, the associated multiple correlation metrics resulted in a decision conflict, since a particular bit sequence may optimise one correlation metric but not the others. GAs can resolve this conflict by invoking the Pareto optimality approach, in order to assist the search for the optimum solution. The Pareto optimality approach is accomplished by identifying the non-dominated individuals in a population by considering the figure of merit from each antenna independently. The corresponding individuals are then placed in the mating pool. This approach is compared to the traditional way of selecting individuals for the mating pool, which contains all non-identical individuals. Simulation results have shown that the Pareto optimality approach performed consistently better in terms of a lower BEP than the traditional approach. Furthermore, a substantial performance gain was achieved, when antenna diversity was invoked compared to no diversity, without incurring additional computational complexity at the multi-user detector.

Finally, we applied the GA-assisted multi-user detector in an asynchronous CDMA system in Chapter 13. In order to reduce the computational complexity involved in detecting the entire bit sequence of all the users, a truncated observation window was invoked, such that the window encompasses at most one complete bit duration of each user. The EEBs were then tentatively detected according to two different strategies. Firstly, the EEBS can be tentatively estimated based on the hard decisions of the matched filter outputs. The GA-assisted multi-user detector then proceeded to search for the desired bits sequence that optimized the correlation metric. Since the estimated EEBs were corrupted by the interference inflicted by other users, the EBEP was high and this limited the overall performance of the GA-assisted multi-user detector. In our second approach, the EEBs and the desired bits were estimated by the GA-assisted multi-user detector simultaneously. As a result, the EEBs benefited from a lower bit error probability and the overall performance of the detector approached the optimum performance.

24.3 Part III: M-ary Single-Carrier CDMA

in Chapter 15 M-ary orthogonal modulation techniques applicable to CDMA were studied. The simulations and numerical results of this chapter demonstrated that for a given number of chips per symbol, increasing M does not significantly improve the BER performance of the system, as illustrated in Figures 15.8 and 15.9, unless the number of users is quite low. By contrast, the bandwidth efficiency of $\eta = \frac{K \log_2 M}{N_c}$ is higher for higher values of M, as shown in Table 15.3. However, for a given processing gain, increasing M gives a better BER performance, regardless of the user capacity, as shown in Figures 15.10 and 15.11. Furthermore, the overall bandwidth efficiency of $\eta = \frac{K \log_2 M}{N_c}$ is also higher for higher values of M due to the fact that the system can accommodate more users and still achieve the required BER, as illustrated in Table 15.4.

Despite its advantages M-ary orthogonal modulation has rarely been employed in practical systems because of the potentially high complexity of the receiver, which requires 2^M number of correlators and hence increases exponentially with M, as becomes explicit in Figure 15.7. However, the employment of a novel modulation technique based on the residue number system (RNS) can be used for overcoming this impediment, as was shown in Chapters 17 and 18. It was demonstrated that the inherent properties of the RNS can be exploited for providing error correction and/or detection capabilities during the detection of the CDMA signals.

In Chapter 16 the performance of RS-coded DS-CDMA systems using 'errors-and-erasures' decoding was investigated and compared to that using 'error-correction-only' decoding over dispersive multipath Rayleigh fading channels, when non-coherent M-ary orthogonal modulation and diversity reception using EGC or SC schemes were considered. Viterbi's RTT technique has been studied, in order to quantify the reliability of the received RS coded symbols. The symbols having a low-confidence were erased before RS 'errors-and-erasures' decoding. The PDFs associated with the RTT under the hypotheses of correct detection and erroneous detection of the M-ary signals were derived and the optimum thresholds were determined. Our numerical results showed that by using 'errors-and-erasures' decoding associated with the RTT, RS codes of a given code rate can achieve a higher coding gain than upon using 'error-correction-only' decoding without erasure information. DS-CDMA systems using the proposed 'errors-and-erasures' decoding could also support more simultaneous users than without erasure information. Moreover, the numerical results showed that the optimum threshold for the RTT over multipath Rayleigh fading channels was around [1.5, 2] for the practical range of average SNR per bit values and for a given number of simultaneous users.

In Chapter 17 we discussed the mathematical models of RNSs and RRNSs as well as the associated principle of error-detection and error-correction in RRNSs by using a range of examples. Moreover, we summarized the basic coding theory of RRNS codes, in order to support our forthcoming deliberations. A RRNS codeword does not change its error-detection and error-correction characteristics after arithmetic operations such as addition, subtraction and multiplication, while conventional linear codes such as Hamming codes, BCH codes and convolutional codes, usually cannot provide their error-control properties after multiplyng codewords with each other. Hence, in conventional communication system design error protection of signals in signal processing and signal transmission procedures are treated separately. However, in most situations in conjunction with the RRNS representation introduced before signal processing, the RRNS can be used not only for the protection of the signals while they are being processed in the transceivers, but also for enhancing the system's error resilience over the communication channels. From this point of view, we can argue that, to some extent communication systems might be simplified by unifying the whole encoding and decoding procedure across global communication systems.

In digital communications, upon exploiting the inherent properties of the RNS arithmetic, an RRNS can be designed for improving the system's bit error rate (BER) performance. In order to fulfil this potential, an RNST unit, which implements the binary-to-residue transformation and an IRNST unit, which carries out the residue-to-binary transformation have to be incorporated in the transmitter and the receiver of the system, respectively.

In Chapter 18 we presented a novel parallel DS-CDMA system that uses a set of orthogonal signals and associated correlators. The output of the correlators was processed according to the principles of the RNS arithmetic algorithms. This technique is amenable to the transmission of signals conveying a high number of bits per symbol, while using a relatively low number of orthogonal signals, and therefore it is attractive in the context of high-bit-rate communications. We discussed the characteristics of the RNS arithmetic and demonstrated how these characteristics affect the communications performance in AWGN using coherent or noncoherent receivers and in dispersive multipath Rayleigh fading channels using MRC, EGC or SC schemes. Based on the independence and on the lack of weighted significance of the residue digits of a RNS, novel receivers were proposed for receiving RNS arithmetic based DS-CDMA signals. In this novel receiver a RRNS code can be decoded by first discarding a number of the lowest-reliability residue digits according to the so-called ratio statistic test, which is based on the ratio of the (first) maximum to the 'second maximum' of the inputs of the RRNS decoder, and then decoding the reduced and simultaneously simplified RRNS code. Concatenated coding techniques were discussed and evaluated in a dispersive multipath fading scenario, and the concatenated coding performance in conjunction with an RS code as the outer code and the RRNS code as the inner code has been evaluated both analytically and numerically. The results throughout the sections associated with RRNS decoding and concatenated RS-RRNS decoding demonstrated that using RRNS decoding with the lowest-reliability inputs of the RRNS decoder being dropped is a high-efficiency, low-complexity decoding technique.

Following a brief classification of the most popular MC-CDMA systems in Chapter 19 we introduced slow frequency hopping for improving the system's performance and flexibility. The proposed SFH/MC DS-CDMA system is capable of efficiently amalgamating the techniques of slow FH, OFDM and DS-CDMA. Nonlinear constant-weight codes were introduced, in order to control the associated FH patterns, and hence to activate a number of subcarriers, in order to support multi-rate services. Hard-detection or blind joint soft-detection were proposed for signal detection. The hard-detection exploits the explicit knowledge of of the FH patter employed, while the soft-detection assumes no apriori knowledge of the FH patterns used. The properties of the constant-weight FH codes were investigated in conjunction with specific system parameters, and the performance of the proposed SFH/MC DS-CDMA systems using either hard-detection or blind joint soft-detection has been evaluated, under the conditions of supporting constant-rate or multi-rate services. From the results we concluded that the blind joint soft-detection techniques are capable of detecting the transmitted information and simultaneously acquiring the FH patterns.

24.4 Part IV: Multi-Carrier CDMA

In Chapter 20 following the brief motivational notes of Section 20.1 in Section 20.2 we compared various broadband CDMA schemes supporting ubiquitous communications and maintaining the target integrity, regardless of the dispersion of the specific propagation environments encountered. Specifically, multicarrier DS-CDMA (MC DS-CDMA) was proposed for satisfying the various requirements of broadband systems and for supporting ubiquitous communications, with the aid of STS-based transmit diversity. We discussed the design of the various system parameters rendering the STS-assisted broadband MC DS-CDMA scheme capable of supporting ubiquitous communications without compromising the achievable BER. We also considered the user capacity extension achievable by STS-assisted broadband MC DS-CDMA with the aid of TF-domain spreading. Our studies suggest that broadband MC DS-CDMA using STS constitutes a promising multiple-access scheme, which is capable of avoiding the various design limitations that are unavoidable, when using single-carrier DS-CDMA or MC-CDMA.

In Section 20.3 we elaborated further and based on the typical delay-spread statistics of the time-varying wireless channels, such as those encountered in indoor, open rural, suburban and urban areas, we demonstrated how the chip duration of the spreading codes can be optimized. More specifically, the effect of the number of data bits involved in the S-P conversion stage of Figure 20.4 has been investigated in the context of frequency-selective Rayleigh fading channels. Furthermore, we considered the potential capacity extension achievable by STS-assisted broadband MC DS-CDMA with the aid of TF-domain spreading. The corresponding BER performance has been investigated in the context of the correlation based single-user detector and the decorrelating multi-user detector for transmissions over frequency-selective Rayleigh fading channels. Our studies suggest that broadband MC DS-CDMA using STS constitutes a promising multiple-access option, which is capable of avoiding various design limitations that are unavoidable, when using single-carrier DS-CDMA or MC-CDMA.

24.5 Part IV: Standards and Networking

In Chapter 22 we presented an overview of the terrestrial radio transmission technology of 3G mobile radio systems proposed by ETSI, ARIB, and TIA. All three proposed systems are based on Wideband-CDMA. Despite the call for a common global standard, there are some differences in the proposed technologies, notably, the chip rates and inter-cell operation. These differences are partly due to the existing 2G infrastructure already in use all over the world, and are specifically due to the heritage of the GSM and the IS-95 systems. Huge capital has been invested in these current 2G mobile radio systems. Therefore, the respective regional standard bodies endeavoured to ensure that the 3G systems are compatible with the 2G systems. Because of the diversified nature of these 2G mobile radio systems, it is not an easy task to reach a common 3G standard that can maintain perfect backwards compatibility.

In Chapter 23 we demonstrated that when the number of active users in a DS-CDMA system is a time-varying random variable and when the conventional matched filter based receiver is employed, an adaptive rate transmission scheme using variable spreading factors (VSF) can be employed for compensating the effects of the time-varying MUI level experienced. The VSF-assisted adaptive rate transmission scheme is capable of significantly increasing the system's effective throughput. Specifically, our results show that the effective throughput may be increased by up to 40%, when compared to that of DS-CDMA systems using constant spreading factors. This increased throughput is achieved without wasting power, without imposing extra interference upon other users and without increasing the BER.

Appendix **A**

Appendix to Chapter 15

A.1 Derivation of the correlator output in Equations 15.27 and 15.28

A.1.1 In-phase channel

The output of the mth correlator of the l_pth branch was given by Equation 15.26, which is repeated here for convenience :

$$Z_I^{(i)}(m, l_p) = \frac{1}{\sqrt{T_s}} \int_{\tau_{l_p}^{(i)}}^{T_s + \tau_{l_p}^{(i)}} d_I^{(i),l_p}(t) V_m^{(i)}(t - \tau_{l_p}^{(i)}) dt. \tag{A.1}$$

Substituting $d_I^{(i),l_p}(t)$ given by Equation 15.24 into Equation A.1, we obtain :

$$\begin{aligned}
Z_I^{(i)}(m,l_p) &= \int_{\tau_{l_p}^{(i)}}^{T_s+\tau_{l_p}^{(i)}} \sqrt{\frac{P_s}{2T_s}} \left|\alpha_{l_p}^{(i)}\right| V_\mu^{(i)}(t-\tau_{l_p}^{(i)}) V_m^{(i)}(t-\tau_{l_p}^{(i)}) \\
&\times a^{(i,i)}(\tau_{l_p}^{(i)},\tau_{l_p}^{(i)}) \cos\theta_{l_p}^{(i)} dt \\
&+ \sum_{\substack{l=1\\ l\neq l_p}}^{L_P} \sqrt{\frac{P_s}{2T_s}} \int_{\tau_{l_p}^{(i)}}^{T_s+\tau_{l_p}^{(i)}} \alpha_l^{(i)} V_\mu^{(i)}(t-\tau_l^{(i)}) V_m^{(i)}(t-\tau_{l_p}^{(i)}) \\
&\times a^{(i,i)}(\tau_l^{(i)},\tau_{l_p}^{(i)}) \cos\theta_l^{(i)} dt \\
&+ \sum_{\substack{k=1\\ k\neq i}}^{K} \sum_{l=1}^{L_P} \sqrt{\frac{P_s}{2T_s}} \int_{\tau_{l_p}^{(i)}}^{T_s+\tau_{l_p}^{(i)}} \alpha_l^{(k)} V_\mu^{(k)}(t-\tau_l^{(k)}) V_m^{(i)}(t-\tau_{l_p}^{(i)}) \\
&\times a^{(k,i)}(\tau_l^{(k)},\tau_{l_p}^{(i)}) \cos\theta_l^{(k)} dt \\
&+ \frac{1}{2\sqrt{T_s}} \int_{\tau_{l_p}^{(i)}}^{T_s+\tau_{l_p}^{(i)}} n_c(t) a^{(i)}(t-\tau_{l_p}^{(i)}) V_m^{(i)}(t-\tau_{l_p}^{(i)}) dt.
\end{aligned} \tag{A.2}$$

From the above expression, it is observed that the correlator output contains the desired signal and various interference terms, when $m=\mu$, and only interference if $m\neq\mu$, which will be elaborated on below. Thus, $Z_I^i(m,l_p)$ can be written as a function of the various interference terms as :

$$Z_I^{(i)}(m,l_p) = \begin{cases} \alpha_{l_p}^{(i)}\sqrt{\frac{\xi_s}{2}}\cos\theta_{l_p}^{(i)} + I_{IM}^{(i,i)}(l_p) + I_{IMAI}^{(i,k)}(l_p) + N_I^{(i)}(l_p), & \text{for } m=\mu \\ I_{IM}^{(i,i)}(l_p) + I_{MAI,I}^{(i,k)}(l_p) + N_I^{(i)}(l_p), & \text{for } m\neq\mu. \end{cases} \tag{A.3}$$

where $\xi_s=\beta\times\xi_b$ is the symbol energy and ξ_b is the bit energy.

$$\begin{aligned}
I_{IM}^{(i,i)}(l_p) &= \sum_{\substack{l=1\\ l\neq l_p}}^{L_P} \sqrt{\frac{P_s}{2T_s}} \alpha_l^{(i)} \int_{\tau_{l_p}^{(i)}}^{T_s+\tau_{l_p}^{(i)}} V_\mu^{(i)}(t-\tau_l^{(i)}) V_m^{(i)}(t-\tau_{l_p}^{(i)}) \\
&\times a^{(i,i)}(\tau_l^{(i)},\tau_{l_p}^{(i)}) \cos\theta_l^{(i)} dt
\end{aligned} \tag{A.4}$$

is the self-interference inflicted by the M signalling sequences due to multipath propagation,

$$I_{MAI,I}^{(i,k)}(l_p) = \sum_{\substack{k=1 \\ k\neq i}}^{K} \sum_{l=1}^{L_P} \sqrt{\frac{P_s}{2T_s}} \alpha_l^{(i)} \int_{\tau_{l_p}^{(i)}}^{T_s+\tau_{l_p}^{(i)}} V_\mu^{(k)}(t-\tau_l^{(k)}) V_m^{(i)}(t-\tau_{l_p}^{(i)}) \times a^{(k,i)}(\tau_l^{(k)}, \tau_{l_p}^{(i)}) \cos\theta_l^{(k)} dt \quad \text{(A.5)}$$

is the multiple access interference (MAI) due to other users, and

$$N_I^{(i)}(l_p) = \frac{1}{2\sqrt{T_s}} \int_{\tau_{l_p}^{(i)}}^{T_s+\tau_{l_p}^{(i)}} n_c(t) a^{(i)}(t-\tau_{l_p}^{(i)}) V_m(t-\tau_{l_p}^{(i)}) dt. \quad \text{(A.6)}$$

is the white Gaussian noise, which remains a noise-induced term.

A.1.2 Quadrature channel

The output of the mth correlator of the l_pth branch is given by

$$Z_Q^{(i)}(m, l_p) = \frac{1}{\sqrt{T_s}} \int_{\tau_{l_p}^{(i)}}^{T_s+\tau_{l_p}^{(i)}} d_Q^{(i),l_p}(t) V_m^{(i)}(t-\tau_{l_p}^{(i)}) dt. \quad \text{(A.7)}$$

Substituting $d_Q^{(i),l_p}(t)$ given by Equation 15.25 into Equation A.7 and breaking up the equation into the various interference terms as described in Subsection A.1.1, we obtain

$$Z_Q^{(i)}(m, l_p) = \begin{cases} -\alpha_{l_p}^{(i)} \sqrt{\frac{\xi_s}{2}} \sin\theta_{l_p}^{(i)} - I_{QM}^{(i,i)}(l) - I_{MAI,Q}^{(i,k)}(l) + N_Q^{(i)}(l), & \text{for } m = \mu \\ -I_{QM}^{(i,i)}(l) - I_{MAI,Q}^{(i,k)}(l) + N_Q^{(i)}(l), & \text{for } m \neq \mu. \end{cases} \quad \text{(A.8)}$$

A.2 Derivation of the unconditional probability in Equation 15.43

Suppose that

$$p_{s^{(i)}(m)}(s_m|c^2) = \frac{1}{2\sigma^2} \left(\frac{s_m}{c^2}\right)^{\frac{L-1}{2}} e^{-\frac{c^2+s_m}{2\sigma^2}} I_{L-1}\left(\frac{\sqrt{c^2 s_m}}{\sigma^2}\right) \quad ; m = \mu. \quad \text{(A.9)}$$

From [90], the so-called characteristic function of the above equation is

$$\psi_{s^{(i)}(m)}(jv) = \frac{1}{(1 - jv2\sigma^2)^L} \exp\left(\frac{jvc^2}{1 - jv2\sigma^2}\right). \tag{A.10}$$

Thus, representing $p_{s^{(i)}(m)}(s_m|c^2)$ in terms of its characteristic function by taking the inverse Fourier transform yields

$$\begin{aligned} p_{s^{(i)}(m)}(s_m|c^2) &= \frac{1}{2\pi}\int_{-\infty}^{\infty} \psi_{s^{(i)}(m)}(jv)e^{-jvs_m}dv \\ &= \frac{1}{2\pi}\int_{-\infty}^{\infty} \frac{1}{(1 - jv2\sigma^2)^L} \exp\left(\frac{jvc^2}{1 - jv2\sigma^2}\right) e^{-jvs_m}dv. \end{aligned} \tag{A.11}$$

But $p_{s^{(i)}(m)}(s_m|c^2)$ is a conditional probability and in order to render it unconditional, we must weight the occurrence of $p_{s^{(i)}(m)}(s_m|c^2)$ by its probability of occurence given by its pdf of $p(c^2)$ and average or integrate it over the range $0 \ldots \infty$, yielding :

$$p_{s^{(i)}(m)}(s_m) = \int_0^{\infty} p_{s^{(i)}(m)}(s_m|c^2)p(c^2)dc^2, \quad m = \mu. \tag{A.12}$$

Following [90], $c^2 = \frac{\xi_s}{2}\sum_{l=1}^{L}(\alpha_l^{(i)})^2$. Since $\alpha_l^{(i)}$ is a Rayleigh distributed random variable, c^2 is a Chi-square distribution with $2L$ degrees of freedom. Thus [90] :

$$p(c^2) = \frac{(c^2)^{L-1}}{(2\sigma_R^2)^L(L-1)!} \exp\left(-\frac{(c^2)}{2\sigma_R^2}\right). \tag{A.13}$$

Therefore, upon substituting Equations A.11 and A.13 in Equation A.12, we have :

$$\begin{aligned} p_{s^{(i)}(m)}(s_m) &= \int_0^{\infty} \frac{1}{2\pi}\int_{-\infty}^{\infty} \frac{1}{(1 - jv2\sigma^2)^L} e^{\left(\frac{jvc^2}{1-jv2\sigma^2}\right)} e^{-jvs_m}dv \; p(c^2)dc^2 \\ &= \frac{1}{2\pi}\int_{-\infty}^{\infty} \frac{e^{-jvs_m}}{(1 - jv2\sigma^2)^L} \\ &\quad \cdot \underbrace{\left\{\int_0^{\infty} e^{\left(\frac{jvc^2}{1-jv2\sigma^2}\right)} e^{-jvs_m} p(c^2)dc^2\right\}}_{\psi\left(\frac{jv}{1-jv2\sigma^2}\right)} dv. \end{aligned} \tag{A.14}$$

Notice that the term $\psi\left(\frac{jv}{1-jv2\sigma^2}\right)$ in Equation A.14 is the characteristic function of $p(c^2)$ [90], which is given by :

$$\begin{aligned}\psi\left(\frac{jv}{1-jv2\sigma^2}\right) &= \left(\frac{1}{1-\frac{jv}{1-jv2\sigma^2}\times 2\sigma_R^2}\right)^L \\ &= \frac{(1-jv2\sigma^2)^L}{(1-jv2\sigma^2-jv2\sigma_R^2)^L}.\end{aligned} \tag{A.15}$$

Hence, substituting Equation A.15 in Equation A.14 yields :

$$\begin{aligned}p_{s^{(i)}(m)}(s_m) &= \frac{1}{2\pi}\int_{-\infty}^{\infty}\frac{e^{-jvs_m}}{(1-jv2\sigma^2)^L}\frac{(1-jv2\sigma^2)^L}{(1-jv2\sigma^2-jv2\sigma_R^2)^L}dv \\ &= \frac{1}{2\pi}\int_{-\infty}^{\infty}\frac{e^{-jvs_m}}{(1-jv2\sigma_T^2)^L},\end{aligned} \tag{A.16}$$

where $\sigma_T^2 = \sigma^2 + \sigma_R^2$ (sum of interference and Rayleigh variance). The term $p_{s^{(i)}(m)}(s_m)$ in Equation A.16 is the inverse Fourier transform of the characteristic function $\frac{1}{(1-jv2\sigma_T^2)^L}$, and it is given by [90] :

$$p_{s^{(i)}(m)}(s_m) = \frac{1}{(2\sigma_T^2)^L(L-1)!}s_m^{L-1}\exp\left(-\frac{s_m}{2\sigma^2}\right). \tag{A.17}$$

Appendix **B**

Appendix to Chapter 16: Derivation of Eq.(16.37)

In this Appendix we derive the correct symbol probability, $P(H_1)$, of the SC scheme used in the context of RS-coded M-ary CDMA. $P(H_1)$ associated with the SC scheme can be expressed in the form of Eq.(16.35) with $f_{U_m^i}(x)$ and $f_{U_j^i}(y)$ given by Eq.(16.28) and Eq.(16.31), respectively. In order to derive a general result, we let $\lambda_1 = 1/(1+\overline{\gamma}_c)$ and $\lambda_2 = 1$. Then, Eq.(16.28) and Eq.(16.31) can be expressed as:

$$f_{U_m^i}(y) = \lambda_2 L \exp(-\lambda_2 y)\left[1-\exp(-\lambda_2 y)\right]^{L-1}, \tag{B.1}$$

$$f_{U_j^i}(y) = \lambda_1 L \exp(-\lambda_1 y)\left[1-\exp(-\lambda_1 y)\right]^{L-1}. \tag{B.2}$$

With the aid of Eq.(B.1), it can be shown that:

$$\int_0^y f_{U_m^i}(x)dx = \left[1-\exp(-\lambda_2 y)\right]^{L}. \tag{B.3}$$

Substituting Eq.(B.2) and Eq.(B.3) into Eq.(16.35), $P(H_1)$ of the SC scheme can be expressed as:

$$\begin{aligned} P(H_1) &= \int_0^\infty \lambda_1 L \exp(-\lambda_1 y)\left[1-\exp(-\lambda_1 y)\right]^{L-1}\left[1-\exp(-\lambda_2 y)\right]^{L(M-1)} dy \\ &= \sum_{k=0}^{L-1}(-1)^k\binom{L-1}{k}\lambda_1 L\int_0^\infty \exp\left[-(1+k)\lambda_1 y\right]\left[1-\exp(-\lambda_2 y)\right]^{L(M-1)} dy. \end{aligned} \tag{B.4}$$

Upon using the relationship of:

$$\int_0^\infty \exp\left[-((1+k)\lambda_1 + n\lambda_2)y\right]\left[1-\exp(-\lambda_2 y)\right]^{L(M-1)-n} dy$$
$$= \frac{[L(M-1)-n]\lambda_2}{(1+k)\lambda_1 + n\lambda_2} \int_0^\infty \exp\left[-((1+k)\lambda_1 + (n+1)\lambda_2)y\right]\left[1-\exp(-\lambda_2 y)\right]^{L(M-1)-n-1} dy, \tag{B.5}$$

the integral in Eq.(B.4) can be expressed as:

$$\int_0^\infty \exp\left[-(1+k)\lambda_1 y\right]\left[1-\exp(-\lambda_2 y)\right]^{L(M-1)} dy$$
$$= \frac{1}{(1+k)\lambda_1} \prod_{n=1}^{L(M-1)} \frac{n}{n+(1+k)\lambda_1/\lambda_2}. \tag{B.6}$$

Substituting the above equation into Eq.(B.4) and letting $m = 1+k$, it can be shown that $P(H_1)$ can be expressed as:

$$P(H_1) = \sum_{m=1}^{L} (-1)^{m+1} \binom{L}{m} \prod_{n=1}^{L(M-1)} \frac{n}{n+m\lambda_1/\lambda_2}. \tag{B.7}$$

Finally, replacing λ_1 by $1/(1+\overline{\gamma}_c)$ and λ_2 by 1 in Eq.(B.7), the closed form of the correct symbol probability, $P(H_1)$ associated with the SC scheme can be obtained as shown in Eq.(16.37).

Appendix C

Appendix to Chapter 18

C.1 Probability Density Function of the λ-th Maximum

In this appendix we derive the probability density function (PDF) of the λ-th maximum of the independent random variables $\{Z_1, Z_2, \ldots, Z_n\}$, which may be expressed as:

$$Y = \max_{\lambda} \{Z_1, Z_2, \ldots, Z_n\}, \tag{C.1}$$

where Z_i for $i = 1, 2, \ldots, n$ follows the PDF of $f_{Z_i}(x)$. The distribution function of Y can be written as:

$$\begin{aligned} P(Y < y) &= P\left(\max_{\lambda} \{Z_1, X_2, \ldots, Z_n\} < y\right) \\ &= \sum_{i=1}^{n} \sum_{Q\binom{n-1}{\lambda-1}} P\left(Z_{i_1} > Z_i, Z_{i_2} > Z_i, \ldots, Z_{i_{\lambda-1}} > Z_i;\right. \\ &\quad \left. Z_{j_1} < Z_i, Z_{j_2} < Z_i, \ldots, Z_{j_{n-\lambda}} < Z_i;\ i_l, j_m \neq i | Z_i = Y < y\right) \\ &= \sum_{i=1}^{n} \sum_{Q\binom{n-1}{\lambda-1}} \int_{-\infty}^{y} P\left(Z_{i_1} > x, Z_{i_2} > x, \ldots, Z_{i_{\lambda-1}} > x;\right. \\ &\quad \left. Z_{j_1} < x, Z_{j_2} < x, \ldots, Z_{j_{n-\lambda}} < x;\ i_l, j_m \neq i | Z_i = x\right) f_{Z_i}(x) dx, \end{aligned} \tag{C.2}$$

where $\sum_{Q\binom{n}{i}}$ represents the sum of different selections of i out of n. Note that at the second step of the above derivation, we assumed that $Z_i = Y$ was the λ-th maximum. Consequently, there were $\lambda - 1$ out of the remaining $(n-1)$ variables, whose values were lower than Z_i, and the values of the remaining $(n - \lambda)$ variables were higher than Z_i. Since $\{Z_i\}$ for

$i = 1, 2, \ldots, n$ are independent random variables, Eq.(C.2) can be expressed as:

$$P(Y < y) = \sum_{i=1}^{n} \sum_{Q\binom{n-1}{\lambda-1}} \int_{-\infty}^{y} \left(\prod_{l=1}^{\lambda-1} P(Z_{i_l} > x) \right) \left(\prod_{m=1}^{n-\lambda} P(Z_{j_m} < x) \right) f_{Z_i}(x) dx. \quad \text{(C.3)}$$

After differentiating the above equation with respect to the variable y, we finally obtained the PDF of Eq.(C.1) as:

$$f_Y(y) = \sum_{i=1}^{n} \sum_{Q\binom{n-1}{\lambda-1}} \left(\prod_{l=1}^{\lambda-1} P(Z_{i_l} > y) \right) \left(\prod_{m=1}^{n-\lambda} P(Z_{j_m} < y) \right) f_{Z_i}(y). \quad \text{(C.4)}$$

Note that when $\lambda = 1$ or $\lambda = n$, and $\{Z_i\}$ obey identical distributions, then $\sum_{Q\binom{n-1}{0}}$ and $\sum_{Q\binom{n-1}{n-1}}$ are equal to 1, and Eq.(C.4) represents the well-known PDF of the distribution $Y = \max\{Z_1, Z_2, \ldots, Z_n\}$ or $Y = \min\{Z_1, Z_2, \ldots, Z_n\}$, where the corresponding PDFs are written as:

$$f_Y(y) = n f_{Z_i}(y) \left[P(Z_j < y) \right]^{n-1}, \quad \text{(C.5)}$$

and

$$f_Y(y) = n f_{Z_i}(y) \left[P(Z_j > y) \right]^{n-1}, \quad \text{(C.6)}$$

respectively.

C.2 The PDFs of the first maximum, the 'second' maximum and their ratio under the hypotheses H_1 of correct detection and H_0 of erroneous detection

In this appendix, we derive the PDFs of the maximum and the 'second' maximum of the decision variables, $\{Z_1, Z_2, \ldots, Z_M\}$ of the demodulator, as well as those of their ratio defined in Eq.(18.113). Both the hypothesis that the demodulator output is correct (H_1) and that it is in error (H_0) are considered. We assume that Z_1 is the output variable matched to the transmitted signal, while $Z_j,\ j > 1$ are the output variables mismatched to the transmitted signal. Moreover, we assume that $Z_j,\ j > 1$ are i.i.d random variables.

C.2.1 PDFs under the hypothesis H_1

Let the maximum and the 'second' maximum of the decision variables be represented as:

$$Y_1 = \max_1 \{Z_1, Z_2, \dots, Z_M\}, \quad \text{(C.7)}$$

$$Y_2 = \max_2 \{Z_1, Z_2, \dots, Z_M\}. \quad \text{(C.8)}$$

Then, the PDF of Y_1 conditioned on H_1 can be expressed as:

$$\begin{aligned} f_{Y_1}(y|H_1) &= \frac{d}{dy} P(Y_1 < y|H_1) \\ &= \frac{1}{P(H_1)} \cdot \frac{d}{dy} P(Y_1 < y, H_1) \\ &= \frac{1}{P(H_1)} \cdot \frac{d}{dy} P(Z_2 < Z_1, Z_3 < Z_1, \dots, Z_M < Z_1; Z_1 = Y_1 < y) \\ &= \frac{1}{P(H_1)} \cdot \frac{d}{dy} \int_0^y f_{Z_1}(x) \left[\int_0^x f_{Z_2}(z) dz \right]^{M-1} dx. \end{aligned} \quad \text{(C.9)}$$

Similarly, the PDF of Y_2 conditioned on H_1 can be expressed as:

$$\begin{aligned} f_{Y_2}(y|H_1) &= \frac{d}{dy} P(Y_2 < y|H_1) \\ &= \frac{1}{P(H_1)} \cdot \frac{d}{dy} P(Y_2 < y, H_1) \\ &= \frac{M-1}{P(H_1)} \cdot \frac{d}{dy} P(Z_1 > Z_i, Z_2 < Z_i, Z_3 < Z_i, \dots, Z_M < Z_i; Z_i = Y_2 < y) \\ &= \frac{M-1}{P(H_1)} \cdot \frac{d}{dy} \int_0^y f_{Z_i}(x) \left[\int_x^\infty f_{Z_1}(z) dz \right] \left[\int_0^x f_{Z_2}(z) dz \right]^{M-2} dx. \end{aligned} \quad \text{(C.10)}$$

Note that, in the third step of Eq.(C.10), we assumed that $Z_i \neq Z_1$ is the 'second' maximum amongst the decision variables, ie $Z_2 = Z_i$.

After obtaining the PDFs of the maximum and the 'second' maximum of the decision variables $\{Z_1, Z_2, \dots, Z_M\}$, the PDF of their ratio defined in Eq.(18.113), ie $\lambda = Y_1/Y_2$, conditioned on the maximum larger than the 'second' maximum, ie $Y_2 < Y_1$ can be derived using the approach of [598](pp.244):

$$\begin{aligned}
f_\lambda(y|H_1) &= \frac{d}{dy}P\left(\lambda = \frac{Y_1}{Y_2} \le y|Y_2 < Y_1, H_1\right) \\
&= \frac{1}{P(Y_2 < Y_1|H_1)}\frac{d}{dy}P\left(\lambda = \frac{Y_1}{Y_2} \le y, Y_2 < Y_1|H_1\right) \\
&= \frac{1}{P(Y_2 < Y_1|H_1)}\frac{d}{dy}\int_0^\infty P\left(y_2 < Y_1 \le y_2 y|Y_2 = y_2, H_1\right) f_{Y_2}(y_2|H_1)dy_2 \\
&= \frac{1}{P(Y_2 < Y_1|H_1)}\frac{d}{dy}\int_0^\infty \left[\int_{y_2}^{y_2 y} f_{Y_1}(y_1|H_1)dy_1\right] f_{Y_2}(y_2|H_1)dy_2 \\
&= \frac{1}{P(Y_2 < Y_1|H_1)}\int_0^\infty y_2 f_{Y_1}(y_2 y|H_1) f_{Y_2}(y_2|H_1)dy_2, \; y \ge 1, \qquad \text{(C.11)}
\end{aligned}$$

where $P(Y_2 < Y_1|H_1)$ represents the probability of $Y_2 < Y_1$ conditioned on the correct detection of H_1. $P(Y_2 < Y_1|H_i)$, $i = 0, 1$ of the probability of $Y_2 < Y_1$ conditioned on the hypothesis H_i can be expressed as:

$$P(Y_2 < Y_1|H_i) = \int_0^\infty \left[\int_0^{y_1} f_{Y_2}(y_2|H_i)dy_2\right] f_{Y_1}(y_1|H_i)dy_1, \qquad \text{(C.12)}$$

where $f_{Y_2}(y_2|H_i)$ and $f_{Y_1}(y_1|H_i)$ were the PDFs of Y_2 and Y_1 under the hypotheses H_i, respectively.

C.2.2 PDFs under the hypothesis H_0

Under the hypothesis H_0, the PDF of Y_1 can be expressed as:

$$\begin{aligned}
f_{Y_1}(y|H_0) &= \frac{d}{dy}P(Y_1 < y|H_0) \\
&= \frac{1}{P(H_0)} \cdot \frac{d}{dy}P(Y_1 < y, H_0) \\
&= \frac{1}{P(H_0)} \cdot \frac{d}{dy}P(Z_1 \ne Y_1 < y) \\
&= \frac{M-1}{P(H_0)} \cdot \frac{d}{dy}P(Z_1 < Z_i, Z_2 < Z_i, \ldots, Z_M < Z_i; Z_i = Y_1 < y) \\
&= \frac{M-1}{P(H_0)} \cdot \frac{d}{dy}\int_0^y f_{Z_i}(x)\left[\int_0^x f_{Z_1}(z)dz\right]\left[\int_0^x f_{Z_2}(z)dz\right]^{M-2} dx. \qquad \text{(C.13)}
\end{aligned}$$

The PDF of $f_{Y_2}(y|H_0)$ conditioned on H_0 can be expressed as:

$$\begin{aligned} f_{Y_2}(y|H_0) &= \frac{d}{dy}P(Y_2 < y|H_0) \\ &= \frac{1}{P(H_0)} \cdot \frac{d}{dy}P(Y_2 < y, H_0) \\ &= \frac{1}{P(H_0)} \cdot \left[\frac{d}{dy}P(Z_1 = Y_2 < y) + \frac{d}{dy}P(Z_1 \neq Y_2 < y)\right] \\ &= \frac{M-1}{P(H_0)} \cdot \left\{\frac{d}{dy}P(Z_i > Z_1, Z_2 < Z_1, \ldots, Z_M < Z_1; Z_1 = Y_2 < y)\right. \\ &\quad \left. + (M-2)\frac{d}{dy}P(Z_i > Z_j, Z_1 < Z_j, \ldots, Z_M < Z_j; Z_j = Y_2 < y)\right\} \\ &= \frac{M-1}{P(H_0)} \cdot \left\{\frac{d}{dy}\int_0^y f_{Z_1}(x)\left[\int_x^\infty f_{Z_i}(z)dz\right]\left[\int_0^x f_{Z_2}(z)dz\right]^{M-2} dx\right. \\ &\quad +(M-2)\frac{d}{dy}\int_0^y f_{Z_j}(x)\left[\int_x^\infty f_{Z_i}(z)dz\right] \\ &\quad \left. \cdot \left[\int_0^x f_{Z_1}(z)dz\right]\left[\int_0^x f_{Z_2}(z)dz\right]^{M-3} dx\right\}, \end{aligned} \tag{C.14}$$

where at the third step of the derivation, the ‘second’ maximum can be Z_1 or any one of the other decision variables. At the fourth step, we assumed that Z_i was the maximum, and $Z_j \neq Z_1, Z_i$ was the ‘second’ maximum, when it was not Z_1.

Again, the distribution of the ratio λ in Eq.(18.113) associated with the RST under the hypothesis H_0 of erroneous demodulation can be derived using Eq.(C.11) under the hypothesis of H_0.

Appendix D

Appendix to Chapter 21

In this Appendix we compute the result of the integral:

$$\Theta(\Omega, b) = \int_0^\infty Q_\mu(r, b) p_R(r) dr, \tag{D.1}$$

where $Q_\mu(r, b)$ is the generalized Marcum's Q function, which is defined as [90]:

$$Q_\mu(r, b) = \int_b^\infty x \left(\frac{x}{r}\right)^{\mu-1} \exp\left(-\frac{x^2 + r^2}{2}\right) I_{\mu-1}(rx) dx \tag{D.2}$$

with $I_\alpha(x)$ defined as [90]:

$$I_\alpha(x) = \sum_{i=0}^\infty \frac{(x/2)^{\alpha+2i}}{i!\Gamma(\alpha + i + 1)}, \ x \geq 0. \tag{D.3}$$

In Eq.(D.1) R is a random variable, which is defined as:

$$R = \sqrt{\sum_{m=1}^{2p} X_m^2}, \tag{D.4}$$

where the X_m, $m = 1, 2, \ldots, 2p$ are statistically independent, identically distributed zero mean Gaussian random variables. The PDF of R is given by [90]:

$$p_R(r) = \frac{2r^{2p-1}}{\Omega^p (p-1)!} \exp\left(-\frac{r^2}{\Omega}\right), \ r \geq 0. \tag{D.5}$$

Substituting Eq.(D.2), (D.3) and (D.5) into Eq.(D.1) we obtain the integral in the form of:

$$\begin{aligned}\Theta(\sigma, b) &= \int_0^\infty Q_\mu(r,b)p_R(r)dr \\ &= \sum_{i=0}^{\infty} \frac{1}{i!(\mu+i-1)!(p-1)!2^{2i+\mu-2}\Omega^p} \\ &\quad \cdot \int_0^\infty r^{2(i+p)-1} \exp\left(-\frac{(\Omega+2)r^2}{2\Omega}\right) dr \\ &\quad \cdot \int_b^\infty x^{2(i+\mu)-1} \exp\left(-\frac{x^2}{2}\right) dx. \end{aligned} \tag{D.6}$$

Upon evaluating the integrals of Eq.(D.6), we finally simplify the integral of Eq.(D.1) to:

$$\Theta(\Omega, b) = \sum_{i=0}^{\infty} \sum_{j=0}^{i+\mu-1} \frac{(b^2)^{i+\mu-j-1}}{(i+\mu-j-1)!2^{i+\mu-p-j-1}} \binom{i+p-1}{i} \frac{\exp(-b^2/2)}{\Omega^p(1+2/\Omega)^{i+p}}. \tag{D.7}$$

Glossary

ACL	Auto-correlation of a sequence
Adaptive-rate	A term applied to techniques that adapt the bit rate according to certain criteria
AQAM	Adaptive Quadrature Amplitude Modulation, a transmission scheme where the modulation mode is adapted according to certain criteria
ARIB	Association of Radio Industries and Businesses in Japan
AWGN	Additive White Gaussian Noise
BCH	Bose-Chaudhuri-Hocquenghem codes constitute a class of forward error correcting codes (FEC)
BER	Bit error rate, the number of the bits received incorrectly
Blind detection	A data or parameter estimation technique that does not require reference sequences to be transmitted or parameter estimation to be carried out separately
BPS	Bits Per Symbol, indicates the throughput performance
BPSK	Binary Phase Shift Keying, a type of data modulation scheme
BS	A common abbreviation for Base Station
CATT	Chinese Academy of Telecommunication Technology
CCL	Cross-correlation, usually of two different sequences
CDMA	Code Division Multiple Access, a multiple access scheme where multiple users transmit simultaneously within the same bandwidth and are separated through the use of a unique spreading code for each user
CIR	Channel Impulse Response
CRAD	Coherent Receiver Antenna Diversity, where the received signals from more than one antenna are coherently combined to obtain signal gain.
DBPSK	Differential Binary Phase Shift Keying, a type of data modulation scheme
Decorrelator	A detector that removes the correlation of all the interferer signals with the signal of the desired user

Diversity	A technique employed to obtain performance gain where different received versions of the same source signal are combined in order to improve the system performance
DS-CDMA	Direct Sequence Code Division Multiple Access, a sub-class of CDMA where each transmitted bit is directly multiplied with a spreading sequence in order to spread its bandwidth.
ETSI	European Telecommunications Standards Institute
FDMA	Frequency Division Multiple Access, a multiple access scheme where different users transmit in different bandwidths in order not to interfere with each other
FFH	Fast Frequency Hopping
FH-CDMA	Frequency Hopping Code Division Multiple Access, a sub-class of CDMA where the carrier frequency of the CDMA user is switched according to a pattern determined by its unique code
IC	Interference Cancellation, a type of multiuser receiver for CDMA where the received signal is regenerated from previous data estimates and cancelled from the composite received signal, in order to provide more reliable estimates after the cancellation stages
IMT-2000	International Mobile Telecommunications 2000
Interleaving	A technique employed to randomize burst errors caused by fading in the mobile channel. The transmitted bits are arranged according to a known order before transmission and at the receiver the received symbols are re-arranged into the pre-transmission order so that the bursty errors can be separated. This helps improve the performance of the channel decoder.
IS-95	Interim Standard 95, the definition of the cellular (800 MHz) CDMA Common Air Interface
ISI	Inter-symbol Interference, interference caused by the time dispersion of the wideband channel where the transmitted symbols interfere with each other
JD	Joint Detection or Joint Detector, a type of multiuser receiver that uses equalization techniques to jointly detect the symbols of multiple users
JD-CDMA	Joint Detection CDMA system, a CDMA system that employs joint detection receivers
LMS	Least Mean Square algorithm, a linear adaptive filtering algorithm that recursively optimizes the filter tap weights in order to obtain the minimum mean square error at the output of the filter
MAI	Multiple Access Interference, the interference caused by multiple users transmitting simultaneously within the same bandwidth and is usually used in the context of CDMA systems
MAP	Maximum A Posteriori, the maximum a posteriori probability criterion maximizes the probability of making a correct decision
Matched filter	A filter that has an impulse response that is matched to the waveform of the desired signal and maximizes the SNR at the output of the filter

MC-CDMA Multi-Carrier Code Division Multiple Access, a sub-class of CDMA where a data symbol is spread with a spreading sequence into say, Q chips and each chip of the spread data symbol is transmitted over a narrowband subcarrier in the frequency domain.

MLSE Maximum Likelihood Sequence Estimation, a sequence estimation technique that produces the most likely transmitted sequence based on a metric that is optimized for a certain criterion

MMSE Minimum Mean Square Error

MMSE-BDFE Minimum Mean Square Error Block Decision Feedback Equalizer, a type of joint detection receiver that minimizes the mean square error and feeds back already detected symbols to improve the reliability of the output estimates

MMSE-BLE Minimum Mean Square Error Block Linear Equalizer, a type of joint detection receiver that linearly minimizes the mean square error

MRC Maximal Ratio Combining, a diversity combining technique where multiple received signals are coherently combined

MS A common abbreviation for Mobile Station

Multipath diversity Multiple versions of the transmitted signal are obtained at the receiver due to the different multipaths in a channel and the signals of these paths can be combined in order to provide performance gain

Multiuser receiver A receiver that employs available knowledge on the properties of all the transmitting users in order to detect the data symbols of all the users

Near-far effect The phenomenon that occurs when the signals from different users arrive at the base station with different signal strengths. The stronger signals swamp out the weaker signals, thus severely degrading the performance of the weaker signals.

PDF Probability Density Function

PIC Parallel Interference Cancellation, an interference cancellation receiver where the received signals of all the interferers are cancelled from the received composite signal at each cancellation stage in order to generate a more reliable signal for the data estimation of the desired user

PN sequence Pseudo-noise sequence, or pseudo-random sequence, which is a generated sequence that exhibits noise-like properties

Power control A technique used to combat the near-far effect where the power control algorithm attempts to regulate the transmitted powers of all the users such that the signals of all the users arrive with similar strengths at the receiver.

PSD Power Spectral Density

PSP Per Survivor Processing, a trellis-decoding algorithm, where the required parameters, for example CIR estimates, are unknown. The parameter estimation is carried out in a "per-survivor" fashion, which means that a parameter estimator is assigned to each surviving data sequence of the trellis.

QAM Quadrature Amplitude Modulation

RAKE A multipath diversity combiner, that inherited its name from the way it "rakes" in all the incoming pulses to form an equalized signal. The signal energy from different multipaths are combined according to the chosen diversity combining technique.

RLS Recursive Least Squares, an adaptive filtering technique where a recursive method is used to adapt the filter tap weights such that the square of the error between the filter output and the desired response is minimized

SFH Slow Frequency Hopping

SIC Successive Interference Cancellation, an interference cancellation receiver where only the received signals of all the interferers that are more reliable than the desired signal are cancelled from the received composite signal in order to generate a more reliable signal for the data estimation of the desired user

SINR Signal to Interference plus Noise ratio, same as signal to noise ratio (SNR) when there is no interference.

SNR Signal to Noise Ratio, noise energy compared to the signal energy

SOVA Soft Output Viterbi Algorithm, a trellis algorithm that generates the most likely sequence in soft decisions according to the constraints of the trellis and the received signal

SSMA Spread Spectrum Multiple Access

TDD Time Division Duplex, a transmission protocol where the uplink and downlink transmissions are carried out in the same frequency but separated in time

TDD-CDMA Time Division Duplex Code Division Multiple Access, a multiple access scheme that combines TDD and CDMA

TDMA Time Division Multiple Access, a multiple access technique where multiple users transmit in the same bandwidth but are separated in time through user-designated timeslots

TH-CDMA Time Hopping Code Division Multiple Access, a sub-class of CDMA where each user transmits in the timeslots determined by its spreading sequence

TIA Telecommunications Industry Association in the USA

UMTS Universal Mobile Telecommunications Systems

UTRA UMTS Terrestrial Radio Access

Viterbi algorithm A trellis algorithm that generates the most likely sequence according to the constraints of the trellis and the received signal

VSF Variable Spreading Factor, an adaptive rate transmission scheme for CDMA, where the bit rate is adapted by varying the spreading factor but keeping the chip rate constant

W-CDMA Wideband Code Division Multiple Access, a high chip-rate and bit-rate CDMA air interface, where the mobile channel bandwidth is very wide and the fading within the channel is frequency-selective. In general, the minimum bandwidth of wideband CDMA is 5 MHz.

WMF Whitening Matched Filter, a filter that whitens the received noise and maximizes the SNR at the output of the filter

ZF-BDFE Zero Forcing Block Decision Feedback Equalizer, a type of joint detection receiver that eliminates all the interference at the expense of noise enhancement and feeds back already detected symbols to improve the reliability of the output estimates

ZF-BLE Zero Forcing Block Linear Equalizer, a type of linear joint detection receiver that eliminates all the interference at the expense of noise enhancement

Bibliography

[1] J. Mitola, "The software radio architecture," *IEEE Communications Magazine*, pp. 26–38, May 1995.

[2] J. Mitola, "Technical challenges in the globalization of software radio," *IEEE Communications Magazine*, pp. 84–89, February 1999.

[3] W. Tuttlebee, ed., *Software Defined Radio, Volumes I and II*. John Wiley, 2002.

[4] L. Hanzo, C. Wong, and M. Yee, *Adaptive Wireless Transceivers*. John Wiley, IEEE Press, 2002. (For detailed contents, please refer to http://www-mobile.ecs.soton.ac.uk.).

[5] L. Hanzo, T. Liew, and B. Yeap, *Turbo Coding, Turbo Equalisation and Space-Time Coding*. John Wiley, IEEE Press, 2002. (For detailed contents, please refer to http://www-mobile.ecs.soton.ac.uk.).

[6] J. Holland, *Adaptation in Natural and Artificial Systems*. Ann Arbor, Michigan: University of Michigan Press, 1975.

[7] T.-J. Lim, L. K. Rasmussen, and H. Sugimoto, "An asynchronous multiuser CDMA detector based on the Kalman filter," *IEEE Journal of Selected Areas in Communications*, vol. 16, pp. 1711–1722, December 1998.

[8] X. Wang and H. Poor, "Adaptive joint multiuser detection and channel estimation in multipath fading CDMA," *Wireless Networks*, vol. 4, pp. 453–470, June 1998.

[9] D. E. Goldberg, *Genetic Algorithms in Search, Optimization, and Machine Learning*. Reading, Massachusetts: Addison-Wesley, 1989.

[10] K. S. Gilhousen, I. M. Jacobs, R. Padovani, A. J. Viterbi, L. A. Weaver, and C. E. Wheatley, "On the capacity of a cellular CDMA system design," *IEEE Transactions on Vehicular Technology*, vol. 40, pp. 303–312, May 1991.

[11] Telecomm. Industry Association (TIA), Washington, DC, *Mobile station - Base station compatibility standard for dual-mode wideband spread spectrum cellular system, EIA/TIA Interim Standard IS-95*, 1993.

[12] J. Holtzman, "A simple, accurate method to calculate spread-spectrum multiple-access error probabilities," *IEEE Transactions on Communications*, vol. 40, pp. 461–464, March 1992.

[13] T. Eng, N. Kong, and L. Milstein, "Comparison of diversity combining techniques for Rayleigh-fading channels," *IEEE Transactions on Communications*, vol. 44, pp. 1117–1129, September 1996.

[14] R. E. Ziemer and R. L. Peterson, *Digital Communications and Spread Spectrum Systems*. New York: Macmillan Publishing Company, 1985.

[15] A. Polydoros and C. Weber, "A unified approach to serial search spread-spectrum code acquisition-Part I: general theory," *IEEE Transactions on Communications*, vol. 32, pp. 542–549, May 1984.

[16] D. Sarwate, "Acquisition of direct-sequence spread-spectrum signals," in *Wireless Communication - TDMA versus CDMA* (S. G. Glisic and P. L. Leppanen, eds.), pp. 121–145, Kluwer Academic Publishers, 1997.

[17] U. Madhow, "MMSE interference suppression for timing acquisition and demodulation in direct-sequence CDMA systems," *IEEE Transactions on Communications*, vol. 46, pp. 1065–1075, August 1998.

[18] K. Chugg, "Blind acquisition characteristics of PSP-based sequence detectors," *IEEE Journal on Selected Areas in Communications*, vol. 16, pp. 1518–1529, October 1998.

[19] R. Rick and L. Milstein, "Optimal decision strategies for acquisition of spread-spectrum signals in frequency-selective fading channels," *IEEE Transactions on Communications*, vol. 46, pp. 686–694, May 1998.

[20] D.-W. Lee and L. Milstein, "Analysis of a multicarrier DS-CDMA code-acquisition system," *IEEE Transactions on Communications*, vol. 47, pp. 1233–1244, August 1999.

[21] S. Bensley and B. Aazhang, "Maximum-likelihood synchronization of a single user for code-division multiple-access communication systems," *IEEE Transactions on Communications*, vol. 46, pp. 392–399, March 1998.

[22] W. R. Braun, "PN acquisition and tracking performance in DS/CDMA systems with symbol-length spreading sequences," *IEEE Transactions on Communications*, vol. 45, pp. 1595–1601, December 1997.

[23] R. Rick and L. Milstein, "Parallel acquisition in mobile DS-CDMA systems," *IEEE Transactions on Communications*, vol. 45, pp. 1466–1476, November 1997.

[24] U. Cheng, "Performance of a class of parallel spread-spectrum code acquisition schemes in the presence of data modulation," *IEEE Transactions on Communications*, vol. 36, pp. 596–604, May 1988.

[25] E. Sourour and S. Gupta, "Direct-sequence spread-spectrum parallel acquisition in nonselective and frequency-selective Rician fading channels," *IEEE Journal on Selected Areas in Communications*, vol. 10, pp. 535–544, April 1992.

[26] P.-T. Sun and C.-Y. Chu, "Hidden preamble detector for acquisition of frequency hopping multiple-access communication system," *IEE Proceedings Communications*, vol. 144, pp. 161–165, June 1997.

[27] Y.-H. You, T.-H. Moon, J.-H. Kim, and C.-E. Kang, "Threshold decision technique for direct sequence code synchronisation in a fading mobile channel," *IEE Proceedings Communications*, vol. 144, pp. 1155–1160, June 1997.

[28] J. Holmes and C.-C. Chen, "Acquisition time performance of PN spread-spectrum systems," *IEEE Transactions on Communications*, vol. 25, pp. 778–784, August 1977.

[29] V. Jovanovic and E. Sousa, "Analysis of non-coherent correlation in DS/BPSK spread spectrum acquisition," *IEEE Transactions on Communications*, vol. 43, pp. 565–573, February/March/April 1995.

[30] A. Polydoros and C. Weber, "A unified approach to serial search spread-spectrum code acquisition-Part II: a matched-filter receiver," *IEEE Transactions on Communications*, vol. 32, pp. 550–560, May 1984.

[31] E. Sourour and S. Gupta, "Direct-sequence spread-spectrum parallel acquisition in a fading mobile channel," *IEEE Transactions on Communications*, vol. 38, pp. 992–998, July 1990.

[32] S. Glisic, T. Poutanen, W. Wu, G. Petrovic, and Z. Stefanovic, "New PN code acquisition scheme for CDMA networks with low signal-to-noise ratios," *IEEE Transactions on Communications*, vol. 47, pp. 300–310, February 1999.

[33] C.-J. Kim, H.-J. Lee, and H.-S. Lee, "Adaptive acquisition of PN sequences for DSSS communications," *IEEE Transactions on Communications*, vol. 46, pp. 993–996, August 1998.

[34] B.-H. Kim and B.-G. Lee, "DSA: a distributed sample-based fast DS/CDMA acquisition technique," *IEEE Transactions on Communications*, vol. 47, pp. 754–765, May 1999.

[35] B.-H. Kim and B.-G. Lee, "Performance analysis of DSA-based DS/CDMA acquisition," *IEEE Transactions on Communications*, vol. 47, pp. 817–822, June 1999.

[36] M. Salih and S. Tantaratana, "A closed-loop coherent PN acquisition system with a pre-loop estimator," *IEEE Transactions on Communications*, vol. 47, pp. 1394–1405, September 1999.

[37] R. Smith and S. Miller, "Acquisition performance of an adaptive receiver for DS-CDMA," *IEEE Transactions on Communications*, vol. 47, pp. 1416–1424, September 1999.

[38] D. Dicarlo and C. Weber, "Statistical performance of single dwell serial synchronization systems," *IEEE Transactions on Communications*, vol. 28, pp. 1382–1388, August 1980.

[39] Y.-T. Su, "Rapid code acquisition algorithms employing PN matched filters," *IEEE Transactions on Communications*, vol. 36, pp. 724–732, June 1988.

[40] U. Madhow and M. Pursley, "Acquisition in direct-sequence spread-spectrum communication networks: an asymptotic analysis," *IEEE Transactions on Information Theory*, vol. 39, pp. 903–912, May 1993.

[41] P. Baier, K. Dostert, and M. Pandit, "A novel spread-spectrum receiver synchronization scheme using a saw-tapped delay line," *IEEE Transactions on Communications*, vol. 30, pp. 1037–1047, May 1982.

[42] C.-D. Chung, "Differentially coherent detection technique for direct-sequence code acquisition in a Rayleigh fading mobile channel," *IEEE Transactions on Communications*, vol. 43, pp. 1116–1126, February/March/April 1995.

[43] D. Dicarlo and C. Weber, "Multiple dwell serial search: performance and application to direct sequence code acquisition," *IEEE Transactions on Communications*, vol. 31, pp. 650–659, May 1983.

[44] B. Ibrahim and A. Aghvami, "Direct sequence spread spectrum matched filter acquisition in frequency-selective Rayleigh fading channels," *IEEE Journal on Selected Areas in Communications*, vol. 12, pp. 885–890, June 1994.

[45] J. Li and S. Tantaratana, "Optimal and suboptimal coherent acquisition schemes for PN sequences with data modulation," *IEEE Transactions on Communications*, vol. 43, pp. 554–564, Fenruary/March/April 1995.

[46] H.-R. Park and B.-J. Kang, "On the performance of a maximum-likelihood code-acquisition technique for preamble search in a CDMA reverse link," *IEEE Transactions on Vehicular Technology*, vol. 47, pp. 65–74, February 1998.

[47] M. Zarrabizadeh and E. Sousa, "A differentially coherent PN code acquisition receiver for CDMA systems," *IEEE Transactions on Communications*, vol. 45, pp. 1456–1465, November 1997.

[48] U. Cheng, W. Hurd, and J. Statman, "Spread-spectrum code acquisition in the presence of doppler shift and data modulation," *IEEE Transactions on Communications*, vol. 38, pp. 241–250, February 1990.

[49] K. K. Chawla and D. V. Sarwate, "Parallel acquiaition of PN sequences in DS/SS systems," *IEEE Transactions on Communications*, vol. 42, pp. 2155–2164, May 1994.

[50] L.-L. Yang and J. Simsa, "Performance evaluation of spread-spectrum code acquisition system using four-state Markov process," in *Proceedings IEEE ISSSTA'98*, (Sun City: South Africa), pp. 848–852, September 1998.

[51] L.-L. Yang and L. Hanzo, "Serial acquisition techniques for DS-CDMA signals in frequency-selective multi-user mobile channels," in *Proceedings of IEEE VTC'99*, (Houston, USA), pp. 2398–2402, May 1999.

[52] M. Katz and S. Glisic, "Modeling of code acquisition process in CDMA networks - quasi-synchronous systems," *IEEE Transactions on Communications*, vol. 46, pp. 1564–1568, December 1998.

[53] W.-H. Sheen, J.-K. Tzeng, and C.-K. Tzou, "Effects of cell correlations in a matched-filter PN code acquisition for direct-sequence spread-spectrum systems," *IEEE Transactions on Vehicular Technology*, vol. 48, pp. 724–732, May 1999.

[54] D. Zheng, J. Li, S. Miller, and E. Strom, "An efficient code-timing estimator for DS-CDMA signals," *IEEE Transactions on Signal Processing*, vol. 45, pp. 82–89, January 1997.

[55] E. Strom, S. Parkvall, S. Miller, and B. Ottersten, "Propagation delay estimation in asynchronous direct-sequence code-division multiple access systems," *IEEE Transactions on Communications*, vol. 44, pp. 84–93, January 1996.

[56] L.-C. Chu and U. Mitra, "Analysis of MUSIC-based delay estimators for direct-sequence code-division multiple-access systems," *IEEE Transactions on Communications*, vol. 47, pp. 133–138, January 1999.

[57] M. Zeng, A. Annamalai, and V. K. Bhargava, "Recent advances in cellular wireless communications," *IEEE Communications Magazine*, pp. 128–138, September 1999.

[58] P. Chaudhury, W. Mohr, and S. Onoe, "The 3GPP proposal for IMT-2000," *IEEE Communications Magazine*, pp. 72–81, December 1999.

[59] M. Progler, C. Evci, and M. Umehira, "Air interface access schemes for broadband mobile systems," *IEEE Communications Magazine*, pp. 106–115, September 1999.

[60] L. Kleinrock, "On some principles of nomadic computing and multi-access communications," *IEEE Communications Magazine*, pp. 46–50, July 2000.

[61] P. Bender, P. Black, M. Grob, R. Padovani, N. Sindhushayana, and A. Viterbi, "CDMA/HDR: A bandwidth-efficient high-speed wireless data service for nomadic users," *IEEE Communications Magazine*, pp. 70–77, July 2000.

[62] J. Chuang and N. Sollenberger, "Beyond 3G: wideband wireless data access based on OFDM and dynamic packet assignment," *IEEE Communications Magazine*, pp. 78–87, July 2000.

[63] N. Dimitriou, R. Tafazolli, and G. Sfikas, "Quality of service for multimedia CDMA," *IEEE Communications Magazine*, pp. 88–94, July 2000.

[64] R. Steele and L. Hanzo, *Mobile Radio Communications*. IEEE Press-John Wiley, 2 ed., 1999.

[65] E. Sourour and M. Nakagawa, "Performance of orthogonal multicarrier CDMA in a multipath fading channel," *IEEE Transactions on Communications*, vol. 44, pp. 356–367, March 1996.

[66] L.-L. Yang and L. Hanzo, "Blind soft-detection assisted frequency-hopping multicarrier DS-CDMA systems," in *Proceedings of IEEE GLOBECOM'99*, (Rio de Janeiro, Brazil), pp. 842–846, December:5-9 1999.

[67] S. Slimane, "MC-CDMA with quadrature spreading for wireless communication systems," *European Transactions on Telecommunications*, vol. 9, pp. 371–378, July–August 1998.

[68] L.-L. Yang and L. Hanzo, "Slow frequency-hopping multicarrier DS-CDMAfor transmission over Nakagami multipath fading channels," *IEEE Journal on Selected Areas in Communications*, vol. 19, no. 7, pp. 1211–1221, 2001.

[69] S. Verdu, "Wireless bandwidth in the making," *IEEE Communications Magazine*, pp. 53–58, July 2000.

[70] S. Nanda, K. Balachandran, and S. Kumar, "Adaptation techniques in wireless packet data services," *IEEE Communications Magazine*, pp. 54–64, January 2000.

[71] A. J. Goldsmith and S. G. Chua, "Variable-rate variable-power MQAM for fading channels," *IEEE Transactions on Communications*, vol. 45, pp. 1218–1230, October 1997.

[72] M. S. Alouini and A. J. Goldsmith, "Capacity of Rayleigh fading channels under different adaptive transmission and diversity-combining techniques," *IEEE Transactions on Vehicular Technology*, vol. 48, pp. 1165–1181, July 1999.

[73] A. Duel-Hallen, S. Hu, and H. Hallen, "Long-range prediction of fading signals," *IEEE Signal Processing Magazine*, pp. 62–75, May 2000.

[74] L.-L. Yang and L. Hanzo, "Blind joint soft-detection assisted slow frequency-hopping multicarrier DS-CDMA," *IEEE Transactions on Communications*, vol. 48, pp. 1520 –1529, September 2000.

[75] R. Prasad and S. Hara, "Overview of multi-carrier CDMA," *IEEE Communications Magazine*, vol. 35, pp. 126–133, Dec. 1997.

[76] L. Hanzo, W. Webb, and T. Keller, *Single- and Multi-carrier Quadrature Amplitude Modulation: Principles and Applications for Personal Communications, WLANs and Broadcasting*. London: IEEE Press, and John Wiley & Sons, 2nd ed., 1999.

[77] L.-L. Yang and L. Hanzo, "Parallel code-acquisition for multicarrier DS-CDMA systems communicating over multipath Nakagami fading channels," in *Proceedings of IEEE GLOBECOM'2000*, (San Francisco, California), November 27 - December 1 2000.

[78] S. Verdú, *Multiuser Detection*. Cambridge University Press, 1998.

[79] C. Berrou and A. Glavieux, "Near optimum error correcting coding and decoding: turbo-codes," *IEEE Transactions on Communications*, vol. 44, pp. 1261–1271, Oct. 1996.

[80] Y. Li and N. R. Sollenberger, "Adaptive antenna arrays for OFDM systems with cochannel interference," *IEEE Transactions on Communications*, vol. 47, pp. 217–229, Feb 1999.

[81] V. Tarokh, N. Seshadri, and A. R. Calderbank, "Space-time codes for high data rate wireless communication: performance criterion and code construction," *IEEE Transactions on Information Theory*, vol. 44, pp. 744–765, March 1998.

[82] M. K. Simon, J. K. Omura, R. A. Scholtz, and B. K. Levitt, *Spread Spectrum Communications Handbook*. McGraw Hill, 1994.

[83] A. J. Viterbi, *CDMA: Principles of Spread Spectrum Communication*. Addison-Wesley Publishing Company, 1995.

[84] L. E. Miller and J. S. Lee, *CDMA Systems Engineering Handbook*. Boston: Artech House, 1998.

[85] J. S. Lee, "Overview of the technical basis of QUALCOMM's CDMA cellular telephone system design: A view of North American TIA/EIA IS-95," in *International Conference on Communications Systems (ICCS)*, (Singapore), pp. 353–358, 1994.

[86] R. Prasad, *CDMA for Wireless Personal Communications*. Artech House, Inc., 1996.

[87] S. G. Glisic and P. A. Leppännen, *Wireless Communications TDMA versus CDMA*. Kluwer Academic Publishers, 1997.

[88] S. Glisic and B. Vucetic, *Spread Spectrum CDMA Systems for Wireless Communications*. Artech House, Inc., 1997.

[89] P. W. Baier, "A critical review of CDMA," in *Proceedings of the IEEE Vehicular Technology Conference (VTC)*, (Atlanta, USA), pp. 6–10, Apr. 28-May 1 1996.

[90] J. G. Proakis, *Digital Communications*. Mc-Graw Hill International Editions, 3rd ed., 1995.

[91] G. Stüber, *Principles of Mobile Communication*. Kluwer Academic Publishers, 1996.

[92] L. Hanzo, W. Webb, and T. Keller, *Single and Multicarrier Quadrature Amplitude Modulation*. John-Wiley IEEE Press, 2000.

[93] R. L. Pickholtz, L. B. Milstein, and D. L. Schilling, "Spread spectrum for mobile communications," *IEEE Transactions on Vehicular Technology*, vol. 40, pp. 313–322, May 1991.

[94] A. J. Viterbi, "Wireless digital communication: a view based on three lessons learned," *IEEE Communications Magazine*, pp. 33–36, Sep. 1991.

[95] A. Klein, B. Steiner, and A. Steil, "Known and novel diversity approaches as powerful means to enhance the performance of cellular mobile radio systems," *IEEE Journal on Selected Areas in Communications*, vol. 14, pp. 1784–1795, Dec. 1996.

[96] A. Baier, U.-C. Fiebig, W. Granzow, W. Koch, P. Teder, and J. Thielecke, "Design study for a CDMA-based third-generation mobile radio system," *IEEE Journal on Selected Areas in Communications*, vol. 12, pp. 733–743, May 1994.

[97] T. Ottosson and A. Svensson, "On schemes for multirate support in DS-CDMA systems," *Wireless Personal Communications (Kluwer)*, vol. 6, pp. 265–287, Mar. 1998.

[98] S. Ramakrishna and J. M. Holtzman, "A comparison between single code and multiple code transmission schemes in a CDMA system," in *Proceedings of the IEEE Vehicular Technology Conference (VTC)*, (Ottawa, Canada), pp. 791–795, May 18-21 1998.

[99] R. E. Blahut, *Theory and Practice of Error Control Codes*. Addison-Wesley Publishing Company, 1983.

[100] M. P. Lötter and L. P. Linde, "A comparison of three families of spreading sequences for CDMA applications," in *Proceedings of IEEE South African Symposium on Communications and Signal Processing (COMSIG)*, (Stellenbosch, South Africa), pp. 68–75, Oct. 4 1994.

[101] R. Gold, "Optimal binary sequences for spread spectrum multiplexing," *IEEE Transactions on Information Theory*, vol. 13, pp. 619–621, Oct. 1967.

[102] T. Kasami, *Combinatorial Mathematics and its Applications*. University of North Carolina Press, 1969.

[103] M. Pursley, "Performance evaluation for phase-coded SSMA communication - part 1: System analysis," *IEEE Transactions on Communications*, vol. 25, pp. 795–799, Aug. 1977.

[104] A. D. Whalen, *Detection of Signals in Noise*. Academic Press, 1971.

[105] M. Failli, "Digital land mobile radio communications COST 207," tech. rep., European Commission, Luxembourg, 1989.

[106] S. Verdú, "Minimum probability of error for asynchronous Gaussian multiple access channels," *IEEE Transactions on Information Theory*, vol. 32, pp. 85–96, Jan. 1986.

[107] J. S. Thompson, P. M. Grant, and B. Mulgrew, "Smart antenna arrays for CDMA systems," *IEEE Personal Communications Magazine*, vol. 3, pp. 16–25, Oct. 1996.

[108] J. S. Thompson, P. M. Grant, and B. Mulgrew, "Performance of antenna array receiver algorithms for CDMA," in *Proceedings of the IEEE Global Telecommunications Conference (GLOBECOM)*, (London, UK), pp. 570–574, Nov. 18-22 1996.

[109] A. F. Naguib and A. Paulraj, "Performance of wireless CDMA with m-ary orthogonal modulation and cell site antenna arrays," *IEEE Journal on Selected Areas in Communications*, vol. 14, pp. 1770–1783, Dec. 1996.

[110] L. C. Godara, "Application of antenna arrays to mobile communications, part I: Performance improvement, feasibility, and system considerations," *Proceedings of the IEEE*, vol. 85, pp. 1031–1060, Aug. 1997.

[111] R. Kohno, H. Imai, M. Hatori, and S. Pasupathy, "Combination of adaptive array antenna and a canceller of interference for direct-sequence spread-specturn multiple-access system," *IEEE Journal on Selected Areas in Communications*, vol. 8, pp. 675–681, May 1998.

[112] R. Lupas and S. Verdú, "Linear multiuser detectors for synchronous code divison multiple access channels," *IEEE Transactions on Information Theory*, vol. 35, pp. 123–136, Jan. 1989.

[113] R. Lupas and S. Verdú, "Near-far resistance of multiuser detectors in asynchronous channels," *IEEE Transactions on Communications*, vol. 38, pp. 509–519, Apr. 1990.

[114] Z. Zvonar and D. Brady, "Suboptimal multiuser detector for frequency selective Rayleigh fading synchronous CDMA channels," *IEEE Transactions on Communications*, vol. 43, pp. 154–157, Feb.-Apr. 1995.

[115] Z. Zvonar and D. Brady, "Differentially coherent multiuser detection in asynchronous CDMA flat Rayleigh fading channels," *IEEE Transactions on Communications*, vol. 43, pp. 1252–1255, Feb-Apr 1995.

[116] Z. Zvonar, "Combined multiuser detection and diversity reception for wireless CDMA systems," *IEEE Transactions on Vehicular Technology*, vol. 45, pp. 205–211, Feb. 1996.

[117] T. Kawahara and T. Matsumoto, "Joint decorrelating multiuser detection and channel estimation in asynchronous CDMA mobile communications channels," *IEEE Transactions on Vehicular Technology*, vol. 44, pp. 506–515, Aug. 1995.

[118] M. Hosseinian, M. Fattouche, and A. B. Sesay, "A multiuser detection scheme with pilot symbol-aided channel estimation for synchronous CDMA systems," in *Proceedings of the IEEE Vehicular Technology Conference (VTC)*, (Ottawa, Canada), pp. 796–800, May 18-21 1998.

[119] M. J. Juntti, B. Aazhang, and J. O. Lilleberg, "Iterative implementation of linear multiuser detection for dynamic asynchronous CDMA systems," *IEEE Transactions on Communications*, vol. 46, pp. 503–508, Apr. 1998.

[120] P.-A. Sung and K.-C. Chen, "A linear minimum mean square error multiuser receiver in Rayleigh fading channels," *IEEE Journal on Selected Areas in Communications*, vol. 14, pp. 1583–1594, Oct. 1996.

[121] A. Duel-Hallen, "Decorrelating decision-feedback multiuser detector for synchronous code-division multiple-access channel," *IEEE Transactions on Communications*, vol. 41, pp. 285–290, Feb. 1993.

[122] G. H. Golub and C. F. van Loan, *Matrix Computations*. North Oxford Academic, 1983.

[123] L. Wei and C. Schlegel, "Synchronous DS-SSMA system with improved decorrelating decision-feedback multiuser detection," *IEEE Transactions on Vehicular Technology*, vol. 43, pp. 767–772, Aug. 1994.

[124] A. Hafeez and W. E. Stark, "Combined decision-feedback multiuser detection/soft-decision decoding for CDMA channels," in *Proceedings of the IEEE Vehicular Technology Conference (VTC)*, (Atlanta, USA), pp. 382–386, Apr. 28-May 1 1996.

[125] S. Haykin, *Adaptive Filter Theory*. Prentice-Hall International, Inc., 1996.

[126] A. Klein and P. W. Baier, "Linear unbiased data estimation in mobile radio sytems applying CDMA," *IEEE Journal on Selected Areas in Communications*, vol. 11, pp. 1058–1066, Sep. 1993.

[127] A. Klein, G. K. Kaleh, and P. W. Baier, "Zero forcing and minimum mean square error equalization for multiuser detection in code division multiple access channels," *IEEE Transactions on Vehicular Technology*, vol. 45, pp. 276–287, May 1996.

[128] J. Blanz, A. Klein, M. Nasshan, and A. Steil, "Performance of a cellular hybrid C/TDMA mobile radio system applying joint detection and coherent receiver antenna diversity," *IEEE Journal on Selected Areas in Communications*, vol. 12, pp. 568–579, May 1994.

[129] P. Jung and J. Blanz, "Joint detection with coherent receiver antenna diversity in CDMA mobile radio systems," *IEEE Transactions on Vehicular Technology*, vol. 44, pp. 76–88, Feb. 1995.

[130] P. Jung, J. Blanz, M. Nasshan, and P. W. Baier, "Simulation of the uplink of the JD-CDMA mobile radio systems with coherent receiver antenna diversity," *Wireless Personal Communications (Kluwer)*, vol. 1, no. 1, pp. 61–89, 1994.

[131] A. Steil and J. J. Blanz, "Spectral efficiency of JD-CDMA mobile radio systems applying coherent receiver antenna diversity," in *Proceedings of the International Symposium on Spread Spectrum Techniques and Applications (ISSSTA)*, (Mainz, Germany), pp. 313–319, Sep. 22-25 1996.

[132] P. Jung, M. Nasshan, and J. Blanz, "Application of turbo codes to a CDMA mobile radio system using joint detection and antenna diversity," in *Proceedings of the IEEE Vehicular Technology Conference (VTC)*, (Stockholm, Sweden), pp. 770–774, June 8-10 1994.

[133] P. Jung and M. Nasshan, "Results on turbo-codes for speech transmission in a joint detection CDMA mobile radio system with coherent receiver antenna diversity," *IEEE Transactions on Vehicular Technology*, vol. 46, pp. 862–870, Nov. 1997.

[134] M. M. Nasshan, A. Steil, A. Klein, and P. Jung, "Downlink cellular radio capacity of a joint detection CDMA mobile radio system," in *Proceedings of the 45th IEEE Vehicular Technology Conference (VTC)*, (Chicago, USA), pp. 474–478, Jul. 25-28 1995.

[135] A. Klein, "Data detection algorithms specially designed for the downlink of CDMA mobile radio systems," in *Proceedings of the IEEE Vehicular Technology Conference (VTC)*, (Phoenix, USA), pp. 203–207, May 4-7 1997.

[136] B. Steiner and P. Jung, "Optimum and suboptimum channel estimation for the uplink of CDMA mobile radio systems with joint detection," *European Transactions on Telecommunications*, vol. 5, pp. 39–50, 1994.

[137] M. Werner, "Multistage joint detection with decision feedback for CDMA mobile radio applications," in *Proceedings of the IEEE International Symposium on Personal, Indoor and Mobile Radio Communications (PIMRC)*, pp. 178–183, 1994.

[138] M. K. Varanasi and B. Aazhang, "Multistage detection in asynchronous code-division multiple-access communications," *IEEE Transactions on Communications*, vol. 38, pp. 509–519, Apr. 1990.

[139] M. K. Varanasi, "Group detection for synchronous Gaussian code-division multiple-access channels," *IEEE Transactions on Information Theory*, vol. 41, pp. 1083–1096, July 1995.

[140] M. K. Varanasi, "Parallel group detection for synchronous CDMA communication over frequency-selective Rayleigh fading channels," *IEEE Transactions on Information Theory*, vol. 42, pp. 116–128, Jan. 1996.

[141] Y. C. Yoon, R. Kohno, and H. Imai, "A SSMA system with cochannel interference cancellation with multipath fading channels," *IEEE Journal on Selected Areas in Communications*, vol. 11, pp. 1067–1075, Sep. 1993.

[142] T. R. Giallorenzi and S. G. Wilson, "Suboptimum multiuser receivers for convolutionally coded asynchronous DS-CDMA systems," *IEEE Transactions on Communications*, vol. 44, pp. 1183–1196, Sep. 1996.

[143] Y. Sanada and M. Nakagawa, "A multiuser interference cancellation technique utilizing convolutional codes and multicarrier modulation for wireless indoor communications," *IEEE Journal on Selected Areas in Communications*, vol. 14, pp. 1500–1509, Oct. 1996.

[144] M. Latva-aho, M. Juntti, and M. Heikkilä, "Parallel interference cancellation receiver for DS-CDMA systems in fading channels," in *Proceedings of the IEEE International Symposium on Personal, Indoor and Mobile Radio Communications (PIMRC)*, (Helsinki, Finland), pp. 559–564, Sep. 1-4 1997.

[145] D. Dahlhaus, A. Jarosch, B. H. Fleury, and R. Heddergott, "Joint demodulation in DS/CDMA systems exploiting the space and time diversity of the mobile radio channel," in *Proceedings of the IEEE International Symposium on Personal, Indoor and Mobile Radio Communications (PIMRC)*, (Helsinki, Finland), pp. 47–52, Sep. 1-4 1997.

[146] D. Divsalar, M. K. Simon, and D. Raphaeli, "Improved parallel interference cancellation for CDMA," *IEEE Transactions on Communications*, vol. 46, pp. 258–267, Feb. 1998.

[147] P. Patel and J. Holtzman, "Analysis of a simple successive interference cancellation scheme in a DS/CDMA system," *IEEE Journal on Selected Area in Communications*, vol. 12, pp. 796–807, June 1994.

[148] A. C. K. Soong and W. A. Krzymien, "A novel CDMA multi-user interference cancellation receiver with reference symbol aided estimation of channel parameters," *IEEE Journal on Selected Areas in Communications*, vol. 14, pp. 1536–1547, Oct. 1996.

[149] A. L. C. Hui and K. B. Letaief, "Successive interference cancellation for multiuser asynchronous DS/CDMA detectors in multipath fading links," *IEEE Transactions on Communications*, vol. 46, pp. 384–391, Mar. 1998.

[150] Y. Li and R. Steele, "Serial interference cancellation method for CDMA," *Electronics Letters*, vol. 30, pp. 1581–1583, Sep. 1994.

[151] T. B. Oon, R. Steele, and Y. Li, "Cancellation frame size for a quasi-single-bit detector in asynchronous CDMA channel," *Electronics Letters*, vol. 33, pp. 258–259, Feb. 1997.

[152] T.-B. Oon, R. Steele, and Y. Li, "Performance of an adaptive successive serial-parallel CDMA cancellation scheme in flat Rayleigh fading channels," in *Proceedings of the IEEE Vehicular Technology Conference (VTC)*, (Phoenix, USA), pp. 193–197, May 4-7 1997.

[153] M. Sawahashi, Y. Miki, H. Andoh, and K. Higuchi, "Pilot symbol-assisted coherent multistage interference canceller using recursive channel estimation for DS-CDMA mobile radio," *IEICE Transactions on Communications*, vol. E79-B, pp. 1262–1270, Sep. 1996.

[154] S. Sun, L. K. Rasmussen, H. Sugimoto, and T. J. Lim, "Hybrid interference canceller in CDMA," in *Proceedings of the IEEE International Symposium on Spread Spectrum Techniques and Applications (ISSSTA)*, (Sun City, South Africa), pp. 150–154, Sep. 2-4 1998.

[155] Y. Cho and J. H. Lee, "Analysis of an adaptive SIC for near-far resistant DS-CDMA," *IEEE Transactions on Communications*, vol. 46, pp. 1429–1432, Nov. 1998.

[156] P. Agashe and B. Woerner, "Interference cancellation for a multicellular CDMA environment," *Wireless Personal Communications (Kluwer)*, vol. 3, no. 1-2, pp. 1–14, 1996.

[157] L. K. Rasmussen, T. J. Lim, and T. M. Aulin, "Breadth-first maximum likelihood detection in multiuser CDMA," *IEEE Transactions on Communications*, vol. 45, pp. 1176–1178, Oct. 1997.

[158] L. Wei, L. K. Rasmussen, and R. Wyrwas, "Near optimum tree-search detection schemes for bit-synchronous multiuser CDMA systems over Gaussian and two-path Rayleigh-fading channels," *IEEE Transactions on Communications*, vol. 45, pp. 691–700, June 1997.

[159] M. Nasiri-Kenari, R. R. Sylvester, and C. K. Rushforth, "Efficient soft-in-soft-out multiuser detector for synchronous CDMA with error-control coding," *IEEE Transactions on Vehicular Technology*, vol. 47, pp. 947–953, Aug. 1998.

[160] J. B. Anderson and S. Mohan, "Sequential coding algorithms: a survey and cost analysis," *IEEE Transactions on Communications*, vol. 32, pp. 169–176, Feb. 1984.

[161] C. Schlegel, S. Roy, P. D. Alexander, and Z.-J. Xiang, "Multiuser projection receivers," *IEEE Journal on Selected Areas in Communications*, vol. 14, pp. 1610–1618, Oct. 1996.

[162] P. D. Alexander, L. K. Rasmussen, and C. B. Schlegel, "A linear receiver for coded multiuser CDMA," *IEEE Transactions on Communications*, vol. 45, pp. 605–610, May 1997.

[163] P. B. Rapajic and B. S. Vucetic, "Adaptive receiver structures for asynchronous CDMA systems," *IEEE Journal on Selected Areas in Communications*, vol. 12, pp. 685–697, May 1994.

[164] G. Woodward and B. S. Vucetic, "Adaptive detection for DS-CDMA," *Proceedings of the IEEE*, vol. 86, pp. 1413–1434, July 1998.

[165] Z. Xie, R. T. Short, and C. K. Rushforth, "Family of suboptimum detectors for coherent multiuser communications," *IEEE Journal on Selected Areas in Communications*, vol. 8, pp. 683–690, May 1990.

[166] T. J. Lim, L. K. Rasmussen, and H. Sugimoto, "An asynchronous multiuser CDMA detector based on the Kalman filter," *IEEE Journal on Selected Areas in Communications*, vol. 16, pp. 1711–1722, Dec. 1998.

[167] P. Seite and J. Tardivel, "Adaptive equalizers for joint detection in an indoor CDMA channel," in *Proceedings of the IEEE Vehicular Technology Conference (VTC)*, (Chicago, USA), pp. 484–488, Jul. 25-28 1995.

[168] S. M. Spangenberg, D. G. M. Cruickshank, S. McLaughlin, G. J. R. Povey, and P. M. Grant, "Advanced multiuser detection techniques for downlink CDMA, version 2.0," tech. rep., Virtual Centre of Excellence in Mobile and Personal Communications Ltd (Mobile VCE), July 1999.

[169] G. J. R. Povey, P. M. Grant, and R. D. Pringle, "A decision-directed spread-spectrum RAKE receiver for fast-fading mobile channels," *IEEE Transactions on Vehicular Technology*, vol. 45, pp. 491–502, Aug. 1996.

[170] H. Liu and K. Li, "A decorrelating RAKE receiver for CDMA communications over frequency-selective fading channels," *IEEE Transactions on Communications*, vol. 47, pp. 1036–1045, Jul. 1999.

[171] Z. Xie, C. K. Rushforth, R. T. Short, and T. K. Moon, "Joint signal detection and parameter estimation in multiuser communications," *IEEE Transactions on Communications*, vol. 41, pp. 1208–1216, Aug. 1993.

[172] N. Seshadri, "Joint data and channel estimation using blind trellis search techniques," *IEEE Transactions on Communications*, vol. 42, pp. 1000–1011, Feb/Mar/Apr 1994.

[173] R. Raheli, A. Polydoros, and C.-K. Tzou, "Per-survivor-processing: A general approach to MLSE in uncertain environments," *IEEE Transactions on Communications*, vol. 43, pp. 354–364, Feb/Mar/Apr 1995.

[174] R. Raheli, G. Marino, and P. Castoldi, "Per-survivor processing and tentative decisions: What is in between?," *IEEE Transactions on Communications*, vol. 44, pp. 127–129, Feb. 1998.

[175] T. K. Moon, Z. Xie, C. K. Rushforth, and R. T. Short, "Parameter estimation in a multi-user communication system," *IEEE Transactions on Communications*, vol. 42, pp. 2553–2560, Aug. 1994.

[176] R. Iltis and L. Mailaender, "Adaptive multiuser detector with joint amplitude and delay estimation," *IEEE Journal on Selected Areas in Communications*, vol. 12, pp. 774–785, June 1994.

[177] U. Mitra and H. V. Poor, "Adaptive receiver algorithms for near-far resistant CDMA," *IEEE Transactions on Communications*, vol. 43, pp. 1713–1724, Feb-Apr 1995.

[178] U. Mitra and H. V. Poor, "Analysis of an adaptive decorrelating detector for synchronous CDMA," *IEEE Transactions on Communications*, vol. 44, pp. 257–268, Feb. 1996.

[179] X. Wang and H. V. Poor, "Blind equalization and multiuser detection in dispersive CDMA channels," *IEEE Transactions on Communications*, vol. 46, pp. 91–103, Jan. 1998.

[180] X. D. Wang and H. V. Poor, "Blind multiuser detection: a subspace approach," *IEEE Transactions on Information Theory*, vol. 44, pp. 677–690, Mar. 1998.

[181] M. Honig, U. Madhow, and S. Verdú, "Blind adaptive multiuser detection," *IEEE Transactions on Information Theory*, vol. 41, pp. 944–960, July 1995.

[182] N. B. Mandayam and B. Aazhang, "Gradient estimation for sensitivity analysis and adaptive multiuser interference rejection in code division multiple access systems," *IEEE Transactions on Communications*, vol. 45, pp. 848–858, July 1997.

[183] S. Ulukus and R. D. Yates, "A blind adaptive decorrelating detector for CDMA systems," *IEEE Journal on Selected Areas in Communications*, vol. 16, no. 8, pp. 1530–1541, 1998.

[184] T. J. Lim and L. K. Rasmussen, "Adaptive symbol and parameter estimation in asynchronous multiuser CDMA detectors," *IEEE Transactions on Communications*, vol. 45, pp. 213–220, Feb. 1997.

[185] J. Miguez and L. Castedo, "A linearly constrained constant modulus approach to blind adaptive multiuser interference suppression," *IEEE Communications Letters*, vol. 2, pp. 217–219, Aug. 1998.

[186] D. N. Godard, "Self-recovering equalization and carrier tracking in two-dimensional data communication systems," *IEEE Transactions on Communications*, vol. 28, pp. 1867–1875, Nov. 1980.

[187] K. Wesolowsky, "Analysis and properties of the modified constant modulus algorithm for blind equalization," *European Transactions on Telecommunications and Related Technologies*, vol. 3, pp. 225–230, May-June 1992.

[188] K. Fukawa and H. Suzuki, "Orthogonalizing matched filtering (OMF) detector for DS-CDMA mobile communication systems," *IEEE Transactions on Vehicular Technology*, vol. 48, pp. 188–197, Jan. 1999.

[189] U. Fawer and B. Aazhang, "Multiuser receiver for code division multiple access communications over multipath channels," *IEEE Transactions on Communications*, vol. 43, pp. 1556–1565, Feb-Apr 1995.

[190] W. H. Press, S. A. Teukolsky, W. T. Vetterling, and B. P. Flannery, *Numerical Recipes in C: The Art of Scientific Computing*. Cambridge University Press, 1993.

[191] Y. Bar-Ness, "Asynchronous multiuser CDMA detector made simpler: Novel decorrelator, combiner, canceller, combiner (dc^3) structure," *IEEE Transactions on Communications*, vol. 47, pp. 115–122, Jan. 1999.

[192] K. Yen and L. Hanzo, "Hybrid genetic algorithm based multi-user detection schemes for synchronous CDMA systems," in *Proceedings of the IEEE Vehicular Technology Conference (VTC)*, (Tokyo, Japan), May 15-18 2000.

[193] W. M. Jang, B. R. Vojčić, and R. L. Pickholtz, "Joint transmitter-receiver optimization in synchronous multiuser communications over multipath channels," *IEEE Transactions on Communications*, vol. 46, pp. 269–278, Feb. 1998.

[194] B. R. Vojčić and W. M. Jang, "Transmitter precoding in synchronous multiuser communications," *IEEE Transactions on Communications*, vol. 46, pp. 1346–1355, Oct. 1998.

[195] R. Tanner and D. G. M. Cruickshank, "Receivers for nonlinearly separable scenarios in DS-CDMA," *Electronics Letters*, vol. 33, pp. 2103–2105, Dec. 1997.

[196] R. Tanner and D. G. M. Cruickshank, "RBF based receivers for DS-CDMA with reduced complexity," in *Proceedings of the IEEE International Symposium on Spread Spectrum Techniques and Applications (ISSSTA)*, (Sun City, South Africa), pp. 647–651, Sep. 2-4 1998.

[197] C. Berrou, P. Adde, E. Angui, and S. Faudeil, "A low-complexity soft-output Viterbi decoder architecture," in *Proceedings of IEEE International Conference on Communications (ICC)*, (Geneva, Switerland), pp. 737–740, May 23-26 1993.

[198] T. R. Giallorenzi and S. G. Wilson, "Multiuser ML sequence estimator for convolutionally coded asynchronous DS-CDMA systems," *IEEE Transactions on Communications*, vol. 44, pp. 997–1008, Aug. 1996.

[199] M. Moher, "An iterative multiuser decoder for near-capacity communications," *IEEE Transactions on Communications*, vol. 46, pp. 870–880, July 1998.

[200] M. Moher and P. Guinaud, "An iterative algorithm for asynchronous coded multiuser detection," *IEEE Communications Letters*, vol. 2, pp. 229–231, Aug. 1998.

[201] P. D. Alexander, A. J. Grant, and M. C. Reed, "Iterative detection in code-division multiple-access with error control coding," *European Transactions on Telecommunications*, vol. 9, pp. 419–426, Sep.-Oct. 1998.

[202] P. D. Alexander, M. C. Reed, J. A. Asenstorfer, and C. B. Schlegel, "Iterative multiuser interference reduction: Turbo CDMA," *IEEE Transactions on Communications*, vol. 47, pp. 1008–1014, Jul. 1999.

[203] M. C. Reed, C. B. Schlegel, P. D. Alexander, and J. A. Asenstorfer, "Iterative multiuser detection for CDMA with FEC: Near-single-user performance," *IEEE Transactions on Communications*, vol. 46, pp. 1693–1699, Dec. 1998.

[204] X. D. Wang and H. V. Poor, "Iterative (turbo) soft interference cancellation and decoding for coded CDMA," *IEEE Transactions on Communications*, vol. 47, pp. 1046–1061, Jul. 1999.

[205] T. Ojanperä, A. Klein, and P.-O. Anderson, "FRAMES multiple access for UMTS," *IEE Colloquium (Digest)*, pp. 7/1–7/8, May 1997.

[206] E. A. Lee and D. G. Messerschmitt, *Digital Communication*. Kluwer Academic Publishers, 1988.

[207] D. F. Mix, *Random Signal Processing*. Prentice-Hall International, Inc., 1995.

[208] H. R. Karimi and N. W. Anderson, "A novel and efficient solution to block-based joint-detection using approximate Cholesky factorization," in *Proceedings of the IEEE International Symposium on Personal, Indoor and Mobile Radio Communications (PIMRC)*, (Boston, USA), pp. 1340–1344, Sep 8-11 1998.

[209] R. Karimi, "Efficient multi-rate multi-user detection for the asynchronous WCDMA uplink," in *Proceedings of the IEEE Vehicular Technology Conference (VTC Fall)*, (Amsterdam, The Netherlands), pp. 593–597, Sep. 19-22 1999.

[210] N. Benvenuto and G. Sostrato, "Joint detection with low computational complexity for hybrid TD-CDMA systems," in *Proceedings of the IEEE Vehicular Technology Conference (VTC Fall)*, (Amsterdam, The Netherlands), pp. 618–622, Sep. 19-22 1999.

[211] P. A. Bello, "Sample size required in error-rate measurement on fading channels," *Proceedings of the IEEE*, vol. 86, July 1998.

[212] A. S. Barbulescu and S. S. Pietrobon, "Interleaver design for turbo codes," *IEE Electronics Letters*, vol. 30, pp. 2107–2108, Dec 1994.

[213] J. Hagenauer and P. Hoeher, "A Viterbi algorithm with soft-decision outputs and its applications," in *Proceedings of IEEE Global Telecommunications Conference*, (Dallas, USA), pp. 1680–1686, Nov. 27-30 1989.

[214] E. Papproth and G. K. Kaleh, "Near-far resistant channel estimation for the DS-CDMA uplink," in *Proceedings of the IEEE International Symposium on Personal, Indoor and Mobile Radio Communications (PIMRC)*, (Toronto, Canada), pp. 758–762, Sep. 27-29 1995.

[215] T. Ojanperä and R. Prasad, *Wideband CDMA for Third Generation Mobile Communications*. Artech House, Inc., 1998.

[216] W. T. Webb and R. Steele, "Variable rate QAM for mobile radio," *IEEE Transactions on Communications*, vol. 43, pp. 2223 – 2230, July 1995.

[217] S. W. Kim, "Adaptive rate and power DS/CDMA communications in fading channels," *IEEE Communications Letters*, vol. 3, pp. 85–87, Apr. 1999.

[218] F. Adachi, K. Ohno, A. Higashi, T. Dohi, and Y. Okumura, "Coherent multicode DS-CDMA mobile radio access," *IEICE Transactions on Communications*, vol. E79-B, pp. 1316–1325, Sep. 1996.

[219] T. Dohi, Y. Okumura, A. Higashi, K. Ohno, and F. Adachi, "Experiments on coherent multicode DS-CDMA," *IEICE Transactions on Communications*, vol. E79-B, pp. 1326–1332, Sep. 1996.

[220] H. D. Schotten, H. Elders-Boll, and A. Busboom, "Adaptive multi-rate multi-code CDMA systems," in *Proceedings of the IEEE Vehicular Technology Conference (VTC)*, (Ottawa, Canada), pp. 782–785, May 18-21 1998.

[221] M. Saquib and R. Yates, "Decorrelating detectors for a dual rate synchronous DS/CDMA channel," *Wireless Personal Communications (Kluwer)*, vol. 9, pp. 197–216, May 1999.

[222] M. J. Juntti, "Multiuser detector performance comparisons in multirate CDMA systems," in *Proceedings of the IEEE Vehicular Technology Conference (VTC)*, (Ottawa, Canada), pp. 36–40, May 18-21 1998.

[223] S. Abeta, S. Sampei, and N. Morinaga, "Channel activation with adaptive coding rate and processing gain control for cellular DS/CDMA systems," in *Proceedings of the IEEE Vehicular Technology Conference (VTC)*, (Atlanta, USA), pp. 1115–1119, Apr. 28-May 1 1996.

[224] M. Hashimoto, S. Sampei, and N. Morinaga, "Forward and reverse link capacity enhancement of DS/CDMA cellular system using channel activation and soft power control techniques," in *Proceedings of the IEEE International Symposium on Personal, Indoor and Mobile Radio Communications (PIMRC)*, (Helsinki, Finland), pp. 246–250, Sep. 1-4 1997.

[225] V. K. N. Lau and S. V. Maric, "Variable rate adaptive modulation for DS-CDMA," *IEEE Transactions on Communications*, vol. 47, pp. 577–589, Apr. 1999.

[226] S. Tateesh, S. Atungsiri, and A. M. Kondoz, "Link adaptive multi-rate coding verification system for CDMA mobile communications," in *Proceedings of the IEEE Global Telecommunications Conference (GLOBECOM)*, (London, UK), pp. 1969–1973, Nov. 18-22 1996.

[227] Y. Okumura and F. Adachi, "Variable data rate transmission with blind rate detection for coherent DS-CDMA mobile radio," *Electronics Letters*, vol. 32, pp. 1865–1866, Sep. 1996.

[228] J. S. Blogh, P. Cherriman, and L. Hanzo, "Adaptive beamforming assisted, power controlled dynamic channel allocation for adaptive modulation," in *Proceedings of the IEEE Vehicular Technology Conference (VTC Fall)*, (Amsterdam, The Netherlands), pp. 2348–2352, Sep. 19-22 1999.

[229] K. Miya, O. Kato, K. Homma, T. Kitade, M. Hayashi, and T. Ue, "Wideband CDMA systems in TDD-mode operation for IMT-2000," *IEICE Transactions on Communications*, vol. E81-B, pp. 1317–1326, July 1998.

[230] O. Kato, K. Miya, K. Homma, T. Kitade, M. Hayashi, and M. Watanabe, "Experimental performance results of coherent wideband DS-CDMA with TDD scheme," *IEICE Transactions on Communications*, vol. E81-B, pp. 1337–1344, July 1998.

[231] I. Jeong and M. Nakagawa, "A novel transmission diversity system in TDD-CDMA," *IEICE Transactions on Communications*, vol. E81-B, pp. 1409–1416, July 1998.

[232] T. Keller and L. Hanzo, "Adaptive orthogonal frequency division multiplexing schemes," in *Proceedings of the ACTS Mobile Communications Summit*, (Rhodes, Greece), pp. 794–799, June 1998.

[233] T. Keller and L. Hanzo, "Blind-detection assisted sub-band adaptive turbo-coded OFDM schemes," in *Proceedings of the IEEE Vehicular Technology Conference (VTC Spring)*, (Houston, USA), pp. 489–493, May 16-20 1999.

[234] S. Sampei, S. Komaki, and N. Morinaga, "Adaptive modulation/TDMA scheme for large capacity personal multimedia communications systems," *IEICE Transactions on Communications*, vol. E77-B, pp. 1096–1103, September 1994.

[235] J. M. Torrance, *Adaptive Full Response Digital Modulation for Wireless Communications Systems.* PhD thesis, University of Southampton, 1997.

[236] M. S. Yee and L. Hanzo, "Multi-level radial basis function network based equalisers for Rayleigh channel," in *Proceedings of the IEEE Vehicular Technology Conference (VTC Spring)*, (Houston, USA), pp. 707–711, May 16-20 1999.

[237] A. J. Goldsmith and S. G. Chua, "Variable rate variable power MQAM for fading channels," *IEEE Transactions on Communications*, vol. 45, pp. 1218 – 1230, October 1997.

[238] C. H. Wong and L. Hanzo, "Upper-bound of a wideband burst-by-burst adaptive modem," in *Proceedings of the IEEE Vehicular Technology Conference (VTC Spring)*, (Houston, USA), pp. 1851–1855, May 16-20 1999.

[239] T. Eyceoz, A. Duel-Hallen, and H. Hallen, "Deterministic channel modeling and long range prediction of fast fading mobile radio channels," *IEEE Communications Letters*, vol. 2, pp. 254–256, Sep. 1998.

[240] F. Adachi, M. Sawahashi, and K. Okawa, "Tree-structured generation of orthogonal spreading codes with different lengths for forward link of DS-CDMA mobile radio," *Electronics Letters*, vol. 33, pp. 27–28, Jan. 1997.

[241] A.-L. Johansson and A. Svensson, "Successive interference cancellation schemes in multi-rate DS/CDMA systems," in *Wireless Information Networks (Baltzer)*, pp. 265–279, 1996.

[242] A. Toskala, J. P. Castro, E. Dahlman, M. Latva-aho, and T. Ojanperä, "FRAMES FMA2 Wideband-CDMA for UMTS," *European Transactions on Telecommunications*, vol. 9, pp. 325–335, Jul.-Aug. 1998.

[243] T. Ue, S. Sampei, and N. Morinaga, "Symbol rate and modulation level controlled adaptive modulation/TDMA/TDD for personal communication systems," in *Proceedings of the IEEE Vehicular Technology Conference (VTC)*, (Chicago, USA), pp. 306–310, Jul. 25-28 1995.

[244] S. M. Alamouti, "A simple transmit diversity technique for wireless communications," *IEEE Journal on Selected Areas in Communications*, vol. 16, pp. 1451–1458, Oct. 1998.

[245] V. Tarokh, N. Seshadri, and A. R. Calderbank, "Space-time codes for high data rate wireless communication: Performance analysis and code construction," *IEEE Transactions on Information Theory*, vol. 44, pp. 744–765, Mar. 1998.

[246] S. R. Kim, J. G. Lee, and H. Lee, "Interference cancellation scheme with simple structure and better performance," *Electronics Letters*, vol. 32, pp. 2115–2117, Nov. 1996.

[247] D. Koulakiotis and A. H. Aghvami, "Evaluation of a DS/CDMA multiuser receiver employing a hybrid form of interference cancellation in Rayleigh-fading channels," *IEEE Communications Letters*, vol. 2, pp. 61–63, Mar. 1998.

[248] A. L. Johansson and A. Svensson, "Multistage interference cancellation in multirate DS/CDMA on a mobile radio channel," in *Proceedings of the IEEE Vehicular Technology Conference (VTC)*, (Atlanta, USA), pp. 666–670, Apr. 28-May 1 1996.

[249] M. C. Reed, C. B. Schlegel, P. D. Alexander, and J. A. Asenstorfer, "Iterative Multiuser Detection for CDMA with FEC: Near single user performance," *IEEE Transaction on Communication*, vol. 46, pp. 1693–1699, Dec. 1999.

[250] Y. Zhang, "Reduced complexity iterative multiuser detection for DS/CDMA with FEC," in *International Conference on Universal Personal Communications*, no. 12, (San Diego, U.S.A), pp. 10–14, Oct. 1997.

[251] P. D. Alexander, A. J. Grant, and M. C. Reed, "Performance analysis of an iterative decoder for code-division mulitple-access," *European Transaction on Telcommunicaiton*, vol. 9, pp. 419–426, Sept./Oct 1998.

[252] M. C. Reed, *Iterative Receiver Techniques for Coded Multiple Access Communication System.* PhD thesis, The University of South Australia, 1999.

[253] G. Ungerböck, "Channel coding with multilevel/phase signals," *IEEE Transactions on Information Theory*, vol. 28, pp. 55–67, Jan 1982.

[254] D. Divsalar and M. K. Simon, "The design of trellis coded MPSK for fading channel: Performance criteria," *IEEE Transactions on Communications*, vol. 36, pp. 1004–1012, Sept 1988.

[255] P. Robertson and T. Wörz, "Bandwidth-Efficient Turbo Trellis-Coded Modulation Using Punctured Component Codes," *IEEE Journal on Selected Areas in Communications*, vol. 16, pp. 206–218, Feb 1998.

[256] C. Douillard, M. Jezequel, C. Berrou, A. Picart, P. Didier, and A. Glavieux, "Iterative correction of intersymbol interference: Turbo-equalization," *European Transactions on Telecommunications and Related Technologies*, vol. 6, pp. 507–511, Sept–Oct 1995.

[257] P. Patel and J. Holtzman, "Analysis of a simple successive interference cancellation scheme in DS/CDMA system," *IEEE Journal of Selected Areas in Communications*, vol. 12, pp. 796–807, June 1994.

[258] P. Robertson, E. Villebrun, and P. Höher, "A comparision of optimal and sub-optimal MAP decoding algorithms operating in the log domain," in *Preceedings of the International Conference on Communications*, (Seattle, Unites States), pp. 1009–1013, 28-22 May 1995.

[259] C. E. Shannon, "A mathematical theory of communication," *The Bell System Technical Journal*, vol. 27, pp. 379–423, July 1948.

[260] R. Gallager, "Low Density Parity Check Codes," *IRE Transactions on Information Theory*, 1962.

[261] R. Gallager, "Low Density Parity Check Codes," *Ph.D thesis,M.I.T,USA*, 1963.

[262] M. R. Tanner, "A Recursive Approach to Low Complexity Codes," *IEEE Transactions on Information Theory*, vol. 27, September 1981.

[263] D.J.C MacKay and R.M. Neal, "Good error-correction codes based on very sparse matrices," *IEEE Transactions on Information Theory*, vol. 45, pp. 399–431, March 1999.

[264] D. J. C Mackay and R. M. Neal, "Near Shannon Limit Performance of Low Density Parity Check Codes," *Electronics Letters*, vol. 33, pp. 457–458, March 1997.

[265] M. E. Rollins and S. J. Simmons, "Simplified per-survivor Kalman processing in fast frequency-selective fading channels," *IEEE Transactions on Communications*, vol. 45, pp. 544–553, May 1997.

[266] W. Koch and A. Baier, "Optimum and sub-optimum detection of coded data disturbed by time-varying inter-symbol interference," in *Proceedings of IEEE Global Telecommunications Conference (GLOBECOM)*, pp. 1679–1684, 1990.

[267] J. A. Erfanian, S. Pasupathy, and G. Gulak, "Reduced complexity symbol detectors with parallel structures for ISI channels," *IEEE Transactions on Communications*, vol. 42, pp. 1661–1671, Feb.-Apr. 1994.

[268] J. Blogh and L. Hanzo, *3G Systems and Intelligent Networking*. John Wiley and IEEE Press, 2002. (For detailed contents, please refer to http://www-mobile.ecs.soton.ac.uk.).

[269] L. Hanzo, L. L. Yang, E. L. Kuan, and K. Yen, *Single- and Multi-Carrier CDMA*. John Wiley and IEEE press, May 2003.

[270] G. J. Foschini, "Layered Space-time architecture for wireless communication in a fading environment when using multi-element antennas," *Bell Labs Tech. J.*, pp. 41–59, 1996.

[271] J. Winters, "Smart antennas for wireless systems," *IEEE Personal Communications*, vol. 5, pp. 23–27, February 1998.

[272] R. Derryberry, S. Gray, D. Ionescu, G. Mandyam, and B. Raghothaman, "Transmit diversity in 3g CDMA systems," *IEEE Communications Magazine*, vol. 40, pp. 68–75, April 2002.

[273] W. Lee, *Mobile Communications Engineering*. New York: McGraw-Hill, 2nd ed., 1998.

[274] B. Hochwald, T. L. Marzetta, and C. B. Papadias, "A transmitter diversity scheme for wideband CDMA systems based on space-time spreading," *IEEE Journal on Selected Areas in Communications*, vol. 19, pp. 48–60, January 2001.

[275] N. Nakagami, "the m-distribution, a general formula for intensity distribution of rapid fading," in *Statistical Methods in Radio Wave Propagation* (W. G. Hoffman, ed.), Oxford, England: Pergamon, 1960.

[276] T. Eng and L. Milstein, "Coherent DS-CDMA performance in Nakagami multipath fading," *IEEE Transactions on Communications*, vol. 43, pp. 1134–1143, Feb./Mar./Apr. 1995.

[277] V. Aalo, O. Ugweje, and R. Sudhakar, "Performance analysis of a DS/CDMA system with noncoherent M-ary orthogonal modulation in Nakagami fading," *IEEE Transactions on Vehicular Technology*, vol. 47, pp. 20–29, February 1998.

[278] M.-S. Alouini and A. Goldsmith, "A unified approach for calculating error rates of linearly modulated signals over generalized fading channels," *IEEE Transactions on Communications*, vol. 47, pp. 1324–1334, September 1999.

[279] M. Simon and M.-S. Alouini, "A unified approach to the probability of error for noncoherent and differentially coherent modulation over generalized fading channels," *IEEE Transactions on Communications*, vol. 46, pp. 1625–1638, December 1998.

[280] M. Simon and M.-S. Alouini, "A unified approach to the performance analysis of digital communication over generalized fading channels," *Proceedings of the IEEE*, vol. 86, pp. 1860–1877, September 1998.

[281] L. E. Millera and J.-S. Lee, *CDMA Systems Engineering Handbook*. Artech House Pubs., 1998.

[282] C. Darwin, *On the Origin of Species*. London: John Murray, 1859.

[283] I. Rechenberg, "Cybernetic solution path of an experimental problem," tech. rep., Ministry of Aviation, Royal Aircraft Establishment, U.K., 1965.

[284] H.-P. Schwefel, *Evolutionsstrategie und numerische Optimierung*. PhD thesis, Technische Universität Berlin, 1975.

[285] L. Fogel, A. J. Owens, and M. J. Walsh, *Artificial Intelligence through Simulated Evolution*. New York: John Wiley, 1966.

[286] T. Bäck, U. Hammel, and H.-P. Schwefel, "Evolutionary computation: Comments on the history and current state," *IEEE Transactions on Evolutionary Computation*, vol. 1, pp. 3–17, April 1997.

[287] M. Mitchell, *An Introduction to Genetic Algorithms*. Cambridge, Massachusetts: MIT Press, 1996.

[288] K. S. Tang, K. F. Man, S. Kwong, and Q. He, "Genetic algorithms and their applications," *IEEE Signal Processing Magazine*, vol. 13, pp. 22–37, November 1996.

[289] D. Whitley, "A genetic algorithm tutorial," *Statistics and Computing*, vol. 4, pp. 65–85, June 1994.

[290] S. Forrest, "Genetic algorithms: Principles of natural selection applied to computation," *Science*, vol. 261, pp. 872–878, August 1993.

[291] H. Mühlenbein, *Foundations of Genetic Algorithms*, ch. Evolution in time and space – The Parallel Genetic Algorithm, pp. 316–337. California, USA: G. Rawlins, ed., Morgan Kaufmann, 1991.

[292] J. J. Grefenstette and J. E. Baker, "How genetic algorithms work: A critical look at implicit parallelism," in *Proceedings of the Third International Conference on Genetic Algorithms* (J. D. Schaffer, ed.), (California, USA), pp. 20–27, Morgan Kaufmann, 1989.

[293] B. L. Miller and D. E. Goldberg, "Genetic algorithms, selection schemes, and the varying effects of noise," *Evolutionary Computation*, vol. 4, pp. 113–131, Summer 1996.

[294] G. Harik, E. Cantú-Paz, D. E. Goldberg, and B. L. Miller, "The gambler's ruin problem, genetic algorithms, and the sizing of populations," in *Proceedings of the 1997 IEEE Conference on Evolutionary Computation* (T. Bäck, ed.), (New York), pp. 7–12, IEEE Press, 1997.

[295] M. D. Vose and G. E. Liepins, "Punctuated equilibria in genetic search," *Complex Systems*, vol. 5, pp. 31–44, January/February 1991.

[296] A. E. Nix and M. D. Vose, "Modeling genetic algorithms with Markov chains," *Annals of Mathematics and Artificial Intelligence*, vol. 5, pp. 79–88, January/February/March 1992.

[297] M. D. Vose, *Foundations of Genetic Algorithms 2*, ch. Modeling Simple Genetic Algorithms, pp. 63–73. California, USA: L. D. Whitley, ed., Morgan Kaufmann, 1993.

[298] A. H. Wright, *Foundations of Genetic Algorithms*, ch. Genetic Algorithms for Real Parameter Optimization, pp. 205–218. California, USA: G. Rawlins, ed., Morgan Kaufmann, 1991.

[299] C. Z. Janikow and Z. Michalewicz, "An experimental comparison of binary and floating point representations in genetic algorithms," in *Proceedings of the Fourth International Conference on Genetic Algorithms* (R. K. Belew and L. B. Booker, eds.), (California, USA), pp. 31–36, Morgan Kaufmann, 1991.

[300] R. Tanese, *Distributed Genetic Algorithms for Function Optimization*. PhD thesis, University of Michigan, 1989.

[301] J. E. Baker, "Adaptive selection methods for genetic algorithms," in *Proceedings of the First International Conference on Genetic Algorithms and Their Applications* (J. J. Grefenstette, ed.), (New Jersey, USA), pp. 101–111, Lawrence Erlbaum Associates, 1985.

[302] T. Blickle and L. Thiele, "A comparison of selection schemes used in evolutionary algorithms," *Evolutionary Computation*, vol. 4, pp. 361–394, Winter 1996.

[303] D. E. Goldberg and K. Deb, *Foundations of Genetic Algorithms*, ch. A Comparative Analysis of Selection Schemes Used in Genetic Algorithms, pp. 69–93. California, USA: G. Rawlins, ed., Morgan Kaufmann, 1991.

[304] L. Eshelman and J. Schaffer, "Preventing premature convergence in genetic algorithms by preventing incest," in *Proceedings of the Fourth International Conference on Genetic Algorithms* (R. K. Belew and L. B. Booker, eds.), (California, USA), pp. 115–122, Morgan Kaufmann, 1991.

[305] G. Syswerda, "Uniform crossover in genetic algorithms," in *Proceedings of the Third International Conference on Genetic Algorithms* (J. D. Schaffer, ed.), (California, USA), pp. 2–9, Morgan Kaufmann, 1989.

[306] W. Spears and K. De Jong, *Foundations of Genetic Algorithms*, ch. An Analysis of Multi-Point Crossover, pp. 301–315. California, USA: G. Rawlins, ed., Morgan Kaufmann, 1991.

[307] J. D. Schaffer, R. A. Caruana, L. J. Eshelman, and R. Das, "A study of control parameters affecting online performance of genetic algorithms for function optimization," in *Proceedings of the Third International Conference on Genetic Algorithms* (J. D. Schaffer, ed.), (California, USA), pp. 51–60, Morgan Kaufmann, 1989.

[308] J. J. Grefenstette, "Optimization of control parameters for genetic algorithms," *IEEE Transactions on Systems, Man and Cybernetics*, vol. SMC-16, pp. 122–128, January 1986.

[309] T. Bäck, "Optimal mutation rates in genetic search," in *Proceedings of the Fifth International Conference on Genetic Algorithms* (S. Forrest, ed.), (California, USA), pp. 2–8, Morgan Kaufmann, 1993.

[310] T. Bäck, "Self adaptation in genetic algorithms," in *Proceedings of the First European Conference on Artificial Life* (F. J. Varela and P. Bourgine, eds.), (Massachusetts, USA), pp. 263–271, MIT Press, 1992.

[311] M. J. Juntti, T. Schlösser, and J. O. Lilleberg, "Genetic algorithms for multiuser detection in synchronous CDMA," in *IEEE International Symposium on Information Theory – ISIT'97*, (Ulm, Germany), p. 492, 1997.

[312] X. F. Wang, W. S. Lu, and A. Antoniou, "A genetic algorithm-based multiuser detector for multiple-access communications," in *IEEE International Symposium on Circuits and System – ISCAS'98*, (Monterey, California, USA), pp. 534–537, 1998.

[313] C. Ergün and K. Hacioglu, "Application of a genetic algorithm to multi-stage detection in CDMA systems," in *Proceedings of the 9th Mediterranean Electrotechnical Conference – MELECON'98*, (Tel-Aviv, Israel), pp. 846–850, 1998.

[314] C. Ergün and K. Hacioglu, "Multiuser detection using a genetic algorithm in CDMA communications systems," *IEEE Transactions on Communications*, vol. 48, pp. 1374–1383, August 2000.

[315] S. Abedi, *Genetic Multiuser Detection for Code Division Multiple Access Systems*. PhD thesis, University of Surrey, 2000.

[316] M. B. Pursley, "Performance evaluation for phase-coded spread-spectrum multiple-access communication-part i: System analysis," *IEEE Transactions on Communications*, vol. COM-25, pp. 795–799, August 1977.

[317] R. Morrow, "Bit-to-bit error dependence in slotted DS/SSMA packet systems with random signature sequences," *IEEE Transactions on Communications*, vol. 37, pp. 1052–1061, October 1989.

[318] J. Holtzman, "A simple, accurate method to calculate spread-spectrum multiple-access error probabilities," *IEEE Transactions on Communications*, vol. 40, pp. 461–464, March 1992.

[319] S. Verdú, *Multiuser Detection*. New York, USA: Cambridge University Press, 1998.

[320] M. Varanasi and B. Aazhang, "Near-optimum detection in synchronous code-division multiple-access systems," *IEEE Transactions on Communications*, vol. 39, pp. 725–736, May 1991.

[321] S. Verdú, "Minimum probability of error for asynchronous Gaussian multiple-access channel," *IEEE Transactions on Communications*, vol. 32, pp. 85–96, January 1986.

[322] S. Abedi and R. Tafazolli, "Genetically modified multiuser detection for code division multiple access systems," *Journal on Selected Areas in Communications*, vol. 20, pp. 463–473, Feb. 2002.

[323] L. Wei, L. K. Rasmussen, and R. Wyrwas, "Near optimum tree-search detection schemes for bit-synchronous multiuser CDMA systems over Gaussian and two-path Rayleigh-fading channels," *IEEE Transactions on Communications*, vol. 45, pp. 691–700, June 1997.

[324] L. K. Rasmussen, T.-J. Lim, and T. M. Aulin, "Breadth-first maximum likelihood detection in multiuser CDMA," *IEEE Transactions on Communications*, vol. 45, pp. 1176–1178, October 1997.

[325] J.-S. Lee and L. E. Miller, *CDMA Systems Engineering Handbook*. Boston, USA: Artech House Publishers, 1998.

[326] J. Cavers, "An analysis of pilot symbol assisted modulation for Rayleigh fading channels," *IEEE Transactions on Vehicular Technology*, vol. 40, pp. 686–693, November 1991.

[327] T. Ojanperä and R. Prasad, *Wideband CDMA for Third Generation Mobile Communications*. Boston, USA: Artech House Publishers, 1998.

[328] Z. Xie, C. Rushforth, R. Short, and T. Moon, "Joint signal detection and parameter estimation in multiuser communications," *IEEE Transactions on Communications*, vol. 41, pp. 1208–1215, August 1993.

[329] U. Fawer and B. Aazhang, "A multiuser receiver for code division multiple access communications over multipath channels," *IEEE Transactions on Communications*, vol. 43, pp. 1556–1565, February/March/April 1995.

[330] T. Kawahara and T. Matsumoto, "Joint decorrelating multiuser detection and channel estimation in asynchronous CDMA mobile communication channels," *IEEE Transactions on Vehicular Technology*, vol. 44, pp. 506–515, August 1995.

[331] S. Chen and Y. Wu, "Maximum likelihood joint channel and data estimation using genetic algorithms," *IEEE Transactions on Signal Processing*, vol. 46, pp. 1469–1473, May 1998.

[332] Z. Zvonar and M. Stojanovic, "Performance of antenna diversity multiuser receivers in CDMA channels with imperfect fading estimation," *Wireless Personal Communications*, vol. 3, no. 1-2, pp. 91–110, 1996.

[333] D. N. Kalofonos, M. Stojanovic, and J. G. Proakis, "Analysis of the impact of channel estimation errors on the performance of a MC-CDMA system in a Rayleigh fading channel," in *IEEE Global Telecommunications Conference*, vol. 4, (Phoenix, Arizona, USA), pp. 213–217, November 1997.

[334] M. Omidi, P. Gulak, and S. Pasupathy, "Parallel structures for joint channel estimation and data detection over fading channels," *IEEE Journal of Selected Areas in Communications*, vol. 16, pp. 1616–1629, December 1998.

[335] M. Stojanovic and Z. Zvonar, "Performance of multiuser detection with adaptive channel estimation," *IEEE Transactions on Communications*, vol. 47, pp. 1129–1132, August 1999.

[336] P. Schramm, "Differentially coherent demodulation for differential bpsk in spread spectrum systems," *IEEE Transactions on Vehicular Technology*, vol. 48, pp. 1650–1656, September 1999.

[337] H. Liu and Z. Siveski, "Differentially coherent decorrelating detector for CDMA single-path time-varying Rayleigh fading channels," *IEEE Transactions on Communications*, vol. 47, pp. 590–597, April 1999.

[338] M. Juntti, *Multiuser Demodulation for DS-CDMA Systems in Fading Channels*. PhD thesis, University of Oulu, 1997.

[339] A. Klein, B. Steiner, and A. Steil, "Known and novel diversity approaches as a powerful means to enhance the performance of cellular mobile radio systems," *IEEE Journal of Selected Areas in Communications*, vol. 14, pp. 1784–1795, December 1996.

[340] P. Diáz and R. Agustí, "The use of coding and diversity combining for mitigating fading effects in a DS/CDMA system," *IEEE Transactions on Vehicular Technology*, vol. 47, pp. 95–102, February 1998.

[341] P. Jung and J. Blanz, "Joint detection with coherent receiver antenna diversity in CDMA mobile radio systems," *IEEE Transactions on Vehicular Technology*, vol. 44, pp. 76–88, February 1995.

[342] A. Naguib and A. Paulraj, "Performance of wireless CDMA with M-ary orthogonal modulation and cell site antenna arrays," *IEEE Journal of Selected Areas in Communications*, vol. 14, pp. 1770–1783, December 1996.

[343] P. van Rooyen, R. Kohno, and I. Oppermann, "DS-CDMA performance with maximum ratio combining and antenna arrays," *Wireless Networks*, vol. 4, pp. 479–488, June 1998.

[344] N. Srinivas and K. Deb, "Multiobjective optimization using nondominated sorting in genetic algorithms," *Evolutionary Computation*, vol. 2, pp. 221–248, Autumn 1994.

[345] E. Zitzler and L. Thiele, "Multiobjective evolutionary algorithms: A comparative case study and the strength Pareto approach," *IEEE Transactions on Evolutionary Computation*, vol. 3, pp. 257–271, November 1999.

[346] J. Panicker and S. Kumar, "Effect of system imperfections on BER performance of a CDMA receiver with multipath diversity combining," *IEEE Transactions on Vehicular Technology*, vol. 45, pp. 622–630, November 1996.

[347] R. Lupas and S. Verdú, "Near-far resistance of multi-user detectors in asynchronous channels," *IEEE Transactions on Communications*, vol. 38, pp. 496–508, April 1990.

[348] F.-C. Zheng and S. K. Barton, "Near-far resistant detection of CDMA signals via isolation bit insertion," *IEEE Transactions on Communications*, vol. 43, pp. 1313–1317, February/March/April 1995.

[349] Z. Xie, R. Short, and C. Rushforth, "A family of suboptimum detectors for coherent multiuser communications," *IEEE Journal of Selected Areas in Communications*, vol. 8, pp. 683–690, May 1990.

[350] S. S. H. Wijayasuriya, G. H. Norton, and J. P. McGeehan, "A sliding window decorrelating receiver for multiuser DS-CDMA mobile radio networks," *IEEE Transactions on Vehicular Technology*, vol. 45, pp. 503–521, August 1996.

[351] M. J. Juntti and B. Aazhang, "Finite memory-length linear multiuser detection for asynchronous CDMA communications," *IEEE Transactions on Communications*, vol. 45, pp. 611–622, May 1997.

[352] J. Shen and Z. Ding, "Edge decision assisted decorrelators asynchronous CDMA channels," *IEEE Transactions on Communications*, vol. 47, pp. 438–445, March 1999.

[353] M. Varanasi and B. Aazhang, "Multistage detection in asynchronous code division multiple-access communications," *IEEE Transactions on Communications*, vol. 38, pp. 509–519, April 1990.

[354] R. Prasad and S. Hara, "Overview of multi-carrier CDMA," in *Proceedings of the IEEE ISSSTA'96*, pp. 107–114, 1996.

[355] S. L. Miller and B. J. Rainbolt, "MMSE detection of multicarrier CDMA," *IEEE Journal on Selected Areas in Communications*, vol. 18, pp. 2356–2362, Nov 2000.

[356] P. Zong, K. Wang, and Y. Bar-Ness, "Partial sampling mmse interference suppression in asynchronous multicarrirer CDMA system," *IEEE Journal on Selected Areas in Communications*, vol. 19, pp. 1605–1613, August 2001.

[357] X. Gui and T.-S. Ng, "Performance of asynchronous orthogonal multicarrier CDMA system in frequency selective fading channel," *IEEE Transactions on Communications*, vol. 47, pp. 1084–1091, July 1999.

[358] M. Schnell and S. Kaiser, "Diversity considerations for MC-CDMA systems in mobile communications," in *Proceedings of IEEE ISSSTA 1996*, pp. 131–135, 1996.

[359] K. Yen and L. Hanzo, "Hybrid Genetic Algorithm Based Multiuser Detection Schemes for Synchronous CDMA Systems," in *Proceedings of the 51st IEEE Vehicular Technology Conference*, (Tokyo, Japan), pp. 1400–1404, 18 May 2000.

[360] K. Yen and L. Hanzo, "Genetic algorithm assisted joint multiuser symbol detection and fading channel estimation for synchronous CDMA systems," *IEEE Journal on Selected Areas in Communications*, vol. 19, pp. 985 – 998, June 2001.

[361] K. Yen and L. Hanzo, "Genetic algorithm assisted multiuser detection in asynchronous CDMA communications," in *Proceedings of the IEEE International Conference on Communications*, vol. 3, pp. 826 –830, 2001.

[362] G. Turin, "The effects of multipath and fading on the performance of direct-sequence CDMA systems," *IEEE Journal on Selected Areas in Communications*, vol. SAC-2, pp. 597–603, July 1984.

[363] M. Kavehrad and B. Ramamurthi, "Direct-sequence spread spectrum with DPSK modulation and diversity for indoor wireless communications," *IEEE Transactions on Communications*, vol. COM-35, pp. 224–236, February 1987.

[364] P. Enge and D. Sarwate, "Spread spectrum multiple access performance of orthogonal code: Linear receivers," *IEEE Transactions on Communications*, vol. COM-35, pp. 1309–1319, December 1987.

[365] K. Pahlavan and M. Chase, "Spread-spectrum multiple-access performance of orthogonal codes for indoor radio communications," *IEEE Transactions on Communications*, vol. 38, pp. 574–577, May 1990.

[366] L. Jalloul and J. Holtzman, "Performance analysis of DS/CDMA with noncoherent M-ary orthogonal modulation in multipath fading channels," *IEEE Journal on Selected Areas in Communications*, vol. 12, pp. 862–870, June 1994.

[367] E. K. Hong, K. J. Kim, and K. C. Whang, "Performance evaluation of DS-CDMA system with M-ary orthogonal signalling," *IEEE Transactions on Vehicular Technology*, vol. 45, pp. 57–63, February 1996.

[368] Q. Bi, "Performance analysis of a CDMA cellular system," in *Proceedings of the IEEE Vehicular Technology Conference*, (Denver, CO), pp. 43–46, May 1992.

[369] Q. Bi, "Performance analysis of a CDMA cellular system in the multipath fading environment," in *Proceedings of the IEEE International Conference on Personal, Indoor and Mobile Radio Communications*, (Boston, MA), pp. 108–111, October 1992.

[370] K. Cheun, "Performance of direct-sequence spread-spectrum RAKE receivers with random spreading sequences," *IEEE Transactions on Communications*, vol. 45, pp. 1130–1143, September 1997.

[371] R. S. Lunayach, "Performance of a direct sequence spread-spectrum system with long period and short period code sequences," *IEEE Transactions on Communications*, vol. COM-31, pp. 412–419, March 1983.

[372] L.-L. Yang and L. Hanzo, "Performance of a residue number system based orthogonal signalling scheme in AWGN channels." Yet to be published.

[373] L.-L. Yang and L. Hanzo, "Performace of a residue number system based orthogonal signalling scheme over frequency-nonselective , slowly fading channel." Yet to be published.

[374] L.-L. Yang and L. Hanzo, "Performance of residue number system based DS-CDMA over multipath fading channels using orthogonal sequences," *European Transactions on Telecommunications*, vol. 9, pp. 525–535, November/December 1998.

[375] L.-L. Yang and L. Hanzo, "Residue number system arithmetic assisted M-ary modulation," *IEEE Communications Letters*, vol. 3, pp. 28–30, February 1999.

[376] L.-L. Yang and L. Hanzo, "Residue number system based multiple code DS-CDMA systems," in *Proceedings of IEEE VTC'99*, (Houston, USA), pp. 1450–1454, May 1999.

[377] L.-L. Yang and L. Hanzo, "Ratio statistic test assisted residue number system based parallel communication systems," in *Proceedings of IEEE VTC'99*, (Houston, USA), pp. 894–898, May 1999.

[378] K. Yen, L.-L. Yang, and L. Hanzo, "Residual number system assisted CDMA – a new system concept," in *Proceedings of 4th ACTS Mobile Communications Summit'99*, (Sorrento, Italy), pp. 177–182, June 8–11 1999.

[379] S. Haykin, *Digital Communications*. New York: John Wiley and Sons, 1988.

[380] R. Pickholtz, D. Schilling, and L. Milstein, "Theory of spread-spectrum communications — a tutorial," *IEEE Transactions on Communications*, vol. COM-30, pp. 855–884, May 1982.

[381] S. Rappaport and D. Grieco, "Spread-spectrum signal acquisition: Methods and technology," *IEEE Communications Magazine*, vol. 22, pp. 6–21, June 1984.

[382] E. Ström, S. Parkvall, S. Miller, and B. Ottersten, "Propagation delay estimation in asynchronous direct-sequence code division multiple access systems," *IEEE Transactions on Communications*, vol. 44, pp. 84–93, January 1996.

[383] R. Rick and L. Milstein, "Optimal decision strategies for acquisition of spread-spectrum signals in frequency-selective fading channels," *IEEE Transactions on Communications*, vol. 46, pp. 686–694, May 1998.

[384] R. D. Gaudenzi, T. Garde, F. Giannetti, and M. Luise, "A performance comparison of orthogonal code division multiple-access techniques for mobile satellite communications," *IEEE Journal on Selected Areas in Communications*, vol. 13, pp. 325–332, February 1995.

[385] M. Chase and K. Pahlavan, "Performance of DS-CDMA over measured indoor radio channels using random orthogonal codes," *IEEE Transactions on Vehicular Technology*, vol. 42, pp. 617–624, November 1993.

[386] S.-W. Kim and W. Stark, "Performance limits of Reed-Solomon coded CDMA with orthogonal signaling in a Rayleigh-fading channel," *IEEE Transactions on Communications*, vol. 46, pp. 1125–1134, September 1998.

[387] S. Lin and J. Costello, *Error Control Coding: Fundamentals and Applications*. Englewood Cliffs, NJ: Prentice-Hall, 1983.

[388] R. E. Blahut, *Fast Algorithms for digital Signal Processing*. Addison Wesley Publishing Company, 1984.

[389] C. Keller and M. Pursley, "Diversity combining for channels with fading and partial-band interference," *IEEE Journal on Selected Areas in Communications*, vol. SAC-5, pp. 248–259, February 1987.

[390] G. Chyi, G. Proakis, and C. M. Keller, "On the symbol error probability of maximum-selection diversity reception schemes over a Rayleigh fading channel," *IEEE Transactions on Communications*, vol. COM-37, pp. 79–83, January 1989.

[391] L.-L. Yang and L. Hanzo, "Performance analysis of m-ary orthogonal signaling using errors-and-erasures decoding over frequency-selective Rayleigh fading channels," *IEEE Journal on Selected Areas of Communications*, vol. 19, pp. 211–221, February 2001.

[392] R. W. Watson and C. W. Hastings, "Self-checked computation using residue arithmetic," *Proceedings of the IEEE*, vol. 54, pp. 1920–1931, December 1966.

[393] N. S. Szabo and R. I. Tanaka, *Residue Arithmetic and Its Applications to Computer Technology*. New York: McGraw-Hill Book Company, 1967.

[394] E. D. Claudio, G. Orlandi, and F. Piazza, "A systolic redundant residue arithmetic error correction circuit," *IEEE Transactions on Computers*, vol. 42, pp. 427–432, April 1993.

[395] H. Krishna, K.-Y. Lin, and J.-D. Sun, "A coding theory approach to error control in redundant residue number systems - Part I: theory and single error correction," *IEEE Transactions Circuits Syst.*, vol. 39, pp. 8–17, January 1992.

[396] J.-D. Sun and H. Krishna, "A coding theory approach to error control in redundant residue number systems - Part II: multiple error detection and correction," *IEEE Transactions Circuits Syst.*, vol. 39, pp. 18–34, January 1992.

[397] H. Krishna and J.-D. Sun, "On theory and fast algorithms for error correction in residue number system product codes," *IEEE Transactions on Computer*, vol. 42, pp. 840–852, July 1993.

[398] W. Jenkins and E. Altman, "Self-checking properties of residue number error checkers based on mixed radix conversion," *IEEE Transactions on Circuit and Systems*, vol. 35, pp. 159–167, February 1988.

[399] F. Barsi and P. Maestrini, "Error correction properties of redundant residue number systems," *IEEE Transactions on Computers*, vol. 22, pp. 307–315, March 1973.

[400] S.-S. Yau and Y.-C. Liu, "Error correction in redundant residue number systems," *IEEE Transactions on Computers*, vol. 22, pp. 5–11, January 1973.

[401] D. Mandelbaum, "Error correction in residue arithmetic," *IEEE Transactions on Computers*, vol. 21, pp. 538–545, June 1972.

[402] M. Etzel and W. Jenkins, "Redundant residue number systems for error detection and correction in digital filters," *IEEE Transactions on Acoustics, Speech, and Signal Processing*, vol. 28, pp. 538–545, October 1980.

[403] W. Jenkins, "The design of error checkers for self-checking residue number arithmetic," *IEEE Transactions on Computers*, vol. 32, pp. 388–396, April 1983.

[404] F. Barsi and P. Maestrini, "Improved decoding algorithms for arithmetic residue codes," *IEEE Transactions on Information Theory*, vol. 24, pp. 640–644, September 1978.

[405] V. Ramachandran, "Single residue error correction in residue number systems," *IEEE Transactions on Computers*, vol. 32, pp. 504–507, May 1983.

[406] L.-L. Yang and L. Hanzo, "Coding theory and performance of redundant residue number system codes." submitted to European Transactions on Telecommunications, 2003.

[407] L.-L. Yang and L. Hanzo, "Performance of residue number system based DS-CDMA over multipath fading channels using orthogonal sequences," *European Transactions on Telecommunications*, vol. 9, pp. 525–536, November - December 1998.

[408] M. A. Soderstrand, W. K. Jenkins, and G. A. Jullien, *Residue Number System Arithmetic: Modern Applications in Digital Signal Processing*. New York, USA: IEEE Press, 1986.

[409] M. A. Soderstrand, "A high-speed, low-cost, recursive digital filter using residue number arithmetic," *Proceeding IEEE*, vol. 65, pp. 1065–1067, July 1977.

[410] W. K. Jenkins and B. J. Leon, "The use of residue number system in the design of finite impulse response filters," *IEEE Transactions on Circuits Systems*, vol. CAS-24, pp. 191–201, April 1977.

[411] R. Krishnan, G. Jullien, and W. Miller, "Complex digital signal processing using quadratic residue number systems," *IEEE Transactions on Acoustics, Speech and Signal Processing*, vol. 34, pp. 166–176, February 1986.

[412] G. Alia and E. Martinelli, "A vlsi modulo m multiplier," *IEEE Transactions on Computers*, vol. 40, pp. 873–878, July 1991.

[413] T. Vu, "Efficient implementations of the Chinese remainder theorem for sign detection and residue decoding," *IEEE Transactions on Computers*, vol. 34, pp. 646–651, July 1985.

[414] R. Cosentino, "Fault tolerance in a systolic residue arithmetic processor array," *IEEE Transactions on Computers*, vol. 37, pp. 886–889, July 1988.

[415] B.-D. Tseng, G. Jullien, and W. Miller, "Implementation of fft structure using the residue number system," *IEEE Transactions on Computers*, vol. 28, pp. 831–844, November 1979.

[416] F. Barsi and P. Maestrini, "Error detection and correction by product codes in residue number system," *IEEE Transactions on Computers*, vol. 23, pp. 915–924, September 1974.

[417] A. P. Shenoy and R. Kumaresan, "Fast base extension using a redundant modulus in rns," *IEEE Transactions on Computers*, vol. 38, pp. 292–296, February 1989.

[418] D. Radhakrishnan and Y. Yuan, "Novel approaches to the design of vlsi rns multipliers," *IEEE Transactions on Circuit and Systems-II*, vol. 39, pp. 52–57, January 1992.

[419] G. Alia and E. Martinelli, "On the lower bound to the vlsi complexity of number conversion from weighted to residue representation," *IEEE Transactions on Computers*, vol. 42, pp. 962–967, August 1993.

[420] G. Alia and E. Martinelli, "A vlsi algorithm for direct and reverse conversion from weighted binary number system to residue number system," *IEEE Transactions on Circuits and Systems*, vol. 31, pp. 1033–1039, December 1984.

[421] K. Elleithy and M. Bayoumi, "Fast and flexible architectures for rns arithmetic decoding," *IEEE Transactions on Circuits and Systems-II*, vol. 39, pp. 226–235, April 1992.

[422] S. G. Glisic and P. A. Leppanen, *Wireless Communications: TDMA versus CDMA*. Kluwer Academic Publishers, Boston, 1997.

[423] P. Enge and D. Sarwate, "Spread spectrum multiple access performance of orthogonal code: impulse noise," *IEEE Transactions on Communications*, vol. COM-36, pp. 98–105, January 1988.

[424] L.-L. Yang and C.-S. Li, "DS-CDMA performance of random orthogonal codes over Nakagami multipath fading channels," in *Proceedings of IEEE ISSSTA'96*, (Mainz, Germany), pp. 68–72, IEEE, Sept 1996.

[425] R. V. Nee and A. D. Wild, "Reducing the peak-to-average power ratio of OFDM," in *Proceedings of IEEE Vehicular Technology Conference (VTC'98)* [599], pp. 2072–2076.

[426] W. G. Jeon, K. H. Chang, and Y. S. Cho, "An adaptive data predistorter for compensation of nonlinear distortion in OFDM systems," *IEEE Transactions on Communications*, vol. 45, pp. 1167–1171, October 1997.

[427] L. B. Milstein, T. S. Rappaport, and R. Barghouti, "Performance evaluation for cellular CDMA," *IEEE Journal on Selected Areas in Communications*, vol. 10, no. 4, pp. 680–689, 1992.

[428] T. Vlachus and E. Geraniotis, "Performance study of hybrid spread-spectrum random-access communications," *IEEE Transactions on Communications*, vol. 39, pp. 975–985, June 1991.

[429] G. D. Forney, "Exponential error bounds for erasure, list, and decision feedback scheme," *IEEE Transactions on Information Theory*, vol. 14, pp. 206–220, March 1968.

[430] A. J. Viterbi, "A robust ratio-threshold technique to mitigate tone and partial band jamming in coded MFSK systems," in *Proceedings of IEEE Military Communications Conferences Rec.*, pp. 22.4.1–22.4.5, IEEE, October 1982.

[431] L.-F. Chang and R. McEliece, "A study of Viterbi's ratio threshold AJ technique," in *Proceedings of IEEE Military Communications Conferences Rec.*, pp. 182–186, IEEE, October 1984.

[432] C. Baum and M. Pursley, "Bayesian methods for erasure insertion in frequency-hop communication system with partial-band interference," *IEEE Transactions on Communications*, vol. 40, pp. 1231–1238, July 1992.

[433] C. Baum and M. Pursley, "A decision-theoretic approach to the generation of side information in frequency-hop multiple-access communications," *IEEE Transactions on Communications*, vol. 43, pp. 1768–1777, February/March/April 1995.

[434] C. Baum and M. Pursley, "Bayesian generation of dependent erasures for frequency-hop communications and fading channels," *IEEE Transactions on Communications*, vol. 44, pp. 1720–1729, December 1996.

[435] C. Baum and M. Pursley, "Erasure insertion in frequency-hop communications with fading and partial-band interference," *IEEE Transactions on Vehicular Technology*, vol. 46, pp. 949–956, November 1997.

[436] G. Forney, *Concatenated Codes*. MIT Press, Cambridge, Massachusetts, 1966.

[437] T. Kasami, T. Takata, T. Fujiwara, and S. Lin, "A concatenated coded modulation scheme for error control," *IEEE Transactions on Communications*, vol. 38, pp. 752–763, June 1990.

[438] T. Kasami, T. Takata, K. Yamashita, T. Fujiwara, and S. Lin, "On bit-error probability of a concatenated coding scheme," *IEEE Transactions on Communications*, vol. 45, pp. 536–543, May 1997.

[439] L. B. Milstein, R. B. Pickholtz, and D. L. Schilling, "Optimization of the processing gain of an fsk-fh system," *IEEE Transactions on Communications*, vol. 28, pp. 1062–1069, July 1980.

[440] C. W. Helstrom, *Probability and Stochastic Processes for Engineering*. New York: Macmillion Publishing Company, 2rd ed., 1991.

[441] *Consultative Committee for Space Data System: Recommendation for Space Data System Standard: Telemetry Channel Coding "Blue Book,"*, May 1984.

[442] S. B. Weinstein and P. Ebert, "Data transmission by frequency-division multiplexing using the discrete Fourier transform," *IEEE Transactions on Communication Technology*, vol. 19, pp. 628–634, October 1971.

[443] J. Bingham, "Multicarrier modulation for data transmission: An idea whose time has come," *IEEE Communications Magazine*, pp. 5–14, May 1990.

[444] I. Kalet, "The multitone channel," *IEEE Transactions on Communications*, vol. 37, pp. 119–124, February 1989.

[445] L.-L. Yang and L. Hanzo, "Slow frequency-hopping multicarrier DS-CDMA," in *International Symposium on Wireless Personal Multimedia Communications (WPMC'99)*, (Amsterdam, The Netherlands), pp. 224–229, September:21–23 1999.

[446] R. Li and G. Stette, "Time-limited orthogonal multicarrier modulation schemes," *IEEE Transactions on Communications*, vol. 43, pp. 1269–1272, February/March/April 1995.

[447] L. Goldfeld and D. Wulich, "Multicarrier modulation system with erasures-correcting decoding for Nakagami fading channels," *European Transactions on Telecommunications*, vol. 8, pp. 591–595, November–December 1997.

[448] E. Sousa, "Performance of a direct sequence spread spectrum multiple access system utilizing unequal carrier frequencies," *IEICE Transactions on Communications*, vol. E76-B, pp. 906–912, August 1993.

[449] B. Saltzberg, "Performance of an efficient parallel data transmission system," *IEEE Transactions on Communication Technology*, vol. 15, pp. 805–811, December 1967.

[450] C. Baum and K. Conner, "A multicarrier transmission scheme for wireless local communications," *IEEE Journal on Selected Areas in Communications*, vol. 14, pp. 512–529, April 1996.

[451] V. Dasilva and E. Sousa, "Multicarrier orthogonal CDMA signals for quasi-synchronous communication systems," *IEEE Journal on Selected Areas in Communications*, vol. 12, pp. 842–852, June 1994.

[452] L. Vandendorpe and O. V. de Wiel, "MIMO DEF equalization for multitone DS/SS systems over multipath channels," *IEEE Journal on Selected Areas in Communications*, vol. 14, pp. 502–511, April 1996.

[453] N. Al-Dhahir and J. Cioffi, "A bandwidth-optimized reduced-complexity equalized multicarrier transceiver," *IEEE Transactions on Communications*, vol. 45, pp. 948–956, August 1997.

[454] P. Jung, F. Berens, and J. Plechinger, "A generalized view on multicarrier CDMA mobile radio systems with joint detection (Part i)," *FREQUENZ*, vol. 51, pp. 174–184, July–August 1997.

[455] S. Hara and R. Prasad, "Design and performance of multicarrier CDMA system in frequency-selective Rayleigh fading channels," *IEEE Transactions on Vehicular Technology*, vol. 48, pp. 1584–1595, September 1999.

[456] V. Tarokh and H. Jafarkhani, "On the computation and reduction of the peak-to-average power ratio in multicarrier communications," *IEEE Transactions on Communications*, vol. 48, pp. 37–44, January 2000.

[457] D. Wulich and L. Goldfied, "Reduction of peak factor in orthogonal multicarrier modulation by amplitude limiting and coding," *IEEE Transactions on Communications*, vol. 47, pp. 18–21, January 1999.

[458] H.-W. Kang, Y.-S. Cho, and D.-H. Youn, "On compensating nonlinear distortions of an OFDM system using an efficient adaptive predistorter," *IEEE Transactions on Communications*, vol. 47, pp. 522–526, April 1999.

[459] Y.-H. Kim, I. Song, S. Seokho, and S. R. Park, "A multicarrier CDMA system with adaptive subchannel allocation for forward links," *IEEE Transactions on Vehicular Technology*, vol. 48, pp. 1428–1436, September 1999.

[460] T. Lok, T. Wong, and J. Lehnert, "Blind adaptive signal reception for MC-CDMA systems in Rayleigh fading channels," *IEEE Transactions on Communications*, vol. 47, pp. 464–471, March 1999.

[461] B. Rainbolt and S. Miller, "Multicarrier CDMA for cellular overlay systems," *IEEE Journal on Selected Areas in Communications*, vol. 17, pp. 1807–1814, October 1999.

[462] S.-M. Tseng and M. Bell, "Asynchronous multicarrier DS-CDMA using mutually orthogonal complementary sets of sequences," *IEEE Transactions on Communications*, vol. 48, pp. 53–59, January 2000.

[463] D. Rowitch and L. Milstein, "Convolutionally coded multicarrier DS-CDMA systems in a multipath fading channel – Part I: Performance analysis," *IEEE Transactions on Communications*, vol. 47, pp. 1570–1582, October 1999.

[464] D. Rowitch and L. Milstein, "Convolutionally coded multicarrier DS-CDMA systems in a multipath fading channel – Part II: Narrow-band interference suppression," *IEEE Transactions on Communications*, vol. 47, pp. 1729–1736, November 1999.

[465] D.-W. Lee and L. Milstein, "Comparison of multicarrier DS-CDMA broadcast systems in a multipath fading channel," *IEEE Transactions on Communications*, vol. 47, pp. 1897–1904, December 1999.

[466] N. Yee, J.-P. Linnartz, and G. Fettweis, "Multi-carrier CDMA in indoor wireless radio network," *IEICE Transactions on Communications*, vol. E77-B, pp. 900–904, July 1994.

[467] S. Kondo and L. Milstein, "On the use of multicarrier direct sequence spread spectrum systems," in *Proceedings of IEEE MILCOM'93*, (Boston, MA), pp. 52–56, Oct. 1993.

[468] V. M. DaSilva and E. S. Sousa, "Performance of orthogonal CDMA codes for quasi-synchronous communication systems," in *Proceedings of IEEE ICUPC'93*, (Ottawa, Canada), pp. 995–999, Oct. 1993.

[469] L. Vandendorpe, "Multitone direct sequence CDMA system in an indoor wireless environment," in *Proceedings of IEEE First Symposium of Communications and Vehicular Technology in the Benelux, Delft, The Netherlands*, pp. 4.1–1–4.1–8, Oct. 1993.

[470] B. Steiner, "Time domain uplink channel estimation in multicarrier-CDMA mobile radio system concepts," in *Multi-Carrier Spread-Spectrum* (K. Fazel and G. P. Fettweis, eds.), pp. 153–160, Kluwer Academic Publishers, 1997.

[471] K.-W. Yip and T.-S. Ng, "Tight error bounds for asynchronous multicarrier CDMA and their application," *IEEE Communications Letters*, vol. 2, pp. 295–297, November 1998.

[472] S. Kondo and L. Milstein, "Performance of multicarrier DS CDMA systems," *IEEE Transactions on Communications*, vol. 44, pp. 238–246, February 1996.

[473] B. Popovć, "Spreading sequences for multicarrier CDMA systems," *IEEE Transactions on Communications*, vol. 47, pp. 918–926, June 1999.

[474] P. Jung, P. Berens, and J. Plechinger, "Uplink spectral efficiency of multicarrier joint detection code division multiple access based cellular radio systems," *Electronics Letters*, vol. 33, no. 8, pp. 664–665, 1997.

[475] D.-W. Lee, H. Lee, and J.-S. Kim, "Performance of a modified multicarrier direct sequence CDMA system," *Electronics and Telecommunications Research Institute Journal*, vol. 19, pp. 1–11, April 1997.

[476] A. Chouly, A. Brajal, and S. Jourdan, "Orthogonal multicarrier techniques applied to direct sequence spread spectrum CDMA systems," in *Proceedings of the IEEE GLOBECOM '93*, (Houston, USA), pp. 1723–1728, November 1993.

[477] L. Rasmussen and T. Lim, "Detection techniques for direct sequence and multicarrier variable rate for broadband CDMA," in *Proceedings of the ICCS/ISPACS '96*, pp. 1526–1530, 1996.

[478] P. Jung, F. Berens, and J. Plechinger, "Joint detection for multicarrier CDMA mobile radio systems-Part II: Detection techniques," in *Proceedings of the IEEE ISSSTA*, vol. 3, (Mainz, Germany), pp. 996–1000, September 1996.

[479] Y. Sanada and M. Nakagawa, "A multiuser interference cancellation technique utilizing convolutional codes and multicarrier modulation for wireless indoor communications," *IEEE Journal on Selected Areas in Communications*, vol. 14, pp. 1500–1509, October 1996.

[480] Q. Chen, E. S. Sousa, and S. Pasupathy, "Multicarrier CDMA with adaptive frequency hopping for mobile radio systems," *IEEE Journal on Selected Areas in Communications*, vol. 14, pp. 1852–1857, December 1996.

[481] N. Yee, J.-P. Linnartz, and G. Fettweis, "Multicarrier CDMA in indoor wireless radio networks," in *Proceedings of PIMRC'93*, pp. 109–113, 1993.

[482] L. Vandendorpe, "Multitone spread spectrum multiple access communications system in a multipath Rician fading channel," *IEEE Transactions on Vehicular Technology*, vol. 44, no. 2, pp. 327–337, 1995.

[483] L.-L. Yang and L. Hanzo, "Overlapping M-ary frequency shift keying spread-spectrum multiple-access systems using random signature sequences," *IEEE Transactions on Vehicular Technology*, vol. 48, pp. 1984–1995, November 1999.

[484] J. Wang and M. Moeneclaey, "Hybrid DS/SFH-SSMA with predetection diversity and coding over indoor radio multipath Rician-fading channels," *IEEE Transactions on Communications*, vol. 40, pp. 1654–1662, October 1992.

[485] A. J. Viterbi and J. K. Omura, *Principle of Digital Communication and Coding*. New York: McGraw-Hill, 1979.

[486] T. R. N. Rao and E. Fujiwara, *Error-Control Coding for Computer Systems*. New Jersey: Prentice Hall, 1989.

[487] S. M. Johnson, "A new upper bound for error-correcting codes," *IRE Transactions on Information Theory*, vol. 8, no. 2, pp. 203–207, 1962.

[488] F. J. Macwilliams and N. J. A. Sloane, *The Theory of Error-Correcting Codes*. New York: North-Holland, 1977.

[489] L.-L. Yang and L. Hanzo, "Performance of generalized multicarrier DS-CDMA over Nakagami-m fading channels," *IEEE Transactions on Communications, (http://www-mobile.ecs.soton.ac.uk/lly)*, vol. 50, pp. 956–966, June 2002.

[490] L.-L. Yang and L. Hanzo, "A space-time spreading assisted broadband multicarrier DS-CDMA scheme: System design and performance analysis," *Submitted for Publication (http://www-mobile.ecs.soton.ac.uk/lly)*, July 2003.

[491] L.-L. Yang and L. Hanzo, "Performance analysis of space-time spreading assisted wideband CDMA systems communicating over multipath Nakagami fading channels," *Submitted for Publication (http://www-mobile.ecs.soton.ac.uk/lly)*, May 2003.

[492] R. Ziemer and R. Peterson, *Digital Communications and Spread Spectrum System*. New York: Macmillan Publishing Company, 1985.

[493] L-L. Yang and H. Wei and L. Hanzo, "Multicarrier code-division multiple-acess using both time-domain and frequency-domain spreading," *Manuscript*, 2003.

[494] P. Fan and L. Hao, "Generalised orthogonal sequences and their applications in synchronous CDMA systems," *IEICE Transactions on Fundamentals*, vol. E83-A, pp. 2054–2069, Nov. 2000.

[495] S. Stańczak, H. Boche, and M. Haardt, "Are LAS-codes a miracle?," in *GLOBECOM '01*, vol. 1, (San Antonio, Texas), pp. 589–593, Nov. 2001.

[496] B. J. Choi and L. Hanzo, "On the Design of the LAS Spreading Codes," in *IEEE VTC 2001 Fall Conference*, (Vancouver, Canada), pp. 2172–2176, Sept. 2001.

[497] C.-C. Tseng and C. L. Liu, "Complementary Sets of Sequences," *IEEE Transactions in Information Theory*, vol. 18, pp. 644–652, Sep. 1972.

[498] R. Sivaswamy, "Multiphase Complementarty Codes," *IEEE Transactions in Information Theory*, vol. 24, pp. 546–552, Sep. 1978.

[499] R. L. Frank, "Polyphase Complementary Codes," *IEEE Transactions on Information Theory*, vol. 26, pp. 641–647, Nov. 1980.

[500] M. Katz, *Code Acquisition in Advanced CDMA Networks*. Acta Universitatis Ouluensis Technica, Oulu, C 175, 2002.

[501] A. J. Viterbi, *CDMA: Principles of Spread Spectrum Communications*. New York: Addison-Wesley Publishing Company, 1995.

[502] R. Ward, "Acquisition of pseudonoise signals by sequential estimation," *IEEE Transactions on Communications Technology*, vol. 13, pp. 475–483, December 1965.

[503] C. Kilgus, "Pseudonoise code acquisition using majority logic decoding," *IEEE Transactions on Communications*, vol. 21, pp. 772–774, June 1973.

[504] R. Ward and K. Liu, "Acquisition of pseudonoise signals by recursion-aided sequential estimation," *IEEE Transactions on Communications*, vol. 25, pp. 784–794, August 1977.

[505] C. Berrou and A. Glavieux and P. Thitimajshima, "Near shannon limit error-correcting coding and decoding: Turbo codes," in *Proceedings of the International Conference on Communications*, (Geneva, Switzerland), pp. 1064–1070, May 1993.

[506] J. Hagenauer, E. Offer, and L. Papke, "Iterative decoding of binary block and convolutional codes," *IEEE Transactions on Information Theory*, vol. 42, pp. 429–445, March 1996.

[507] L. Milstein, J. Gevargiz, and P. Das, "Rapid acquisition for direct sequence spread-spectrum communications using parallel saw convolvers," *IEEE Transactions on Communications*, vol. 33, pp. 593–600, July 1985.

[508] D. Chase, "A class of algorithms for decoding block codes with channel measurement information," *IEEE Transactions on Information Theory*, vol. IT-18, pp. 170–182, January 1972.

[509] L.-L. Yang and L. Hanzo, "Serial acquisition of DS-CDMA signal in multipath fading mobile channels," *IEEE Transactions on Vehicular Technology*, vol. 50, pp. 617 – 628, March 2001.

[510] K. Higuchi, M. Sawahashi, and F. Adachi, "Fast cell search algorithm in inter-cell asynchronous DS-CDMA mobile radio," *IEICE Transactions on Communications*, vol. E81-B, pp. 1527–1534, July 1998.

[511] P.-G. Andermo and L.-M. Ewerbring, "A CDMA-based radio access design for UMTS," *IEEE Personal Communications*, vol. 2, pp. 48–53, February 1995.

[512] J. Rapeli, "UMTS: Targets, system concept, and standardization in a global framework," *IEEE Personal Communications*, vol. 2, pp. 20–28, February 1995.

[513] M. Callendar, "Future public land mobile telecommunication systems," *IEEE Personal Communications*, vol. 12, no. 4, pp. 18–22, 1994.

[514] *The UMTS Forum website*. http://www.umts-forum.org/.

[515] *The 3GPP1 website*. http://www.3gpp.org.

[516] *The 3GPP2 website*. http://www.3gpp2.org.

[517] T. Ojanperä and R. Prasad, *Wideband CDMA for Third Generation Mobile Communications*. London: Artech House, 1998.

[518] B. H. Walke, M. P. Althoff, and P. Seidenberg, *UMTS - The Fundamentals*. Chichester, UK: John Wiley, 2002.

[519] H. Holma and A. Toskala, *WCDMA for UMTS*. Chichester, UK: John Wiley, 2002.

[520] K. Tachikawa, *W-CDMA Mobile Communications System*. Chichester, UK: John Wiley, 2002.

[521] J. Korhonen, *Introduction to 3G Mobile Communications*. Artech House, 2001.

[522] E. Dahlman, B. Gudmundson, M. Nilsson, and J. Sköld, "UMTS/IMT-2000 based on wideband CDMA," *IEEE Communications Magazine*, vol. 36, pp. 70–80, Sep. 1998.

[523] T. Ojanperä, "Overview of research activities for third generation mobile communications," in Glisic and Leppännen [87], ch. 2 (Part 4), pp. 415–446.

[524] European Telecommunications Standards Institute, *The ETSI UMTS Terrestrial Radio Access (UTRA) ITU-R RTT Candidate Submission*, June 1998. ETSI/SMG/SMG2.

[525] Association of Radio Industries and Businesses, *Japan's Proposal for Candidate Radio Transmission Technology on IMT-2000: W-CDMA*, June 1998.

[526] F. Adachi, M. Sawahashi, and H. Suda, "Wideband DS-CDMA for next-generation mobile communications systems," *IEEE Communications Magazine*, vol. 36, pp. 56–69, September 1998.

[527] F. Adachi and M. Sawahashi, "Wideband wireless access based on DS-CDMA," *IEICE Transactions on Communications*, vol. E81-B, pp. 1305–1316, July 1998.

[528] A. Sasaki, "Current situation of IMT-2000 radio transmission technology study in Japan," *IEICE Transactions on Communications*, vol. E81-B, pp. 1299–1304, July 1998.

[529] P. Baier, P. Jung, and A. Klein, "Taking the challenge of multiple access for third-generation cellular mobile radio systems — a European view," *IEEE Communications Magazine*, vol. 34, pp. 82–89, February 1996.

[530] E. Berruto, M. Gudmundson, R. Menolascino, W. Mohr, and M. Pizarroso, "Research activities on UMTS radio interface, network architectures, and planning," *IEEE Communications Magazine*, vol. 36, pp. 82–95, February 1998.

[531] T. Ojanperä and R. Prasad, ed., *Wideband CDMA for 3rd Generation Mobile Communications*. Artech House Publishers, 1998.

[532] J. Schwarz da Silva, B. Barani, and B. Arroyo-Fernández, "European mobile communications on the move," *IEEE Communications Magazine*, vol. 34, pp. 60–69, February 1996.

[533] E. Nikula, A. Toskala, E. Dahlman, L. Girard, and A. Klein, "FRAMES multiple access for UMTS and IMT-2000," *IEEE Personal Communications Magazine*, vol. 5, pp. 16–25, Apr. 1998.

[534] F. Ovesjö, E. Dahlman, T. Ojanperä, A. Toskala, and A. Klein, "FRAMES multiple access mode 2 - wideband CDMA," in *Proceedings of the IEEE International Symposium on Personal, Indoor and Mobile Radio Communications (PIMRC)*, (Helsinki, Finland), pp. 42–48, Sep. 1-4 1997.

[535] E. L. Kuan, C. H. Wong, and L. Hanzo, "Burst-by-burst adaptive joint detection CDMA," in *Proceedings of the IEEE Vehicular Technology Conference (VTC Spring)*, (Houston, USA), pp. 1628–1632, May 16-20 1999.

[536] M. Sunay, Z.-C. Honkasalo, A. Hottinen, H. Honkasalo, and L. Ma, "A dynamic channel allocation based TDD DS CDMA residential indoor system," in *IEEE 6th International Conference on Universal Personal Communications, ICUPC'97*, (San Diego, CA), pp. 228–234, October 1997.

[537] J. Lee, *CDMA Systems Engineering Handbook*. London: Artech House Publishers, 1998.

[538] L. Hanzo, C. Wong, and P. Cherriman, "Channel-adaptive wideband video telephony," *IEEE Signal Processing Magazine*, vol. 17, pp. 10–30, July 2000.

[539] P. Cherriman and L. Hanzo, "Programmable H.263-based wireless video transceivers for interference-limited environments," *IEEE Transactions on Circuits and Systems for Video Technology*, vol. 8, pp. 275–286, June 1998.

[540] A. Fujiwara, H. Suda, and F. Adachi, "Turbo codes application to DS-CDMA mobile radio," *IEICE Transactions on Communications*, vol. E81A, pp. 2269–2273, November 1998.

[541] M. Juntti, "System concept comparison for multirate CDMA with multiuser detection," in *Proceedings of IEEE Vehicular Technology Conference (VTC'98)* [599], pp. 18–21.

[542] T. Kasami, *Combinational Mathematics and its Applications*. University of North Carolina Press, 1969.

[543] L. Hanzo, P. Cherriman, and J. Streit, *Wireless Video Communications: From Second to Third Generation Systems, WLANs and Beyond.* IEEE Press and John Wiley, 2001. (For detailed contents please refer to http://www-mobile.ecs.soton.ac.uk.).

[544] A. Brand and A. Aghvami, "Multidimensional PRMA with prioritized Bayesian broadcast — a MAC strategy for multiservice traffic over UMTS," *IEEE Transactions on Vehicular Technology*, vol. 47, pp. 1148–1161, November 1998.

[545] R. Ormondroyd and J. Maxey, "Performance of low rate orthogonal convolutional codes in DS-CDMA," *IEEE Transactions on Vehicular Technology*, vol. 46, pp. 320–328, May 1997.

[546] E. L. Kuan and L. Hanzo, "Joint detection CDMA techniques for third-generation transceivers," in *Proceedings of the ACTS Mobile Communications Summit*, (Rhodes, Greece), pp. 727–732, June 1998.

[547] A. Chockalingam, P. Dietrich, L. Milstein, and R. Rao, "Performance of closed-loop power control in DS-CDMA cellular systems," *IEEE Transactions on Vehicular Technology*, vol. 47, pp. 774–789, August 1998.

[548] R. Gejji, "Forward-link-power control in CDMA cellular-systems," *IEEE Transactions on Vehicular Technology*, vol. 41, pp. 532–536, November 1992.

[549] K. Higuchi, M. Sawahashi, and F. Adachi, "Fast cell search algorithm in DS-CDMA mobile radio using long spreading codes," in *Proceedings of IEEE VTC'97*, vol. 3, (Phoenix, AZ), pp. 1430–1434, IEEE, 4–7 May 1997.

[550] M. Golay, "Complementary series," *IRE Transactions on Information Theory*, vol. IT-7, pp. 82–87, 1961.

[551] V. Tarokh, H. Jafarkhani, and A. Calderbank, "Space-time block codes from orthogonal designs," *IEEE Transactions on Information Theory*, vol. 45, pp. 1456–1467, May 1999.

[552] W. Lee, *Mobile Communications Engineering*. New York: McGraw-Hill, 2nd ed., 1997.

[553] H. Wong and J. Chambers, "Two-stage interference immune blind equaliser which exploits cyclostationary statistics," *Electronics Letters*, vol. 32, pp. 1763–1764, September 1996.

[554] C.-C. Lee and R. Steele, "Effect of soft and softer handoffs on CDMA system capacity," *IEEE Transactions on Vehicular Technology*, vol. 47, pp. 830–841, August 1998.

[555] M. Gustafsson, K. Jamal, and E. Dahlman, "Compressed mode techniques for inter-frequency measurements in a wide-band DS-CDMA system," in *Proceedings of IEEE International Symposium on Personal, Indoor and Mobile Radio Communications, PIMRC'97*, (Marina Congress Centre, Helsinki, Finland), pp. 231–235, IEEE, 1–4 September 1997.

[556] D. Knisely, S. Kumar, S. Laha, and S. Nanda, "Evolution of wireless data services: IS-95 to cdma2000," *IEEE Communications Magazine*, vol. 36, pp. 140–149, October 1998.

[557] Telecommunications Industry Association (TIA), *The cdma2000 ITU-R RTT Candidate Submission*, 1998.

[558] D. Knisely, Q. Li, and N. Rames, "cdma2000: A third generation radio transmission technology," *Bell Labs Technical Journal*, vol. 3, pp. 63–78, July–September 1998.

[559] S. Willenegger, "Overview of cdma2000 physical layer," *Journal of Communications and Networks*, vol. 2, pp. 5–17, March 2000.

[560] Y. Okumura and F. Adachi, "Variable-rate data transmission with blind rate detection for coherent DS-CDMA mobile radio," *IEICE Transactions on Communications*, vol. E81B, pp. 1365–1373, July 1998.

[561] M. Raitola, A. Hottinen, and R. Wichman, "Transmission diversity in wideband CDMA," in *Proceeding of VTC'99 (Spring)* [600], pp. 1545–1549.

[562] J. Liberti and T. Rappaport, "Analytical results for capacity improvements in CDMA," *IEEE Transactions on Vehicular Technology*, vol. 43, pp. 680–690, August 1994.

[563] S.Moshavi, "Multiuser detection for DS-CDMA communications," *IEEE Communications Magazine*, vol. 34, pp. 124–136, Oct. 1996.

[564] T. Lim and S. Roy, "Adaptive filters in multiuser (MU) CDMA detection," *Wireless Networks*, vol. 4, pp. 307–318, June 1998.

[565] L. Wei, "Rotationally-invariant convolutional channel coding with expanded signal space, part I and II," *IEEE Transactions on Selected Areas in Comms*, vol. SAC-2, pp. 659–686, September 1984.

[566] T. Lim and M. Ho, "LMS-based simplifications to the Kalman filter multiuser CDMA detector," in *Proceedings of IEEE Asia-Pacific Conference on Communications/International Conference on Communication Systems*, (Singapore), November 1998.

[567] D. You and T. Lim, "A modified blind adaptive multiuser CDMA detector," in *Proceedings of IEEE International Symposium on Spread Spectrum Techniques and Application (ISSSTA'98)* [601], pp. 878–882.

[568] S. Sun, L. Rasmussen, T. Lim, and H. Sugimoto, "Impact of estimation errors on multiuser detection in CDMA," in *Proceedings of IEEE Vehicular Technology Conference (VTC'98)* [599], pp. 1844–1848.

[569] Y. Sanada and Q. Wang, "A co-channel interference cancellation technique using orthogonal convolutional codes on multipath Rayleigh fading channel," *IEEE Transactions on Vehicular Technology*, vol. 46, pp. 114–128, February 1997.

[570] P. Tan and L. Rasmussen, "Subtractive interference cancellation for DS-CDMA systems," in *Proceedings of IEEE Asia-Pacific Conference on Communications/International Conference on Communication Systems*, (Singapore), November 1998.

[571] K. Cheah, H. Sugimoto, T. Lim, L. Rasmussen, and S. Sun, "Performance of hybrid interference canceller with zero-delay channel estimation for CDMA," in *Proceeding of Globecom'98*, (Sydney, Australia), pp. 265–270, IEEE, 8–12 November 1998.

[572] S. Sun, L. Rasmussen, and T. Lim, "A matrix-algebraic approach to hybrid interference cancellation in CDMA," in *Proceedings of IEEE International Conference on Universal Personal Communications '98*, (Florence, Italy), pp. 1319–1323, October 1998.

[573] A. Johansson and L. Rasmussen, "Linear group-wise successive interference cancellation in CDMA," in *Proceedings of IEEE International Symposium on Spread Spectrum Techniques and Application (ISSSTA'98)* [601], pp. 121–126.

[574] D. Guo, L. Rasmussen, S. Sun, T. Lim, and C. Cheah, "MMSE-based linear parallel interference cancellation in CDMA," in *Proceedings of IEEE International Symposium on Spread Spectrum Techniques and Application (ISSSTA'98)* [601], pp. 917–921.

[575] L. Rasmussen, D. Guo, Y. Ma, and T. Lim, "Aspects on linear parallel interference cancellation in CDMA," in *Proceedings of IEEE International Symposium on Information Theory*, (Cambridge, MA), p. 37, August 1998.

[576] L. Rasmussen, T. Lim, H. Sugimoto, and T. Oyama, "Mapping functions for successive interference cancellation in CDMA," in *Proceedings of IEEE Vehicular Technology Conference (VTC'98)* [599], pp. 2301–2305.

[577] S. Sun, T. Lim, L. Rasmussen, T. Oyama, H. Sugimoto, and Y. Matsumoto, "Performance comparison of multi-stage SIC and limited tree-search detection in CDMA," in *Proceedings of IEEE Vehicular Technology Conference (VTC'98)* [599], pp. 1854–1858.

[578] H. Sim and D. Cruickshank, "Chip based multiuser detector for the downlink of a DS-CDMA system using a folded state-transition trellis," in *Proceeding of VTC'99 (Spring)* [600], pp. 846–850.

[579] G. L. Turin, "Introduction to spread-spectrum antimultipath techniques and their application to urban digital radio," *Proceedings of IEEE*, vol. 68, pp. 328–353, March 1980.

[580] B. Lu and X. D. Wang, "Iterative receivers for multiuser space-time coding systems," *IEEE Journal on Selected Areas in Communications*, vol. 18, pp. 2322–2335, November 2000.

[581] D. Bertseka and R. Gallager, *Data Networks*. Englewood Cliffs, N.J.: Prentice Hall, 2nd ed., 1992.

[582] E. L. Kuan and L. Hanzo, "Comparative study of adaptive-rate CDMA transmission employing joint-detection and interference cancellation receivers," in *Proceedings of the IEEE Vehicular Technology Conference (VTC) 2000, Spring conference*, (Tokyo, Japan), May 15-18 2000.

[583] E. L. Kuan, C. H. Wong, and L. Hanzo, "Burst-by-burst adaptive joint detection CDMA," in *Proceedings of the IEEE Vehicular Technology Conference (VTC Spring)*, (Houston, USA), pp. 1628–1632, May 1999.

[584] L. Hanzo, P. Cherriman, and E. Kuan, "Interactive cellular and cordless video telephony: State-of-the-art, system design principles and expected performance," *Proceedings of the IEEE*, pp. 1388–1413, September 2000.

[585] J. Laiho-Steffens, A. Wacker, and P. Aikio, "The Impact of the Radio Network Planning and Site Configuration on the WCDMA Network Capacity and Quality of Service," in *IEEE Proceedings of Vehicular Technology Conference*, (Tokyo, Japan), pp. 1006–1010, 2000.

[586] R. D. K. Hiltunen, "WCDMA downlink capacity estimation," in *IEEE Proceedings of Vehicular Technology Conference*, (Tokyo, Japan), pp. 992–996, 2000.

[587] K. Sipilä, Z.-C. Honkasalo, J. Laiho-Steffens, and A. Wacker, "Estimation of capacity and required transmission power of wcdma downlink based on a downlink pole equation," in *IEEE Proceedings of Vehicular Technology Conference*, (Tokyo, Japan), pp. 1002–1005, 2000.

[588] L. Wang and A. Aghvami, "Optimal power allocation based on QoS balance for a multi-rate packet CDMA system with multi-media traffic," *Proceedings of Globecom'99*, pp. 2778–2782, December 1999.

[589] D. Koulakiotis and A. Aghvami, "Data detection techniques for DS/CDMA mobile systems: A review," *IEEE Personal Communications*, pp. 24–34, June 2000.

[590] M. Cheng and J.-I. Chuang, "Performance evaluation of distributed measurement-based dynamic channel assignment in local wireless communications," *IEEE Journal on Selected Areas of Communications*, vol. 14, pp. 698–710, May 1996.

[591] R. Owen, P. Jones, S. Dehgan, and D. Lister, "Uplink WCDMA capacity and range as a function of inter-to-intra cell interference: theory and practice," *Proceedings of IEEE Vehicular Technology Conference*, pp. 298–303, 2000.

[592] J. Liberti and T. Rappaport, "A geometrically based model for line-of-sight multipath radio channels," *Proceedings of IEEE Vehicular Technology Conference*, pp. 844–848, 1996.

[593] R. Ertel, P. Cardieri, K. Sowerby, T. Rappaport, and J. Reed, "Overview of Spatial Channel Models for Antenna Array Communications Systems," *IEEE Personal Communications*, pp. 10–22, February 1998.

[594] R. Kohno, *Chapter 1: Spatial and Temporal Communication Theory using Software Antennas for Wireless Communications*. Kluwer Academic Publishers, 1997.

[595] Y. Ogawa and T. Ohgane, "Adaptive Antennas for Future Mobile Radio," *IEICE Transactions on Fundamentals*, vol. E79-A, pp. 961–967, July 1996.

[596] X. Wu, L.-L. Yang, and L. Hanzo, "Uplink capacity investigations of tdd/cdma," *Proceedings of the IEEE Vehicular Technology Conference*, pp. 997–1001, May 2002.

[597] J. M. Torrance and L. Hanzo, "On the upper bound performance of adaptive QAM in a slow Rayleigh fading channel," *IEE Electronics Letters*, pp. 169 – 171, April 1996.

[598] H. J. Larson and B. O. Shubert, *Probabilitic Models in Engineering Sciences, Volume I: Random Variables and Stochastic Processes*. New York: John Wiley & Sons, 1979.

[599] IEEE, *Proceedings of IEEE Vehicular Technology Conference (VTC'98)*, (Ottawa, Canada), 18–21 May 1998.

[600] IEEE, *Proceeding of VTC'99 (Spring)*, (Houston, TX), 16–20 May 1999.

[601] IEEE, *Proceedings of IEEE International Symposium on Spread Spectrum Techniques and Application (ISSSTA'98)*, (Sun City, South Africa), September 1998.

Subject Index

Symbols

A

D

E

F

G

H

I

J

Q

R

S

T

U

V

Author Index

B

C

D

E

H

I

J

L

M

R

S

T

U

X

Y

Z